Human Anatomy

Human Anatomy

Elaine N. Marieb, R.N., Ph.D.

Holyoke Community College

Jon Mallatt, Ph.D.

Washington State University

The Benjamin/Cummings Publishing Company, Inc.

Redwood City, California • Menlo Park, California
Reading, Massachusetts • New York • Don Mills, Ontario
Wokingham, U.K. • Amsterdam • Bonn • Sydney
Singapore • Tokyo • Madrid • San Juan

Sponsoring editor: Melinda Adams

Developmental editor: Laura Bonazzoli

Production editor: Wendy Earl

Production consultant: Anne Friedman

Art supervisor: Kathie Minami

Art coordinators: Lynn Sanchez, Kay Brown, Betty Gee

Manuscript editor: Sally Peyrefitte

Text designer: Gary Head

Cover designer: Mark Ong

Illustrators: Martha Blake, Raychel Ciemma, Barbara Cousins, Charles W. Hoffman, Georg Klatt, Jeanne Koelling, Stephanie McCann, Linda McVay, Ken Miller, Elizabeth Morales-Denney, Laurie O'Keefe, Carla Simmons, Nadine Sokol, Cyndie Wooley

Production artists: Shirley Bortoli, Linda Harris, Carol Verbeeck

Dummy artist: Brenn Lea Pearson

Photo editor: Cecilia Mills

Photo researcher: Kelli West

Editorial assistant: Mark Thomas Childs

Production assistant: Frank Edward Vaughn

Proofreader: Jenny Pulsipher

Indexer: Elinor Lindheimer

Compositor: Graphic Typesetting Service Inc.

Color separator: Color Response, Inc.

Manufacturing supervisor: Casimira Kostecki

Text printer and binder: Rand McNally

Library of Congress Cataloging-in-Publication Data

Marieb, Elaine Nicpon
    Human anatomy / Elaine N. Marieb, Jon Mallatt.
        p.   cm. — (The Benjamin/Cummings series in life
    sciences)
    Based on her larger work titled: Human anatomy and
    physiology.
    Includes index.
    ISBN 0-8053-4060-2
    1. Human anatomy.   I. Mallatt, Jon.   II. Marieb,
    Elaine Nicpon, 1936–
Human anatomy and physiology.   III. Title.   IV. Series.
QM23.2.M348   1992
611—dc20   91-36531
                                                    CIP

            ISBN 0-8053-4060-2

    2 3 4 5 6 7 8 9 10–RN–95 94 93 92

The Benjamin/Cummings Publishing Company, Inc.
390 Bridge Parkway
Redwood City, California 94065

About the Authors

When Elaine Marieb and Jon Mallatt walk into their classrooms, they face the same challenges all anatomy instructors do. They must communicate a vast amount of information to students in a way that stimulates their interest, not dampens it. Thirty-eight combined years of teaching experience in both the laboratory and the classroom have enhanced their sensibilities about pedagogy and presentation. The insights gained from this experience are distilled in *Human Anatomy*—a book that truly addresses the problems students encounter in this course.

Dr. Marieb started her teaching career at Springfield College, where she taught anatomy and physiology to physical education majors. She then joined the faculty of the Biological Sciences Division of Holyoke Community College in 1969 after receiving her Ph.D. in Zoology from the University of Massachusetts at Amherst. As a result of her contact with students at Holyoke Community College, many of whom were pursuing degrees in nursing, she developed a desire to better understand the relationship of the scientific study of the human body to the clinical aspects of nursing practice. While continuing to teach full-time, Dr. Marieb earned an associate degree in nursing in 1980 from Holyoke Community College. She continued her nursing education, receiving a bachelor's degree in nursing from Fitchburg State College and then a master's degree in gerontological nursing from the University of Massachusetts in 1985. Dr. Marieb was honored with the "Outstanding Educator Award" for excellence in classroom instruction at Holyoke Community College. She has also enjoyed the honors of being the keynote speaker at the 1990 *Human Anat-*omy *and Physiology Society* meeting in Madison, Wisconsin and serving on the Instrumentation Committee of the National Science Foundation.

Dr. Marieb is an accomplished author, with several successful textbooks and laboratory manuals to her credit. These publications are listed on the page facing the title page.

The perfect complement to Dr. Marieb's nursing background, Dr. Mallatt has a Ph.D. in Anatomy from the University of Chicago. He is currently an Associate Professor of Zoology at Washington State University, where he has been teaching human anatomy to undergraduates of all backgrounds for 12 years. He is also a member of the department of Basic Medical Sciences, where he teaches courses in Histology and Anatomy of the Trunk in the WAMI Medical Program. Additionally, Dr. Mallatt holds a position as adjunct Associate Professor in the department of Biological Structure at the University of Washington. His particular areas of expertise in the study of anatomy are histology, comparative anatomy, embryology, and anatomical drawing. As a trained anatomist with a talent in anatomical illustration, Dr. Mallatt is meticulous about anatomical accuracy, detail, and currency. The superlative art in this text reflects his expertise.

Dr. Mallatt is an accomplished researcher with 20 scientific publications in the field of anatomy to his credit.

Both authors began their careers in publishing over 10 years ago as reviewers for textbooks in Human Anatomy and Human Anatomy and Physiology.

Preface to the Instructor

As a teacher, you know that teaching anatomy is not just the presentation of facts. You must provide information in a framework that encourages genuine understanding, devise new presentations to help students remember large amounts of material, and help students apply what they have learned to new situations. All the while you hope that you inspire in the students a real love of the subject.

After many years of teaching human anatomy, we became convinced that new approaches to the subject could excite and challenge the students' natural curiosity, which is why we decided to write this book. We were fortunate to have collaborated with Benjamin/Cummings, a publisher that shared our goal: to set a new standard for pedagogical and visual effectiveness in an anatomy text.

Intended Audience and Objectives of the Book

This book is designed for one-semester or one-quarter introductory anatomy courses that serve students in prenursing, allied health, physical education, premedical, prephysical therapy, and other fields. We present the basic concepts of anatomy—gross, microscopic, developmental, and clinical—in a manner that is clearly written, effectively organized, up-to-date, and well illustrated. We realize that anatomy involves gargantuan amounts of factual material, and have tried to make this material as logical and accessible as possible. To this end, we present anatomy as a "story" that can be explained and understood—convincing the students that the structure of the body makes sense. We have done this primarily by using a functional approach, explaining wherever possible how the shape and composition of the anatomical structures allow them to perform their functions. Such functional anatomy is not physiology (which focuses on biological mechanisms), but more akin to "design analysis." This approach is not used by other texts at this level. To further encourage understanding, this interactive book asks students to relate the material to their life experiences and to answer conceptual questions as they read along. The book also includes extensive, easy-to-use pedagogical devices, color coding, and art that functions as a teaching tool. These features will especially benefit students who may be less thoroughly or less recently prepared for a rigorous science course.

Organization

This book organizes the human body by organ systems. Chapter 1 introduces the terms and presents fundamental concepts of anatomy and the basic organization of the body. Chapter 2 covers the cell and its organelles. Chapter 3 presents the basics of embryology. Chapter 4 introduces the types of tissues. Chapter 5 covers the skin and its appendages, as an example of a simple organ system. Chapters 6 and 7 explain the skeleton, Chapter 8, the joints, and Chapters 9–10 the muscles. Chapters 11–15 present the nervous system: Chapter 11 introduces nervous tissue and also explains the basic organization of the nervous system; Chapter 12 covers the central nervous system; Chapter 13, the peripheral nervous system; Chapter 14, the autonomic nervous system; and Chapter 15, the special senses. Chapters 16–19 present the circulatory system: Chapter 16 deals with blood; Chapter 17, the heart; Chapter 18, blood vessels; and Chapter 19, the lymphatic system. Chapters 20–24 cover the visceral organ systems: Chapter 20 discusses the respiratory system; Chapter 21, the digestive system; Chapter 22, the urinary system; Chapter 23, the reproductive system; and Chapter 24, the endocrine system and hormones. The final chapter, Chapter 25, deals with surface anatomy, encouraging the students to review as much anatomy as possible by examining their own bodies.

This organization scheme is basically traditional; however, there are several unique aspects. First, among the many anatomy books that have appeared in the last thirty years, this book alone covers the basics of embryology early in the book (Chapter 3); this presentation provides the students with the background they need to understand the development of the specific organ systems in later chapters. Second, we place the endocrine chapter (Chapter 24) near the end of the book—after, rather than before, the chapters on the visceral organ systems (digestive, urinary, reproductive). We did this because we feel that students must learn what the visceral functions are before they can truly remember the numerous and varied roles played by hormones produced by the endocrine organs. Nonetheless, we realize that many anatomy courses teach the endocrine system with the nervous system, so we designed the endocrine chapter so that it can also be interjected and understood at that earlier point of the course. Finally, by placing our surface anatomy chapter (Chapter 25) at the end of the book, we can cover more of the clinical information related to surface body features. Other books place this chapter earlier, after bones and muscles, thereby forfeiting the opportunity to discuss the surface anatomy of blood vessels (pulse points), lymph nodes, nerves, and the visceral organs. Because it is in the last chapter, our coverage of surface anatomy provides the students with a unique summary of the entire body.

Development of this Book

In order to provide a book that is both original and responsive to the needs of the market, this text underwent extensive development. Its two drafts were examined for content and presentation by 38 different reviewers. In the second round of reviews, the manuscript was compared to the major undergraduate anatomy texts currently available. This was done to assure that our book is an improvement over other books published for this market. During a series of special art reviews, a faculty survey tested the clarity and pedagogical effectiveness of the cadaver photos in the book vis à vis the other anatomy texts. A separate survey evaluated our muscle art (pp. 244–291) for accuracy, effectiveness, and beauty compared to that in other books.

This book, like all major anatomy books in this market, is based on a larger anatomy and physiology text—in this case, *Human Anatomy and Physiology* by Elaine N. Marieb. However, special effort was made to assure that this is not just a "spin-off" but an entirely new product, specifically written and developed for the anatomy market. For example, there are over 100 new pieces of art and over 80 new color photos that are not present in the parent book. The embryology and surface anatomy chapters are completely new, and the chapters on the autonomic nervous system, lymphatic system, digestive system, and endocrine system have been extensively reworked from an anatomical perspective. New sections that provide overviews of the organ systems have been added to

the chapter on muscles, (pp. 241–243) and the introductory nervous system chapter (pp. 310–312). Given the clinical importance of the heart, an expanded discussion of the four heart chambers is presented in Chapter 17 (pp. 463–466). In keeping with the functional anatomy emphasis of this book, new sections have been added on the design characteristics of long bones (p. 124), the functional morphology of cartilage tissue (p. 120), and the structural basis of capillary permeability (p. 486). Presentations of microscopic anatomy have been refined and expanded throughout the text. Finally, it is not an exaggeration that virtually every paragraph of the book has been scrutinized, edited, and where deemed necessary, rewritten to meet the special needs of the anatomy market.

Unique Approach to Anatomy

We have worked diligently to distinguish this book from the many other anatomy books currently available. First, we emphasize strongly the *functional anatomy theme*, giving careful consideration to the adaptive characteristics of the anatomical structures of the body (see pp. 120, 124, and 486 for examples).

Second, this book is as *current* as possible. We believe that the undergraduate anatomy texts presently available are about fifteen years out-of-date in light of modern anatomical knowledge. We attempt to close this gap, while keeping our book readable and accessible to the undergraduate audience.

Third, the production team has developed a *superior art program*, using step-by-step diagrams of developmental processes (see Figures 3.11, 12.3, 17.16, 24.13 and others), many figures with insets and orientational diagrams to guide students to key structures (e.g., see Chapters 2 and 3), and placement of figures as close as possible to their text references. We are especially proud of the renderings of the skeletal muscles in Chapter 10 and the color photos of surface anatomy in Chapter 25.

Fourth, we provide an effective treatment of *microscopic anatomy*. Many undergraduate texts treat histology as a specialized and minor subfield of anatomy that takes a back seat to gross anatomy. This is unfortunate, because most physiological and disease processes take place at the cell and tissue level, and most allied health students require a solid background in histology to prepare them for their physiology courses.

Fifth, we provide an anatomically valid treatment of *embryology*. An in-depth understanding of anatomy is enriched by a simultaneous presentation of how the structures develop. Currently, all major anatomy texts present separate information on the embryology with each organ system—as does this text. However, none of the competing texts presents the basic facts of early development that students need to properly approach and understand embryology until the very end of the book (with the female reproductive organs). We are convinced that such basics should be presented early on, so we wrote Chapter 3 as an introduction to embryology. Because a comprehensive presentation of development early in the book could be intimidating to students, we have used a "velvet glove" approach, providing only the most important concepts in a concise, understandable way (see pp. 48–63).

Sixth, we present more *regional* anatomy, carefully describing the locations of the various organs. For example, the vertebrae in the spinal column are used as a yardstick to indicate the positions of various visceral organs in the trunk, and the courses of the vessels and nerves of the limbs are described in relation to nearby muscle groups. This not only helps students locate structures in the laboratory, but also allows them to appreciate the effects of injuries that are confined to limited body regions. The surface anatomy chapter (Chapter 25) provides the best demonstration of the use of regional anatomy.

Finally, this book is an *interactive learning tool* that encourages understanding, rather than an encyclopedia that encourages rote memorization. Whereas many scientific textbooks read like long lists of terms, this book reads like an instructor talking to, explaining to, and challenging the student. Many of the chapters include special topic boxes that present interesting information in a lighter, more inviting format. For those organs and systems that include large numbers of parts (e.g., skeletal muscles, cranial nerves, sensory receptors), we present these parts in illustrated tables that organize the information in a logical way (pp. 246–292; 366–367; 371–377). In-text Critical Thinking questions and Clinical Application sections challenge the student and encourage synthesis of information. The complex terminology of anatomy is always the most difficult aspect of the subject to make interesting and accessible. To this end, we have included the Latin or Greek translations of virtually every term as it is introduced in the text. Our anatomy book presents the definitions directly where the terms are introduced in the text.

Special Features

Art and Photo Program

Art plays a critical role in helping students visualize anatomical structures and concepts. Writing this text has been a process of translating our knowledge of anatomy into words and pictures. Our team of medical illustrators read the manuscript, did additional research when necessary, and provided highly accurate figures to depict the anatomy. Each illustration was carefully reviewed at every stage of its development to ensure accuracy.

The art program is full color throughout. The color is used not only for aesthetic purposes but also in a functional sense. For example, the color protocol for the embryonic germ layers seen in Chapter 3 (ectoderm in blue, mesoderm in red, endoderm in yellow) is used throughout the later embryology figures of this book. Additionally, throughout the nervous system unit, sensory neurons are always shown in blue (and green), motor neurons in red (and yellow). Consistency of this nature encourages automatic learning by the student and continually reinforces their memory. In many figures, the body location or general structure of a body part is first illustrated in a simple diagram for orientation, and then the structures of interest are shown in enlarged, detailed view. Not only are light and scanning electron micrographs used abundantly, but images produced by modern medical scanning techniques (CT, PET, and MRI) are inserted where appropriate to enhance the students' interest and grasp of structure. Additionally, the highly acclaimed Bassett photographs of dissected cadavers are sprinkled throughout the book. These photos should especially benefit the many anatomy courses in which human cadavers are unavailable.

Illustrated Tables

The use of illustrations within summary tables is a highly effective and efficient method of helping students learn important and detailed information. Thus, many chapters include illustrated summary tables. These tables run the gamut in level of complexity from the exceptionally complete tissue tables in Chapter 4 (presented as Figures 4.3, 4.15, 4.17, and 4.18) to the simple but highly effective table showing bone fractures in Chapter 6 (p. 132). In particular, the tables illustrating the skeletal muscles in Chapter 10 are noteworthy, because, in addition to doing what all muscle tables do (summarizing the locations, attachments, and functions of the muscles), every table includes an introductory essay explaining general concepts about a functional group of muscles. Other illustrated tables cover the stages of the fetal period (Table 3.2), various bones of the skeleton (Tables 7.3–7.5), the synovial joints of the body (Table 8.1), the classes of sensory receptors (Table 13.1), the cranial nerves (Table 13.2), and more.

Color-Coded Pages

Each body system has been assigned a specific color. For example, the chapters on the cardiovascular system, Chapters 16–18, are identified by a red color tab at the top corner of each page. Thus, if you wish to flip quickly to the chapters on this system, you simply open the book to the red-edged pages. This makes the book much easier to use, greatly simplifying the task of looking up information. The Brief Contents on p. xvii provides a guide to the color-coding scheme.

Closer Look Boxes

Many chapters of this book have special topic boxes ("A Closer Look"), generally of a clinical nature. Some of these boxes deal with timely topics, such as artificial joints (p. 210), bone transplants (p. 184), the effects of steroid use on the physique of athletes (p. 233), and AIDS (p. 522). Other boxes address timeless topics that interest students, such as the general aspects of cancer (p. 42), problems with visual focusing (p. 426), atherosclerosis and coronary bypass surgery (p. 484), and lung cancer and smoking (p. 554). Special properties of cartilage (p. 120), and porphyria, a vampire-like disorder of the skin (p. 112), are topics of other boxes.

Life Span Approach

Most chapters close with a "Throughout Life" section that summarizes the embryonic development of organs of the system and then examines how these organs change throughout the life span. Diseases particularly common during certain periods of life are pointed out, and effects of aging are considered. The implications of aging are particularly important to students in the health-related curricula because many of their patients will be in the senior age group.

Clinical Applications Throughout the Text

Clinical material that applies anatomical concepts is always of interest to students in the allied health and physical education fields. Such material is included in each chapter, set off by a **red cross symbol.** Furthermore, most of the Closer Look boxes and Critical Thinking questions (described below) contain clinical material, and lists of Related Clinical Terms are provided at the end of each chapter.

Critical Thinking Questions Throughout the Text

As part of the interactive nature of this book, a number of Critical Thinking questions are included in the text portion of each chapter. These questions are indicated by a symbol that resembles a **square puzzle,** and answers are provided directly below each question. They ask the student to synthesize information from previous paragraphs, or to call

on their own life experiences. "Learning by doing" is an effective learning tool, and we believe the students will enjoy the challenge of these questions. Although some of the Critical Thinking questions are purely anatomical, most are clinical in nature. Additional Critical Thinking questions are included in the Review Questions at the end of each chapter. There are approximately 10 Critical Thinking questions per chapter.

In-Text Learning Aids

Numerous pedagogical devices have been incorporated into *Human Anatomy* to enhance its utility as a learning tool.

Chapter Outlines and Student Objectives

Each chapter begins with an outline of the major topics of the chapter, followed directly by a list of specific objectives that the student should strive to master. Page references allow students to relate the outline and objectives to specific text passages.

Word Pronunciations and Derivations

The introduction of each technical term in the text is directly followed by its pronunciation and, when helpful, by the literal meaning of the term (usually its Greek or Latin derivation). The pronunciations are adopted from the *Encyclopedia and Dictionary of Medicine, Nursing, and Allied Health*, Fourth Edition, by B. Miller and C. Keane, 1987, Philadelphia, W.B. Saunders. (The pronunciation scheme is explained in the Preface to the Student on pp. xv–xvi.)

Related Clinical Terms

A list of clinical terms that are relevant to the content of the chapter is provided at the end of each chapter.

Chapter Summary

Thorough chapter summaries in a study outline format with page references will help students review the chapter they have just read.

Review Questions

Numerous end-of-chapter questions are provided. These questions are grouped into several sections: short, objective questions ("Multiple Choice and Matching"), longer questions ("Short Answer and Essay"), and finally, "Critical Thinking and Clinical

Application Questions." The latter, which are identified by the **square puzzle,** challenge the students to synthesize information and solve new problems. There are approximately 30 review questions per chapter.

Appendices

The appendices offer several useful references for the student. They include: Appendix A: The Metric System; Appendix B: Answers to Multiple Choice and Matching Questions from the individual chapters; and Appendix C: Word Roots, Prefixes, Suffixes, and Combining Forms.

Glossary

At the end of this book, students will find an extensive glossary that includes both the definitions and phonetic pronunciations of key terms.

Teaching Package

The teaching package accompanying *Human Anatomy* has been developed to help instructors and students derive the greatest benefit from this text.

Laboratory Manual

The *Human Anatomy Laboratory Manual with Cat Dissection* by Elaine Marieb (1992) that accompanies this text contains 31 gross anatomy and histology exercises for all major body systems. Illustrated in full color, with a convenient spiral binding, this lab manual has an accompanying *Instructor's Guide* by Linda Kollett of Massasoit Community College.

Instructor's Guide/Test Bank

An *Instructor's Guide and Test Bank* has been prepared for this textbook by Jon Mallatt and Elaine N. Marieb. It provides detailed annotated lecture outlines, suggested class activities and discussion topics, additional clinical terminology, lists of suggested readings, recommended audiovisual aids, and the answers to all the subjective questions in the textbook (i.e., to the Short Answer and Essay Questions, and the Critical Thinking and Clinical Application Questions). The Test Bank section of the manual offers approximately 2,000 new multiple choice questions for use in course exams. It contains two different Tests per book chapter (Tests A and B), each of which consists of 25–30 questions. At the end of the test bank are two Final Exams, each consisting of over 200 ques-

tions. The instructor may use the tests intact, or may choose only those questions that are most appropriate. The questions in the Test Bank are entirely new and different from those provided in the textbook itself. This Test Bank is available on a microcomputer test generation program for the IBM PC, IBM PS/2, and Macintosh computers. This software is available to qualified adopters of *Human Anatomy* through the publisher or your local representative.

Overhead Acetate Transparencies

A package of 200 acetates, for use with overhead projectors during lecture presentations, is available. This package includes over 100 of the most important illustrations from *Human Anatomy* and additional acetates from sources other than this textbook. On all acetates, the type has been enlarged for easy viewing in larger lecture halls.

Photo Atlas

Authored by Robert A. Chase, M.D., Professor of Anatomy at Stanford University, *The Bassett Atlas of Human Anatomy* offers an extraordinary collection of full-color dissection photos. An inexpensive alternative to the costly collections now available, it will be extremely useful to your students in their anatomy laboratory.

Human Dissection Slides

Qualified adopters of this *Human Anatomy* text will be eligible for a set of 85 complimentary dissection slides taken from *The Bassett Atlas of Human Anatomy.*

Coloring Atlas

A *Coloring Atlas of Human Anatomy* has been prepared by Steve Langjahr and Bob Brister of Antelope Valley College. Accompanying Macintosh software, keyed to this atlas, is available.

The Anatomy Tutor

Programmed by Dr. Marvin Branstrom of Cañada College, these student tutorials include a module for each organ system on a total of four disks for the Macintosh computer and utilize the Hypercard program. Each module, class tested and reviewed for accuracy and ease of use, is designed to help the student review the anatomy of the body systems.

The Human Musculature Videotape

Scripted, developed, and produced by Rose Leigh Vines, Ph.D. and Allan Hinderstein of California State

University, Sacramento, this highly acclaimed video takes the student on an 18-minute tour of 38 muscles on a prosected cadaver. Ideal for the anatomy lab, the origin and insertion of each muscle are also shown.

The Human Nervous System Videotape

Produced by the same team of anatomy instructors and media specialists from California State University, Sacramento, this new 20-minute video shows regions of the brain and cranial nerves, and regions of the spinal cord and spinal nerves. Because it shows structures on a prosected cadaver that are usually difficult to see with mere models or drawings, this video is also ideally suited for the laboratory.

Acknowledgments

The authors would like to thank the many dedicated and skilled people who made this book possible. Connie Spatz, the biology acquisitions editor at Benjamin/Cummings from 1987–1989, enthusiastically initiated the project and brought us together. Melinda Adams, the current acquisitions editor, has been extremely supportive and fair in guiding the project through to completion. She did a masterful job of motivating and inspiring the authors, and has kept us in touch with market conditions during these past two years. Our most special thanks are extended to Laura Bonazzoli, the developmental editor who labored tirelessly to make our presentation clear and concise. Her intelligent insights and literary talents were invaluable, and she gave generously of her time, spending endless hours on the phone to help us work through the rough spots in the manuscript. Laura was the stable center around which the project revolved during its first two years. Sally Peyrefitte, the manuscript editor, picked up the torch from Laura and proved remarkably competent in polishing the final stages of the manuscript. We also appreciate the efforts of editorial assistant Mark Childs, who coordinated the seemingly endless number of reviews, edited the *Instructor's Manual and Test Bank*, and remained enthusiastic throughout the project. Cecilia Mills, the head of photo research for this book, did an excellent job finding the wide variety of photographs; her work is much appreciated. Kathie Minami and Lynn Sanchez, the art supervisor and coordinator, faced the difficult task of coordinating the artwork and commissioning over 100 new pieces of art. This was a remarkable accomplishment, considering the tight publication schedule. Wendy Earl, the production editor, did a beautiful job of assembling the book. Wendy's competence and organizational skills were evident from the moment she joined the project. She removed the stress of a tight production schedule from the shoulders of the authors, providing us with peace of mind, for which we are grateful. Anne Friedman kindly acted, in an advisory capacity, as production consultant. We would also like to thank Stacy Treco, the marketing manager, who kept us aware of the needs of modern anatomy students. We also appreciate the contribution of Jamie Redinius, who assembled the pronunciations of technical terms.

Our artists, whose names are listed on page iv, have been superb. There is little doubt that their illustrations will imbue the students with lasting, colorful imagery from their anatomy studies.

Last but not least, we thank our spouses, "Zeb" Marieb and Marisa de los Santos, for their loving and patient support of our work on this project over the last 36 months.

Elaine N. Marieb

Jon Mallatt

Text, Art, and Photo Reviewers

Ann Allworth, *Quinsigamond Community College*

Jane Aloi, *Saddleback College*

Robert Anthony, *Triton College*

Carlton Baker, *St. Philip's College*

William Becker, *American River College*

Leann Blem, *Virginia Commonwealth University*

Marvin Branstrom, *Cañada College*

Ruth Ebeling, *Biola University*

Victor Eroschenko, *University of Idaho*

Mark Frasier, *Colorado State University*

Mary Jo Fourier, *Johnson County Community College*

Larry Ganion, *Ball State University*

John Hein, *University of Wisconsin at Oshkosh*

Nicky Huls, *West Valley College*

Jackson E. Jeffrey, *Virginia Commonwealth University*

Anne Johnson, *Southern Illinois University at Carbondale*

David A. Kaufmann, *University of Florida*

Andrew J. Kuntzman, *Wright State University*

Robert J. Laird, *University of Central Florida*

Sue Y. Lee, *Humboldt State University*

Robert Lewke, *Chippewa Valley Technical College*

Donn D. Martin, *Texas Wesleyan University*

Randall M. McKee, *University of Wisconsin; Parkside*

Lewis M. Milner, *North Central Technical College*

Sherwin Mizell, *Indiana University*

Patricia Munn, *Longview Community College*

Lowell Neudeck, *Northern Michigan University*

Jamie L. Redinius, *University of Alaska*

Ralph Reiner, *College of the Redwoods*

Gary Resnick, *Irvine Valley College*

Ernest Riley, *Sierra College*

Roscoe Root, *Lansing Community College*

William Saltarelli, *Central Michigan University*

Greg Smith, *Saint Mary's College of California*

Mary Lynne Stephanou, *Santa Monica College*

James Waters, *Humboldt State University*

Richard Welton, *Southern Oregon State College*

Laura S. Zambrano, *Hudson Valley Community College*

Preface to the Student

This book is written for you. In a way, it was written by our students, because it incorporates their suggestions, the answers to questions they most often ask, and explanatory approaches that have been most successful in helping them learn about the human body. Together, we have accumulated 38 years of experience in teaching human anatomy (we don't feel that old!), and we have listened to our students all along.

Human anatomy is more than just an important and intellectually challenging field—it is interesting. To help you get involved in the study of this exciting subject, a number of special features are incorporated throughout the book.

We want you to enjoy your learning experience, so the tone of this book is intentionally informal and unintimidating. The book is a guide to the understanding of your own body, not an encyclopedia of human anatomy. Certainly you will be learning many new names and structures, but we will show you that all of this information makes sense: The anatomy of body structures reflects their functions in a logical way, and the complex scientific names of body structures actually have simple meanings (for example, the *brachiocephalic* artery, literally means "arm-head" artery, and it supplies blood to precisely those body parts). Throughout the book, we will point out various features that make anatomy understandable and easier to remember. We explain anatomical concepts thoroughly, using abundant analogies and we present examples from familiar events whenever possible.

The illustrations and tables are designed with your learning needs in mind. Most of the tables are summaries of important information in the text and should be valuable resources for reviewing for exams. In all cases, the figures have been placed as close as possible to their text references. Special topic boxes, entitled "A Closer Look," alert you to advances in medicine or new scientific discoveries that can be applied to your daily life. Shorter pieces of clinical information that relate to anatomy are interspersed throughout the text, and are identified by a **red cross symbol.**

Quiz shows on televison are interesting and challenging entertainment, and this book contains its own "quiz show" in the form of Critical Thinking questions. Several times in each chapter you will encounter this **square puzzle,** which indicates that a question will follow. Most of these questions relate to clinical anatomy, although others test your understanding of purely anatomical concepts. Your job is to deduce the answer, based on the information ("clues") presented in the material you have been reading to that point. The answer is written directly after each question, so you will probably want to cover the answer with your hand as you are thinking. Do not be frustrated if you cannot answer the question right away, but do try to figure out the essential point. We think you will be surprised at how well you perform, and at how effective these quiz questions are in keeping your mind focused on the subject matter you are reading.

To help you organize your reading and studying, each chapter of this book begins with an outline of its major topics, with specific learning objectives listed for each topic. Furthermore, all chapters end with a comprehensive summary and several dozen review questions to help you review your understanding of the material. The last set of review questions in each chapter are of the Critical Thinking type—similar to those interspersed throughout the text—which test your ability to synthesize and apply your knowledge to clinical situations.

Anyone taking an anatomy course must learn several thousand new names, most of which are in the form of long Greek and Latin words—learning anatomy has been compared to learning a foreign language. To help you learn the vocabulary, important terms within the text are highlighted in bold or italic type. They are followed by a literal translation of the word meaning, and for hard-to-pronounce terms, a phonetic spelling.

Pronunciation Key

The pronunciations in this book are very easy to use *if you know the rules*. Our pronunciation key takes advantage of the fact that, in scientific terms, long vowels usually occur at the end of syllables, whereas short vowels usually occur at the beginning or in the middle of syllables. Therefore:

1. When vowels are *unmarked*, they have a long sound when they occur at the end of a syllable, and a short sound when they are at the beginning or in the middle of a syllable. For example, in the word *kidney* (kid′ne), you automatically know that the middle 'i' is short and the terminal 'e' sound is long. In the word *renal* (re′nal), the 'e' is long, and the 'a' is short.

2. When a vowel sound violates the above rule, it is marked with a short or long symbol. Only those short vowels that come at the end of a syllable are marked with the short symbol, called a breve (˘); and only those long vowels that occur at the start or middle of a syllable are marked with a long symbol (¯). For example, in the word *pelvirectal* (pel″vĭ-rek′tal), all four vowels are short, but only the 'i' is marked as short, because it is the only vowel that falls at the end of a syllable. In *methane* (meth′ān), the long 'a' is marked as such because it does not fall at the end of a syllable.

3. Short 'a' is *never* marked with a breve. Instead, short 'a' at the end of a syllable is indicated by 'ah'. For example, in the word *papilla* (pah-pil′ah), each 'a' is short and pronounced as 'ah'.

4. In words that have more than one syllable, the syllable with the strongest accent is followed by a prime (′) mark, and syllables with the second strongest accent, where present, are indicated with a double prime (″). The unaccented syllables are followed by dashes. An example of these principles is the word *anesthetic* (an″es-thet′ik), in which thet′ is emphasized most strongly, an″ has a secondary emphasis, and the other two syllables are not spoken with any emphasis.

We would appreciate hearing from you about your experiences with this textbook or suggestions for improvements in future editions. Please address your comments to the authors, Elaine N. Marieb and Jon Mallatt, c/o Benjamin/Cummings Publishing Company, 390 Bridge Parkway, Redwood City, CA 94065.

We hope that you enjoy *Human Anatomy* and that this book makes learning about the body's structure an exciting and rewarding experience. Perhaps the best bit of advice we can give you is that memory depends on understanding. Thus, if you try to achieve understanding instead of rote memorization, your memory will not fail you very often.

Elaine N. Marieb and Jon Mallatt

Brief Contents

Detailed Contents

continued

continued

continued

continued

continued

continued

continued

A Brief Guided Tour to *Human Anatomy*

In so many ways, Marieb and Mallatt make anatomy clear, memorable, and lively to your students.

Marieb and Mallatt include the most current anatomical research.

Anatomy is presented with a functional approach.

Clinical applications are integrated into the text and distinguished by a red cross icon.

Extensive cross-referencing ties concepts together and helps students see relationships among different body systems.

By recognizing scientific controversy, Marieb and Mallatt present anatomy as a "live" science.

Critical-thinking questions—and answers—are integrated into the text and flagged with a puzzle icon.

Supporting Cells

All neurons associate closely with non-nervous **supporting cells**, of which there are six types. Four of these are in the CNS, and two are in the PNS (Figure 11.11). Each type of supporting cell has a unique function, but in general, these cells provide a supportive scaffolding for the neurons. Furthermore, supporting cells cover all parts of the neurons that are not involved in synapses. Such masking insulates the neurons and prevents the electrical activities of adjacent neurons from interfering with each other.

The importance of supporting cells in insulating nerve fibers from one another is illustrated in the painful disorder called **tic douloureux** (tik doo"loo-roo'; "wincing in pain"). In this condition, the supporting cells around the sensory nerve fibers in the main nerve of the face (the trigeminal nerve) degenerate and are lost. As a result, impulses in nerve fibers that carry *touch* sensations proceed to influence and stimulate the uninsulated *pain* fibers in the same nerve, leading to a perception of pain by the brain. Because of this crossover, the softest touch to the face can produce agonizing pain. For more information on tic douloureux, see Table 13.2, page 371. ■

Supporting Cells in the CNS

The supporting cells in the CNS are collectively called the **neuroglia** (nu-rog'le-ah; "nerve glue") or **glial** (gle'al) **cells**. (Most authorities restrict the name *neuroglia* to the supporting cells in the CNS, but others consider all supporting cells neuroglia, including those in the PNS.) Like neurons, most glial cells have branching processes and a central cell body (Figure 11.11a–d). Neuroglia can be distinguished from neurons, however, by their much smaller size and by their darker-staining nuclei (Figure 11.4). Neuroglia outnumber neurons in the CNS by as much as 50 to 1, and they make up about half the mass of the brain. Unlike neurons, glial cells can divide throughout life.

Can you deduce which of the two cell types in neural tissue—neurons or neuroglia—gives rise to more brain tumors?

The glial cells do. Since glial cells can divide regularly, they accumulate the "mistakes" in DNA replication that may transform them into neoplastic cells. This does not occur in neurons, which do not divide. Therefore, most tumors that originate in the brain (60%) are **gliomas** (tumors formed by uncontrolled proliferation of glial cells). ■

Star-shaped **astrocytes** (as'tro-sītz; "star cells") are the most abundant glial cells (Figure 11.11a). They have many radiating processes with bulbous ends.

Some of these bulbs cling to neurons, whereas others cling to capillaries. Because of these connections, some scientists believe that astrocytes transfer nutrients from the capillary blood to the neurons, thereby "nursing" the nerve cells. While their nutritive function is still disputed, most agree that astrocytes help control the ionic environment around neurons: The concentrations of various ions outside the axons must be kept within narrow limits for nerve impulses to be generated and conducted. Additionally, astrocytes recapture (and recycle) released neurotransmitters.

Microglia are the smallest and least abundant of the neuroglia (Figure 11.11b). They have elongated cell bodies and cell processes with many pointed projections, like a thorny bush. The microglia are phagocytes, the macrophages of the CNS. They engulf invading microorganisms and injured or dead neurons. The origin of microglia is controversial. Some authorities believe they originate, like the other macrophages of the body, from a type of blood cell called a monocyte. Others claim that microglia derive from the ectoderm of the embryonic neural tube, as do the other neuroglial cells.

You will recall from Chapter 3 (p. 55) that the CNS originates in the embryo as a hollow neural tube and retains a central cavity throughout life. **Ependymal cells** (ĕ-pen'dĭ-mal; "wrapping garment") form a simple epithelium that lines the central cavity of the spinal cord and brain (Figure 11.11c). Here, these cells provide a fairly permeable layer between the cerebrospinal fluid that fills this cavity and the tissue fluid that bathes the cells of the CNS. Ependymal cells bear cilia that help circulate the cerebrospinal fluid.

Oligodendrocytes (ol"ĭ-go-den'dro-sītz) (Figure 11.11d) have fewer branches than astrocytes. Indeed, their name means "few-branch cells." Oligodendrocytes line up in small groups along the thicker axons in the CNS. They wrap their cell processes around these axons, producing insulating coverings called *myelin sheaths* (discussed in detail below).

Supporting Cells in the PNS

The two kinds of supporting cells in the PNS are *satellite cells* and *Schwann cells*. These very similar cell types differ mainly in location. **Satellite cells** surround neuron cell bodies within ganglia (Figure 11.11e). Their name comes from a fancied resemblance to the moons (satellites) around a planet. **Schwann cells** (also called *neurolemmocytes*) surround all axons in the PNS and form myelin sheaths around many of these axons.

Myelin

Myelin (mi'ĕ-lin), a lipoprotein, is a fatty substance that surrounds the thicker axons of the body. It takes

Boldfaced key terms accompanied by their pronunciation and word derivations, as well as definitions, are easier to remember.

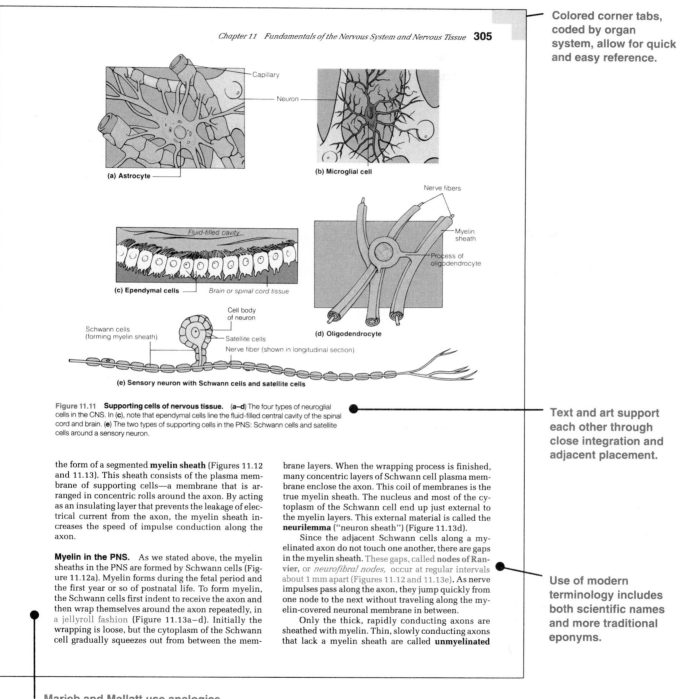

Figure 11.11 **Supporting cells of nervous tissue.** **(a–d)** The four types of neuroglial cells in the CNS. In **(c)**, note that ependymal cells line the fluid-filled central cavity of the spinal cord and brain. **(e)** The two types of supporting cells in the PNS: Schwann cells and satellite cells around a sensory neuron.

the form of a segmented **myelin sheath** (Figures 11.12 and 11.13). This sheath consists of the plasma membrane of supporting cells—a membrane that is arranged in concentric rolls around the axon. By acting as an insulating layer that prevents the leakage of electrical current from the axon, the myelin sheath increases the speed of impulse conduction along the axon.

Myelin in the PNS. As we stated above, the myelin sheaths in the PNS are formed by Schwann cells (Figure 11.12a). Myelin forms during the fetal period and the first year or so of postnatal life. To form myelin, the Schwann cells first indent to receive the axon and then wrap themselves around the axon repeatedly, in a jellyroll fashion (Figure 11.13a–d). Initially the wrapping is loose, but the cytoplasm of the Schwann cell gradually squeezes out from between the mem-

brane layers. When the wrapping process is finished, many concentric layers of Schwann cell plasma membrane enclose the axon. This coil of membranes is the true myelin sheath. The nucleus and most of the cytoplasm of the Schwann cell end up just external to the myelin layers. This external material is called the **neurilemma** ("neuron sheath") (Figure 11.13d).

Since the adjacent Schwann cells along a myelinated axon do not touch one another, there are gaps in the myelin sheath. These gaps, called **nodes of Ranvier,** or *neurofibral nodes,* occur at regular intervals about 1 mm apart (Figures 11.12 and 11.13e). As nerve impulses pass along the axon, they jump quickly from one node to the next without traveling along the myelin-covered neuronal membrane in between.

Only the thick, rapidly conducting axons are sheathed with myelin. Thin, slowly conducting axons that lack a myelin sheath are called **unmyelinated**

1

The Human Body: An Orientation

Student Objectives

An Overview of Anatomy (pp. 2–6)

1. Define anatomy and physiology, and describe the subdivisions of anatomy.

2. Name (in order of increasing complexity) the different levels of structural organization that make up the human body, and explain their relationships.

3. List the organ systems of the body, and briefly state the main functions of each system.

4. Classify the levels of structural organization in the body according to relative and actual size.

5. Give an example of anatomical variation.

Gross Anatomy: An Introduction (pp. 7–15)

6. Define the anatomical position.

7. Use anatomical terminology to describe body directions, regions, and body planes.

8. Describe the basic structures that humans share with other vertebrates.

9. Locate the major body cavities and their subdivisions.

10. Name the nine regions and four quadrants of the abdomen, and name the visceral organs that relate to these regions.

Microscopic Anatomy: An Introduction (pp. 15–16)

11. Explain how human tissue is prepared and examined for its microscopic structure.

12. Distinguish tissue viewed by light microscopy from that viewed by electron microscopy.

Clinical Anatomy: Introduction to Medical Imaging Techniques (pp. 17–20)

13. Describe six medical imaging techniques that reveal the anatomy of living patients.

As you make your way through this book, you will be learning about one of the most fascinating subjects possible—your own body. Such a study is not only interesting and highly personal, but timely as well. The current information blizzard brings news of some medical advance almost daily. Learning how your body is built and how it works enables you to appreciate new techniques for detecting and treating disease and to apply guidelines for staying healthy. For those of you preparing for a career in the health sciences, the study of anatomy provides the strong foundation needed to support your clinical experiences.

An Overview of Anatomy

Anatomy is the study of the structure of the human body. It is also called **morphology** (mor"fol'o-je), the science of form. An old and proud science, anatomy has been a field of serious intellectual investigation for at least 2300 years. It was the most prestigious biological discipline of the 1800s and is still a dynamic and growing science.

Anatomy is closely related to **physiology**, the study of body function. Although you may be studying anatomy and physiology in separate courses, the two are truly inseparable, because structure tends to reflect function. For example, the lens of the eye is transparent and curved; it could not perform its function of focusing light if it were opaque and uncurved. Likewise, the thick, long bones in our legs could not support our weight if they were soft and thin. This textbook stresses the intimacy of the relationship between structure and function. In almost all cases, a description of the anatomy of a body part is accompanied by an explanation of its function, emphasizing the structural characteristics that contribute to that function. This approach is called functional anatomy.

Anatomical Terminology

Most anatomical terms are based on ancient Greek and Latin words. For example, the arm is the brachium (Greek for "arm"), and the thigh bone is the femur (Latin for "thigh"). This terminology, which came into use when Latin was the official language of science, provides a standardized set of names that scientists can use worldwide, no matter what language they speak. Learning anatomical terminology can be

difficult, but we will help by translating many of the names for you as we present them in the text. For further help, see the Glossary on p. 717 and the list of word roots in Appendix C (p. 714).

Subdivisions of Anatomy

Gross Anatomy

Anatomy is a broad science with many subdivisions. **Gross anatomy** is the study of those body structures visible to the naked eye (*gross* = large). Such structures include the bones, lungs, and muscles. An important technique for studying gross anatomy is **dissection** (dĭ-sek'shun; "cut apart"). In dissection, connective tissue is removed from between the body organs so that the organs can be seen more clearly. The term *anatomy* is derived from Greek words meaning "I cut apart" ("I dissect").

Studies of gross anatomy can be approached in several different ways. In **regional anatomy**, all structures in a region of the body, such as the abdomen or head, are examined as a group. In **systemic** (sistem'ik) **anatomy**, by contrast, all the organs with related functions are studied together. For example, when studying the muscular system, you examine the muscles of the entire body. The systemic approach to anatomy is best for relating structure to function. Therefore, it is the approach taken in most college anatomy courses, and in this book. Medical schools, however, favor regional anatomy because many injuries and diseases involve specific body regions (sprained ankle, sore throat, heart disease); furthermore, surgeons must know each body region in detail.

Another subdivision of gross anatomy is **surface anatomy**, the study of internal body structures as they relate to the overlying skin surface. This knowledge is used to identify the muscles that bulge beneath the skin in weight lifters, and clinicians use it to locate blood vessels for placing catheters, feeling pulses, and drawing blood.

Microscopic Anatomy

Microscopic anatomy, or **histology** (his-tol'o-je; "tissue study"), is the study of structures too small to be seen without the help of a microscope. These include cells and cell parts; groups of cells, called tissues; and the microscopic details of the organs of the body (stomach, spleen, and so on). A knowledge of microscopic anatomy is important because many physiological and disease processes occur at the cellular level.

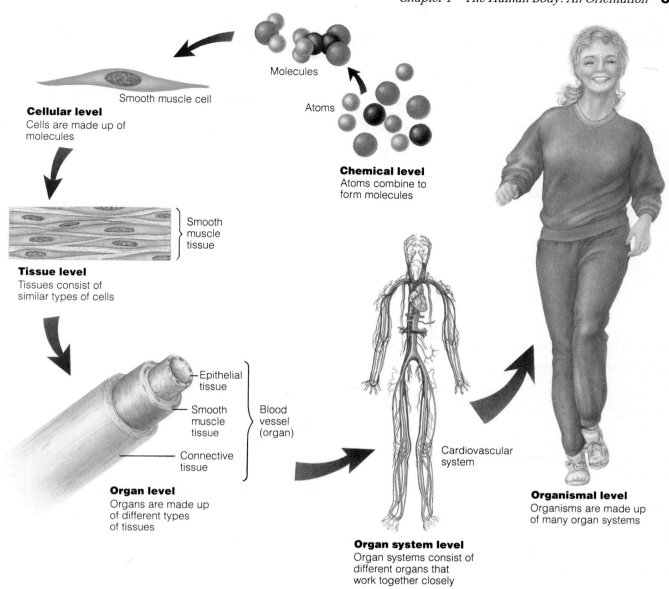

Cellular level
Cells are made up of molecules

Smooth muscle cell

Molecules

Atoms

Chemical level
Atoms combine to form molecules

Smooth muscle tissue

Tissue level
Tissues consist of similar types of cells

Epithelial tissue

Smooth muscle tissue

Connective tissue

Blood vessel (organ)

Organ level
Organs are made up of different types of tissues

Cardiovascular system

Organ system level
Organ systems consist of different organs that work together closely

Organismal level
Organisms are made up of many organ systems

Figure 1.1   Levels of structural complexity.   In this diagram, components of the cardiovascular system illustrate the various levels of complexity in the human body, from the chemical to the organismal.

Other Branches of Anatomy

Developmental anatomy traces the structural changes that occur in the body throughout the life span. **Embryology** concerns developmental changes that occur before birth. A knowledge of embryological anatomy provides a rational explanation for the complex design of the adult human body and helps to explain birth defects, which are deviations from the normal developmental pathway. Developmental anatomy also includes the effects of aging.

Some specialized branches of anatomy are used primarily for medical diagnosis and scientific research. **Pathological** (pah-tho-loj'i-kal) **anatomy** deals with the structural changes in cells, tissues, and organs caused by disease (**pathology** is the study of disease). **Radiographic** (ra"de-o-graf'ic) **anatomy** is the study of internal body structures by means of X rays and other forms of radiation (see pp. 17–20). **Functional morphology** emphasizes the functional properties of body structures and assesses the efficiency of their design.

The Hierarchy of Structural Organization

The human body has many levels of structural complexity (Figures 1.1 and 1.2). At the **chemical level**, *atoms,* the tiny building blocks of matter, combine to form small *molecules,* such as carbon dioxide (CO_2) and water (H_2O), and larger *macromolecules* (*macro* = big). There are four classes of macromolecules in

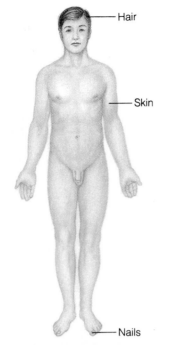

○ **(a) Integumentary system**

Forms the external body covering; protects deeper tissues from injury; synthesizes vitamin D; site of cutaneous (pain, pressure, etc.) receptors, and sweat and oil glands.

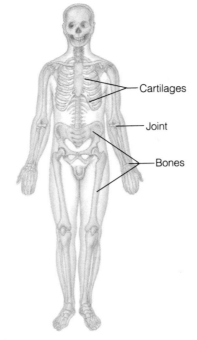

○ **(b) Skeletal system**

Protects and supports body organs; provides a framework the muscles use to cause movement; blood cells are formed within bones; stores minerals.

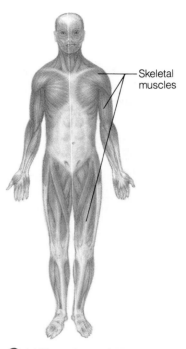

○ **(c) Muscular system**

Allows manipulation of the environment, locomotion, and facial expression; maintains posture; produces heat.

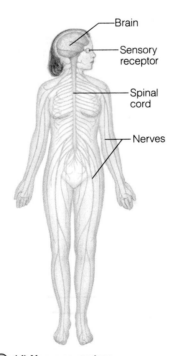

○ **(d) Nervous system**

Fast-acting control system of the body; responds to internal and external changes by activating appropriate muscles and glands.

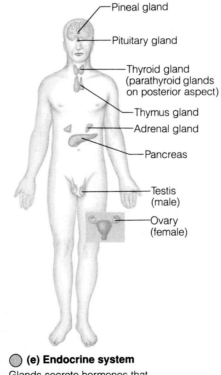

○ **(e) Endocrine system**

Glands secrete hormones that regulate processes such as growth, reproduction, and nutrient use (metabolism) by body cells.

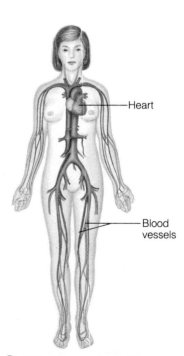

○ **(f) Cardiovascular system**

Blood vessels transport blood which carries oxygen, carbon dioxide, nutrients, wastes, etc.; the heart pumps blood.

Figure 1.2   Summary of the body's organ systems.   The structural components of each organ system are illustrated, with the main functions of the system described below each illustration. Note that the blood vessels (cardiovascular system) and the lymph vessels (lymphatic system) can be grouped together as the circulatory system. The immune system is sometimes considered a functional system rather than a true organ system.

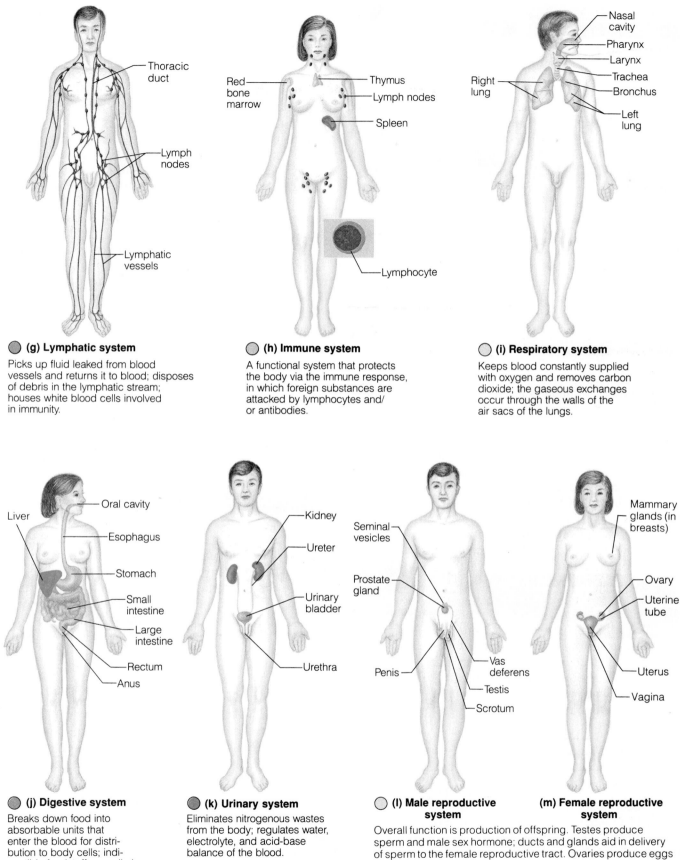

(g) Lymphatic system

Picks up fluid leaked from blood vessels and returns it to blood; disposes of debris in the lymphatic stream; houses white blood cells involved in immunity.

Labels: Thoracic duct, Lymph nodes, Lymphatic vessels

(h) Immune system

A functional system that protects the body via the immune response, in which foreign substances are attacked by lymphocytes and/or antibodies.

Labels: Red bone marrow, Thymus, Lymph nodes, Spleen, Lymphocyte

(i) Respiratory system

Keeps blood constantly supplied with oxygen and removes carbon dioxide; the gaseous exchanges occur through the walls of the air sacs of the lungs.

Labels: Nasal cavity, Pharynx, Larynx, Trachea, Bronchus, Left lung, Right lung

(j) Digestive system

Breaks down food into absorbable units that enter the blood for distribution to body cells; indigestible foodstuffs are eliminated as feces.

Labels: Oral cavity, Liver, Esophagus, Stomach, Small intestine, Large intestine, Rectum, Anus

(k) Urinary system

Eliminates nitrogenous wastes from the body; regulates water, electrolyte, and acid-base balance of the blood.

Labels: Kidney, Ureter, Urinary bladder, Urethra

(l) Male reproductive system

Labels: Seminal vesicles, Prostate gland, Penis, Vas deferens, Testis, Scrotum

(m) Female reproductive system

Overall function is production of offspring. Testes produce sperm and male sex hormone; ducts and glands aid in delivery of sperm to the female reproductive tract. Ovaries produce eggs and female sex hormones; remaining structures serve as sites for fertilization and development of the fetus. Mammary glands of female breasts produce milk to nourish the newborn.

Labels: Mammary glands (in breasts), Ovary, Uterine tube, Uterus, Vagina

the body: carbohydrates (sugars), lipids (fats), proteins, and nucleic acids (DNA, RNA). These form the building blocks of the structures at the **cellular level**: the *cells* and their functional subunits, called *cellular organelles.* Cells are the smallest living things in our body.

The next level is the **tissue level**. Tissues are groups of cells that together perform a common function. Only four tissue types together make up all organs of the human body: epithelial tissue (epithelium), connective tissue, muscle tissue, and nervous tissue. Each tissue plays a characteristic role in the body (see Chapter 4). Briefly, epithelium (ep″ĭ-the′le-um) covers the body surface and lines its cavities; connective tissue supports the body and protects its organs; muscle tissue allows for movement; and nervous tissue provides a means of rapid internal communication by transmitting electrical impulses.

Extremely complex physiological processes occur at the **organ level**. An organ is a discrete structure made up of more than one tissue. In fact, most organs contain all four tissues. The liver, the kidney, and the heart are good examples. You can think of each organ in the body as a highly specialized functional center responsible for an activity that no other organ can perform.

Organs that work closely together to accomplish a common purpose make up an **organ system**, the next level. For example, organs of the respiratory system—the lungs, trachea, bronchi, and others—ensure that air flows in and out of the lungs so that oxygen can enter the blood and carbon dioxide can be removed. Organs of the digestive system—the mouth, esophagus, stomach, intestine, and so forth—break down the food we eat so that the nutrients can be absorbed into the blood. The body's organ systems are the *integumentary* (skin), *skeletal, muscular, nervous, endocrine, circulatory (cardiovascular* plus *lymphatic), immune, respiratory, digestive, urinary,* and *reproductive* systems. Figure 1.2 provides a brief overview of the organ systems and their basic functions.

The highest level of organization is the **organism**, exemplified by a living human being (Figure 1.1). The organismal level represents the sum total of all levels working in unison to promote life.

Scale: What Size Is It?

To describe the dimensions of cells, tissues, and organs, anatomists need a precise system of measurement. The **metric system** provides such precision. Familiarity with this system enables you to understand the sizes, weights, and volumes of body structures. Metric units are described in detail in Appendix A on page 711, and you will find yourself referring to that appendix often as you read through this book.

An important unit of *length* is the **meter** (m), which is a little longer than a yardstick. If you are 6 feet tall, your metric height is 1.83 meters. Most adults are between 1.5 and 2 meters tall. A **centimeter** (cm) is a hundredth of a meter (*cent* = hundred). Many of our organs are several centimeters in height, length, and width. (It helps to remember that a nickel is about 2 cm in diameter.) A **micrometer** (μm) is a millionth of a meter (*micro* = small or millionth). Cells, organelles, and tissues are measured in micrometers. Human cells average about 10 μm in diameter, although they range from 5 μm to 100 μm. The human cell with the largest diameter, the egg cell (ovum), is about the size of the tiniest dot you could make on this page with a pencil.

The metric system also measures *volume* and *weight* (mass). A **liter** (L) is a volume slightly larger than a quart (think of a quart container of milk). A **milliliter** (ml) is one thousandth of a liter (*milli* = thousandth). A **kilogram** is a mass equal to about 2.2 pounds, and a **gram** (g) is a thousandth of a kilogram (*kilo* = thousand).

Can you calculate your own weight in kilograms?

To do this, just divide your weight in pounds by 2.2. For example, if you weigh 150 pounds, then you weigh 68 kg. ■

Anatomical Variability

We all know from looking at the faces and body shapes of the people around us that humans differ in their external anatomy. The same kind of variability holds for our "insides" as well. Thus, not every structural detail described in an anatomy book is true of all people or of all the cadavers (dead bodies) you observe in the anatomy lab. In some bodies, for example, a certain blood vessel may branch off higher than usual, a nerve or vessel may be somewhat "out of place," or a small muscle may be missing. Such variability does not cause confusion, however, since well over 90% of all structures present within any human body match the textbook descriptions. Extreme variations are seldom seen, because excessive variability is incompatible with life. For example, no living person could be missing the blood vessels to the brain.

Can you think of some other easily observed traits that vary noticeably among different people?

Your examples may include eye color, curly or straight hair, presence or absence of ear lobes, presence or absence of wisdom teeth, and the individual nature of fingerprints. ■

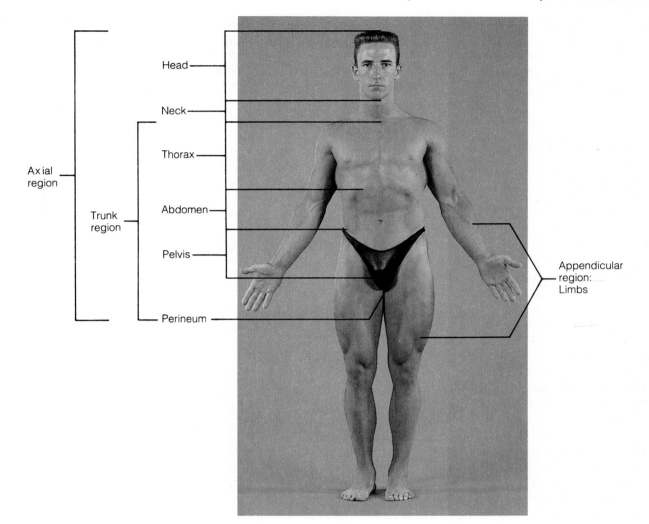

Head

Neck

Thorax

Abdomen

Pelvis

Perineum

Axial region

Trunk region

Appendicular region: Limbs

Figure 1.3   The anatomical position.   The largest divisions of the body are indicated.

Gross Anatomy: An Introduction

The Anatomical Position

To describe the various body parts and their locations, we need an initial reference point. This reference point is the **anatomical position** (Figure 1.3). In the anatomical position, an individual stands erect with feet together and eyes forward. The palms face anteriorly with the thumbs pointed away from the body. It is essential to learn the anatomical position because most of the directional terminology used in anatomy refers to the body in this position. Additionally, the terms *right* and *left* always refer to those sides of the person or cadaver being viewed—not to the right and left sides of the viewer.

Directional and Regional Terms

Medical personnel and anatomists use certain directional terms to explain precisely where one body structure lies in relation to another. For example, we could describe the relationship between the eyebrows and the nose informally by stating, "The eyebrows are located on each side of the face to the right and left of the nose and higher than the nose." In anatomical terminology, this condenses to "The eyebrows are lateral and superior to the nose." Clearly, the anatomical terminology is less wordy and confusing. The standardized terms of direction are defined and illustrated in Table 1.1. Most often used are the paired terms **superior/inferior, anterior (ventral)/posterior (dorsal), medial/lateral,** and **superficial/deep.**

Regional terms are the names of specific body areas. These areas are shown in Figures 1.3 and 1.4.

Table 1.1 Orientation and Directional Terms

Term	Definition	Illustration	Example
Superior (cranial or cephalad)	Toward the head end or upper part of a structure or the body; above		The forehead is superior to the nose
Inferior (caudal)	Away from the head end or toward the lower part of a structure or the body; below		The navel is inferior to the breastbone
Anterior (ventral)*	Toward or at the front of the body; in front of		The breastbone is anterior to the spine
Posterior (dorsal)*	Toward or at the back of the body; behind		The heart is posterior to the breastbone
Medial	Toward or at the midline of the body; on the inner side of		The heart is medial to the arms
Lateral	Away from the midline of the body; on the outer side of		The eye is lateral to the bridge of the nose
Intermediate	Between a more medial and a more lateral structure		The collarbone is intermediate between the breastbone and shoulder
Proximal	Closer to the origin of the body part or the point of attachment of a limb to the body trunk		The elbow is proximal to the wrist
Distal	Farther from the origin of a body part or the point of attachment of a limb to the body trunk		The knee is distal to the thigh
Superficial	Toward or at the body surface		The skin is superficial to the skeleton
Deep	Away from the body surface; more internal		The lungs are deep to the skin

*Whereas the terms *ventral* and *anterior* are synonymous in humans, this is not the case in four-legged animals. *Ventral* specifically refers to the "belly" of a vertebrate animal and thus is the inferior surface of four-legged animals. Likewise, although the dorsal and posterior surfaces are the same in humans, the term *dorsal* specifically refers to an animal's back. Thus, the dorsal surface of four-legged animals is their superior surface. Additionally, the superior/inferior term pair mean toward and away from the head end respectively in humans, but the term pair anterior/posterior conveys this meaning in four-legged animals.

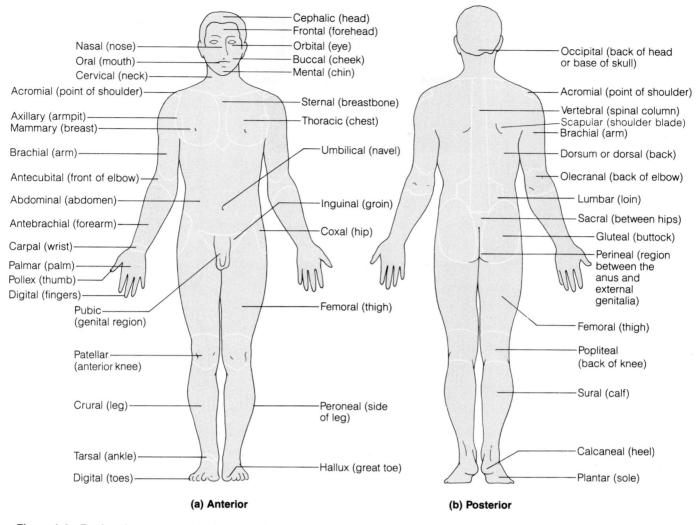

Figure 1.4   Regional terms: names of specific body areas.

The most fundamental divisions of the body (Figure 1.3) are the *axial* and *appendicular* (ap″en-dik′u-lar) *regions.* The **axial region**, so named because it makes up the main axis of the body, consists of the *head, neck,* and *trunk.* The trunk, in turn, is divided into the *thorax* (chest), *abdomen,* and *pelvis;* the trunk also includes the region around the anus and external genitals, called the *perineum* (per″ĭ-ne′um; "around the anus"). The **appendicular region** of the body consists of the limbs (*appendage* can mean limb), also called the extremities.

The fundamental divisions of the body are further divided into smaller regions. See Figure 1.4.

As you study Figure 1.4, can you determine precisely where you would be injured if you pulled a muscle in your *axillary* region, cracked a bone in your *occipital* region, or received a cut on a *digit*?

These injuries would be to the armpit, base of the skull (posteriorly), and a finger or toe, respectively. ■

Body Planes and Sections

In the study of anatomy, the body is often *sectioned* (cut) along a flat surface called a plane. The most frequently used body planes are sagittal, frontal, and transverse planes, which lie at right angles to one another (Figure 1.5). A section bears the name of the plane along which it is cut. Thus, a cut along a sagittal plane produces a sagittal section.

A **sagittal plane** (saj′ĭ-tal; "arrow") lies vertically and divides the body into right and left parts. The specific sagittal plane that lies exactly in the midline is the **median plane**, or **midsagittal plane** (Figure 1.5a). All other sagittal planes, offset from the midline, are **parasagittal planes** (*para* = near).

Frontal planes, like sagittal planes, lie vertically. Frontal planes, however, divide the body into anterior and posterior parts (Figure 1.5b). A frontal plane is also called a **coronal plane** (kŏ-ro′nal; "crown").

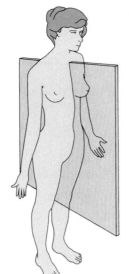

(a) Median sagittal plane

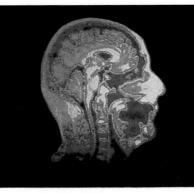

Median sagittal
section through head

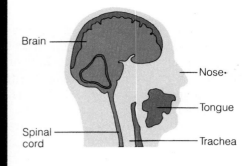

Brain

Nose

Tongue

Spinal
cord

Trachea

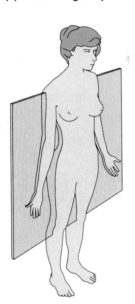

(b) Frontal (coronal) plane

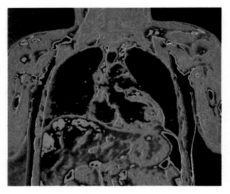

Frontal section through torso

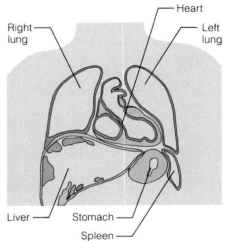

Heart

Left
lung

Right
lung

Liver

Stomach

Spleen

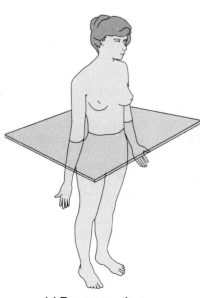

(c) Transverse plane

Posterior

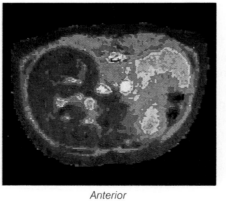

Anterior

Transverse section through torso
(Superior view)

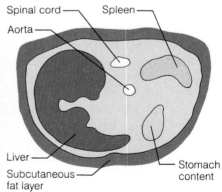

Spinal cord

Spleen

Aorta

Liver

Subcutaneous
fat layer

Stomach
content

Figure 1.5 Planes of the body. The left side of the figure illustrates the three major types of planes (sagittal, frontal, and transverse) relative to the anatomical position. In the center are sections through each type of plane, as visualized by the medical imaging technique called magnetic resonance imaging (MRI). At right, the organs in the images are identified in simplified diagrams.

A **transverse**, or **horizontal, plane** runs horizontally from right to left, dividing the body into superior and inferior parts (Figure 1.5c). There are, of course, many different transverse planes existing at every possible level from head to foot. A transverse section is also called a **cross section**.

Cuts made along any plane that lies diagonally between the horizontal and the vertical are called **oblique sections**. Such sections are often confusing and difficult to interpret and are not used very much.

The ability to interpret sections through the body, especially transverse sections, is increasingly important in the clinical sciences. New medical imaging devices (described on pp. 17–20) produce sectional images rather than three-dimensional images. It can be difficult to decipher an object's overall shape from sectioned material. A cross section of a banana, for example, looks like a circle and gives no indication of the whole banana's crescent shape. It will take practice, but you will gradually learn to relate two-dimensional sections to three-dimensional shapes.

The Human Body Plan

Humans belong to the group of animals called *vertebrates*. This group also includes cats, rats, birds, lizards, frogs, and fish. All vertebrates share the following basic features (Figure 1.6):

1. Tube-within-a-tube body plan. Our digestive organs form a tube that runs through the axial region of the body, which can be viewed as forming a larger tube.

2. Bilateral symmetry. The left half of our body is essentially a mirror image of the right half. Most body structures occur in pairs (the right and left hands are examples). Structures in the median plane are unpaired, but they tend to have identical right and left sides (the nose is an example).

3. Dorsal hollow nerve cord. All vertebrate embryos have a hollow nerve cord running through the dorsal

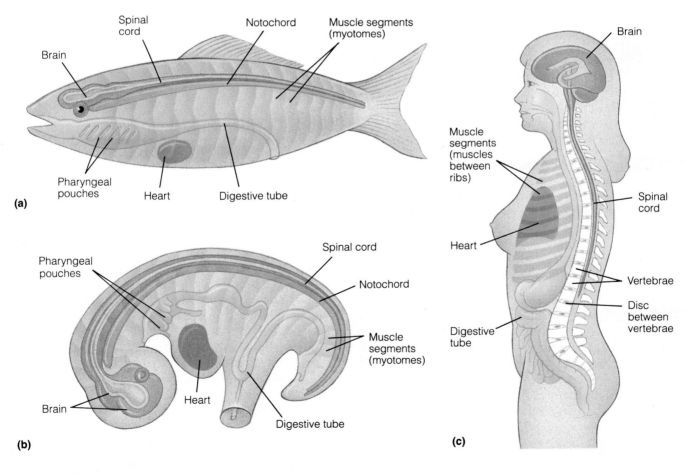

Figure 1.6   Basic human body plan, indicated by structures shared with all vertebrates.   The bodies are shown as semi-transparent, in order to reveal the internal organs. **(a)** Generalized vertebrate, as represented by a primitive fish (highly simplified). **(b)** Human embryo 5 weeks after conception; note the features shared with the fish. **(c)** Adult human.

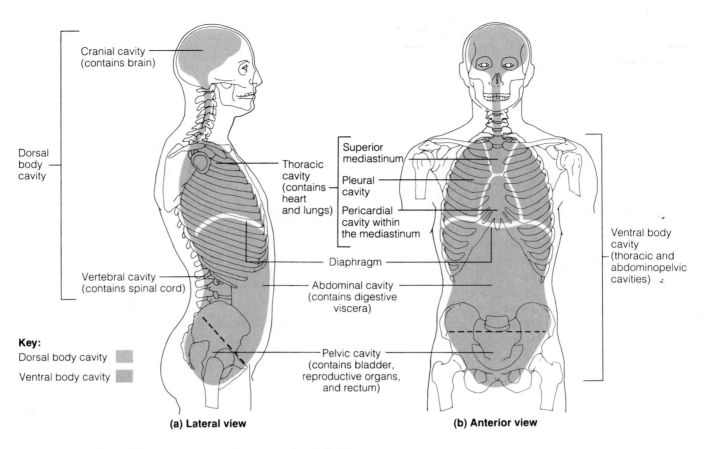

Cranial cavity
(contains brain)

Dorsal
body
cavity

Thoracic
cavity
(contains
heart
and lungs)

Superior
mediastinum

Pleural
cavity

Pericardial
cavity within
the mediastinum

Ventral body
cavity
(thoracic and
abdominopelvic
cavities)

Diaphragm

Vertebral cavity
(contains spinal cord)

Abdominal cavity
(contains digestive
viscera)

Key:
Dorsal body cavity
Ventral body cavity

Pelvic cavity
(contains bladder,
reproductive organs,
and rectum)

(a) Lateral view

(b) Anterior view

Figure 1.7   Dorsal and ventral body cavities and their subdivisions.

region of the body in the median plane. In adults, this develops into the brain and spinal cord.

4. Notochord and vertebrae. The **notochord** (no'to-kord; "back string") is a stiffening rod in the back just deep to the spinal cord. In humans, a complete notochord forms in the embryo, though most of it is quickly replaced by the development of **vertebrae** (ver'tĕ-bre), the bony pieces of the vertebral column, or backbone. Still, some of the notochord persists throughout life as the cores of the discs between the vertebrae.

5. Segmentation. The "outer tube" of our body shows evidence of segmentation. Segments are repeating units of similar structure that run from the head along the full length of the trunk. In humans, the ribs and the muscles between the ribs are evidence of segmentation, as are the many nerves branching off our spinal cord. The bony vertebral column, with its repeating vertebrae, is also segmented.

6. Pharyngeal pouches. Pharyngeal (far-rin'je-al) pouches form the clefts between the gills of fish. In humans, the embryonic pharyngeal pouches give rise to various structures in the head and neck, including the middle ear cavity (defined in the next section).

Body Cavities and Membranes

Within the body are two large cavities called the dorsal and ventral cavities (Figure 1.7). These are closed to the outside, and each contains internal organs.

Dorsal Body Cavity

The **dorsal body cavity** is subdivided into a **cranial cavity**, which lies in the skull and encases the brain, and a **vertebral cavity**, which runs through the vertebral column to enclose the spinal cord. Since the spinal cord issues from and is essentially a continuation of the brain, the cranial and spinal cavities are continuous with one another. The vital and fragile nervous organs housed by the dorsal body cavity are protected by the hard, bony walls of the dorsal cavity.

Ventral Body Cavity

The more anterior and larger of the closed body cavities is the **ventral body cavity**. It has two main subdivisions, the *thoracic cavity* and the *abdominopelvic cavity.* The ventral body cavity houses a group of organs that are collectively called the **viscera** (vis'er-ah), or **visceral organs**.

The more superior subdivision, the **thoracic cavity**, is surrounded by the ribs and muscles of the chest. The thoracic cavity contains the lungs, each surrounded by a **pleural cavity** (ploo′ral; "the side, a rib"). The thoracic cavity also contains a thick band of organs between the lungs called the **mediastinum** (me″de-ah-sti′num; "in the middle"). The mediastinum contains the heart surrounded by a **pericardial cavity** (per″ĭ-kar′de-al; "around the heart"). It also houses other major thoracic organs, such as the esophagus (gullet) and trachea (windpipe).

The thoracic cavity is separated from the more inferior **abdominopelvic** (ab-dom″ĭ-no-pel′vic) **cavity** by the diaphragm, a dome-shaped muscle important in breathing. The abdominopelvic cavity, as its name suggests, has two regions. The superior region, called the **abdominal cavity**, contains the stomach, intestines, liver, and other organs. The inferior region, or **pelvic cavity**, contains the bladder, some reproductive organs, and the rectum. These two regions are continuous, not being separated from each other by any muscular or membranous partition.

Can you deduce which basic division of the trunk—thorax, abdomen, or pelvis—is most likely to sustain damage if subjected to heavy blows or other types of physical trauma?

When the trunk is subjected to physical trauma, as often happens in automobile accidents, the most vulnerable organs are those within the abdominal cavity. This is because the walls of the abdominal cavity are formed only by muscles and are not reinforced by bone. The bony pelvis gives the pelvic organs some protection. The thoracic organs, within the bony rib cage, are the best protected. ✖

Membranes in the Ventral Body Cavity

The inner surface of the walls of the ventral cavity, and the outer surfaces of most organs in this cavity, are covered by a thin membrane called a **serous** (se′rus) **membrane**, or **serosa** (se-ro′sah) (Figure 1.8). The part of the membrane that lines the cavity walls is generally called the **parietal** (pah-ri′ĕ-tal) **serosa** (*paries* = a wall). The parietal serosa is continuous with the **visceral serosa**, which covers the visceral organs in the cavity. You can visualize this relationship by pushing your fist into a limp balloon (Figure 1.8a). The part of the balloon that clings closely to your fist represents the visceral serosa clinging to the organ's outer surface. The outer wall of the balloon represents the parietal serosa that lines the walls of the cavity. The two serous layers are separated not by air but by a thin sheet of lubricating fluid called **serous fluid** (*serous* = watery). This fluid is secreted by both serous membranes. Although there is a potential space between the two membranes, they tend to lie very close to each other.

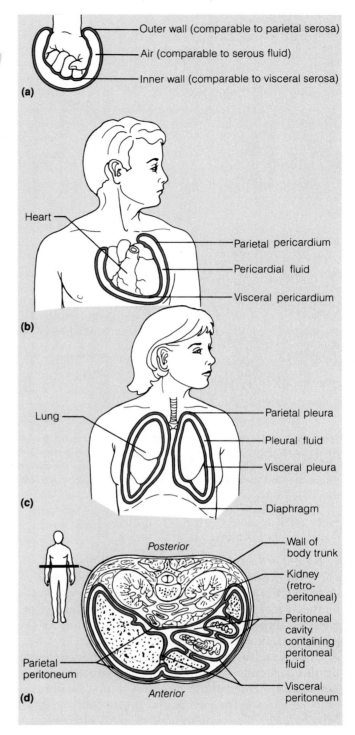

Figure 1.8 The serous membranes of the ventral body cavity. Serous membranes are shown in blue. (**a**) A fist is thrust into a flaccid balloon to show the relationship between the parietal and visceral layers of serous membrane. (**b**) Serous membranes associated with the heart: The parietal layer is the outer layer; the visceral layer clings to the heart. (**c**) Serous membranes around the lungs: The visceral pleura covers each lung; the parietal pleura covers the inner walls of the thoracic cavity. (**d**) Transverse section through the abdomen at the level of the liver and kidneys. The parietal peritoneum covers the inner wall of the abdominopelvic cavity; the visceral peritoneum covers the outer surfaces of most organs in that cavity. However, some abdominal viscera, such as the kidneys, are retroperitoneal (behind the peritoneum).

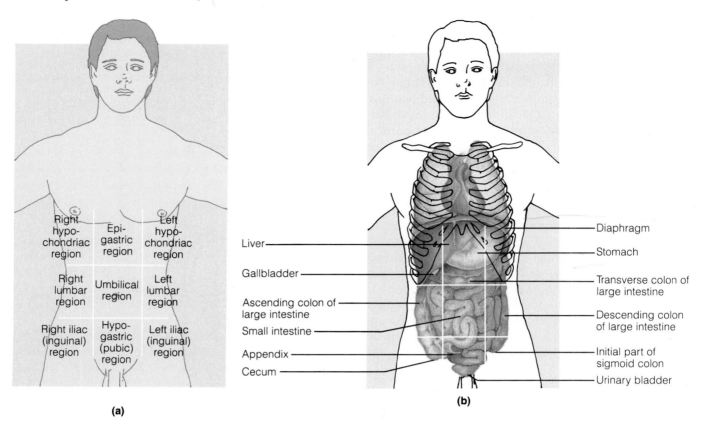

(a)

(b)

Figure 1.9 The nine abdominal regions. (a) Division of the abdomen into nine regions delineated by four planes. The superior horizontal plane is just inferior to the rib cage; the inferior horizontal plane is just superior to the hip bones. The two parasagittal planes lie just medial to each nipple. (b) Anterior view of the abdominopelvic cavity showing many of its visceral organs.

The slippery serous fluid allows the visceral organs to slide without friction across the cavity walls as they carry out their routine functions. This is extremely important for mobile organs, such as the pumping heart and the churning stomach.

The specific name of a serous membrane depends on the particular cavity and organs with which it is associated. For example, the serous **pericardium** is associated with the pericardial cavity and the heart (Figure 1.8b). The parietal and visceral **pleurae** (singular: **pleura**) cover the inner thoracic walls and the lungs (Figure 1.8c). The parietal and visceral **peritonea** (singular: **peritoneum** [per″ĭ-to-ne′um]) are the serous membranes that line the abdominopelvic cavity and cover its organs.

Other Cavities

In addition to the large, closed body cavities, there are several smaller body cavities. Many of these are in the head, and most open to the body exterior:

1. Oral cavity. The oral cavity, commonly called the mouth, contains the tongue and teeth. The oral cavity is continuous with the rest of the digestive tube, which opens to the exterior at the anus.

2. Nasal cavity. Located within and posterior to the nose, the nasal cavity is part of the passages of the respiratory system.

3. Orbital cavities. The orbital cavities (orbits) house the eyes and present them in an anterior position.

4. Middle ear cavities. Each middle ear cavity lies just medial to an ear drum, and is carved into the bone of the skull. These cavities contain tiny bones that transmit sound vibrations to the organ of hearing in the inner ears.

5. Synovial (sĭ-no′ve-al) **cavities.** Synovial cavities are joint cavities: They are enclosed within fibrous capsules that surround the freely movable joints of the body (such as the knee joint and hip joint). Like the serous membranes of the ventral body cavity, mem-

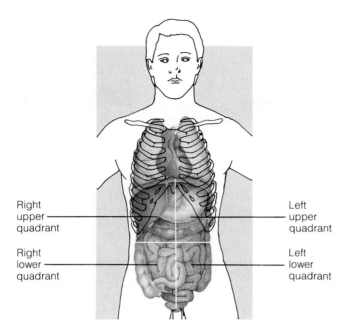

Figure 1.10   The four abdominal quadrants. In this scheme, two planes divide the abdominopelvic cavity into four quarters.

Right upper quadrant

Left upper quadrant

Right lower quadrant

Left lower quadrant

branes lining the synovial cavities secrete a lubricating fluid that reduces friction as the bones move across one another.

Abdominal Regions and Quadrants

Since the abdominopelvic cavity is large and contains many organs, it is helpful to divide it into smaller areas for study. In one method of division, two transverse planes and two parasagittal planes divide the abdomen into nine **regions** (Figure 1.9a): the right and left hypochondriac regions, the epigastric region, the right and left lumbar regions, the umbilical region, the right and left iliac regions, and the hypogastric region. These nine regions are discussed in more detail in Chapter 21 (pp. 563–564). An alternative scheme is to divide the abdomen into just four **quadrants** ("quarters") by drawing one vertical and one horizontal plane through the navel (see Figure 1.10). Finally, you can use Figure 1.9b to become familiar with the locations of some of the major viscera (the stomach, liver, intestines, and so on).

As shown in Figures 1.7b and 1.9b, the most superior part of the abdominal cavity is protected by the rib cage. This protection is especially important for a large visceral organ on the right side of the body, which could bleed profusely if ruptured by a blow. Can you deduce which organ we are referring to?

It is the liver. Similarly, the spleen, another abdominal organ subject to rupture, is protected by the rib cage on the *left* (see Figure 1.5c). ■

Microscopic Anatomy: An Introduction

Light and Electron Microscopy

Microscopy is the examination of small structures by means of a microscope. When microscopes were introduced in the early 1600s, they opened up a tiny new universe whose existence was unsuspected before that time. Two main types of microscopes are now used to investigate the fine structure of organs, tissues, and cells: the **light microscope (LM)** and the **transmission electron microscope (TEM** or just **EM)**. Light microscopy illuminates body tissue with a beam of light, whereas electron microscopy uses a beam of electrons. LM is used for lower magnification viewing, EM for higher magnification (Figure 1.11a and b). Light microscopy can produce sharp images of tissues and cells, but not of the small structures within cells (organelles). Most such structures are too small to alter the path of light waves enough to appear on the image. Since electron beams behave like light waves of much smaller wavelength, EM can produce sharp images at much greater magnification. The fine details of cells and tissues, as viewed by electron microscopy, are said to comprise their **ultrastructure**.

Preparing Human Tissue For Microscopy

Elaborate steps are taken to prepare human or animal tissue for microscopic viewing. The specimen must be **fixed** (preserved), and then cut into **sections** (slices) thin enough to transmit light (or electrons). Finally, the specimen must be **stained** to enhance contrast. The stains used in light microscopy are beautifully colored organic dyes, most of which were originally invented by the textile industry in the mid 1800s. These dyes helped to usher in the golden age of microscopic anatomy from 1860 to 1900. The stains come in almost all colors of the rainbow. Many consist of charged dye molecules (negative or positive molecules) that bind within the tissue to macromolecules of the opposite charge. This electrical attraction is the basis of the staining ability. Stains with negatively charged dye molecules are called **acidic stains**. Positively charged dyes, by contrast, are called **basic stains**. Since different parts of cells and tissues take

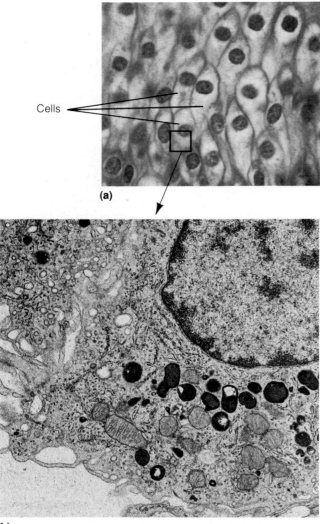

Cells

(a)

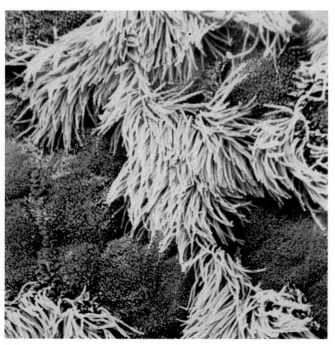

(b)

up different dyes, the stains visually demarcate different anatomical structures.

For transmission electron microscopy, tissue sections are stained with heavy-metal salts. These metals deflect electrons in the beam to different extents, thus providing contrast in the image. Electron microscope images contain only shades of gray (Figure 1.11b), because color is a property of light, not of electron waves.

Scanning Electron Microscopy

The types of microscopy introduced so far are used to view cells and tissue that have been sectioned. Another kind of electron microscopy, **scanning electron microscopy (SEM)**, provides three-dimensional pictures of whole, unsectioned surfaces with striking clarity (Figure 1.11c). First, the specimen is preserved and coated with fine layers of carbon and gold dust. Then, an electron beam scans the specimen, causing other, secondary, electrons to be emitted from its surface. A detector captures these emitted electrons and assembles them into a three-dimensional image on a video screen (cathode ray tube), based on the principle that more electrons are produced by the high points on the specimen surface than by the lower points. Although artificially constructed, the SEM image is accurate and looks very real.

Artifacts

The tissue we see under the microscope has been exposed to many procedures that alter its original condition. Because each preparatory step introduces minor distortions, called **artifacts**, most microscopic structures viewed by anatomists are not exactly like those in living tissue. Furthermore, the human and animal corpses studied in the anatomy laboratory have also been preserved, so their organs have a drabber color and a texture different from that of living organs. Keep these principles in mind as you look at the micrographs (microscope pictures) and the photos of human cadavers in this book.

(c)

Figure 1.11   Cells viewed by three types of microscopy.
(**a**) Light micrograph (cells from the epithelium lining the bladder). (**b**) Transmission electron micrograph (enlarged area of the cell region that is indicated in a box in part a). (**c**) Scanning electron micrograph: surface view of cells lining the trachea, or windpipe. The long, grass-like structures on the surfaces of these cells are cilia, and the tiny knob-like structures are microvilli (see Chapter 4); this scanning electron micrograph is artificially colored.

Clinical Anatomy: An Introduction to Medical Imaging Techniques

Physicians have long sought ways to examine the body's internal organs for evidence of disease without subjecting the patient to the shock and pain of exploratory surgery. Physicians can identify some diseases and injuries by feeling the patient's deep organs through the skin or by using traditional X rays—but these are relatively crude methods. Modern medical science is developing powerful new techniques for viewing the internal anatomy of living people. These imaging techniques not only reveal the structure of our functioning internal organs, but also can extract information about the workings of our molecules. The new techniques all rely on powerful computers to construct images from raw data transmitted by electrical signals.

X Rays

Before considering the new techniques, we must discuss traditional X rays, for these still play the major role in medical diagnosis (Figure 1.12). X rays are best for visualizing bones and locating abnormal dense structures, such as some tumors and tuberculosis nodules in the lungs. Discovered in 1895 and used in medicine ever since, X rays are electromagnetic waves of very short wavelength. When X rays are directed at the body, some are absorbed. The amount of absorption depends on the density of the matter encountered. X rays that pass through the body expose a piece of film behind the patient. The resulting radiograph is a negative: The darker, exposed areas on the film represent soft organs that are easily penetrated, whereas light, unexposed areas correspond to denser structures, like bones, that absorb most X rays.

Exposure to high or prolonged levels of X radiation can damage tissue, cause hereditary disorders, or predispose one to cancer. Therefore, modern X-ray techniques attempt to minimize the exposure of the patient to radiation.

In a variation of radiography, X-ray images are viewed on a fluorescent screen, or **fluoroscope**, as they are being generated. Although not as sharp as the film image, the fluoroscope image enables us to view internal organs as they move. Alternatively, movements can be recorded with X-ray cinema film, a technique known as **cineradiography**.

Conventional X rays have several limitations that have prompted clinicians to seek more ad-

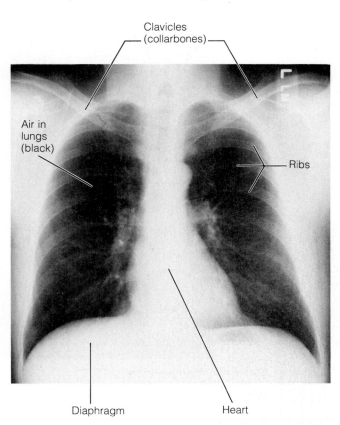

Clavicles (collarbones)

Air in lungs (black)

Ribs

Diaphragm   Heart

Figure 1.12   An X-ray image (radiograph) of the chest.

vanced imaging techniques. Looking at Figure 1.12, can you deduce some of the problems with standard X-ray images?

Conventional radiography has two main limitations. First, it produces blurry images of soft tissues. For hollow organs, we can partly overcome this problem by filling them with a liquid called **contrast medium**, which contains heavy elements like barium or iodine. Second, conventional radiographs flatten the three-dimensional structure of the body onto a two-dimensional image. Therefore, organs appear to overlap, and denser organs obscure less dense structures behind them. To avoid the confusion of such superimposition, most of the new imaging techniques produce pictures of body *sections,* in which organs do not overlap. ■

CT and DSA: Advanced X-Ray Techniques

The best known of the new imaging techniques is a refined X-ray technology called **computed tomography (CT)**, or **computed axial tomography (CAT)** (Figure 1.13). A CT scanner is shaped like a square metal nut (as in "nuts and bolts") standing on its side. The

Anterior

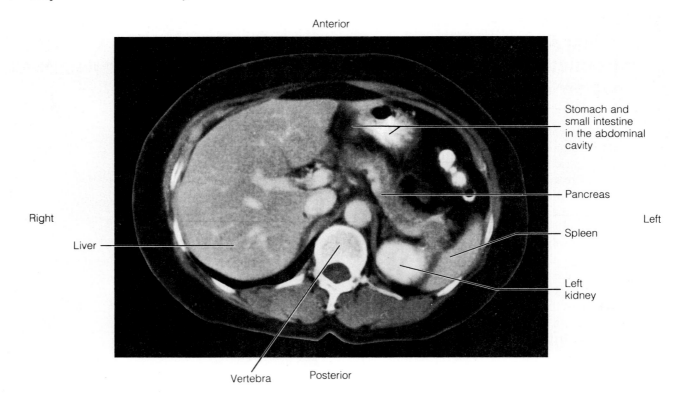

Stomach and small intestine in the abdominal cavity

Pancreas

Right

Left

Liver

Spleen

Left kidney

Vertebra   Posterior

Figure 1.13   Computed tomography (CT).   CT scan through the superior abdomen. CT sections are conventionally oriented as if viewed from an inferior direction, with the posterior surface of the body directed toward the inferior part of the picture; therefore, the patient's right side is at the left side of the picture.

patient lies in the central hole, situated between an X-ray tube and a recorder, both of which are in the scanner. The tube and recorder rotate to take about 12 successive X rays around the person's full circumference. The fan-shaped X-ray beam is thin, so all pictures are confined to a single transverse body plane about 0.3 cm thick. Information is obtained from all possible angles so that every organ is recorded from its best angle, with the fewest structures blocking it. The computer translates all the recorded information into a detailed picture of the body section, which it displays on a viewing screen. Soft structures are represented more clearly than in conventional radiographs, because computerized image-enhancement techniques are used.

The most sophisticated CT systems can use information from many transverse sections, taken one after another, to assemble accurate three-dimensional images of entire organs. For example, a patient's diseased hip joint can be imaged in detail, so that a replacement joint of correct dimensions can be manufactured prior to surgery. Other advanced CT systems can record moving organs, like the heart, and reconstruct these as three-dimensional moving images.

Another computer-assisted X-ray technique is **digital subtraction angiography (DSA)** (*angiography*

= "vessel pictures"). This technique provides an unobstructed view of diseased blood vessels. Its principle is simple: Conventional radiographs are taken before and after a contrast medium is injected into an artery. Then the computer subtracts the "before" image from the "after" image, eliminating all traces of body structures that obscure the vessel. DSA is often used to identify blockage of the arteries that supply the heart wall and the brain.

PET

Positron emission tomography (PET) (Figure 1.14) is an advanced procedure for following radioactive isotopes that are injected into the body. The special advantage of PET is that its images contain messages about the chemical processes of life. For example, radioactively tagged sugar or water molecules are injected into the bloodstream and followed to the body areas that take them up most avidly. This procedure locates our body's most active cells and pinpoints the body regions that receive the greatest supply of blood. As the isotopes decay, they emit particles called positrons, which indirectly lead to the production of gamma rays. Sensors within the doughnut-shaped scanner pick up the emitted gamma rays, which are

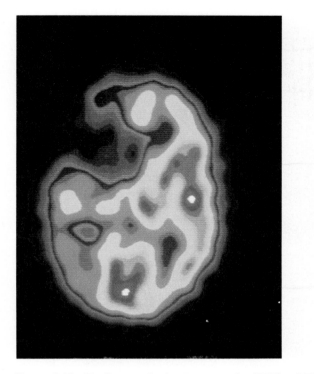

Figure 1.14 Positron emission tomography (PET). A PET scan of the brain, in transverse section, shows an area with no neural activity (the frontal region of the brain at upper left). This area was destroyed by a stroke.

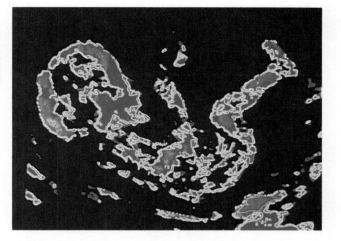

Figure 1.15 Ultrasound image of a fetus in the uterus.

translated into electrical impulses and sent to the computer. A picture of the isotope's location is then constructed on the viewing screen in vivid colors (Figure 1.14).

The most dramatic use of PET scans has been to map sugar consumption in the brain, revealing minimally active brain areas damaged by a stroke or Alzheimer's disease. Furthermore, by determining which areas of the normal healthy brain are most active during certain tasks (such as speech, seeing, comprehension), we are finally obtaining direct evidence for the functions of specific brain regions.

Sonography

Sonography, or **ultrasound imaging** (Figure 1.15), has some distinct advantages over the techniques explained so far. First, the equipment is inexpensive. Second, ultrasound seems to be safer than ionizing forms of radiation, with fewer harmful effects on living tissues. The body is probed with pulses of sound waves, which, when reflected and scattered to different extents by different body tissues, cause echoes. A computer analyzes these echoes to construct sectional images of the outlines of organs. A single hand-held device is used both to emit the sound and to pick up the echoes. This device is easy to move around the body, so sections can be scanned from many different

body planes. Because of its safety, ultrasound is the imaging technique of choice for determining the age and health of a developing fetus. Sonography is of little value for viewing air-filled structures (lungs) or structures surrounded by bone (brain and spinal cord) because sound waves do not penetrate hard objects well and rapidly dissipate in air. Ultrasound images are also somewhat blurry, although their sharpness is being improved.

MRI

Magnetic resonance imaging (MRI) is a technique with tremendous appeal because it produces high-contrast images of our *soft* tissues (Figure 1.16), an area in which X-ray imaging is weak. In its present form, MRI primarily maps the element hydrogen in our body, most of which is in water. Thus, MRI tends to distinguish body tissues from one another on the basis of differences in water content. For example, MRI can distinguish the fatty white matter from the more watery gray matter of the brain. Bones, which contain less water than other tissues, do not show up at all, so MRI peers easily through the skull to reveal the brain. Many tumors show up distinctly, and MRI has even revealed tumors that had been missed by direct observation during exploratory surgery!

The technique subjects the patient to magnetic fields up to 60,000 times stronger than that of the earth. The patient lies within a chamber, with his or her body surrounded by a huge magnet. When the magnet is turned on, the nuclei of the body's hydrogen atoms—single protons that spin like tops—line up parallel to the strong magnetic field. The patient is then exposed to radio waves, which knock the spinning protons out of alignment. When the radio waves are turned off, the protons return to their alignment in

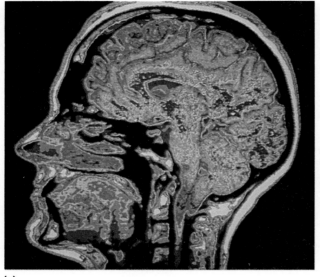

(a)

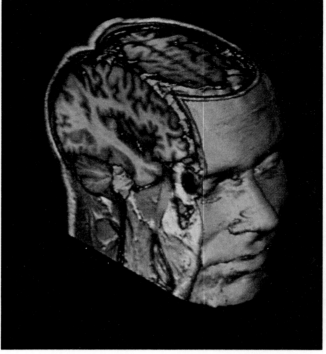

(b)

Figure 1.16   Magnetic resonance image (MRI) of a sagittal section of the head. **(a)** The MRI technique produces great clarity and contrast, in (artificial) colors. **(b)** Volume rendering of an MRI of the head. The flat surfaces show the original MRI data.

the magnetic field, emitting their own faint radio waves in the process. Sensors detect these waves, and the computer translates them into images.

A new variation of this technique, called **magnetic resonance spectroscopy (MRS)**, maps the body distribution of elements other than hydrogen to reveal more about how disease changes body chemistry. Furthermore, computer scientists are now using advanced volume rendering techniques to display MRI scans in three dimensions (Figure 1.16b).

Despite the advantages of MRI, its powerful magnets present problems. They can "suck" metal objects, like implanted pacemakers and loose tooth fillings, through and from the body. Moreover, there is no convincing evidence that such strong magnetic fields can be used without risk to the body, although the procedure is currently considered safe.

Although they are quite stunning, the pictures produced by the new imaging devices are abstractions assembled within the "mind" of a computer. They are artificially enhanced for sharpness and artificially colored to increase contrast (all colors are "phoney"). The images are not inaccurate; they are, however, several steps removed from direct visual observation.

In conclusion, medical science has some remarkable new diagnostic tools, and there is little question that they have made their mark. CT and PET scans now account for about 25% of imaging. Ultrasonography, because of its safety and low cost, has become the most widespread of the new techniques. The use of MRI scans is increasing rapidly. Even so, conventional X rays remain the workhorse of diagnostic imaging techniques and still account for over half of all imaging currently done.

Chapter Summary

AN OVERVIEW OF ANATOMY (pp. 2–6)

1. Anatomy is the study of body structure. In this book, structures are considered in terms of their function.

Anatomical Terminology (p. 2)

2. Because most structures in the body have formal Greek and Latin names, learning anatomy is similar to learning a new language.

Subdivisions of Anatomy (pp. 2–3)

3. Subdivisions of anatomy include gross anatomy, microscopic anatomy (histology), and developmental anatomy.

The Hierarchy of Structural Organization (pp. 3–6)

4. The levels of structural organization of the body, from simplest to most complex, are chemical, cellular, tissue, organ, organ system, and organismal.

5. The organ systems in the body are the integumentary (skin), skeletal, muscular, nervous, endocrine, circulatory, immune, respiratory, digestive, urinary, and reproductive systems.

Scale: What Size Is It? (p. 6)

6. Important units of length measurement are meters (m) for the organism, centimeters (cm) for the organs, and micrometers (μm) for cells.

Anatomical Variability (p. 6)

7. Individual bodies vary, so anatomical structures do not always match textbook descriptions.

GROSS ANATOMY: AN INTRODUCTION (pp. 7–15)

The Anatomical Position (p. 7)

1. In the adult anatomical position, the body stands erect, facing forward. The arms are at the sides, with the palms forward.

Directional and Regional Terms (pp. 7–9)

2. Directional terms allow anatomists to describe the location of body structures with precision. Important terms include superior/inferior; anterior/posterior; ventral/dorsal; medial/lateral; proximal/distal; and superficial/deep (see Table 1.1).

3. Regional terms are used to designate specific areas of the body (see Figures 1.3 and 1.4).

Body Planes and Sections (pp. 9–11)

4. The body or its organs may be cut along planes to produce different types of sections. Frequently used are sagittal, frontal, and transverse planes.

The Human Body Plan (pp. 11–12)

5. The basic structures we share with all other vertebrate animals are the tube-within-a-tube body plan, bilateral symmetry, a dorsal hollow nerve cord, notochord and vertebrae, segmentation, and pharyngeal pouches.

Body Cavities and Membranes (pp. 12–15)

6. The body contains two major closed cavities: the dorsal cavity, subdivided into the cranial and vertebral cavities; and the ventral body cavity, subdivided into the thoracic cavity and the abdominopelvic cavity.

7. The walls of the ventral body cavity and the surfaces of the visceral organs it contains are covered with thin membranes, the parietal and visceral serosae, respectively. The serosae produce a thin layer of lubricating fluid that decreases friction between moving organs.

8. There are several smaller body cavities: oral, nasal, orbital, middle ear, and synovial.

Abdominal Regions and Quadrants (p. 15)

9. To map the visceral organs in the abdominopelvic cavity, physicians divide the abdomen into nine regions or four quadrants.

MICROSCOPIC ANATOMY: AN INTRODUCTION (pp. 15–16)

Light and Electron Microscopy (p. 15)

1. To illuminate cells and tissues, the light microscope (LM) uses light beams and the transmission electron microscope (TEM or EM) uses electron beams. EM produces sharper images than LM at higher magnification.

Preparing Human Tissues for Microscopy (pp. 15–16)

2. The preparation of tissues for microscopy involves preservation (fixation), sectioning, and staining. Stains for LM are colored dyes, while stains for TEM are heavy-metal salts.

Scanning Electron Microscopy (p. 16)

3. Scanning electron microscopy (SEM) provides sharp, three-dimensional images at high magnification.

Artifacts (p. 16)

4. All steps in the preparation of tissue for microscopy produce minor distortions called artifacts.

CLINICAL ANATOMY: AN INTRODUCTION TO MEDICAL IMAGING TECHNIQUES (pp. 17–20)

1. In conventional radiographs, X rays are used to produce negative images of internal body structures. Denser structures in the body appear lighter (whiter) on the X-ray film.

2. CT produces improved X-ray images that are computer enhanced for clarity, and are taken in cross section to avoid overlapping images of adjacent organs. DSA produces sharp X-ray images of blood vessels injected with a contrast medium.

3. PET scans track radioisotopes in the body, locating areas of high energy consumption and high blood flow.

4. Ultrasonography provides sonar images of developing fetuses and internal body structures.

5. MRI subjects the body to strong magnetic fields and radio waves, producing high-contrast images of soft body structures.

Review Questions

Multiple Choice and Matching Questions

1. The correct sequence of levels forming the body's structural hierarchy is (a) organ, organ system, cellular, chemical, tissue; (b) chemical, tissue, cellular, organismal, organ, organ system; (c) chemical, cellular, tissue, organ, organ system, organismal; (d) organismal, organ system, organ, chemical, cellular, tissue.

2. Using what you learned about scale in this chapter, match each structure at left with the letter indicating its correct diameter or height, at right.

__d__	**(1)** white blood cell	**(a)**	2 m
__b__	**(2)** stomach	**(b)**	25 cm
__a__	**(3)** professional basketball player	**(c)**	1 μm
		(d)	10 μm

3. For each part (a–e), circle the structure or organ that matches the given directional term.

(a) distal: the elbow/the wrist
(b) lateral: the hip bone/the navel
(c) superior: the nose/the chin
(d) anterior: the toes/the heel
(e) superficial: the scalp/the skull

4. Which of these organs would not be cut by a section through the midsagittal plane of the body? (Hint: See Figure 1.9.) (a) urinary bladder, (b) gallbladder, (c) small intestine, (d) heart.

5. Relate each of the following conditions or statements to either the dorsal body cavity (D) or the ventral body cavity (V).

(a) surrounded by the skull and the vertebral column
(b) includes the thoracic and abdominopelvic cavities
(c) contains the brain and spinal cord
(d) located more anteriorly
(e) contains the heart, lungs, and many digestive organs

6. Radiographs. List the following structures, from darkest (black) to lightest (white), as they would appear on an X-ray film. Number the darkest one 1, the next darkest 2, and so on.

__2__ **(a)** soft tissue
__3__ **(b)** femur (bone of the thigh)
__1__ **(c)** air in lungs
__4__ **(d)** gold (metal) filling in a tooth

7. A large organ in the right upper quadrant of the abdomen is the (a) stomach, (b) liver, (c) descending colon, (d) heart.

8. The ventral surface of the body is the same as its _____ surface. (a) medial, (b) superior, (c) superficial, (d) anterior, (e) distal.

9. Two of the fingers on Sally's left hand are fused together by a web of skin. Twenty-year-old Joey has a small streak of white hair on the top of his head, although the rest of his hair is black. Both Sally and Joey exhibit (a) rare but minor diseases, (b) serious birth defects, (c) examples of anatomical variation.

10. Which of the following relationships is incorrect? (a) visceral peritoneum/outer surface of small intestine, (b) parietal part of pericardium/outer surface of heart, (c) parietal pleura/wall of thoracic cavity.

11. Which microscopic technique provides sharp pictures of three-dimensional structures at high magnification? (a) light microscopy, (b) X-ray microscopy, (c) scanning electron microscopy, (d) transmission electron microscopy.

12. Histology is the same as (a) pathological anatomy, (b) ultrastructure, (c) functional morphology, (d) surface anatomy, (e) microscopic anatomy.

Short Answer and Essay Questions

13. Construct a table that lists the organ systems of the body. Name two organs belonging to each system, and describe the overall or major function of each system.

14. Describe the anatomical position, and then assume this position.

15. Describe two ways you could distinguish a transmission electron micrograph from a light micrograph.

16. (a) Define bilateral symmetry. (b) Give an example of segmentation in the human body. (c) Define dissection.

17. Insects have a ventral solid nerve cord. In what two ways does this nerve cord differ from the human spinal cord?

18. The following advanced imaging techniques are discussed in the text: CT, DSA, PET, sonography, and MRI. (a) Which of these techniques uses X rays? (b) Which uses radio waves and

magnetic fields? (c) Which uses radioisotopes? (d) Which displays body regions in sections? You may have more than one answer to each question.

19. Give the formal regional term for each of these body areas: (a) arm, (b) chest, (c) groin, (d) armpit.

20. Felix dreaded his first day of anatomy lab because he knew he would have to look at a cadaver. When he saw the cadaver, however, he decided it was not so bad because it did not look quite real. How does a preserved cadaver differ from a living body in appearance?

Critical Thinking and Clinical Application Questions

21. At the clinic, Harry was told that blood would be drawn from his antecubital region. What body part was Harry asked to hold out? Later, the nurse came in and gave Harry a shot of penicillin in his gluteal region. Did Harry take off his shirt or lower his pants to receive the injection? Before Harry left, the nurse noticed that he had a nasty bruise on his cervical region. What part of his body was bruised?

22. Rebecca is 3 months pregnant, and her physician wants to measure the size of the head of the fetus in her womb to make sure it is developing normally. Which medical imaging technique will the physician probably use?

23. Dominic is behaving abnormally, and his doctors strongly suspect he has a tumor in his brain. Which of the following medical imaging devices—conventional X ray, DSA, PET, sonography, or MRI—would probably be best for precisely locating the tumor within Dominic's brain?

24. The Nguyen family was traveling in their van and had a minor accident. The children in the back seat were wearing lap belts, but they still sustained bruises around the abdomen and had some internal injuries. Why is the abdominal area more vulnerable to damage than others?

25. Felicity was staining some tissue sections for light microscopy and spilled the colored stain on her lab coat. Is this stain likely to wash out easily? Explain your answer.

2

Cells: The Living Units

Student Objectives

Introduction to Cells (p. 24)

1. Define a cell, its basic activities, and its three major regions.

The Plasma Membrane (pp. 24–28)

2. Describe the composition and basic functions of the plasma membrane.

3. Explain how various molecules move across the plasma membrane.

The Cytoplasm (pp. 28–35)

4. Describe the cytosol (cytoplasmic matrix).

5. Discuss mitochondria.

6. Compare the structure and function of ribosomes and endoplasmic reticulum.

7. Describe the Golgi apparatus, and explain how it relates to the rough endoplasmic reticulum.

8. Compare lysosomes with peroxisomes.

9. Describe the cytoskeleton.

10. Describe centrioles, and explain how they relate to microtubules.

11. Explain the structure of glycogen granules and lipid droplets.

The Nucleus (pp. 35–39)

12. Outline the structure and function of the nuclear envelope, chromatin, and the nucleolus.

The Cell Life Cycle (pp. 39–42)

13. List the phases of the cell life cycle, including mitosis, and describe a key event of each phase.

Developmental Aspects of Cells (pp. 43–44)

14. Present some theories of cell differentiation and aging.

All living organisms are cellular in nature, from one-celled "generalists" like amoebas to complex multicellular organisms such as trees, dogs, and humans. Just as bricks and timbers are the structural units of a house, cells are structural units of all living things. The human body has about *50 trillion* cells.

This chapter focuses on the structures and functions that are shared by the different cells of the body. Specialized cells and their unique functions are addressed in later chapters.

Introduction to Cells

The English scientist Robert Hooke first observed plant cells with a crude microscope in the late 1600s. However, it was not until the 1830s that two German scientists, Matthias Schleiden and Theodor Schwann, were bold enough to assert that all living things are composed of cells. Shortly thereafter, the German pathologist Rudolf Virchow extended this idea by contending that cells arise only from other cells. Virchow's proclamation was a landmark in the history of biology because it challenged the widespread theory of spontaneous generation, which held that organisms arise from garbage or other nonliving matter. Since the late 1800s, cell research has been exceptionally fruitful. Currently, scientific knowledge about the cell doubles every 7 years or so!

Cells are the smallest living units in our body. Each cell performs all the functions necessary to sustain life: It can obtain nutrients and other essential substances from the surrounding body fluids and use these nutrients to manufacture the molecules it needs to survive. Each cell can also dispose of its wastes and maintain its shape and integrity. Finally, cells can replicate themselves. All these functions are divided among the cell's numerous subunits, most of which are called cellular **organelles** (literally, "little organs"). Most chemical reactions in cells are directed and controlled by **enzymes** (en'zīmz), catalyst molecules made of protein. The sum of all the chemical reactions in the cell (and thus in the whole body) is called **metabolism** (mě-tab'o-lizm; "a state of change").

The trillions of cells in the human body are made up of about 200 different types that vary greatly in size, shape, and function. They include sphere-shaped fat cells, disc-shaped red blood cells, branching nerve cells, and the cube-like cells of kidney tubules. Although different cell types perform different functions, virtually all animal cells contain the same basic parts and can be described in terms of a generalized cell (Figure 2.1).

Human cells have three main parts: the plasma membrane, the cytoplasm, and the nucleus. The *plasma membrane* is the outer boundary. Internal to this membrane is the *cytoplasm* (si'to-plazm), which makes up the bulk of the cell and surrounds the nucleus. The *nucleus* (nu'kle-us) controls cell activities and lies near the cell's center. These cell structures are discussed next, and are summarized in Table 2.1 on page 38.

The Plasma Membrane

The outer cell membrane (Figure 2.2) is called the **plasma membrane** or **plasmalemma** (plaz"mah-lem'ah; *lemma* = sheath, husk). This thin, flexible layer defines the extent of the cell and acts as a fragile barrier.

Structure

The *fluid mosaic model* of membrane structure depicts the plasma membrane as a double layer, or bilayer, of lipid molecules with protein molecules dispersed within it (Figure 2.2a). Its most abundant lipids are phospholipids. Each phospholipid molecule has a polar "head" that is charged, and an uncharged, nonpolar "tail" made of two chains of fatty acids. The polar heads are attracted to water—the main constituent of both the cytoplasm and the fluid external to the cell—so they lie on both the inner and outer faces of the membrane. The nonpolar tails, by contrast, avoid water and line up in the center of the membrane. The result is two parallel sheets of phospholipid molecules lying tail-to-tail, forming the membrane's basic bilayered structure.

The inner and outer layers of the membrane differ somewhat in the specific kinds of lipids they contain. Sugar groups are attached to about 10% of the outer lipid molecules, making them "sugar-fats," or glycolipids (gli"ko-lip'ids). The plasma membrane also contains substantial amounts of cholesterol, another lipid. Cholesterol helps stabilize the membrane and keep it fluid.

Even though we know that the plasma membrane is bilayered, in electron micrographs it appears to have *three* layers—two dark layers sandwiching a

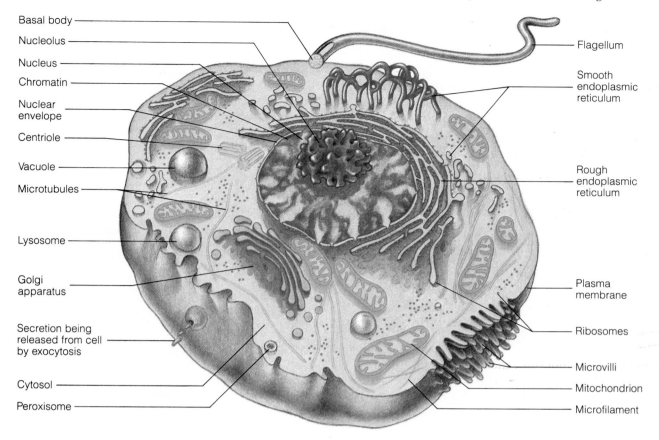

Basal body
Nucleolus
Nucleus
Chromatin
Nuclear envelope
Centriole
Vacuole
Microtubules
Lysosome
Golgi apparatus
Secretion being released from cell by exocytosis
Cytosol
Peroxisome

Flagellum
Smooth endoplasmic reticulum
Rough endoplasmic reticulum
Plasma membrane
Ribosomes
Microvilli
Mitochondrion
Microfilament

Figure 2.1   Structure of a generalized cell.   No cell type is exactly like this one, but virtually all the features shown occur in every animal cell. Exceptions are the microvilli and flagellum, which occur only in certain cells (discussed in Chapter 4). In this figure, the organelles are not drawn to scale.

light one (Figure 2.2b). How can this discrepancy between molecular structure and visualized structure be explained? Evidently, the osmium ions of the stain used for electron microscopy (osmium tetroxide) attach to the polar heads of the phospholipid molecules, staining both surfaces of the membrane but leaving the central part of the membrane (the tails) unstained.

Proteins make up about half of the plasma membrane by weight. The membrane proteins are of two distinct types: integral and peripheral (Figure 2.2a). **Integral proteins** are firmly embedded in the lipid bilayer. Some integral proteins protrude from one side of the membrane only, but most are *transmembrane proteins* that span the whole width of the membrane and protrude from both sides (*trans* = across). **Peripheral proteins**, by contrast, are not embedded in the lipid bilayer at all. Instead, they attach to the membrane surface, usually the inner surface. The peripheral proteins include a network of filaments that helps support the membrane from its cytoplasmic side. Without this supportive network, the plasma membrane would break up into many tiny sacs.

Short chains of carbohydrate molecules attach to the integral proteins. These sugars project from the external cell surface as part of the *cell coat*, or *glycocalyx* (gli″ko-kal′iks; "sugar covering"). The sugars of the membrane's glycolipids are also part of the glycocalyx. You can therefore think of your cells as "sugar-coated." The glycocalyx is sticky and may help adjacent cells bind together. Because every cell type has a different pattern of sugars in its glycocalyx, the glycocalyx is also a distinctive biological marker by which cells recognize each other. For example, a sperm recognizes the ovum (egg cell) by the distinctive nature of the ovum's cell coat.

Functions

Besides acting as the cell's protective boundary, the plasma membrane determines which substances can enter and leave the cell. The membrane is a selectively permeable barrier; that is, it allows some substances to pass while excluding others. Only small, uncharged molecules like oxygen, carbon dioxide,

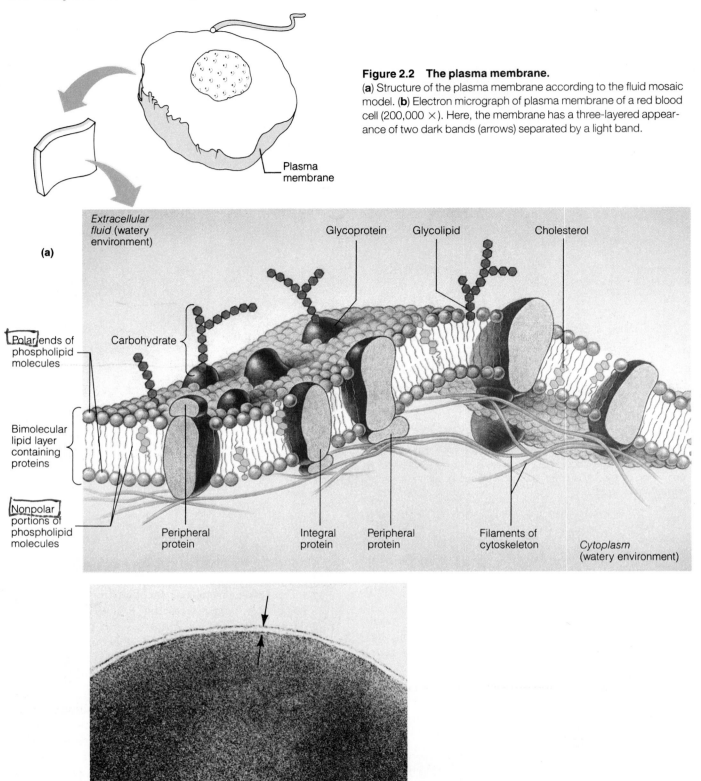

Figure 2.2 The plasma membrane.
(**a**) Structure of the plasma membrane according to the fluid mosaic model. (**b**) Electron micrograph of plasma membrane of a red blood cell (200,000 ×). Here, the membrane has a three-layered appearance of two dark bands (arrows) separated by a light band.

Plasma membrane

(a)

Extracellular fluid (watery environment)

Glycoprotein

Glycolipid

Cholesterol

Polar ends of phospholipid molecules

Carbohydrate

Bimolecular lipid layer containing proteins

Nonpolar portions of phospholipid molecules

Peripheral protein

Integral protein

Peripheral protein

Filaments of cytoskeleton

Cytoplasm (watery environment)

(b)

and fat-soluble drugs can pass freely through the lipid bilayer. This free passage of molecules is accomplished through the process called diffusion. *Diffusion* is the tendency of molecules in a solution to move from a region of their greater concentration to a region of their lesser concentration. Water, like other molecules, diffuses down its concentration gradient.

The diffusion of water molecules across a membrane is called *osmosis* (oz-mo′sis).

The lipid bilayer excludes water soluble or charged molecules, such as sugar molecules and ions; such substances can cross the plasma membrane only by means of specific *transport mechanisms.* These mechanisms carry or pump molecules across the

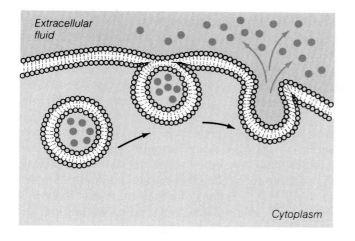

Figure 2.3   Exocytosis. A membrane-bound vesicle containing the substance to be secreted migrates to the plasma membrane, and the two membranes fuse. The fused site opens and releases the contents of the secretory vesicle into the extracellular space.

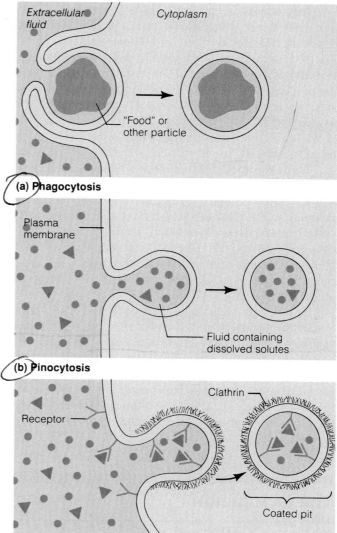

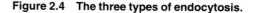

(c) **Receptor-mediated endocytosis**

Figure 2.4   The three types of endocytosis.

membrane or form channels through which specific molecules pass. All these transport mechanisms are formed by the intrinsic membrane proteins.

The largest molecules (macromolecules) and large solid particles are transported through the plasma membrane by another set of processes, called *bulk transport.* Knowledge of the two general types of bulk transport, *exocytosis* and *endocytosis,* is essential to the understanding of basic functional anatomy.

Exocytosis (ek″so-si-to′sis; "out of the cell") is a mechanism that moves substances from the cytoplasm to the outside of the cell (Figure 2.3). Exocytosis accounts for most secretion processes, such as the release of mucus or protein hormones from various gland cells of the body. In exocytosis, the substance or cell product to be released is first enclosed within a membrane-lined sac called a *vesicle.* The vesicle migrates to the plasma membrane and fuses with it. The fused area then ruptures, releasing the contents of the sac into the space outside the cell.

Endocytosis (en″do-si-to′sis; "into the cell") is the mechanism by which large particles and macromolecules *enter* cells (Figure 2.4). The substance to be taken into the cell is enclosed by an infolding portion of the plasma membrane. Once a membranous vesicle is formed, it pinches off from the plasma membrane and moves into the cytoplasm, where its contents are digested. Three types of endocytosis are recognized: phagocytosis, pinocytosis, and receptor-mediated endocytosis.

Phagocytosis (fag″o-si-to′sis) is literally "cell eating." In this process, parts of the plasma membrane and cytoplasm protrude and flow around some relatively large material, such as a clump of bacteria or cell debris, and engulf it (Figure 2.4a). The membranous vesicle thus formed is called a **phagosome**

(fag′o-sōm; "eaten body"). In most cases, the phagosome then fuses with *lysosomes* (li′so-sōmz), organelles containing digestive enzymes that break down the contents of the phagosome. Some cells—most white blood cells, for example—are experts at phagocytosis. Such cells help to police and protect the body by ingesting bacteria, viruses, and other foreign substances. They also "eat" the body's dead and diseased cells.

Phagocytic cells gather in the air spaces of the lungs—especially in the lungs of smokers. Can you explain this connection?

These phagocytic cells serve to remove bacteria and dust particles from the inhaled air. In smokers, these cells are packed with indigestible particles of carbon. ■

Just as we can say that cells eat, we can also say that they drink. **Pinocytosis** (pin"o-si-to'sis) is "cell drinking" (Figure 2.4b). In pinocytosis, a bit of infolding plasma membrane surrounds a tiny droplet of external fluid containing dissolved molecules. This droplet enters the cell in a tiny membranous *pinocytotic vesicle.* Unlike phagocytosis, pinocytosis is a routine activity of most cells. It is particularly important in cells that function in nutrient absorption, such as cells that line the intestines.

Unlike pinocytosis, which is a nonspecific ingestion process, **receptor-mediated endocytosis** is exquisitely selective (Figure 2.4c). Membrane **receptors** are proteins in the plasma membrane that bind only with certain molecules. Once such binding occurs, both the receptors and the attached molecules are internalized in a small vesicle called a **coated pit**. This term refers to a coat of *clathrin* (klath'rin; "lattice") protein around the vesicle.

Substances taken up by receptor-mediated endocytosis include signal molecules that carry messages to the cell from other regions of the body. Protein hormones are examples of such signal molecules. When the coated pit combines with a lysosome in the cytoplasm, the endocytosed substance is digested or released into the cell. Then the membranes containing the receptors rejoin the plasma membrane.

The Cytoplasm

Cytoplasm, literally "cell-forming material," is the part of the cell that lies internal to the plasma membrane and external to the nucleus. Most cellular activities are carried out in the cytoplasm. It consists of three major elements: *cytosol (cytoplasmic matrix), organelles,* and *inclusions* (see Table 2.1 on p. 38).

Cytosol (Cytoplasmic Matrix)

The **cytosol** (si'to-sol), or **cytoplasmic matrix**, is the jelly-like substance within which the other cytoplasmic elements are suspended (Figure 2.1). It consists of water, ions, and many enzymes. Some of these enzymes start the breakdown of foodstuffs (sugars, amino acids, lipids) that provide the raw material and energy for cell activities. In many cell types, the cytosol makes up about half the volume of the cytoplasm.

Cytoplasmic Organelles

Typically, the cytoplasm contains about nine types of organelles. Each has a different function, and each is essential for the survival of the cell. The organelles include mitochondria, ribosomes, rough and smooth endoplasmic reticulum, Golgi apparatus, lysosomes, peroxisomes, cytoskeleton, and centrioles (Figure 2.1). As we will explain, most organelles are bounded by a membrane that is similar in composition to the plasma membrane but lacks a glycocalyx. With very few exceptions, all cells of the human body share the same kinds of organelles. However, when a cell type performs a special body function, the particular organelles that contribute to that function are especially abundant in that cell. Thus, certain organelles are better developed in some cells than in others. Examples of this principle will be provided as we explore the organelles and their roles.

Mitochondria

Mitochondria (mi"to-kon'dre-ah) are the main "power plants" of the cells. These organelles usually are diagrammed as bean-shaped, reflecting their appearance in thin sections (Figure 2.5), but actually they are long and thread-like (*mitos* = thread). In living cells, mitochondria squirm and change shape as they move through the cytoplasm. Most organelles are surrounded by a membrane, but mitochondria are actually enclosed by *two* membranes: The outer membrane is smooth and featureless, but the inner membrane folds inward to produce shelf-like **cristae** (krĭ'ste; "crests"). These protrude into the **matrix**, the jelly-like substance within the mitochondrion. The matrix also contains scattered, dark spheres of calcium phosphate, the **matrix granules** (Figure 2.5b).

Mitochondria generate most of the energy the cell uses to carry out work. That is, they use food energy to make most of the cell's **ATP (adenosine triphosphate)**, the high-energy molecules that power most chemical reactions in the cell. Most ATP is produced on the mitochondrial cristae—and those cell types with the highest energy requirements have the largest number of cristae in their mitochondria.

Mitochondrial matrix granules do not function in energy production. Rather, they bind calcium ions and release these ions into the cytoplasm to trigger various cellular events (including muscle contraction, some kinds of secretion, and cell proliferation).

Mitochondria are far more complex than any other organelle. They even contain some genetic material (DNA), as if they were miniature cells themselves. Intriguingly, mitochondria are very similar to a group of bacteria (purple bacteria phylum). It is now widely believed that mitochondria arose from bacteria that invaded the ancient ancestors of animal and plant cells.

If a particular cell type has a large number of mitochondria in its cytoplasm, what can you say about the activity level of that cell?

The cell performs some process that requires a great deal of energy and oxygen. The muscle cell,

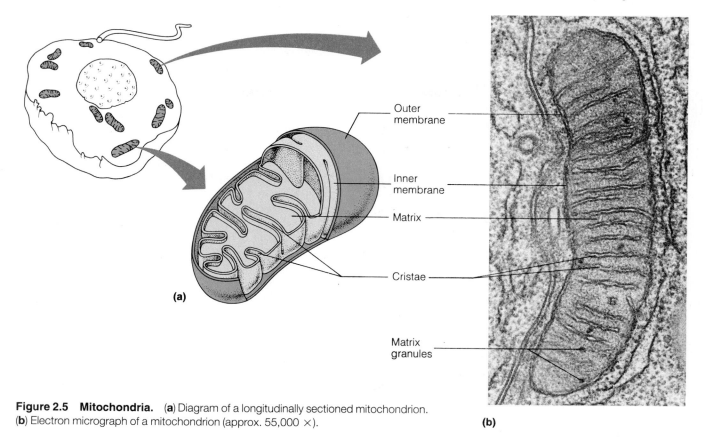

Outer
membrane

Inner
membrane

Matrix

Cristae

Matrix
granules

(a)

(b)

Figure 2.5   Mitochondria.   (a) Diagram of a longitudinally sectioned mitochondrion.
(b) Electron micrograph of a mitochondrion (approx. 55,000 ×).

which requires large amounts of energy for contraction, is an example of a mitochondria-rich cell. ■

Ribosomes

Ribosomes (ri'bo-sōmz) are small, dark-staining granules (Figure 2.6). Unlike most organelles, they are not
surrounded by a membrane, but are constructed of
proteins plus **ribosomal RNA** (RNA = ribonucleic
acid). Each ribosome consists of two subunits that fit
together like the body and cap of an acorn (Figure
2.6a).

Ribosomes are the sites of protein synthesis

within the cell. On the ribosomes, building blocks
called amino acids are linked together to form protein
molecules. This assembly is dictated by the genetic
material (DNA) in the cell nucleus, whose instructions are carried to the ribosomes by messenger molecules called **messenger RNA (mRNA).**

Many ribosomes float freely within the cytosol.
Such **free ribosomes** (Figure 2.6b) manufacture the
soluble proteins that function within the cytosol itself. By contrast, ribosomes attached to the membranes of the rough endoplasmic reticulum synthesize proteins with somewhat different destinies, as
described shortly.

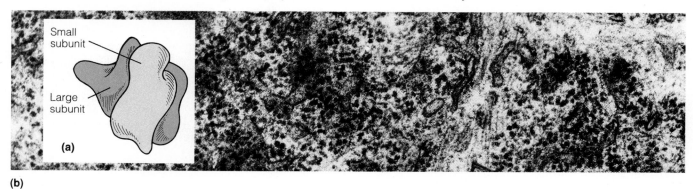

Small
subunit

Large
subunit

(a)

(b)

Figure 2.6   Ribosomes.   (a) Diagram of a ribosome, showing the large and small subunits. **(b)** Electron micrograph of free ribosomes (dark dots) in the cytosol (100,000 ×).

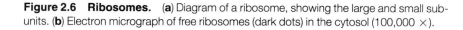

Endoplasmic Reticulum

Endoplasmic reticulum (en"do-plaz'mik ret-tik'u-lum), or **ER**, is literally the "network within the cytoplasm." As shown in Figure 2.7, this is an extensive system of membrane-lined tubes and envelopes that twists through the cytoplasm. There are two distinct varieties of ER: rough ER and smooth ER. Either variety may predominate in a given cell type, depending on the specific functions of the cell.

Rough Endoplasmic Reticulum. The **rough endoplasmic reticulum (rough ER)** consists mainly of stacked envelopes called **cisternae** (sis-ter'ne; "fluid-filled cavities"). Ribosomes stud the external face of both membranes of the rough ER. Proteins assembled on these ribosomes thread their way into the fluid-filled interior of the ER cisternae (Figure 2.7c).

The rough ER has several functions. Its ribosomes manufacture all proteins that are secreted from cells; thus, rough ER is especially well-developed in gland cells that secrete a large amount of protein (mucous cells, for example). More significantly, the rough ER manufactures both the integral proteins and the phospholipid molecules of all the cell's membranes. The rough ER can therefore be considered the cell's "membrane factory."

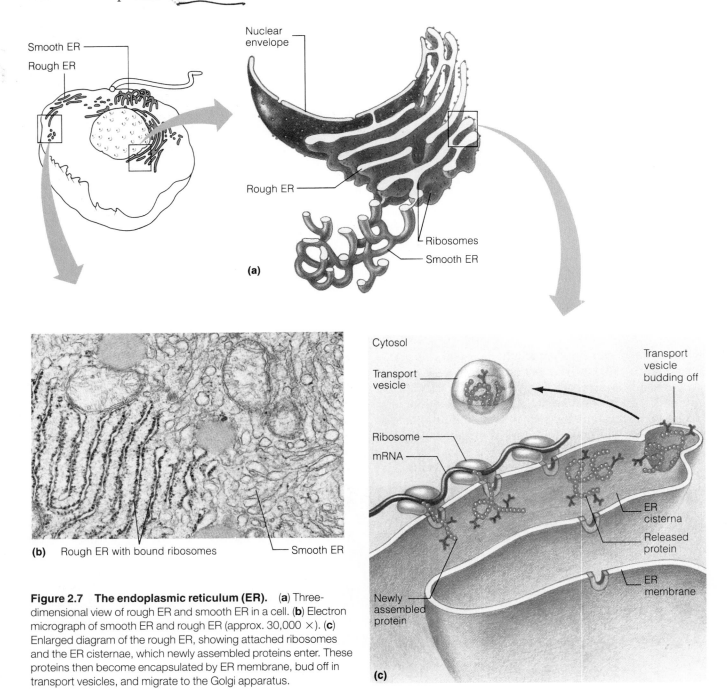

Figure 2.7 The endoplasmic reticulum (ER). **(a)** Three-dimensional view of rough ER and smooth ER in a cell. **(b)** Electron micrograph of smooth ER and rough ER (approx. 30,000 ×). **(c)** Enlarged diagram of the rough ER, showing attached ribosomes and the ER cisternae, which newly assembled proteins enter. These proteins then become encapsulated by ER membrane, bud off in transport vesicles, and migrate to the Golgi apparatus.

Smooth Endoplasmic Reticulum.   The **smooth endoplasmic reticulum (smooth ER)** is continuous with the rough ER (Figure 2.7a and b). It consists of tubules arranged in a branching network. No ribosomes attach to its membranes, and it makes no proteins. Instead, the smooth ER participates in lipid metabolism (making or breaking down *fats*) and stores calcium ions. All functions of the smooth ER are performed through the action of enzymes (integral proteins) in its membrane.

In addition to metabolizing lipids, the smooth ER detoxifies lipid-soluble drugs that enter the cell (such as alcohol and the sedative phenobarbital). This detoxifying function is especially well-developed in liver cells. Can you deduce what happens to the smooth ER in liver cells of people who use such drugs?

The amount of smooth ER increases markedly as the liver becomes more efficient at detoxifying the drug. Therefore, we gradually gain resistance or tolerance to many drugs with increased use. ■

Golgi Apparatus

The **Golgi** (gol′je) **apparatus** is a stack of three to ten disc-shaped envelopes (cisternae), each lined by a membrane (Figure 2.8). It resembles a stack of hollow saucers, one cupped inside the next. The convex face of the Golgi stack, the *cis* face, receives spherical membranous transport vesicles from the nearby rough ER. Other vesicles bud off the concave, or *trans*, face of the Golgi apparatus. The products of the rough ER move through the Golgi stack from the cis to the trans side.

The function of the Golgi apparatus is to sort and package the proteins and membranes made by the rough ER (Figure 2.9). For example, the Golgi apparatus distinguishes which newly manufactured membranes will become part of the mitochondria, which membranes are destined for the plasma membrane, and so on. It then sends these membranes to their correct destinations via the vesicles that leave the trans face (see destination #2 in Figure 2.9, for example). Thus, the Golgi apparatus is like a post office: It receives "letters" (products of the rough ER) that are "addressed" to different locations within the cell and ensures that they get to their destinations.

The Golgi apparatus also plays a role in secretion. Gland cells that secrete proteins (mucous cells, for example) have a well-developed Golgi apparatus. In such cells, the rough ER makes the product, and the Golgi apparatus packages it; the product then leaves the trans face in vesicles called **secretory granules** (see destination #1 in Figure 2.9). These granules ultimately release their contents by exocytosis.

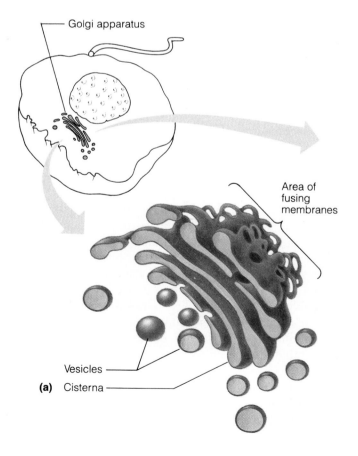

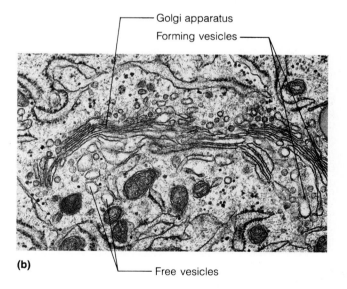

Figure 2.8   Golgi apparatus.   (**a**) Three-dimensional view of a Golgi apparatus that has been cut in half. The concave surface at lower left, near the vesicles, is the trans face; the surface at upper right (area of fusing membranes) is the cis face. (**b**) Electron micrograph of the Golgi apparatus. The cis face (above) receives small transport vesicles. Larger vesicles (forming vesicles and free vesicles) leave the trans face (approx. 85,000 ×).

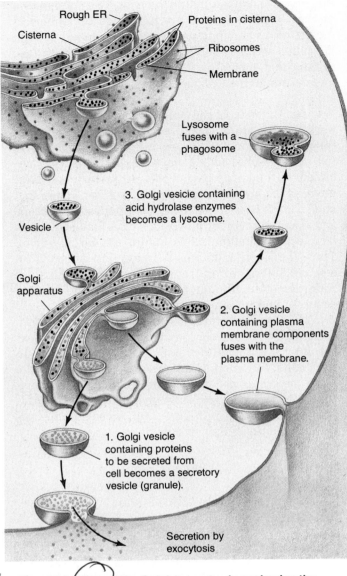

Figure 2.9 **Role of the Golgi apparatus in packaging the products of the rough ER for use in the cell and for secretion.** Protein-containing transport vesicles pinch off the rough ER and migrate to fuse with the cis face of the Golgi apparatus. As it passes through the Golgi apparatus, the product is sorted (and its chemistry is slightly modified). The product is then packaged within vesicles, which leave the trans face of the Golgi apparatus and head for various destinations (#1–3), as shown.

Another function of the Golgi apparatus is packaging digestive enzymes into membranous sacs called lysosomes that stay within the cell (destination #3 in Figure 2.9).

Lysosomes

Lysosomes (literally, "disintegrator bodies") are spherical, membrane-bound sacs containing a large variety of digestive enzymes (Figure 2.10). These enzymes, called acid hydrolases, can digest virtually all varieties of large biological molecules. Lysosomes can be considered the cell's "demolition crew," for they digest unwanted substances. Within lysosomes, digestion can proceed safely, because the enclosing membrane keeps the destructive enzymes inside the lysosome and away from other cell components.

As mentioned earlier, lysosomes break down substances that enter the cell by endocytosis (p. 27). Phagocytic cells, for example, contain many lysosomes and use them to degrade ingested bacteria and viruses. Lysosomes also digest the cell's own worn-out membranes and organelles as part of a normal recycling process in which aging organelles are replaced. When an entire cell wears out or is seriously injured, its lysosomes release their enzymes into the cytosol, causing the whole cell to self-digest. For this reason, lysosomes have been called "suicide sacs."

In an inherited condition called Tay-Sachs disease, an infant's lysosomes lack a specific enzyme that breaks down certain glycolipids in the normal recycling of worn-out cellular membranes. Such glycolipids are especially abundant in the membranes of nerve cells.

Children with Tay-Sachs disease inevitably die within a year and a half. Can you deduce why?

The undigested glycolipids build up within the lysosomes of nerve cells. Soon their accumulation interferes with the function of the nerve cells. Neurological deterioration leads to mental retardation, blindness, spastic movements, and death of the child.

Peroxisomes

Peroxisomes (pĕ-roks'ĭ-sōmz; "peroxide bodies") are membrane-lined sacs that resemble small lysosomes in typical microscopic sections (Figure 2.1). Like lysosomes, peroxisomes contain enzymes; however, the most important role of peroxisome enzymes is to "disarm" dangerous **free radicals** by converting them to hydrogen peroxide, which in turn is converted to harmless water and oxygen. Free radicals are normal byproducts of cellular metabolism, but if allowed to accumulate, they can have a devastating effect on the cells. In addition, peroxisomes use highly reactive hydrogen peroxide molecules to destroy certain toxic chemicals that have entered the cells. Peroxisomes are numerous in liver and kidney cells, which are very active in such detoxification.

Cytoskeleton

The **cytoskeleton**, literally "cell skeleton," is an elaborate series of rods running through the cytosol (Fig-

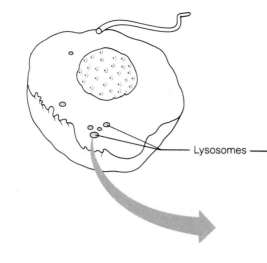

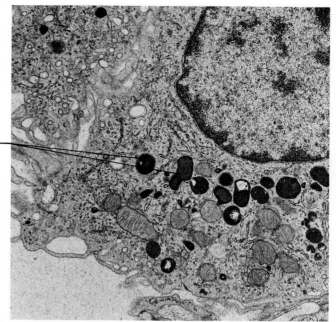

Lysosomes

Figure 2.10   Lysosomes. Electron micrograph of a cell containing lysosomes. Lysosomes often stain darkly as shown (10,000 ×).

ure 2.11a and b). This network acts as a cell's "bones" and "muscles" by supporting cellular structures and generating various cell movements. The three types of rods in the cytoskeleton are *microtubules, microfilaments,* and *intermediate filaments*. None of these is covered by a membrane.

Microtubules, the elements with the largest diameter, are hollow tubes made of spherical protein subunits called tubulins. All microtubules radiate from a small region of cytoplasm near the nucleus called the *centrosome* or *cell center*. This radiating pattern of stiff microtubules determines the overall shape of the cell, as well as the distribution of cellular organelles. Mitochondria, lysosomes, and secretory granules attach to the microtubules like ornaments hanging from the limbs of a Christmas tree. These attached organelles are continually pulled along the microtubules by small **motor proteins** that act like train engines on the microtubular "railroad tracks." Microtubules are remarkably dynamic organelles, constantly growing out from the cell center, disassembling, then reassembling again.

Microfilaments, the thinnest elements of the cytoskeleton, are strands of the contractile protein **actin** (ak'tin). They concentrate most heavily in the cell periphery, on the cytoplasmic side of the plasma membrane. Microfilaments interact with another protein called **myosin** (mi'o-sin) to generate contractile forces within the cell. Some cells send out then retract extensions called *pseudopods* (soo'do-pods; "false feet"), which allow the cells to crawl along. Microfilaments are responsible for this crawling action, called **amoeboid motion** (ah-me'boid; "changing shape").

Except in muscle cells, where they are stable and permanent, microfilaments are labile. That is, they are constantly breaking down and re-forming from smaller subunits.

Intermediate filaments are tough, insoluble protein fibers, with a diameter between those of microfilaments and microtubules. Intermediate filaments are the most stable and permanent of the cytoskeletal elements. Their most important property is high tensile strength. That is, they act as strong guy wires to resist *pulling* forces that are placed on the cell.

Some researchers believe that an additional fine, web-like element weaves its way through the cytosol, interconnecting the other cytoskeletal elements. They have named this web the *microtrabecular lattice* (mi"kro-trah-bek'u-lar; "small-stranded"), and they believe it gives the cytosol its jelly-like consistency. However, the existence of this lattice has been difficult to demonstrate with certainty and is still hotly disputed.

Centrioles

Centrioles (sen'tre-ōlz) are barrel-shaped structures in the centrosome (Figure 2.12). They occur in pairs, with one centriole perpendicular to the other. The wall of each centriole consists of nine groups of three microtubules; it is not membrane-lined. Unlike most other microtubules, those in centrioles are stable and do not disassemble. As we will discuss below, centrioles participate in cell division by organizing the mitotic spindle (see Figure 2.18, p. 40).

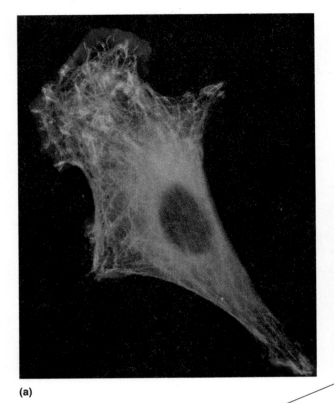

(a)

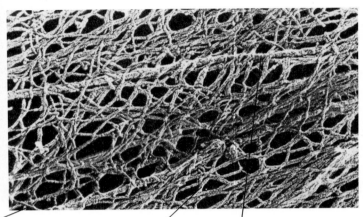

(b)

Figure 2.11 The cytoskeleton. (**a**) Micrograph of the cytoskeleton (approx. 1200 ×). Microtubules appear green and actin microfilaments appear blue. Intermediate filaments form most of the rest of the network. (**b**) Electron micrograph of the cytoskeleton (51,400 ×). Below are diagrams of a microfilament, intermediate filament, and a microtubule, showing the molecular subunits of each element.

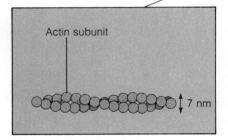

Microfilament

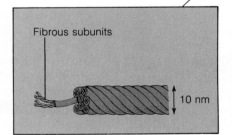

Intermediate filament

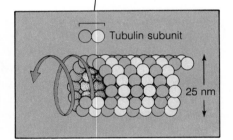

Microtubule

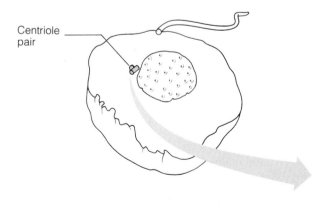

Figure 2.12 Centrioles. Three-dimensional view of a pair of centrioles, oriented at right angles to one another as they are usually seen in the cell.

Cytoplasmic Inclusions

Inclusions are structures in the cytoplasm that are not permanent, not essential, and not present in all cell types. Specifically, inclusions are pigments, crystals of protein, and food stores. The food stores, by far the most important inclusions, are *glycogen granules* and *lipid droplets*. Glycogen (gli′ko-gen; "sugar-generating") is a long, branching chain of glucose molecules (glucose is the cell's main sugar and energy source). Structurally, glycogen takes the form of dense, spherical glycogen granules. These granules resemble ribosomes when viewed by electron microscopy, but their diameter is twice as large. Muscle cells, which experience rapid demands for energy, contain many glycogen granules. Lipid droplets are spherical drops of stored fat. They can have the same size and appearance as lysosomes but can be distinguished by their lack of a surrounding membrane. Only a few cell types contain lipid droplets: Small lipid droplets are found in liver cells, large ones in fat cells.

The Nucleus

The **nucleus**, literally a "central core" or "kernel," is the control center of the cell. Its genetic material, **deoxyribonucleic acid (DNA)**, directs the cell's activities by providing the instructions for protein synthesis. The nucleus can be compared to a central genetic library, design department, construction superintendent, and board of directors—all rolled into one. Whereas most cells have only one nucleus, some, including skeletal muscle cells, are *multinucleate* (mul″tĭ-nu′kle-āt); that is, they have many nuclei (*multi* = many). The presence of more than one nucleus usually signifies that a cell has a larger-than-usual amount of cytoplasm to regulate. One cell type in our body, the mature red blood cell, has no nucleus at all (is *anucleate*), for its nucleus is ejected when this cell first enters the bloodstream.

The nucleus, which averages 5 μm in diameter, is larger than any of the organelles. Although it is usually spherical or oval, it generally conforms to the shape of the cell. If a cell is elongated, for example, the nucleus may also be elongated. The main parts of the nucleus are the *nuclear envelope, chromatin and chromosomes,* and *nucleoli.*

Nuclear Envelope

The nucleus is surrounded by a **nuclear envelope** that consists of two parallel nuclear membranes separated by a fluid-filled space (Figure 2.13a). The envelope is continuous with the rough ER and has ribosomes on its outer face. Evidently, it is a specialized part of the rough ER.

At various points, the two layers of the nuclear envelope fuse, and **nuclear pores** penetrate the fused regions. There are several thousand pores per nucleus. Like other cellular membranes, the membranes of the nuclear envelope are selectively permeable, but large molecules pass easily through the nuclear pores. Protein molecules imported from the cytoplasm and RNA molecules exported from the nucleus routinely travel through the pores.

The nuclear envelope encloses a jelly-like fluid called *nucleoplasm* (nu′kle-o-plazm″), within which the chromatin and nucleoli are suspended. Like the cytosol, the nucleoplasm contains salts, nutrients, and other essential chemicals. It even contains a supportive nucleoskeleton comparable to the cytoskeleton.

Chromatin and Chromosomes

Most cells undergo brief stages of cell division, followed by longer stages during which they carry out normal cellular activities. When cells in the nondividing stage are examined by electron microscopy, some regions of the nucleus stain darkly (Figure 2.13b). These dark regions, called **condensed chromatin** (kro′mah-tin; "colored substance"), contain tightly coiled strands of DNA. Other regions of the nucleus stain poorly. These light regions, called **extended chromatin**, contain fine uncoiled strands of DNA. While the DNA of condensed chromatin is always inactive, that of extended chromatin directs the synthetic activities of the cell during a cell's nondividing stage. Understandably, the most active cells of the body tend to have a large amount of extended chromatin and little condensed chromatin. In general, most of the condensed chromatin is near the periphery of the nucleus, just internal to the nuclear envelope.

The strands of chromatin in the nucleus each contain an exceptionally long molecule of DNA, which is a double helix that resembles a spiral staircase (Figure 2.14). This double helix is in turn composed of four kinds of subunits called *nucleotides,* each of which contains a distinct base. These bases—thymine (T), adenine (A), cytosine (C), and guanine (G)—bind to form the "stairs" of the "staircase" and to hold the DNA helix together.

The long DNA helix wraps around clusters of eight spherical protein molecules called **histones** (his′tōnz). These DNA-histone clusters, each of which is called a **nucleosome** (nu′kle-o-sōm″; "nuclear body"), appear under the electron microscope as

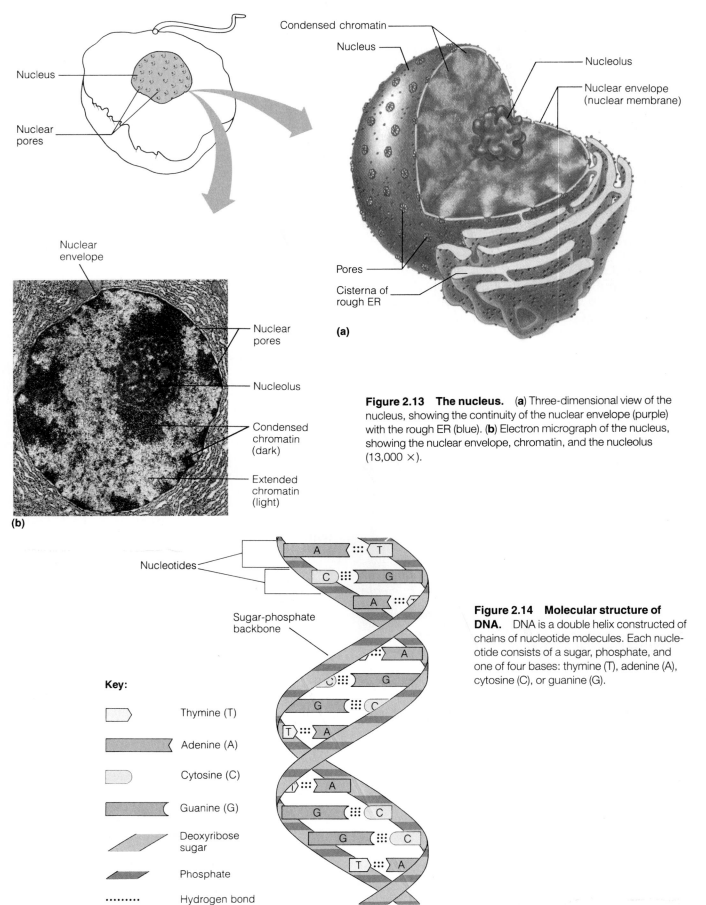

Nucleus

Nuclear pores

Condensed chromatin

Nucleus

Nucleolus

Nuclear envelope (nuclear membrane)

Pores

Cisterna of rough ER

(a)

Nuclear envelope

Nuclear pores

Nucleolus

Condensed chromatin (dark)

Extended chromatin (light)

(b)

Figure 2.13 The nucleus. (**a**) Three-dimensional view of the nucleus, showing the continuity of the nuclear envelope (purple) with the rough ER (blue). (**b**) Electron micrograph of the nucleus, showing the nuclear envelope, chromatin, and the nucleolus (13,000 ×).

Nucleotides

Sugar-phosphate backbone

Key:

Thymine (T)

Adenine (A)

Cytosine (C)

Guanine (G)

Deoxyribose sugar

Phosphate

Hydrogen bond

Figure 2.14 Molecular structure of DNA. DNA is a double helix constructed of chains of nucleotide molecules. Each nucleotide consists of a sugar, phosphate, and one of four bases: thymine (T), adenine (A), cytosine (C), or guanine (G).

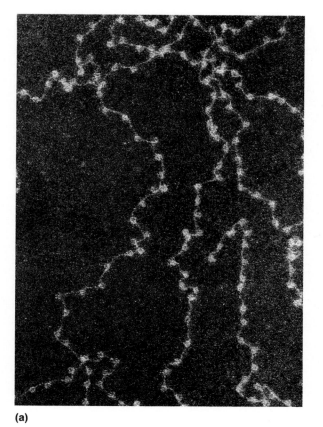

(a)

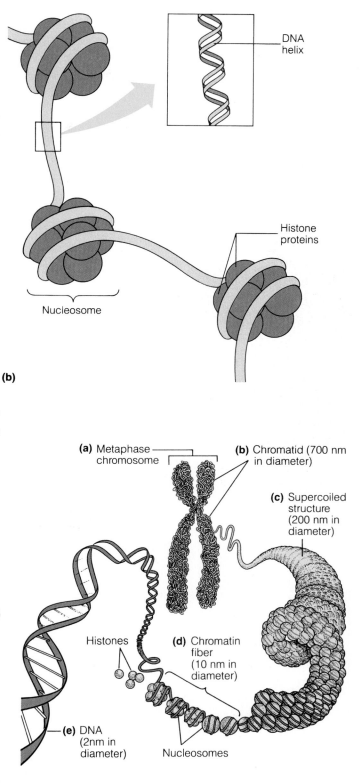

(b)

Figure 2.15   Chromatin.   **(a)** High-power electron micrograph of extended chromatin fibers, which have the appearance of beads on a string (216,000 ×). **(b)** A model for the structure of a chromatin fiber. The DNA molecule, a long double helix (inset), winds around spherical particles called nucleosomes. Each nucleosome is composed of eight histone proteins.

beads on a string (Figure 2.15). Chromatin is thus made up of strands of DNA plus their associated nucleosomes.*

Within a dividing cell, the chromatin coils in a complex manner, forming thick rods called **chromosomes** (kro'mo-sōmz; "colored bodies") (Figure 2.16). Chromosomes are moved extensively during cell division (p. 40), and their compact nature avoids entanglement and breakage of the delicate chromatin strands during such movements.

Nucleoli

The cell nucleus contains one or several dark-staining bodies called **nucleoli** (nu-kle'o-li; singular, **nucleo-**

---

*The term *chromatin,* which technically means "DNA and nucleosomes (or histones)," can also be used as an abbreviation for condensed chromatin (see Figure 2.13a).

Figure 2.16   Chromosome structure.   The chromosome at the center (a) is shown in its maximally coiled state, in the metaphase stage of mitosis (cell division). Parts (b) through (e) indicate progressively smaller subunits of the chromosome. Much of the chromatin uncoils (extends) in nondividing cells, for DNA must be uncoiled to dictate normal cellular activities. This diagram is simplified, and even more complex levels of coiling are present in actual chromosomes.

Table 2.1 Parts of the Cell: Structure and Function

Cell part	Structure	Functions
Plasma membrane (Figure 2.2)	Membrane composed of a double layer of lipids (phospholipids, cholesterol) within which proteins are embedded; proteins may extend entirely through the lipid bilayer or protrude on only one face; externally facing proteins and some lipids have attached sugar groups	Serves as an external cell barrier; acts in transport of substances into or out of the cell; externally facing proteins act as receptors (for hormones, neurotransmitters) and in cell-to-cell recognition
Cytoplasm	Cellular region between the nuclear and plasma membranes; consists of fluid *cytosol,* containing dissolved solutes, *inclusions* (stored nutrients, pigment granules), and *organelles,* the metabolic machinery of the cytoplasm	
Cytoplasmic organelles · Mitochondria (Figure 2.5)	Rod-like, double-membrane structures; inner membrane folded into projections called cristae	Site of ATP synthesis; powerhouse of the cell
· Ribosomes (Figure 2.6)	Dense particles consisting of two subunits, each composed of ribosomal RNA and proteins; free or attached to rough ER	The sites of protein synthesis
· Rough endoplasmic reticulum (rough ER) (Figure 2.7)	Membrane system enclosing a cavity, the cisterna, and coiling through the cytoplasm; externally studded with ribosomes	Makes proteins that are secreted from the cell; makes the cell's membranes
· Smooth endoplasmic reticulum (smooth ER) (Figure 2.7)	Membranous system of tubules; free of ribosomes	Site of lipid metabolism
· Golgi apparatus (Figure 2.8)	A stack of smooth membrane sacs close to the nucleus	Packages, modifies, and segregates proteins for secretion from the cell and inclusion in lysosomes; sends membranes from rough ER to their destinations
· Lysosomes (Figure 2.10)	Membranous sacs containing acid hydrolases	Sites of intracellular digestion
· Peroxisomes	Membranous sacs of oxidase enzymes	The enzymes detoxify a number of toxic substances; the most important enzyme, catalase, breaks down hydrogen peroxide
· Microfilaments (Figure 2.11)	Fine filaments of the contractile protein actin	Involved in muscle contraction and other types of intracellular movement; help form the cell's cytoskeleton
· Intermediate filaments (Figure 2.11)	Protein fibers; composition varies	The stable cytoskeletal elements; resist tensile forces acting on the cell
· Microtubules (Figure 2.11)	Cylindrical structures composed of tubulin proteins	Support the cell and give it shape; involved in intracellular and cellular movements, form centrioles
· Centrioles (Figure 2.12)	Paired cylindrical bodies, each composed of nine triplets of microtubules	Organize a microtubule network during mitosis to form the spindle; form the bases of cilia and flagella
Nucleus (Figure 2.13)	Surrounded by the nuclear membrane; contains fluid nucleoplasm, nucleoli, and chromatin	Control center of the cell; responsible for transmitting genetic information and providing the instructions for protein synthesis
· Nuclear membrane (Figure 2.13)	Double bilipid membrane containing proteins; pierced by pores; continuous with the cytoplasmic ER	Separates the nucleoplasm from the cytoplasm and regulates passage of substances to and from the nucleus
· Nucleoli (Figure 2.13)	Dense spherical (non-membrane-bounded) bodies	Site of ribosome subunit manufacture
· Chromatin (Figure 2.13)	Granular, threadlike material composed of DNA and histone proteins	DNA constitutes the genes

lus: "little nucleus") (Figure 2.13). Nucleoli contain parts of several different chromosomes and serve as the cell's "ribosome-producing machines": Specifically, the large and small subunits of ribosomes are assembled at the nucleoli. From there, these subunits leave the nucleus through the nuclear pores and join within the cytoplasm to form complete ribosomes.

The Cell Life Cycle

The cell life cycle is the series of changes a cell undergoes from the time it forms until it reproduces itself. This cycle can be divided into two major periods: *interphase*, in which the cell grows and carries on its usual activities; and *cell division*, or the *mitotic phase*, during which it divides into two cells (Figure 2.17).

Interphase

In addition to carrying on its life-sustaining activities, an interphase cell prepares for the next cell division. Interphase is divided into G₁, S, and G₂ subphases. During **G₁ (growth 1)**, the first part of interphase, cells are metabolically active, synthesize proteins rapidly, and grow vigorously. This is the most variable phase in terms of duration. In cells with rapid division rates, G₁ lasts several hours; in cells that divide slowly, it can last days or even years.

For most of G₁, no cell activities directly related to cell division occur. However, near the end of G₁, the centrioles start to replicate in preparation for cell division. During the next stage, the **S (synthetic)** phase, DNA replicates itself, ensuring that the two future cells receive identical copies of the genetic material.

The final part of interphase, called **G₂ (growth 2)**, is brief. In this period, enzymes needed for the division process are synthesized. Centrioles finish copying themselves at the end of G₂. Throughout the S and G₂ phases, the cell continues to grow and to carry out its normal metabolic activities.

Cell Division

Cell division is essential for body growth and tissue repair. Short-lived cells that continuously wear away, such as cells of the skin and the intestinal lining, reproduce themselves almost continuously. Others, such as liver cells, reproduce slowly (replacing those cells that gradually wear out), but can divide quickly if the organ is damaged. Cells of nervous tissue, skeletal muscle, and heart muscle are totally unable to divide once they are fully mature; repair is carried out

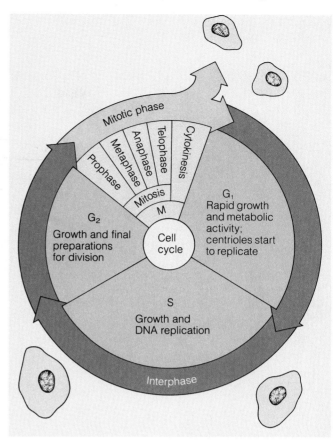

Figure 2.17   The cell cycle. The two basic phases in the life and reproduction of each cell are interphase and the mitotic (M) phase. The length of the cell cycle varies in different cell types, but the G₁ stage of interphase tends to be the longest and the most variable in duration. See the text for details.

by scar tissue (a fibrous type of connective tissue).

Cells divide in the **M (mitotic) phase** of their life cycle, which follows interphase (Figure 2.17). In most cell types, division involves two distinct events: *mitosis* (mi-to′sis), or division of the nucleus, and *cytokinesis* (si″to-ki-ne′sis), or division of the cytoplasm.

Mitosis

Mitosis is the series of events that parcel out the replicated DNA of the original cell to two new cells, culminating in the division of the nucleus. Throughout these events, the chromosomes are evident as thick rods or threads (*mitosis* literally means "the stage of threads"). Mitosis is said to have four consecutive phases: prophase, metaphase, anaphase, and telophase. However, mitosis is actually a continuous process, with each phase merging smoothly into the next. Its duration varies according to cell type, but mitosis typically lasts about 2 hours. Mitosis is described in detail in Figure 2.18.

Figure 2.18 The stages of mitosis. The cells shown are from an early embryo of a whitefish. (Micrographs approx. 600 ×.)

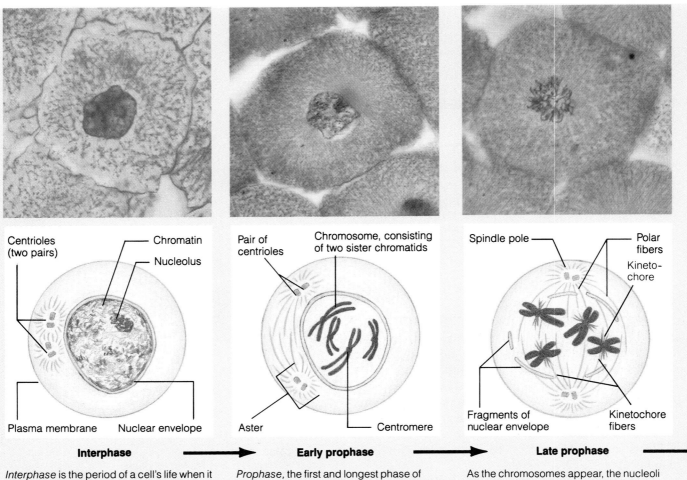

Interphase

Interphase is the period of a cell's life when it is carrying out its normal metabolic activities and growing. During various periods of this phase, the centrioles begin replicating (G_1 through G_2), DNA is replicated (S), and the final preparations for mitosis are completed (G_2). The centriole pair finishes replicating into two pairs during G_2. During interphase, the chromosomal material is seen in the form of diffuse (extended) chromatin, and the nuclear membrane and nucleolus are intact and visible.

Early prophase

Prophase, the first and longest phase of mitosis, begins when the chromatin threads start to coil and condense, forming barlike *chromosomes* that are visible with a light microscope. Since DNA replication has occurred during interphase, each chromosome is actually made up of two identical chromatin threads, now called *chromatids*. The chromatids of each chromosome are held together by a small, buttonlike body called a *centromere*. Chromatids can be thought of as half-chromosomes; after they separate, each will be considered a new chromosome.

Late prophase

As the chromosomes appear, the nucleoli disappear, the cytoskeletal microtubules disassemble, and the centriole pairs separate from one another. The centrioles act as focal points for growth of a new assembly of microtubules called the **mitotic spindle.** As these microtubules lengthen, the centrioles are pushed farther and farther apart, toward opposite ends (poles) of the cell, until the spindle finally extends from pole to pole. While the centrioles are still moving away from each other, the nuclear membrane breaks up into fragments, allowing the spindle to occupy the center of the cell and to interact with the chromosomes. The centrioles also direct the formation of *asters,* systems of microtubules that radiate outward from the ends of the spindle and anchor it to the plasma membrane. Meanwhile, some of the growing spindle microtubules attach to special protein complexes, called *kinetochores (ki-ne'to-korz),* on each chromosome's centromere. Such microtubules are called *kinetochore fibers.* The remaining spindle microtubules, which do not attach to any chromosomes, are called *polar fibers.* The kinetochore fibers pull on each chromosome from both poles, resulting in a tug-of-war that ultimately draws the chromosomes to the middle of the cell.

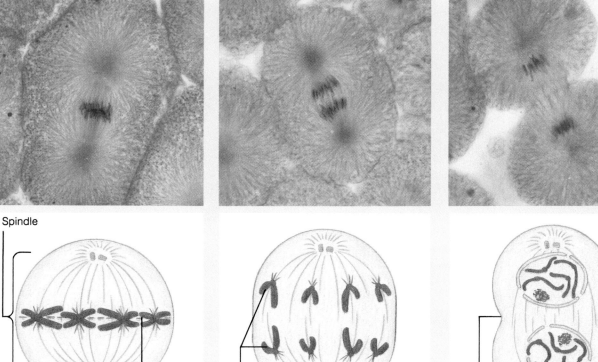

Spindle

Metaphase plate

Daughter
chromosomes

Cleavage
furrow

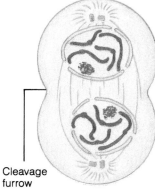

Metaphase

Metaphase is the second phase of mitosis. The chromosomes cluster at the middle of the cell, with their centromeres precisely aligned at the exact center, or *equator,* of the spindle. This arrangement of the chromosomes along a plane midway between the poles is called the *metaphase plate*.

Anaphase

Anaphase, the third phase of mitosis, begins abruptly as the centromeres of the chromosomes split, and each chromatid now becomes a chromosome in its own right. The kinetochore fibers shorten and, like a recoiling rubber band, gradually pull each chromosome toward the pole it faces. The polar fibers, by contrast, lengthen and push the two poles of the cell apart. Anaphase is easy to recognize because the moving chromosomes look V-shaped. The centromeres, which are attached to the kinetochore fibers, lead the way, and the chromosomal "arms" dangle behind them. Anaphase is the shortest stage of mitosis; it typically lasts only a few minutes.

This process of moving and separating the chromosomes is helped by the fact that the chromosomes are short, compact bodies. Diffuse chromatin threads would tangle, trail, and break, which would damage the genetic material and result in its imprecise "parceling out" to the daughter cells.

Telophase and cytokinesis

Telophase begins as soon as chromosomal movement stops. This final phase is like prophase in reverse. The identical sets of chromosomes at the opposite poles of the cell uncoil and resume their threadlike extended chromatin form. A new nuclear membrane, derived from the rough ER, reforms around each chromatin mass. Nucleoli reappear within the nuclei, and the spindle breaks down and disappears. Mitosis is now ended. The cell, for just a brief period, is a binucleate cell (has two nuclei) and each new nucleus is identical to the original mother nucleus.

As a rule, as mitosis draws to a close, *cytokinesis* occurs, completing the division of the cell into two daughter cells. Cytokinesis occurs as a ring of peripheral actin and myosin filaments (not shown) contracts at the *cleavage furrow* and squeezes the cells apart.

A CLOSER LOOK Cancer—The Intimate Enemy

The clinical definition of cancer—the uncontrolled division and spread of cells within the body—gives no hint of its status as one of the most universally feared diseases. Our fears seem to arise instead from cancer's overwhelming prevalence in our lives: Almost *half* of all Americans develop cancer during their lifetime, and a fifth will die from it. Cancer can arise from almost any cell type, but the most common cancers arise in the skin, lung, large intestine, breasts, the prostate gland below the male bladder, and the stomach (in that order).

Cancer is a variety of *neoplasm* (ne'o-plazm; "new growth"). A neoplasm, also called a *tumor*, is defined as an abnormal mass of proliferating cells. Neoplasms are classified as benign ("kindly, favorable") or malignant ("bad"). A *benign neoplasm* (*benign tumor*) is strictly localized: Its cells remain compacted, and it is often encapsulated. Benign tumors tend to grow in a leisurely way and, if surgically removed before they compress vital organs, seldom kill their hosts. By contrast, cancers are *malignant neoplasms*, nonencapsulated "killers" that grow more relentlessly. Their cells often resemble immature (undifferentiated) cells as they invade their surroundings. This invasive property is reflected in the word *cancer*, from the Latin meaning a creeping "crab." The most devastating property of cancer cells is their ability to break away from the parent mass, the primary tumor, and

enter the blood or lymph vessels to travel to other body organs, where they form secondary growths (satellite tumors). This capacity for travel is called *metastasis* (mě-tas'tah-sis; "beyond standing"). As you can see, what distinguishes malignant cancer from a benign neoplasm are the invasive and metastatic properties of cancer cells. Cancer cells release chemicals and enzymes that eat through blood vessels and allow these cells to burrow through tissues.

Malignancies can kill their victims by blocking, pressing on, or disrupting the functions of vital organs. Cancer cells are parasites that consume an exceptional amount of the body's nutrients and thus cause a tremendous loss of body weight. Such wasting away severely weakens the patient and is the main contributor to death in many cases.

Cancer begins with *just one cell* that has accumulated damaging alterations in certain genes through exposure to carcinogenic (cancer-causing) chemicals or other factors. These damaged genes, called *oncogenes* (*onco* = tumor, cancer), dictate the uncontrolled division of cancer cells. Oncogenes are mutated versions of normal genes that regulate natural cell growth and multiplication. Furthermore, cancer can also result from damage to various genes that normally suppress uncontrolled cell division (tumor-suppressor genes). The discovery of oncogenes and tumor-suppressor genes was the most brilliant outcome of

the famed "war on cancer" funded by the United States and other governments in the 1970s and 1980s. Sadly, zeroing in on cancer's cause has not led to the long-sought "cure for cancer." Although the incidence of some cancers is down (stomach and cervical cancer), the incidence of more cancers is up (lung, breast, and brain cancers, for example), and the overall rate of cancer deaths has actually increased over the past few decades. This is especially frustrating in light of the fact that the amount of money spent treating cancer in the United States has tripled in the last 10 years.

Many forms of cancer are preceded by observable structural changes in the tissue, called *preneoplastic lesions*, in which the appearance of the cells changes. An example is the development of *leukoplakia* (loo"ko-pla'ke-ah; "white plates"), white patches in the mouth (on the gums, tongue). These patches may result from heavy smoking or from long-term irritation due to badly fitting dentures. Although preneoplastic lesions sometimes progress quickly to cancer, they may remain stable for a long time and even revert to a normal condition if the environmental stimulus is removed.

The American Medical Association makes the following rough estimates of the most common cancer-causing agents and behaviors: Natural chemicals in food may cause 35% of all cancers; tars and other carcinogens in tobacco (smoking),

Cytokinesis

The division of the cytoplasm that completes the separation of one cell into two is called **cytokinesis,** literally "cells moving (apart)." It begins during the final phases of mitosis and is completed after mitosis ends (see Figure 2.18). Essentially, a ring of contractile microfilaments in the center of the original cell constricts to pinch that cell in two. The two new cells, called daughter cells, then enter the interphase part of their life cycle.

After studying Figure 2.18, try to answer the following question: The drug *vinblastine*, which blocks mitosis, is used in cancer therapy to stop the runaway division of cancer cells. Vinblastine inhibits the assembly and growth of microtubules. Can you explain exactly how that prevents mitosis?

Without microtubules, the mitotic spindle cannot form, and without the mitotic spindle, mitosis and cell division are impossible. ∎

30%; infections, 10%; sexual and reproductive behavior, 7% (for example, sex with many partners is associated with cancer of the uterus); occupational hazards, 4%; and alcohol, 3%. Other, less common factors account for the remaining 11%.

Cancer management techniques vary, but the usual sequence is as follows:

1. Diagnosis. Screening procedures, such as examining one's breasts or testicles for lumps, can detect some cancers very early and have saved many lives. Even so, most tumors are recognized only after they cause symptoms. A *biopsy* (bi'op-se; "life viewing") is another diagnostic procedure used to identify cancers: A sample of the primary tumor is surgically removed and examined under the microscope for structural changes typical of malignant cells. Tumors may also be recognized by the medical imaging techniques (X rays, MRI, CT) that are described in Chapter 1.

2. Staging. Several methods (physical examination, histological examination, laboratory tests, and imaging techniques) determine the extent of the disease (size of the neoplasm, degree of metastasis). The patient's probability of cure is designated according to stage: Stage 1 has the greatest probability, stage 4 the least.

3. Treatment. Most cancers are removed by surgery. To destroy the metastasized cells, surgery is commonly followed by radiation therapy (X irradiation of the primary

President Nixon signs the federal Cancer Act in 1971, beginning the war on cancer. Despite advances in research and treatment, cancer deaths have actually increased.

tumor and treatment with radioactive isotopes) and chemotherapy (treatment with anticancer drugs, usually administered intravenously or orally). Anticancer drugs have unpleasant side effects, because most of them target all the rapidly dividing cells in our body, including normal ones. The side effects of chemotherapy—nausea, vomiting, and loss of hair—are extremely distressing to the patient.

Treatment can arrest and cure about half of all cancer cases. Many

cancers, however, have very low survival rates. The deadliest cancers include lung cancer, cancers of the digestive organs, ovarian cancer, and a cancer of the pigment cells of the skin, called melanoma.

Many new treatments for cancer are under development. The most promising methods focus on delivering anticancer drugs more exclusively and precisely to the cancer cells and on increasing the effectiveness of the body's own immune system in the fight against cancer cells.

Developmental Aspects of Cells

Youth

We all begin life as a single cell, the fertilized egg, from which arise all the cells of our bodies. Early in embryonic development, the cells begin to specialize: Some become liver cells; some become nerve cells;

others become the transparent lens of the eye. Every cell in the body carries the same genes. (A *gene*, simply speaking, is a segment of DNA that encodes for a specific protein and therefore dictates a specific cell function.) If all cells have identical genes, how can one cell be so different from another? This question is currently the subject of extensive research. Apparently, cells in various regions of the developing embryo are exposed to different chemical signals that channel the cells into specific pathways of development. In a young embryo that consists of just a few cells, slight differences in oxygen and carbon dioxide concentrations between the more superficial and the deeper cells may be the major signals. But as development continues, cells begin to release chemicals that influence the development of neighboring cells by switching some of their genes "on" or "off." Some genes are active in all our cells; for example, all cells must carry out protein synthesis and make ATP. However, the genes for enzymes that catalyze the synthesis of specialized proteins such as hormones or mucus are activated only in certain cell populations. Hence, the secret of cell specialization lies in the kinds of proteins made and reflects differential gene activation in the different cell types.

Cell specialization leads to *structural* variation among the cell types in the body. Different organelles come to predominate in different cells. For example, muscle cells make tremendous quantities of actin and myosin protein, and their cytoplasm fills with actin microfilaments. High-energy cells, like those in the liver, make more enzymes for the manufacture of ATP and have more mitochondria than other cells. Phagocytic cells produce more lysosomal enzymes, and contain many lysosomes. Through such events arise the 200 or so different cell types of our body. The development of specific and distinctive features within cells is called **cell differentiation**.

Most organs are well-formed and functional long before birth, but the body continues to grow by forming more cells throughout childhood and adolescence. Once adult size is reached, cell division slows considerably and occurs primarily to replace short-lived cell types and to repair wounds.

As you can see, cell proliferation and differentiation are precisely orchestrated events. This precise coordination is disrupted in cancer, a disease in which cells multiply uncontrollably and lose some of their differentiated features. Cancer is discussed in the box on pages 42–43. ■

Aging

There is no doubt that cellular aging occurs and that it accounts for most problems associated with old age. While the process of aging is very complex, most theories recognize two main factors: Some researchers believe that aging is due to progressive "wear and tear" on the body, whereas others believe that aging is "programmed" into our genes.

Advocates of the *wear-and-tear theory* attribute aging to little chemical insults that build up and assault our cells throughout life. For example, environmental toxins such as pesticides, alcohol, carbon monoxide, and bacterial toxins may damage cell membranes, poison enzyme systems, or cause "mistakes" in DNA replication. Our cells fight to detoxify the destructive chemicals and repair the damaged DNA, but (according to the theory) these protective mechanisms are eventually overwhelmed. A specific wear-and-tear theory postulates that free radicals (p. 32), the highly reactive byproducts of normal metabolism, build up and progressively damage the essential molecules of our cells. Another theory holds that glucose (blood sugar) progressively links together the proteins in and between our cells, disrupting the protein functions that are vital to life. Supporting this cross-linking theory is the fact that people with diabetes, who have high levels of blood sugar, usually age faster and die sooner than average.

Health statistics also suggest that obese people have shortened life spans. This holds true even when all the life-threatening diseases that accompany obesity such as heart disease and diabetes have been factored out. In nutrition tests, slightly *undernourished* rats live longer than normal.

The *genetic theories of aging* propose that aging is programmed into our genes. These theories originated from the observation that the body structure ages in a predictable, patterned way, as if aging were a normal part of human development (and development is known to be controlled by genes). Rats and fruit flies can be bred to live longer than usual, and genes that increase and decrease longevity have been identified in animals.

A third group of theories combines the wear-and-tear theories with the genetic theories. One of these combined theories proposes that the buildup of free radicals or other harmful molecules over time activates certain genes that accelerate aging.

Finally, there are some theories of aging that are difficult to classify with the above types. One such theory proposes that aging occurs because errors in DNA replication accumulate over time as our cells divide. Another theory attributes aging to a declining ability of the immune system to destroy microorganisms or tumors.

Related Clinical Terms

Anaplasia (an″ah-pla′ze-ah) (*an* = without, not; *plas* = form) An abnormality in cell structure. For example, some cancer cells resemble not their parent cells but undifferentiated or embryonic cells.

Dysplasia (dis-pla′ze-ah) (*dys* = abnormal) A change in cell size, shape, or arrangement due to long-term irritation or inflammation (from infections, for example).

Gerontology (jer″on-tol′o-je) (*geron* = old man; *logos* = study of) The study of the aging process and the elderly.

Hyperplasia (hi″per-pla′ze-ah; "excess shape") Excessive cell proliferation; hyperplastic cells, however, do retain their normal form and arrangement within tissues.

Hypertrophy (hi-per′tro-fe; "excess growth") Growth of an organ or tissue due to an increase in the size of its cells. Hypertrophy is a normal response of skeletal muscle cells to exercise. Hypertrophy differs from hyperplasia, the condition in which cells increase in number but not in size.

Necrosis (ne-kro′sis) (*necros* = death; *osis* = process; condition) Death of a cell or group of cells due to injury or disease.

Chapter Summary

INTRODUCTION TO CELLS (p. 24)

1. Cells are the basic structural and functional units of life.

2. There are about 200 cell types in the human body, which vary widely in size, shape, and function. In this chapter, however, the features shared by all cells are emphasized.

3. Each cell has three main regions: plasma membrane, cytoplasm, and nucleus.

THE PLASMA MEMBRANE (pp. 24–28)

Structure (pp. 24–25)

1. The plasma membrane defines the cell's outer boundary. The fluid mosaic model interprets this membrane as a flexible bilayer of lipid molecules (phospholipids, cholesterol, and glycolipids), within which are proteins. When viewed by electron microscopy, the membrane has two dark layers (phospholipid heads) that sandwich an inner, light layer (phospholipid tails).

2. Most proteins in the membrane are integral proteins and extend entirely through the membrane. Peripheral proteins, by contrast, help support the membrane from its cytoplasmic side.

3. Sugar groups of membrane glycoproteins and glycolipids project from the cell surface and contribute to the cell coat (glycocalyx), which functions in cell-to-cell adherence and recognition.

Functions (pp. 25–28)

4. Besides acting as a fragile barrier to protect the cell contents, the plasma membrane determines what enters and leaves the cell. Only small, uncharged molecules may diffuse freely through the membrane; larger or charged molecules must pass by transport mechanisms that involve the integral proteins. Some membrane proteins also serve as receptors for extracellular molecules that bind to the plasma membrane.

5. Large particles and macromolecules pass through the membrane by bulk transport (exocytosis and endocytosis). In exocytosis, membrane-lined cytoplasmic vesicles fuse with the plasma membrane and release their contents to the outside of the cell.

6. Endocytosis brings large substances into the cell, as packets of plasma membrane fold in to form cytoplasmic vesicles. If the substance is a particle, the process is called phagocytosis; if the substance is dissolved molecules in the extracellular fluid, the process is known as pinocytosis. Receptor-mediated endocytosis is selective: Specific molecules attach to receptors on the membrane before being taken into the cell.

THE CYTOPLASM (pp. 28–35)

Cytosol (Cytoplasmic Matrix) (p. 28)

1. The cytosol, or cytoplasmic matrix, is the viscous substance in which cytoplasmic organelles and inclusions are embedded.

Cytoplasmic Organelles (pp. 28–34)

2. Each organelle performs its own specific functions. The various cell types in the body have different numbers of each organelle type.

3. Mitochondria are thread-like organelles covered by two membranes, the inner of which forms shelf-like cristae. Mitochondria are the main sites of ATP synthesis.

4. Ribosomes are dark-staining granules that consist of two subunits, each made of protein and ribosomal RNA. Ribosomes are the sites of protein synthesis. Free ribosomes make proteins used in the cytosol.

5. The rough endoplasmic reticulum is a ribosome-studded system of membrane-lined envelopes (cisternae). Its ribosomes make proteins, which enter the cisternae, and which may ultimately be secreted by the cell. The rough ER also makes all the cell's membranes.

6. The smooth endoplasmic reticulum, a network of membrane-lined tubes containing no ribosomes, is involved in the metabolism of lipids.

7. The Golgi apparatus is a stack of disc-shaped envelopes that has a cis (convex) and a trans (concave) face. It sorts the products of the rough endoplasmic reticulum and then sends these products, in membrane-bound vesicles, to their proper destination. Lysosomes and secretory granules also arise from the Golgi apparatus.

8. Lysosomes are spherical, membrane-lined sacs of digestive enzymes. They digest deteriorated organelles and endocytosed substances. Dying cells are killed and digested by enzymes released from their own lysosomes.

9. Peroxisomes are membranous, enzyme-containing sacs that protect the cell from free radicals and hydrogen peroxide. Peroxisomes also use hydrogen peroxide to break down some organic poisons and carcinogens.

10. The cytoskeleton includes protein rods of three distinct types—microtubules, actin microfilaments, and intermediate filaments—all in the cytosol. Microtubules, which radiate out from the centrosome, give the cell shape; they also organize the distribution and the transport of various organelles within the cytoplasm. Actin microfilaments interact with myosin to produce contractile forces. Both microtubules and microfilaments tend to be labile, but intermediate filaments, which act to resist tension placed on the cell, are stable.

11. Centrioles are paired, barrel-shaped organelles made up of microtubules. They lie in the centrosome. The centrosome anchors the microtubules of the cytoskeleton, and it anchors the mitotic spindle during cell division.

Cytoplasmic Inclusions (p. 35)

12. Inclusions are impermanent structures in the cytoplasm. Examples include food stores, such as glycogen granules and lipid droplets.

THE NUCLEUS (pp. 35–39)

1. The nucleus contains genetic material (DNA) and is the control center of the cell. Most cells have one centrally located nucleus shaped like a sphere or an egg.

Nuclear Envelope (p. 35)

2. The nucleus is surrounded by a nuclear envelope, which is penetrated by nuclear pores. The nuclear envelope is continuous with the rough endoplasmic reticulum.

Chromatin and Chromosomes (pp. 35–37)

3. Chromatin is strand-like material (DNA and histones) in the nucleus. This chromatin is distributed in chromosomes. During cell division, all chromatin is highly coiled, making the chromosomes appear as thick rods. In nondividing cells, the chromatin is a mixture of dark, coiled regions (condensed chromatin) and light, uncoiled regions (extended chromatin).

4. The DNA molecule is a double helix consisting of four types of nucleotides (with bases of thymine, adenine, cytosine, and guanine).

5. A structural subunit of chromatin is the nucleosome, a cluster of eight histone proteins around which DNA coils.

Nucleoli (pp. 37–39)

6. A nucleolus is a dark-staining body within the nucleus, associated with parts of several chromosomes. Nucleoli manufacture the subunits of ribosomes.

THE CELL LIFE CYCLE (pp. 39–42)

1. The cell life cycle is the series of changes a cell experiences from the time it forms until it divides.

Interphase (p. 39)

2. Interphase is the nondividing phase of the cell life cycle. It consists of G_1, S, and G_2 subphases. During G_1, the cell grows vigorously; during S, DNA replicates; and during G_2, the final preparations for division are made.

Cell Division (pp. 39–42)

3. Cell division, essential for body growth and repair, occurs during the M (mitotic) phase. Cell division has two distinct aspects: mitosis and cytokinesis.

4. Mitosis, the division of the nucleus, consists of prophase, metaphase, anaphase, and telophase. Mitosis parcels out the replicated chromosomes to two daughter nuclei. Cytokinesis, completed after mitosis, is the division of the components of the cytoplasm.

DEVELOPMENTAL ASPECTS OF CELLS (pp. 43–44)

Youth (pp. 43–44)

1. The first cell of a human is the fertilized egg. Cell differentiation begins early in development and is thought to reflect differential gene activation. Different organelles come to predominate in different cell types.

2. During adulthood, cell numbers remain fairly constant, and cell division occurs primarily to replace lost cells.

Aging (p. 44)

3. Aging of cells (and of the whole body) may reflect wear and tear from chemical insults, and/or it may be a genetically pro-

grammed event. Aging may reflect progressive disorders of immunity or accumulated "mistakes" in gene copying prior to cell divisions.

Review Questions

Multiple Choice and Matching Questions

1. The endocytotic process in which particulate matter is brought into the cell is called (a) phagocytosis, (b) pinocytosis, (c) exocytosis.

2. The nuclear substance composed of histone proteins and DNA is (a) chromatin, (b) the nuclear envelope, (c) nucleoplasm, (d) nuclear pores.

3. Final preparations for cell division are made during this stage of the cell life cycle: (a) G_1, (b) G_2, (c) M, (d) S.

4. The fundamental bilayered structure of the plasma membrane is determined almost exclusively by (a) phospholipid molecules, (b) peripheral proteins, (c) cholesterol molecules, (d) integral proteins.

5. Identify the cell structure or organelle described by each of the following statements.

(a) This is a stack of 3–10 membrane discs, associated with vesicles. *Golgi apparatus*
(b) A continuation of the nuclear envelope forms this cytoplasmic organelle. *RER*
(c) In nondividing cells, the highly coiled parts of chromosomes form this type of chromatin. *condensed*
(d) This type of endoplasmic reticulum is a network of hollow tubes, not envelopes. *smooth*
(e) This is a spherical cluster of eight histone molecules wrapped by a DNA strand. *nucleosome*
(f) These are the cytoskeletal rods with the thickest diameter (choose from microtubules, microfilaments, intermediate filaments).
(g) This is the only organelle with DNA, cristae, and matrix granules. *mitochondria*

6. Circle the false statement about centrioles. (a) They start to duplicate in G_1; (b) they lie in the centrosome; (c) they are made of microtubules; (d) they are membrane-lined barrels lying parallel to each other.

7. The trans face of the Golgi apparatus (a) is its convex face (b) is where products leave the Golgi apparatus in vesicles, (c) receives transport vesicles from the rough ER, (d) is in the very center of the Golgi stack.

8. Circle the false statement about lysosomes (a) They have the same structure and function as peroxisomes; (b) they form by budding off the Golgi apparatus; (c) lysosomal enzymes do not occur freely in the cytosol in healthy cells; (d) they are abundant in phagocytic cells.

9. Name the appropriate stage of mitosis (telophase, anaphase, metaphase, or prophase) for each of the following.

M (a) the chromosomes are lined up in the middle of the cell
P (b) the nuclear membrane fragments
T (c) the nuclear membrane re-forms
P (d) the mitotic spindles form
A (e) the chromosomes (chromatids) are V-shaped

10. Name the proper cytoskeletal element (microtubules, actin microfilaments, or intermediate filaments) for each of the following.

M (a) give the cell its shape
I (b) resist tension placed on a cell
M (c) radiate from the cell center
A (d) interact with myosin to produce contraction force
I (e) are the most stable

11. Different organelles are abundant in different cell types. Match the cell types with their abundant organelles by placing the correct letter from the column at right into each blank at left. Follow the hints provided in parentheses.

Cell Type	Abundant Organelle
b (1) cell lining the small intestine (assembles fats)	**(a)** mitochondria
α (2) white blood cell (phagocytic)	**(b)** smooth ER
c/b (3) liver cell (detoxifies carcinogens)	**(c)** peroxisomes
d (4) muscle cell (highly contractile)	**(d)** microfilaments
e (5) mucous cell (secretes protein product)	**(e)** rough ER
f (6) cell in outer layer of skin (withstands tension)	**(f)** intermediate filaments
α (7) kidney tubule cell (makes and uses large amounts of ATP)	**(g)** lysosomes

12. Some cellular organelles and structures resemble one another in electron micrographs and are easy to confuse. Which pair is least likely to be confused with one another? (a) glycogen granules and ribosomes, (b) mitochondria and nuclei, (c) smooth ER and Golgi apparatus, (d) lysosomes and fat droplets, (e) intermediate filaments and microfilaments.

13. Cancer is the same as (a) all tumors, (b) all neoplasms, (c) all malignant neoplasms, (d) benign tumors, (e) AIDS.

14. All enzymes are made (a) in mitochondria, (b) in the nucleus by DNA, (c) in lysosomes, (d) on ribosomes.

Short Answer and Essay Questions

15. Some of the intrinsic proteins and lipids in the plasma membrane have sugar chains attached to their external side. What role does this sugar coat play in the life of the cell?

16. Name two differences in structure between the smooth ER and the rough ER.

17. List all the cytoplasmic organelles that are composed (at least in part) of lipid-bilayer membranes. Then list all the cytoplasmic organelles that are not membranous.

18. What is the difference between the cytoplasm and the cytosol (cytoplasmic matrix)?

19. In this chapter, many cell structures are given descriptive nicknames to help you remember their functions and forms.

Write the name of the organelle or structure to which each of the following nicknames refers, then briefly explain why each nickname is appropriate.

(a) membrane factory
(b) a cell's "bones and muscles"
(c) demolition crew (also called suicide sacs)
(d) central post office
(e) the cell's control center
(f) a ribosome-making machine
(g) power plants
(h) "beer barrels" with walls of microtubules
(i) the railroad tracks of the cytoskeleton

20. Thelma, an anatomy student, said the Golgi apparatus reminded her of a drooping stack of pancakes with drops of syrup flowing off. Explain what she meant by this analogy.

21. Martin missed a point on his anatomy test because he thought *nucleus* and *nucleolus* were the same word and the same structure. Distinguish the nucleus from the nucleolus.

22. (a) What is chemotherapy for cancer (as currently employed)? (b) How does it work? (c) What are some of its side effects?

Critical Thinking and Clinical Application Questions

23. Chrissy, who is 30 years old, stays 5 kg (11 pounds) underweight because she thinks being thin will slow her aging process. After warning her not to lose any more weight or to become anorexic, her doctor admitted that there is at least some scientific evidence to support her view. What is that evidence?

24. After studying the information box on cancer in this chapter, deduce which of the various types of cancer causes the most deaths in the United States. (Hint: Most skin cancers are easily cured.)

25. Kareem had a nervous habit of chewing on the inner lining of his lip with his front teeth. The lip grew thicker and thicker from years of continual irritation. Kareem's dentist noticed his greatly thickened lip, then told him to have it checked to see if the thickening was a tumor. A biopsy revealed hyperplasia and scattered areas of dysplasia, but no evidence of neoplasia. What do these terms mean? Did Kareem have cancer of the mouth?

3

Basic Embryology

Student Objectives

1. List the practical and clinical reasons for studying embryology.

Introduction to Embryology (pp. 49–50)

2. Distinguish the embryonic period from the fetal period of development.

3. Sketch the basic structural plan of the adult body, which is established during the embryonic period.

The Embryonic Period (pp. 50–58)

4. Describe the earliest stage of development, from zygote to blastocyst (week 1).

5. Describe how the embryo becomes a two-layered disc (week 2).

6. Explain the formation of the three germ layers (week 3).

7. Discuss how the body folds from a flat disc into its three-dimensional, tubular shape (week 4).

8. List the main derivatives of each germ layer.

9. Describe the main events of the second month of development.

The Fetal Period (pp. 58–62)

10. List the major events of fetal development.

This chapter describes the **prenatal period** (pre-na'tal; "before birth"), the time between conception and birth. In just 38 weeks, a single fertilized egg cell develops into a fully formed human being, the newborn infant. The body does not change this much again during the entire life span of 70–90 years. In this chapter, we describe how an embryo bridges the vast structural gulf between a single cell and the human body.

A knowledge of embryology is valuable for two reasons. By studying the body in its most basic form, we can better understand (1) the more complex anatomy of the adult and (2) the origin of birth defects (see the box on pp. 60–61). Since about 6% of children have some significant birth defect, this information is important indeed, particularly for prospective parents and those who counsel them.

Introduction to Embryology

The prenatal period is divided into two stages (Figure 3.1): the **embryonic period,** which spans the first 8 weeks, and the **fetal** (fe'tal) **period,** which encompasses the remaining 30 weeks. The embryonic period is an exceptionally "busy" one: By its end, all major organs are in place, and the body looks distinctly human. In the longer fetal period that follows, the organs mostly grow larger and more complex.

During the fetal period, the individual is called a **fetus** (fe'tus; "the young in the womb"). The definition of the term **embryo,** however, is not so straightforward. In a loose sense, *embryo* applies to the developing individual throughout the embryonic

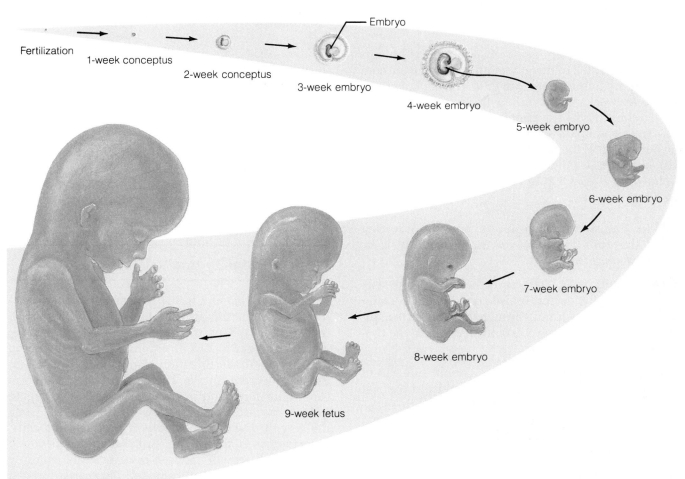

Fertilization

1-week conceptus

2-week conceptus

3-week embryo

Embryo

4-week embryo

5-week embryo

6-week embryo

7-week embryo

8-week embryo

9-week fetus

12-week fetus

Figure 3.1 Diagram showing the actual size of a human conceptus from fertilization to the early fetal stage (kun-sep'tus; "that which is conceived").

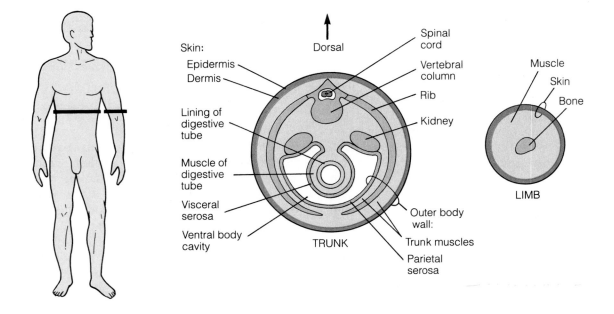

Figure 3.2 The adult human body plan (simplified cross section). The development of this plan is traced in this chapter.

period. More correctly, however, the term is not applied until week 2 of development, and some authors do not apply it until week 3 or 4. In this book, *embryo* is used for weeks 2–8.

To simplify the treatment of developmental anatomy in this chapter, we will limit our discussion to the derivation of the basic adult body plan. This plan, which can be visualized by month 2 of development, is shown in Figure 3.2. In the figure, note the following adult structures:

1. Skin. The skin has two layers: an outer covering called the *epidermis* and an inner, leathery layer called the *dermis.*

2. Outer body wall. The outer body wall consists mostly of trunk muscles. It also contains the dorsally located vertebral column, through which runs the spinal cord. Ribs attach to each bony vertebra in the thorax region of the trunk wall.

3. Body cavity and digestive tube. The slit-like part of the ventral body cavity, lined by visceral and parietal serosae, surrounds the digestive tube (stomach, intestines, and so on) in the trunk region. The digestive tube has a muscular wall and is lined internally by a sheet of cells. This lining is identified in the figure.

4. Kidneys and gonads. The kidneys lie directly deep to the dorsal body wall, in the lumbar region. The gonads (male testes and female ovaries) originate in a

similar position but migrate to other parts of the body during the fetal period.

5. Limbs. The limbs consist mostly of bone, muscle, and skin.

We can derive this adult body plan by following the events of month 1 of human development.

The Embryonic Period

Week 1: From Zygote to Blastocyst

Each month, one of a fertile woman's two ovaries releases an immature egg, called an *oocyte* ('Ovulation' in Figure 3.3). This cell then enters a *uterine tube,* which provides a direct route to the uterus (womb). **Conception**, the fertilization of an oocyte by a sperm, generally occurs in the lateral third of the uterine tube. The fertilized oocyte, now called a **zygote** (zi'gōt; "a union"), moves toward the uterus. Along the way, it divides repeatedly to produce two cells, then four cells, and so on. Since there is not much time for cell growth between divisions, the resulting cells become smaller and smaller. This early division sequence, called **cleavage**, provides the large number of cells needed as building blocks for the embryo.

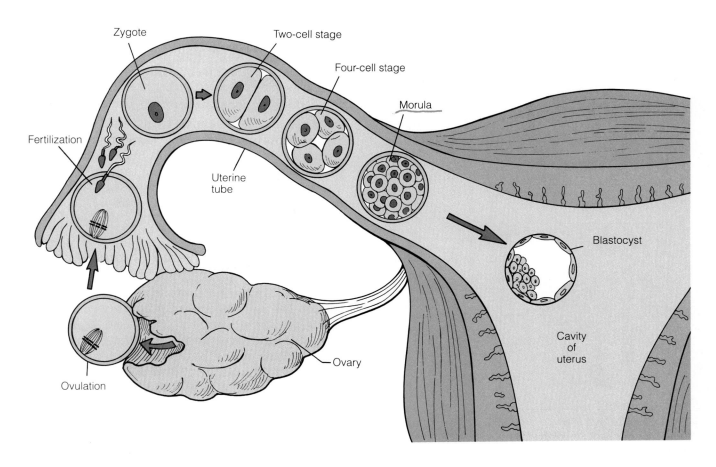

Figure 3.3   Fertilization and the events of the first 6 days of development. The ovary, uterus, and uterine tubes, which lie in the woman's pelvis, are shown in posterior view.

About 72 hours after fertilization, the cleavage divisions have generated a solid cluster of 12–16 cells called a **morula** (mor′u-lah; "mulberry"). During day 4, the morula—now consisting of about 60 cells—enters the uterus. Fluid appears between the cells and gathers into a central cavity. This new fluid-filled structure is called a **blastocyst** (blas′to-sist), a name that roughly means "hollow bag of germ cells" (*blast* = bud or sprout, germinator, or generator).

The blastocyst contains two distinct types of cells; in fact, this is the earliest stage where cell differentiation is evident (Figure 3.4a). The cluster of cells on one side of the cavity is called the **inner cell mass**, while the layer of cells around the outside of the cavity is called the **trophoblast** (trof′o-blast; "nourishment generator"). The inner cell mass will form the embryo itself, while the trophoblast will help form the placenta, the structure by which the fetus obtains nutrients from the mother. In this chapter, we will focus on the inner cell mass and the embryo, leaving the trophoblast and placenta for the discussion of the female reproductive system (Chapter 23).

The blastocyst stage lasts about 3 days, from day 4 to day 7.* For most of this time, the blastocyst floats freely in the cavity of the uterus, but on day 6, it starts to burrow into the wall of the uterus (Figure 3.4b). This process, called **implantation**, takes about a week to complete.

Try to answer the following questions about week 1 of development.

Sometimes a woman's ovaries release two eggs at once that are fertilized by two different sperm. Can you deduce the outcome? (Answer is at the top of p. 52).

---

*All dates given for the developmental events represent *average* times. Actual dates vary by 1–2 days or more among different pregnancies.

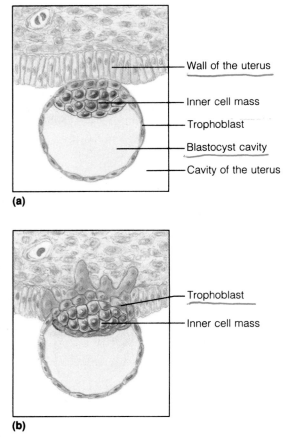

(a)

- Wall of the uterus
- Inner cell mass
- Trophoblast
- Blastocyst cavity
- Cavity of the uterus

(b)

- Trophoblast
- Inner cell mass

Figure 3.4 Implantation of the blastocyst during week 2 of development. **(a)** Blastocyst that has just adhered to the wall of the uterus (day 6). **(b)** Implanting embryo (day 7). The embryos in this figure are in cross-sectional view.

The result is nonidentical, or fraternal, twins, who resemble one another no more closely than typical brothers and sisters. Fraternal twins are also called dizygotic twins (*dizygotic* = from two zygotes).

In some pregnancies, the inner cell mass of a single blastocyst splits into two at the end of week 1. Can you deduce the consequence of this occurrence?

This produces identical twins, also called monozygotic twins (*monozygotic* = from one zygote). Less often, monozygotic twins result from a splitting that occurs earlier in week 1, during the early stages of cleavage. Overall, twins account for 1 in 90 births in the United States, and identical twins are about half as common as fraternal twins. ▪

Week 2: The Two-Layered Embryo

Figure 3.5 shows the embryo about 9 days after fertilization. The inner cell mass has now divided into two sheets of cells, the **epiblast** (ep′ĭ-blast; "upper germ layer") and the **hypoblast** (hi′po-blast; "lower germ layer"). Extensions of these cell sheets form two fluid-filled sacs. This arrangement resembles two balloons touching one another, with the epiblast and hypoblast at the area of contact. Together, the epiblast and hypoblast make up the **bilaminar** ("two-layered") **embryonic disc,** which will give rise to the whole body.

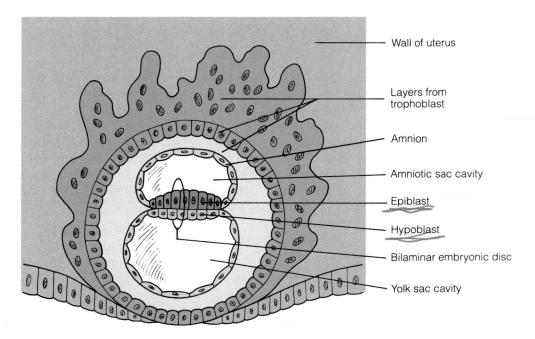

- Wall of uterus
- Layers from trophoblast
- Amnion
- Amniotic sac cavity
- Epiblast
- Hypoblast
- Bilaminar embryonic disc
- Yolk sac cavity

Figure 3.5 Embryo at day 9. Embryo is almost completely implanted in the wall of the uterus. It is now a bilaminar embryonic disc of epiblast and hypoblast situated between an amniotic sac and a yolk sac.

The sac adjacent to the epiblast is the *amniotic* (am"ne-ot'ik) *sac,* and the sac adjacent to the hypoblast is the *yolk sac.* The **amniotic sac** is defined by an external membrane called the **amnion,** and its internal **amniotic sac cavity** is filled with *amniotic fluid.* This fluid buffers the embryo and fetus from physical shock throughout pregnancy, allowing the mother to withstand jostling and bumping without harm to the fetus. The **yolk sac** holds a very small amount of yolk, an insignificant food source. Despite its insufficiency as a food sac, however, the human yolk sac is important because the digestive tube forms from part of it (see p. 56). Furthermore, tissue around the yolk sac gives rise to the earliest blood cells and blood vessels (see Chapter 16, p. 456).

You may have heard the expression "The mother's water broke just before she gave birth." Based on the information in the previous paragraph, can you guess what that means?

It means that the amniotic sac ruptures and its fluid is released, just before or during labor. ■

Week 3: The Three-Layered Embryo

The Primitive Streak and the Three Germ Layers

During week 3, the embryo changes from a two-layered disc to a three-layered disc. In so doing, it establishes the three *primary germ layers*—ectoderm, endoderm, and mesoderm—from which all body tissues develop. To provide a brief preview, the *ectoderm* ("outer skin") becomes the epidermis of the skin, as well as the brain and spinal cord; the *endoderm* ("inner skin") becomes the lining of the digestive tube; and the *mesoderm* ("middle skin") gives rise to many structures, especially the muscles and skeleton.

Germ-layer formation begins about day 15, when a raised groove called the **primitive streak** appears on the dorsal surface of the epiblast (Figure 3.6). Epiblast cells move inward (invaginate) at this streak to form a new germ layer between the epiblast and hypoblast (Figure 3.7). This new, middle layer is the **mesoderm.** Some of the invaginating epiblast cells enter the underlying hypoblast layer and mingle with hypoblast cells to form the **endoderm.** The epiblast cells that stay on the embryo's dorsal surface make up the **ectoderm.** In this way, the three primary germ layers of the body are established. With the appearance of the mesoderm, the bilaminar embryonic disc has become *trilaminar* (three-layered).

The three germ layers differ in the arrangement of their cells. In other words, they are different types of *tissue.* Ectoderm and endoderm are epithelial tissues, or epithelia—sheets of tightly joined cells (recall p. 6). Mesoderm, by contrast, is a *mesenchyme* tissue (mes'eng-kīm; "poured into the middle [of the embryo]"). A mesenchyme is any embryonic tissue with star-shaped cells that do not attach to one another. Thus, mesenchyme cells are free to migrate widely within the embryo.

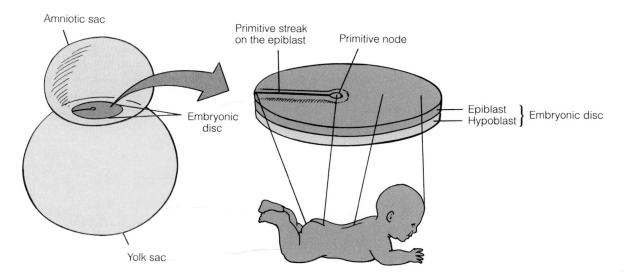

Figure 3.6   The primitive streak stage.   The primitive streak appears on the dorsal surface of the embryonic disc on about day 15. The inset below shows the relation of the primitive streak and embryonic disc to the body that forms later. Ultimately, the primitive streak is reduced to a strip of skin epidermis over the tailbone (coccyx) in the cleft between the buttocks.

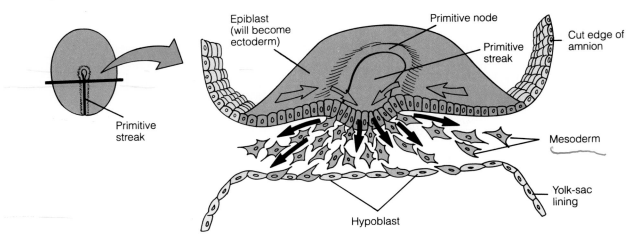

Figure 3.7   Transverse section through the embryonic disc at the primitive streak.
(Inset at upper left shows the orientation of the disc).   Epiblast cells migrate medially into the primitive streak to become mesoderm. Some invaginating epiblast cells contribute to the hypoblast layer to produce endoderm. The epiblast that remains on the surface is the ectoderm.

Figure 3.8   Formation of the mesoderm and the notochord in a 16-day embryo.
(**a**) Dorsal surface view of the embryonic disc. Orientation is the same as in Figure 3.7, with the future head region at the top of the diagram (see the inset showing the baby at upper right). The solid blue lines show the path taken by surface epiblast cells as they migrate toward the primitive streak. The broken pink arrows show the path of the same cells as they migrate as mesoderm beneath the surface cell layer. The epiblast cells that enter at the primitive node contribute to the notochord, which defines the body axis. (**b**) Cross-sectional view of the embryonic disc, through the primitive streak. (**c**) Cross section taken anterior to the primitive streak, in the future thorax. This is the same view as that of the adult body plan in Figure 3.2.

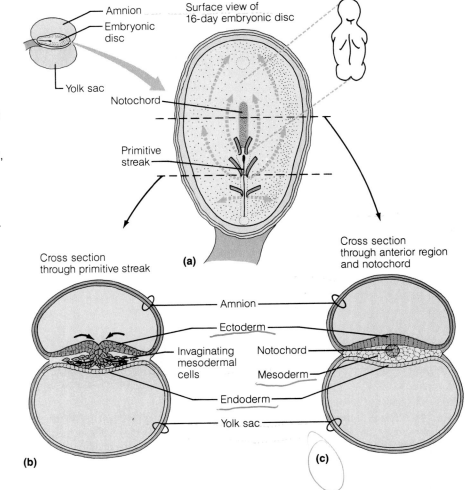

The Notochord

At one end of the primitive streak is a swelling called the **primitive node** (Figures 3.6 through 3.8a). The epiblast cells that invaginate at this node migrate straight anteriorly, and together with a few cells from the underlying endoderm, form a rod called the **notochord** (Figure 3.8a). The notochord defines the body *axis* (the midline that divides the left and right sides of the body) and is the site of the future vertebral column. The notochord appears on day 16, and by day 18 it reaches the future head region.

Neurulation

As the notochord develops, it signals the overlying ectoderm to start forming the spinal cord and brain (Figures 3.9 and 3.10). This event is called **neurulation** (nu″roo-la′shun). Specifically, the ectoderm in the dorsal midline thickens into a **neural plate,** and then starts to fold inward as a **neural groove.** This groove deepens until a hollow **neural tube** pinches off into the body. Closure of the neural tube begins in the neck region at the end of week 3, and then proceeds both cranially and caudally. Closure is completed by the end of week 4. The cranial part of this neural tube becomes the brain, and the rest becomes the spinal cord.

As shown in Figure 3.9, **neural crest** cells are pulled into the body along with the invaginating neural tube. The neural crest cells originate from ectodermal cells just lateral to the future neural tube, and they lie just external to that tube once its closure is completed. The neural crest forms the sensory nerve cells and some other important structures, as described below.

The Mesoderm Begins to Differentiate

In the middle of week 3, the mesoderm lies lateral to the notochord on both sides (Figure 3.10a). By the end of this week, the mesoderm on each side has divided into three regions (Figure 3.10b–d):

1. Somites (so′mītz; "bodies"). Somites are the blocks of mesoderm immediately lateral to the notochord. They represent our first body segments. About 40 pairs of somites are present by the end of week 4. These run along the future back region from *head to tail* (crown to rump).

2. Intermediate mesoderm. The intermediate mesoderm forms a spherical mass just lateral to each somite.

3. Lateral plate. Initially, the lateral plate—the most lateral unsegmented part of the mesoderm—forms a continuous layer from head to tail. Soon, a wedge-like

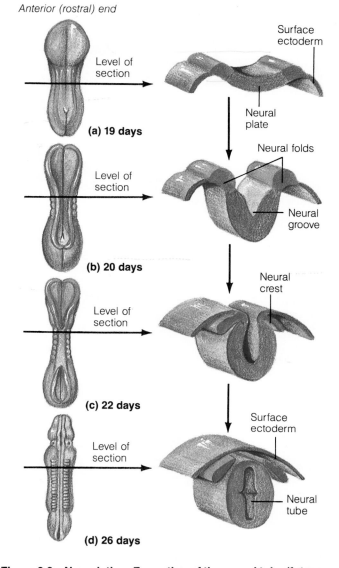

Anterior (rostral) end

Surface ectoderm

Level of section

Neural plate

(a) 19 days

Neural folds

Level of section

Neural groove

(b) 20 days

Neural crest

Level of section

(c) 22 days

Surface ectoderm

Level of section

Neural tube

(d) 26 days

Figure 3.9 Neurulation: Formation of the neural tube (future spinal cord and brain). At left, embryos are shown in dorsal surface view, with head above and tail below. The figures at right are transverse sections of the ectoderm. Ectoderm folds inward to form the neural tube. Also note the neural crest.

space, called the **coelom** (se′lom; "cavity"), divides the lateral plate into two layers: the **somatic mesoderm** (so-mat′ik; "body"), apposed to the ectoderm, and the **splanchnic mesoderm** (splangk′nik; "viscera"), apposed to the endoderm. The coelom that intervenes between the splanchnic and somatic mesoderm will become the ventral body cavity.

When you compare the cross sections in Figure 3.10 with the adult body plan in Figure 3.2, you can begin to see a few similarities, especially if you match structures according to their colors. The main difference between the 3-week embryo and the adult body

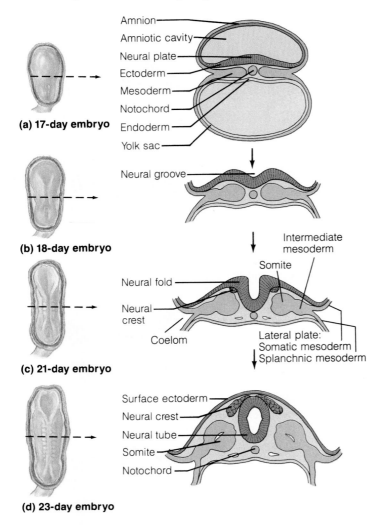

(a) 17-day embryo

Amnion
Amniotic cavity
Neural plate
Ectoderm
Mesoderm
Notochord
Endoderm
Yolk sac

Neural groove

(b) 18-day embryo

Intermediate mesoderm
Somite
Neural fold
Neural crest
Coelom
Lateral plate:
Somatic mesoderm
Splanchnic mesoderm

(c) 21-day embryo

Surface ectoderm
Neural crest
Neural tube
Somite
Notochord

(d) 23-day embryo

Figure 3.10   Changes in the mesoderm near the end of week 3 of development. Diagrams on the left show embryos in dorsal surface view. Diagrams on the right are cross sections. The mesoderm forms somites, intermediate mesoderm, and somatic and splanchnic mesoderm. The formation of the neural tube by ectoderm is also shown.

plan is that the embryo is still a flat disc. The three-dimensional, cylindrical body shape forms during week 4.

Week 4: The Body Takes Shape

Folding

The embryo achieves a cylindrical body shape when its sides fold medially and it lifts off the yolk sac by protruding into the amniotic cavity (Figure 3.11). In the simplest sense, this resembles three stacked sheets of paper folding into a tube. At the same time, the head and tail regions fold under, as shown on the right side of Figure 3.11a. As a result of this folding, the embryo acquires a tadpole shape by day 24. As the

embryo becomes cylindrical, it encloses a tubular part of the yolk sac, called the **primitive gut** (future digestive tube). This tube is lined by endoderm. The embryo remains attached to the yolk sac below by a narrow duct located at the future navel. This duct becomes incorporated into the umbilical cord.

Derivatives of the Germ Layers

At this point, the basic human body plan has been attained. Figure 3.12b is a cross section through the trunk of a 1-month-old embryo that can be compared directly to the adult body section in Figure 3.2. We can use such a comparison to explain the adult derivatives of the germ layers. These derivatives are summarized in Table 3.1 on page 58.

Derivatives of Ectoderm. The ectoderm becomes the brain, spinal cord, and epidermis of the skin. The early epidermis, in turn, produces the hair, fingernails, toenails, sweat glands, and the skin's oil glands. Neural crest cells, from ectoderm, give rise to the sensory nerve cells. Furthermore, much of the neural crest breaks up into a mesenchyme tissue, which wanders widely through the embryonic body. These wandering neural crest derivatives produce such varied structures as the pigment cells and the bones of the face.

Derivatives of Endoderm. The endoderm becomes the inner epithelial lining of the digestive tube and its derivatives (respiratory tubes, bladder). It also gives rise to the secretory cells of glands that develop from gut-lining epithelium (the liver and pancreas, for example).

Derivatives of the Mesoderm and Notochord. The mesoderm has a complex fate, so you may wish to start by reviewing its basic parts in Figure 3.12: the somites, intermediate mesoderm, and somatic and splanchnic mesoderm. Also note the location of the notochord.

The *notochord* gives rise to an important part of our spinal column, the springy cores of the discs between our vertebrae. These spherical centers, each called a *nucleus pulposus* (pul-po′sus), give our vertebral column some bounce as we walk.

Each of the *somites* divides into three parts. One part is the **sclerotome** (skle′ro-tōm; "hard piece"). Its cells migrate medially, gather around the notochord and the neural tube, and produce the vertebra and rib at the associated level. The lateralmost part of each somite is a **dermatome** ("skin piece"). Its cells migrate externally until they lie directly deep to the ectoderm, where they form the dermis of the skin in the dorsal part of the body. The third part of each somite is the **myotome** (mi′o-tōm; "muscle piece"), which stays be-

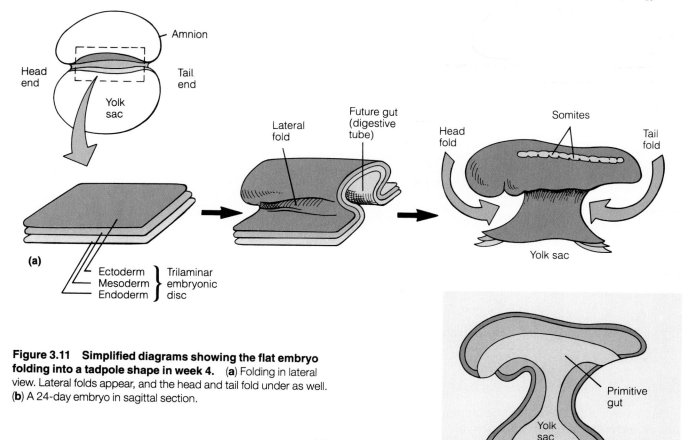

(a)

Ectoderm ⎫
Mesoderm ⎬ Trilaminar embryonic disc
Endoderm ⎭

Figure 3.11 Simplified diagrams showing the flat embryo folding into a tadpole shape in week 4. (**a**) Folding in lateral view. Lateral folds appear, and the head and tail fold under as well. (**b**) A 24-day embryo in sagittal section.

(b)

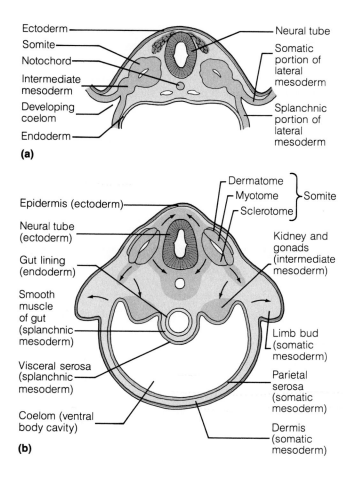

(a)

(b)

hind after the sclerotome and dermatome migrate away. Myotomes form the segmented trunk musculature of the body wall (Figures 1.6b and 3.13). Every myotome and dermatome gains its own segmental nerve supply, so the derivatives of the somites define the fundamental pattern of segmentation of the human body.

The *intermediate mesoderm,* lateral to each somite, forms the kidneys and the gonads. By comparing Figures 3.12 and 3.2, you can see that the intermediate mesoderm lies in the same relative location as the adult kidneys.

We now turn to the splanchnic and somatic mesoderm which are separated by the coelom body cavity. By now, the *splanchnic mesoderm* surrounds the gut (Figure 3.12b). It gives rise to the entire wall of the digestive tube, except the inner epithelium layer; that is, it forms the musculature and the slippery visceral

Figure 3.12 The germ layers in week 4 and their adult derivatives. These cross-sectional views should be compared to Figure 3.2. (**a**) The flat embryo on day 22, just before it begins to fold into a tube. (**b**) The cylindrical human body plan on day 28. This cross section is taken from the trunk region, where the posterior limb buds (future legs) attach.

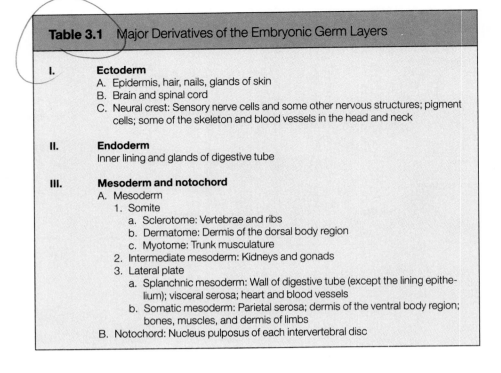

Table 3.1   Major Derivatives of the Embryonic Germ Layers

I.   Ectoderm
   A. Epidermis, hair, nails, glands of skin
   B. Brain and spinal cord
   C. Neural crest: Sensory nerve cells and some other nervous structures; pigment cells; some of the skeleton and blood vessels in the head and neck

II.   Endoderm
   Inner lining and glands of digestive tube

III.   Mesoderm and notochord
   A. Mesoderm
      1. Somite
         a. Sclerotome: Vertebrae and ribs
         b. Dermatome: Dermis of the dorsal body region
         c. Myotome: Trunk musculature
      2. Intermediate mesoderm: Kidneys and gonads
      3. Lateral plate
         a. Splanchnic mesoderm: Wall of digestive tube (except the lining epithelium); visceral serosa; heart and blood vessels
         b. Somatic mesoderm: Parietal serosa; dermis of the ventral body region; bones, muscles, and dermis of limbs
   B. Notochord: Nucleus pulposus of each intervertebral disc

serosa of the digestive tube. Splanchnic mesoderm also gives rise to the heart and most blood vessels.

Somatic mesoderm, just external to the coelom, produces the parietal serosa and the dermis of the skin in the ventral body region. Its cells migrate into the forming limbs and produce the bone, muscle, and dermis of each limb. Somatic mesoderm thus forms all components of the limbs except the epidermis of the skin, which arises from ectoderm.

Weeks 5–8: The Second Month of Embryonic Development

Figure 3.13 shows an embryo near the beginning of month 2. At this time, the embryo is only about a half

centimeter long! Around day 28, the first rudiments of the limbs appear as **limb buds**. The upper limb buds appear slightly earlier than the lower limb buds.

One can think of month 2 as the time when the body changes from tadpole-like to recognizably human in form (Figure 3.1). The limbs grow from rudimentary buds to fully formed extremities with fingers and toes. The head enlarges quickly and occupies almost half the volume of the entire body. The eyes, ears, and nose appear, and the face gains a human appearance as the embryonic period draws to a close. The protruding tail of the 1-month-old embryo disappears at the end of week 8. All major organs are in place by the end of month 2, at least in rudimentary form. In successive chapters we discuss the development of each organ system, so we will return to the events of month 2 often.

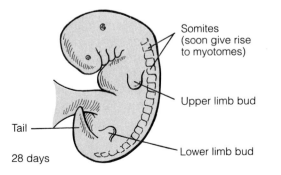

Somites (soon give rise to myotomes)

Upper limb bud

Tail

28 days

Lower limb bud

Figure 3.13   A 28-day embryo.   Note the limb buds. Also note the segmented somites, which soon give rise to segmented myotomes.

The Fetal Period

The main events of the fetal period—weeks 9 through 38—are listed chronologically in Table 3.2. As we stated earlier, the fetal period is a time of rapid growth of the body structures that were established in the embryo. The fetal period is more than just a time of growth, however. During the first half of this period, cells are still differentiating into specific cell types to form the body's distinctive tissues and are completing the fine details of body structure. Color photos of

(text continues on page 62)

Table 3.2 Developmental Events of the Fetal Period

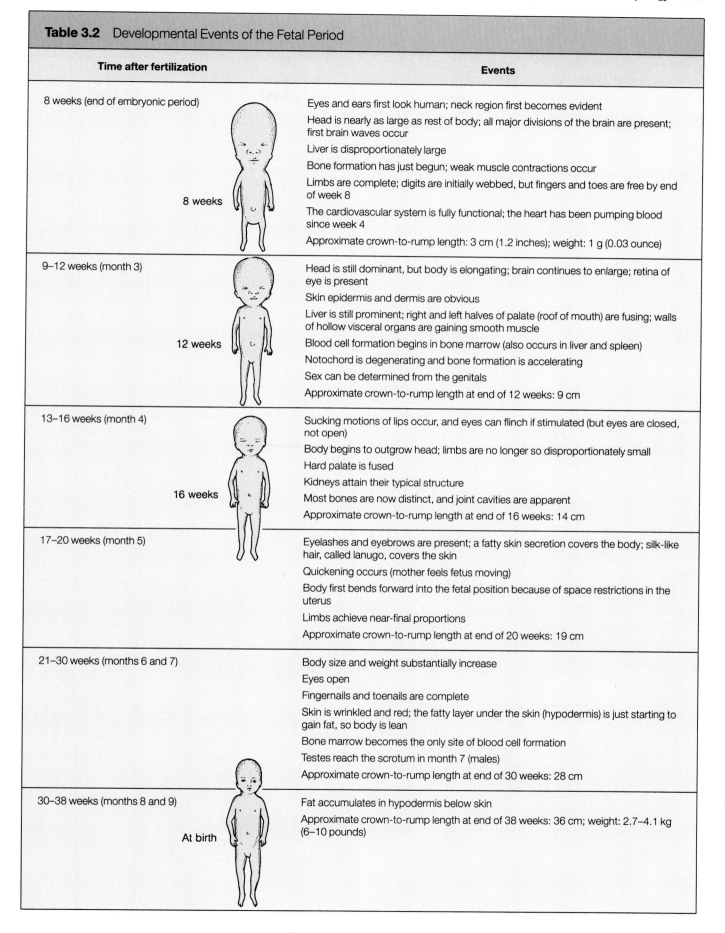

Time after fertilization	Events
8 weeks (end of embryonic period) 8 weeks	Eyes and ears first look human; neck region first becomes evident Head is nearly as large as rest of body; all major divisions of the brain are present; first brain waves occur Liver is disproportionately large Bone formation has just begun; weak muscle contractions occur Limbs are complete; digits are initially webbed, but fingers and toes are free by end of week 8 The cardiovascular system is fully functional; the heart has been pumping blood since week 4 Approximate crown-to-rump length: 3 cm (1.2 inches); weight: 1 g (0.03 ounce)
9–12 weeks (month 3) 12 weeks	Head is still dominant, but body is elongating; brain continues to enlarge; retina of eye is present Skin epidermis and dermis are obvious Liver is still prominent; right and left halves of palate (roof of mouth) are fusing; walls of hollow visceral organs are gaining smooth muscle Blood cell formation begins in bone marrow (also occurs in liver and spleen) Notochord is degenerating and bone formation is accelerating Sex can be determined from the genitals Approximate crown-to-rump length at end of 12 weeks: 9 cm
13–16 weeks (month 4) 16 weeks	Sucking motions of lips occur, and eyes can flinch if stimulated (but eyes are closed, not open) Body begins to outgrow head; limbs are no longer so disproportionately small Hard palate is fused Kidneys attain their typical structure Most bones are now distinct, and joint cavities are apparent Approximate crown-to-rump length at end of 16 weeks: 14 cm
17–20 weeks (month 5)	Eyelashes and eyebrows are present; a fatty skin secretion covers the body; silk-like hair, called lanugo, covers the skin Quickening occurs (mother feels fetus moving) Body first bends forward into the fetal position because of space restrictions in the uterus Limbs achieve near-final proportions Approximate crown-to-rump length at end of 20 weeks: 19 cm
21–30 weeks (months 6 and 7)	Body size and weight substantially increase Eyes open Fingernails and toenails are complete Skin is wrinkled and red; the fatty layer under the skin (hypodermis) is just starting to gain fat, so body is lean Bone marrow becomes the only site of blood cell formation Testes reach the scrotum in month 7 (males) Approximate crown-to-rump length at end of 30 weeks: 28 cm
30–38 weeks (months 8 and 9) At birth	Fat accumulates in hypodermis below skin Approximate crown-to-rump length at end of 38 weeks: 36 cm; weight: 2.7–4.1 kg (6–10 pounds)

A CLOSER LOOK **Focus on Birth Defects**

About 3% of all newborns exhibit some significant birth defect or **congenital abnormality** (*congenital* = present at birth). This figure increases to 6% by age 1, when initially unrecognized defects become evident. Many people are startled to hear that birth defects are this common. The 12 most common and serious defects (listed on the facing page) affect almost 3 children per 100 births. We will consider most of these specific birth defects in later chapters. Here, we provide some general information on congenital abnormalities and explore their causes.

Birth defects result from many factors, including defective chromosomes or genes, exposure to chemicals or viruses passed from the mother to the embryo, and exposure to destructive forms of radiation. Chemical, physical, or biological factors that disrupt developmental events and cause birth defects are called **teratogens** (ter'ah-to-jenz; "monster formers"). Teratogens include alcohol, nicotine, many drugs, X rays, radioactivity, and maternal infections (particularly German measles).

When a woman drinks alcohol, she intoxicates her fetus. Excessive alcohol consumption results in **fetal alcohol syndrome (FAS),** a condition typified by microcephaly (small head and jaw), heart abnormalities, joint defects, and deficient growth. FAS is now recognized as the most common cause of mental retardation in the United States. It is not known at this time whether even small amounts of alcohol might cause developmental problems, so many physicians advise pregnant women against drinking any alcohol at all. Similarly, infants born to mothers addicted to illegal drugs, such as crack cocaine, have severe mental, social, and emotional problems.

Nicotine from heavy tobacco smoking constricts the blood vessels that supply the uterus. The chemical substances in tobacco smoke also lessen the ability of fetal blood to carry oxygen. The subsequent oxygen deficiency impairs the growth and development of the fetus. Therefore, children born to heavy smokers tend to weigh less than average, and their mental development may be affected.

Thalidomide (thah-lid'o-mīd) was a sedative used by thousands of pregnant women in Europe from about 1958 to 1962 to alleviate morning sickness. Use of this drug during the period of limb differentiation (days 24–36) sometimes resulted tragically in the development of short, flipper-like limbs. Environmental toxins, such as heavy metals and pesticides, can also act as teratogens. Embryos and fetuses (and even children) are especially sensitive to the effects of these toxins.

An expectant mother must be particularly careful of her health and her chemical intake during the third to eighth weeks after conception, for most teratogen-caused birth defects originate at this time. It is easy to understand why weeks 3–8 are the "danger period": Insults delivered before week 3 tend to destroy critical early cell lines so that

A boy with fetal alcohol syndrome (FAS), shown in infancy and childhood. Typical features are unusually shaped lips, low body weight, and poorly aligned teeth. Behavioral traits of FAS often include difficulty in concentrating, impulsiveness, and mental retardation.

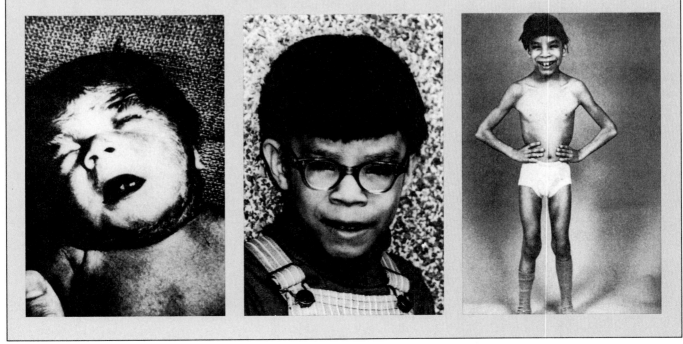

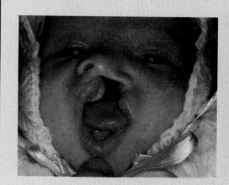

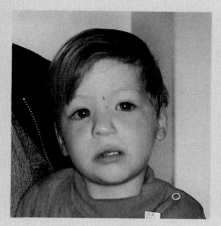

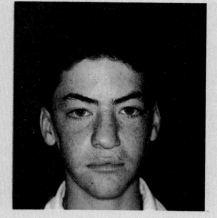

A boy born with a cleft lip and palate (above), which was repaired by surgery during infancy. As the boy grew to adolescence, there was little evidence of the original defect.

development cannot proceed and a natural abortion occurs. After week 8, by contrast, almost all organs are already in place, so teratogens may produce only minor structural malformations. The later, fetal period is not entirely risk-free, however: Exposure of the fetus to teratogens can affect body growth and organ functions, produce structural abnormalities of the eyes and brain, and lead to mental retardation.

Even though many different teratogens have been identified, it must be emphasized that most birth defects cannot be traced to any known teratogen; that is, for any individual deformed baby, it is usually impossible to determine what caused the defects. Most birth defects can be caused by many factors.

Birth defects are not the only evidence that the pathway from conception to birth is a difficult and dangerous one. Though it seems unbelievable, *most* human embryos die within the womb. About 15%–30% of all fertilized zygotes never implant in the uterine wall, either because they have lethal chromosomal defects or because the uterine wall is unprepared to receive them. Of the embryos that do implant, about one-third are so malformed that they die, most often during week 2 or 3. Smaller numbers are miscarried later in development.

Overall, at least 50% of fertilized human ova fail to produce living infants—and some studies put this number as high as 75%! We fail to appreciate how many embryos die before birth because embryos lost during the first 2 weeks are shed in the woman's next menstrual flow, before the woman ever learns she is pregnant. Some scientists speculate that this great loss of embryos is nature's screening process, which assures that most surviving babies are born healthy.

Most Common Birth Defects*

Birth defect	Number per 1000 births
1. Heart defects	7
2. Mental retardation	3
3. Pyloric stenosis (a valve in the stomach is too narrow)	3
4. Anencephaly (most of the brain is absent)	2
5. Spina bifida (vertebrae are incomplete, leading to damaged spinal cord)	1.5
6. Down syndrome[†] (child has distinct facial appearance, and mild mental retardation)	1.5
7. Cleft palate and cleft lip	1.5
8. Hypospadias (urine leaks from slit in underside of penis)	1.5
9. Clubfoot (feet turn inward)	1
10. Congenital dislocation of hip	1
11. Congenital deafness	0.7
12. Cystic fibrosis[†] (thick mucus and other factors cause severe health problems)	0.5

*Data from Clayman C.B. (editor): *American Medical Association Encyclopedia of Medicine.* New York: Random House, 1989.
[†]These two abnormalities result from chromosomal and genetic defects. The other birth defects on the list have many different causes, which may include exposure to teratogens.

Figure 3.14 Photographs of a developing fetus. The major events of fetal development are growth and tissue specialization. **(a)** Fetus in month 3, about 6 cm long. **(b)** Fetus late in month 5, about 19 cm long. By birth, the fetus is typically 35 cm long from crown to rump.

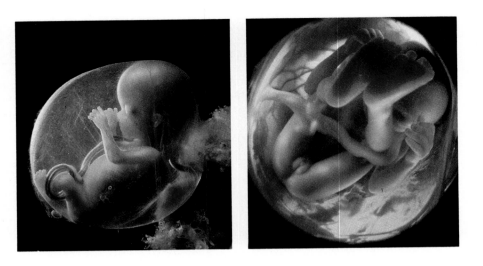

some early fetuses are presented in Figure 3.14.

Normal births typically occur 38 weeks after conception. A **premature birth** is one that occurs before that time. Infants born as early as week 30 (7 months) usually survive without requiring life-saving measures. Modern technology can save many fetuses born earlier, sometimes as early as 22 weeks (in month 6). However, the lungs of such premature infants are so immature that death may result from respiratory distress (see p. 552 for more details).

Related Clinical Terms

Abortion (*abort* = born prematurely) Premature removal of the embryo or fetus from the uterus. Abortions can be spontaneous (resulting from fetal death) or induced (most legal abortions are performed by suction).

Ectopic pregnancy (ek-top′ik) (*ecto* = outside) A pregnancy in which the embryo implants in any site other than the uterus. Ectopic pregnancy usually occurs in one of the uterine tubes, a condition known as a tubal pregnancy. Since the uterine tube is unable to establish a placenta or accommodate growth, it ruptures unless the condition is diagnosed early. Once rupture occurs, internal bleeding may threaten the woman's life. Sometimes, an embryo implants in the wall of the ventral body cavity; these pregnancies are also dangerous. For obvious reasons, ectopic pregnancies almost never result in live births.

Teratology (ter″ah-tol′o-je) (*terato* = a monster; *logos* = study of) The scientific study of birth defects and of fetuses with congenital deformities.

Ultrasonography Noninvasive technique that uses ultrasound echoes to visualize the position and size of the fetus and placenta (see p. 19).

Chapter Summary

1. The study of the 38-week prenatal period is called embryology. Embryology helps one understand the basic organization of the human body, as well as the origin of birth defects (see the box on pp. 60–61).

INTRODUCTION TO EMBRYOLOGY (pp. 49–50)

1. The embryonic period, during which the basic body plan is established, is the first 8 weeks of prenatal development. The fetal period, primarily a time of growth and maturation, is the remaining 30 weeks.

2. Basic features of the adult body that are established in the embryonic period include the skin, trunk muscles, vertebral column, spinal cord and brain, digestive tube, ventral body cavity, heart, kidneys, gonads, and limbs (see Figure 3.2).

THE EMBRYONIC PERIOD (pp. 50–58)

Week 1: From Zygote to Blastocyst (pp. 50–52)

1. After fertilization occurs in the uterine tube, the zygote undergoes cleavage (early cell division without growth) to produce a solid ball of cells called the morula. During day 4, the morula enters the uterus.

2. The morula gains an internal blastocyst cavity and becomes a blastocyst, which consists of an inner cell mass and a trophoblast. The inner cell mass becomes the embryo, and the trophoblast forms part of the placenta. The blastocyst implants in the uterine wall from about day 6 to day 13.

Week 2: The Two-Layered Embryo (pp. 52–53)

3. By day 7, the inner cell mass has formed two touching sheets of cells, the epiblast and hypoblast. These layers form the bilaminar embryonic disc.

4. The amniotic sac, which is adjacent to the epiblast, contains amniotic fluid that buffers the growing embryo from physical trauma. The yolk sac, adjacent to the hypoblast, later contributes to the digestive tube.

Week 3: The Three-Layered Embryo (pp. 53–56)

5. At the start of week 3, the primitive streak appears on the posterior part of the epiblast layer. Epiblast cells invaginate here to form the mesoderm, which fills the space between the other two layers. The embryonic disc now has three layers: ectoderm (the remaining epiblast), mesoderm, and endoderm (a mixture of former hypoblast and invaginated epiblast cells).

6. Ectoderm and endoderm are epithelial tissues (sheets of closely attached cells). Mesoderm, by contrast, is a mesenchyme tissue (embryonic tissue with star-shaped, migrating cells that remain unattached to one another).

7. The primitive node, at the anterior end of the primitive streak, is the site where the notochord originates. The notochord defines the body axis and later forms each nucleus pulposus of the vertebral column.

8. The notochord signals the overlying ectoderm to fold into a neural tube, the future spinal cord and brain. This infolding carries other ectoderm cells, neural crest cells, into the body.

9. By the end of week 3, the mesoderm on both sides of the notochord has condensed into three parts. These are, from medial to lateral, somites (segmented blocks), intermediate mesoderm (segmented balls), and the lateral plate (an unsegmented layer). The coelom (future body cavity) divides the lateral plate into somatic and splanchnic mesoderm layers.

Week 4: The Body Takes Shape (pp. 56–58)

10. The flat embryo folds inward at its sides and at its head and tail, taking on a tadpole shape. As this occurs, a tubular part of the yolk sac becomes enclosed in the body as the primitive gut.

11. The basic body plan is attained as the body achieves this tadpole shape. This development can be demonstrated by comparing Figures 3.2 and 3.12. The major adult structures derived from ectoderm, endoderm, and mesoderm are summarized in Table 3.1.

12. The limb buds appear at the end of week 4.

Weeks 5–8: The Second Month of Embryonic Development (p. 58)

13. All major organs are present by the end of month 2. During this month, the embryo changes from tadpole-like to human-like as the eyes, ears, nose, and limbs form and take on their human shape.

THE FETAL PERIOD (pp. 58–62)

1. Cell and tissue differentiation is completed in the first months of the fetal period. The fetal period is a time of growth and maturation of the body.

2. The main events of the fetal period are summarized in Table 3.2.

Review Questions

Multiple Choice and Matching Questions

1. Indicate whether each of the following describes (a) cleavage or (b) events at the primitive streak.

_____ **(1)** period when a morula forms

_____ **(2)** period when the notochord forms

_____ **(3)** period when the three embryonic germ layers appear

_____ **(4)** period when individual cells become markedly smaller

2. The outer layer of the blastocyst, which attaches to the uterus wall, is the (a) yolk sac, (b) trophoblast, (c) amnion, (d) inner cell mass.

3. Most birth defects can be traced to disruption of the developmental events during this part of the prenatal period: (a) first 2 weeks, (b) second half of month 1 and all of month 2, (c) month 3, (d) end of month 4, (e) months 8 and 9.

4. The primary germ layer that forms the trunk muscles, heart, and skeleton is (a) ectoderm, (b) endoderm, (c) mesoderm.

5. Match each embryonic structure in column A with its adult derivative in column B.

Column A	Column B
_____ **(1)** notochord	**(a)** kidney
_____ **(2)** ectoderm (not neural tube)	**(b)** ventral body cavity
_____ **(3)** intermediate mesoderm	**(c)** pancreas secretory cells
_____ **(4)** splanchnic mesoderm	**(d)** parietal serosa
_____ **(5)** sclerotome	**(e)** nucleus pulposus
_____ **(6)** coelom	**(f)** visceral serosa
_____ **(7)** neural tube	**(g)** hair and epidermis
_____ **(8)** somatic mesoderm	**(h)** brain
_____ **(9)** endoderm	**(i)** ribs

6. Match each date in column A (approximate time after conception) with a developmental event or stage in column B.

Column A	Column B
_____ **(1)** 38 weeks	**(a)** blastocyst
_____ **(2)** 15 days	**(b)** implantation
_____ **(3)** 4–7 days	**(c)** first few somites
_____ **(4)** about 21 days	**(d)** flat embryo becomes cylindrical
_____ **(5)** about 21–24 days	**(e)** embryo/fetus boundary
_____ **(6)** about 60 days	**(f)** birth
_____ **(7)** days 6–13	**(g)** primitive streak appears

7. The most common birth defect is (a) clubfoot, (b) cleft palate, (c) heart defects, (d) deafness.

8. Teratology is (a) the study of birth defects, (b) cancer in the embryo, (c) a very early stage of embryonic development, (d) the rarest kind of congenital abnormality, (e) another name for fetal alcohol syndrome.

9. The youngest premature babies that can currently be saved by modern technology are born in week _____ of development. (a) 8, (b) 16, (c) 22, (d) 38, (e) 2.

10. Somites are evidence of (a) a structure from ectoderm, (b) a structure from endoderm, (c) a structure from lateral plate mesoderm, (d) segmentation in the human body.

11. Circle the incorrect statement about the neural crest. (a) It comes from ectoderm. (b) It produces some nervous structures. (c) It is inside the neural tube. (d) Part of it becomes a wandering mesenchyme.

12. Which of the following embryonic structures are segmented? (More than one is correct.) (a) somites, (b) lateral plate, (c) endoderm, (d) myotomes.

13. Deduce what the primitive streak becomes in the adult body: (a) the navel, (b) the mouth opening, (c) skin epidermis over the tailbone, (d) the cleft between the index finger and thumb.

14. Endoderm gives rise to (a) the entire digestive tube, (b) the skin, (c) the neural tube, (d) the kidney, (e) none of the above.

15. When it is 1½ months old, an average embryo is the size of (a) Lincoln on a penny, (b) an adult fist, (c) its mother's nose, (d) the head of a common pin.

Short Answer and Essay Questions

16. What important event occurs at the primitive streak?

17. What important event occurs at the primitive node?

18. What is the function of the amniotic sac and the fluid it contains?

19. The epiblast forms all the ectoderm. Does the hypoblast form all the endoderm? Explain your answer.

20. (a) What is mesenchyme? (b) How does it differ from an epithelium?

21. Explain how the flat embryonic disc takes on the cylindrical shape of a tadpole.

22. While Mortimer was cramming for his anatomy test, he read that some parts of the mesoderm become segmented. He suddenly realized that he could not remember what segmentation is. Define segmentation, and give two examples of segmented structures in the embryo. (You may need to refer to Chapter 1.)

Critical Thinking and Clinical Application Questions

23. Mary, a heavy smoker, has ignored a friend's advice to quit smoking during her pregnancy. Describe how Mary's smoking might affect her fetus.

24. A friend in your dormitory, a freshman, tells you she just discovered she is 3-months pregnant. She recently bragged that since she came to college she has been drinking alcohol heavily and experimenting with every kind of recreational drug she can find. Circle the best advice you could give her, and explain your choice. (a) She must stop taking the drugs, but they could not have affected her fetus during these first few months of her pregnancy. (b) Harmful substances usually cannot pass from mother to embryo, so she can keep using drugs. (c) There could be defects in the fetus, so she should stop using drugs and visit a doctor as soon as possible. (d) If she has not taken any drugs in the last week, she is okay.

25. The citizens of Nukeville brought a class-action lawsuit against the local power company, which ran a nuclear power plant on the outskirts of town. The people claimed that radiation leaks were causing birth defects. As evidence, they showed that of the 998 infants born in town since the plant opened, 20 had congenital abnormalities. Of these 20 infants, 6 had heart defects, 3 were mentally retarded, 2 had pyloric stenosis, 2 were born without a brain, 2 had spina bifida, 2 had Down's syndrome, and 3 had cleft palates. Did the citizens win or lose their case? Explain.

26. Before Delta studied embryology in her anatomy course, she imagined that a developing human is a shapeless mass of indistinct tissues until about halfway through pregnancy. Was Delta correct? At what stage does the embryo or fetus really start to look like a developing human?

4

Tissues

Student Objectives

1. Define tissue, and list the four main types of tissue in the body.

Epithelia and Glands (pp. 66–78)

2. List several functional and structural characteristics of epithelial tissue.

3. Classify the different epithelia of the body.

4. Describe the surface features of epithelia and epithelial cells.

5. Define exocrine and endocrine glands.

6. Explain how multicellular exocrine glands are classified.

Connective Tissue (pp. 78–83)

7. List the structural elements of loose areolar connective tissue, and describe each element.

8. Describe the types of connective tissue, and note their main functions.

Covering and Lining Membranes (pp. 83–90)

9. Discuss the structure and function of mucous, serous, and cutaneous membranes.

Muscle Tissue (p. 91)

10. Briefly describe the three types of muscle tissue.

Nervous Tissue (p. 91)

11. Note the general characteristics of nervous tissue.

Tissue Response to Injury (pp. 91–94)

12. Describe the inflammatory and repair processes by which tissues recover from injury.

The Tissues Throughout Life (pp. 94–95)

13. Indicate the embryonic derivation of each tissue class.

14. Briefly describe changes that occur in tissues with age.

The cells of the human body do not operate independently of one another. Instead, related cells live and work together in cell communities called **tissues.** A tissue is more precisely defined as a group of closely associated cells that perform related functions and are similar in structure. However, tissues do not consist entirely of cells: Between the cells is nonliving *extracellular material.*

The word *tissue* derives from the Old French word meaning "to weave," reflecting the fact that the different tissues weave together to form the "fabric" of the human body. The four basic types of tissue are epithelium, connective tissue, muscle tissue, and nervous tissue. If we were to assign a single term to each basic tissue that would best describe its general role, the terms would most likely be *covering* (epithelium), *support* (connective), *movement* (muscle), and *control* (nervous). However, these terms reflect only a fraction of the functions that each tissue performs.

As we explained in Chapter 1, tissues are the building blocks of the body's organs. Since most organs contain all four tissue types, learning about tissues will make it easier for you to understand the structure of the organs discussed in the remaining chapters of this book.

Epithelia and Glands

An **epithelium** (ep″ĭ-the′le-um; "covering") is a sheet of cells that covers a body surface or lines a body cavity. With minor exceptions, epithelia cover all of the outer and inner surfaces of the body. Examples of epithelia include the outer layer of the skin, the inner lining of all hollow viscera (such as the stomach and respiratory tubes), the lining of the ventral body cavity, and the inner lining of all blood vessels. Epithelia also form most of the body's glands.

Epithelia occur at the interfaces between two different environments (the epidermis of the skin, for example, lies between the inside and outside of the body). Therefore, all functions of epithelia reflect their role as interface tissues: New stimuli, including harmful ones, are experienced at body interfaces, so epithelia both *protect* the underlying tissues and contain nerve endings for *sensory reception.* Nearly all substances that are received or given off by the body must pass across an epithelium, so epithelia function in *secretion* (the release of molecules from cells), *absorption* (bringing small molecules into the body),

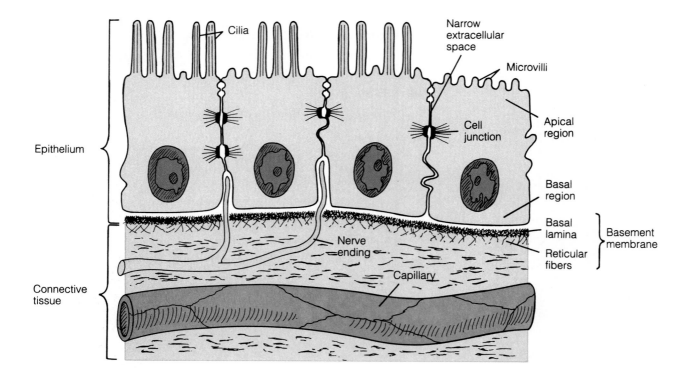

Figure 4.1 Special characteristics of epithelium.   A sheet of closely joined epithelial cells rests on connective tissue proper. Epithelia contain nerve endings but no blood vessels. Note the special features on the epithelial cell surfaces: cilia, microvilli, cell junctions, and basal lamina.

66

and *ion transport* (moving ions across the interface). Furthermore, body fluids can be *filtered* across thin epithelia.

Special Characteristics of Epithelia

Epithelial tissues have many characteristics that distinguish them from other tissue types (Figure 4.1):

1. Cellularity. Epithelia are composed almost entirely of cells. A tiny amount of extracellular material is confined to the narrow spaces between the cells.

2. Specialized contacts. Adjacent epithelial cells are held together at many points by special cell junctions.

3. Polarity. All epithelia have a free upper (apical) surface, and a lower (basal) surface. They exhibit polarity, a term meaning that the cell regions near the apical surface differ from those near the basal surface. As shown in Figure 4.1, the basal surface of an epithelium lies on a thin sheet called a *basal lamina,* which is part of a *basement membrane* (this is explained further on p. 76).

4. Support by connective tissue. All epithelial sheets in the body are supported by an underlying layer of connective tissue.

5. Avascular but innervated. Whereas most tissues in the body are *vascularized* (contain blood vessels), epithelium is *avascular* (a-vas'ku-lar) (lacks blood vessels). Epithelial cells receive their nutrients from capillaries in the underlying connective tissue. Although blood vessels do not penetrate epithelial sheets, nerve endings do; that is, epithelium is *innervated.*

6. Regeneration. Epithelial tissue has a high regenerative capacity. Some epithelia are exposed to friction, and their surface cells rub off. Others are destroyed by hostile substances in the external environment (bacteria, acids, smoke). As long as epithelial cells receive adequate nutrition, they can replace lost cells quickly by cell division.

Classification of Epithelia

There are many varieties of epithelia in the body. In classifying them, each epithelium is given two names. The first name indicates the number of cell layers in the epithelium, and the last name describes the shape of the cells (Figure 4.2). In classifying by cell layers, an epithelium is called **simple** if it has just one cell layer or **stratified** if it has more than one layer. In clas-

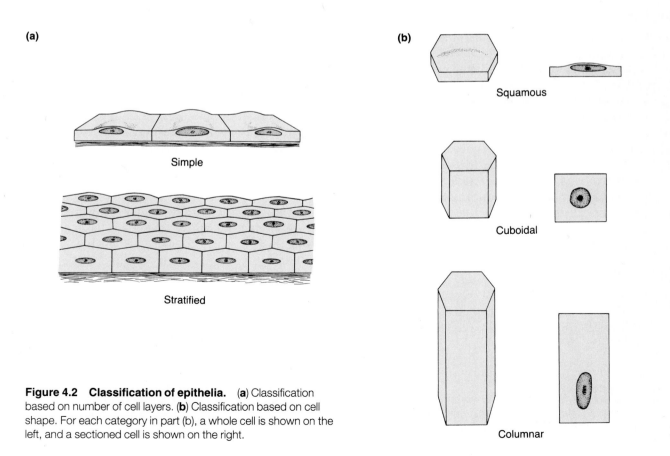

(a)

Simple

Stratified

(b)

Squamous

Cuboidal

Columnar

Figure 4.2 Classification of epithelia. (a) Classification based on number of cell layers. **(b)** Classification based on cell shape. For each category in part (b), a whole cell is shown on the left, and a sectioned cell is shown on the right.

sifying epithelia by cell shape, the cells are called either squamous, cuboidal, or columnar. **Squamous** (sqwa'mus; "plate-like") cells are wider than they are tall; **cuboidal** cells are about as tall as they are wide (like cubes); and **columnar** cells are taller than they are wide (like columns). In each case, the shape of the nucleus conforms to that of the cell: The nucleus of a squamous cell is disc-shaped; that of a cuboidal cell is spherical; and that of a columnar cell is an oval elongated from top to bottom. The shape of the nucleus is an important feature to keep in mind when distinguishing epithelial types.

Simple epithelia are easy to classify by cell shape because all cells in the layer usually have the same shape. In stratified epithelia, however, cell shapes vary among the different cell layers. To avoid ambiguity, stratified epithelia are named according to the shape of the cells in the *apical* layer. This naming system will become clearer as we explore the specific epithelial types.

As you read about the epithelial classes, follow Figure 4.3. Using the photomicrographs, try to pick out the individual cells within each epithelium. This is not always easy, because the boundaries between epithelial cells often are indistinct. Furthermore, the nucleus of a particular cell may or may not be visible, depending on the precise plane of the cut made to prepare the tissue slides.

Simple Epithelia

Simple Squamous Epithelium (Figure 4.3a). A simple squamous epithelium is a single layer of flat cells. From a superior view, the closely fitting cells resemble a tiled floor. From a cross-sectional view, the cells resemble fried eggs seen from the side. Thin and often permeable, this type of epithelium occurs wherever small molecules must pass through a membrane quickly. The walls of capillaries consist exclusively of this epithelium, whose exceptional thinness encourages efficient exchange of nutrients and wastes between the bloodstream and surrounding tissue cells. In the lungs, this epithelium forms the thin walls of the air sacs, where gas exchange occurs.

Two simple squamous epithelia in the body have special names that reflect their locations. **Endothelium** (en"do-the'le-um; "inner covering") provides a slick, friction-reducing lining in all hollow organs of the circulatory system—blood vessels, lymphatic vessels, and the heart (Figure 4.4 on p. 73). **Mesothelium** (mez"o-the'le-um; "middle covering") is the epithelium that lines the ventral body cavity and covers the organs in this cavity. It forms the slick surface layer of serous membranes (serosae: see p. 90).

Simple Cuboidal Epithelium (Figure 4.3b). Simple cuboidal epithelium consists of a single layer of cells as tall as they are wide. This epithelium forms the walls of the smallest ducts of glands, and of many tubules in the kidney. Its functions are the same as those of simple columnar epithelium.

Simple Columnar Epithelium (Figure 4.3c). Simple columnar epithelium is a single layer of tall cells aligned like soldiers in a row. This epithelium lines the digestive tube from the stomach to the anal canal. It functions in absorption, secretion, and ion transport. The structure of simple columnar epithelium is ideal for these functions: It is thin enough to allow large numbers of molecules to pass through it quickly, yet thick enough to house the cellular machinery needed to perform the complex processes of molecular transport.

Some simple columnar epithelia bear *cilia* (sil'e-ah; "eyelashes"), whip-like bristles on the apex of epithelial cells (Figure 4.1) that beat rhythmically to move substances across certain body surfaces. For example, a simple ciliated columnar epithelium lines the inside of the uterine tube. Its cilia help move the ovum to the uterus, a journey we traced in Chapter 3. Cilia are considered in detail later in this chapter.

Pseudostratified Columnar Epithelium (Figure 4.3d). The cells of pseudostratified (soo-do-strat'ĭ-fīd) columnar epithelium are varied in height. While all of its cells rest on the basement membrane, only the tall cells reach the apical surface of the epithelium. The short cells are undifferentiated and continuously give rise to the tall cells. The cell nuclei lie at several different levels, giving the false impression that this epithelium is stratified (*pseudostratified* = false stratified).

Pseudostratified columnar epithelium, like simple columnar epithelium, functions in secretion or absorption. A ciliated type lines the interior of the respiratory tubes. Here, the cilia propel sheets of dust-trapping mucus out of the lungs.

Stratified Epithelia

Stratified epithelia contain two or more layers of cells. They regenerate from below; that is, the basal cells divide and push apically to replace the older surface cells. Stratified epithelia are more durable than simple epithelia, and their major (but not their only) role is protection.

Stratified Squamous Epithelium (Figure 4.3e). As you might expect, stratified squamous epithelium consists of many cell layers whose surface cells are squamous in shape. In the deeper layers, the cells are cuboidal or columnar. Of all the epithelial types, this is thickest and best adapted for protection. It covers the often-abraded surfaces of our body, forming the epidermis of the skin and the inner lining of the mouth, esophagus, and vagina. To avoid memorizing

(text continues on p. 73)

Figure 4.3 Epithelial tissues. (**a**) to (**d**), Simple epithelia.

(a) Simple squamous epithelium

Description: Single layer of flattened cells with disk-shaped central nuclei and sparse cytoplasm; the simplest of the epithelia.

Location: Air sacs of lungs; kidney glomeruli; lining of heart, blood vessels, and lymphatic vessels; lining of ventral body cavity (serosae).

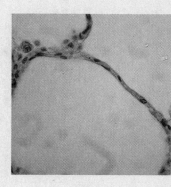

Function: Allows passage of materials by diffusion and filtration in sites where protection is not important; secretes lubricating substances in serosae.

Photomicrograph: Simple squamous epithelium forming walls of alveoli (air sacs) of the lung (280×).

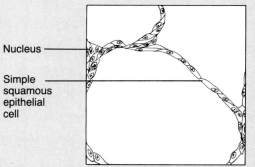

Nucleus

Simple squamous epithelial cell

(b) Simple cuboidal epithelium

Description: Single layer of cubelike cells with large, spherical central nuclei.

Location: Kidney tubules; ducts and secretory portions of small glands; ovary surface.

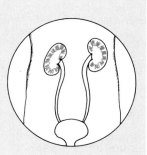

Function: Secretion and absorption.

Photomicrograph: Simple cuboidal epithelium in kidney tubules (260X)

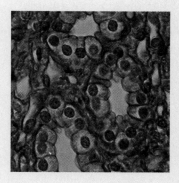

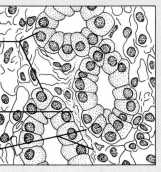

Simple cuboidal epithelial cells

Basement membrane

Connective tissue

Figure 4.3   (continued)

(c) Simple columnar epithelium

Description: Single layer of tall cells with *oval* nuclei; some cells bear cilia; layer may contain mucus-secreting glands (goblet cells).

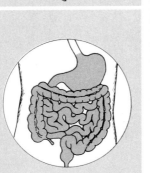

Location: Nonciliated type lines most of the digestive tract (stomach to anal canal), gallbladder and excretory ducts of some glands; ciliated variety lines small bronchi, uterine tubes, and some regions of the uterus.

Function: Absorption; secretion of mucus, enzymes, and other substances; ciliated type propels mucus (or reproductive cells) by ciliary action.

Photomicrograph: Simple columnar epithelium of the stomach mucosa (280 ×).

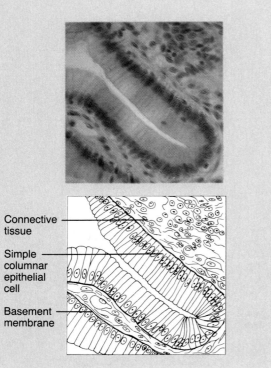

Connective tissue

Simple columnar epithelial cell

Basement membrane

(d) Pseudostratified epithelium

Description: Single layer of cells of differing heights, some not reaching the free surface; nuclei seen at different levels; may contain goblet cells and bear cilia.

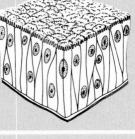

Location: Nonciliated type in ducts of large glands, parts of male urethra; ciliated variety lines the trachea, most of the upper respiratory tract.

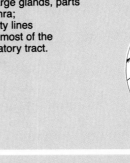

Function: Secretion, particularly of mucus; propulsion of mucus by ciliary action.

Photomicrograph: Pseudostratified ciliated columnar epithelium lining the human trachea (430 ×).

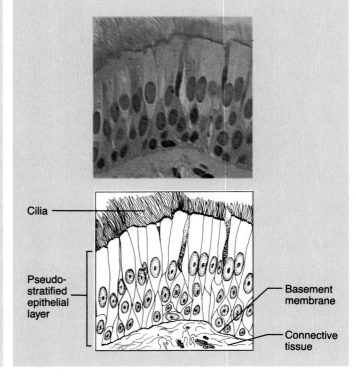

Cilia

Pseudo-stratified epithelial layer

Basement membrane

Connective tissue

Figure 4.3 (continued) (e) to (h), Stratified epithelia.

(e) Stratified squamous epithelium

Description: Thick membrane composed of several cell layers; basal cells are cuboidal or columnar and metabolically active; surface cells are flattened (squamous); in the keratinized type, the surface cells are full of keratin and dead; basal cells are active in mitosis and produce the cells of the more superficial layers.

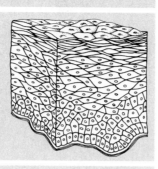

Location: Nonkeratinized type forms the moist linings of the esophagus, mouth, and vagina; keratinized variety forms the epidermis of the skin, a dry membrane.

Function: Protects underlying tissues in areas subjected to abrasion.

Photomicrograph: Stratified squamous epithelium lining of the esophagus (173 ×).

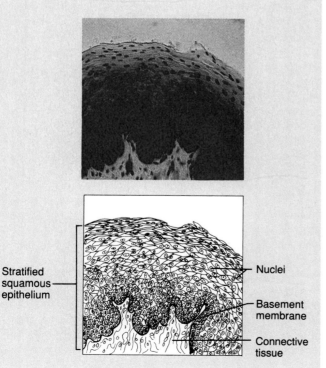

Stratified squamous epithelium

Nuclei

Basement membrane

Connective tissue

(f) Stratified cuboidal epithelium

Description: Generally two layers of cube-like cells.

Location: Largest ducts of sweat glands, mammary glands, and salivary glands.

Function: Protection.

Photomicrograph: Stratified cuboidal epithelium forming a salivary gland duct (400 ×).

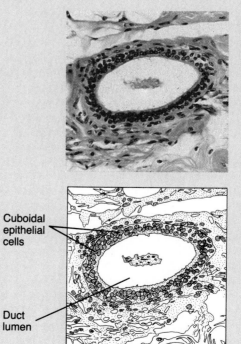

Cuboidal epithelial cells

Duct lumen

Figure 4.3   (continued)

(g) Stratified columnar epithelium

Description: Several cell layers; basal cells usually cuboidal; superficial cells elongated and columnar.

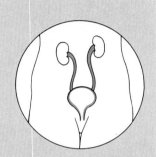

Location: Rare in the body; small amounts in male urethra and in large ducts of some glands.

Function: Protection; secretion.

Photomicrograph: Stratified columnar epithelium lining of the male urethra (360×).

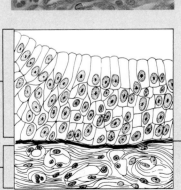

Stratified columnar epithelium

Underlying connective tissue

Basement membrane

(h) Transitional epithelium

Description: Resembles both stratified squamous and stratified cuboidal; basal cells cuboidal or columnar; surface cells dome-shaped or squamous-like, depending on degree of organ stretch.

Location: Lines the ureters, bladder, and part of the urethra.

Function: Stretches readily and permits distension of urinary organ by contained urine.

Photomicrograph: Transitional epithelium lining of the bladder, relaxed state (170×); note the bulbous, or rounded, appearance of the cells at the surface; these cells flatten and become elongated when the bladder is filled with urine.

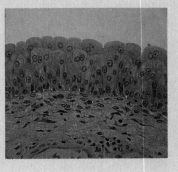

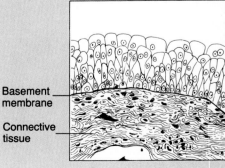

Basement membrane

Connective tissue

Transitional epithelium

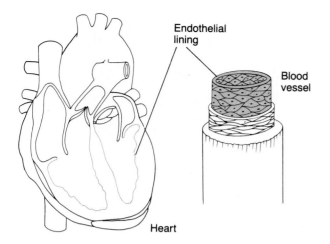

Figure 4.4   Endothelium. Endothelium is a simple squamous epithelium that lines the interior of the circulatory vessels and the heart.

all its locations, simply remember that this epithelium covers the skin and extends a certain distance into every body opening that is directly continuous with the skin.

The epidermis of the skin is *keratinized,* meaning that its surface cells contain an especially tough protective protein called *keratin.* The other stratified squamous epithelia of the body lack keratin and are *nonkeratinized.* Keratin is explained in detail in Chapter 5 (p. 101).

Stratified Cuboidal and Columnar Epithelia (Figure 4.3f–g). Stratified cuboidal and stratified columnar epithelia are rare types of tissue, forming little else than the large ducts of some glands.

Transitional Epithelium (Figure 4.3h). Transitional epithelium lines the inside of the hollow urinary organs. Such organs (the urinary bladder, for example), stretch as they fill with urine. As the transitional epithelium stretches, it thins from about six cell layers to three, and its apical cells flatten from a rounded shape to a squamous shape. Thus, this epithelium undergoes "transitions" in shape. It also forms an impermeable barrier that keeps urine from permeating the wall of the bladder.

Epithelial Surface Features

The apical, lateral, and basal cell surfaces of epithelia have special features.

Apical Surface Features: Microvilli and Cilia

Microvilli (mi″cro-vĭ′li; "little shaggy hairs") are finger-like extensions of the plasma membrane of apical epithelial cells (see Figure 4.1 and Figure 1.11c on p. 16). They occur on almost every epithelium in the body but are longest and most abundant on epithelia that absorb nutrients (in the small intestine) or transport ions (in the kidney). In such epithelia, microvilli maximize the surface area across which small molecules enter or leave cells. Microvilli are also abundant on epithelia that secrete mucus, where they help anchor the mucous sheets to the epithelial surface. Finally, microvilli may act as "stiff knobs" to resist abrasion.

Cilia, you will recall, are whip-like, highly motile extensions of the apical surface membranes of certain epithelial cells (Figure 4.1). Each cilium contains a core of microtubules, nine pairs of which encircle one middle pair (Figure 4.5). Each pair of microtubules is called a *doublet.* Cilia movement is generated when adjacent doublets grip one another with side arms and start to crawl along each other's length, like centipedes running over each other's backs. This action causes the cilium to bend.

The microtubules in cilia are arranged in much the same way as in *centrioles* (compare Figure 4.5b to 4.5c). Indeed, cilia originate as their microtubules assemble around centrioles that have migrated from the centrosome to the apical plasma membrane. The centriole at the base of each cilium is called a **basal body** (Figure 4.5a).

The cilia on an epithelium bend and move in coordinated waves, like the waves that pass across a field of grass on a windy day. These waves push mucus and other substances over the epithelial surface. Each cilium executes a propulsive *power stroke,* followed by a nonpropulsive *recovery stroke* (see Figure 4.6). This sequence ensures that fluid is moved in one direction only.

An extremely long, isolated cilium is called a *flagellum* (flah-jel′um; "whip"). The only flagellated cells in the human body are sperm, which use their flagella to swim through the female reproductive tract.

✚ *Kartagener's syndrome,* a type of *immotile cilia syndrome,* is an inherited disease in which the side arms within the cilia fail to form. This condition leads to frequent respiratory infections, because the nonfunctional cilia cannot sweep inhaled bacteria out of the respiratory tubes.

✚ All males with immotile cilia syndrome are sterile. Can you guess why?

The flagella in their sperm do not function, so the sperm cannot swim to meet and fertilize an ovum. ∎

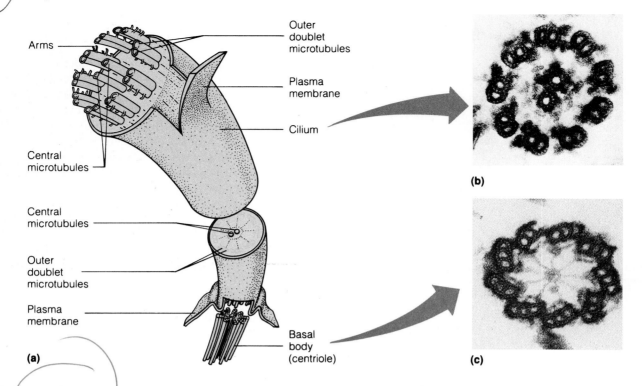

Figure 4.5   A cilium.   (a) Three-dimensional view of a cilium, sectioned to show the arrangement of microtubules in its core. Note the basal body (centriole) below, which lies in the apical cytoplasm of the cell. (b) Electron micrograph of a cilium in cross section (approx. 150,000 ×). (c) Electron micrograph of the basal body in cross section (approx. 150,000 ×).

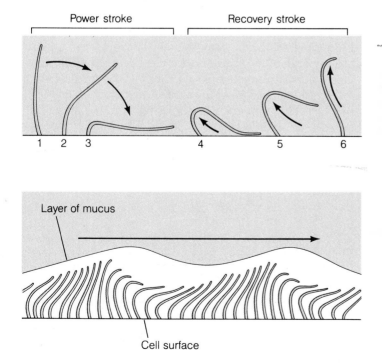

Figure 4.6   Cilia movement.   In the initial power stroke, a cilium moves in an arc. This movement is followed by a recovery stroke, in which the cilium bends and pulls back, keeping its tip low.

Lateral Surface Features: Cell Junctions

Three factors act to bind epithelial cells to one another: (1) adhesive proteins in the scant material between the cells; (2) the wavy contour of lateral plasma membranes of adjacent epithelial cells, which enables the cells to fit together in tongue-and-groove fashion; and (3) special cell junctions (Figures 4.7–4.9). Cell junctions, the most important of the factors, are most characteristic of epithelial tissue but are occasionally found in other tissue types as well.

Desmosomes.   The junctions that act to bind cells together are called **desmosomes** (dez′mo-sōmz; "binding bodies"). These adhesive spots are scattered like rivets along the abutting sides of adjacent cells (Figure 4.7). Desmosomes have a complex structure: On the cytoplasmic face of each plasma membrane is a circular plaque. The plaques of neighboring cells are joined by thin *linker proteins* that span the narrow extracellular space. In addition, intermediate filaments (the cytoskeletal elements that resist tension) insert into each plaque from its inner, cytoplasmic side. Bundles of these filaments extend across the cytoplasm and anchor at other desmosomes on the op-

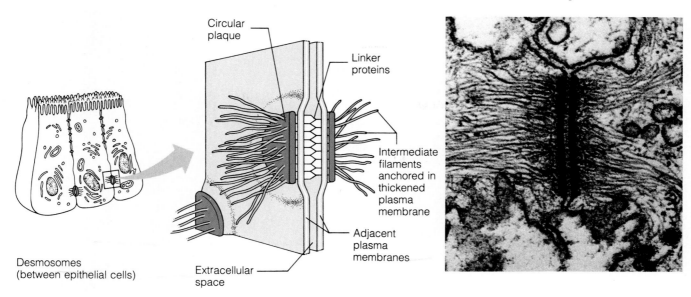

Figure 4.7 Desmosome. The diagrams at left and center show the location and structure of a desmosome, a spot junction that connects adjacent epithelial cells. At right is a desmosome viewed by electron microscopy (58,000 ×).

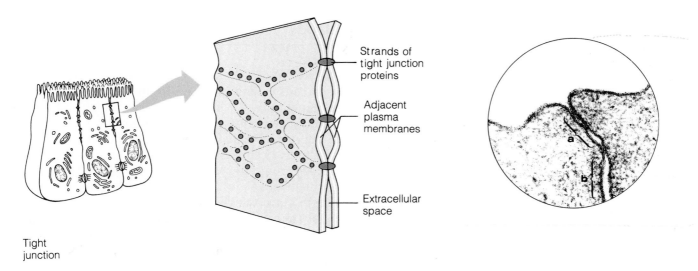

Figure 4.8 Tight junction between epithelial cells. Drawing at left and electron micrograph at right (approx. 40,000 ×). In the micrograph, the structure labeled a is the tight junction and the structure labeled b shows the normal width of the intercellular space.

posite side of the same cell. This arrangement not only holds adjacent cells together but also interconnects all intermediate filaments of the entire epithelium into one continuous network of strong "guy wires." The epithelium is thus less likely to tear when pulled upon, because the pulling forces are distributed evenly throughout the sheet.

Tight Junctions. In the apical region of most epithelial types, a belt-like junction extends around the periphery of each cell (Figure 4.8). This junction is called a **tight junction** or a *zonula occludens* (zōn'u-lah o-klood'enz; "belt that shuts off"). At tight junctions, the adjacent cells are so close that some proteins in their plasma membranes are fused. This fusion forms a tight seal that closes off the extracellular space. Thus, tight junctions prevent molecules from passing between the cells of epithelial tissue. For example, the tight junctions in the epithelium lining the digestive tract keep digestive enzymes, ions, and microorganisms in the intestine from seeping into the bloodstream.

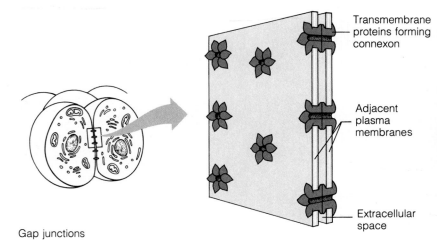

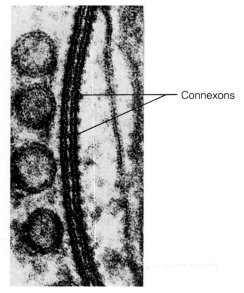

Figure 4.9 Gap junction between epithelial cells. Drawing (left) and electron micrograph (right; approx. 200,000 ×).

Gap Junctions. A **gap junction,** or *nexus* (nek'sus; "bond"), is a spot-like junction that can occur anywhere along the lateral membranes of adjacent cells (Figure 4.9). Gap junctions let small molecules move directly between neighboring cells. At gap junctions, the adjacent plasma membranes are very close, and the cells are connected by hollow cylinders of protein (connexons). Ions, simple sugars, and other small molecules pass through these cylinders from one cell to the next.

Basal Feature: The Basal Lamina

At the border between the epithelium and the connective tissue deep to it is a supporting sheet called the **basal lamina** (lam'ĭ-nah; "sheet") (recall Figure 4.1). This thin, noncellular sheet consists of proteins secreted by the epithelial cells. Functionally, the basal lamina acts as a selective filter; that is, it determines which molecules from capillaries in the underlying connective tissue will be allowed to enter the epithelium. The basal lamina also acts as scaffolding along which regenerating epithelial cells can migrate. Luckily, infections and toxins that destroy epithelial cells usually leave the basal lamina in place, for without this lamina, epithelial regeneration is difficult or impossible.

Directly deep to the basal lamina is a layer of reticular fibers (defined shortly) belonging to the underlying connective tissue. These layers together form the **basement membrane** (Figure 4.1). The thin basal lamina can be seen only by electron microscopy, but the thicker basement membrane is visible by light microscopy (Figure 4.3).

Glands

Many epithelial cells make and secrete a product. Such cells constitute **glands.** The products of glands are aqueous (water-based) fluids that usually contain proteins. *Secretion* is the process whereby gland cells obtain needed substances from the blood and transform them chemically into a product that is then discharged from the cell. More specifically, the protein product is made in the rough ER, then packaged into secretory granules by the Golgi apparatus and ultimately released from the cell by exocytosis (recall p. 31 and Figure 2.9). Gland cells that secrete proteins almost always have a well-developed rough ER, a large Golgi apparatus, and secretory granules in their apical cytoplasm.

Glands are classified as **exocrine** (ek'so-krin; "external secretion") or **endocrine** (en'do-krin; "internal secretion"), depending on where they release their product, and as **unicellular** ("one-celled") or **multicellular** ("many-celled") on the basis of cell number. Unicellular glands are scattered within epithelial sheets, whereas most multicellular glands develop by invagination of an epithelial sheet into the underlying connective tissue.

Exocrine Glands

Exocrine glands are numerous, and many of their products are familiar ones. All exocrine glands secrete their products onto body surfaces (skin) or into body cavities (like the digestive tube), and multicellular exocrine glands have **ducts** that carry their product to the epithelial surfaces. Exocrine glands are a diverse group: They include glands that secrete mucus, the sweat glands and oil glands of the skin, salivary glands of the mouth, the liver (which secretes bile), the pancreas (which secretes digestive enzymes), mammary glands (which secrete milk), and many others.

Unicellular Exocrine Glands. The only important example of a one-celled exocrine gland is the **goblet cell** (Figure 4.10). True to its name, a goblet cell is indeed shaped like a goblet (a drinking glass with a stem). Goblet cells are scattered within the epithelial lining of the intestines and respiratory tubes, between columnar cells with other functions. They produce **mucin** (mu'sin), a glycoprotein (sugar-protein) that dissolves in water when secreted. The resulting complex of mucin and water is viscous **mucus.** Mucus covers, protects, and lubricates many internal body surfaces.

Multicellular Exocrine Glands. Each multicellular exocrine gland has two basic parts: an epithelium-lined *duct* and a *secretory unit* consisting of the secretory epithelium (Figure 4.11). Also, in all but the simplest glands, a supportive connective tissue surrounds the secretory unit and supplies it with blood vessels and nerve fibers. Often, the connective tissue forms a *fibrous capsule* that extends into the gland proper and partitions the gland into subdivisions called *lobes.*

Multicellular glands are classified on the basis of the structure of their ducts (Figure 4.11). **Simple** glands have an unbranched duct, whereas **compound** glands have a branched duct. The glands are further categorized according to the structure of their secretory units: Glands are **tubular** if their secretory cells form tubes and **alveolar** (al-ve'o-lar) if the secretory cells form spherical sacs (an *alveolus* is literally a small hollow cavity). Furthermore, some glands are **tubuloalveolar;** that is, they contain both tubular and alveolar units. It should also be noted that another word for *alveolar* is *acinar* (as'ĭ-nar) (an *acinus* is literally a grape or berry).

Endocrine Glands

Endocrine glands have no ducts, so they are often referred to as *ductless glands.* They secrete directly into the bloodstream rather than onto an epithelial surface. More specifically, endocrine glands produce messenger molecules called **hormones** (hor'mōnz; "exciters"), which they release into the extracellular space. These hormones then enter nearby capillaries and travel through the bloodstream to specific *target organs.* Each hormone signals its target organs to respond in some characteristic way. For example, en-

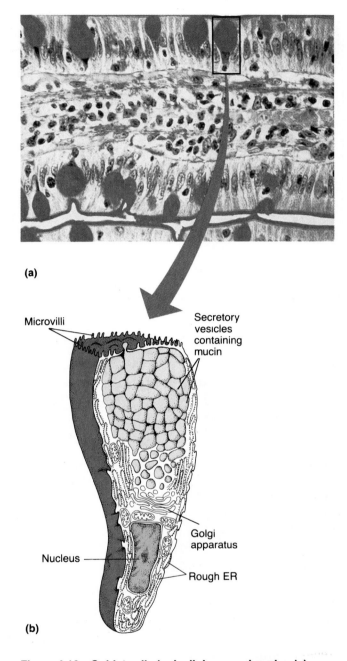

(a)

(b)

Microvilli

Secretory vesicles containing mucin

Golgi apparatus

Nucleus

Rough ER

Figure 4.10   Goblet cells (unicellular exocrine glands).
(**a**) Photomicrograph of mucus-secreting goblet cells in the simple columnar epithelium that lines the small intestine (this epithelium extends along the entire upper border of the photo). The goblet cells stain magenta red (approx. 300 ×). (**b**) Diagram of the fine structure of a goblet cell. Note the many secretory vesicles (secretory granules) and the well-developed rough ER and Golgi apparatus.

docrine cells in the intestine secrete a hormone that targets the pancreas, signaling the pancreas to release the enzymes that help digest a meal.

Although most endocrine glands derive from epithelia, some derive from other tissues. The endocrine system is discussed in detail in Chapter 24.

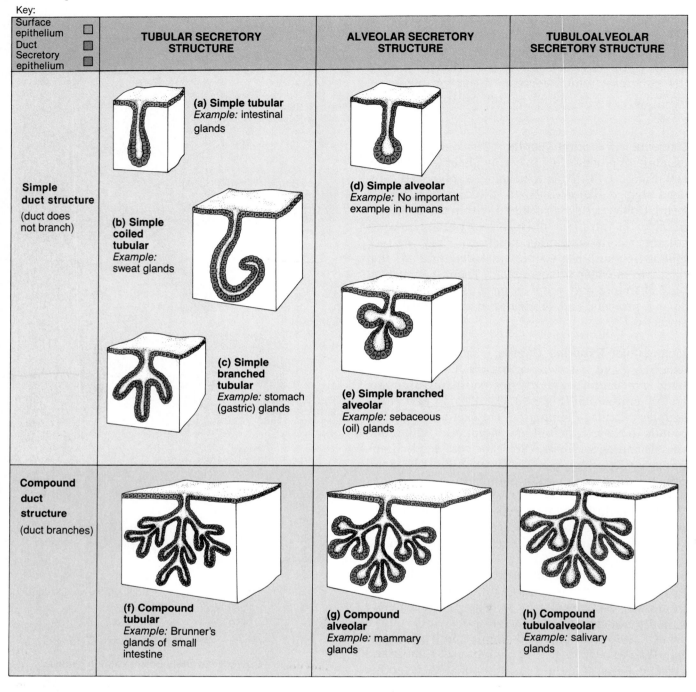

Figure 4.11 Types of multicellular exocrine glands. These glands are classified according to their ducts (simple or compound) and the structure of their secretory units (tubular, alveolar, tubuloalveolar). The largest compound glands, such as the mammary glands and saliva-producing glands, contain many copies of the gland structures illustrated here.

Connective Tissue

Connective tissue is the most diverse type of tissue. There are four main classes of connective tissue and many subclasses (Figure 4.12). The main classes are: (1) *connective tissue proper,* familiar examples of which are fat tissue and the fibrous tissue of ligaments; (2) *cartilage;* (3) *bone tissue;* and (4) *blood.* Connective tissues do far more than just connect the cells and organs of the body together. They also form the basis of the skeleton (bone and cartilage), store and carry nutrients (fat tissue and blood), surround all

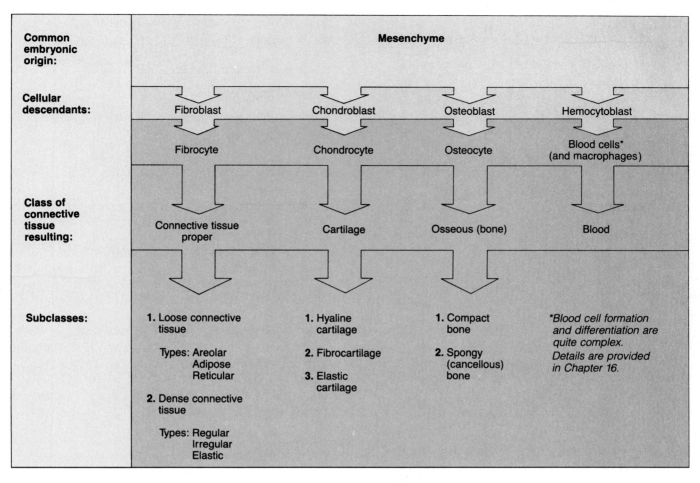

Figure 4.12   Classes of connective tissue.   All connective tissues arise from the same embryonic tissue, mesenchyme.

the blood vessels and nerves of the body (connective tissue proper), and lead the body's fight against infection. The great variety of connective tissues may seem bewildering at first, but all connective tissues share the same simple structural plan (Figure 4.13): Their cells are always separated from one another by a large amount of material called the **extracellular matrix** (ma′triks; "womb"). This differs markedly from epithelial tissue, in which the cells are crowded closely together. Another feature that unifies the many connective tissues is that they all originate from the embryonic tissue called mesenchyme. Mesenchyme is shown in Figure 4.15a on p. 84.

Connective Tissue Proper

Connective tissue proper has two subclasses: *loose connective tissue* and *dense connective tissue*. Along with the functions and locations listed below, connective tissue proper forms the supporting framework, or

stroma (stro′mah; "mattress"), of many large organs in the body.

Areolar Tissue: A Model Connective Tissue

The connective tissues are so diverse that we need a "model" type, or "prototype," against which the other connective tissues can be compared. For our prototype, we will use the most abundant type of connective tissue proper, called loose **areolar** (ah-re′o-lar) **connective tissue** (Figures 4.14 and 4.15b). This is the connective tissue that underlies almost all the epithelia of the body and surrounds almost all the small nerves and blood vessels, including the capillaries. Since no cell can lie very far from a nutrient-carrying capillary, no cell can lie far from areolar connective tissue. In fact, one might say that all other tissues and cells in the body either border areolar connective tissue or are embedded in it.

We will describe the structure of areolar connective tissue in terms of its basic functions. These func-

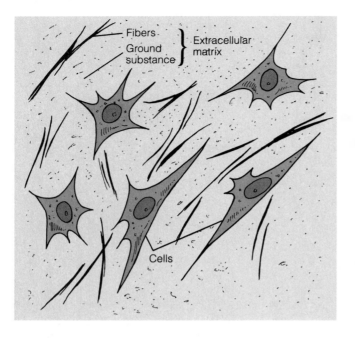

Figure 4.13   The basic organization of connective tissues (simplified).   Cells are separated by an abundant extracellular matrix. The matrix consists of fibers and a jelly-like ground substance (except in blood tissue).

tions, which are shared by many other types of connective tissue, are as follows:

1. Supporting and binding other tissues

2. Holding body fluids

3. Defending the body against infection

4. Storing nutrients as fat

Each of these functions is performed by a different structural part of the areolar tissue.

Fibers Provide Support.   Areolar connective tissue has three types of **fibers** in its extracellular matrix that give this tissue its supportive properties (Figure 4.14): *collagen, reticular,* and *elastic fibers.* **Collagen fibers,** the strongest and most abundant type, allow connective tissue to withstand tension (pulling forces). Pulling tests show that collagen fibers are stronger than *steel* fibers of the same size! The thick collagen fibers that one sees with the light microscope are bundles of thinner, striped threads called *unit fibrils,* which consist of still thinner strands that are strongly cross-linked to one another. This cross-linking is the source of collagen's great tensile strenth.

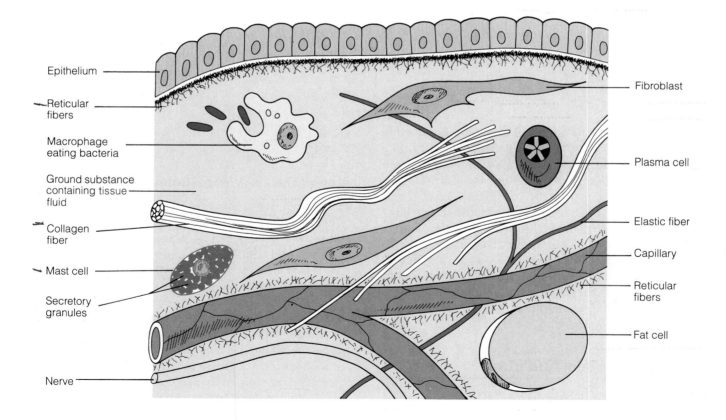

Figure 4.14   Areolar connective tissue: A model connective tissue.   This tissue underlies epithelia and surrounds capillaries. Note the various cell types and the three classes of fibers (collagen, reticular, elastic) embedded in the ground substance.

Reticular (re-tik′u-lar) **fibers** are a special type of collagen unit fibrils. These short fibers cluster into delicate networks (*reticulum* = network) that cover and support all structures bordering the connective tissue. Capillaries are coated with fuzzy nets of reticular fibers, and reticular fibers also border the nearby epithelia (Figure 4.14). Individual reticular fibers glide freely across one another whenever the network is pulled, so they allow more "give" than do collagen fibers. Thus, capillaries can expand a bit without being choked by their collar of reticular fibers.

Areolar connective tissue also contains **elastic fibers.** Long and thin, these fibers branch to form wide networks within the extracellular matrix. Elastic fibers contain a rubber-like protein called **elastin** (e-las′tin), which allows them to function like rubber bands. Connective tissue can stretch only so much before its thick, rope-like collagen fibers become taut. The elastic fibers then recoil so that the stretched tissue springs back to shape.

A single kind of cell makes and secretes all the fibers of areolar connective tissue: the **fibroblast** (fi′bro-blast; "fiber-generator"). Fibroblasts are the most abundant cell type in connective tissue proper. They are shaped either like spindles (elongated to a point at both ends) or like stars. Resting fibroblasts that are not actively secreting their product are sometimes called *fibrocytes* (fi′bro-sīts; "fiber cells").

✚ Vitamin C, which is abundant in citrus fruits (such as oranges and lemons), is necessary for the proper cross-linking of the molecules that make up collagen fibers. A deficiency of this vitamin in the diet can lead to **scurvy,** a weakening of collagen and connective tissue throughout the body. As we will explain later in this text, strong collagen is necessary for holding teeth in their sockets, for reinforcing the walls of blood vessels, and for healing wounds and forming scar tissue.

▨ Given the above information, can you list the major symptoms of scurvy?

The teeth fall out, blood vessels rupture, and wounds fail to heal. ■

Ground Substance Holds Fluid. Recall that areolar connective tissue surrounds the capillaries of the body (Figure 4.14). Fluid continually pours out of capillaries and enters the extracellular matrix. This clear fluid, called **tissue fluid** (or *interstitial fluid*), derives from blood and contains all the small molecules of blood. All cells that border the connective-tissue matrix—essentially, all cells of the body—obtain their blood-derived oxygen and nutrients from the tissue fluid and release their wastes into this fluid as well. The waste molecules then diffuse back into the capillaries, to be carried away by the blood. The tissue fluid that occupies areolar connective tissue is

continually recycled, and it normally returns to the capillaries at the same rate that it forms.

The part of the extracellular matrix that holds this tissue fluid is called **ground substance.** This jelly-like material consists of large sugar and sugar-protein molecules that soak up fluid like a sponge. The molecules of ground substance are made and secreted by the nearby fibroblasts.

So far, we have established that the extracellular matrix of connective tissue is a combination of (1) fibers and (2) ground substance. Tissue fluid can be considered another basic component of the extracellular matrix.

Defense Cells Provide Protection. Areolar connective tissue is the main battlefield in the body's war against infectious microorganisms, such as bacteria, viruses, fungi, and parasites. It is easy to see why this is so, for microorganisms that invade the body always enter areolar tissue after penetrating the body's epithelial surfaces. Every effort is made to destroy the microorganisms at their entry site in the areolar tissue—otherwise they enter the capillaries and use the circulatory vessels to spread throughout the body. Areolar connective tissue contains a variety of defense cells, all of which originate as blood cells then migrate to the connective tissue by leaving the capillaries. These defenders gather at infection sites in large numbers, where each cell type uses its own strategy to destroy the invaders. The defense cells include the following:

1. Macrophages (mak′ro-fāj-es). These "big eaters" are oval cells whose surface is ruffled by pseudopods (Figure 4.14). Macrophages are the nonspecific phagocytic cells of our body; that is, they engulf and devour a wide variety of foreign materials, ranging from whole bacteria to foreign molecules to dirt particles. Macrophages also dispose of dead tissue cells.

2. Plasma cells. These egg-shaped cells secrete protein molecules called *antibodies.* Antibodies bind to foreign molecules and microorganisms, marking them for destruction. Antibodies and plasma cells are discussed in more detail in Chapter 16 (pp. 450–451).

3. Mast cells. Also oval in shape, **mast cells** possess large secretory granules that contain molecules of *heparin* and *histamine.* In the presence of infectious organisms, the contents of the granules are released quickly and affect nearby blood vessels almost immediately. Histamine (his′tah-mēn) increases the permeability of the capillary walls, causing the infected areolar tissue to swell with excess tissue fluid. This process is a part of inflammation, a defense process described later in the chapter. Heparin is an anticlotting agent when free in the bloodstream, but the function of the heparin from human mast cells is unknown.

4. Neutrophils, lymphocytes, and eosinophils. These are white blood cells that fight infection (Chapter 16, p. 448). *Neutrophils* (nu′tro-filz) are experts at phagocytizing bacteria.

 Cellular defenses are not the only means by which areolar connective tissue fights infection. Both the viscous ground substance and the dense networks of collagen fibers in the extracellular matrix slow the progress of invading microorganisms. Some bacteria, however, secrete enzymes that rapidly break down ground substance or collagen.

What can you deduce about the virulence of such bacteria?

These particular bacteria are highly invasive; that is, they spread rapidly through the connective tissues and are especially difficult for the body's defenses to control. An example of a bacterium that degrades ground substance is the *Streptococcus* responsible for strep throat. ■

Fat Cells Store Nutrients.   A minor function of areolar connective tissue is to store energy reserves as fat. The large, fat-storing cells are **fat cells** (also called *adipose* [ad′ĭ-pōs] *cells*, or *adipocytes*) (Figure 4.14). Fat cells are egg-shaped, and their cytoplasm is dominated by a giant lipid droplet that flattens the nucleus and cytoplasm at one end of the cell. Mature fat cells are among the largest cells in the body and are incapable of division. In areolar connective tissue, fat cells occur singly or in small groups.

Now that we have established the structure of a "model" (areolar) connective tissue, we can consider the other connective tissues. We begin with those that most closely resemble areolar connective tissue, then proceed to those that are less similar.

Other Loose Connective Tissues

Two connective tissues are specialized variations of areolar tissue: adipose (fat) tissue and reticular tissue. Areolar, adipose, and reticular tissues make up the loose connective tissues.

Adipose Tissue (Figure 4.15c).   **Adipose tissue** is similar to areolar connective tissue in structure and function, but its nutrient-storing function is much greater. Correspondingly, adipose tissue is crowded with fat cells, which account for 90% of its mass. These fat cells are grouped into large clusters called lobules. Adipose tissue is richly vascularized, reflecting its high metabolic activity. Without the fat stores in our adipose tissue, we could not live for more than a few days without eating. Adipose tissue is certainly abundant: It constitutes 18% of an average person's body weight (15% in men and 22% in women). In-

deed, a chubby person's body can be 50% fat without being morbidly obese.

Much of the body's adipose tissue occurs in the layer beneath the skin called the *hypodermis.* Adipose tissue also is abundant in the mesenteries (sheets of serous membranes that hold the stomach and intestines in place), around the kidney, and in the orbit posterior to the eyeball.

Reticular Connective Tissue (Figure 4.15d).   **Reticular connective tissue** resembles areolar tissue, but the only fibers in its matrix are reticular fibers. These fine fibers form a broad, three-dimensional network, a labyrinth of caverns that hold many free cells. The bone marrow, spleen, and lymph nodes, which house many free blood cells outside of their capillaries, consist largely of reticular connective tissue. Fibroblasts called **reticular cells** lie along the reticular network of this tissue. Reticular tissue is discussed further in Chapters 16 (p. 453) and 19 (p. 524).

Dense Connective Tissue

Dense connective tissue, or *fibrous connective tissue,* contains more collagen than does areolar connective tissue but is otherwise built on the same plan. With its thick collagen fibers, it can resist extremely strong pulling (tensile) forces. The two main kinds of dense connective tissue are *irregular* and *regular.*

Dense Irregular Connective Tissue (Figure 4.15e). **Dense irregular connective tissue** is similar to areolar tissue, but its collagen fibers are much thicker. These fibers run in many different planes, allowing this tissue to resist strong tensions from many different directions. This tissue dominates the leathery dermis of the skin, which is stretched, pulled, and hit from various angles. This tissue also makes up the *fibrous capsules* that surround certain organs in the body (kidney, lymph nodes, bones).

Dense Regular Connective Tissue (Figure 4.15f). The collagen fibers in **dense regular connective tissue** usually run in the same direction, parallel to the direction of pull. Crowded between the collagen fibers are rows of fibroblasts, which continuously manufacture the fibers and a scant ground substance. When this tissue is not under tension, its collagen fibers are slightly wavy. Unlike areolar connective tissue, dense regular connective tissue is poorly vascularized and contains no fat cells or defense cells.

With its enormous tensile strength, dense regular connective tissue is the main component of *ligaments,* bands or sheets that bind bones to one another. It also forms *tendons,* which are cords that attach muscles to bones, and *aponeuroses* (ap″o-nu-ro′sēs), which resemble sheet-like tendons.

Dense regular connective tissue also forms **fascia** (fash′e-ah; "a band"), a fibrous membrane that wraps around muscles, around groups of muscles, and around large vessels and nerves. Many sheets of fascia occur throughout the body, binding structures together like plastic sandwich wrap. When the word *fascia* is used alone, it is understood to mean *deep fascia. Superficial fascia,* something else entirely, is the fatty hypodermis below the skin.

Elastic Connective Tissue (Figure 4.15g).

The vocal cords and some ligaments connecting adjacent vertebrae are dominated by thick, parallel elastic fibers. These structures combine strength with great elasticity. The dense tissue they contain is called **elastic connective tissue.**

Other Connective Tissues: Cartilage, Bone, and Blood

As we have seen, connective tissue proper has the ability to resist tension (pulling). Cartilage and bone are the firm connective tissues that resist *compression* (pressing) as well as tension. Like areolar tissue, they consist of cells separated by a matrix consisting of fibers, ground substance, and tissue fluid. However, these skeletal tissues exaggerate the supportive functions of connective tissue and play no role in fat storage or defense against disease. After discussing cartilage and bone, we will introduce blood, the most unusual connective tissue.

Cartilage

Cartilage, a firm but flexible tissue, makes up several parts of the skeleton. For example, it forms the supporting rings of the trachea (windpipe) and gives shape to the nose and ears. Like all connective tissues, cartilage consists of cells separated by an abundant extracellular matrix (Figure 4.15h). This matrix contains fine collagen fibrils, a ground substance, and an exceptional quantity of tissue fluid; in fact, cartilage consists of up to 80% water! The arrangement of water in its matrix enables cartilage to spring back from compression, as we will explain in Chapter 6.

Cartilage is simpler than other connective tissues: It contains no blood vessels or nerves and just one kind of cell, the **chondrocyte** (kon′dro-sīt; "cartilage cell"). Each chondrocyte resides in a cavity (lacuna) surrounded by the firm matrix. Immature chondrocytes are *chondroblasts,* cells that actively secrete the matrix during cartilage growth. Cartilage is found in three varieties, each dominated by a particular fiber type: hyaline cartilage, elastic cartilage, and fibrocartilage (Figure 4.15 h–j). These are discussed in depth in Chapter 6.

Bone Tissue

Because of its rock-like hardness, **bone tissue** has a tremendous ability to support and protect body structures (Figure 4.15k). Bone matrix contains inorganic calcium salts (bone salts), which enable bone to resist compression, and an abundance of collagen fibers, which allow bone to withstand strong tension.

Immature bone cells, called *osteoblasts* (os″te-o-blasts′; "bone formers"), secrete the collagen fibers and ground substance of the bone matrix. Then, bone salts precipitate on and between the collagen fibers, hardening the matrix. The mature bone cells, called **osteocytes,** inhabit cavities (lacunae) in this hardened matrix. Bone is a living and dynamic tissue well-supplied with blood vessels. It is discussed further in Chapter 6.

Blood

Blood, the fluid in the blood vessels, is the most atypical connective tissue. It is classified as a connective tissue because it consists of blood *cells* surrounded by a nonliving matrix, the liquid blood *plasma* (Figure 4.15l). Its cells and matrix are very different from those in other connective tissues, however. Blood functions as the transport vehicle for the cardiovascular system, carrying nutrients, wastes, respiratory gases, and many other substances throughout the body. Blood is discussed in detail in Chapter 16.

Covering and Lining Membranes

Now that we have discussed connective and epithelial tissues, we can consider the **covering and lining membranes** (Figure 4.16, p. 90) that combine these two tissue types. These membranes, which cover broad areas within the body, consist of an epithelial sheet plus the underlying layer of connective tissue proper. These membranes are of three types: the cutaneous membrane, mucous membranes, and serous membranes.

The **cutaneous** (ku-ta′ne-us) **membrane,** or skin (*cutis* = skin), covers the outer surface of the body (Figure 4.16a). Its outer epithelium is the thick epidermis, and its inner connective tissue is the dense dermis. It is a dry membrane. The skin is discussed further in Chapter 5.

A **mucous membrane,** or **mucosa** (mu-ko′sah), lines the inside of every hollow internal organ that opens to the outside of the body. More specifically, mucous membranes line the tubes of the respiratory, digestive, reproductive, and urinary systems (Figure *(text continues on p. 90)*

Figure 4.15 Connective tissues.

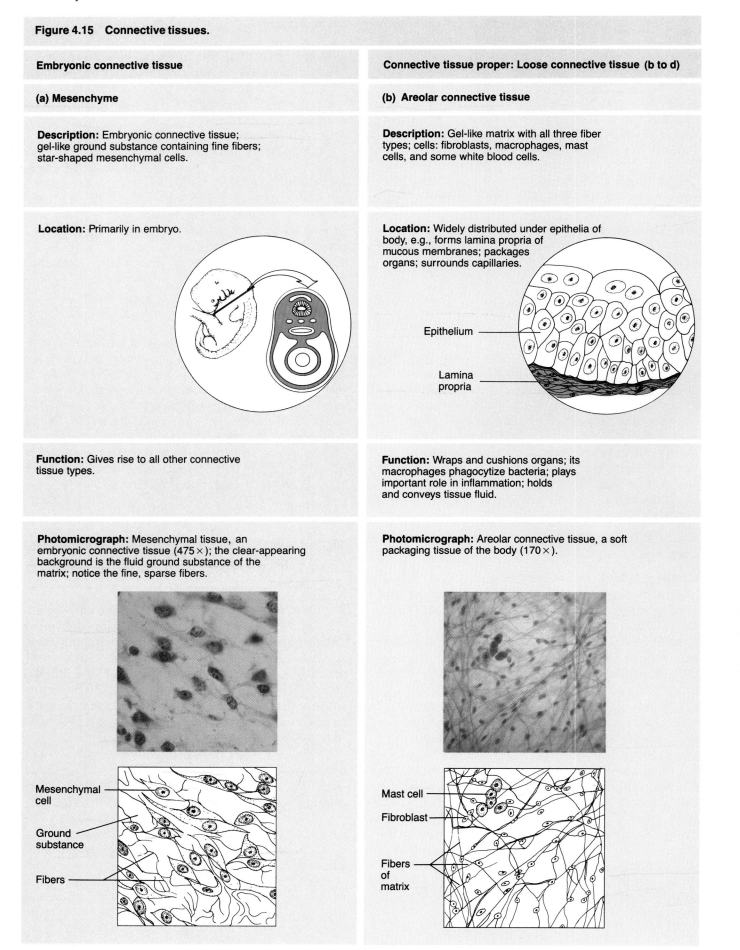

Embryonic connective tissue

(a) Mesenchyme

Description: Embryonic connective tissue; gel-like ground substance containing fine fibers; star-shaped mesenchymal cells.

Location: Primarily in embryo.

Function: Gives rise to all other connective tissue types.

Photomicrograph: Mesenchymal tissue, an embryonic connective tissue (475×); the clear-appearing background is the fluid ground substance of the matrix; notice the fine, sparse fibers.

Mesenchymal cell

Ground substance

Fibers

Connective tissue proper: Loose connective tissue (b to d)

(b) Areolar connective tissue

Description: Gel-like matrix with all three fiber types; cells: fibroblasts, macrophages, mast cells, and some white blood cells.

Location: Widely distributed under epithelia of body, e.g., forms lamina propria of mucous membranes; packages organs; surrounds capillaries.

Epithelium

Lamina propria

Function: Wraps and cushions organs; its macrophages phagocytize bacteria; plays important role in inflammation; holds and conveys tissue fluid.

Photomicrograph: Areolar connective tissue, a soft packaging tissue of the body (170×).

Mast cell

Fibroblast

Fibers of matrix

Figure 4.7 (continued)

(c) Adipose tissue

Description: Matrix as in areolar, but very sparse; closely packed adipocytes, or fat cells, have nucleus pushed to the side by large fat droplet.

Location: Under skin; around kidneys and eyeballs; in bones and within abdomen; in breasts.

Function: Provides reserve food fuel; insulates against heat loss; supports and protects organs.

Photomicrograph: Adipose tissue from the subcutaneous layer under the skin (500×).

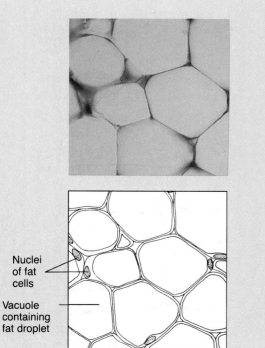

Nuclei of fat cells

Vacuole containing fat droplet

(d) Reticular connective tissue

Description: Network of reticular fibers in a typical loose ground substance; reticular cells predominate.

Location: Lymphoid organs (lymph nodes, bone marrow, and spleen).

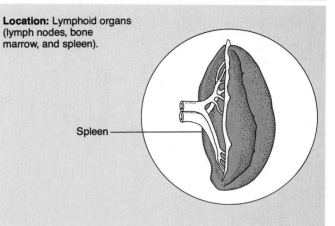

Spleen

Function: Fibers form a soft internal skeleton that supports other cell types.

Photomicrograph: Dark-staining network of reticular connective tissue fibers forming the internal skeleton of the spleen (625×).

Reticular cell

Blood cells

Reticular fibers

Figure 4.15 (continued)

Connective tissue proper: Dense connective tissue (e to g)

(e) Dense irregular connective tissue

Description: Primarily irregularly arranged collagen fibers; some elastic fibers; major cell type is the fibroblast.

Location: Dermis of the skin; submucosa of digestive tract; fibrous capsules of organs and of joints.

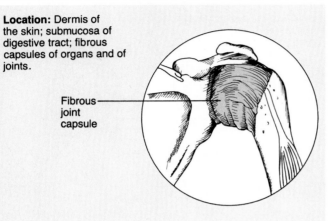

Fibrous joint capsule

Function: Able to withstand tension exerted in many directions; provides structural strength.

Photomicrograph: Dense irregular connective tissue from the dermis of the skin (475×).

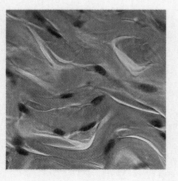

Collagen fibers

Nuclei of fibroblasts

(f) Dense regular connective tissue

Description: Primarily parallel collagen fibers; a few elastin fibers; major cell type is the fibroblast.

Location: Tendons, most ligaments, aponeuroses.

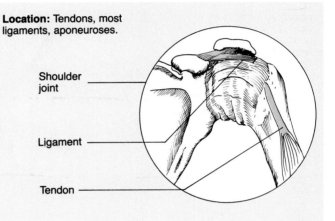

Shoulder joint

Ligament

Tendon

Function: Attaches muscles to bones or to muscles; attaches bones to bones; withstands great tensile stress when pulling force is applied in one direction.

Photomicrograph: Dense regular connective tissue from a tendon (200×).

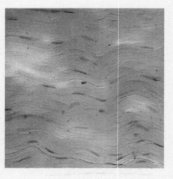

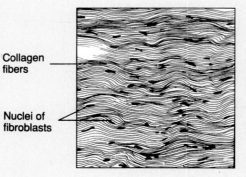

Collagen fibers

Nuclei of fibroblasts

Figure 4.15 (continued)

Cartilage: (h to j)

(g) Elastic connective tissue

(h) Hyaline cartilage

Description: Same as for the other dense connective tissues, but predominant fiber type is elastin.

Description: Amorphous but firm matrix; collagen fibers form an imperceptible network; chondroblasts produce the matrix and when mature (chondrocytes) lie in lacunae.

Location: Walls of the aorta, some parts of trachea and bronchi; forms the vocal cords and the ligamenta flava connecting the vertebrae.

Vocal cords

Location: Forms most of the embryonic skeleton; covers the ends of long bones in joint cavities; forms costal cartilages of the ribs; cartilages of the nose, trachea, and larynx.

Costal cartilages

Function: Provides durability with stretch.

Function: Supports and reinforces; has resilient cushioning properties; resists compressive stress.

Photomicrograph: Elastic connective tissue in a wall of the aorta (190×); notice the wavy appearance of the elastin fibers.

Photomicrograph: Hyaline cartilage from the trachea (475×).

Fibroblast

Elastin fiber

Chondrocyte in lacuna

Matrix

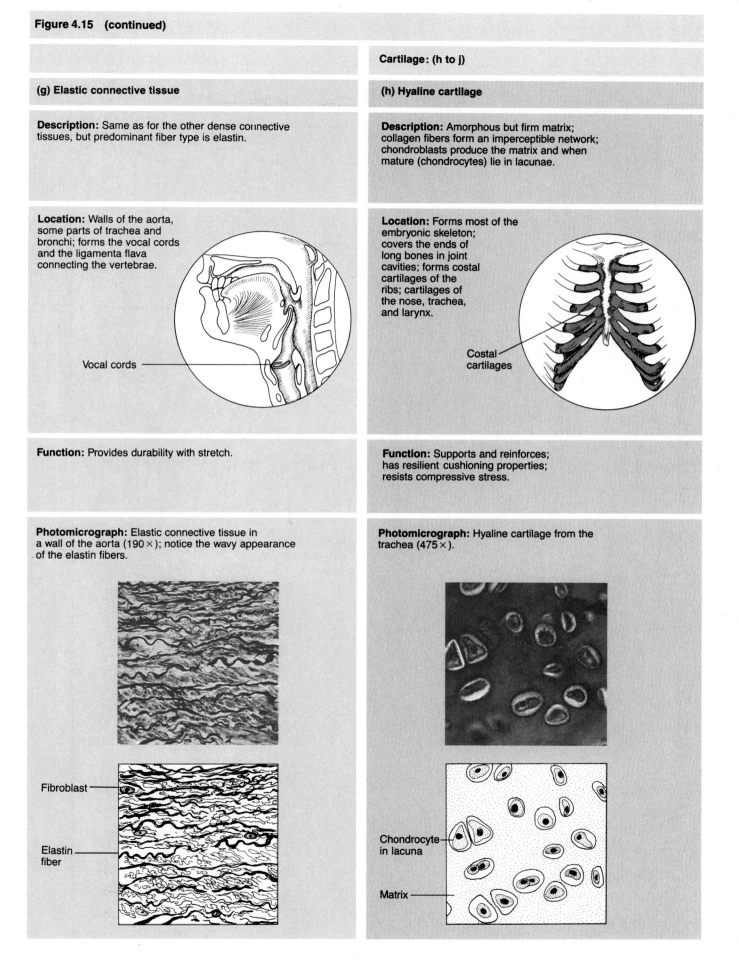

Figure 4.15 (continued)

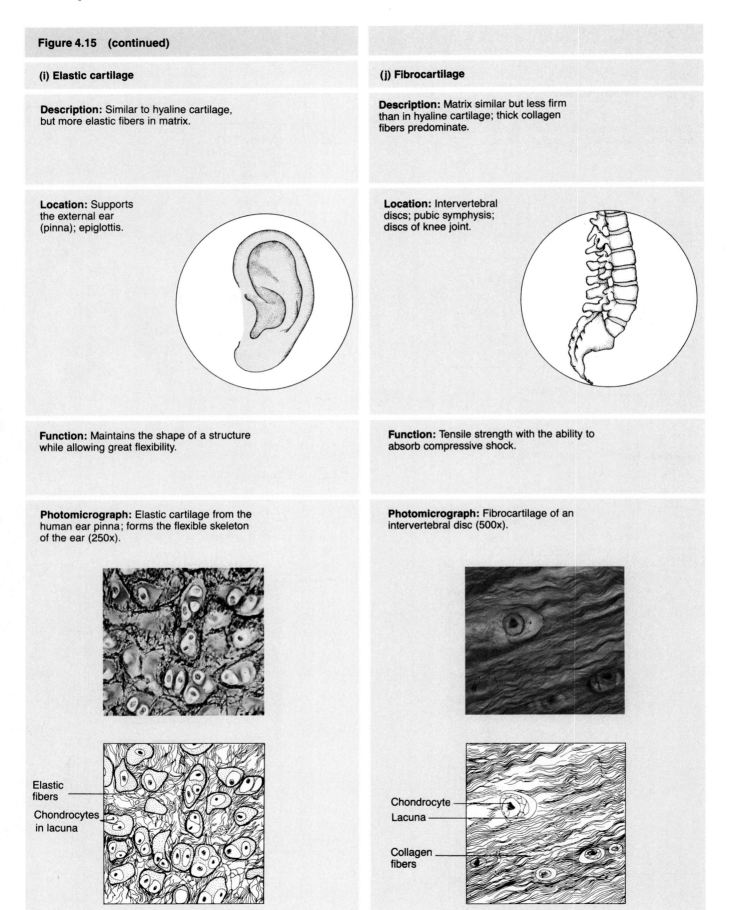

(i) Elastic cartilage

Description: Similar to hyaline cartilage, but more elastic fibers in matrix.

Location: Supports the external ear (pinna); epiglottis.

Function: Maintains the shape of a structure while allowing great flexibility.

Photomicrograph: Elastic cartilage from the human ear pinna; forms the flexible skeleton of the ear (250x).

Elastic fibers

Chondrocytes in lacuna

(j) Fibrocartilage

Description: Matrix similar but less firm than in hyaline cartilage; thick collagen fibers predominate.

Location: Intervertebral discs; pubic symphysis; discs of knee joint.

Function: Tensile strength with the ability to absorb compressive shock.

Photomicrograph: Fibrocartilage of an intervertebral disc (500x).

Chondrocyte

Lacuna

Collagen fibers

Figure 4.15   (continued)

Others: (k and l)

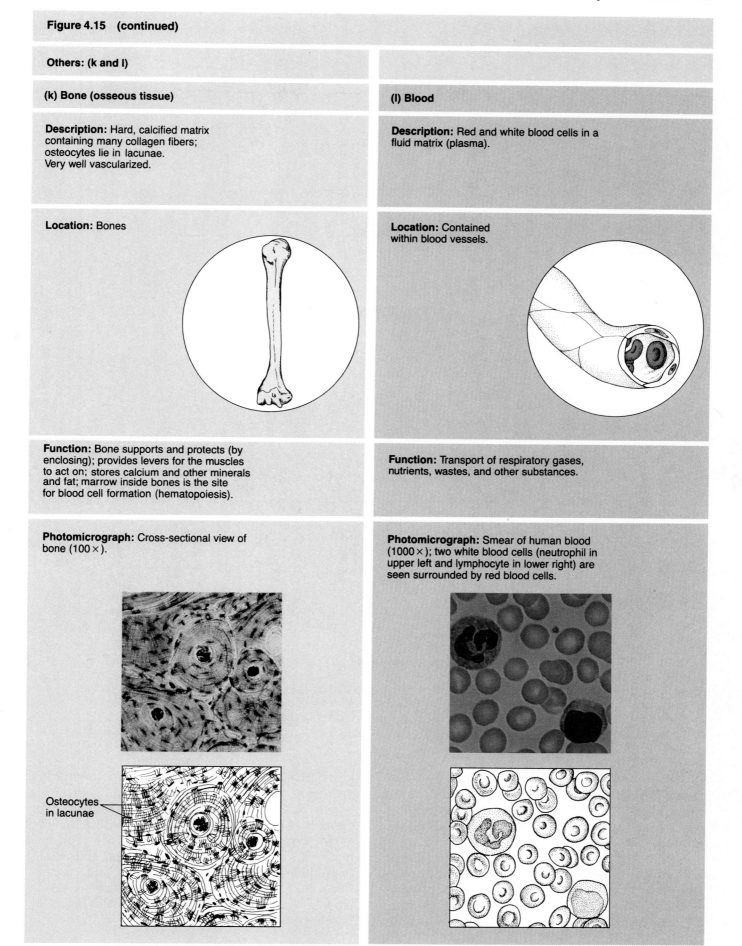

(k) Bone (osseous tissue)

Description: Hard, calcified matrix containing many collagen fibers; osteocytes lie in lacunae. Very well vascularized.

Location: Bones

Function: Bone supports and protects (by enclosing); provides levers for the muscles to act on; stores calcium and other minerals and fat; marrow inside bones is the site for blood cell formation (hematopoiesis).

Photomicrograph: Cross-sectional view of bone (100×).

Osteocytes in lacunae

(l) Blood

Description: Red and white blood cells in a fluid matrix (plasma).

Location: Contained within blood vessels.

Function: Transport of respiratory gases, nutrients, wastes, and other substances.

Photomicrograph: Smear of human blood (1000×); two white blood cells (neutrophil in upper left and lymphocyte in lower right) are seen surrounded by red blood cells.

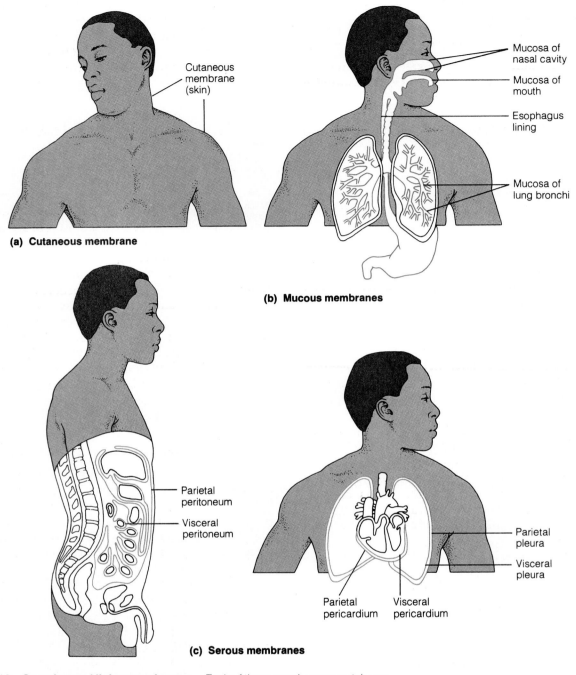

(a) Cutaneous membrane

(b) Mucous membranes

- Mucosa of nasal cavity
- Mucosa of mouth
- Esophagus lining
- Mucosa of lung bronchi

(c) Serous membranes

- Parietal peritoneum
- Visceral peritoneum
- Parietal pleura
- Visceral pleura
- Parietal pericardium
- Visceral pericardium

Figure 4.16 Covering and lining membranes.   Each of these membranes contains an epithelium and an underlying connective tissue. (**a**) The cutaneous membrane is the skin. (**b**) Mucous membranes line body cavities that are open to the exterior. (**c**) Serous membranes line ventral body cavities that are closed to the exterior.

4.16b). Although different mucous membranes vary widely in the types of epithelia they contain, all are wet or moist. As their name implies, many mucous membranes secrete mucus. Not all of them do so, however.

All mucous membranes consist of an epithelial sheet directly underlain by a layer of loose connective tissue called the *lamina propria* (lam′ĭ-nah pro′pre-ah; "one's own layer"). In some mucous membranes, the lamina propria rests on a third layer of smooth muscle cells. We will discuss the specific mucous membranes of the body in later chapters.

Serous membranes or **serosae,** introduced in Chapter 1, are the slippery membranes that line the closed ventral body cavity (Figure 4.16c). A serous membrane consists of a simple squamous epithelium (mesothelium) lying on a thin layer of areolar connective tissue. The lubricant produced by serous membranes, *serous fluid,* exudes from the capillaries in the connective tissue.

Muscle Tissue

The two remaining tissue types are muscle and nervous tissues. These are sometimes called *composite tissues* because, along with their own muscle or nerve cells, they contain small amounts of areolar connective tissue. (Areolar connective tissue surrounds all blood vessels, and both muscle and nervous tissue are richly vascularized.)

Muscle tissues (Figure 4.17) bring about most kinds of body movements. Muscle cells are called **muscle fibers.** They have an elongated shape and contract forcefully as they shorten. These fibers contain many myofilaments (mi″o-fil′ah-ments; "muscle filaments"), elaborate versions of the actin and myosin filaments that bring about contraction in all cell types (p. 33). There are three kinds of muscle tissue: skeletal, cardiac, and smooth.

Skeletal muscle tissue (Figure 4.17a) is the major component of organs called skeletal muscles, which pull on bones to cause body movements. Skeletal muscle fibers are long, large cylinders that contain many nuclei. Their obvious banded, or striated, appearance reflects a highly organized arrangement of their myofilaments.

Cardiac muscle tissue (Figure 4.17b) occurs in the wall of the heart. It contracts to propel blood through the blood vessels. Like skeletal muscle fibers, cardiac muscle fibers are striated. However, they differ in two ways: (1) each cardiac cell has just one nucleus, and (2) cardiac cells branch and join at unique cellular junctions called intercalated (in-ter′kah-la″ted) discs.

Smooth muscle tissue (Figure 4.17c) is so named because there are no visible striations in its fibers. These fibers are spindle-shaped and contain one centrally located nucleus. Smooth muscle primarily occurs in the walls of hollow organs (the digestive and urinary organs, uterus, and blood vessels). It generally acts to squeeze substances through these organs by alternately contracting and relaxing. Muscle tissue is described in detail in Chapter 9.

Nervous Tissue

Nervous tissue is the main component of the nervous organs—the brain, spinal cord, and nerves—which regulate and control body functions. Nervous tissue contains two types of cells, neurons and supporting cells. **Neurons** are the highly specialized nerve cells (Figure 4.18) that generate and conduct electrical impulses. Neurons have extensions, or processes, that allow them to transmit impulses over substantial distances within the body. **Supporting cells** are nonconducting cells that nourish, insulate, and protect the delicate neurons. A more complete discussion of nervous tissue appears in Chapter 11.

Tissue Response to Injury

The body has many mechanisms for protecting itself from injury and invading microorganisms. Intact epithelia act as a physical barrier, but once that barrier has been penetrated, protective responses are activated in the underlying connective tissue proper. These are the *inflammatory* and *immune responses.* Inflammation is a nonspecific, local response that develops quickly and limits the damage to the injury site. The immune response, by contrast, takes longer to develop and is highly specific. It destroys particular infectious microorganisms and foreign molecules at the site of infection and throughout the body. We will concentrate on inflammation here, saving the immune response for Chapter 19.

Inflammation

Almost any injury or infection will lead to an inflammatory response. For example, let us assume the skin is cut by a dirty piece of glass or that it has an infected pimple. As short-term or acute inflammation develops in the connective tissue, it produces four symptoms: *heat, redness, swelling,* and *pain.* Let us trace the source of each symptom.

The initial insult induces the release of *inflammatory chemicals* into the nearby tissue fluid. Injured tissue cells, macrophages, mast cells, and proteins from blood all serve as sources of these inflammatory mediators. These chemicals signal nearby blood vessels to dilate (enlarge), thus increasing the flow of blood to the injury site. The increase in blood flow is the source of the *heat* and *redness* of inflammation. Certain inflammatory chemicals, such as histamine (p. 81), increase the permeability of the capillaries, causing large amounts of tissue fluid to leave the bloodstream. The resulting accumulation of fluid in the connective tissue, called **edema** (ĕ-de′mah; "swelling") causes the *swelling* of inflammation. The excess fluid presses on nerve endings, contributing to the sensation of *pain.* Some of the inflammatory chemicals also cause pain by affecting the nerve endings directly.

Although at first glance inflammatory edema seems detrimental, it is actually beneficial. The entry of blood-derived fluid into the injured connective tis-

(text continues on p. 94)

Figure 4.17   Muscle tissues.

(a) Skeletal muscle

Description: Long, cylindrical, multinucleate cells; obvious striations.

Location: In skeletal muscles attached to bones or occasionally to skin.

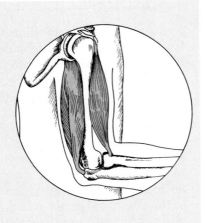

Function: Voluntary movement; locomotion; manipulation of the environment; facial expression. Voluntary control.

Photomicrograph: Skeletal muscle (approx. 30x). Notice the obvious banding pattern and the fact that these large cells are multinucleate.

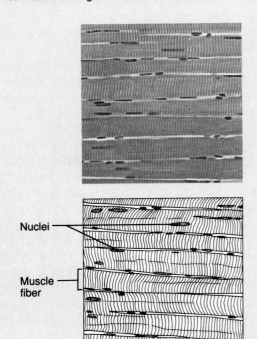

Nuclei

Muscle fiber

(b) Cardiac muscle

Description: Branching, striated, generally uninucleate cells that interdigitate at specialized junctions (intercalated discs).

Location: The walls of the heart.

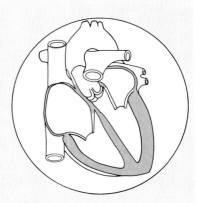

Function: As it contracts, it propels blood into the circulation; involuntary control.

Photomicrograph: Cardiac muscle (250x); notice the striations, branching of fibers, and the intercalated discs.

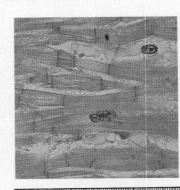

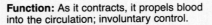

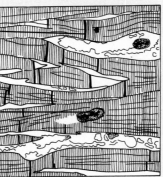

Inter-calated disc

Nucleus

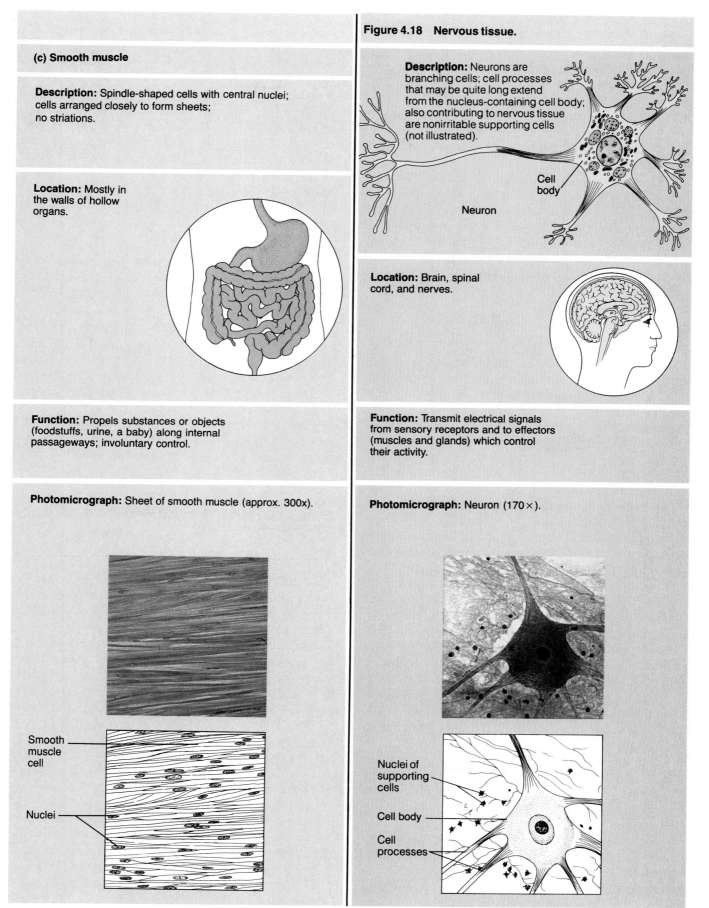

(c) Smooth muscle

Description: Spindle-shaped cells with central nuclei; cells arranged closely to form sheets; no striations.

Location: Mostly in the walls of hollow organs.

Function: Propels substances or objects (foodstuffs, urine, a baby) along internal passageways; involuntary control.

Photomicrograph: Sheet of smooth muscle (approx. 300x).

Smooth muscle cell

Nuclei

Figure 4.18 Nervous tissue.

Description: Neurons are branching cells; cell processes that may be quite long extend from the nucleus-containing cell body; also contributing to nervous tissue are nonirritable supporting cells (not illustrated).

Cell body

Neuron

Location: Brain, spinal cord, and nerves.

Function: Transmit electrical signals from sensory receptors and to effectors (muscles and glands) which control their activity.

Photomicrograph: Neuron (170×).

Nuclei of supporting cells

Cell body

Cell processes

sue (1) helps to dilute toxins secreted by bacteria, (2) brings in oxygen and nutrients from the blood, necessary for tissue repair, and (3) brings in antibodies from the blood to fight infection. In very severe infections and in all wounds that sever blood vessels, the fluid leaking from the capillaries contains *clotting* proteins. In these cases, clotting occurs in the connective tissue matrix. The fibrous clot isolates the injured area and "walls in" the infectious microorganisms, preventing their spread.

The next stage in inflammation is *stasis* ("standing"). This is the slowdown in local blood flow that necessarily follows a massive exit of fluid from the capillaries. At this stage, white blood cells begin to leave the small vessels. First to appear at the infection site are neutrophils, then macrophages. These cells devour the infectious microorganisms and the damaged tissue cells as well.

Repair

Even as inflammation proceeds, repair begins. Tissue repair can occur in two major ways: by regeneration and by fibrosis. **Regeneration** is the replacement of a destroyed tissue by the same kind of tissue, whereas **fibrosis** involves the proliferation of a fibrous connective tissue called **scar tissue.** The process of tissue repair in a skin wound involves both regeneration and fibrosis. Once the blood within the cut has clotted, the surface part of the clot dries to form a scab (Figure 4.19a and b). At this point, tissue repair begins with a step called organization.

Organization is the process by which the clot is replaced by granulation tissue (Figure 4.19b). **Granulation tissue** is a delicate pink tissue made of several elements. It contains capillaries that grow in from nearby areas, as well as proliferating fibroblasts that produce new collagen fibers to bridge the gash. As organization proceeds, macrophages digest the original clot, and the deposit of collagen continues. As more collagen is made, the granulation tissue gradually transforms into fibrous scar tissue.

During organization, the surface epithelium begins to *regenerate,* growing under the scab until the scab falls away. The end result is a fully regenerated epithelium and an underlying area of scar.

The above describes the healing of a wound (a cut, scrape, puncture). In pure *infections* (a pimple or sore throat), by contrast, there is usually no clot formation or scarring. Only severe infections lead to scarring.

The capacity for regeneration varies widely among the different tissues. Epithelia regenerate extremely well, as do bone, areolar connective tissue, dense irregular connective tissue, and blood-forming tissue. Smooth muscle and dense regular connective tissue have a weak to moderate capacity for regeneration, but skeletal muscle and cartilage regenerate poorly, if at all. Cardiac muscle and the nervous tissue in the brain and spinal cord have essentially no regenerative capacity. They can be replaced only by scar tissue.

In nonregenerating tissues and in exceptionally severe wounds, fibrosis totally replaces the lost tissue. The resulting scar appears as a pale, often shiny area and shrinks during the months after it first forms. A scar consists mostly of collagen fibers and contains few cells or capillaries. Although it is very strong, it lacks the flexibility and elasticity of most normal tissues. Nor can it perform the normal functions of the tissue it has replaced.

Scar tissue that forms in the wall of the urinary bladder, the heart, or another hollow organ may severely hamper the function of that organ. Why do you think this is so?

The normal shrinking of the scar reduces the internal volume of the organ, possibly hindering or blocking the movement of substances through that organ. Irritation of visceral organs can cause them to adhere to one another or to the body wall as they scar. Such **adhesions** prevent the normal churning actions of loops of the intestine, dangerously halting the movement of food through the digestive tube. Adhesions can also restrict the movement of the heart and lungs and immobilize the joints. ■

The Tissues Throughout Life

The embryonic derivations of the four basic tissues are as follows: Connective and muscle tissues derive from mesenchyme, mostly of the mesoderm germ layer. Most epithelial tissues develop from embryonic epithelium, of the ectoderm and endoderm layers. A few epithelia, however, derive from mesodermal mesenchyme, namely the endothelium (lining the vessels) and mesothelium (lining the ventral body cavity). The nervous tissue in the brain and spinal cord derives from ectodermal epithelium (recall Chapter 3, p. 56).

By the end of the second month of development, the primary tissues have appeared, and all major organs are in place. In virtually all tissues, the cells divide throughout the prenatal period, providing the rapid growth that occurs before birth. The division of *nerve* cells, however, stops permanently during the fetal period. After birth, the cells of most other tissues continue to divide until adult body size is reached. Cellular division then slows greatly, although many tissues retain a regenerative capacity. In adulthood,

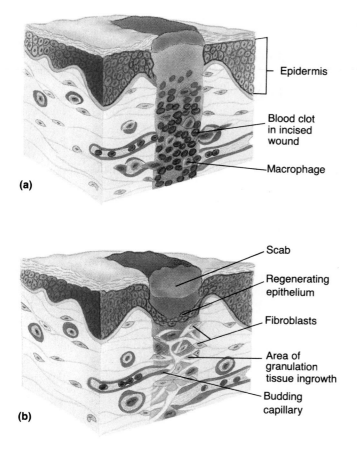

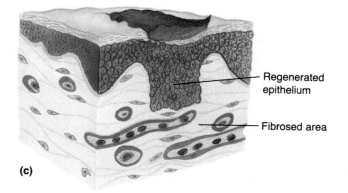

(a)

- Epidermis
- Blood clot in incised wound
- Macrophage

(c)

- Regenerated epithelium
- Fibrosed area

(b)

- Scab
- Regenerating epithelium
- Fibroblasts
- Area of granulation tissue ingrowth
- Budding capillary

Figure 4.19   Tissue repair of a skin wound. (**a**) At the time of injury, inflammatory chemicals are released, causing local blood vessels to dilate and become more permeable. Fluid, white blood cells, and blood proteins can then enter the injured site. Cut vessels bleed, blood clots within the wound, and the surface of the clot dries to become a scab. (**b**) Organization, or the formation of granulation tissue. Capillary buds invade the clot, restoring a vascular supply. Fibroblasts enter the region and secrete collagen in abundance to bridge the wound. Macrophages dispose of dead cells and other debris. Surface epithelial cells proliferate and migrate under the scab to cover the granulation tissue. (**c**) About a week after injury, the fibrosed area (scar) is contracting, and regeneration of the epithelium is almost complete. Depending on the severity of the original wound, the scar may or may not be visible beneath the epidermis.

only the epithelial tissues and blood cell-forming tissues are highly mitotic.

Some tissues that regenerate throughout life do so through the division of their mature, differentiated cells (the gland cells in the liver and endothelial cells are examples). However, many tissues contain populations of **stem cells,** relatively undifferentiated cells that divide to produce new tissue cells as needed. Stem cells occur in the epidermis of the skin, the lining of the digestive tube, in blood-forming tissue, and elsewhere.

Given good nutrition, good circulation, and relatively infrequent wounds and infections, our tissues normally function well through youth and middle age. The importance of nutrition for tissue health cannot be overemphasized. For example, vitamin A is

needed for the normal regeneration of epithelium (liver and carrots are rich in this vitamin). Since proteins are the structural material of the body, an adequate intake of protein is essential for the tissues to retain their structural integrity.

With increasing age, the epithelia thin and are more easily breached. The amount of collagen in the body declines, making tissue repair less efficient. Bone, muscle, and nervous tissues begin to atrophy. These events are due in part to a decrease in circulatory efficiency, which reduces the delivery of nutrients to the tissues, but in some cases diet is a contributing factor. As their income declines or as chewing becomes more difficult, older people tend to eat soft foods, which may be low in protein and vitamins. As a result, the health of the tissues suffers.

✚ Related Clinical Terms

Adenoma (ad"ĕ-no′mah) (*aden* = gland; *oma* = tumor) Any neoplasm of glandular epithelium, benign or malignant. The malignant type is more specifically called *adenocarcinoma.*

Carcinoma (kar"sĭ-no′mah) (*karkinos* = crab, cancer) Cancer arising in an epithelium.

Healing by first intention The simplest type of wound healing, this occurs when the edges of the wound are brought together by sutures, staples, or other means to close surgical incisions. Only small amounts of granulation tissue usually form.

Healing by second intention The edges of the wound remain separated, and the gap is bridged by large amounts of granulation tissue. This is the manner by which unattended wounds heal. It is slower than healing by first intention and leads to larger scars.

-itis (i'tis) Suffix meaning "inflammation of." For example, appendic*itis* is inflammation of the appendix.

Lesion (le'zhun; "wound") Any injury, wound, or infection that affects tissue over an area of a definite size (as opposed to widely spread throughout the body).

Pus A collection of tissue fluid, dead tissue cells, bacteria, and white blood cells in an infected area. Bacteria-fighting neutrophils are an important component of pus.

Sarcoma (sar-ko'mah) (*sarkos* = flesh; *oma* = tumor) Cancer arising in the mesenchyme-derived tissues, that is, in connective tissues and muscle.

Septicemia (sep"tĭ-se'me-ah) (*septikos* = rotting; *hemia* = blood) An infection in which bacteria enter the blood from the original infection site and thus spread throughout the body. Many body defenses act to prevent local infections from leading to septicemia.

Suction lipectomy (*lip* = fat; *ectomy* = removal) Surgery in which the fat of the hypodermis is suctioned from places where it is most abundant, such as the thighs, abdomen, and breasts. This is a cosmetic procedure that does not result in a permanent reduction of fat.

Chapter Summary

1. Tissues are collections of structurally similar cells with related functions. Tissues also contain nonliving material between their cells. The four primary tissues are epithelium, connective tissue, muscle, and nervous tissue.

EPITHELIA AND GLANDS (pp. 66–78)

1. Epithelia are sheets of cells that cover body surfaces and line body cavities. Their functions include protection, sensory reception, absorption, ion transport, and filtration. Glandular epithelia also function in secretion.

Special Characteristics of Epithelia (p. 67)

2. Epithelia exhibit a high degree of cellularity, little extracellular material, specialized cell junctions, polarity, avascularity, and the ability to regenerate. They are underlain by loose connective tissue.

Classification of Epithelia (pp. 67–73)

3. Epithelia are classified by cell shape as squamous, cuboidal, or columnar, and by the number of cell layers as simple (one layer) or stratified (more than one layer). Stratified epithelia are named according to the shape of their apical cells.

4. Simple squamous epithelium is a single layer of flat cells. Its thinness allows molecules to pass through it rapidly. It lines the air sacs of the lungs, the interior of blood vessels (endothelium), and the ventral body cavity (mesothelium).

5. Simple cuboidal epithelium occurs in kidney tubules and the small ducts of glands.

6. Simple columnar epithelium lines the stomach and intestines, and a ciliated version lines the uterine tubes. Its cells are active in secretion, absorption, and ion transport.

7. Pseudostratified columnar epithelium is a simple epithelium that contains both short and tall cells. A ciliated version lines most of the respiratory passages.

8. Stratified squamous epithelium is multilayered and thick. Its apical cells are flat, and it resists abrasion. Examples are the epidermis and the lining of the mouth, esophagus, and vagina.

9. Transitional epithelium is a stratified epithelium that thins when it stretches. It lines the hollow urinary organs.

Epithelial Surface Features (pp. 73–76)

10. Features on apical epithelial surfaces: Microvilli occur on most epithelia. They increase the epithelial surface area and may anchor sheets of mucus. Cilia are whip-like projections that beat to move fluid (usually mucus). Microtubules in the cores of cilia generate ciliary movement.

11. Features on lateral cell surfaces: The main types of cell junctions are desmosomes that bind cells together, tight junctions that close off the extracellular spaces, and gap junctions that allow passage of small molecules between cells.

12. Feature on the basal epithelial surface: Epithelial cells lie on a protein sheet called the basal lamina. This acts as a filter and a scaffolding on which regenerating epithelial cells can grow. The basal lamina, plus some underlying reticular fibers, form the thicker basement membrane.

Glands (pp. 76–77)

13. A gland is one or more cells specialized to secrete a product. Most glandular products are proteins released by exocytosis.

14. Exocrine glands secrete onto body surfaces or into body cavities. Mucus-secreting goblet cells are unicellular exocrine glands. Multicellular exocrine glands are classified by the structure of their ducts as simple or compound and by the structure of their secretory units as tubular, alveolar (acinar), or tubuloalveolar.

15. Endocrine (ductless) glands secrete hormones, which enter the circulatory vessels and travel to target organs, from which they signal a response.

CONNECTIVE TISSUE (pp. 78–83)

1. Connective tissue is the most diverse class of tissues. Its four basic subtypes are connective tissue proper (loose and dense), cartilage, bone, and blood. Despite their diversity, all derive from mesenchyme and consist of cells separated by abundant extracellular matrix. In all connective tissues except blood, the matrix consists of fibers, ground substance, and tissue fluid.

Connective Tissue Proper (pp. 79–83)

2. Loose areolar connective tissue can be considered a prototype connective tissue against which the other types are compared. Areolar connective tissue surrounds capillaries and underlies most epithelia. Its main functions are to (1) support and bind other tissues with its fibers (collagen, reticular, elastic), (2) hold tissue fluid in its jelly-like ground substance, (3) fight infection with its many blood-derived defense cells, and (4) store nutrients in fat cells.

3. The fibroblast, the most abundant cell type in connective tissue proper, produces both the fibers and the ground substance of the extracellular matrix. The function of collagen fibers is to resist tension placed on the tissue.

4. Adipose connective tissue is similar to areolar connective tissue but is dominated by fat cells. It is plentiful in the hypodermis below the skin.

5. Reticular connective tissue resembles areolar connective tissue, but its only fibers are reticular fibers. These form networks to hold free blood cells. This tissue occurs in bone marrow, lymph nodes, and the spleen.

6. Dense connective tissue contains exceptionally thick collagen fibers and resists tremendous pulling forces. In dense irregular connective tissue, the collagen fibers run in various directions. This tissue occurs in the dermis and in organ capsules. Dense regular connective tissue contains parallel bundles of

collagen fibers separated by rows of fibroblasts. This tissue, which is subject to high tension from a single direction, forms tendons, ligaments, and fascia.

Other Connective Tissues: Cartilage, Bone, and Blood (p. 83)

7. Cartilage and bone have the basic structure of connective tissue (cells and matrix), but their stiff matrix allows them to resist compression. Cartilage is springy and avascular. Its matrix contains mostly water.

8. Bone tissue has a firm, collagen-rich matrix embedded with calcium salts. This mineral gives bone compressive strength.

9. Blood consists of red and white blood cells in a fluid matrix called plasma. It is the most atypical connective tissue.

COVERING AND LINING MEMBRANES (pp. 83–90)

1. Membranes, each consisting of an epithelium plus an underlying layer of connective tissue, cover broad surfaces in the body. The cutaneous membrane (skin), which is dry, covers the body surface. Mucous membranes, which are moist, line the hollow internal organs that open to the body exterior. Serous membranes, which are slippery, line the ventral body cavity.

MUSCLE TISSUE (p. 91)

1. Muscle tissue consists of long muscle cells (fibers) specialized to contract and generate movement. A scant extracellular matrix separates the muscle fibers.

2. There are three types of muscle tissue:
Skeletal muscle, which forms muscles that move the skeleton. Its cells are cylindrical and striated.
Cardiac muscle, located in the wall of the heart, pumps blood. Its cells branch and have striations.
Smooth muscle, located in the walls of hollow organs, usually propels substances through these organs. Its cells are spindle-shaped and lack striations.

NERVOUS TISSUE (p. 91)

1. Nervous tissue, the main tissue of the nervous system, is composed of neurons and supporting cells.

2. Neurons are branching cells that receive and transmit electrical impulses. Nervous tissue regulates body functions.

TISSUE RESPONSE TO INJURY (pp. 91–94)

Inflammation (pp. 91–94)

1. Inflammation is a response to tissue injury and infection. It is localized to the connective tissue and the vessels of the injury site. Inflammation begins with dilation of blood vessels (causing redness and heat), followed by edema (causing swelling and pain). Stasis results, and white blood cells migrate into the injured tissue.

Repair (p. 94)

2. Tissue repair begins during inflammation and may involve tissue regeneration, fibrosis (scarring), or both. Repair of a cut begins with organization, during which the clot is replaced with granulation tissue. Collagen deposition replaces granulation tissue with scar tissue.

3. Certain tissues, like cardiac muscle and most nervous tissue, do not regenerate and are replaced only by scar tissue.

THE TISSUES THROUGHOUT LIFE (pp. 94–95)

1. Most epithelial and nervous tissues develop from embryonic epithelia, the ectoderm and endoderm. Exceptions are endothelium and mesothelium, which arise from mesodermal mesenchyme. Connective and muscle tissues develop from mesenchyme.

2. The decrease in mass and viability seen in most tissues during old age partially reflects circulatory deficits or poor nutrition.

Review Questions

Multiple Choice and Matching Questions

1. Use the key to match each basic tissue type with a description below.

Key: **(a)** connective tissue   **(b)** epithelium
     (c) muscle tissue   **(d)** nervous tissue

_a_ **(1)** composed largely of nonliving extracellular matrix; important in protection, support, defense, and holding tissue fluid

_c_ **(2)** the tissue immediately responsible for body movement

_d_ **(3)** the tissue that provides an awareness of the external environment and enables us to react to it

_b_ **(4)** the tissue that lines body cavities and covers surfaces

_b_ **(5)** the tissue that includes most glands

2. An epithelium that has several cell layers, with flat cells in the apical layer, is called (choose all that apply): (a) ciliated, (b) columnar, (c) stratified, (d) simple, (e) squamous.

3. Match the epithelial type named in column B with the appropriate location in column A.

	Column A		Column B
_b_	**(1)** lines inside of stomach and most of intestines	**(a)**	pseudostratified ciliated columnar
_f_	**(2)** lines inside of mouth	**(b)**	simple columnar
		(c)	simple cuboidal
_a_	**(3)** lines much of respiratory tract (including the trachea)	**(d)**	simple squamous
		(e)	stratified columnar
_d_	**(4)** endothelium and mesothelium	**(f)**	stratified squamous
_g_	**(5)** lines inside of urinary bladder	**(g)**	transitional

4. Match the epithelium named in column B with the appropriate function in column A.

	Column A		Column B
_d_	**(1)** protection	**(a)**	endothelium
_a_	**(2)** small molecules pass through most rapidly	**(b)**	simple columnar
		(c)	a type of ciliated epithelium
_c_	**(3)** propel sheets of mucus	**(d)**	stratified squamous
_b_	**(4)** absorption, secretion, or ion transport	**(e)**	transitional
_e_	**(5)** stretches		

5. The type of gland that secretes products such as milk, saliva, bile, or sweat through a duct is (a) an endocrine gland, (b) an exocrine gland, (c) a goblet cell.

6. Circle the one false statement about mucous and serous membranes. (a) The epithelial type is the same in all serous membranes, but different epithelial types occur among different mucous membranes. (b) Serous membranes line closed body cavities, whereas mucous membranes line body cavities that open to the outside. (c) Serous membranes always produce serous fluid, and mucous membranes always secrete mucus. (d) Both membranes contain an epithelium plus a layer of loose connective tissue.

7. In connective tissue proper, the cell type that secretes the fibers and ground substance is the (a) fibroblast, (b) plasma cell, (c) mast cell, (d) macrophage, (e) chondrocyte.

8. Identify the cell surface features described below.

(1) whip-like extensions that move fluids across epithelial surfaces *cilia*
(2) little "fingers" on apical epithelial surfaces that increase cell surface area and anchor mucus *microvilli*
(3) the spot-like cell junction that holds cells together so that they are not pulled apart *desmosome*
(4) a cell junction that closes off the extracellular space *tight junction*
(5) of the basement membrane and the basal lamina, the one that can be seen by light microscopy

9. Scar tissue is a variety of (a) epithelium, (b) connective tissue, (c) muscle tissue, (d) nerve tissue, (e) all of these.

Short Answer and Essay Questions

10. Define tissue.

11. Explain the classification of multicellular exocrine glands, and supply an example of each class.

12. Name the specific type of connective tissue being described: (1) around the capillaries; (2) forms the vocal cords; (3) the original, embryonic connective tissue; (4) hardest tissue of the skull; (5) main tissue in ligaments; (6) dominates the dermis; (7) dominates the hypodermis.

13. Name four functions of areolar connective tissue, and relate each function to a specific structural part of this tissue.

14. The matrix of connective tissue is extracellular. How does the matrix arrive at its characteristic position?

15. (a) Where does tissue fluid come from? (b) What is the function of tissue fluid?

16. (a) How do collagen and elastic fibers differ in function? (b) How do cartilage and bone differ from dense connective tissue proper in their basic function?

17. Compare and contrast skeletal, cardiac, and smooth muscle tissues relative to their body locations.

18. What is the function of macrophages?

19. Name the four classic symptoms of inflammation, and explain what each symptom represents in terms of changes in the injured tissue.

20. How is granulation tissue related to scar tissue?

21. Define endocrine gland, and explain the basic function of hormones.

22. Name two specific tissues that regenerate well and two that regenerate poorly.

23. Consuelo was studying the blood vessels in the circulatory system chapter of her anatomy book, and this chapter kept mentioning endothelium. She realized that she had forgotten what endothelium is. Can you help her?

24. Of the four basic tissue types, which two derive from mesenchyme?

25. On his anatomy test, Bruno answered three questions incorrectly. He confused a gap junction with a tight junction, a basal lamina with a basement membrane, and a mucous membrane with a sheet of mucus. What are the differences between each of these sound-alike pairs of structures?

26. The epidermis (epithelium of the skin) is a keratinized stratified squamous epithelium. Discuss why such an epithelium is better able to protect the body's external surface than a simple columnar epithelium would be.

27. What is fascia?

Critical Thinking and Clinical Application Questions

28. Systemic lupus erythematosus, or lupus, is a condition that sometimes affects young women. It is a chronic (persistent) inflammation that affects all or most of the connective tissue proper in the body. Suzy is told by her doctor that she has lupus, and she asks if it will have widespread or merely localized effects within the body. What would the physician answer?

29. Sailors who made long ocean journeys in the time of Christopher Columbus ate only bread, water, and salted meat on their journeys. They often suffered from scurvy. Eventually the problem was solved, and ocean sailors no longer developed scurvy. Try to deduce how the problem was solved. (Hint: It was not by inventing faster boats!)

30. Three patients in an intensive care unit have sustained damage and widespread tissue death in three different organs. One patient has brain damage from a stroke, another had a heart attack that destroyed cardiac muscle, and the third injured much of her liver (a gland) in a crushing car accident. All three patients have stabilized and will survive, but only one will gain full functional recovery through tissue regeneration. Which one, and why?

31. In adults, over 90% of all cancers are either adenomas (adenocarcinomas) or carcinomas. In fact, cancers of the skin, lung, colon, breast, and prostate are all in these categories. Which one of the four basic tissue types gives rise to most cancers?

5

The Integumentary System

Student Objectives

The Skin and the Hypodermis (pp. 100–104)

1. Name the tissue types that compose the epidermis and dermis. List the major layers of each, and describe the functions of each layer.

2. Describe the structure and function of the hypodermis.

3. Describe the factors that contribute to skin color.

Appendages of the Skin (pp. 104–110)

4. List the parts of a hair and a hair follicle, explaining the function of each part.

5. Explain the basis of hair color. Describe the distribution, growth, and replacement of hairs and how hair changes throughout life.

6. Compare the structure and location of oil and sweat glands.

7. Compare eccrine and apocrine sweat glands.

8. Describe the structure of nails.

Clinical Conditions: Burns and Cancer (pp. 110–113)

9. Explain why serious burns are life-threatening and how burns are treated. Differentiate first-, second-, and third-degree burns.

10. Summarize the characteristics and warning signs of skin cancers.

The Skin Throughout Life (p. 113)

11. Briefly explain the changes that occur in the skin from birth to old age.

Would you be enticed by an advertisement for a coat that is waterproof, stretchable, washable, and permanent-press, that automatically repairs small rips and burns, and is guaranteed to last a lifetime with reasonable care? Sounds too good to be true, but you already have such a coat—your skin. The skin and its appendages (sweat glands, oil glands, hair, and nails) serve a number of functions. Together, these organs make up the **integumentary system** (in-teg"u-men'tar-e; "covering").

The Skin and the Hypodermis

The **skin (integument)** and its appendages are the first *organs* discussed in this book. The structure of the skin is shown in Figure 5.1. Recall from Chapter 1 that an organ consists of tissues working together to perform certain functions. Although the skin is less complex than most other organs, it is still an architectural

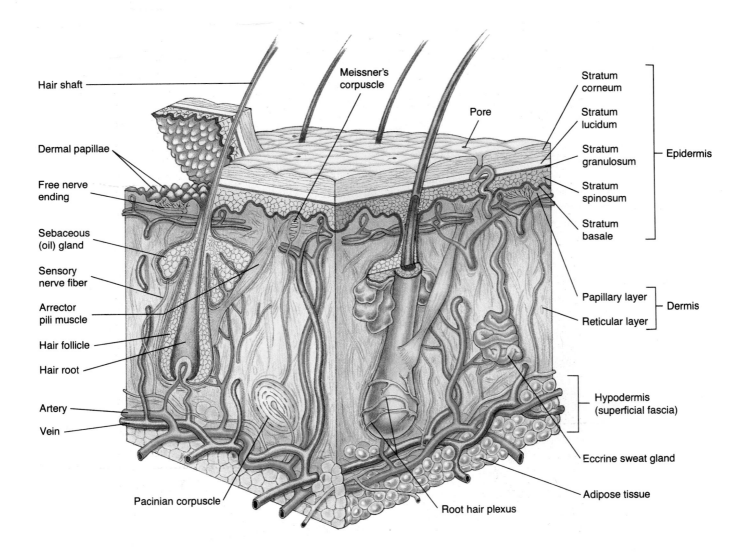

Hair shaft

Dermal papillae

Free nerve ending

Sebaceous (oil) gland

Sensory nerve fiber

Arrector pili muscle

Hair follicle

Hair root

Artery

Vein

Pacinian corpuscle

Meissner's corpuscle

Pore

Root hair plexus

Stratum corneum

Stratum lucidum

Stratum granulosum

Stratum spinosum

Stratum basale

Epidermis

Papillary layer

Reticular layer

Dermis

Hypodermis (superficial fascia)

Eccrine sweat gland

Adipose tissue

Figure 5.1 Skin structure.   Three-dimensional view of the skin and the underlying hypodermis. The epidermis and dermis have been pulled apart at the left corner to reveal the dermal papillae. The various nerve endings are sensory receptors, discussed in Chapter 13.

marvel. It covers the entire body and accounts for about 7% of our total weight, making it the largest organ. It has been estimated that every square centimeter of the skin contains 70 cm of blood vessels, 55 cm of nerves, 100 sweat glands, 15 oil glands, 230 sensory receptors, and about 500,000 cells that are constantly dying and being replaced.

The skin, which varies in thickness from 1.5 to 4 mm or more in different regions of the body, has two distinct layers. The superficial layer is the *epidermis*, a thick epithelium. Deep to the epidermis is the *dermis*, a fibrous connective tissue. Just deep to the skin lies a fatty layer called the *hypodermis*. Although the hypodermis is not part of the integumentary system, it shares some of the skin's functions and is considered in this chapter.

The skin performs many functions, most but not all of which are protective. It cushions and insulates the deeper body organs and protects the entire body from mechanical damage (bumps and cuts). The skin also offers protection from harmful chemicals, thermal damage (heat and cold), and invading bacteria. The epidermis is waterproof, preventing unnecessary loss of water across the body surface. The skin's rich capillary networks and sweat glands regulate the loss of heat from the body, helping to control body temperature. The skin also acts as a mini-excretory system: Urea, salts, and water are lost as we sweat. Skin screens out harmful ultraviolet (UV) rays from the sun, and its epidermal cells use these UV rays to synthesize vitamin D. Finally, the skin contains sense organs called *sensory receptors* that are associated with nerve endings. By sensing touch, pressure, temperature, and pain, these receptors keep us aware of what is happening at the body surface. As we describe the anatomy of the skin, we will explore its functions in more detail.

Epidermis

The **epidermis** (ep″ĭ-der′mis; "on the skin") is a keratinized stratified squamous epithelium (Figure 5.2). It contains four distinct types of cells: *keratinocytes, melanocytes, Merkel cells,* and *Langerhans cells.* Keratinocytes are by far the most abundant of these, so we will consider them first. We will discuss the other cell types later when we examine the various layers of the epidermis.

Keratinocytes

The chief role of **keratinocytes** (ke-rat′ĭ-no-sīts; "keratin cells") is to produce **keratin,** a tough fibrous protein that gives the epidermis its protective properties. Tightly connected to one another by a large number of desmosomes, the keratinocytes arise in the deepest

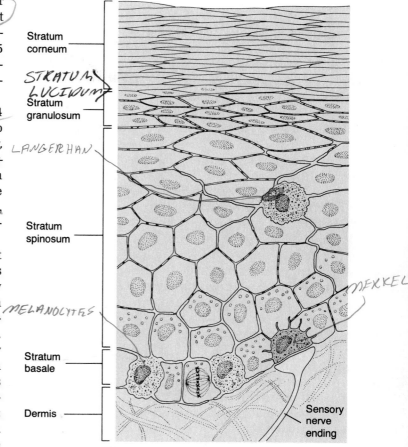

Figure 5.2   Layers and cells of the epidermis of thin skin.   The four cell types are keratinocytes (pink), melanocytes (grey), Langerhans cell (light blue), and Merkel cell (purple).

part of the epidermis from cells that undergo almost continuous mitosis. As these cells are pushed toward the skin surface by the production of new cells beneath them, they manufacture the keratin that eventually fills their cytoplasm. By the time the keratinocytes approach the skin surface, they are dead, flat sacs completely filled with keratin. Millions of these dead cells rub off every day, giving us an entirely new epidermis every 35 to 45 days—the average time from the "birth" of a keratinocyte to its final wearing away. In the normal, healthy epidermis, the production of new cells balances the loss at the surface. Where the skin experiences friction, both cell production and keratin formation are accelerated.

✚   Persistent friction (from a poorly fitting shoe, for example), causes an excessive thickening of the epidermis called a **callus.** Short-term but severe friction (from wielding a hoe, for example), can cause a **blister,** the separation of the epidermis and dermis by a fluid-filled pocket. A large blister is called a **bulla** (bul′ah; "bubble"). ■

Layers of the Epidermis

In **thick skin,** which covers the palms and soles, the thickened epidermis consists of five layers, or *strata* (stra'tah; "bed sheets") (Figures 5.1 and 5.3). In **thin skin,** which covers the rest of the body, only four strata are present (Figure 5.2).

Stratum Basale (Basal Layer). The **stratum basale** (stra'tum ba-sal'e), the deepest epidermal layer, is firmly attached to the underlying dermis along a wavy borderline. Also called the *stratum germinativum* (jer-mĭ-na"te'vum; "germinating layer"), this stratum consists of a single row of cells, mostly stem cells representing the youngest keratinocytes (Figure 5.2). These cells divide rapidly, and many mitotic nuclei are visible. Occasional **Merkel cells** are seen amid the keratinocytes. Each hemisphere-shaped Merkel cell is intimately associated with a disc-like sensory nerve ending and may serve as a receptor for touch.

Between 10% and 25% of the cells in the stratum basale are **melanocytes** (mel'ah-no-sīts; "melanin cells"). These make the dark skin pigment **melanin** (mel'ah-nin; "black"). The spider-shaped melanocytes have many branching processes that reach and touch all the keratinocytes in the basal layer. Melanin is made in membrane-lined granules and then transferred through the cell processes to nearby keratinocytes. Consequently, the basal keratinocytes contain more melanin than do the melanocytes themselves. The melanin granules accumulate on the superficial side of each keratinocyte, forming a shield of pigment over the nucleus. This shield protects against UV rays, which can damage DNA and cause cancer (see p. 110). In Caucasians, the melanin disappears a short distance above the basal layer, where it is digested by lysosomes in the keratinocytes. In black people, no such digestion occurs, so melanin occupies keratinocytes throughout the epidermis. Although black people have a darker melanin and more pigment in each melanocyte, they do *not* have more melanocytes in their skin. In all but the darkest people, melanin builds up in response to the sun's UV rays, the protective response we know as suntanning.

Stratum Spinosum (Spiny Layer). The **stratum spinosum** (spi-no'sum) is several cell layers thick (Figure 5.2). Mitosis occurs here, but less frequently than in the basal layer. In typical histological slides, the keratinocytes in this layer have many spine-like extensions. However, these spines do not exist in living cells: They arise during tissue preparation, when the cells shrink while holding tight at their many desmosomes. Cells of the stratum spinosum also contain thick bundles of intermediate filaments, called *tonofilaments* ("tension filaments"). Tonofilaments consist of a tension-resisting protein called pre-keratin.

Scattered among the keratinocytes of the stratum spinosum are **Langerhans cells.** These star-shaped cells are macrophages that help activate the immune system. (Most macrophages occur within connective tissue, as we discussed on page 81, but a few occur in epithelia as well.)

Stratum Granulosum (Granular Layer). The thin **stratum granulosum** (gran"u-lo'sum) consists of three to five layers of flattened keratinocytes. Along with abundant tonofilaments, these cells contain *keratohyaline granules* and *lamellated granules.* The keratohyaline (ker"ah-to-hi'ah-lin) granules contribute to the formation of keratin in the higher strata, as described below. The lamellated granules (lam'ĭ-la-ted; "plated") contain a waterproofing glycolipid that is secreted into the extracellular space and is the major factor for slowing water loss across the epidermis. Furthermore, the plasma membranes of the cells thicken, so that they become more resistant to destruction. You might say that the keratinocytes are "toughening up" to make the outer strata the strongest skin region.

Like all epithelia, the epidermis relies on capillaries in the underlying connective tissue (the dermis) for its nutrients. Above the stratum granulosum, the epidermal cells are too far from the dermal capillaries, so they die. This is a completely normal sequence of events.

Stratum Lucidum (Clear Layer). The **stratum lucidum** (lu'sĭ-dum) occurs in thick skin (Figure 5.3), but not in thin skin. Appearing through the light microscope as a thin translucent band, this layer consists of a few rows of flat dead keratinocytes. Electron microscopy shows that its cells are identical to cells at the bottom of the stratum corneum, the next layer.

Stratum Corneum (Horny Layer). The most external part of the epidermis, the **stratum corneum** (kor'ne-um), is many cells thick. This stratum is far thicker in thick skin than in thin skin. Its dead cells are flat sacs completely filled with keratin, because their nuclei and organelles were digested away by lysosome enzymes upon cell death. Keratin consists of tonofilaments embedded in a thick "glue" from the keratohyaline granules. Both the keratin and the thickened plasma membranes of cells in the stratum corneum protect the skin against abrasion and penetration. Additionally, the glycolipid between its cells keeps this layer waterproof. It is amazing that a layer of dead cells can still perform so many functions!

The cells of the stratum corneum are referred to as *cornified,* or *horny* cells (*cornu* = horn). These cells are the dandruff shed from the scalp and the flakes that come off dry skin. The average person sheds 18 kg (40 pounds) of these skin flakes in a lifetime. The common saying "Beauty is only skin deep" is especially interesting in light of the fact that nearly everything we see when we look at someone is dead!

Dermis

The **dermis,** the second major layer of the skin, is a strong, flexible connective tissue. The cells in the dermis are those of any connective tissue proper: fibroblasts, macrophages, mast cells, and scattered white blood cells (see p. 81). The fiber types—collagen, elastic, and reticular fibers—also are typical. The dermis binds the entire body together like a body stocking. This layer is your "hide" and corresponds to animal hides used to make leather products.

The dermis is richly supplied with nerve fibers and blood vessels. The nerve supply of the dermis (and epidermis) is discussed on p. 364 of Chapter 13. The dermal blood vessels are so extensive that the dermis can hold 5% of all blood in the body. When organs, such as exercising muscles, need more blood, the nervous system constricts the dermal blood vessels. This shunts more blood into the general circulation, making it available to the muscles and other organs. Conversely, the dermal vessels engorge with warm blood on hot days, allowing heat to radiate from the body—a cooling effect.

The dermis has two layers: papillary and reticular (Figure 5.1). The **papillary** (pap′ĭ-lar-e) **layer,** the superficial 20% of the dermis, is areolar connective tissue. It includes the **dermal papillae** (pah-pil′le; "nipples"), finger-like pegs that indent the overlying epidermis. On the palms of the hands and the soles of the feet, the dermal papillae lie atop larger mounds called *dermal ridges.* These dermal ridges elevate the overlying epidermis into *epidermal ridges (friction ridges),* which are responsible for fingerprints and footprints. Epidermal ridges increase friction and enhance the gripping ability of the hands and feet. Patterns of these ridges are genetically determined and unique to each person. Because *sweat pores* open along the crests of the epidermal ridges, they leave distinct *fingerprints* on almost anything they touch. Thus, fingerprints are "sweat films."

The deeper, **reticular layer,** which accounts for about 80% of the thickness of the dermis, is dense irregular connective tissue. Its extracellular matrix contains thick bundles of interlacing collagen fibers that run in many different planes. However, most run parallel to the skin surface. The reticular layer is so called because of these networks of collagen fibers (*reticulum* = network), not because it has any special abundance of reticular fibers. Separations, or less dense regions, between the collagen bundles form the *lines of cleavage* or *tension lines of the skin.* These invisible lines occur over the entire body: They run longi-

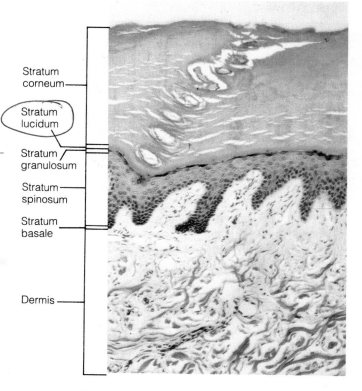

Stratum corneum

Stratum lucidum

Stratum granulosum

Stratum spinosum

Stratum basale

Dermis

Figure 5.3 The epidermis of thick skin. This photomicrograph is magnified 150 ×.

tudinally in the skin of the limbs and head and in circular patterns around the neck and trunk. A knowledge of cleavage lines is important to surgeons: When a surgical incision is made *parallel* to these lines, the skin tends to gape less and heals more readily than when the incision is made *across* cleavage lines.

The collagen fibers of the dermis give skin its strength and resilience. Thus, many jabs and scrapes do not penetrate the tough dermis. Furthermore, elastic fibers in the dermis provide the skin with stretch-recoil properties.

The deep part of the dermis is responsible for markings on our skin surface called *flexure lines.* Observe, for example, the deep skin creases on your palm. Flexure lines result from a continual folding of the skin, often over joints, where the dermis attaches tightly to underlying structures. Flexure lines are also visible on the wrists, soles, fingers, and toes.

Can you guess how stretch marks on the skin, commonly seen when one gains too much weight, relate to the dermis?

Extreme stretching of the skin, which occurs in obesity and pregnancy, can tear the dermis. Such dermal tearing is indicated by silvery white scars termed *striae* (stri′e; "streaks"), the frequently observed "stretch marks." ■

Hypodermis

Just deep to the skin is the fatty **hypodermis** (hi″po-der′mis; "below the skin" in Greek) (see Figure 5.1). This layer is also called the **superficial fascia** and the *subcutaneous layer* (sub″ku-ta′ne-us; "below the skin" in Latin). It consists of both areolar and adipose connective tissue, although the adipose tissue normally dominates. Besides storing fat, the hypodermis anchors the skin to the underlying structures (mostly to muscles) and allows the skin to slide relatively freely over those structures. Sliding skin protects us by ensuring that many blows just glance off our bodies. The hypodermis is also an insulator: Since fat is a poor conductor of heat, it helps prevent heat loss from the body. The hypodermis thickens markedly when one gains weight, but this thickening occurs in different body areas in the two sexes. In females, subcutaneous fat accumulates first in the thighs and breasts, whereas in males it first accumulates in the anterior abdomen (as a "beer belly").

Skin Color

Three pigments contribute to skin color: melanin, carotene, and hemoglobin. **Melanin,** the most important pigment, is made from an amino acid called tyrosine. Melanin ranges in color from yellow to reddish to brown to black. Its manufacture depends on an enzyme in melanocytes called *tyrosinase* (ti-ro′sĭ-nās). As we noted earlier, melanin passes from melanocytes to keratinocytes in the stratum basale of the epidermis. Freckles and pigmented moles are localized accumulations of melanin.

Carotene (kar′o-tēn) is a yellow to orange pigment derived from certain plant products, such as carrots. It tends to accumulate in the stratum corneum of the epidermis and in fat tissue of the hypodermis. Its color is most obvious in the palms and soles, where the stratum corneum is thickest. It should be noted that the yellowish tinge of the skin of Asian peoples is due to variations in melanin, not to carotene.

The pink hue of Caucasian skin reflects the crimson color of oxygenated **hemoglobin** (he′mo-glo″bin) in the capillaries of the dermis. Since Caucasian skin contains little melanin, the epidermis is nearly transparent and allows blood's color to show through.

When hemoglobin is poorly oxygenated, both the blood and the skin of Caucasians appear blue, a condition called **cyanosis** (si″ah-no′sis; "dark blue condition"). Skin often becomes cyanotic during heart failure or severe respiratory disorders. In black people, the skin is too dark to reveal the color of the underlying vessels, but cyanosis is apparent in the mucous membranes and nail beds.

 Can you imagine what bruises are?

Black-and-blue marks represent discolored blood that is visible through the skin. Such bruises, usually caused by blows, reveal sites where blood has escaped from the circulation and clotted below the skin. The general term for a clotted mass of escaped blood anywhere in the body is a **hematoma** (he″mah-to′mah; "blood swelling"). ■

Have you ever wondered why the skin colors of different human populations vary from dark to light? According to one theory, this variation reflects the fact that the sun's UV rays are both dangerous and essential. On the one hand, the intense rays of the tropical sun pose the greatest threat of skin cancer, and the inhabitants of equatorial regions developed a thick screen of dark melanin that eliminates this danger. (Skin cancer is almost nonexistent among the native peoples of Africa.) On the other hand, UV rays stimulate the deep epidermis to produce vitamin D, a vital hormone that is required for the uptake of calcium from the diet and is essential for healthy bones. Producing vitamin D poses no problem in the sunny tropics, but Caucasians in northern Europe receive very little sunlight in the long, dim winter. Therefore, their colorless epidermis ensures enough UV penetration for vitamin D production. The majority of humans are native to intermediate latitudes (China, the Middle East, and so on) and have moderately brown skin that is neither "white" nor "black." Such skin is dark enough to be a sunscreen in the summer (especially when it tans), yet light enough to allow vitamin D production in the moderate winters of intermediate latitudes.

Appendages of the Skin

Along with the skin itself, the integumentary system includes several derivatives of the epidermis. These **skin appendages** include the hair and hair follicles, sebaceous (oil) glands, sweat glands, and nails. Although these appendages derive from the epidermis, they all extend into the dermis.

Hair and Hair Follicles

Hairs associate with hair follicles to form complex structural units (Figures 5.1, 5.4, and 5.5a). In these units, the **hairs** are the long filaments, and the **hair follicles** are tubular invaginations of the epidermis from which the hairs grow.

Hair

Although hair serves other mammals by keeping them warm, human body hair is far less luxuriant and less useful. Even so, hair is distributed over all our skin surface, except on the palms, soles, nipples, and parts of the external genitalia (the head of the penis, for example). The main function of our sparse body hair is to sense insects on the skin before they sting us. The hair on the head protects the head against direct sunlight in summer and against heat loss on cold days. Eyelashes shield the eyes, and nose hairs filter large particles like insects and lint from the air we inhale.

A hair is a flexible strand made of keratinized cells. The **hard keratin** that dominates hairs confers two advantages over the **soft keratin** that is found in typical epidermal cells: (1) It is tougher and more durable, and (2) its individual cells do not flake off. The chief parts of a hair are the **shaft,** the part that projects above the skin surface, and the **root,** the part embedded in the skin (Figure 5.5a). The shape of its shaft determines whether a hair is straight or curly. If the shaft is flat and ribbon-like in cross-section, the hair is kinky; if oval in section, the hair is wavy; if perfectly round, the hair is straight. Of these three types, straight hair is the strongest.

A hair consists of three concentric layers of keratinized cells (Figure 5.5b). Its central core, the **medulla** (mĕ-dul′ah; "middle"), consists of large cells that are partially separated by air spaces. The medulla is absent in fine hairs. The **cortex,** a bulky layer surrounding the medulla, consists of several layers of flattened cells. The outermost **cuticle** is a single layer of cells that overlap one another from below like shingles on a roof (Figure 5.4). This shingle pattern helps to keep neighboring hairs apart so that hair does not mat. The cuticle is the most heavily keratinized part of the hair, providing strength and keeping the inner layers tightly compacted.

Because it is subjected to the most abrasion, the oldest part of the cuticle tends to wear away at the tip of the hair shaft. Can you guess the common name for this phenomenon?

The abrasion allows the keratin fibrils in the cortex and medulla to frizz out, a phenomenon called "split ends." ■

Hair pigment is made by melanocytes at the base of the hair follicle and is transferred to the cells of the hair root. Different proportions of two types of melanin (black-brown colored and yellow-rust colored) combine to produce all the common hair colors—black, brown, red, and blond. Graying or whitening of hair results from a decrease in the production of melanin (signaled by genes that usually do not act until one is about 40 years old) and from the replacement of melanin by colorless air bubbles in the hair shaft.

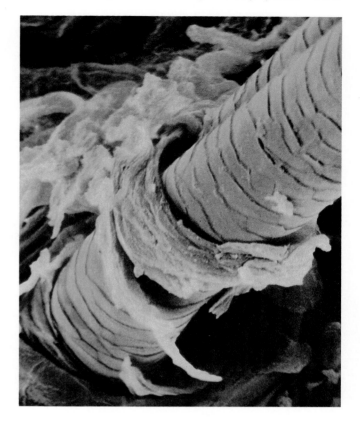

Figure 5.4   Scanning electron micrograph showing a hair shaft emerging from a follicle at the epidermal surface.   Note how the scale-like cells of the cuticle overlap one another (1500 ×; artificially colored).

Hair Follicles

Hair follicles extend from the epidermal surface into the dermis. In the scalp, they may even extend into the hypodermis. The deep end of the follicle is expanded, forming a **hair bulb** (Figure 5.5c). A knot of sensory nerve endings wraps around each hair bulb to form a **root hair plexus.** Bending the hair shaft stimulates these nerve endings. Hairs are thus exquisite touch receptors. (Feel the tickle as you run your hand over the hairs on your forearm.) A nipple-like bit of the dermis, the **connective tissue papilla** *(hair papilla),* protrudes into each hair bulb. This papilla contains a knot of capillaries that supply nutrients to the growing hair. If the hair papilla is destroyed by trauma, the follicle permanently stops producing hair. The epithelial cells in the hair bulb just above the papilla make up the **hair matrix.** These cells multiply to produce the growing hair.

The wall of a hair follicle is actually a compound structure composed of an outer **connective tissue root sheath,** derived from the dermis, and an inner **epithelial root sheath,** from the epidermis (Figure 5.5c and d). The epithelial root sheath is composed of external and internal parts: Of these, the **internal root sheath** derives from the matrix cells, as does the hair root,

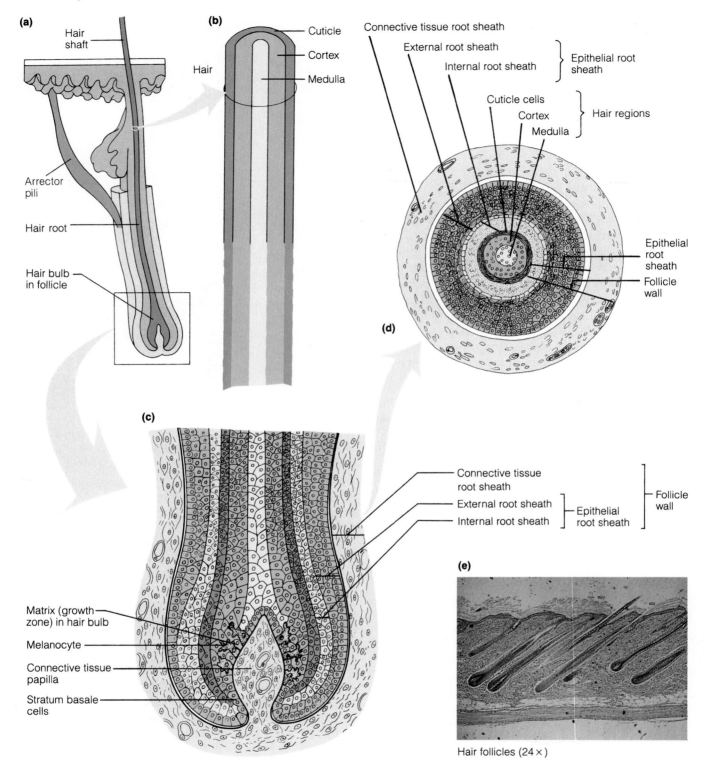

(a) Hair shaft
Hair
Arrector pili
Hair root
Hair bulb in follicle

(b) Cuticle
Cortex
Medulla
Hair

(d) Connective tissue root sheath
External root sheath
Internal root sheath
Epithelial root sheath
Cuticle cells
Cortex
Medulla
Hair regions
Epithelial root sheath
Follicle wall

(c) Connective tissue root sheath
External root sheath
Internal root sheath
Epithelial root sheath
Follicle wall
Matrix (growth zone) in hair bulb
Melanocyte
Connective tissue papilla
Stratum basale cells

(e) Hair follicles (24×)

Figure 5.5 Structure of a hair and hair follicle. (**a**) Longitudinal section of a hair within its follicle. (**b**) Enlarged longitudinal section of a hair. (**c**) Enlarged longitudinal section of the hair bulb at the base of the follicle. This bulb encloses the hair matrix, the actively dividing epithelial cells that produce the hair. (**d**) Cross section of a hair and hair follicle. (**e**) Photomicrograph of scalp tissue showing numerous hair follicles.

which it surrounds. The **external root sheath,** by contrast, is a direct continuation of the skin epidermis.

Associated with each hair follicle (Figure 5.5a) is a bundle of smooth muscle cells called an **arrector pili** muscle (ah-rek′tor pi′li; "raiser of the hair"). As you can see in Figure 5.5e, most hair follicles lie at an oblique angle to the skin surface. Each arrector pili muscle extends from the most superficial part of the dermis to a deep-lying hair follicle. The contraction of this muscle causes the hair to stand erect and dimples the skin surface to produce goose bumps in response to cold or fear. This is the basis for the expression "a hair-raising experience." This response helps keep furry animals warm in the winter by trapping a layer of insulating air in their fur. It also makes a frightened animal with its hair on end look larger and more formidable to its enemy. Obviously, hair raising is not very useful to people, because our body hair is so short and sparse.

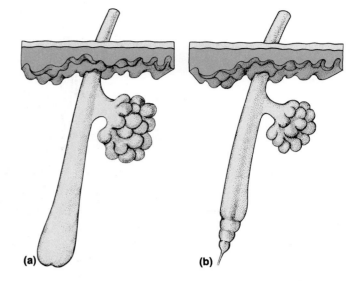

Figure 5.6   Overall shape of (a) an active and (b) a resting hair follicle.   The hair is shed during or just after each resting phase.

Types and Growth of Hair

Hairs come in various sizes and shapes, but as a rule, they can be classified as either **vellus** (vel′us: *vell* = wool, fleece) or **terminal.** The body hair of children and women is of the fine, short vellus variety. The longer, coarser hair of everyone's scalp is terminal hair. Terminal hairs appear at puberty in the axillary (armpit) and pubic regions of both sexes and on the face, chest, arms, and legs of males. These terminal hairs grow under the influence of male sex hormones called *androgens* (of which *testosterone* is the most important type).

In women, small amounts of androgens are normally produced by both the ovaries and the adrenal glands. Excessive hairiness, or **hirsutism,** (her′soot-izm; "shaggy state") may result from a tumor of the adrenal gland that secretes androgens in excess. Since few women want a beard or hairy chest, such tumors are surgically removed as soon as possible.

Hair growth in certain body areas that is seen as undesirable (such as hair on the upper lip of a female) is sometimes treated with *electrolysis.* This procedure uses electricity to destroy the hair root and hair matrix, permanently halting hair growth at the treated follicle. ■

Hairs grow an average of 2 mm per week, although this rate varies widely among body regions and with sex and age. Each follicle goes through growth cycles (Figure 5.6). In each cycle, an active growth phase is followed by a resting phase, in which the hair matrix is inactive and the follicle atrophies somewhat. In each resting phase, the hair root detaches from its matrix and eventually falls out. After the resting phase, the matrix proliferates and forms a new hair. The life span of hairs varies: On the scalp, the follicles stay active for an average of 4 years before becoming inactive for several months. On the eyebrows, by contrast, the follicles remain active for only a few months, a fact that explains why the eyebrows never grow very long. When people are told that hair grows in cycles, they naturally ask why all the hairs on the head do not fall out simultaneously every few years. The answer is that the cycles of adjacent hair follicles are not in synchrony, so only a small percentage of the hairs on our head sheds at any one time. We typically shed about 90 scalp hairs daily, so there is no need to worry about baldness just because a few hairs come off on a hair brush.

Hair Thinning and Baldness

Given ideal conditions, hair grows fastest from the teen years to the 40s. When hairs are no longer replaced as quickly as they are shed, the hair thins. By age 60 to 65, both sexes usually experience some degree of balding. Coarse terminal hairs are replaced by vellus hairs, and the hair becomes increasingly wispy.

True baldness is different. The most common type, **male pattern baldness,** is a genetically determined, gender-influenced condition. It is thought to be caused by a gene that does not act until adulthood, at which time it changes the response of hair follicles to androgens. Now the hairs respond to androgens by shortening their growth cycles. The cycles become so short that many hairs never even emerge from their follicles before shedding, and those that do emerge are fine vellus hairs that look like peach fuzz in the

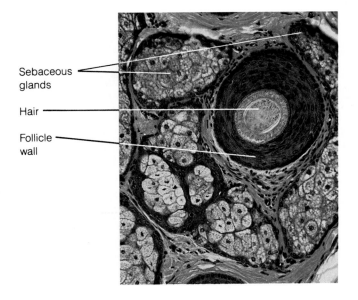

Sebaceous glands

Hair

Follicle wall

Figure 5.7 Sebaceous glands. In this cross section, sebaceous glands surround a hair follicle into which the sebum is secreted. The tissue colored blue-green is the connective tissue of the dermis (150 ×).

"bald" area. Until recently, the only effective treatment for male pattern baldness was drugs that inhibit the production of male sex hormones. However, these drugs also diminish sex drive—an unacceptable solution to most men and their partners. Quite by accident, it was discovered that minoxidil, a drug used to reduce high blood pressure, stimulates hair growth in some bald men. This drug works by dilating blood vessels, thereby increasing blood flow to the skin and the hair follicles. Minoxidil is available in ointment form for application to the scalp.

➕ Hair thinning can be induced by a number of other factors that lengthen follicular resting periods and upset the balance between the rates of hair loss and replacement. For example, hair loss can be prompted by certain stressors, such as high fever, surgery, or emotional trauma. Chemotherapy drugs used to treat cancer target the most rapidly dividing cells in the body, thus destroying many matrix cells and causing hair loss. In all of these cases, the hair can recover and regrow. However, hair loss due to severe burns, excessive radiation, or genetic factors is permanent. ◼

Sebaceous Glands

The **sebaceous glands** (se-ba′shus; "greasy") are the skin's oil glands (Figures 5.1 and 5.7). They occur over the entire body, except on the palms and soles. Sebaceous glands are simple alveolar glands with several alveoli opening into a single duct (see Figure 4.11e on

p. 78). Their oily product, called **sebum** (se′bum; "animal fat"), is secreted in a most unusual way: The central cells in the alveoli accumulate oily lipids until they become engorged and burst apart. This is called **holocrine secretion** (hol′o-krin; *holos* = whole), because *whole* cells break up to form the product. Most sebaceous glands are associated with hair follicles, emptying their sebum into the upper third of the follicle. From the follicle, the sebum flows superficially to cover the skin. In addition to making our skin and hair greasy, sebum collects dirt, so one might say its function is to support the soap and shampoo industry! More seriously, sebum softens and lubricates the hair and skin, prevents hair from becoming brittle, and keeps the epidermis from cracking. It also helps to slow the loss of water across the skin and has a bactericidal (bacterium-killing) action.

The secretion of sebum is stimulated by hormones, especially androgens. The sebaceous glands are relatively inactive during childhood but are activated in both sexes during puberty, when the production of androgens begins to rise.

➕ In some teenagers, so much sebum is made that it cannot be ducted from the glands quickly enough. When a sebaceous gland becomes blocked by sebum, a whitehead appears on the skin surface. If the material oxidizes and dries, it darkens to form a blackhead (note that the black color of a blackhead is *not* due to dirt). Blocked sebaceous glands are likely to be infected by bacteria, particularly staphylococcus, which feed on the nutrient-rich sebum. This infection produces pimples. The acne that results can be mild or extremely severe, leading to permanent scarring.

Seborrhea (seb″o-re′ah; "fast-flowing sebum") is another condition caused by overactivity of sebaceous glands. It begins on the scalp as an oily crust from which scales flake off, forming a greasy type of dandruff. Probably stimulated by androgens, seborrhea is seen most often in adolescent males. Seborrhea in infants is known as "cradle cap." Careful washing to remove the excess oil usually helps. ◼

Sweat Glands

Humans have more than 2.5 million **sweat glands (sudoriferous glands).** Sweat glands are distributed over the entire skin surface, except on the nipples and parts of the external genitalia. We normally produce about 500 ml of sweat per day, but this amount can increase to 12 liters on hot days during vigorous exercise. That is over 3 gallons of sweat! Sweating prevents overheating of the body, since evaporation of sweat from the skin is a cooling process. Of all the species in the animal kingdom, humans have the most sweat glands, so we can run and work in the hot sun better than any other mammal. Hair would interfere with the evaporation of sweat, so our exceptional sweating ability seems to explain our near absence of body hair. Heat-induced sweating begins on the forehead (cooling the important brain region) and then spreads inferiorly over the rest of the body. Emotionally induced sweating—the "cold sweat" brought on by fright, embarrassment, or nervousness— begins on the palms, soles, and axillae and then spreads to other body areas.

There are two types of sweat glands: eccrine and apocrine. **Eccrine glands** (ek'rin; "secreting") are by far the more numerous type (Figure 5.1). They are most abundant on the palms, soles, and forehead. Each is a simple coiled tubular gland (see Figure 4.11b on p. 78). The coiled secretory base lies in the deep dermis and hypodermis, while the duct runs superficially to open at the skin surface through a funnel-shaped **pore.** (Although most pores on the skin surface are sweat pores, the "pores" seen on the face are openings of hair follicles.)

Sweat is an unusual secretory product in that it is primarily a filtrate of the blood, which passes through the secretory cells of the sweat glands and is released by exocytosis. Sweat is 99% water, with some salts (mostly sodium chloride) and traces of metabolic wastes (urea, ammonia, uric acid). Sweat is acidic, so it retards the growth of bacteria on the skin.

Apocrine (ap'o-krin) *glands* are largely confined to the axillary, anal, and genital areas. They are larger than eccrine glands, and their ducts open into hair follicles. The product of apocrine glands contains the same components as true sweat, plus fatty substances and proteins. Consequently, it is viscous and sometimes has a milky or yellow color. The secretion is odorless, but when its organic molecules are decomposed by bacteria on the skin, it takes on a musky smell. This is the basis of body odor.

Apocrine glands start to function at puberty under the influence of androgens. Although their secretion certainly increases on hot days, apocrine glands play little role in cooling the body. Since their activity is increased by sexual foreplay and they enlarge and recede with the phases of a woman's menstrual cycle, they may be analogous to the sexual scent glands of other animals.

The skin forms several types of modified sweat glands. *Ceruminous glands* (sĕ-roo'mĭ-nus; "pertaining to wax") are modified apocrine glands in the lining of the external ear canal. Their product is a component of earwax (see p. 430). The *mammary glands* are specialized sweat glands highly modified to secrete milk. Although they are part of the integumentary system, we will discuss the mammary glands as part of the female reproductive system (Chapter 23).

Nails

A **nail** (Figure 5.8) is a scale-like modification of the epidermis. It corresponds to the hoof or claw of other mammals. Like hair, nails are made of hard keratin. Nails are built-in finger tools that enable us to pick up small objects and scratch our skin when it itches. Each nail has a distal **free edge**, a **body** (the visible attached part), and a **root** (the proximal part embedded in the skin). The nail rests on a bed of epidermis called the **nail bed.** This bed contains only the deeper layers of the epidermis, for the nail itself corresponds to the superficial keratinized layers.

Nails look pink because of the rich network of capillaries in the underlying dermis. At the root and the proximal end of the nail body, the nail bed thickens to form the **nail matrix,** the actively growing part of the nail. The nail matrix is thick enough that the pink dermis cannot show through it. Instead, we see a white crescent, the **lunula** (lu'nu-la; "little moon"),

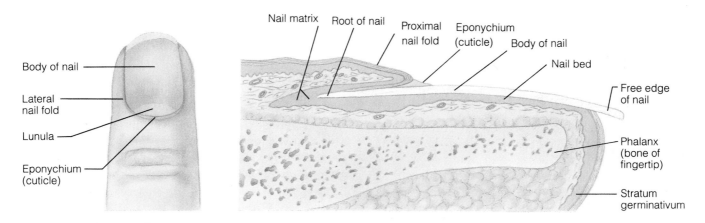

Figure 5.8   Structure of a nail.   A longitudinal section through a fingertip is shown at right.

under the nail's proximal region. The lateral and proximal borders of the nail are overlapped by skin folds, called **nail folds.** The proximal nail fold projects onto the nail body as the cuticle or **eponychium** (ep″o-nik′e-um; "on the nail").

 Can you guess the most common cause of an ingrown toenail?

An *ingrown toenail* is just what its name implies, a nail whose growth pushes it painfully into the lateral nail fold. This happens when the nail grows crookedly, most commonly from the pressure of a badly fitting shoe. ■

✚ Clinical Conditions: Burns and Cancer

It is difficult to scoff at anything that goes wrong with the skin, for when the skin rebels, the outbreak is quite visible. Skin can develop more than 1000 different conditions and ailments. The most common skin disorders are bacterial, viral, or yeast infections, some of which are described in the Clinical Terms at the end of this chapter (also see the Closer Look box on p. 112). The most severe threats to the skin are burns and skin cancer, considered next.

Burns

Burns are a devastating threat to the body, primarily because of their effects on the skin. A burn is tissue damage inflicted by heat, electricity, radiation, extreme friction, or certain harmful chemicals. Burns are the major cause of death in people under 40 years of age and the third leading cause of death in all age groups.

The immediate threat to life from serious burns is a catastrophic loss of body fluids. Inflammatory edema (p. 91) is severe. As fluid seeps from the burned surfaces, the body quickly dehydrates and loses essential salts. This in turn leads to fatal circulatory shock (inadequate blood circulation caused by the reduction in blood volume). To save the patient, the lost fluids must be replaced immediately. After the initial crisis has passed, infection becomes the main threat. Bacteria, fungi, and other pathogens can easily invade areas where the skin barrier is destroyed.

Burns are classified by their severity (depth) as first-, second-, or third-degree burns. Third-degree burns are the *most* severe, not the least, as one might expect. In **first-degree burns,** only the epidermis is damaged. Symptoms include redness, swelling, and pain. Generally, first-degree burns heal in a few days without special attention. Sunburn is usually a first-

degree burn. **Second-degree burns** involve injury to the epidermis and upper part of the dermis. Symptoms resemble those of first-degree burns, but blisters also appear. The skin regenerates with little or no scarring in 3 to 4 weeks if care is taken to prevent infection. First- and second-degree burns are called **partial-thickness burns.**

Third-degree burns consume the entire thickness of the skin. They are **full-thickness burns.** The burned area appears white, cherry red, or blackened. Although the skin might eventually regenerate, it is usually impossible to wait for regeneration because of fluid loss and infection. Therefore, skin from other parts of the patient's body must be grafted onto the burned area.

In general, burns are considered critical if any of the following conditions exists: (1) Over 10% of the body has third-degree burns; (2) 25% of the body has second-degree burns; (3) there are third-degree burns on the face, hands, or feet. A quick way to estimate how much surface is affected by a burn is to use the **rule of nines** (Figure 5.9). This method divides the body surface into 11 regions, each accounting for 9% (or a multiple of 9%) of total body area. This method is only roughly accurate, so special tables are used when greater accuracy is desired.

Skin Cancer

Many types of tumors arise in the skin. Most are benign (warts, for example), but some are malignant. Skin cancer is the most common type of cancer, with about half a million new cases appearing each year in the United States. As mentioned earlier, the most important risk factor for skin cancer is overexposure to the UV rays in sunlight. The important types of skin cancer are pictured in Figure 5.10.

Basal Cell Carcinoma

Basal cell carcinoma is the least malignant and most common of the skin cancers. Cells of the stratum basale proliferate, invading the dermis and hypodermis and causing tissue erosions there. The cancer lesions appear most often as dome-shaped, shiny nodules on sun-exposed areas of the face. These nodules later develop a central ulcer and a "pearly," beaded edge. They often bleed. Basal cell carcinoma grows relatively slowly, and metastasis seldom occurs before the carcinoma is noticed. Full cure by surgical excision is the rule in 99% of cases.

Squamous Cell Carcinoma

Squamous cell carcinoma arises from the keratinocytes of the stratum spinosum. The lesion appears as a scaly, reddened papule (small, rounded elevation)

(text continues on p. 113)

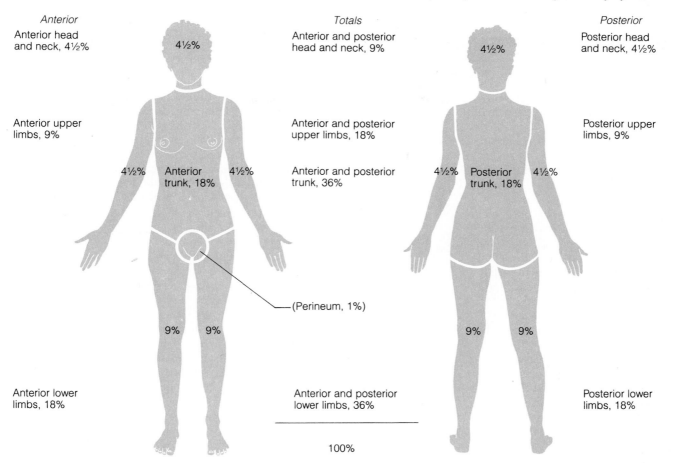

Anterior

Anterior head
and neck, 4½%

Anterior upper
limbs, 9%

Anterior
trunk, 18%

4½%   4½%

4½%

9%   9%

Anterior lower
limbs, 18%

Totals

Anterior and posterior
head and neck, 9%

Anterior and posterior
upper limbs, 18%

Anterior and posterior
trunk, 36%

(Perineum, 1%)

Anterior and posterior
lower limbs, 36%

100%

Posterior

Posterior head
and neck, 4½%

Posterior upper
limbs, 9%

Posterior
trunk, 18%

4½%   4½%

4½%

9%   9%

Posterior lower
limbs, 18%

Figure 5.9   Estimating the extent of burns using the rule of nines.

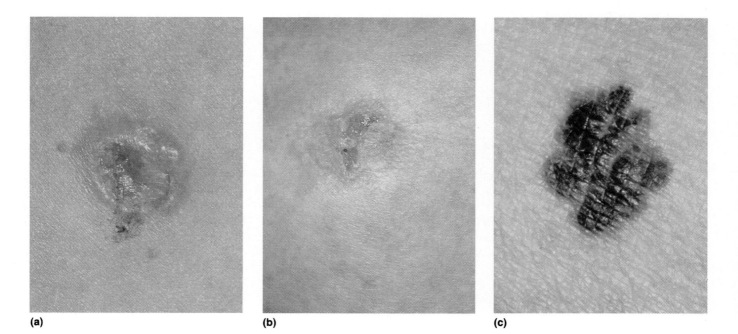

(a)   **(b)**   **(c)**

Figure 5.10   Photographs of skin cancers.   (**a**) Basal cell carcinoma. (**b**) Squamous cell
carcinoma. (**c**) Malignant melanoma. This begins as a multicolored dark lesion with an irregular shape.

A CLOSER LOOK Vampires and Touch-Me-Nots

Stories of vampires and were-wolves are firmly entrenched in the folklore of many countries. Although many aspects of vampire legends derive from the "behavior" of bloating corpses that shift in their graves (the "undead"), other aspects of these legends may be based on living people with skin conditions. One such condition is **porphyria** (por-fēr'e-a; "purple"), a rare, inherited disease. People with this condition cannot make the heme part of hemoglobin, the oxygen-carrying molecule of red blood cells. Unfortunate victims who were once shunned as vampires or werewolves may actually have been afflicted by this disease, which affects about 1 of every 25,000 people.

The cause of porphyria is a deficiency in some enzymes that catalyze the biochemical pathway leading to heme formation. Without these enzymes, the intermediate products of this pathway build up, spill into the circulation, and eventually cause lesions throughout the body—especially when exposed to rays of violet light from sunlight. These intermediate products are called **porphyrins,** after which porphyria is named. Actually a group of diseases, the porphyrias are characterized by five "P's": (1) *puberty* (the approximate time that symptoms appear), (2) *psychiatric* abnormalities, (3) *pain,* (4) *polyneuropathy* (different neurological defects), and (5) *photosensitivity* (sensitivity to light) in many cases. Symptoms tend to wax and wane and are aggravated by exposure to many types of drugs, including the chemicals in garlic.

Sunlight creates all kinds of "nasty damage" to some porphyria victims (a reason, perhaps, why

vampires were said to hide in dark basements and coffins during the daylight hours). When exposed to sunlight, the skin becomes lesioned and scarred, and the fingers, toes, and nose are often mutilated. The teeth grow prominent as the gums degenerate (the basis of large vampire "fangs"?). Rampant growth of hair causes the sufferer's face to become wolf-like and the hands to resemble paws. One treatment for porphyria is the injection of normal heme molecules extracted from healthy red blood cells. Since heme injections were not available in the Middle Ages, the next best thing would have been to drink blood, as vampires were said to do. The claim that garlic keeps vampires away may stem from the fact that garlic severely aggravates porphyria symptoms.

Epidermolysis bullosa (literally "epidermal breakdown with blisters"), or **EB,** is a group of hereditary disorders marked by an inadequate or faulty synthesis of collagen or by the premature breakdown of collagen fibrils. All of these disorders lead to a lack of cohesion between the layers of the skin and of mucous membranes. As a result, even a simple touch can cause these layers to separate and blister. For this reason, people suffering from EB have been called "touch-me-nots."

EB produces many symptoms, as might be expected when a body-wide structural protein like collagen is involved. In some cases, the skin blisters and erosions are limited to the feet and hands. The ends of the digits may fuse together with scarred skin, causing the hands or feet to resemble cocoons. In other cases, the mucous membranes of the digestive, respiratory, and urinary organs are also affected. In the most severe cases, generalized and fatal blistering occurs in major vital organs.

The main problem with EB occurs when the blisters break, giving pathogens (bacteria, for example) an easy route of entry to the deeper body tissues. Therefore, touch-me-nots suffer frequent infections. Since there is no known cure for EB, treatments have been aimed at relieving the symptoms and preventing infection: administering antibiotics, dressing the wounds, and ensuring proper nutrition. However, there is now some hope for a more specific drug therapy. Phenytoin (Dilantin), a drug used to manage epilepsy, is known to inhibit the breakdown of collagen. In clinical trials, phenytoin is showing promise against some forms of EB.

that tends to grow rapidly and metastasize if not removed. If it is caught early, the chance of a complete cure is good. The carcinoma can be removed by radiation therapy, by surgery, or by skin creams containing anticancer drugs.

Malignant Melanoma

Malignant melanoma (mel″ah-no′mah) is a cancer of melanocytes. This is the most dangerous kind of skin cancer. It accounts for only 5% of skin cancers, but its incidence is increasing rapidly. Melanomas can begin wherever there is pigment, but about one-third of them originate in existing moles. Melanoma usually appears as an expanding dark patch whose cells metastasize rapidly into surrounding circulatory vessels. The chances for surviving melanoma are not high, but early detection helps. The usual therapy for malignant melanoma is wide surgical excision accompanied by chemotherapy. Even so, melanoma is notoriously unresponsive to chemotherapy.

The American Cancer Society suggests that sun worshippers periodically examine their skin for moles and new pigment spots, applying the **ABCD rule** for recognizing melanoma: **A.** *Assymetry:* The two halves of the spot or mole do not match; **B.** *Border irregularity:* The borders are not smooth but show indentations and notches; **C.** *Color:* The pigment spot contains several colors, including blacks, browns, tans, and sometimes blues and reds; **D.** *Diameter* is larger than 6 mm (larger than a pencil eraser). ■

The Skin Throughout Life

The epidermis develops from embryonic ectoderm, and the dermis and hypodermis develop from mesoderm (Chapter 3, p. 56). Melanocytes, however, develop from neural crest cells that migrate into the ectoderm during the first 3 months of prenatal development. By the end of the fourth month, the skin is fairly well formed: The epidermis has all its strata, dermal papillae are obvious, and the epidermal derivatives are present in rudimentary form. During the fifth and sixth months, the fetus is covered with a downy coat of delicate hairs called the *lanugo* (lah-nu′go; "wool or down"). This hairy cloak is shed by the seventh month, and vellus hairs make their appearance.

When a baby is born, its skin is covered with *vernix caseosa* (ver′niks ka′se-o-sah; "varnish of cheese"), a cheesy-looking substance produced by sebaceous glands. It protects the skin of the fetus within the water-filled amnion. A newborn's skin is very thin, but it thickens during infancy and childhood.

With the onset of adolescence, acne may appear. Acne generally subsides in early adulthood, and the skin reaches its optimal appearance in the 20s and 30s. Thereafter, the skin starts to show the effects of continued environmental assaults (abrasion, wind, sun, chemicals). Scaling and various kinds of skin inflammation, called **dermatitis** (der″mah-ti′-tis), become more common.

In middle age, the lubricating and softening substances produced by the glands start to become deficient. As a result, the skin becomes dry and itchy. People with naturally oily skin may avoid this dryness, so their skin often ages well. Therefore, young people who suffer the humiliation of oily skin and acne may be rewarded later in life by looking young longer. With age, the elastic fibers in the dermis begin to break and degenerate, collagen fibers link together, and the skin loses its suppleness. These alterations of dermal fibers are hastened by prolonged exposure to sun and wind. The loss of skin elasticity, along with a loss of subcutaneous tissue, leads to wrinkling. Melanocytes also decrease in number and activity, resulting in less protection from UV radiation and a higher incidence of skin cancer in senior citizens. As a general rule, redheads and fair-haired individuals, who have less melanin to begin with, show age-related changes at a younger age than do people with darker skin and hair. For the same reason, black people look youthful much longer than Caucasians. As old age approaches, the replacement rate of epidermal cells slows, the skin thins, and its susceptibility to bruises and other types of injury increases.

Related Clinical Terms

Albinism (al′bĭ-nizm; "white condition")   Inherited condition in which a person's melanocytes do not make melanin, usually because the enzyme tyrosinase is absent. An albino's skin is pink, and the hair is pale yellow or white.

Athlete's foot   Itchy, red, peeling condition of the skin between the toes resulting from fungus infection.

Boils and carbuncles (kar′bung″k′ls; "little glowing embers")   Infection and inflammation of hair follicles and sebaceous glands, in which the infection has spread into the underlying hypodermis. Most people liken a boil to a giant pimple. Carbuncles are composite boils. The common cause is bacterial infection (often *Staphylococcus aureus*).

Cold sores (fever blisters)   Small, fluid-filled blisters that sting and hurt. They usually occur around the lips and in the mucosa inside the mouth. They are caused by the herpes simplex virus. This virus localizes in nerve cells that supply the skin, where it remains dormant until activated by emotional upset, fever, or UV radiation.

Decubitus ulcer (de-ku'bĭ-tus ul'ser; "lying down open sore")   A bedsore. More specifically, this is a localized breakdown of skin due to the pressure from prolonged confinement in bed and the subsequent reduction in blood supply. A decubitus ulcer usually occurs over a bony prominence, such as the hip or the heel.

Dermatology   The branch of medicine that studies and treats disorders of the skin.

Impetigo (im″pĕ-ti'go; "an attack")   Pink, water-filled, raised lesions (common around the mouth and nose) that develop a yellow crust and eventually rupture. This is a contagious condition caused by a staphylococcus infection and is common in school-age children.

Keloid (ke'loid; "growth-forming")   Excessive and prolonged proliferation of connective tissue during healing of skin wounds. It results in a large, unsightly mass of scar tissue at the skin surface. Keloids are more common in blacks than in whites.

Mongolian spot   Blue-black spot in the skin of the buttocks region of most Asian and Native American children. It results from a collection of melanocytes in the dermis that fail to migrate to the epidermis and disappears by age 3 or 4.

Psoriasis (so-ri'ah-sis; "an itching")   A chronic condition characterized by reddened epidermal papules covered with dry, silvery scales. Its cause is unknown, but this condition may be hereditary in some cases. Attacks recur and are often triggered by trauma, infection, hormonal changes, and stress.

Vitiligo (vit″il-i'go; "blemish")   An abnormality of skin pigmentation characterized by light spots surrounded by areas of normally pigmented skin. It can be cosmetically disfiguring, especially in people with dark skin.

Chapter Summary

1. The integumentary system consists of the skin and its appendages (hair, sebaceous glands, sweat glands, nails). The hypodermis, although not part of the integumentary system, is also considered in this chapter.

THE SKIN AND THE HYPODERMIS (pp. 100–104)

Epidermis (pp. 101–103)

1. The epidermis is a keratinized stratified squamous epithelium. Its main cell type is the keratinocyte. From deep to superficial, its strata are the basale, spinosum, granulosum, lucidum, and corneum. The stratum lucidum occurs only in thick skin (on palms and soles), not in thin skin (rest of the body).

2. The dividing cells of the stratum basale are the source of new keratinocytes for the growth and renewal of the epidermis.

3. In the stratum spinosum, the keratinocytes contain strong prekeratin tonofilaments. In the stratum granulosum, the cells contain granules of keratohyaline, which combine with the tonofilaments to form the tension-resisting keratin of the more superficial layers. Stratum granulosum cells also secrete a waterproofing substance that seals the spaces between cells in the superficial layers. Keratinized cells of the stratum corneum protect the skin. They are shed as skin flakes and dandruff.

4. Scattered among the keratinocytes in the deepest layers of the epidermis are pigment-producing melanocytes, sensory Merkel cells, and phagocytic Langerhans cells.

Dermis (p. 103)

5. The leathery dermis, composed of connective tissue proper, is well supplied with vessels and nerves. Also located in the dermis are glands and hair follicles.

6. The superficial papillary layer consists of areolar connective tissue, whereas the deep reticular layer is a dense irregular connective tissue. The papillary layer is responsible for the epidermal friction ridges that produce fingerprints. In the reticular layer, less dense regions between the bundles of collagen produce cleavage lines.

Hypodermis (p. 104)

7. The hypodermis underlies the skin. Composed primarily of adipose tissue, the hypodermis stores fat, insulates the body against heat loss, and absorbs and deflects blows. When one gains weight, this layer thickens most markedly in the thighs and breasts (females) or in the anterior abdominal wall (males).

Skin Color (p. 104)

8. Skin color reflects the amount of pigments (melanin and/or carotene) in the skin and the oxygenation level of hemoglobin in the dermal blood vessels.

9. Melanin, made by melanocytes and transferred to keratinocytes, protects the keratinocyte nuclei from the damaging effects of UV radiation. The epidermis also produces vitamin D under the stimulus of UV radiation.

APPENDAGES OF THE SKIN (pp. 104–110)

1. Skin appendages, which arise from the epidermis but project into the dermis, include hairs and hair follicles, sebaceous glands, sweat glands, and nails.

Hair and Hair Follicles (pp. 104–108)

2. A hair is produced by a tube-shaped hair follicle and made of heavily keratinized cells. Each hair has an outer cuticle, a cortex, and usually a central medulla. Hairs have roots and shafts. Hair color reflects the amount and variety of melanin present.

3. A hair follicle consists of an inner, epithelial root sheath that includes the hair matrix (region of the hair bulb that produces the hair) and an outer, connective tissue root sheath from the dermis. Capillaries in the hair papilla nourish the hair, and a nerve knot surrounds the hair bulb. Arrector pili muscles pull the hairs erect and produce goose bumps.

4. Fine and coarse hairs are called vellus and terminal hairs, respectively. At puberty, under the influence of androgens, terminal hairs appear in the axilla and on the anal-genital skin.

5. Hairs grow in cycles that consist of growth and resting phases. Hair thinning results from factors that lengthen follicular resting phases. Examples of such factors are the natural atrophy of hair follicles with age and a delayed-action gene for male pattern baldness.

Sebaceous Glands (p. 108)

6. Sebaceous glands are simple alveolar glands that usually empty into hair follicles. Their oily holocrine secretion is called sebum.

7. Sebum lubricates the skin and hair, slows water loss across the skin, and acts as a bactericidal agent. Sebaceous glands secrete increased amounts of sebum at puberty under the influence of androgens. Blocked and infected sebaceous glands are the basis of pimples and acne.

Sweat Glands (pp. 108–109)

8. Eccrine sweat glands are simple coiled tubular glands that secrete sweat, a modified filtrate of the blood. Evaporation of sweat cools the skin and the body.

9. Apocrine sweat glands, which may function as scent glands, occur primarily in the axillary, anal, and genital regions. Their secretion is similar to eccrine sweat but also contains proteins and fatty acids on which bacteria thrive. Bacterial action causes odor.

Nails (pp. 109–110)

10. A nail is a scale-like modification of the epidermis that covers the dorsal tip of a finger or toe. The actively growing region is the nail matrix.

CLINICAL CONDITIONS: BURNS AND CANCER (pp. 110–113)

1. The most common skin disorders result from infections.

Burns (p. 110)

2. In severe burns, the initial threat is loss of body fluids leading to circulatory collapse. The secondary threat is overwhelming infection.

3. The extent of a burn may be estimated by using the rule of nines. The severity of burns is indicated by the terms first-degree, second-degree, and third-degree. Third-degree burns harm the full thickness of the skin and require grafting for successful recovery.

Skin Cancer (pp. 110–113)

4. The major cause of skin cancer is exposure to the UV radiation of sunlight.

5. Basal cell carcinoma and squamous cell carcinoma are the most common types of skin cancer and usually are curable if detected early. Malignant melanoma, a cancer of melanocytes, is rare but usually fatal. The ABCD rule can be used to determine whether a mole or spot has melanoma.

THE SKIN THROUGHOUT LIFE (p. 113)

1. Epidermis develops from embryonic ectoderm, dermis from mesoderm, and melanocytes from neural crest.

2. The fetus exhibits a downy lanugo coat. Fetal sebaceous glands produce vernix caseosa, which helps protect the fetus within the amniotic sac.

3. In old age, the replacement rate of epidermal cells declines, and the skin and hair thin. Skin glands become less active. Loss of elastin fibers and of subcutaneous fat lead to skin wrinkling.

Review Questions

Multiple Choice and Matching Questions

1. Which cell type in the epidermis is most numerous? (a) keratinocyte, (b) melanocyte, (c) Langerhans cell, (d) Merkel cell.

2. Which epidermal cell type is probably a macrophage? (a) keratinocyte, (b) melanocyte, (c) Langerhans cell, (d) Merkel cell.

3. Match the names of epidermal layers provided in the key with the appropriate descriptions given below.

Key: **(a)** stratum basale   **(d)** stratum lucidum
      (b) stratum corneum   **(e)** stratum spinosum
      (c) stratum granulosum

_____ **(1)** desmosomes and shrinkage artifacts give its cells "prickly" projections
_____ **(2)** its cells are flat, dead bags of keratin
_____ **(3)** its cells divide, and it is also called the stratum germinativum
_____ **(4)** contains keratohyaline and lamellated granules
_____ **(5)** present only in thick skin

4. The ability of the epidermis to resist rubbing and abrasion is largely due to the presence of (a) melanin, (b) carotene, (c) collagen, (d) keratin.

5. Skin color is determined by (a) melanin, (b) carotene, (c) oxygenation level of the blood, (d) all of these.

6. What is the major factor accounting for the waterproof nature of the skin? (a) desmosomes in stratum corneum, (b) glycolipid between stratum corneum cells, (c) the thick insulating fat of the hypodermis, (d) the dermis.

7. Circle the false statement about the papillary layer of the dermis: (a) It produces the patterns for fingerprints. (b) It is responsible for the toughness of the skin. (c) It contains nerve endings that respond to stimuli. (d) It contains capillaries that fill with blood to radiate heat on warm days.

8. Circle the false statement about vitamin D: (a) Dark-skinned people make no vitamin D. (b) If the production of vitamin D is inadequate, one may develop weak bones. (c) If the skin is not exposed to sunlight, one may develop weak bones. (d) Vitamin D is needed for the uptake of calcium from food in the intestine.

9. Which of the following is *not* an epidermal appendage? (a) hair, (b) sweat gland, (c) dermis, (d) sebaceous gland.

10. Use logic to deduce the answer this question. Given what you know about skin color and skin cancer, the highest rate of skin cancer on earth occurs among (a) blacks in tropical Africa, (b) scientists in research stations in Antarctica, (c) whites in northern Australia, (d) Norwegians in the United States, (e) blacks in the United States.

11. An arrector pili muscle (a) is associated with each sweat gland, (b) causes the hair to stand up straight, (c) enables each hair to be stretched when wet, (d) squeezes hair upward so it can grow out.

12. This is the type of sweat gland whose secretion produces an odor through bacterial action: (a) apocrine gland, (b) eccrine gland, (c) sebaceous gland, (d) cutaneous gland.

13. Sebum (a) lubricates the skin and hair, (b) consists of dead cells and fatty substances, (c) collects dirt, so hair has to be washed, (d) all of these.

14. The "rule of nines" is helpful in (a) diagnosing skin cancer, (b) estimating the extent of a burn, (c) estimating how serious a burn is, (d) preventing acne.

15. An eccrine sweat gland is a _____ _____ gland: (a) compound tubular, (b) compound tubuloacinar, (c) simple squamous, (d) simple tubular.

Short Answer and Essay Questions

16. Is a bald man really hairless? Explain.

17. Can you explain why thin skin is also called hairy skin and why thick skin is called hairless skin? (Hint: Check their locations.)

18. Distinguish first-, second-, and third-degree burns.

19. What is cyanosis, and what does it reflect?

20. Why does skin wrinkle, and what factors accelerate the wrinkling process?

21. A man has first-degree burns over 25% of his body. Is this serious? Is he likely to die?

22. Why are there no skin cancers that originate from stratum corneum cells?

23. Explain each of these familiar phenomena in terms of what you learned in this chapter: (a) pimples and blackheads, (b) goose bumps, (c) suntanning, (d) dandruff, (e) greasy hair and "shiny nose," (f) stretch marks from gaining weight, (g) freckles, (h) turning blue from holding your breath, (i) leaving fingerprints, (j) pores on the face, (k) the almost hairless body of humans, (l) getting gray hairs, (m) blisters, (n) bruises.

24. Eric, an anatomy student, said it was so hot one day that he was "sweating like a pig." His professor overheard him and remarked, "That is a stupid expression, Eric. No pig ever sweats nearly as much as a person does!" Explain her remark.

25. List three functions of the subcutaneous layer.

Critical Thinking and Clinical Application Questions

26. Victims of severe burns demonstrate the vital functions performed by the skin. What are the two most important problems encountered clinically with burn patients? Explain each in terms of the absence of skin. *Fluid → out body*
Bacteria → in body

27. Mrs. Fawcett, a Cherokee Indian, worries about the bluish spots on her infant daughter's buttocks, but her husband tells her "they are just harmless spots." She is unconvinced and asks a doctor about them. What does she learn?

28. Dean, a 40-year-old beach boy, complains to you that although his suntan made him popular when he was young, now his face is all wrinkled, and he has several moles that are growing rapidly and are as large as coins. He shows you the moles, and immediately you think "ABCD." What does that mean?

Asymmetry
Border Irregularity
Color
Diameter

29. Mrs. Ibañez volunteered to help at a hospital for children with cancer. When she first entered the cancer ward, she was upset by the fact that most of the children had no hair. What is the explanation for their baldness?

30. Patients in hospital beds are rotated every 2 hours to prevent bedsores. Exactly why is this effective? (Hint: See the Clinical Terms.)

31. Carmen slipped on some ice and split open the skin of her chin on the sidewalk. As the physician in the emergency room was giving her six stitches, he remarked that the split was straight along a cleavage line. How cleanly is her wound going to heal, and is major scarring likely to occur?

32. Count Dracula, the most famous vampire of legends, was based on a real person who lived in eastern Europe about 600 years ago. He killed at least 200,000 people in the region he ruled. He was indeed a "monster," even though he was not a real vampire. Based on the Closer Look box in this chapter, the historical Count Dracula may have suffered from which of the following? (a) porphyria, (b) EB, (c) halitosis, (d) vitiligo. Explain your answer.

33. Monroe was attacked with a knife by a mugger and received several slashes on his shoulders. Over time, these cuts developed into keloids. When they started to itch, he sought help. His physician told him that it is useless to remove keloids surgically because they will only come back worse than before. Explain why this is so. (Hint: See the Clinical Terms.)

6

Bones and Skeletal Tissues

TEST #1
1ST EXM

Student Objectives

Cartilages (pp. 118–121)

1. Locate the major cartilage elements of the adult human body, and explain the functional properties of cartilage tissue.

2. Compare the structure, functions, and locations of the three kinds of cartilage tissue.

3. Explain how cartilage grows.

Bones (pp. 121–134)

4. Explain why bones can be considered organs.

5. Describe the main functions of the bony skeleton.

6. Describe the gross anatomy of a typical long bone and flat bone.

7. Explain how bones withstand tension and compression.

8. Describe the histology of compact and spongy bone.

9. Discuss the chemical composition of bone tissue and the functions of its organic and inorganic parts.

10. Compare and contrast the two types of bone formation: intramembranous and endochondral ossification.

11. Describe how endochondral bones grow at their epiphyseal plates.

12. Discuss how bone tissue is remodeled within the skeleton.

13. Explain the steps in the healing of bone fractures.

14. Describe osteoporosis, osteomalacia, Paget's disease, osteomyelitis, and osteosarcoma.

The Skeleton Throughout Life (pp. 134–135)

15. Describe how bone architecture and bone mass change from the embryonic period to old age.

All of us have heard expressions like "bone-tired," "dry as a bone," and "bag of bones"—pretty unflattering and inaccurate images of our main skeletal elements. Our brains, not our bones, convey feelings of fatigue, and bones are far from dry. As for "bag of bones," they are indeed more prominent in some of us, but without bones to form our internal skeleton we would creep along like slugs, lacking a defined shape. Bones conjure up images of graveyards and death, but they are very much alive. As we will explain, bone is one of our most dynamic tissues.

Along with its bones, the skeleton contains cartilages, which are more flexible and resilient than bones, though not as strong. In this chapter, we discuss the structure, function, and growth of cartilage and bone tissue. The individual bones of the skeleton are considered in Chapter 7.

Cartilages

Location and Basic Structure

The distribution of cartilage in the adult body is shown in Figure 6.1. These cartilages include (1) **articular cartilages,** which cover the ends of most bones at the movable joints; (2) **costal cartilages,** which connect the ribs to the sternum (breastbone); (3) cartilages that hold open the air tubes of the respiratory system; (4) cartilages in the larynx (voice box), including the *epiglottis,* a flap that keeps food from entering the larynx and the lungs; (5) cartilage in the discs between the vertebrae; (6) cartilages in the nose; and (7) cartilage in the external ear.

Cartilage is far more abundant in the embryo than in the adult: Most of the skeleton is first formed in cartilage, which is subsequently replaced by bone tissue in the fetal and childhood periods (pp. 127–129).

A typical piece of cartilage in the skeleton is composed of cartilage tissue, which contains no nerves or blood vessels. The cartilage is surrounded by a layer of dense connective tissue, the *perichondrium* (per″ĭ-kon′dre-um; "around the cartilage"). This strong layer acts like a girdle to resist outward expansion when the cartilage is subjected to pressure. The perichondrium also functions in the growth and repair of cartilage.

Cartilage tissue consists primarily of water and is very resilient. (Resilience is the ability to spring back to the original shape after being compressed.) For more information on the unique properties of cartilage tissue, see the box on p. 120.

Classification of Cartilage

Three types of cartilage tissue occur within the body: *hyaline cartilage, elastic cartilage,* and *fibrocartilage* (Figure 6.1). While reading about these, keep in mind that cartilage is a connective tissue that consists of cells called *chondrocytes* and an extracellular matrix containing a jelly-like ground substance and fibers (recall Chapter 4, p. 83).

Hyaline Cartilage

Hyaline (hi′ah-līn) **cartilage** looks like frosted glass when viewed by the unaided eye (*hyalin* = glass). It is the most abundant kind of cartilage (Figure 6.1). When viewed under the light microscope, its chondrocytes appear spherical (Figure 4.15h on p. 87). Each chondrocyte occupies a cavity in the matrix called a *lacuna* (lah-ku′nah; literally a "pit" or "cavity"). The only type of fiber in the matrix is a kind of collagen unit fibril, which forms networks that are too thin to be seen under the light microscope. Hyaline cartilage provides support with flexibility and resilience.

Elastic Cartilage

Elastic cartilage is similar to hyaline cartilage, but its matrix contains many elastic fibers (Figure 4.15i, p. 88). This cartilage is more elastic than hyaline cartilage and better able to tolerate repeated bending. The epiglottis, which bends down to cover the larynx each time we swallow, is made of elastic cartilage, as is the cartilage that supports the outer ear.

Fibrocartilage

Fibrocartilage is often found where hyaline cartilage meets a true ligament or a tendon. Fibrocartilage is a perfect structural intermediate between hyaline cartilage and dense regular connective tissues: Microscopically, it consists of rows of chondrocytes (a cartilage feature) alternating with rows of thick collagen fibers (a feature of dense regular connective tissue) (Figure 4.15j, p. 88). Fibrocartilage occurs where both tensile strength and the ability to withstand heavy pressure are required. For example, the intervertebral discs (resilient cushions between the bony vertebrae) contain fibrocartilage (Figure 6.1), as does the knee joint.

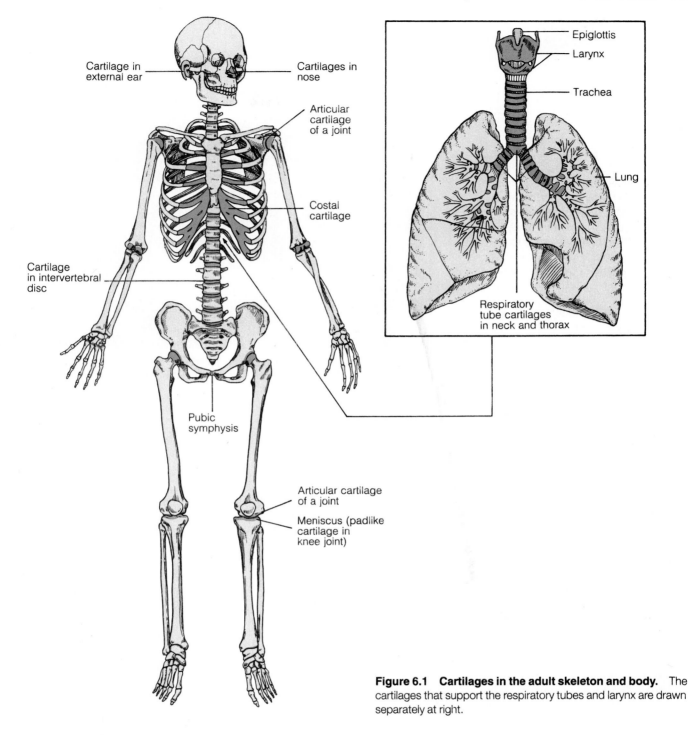

Figure 6.1   Cartilages in the adult skeleton and body.   The cartilages that support the respiratory tubes and larynx are drawn separately at right.

Growth of Cartilage

Cartilage grows in two ways. In **appositional** (ap″o-zish′un-al) **growth,** a "growth from outside," carti-lage-forming cells (chondroblasts) in the surrounding perichondrium produce the new cartilage tissue. In **interstitial** (in″ter-stish′al) **growth,** a "growth from within," the chondrocytes within the cartilage divide and secrete new matrix. Cartilage stops growing in the

late teens when the skeleton itself stops growing, and chondrocytes do not divide again.

It may surprise you that cartilage tissue, which grows so rapidly during youth, has little capac-ity for regeneration and healing in adults. Can you ex-plain why injured adult cartilage regenerates so poorly?

Adult cartilage regenerates poorly because the chondrocytes have lost all capacity for division. The

A CLOSER LOOK **Cartilage—Strong Water**

It's match point: Avery tosses the tennis ball in the air, leans back, then smashes it over the net into the opposing court. Brazinsky runs backwards, then leaps up to return the shot. Tennis anyone? Not without healthy cartilage, the tissue that forms the framework for building our bones and fashions the spongy cushions of our movable joints.

Cartilage is resilient: If you push on it and then ease the pressure, it bounces back. This resilience allows the articular cartilages at the ends of our bones to cushion the impact of various activities, such as lifting heavy objects, striking a tennis ball, sprinting, jogging, or jumping. Once an action is complete, the cartilage tissue springs back to await the next movement.

Another important feature of cartilage tissue is its rapid growth, which enables it to keep pace with the rapid growth of the embryo. Indeed, most of the embryo's "bones" originate as cartilage models that are only gradually replaced by bone tissue. Throughout youth, growth plates of cartilage remain in most bones and bring about most growth of the skeleton. One would expect cartilage to require a rich blood supply for such rapid growth, yet cartilage contains no blood vessels. Instead, it receives all its oxygen and nutrients from vessels in the perichondrium layer surrounding it. In fact, cartilage and blood vessels are strictly incompatible: Cartilage secretes chemicals that prevent blood vessels from growing into it, and cartilage cannot develop in the parts of the embryo that are well supplied with blood vessels or oxygen.

The explanation for both the resilience of cartilage and its remarkable growth properties lies in its ability to hold tremendous amounts of water. Like all connective tissues, cartilage consists of cells, called chondrocytes, separated by an extracellular matrix of jelly-like ground substance, collagen fibers, and watery tissue fluid (Chapter 4, p. 83,

The cartilages of professional tennis players should be well nourished and well hydrated because of the great physical demands this sport places on the joints.

pp. 87–88). The ground substance of every connective tissue contains long sugar molecules (glycosaminoglycans) bearing many negative charges. In cartilage, however, there are far more of these negative charges than in other connective tissues. These charges strongly attract water molecules, which form many layers of "water shells" around each charged site. In fact, water constitutes 60% to 80% of cartilage by weight. This attraction is the source of cartilage's resilience: When compression forces are applied to the cartilage, water is forced away from the charged sites. As the pressure increases, the negative charges are pushed closer together and repel each other, resisting further compression. Then, when the pressure is released, the water molecules rush back to their original sites, causing the cartilage to spring back forcefully to its original shape. These forces also play a vital role in the nourishment of cartilage: The resulting flow of liquids carries nutrients to the cartilage cells. For this reason, long periods of inactivity can weaken the cartilages of our joints.

The rapid growth of cartilage is another result of its capacity to hold fluid. Because cartilage matrix is mostly water, nutrients from distant capillaries can diffuse through it quickly enough to supply the high metabolic needs of the rapidly dividing cartilage cells (chondroblasts). There is no need for a slow, time-consuming process of growing new capillary beds within the cartilage tissue itself. Furthermore, the organic matrix secreted by the chondroblasts attracts so much water that the cartilage expands (grows) quickly with little expenditure of materials.

Another important feature of cartilage tissue is its ability to resist tension (pulling forces). This property is attributable not to its high water content, but to its collagen fibrils, which run in a network throughout the extracellular matrix. Although cartilage is strong in resisting compression and tension, it is weak in resisting shear forces (twisting and bending). Because of this weakness, torn cartilages are a common sports injury.

little healing that does occur within adult cartilage is due to the ability of the surviving chondrocytes to secrete more extracellular matrix. ∎

Under certain conditions, crystals of calcium phosphate precipitate in the matrix of cartilage. Such calcification is both a sign of aging and a normal stage in the growth of most bones (p. 127). Note, however, that this **calcified cartilage** is not bone: Bone and cartilage are always distinct tissues.

Bones

The bones of the skeleton are *organs.* Recall that an organ contains several different tissues. Although bones are dominated by bone tissue, they also contain nervous tissue in their nerves, blood tissue in their blood vessels, cartilage in their articular cartilages, and epithelial tissue lining the blood vessels.

Functions of Bones

Besides contributing to body shape, bones perform other important functions:

1. Support. Bones provide a hard framework that supports the body and cradles its soft organs. For example, the bones of the legs are pillars that support the body trunk when we stand, and the bony pelvis acts like a bowl to hold the pelvic organs.

2. Protection. The bones of the skull form a protective case for the brain. The vertebrae surround the spinal cord, and the rib cage helps protect the organs of the thorax.

3. Movement. Skeletal muscles, which attach to the bones by tendons, use the bones as levers to move the body and its parts. As a result, we can walk, grasp objects, and move our rib cage to breathe. The arrangement of the bones and the structure of the joints determine the types of movement that are possible.

4. Mineral storage. Bone serves as a reservoir for minerals, the most important of which are calcium and phosphate. The stored minerals can be released into the bloodstream as ions for distribution to all parts of the body as needed.

5. Blood cell formation. Bones contain red and yellow *bone marrow.* Red bone marrow manufactures the blood cells. Bone marrow and the formation of blood cells are described in detail in Chapter 16 (p. 453).

Classification of Bones

Bones come in many sizes and shapes. For example, the tiny pisiform bone of the wrist is the size and shape of a pea, while the femur (thighbone) is large and elongated. The shape of each bone fills a particular need. The femur, for example, must withstand great weight and pressure, and its hollow cylindrical design provides maximum strength with minimum weight.

Bones are classified by their shape as long, short, flat, or irregular (Figure 6.2).

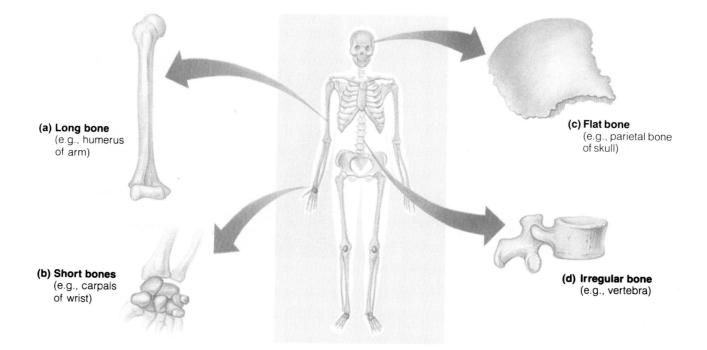

(a) Long bone
(e.g., humerus of arm)

(b) Short bones
(e.g., carpals of wrist)

(c) Flat bone
(e.g., parietal bone of skull)

(d) Irregular bone
(e.g., vertebra)

Figure 6.2   Classification of bones on the basis of shape.

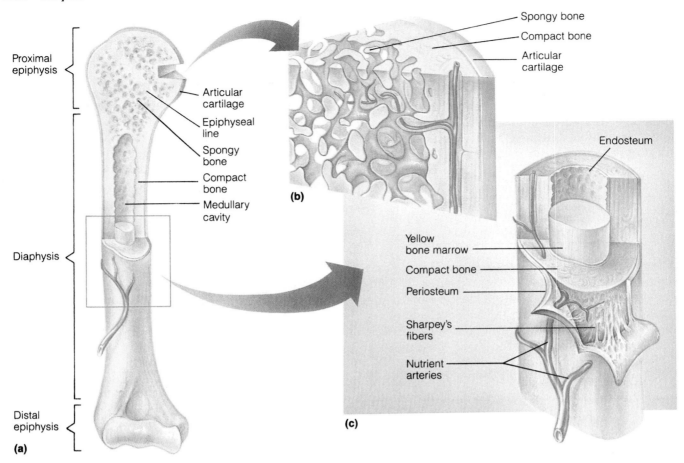

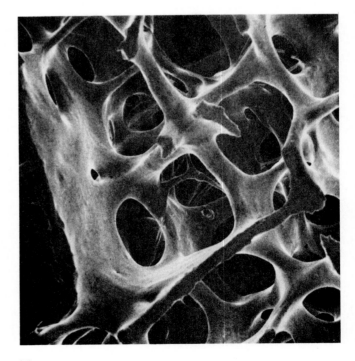

Figure 6.3   The structure of a long bone (humerus of arm). (a) Anterior view, with the superior half of the bone sectioned frontally to show the interior. (b) Enlargement of (a), showing the spongy bone and compact bone of the epiphysis. (c) Enlargement of the diaphysis from (a). (d) Scanning electron micrograph of spongy bone, showing a network of bone trabeculae (20 ×).

1. Long bones. As their name suggests, long bones are considerably longer than they are wide. A long bone has a shaft plus two ends. Most bones in the limbs are long bones. Note that this bone classification reflects the elongated shape of these bones, not their overall size: The three bones within each finger are long bones, even though they are very small.

2. Short bones. Short bones are roughly cube-shaped. Bones of the wrist are examples.

Sesamoid (ses′ah-moid) *bones* are a special type of short bone that form within a tendon. The best example is the kneecap, or patella. Sesamoid bones vary in size and number in different people. Some clearly act to alter the direction of pull of a tendon, but the function of others is unknown. The name *sesamoid* means "shaped like a sesame seed."

3. Flat bones. Flat bones are thin, flattened, and usually somewhat curved. Most skull bones are flat, as are the ribs and sternum.

4. Irregular bones. Irregular bones have various shapes that do not fit into any of the above categories. Examples are the vertebrae and the hip bones.

Gross Anatomy of Bones

Compact and Spongy Bone

Every bone of the skeleton has a dense outer layer that looks smooth and solid to the naked eye. This external layer is **compact bone** (Figures 6.3 and 6.4). Internal to this is **spongy bone** (also called *cancellous bone*), a honeycomb of small needle-like or flat pieces called *trabeculae* (trah-bek′u-le; "little beams"). In this network, the open spaces between the trabeculae are filled with red or yellow bone marrow.

Structure of a Typical Long Bone

With few exceptions, all the long bones in the body have the same general structure (Figure 6.3).

Diaphysis and Epiphyses. The tubular **diaphysis** (di-af′ĭ-sis), or shaft, forms the long axis of a long bone. The **epiphyses** (e-pif′ĭ-sēz) are the bone ends. In many cases, the epiphyses are more expanded than the diaphysis. The joint surface of each epiphysis is covered with a thin layer of articular cartilage. Between the diaphysis and each epiphysis of an adult long bone is an **epiphyseal line.** This line is a remnant of the epiphyseal plate, a disc of hyaline cartilage that grows during childhood to lengthen the bone (p. 129).

Blood Vessels. Unlike cartilage, bone tissue is very well vascularized. The main vessels supplying the di-

aphysis are the *nutrient artery* (Figure 6.3c) and *nutrient vein.* Together these run through a hole in the wall of the diaphysis, the *nutrient foramen* (fora′men; "opening"). The nutrient artery runs internally to supply the bone marrow and the spongy bone, and it sends branches externally to help supply the compact bone. Several *epiphyseal arteries* and *veins* (not illustrated) supply each epiphysis in the same way.

The Medullary Cavity. The interior of long bones is composed largely of spongy bone. The very center of the diaphysis of long bones, however, contains no bone tissue at all and is called the **medullary cavity** (med′u-lar-e; "middle") or *marrow cavity.* As its name implies, this cavity is filled with yellow bone marrow. (Recall that the spaces between the trabeculae of spongy bone are also filled with marrow.)

Membranes. A white membrane of connective tissue called the **periosteum** (per″e-os′te-um; "around the bone") covers the entire outer surface of each bone, except the region of the epiphyses where articular cartilage occurs. This membrane consists of a superficial layer of dense irregular connective tissue and a deep layer that contacts the compact bone. This deep layer contains bone-depositing cells called *osteoblasts* (os′te-o-blasts″; "bone generators") and bone-destroying cells called *osteoclasts* ("bone breakers"). These cells remodel bone surfaces throughout our lives. The periosteum is richly supplied with nerves (which is why broken bones are painful) and with blood vessels (which bleed profusely when a bone is fractured). The vessels that supply the periosteum are branches from the nutrient and epiphyseal vessels. The periosteum is secured to the underlying bone by **Sharpey's fibers** *(perforating fibers),* thick bundles of collagen that run from the periosteum into the bone matrix (Figure 6.3c). The periosteum also provides an insertion point for tendons and ligaments. At these points, Sharpey's fibers are exceptionally dense.

Whereas periosteum covers the external surface of bones, the *internal* bone surfaces are covered by a much thinner membrane of connective tissue called **endosteum** (en-dos′te-um; "within the bone"). Specifically, endosteum covers the trabeculae of spongy bone. Endosteum also lines the central canals of osteons (see p. 124). Like periosteum, endosteum contains both osteoblast and osteoclast cells.

Structure of Short, Irregular, and Flat Bones

Short, irregular, and flat bones have much the same composition as long bones: periosteum-covered compact bone externally and endosteum-covered spongy

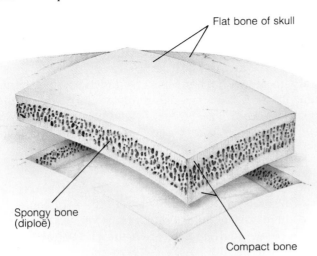

Flat bone of skull

Spongy bone
(diploë)

Compact bone

Figure 6.4   Structure of a flat bone of the skull (in section).   As in other bones, a thick periosteum covers the outer surfaces of the compact bone, and a thin endosteum lines the network of spongy bone. For simplicity, however, these membranes are not illustrated.

bone internally. These bones are not cylindrical, so they have no diaphysis or epiphyses. They contain bone marrow (between the trabeculae of their spongy bone), but no marrow cavity is present. Figure 6.4 shows a typical flat bone of the skull. In flat bones, the internal spongy bone is called **diploë** (dip′lo-e; "folded"). The whole arrangement of a flat bone resembles a stiffened sandwich.

Bone Design and Stress

The anatomy of each bone reflects the stresses most commonly placed on it. Bones are subjected to compression as weight bears down on them or as muscles pull on them. The loading usually is off center, however, and tends to *bend* the bone (Figure 6.5a). Bending compresses the bone on one side and stretches it (subjects it to tension) on the other. Both compression and tension are greatest at the external bone surfaces. Therefore, strong, compact bone tissue occurs in the external region of the bone. Internal to this region, however, tension and compression forces tend to cancel each other out, decreasing overall stress. Solid bone does not form here; spongy bone is sufficient. Because no stress occurs at the bone's center, the lack of bone tissue in the central medullary cavity does not impair the strength of long bones. Furthermore, the presence of spongy bone and marrow cavities serves to lighten the heavy skeleton and to provide room for the bone marrow.

The trabeculae of spongy bone align precisely along stress lines. Spongy bone is not a random network of bone fragments, but an organized pattern of tiny struts (Figure 6.5b and c) as carefully positioned as the flying buttresses that support the walls of a Gothic cathedral.

Microscopic Structure of Bone

Like other connective tissues, bone tissue consists of cells separated by an extracellular matrix (see the enlargement in Figure 6.6a). In addition to the usual matrix components of collagen fibers and ground substance, however, bone matrix contains mineral crystals. It also contains a small amount of tissue fluid, although bone contains less water than other connective tissues do.

Compact Bone

Viewed by the unaided eye, compact bone looks solid. However, microscopic examination reveals that it is riddled with passageways for blood vessels, lymphatic vessels, and nerves (Figure 6.6a). An important structural component of compact bone is the **osteon** (os′te-on; "bone"), or **Haversian** (ha-ver′shan) **system.** Osteons are long, cylinder-shaped structures oriented parallel to the long axis of the bone and the main compression stresses. Functionally, osteons can be viewed as miniature weight-bearing pillars. Structurally, an osteon is a group of concentric tubes resembling the rings in a cross section of a tree trunk (Figure 6.7). Each of the tubes is a bone **lamella** (lah-mel′ah; "little plate"). More precisely, a lamella is a layer of bone matrix in which all the collagen fibers run in a single direction. However, the fibers of adjacent lamellae always run in opposite directions. This alternating pattern is optimal for withstanding torsion stresses—the adjacent lamellae in an osteon reinforce one another to resist twisting.

Through the core of each osteon runs a canal called the **central canal,** or **Haversian canal.** Like all internal bone cavities, it is lined by endosteum. The central canal contains its own blood vessels, which supply nutrients to the bone cells of the osteon, and its own nerve fibers. **Perforating canals,** also called **Volkmann's** (fōlk′mahnz) **canals,** also occur in compact bone (Figure 6.6a). These lie at right angles to the central canals and connect the blood and nerve supply of the periosteum to that of the central canals and the marrow cavity.

The mature bone cells, or **osteocytes,** are spider-shaped cells (Figure 6.6a). They occupy small cavities in the solid matrix called **lacunae.** Thin tubes called **canaliculi** (kan″ah-lik′u-li; "little canals") run through the matrix, connecting neighboring lacunae to one another and to the central canals. The long, arm-like processes of neighboring osteocytes touch each other within these canaliculi, forming gap junctions (p. 76). Nutrients diffusing from capillaries in the central canal are relayed across these gap junctions, from one osteocyte to the next, throughout the entire osteon. Such a cell-to-cell relay is necessary because the intervening bone matrix is solid and imper-

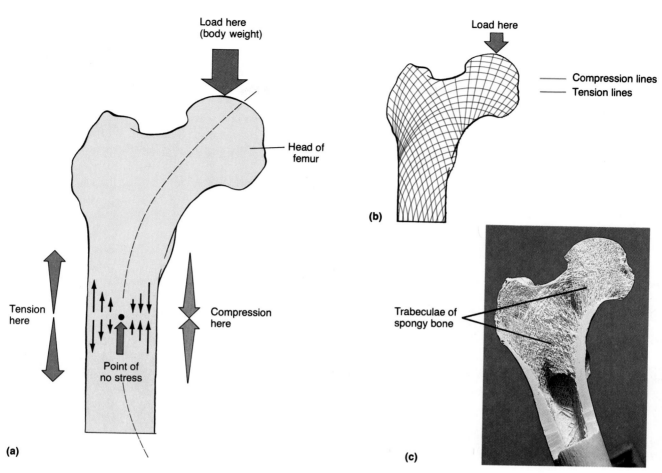

Figure 6.5   Bone anatomy and stress.   **(a)** Because the load placed on most bones is off center, bones are subjected to bending stress. Body weight is transmitted to the head of the femur and threatens to bend the bone along the indicated arc. This bending compresses the bone at one side (converging arrows) and pulls it in tension at the other side (diverging arrows). Because these two forces cancel each other internally (at the point of no stress), much less bone material is needed internally than peripherally. **(b)** A diagram of the stress lines (stress trajectories) experienced by the loaded femur. **(c)** The actual pattern of trabeculae of spongy bone within a femur. A comparison of the two patterns in (b) and (c) shows that they match closely, illustrating that spongy bone provides support along the stress lines.

meable to nutrients. The function of osteocytes is to maintain the bone matrix: If osteocytes die, the surrounding matrix is reabsorbed.

Not all lamellae in compact bone occur within osteons. **Circumferential lamellae** occur in the external and internal surfaces of the layer of compact bone. Each of these lamellae extends around the entire circumference of the diaphysis (Figure 6.6a). Circumferential lamellae effectively resist twisting of the entire long bone.

Spongy Bone

The microscopic anatomy of spongy bone is less complex than that of compact bone. Although each trabecula of spongy bone contains several layers of lamellae

and osteocytes, the trabeculae are too small to contain osteons or blood vessels of their own. Nutrients reach the osteocytes within the trabeculae through diffusion from capillaries in the surrounding endosteum.

Chemical Composition of Bone Tissue

Bone tissue has both organic and inorganic components. The organic components—cells, fibers, and ground substance—account for 35% of the tissue mass. These organic substances, particularly collagen, contribute the flexibility and tensile strength that allow bone to resist stretching and twisting. Collagen is remarkably abundant in bone tissue.

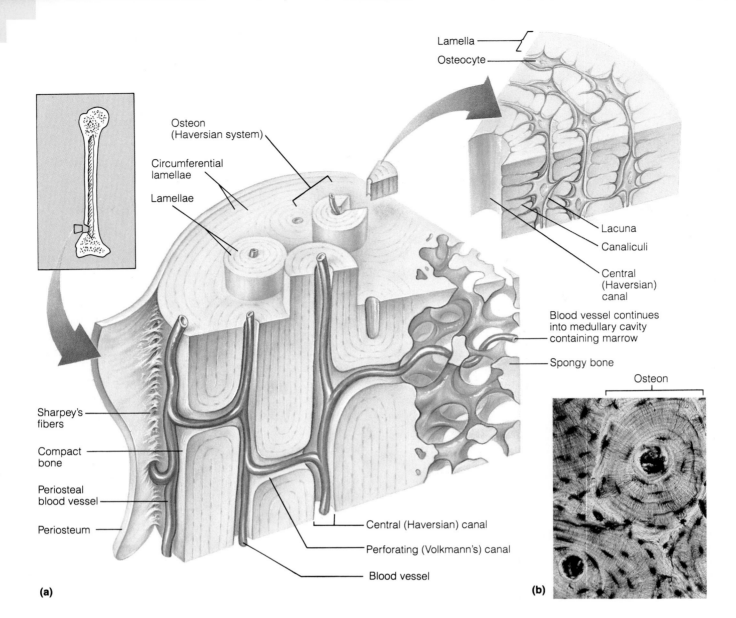

Figure 6.6 Microscopic structure of compact bone. (**a**) A pie-shaped segment of compact bone from the wall of the humerus (the bone of the arm). Note the pillar-shaped osteons and the passageways holding blood vessels. The enlargement at upper right is a more highly magnified view of a part of one osteon. (**b**) Photomicrograph of a cross section of one osteon and parts of others (90 ×).

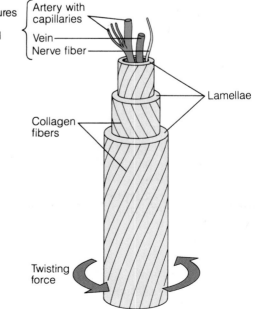

Figure 6.7 Diagram of an isolated osteon. The osteon is drawn as if it were pulled out like a telescope to show the lamellae that comprise it. Note the different directions of the collagen fibers in the three different lamellae. The contents of a central canal are also shown. The arrows below indicate twisting forces placed on the osteon.

The balance of bone tissue, 65% by mass, consists of inorganic hydroxyapatites (hi-drok″se-ap′ah-tītz), or mineral salts, primarily of calcium phosphate. These mineral salts are present in the form of tiny crystals that lie in and around the collagen fibers in the extracellular matrix. The crystals pack tightly into a highly ordered pattern, thereby providing bone with its exceptional hardness, which allows it to resist compression.

The proper combination of organic and inorganic elements allows bones to be exceedingly durable and strong without being brittle. It is always surprising to hear that healthy bone is half as strong as steel in resisting compression and fully as strong as steel in resisting tension.

Because of bone salts, bones can last long after death and provide an enduring "monument." In fact, skeletal remains many centuries old have revealed the shapes, sizes, and racial characteristics of ancient peoples, the kinds of work they did, and many of the ailments they suffered (arthritis, for example).

A long bone that is taken from a fresh cadaver and soaked in a solution of weak acid for several weeks will maintain its original form but will look leathery and can be easily tied into a knot. Can you explain this?

The acid dissolves away the bone's mineral, leaving only the organic component (mainly collagen). The fact that a demineralized bone can be knotted demonstrates the great flexibility provided by the collagen within bone. This fact also confirms that without its mineral content, bone bends too easily to support weight. ■

Bone Development

Osteogenesis (os″te-o-jen′ĕ-sis) and **ossification** are both names for the process of bone tissue formation. Osteogenesis begins in the embryo, proceeds through childhood and adolescence as the skeleton grows, and occurs at a slower rate in the adult as part of a continual remodeling of the full-grown skeleton.

Before week 8, the skeleton of the human embryo consists only of hyaline cartilage and some membranes of mesenchyme. Bone tissue first appears in week 8 and eventually replaces all the mesenchyme and most of the cartilage in the skeleton. Some bones, called **membrane bones,** develop from a mesenchymal membrane through a process called **intramembranous ossification.** Other bones develop by replacing a piece of hyaline cartilage through a process called **endochondral ossification** (*endo* = within; *chondro* = cartilage). These bones are called **endochondral** or **cartilage bones.**

Intramembranous Ossification

Membrane bones form directly from mesenchyme without first being modeled in cartilage. All bones of the skull, except a few at the base of the skull, are in this category. The clavicles (collarbones) are the only bones formed by intramembranous ossification that are not in the skull.

Intramembranous ossification proceeds in the following way: During week 8 of embryonic development, certain mesenchyme cells cluster and become bone-forming **osteoblasts** (see the top diagram in Figure 6.8). These cells begin secreting the organic part of bone matrix, called **osteoid** (os′te-oid; "bone-like"), which then becomes mineralized. Once surrounded by their own matrix, the osteoblasts are called osteocytes (Figure 6.8, step 1). The new bone tissue forms between embryonic blood vessels, which are woven in a random network. The result is **woven bone tissue,** with bone trabeculae that run in networks (Figure 6.8, step 2). This embryonic tissue lacks the lamellae that occur in mature spongy bone. During this same stage, more mesenchyme condenses just external to the developing membrane bone and becomes the periosteum.

To complete the development of a membrane bone, the trabeculae at the periphery grow thicker until plates of compact bone are present on both surfaces (Figure 6.8, step 3). In the center of the membrane bone, the trabeculae remain distinct, and spongy bone results. The final pattern is that of the flat bone pictured in Figure 6.4.

Endochondral Ossification

Most bones of the skeleton inferior to the skull are endochondral bones. They are first modeled in hyaline cartilage, which is gradually replaced by bone tissue. Endochondral ossification begins late in the second month of development and is not completed until the skeleton stops growing in early adulthood. Growing endochondral bones increase both in length and in width, but we will initially consider only the increase in length, using a large long bone as an example (Figure 6.9).

Stage 1: A bone collar forms around the diaphysis. In the late embryo (week 8), the endochondral bone begins as a piece of cartilage called a *cartilage model.* Like all cartilages, it is surrounded by a perichondrium. Then, the perichondrium surrounding the diaphysis becomes a bone-forming periosteum. Osteoblasts in this new periosteum lay down a collar of bone tissue around the diaphysis.

Stage 2: Cartilage calcifies in the center of the diaphysis. As the bone collar forms, the chondrocytes in the

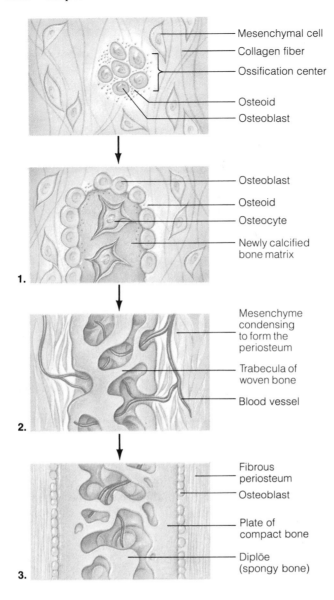

Figure 6.8 Intramembranous ossification: Development of a flat bone of the skull in the fetus. The unnumbered top view shows mesenchyme cells becoming osteoblasts. The events depicted in diagrams 1 through 3 are described on p. 127. The two lower views are at much lower magnification than the two upper ones.

center of the diaphysis enlarge and signal the surrounding cartilage matrix to calcify. The matrix of this calcified cartilage is impermeable to diffusing nutrients. Cut off from all nutrients, the chondrocytes die. No longer kept healthy by chondrocytes, the cartilage matrix starts to deteriorate. Although this weakens the diaphysis of the hyaline cartilage model, it is well stabilized by the bone collar around it. These changes affect only the center of the diaphysis. Elsewhere, the cartilage remains healthy and continues to grow, causing the entire endochondral bone to elongate.

Stage 3: The periosteal bud invades the diaphysis, and the first bone trabeculae form. The cavities within the diaphysis are immediately invaded by a collection of elements called the **periosteal bud.** This bud consists of a nutrient artery and vein (p. 123), along with the cells that will form the bone marrow. Most important, the bud contains bone-forming and bone-destroying cells (pre-osteoblasts and osteoclasts). The entering osteoblasts start to secrete osteoid around the remaining fragments of calcified cartilage matrix, forming bone-covered trabeculae. In this way, the earliest version of spongy bone appears within the diaphysis.

By the third month of development, bone tissue has begun to appear both around the diaphysis and in its center. This bone tissue of the diaphysis makes up the **primary ossification center.**

Stage 4: The bone epiphyses organize for growth, and the medullary cavity forms. While changes are occurring in the diaphysis, the cartilage in the epiphyses stays healthy and grows briskly. In fact, each epiphyseal region nearest the diaphysis organizes into a pattern that allows exceptionally rapid and efficient growth (Figure 6.10). The cartilage cells here form tall columns, like coins in a stack. The chondroblasts at the "top" of the stack (zone 1 in Figure 6.10) divide quickly, pushing the epiphysis away from the diaphysis, causing the entire long bone to lengthen. Meanwhile, the older chondrocytes deeper in the stack enlarge, signal the surrounding matrix to calcify (zone 2), and then die and disintegrate (zone 3). This process leaves long spicules (trabeculae) of calcified cartilage on the diaphysis side of the epiphysis-diaphysis junction (zone 4), hanging like stalactites from the roof of a cave. These spicules are quickly covered with bone tissue by osteoblasts and are eventually digested away by osteoclasts. Since the bone spicules are removed from within the diaphysis at the same rate that they form at the epiphysis, they remain a constant length. Because the tips of the spicules are continually reabsorbed, a boneless medullary cavity forms in the center of the diaphysis and enlarges as the long bone lengthens.

Throughout the fetal period, the rapidly growing epiphyses consist only of cartilage.

Stage 5: The epiphyses ossify. Shortly before or after birth, the epiphyses gain bone tissue (Figure 6.9d): First, the cartilage in the center of each epiphysis calcifies and degenerates. Then, an osteogenic bud containing the epiphyseal vessels invades each epiphysis, and bone trabeculae appear, just as they appeared earlier in the primary ossification center. The areas of bone formation in the epiphyses are called **secondary ossification centers.** The larger long bones of the body can have several ossification centers in each epiphysis.

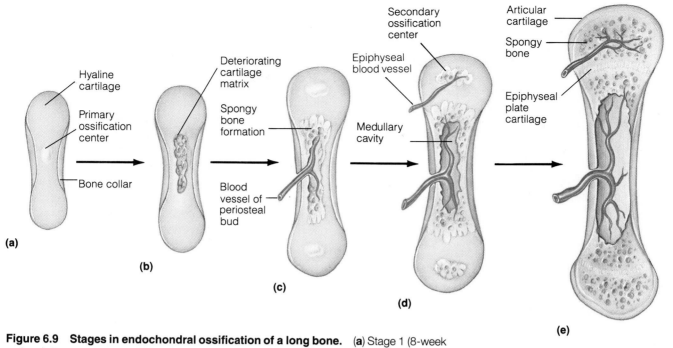

Figure 6.9   Stages in endochondral ossification of a long bone.   (**a**) Stage 1 (8-week embryo): Formation of a bone collar around the diaphysis of the hyaline cartilage model. (**b**) Stage 2 (8-week embryo): Cartilage calcification in the center of the diaphysis and appearance of cavities. (**c**) Stages 3 and 4 (week 9 to birth): Invasion of the center of the diaphysis by the periosteal bud and the formation of the first trabeculae of spongy bone. The epiphyses then organize for efficient growth, and the medullary cavity forms. (**d**) Stage 5 (time of birth): Ossification of the epiphyses. (**e**) Bone growth in the child and adolescent. Note the epiphyseal plate cartilage and the articular cartilage.

Once the secondary ossification centers have appeared, hyaline cartilage remains at only two places (Figure 6.9e): (1) on the epiphyseal surfaces, where it forms the articular cartilages; and (2) between the diaphysis and epiphyses, where it forms the **epiphyseal plates** (also called *epiphyseal discs* or *growth plates*).

By this time, the baby has been born, and the endochondral bones continue to elongate without interruption.

Postnatal Growth of Endochondral Bones

During childhood and adolescence, the endochondral bones lengthen entirely by growth of the epiphyseal plates. The epiphyseal plate retains all the cartilage zones pictured in Figure 6.10. Because its cartilage is replaced with bone tissue on the diaphysis side about as quickly as it grows, the epiphyseal plate maintains a constant thickness while the whole bone lengthens.

As adolescence draws to an end, the chondroblasts in the epiphyseal plates divide less often, and the plates become thinner. Eventually, they exhaust their supply of mitotically active cartilage cells and are replaced by bone tissue. Long bones stop length-ening when the bone of the epiphyses and diaphysis fuses. This process, called *closure of the epiphyseal plates,* occurs at about 18 years of age in females and 21 years of age in males. After our epiphyseal plates close, we can grow no taller.

Our discussion so far has revealed the surprising fact that *bone* tissue is not responsible for the increase in length of growing endochondral bones. Can you determine what other tissue is?

Hyaline cartilage does the growing. Bone tissue merely replaces the cartilage. ■

Growing bones must widen as they lengthen. How do they widen? Simply, osteoblasts in the periosteum add bone tissue to the external face of the diaphysis as osteoclasts in the endosteum remove bone from the internal surface of the diaphysis wall. These two processes occur at about the same rate, so that the circumference of the long bone expands and the bone widens. Growth of a bone by the addition of bone tissue to its surfaces is called **appositional growth.**

We have been discussing the growth and development of large long bones. The other types of endochondral bones grow in slightly different ways. Short bones, such as those in the wrist, arise from only a single ossification center. Most of the irregular bones,

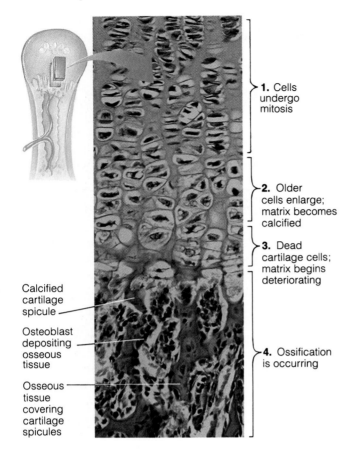

1. Cells undergo mitosis

2. Older cells enlarge; matrix becomes calcified

3. Dead cartilage cells; matrix begins deteriorating

4. Ossification is occurring

Calcified cartilage spicule

Osteoblast depositing osseous tissue

Osseous tissue covering cartilage spicules

Figure 6.10   Organization of the cartilage within the epiphysis of a long bone (250 ×).   Cartilage cells are stacked in vertical columns like coins. This neat arrangement first appears in the early fetus, and after birth it remains in the epiphyseal plate. Four zones are present: (1) Chondroblasts undergo mitosis, multiplying vigorously. (2) Chondrocytes enlarge, then their matrix calcifies. (3) Chondrocytes die and disintegrate, leaving ragged pieces (spicules) of cartilage matrix in zone 4. (4) Bone matrix (osseous tissue) covers the cartilage spicules, forming spicules (trabeculae) of bone.

such as the hip bone and vertebrae, develop from several distinct ossification centers. Small long bones, such as those in the palm and fingers, form from a primary ossification center (diaphysis) plus a single secondary center (that is, they have just one epiphysis).

Hormonal Regulation of Bone Growth

The growth of bones that occurs throughout youth is exquisitely controlled by several hormones acting in concert. During infancy and childhood, the most important stimulator of epiphyseal plate activity is *growth hormone.* This hormone is released by the pituitary gland, which lies on the inferior surface of the brain. Growth hormone has multiple effects on body growth; its ultimate effect on the skeleton is to stimulate the epiphyseal plate cartilages to grow. Hormones

from the thyroid gland (an endocrine gland in the neck) modulate the activity of growth hormone, ensuring that the skeleton keeps its proper proportions as it grows. At puberty, the male and female sex hormones (*androgens* and *estrogens,* respectively) are released in increasing amounts. These hormones initially promote the growth spurt of the adolescent and later induce the epiphyseal plates to close and growth to cease. Sex hormones also promote the masculinization or feminization of specific parts of the skeleton (broader hip bones in females, for example). Excesses or deficits of any of these hormones can cause abnormal growth of the skeleton. For example, an excess of growth hormone results in gigantism (abnormal tallness) and deficits of growth hormone or thyroid hormone produce various kinds of dwarfism (p. 676).

Bone Remodeling

Bones appear to be the most lifeless of body organs, and once they are formed, they seem set for life. Nothing could be further from the truth. Bone is a very dynamic and active tissue. Large amounts of bone matrix and thousands of osteocytes are continuously being removed and replaced within the skeleton, and the small-scale architecture of our bones constantly changes. As much as half a gram of calcium may enter or leave the adult skeleton each day!

In the adult skeleton, bone deposit and bone removal occur at many periosteal and endosteal surfaces. Together, the two processes constitute **bone remodeling.** They are coordinated by "packets" of osteoblasts and osteoclasts called remodeling units. In healthy young adults, the total mass of bone in the body stays constant, an indication that the rates of bone deposit and reabsorption are essentially equal. The remodeling process is not uniform, however. Some bones (or bone parts) are very heavily remodeled, but others are not. For example, the distal region of the femur is fully replaced every 5 to 6 months, whereas the diaphysis of the femur changes much more slowly.

Bone deposit is accomplished by *osteoblasts,* hemisphere-shaped cells with short cellular processes. These cells lay down organic osteoid on bone surfaces, and calcium salts crystallize within this osteoid. As we stated earlier, the osteoblasts transform into osteocytes when they become surrounded by bone matrix.

Bone reabsorption is accomplished by **osteoclasts** (Figure 6.11). Each of these giant cells has dozens of nuclei. Osteoclasts crawl along bone surfaces, breaking down the bone tissue. The part of their plasma membrane that touches the bone surface is highly folded (ruffled). Lysosomal enzymes are re-

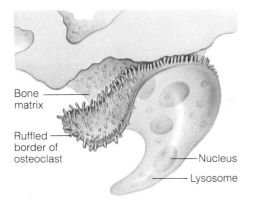

Bone matrix

Ruffled border of osteoclast

Nucleus

Lysosome

Figure 6.11 Osteoclasts are bone-degrading cells. Osteoclasts are large and multinucleate and have a ruffled border on the surface that degrades the bone.

leased across this expanded membrane to digest the organic part of the bone matrix. Osteoclasts also secrete acid, which dissolves the mineral part of the matrix. The liberated calcium ions (Ca^{2+}) and phosphate ions (PO_4^{3-}) enter the tissue fluid and the bloodstream.

Bone-forming osteoblasts derive from mesenchyme cells or, in the adult, from undifferentiated cells that retain mesenchyme-like properties. The origin of osteoclasts, on the other hand, has long been controversial. Recent studies indicate that osteoclasts arise from immature blood cells called *hematopoietic stem cells* (p. 455), and that they may be related to macrophages. Many stem cells fuse together to form each osteoclast.

Control of Remodeling

Why is the bone of the skeleton remodeled so extensively?

First, bone remodeling helps maintain constant concentrations of Ca^{2+} and PO_4^{3-} in body fluids. When the concentration of Ca^{2+} in body fluids starts to fall, a hormone is released by the parathyroid (par″ah-thi′roid) glands of the neck. This *parathyroid hormone* stimulates osteoclasts to reabsorb bone, a process that releases more Ca^{2+} into the blood.

Second, bone is remodeled in response to the mechanical stress it experiences. For example, both the osteons of compact bone and the trabeculae of spongy bone are constantly replaced by new osteons and trabeculae that are more precisely parallel to newly experienced compressive stresses. Furthermore, bone grows thicker in response to the forces of exercise and the force of gravity.

Given that bones grow stronger with use, can you deduce the effects of disuse on bones?

In the absence of mechanical stress, bone tissue is lost. This fact explains why the bones of bedridden people atrophy. A loss of bone under low-gravity conditions is the main obstacle to long missions in outer space. To slow the loss of bone, astronauts perform isometric (weight-bearing) exercises during space missions. ∎

Repair of Bone Fractures

Despite their strength, bones are susceptible to **fractures,** or breaks. In young people, most fractures result from trauma (sports injuries, falls, and car accidents, for example) that twists or smashes the bones. In old age, bones thin and weaken, and fractures occur more often. The common types of fractures are explained in Table 6.1.

A fracture is treated by **reduction,** the realignment of the broken bone ends. In **closed reduction,** the bone ends are coaxed back into their normal position by the physician's hands. In **open reduction,** the bone ends are secured together surgically with pins or wires. After the broken bone is reduced, it is immobilized by a cast or traction to allow the healing process to begin. Healing time is about 6 to 8 weeks for a simple fracture, but it is longer for large, weight-bearing bones and for the bones of elderly people.

The healing of a simple fracture involves several phases (Figure 6.12).

1. Hematoma formation. The fracture is usually accompanied by hemorrhaging. Blood vessels break both in the periosteum and inside the bone, releasing blood that clots to form a hematoma. The stages of inflammation, described in Chapter 4 (p. 91), are evident in and around the clot.

2. Fibrocartilaginous callus formation. Within a few days, new blood vessels grow into the clot. The periosteum and endosteum near the fracture site show a proliferation of bone-forming cells. These cells invade the clot, filling it with repair tissue called *soft callus* (kal′us; "hard skin"). Initially, the soft callus is a fibrous granulation tissue (p. 94). This tissue then transforms into a dense connective tissue containing fibrocartilage and hyaline cartilage. At this point, the soft callus is also called a **fibrocartilaginous callus.**

3. Bony callus formation. Within a week, trabeculae of new bone begin to form in the callus, mostly by endochondral ossification. These trabeculae span the width of the callus and unite the two fragments of the broken bone. The callus is now called a **bony callus,** or *hard callus,* and its trabeculae grow progressively thicker and stronger and become firm by about 2 months after the injury.

Table 6.1 Common Types of Fractures

Fracture type	Illustration	Description	Comments
Simple	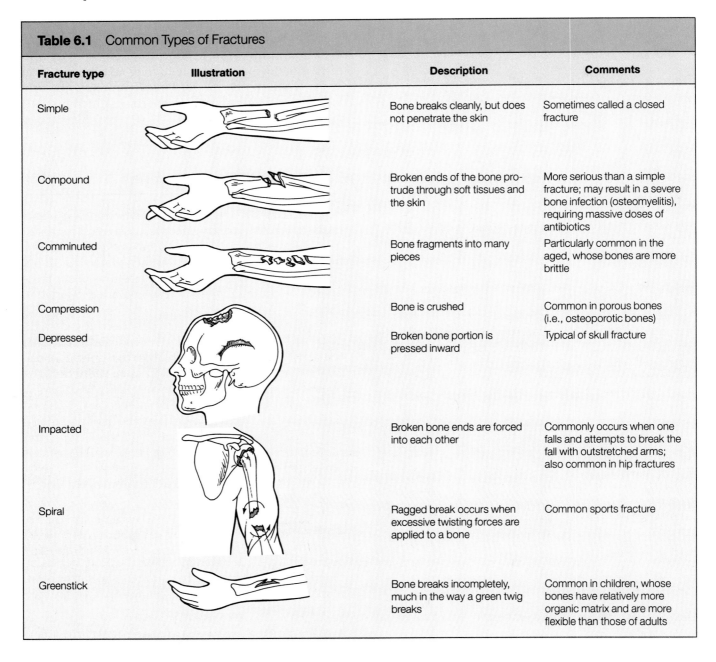	Bone breaks cleanly, but does not penetrate the skin	Sometimes called a closed fracture
Compound		Broken ends of the bone protrude through soft tissues and the skin	More serious than a simple fracture; may result in a severe bone infection (osteomyelitis), requiring massive doses of antibiotics
Comminuted		Bone fragments into many pieces	Particularly common in the aged, whose bones are more brittle
Compression		Bone is crushed	Common in porous bones (i.e., osteoporotic bones)
Depressed		Broken bone portion is pressed inward	Typical of skull fracture
Impacted		Broken bone ends are forced into each other	Commonly occurs when one falls and attempts to break the fall with outstretched arms; also common in hip fractures
Spiral		Ragged break occurs when excessive twisting forces are applied to a bone	Common sports fracture
Greenstick		Bone breaks incompletely, much in the way a green twig breaks	Common in children, whose bones have relatively more organic matrix and are more flexible than those of adults

4. Bone remodeling. Over a period of many months, the bony callus is remodeled. The excess bony material is removed from the exterior of the bone shaft and the interior of the medullary cavity. Compact bone is laid down to reconstruct the shaft walls. The repaired area resembles the original unbroken bone region, since it is responding to the same set of mechanical stresses. ■

✚ Bone Diseases

Osteoporosis

Osteoporosis (os″te-o-po-ro′sis; "bone-porous-condition") refers to a group of diseases in which bone reabsorption outpaces bone deposition. Osteoporotic bones become porous and light. The compact bone looks thinner and less dense than normal, and the spongy bone has fewer trabeculae (Figure 6.13). Although the mass of the bone declines, the chemical composition of the bone matrix remains normal. The loss of bone mass often leads to fractures. Even though osteoporosis affects the whole skeleton, the vertebral column is most vulnerable, and compression fractures of the vertebrae are frequent. The femur, especially its neck near the hip joint, is also very susceptible to fracture. Such a fracture, called a broken hip, is a common problem in people with osteoporosis.

Osteoporosis occurs most often in the aged, particularly in women who have gone through menopause. The hormone estrogen helps to maintain the health and normal density of a woman's skeleton.

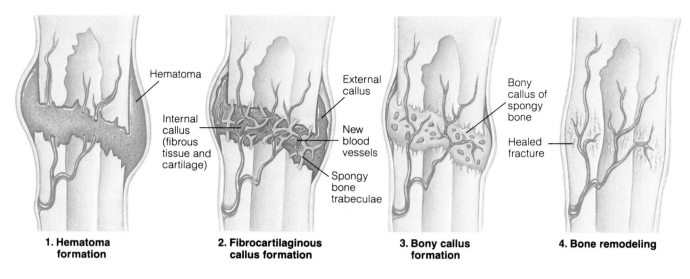

1. **Hematoma formation**

2. **Fibrocartilaginous callus formation**

3. **Bony callus formation**

4. **Bone remodeling**

Figure 6.12   Stages in the healing of a bone fracture.

After menopause, however, the secretion of estrogen wanes, and estrogen deficiency is strongly implicated in osteoporosis in older women. Additional factors that may contribute to age-related osteoporosis include insufficient exercise to stress the bones, a diet poor in calcium and protein, and many other factors. Osteoporosis can develop at any age in an immobile or bedridden person. The condition is usually treated by supplemental calcium and vitamin D, increased exercise, and in selected cases, estrogen replacement. Frustratingly, these treatments can only slow the loss of bone, not reverse it. A new drug called etidronate, which suppresses the activity of osteoclasts, shows promise in reversing osteoporosis in the vertebral column.

Osteomalacia and Rickets

The term **osteomalacia** (os″te-o-mah-la′she-ah), literally "soft bones," applies to a number of disorders in adults in which the bones are inadequately mineralized. The osteoid matrix is produced, but calcification does not occur, so the bones soften and weaken. The lower limbs bow as their bones are deformed by body weight. The main symptom is pain when weight is put on the affected bone.

Rickets, the analogous disease in children, is accompanied by many of the same signs and symptoms. Since the bones are still growing rapidly, however, rickets is more severe than adult osteomalacia. Along with bowed legs, malformities of the child's head and rib cage are common. Because the epiphyseal plates cannot be replaced with calcified bone, they grow abnormally thick, and the epiphyses of the long bones become abnormally long.

Osteomalacia and rickets are caused by inadequate amounts of vitamin D or calcium phosphate in the diet. The role of vitamin D in the uptake of calcium from food is discussed on p. 104 of Chapter 5.

Can you guess how osteomalacia and rickets are usually cured? Additionally, can you deduce why repeated pregnancies may cause a woman to develop osteomalacia?

Drinking vitamin D-fortified milk and exposing the skin to sunlight usually provides a full cure for osteomalacia and rickets.

Pregnancy leads to a transfer of large amounts of calcium from a woman's skeleton to the growing skeleton of her fetus. Therefore, it is especially important for women to eat a calcium-rich diet during pregnancy (and between pregnancies as well). ■

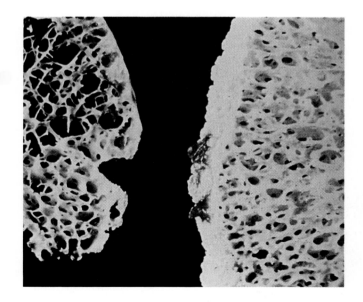

Figure 6.13   Osteoporosis: The contrasting architecture of osteoporotic bone (left) and normal bone (right).

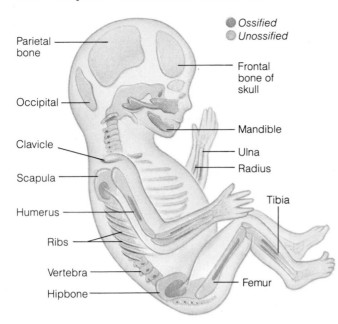

● *Ossified*
● *Unossified*

Parietal bone

Frontal bone of skull

Occipital

Mandible

Clavicle

Ulna

Radius

Scapula

Tibia

Humerus

Ribs

Vertebra

Hipbone

Femur

Figure 6.14 Primary ossification centers in the skeleton of a 10-week-old fetus.

Paget's Disease

Paget's (paj′ets) **disease** is characterized by excessive rates of bone deposition and bone reabsorption. The newly formed bone, called *Pagetic bone,* has an abnormally high ratio of immature woven bone to mature compact bone. This, along with reduced mineralization, makes the bones soft and weak. Late in the course of the disease, the activity of osteoclasts wanes while osteoblasts stay active. Therefore, the bones can thicken in an irregular manner, and the medullary cavity may become filled with bone. Paget's disease may affect many parts of the skeleton but is usually a localized and intermittent condition. The condition rarely occurs before the age of 40 years and affects about 3% of all elderly people. It progresses slowly, often produces no symptoms, and is seldom life-threatening. Its cause is unknown, but it may be initiated by a virus. Its treatment is complex and beyond the scope of this book.

Osteomyelitis

Bacterial infection of the bone and bone marrow is called **osteomyelitis** (os″te-o-mi-ĕ-li′tis; "bone and marrow inflammation"). It can enter bones from infections in surrounding tissues or through the bloodstream, or it can follow a compound bone fracture. Bone infections are difficult to treat, even with antibiotics, and recovery can be slow and painful.

Osteosarcoma

Given what you have learned about the terminology of neoplasms, you can deduce what an **osteosarcoma** is. Recall that a *sarcoma* is any cancer arising from a connective tissue cell or muscle cell and that *osteo* means "bone." Therefore, osteosarcoma is a form of bone cancer.

Osteosarcoma primarily affects young people between 10 and 25 years of age. It usually originates in a long bone of the upper or lower limb, with 50% of cases arising near the knee. The cancer cells derive from osteoblast-like cells of mesenchymal origin in the parts of the diaphyses nearest the epiphyses. Secreting osteoid and growing quickly, the tumor alters the affected bone by eroding the medullary cavity internally and the compact bone externally. The tumor metastasizes to the lungs and other organs, and most deaths result from the secondary tumors in the lungs. Most osteosarcomas are recognized by the pain and the visible swelling they produce in the affected bone, and the diagnosis is confirmed by X rays or other medical imaging techniques. Treatment is by amputation of the bone or limb, followed by chemotherapy and surgical removal of any metastases in the lung. The survival rate is about 50% if the disease is detected early. ■

The Skeleton Throughout Life

It can be said that the bones are on a timetable from the time they form until death. The mesoderm germ layer gives rise to embryonic mesenchyme cells, which in turn produce the membranes and the cartilages that form most of the embryonic skeleton. These structures then ossify according to a predictable schedule. Although each bone of the skeleton has its own developmental schedule, most long bones start ossifying by week 8 and have obvious primary ossification centers by week 12 (Figure 6.14). So precise is the ossification timetable that the age of a fetus in the womb can be determined from X-ray films or sonograms of the fetal skeleton.

At birth, all bones are relatively smooth and featureless. However, as the child uses its muscles more and more, the bones develop projections and other markings. Children's bones are not particularly weak, but the cartilage of their epiphyseal plates is not as strong as bone. Thus, childhood injuries often knock the epiphyses off the diaphysis. To treat such injuries, the bone parts are manipulated back into place, then stabilized with metal pins. As we mentioned earlier, the skeleton keeps growing until the age of 18 to 21 years. In children and adolescents, the rate of bone formation exceeds the rate of bone reabsorption. In young adults, these two processes are in balance. In old age, reabsorption predominates. Beginning in the fourth decade of life, the mass of both compact and spongy bone declines. Among young adults, skeletal mass is generally greater in males than in females, and in blacks than in whites. With age, the rate of bone loss is faster in whites than in blacks, and in females than in males.

As bone mass declines with age, other changes occur. An increasing number of osteons remain incompletely formed, and mineralization is less complete. There is also an increased amount of non-living compact bone, reflecting a diminished blood supply to the bones in old age.

✚ Related Clinical Terms

Achondroplasia (a-kon″dro-pla′ze-ah) (*a* = without; *chondro* = cartilage; *plasia* = shape)   A congenital condition involving defective growth of cartilage and defective endochondral ossification. Achondroplasia results in typical dwarfism, in which the limbs are short but the membrane bones are of normal size.

Bone graft   Transplanting a piece of bone from one part of a person's skeleton to another part where bone has been damaged or lost. The graft, often taken from the crest of the iliac bone of the hip, encourages regrowth of lost bone.

Bony spur   An abnormal projection from a bone due to bone overgrowth. Common in aging bones.

Metastatic calcification   Deposit of bone salts in those tissues that do not normally calcify. The salts precipitate as a result of abnormally high concentrations of calcium in the body fluids that in turn can result from an excessive secretion of parathyroid hormone.

Ostealgia (os″te-al′je-ah) (*algea* = pain)   Pain in a bone.

Pathologic fracture   Fracture occurring in a diseased bone and involving slight or no physical trauma. An example is a broken hip caused by osteoporosis: The hip breaks first, causing the person to fall.

Traction ("pulling")   Placing a sustained tension on a region of the body in order to keep the parts of a fractured bone in the proper alignment. The traction also keeps the bone immobilized as it heals. A broken femur is often elevated and placed in traction, with the patient confined to bed until the fracture heals. Without such traction, strong spasms of the large muscles of the thigh would separate the fracture. Traction is also used to immobilize fractures of the vertebral column, since any movement there could crush the spinal cord.

Chapter Summary

CARTILAGES (pp. 118–121)

Location and Basic Structure (p. 118)
1. Important cartilages in the adult body are articular cartilages, costal cartilages, cartilages of the respiratory tubes, the epiglottis, and cartilage in intervertebral discs. Cartilage makes up most of the embryonic skeleton.

2. Perichondrium is the girdle of dense connective tissue that surrounds a piece of cartilage.

3. Cartilage is resilient: Water squeezed out of its matrix by compression rushes back in as the compression eases, causing the cartilage to spring back.

4. Growing cartilage enlarges quickly because the small amount of matrix it manufactures attracts a large volume of water. Cartilage is avascular and is weak in resisting shear stress.

Classification of Cartilage (p. 118)
5. Locations of hyaline, elastic, and fibrocartilage are shown in Figure 6.1.

6. Hyaline cartilage is the most common type. Its matrix contains fine collagen fibrils. Elastic cartilage resembles hyaline cartilage, but its matrix contains elastic fibers that make it pliable. Fibrocartilage contains thick collagen fibers and can resist both compression and extreme tension.

Growth of Cartilage (pp. 119–121)
7. Cartilages grow from within (interstitial growth) and by the addition of new cartilage tissue at the periphery (appositional growth). In adults, damaged cartilage regenerates poorly. In the growing and aged skeleton, cartilage calcifies.

BONES (pp. 121–134)
1. A bone in the skeleton, such as the femur, is an organ.

Functions of Bones (p. 121)
2. Bones protect and support soft organs, serve as levers for muscles to pull on, store calcium, and contain bone marrow that manufactures blood cells.

Classification of Bones (pp. 121–123)
3. Bones are classified on the basis of their shape as long, short, flat, or irregular.

Gross Anatomy of Bones (pp. 123–124)
4. Bone organs have an external layer of compact bone and are filled internally with spongy bone, in which trabeculae are arranged in networks.

5. A long bone is composed of a diaphysis (shaft) and epiphyses (ends). Epiphyseal vessels supply each epiphysis, and nutrient vessels supply the diaphysis. Bone marrow occurs within the spongy bone and in a central medullary (marrow) cavity. A periosteum covers the outer surface of each bone, whereas an endosteum covers the inner bone surfaces.

6. Flat bones consist of two plates of compact bone separated by a layer of spongy bone.

7. Both the density of bone material and the magnitude of bending stresses decline from the superficial to the deep regions of the bones. The spaces within bones lighten the skeleton and hold bone marrow.

8. The trabeculae of spongy bone are arranged along the dominant lines of stress experienced by the bone.

Microscopic Structure of Bone (pp. 124–125)

9. An important structural unit in compact bone is the osteon, a pillar consisting of a central canal surrounded by concentric lamellae. Osteocytes, embedded in lacunae, are connected to each other and to the central canal by canaliculi.

10. Bone lamellae are sheets of bone matrix. The collagen fibers in adjacent lamellae run in different directions. This arrangement gives bone tissue great strength in resisting torsion (twisting).

Chemical Composition of Bone Tissue (pp. 125–127)

11. Bone consists of cells (osteocytes, osteoblasts, and osteoclasts) plus an extracellular matrix. The matrix contains organic substances secreted by osteoblasts, including collagen, which gives bone tensile strength. Crystals of calcium phosphate salts, which precipitate in this matrix, make bone hard and able to resist compression.

Bone Development (pp. 127–129)

12. Flat bones of the skull form by intramembranous ossification of embryonic mesenchyme. A network of bone tissue woven around capillaries appears first and is then remodeled into a flat bone.

13. Before it mineralizes, the organic part of bone matrix is called osteoid.

14. Most bones develop by endochondral ossification of a hyaline cartilage model, starting in the late embryonic period (week 8). The stages of development of a long bone are (1) formation of a bone collar around the diaphysis; (2) calcification of cartilage in the center of the diaphysis; (3) growth of a periosteal bud into the center of the shaft, where the first bone trabeculae form; (4) organization of the cartilage in the epiphysis for efficient growth; and (5) ossification in the epiphyses. The primary ossification center is in the diaphysis. The secondary ossification centers are in the epiphyses.

Postnatal Growth of Endochondral Bones (pp. 129–130)

15. Endochondral bones lengthen during youth through the growth of the epiphyseal plate cartilages, which close in early adulthood.

16. Bones increase in width through appositional growth.

Bone Remodeling (pp. 130–131)

17. New bone tissue is continuously deposited and reabsorbed in response to hormonal and mechanical stimuli. Together, these processes are called bone remodeling.

18. Osteoid is secreted by osteoblasts at areas of bone deposit. Calcium salt is then deposited in the osteoid.

19. Osteoclasts break down bone tissue by secreting digestive enzymes and acid onto bone surfaces. This process releases Ca^{2+} and PO_4^{3-} into the blood. Parathyroid hormone stimulates this reabsorption of bone.

20. Compressive forces and gravity acting on the skeleton help maintain bone strength, as bones thicken at sites of stress.

Repair of Bone Fractures (pp. 131–132)

21. Fractures are treated by open or closed reduction. Healing involves the formation of a hematoma, a fibrocartilaginous callus, then a bony callus, and a remodeling of the callus into the original bone pattern.

Bone Diseases (pp. 132–134)

22. Osteoporosis is any condition in which bone breakdown outpaces bone formation, causing bones to weaken. Postmenopausal women are most susceptible.

23. Osteomalacia and rickets occur when bones are inadequately mineralized. The bones become soft and deformed. The most common cause is inadequate intake of vitamin D.

24. Paget's disease is characterized by excessive and abnormal remodeling of bone.

25. Osteomyelitis is bacterial infection of a bone and its marrow. Osteosarcoma is the most common form of bone cancer.

THE SKELETON THROUGHOUT LIFE (pp. 134–135)

1. Bone formation in the fetus occurs in a predictable and precisely timed manner.

2. The mass of the skeleton increases dramatically during puberty and adolescence, when bone formation exceeds reabsorption.

3. Bone mass is constant in young adulthood, but beginning in the 40s, bone reabsorption exceeds formation.

Review Questions

Multiple Choice and Matching Questions

1. Which is a function of the skeletal system? (a) support, (b) blood cell formation, (c) mineral storage, (d) providing levers for muscle activity, (e) all of these.

2. Articular cartilages are located (a) at the ends of bones, (b) between the ribs and the sternum (breastbone), (c) between the epiphysis and diaphysis, (d) in the nose.

3. Cartilages relate to their perichondrium just as bones relate to their (a) articular cartilages, (b) spongy bone layer, (c) osteons, (d) marrow, (e) periosteum.

4. Bone versus cartilage tissue. Indicate whether each of the following statements is true (T) or false (F).

_____ **(1)** Cartilage is more resilient than bone.

_____ **(2)** Cartilage is especially strong in resisting shear (bending and twisting) forces.

_____ **(3)** Cartilage can grow faster than bone in the growing skeleton.

_____ **(4)** In the adult skeleton, cartilage regenerates faster than bone when damaged.

_____ **(5)** Neither bone nor cartilage contains capillaries.

_____ **(6)** Bone tissue contains very little water compared to other connective tissues, while cartilage tissue contains a large amount of water.

_____ **(7)** Nutrients diffuse quickly through cartilage matrix but very poorly through the solid bone matrix.

5. A bone that has essentially the same width, length, and height is most likely (a) a long bone, (b) a short bone, (c) a flat bone, (d) an irregular bone.

6. The shaft of a long bone is properly called the (a) epiphysis, (b) periosteum, (c) diaphysis, (d) compact bone.

7. The osteon exhibits (a) a central canal containing blood vessels, (b) concentric lamellae of matrix, (c) osteocytes in lacunae, (d) all of these.

8. The organic components of bone matrix are important in providing all but (a) tensile strength, (b) hardness, (c) ability to resist twisting, (d) flexibility.

9. The flat bones of the skull develop from (a) areolar tissue, (b) hyaline cartilage, (c) mesenchyme membranes, (d) compact bone.

10. The following events apply to the endochondral ossification process as it occurs in the primary ossification center. Put these events in their proper order by assigning each a number (1–6).

_____ **(a)** Cartilage in the diaphysis calcifies and contains cavities.

_____ **(b)** A collar of bone is laid down around the hyaline cartilage model just deep to the periosteum.

_____ **(c)** Periosteal bud invades the center of the diaphysis.

_____ **(d)** Perichondrium becomes more vascularized and becomes a periosteum.

_____ **(e)** Osteoblasts first deposit bone tissue around the cartilage spicules within the diaphysis.

_____ **(f)** Osteoclasts remove the bone from the center of the diaphysis, leaving a medullary cavity that then houses marrow.

11. Where within an epiphyseal plate are dividing chondrocytes located? (a) nearest the diaphysis, (b) in the medullary cavity, (c) farthest from the diaphysis, (d) in the primary ossification center, (e) all of these.

12. The remodeling of bone tissue is a function of which cells? (a) chondrocytes and osteocytes, (b) osteoblasts and osteoclasts, (c) chondroblasts and osteoclasts, (d) osteoblasts and osteocytes.

13. Bone growth during childhood and in adults is regulated and directed by (a) growth hormone, (b) thyroid hormone, (c) sex hormones, (d) mechanical stress, (e) all of these.

14. In fracture repair, formation of a bony callus is followed by (a) hematoma formation, (b) first appearance of the soft callus, (c) bone remodeling, (d) formation of granulation tissue.

15. A fracture in which the bone ends penetrate soft tissue and skin is (a) greenstick, (b) compound, (c) simple, (d) comminuted, (e) compression.

16. The disorder in which bones are porous and thin but the chemistry of the bone matrix remains normal is (a) osteomalacia, (b) osteoporosis, (c) osteomyelitis, (d) Paget's disease.

Short Answer and Essay Questions

17. Explain (a) why cartilages are resilient and (b) why cartilage can grow so quickly in the developing skeleton.

18. Compare compact and spongy bone in their macroscopic appearance (how they look to the unaided eye) and in their microscopic structure.

19. Some anatomy students are joking between classes, imagining what bone organs would look like if they had spongy bone on the outside and compact bone on the inside, instead of the other way around. You tell the students that such imaginary bones would be of poor mechanical design and would break very easily. Explain why this is so.

20. As we grow, our long bones increase in diameter, but the thickness of the compact bone of the shaft remains relatively constant. Explain this phenomenon.

21. When and why do the epiphyseal plates close?

22. (a) During what period of life does skeletal mass increase dramatically? Begin to decline? (b) Why are fractures most common in elderly people?

Critical Thinking and Clinical Application Questions

23. Following a motorcycle accident, 22-year-old Ruby was rushed to the emergency room. X-ray films revealed a spiral fracture of her right tibia (main bone of the leg). When more X rays were taken 2 months later, she was told that bony callus was forming normally. What is bony callus?

24. A father brought his 4-year-old daughter to the doctor, complaining the the child didn't "look right." Her forehead was enlarged, her rib cage was knobby, and her lower limbs were bent and deformed. X-ray films revealed very thick epiphyseal plates. The parent was advised to have the child drink milk fortified with vitamin D and to send her outdoors to play in the sun more often. Explain these instructions, and tell what disease the child had.

25. Explain why people in wheelchairs with paralyzed lower limbs have thin, weak bones in their thighs and legs.

26. While walking home from a meeting of his adult singles support group, Ike broke a bone and damaged a knee cartilage in a fall. Assuming no special tissue grafts are made, which will probably heal faster, the bone or the cartilage? Why?

27. Jock went to weight-lifting camp in the summer between seventh and eighth grade. He noticed that the camp trainer put tremendous pressure on him and his friends to improve their strength. After an especially vigorous workout, Jock's arm felt extremely sore and weak around the elbow. He went to the camp doctor, who took X rays and then told him that the injury was serious, for the "end of his upper arm bone was starting to twist off." What had happened? Could the same thing happen to Jock's 23-year-old sister, Trixie, who was also starting a program of weight lifting?

7

Bones of the Skeleton

Student Objectives

1. Define the axial and appendicular skeletons, and describe their basic functions.
2. Describe the kinds of markings on bones.

PART 1: THE AXIAL SKELETON (pp. 142–166)

The Skull (pp. 142–156)

3. Name and describe the bones of the skull. Identify their important markings.
4. Compare the functions of the cranial and the facial skeleton.
5. Define the bony boundaries of the orbit, nasal cavity, and paranasal sinuses.

The Vertebral Column (pp. 156–164)

6. Describe the general structure of the vertebral column, and list its components.
7. Name a function performed by both the spinal curvatures and the intervertebral discs.
8. Discuss the structure of a typical vertebra, and describe the special features of cervical, thoracic, and lumbar vertebrae.

The Bony Thorax (pp. 164–166)

9. Describe the ribs and sternum.
10. Differentiate true from false ribs, and explain how they relate to floating ribs.

| **Chapter Outline** | **Student Objectives** |

PART 2: THE APPENDICULAR SKELETON (pp. 167–182)

The Pectoral Girdle (pp. 167–169)

11. Identify the bones that form the pectoral girdle, and explain their functions.

12. Identify the important bone markings on the pectoral girdle.

Bones of the Free Upper Limb (pp. 169–173)

13. Describe the bones of the arm, forearm, wrist, and hand.

The Pelvic Girdle (pp. 173–177)

14. Name the bones contributing to the hip bone, and relate the strength of the pelvic girdle to its function.

15. Compare and contrast the male and female pelvis.

Bones of the Free Lower Limb (pp. 177–182)

16. Identify the bones of the lower limb and their important markings.

17. Name the three supporting arches of the foot, and explain their importance.

PART 3: AXIAL AND APPENDICULAR SKELETON THROUGHOUT LIFE (pp. 182–185)

18. Describe how skeletal proportions change as we grow.

19. Compare and contrast the skeleton of an aged person with that of a young adult. Discuss age-related changes that may affect health.

The word **skeleton** comes from a Greek word meaning "dried-up body" or "mummy," a rather disparaging description. Nonetheless, the internal human framework is a triumph of design and engineering that would put any skyscraper to shame. The skeleton is strong yet light, and it is well adapted for the locomotor, protective, and manipulative functions it performs.

The skeleton accounts for about 20% of our body weight. It consists of *bones, cartilages, joints,* and *ligaments.* Bones make up most of the skeleton. Joints, also called *articulations,* are the junctions between the bones. Ligaments connect bones and reinforce most joints. Joints and ligaments are considered in detail in Chapter 8.

For descriptive purposes, the 206 named bones of the human skeleton are grouped into the axial and appendicular skeletons (Figure 7.1). The **axial skeleton,** which forms the long axis of the body, includes the skull, vertebral column, and rib cage. The **appendicular** (ap"en-dik'u-lar) **skeleton** consists of the bones of the upper and lower limbs, including the girdles (bones of the shoulder and hip) that attach the limbs to the axial skeleton.

The outer surfaces of bones are rarely smooth and featureless. Instead, they display bulges, depressions, and holes called **bone markings.** These markings include (1) attachment sites for muscles, tendons, and ligaments; (2) the regions of bones that are near joint surfaces; and (3) tunnels for vessels and nerves. Bone markings provide a wealth of information about the function of bones and muscles and the relationship of bones to their associated soft structures. The main types of bone markings are described in Table 7.1. You should familiarize yourself with these, because you will encounter them as we describe the bones.

Table 7.1 Bone Markings

Name of bone marking	Description	Illustration
Projections that are sites of muscle and ligament attachment		
Tuberosity (too"bĕ-ros'ĭ-te)	Large rounded projection; may be roughened	
Crest	Narrow ridge of bone; usually prominent	
Trochanter (tro-kan'ter)	Very large, blunt, irregularly shaped process (The only examples are on the femur.)	
Line	Narrow ridge of bone; less prominent than a crest	
Tubercle (too'ber-kl)	Small rounded projection or process	
Epicondyle (ep"ĭ-kon'dīl)	Raised area on or above a condyle	
Spine	Sharp, slender, often pointed projection	
Projections that help to form joints		
Head	Bony expansion carried on a narrow neck	
Facet	Smooth, nearly flat articular surface	
Condyle (kon'dīl)	Rounded articular projection	
Ramus (ra'mus)	Arm-like bar of bone	
Depressions and openings allowing blood vessels and nerves to pass		
Meatus (me-a'tus)	Canal-like passageway	
Sinus	Cavity within a bone, filled with air and lined with mucous membrane	
Fossa (fos'ah)	Shallow, basin-like depression in a bone, often serving as an articular surface	
Groove	Furrow	
Fissure	Narrow, slit-like opening	
Foramen (fo-ra'men)	Round or oval opening through a bone	

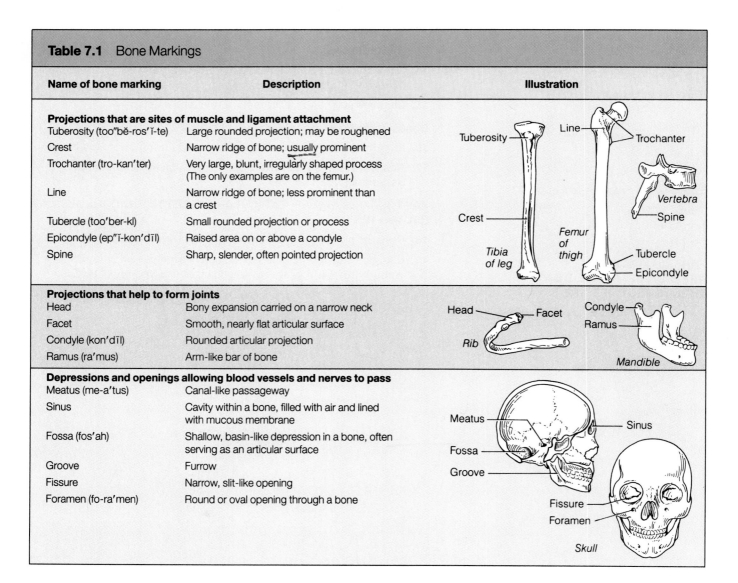

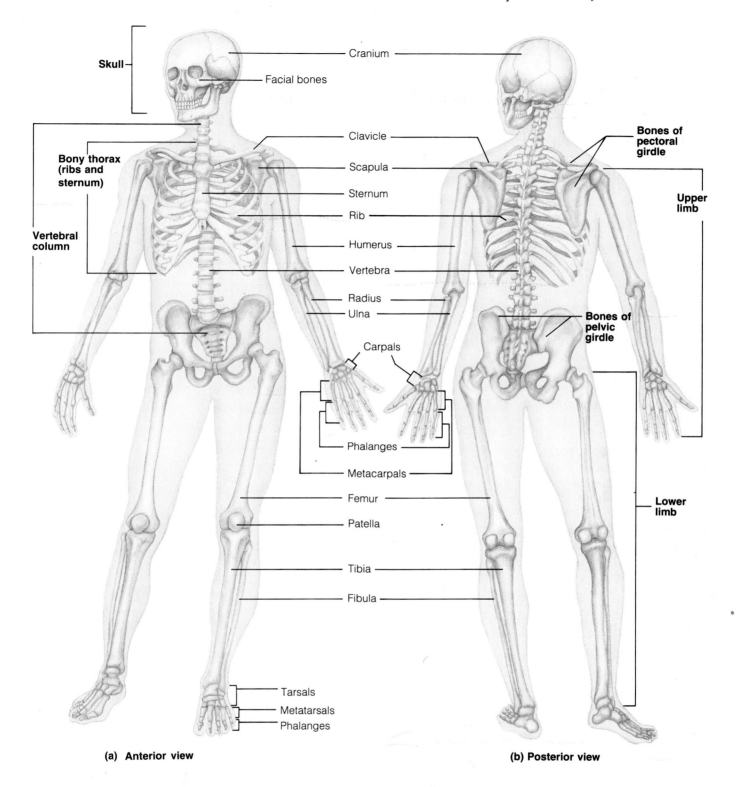

Skull
Cranium
Facial bones

Bony thorax (ribs and sternum)

Vertebral column

Clavicle
Scapula
Sternum
Rib
Humerus
Vertebra
Radius
Ulna

Bones of pectoral girdle

Upper limb

Bones of pelvic girdle

Carpals

Phalanges
Metacarpals

Femur
Patella

Lower limb

Tibia
Fibula

Tarsals
Metatarsals
Phalanges

(a) Anterior view

(b) Posterior view

Figure 7.1 The human skeleton. Bones of the axial skeleton are colored green. Bones of the appendicular skeleton are gold.

PART 1: THE AXIAL SKELETON

The axial skeleton contains 80 named bones, which are arranged into three major regions: the skull, vertebral column, and bony thorax (Figure 7.1). This part of the skeleton supports the head, neck, and trunk, and protects the brain, spinal cord, and the organs in the thorax.

The Skull

The **skull** is the body's most complex bony structure (Figures 7.2 to 7.4). It is formed by two sets of bones, the cranial bones and the facial bones. The *cranial bones,* or **cranium** (kra′ne-um), enclose and protect the brain and provide sites for the attachment of some head and neck muscles. The *facial bones* (1) form the framework of the face; (2) provide cavities for the sense organs of sight, taste, and smell; (3) provide openings for the passage of air and food; (4) hold the teeth; and (5) anchor the muscles of the face.

Most skull bones are flat bones and are firmly united by interlocking joints called **sutures** (soo′cherz; "seams"). The suture lines have an irregular, saw-toothed appearance that is very obvious on the external surfaces of the skull. The longest and most important skull sutures (the *coronal, sagittal, squamous,* and *lambdoidal sutures*) connect the cranial bones (Figure 7.3a). Most other skull sutures connect facial bones and are named according to the specific bones they connect.

Overview of Skull Geography

It is worth spending a few moments on basic skull "geography" before describing the individual bones. With the lower jaw removed, the skull resembles a lopsided, hollow, bony sphere. The facial bones form its anterior aspect, and the cranium forms the rest of the skull. The cranium can be divided into a *vault* and a *base.* The *cranial vault,* also called the *calvaria* (kal-va′re-ah; "bald part of skull") or skullcap, forms the superior, lateral, and posterior aspects of the skull, as well as the forehead region. The *cranial base* or *floor,* is the inferior portion of the skull. Internally, prominent bony ridges divide the skull base into three distinct "steps," or fossae—the anterior, middle, and posterior *cranial fossae* (Figure 7.4c). The brain sits snugly in these cranial fossae and is completely enclosed by the cranial vault. Overall, the brain is said to occupy the *cranial cavity.*

In addition to its large cranial cavity, the skull contains many smaller cavities. These include the middle ear and inner ear cavities (carved into the lateral aspects of the cranial base), plus the nasal cavity and the orbits. The nasal cavity lies in and posterior to the nose, and the orbits house the eyeballs. Air-filled sinuses occur within several bones around the nasal cavity. These are the paranasal sinuses.

Moreover, the skull has about 85 named openings (foramina, canals, fissures). The most important of these provide passageways for the spinal cord, the major blood vessels serving the brain, and the 12 pairs of *cranial nerves,* which conduct impulses to and from the brain. Cranial nerves are discussed in Chapter 13 (p. 369), but we should note here that they are classified by number, using the Roman numerals I–XII.

As you read about the skull, you should locate each bone in Figures 7.2 to 7.4. The skull bones and their markings are also summarized in Table 7.2 on pp. 154–155. The colored dot beside each bone's name in the table corresponds to the color of that bone in the figures.

Cranium

The eight large bones of the cranium are the paired parietal and temporal bones and the unpaired frontal, occipital, sphenoid, and ethmoid bones. Together these form the brain's protective "shell." Because its superior aspect is curved, the cranium is self-bracing. This allows the bones to be thin, and, like an eggshell, the cranium is remarkably strong for its weight.

Frontal Bone

The **frontal bone** forms the forehead. Just superior to the orbits, it protrudes slightly to form *superciliary* (soo″per-sil′e-a-re) *arches.* These "brow ridges" lie just deep to our eyebrows (*supercilium* = eyebrow). The **supraorbital margin** (superior margin of each orbit) is pierced by a **supraorbital foramen,** or *supraorbital notch* (Figure 7.2a). This opening transmits the supraorbital artery and nerve, which supply the forehead. The smooth part of the frontal bone between the superciliary arches in the midline is the **glabella** (glah-bel′ah; "smooth, without hair"). Just inferior to this, the frontal bone meets the nasal bones at the *frontonasal suture.* The regions of the frontal bone lateral to the glabella contain the air-filled **frontal sinuses** (Figures 7.3b).

The frontal bone also forms the roof of each orbit and contributes to the **anterior cranial fossa** (Figure 7.4b and 7.4c). This fossa holds the large frontal lobes of the brain.

(text continues on p. 146)

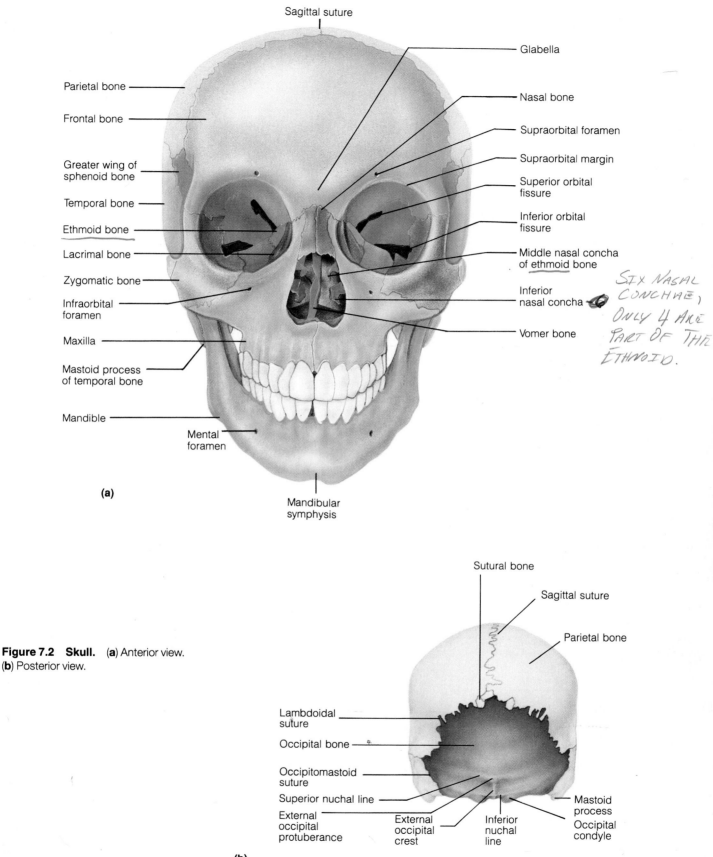

Sagittal suture

Glabella

Parietal bone

Frontal bone

Greater wing of
sphenoid bone

Temporal bone

Ethmoid bone

Lacrimal bone

Zygomatic bone

Infraorbital
foramen

Maxilla

Mastoid process
of temporal bone

Mandible

Mental
foramen

Nasal bone

Supraorbital foramen

Supraorbital margin

Superior orbital
fissure

Inferior orbital
fissure

Middle nasal concha
of ethmoid bone

Inferior
nasal concha

Vomer bone

SIX NASAL
CONCHAE,
ONLY 4 ARE
PART OF THE
ETHNOID.

(a)

Mandibular
symphysis

Figure 7.2   Skull.   (**a**) Anterior view.
(**b**) Posterior view.

Sutural bone

Sagittal suture

Parietal bone

Lambdoidal
suture

Occipital bone

Occipitomastoid
suture

Superior nuchal line

External
occipital
protuberance

External
occipital
crest

Inferior
nuchal
line

Mastoid
process

Occipital
condyle

(b)

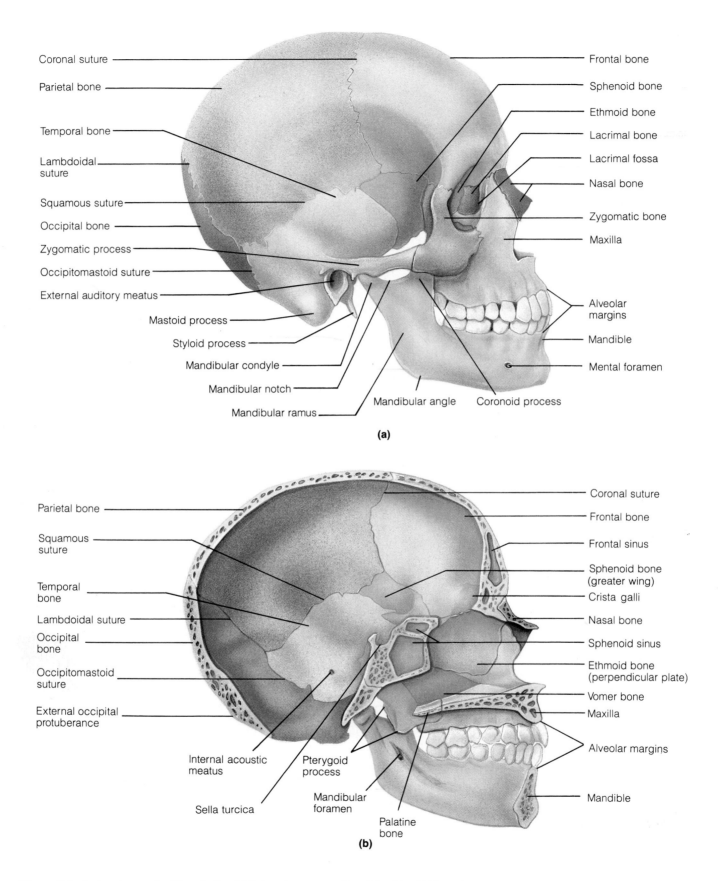

Coronal suture
Parietal bone
Temporal bone
Lambdoidal suture
Squamous suture
Occipital bone
Zygomatic process
Occipitomastoid suture
External auditory meatus
Mastoid process
Styloid process
Mandibular condyle
Mandibular notch
Mandibular ramus

Frontal bone
Sphenoid bone
Ethmoid bone
Lacrimal bone
Lacrimal fossa
Nasal bone
Zygomatic bone
Maxilla
Alveolar margins
Mandible
Mental foramen

Mandibular angle   Coronoid process

(a)

Parietal bone
Squamous suture
Temporal bone
Lambdoidal suture
Occipital bone
Occipitomastoid suture
External occipital protuberance

Internal acoustic meatus
Sella turcica
Pterygoid process
Mandibular foramen
Palatine bone

Coronal suture
Frontal bone
Frontal sinus
Sphenoid bone (greater wing)
Crista galli
Nasal bone
Sphenoid sinus
Ethmoid bone (perpendicular plate)
Vomer bone
Maxilla
Alveolar margins
Mandible

(b)

Figure 7.3   Lateral aspect of the skull. (**a**) External anatomy of the right side. (**b**) View of a skull cut through the midline, showing the internal anatomy of the left half.

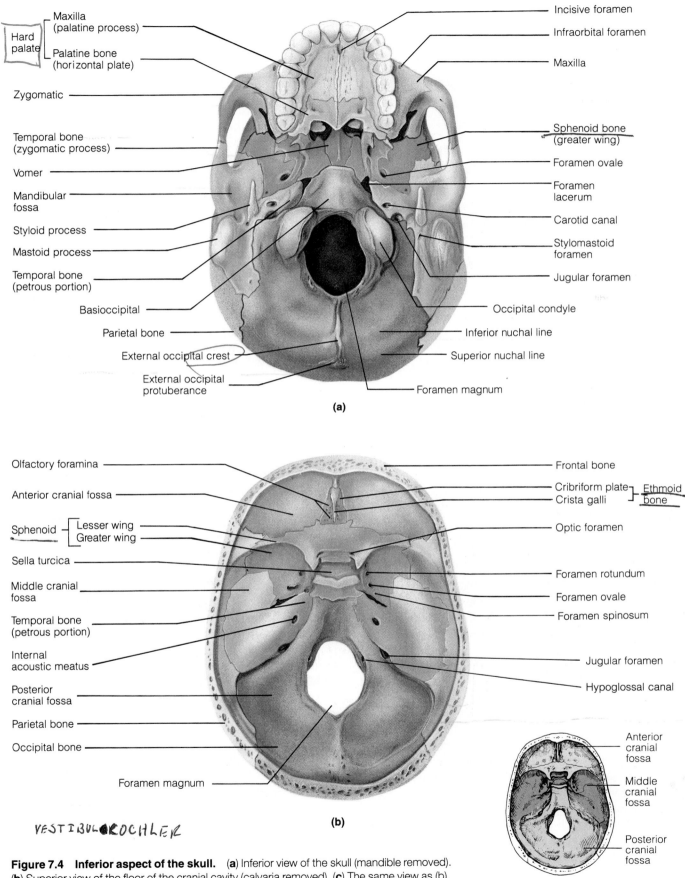

Maxilla (palatine process)
Hard palate
Palatine bone (horizontal plate)
Zygomatic
Temporal bone (zygomatic process)
Vomer
Mandibular fossa
Styloid process
Mastoid process
Temporal bone (petrous portion)
Basioccipital
Parietal bone
External occipital crest
External occipital protuberance

Incisive foramen
Infraorbital foramen
Maxilla
Sphenoid bone (greater wing)
Foramen ovale
Foramen lacerum
Carotid canal
Stylomastoid foramen
Jugular foramen
Occipital condyle
Inferior nuchal line
Superior nuchal line
Foramen magnum

(a)

Olfactory foramina
Anterior cranial fossa
Sphenoid — Lesser wing / Greater wing
Sella turcica
Middle cranial fossa
Temporal bone (petrous portion)
Internal acoustic meatus
Posterior cranial fossa
Parietal bone
Occipital bone
Foramen magnum

Frontal bone
Cribriform plate ⎤ Ethmoid
Crista galli ⎦ bone
Optic foramen
Foramen rotundum
Foramen ovale
Foramen spinosum
Jugular foramen
Hypoglossal canal

VESTIBULOCOCHLER

(b)

Anterior cranial fossa
Middle cranial fossa
Posterior cranial fossa

(c)

Figure 7.4   Inferior aspect of the skull.   **(a)** Inferior view of the skull (mandible removed). **(b)** Superior view of the floor of the cranial cavity (calvaria removed). **(c)** The same view as (b), emphasizing the three cranial fossae. The anterior fossa lies superior to the middle fossa, which lies superior to the posterior fossa.

Parietal Bones and the Major Sutures

The two large **parietal bones,** which are shaped like curved rectangles, make up the bulk of the cranial vault; that is, they form most of the superior part of the skull, as well as its lateral walls (*parietal* = wall). The four largest sutures occur where the parietal bones articulate (form a joint) with other cranial bones.

1. The **coronal suture,** running in the coronal plane, occurs where the parietal bones meet the frontal bone anteriorly (Figure 7.3).

2. The **squamous suture** occurs where each parietal bone meets a temporal bone inferiorly, on the lateral side of the skull (Figure 7.3).

3. The **sagittal suture** occurs where the two parietal bones meet superiorly in the midline of the cranium (Figure 7.2b).

4. The **lambdoid suture** (also called **lambdoidal suture**) occurs where the parietal bones meet the occipital bone posteriorly (Figure 7.2b). This suture is so named because of its resemblance to the Greek letter lambda (Λ).

Sutural Bones

Sutural bones are small bones that occur *within* the sutures, especially within the lambdoidal suture (Figure 7.2b). These bones are irregular in shape, size, and location, and not all people have them. Functionally insignificant, they develop between the major cranial bones during the fetal period.

Occipital Bone

The **occipital bone** (ok-sip′ĭ-tal; "back of the head") makes up the posterior wall of the cranium and cranial base. It articulates with the parietal bones at the lambdoidal suture and with the temporal bones at the *occipitomastoid sutures* (Figure 7.3). Internally, the occipital bone forms the walls of the **posterior cranial fossa** (Figure 7.4b and 7.4c), which holds a large part of the brain called the cerebellum. In the base of the occipital bone is the **foramen magnum,** literally "large hole." Through this opening, the inferior part of the brain connects with the spinal cord. The foramen magnum is flanked laterally by two rocker-like **occipital condyles** (Figure 7.4a). These condyles articulate with the first vertebra of the vertebral column in a way that enables the head to nod. Hidden medial and superior to each occipital condyle is a **hypoglossal** (hi″po-glos′al) **canal,** through which runs cranial nerve XII (the hypoglossal nerve). Anterior to the foramen magnum, the occipital bone joins the sphenoid bone via a band of bone called the **basioccipital** (ba″se-ok-sip′ĭ-tal; "base of the occipital").

Several markings occur on the external surface of the occipital bone (Figures 7.2b and 7.4a). The **external occipital protuberance** is a knob in the midline, at the junction of the base and the posterior wall of the skull. The *external occipital crest* extends anteriorly from the protuberance to the foramen magnum. This crest secures the *ligamentum nuchae* (nu′ke; "of the neck"), a sheet-like ligament that lies in the median plane of the posterior neck and connects the neck vertebrae to the skull. Extending laterally from the occipital protuberance are the **superior nuchal** (nu′kal) **lines,** and running laterally from the occipital crest are the **inferior nuchal lines.** The nuchal lines and the bony regions between them anchor many muscles of the neck and back. The superior nuchal line marks the upper limit of the neck.

Temporal Bones

The **temporal bones** are best viewed on the lateral surface of the skull (Figure 7.5). They lie inferior to the parietal bones and form the inferolateral region of the skull and parts of the cranial floor. The terms *temporal* and *temple,* from the Latin word for "time," refer to the fact that gray hairs, a sign of time's passing, appear first in the temple areas.

Each temporal bone has an intricate shape and is described in terms of its four major regions: the *squamous, tympanic, mastoid,* and *petrous regions.* The plate-shaped **squamous region** (Figure 7.5) abuts the squamous suture. It has a bar-like **zygomatic process** (zi″go-mat′ik; "cheek") that projects anteriorly to meet the zygomatic bone of the face. Together, these two bony structures form the *zygomatic arch,* which defines the projection of the cheek. The small, oval **mandibular** (man-dib′u-lar) **fossa** on the inferior surface of the zygomatic process receives the mandible (lower jawbone), forming the freely movable *temporomandibular joint* (jaw joint).

The **tympanic region** (tim-pan′ik; "eardrum") of the temporal bone surrounds the **external acoustic meatus** (also called the **external auditory meatus**), or external ear canal. It is through this canal that sound enters the ear. The external acoustic meatus and the tympanic membrane (eardrum) at its deep end are part of the external ear structures. In a dried skull, the tympanic membrane has been removed. Thus, part of the *middle ear cavity* deep to the tympanic region may be visible through the meatus. Projecting inferiorly from the tympanic region is the needle-like **styloid process** (sti′loid; "stake-like"). This process is an attachment point for some muscles of the tongue and pharynx and for a ligament that connects the skull to the hyoid bone of the neck.

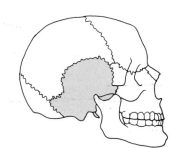

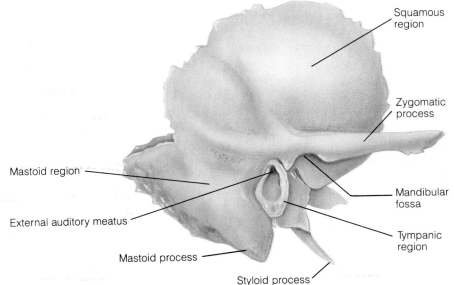

Squamous region

Zygomatic process

Mastoid region

Mandibular fossa

External auditory meatus

Tympanic region

Mastoid process

Styloid process

Figure 7.5   The temporal bone.   Right superficial view.

The **mastoid region** (mas'toid; "breast-shaped") of the temporal bone exhibits a prominent **mastoid process,** an anchoring site for some neck muscles. The mastoid processes can be felt as a lump just posterior to each ear. The **stylomastoid foramen** (Figure 7.4a) is located between the styloid and mastoid processes. Cranial nerve VII (the facial nerve) leaves the skull through the stylomastoid foramen.

The mastoid process is full of air sinuses called *mastoid air cells.* These lie just posterior to the middle ear cavity. Infections can spread from the throat to the middle ear to the mastoid cells. Such an infection, called *mastoiditis,* is difficult to treat.

Before the late 1940s, when antibiotics became available, severe mastoiditis was remarkably common. Many people had their mastoid processes removed surgically to prevent brain infections. Can you deduce why mastoiditis can involve the brain?

The mastoid air cells are separated from the overlying brain by only a thin roof of bone, through which infections can spread to the brain. ■

The **petrous** (pet'rus) **region** of the temporal bone contributes to the cranial base. It appears as a bony wedge between the occipital bone posteriorly and the sphenoid bone anteriorly (Figure 7.4a). The bone in this region is very thick (*petrous* = rocky), and from within the cranial cavity it looks like a mountain ridge. The posterior slope of this ridge lies in the posterior cranial fossa, while the anterior slope is in the **middle cranial fossa,** the fossa that supports the temporal lobes of the brain (Figure 7.4c). Housed within the petrous region are the cavities of the middle and inner ear, which contain the sensory receptors for hearing and balance.

Several foramina penetrate the bone of the petrous region (Figure 7.4a). The large **jugular foramen** appears where the petrous temporal bone joins the occipital bone. Through the jugular foramen pass the largest vein of the head, called the internal jugular vein, and cranial nerves IX, X, and XI. The **carotid** (ka-rot′id) **canal** opens on the skull's inferior aspect just anterior to the jugular foramen. It transmits the internal carotid artery (the main artery to the brain) into the cranial cavity. The **foramen lacerum** (la′ser-um; "lacerated") is a jagged opening between the medial tip of the petrous temporal bone and the sphenoid bone. It serves as a passageway for a few small, insignificant nerves. This foramen is relatively unimportant and is almost completely closed by cartilage in a living person. However, it is conspicuous in a dried skull, and students usually ask its name. The **internal acoustic meatus** lies in the cranial cavity on the posterior face of the petrous ridge (Figure 7.4b). It transmits cranial nerves VII and VIII.

Sphenoid Bone

The **sphenoid bone** (sfe′noid; "wedge-shaped") spans the width of the cranial floor (Figure 7.4). The sphenoid is considered the keystone of the cranium because it forms a central wedge, articulating with every other cranial bone. As shown in Figure 7.6, it consists of a central **body** and three pairs of processes: the *greater wings*, *lesser wings*, and *pterygoid* (ter′ĭ-goid) *processes*. The superior surface of the body bears a saddle-shaped prominence, the **sella turcica** (sel′ah ter′sik-ah; "Turkish saddle"). The seat of this saddle, called the *hypophyseal* (hi″po-fiz′e-al) *fossa*, holds the pituitary gland (the hypophysis). Within the sphenoid body are the paired **sphenoid sinuses** (Figure 7.3b). The **greater wings** project laterally from the sphenoid body, forming parts of the middle cranial

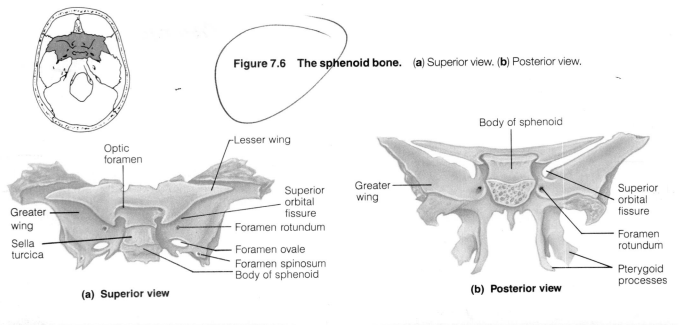

Figure 7.6   The sphenoid bone.   (**a**) Superior view. (**b**) Posterior view.

Optic foramen
Lesser wing
Greater wing
Sella turcica
Superior orbital fissure
Foramen rotundum
Foramen ovale
Foramen spinosum
Body of sphenoid

(a) Superior view

Body of sphenoid
Greater wing
Superior orbital fissure
Foramen rotundum
Pterygoid processes

(b) Posterior view

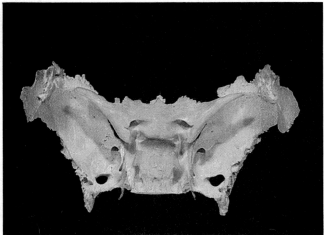

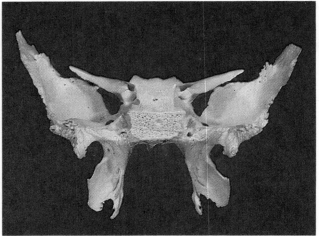

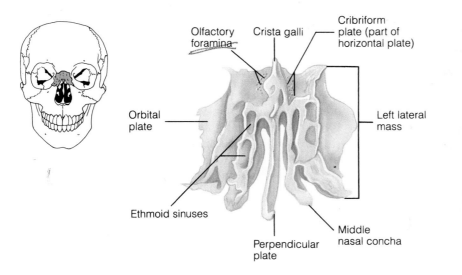

Figure 7.7   The ethmoid bone.   Anterior view.

fossa (Figure 7.4b), the orbit, and the lateral wall of the skull, where the greater wing appears as a flag-shaped area medial to the zygomatic arch (Figure 7.3a). The horn-shaped **lesser wings** form part of the floor of the anterior cranial fossa (Figure 7.4b) and a part of the orbit. The trough-shaped **pterygoid** ("wing-like") **processes** project inferiorly from the sphenoid body (Figures 7.3b and 7.6b). The pterygoid muscles, important in closing the jaw, attach to these pterygoid processes.

The sphenoid bone has five important openings (Figure 7.6). The **optic foramina** lie just anterior to the sella turcica. Cranial nerves II (optic nerves) pass through these holes from the orbits into the cranial cavity. Just lateral to the sphenoid body, on each side, is a crescent-shaped row of four openings. The anterior of these, the **superior orbital fissure,** is a long slit between the greater and lesser wings. It transmits several structures to and from the orbit, including the cranial nerves that control eye movements (III, IV, VI). This fissure is best seen in an anterior view of the orbit (Figure 7.9, p. 152). The **foramen rotundum** lies in the medial part of the greater wing. It is usually oval, despite its name, which means "round opening." The **foramen ovale** (o-val′e) is an oval hole posterior to the foramen rotundum. The foramen rotundum and foramen ovale are passageways through which two large branches of cranial nerve V enter the face. Posterior and lateral to the foramen ovale lies the small **foramen spinosum** (spi-no′sum; "of the spine"), named for a short spine that projects inferiorly from its margins on the inferior aspect of the skull. The foramen spinosum transmits the middle meningeal artery, which supplies blood to the broad inner faces of the parietal and the squamous temporal bones.

Ethmoid Bone   *CRANIAL*

The **ethmoid bone** (Figure 7.7) is the most deeply situated bone of the skull. Lying between the sphenoid bone posteriorly and the nasal bones anteriorly, it forms most of the medial bony area between the nasal cavity and the orbits. The ethmoid bone is remarkably thin-walled and delicate.

The superior surface of the ethmoid bone is formed by paired, horizontal **cribriform** (krib′rĭ-form) **plates.** These plates contribute to the roof of the nasal cavities and the floor of the anterior cranial fossa (Figure 7.4b). The cribriform plates are perforated by tiny holes called olfactory foramina (*cribriform* = perforated like a sieve). The fibers of cranial nerves I (the olfactory nerves) pass through these holes to the brain from smell receptors in the nasal cavities. Between the two cribriform plates, in the midline, is a superior projection called the **crista galli** (kris′tah gal′li; "rooster's comb"). A fibrous membrane called the falx cerebri attaches to the crista galli and helps to secure the brain within the cranial cavity.

The **perpendicular plate** of the ethmoid bone projects inferiorly in the median plane. It forms the superior part of the nasal septum, the vertical sheet that divides the nasal cavity into right and left halves. Flanking the perpendicular plate on each side is a **lateral mass** riddled with the **ethmoid sinuses.** The ethmoid bone is named for these sinuses (*ethmos* = 'sieve'). Extending medially from the lateral masses are the delicate **superior** and **middle nasal conchae** (kong′ke; "shells"), which protrude into the nasal cavity (see Figure 7.10a). The chonchae are curved like scrolls and are named after the conch shells one finds on an ocean beach. The lateral surfaces of the ethmoid's lateral masses are called **orbital plates** because they contribute to the medial walls of the orbits.

Facial Bones

The skeleton of the face consists of 14 bones (Figures 7.2a and 7.3). These include the unpaired mandible and the vomer, plus the paired maxillae, zygomatics, nasals, lacrimals, palatines, and inferior conchae. As a rule, the facial skeleton of men is longer than that of women. Therefore, women's faces are usually rounder and less angular.

Mandible

The U-shaped **mandible** (man'dĭ-bl), or lower jawbone, is the largest, strongest bone in the face. It has a horizontal *body,* which forms the inferior jawline, and two upright *rami* (ra'mi; "branches") (Figure 7.8a). Each **ramus** meets the body posteriorly at a **mandibular angle.** At the superior margin of each **ramus** are *coronoid* and *condylar* processes. The anterior **coronoid process** (kor'o-noid; "crown-shaped") is a flat, triangular projection. It provides an insertion for the temporalis muscle that elevates the lower jaw during chewing. The posterior **condylar process** enlarges superiorly to form the **mandibular condyle,** or *head of the mandible.* This articulates with the temporal bone to form the temporomandibular joint. The coronoid and condylar processes are separated by the **mandibular notch.**

The **body** of the mandible anchors the lower teeth and forms the chin. Its superior border is the **alveolar** (al-ve'o-lar) **margin.** The tooth sockets (called *alveoli*) open onto this margin. Anteriorly, the line of fusion between the two halves of the mandible is called the **mandibular symphysis** or *symphysis menti* (sim'-fi-sis men'ti), literally a "growing together at the chin" (Figure 7.2a).

Several openings pierce the mandible. On the medial surface of each ramus is a **mandibular foramen** (Figure 7.8a), through which the nerve responsible for tooth sensation (inferior alveolar nerve) enters the mandibular body and supplies the roots of the lower teeth. Dentists inject anesthetic into this foramen for working on the lower teeth. The **mental** ("chin") **foramen** is an opening on the anterolateral side of the mandibular body. Blood vessels and nerves pass through the mental foramen to the lower lip and the skin of the chin.

Maxillary Bones

The **maxillary bones,** or **maxillae** (mak-sil'e; "jaws"), form the upper jaw and the central part of the facial skeleton (Figures 7.2a and 7.8b). All other facial bones except the mandible articulate with them. Hence, the maxillae are considered the keystone bones of the facial skeleton.

Like the mandible, the maxillae have an **alveolar margin** that contains teeth in alveoli. The **palatine** (pal'ah-tēn) **processes** of the maxillae project medially from the alveolar margins to form the anterior region of the **hard palate,** or bony roof of the mouth (Figure 7.4a). The **frontal processes** of the maxillae extend superiorly to reach the frontal bone, forming part of the lateral aspect of the bridge of the nose (Figure 7.8b). The maxillary bones lie just lateral to the nasal cavity and contain the **maxillary sinuses** (Figure 7.11). These sinuses, the largest of the paranasal air sinuses, extend from the orbit to the roots of the upper teeth. Laterally, the maxillae articulate with the zygomatic bones via their **zygomatic processes** (Figure 7.8b).

The **inferior orbital fissure** is located deep in the floor of the orbit (Figure 7.9). Several bones contribute to its borders, especially the maxilla and the greater wing of the sphenoid. Several vessels and nerves pass anteriorly through this fissure (Table 7.2), including the maxillary nerve or the infraorbital nerve. The nerve continues anteriorly through an **infraorbital foramen** in the maxilla to enter the face (Figure 7.8b).

Zygomatic Bones

The irregularly shaped **zygomatic bones** are commonly called the cheekbones (*zygoma* = cheekbone). Each joins the zygomatic process of a temporal bone posteriorly, the zygomatic process of the frontal bone superiorly, and the zygomatic process of the maxilla anteriorly (Figure 7.3a). The zygomatic bones form the prominences of the cheeks and define part of the margin of each orbit.

Nasal Bones

The paired, rectangular **nasal bones** join medially to form the bridge of the nose (Figure 7.2a). They articulate with the frontal bone superiorly, the maxillary bones laterally, and the perpendicular plate of the ethmoid bone posteriorly. Inferiorly, they attach to the cartilages that form most of the skeleton of the external nose (see Figure 6.1 on p. 119).

Lacrimal Bones

The delicate, fingernail-shaped **lacrimal** (lak'ri-mal) **bones** are pictured in Figures 7.2a, 7.8b, and 7.9. Each lies in an orbit and contributes to the orbit's medial wall. They articulate with the frontal bone superiorly, the ethmoid bone posteriorly, and the maxilla anteriorly (Figure 7.9). Each lacrimal bone contains a deep groove that contributes to a *lacrimal fossa.* This fossa contains a *lacrimal sac* that gathers tears, allowing

Figure 7.8   Detailed anatomy of some facial bones. (**a**) The mandible. (**b**) The maxilla (nasal and lacrimal bones also shown). (**c**) Palatine bones. Note that the bones in (a) through (c) are not drawn in proportion to one another.

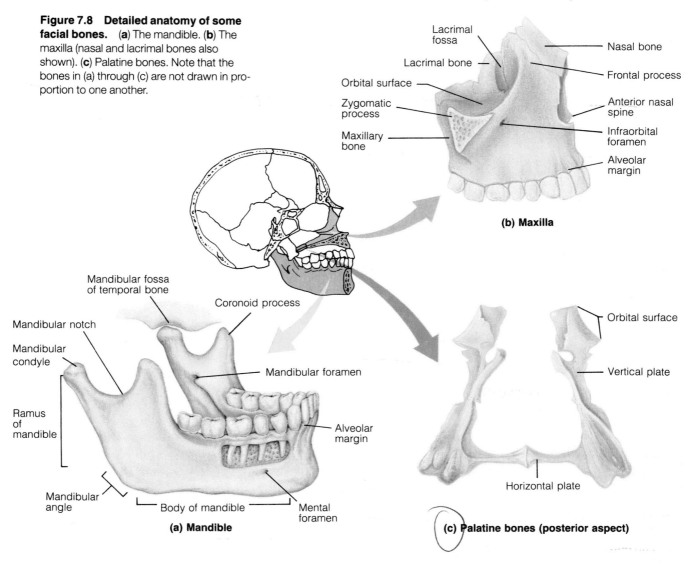

(a) Mandible

(b) Maxilla

(c) Palatine bones (posterior aspect)

the fluid to drain from the eye surface into the nasal cavity (*lacrima* means "tears").

Palatine Bones

The **palatine bones** lie posterior to the maxilla (Figure 7.8). These paired, L-shaped bones are interconnected at their inferior **horizontal plates** (Figure 7.8c), which complete the posterior part of the hard palate (Figure 7.4a). The **vertical plates** form the posterior part of the lateral walls of the nasal cavity and a small part of the orbits (Figure 7.9).

Vomer

The slender, plow-shaped **vomer** (vo'mer; "plowshare") is located within the nasal cavity, where it forms the inferior part of the nasal septum (Figures 7.2a and 7.3b). It is discussed below in connection with the nasal cavity.

Inferior Nasal Conchae

The paired **inferior nasal conchae** are thin, curved bones in the nasal cavity (Figures 7.2a and 7.10a). They project medially from the lateral walls of the nasal cavity, just inferior to the middle nasal conchae of the ethmoid bone. They are the largest of the three pairs of conchae.

Special Characteristics of the Orbits and Nasal Cavity

Two restricted regions of the skull, the orbits and the nasal cavity, are formed from an amazing number of bones. In fact, it is sometimes difficult to comprehend how the many parts form the whole. Thus, even though the bones of these regions have been described, a brief summary is provided here.

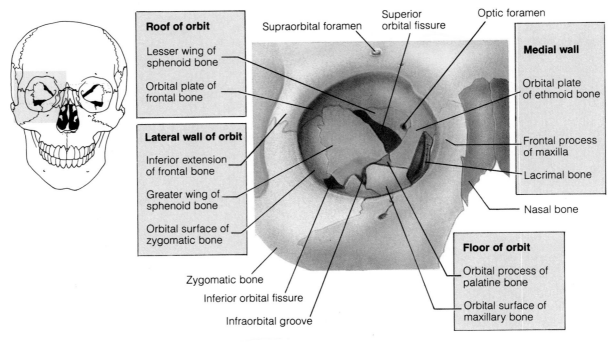

Figure 7.9   Bones of the orbit.

Orbits

The **orbits** are cone-shaped bony cavities that hold the eyes. Also housed in the orbits are the muscles that move the eyes, some fat, and the tear-producing glands. The walls of each orbit are formed by parts of seven bones—the frontal, sphenoid, zygomatic, maxilla, palatine, lacrimal, and ethmoid. Their relationships are shown in Figure 7.9. The superior and inferior orbital fissures and the optic foramen (described earlier) are also present in the orbit.

Nasal Cavity

The **nasal cavity** (Figure 7.10) is constructed of bone and cartilage. The *roof* of the nasal cavity is the ethmoid's cribriform plates. The *floor* is formed by the palatine processes of the maxillae and the palatine bones. The *lateral walls* are shaped by the superior and middle conchae of the ethmoid, the inferior nasal conchae, a part of the maxilla, and the vertical plates of the palatine bones (Figure 7.10a). On these lateral walls, each of the three conchae roofs over a groove-shaped air passageway called a *meatus* (me-a'tus; "a passage"). Therefore, there are *superior, middle,* and *inferior meatuses.*

Recall that the nasal cavity is divided into right and left halves by the nasal septum (Figure 7.10b). The bony part of this septum is formed by the vomer inferiorly and by the perpendicular plate of the ethmoid superiorly. A sheet of cartilage, called the septal cartilage, completes the septum anteriorly.

The walls of the nasal cavity are covered with a mucosa that moistens and warms the entering air. This membrane also secretes mucus to trap dust and therefore cleanse the air of debris.

Can you guess the function of the three pairs of nasal conchae that project into the nasal cavity?

These scroll-shaped bones increase the turbulence of air flowing through the nasal cavity. This swirling increases the contact of inhaled air with the mucosa throughout the nasal cavity, thereby increasing the efficiency with which the air is warmed, moistened, and filtered. ■

Paranasal Sinuses

Several skull bones—the frontal, ethmoid, sphenoid, and both maxillary bones—contain air-filled sinuses that cause them to look rather moth-eaten in an X-ray image. These sinuses are called **paranasal sinuses** (*para* = near) because they cluster around the nasal cavity (Figure 7.11). In fact, they are extensions of the nasal cavity, lined by the same mucous membrane and probably serving the same function of warming, moistening, and filtering inhaled air. The paranasal sinuses also lighten the skull. They connect to the nasal cavity through small openings, most of which occur at the meatuses inferior to the conchae.

(text continues on p. 156)

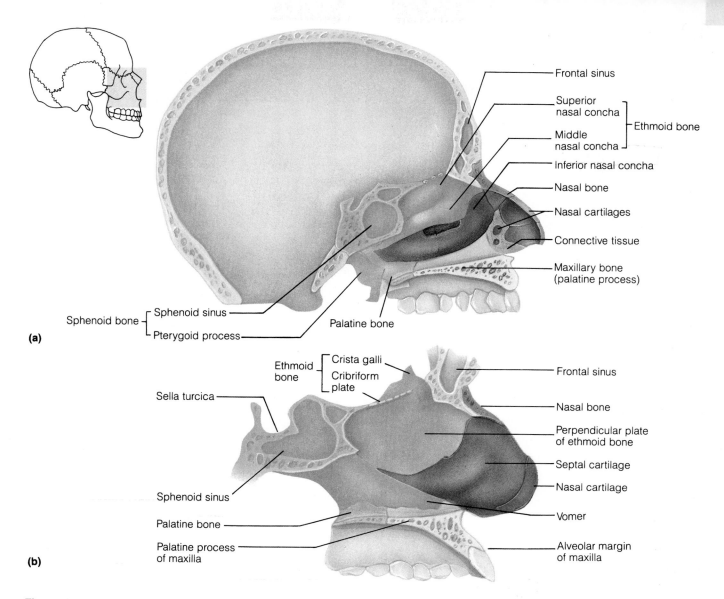

Figure 7.10   Bones of the nasal cavity.   (**a**) Bones forming the left lateral wall of the nasal cavity (nasal septum removed). (**b**) Nasal cavity with nasal septum in place, showing how the ethmoid bone, septal cartilage, and the vomer make up the septum.

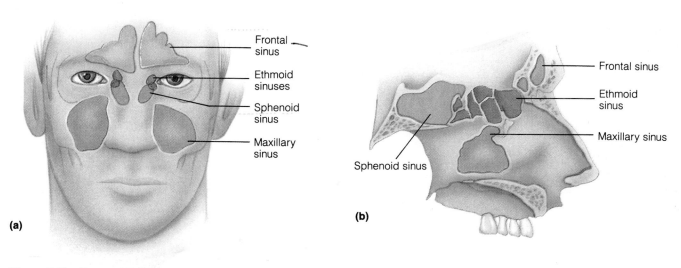

Figure 7.11   Paranasal air sinuses.

Table 7.2 Bones of the Skull

Bone*	Comments	Important markings
CRANIAL BONES		
○ **Frontal** (1) (Figures 7.2a, 7.3, and 7.4b)	Forms forehead; superior part of orbits and anterior cranial fossa; contains sinuses	**Supraorbital foramina:** allow the supraorbital arteries and nerves to pass
○ **Parietal** (2) (Figures 7.2 and 7.3)	Forms most of the superior and lateral aspects of the skull	
● **Occipital** (1) (Figures 7.2b, 7.3, and 7.4)	Forms posterior aspect and most of the base of the skull	**Foramen magnum:** allows the spinal cord to issue from the brain stem to enter the vertebral canal
		Hypoglossal canal: allows passage of the hypoglossal nerve (cranial nerve XII)
		Occipital condyles: articulate with the atlas (first vertebra)
		External occipital protuberance and **nuchal lines:** sites of muscle attachment
		External occipital crest: attachment site of ligamentum nuchae
○ **Temporal** (2) (Figures 7.3, 7.4, and 7.5)	Forms inferolateral aspects of the skull and contributes to the middle cranial fossa; has squamous, mastoid, tympanic, and petrous regions	**Zygomatic process:** helps to form the zygomatic arch, which forms the prominence of the cheek
		Mandibular fossa: articular point of the mandibular condyle
		External acoustic meatus: canal leading from the external ear to the eardrum
		Styloid process: attachment site for several neck and tongue muscles
		Mastoid process: attachment site for several neck muscles
		Stylomastoid foramen: allows cranial nerve VII (facial nerve) to pass
		Jugular foramen: allows passage of the jugular vein and cranial nerves IX, X, and XI
		Internal acoustic meatus: allows passage of cranial nerves VII and VIII
		Carotid canal: allows passage of the internal carotid artery
○ **Sphenoid** (1) (Figures 7.2a, 7.3, 7.4b, and 7.6)	Keystone of the cranium; contributes to the middle cranial fossa and orbits; main parts are the body, greater wings, lesser wings, and pterygoid processes	**Sella turcica:** seat of the pituitary gland
		Optic foramina: allow passage of cranial nerve II and the ophthalmic arteries
		Superior orbital fissures: allow passage of cranial nerves III, IV, VI, and part of V (ophthalmic division)
		Foramen rotundum (2): allows passage of the maxillary division of cranial nerve V
		Foraman ovale (2): allows passage of the mandibular division of cranial nerve V
		Foramen spinosum (2): allows passage of the middle meningeal artery

*The color-coded dot beside each bone name corresponds to the bone's color in Figures 7.2 to 7.10. The number in parentheses ( ) following each bone name denotes the total number of such bones in the body.

Table 7.2   (continued)

Bone*	Comments	Important markings
Ethmoid (1) (Figures 7.2a, 7.3, 7.4b, 7.7, and 7.10)	Helps to form the anterior cranial fossa; forms part of the nasal septum and the lateral walls and roof of the nasal cavity; contributes to the medial wall of the orbit	**Crista galli:** attachment point for the falx cerebri, a dural membrane fold **Cribriform plates:** allow passage of nerve fibers of the olfactory (cranial nerve I) nerves **Superior** and **middle nasal conchae:** form part of lateral walls of nasal cavity; increase turbulence of air flow
Ear ossicles (malleus, incus, and stapes) (2 each)	Found in middle ear cavity; involved in sound transmission; see Chapter 15	

FACIAL BONES

Bone*	Comments	Important markings
Mandible (1) (Figures 7.2a, 7.3, and 7.8a)	The lower jaw	**Coronoid processes:** insertion points for the temporalis muscles **Mandibular condyles:** articulate with the temporal bones in freely movable joints (temporomandibular joints) **Mandibular symphysis:** medial fusion point of the mandibular bones **Alveoli:** sockets for the teeth **Mandibular foramina:** permit the alveolar nerves to pass **Mental foramina:** allow blood vessels and nerves to pass to the chin and lower lip
Maxilla (2) (Figures 7.2a, 7.3, 7.4, and 7.8b)	Keystone bones of the face; form the upper jaw and parts of the hard palate, orbits, and nasal cavity walls	**Alveoli:** sockets for the teeth **Zygomatic processes:** help form the zygomatic arches **Palatine processes:** form the anterior hard palate **Frontal processes:** form part of lateral aspect of bridge of nose **Inferior orbital fissures:** permit maxillary branch of cranial nerve V, the zygomatic nerve, and blood vessels to pass **Infraorbital foramen:** allows passage of infraorbital nerve to skin of face
Zygomatic (2) (Figures 7.2a, 7.3a, and 7.4a)	Form the cheek and part of the orbit	
Nasal (2) (Figures 7.2a and 7.3)	Construct the bridge of the nose	
Lacrimal (2) (Figures 7.2a and 7.3a)	Form part of the medial orbit wall	**Lacrimal fossa:** houses the lacrimal sac which helps to drain tears into the nasal cavity
Palatine (2) (Figures 7.3b, 7.4a, and 7.8c)	Form posterior part of the hard palate and a small part of nasal cavity and orbit walls	
Vomer (1) (Figures 7.2a and 7.10)	Part of the nasal septum	
Inferior nasal concha (2) (Figures 7.2a and 7.10a)	Form part of the lateral walls of the nasal cavity	

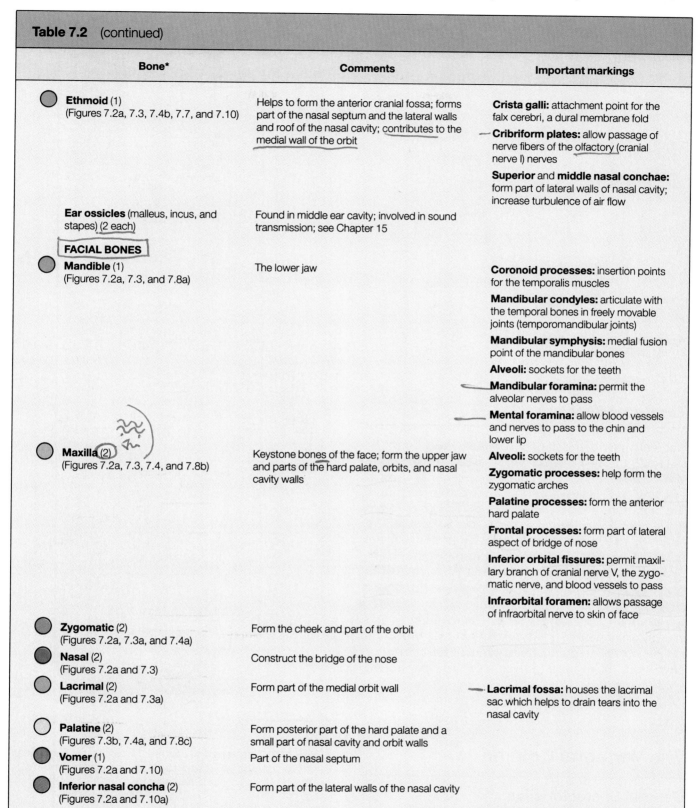

*The color-coded dot beside each bone name corresponds to the bone's color in Figures 7.2 to 7.10. The number in parentheses ( ) following each bone name denotes the total number of such bones in the body.

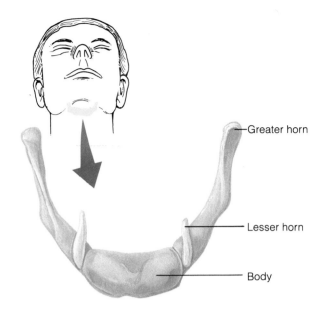

Figure 7.12 The hyoid bone.

The Hyoid Bone

Not really part of the skull, the **hyoid bone** (hi'oid; "U-shaped") lies just inferior to the mandible in the anterior neck (Figure 7.12). This bone resembles an archer's bow, but it also looks like a miniature version of the mandible. It has a **body** and two pairs of **horns** (each horn is also called a *cornu,* the Latin word for "horn"). The hyoid is the only bone in the skeleton that does not articulate directly with any other bone. Instead, its lesser horns are tethered by narrow *stylo-hyoid ligaments* to the styloid processes of the temporal bones. Other ligaments connect the hyoid to the larynx (voice box) inferior to it. The hyoid bone acts as a movable base for the tongue, and its body and greater horns are points of attachment for neck muscles that raise and lower the larynx during swallowing.

The Vertebral Column

General Characteristics

Some people think of the **vertebral column** (Figure 7.13) as a rigid supporting rod, but this picture is inaccurate. The vertebral column, also called the *spinal column* or *spine,* is formed from 26 irregular bones connected into a flexible, curved structure. Serving as the axial support of the trunk, the vertebral column extends from the skull to the pelvis, where it trans-

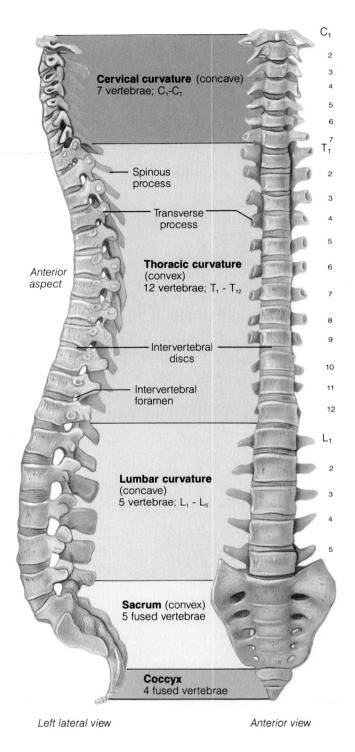

Left lateral view   *Anterior view*

Figure 7.13 The vertebral column. Note the four curvatures in the lateral view at left. The terms convex and concave mean convex *posteriorly* and concave *posteriorly.*

mits the weight of the trunk to the lower limbs. The vertebral column surrounds and protects the delicate spinal cord and provides attachment points for the ribs and for muscles of the neck and back. In the fetus and infant, the vertebral column consists of 33 separate bones, or **vertebrae** (ver'tĕ-bre). Inferiorly, nine of these eventually fuse to form two composite bones,

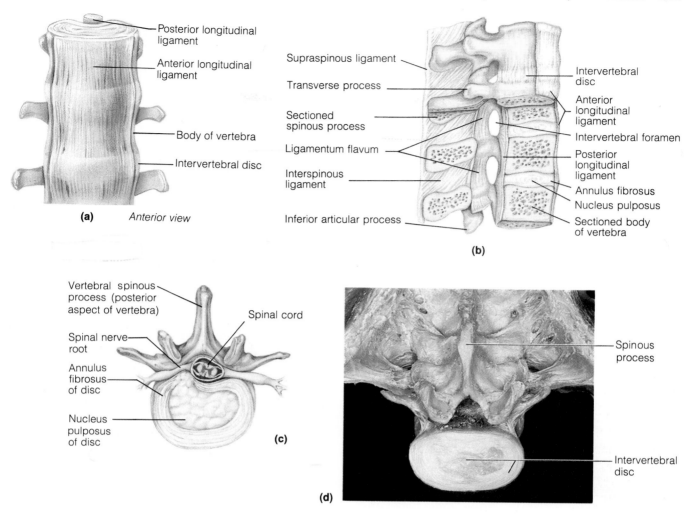

Figure 7.14   Ligaments and intervertebral discs of the spine.   (a) Anterior view of part of the spinal column, showing the anterior longitudinal ligaments. (b) Midsagittal section of the spine, showing other ligaments that interconnect the vertebrae. Some of these ligaments are quite elastic (ligamentum flavum). Also note the parts of the intervertebral discs. (c) A herniated intervertebral disc. (d) Photograph of an intervertebral disc from a cadaver.

the sacrum and the tiny coccyx (tailbone). The remaining 24 bones persist as individual vertebrae separated by intervertebral discs.

Like a tremulous television transmitting tower, the vertebral column cannot possibly stand upright by itself. It must be held in place by an elaborate system of supports. Assuming this role are the strap-like ligaments of the back and the muscles of the trunk. The trunk muscles are discussed in Chapter 10. The major supporting ligaments are the **anterior** and **posterior longitudinal ligaments** (Figure 7.14a). These run vertically along the anterior and posterior surfaces of the spine, from the neck to the sacrum. The anterior longitudinal ligament is wide and attaches strongly to both the bony vertebrae and the intervertebral discs. Along with its supporting role, this thick anterior ligament prevents hyperextension of the back (bending too far backward). The posterior longitudinal ligament (Figure 7.14b) is narrow and relatively weak. It attaches only to the intervertebral discs. This ligament helps to prevent hyperflexion (bending the vertebral column too sharply forward).

Several other posterior ligaments connect each vertebra to those immediately superior and inferior (Figure 7.14b). Of these, the **ligamentum flavum** (fla′vum; "yellow"), which contains elastic connective tissue, is especially strong: It stretches as we bend forward, then recoils as we straighten to an erect position.

Intervertebral Discs

Each **intervertebral disc** is a cushion-like pad (Figure 7.14b and 7.14d) composed of an inner sphere, the **nucleus pulposus** (pul-po′sus; "pulp"), and an outer collar of about 12 concentric rings, the **anulus fibrosus** (an′u-lus fi-bro′sus; "fibrous ring"). Each nucleus pulposus acts like a rubber ball, enabling the spine to absorb compressive stress. In the anulus fibrosus, the outer rings consist of ligament, and the inner ones consist of fibrocartilage. The main purpose of these rings is to contain the nucleus pulposus, limiting its expansion when the spine is compressed. However, these rings also bind the successive vertebrae together, resist tension on the spine, and absorb compressive forces themselves. Collagen fibers in adjacent rings in the anulus cross like an X, allowing the spine to withstand twisting. (This is the same anti-twisting design provided by bone lamellae in osteons: See Figure 6.7 on p. 126).

The intervertebral discs act as shock absorbers during walking, jumping, and running and allow the spine to bend. At points of compression, the discs flatten and bulge out a bit between the vertebrae. The discs are thickest in the lumbar (lower back) and cervical (neck) regions of the vertebral column, a property that enhances the flexibility of these regions.

Collectively, the intervertebral discs make up about 25% of the height of the vertebral column. They flatten somewhat by the end of each day, so our height is always a few centimeters less at night than when we awake in the morning.

Severe or sudden physical trauma to the spine— for example, from lifting a heavy object—may result in herniation of one or more discs. A **herniated disc** is also called a **prolapsed disc** or, in common terms, a slipped disc. It usually involves a rupture of the nucleus pulposus through the anulus fibrosus. The anulus is thinnest posteriorly, so the nucleus usually herniates in that direction. Actually, the posterior longitudinal ligament prevents the herniation from proceeding directly posterior. Thus, the rupture proceeds postero*laterally*—toward the spinal nerve roots exiting from the spinal cord (Figure 7.14c). The resulting pressure on these nerve roots leads to pain or numbness. Herniated discs are treated with bed rest, traction, and pain killers. If this treatment fails, the protruding disc may have to be surgically removed. ■

Divisions and Curvatures

The vertebral column is about 70 cm (28 inches) long in an average adult. It has five major divisions (Figure 7.13). The 7 vertebrae of the neck are the **cervical vertebrae,** the next 12 are the **thoracic vertebrae,** and the 5 that support the lower back are the **lumbar vertebrae.** Remembering the common meal times of 7 A.M., 12 noon, and 5 P.M. will help you recall the number of vertebrae in these three regions of the vertebral column. The vertebrae become progressively larger from the cervical to the lumbar region, as they must support progressively more weight. Inferior to the lumbar vertebrae is the **sacrum** (sa'krum), which articulates with the hip bones of the pelvis. The most inferior part of the vertebral column is the tiny **coccyx** (kok'siks).

All people have the same number of cervical vertebrae. Variations in numbers of vertebrae in the other regions occur in about 5% of people.

Can you deduce which region of the vertebral column (cervical, thoracic, or lumbar) is most likely to experience a herniated disc?

Most herniated discs occur in the lumbar region, because these vertebrae continually experience the largest compressive stresses. Lumbar vertebrae must support the weight of the entire upper body, plus whatever heavy objects one lifts with the upper body. ■

From a lateral view, the four curvatures that give the vertebral column an S-shape are visible (Figure 7.13, left). The **cervical** and **lumbar curvatures** are concave posteriorly, whereas the **thoracic** and **sacral curvatures** are convex posteriorly. These curvatures increase the resilience of the spine, allowing it to function like a spring rather than a straight, rigid rod.

There are several types of abnormal spinal curvatures. Some are congenital (present at birth), while others result from disease, poor posture, or unequal pull of muscles on the spine. **Scoliosis** (sko"le-o'sis; "twisted disease") is an abnormal *lateral* curvature that occurs most often in the thoracic region. It is fairly common during late childhood, particularly in girls, for some unknown reason. The most severe cases result from abnormally structured vertebrae, lower limbs of unequal length, or muscle paralysis. If muscles on one side of the back are nonfunctional, those on the opposite side pull unopposed and move the spine out of alignment. Severe cases of scoliosis must be treated (with body braces or by surgery) before the child stops growing to prevent permanent deformity. Severe scoliosis can compress a lung, causing difficulty in breathing.

Kyphosis (ki-fo'sis; "humped disease"), or hunchback, is an exaggerated thoracic curvature. It is particularly common in aged women because it often results from spinal fractures that follow osteoporosis. Kyphosis may also result from tuberculosis of the spine or osteomalacia.

Lordosis (lor-do'sis; "bent-backward disease"), or swayback, is an accentuated lumbar curvature. It too can result from spinal tuberculosis or osteomalacia. Temporary lordosis is common in people carrying a "large load up front," such as obese men and pregnant women. These people automatically "throw back their shoulders" in an attempt to preserve their center of gravity. ■

General Structure of Vertebrae

The vertebrae share a common structural pattern (Figure 7.15). A vertebra consists of a **body**, or **centrum**, anteriorly and a **vertebral arch** posteriorly. The disc-shaped body is the weight-bearing region. Together, the body and the vertebral arch enclose an opening called the **vertebral foramen.** Successive vertebral foramina of the articulated vertebrae form the long *vertebral canal,* through which the spinal cord passes.

The vertebral arch is a composite structure formed by two pedicles and two laminae. The **pedicles** (ped'ĭ-k'lz; "little feet"), short bony cylinders that project posteriorly from the vertebral body, form the sides of the arch. The **laminae** (lam'ĭ-ne; "sheets") are flattened roof plates that complete the arch posteriorly. Seven different processes project from each vertebral arch. The **spinous process** is a single posterior projection arising at the junction of the two laminae. A **transverse process** projects laterally from each side of the vertebral arch. Both the spinous process and the transverse processes are attachment sites for muscles that move the vertebral column and for ligaments that stabilize it. The **superior** and **inferior articular processes** are paired processes that protrude superiorly and inferiorly, respectively, from the pedicle-lamina junctions. The inferior articular processes of each vertebra form movable joints with the superior articular processes of the vertebra immediately inferior. Thus, successive vertebrae are joined both at their bodies and at their articular processes. The smooth joint surfaces of these processes are *facets* ("little faces").

The pedicles have notches on their superior and inferior borders, forming lateral openings between adjacent vertebrae called **intervertebral foramina** (Figure 7.13). The spinal nerves issuing from the spinal cord pass through these foramina.

Regional Vertebral Characteristics

The different regions of the spine perform slightly different functions, so vertebral structure shows some regional variation. In general, the types of movements that can occur between vertebrae are (1) flexion and extension (anterior bending and posterior straightening of the spine), (2) lateral flexion (bending the spine to the right or left), and (3) rotation (in which the vertebrae rotate on one another in the long axis of the vertebral column). Refer to Table 7.3 on p. 163 while reading this section.

Cervical Vertebrae

The seven **cervical vertebrae** are identified as C_1–C_7. These are the smallest, lightest vertebrae. The first

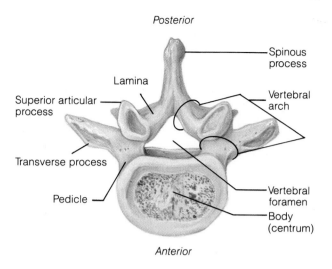

Figure 7.15 Structure of a typical vertebra. Superior view.

two, C_1 and C_2, are unusual, so we will skip them for the moment. The "typical" cervical vertebrae, C_3–C_7, have the following distinguishing features:

1. The body is broader laterally than in the anteroposterior dimension.

2. Except in C_7, the spinous process is short, projects directly posteriorly, and is bifid (bi'fid; "cleaved in two"), that is, split at its tip.

3. The vertebral foramen is large and generally triangular.

4. Each transverse process contains a hole, a **transverse foramen,** through which the vertebral blood vessels pass. These vessels ascend and descend through the neck to help supply the brain.

5. The superior articular facets face superoposteriorly, while the inferior articular facets face inferoanteriorly. The orientation of these articulations allows the neck to carry out an extremely wide range of movements: flexion and extension, lateral flexion, and rotation.

The spinous process of C_7 is not bifid and is much larger than those of the other cervical vertebrae (Figure 7.17a). Because its spinous process is visible through the skin, C_7 is used as a landmark for counting the vertebrae in living people. It is called the **vertebra prominens** ("prominent vertebra").

The first two cervical vertebrae are the *atlas* (C_1) and the *axis* (C_2) (Figure 7.16). No intervertebral disc lies between them, and they have unique structural and functional features: The **atlas** lacks a body and a spinous process. Essentially, the atlas is a ring of bone that consists of **anterior** and **posterior arches,** plus a

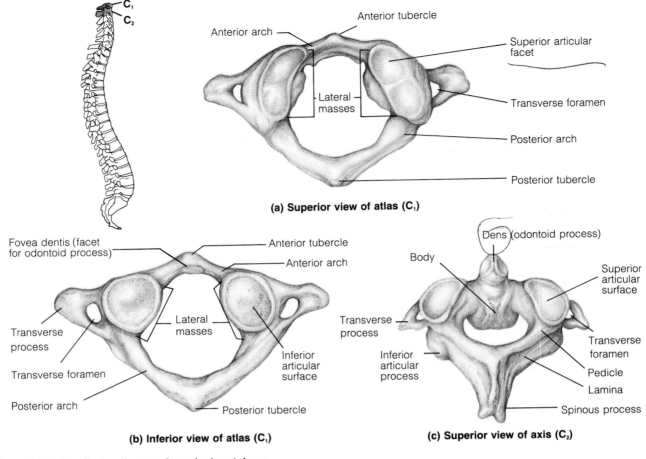

(a) **Superior view of atlas (C₁)**

(b) **Inferior view of atlas (C₁)**

(c) **Superior view of axis (C₂)**

Figure 7.16 The first and second cervical vertebrae.

lateral mass on each side. Each lateral mass has articular facets on both its superior and inferior surfaces. The superior articular facets receive the occipital condyles of the skull. Thus, they "carry" the skull, just as the giant Atlas supported the heavens in Greek mythology. These joints enable you to nod "yes." The inferior articular facets form joints with the axis.

The **axis,** which has a body, spine, and the other typical vertebral processes, is not as specialized as the atlas. In fact, its only unusual feature is the knob-like **dens** (dens; "tooth"), or **odontoid process** (o-don'toid; "tooth-like"), projecting superiorly from its body. The dens is actually the "missing" body of the atlas, which fuses with the axis during embryonic development. Cradled within the anterior arch of the atlas (Figure 7.17a), the dens acts as a pivot for the rotation of the atlas and skull. Hence, this joint enables you to rotate your head from side to side to indicate "no." *Axis* is a good name for the second cervical vertebra because its dens allows the head to rotate on the neck's *axis.*

In some cases of severe head trauma, the skull is driven inferiorly toward the spine. When this happens, the dens may be forced into the brain stem, causing death. If the neck is jerked forward (as in an automobile collision), the dens may move posteriorly into the cervical spinal cord. This injury is also fatal.

Thoracic Vertebrae

Thoracic vertebrae are illustrated in Figure 7.17b and Table 7.3. The 12 thoracic vertebrae, T_1–T_{12}, all articulate with ribs. Other unique characteristics of the thoracic vertebrae are the following:

1. From a superior view, the vertebral body is roughly heart-shaped. Laterally, the vertebral body bears facets for the heads of the ribs. More specifically, there are two half facets, or **demifacets** (dem'e-fas"ets) on each side of the body, one at the superior edge and the other at the inferior edge. Vertebra T_1 differs from this general pattern in that its body bears a full facet for the first rib and a demifacet for the second rib; furthermore, the bodies of T_{10}–T_{12} have only single facets to receive their respective ribs.

2. The vertebral foramen is circular.

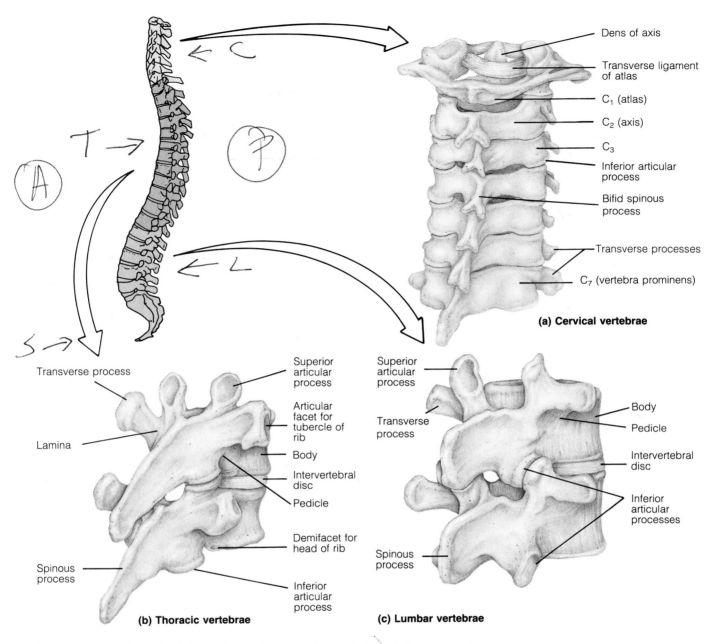

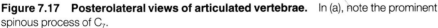

(a) Cervical vertebrae

- Dens of axis
- Transverse ligament of atlas
- C_1 (atlas)
- C_2 (axis)
- C_3
- Inferior articular process
- Bifid spinous process
- Transverse processes
- C_7 (vertebra prominens)

(b) Thoracic vertebrae

- Transverse process
- Lamina
- Spinous process
- Superior articular process
- Articular facet for tubercle of rib
- Body
- Intervertebral disc
- Pedicle
- Demifacet for head of rib
- Inferior articular process

(c) Lumbar vertebrae

- Superior articular process
- Transverse process
- Spinous process
- Body
- Pedicle
- Intervertebral disc
- Inferior articular processes

Figure 7.17   Posterolateral views of articulated vertebrae.   In (a), note the prominent spinous process of C_7.

3. The spinous process is long and points inferiorly.

4. With the exception of T_{11} and T_{12}, the transverse processes have facets that articulate with the tubercles of the ribs (Figure 7.17b).

5. The articular facets lie mainly in the frontal plane; that is, the superior facets face posteriorly, while the inferior facets face anteriorly. Such articulations essentially prohibit the movements of flexion and extension, but they allow rotation between successive vertebrae. Much of the ability to rotate the trunk comes from the thoracic part of the vertebral column. Lateral flexion is also possible but is limited by the ribs.

Lumbar Vertebrae

The lumbar region of the vertebral column, the area commonly referred to as the small of the back, receives the most stress. The enhanced weight-bearing function of the five **lumbar vertebrae** (L_1–L_5) is reflected in their sturdy structure. Their bodies are mas-

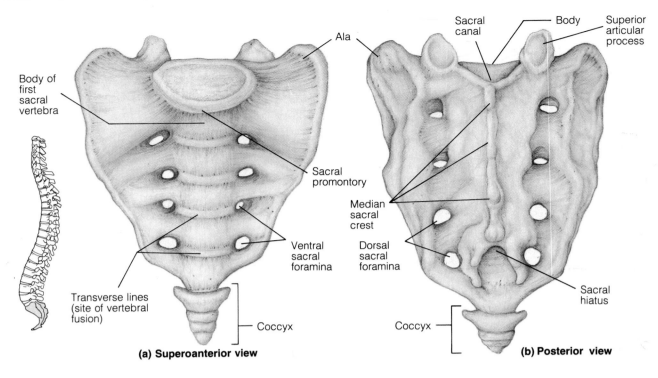

Figure 7.18 The sacrum and coccyx.

sive and appear kidney-shaped from a superior view (Table 7.3 and Figure 7.17c). Other characteristics of the lumbar vertebrae include the following:

1. The pedicles and laminae are shorter and thicker than those of other vertebrae.

2. The spinous processes are short, flat, and hatchet-shaped. They project straight posteriorly. These processes are robust, for large muscles of the back attach to them.

3. The vertebral foramen is triangular.

4. The superior articular facets face posteromedially (or medially), while the inferior articular facets face anterolaterally (or laterally). Such articulations provide stability by preventing rotation between the lumbar vertebrae. Flexion and extension are possible, however. The lumbar region bends, for example, when you bend forward to pick up a coin from the ground. Additionally, lateral flexion is allowed by this spinal region.

In your anatomy course, you may be handed an isolated vertebra and asked to determine whether it is cervical, thoracic, or lumbar. Based on what you have read above, can you think of a reliable scheme for distinguishing these three types of vertebrae?

You are encouraged to devise your own scheme, but here is one that works: First, look for a transverse foramen in each transverse process (if this is present, you have a cervical vertebra). Second, examine the vertebral body for evidence of demifacets or facets for ribs (if these are present, you have a thoracic vertebra). Third, if neither of the first two features is present, look for the distinctive features of lumbar vertebrae (massive body, hatchet-shaped spinous process, and so on). ■

Sacrum

The curved, triangular **sacrum** (Figure 7.18) shapes the posterior wall of the pelvis. It is formed by five fused vertebrae (S_1–S_5). Superiorly, the sacrum articulates with L_5 through a pair of **superior articular processes** and an intervertebral disc. Inferiorly, it joins the coccyx. On each side is a flaring **ala** (ah′lah; "wing"), the fused remnant of the transverse processes of S_1–S_5. The alae articulate with the two hip bones to form the *sacroiliac* (sa″kro-il′e-ak) *joints* of the pelvis.

The anterosuperior margin of the first sacral vertebra bulges anteriorly into the pelvic cavity as the **sacral promontory** (prom′on-tor″e; "a high point of land projecting into the sea"). The body's center of gravity lies about 1 cm posterior to this landmark. Four **transverse lines** cross the anterior surface of the sacrum,

Table 7.3 Regional Characteristics of Cervical, Thoracic, and Lumbar Vertebrae

Characteristic	Cervical (3–7)	Thoracic	Lumbar
Body	Small, wide side to side	Larger than cervical; heart shaped; bears two costal demifacets	Massive; kidney shaped
Spinous process	Short; bifid; projects directly backward	Long; sharp; projects downward	Short; blunt; projects directly backward
Vertebral foramen	Triangular	Circular	Triangular
Transverse processes	Contain foramina	Bear facets for ribs (except T_{11} and T_{12})	No special features
Superior and inferior articulating processes	Superior facets directed superoposteriorly	Superior facets directed posteriorly	Superior facets directed postero-medially (or medially)
	Inferior facets directed inferoanteriorly	Inferior facets directed anteriorly	Inferior facets directed anterola-terally (or laterally)
Movements allowed	Flexion and extension of spine; lateral flexion; rotation	Rotation; some lateral flexion	Flexion and extension; some lateral flexion

Superior view

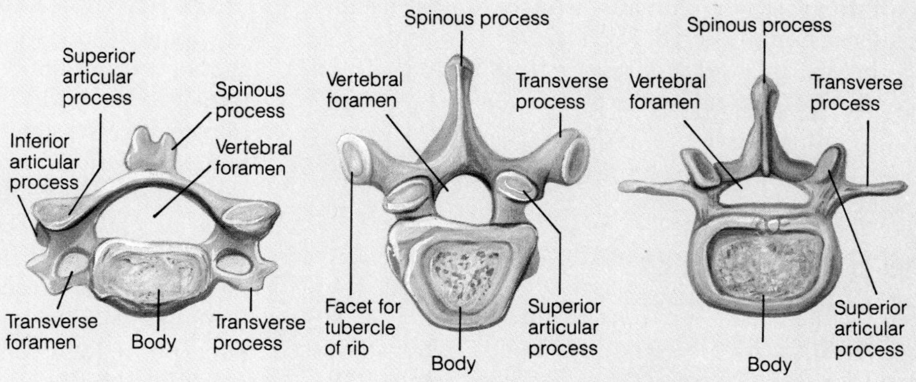

Right lateral view

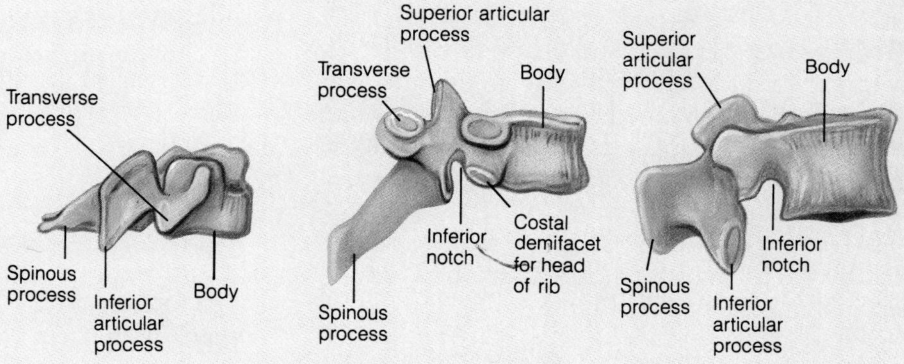

marking the lines of fusion of the sacral vertebrae. The **sacral foramina** transmit vessels and nerves and correspond to intervertebral foramina.

In the dorsal midline, the surface of the sacrum is roughened by the **median sacral crest,** the fused spinous processes of the sacral vertebrae. The vertebral canal continues within the sacrum as the **sacral canal.** The laminae of the fifth (and sometimes the fourth) sacral vertebrae fail to fuse medially, leaving an enlarged external opening at the inferior end of the sacral canal. This opening is the **sacral hiatus** (hi-a′tus; "gap").

Coccyx

The **coccyx,** or tailbone, is small and triangular (Figure 7.18). The name *coccyx* is from a Greek word for "cuckoo," and the bone was so named because of a fancied resemblance to a bird's beak. The coccyx consists of four (or in some cases three or five) vertebrae fused together. Except for the slight support the coccyx affords the pelvic organs, it is an almost useless bone. Occasionally, a baby is born with an unusually long coccyx. In most such cases, this bony "tail" is discreetly snipped off by the physician.

The Bony Thorax

The thorax is the chest, and its bony framework is called the **bony thorax** or **thoracic cage** (Figure 7.19). The bony thorax is roughly cone-shaped and includes the thoracic vertebrae posteriorly, the ribs laterally, and the sternum and costal cartilages anteriorly. The bony thorax forms a protective cage around the heart, lungs, and other thoracic organs. It also supports the shoulder girdles and upper limbs and provides attachment points for many muscles of the back, neck, chest, and shoulders. In addition, the *intercostal spaces* (*inter* = between; *costa* = the ribs) are occupied by the intercostal muscles, which lift and depress the thorax during breathing.

Sternum

The **sternum** (breastbone) lies in the anterior midline of the thorax. Vaguely resembling a dagger, it is a flat bone about 15 cm long consisting of three sections: the manubrium, body, and xiphoid process. The **manubrium** (mah-nu′bre-um; "knife handle"), the superior section, is shaped like the knot in a necktie. It articulates at its *clavicular notches* with the clavicles

(collarbones) laterally. The manubrium also articulates laterally with the first and second ribs. The **body,** or midportion, makes up the bulk of the sternum. It is formed from four separate bones, one inferior to the other, that fuse after puberty. The sides of the sternal body are notched where it articulates with the costal cartilages of the second to seventh ribs. The **xiphoid process** (zif′oid; "sword-like") forms the inferior end of the sternum. This tongue-shaped process is a plate of hyaline cartilage in youth, and it does not fully ossify until about age 40.

✚ In some people, the xiphoid process projects dorsally. Blows to the chest can push such a xiphoid into the underlying heart or liver, causing massive hemorrhage. ■

The sternum has three important anatomical landmarks: the jugular notch, the sternal angle, and the xiphisternal joint (Figure 7.19a). The **jugular notch** (also called the *suprasternal notch*) is the central indentation in the superior border of the manubrium. This notch generally lies in the same horizontal plane as the disc between the second and third thoracic vertebrae (Figure 7.19b). The manubrium is joined to the body of the sternum by a fibrocartilaginous joint that acts as a hinge, allowing the sternal body to swing anteriorly when we inhale. This joint forms the **sternal angle,** a horizontal ridge across the anterior surface of the sternum. The sternal angle is in line with the disc between the fourth and fifth thoracic vertebrae. Anteriorly, the sternal angle lies at the level of the second pair of ribs. It forms a handy reference point for finding the second rib and thus for counting the ribs during a physical examination. The **xiphisternal** (zif″ĭ-ster′nal) **joint** is the point where the sternal body and xiphoid process fuse. It lies at the level of the ninth thoracic vertebra.

Ribs

Twelve pairs of **ribs** (Latin: *costa*) form the flaring sides of the thoracic cage (Figure 7.19a). All ribs attach posteriorly to the thoracic vertebrae and curve inferiorly toward the anterior part of the chest. The superior seven pairs of ribs attach directly to the sternum by their costal cartilages. These are the **true ribs,** or *vertebrosternal* (ver″tĕ-bro-ster′nal) *ribs.* The inferior five pairs of ribs are called **false ribs** because they attach to the sternum either indirectly or not at all. Rib pairs 8–10 attach to the sternum indirectly by joining each other via the costal cartilages immediately superior. These three pairs of ribs are called *vertebrochondral* (ver″tĕ-bro-kon′dral) *ribs.* Rib pairs 11 and 12 are called **floating ribs** or *vertebral ribs* because they

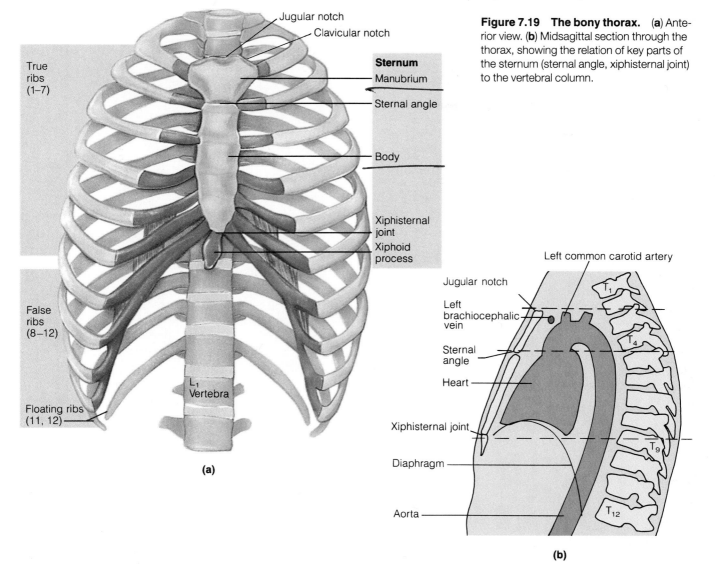

Figure 7.19 The bony thorax. **(a)** Anterior view. **(b)** Midsagittal section through the thorax, showing the relation of key parts of the sternum (sternal angle, xiphisternal joint) to the vertebral column.

have no anterior attachments. Instead, their costal cartilages lie embedded in the muscles of the lateral body wall. The ribs increase in length from pair 1 to 7, then decrease in length from pair 8 to 12. The inferior margin of the rib cage, or *costal margin,* is formed by the costal cartilages of ribs 7 to 10. The right and left costal margins diverge from the region of the xiphisternal joint, where they form the *infrasternal angle* (*infra* = below).

A typical rib is a bowed flat bone (Figure 7.20a–c). The bulk of a rib is simply called the **shaft.** Its superior border is smooth, but its inferior border is sharp and thin and features a **costal groove** on its inner face. This groove lodges the intercostal nerves and vessels. In addition to the shaft, each rib has a head, neck, and tubercle. The wedge-shaped **head** articulates with the vertebral column by two facets: One facet joins the body of the thoracic vertebra of the same number. The other joins the body of the vertebra immediately su-

perior. The **neck** of a rib is the short, constricted region that lies just lateral to the head. Just lateral to the neck, the knob-like **tubercle** articulates with the transverse process of the thoracic vertebra of the same number. Lateral to the tubercle, the shaft angles sharply anteriorly (at the **angle** of the rib) and extends to the costal cartilage anteriorly. The costal cartilages provide secure but flexible attachments of ribs to the sternum. These hyaline cartilages contribute to the elasticity of the thoracic cage.

The *first rib* is atypical (Figure 7.20d). It is flattened from superior to inferior and is quite broad. It forms a horizontal table to support the subclavian vessels (the large artery and vein servicing the upper limb), which groove its superior surface. There are other exceptions to the typical rib pattern: Rib 1 and ribs 10–12 articulate with only one vertebral body, and ribs 11 and 12 do not articulate with a vertebral transverse process.

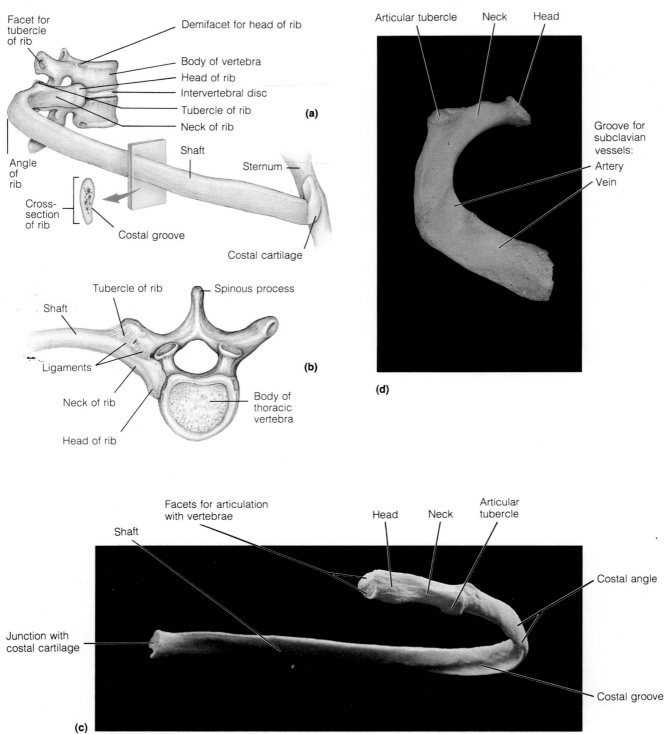

Figure 7.20 Ribs. (a) Lateral view of a typical true rib (ribs 2 through 7), showing its articulations with the vertebral column and sternum. (b) Superior view of rib articulating with a thoracic vertebra. (c) A typical rib (rib 6, right), posterior view. (d) Rib 1, superior view.

PART 2: THE APPENDICULAR SKELETON

Limb bones and their girdles are *appended* to the axial skeleton. Thus, they are collectively called the **appendicular skeleton** (Figure 7.1). The pectoral girdles (pek'tor-al; "chest") attach the upper limbs to the trunk, while the pelvic girdle secures the lower limbs. Although the bones of the upper and lower limbs differ in their functions, they share the same basic structural plan: Each limb is composed of a girdle and three segments connected by movable joints.

The appendicular skeleton enables us to carry out the movements typical of our freewheeling and manipulative life-style. Each time we take a step, throw a ball, or write with a pencil, we use our appendicular skeleton.

As you read about the limb bones, refer to Table 7.4 on page 174, which summarizes the appendicular skeleton.

The Pectoral Girdle

The **pectoral girdle,** or *shoulder girdle,* consists of a *clavicle* (klav'ĭ-k'l) anteriorly and a *scapula* (skap'u-lah) posteriorly (Figure 7.21). The paired pectoral girdles and their associated muscles form the shoulders. The term *girdle* implies a belt completely circling the body, but the pectoral girdles do not quite satisfy this description: Anteriorly, the medial end of each clavicle joins the sternum and first rib, and the lateral ends of the clavicles join the scapulae at the shoulder. However, the two scapulae fail to complete the ring posteriorly, because their medial borders do not join each other or the axial skeleton.

Besides attaching the upper limb to the trunk, the pectoral girdle provides points of insertion for many muscles that move the limb. This girdle is light and allows the upper limbs to be quite mobile. This mobility springs from the following factors:

1. Since only the clavicle attaches to the axial skeleton, the scapula can move quite freely across the thorax, allowing the arm to move with it.

2. The socket of the shoulder joint (the scapula's glenoid cavity) is shallow, so it does not restrict the movement of the humerus (arm bone). Although this arrangement is good for flexibility, it is bad for stability: Shoulder dislocations are fairly common.

Clavicles

The **clavicles** ("little keys"), or collarbones, are slender, slightly curved long bones (Figure 7.21b and c) that extend horizontally across the superior thorax, just deep to the skin. Each clavicle is cone-shaped at its medial **sternal end,** which attaches to the manubrium, and flattened at its lateral **acromial** (ah-kro'me-al) **end,** which articulates with the scapula. The medial two-thirds of the clavicle is convex anteriorly, while the lateral third is concave anteriorly. The superior surface is almost smooth, but the inferior surface is ridged and grooved by the ligaments and muscles that attach to it. Many of these act to bind the clavicle to the rib cage and scapula.

The clavicles perform several functions. Besides providing attachment sites for muscles, they act as braces; that is, they hold the scapulae and arms out laterally from the thorax. This function becomes obvious when a clavicle is fractured: The entire shoulder region collapses medially. The clavicles also transmit compression forces from the upper limbs to the axial skeleton, as when someone reaches forward and pushes a car to a gas station.

➕ The clavicles are not very strong, and they often fracture when a person falls on a shoulder or uses outstretched arms to break a fall. The curves in the clavicle ensure that it fractures anteriorly (outward) at its middle third. If it were to fracture posteriorly (inward), bone splinters would pierce the main blood vessels to the arm (the subclavian vessels), which pass just deep to the clavicle. ■

Scapulae

The **scapulae,** or shoulder blades, are thin, triangular flat bones (Figure 7.21d–f). Their name derives from a word meaning "spade" or "shovel," for ancient cultures made spades from the shoulder blades of animals. The scapulae lie on the dorsal surface of the rib cage, between rib 2 superiorly and rib 7 inferiorly. Each scapula has three borders. The **superior border** is the shortest, sharpest border. The **medial border,** or *vertebral border,* parallels the vertebral column. The **lateral border,** or *axillary border,* abuts the axilla and ends superiorly in a shallow fossa, the **glenoid cavity** (gle'noid; "pit-shaped"). The glenoid cavity articulates with the humerus of the arm, forming the shoulder joint.

Like all triangles, the scapula has three corners, or *angles.* The glenoid cavity lies at the scapula's **lateral angle.** The **superior angle** forms where the superior and medial borders meet, and the **inferior angle** forms at the junction of the medial and lateral borders.

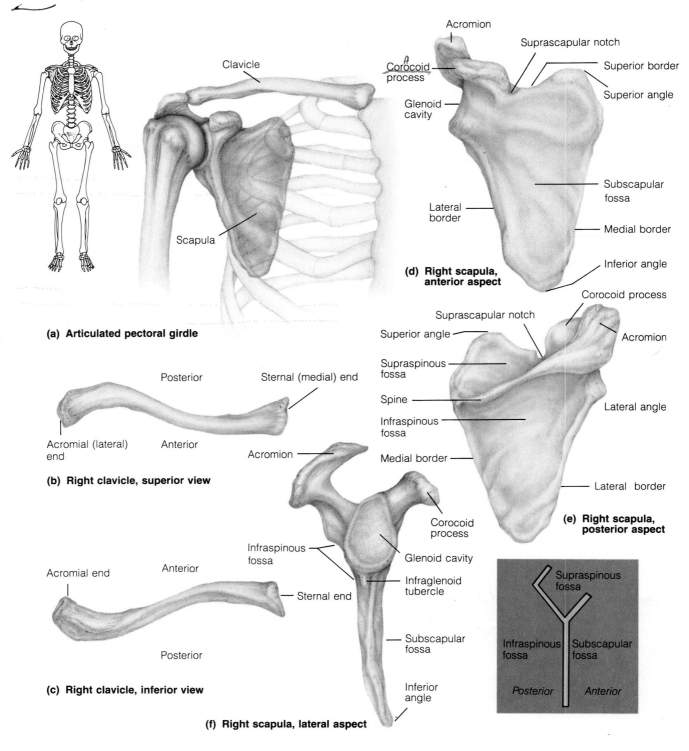

(a) Articulated pectoral girdle

(b) Right clavicle, superior view

(c) Right clavicle, inferior view

(d) Right scapula, anterior aspect

(e) Right scapula, posterior aspect

(f) Right scapula, lateral aspect

Figure 7.21   Bones of the right pectoral girdle.

The inferior angle moves as the arm is raised and lowered and is an important landmark for studying scapular movements.

The anterior, or costal, surface of the scapula is slightly concave and relatively featureless. The posterior surface, by contrast, bears a prominent **spine** that is easily felt through the skin. The spine ends lat-

erally in a flat, triangular projection called the **acromion** (ah-kro′me-on; "apex of shoulder"). The acromion articulates with the acromial end of the clavicle.

The **coracoid** (kor′ah-coid) **process** projects anteriorly from the lateral part of the superior scapular border. The word *corac* means "beak-like," but the coracoid process looks more like a bent finger. The coracoid is an attachment point for the biceps muscle of the arm. Strong ligaments also bind the coracoid to the clavicle. Just medial to the coracoid process lies the **suprascapular notch** (a nerve passage), and just lateral to it lies the glenoid cavity.

Several large but shallow depressions, or fossae, occur on both surfaces of the scapula and are named according to location. The **infraspinous** and **supraspinous fossae** lie inferior and superior to the spine, respectively. The **subscapular fossa** is the shallow concavity formed by the entire anterior surface of the scapula.

Bones of the Free Upper Limb

Thirty bones form the free portion of the skeleton of the upper limb (Figures 7.22 to 7.25). They can be grouped into bones of the arm, forearm, and hand.

Arm

Anatomists use the term *arm* to designate the portion of the upper limb between the shoulder and elbow only. The **humerus** (hu′mer-us) is the only bone of the arm (Figure 7.22). The largest and longest bone in the upper limb, it articulates with the scapula at the shoulder and with the radius and ulna (forearm bones) at the elbow.

At the proximal end of the humerus is the hemispherical **head,** which fits into the glenoid cavity of the scapula. Just inferior to the head is a slight constriction, the **anatomical neck.** Just inferior to the anatomical neck are the lateral **greater tubercle** and the medial **lesser tubercle.** These tubercles are separated by the **intertubercular groove,** or *bicipital* (bi-sip′ĭ-tal) *groove.* The tubercles are sites where muscles attach. The intertubercular groove guides a tendon of the biceps muscle to its attachment point at the rim of the glenoid cavity. Just inferior to the tubercles is the **surgical neck,** so named because it is the most fre-

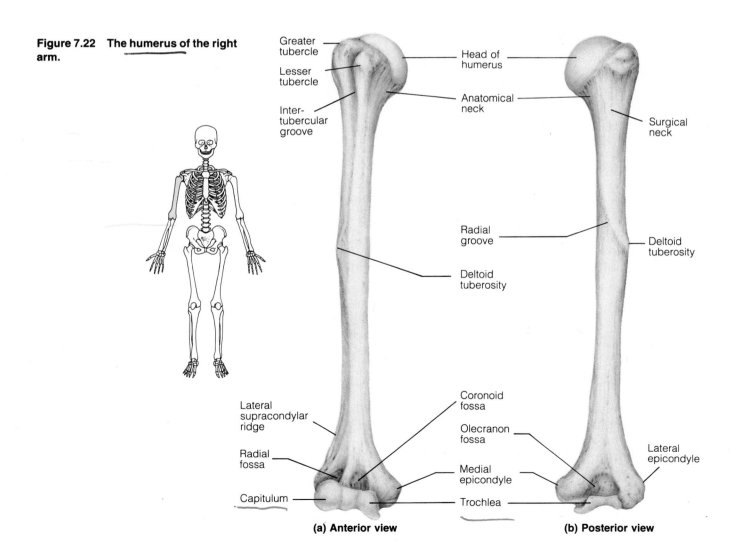

Figure 7.22   The humerus of the right arm.

Greater tubercle

Lesser tubercle

Intertubercular groove

Head of humerus

Anatomical neck

Surgical neck

Radial groove

Deltoid tuberosity

Deltoid tuberosity

Lateral supracondylar ridge

Radial fossa

Capitulum

Coronoid fossa

Olecranon fossa

Medial epicondyle

Trochlea

Lateral epicondyle

(a) Anterior view

(b) Posterior view

quently fractured region of the humerus. About mid-way down the shaft, on the lateral side, is the **deltoid tuberosity.** This roughened area is the attachment site of the deltoid muscle of the shoulder. Near the deltoid tuberosity, a **radial groove** descends obliquely along the posterior surface of the shaft. This groove marks the course of the radial nerve, an important nerve of the upper limb.

At the distal end of the humerus are two condyles, a medial **trochlea** (trok′le-ah; "pulley") and a lateral **capitulum** (kah-pit′u-lum; "small head"). The trochlea looks like an hourglass turned on its side, and the capitulum is shaped like half a ball. These condyles articulate with the ulna and the radius, respectively. They are flanked by the **medial** and **lateral epicondyles** (Figure 7.22), which serve as attachment sites for muscles of the forearm. Superior to the trochlea, on the posterior surface of the humerus, is the deep **olecranon** (o-lek′rah-non) **fossa.** In this position on the anterior surface is a shallower **coronoid** (kor′o-noid) **fossa.** A small **radial fossa** lies lateral to the coronoid fossa.

Forearm

Forming the skeleton of the forearm are two parallel long bones, the radius and ulna (Figures 7.23 and 7.24). Their proximal ends articulate with the humerus, while their distal ends reach the wrist. The radius and ulna articulate with each other both proximally and distally at the small *radioulnar* (ra″de-o-ul′nar) *joints.* Furthermore, they are interconnected along their entire length by a flat ligament called the **interosseous membrane** (in″ter-os′e-us; "between the bones"). In the anatomical position, the radius lies laterally (on the thumb side), and the ulna medially. However, when the palm faces posteriorly, the distal end of the radius crosses over the ulna, and the two bones form an X (see Figure 8.6a on p. 199).

Ulna

The **ulna** (ul′nah; "elbow") is slightly longer than the radius. It has the main responsibility for forming the elbow joint with the humerus. The ulna looks much like a monkey wrench. At its proximal end are two prominent projections, the **olecranon** ("elbow") **process** and the **coronoid** ("crown-shaped") **process,** separated by a deep concavity, the **trochlear notch.** Together, these two processes grip the trochlea of the humerus, forming a stable hinge joint that allows the forearm to be bent upon the arm (flexed), then straightened again (extended). When the forearm is fully extended (Figure 7.24b), the olecranon process "locks" into the olecranon fossa of the humerus. When the forearm is flexed, the coronoid process of the ulna fits into the coronoid fossa of the humerus.

On the lateral side of the coronoid process of the ulna is a smooth depression, the **radial notch,** which articulates with the head of the radius.

Distally, the shaft of the ulna narrows and ends in a knob-like **head.** Medial to this head is a **styloid** ("stake-shaped") **process,** from which a ligament runs to the wrist. The head of the ulna is separated from the bones of the wrist by a disc of fibrocartilage and plays little or no role in hand movements.

Radius

The **radius** ("rod") is thin at its proximal end and widened at its distal end—the opposite of the ulna. The proximal **head** of the radius is shaped like the end of a spool of thread (Figure 7.23). The superior surface of this head is concave, and it articulates with the capitulum of the humerus in the elbow joint. Medially, the head of the radius articulates with the radial notch of the ulna at the *proximal radioulnar joint.* Just distal to the head is a rough bump, the **radial tuberosity.** This is a site of attachment of the biceps muscle of the arm. The distal part of the radius is flared. It has a medial **ulnar notch,** which articulates with the ulna at the *distal radioulnar joint,* and a lateral **styloid process.** Like the styloid process of the ulna, this little knob anchors a ligament that extends to the wrist. The extreme distal end of the radius is concave, and it articulates with carpal bones of the wrist. While the ulna contributes heavily to the elbow joint, the radius contributes heavily to the wrist joint. When the radius moves, the hand moves with it.

Hand

The skeleton of the hand includes the bones of the *carpus,* or wrist; the bones of the *metacarpus,* or palm; and the *phalanges,* or bones of the fingers (Figure 7.25).

Carpus

A "wrist" watch is actually worn on the distal forearm, not on the wrist at all. The true wrist, or **carpus** (kar′pus), is the proximal region of the hand, just distal to the wrist joint. The carpus contains a group of eight marble-sized short bones, or **carpals** (kar′palz), closely united by ligaments. Gliding movements occur between the carpals, making the wrist rather flexible. The carpals are arranged in two irregular rows of four bones each (Figure 7.25). In the proximal row, from lateral (thumb side) to medial, are the **scaphoid** (skaf′oid; "boat-shaped"), **lunate** (lu′nāt; "moon-like"), **triquetral** (tri-kwet′ral; "triangular"), and **pisiform** (pi′sĭ-form; "pea-shaped"). Only the scaphoid and lunate bones articulate with the radius to form the wrist joint. The carpals of the distal row,

Figure 7.23   Radius and ulna of the right forearm.

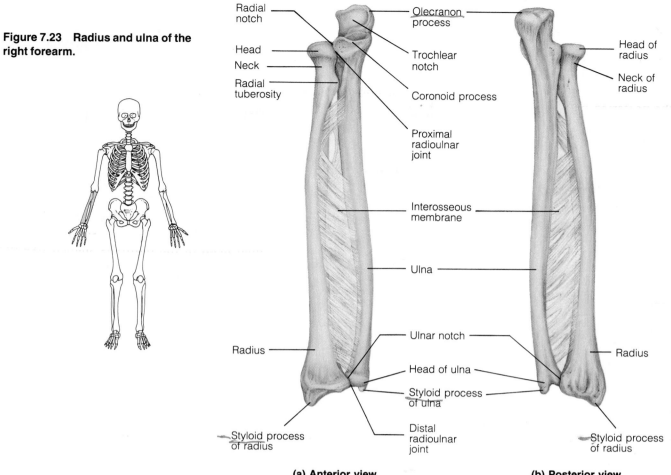

Radial notch
Head
Neck
Radial tuberosity
Olecranon process
Trochlear notch
Coronoid process
Proximal radioulnar joint
Head of radius
Neck of radius
Interosseous membrane
Ulna
Radius
Ulnar notch
Head of ulna
Styloid process of ulna
Radius
Styloid process of radius
Distal radioulnar joint
Styloid process of radius

(a) Anterior view　　　　　　**(b) Posterior view**

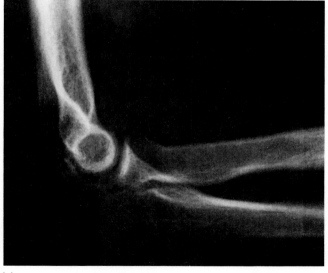

(a)

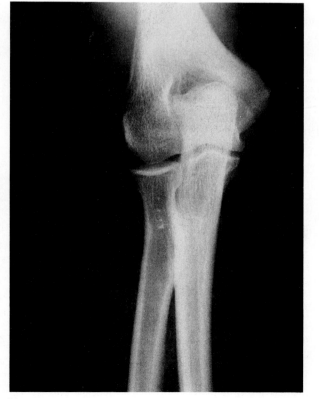

(b)

Figure 7.24   X-ray images of the left elbow joint, formed by articulation of the humerus, ulna, and radius.   (a) Elbow flexed, lateral view. (b) Elbow extended, posterior view.

171

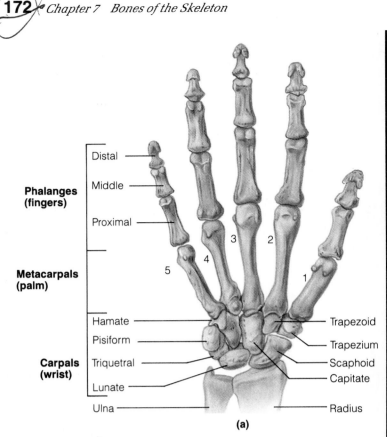

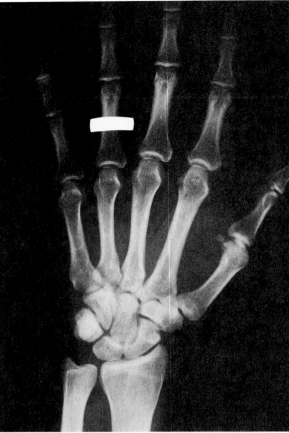

(b)

Figure 7.25   Bones of the hand.   (a) Ventral view of the right hand, illustrating the carpals, metacarpals, and phalanges. **(b)** X-ray image of the right hand. Note the white bar on the proximal phalanx of the ring finger, showing where a ring would be worn.

again from lateral to medial, are the **trapezium** (trah-pe′ze-um; "little table"), **trapezoid** ("four-sided"), **capitate** (kap′i-tāt; "head-shaped"), and **hamate** (ham′āt; "hooked"). Here is a "memory phrase" to help you recall the carpals in the order given above: **S**ally **l**eft **t**he **p**arty **t**o **t**ake **C**athy **h**ome. As in all such memory jogs, the first letter of each word is the first letter of the term you need to remember.

 The arrangement of carpal bones makes the carpus concave anteriorly. A ligamentous band covers this concavity superficially, forming the *carpal tunnel.* Through this narrow tunnel run many long muscle tendons, which extend from the forearm to the fingers. Also crowded in the tunnel is the median nerve, which (roughly speaking) innervates the lateral half of the hand, including the muscles that move the thumb. Inflammation of any element in the carpal tunnel, such as tendons swollen from overuse, can compress the median nerve. This nerve impairment, which affects many workers who type at keyboards all day, is called **carpal tunnel syndrome.**

Can you guess some symptoms of carpal tunnel syndrome?

As the nerve is impaired, the skin of the lateral part of the hand tingles or becomes numb, and movements of the thumb weaken. This condition can be treated by resting the hand in a splint at night, by anti-inflammatory drugs, or by surgery. ■

Metacarpus

Five **metacarpals** radiate distally from the wrist to form the palm of the hand (*meta* = beyond). These small long bones are not given individual names but instead are numbered 1 to 5, from lateral (thumb side) to medial (little finger side). The *bases* of the metacarpals articulate with the carpals proximally, and with each other laterally and medially. The bulbous *heads* of the metacarpals articulate with the proximal phalanges of the fingers to form the knuckles. Metacarpal 1, associated with the thumb, is the shortest and most mobile.

Phalanges

The digits, or fingers, are numbered 1 to 5 beginning with the thumb, or *pollex* (pol′eks). The fingers contain miniature long bones called **phalanges** (fah-lan'jēz). The singular of this term is *phalanx* (fa'langks; "a closely knit row of soldiers"). In most people, the third finger is the longest. With the exception of the thumb, each finger has three phalanges: *proximal, middle,* and *distal.* The thumb has no middle phalanx.

The Pelvic Girdle

The **pelvic girdle,** or *hip girdle,* attaches the lower limbs to the spine and supports the visceral organs of the pelvis (Table 7.4 and Figure 7.26). The full weight of the upper body passes through the pelvic girdle to the free lower limbs. While the pectoral girdle is barely attached to the thoracic cage, the pelvic girdle is attached to the axial skeleton by some of the strongest ligaments in the body. Furthermore, whereas the glenoid cavity of the scapula is shallow, the corresponding socket in the pelvic girdle is a deep cup that secures the head of the femur firmly in place. Consequently, the lower limb and its girdle have less freedom of movement than the upper limb but are much more stable.

The pelvic girdle consists of the paired **hip bones** (Figure 7.26). A hip bone is also called a **coxal** (kok' sal) **bone,** an **os coxae** (*os* = bone; *coxa* = hip), and an **innominate** ("no name") **bone.** Each hip bone unites with its partner anteriorly and with the sacrum posteriorly. The deep, basin-like structure formed by the hip bones, sacrum, and coccyx is the **bony pelvis** (Figure 7.26a).

The hip bone is large and irregularly shaped (Figure 7.26b and c). During childhood, it consists of three separate bones: the *ilium, ischium,* and *pubis.* In adults, these bones are fused and their boundaries indistinguishable. Their names are retained, however, to refer to different regions of the composite hip bone. At the point of fusion of the ilium, ischium, and pubis is a deep hemispherical socket, the **acetabulum** (as"ĕ-tab'u-lum), on the lateral pelvic surface (Figure 7.26b). *Acetabulum* means "vinegar cup" in Latin, a good description. The acetabulum articulates with the head of the femur.

Ilium

The **ilium** (il'e-um; "flank") is a large, flaring bone that forms the superior region of the hip bone. It consists of an inferior **body** and a superior wing-like **ala** ("wing"). The thickened superior margin of the ala is the **iliac crest.** Many muscles attach to this crest, which is thickest about halfway along its length, at the *tubercle of the iliac crest.* Each iliac crest ends anteriorly in a blunt **anterior superior iliac spine** and pos-

teriorly in a sharp **posterior superior iliac spine.** Located inferior to these are the **anterior** and **posterior inferior iliac spines.** The anterior superior iliac spine is an especially prominent anatomical landmark (p. 695) and is easily felt through the skin.

Posteriorly, just inferior to the posterior inferior iliac spine, the ilium is deeply indented to form the **greater sciatic notch** (si-at'ik; "of the hip") (Figure 7.26b). The sciatic nerve, the largest nerve in the body, passes through this notch to enter the thigh. The broad posterolateral surface of the ilium, the *gluteal surface* (glu'te-al; "buttocks"), is crossed by three ridges: the **posterior, anterior,** and **inferior gluteal lines.** These lines define the attachment sites of the gluteal (buttocks) muscles.

The internal surface of the iliac ala is concave. This broad concavity is called the **iliac fossa** (Figure 7.26c). Posterior to this fossa lies a roughened **auricular surface** (aw-rik'u-lar; "ear-shaped"), which articulates with the sacrum, forming the *sacroiliac joint.* The weight of the body is transmitted from the vertebral column to the pelvis through this joint. Running anteriorly and inferiorly from the auricular surface is a robust ridge called the **arcuate line** (ar'ku-āt; "bowed"), which helps define the superior boundary of the true pelvis (p. 177). The inferior part of the ilium joins the ischium and pubic bones.

Ischium

The **ischium** (is'ke-um; "hip") forms the posteroinferior region of the hip bone (Figure 7.26b). Shaped roughly like an L or an arc, it has a thicker, superior **body** and a thinner, inferior **ramus** (*ramus* = branch). Anteriorly, the ischial ramus joins the pubis. The **ischial spine** lies posterior to the acetabulum and projects medially. This triangular spine is an attachment point for a ligament from the sacrum (the sacrospinous ligament). Just inferior to the ischial spine is the **lesser sciatic notch.** Some important nerves and vessels pass through this notch to the perineum (area around the anus and external genitals). The inferior surface of the ischial body is rough and thickened as the **ischial tuberosity** (Figure 7.26b). When we sit, our weight is borne entirely by the ischial tuberosities, which are the strongest parts of the hip bones. A massive ligament runs from the sacrum to each ischial tuberosity. This *sacrotuberous ligament* (not illustrated) helps hold the pelvis together.

Pubis

The **pubis** (pu'bis; "sexually mature"), or *pubic bone,* forms the anterior region of the hip bone (Figure 7.26). In the anatomical position, the pubis lies nearly horizontally, and the bladder rests upon it. Essentially, the

(text continues on p. 176)

Table 7.4   Bones of the Appendicular Skeleton

Body region	Bones*	Illustration	Location	Markings
PART I: Bones of the Pectoral Girdle and Upper Limb				
Pectoral girdle (Figure 7.21)	Clavicle (2)		Superoanterior thorax; articulates medially with sternum and laterally with scapula	Acromial end; sternal end
	Scapula (2)		Posterior thorax; forms part of the shoulder; articulates with humerus and clavicle	Glenoid cavity; spine; acromion; coracoid process; infraspinous, supraspinous, and subscapular fossae
Free upper limb Arm (Figure 7.22)	Humerus (2)		Sole bone of arm; between scapula and elbow	Head; greater and lesser tuberosities; intertubercular groove; deltoid tuberosity; trochlea; capitulum; coronoid and olecranon fossae; radial groove; epicondyles
Forearm (Figure 7.23)	Ulna (2)		Medial bone of forearm between elbow and wrist; forms elbow joint	Coronoid process; olecranon process; radial notch; trochlear notch; styloid process; head
	Radius (2)		Lateral bone of forearm; carries wrist	Radial tuberosity; styloid process; head
Hand (Figure 7.25)	8 Carpals (16) (scaphoid, lunate, triquetral, pisiform, trapezium, trapezoid, capitate, hamate)		Form a bony crescent at the wrist; arranged in two rows of four bones	
	5 Metacarpals (10)		Form the palm; one in line with each digit	
	14 Phalanges (28) (distal, middle, proximal)		Form the fingers; three in digits 2–5; two in digit 1 (the thumb)	

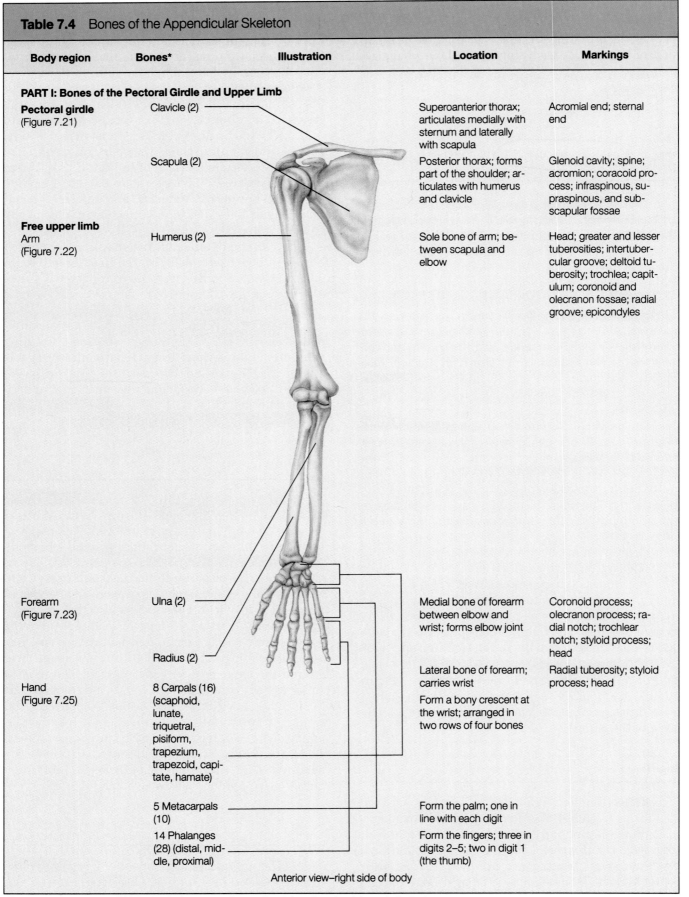

Anterior view–right side of body

*The number in parentheses ( ) following the bone name denotes the total number of such bones in the body.

Table 7.4 (continued)

Body region	Bones*	Illustration	Location	Markings
PART II: Bones of the Pelvic Girdle and Lower Limb				
Pelvic girdle (Figure 7.26)	Coxal (2)		Each coxal bone is formed by the fusion of an ilium, ischium, and pubic bones; the coxal bones fuse anteriorly at the pubic symphysis and form sacroiliac joints with the sacrum posteriorly; girdle consisting of both coxal bones is basin-like	Iliac crest; anterior and posterior iliac spines; auricular surface; greater and lesser sciatic notches; obturator foramen; ischial tuberosity and spine; acetabulum; pubic arch; pubic crest; pubic tubercle
Free lower limb Thigh (Figure 7.27b)	Femur (2)		Sole bone of thigh; between hip joint and knee; largest bone of the body	Head; greater and lesser trochanters; neck; lateral and medial condyles and epicondyles; gluteal tuberosity; linea aspera
Kneecap (Figure 7.27a)	Patella (2)		A sesamoid bone lodged in the tendon of the quadriceps (anterior thigh) muscles	
Leg (Figure 7.28)	Tibia (2)		Larger and more medial bone of leg; between knee and foot	Medial and lateral condyles; tibial tuberosity; anterior crest; medial malleolus
	Fibula (2)		Lateral bone of leg; sticklike	Head; lateral malleolus
Foot (Figure 7.29)	7 Tarsals (14) (talus, calcaneus, navicular, cuboid, and lateral, intermediate, and medial cuneiforms)		Seven bones forming the proximal part of the foot; the talus articulates with the leg bones at the ankle joint; the calcaneus, the largest tarsal, forms the heel	
	5 Metatarsals (10)		Five bones forming each sole	
	14 Phalanges (28) (distal, middle, proximal)		Form the toes; three in digits 2–5, two in digit 1 (the great toe)	

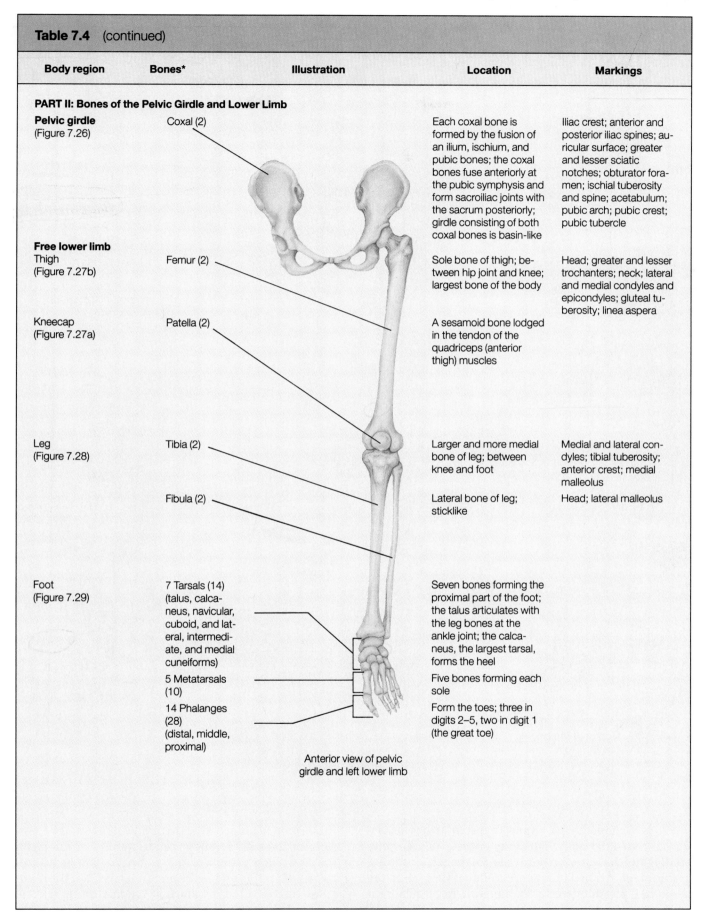

Anterior view of pelvic girdle and left lower limb

*The number in parentheses ( ) following the bone name denotes the total number of such bones in the body.

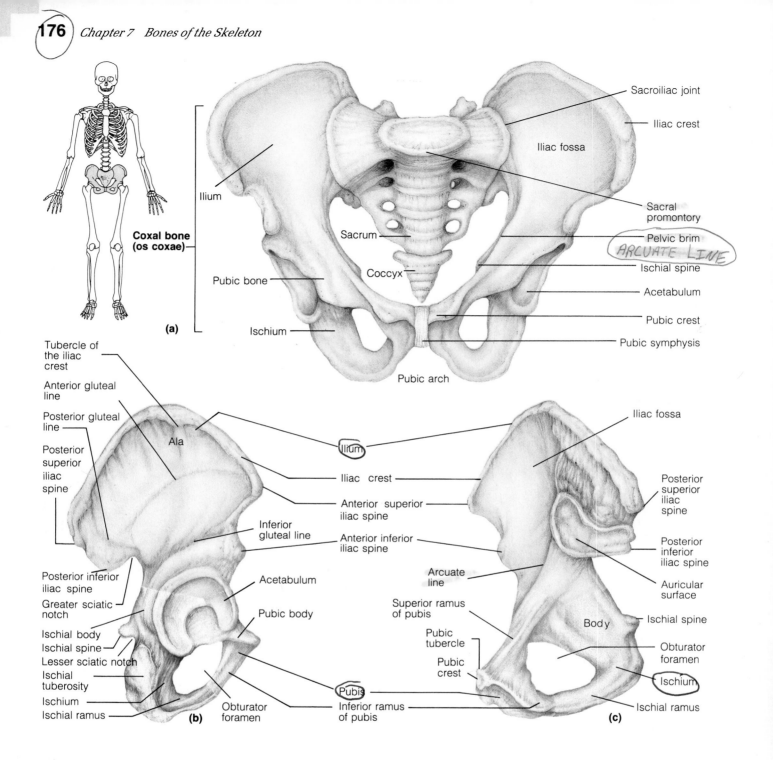

Figure 7.26 Bones of the pelvis. (a) Bony pelvis showing the two hip bones (coxal bones) and the sacrum. (b) Lateral view of the right hip bone, showing the ilium, ischium, and pubic bones (anterior is to the right). (c) Medial view of the right hip bone (anterior is to the left).

pubis is V-shaped, with **superior** and **inferior rami** issuing from a flat **body.** The body of the pubis lies medially, and its anterior border is thickened to form a **pubic crest.** At the lateral end of the pubic crest is the knob-like **pubic tubercle,** an attachment point for an important ligament, the inguinal ligament. The two rami of the pubic bone extend laterally to join with the body and ramus of the ischium. The space between the pubis and ischium is a large hole, the **obturator** (ob"tu-ra'tor) **foramen.** Students often ask the function of this foramen, but the answer is disappointing: Although a few vessels and nerves pass through it, the obturator foramen is almost completely closed by a fibrous membrane, the obturator membrane. In fact, the word *obturator* literally means "closed up."

In the midline, the bodies of the two pubic bones are joined by a disc of fibrocartilage. This joint is the *pubic symphysis (symphysis pubis)*. Inferior to this joint, the inferior pubic rami and the ischial rami form an arch shaped like an inverted V, the **pubic arch** (Figure 7.26a). The acuteness of this arch helps to distinguish the male from the female pelvis (Table 7.5).

True and False Pelvis

The bony pelvis is divided into two parts, the *false (greater) pelvis* and the *true (lesser) pelvis.* These are separated by the **pelvic brim,** a continuous oval ridge that runs from the pubic crest through the arcuate line and the sacral promontory (Figure 7.26a). The **false pelvis,** superior to the pelvic brim, is bounded by the alae of the iliac bones. It is actually part of the abdomen and contains abdominal organs. The **true pelvis** lies inferior to the pelvic brim. It forms a deep "bowl" containing the pelvic organs.

Pelvic Structure and Childbearing

The major differences between the "typical" male and female pelvis are summarized and illustrated in Table 7.5. So consistent are these differences that an anatomist can determine the gender of a skeleton with a 90% degree of certainty merely by examining the pelvis. The female pelvis is adapted for childbearing: It tends to be wider, shallower, and lighter than that of a male. These features provide more room in the true pelvis, which must be wide enough for an infant's head to pass at birth. The size of the true pelvis, particularly the size of its inlet and outlet, is critical to the uncomplicated delivery of a baby. For this reason, obstetricians measure it carefully.

The *pelvic inlet* is the pelvic brim. The largest diameter of this inlet is from side-to-side (Table 7.5). As labor begins, the infant's head enters this inlet, its forehead facing one ilium and its occiput facing the other. If the mother's sacral promontory is too large, it can block the entry of the infant into the true pelvis. The *pelvic outlet* is the inferior margin of the true pelvis (see the photos at bottom of Table 7.5). The outlet's anterior boundary is the pubic arch; its lateral boundaries are the ischial tuberosities, and its posterior boundary is the sacrum and coccyx. Both the coccyx and the ischial spines protrude into the outlet, so a

sharply angled coccyx or unusually large ischial spines can interfere with delivery. The largest dimension of the pelvic outlet is the anteroposterior diameter. Generally, after the infant's head passes through the inlet, it rotates so that the forehead faces posteriorly and the occiput anteriorly. This is the usual position of the infant's head as it leaves the mother's body (see Figure 23.27 on p. 653). Thus, during birth, the infant's head makes a quarter turn to follow the widest dimensions of the true pelvis.

Bones of the Free Lower Limb

The lower limbs carry the entire weight of the erect body and are subjected to exceptional forces when we jump or run. Thus, it is not surprising that the bones of the lower limbs are thicker and stronger than the comparable bones of the upper limbs. The three segments of the free lower limb are the thigh, the leg, and the foot (see Table 7.4 on p. 175)

Thigh

The **femur** (fe′mur; "thigh") is the single bone of the thigh (Figure 7.27b). It is the largest, longest, strongest bone in the body. Its durable structure reflects the fact that the stress on this bone can reach 280 kg per cm^2, or two tons per square inch! The femur courses medially as it descends toward the knee.

Can you propose a functional explanation for the medial course of the femur?

The medial course places the knee joints closer to the body's center of gravity in the midline and thus provides for better balance. The medial course of the femur is more pronounced in women because of their wider pelvis. Some orthopedists think this helps explain the greater incidence of knee problems in female athletes. ■

The ball-like **head** of the femur has a small central pit called the **fovea capitis** (fo′ve-ah cap′ĭ-tis; "pit of the head"). A short ligament, the *ligament of the head of the femur,* runs from this pit to the acetabulum of the hip bone. The head of the femur is carried on a **neck.** The neck does not descend straight vertically but angles laterally to join the shaft. This angled course reflects the fact that the femur articulates with the lateral aspect (rather than the inferior region) of the pelvis. The neck is the weakest part of the femur and is often fractured, an injury commonly called a broken hip. At the junction of the shaft and neck are

Table 7.5   Comparison of the Male and Female Pelves

Characteristic	Female	Male
General structure and functional modifications	Tilted forward; adapted for childbearing; true pelvis defines the birth canal; cavity of the true pelvis is broad, shallow, and has a greater capacity	Tilted less far forward; adapted for support of a male's heavier build and stronger muscles; cavity of the true pelvis is narrow and deep
Bone thickness	Less; bones lighter, thinner, and smoother	Greater; bones heavier and thicker, and markings are more prominent
Acetabula	Smaller; farther apart	Larger; closer
Pubic angle/arch	Broader (80° to 90°); more rounded	Angle is more acute (50° to 60°)
Anterior view		

Pelvic brim

Sacrum	Wider; shorter; sacral curvature is accentuated	Narrow; longer; sacral promontory more ventral
Coccyx	More movable; straighter	Less movable; curves ventrally
Left lateral view		

Pelvic inlet (brim)	Wider; oval from side to side	Narrow; basically heart shaped
Pelvic outlet	Wider; ischial tuberosities shorter, farther apart and everted	Narrower; ischial tuberosities longer, sharper, and point more medially
Posteroinferior view		

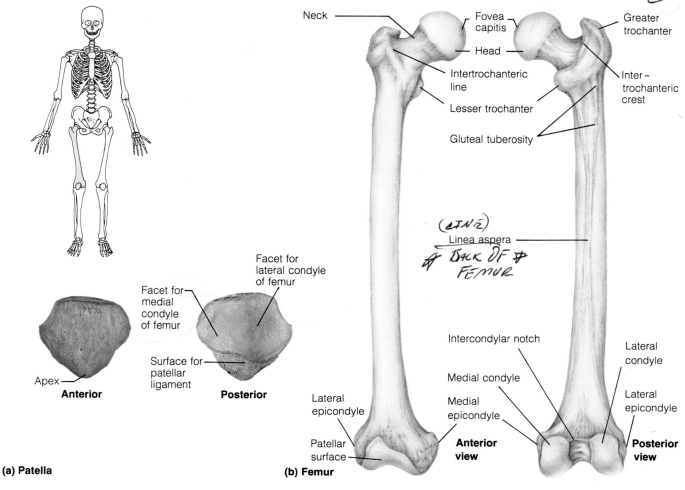

Figure 7.27 The right patella (a) and femur (b).

the lateral **greater trochanter** and medial **lesser trochanter.** Various muscles attach to these robust projections. The two trochanters are interconnected by the *intertrochanteric line* anteriorly and by the prominent *intertrochanteric crest* posteriorly. Inferior to the intertrochanteric crest on the posterior surface of the shaft is the **gluteal tuberosity.** This tuberosity blends into a long vertical ridge, the **linea aspera** (lin'e-ah as'per-ah), literally the "rough line." Both these markings are sites of muscle attachment. Except for its linea aspera, the entire shaft of the femur is smooth and rounded.

Distally, the femur broadens to end in **lateral** and **medial condyles** shaped like wide wheels. The most raised points on the sides of these condyles are the **lateral** and **medial epicondyles,** to which ligaments attach. Posteriorly, the two condyles are separated by a deep **intercondylar notch.** Anteriorly they are separated by a smooth **patellar surface,** which articulates with the kneecap, or patella.

The **patella** (pah-tel'ah; "small pan") is a triangular sesamoid bone enclosed in the tendon that secures the anterior muscles of the thigh to the tibia (Figure 7.27a). It protects the knee joint anteriorly and improves the leverage of the thigh muscles acting across the knee.

Leg

Anatomists use the term *leg* to refer to the part of the lower limb between the knee and the ankle. Two parallel bones, the *tibia and fibula,* form the skeleton of the leg (Figure 7.28). The tibia is more massive than the stick-like fibula and lies medial to it. These two bones articulate with each other both proximally and distally. However, unlike the joints between the radius and ulna of the forearm, the *tibiofibular* (tib"e-o-fib'u-lar) *joints* allow almost no movement. Thus, the two leg bones do not cross one another when the leg

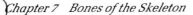

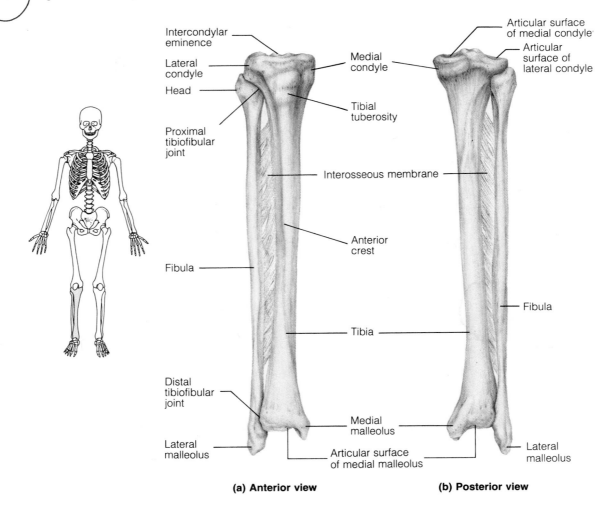

(a) Anterior view **(b) Posterior view**

Figure 7.28   The tibia and fibula of the right leg.

rotates. An **interosseous membrane** connects the tibia and fibula along their entire length. The tibia articulates with the femur to form the knee joint and with the talus bone of the foot at the ankle joint. The fibula, by contrast, does not contribute to the knee joint and merely helps stabilize the ankle joint.

Tibia

The **tibia** (tib′e-ah; "shinbone") receives the weight of the body from the femur and transmits it to the foot. It is second only to the femur in size and strength. At its broad proximal end are the **medial** and **lateral condyles.** Superiorly, the articular surfaces of these condyles are slightly concave and are separated by an irregular projection, the **intercondylar eminence.** The tibial condyles articulate with the corresponding condyles of the femur. The inferior part of the lateral tibial condyle bears a facet that joins the fibula at the *proximal tibiofibular joint.* Just inferior to the condyles, the tibia's anterior surface displays a rough **tibial tu-**berosity. A strong ligament from the patella (the patellar ligament) attaches to the tibial tuberosity.

The shaft of the tibia is triangular in cross section. Its anterior border is the sharp, subcutaneous **anterior crest.** Distally, the end of the tibia is flat where it articulates with the talus of the foot. Medial to this joint surface, the tibia has an inferior projection called the **medial malleolus** (mah-le′o-lus; "little hammer"). This forms the medial bulge of the ankle. The *fibular notch* occurs on the lateral side of the distal tibia and participates in the *distal tibiofibular joint.*

Fibula

The **fibula** (fib′u-la; "pin") is a thin long bone with two expanded ends. Its superior end is its **head,** and its inferior end is the **lateral malleolus.** The lateral malleolus forms the lateral bulge of the ankle and articulates with the talus bone of the foot. The shaft of the fibula is heavily ridged and appears to have been

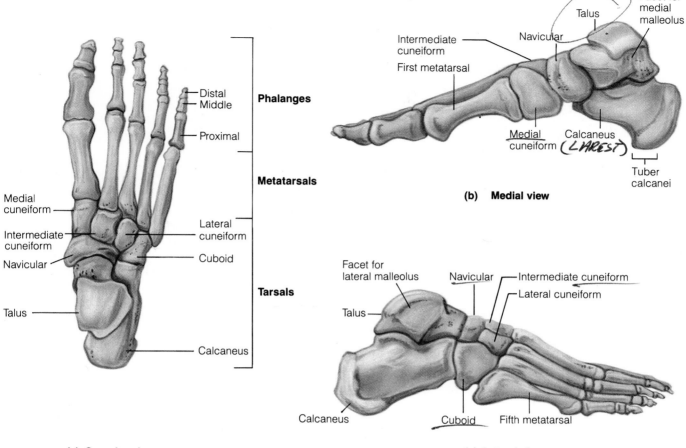

(a) **Superior view**

(b) **Medial view**

(c) **Lateral view**

Figure 7.29   Bones of the right foot.

twisted a quarter turn. The fibula does not bear weight, but several muscles originate from it.

Foot

The skeleton of the foot includes the bones of the *tarsus*, the bones of the *metatarsus*, and the *phalanges*, or toe bones (Figure 7.29). The foot has two important functions: It supports our body weight and acts as a lever to propel the body forward when we walk and run. A single bone could serve both these purposes but would adapt poorly to uneven ground. Segmentation makes the foot pliable, eliminating this problem.

Tarsus

The **tarsus** (tar′sus) makes up the posterior half of the foot and contains seven bones called **tarsals.** The tarsus is comparable to the carpus of the hand. The

weight of the body is borne primarily by the two largest, most posterior tarsal bones: the **talus** (ta′lus; "ankle"), which articulates with the tibia and fibula superiorly, and the strong **calcaneus** (kal-ka′ne-us; "heel bone"), which forms the heel of the foot. The calcaneus carries the talus on its superior surface. The thick tendon of the calf muscles attaches to the posterior surface of the calcaneus. The part of the calcaneus that touches the ground is the **tuber calcanei.** The remaining tarsal bones are the lateral **cuboid** (ku′boid; "cube-shaped"), the medial **navicular** (nah-vik′u-lar; "boat-like"), and the anterior **medial, intermediate,** and **lateral cuneiforms** (ku-ne′ĭ-form; "wedge-shaped").

Metatarsus

The **metatarsus** of the foot, which corresponds to the metacarpus of the hand, consists of five small long bones called **metatarsals.** These are numbered 1 to 5 beginning with the great toe on the medial side of the

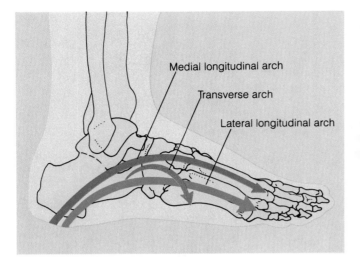

Figure 7.30 Arches of the foot.

foot. The first metatarsal is the largest, and it plays an important role in supporting the weight of the body. The metatarsals are more nearly parallel to one another than are the metacarpals in the palm. Distally, where the metatarsals articulate with the proximal phalanges of the toes, their enlarged heads form the "ball" of the foot.

Phalanges

The 14 phalanges of the toes are smaller than those of the fingers, and thus are less nimble. But their general structure and arrangement are the same. There are three phalanges in each digit except the great toe (the *hallux*), which has only two phalanges. As in the hand, these toe bones are named *distal, middle,* and *proximal phalanges.*

Arches of the Foot

A segmented structure can support weight only if it is arched. The foot has three arches: two *longitudinal* arches (the medial and lateral) and one *transverse* arch (Figure 7.30). These arches are maintained by the interlocking shapes of the foot bones, by strong ligaments, and by the pull of some tendons during muscle activity. The ligaments and muscle tendons provide a certain amount of resilience. In general, the arches "give" when weight is applied to the foot, then they spring back when the weight is removed.

If you examine your wet footprints, you will see that the foot's medial margin, from the heel to the distal end of the first metatarsal, leaves no print. The reason is that the **medial longitudinal arch** curves well above the ground. The talus bone serves as the keystone of the medial longitudinal arch, which originates at the calcaneus, rises to the talus, and then descends to the three medial metatarsals. The **lateral longitudinal arch** is very low. It elevates the lateral edge of the foot just enough to redistribute some of the body weight to the calcaneus and to the head of the fifth metatarsal (that is, to the two ends of the arch). The cuboid bone is the keystone of the lateral longitudinal arch. The two longitudinal arches serve as pillars for the **transverse arch,** which runs obliquely from one side of the foot to the other, following the line of the joints between the tarsals and metatarsals. Together, the three arches of the foot form a half dome that distributes about half our standing and walking weight to the heel bones and half to the heads of the metatarsals.

As we mentioned above, various muscle tendons run inferior to the foot bones and help support the arches of the foot. Such muscles are less active during standing than walking. Therefore, people who stand all day at their jobs may develop fallen arches or "flat feet." Running on hard surfaces can also cause arches to fall, unless one wears shoes that give proper arch support. ■

PART 3: AXIAL AND APPENDICULAR SKELETON THROUGHOUT LIFE

The membrane bones of the skull begin to ossify late in the second month of development. In these flat bones, bone tissue grows outward from **ossification centers** within the mesenchyme membranes. Some of these ossification centers are indicated in Figure 7.31. At birth, the skull bones are separated by still-unossified remnants of the membranes, called **fontanels** (fon″tah-nelz′). The major fontanels are the **anterior, posterior, mastoid,** and **sphenoidal.** The fontanels allow the skull to be compressed slightly as the infant passes through the narrow birth canal, and they accommodate brain growth in the baby. A baby's pulse can be felt in the fontanels like the gushing of a fountain (*fontanel* means "little fountain"). The large, diamond-shaped anterior fontanel can be felt for 1.5 to 2 years after birth. The others usually are replaced by bone by the end of the first year.

You have probably heard of and seen the "soft spots" on an infant's head. Can you state how soft spots relate to the fontanels discussed above?

The soft spots *are* the fontanels covered by the skin of the head. ■

At birth, the skull bones are thin. The frontal bone and the mandible begin as paired bones that fuse medially during childhood. The tympanic part of the temporal bone is merely a C-shaped ring in the newborn.

Several congenital abnormalities may distort the skull and axial skeleton. The most common of these is **cleft palate** (p. 61), a condition in which the right and left halves of the palate fail to join medially. This leaves an opening between the mouth and the nasal cavities that interferes with sucking and thus the baby's ability to nurse. Furthermore, cleft palate can lead to aspiration (inhalation) of food into the nasal cavity and lungs, resulting in pneumonia. Cleft palate is repaired surgically.

The skull changes throughout life, but the changes are most dramatic during childhood. At birth, the baby's cranium is huge relative to its small face (Figure 7.31b). The maxillae and mandible are comparatively tiny, and the contours of the face are flat. The rapid growth of the cranium before and after birth is closely related to the growth of the brain. By 9 months of age, the skull is already half its adult size. By 2 years, it is three-quarters its adult size, and by 8 to 9 years, it has reached almost adult proportions. However, between the ages of 6 and 11, the head appears to enlarge substantially because the face literally grows out from the skull. The jaws, cheekbones, and nose become more prominent. The enlargement of the face correlates with an expansion of the nose and paranasal sinuses and with the development of the large permanent teeth.

In the vertebral column, only the thoracic and sacral curvatures are present at birth (recall Figure 7.13). Both of these *primary curvatures* are convex posteriorly, so that an infant's spine arches like that of a four-legged animal. The *secondary curvatures,* the cervical and lumbar curvatures, are convex anteriorly and develop during the first 2 years of childhood. They result from a reshaping of the intervertebral discs rather than from modifications of the vertebrae themselves. The cervical curvature develops when the baby begins to lift its head (at 3 months), and the lumbar curvature develops when the baby begins to walk (at about 1 year). The lumbar curvature positions the weight of the upper body over the lower limbs, providing optimal balance during standing.

Problems with the vertebral column, such as lordosis or scoliosis, may appear during the early school years, when rapid growth of the long limb bones stretches many muscles. Children normally have a slight lordosis because holding the abdomen anteriorly helps to counterbalance the weight of their relatively large heads, held slightly posteriorly. During childhood, the thorax grows wider, but a true adult posture (head erect, shoulders back, abdomen in, and chest out) does not develop until adolescence. Vertebral curvatures and posture are not under voluntary control. Instead, a person instinctively adopts the

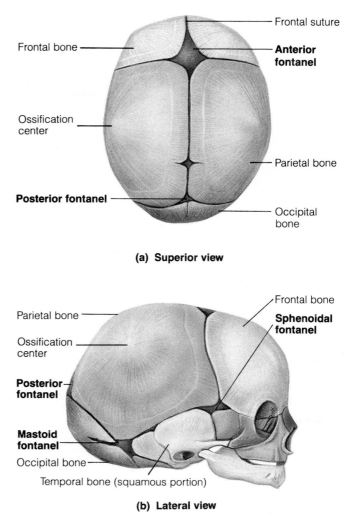

(a) Superior view

(b) Lateral view

Figure 7.31 Skull of a newborn. Note the fontanels and ossification centers.

posture that keeps the body in proper balance and minimizes the likelihood of falling.

Like the axial skeleton, the appendicular skeleton can experience congenital abnormalities. A common birth defect is **dysplasia** (dis-pla′ze-ah; "bad formation") of the hip. In this condition, the acetabulum does not form completely, and the head of the femur tends to slip out of it. Early diagnosis and treatment are essential to prevent permanent crippling.

During youth, the growth of the skeleton not only increases the body's height but also changes the body's proportions; that is, the **upper-lower (UL) body ratio** changes with age. In this ratio, the *lower body segment* (L) is the distance from the top of the pelvic girdle to the ground, while the *upper body segment* (U) is the difference between the lower segment's height and the person's total height. At birth, the UL ratio is about 1.7 to 1. Thus, the head and trunk are

Them Bones, Them Bones Goin' to Walk Around—Clinical Advances in Bone Repair

Although bones have remarkable self-regenerative powers, some conditions frustrate even their most heroic efforts. Outstanding examples include extensive shattering (as in automobile accidents), inadequate circulation in old bones, severe bone atrophy, and certain birth defects. However, recent medical advances are allowing us to deal with some problems that bones cannot handle themselves.

Large fractures and fractures of old bones tend to heal very slowly and often poorly. **Electrical stimulation of fracture sites** has dramatically increased the rapidity and completeness of bone healing in these cases, but just how electricity accomplishes this feat was not known until very recently. It now appears that the electrical fields prevent parathyroid hormone (PTH) from stimulating osteoclasts at the fracture site, thus allowing the formation of bony tissue to increase and accumulate.

A delicate, experimental procedure called the **free vascular fibular graft technique** uses pieces of the fibula (the thin, stick-like bone of the leg) to replace missing or severely damaged bone areas. The fibula is a nonessential bone that does not bear weight in humans; it serves primarily to stabilize the ankle. In the past, extensive bone grafts usually failed because an adequate blood supply could not reach their interior, necessitating eventual amputation. But this new technique, which transplants normal blood vessels along with the bone section, is being used successfully to replace large bone segments. Free vascular fibular grafts have been used to construct a radius (forearm bone) for a child born without one and to replace long bones destroyed by osteomalacia or accidental trauma. Subsequent remodeling of the bones leads to an almost perfect replica of the normally present bone.

Illustration (a) shows the apparatus (external fixator) used to stabi-

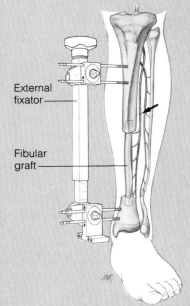

External fixator ─

Fibular graft ─

lize an implanted fibular graft. The arrow points to the site of the anastomosis (joining) of the blood vessels of the patient's tibia and the graft. Illustration (b) is an X-ray image of a healed fibular graft transplant.

Research on bone replacement materials has produced **artificial bone** made of a biodegradable ceramic substance known as TCP. This product is soft enough to be carved into the desired shape, but it is not very strong. Its main application has been to replace parts of non-weight-bearing bones, such as skull bones. Since the material is porous, osteoblasts enter it and lay down bone matrix, and in most instances, the TCP has been completely degraded and replaced by new bone within 2 years.

Another exciting technique, using **crushed bone** from human cadavers, also induces the body to form new bone of its own. This means that surgeons can mold bones where none previously existed and avoid the often painful and time-consuming procedure of bone transplant or grafting. The bone powder is mixed with water to form a paste that can be molded to the shape of the desired bone. When implanted, each bone speck becomes surrounded by connective tissue cells, which form cartilage. Later, bone replaces the cartilage, incorporating the product into the new bony matrix. The major use of this technique thus far has been to treat children born with skeletal defects such as a misshapen skull, missing nose, or cleft palate. However, its widest application will probably be to treat periodontal disease, a condition of bone loss around the teeth, which often results in tooth loss. Advantages of crushed bone include ease of use, ability to pack it into small, difficult-to-reach areas, and the fact that it can be stored (and thus available for immediate use in victims of bone trauma).

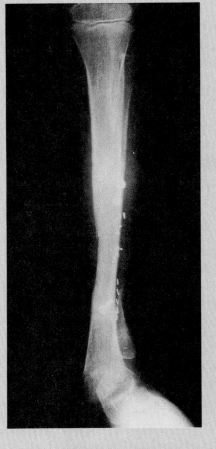

(a) A leg brace for a healing bone. (b) X-ray image of a leg showing a healed fibular graft transplant.

more than 1½ times as long as the lower limbs. The lower limbs grow more rapidly than the trunk from this time on, however, and by the age of 10, the UL ratio is about 1 to 1. It changes little thereafter. During puberty, the female pelvis broadens in preparation for childbearing, and the entire male skeleton becomes more robust. Once adult height is reached, a healthy skeleton changes very little until late middle age.

Old age affects many parts of the skeleton, especially the spine. The water content of the intervertebral discs declines. As the discs become thinner and less elastic, the risk of disc herniation increases. By 55 years, a loss of several centimeters from a person's height is common. Further shortening of the trunk can be produced by osteoporosis of the spine that leads to kyphosis (called "dowager's hump" in the elderly).

Recall from p. 132 that the vertebrae and the neck of the femur often fracture in people with osteoporosis.

The thorax becomes more rigid with increasing age, largely because the costal cartilages ossify. This loss of elasticity of the rib cage leads to shallow breathing, which in turn leads to less efficient gas exchange in the lungs.

All bones, you will recall, lose mass with age. Skull bones lose less mass than most, but they lose enough to produce changes in the facial contours of the elderly. As the bony tissue of the mandible and maxilla declines, the jaws come to look small and child-like once again. If an elderly person loses his or her teeth, this loss of bone from the jaws is accelerated, because the bone of the alveolar region (tooth sockets) is reabsorbed.

Related Clinical Terms

Clubfoot   A birth defect in which the soles of the feet face medially and the toes point inferiorly. Clubfoot occurs in about 1 of every 1000 births. It may be genetically caused or caused by folding of the foot up against the chest during fetal development.

Craniotomy ("cutting the skull")   Surgery to remove part of the cranium, usually done to reach the brain (for example, for removing a brain tumor, a blood clot, or a sample of brain tissue for a biopsy). The piece of skull is removed by drilling a series of round holes through the bone at regular intervals to outline a square and then cutting between the holes with a string-like saw. At the end of the operation, the square piece of bone is replaced, and it heals normally.

Knock-knee   A deformity in which the two knees rub or knock together during walking. Knock-knee usually occurs in children as a result of irregular growth of the bones of the lower limb, injury to the ligaments, or injury to the bone ends at the knee.

Laminectomy (lam″ĭ-nek′to-me; "lamina-cutting")   Surgical removal of a vertebral lamina. Laminectomies are usually done to reach and relieve a herniated disc.

Orthopedist or **orthopedic surgeon** (or″tho-pe′dist)   A physician who specializes in restoring lost function of or repairing damage to bones and joints.

Pelvimetry (pel-vim′ĕ-tre; "pelvis measure")   Measurement of the dimensions of the inlet and outlet of the pelvis, often performed to determine whether the pelvis is large enough to allow normal delivery of a baby.

Spina bifida (spi′nah bĭ′fĭ-dah; "cleft spine")   A common birth defect in which one or more vertebral arches are not complete. Vertebral laminae and spinous process are absent. Cases of spina bifida can be inconsequential to severe. Severe cases can impair the function of the spinal cord and brain. See Chapter 12 (p. 356) for more information.

Spinal fusion   Surgical procedure involving insertion of bone chips to stabilize a region of the vertebral column, particularly after the fracture of a vertebra or the prolapse of a disc.

Chapter Summary

1. The axial skeleton forms the long axis of the body. Its parts are the skull, vertebral column, and bony thorax. The appendicular skeleton consists of the pectoral and pelvic girdles and the long bones of the limbs.

2. Bone markings are landmarks that represent sites of muscle attachment, points of articulation, and sites of passage of blood vessels and nerves (see Table 7.1, p. 140).

▐ PART 1: THE AXIAL SKELETON

THE SKULL (pp. 142–156)

1. The cranium forms the vault and base of the skull and protects the brain. The facial skeleton provides openings for the eyes, openings for respiratory and digestive passages, and areas of attachment for facial muscles.

2. Almost all bones of the skull are joined by immobile sutures.

3. Cranium. The eight bones of the cranium are the paired parietal and temporal bones and the unpaired frontal, occipital, ethmoid, and sphenoid bones (see Table 7.2, pp. 154–155).

4. Facial Bones. The 14 bones of the face are the paired maxillae, zygomatics, nasals, lacrimals, palatines, and inferior conchae, plus unpaired mandible and vomer bones (Table 7.2).

5. Orbits and Nasal Cavity. The orbits and nasal cavity are complicated regions of the skull, each formed by many bones (see Figures 7.9 and 7.10).

6. Paranasal Sinuses. These air-filled sinuses occur in the frontal, ethmoid, sphenoid, and maxillary bones. Paranasal sinuses are extensions of the nasal cavity, to which they are connected.

7. Hyoid Bone. This bowed bone, supported in the anterior neck by muscles and ligaments, serves as a movable base for the tongue, among other functions. Technically, the hyoid bone is not part of the skull.

THE VERTEBRAL COLUMN (pp. 156–164)

1. General Characteristics. The vertebral column includes 24 vertebrae (7 cervical, 12 thoracic, and 5 lumbar) plus the sacrum and coccyx.

2. Intervertebral Discs. These discs, with their nucleus pulposus cores and anulus fibrosus rings, act as shock absorbers and give flexibility to the spine. Herniated lumbar discs are common after middle age.

3. Divisions and Curvatures. The primary curvatures of the vertebral column are the thoracic and sacral, whereas the secondary curvatures are the cervical and lumbar (also see p. 183). Exaggerations of these curves, usually pathological, include kyphosis and lordosis. Scoliosis is an abnormal lateral curvature of the spine.

4. General Structure of Vertebrae. With the exception of C_1, all vertebrae have a body, two transverse processes, superior and inferior articular processes, a spinous process, and a vertebral arch.

5. Regional Vertebral Characteristics. Special characteristics distinguish the cervical, thoracic, and lumbar vertebrae from one another (see Table 7.3, p. 163). The orientation of the articular facets determines the movements possible in the different regions of the spinal column. The sacrum and coccyx, groups of fused vertebrae, contribute to the bony pelvis.

6. The atlas and axis (C_1 and C_2) are atypical vertebrae. The ringlike atlas supports the skull and makes nodding movements possible. The axis has a dens that enables the head to rotate.

THE BONY THORAX (pp. 164–166)

1. The bones of the thorax are the 12 pairs of ribs, the sternum, and the thoracic vertebrae.

2. Sternum. The sternum consists of the manubrium, body, and xiphoid process.

3. Ribs. The first seven pairs of ribs are called true ribs. The rest are false ribs. The eleventh and twelfth are floating ribs. A typical rib consists of a head with facets, a neck, tubercle, and shaft. A costal cartilage occurs on the ventral end of each rib.

PART 2: THE APPENDICULAR SKELETON

THE PECTORAL GIRDLE (pp. 167–169)

1. The pectoral girdles are specialized for mobility. Each pectoral girdle consists of a clavicle and scapula and attaches an upper limb to the bony thorax.

2. Clavicles. These bones hold the arms laterally away from the thorax and transmit pushing forces from the upper limbs to the thorax.

3. Scapulae. Each triangular scapula articulates with a clavicle and a humerus. The borders, angles, and markings of the scapula are summarized in Table 7.4 on pp. 174–175.

BONES OF THE FREE UPPER LIMB (pp. 169–173)

1. Each free upper limb consists of 30 bones and is specialized for mobility.

2. Arm, Forearm, and Hand. The skeleton of the arm consists solely of the humerus. The skeleton of the forearm consists of the radius and ulna. The skeleton of the hand consists of carpals, metacarpals, and phalanges.

THE PELVIC GIRDLE (pp. 173–177)

1. The pelvic girdle, specialized for bearing weight, is composed of two hip bones that bind the lower limbs to the vertebral column. Together with the sacrum, the hip bones form the basin-like bony pelvis.

2. Each hip bone (coxal bone) consists of an ilium, ischium, and a pubis fused together. The acetabulum is at the point of fusion of these three bones.

3. Ilium, Ischium, and Pubis. The ilium is the superior flaring part of the hip bone. Each ilium forms a secure joint with the sacrum. The ischium is a curved bar of bone; we sit on the ischial tuberosities. The V-shaped pubic bones join anteriorly at the pubic symphysis.

4. True and False Pelvis. The pelvic brim is an oval ridge that includes the pubic crest, arcuate line of the ilium, and sacral promontory. It separates the superior false pelvis from the inferior true pelvis.

5. Pelvic Structure and Childbearing. The male pelvis is deep and narrow, with larger, heavier bones than those of the female. The female pelvis, which forms the birth canal, is shallow and wide (see Table 7.5, p. 178).

BONES OF THE FREE LOWER LIMB (pp. 177–182)

1. The free lower limb consists of the thigh, leg, and foot and is specialized for weight bearing and locomotion.

2. Thigh. The long, thick femur is the only bone of the thigh. Its ball-shaped head articulates with the acetabulum.

3. Leg. The bones of the leg are the tibia (which participates in both the knee and ankle joints) and the slender fibula.

4. Foot. The bones of the foot are the tarsals, metatarsals, and phalanges. The most important tarsals are the calcaneus (heel bone) and the talus. The talus articulates with the leg bones at the ankle joint.

5. The foot is supported by three arches that distribute the weight of the body to the heel and ball of the foot.

PART 3: AXIAL AND APPENDICULAR SKELETON THROUGHOUT LIFE (pp. 182–185)

1. Fontanels are membrane-covered regions between the cranial bones in the infant skull. Fontanels allow growth of the brain and ease birth passage. Growth of the cranium early in life is closely related to the rapid growth of the brain. An increase in the size of the face in late childhood follows tooth development and enlargement of the respiratory passageways.

2. The vertebral column is C-shaped at birth. The cervical and lumbar curvatures form when the baby begins to lift its head and to walk, respectively.

3. As the skeleton grows, the upper-lower (UL) body ratio changes from 1.7:1 (at birth) to 1:1 (10 years and beyond).

4. In old age, the intervertebral discs thin. This, along with osteoporosis, leads to a gradual decrease in height. Loss of bone mass predisposes elderly people to fractures.

Review Questions

Multiple Choice and Matching Questions

1. Using the letters from column B, match the bone descriptions in column A. (Some will require more than one choice.)

Column A	Column B
_b, a_ **(1)** connected by the coronal suture	**(a)** ethmoid
_h, g_ **(2)** keystone bone of the cranium	**(b)** frontal
_d_ **(3)** keystone bone of the face	**(c)** mandible
_d, f_ **(4)** form the hard palate	**(d)** maxillary
_e_ **(5)** foramen magnum	**(e)** occipital
_c_ **(6)** forms the chin	**(f)** palatine
_a, b, d_ **(7)** contain paranasal sinuses	**(g)** parietal
_i_ **(8)** contains mastoid sinus	**(h)** sphenoid
	(i) temporal

2. Match the terms in the key with the bone descriptions that follow:

Key: **(a)** clavicle **(b)** ilium **(c)** ischium **(d)** pubis
 (e) sacrum **(f)** scapula **(g)** sternum

_g_ **(1)** bone of the axial skeleton to which the pectoral girdle attaches

_f_ **(2)** its markings include the glenoid cavity and acromion process

_b_ **(3)** its features include the ala, crest, and greater sciatic notch

_a_ **(4)** membrane bone that transmits forces from upper limb to bony thorax

_b_ **(5)** bone of pelvic girdle that articulates with the axial skeleton

_c_ **(6)** the "sit-down" bone

_d_ **(7)** anteriormost bone of the pelvic girdle (inferiorly)

_e_ **(8)** a part of the vertebral column

3. Match the choices in the key with the bone descriptions that follow:

Key: **(a)** carpals **(b)** femur **(c)** fibula **(d)** humerus
 (e) radius **(f)** tarsals **(g)** tibia **(h)** ulna

_b_ **(1)** articulates with the acetabulum and the tibia

_c_ **(2)** its malleolus forms the lateral aspect of the ankle

_e_ **(3)** bone that "carries" the hand and wrist

_a_ **(4)** the wrist bones

_h_ **(5)** bone shaped like a monkey wrench

_e_ **(6)** articulates with the capitulum of the humerus

_f_ **(7)** largest bone is the calcaneus

4. From what you have learned about bone markings in this chapter, a facet is (a) a sharp, almost pointed projection, (b) a smooth, nearly flat articular surface, (c) a canal-like passageway, (d) the same as a condyle.

5. Herniated intervertebral discs tend to herniate (a) superiorly, (b) anterolaterally, (c) posterolaterally, (d) laterally.

Short Answer and Essay Questions

6. In the fetus, how do the relative proportions of the cranium and face compare with those in the skull of a young adult?

7. Name and diagram the four normal vertebral curvatures. Which are primary curvatures, and which are secondary curvatures?

8. List two specific structural characteristics each for cervical, thoracic, and lumbar vertebrae that would enable anyone to identify each type correctly.

9. (a) What is the function of intervertebral discs? (b) Distinguish the anulus fibrosus from the nucleus pulposus regions of a disc. (c) Which herniates in the condition called prolapsed disc?

10. Name all the bones of the thorax.

11. Is a floating rib a true rib or a false rib? Explain.

12. The major function of the pectoral girdle is mobility. (a) What is the major function of the pelvic girdle? (b) Relate these functional differences to the anatomical differences between these girdles.

13. List three differences between the male and female pelvis.

14. Describe the function of the arches of the foot.

15. Briefly describe the anatomical characteristics and impairment of function seen in cleft palate and hip dysplasia.

16. Compare the skeleton of a young adult to that of an 85-year-old person, with respect to bone mass in general and the basic bony structure of the skull, thorax, and vertebral column.

17. (a) Define pelvic girdle and pectoral girdle. (b) Define true pelvis and false pelvis.

18. Identify what types of movement are allowed by the lumbar region of the vertebral column, and compare these with the movements allowed by the thoracic region.

19. Lance was a bright anatomy student, but he sometimes called the leg bones *fibia* and *tibula*. Can you correct this common mistake?

20. What regions do anatomists call the *leg* and the *arm*?

21. Draw the scapula in posterior view, and label all the borders, angles, fossae, and important markings visible from this view.

22. Identify the associated bone and the location of each of these bone markings: (a) greater trochanter, (b) linea aspera, (c) trochlea, (d) coronoid process (in the appendicular skeleton), (e) deltoid tuberosity, (f) greater tubercle, (g) greater sciatic notch, (h) demifacets.

23. List the bones in each of the three cranial fossae.

Critical Thinking and Clinical Application Questions

24. Antonio was hit in the face with a football during practice. An X-ray film revealed multiple fractures of the bones around an orbit. Name the bones that form the margins of an orbit.

25. Malcolm was rushed to the emergency room after trying to break a fall with outstretched arms. The physician took one look at his shoulder and saw that Malcolm had a broken clavicle (and no other injury). Describe the position of Malcolm's shoulder. What part of the clavicle was most likely broken? Malcolm was worried about injury to the main blood vessels to his arm (the subclavian vessels), but he was told such injury was unlikely. Why could the doctor predict this with assurance?

26. Lindsey had polio as a child and was partly paralyzed in one lower limb for over a year. Although she is no longer paralyzed, she now has a severe lateral curvature of the lumbar spine. Explain what has happened, and name her condition.

27. Kelsey's grandmother slipped on a rug and fell heavily to the floor. Her left lower limb was laterally rotated and shorter than the right, and when she tried to get up, she grimaced with pain. Kelsey surmised that her grandmother might have "fractured her hip," which proved to be true. What bone was probably fractured and in what part? Why is a "fractured hip" a common type of fracture in the elderly?

28. Racheal, a hairdresser, developed flat feet. Explain how this may have come about.

29. Mr. Chester, a heavy beer drinker with a large potbelly, complained of severe lower back pains. After X rays were taken, he was found to have displacement of his lumbar vertebrae. What would this condition be called, and what would cause it?

8

Joints

oints, or **articulations,** are places where the rigid elements of the skeleton meet. Most articulations join bone to bone, but some join bone to cartilage, or teeth to their bony sockets. Joints both hold the skeleton together and enable it to be mobile. The graceful movements of ballet dancers and gymnasts attest to the great variety of motions that joints allow. Joints are always the weakest points of any skeleton, yet their structure enables them to resist crushing, tearing, and the various forces that would drive them out of alignment.

Classification of Joints

Joints are classified by structure and by function. The *structural classification* is based on the material that binds the bones together and on the presence or absence of a joint cavity. Structurally, joints are classified as *fibrous, cartilaginous,* and *synovial.* The *functional classification* focuses on the amount of movement allowed at the joint: **Synarthroses** (sin"ar-thro'sēz) are immovable joints; **amphiarthroses** (am"fe-ar-thro'sēz) are slightly movable joints; and **diarthroses** (di"ar-thro'sēz) are freely movable joints. (*Arthro* is Greek for "joint.") Freely movable joints predominate in the limbs, whereas immovable and slightly movable joints are largely restricted to the axial skeleton.

In this chapter, we follow the structural classification of joints, considering the functional properties where appropriate. As you read this discussion, refer to Table 8.1 on pp. 192–193, which summarizes the main joints of the body.

Fibrous Joints

In **fibrous joints** (Figure 8.1), the bones are connected by fibrous tissue (dense regular connective tissue). No joint cavity is present. Most fibrous joints are immovable or only slightly movable. The kinds of fibrous joints are: *sutures, syndesmoses,* and *gomphoses.*

Sutures

In **sutures** ("seams"), the bones are tightly bound by a minimal amount of fibrous tissue (Figure 8.1a). Sutures occur only between the bones of the skull. The

edges of the joining bones are wavy and interlock. Sutures not only knit the bones together but also allow the bones to grow at their edges so that the skull can expand with the brain during childhood. During middle age, the fibrous tissue at sutures becomes ossified, and the skull bones fuse together. At this stage, the closed sutures are more precisely called *synostoses* (sin"os-to'sēz; "bony junctions"). Since movement of the cranial bones would damage the brain, the immovable nature of sutures is a protective adaptation.

Syndesmoses

In **syndesmoses** (sin"des-mo'sēz), the bones are connected by ligaments, bands of fibrous tissue longer than those that occur in sutures (Figure 8.1b). In fact, *syndesmosis* is the Greek word for "ligament." The amount of movement allowed at a syndesmosis depends on the length of the connecting fibers. For ex-

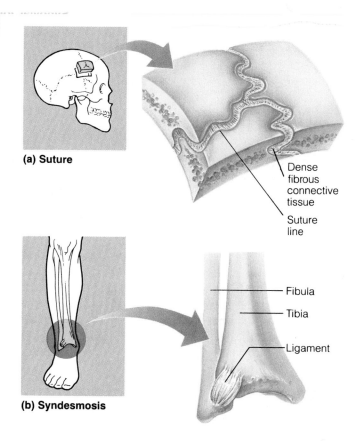

(a) Suture

Dense fibrous connective tissue

Suture line

Fibula

Tibia

Ligament

(b) Syndesmosis

Figure 8.1   Fibrous joints.   (a) Sutures of the skull are fibrous joints with very short connecting fibers. **(b)** In a syndesmosis, the fibrous tissue (ligament) is longer than that in sutures.

189

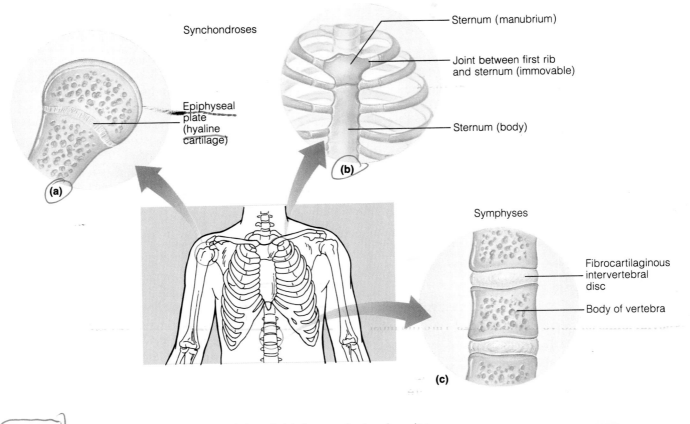

Synchondroses

Sternum (manubrium)

Joint between first rib
and sternum (immovable)

Epiphyseal
plate
(hyaline
cartilage)

Sternum (body)

(b)

(a)

Symphyses

Fibrocartilaginous
intervertebral
disc

Body of vertebra

(c)

Figure 8.2   Cartilaginous joints.   **(a)** The epiphyseal plate in a growing long bone is a
synchondrosis (hyaline cartilage joint); recall that epiphyseal plates are only temporary joints
that close when bones stop growing. **(b)** Another synchondrosis is the joint between rib 1 and
the manubrium of the sternum. **(c)** The intervertebral discs are symphyses (fibrocartilage
joints).

ample, the ligament between the distal ends of the
tibia and fibula is short (Figure 8.1b), allowing only
slightly more movement than a suture does. Such
slight movement is best described as "give." True
movement is still prevented, so this tibiofibular joint
is classified as an immovable joint, or synarthrosis.
(Note, however, that some authorities prefer to call
this joint an amphiarthrosis.) The ligament-like *interosseous membrane* that connects the radius and ulna
along their length is also a syndesmosis, but one with
longer fibers (see Figure 7.23a). This membrane forms
a freely movable joint (diarthrosis) that permits the rotation of the radius around the ulna.

Gomphoses

A **gomphosis** (gom-fo′sis; "bolt") is a peg-in-socket
joint. The only example is the articulation of a tooth
with its socket. In this case, the connecting ligament
is the short *periodontal ligament* (see p. 573 and Figure 21.12).

Cartilaginous Joints

In **cartilaginous** (kar″tĭ-laj′ĭ-nus) **joints,** the articulating bones are united by cartilage (Figure 8.2). They
lack a joint cavity and are not highly movable. The
two types of cartilaginous joints are *synchondroses*
and *symphyses.*

Synchondroses

A joint where *hyaline* cartilage unites the bones is a
synchondrosis (sin″kon-dro′sis; "junction of cartilage"). The epiphyseal plates are synchondroses (Figure 8.2a). Functionally, these plates are classified as
immovable synarthroses. Another example of a synchondrosis is the immovable joint between the first
rib's costal cartilage and the manubrium of the sternum (Figure 8.2b).

Symphyses

A joint where *fibrocartilage* unites the bones is a **symphysis** (sim'fĭ-sis; "growing together") (Figure 8.2c). Examples include the intervertebral discs and the pubic symphysis of the pelvis. While fibrocartilage is the main element of a symphysis, hyaline cartilage is also present in the form of articular cartilages on the bony surfaces. We noted in Chapter 6 that fibrocartilage resists both tension and compression stresses and can act as a resilient shock absorber. Symphyses, then, are slightly movable joints (amphiarthroses) that provide strength with flexibility.

Synovial Joints ③

Synovial joints (sĭ-no've-al; "joint eggs") are the most movable joints of the body, and all are diarthroses. In fact, many authorities equate the terms *synovial joint* and *diarthrosis.* Each synovial joint contains a fluid-filled *joint cavity.* Most joints of the body are in this class, especially those in the limbs (Table 8.1).

General Structure

The basic features of synovial joints (Figure 8.3a) include the following:

1. Articular cartilage. Articular cartilages (p. 118) composed of hyaline cartilage cover the ends of the opposing bones. These spongy cushions absorb compression placed on the joint and thereby keep the bone ends from being crushed.

2. Joint cavity (synovial cavity). A feature unique to synovial joints, the joint cavity is really just a potential space that holds a small amount of fluid.

3. Articular capsule. The joint cavity is enclosed by a two-layered articular capsule (joint capsule). The outer layer is a **fibrous capsule** of dense irregular connective tissue. It is continuous with the periosteum layer of the joining bones. The fibrous capsule strengthens the joint so that the bones are not pulled apart.

The inner layer of the joint capsule is a **synovial membrane** composed of loose connective tissue. Be-

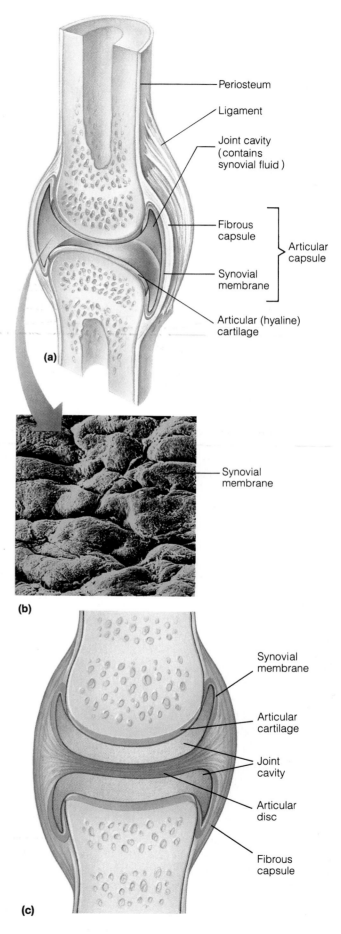

Figure 8.3   Synovial joints.   (a) A typical synovial joint between the ends of two bones. **(b)** Scanning electron micrograph of the surface of a synovial membrane from the knee joint (20 ×). **(c)** A synovial joint that contains an articular disc.

Table 8.1 Structural and Functional Characteristics of Body Joints

Illustration	Joint	Articulating bones	Structural type*	Functional type; movements allowed
	Skull	Cranial and facial bones	Fibrous; suture	Synarthrotic; no movement
	Temporomandibular	Temporal bone of skull and mandible	Synovial; modified hinge (contains articular disc)	Diarthrotic; gliding and uniaxial rotation; slight lateral movement, elevation, depression, protraction, retraction
	Atlanto-occipital	Occipital bone of skull and atlas	Synovial; condyloid	Diarthrotic; biaxial; flexion, extension, abduction, adduction, circumduction
	Atlantoaxial	Atlas (C_1) and axis (C_2)	Synovial; pivot	Diarthrotic; uniaxial; rotation of the head
	Intervertebral	Between adjacent vertebral bodies	Cartilaginous; symphysis	Amphiarthrotic; slight movement
	Intervertebral	Between articular processes	Synovial; plane	Diarthrotic; gliding
	Vertebrocostal	Vertebrae (transverse processes or bodies) and ribs	Synovial; plane	Diarthrotic; gliding
	Sternoclavicular	Sternum and clavicle	Synovial; shallow saddle (contains articular disc)	Diarthrotic; multiaxial (allows clavicle to move in all axes)
	Sternocostal	Sternum and rib 1	Cartilaginous; synchondrosis	Synarthrotic; no movement.
	Sternocostal	Sternum and ribs 2–7	Synovial; double plane	Diarthrotic; gliding
	Acromioclavicular	Acromion process of scapula and clavicle	Synovial; plane	Diarthrotic; gliding; elevation, depression, protraction, retraction
	Shoulder (glenohumeral)	Scapula and humerus	Synovial; ball and socket	Diarthrotic; multiaxial; flexion, extension, abduction, adduction, circumduction, rotation
	Elbow	Ulna (and radius) with humerus	Synovial; hinge	Diarthrotic; uniaxial; flexion, extension
	Radioulnar (proximal)	Radius and ulna	Synovial; pivot	Diarthrotic; uniaxial; rotation around the long axis of forearm to allow pronation and supination
	Radioulnar (distal)	Radius and ulna	Synovial; pivot (contains articular disc)	Diarthrotic; uniaxial; rotation (convex head of ulna rotates in ulnar notch of radius)
	Wrist (radiocarpal)	Radius and proximal carpals	Synovial; condyloid	Diarthrotic; biaxial; flexion, extension, abduction, adduction, circumduction
	Intercarpal	Adjacent carpals	Synovial; plane	Diarthrotic; gliding
	Carpometacarpal of digit 1 (thumb)	Carpal (trapezium) and metacarpal 1	Synovial; saddle	Diarthrotic; biaxial; flexion, extension, abduction, adduction, circumduction, opposition
	of digits 2–5	Carpal(s) and metacarpal(s)	Synovial; plane	Diarthrotic; gliding
	Knuckle (metacarpophalangeal)	Metacarpal and proximal phalanx	Synovial; condyloid	Diarthrotic; biaxial; flexion, extension, abduction, adduction, circumduction
	Finger (interphalangeal)	Adjacent phalanges	Synovial; hinge	Diarthrotic; uniaxial; flexion, extension

*Fibrous joints** indicated by orange circles; **cartilaginous joints** indicated by blue circles; **synovial joints** indicated by purple circles.

Table 8.1 (continued)

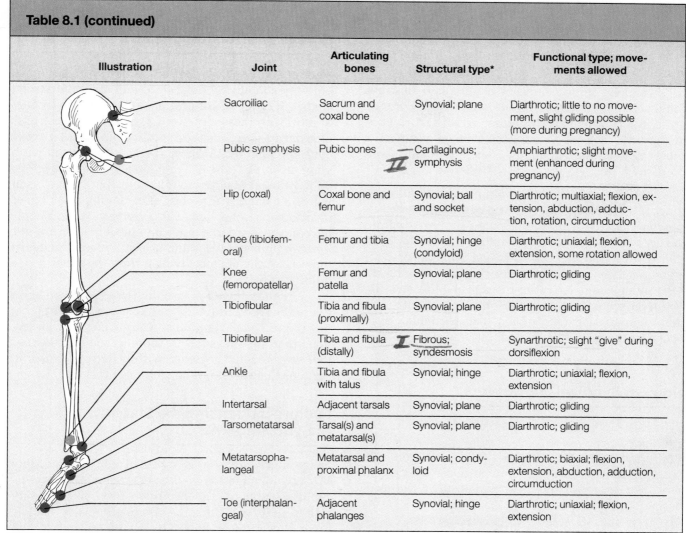

Illustration	Joint	Articulating bones	Structural type*	Functional type; movements allowed
	Sacroiliac	Sacrum and coxal bone	Synovial; plane	Diarthrotic; little to no movement, slight gliding possible (more during pregnancy)
	Pubic symphysis	Pubic bones	Cartilaginous; symphysis	Amphiarthrotic; slight movement (enhanced during pregnancy)
	Hip (coxal)	Coxal bone and femur	Synovial; ball and socket	Diarthrotic; multiaxial; flexion, extension, abduction, adduction, rotation, circumduction
	Knee (tibiofemoral)	Femur and tibia	Synovial; hinge (condyloid)	Diarthrotic; uniaxial; flexion, extension, some rotation allowed
	Knee (femoropatellar)	Femur and patella	Synovial; plane	Diarthrotic; gliding
	Tibiofibular	Tibia and fibula (proximally)	Synovial; plane	Diarthrotic; gliding
	Tibiofibular	Tibia and fibula (distally)	Fibrous; syndesmosis	Synarthrotic; slight "give" during dorsiflexion
	Ankle	Tibia and fibula with talus	Synovial; hinge	Diarthrotic; uniaxial; flexion, extension
	Intertarsal	Adjacent tarsals	Synovial; plane	Diarthrotic; gliding
	Tarsometatarsal	Tarsal(s) and metatarsal(s)	Synovial; plane	Diarthrotic; gliding
	Metatarsophalangeal	Metatarsal and proximal phalanx	Synovial; condyloid	Diarthrotic; biaxial; flexion, extension, abduction, adduction, circumduction
	Toe (interphalangeal)	Adjacent phalanges	Synovial; hinge	Diarthrotic; uniaxial; flexion, extension

*__Fibrous joints__ indicated by orange circles; __cartilaginous joints__ indicated by blue circles; __synovial joints__ indicated by purple circles.

sides lining the joint capsule, this membrane covers all the internal joint surfaces not covered by cartilage.

4. Synovial fluid. The liquid inside the joint cavity is called synovial fluid. Its name, which means "joint egg fluid," refers to its resemblance to raw egg white. Synovial fluid is primarily a filtrate of blood, arising from capillaries in the synovial membrane. Synovial fluid also contains special glycoprotein molecules secreted by the fibroblasts in the synovial membrane. These molecules make synovial fluid a slippery lubricant that eases the movement at the joint. Synovial fluid not only occupies the joint cavity but also occurs *within* the articular cartilages. The pressure placed on joints during their normal movements squeezes the synovial fluid into and out of the articular cartilages, nourishing the cells in these cartilages and lubricating their free surfaces.

5. Reinforcing ligaments. Synovial joints are reinforced and strengthened by a number of band-like ligaments. Most often, the ligaments are intrinsic, or capsular; that is, they are thickened parts of the fibrous capsule itself. In other cases, they are distinct ligaments located just external to the capsule (extracapsular ligaments) or internal to it (intracapsular ligaments). Intracapsular ligaments are covered with a synovial membrane that separates them from the joint cavity through which they run.

6. Nerves and vessels. Synovial joints are richly supplied with sensory nerve fibers. These fibers innervate the articular capsule. Some of these fibers measure pain, but most monitor how much the capsule is being stretched. This monitoring of joint stretch is one of several ways by which the nervous system senses posture and body movements (p. 365). Synovial joints also have a rich blood supply. Most of the blood vessels supply the synovial membrane, where extensive capillary beds produce the blood filtrate that is the basis of synovial fluid.

Each synovial joint in the body is served by branches from *several* major nerves and blood vessels. These branches come from many different directions but supply overlapping areas of the joint capsule.

193

Can you propose a functional explanation for this multiple, overlapping supply of nerves and vessels?

When an injury damages a joint and destroys some vessels and nerves, others survive and keep the joint functioning. Furthermore, when the normal movements at a joint compress a blood vessel, the other vessels stay open and keep the joint well nourished. ■

Certain synovial joints contain a disc of fibrocartilage (Figure 8.3c). Such an **articular disc**, or **meniscus** (mĕ-nis′kus; "crescent"), extends internally from the capsule and divides the joint cavity in two. Articular discs improve the fit between articulating bone ends that have slightly different shapes, making the joint more stable. Articular discs also allow two different movements at the same joint—a distinct movement across each face of the disc. Articular discs occur in the jaw joint, knee joint, and a few others (Table 8.1).

How Do Synovial Joints Work?

Synovial joints are elaborate lubricating devices that allow joining bones to move across one another with a minimum of friction. Without this lubrication, rubbing would wear away the joint surfaces, and excessive friction could overheat and destroy the joint tissues.

Synovial joints are routinely subjected to *compressive* forces as the muscles that move bones pull the bone ends together. As the opposing articular cartilages touch, synovial fluid is squeezed out of them, producing a film of lubricant between the cartilage surfaces. The two moving cartilage surfaces ride on this slippery film, not on each other. Synovial fluid is such an effective lubricant that it yields less friction than ice sliding on ice. When the pressure on the joint ceases, the synovial fluid rushes back into the articular cartilages like water into a sponge, ready to be squeezed out again the next time a load is placed on the joint. This mechanism is called **weeping lubrication.**

You may have heard your knees pop during deep knee bends, or you may know people who like to crack their knuckles by pulling on their fingers. Can you imagine the source of these joint sounds?

The sounds are produced by the popping of gas bubbles within the synovial fluid. More specifically, when one pulls on a synovial joint, the suction draws respiratory gases out of the capillaries of the synovial membrane. The gases coalesce into bubbles in the synovial fluid, which then burst within the joint cavities. Habitually cracking one's knuckles may cause them to enlarge over time, but there is no evidence that this habit damages the joints. ■

Bursae and Tendon Sheaths

Bursae and *tendon sheaths* are not synovial joints, but they contain synovial fluid and are often associated with synovial joints (Figure 8.4). Essentially bags of lubricant, these structures act like "ball bearings" to reduce friction between body elements that move over one another. A **bursa** (ber′sah), from the Latin word meaning "purse," is a flattened fibrous bag lined by a synovial membrane (Figure 8.4b). Bursae occur where ligaments, muscles, skin, tendons, or bones overlie each other and rub together. Many people have never heard of a bursa, but most have heard of a *bunion* (bun′yun). A bunion is an enlarged bursa at the base of the big toe, swollen from the rubbing of a tightly fitting shoe.

A **tendon sheath** is essentially an elongated bursa that wraps around a tendon like a bun around a hot dog (Figure 8.4c). Tendon sheaths occur only on tendons that are subjected to friction, such as tendons that are crowded together within narrow canals or tendons that travel through joint cavities.

Factors Influencing the Stability of Synovial Joints

Since joints are regularly pulled and pushed, they must be stabilized so that they do not dislocate (come out of alignment). The stability of a synovial joint depends on three factors: the nature of the articular surfaces, the number and position of stabilizing ligaments, and muscle tone. We will now explore these factors individually.

Articular Surfaces

The articular surfaces of adjacent bones fit together in a complementary manner. Although the shapes of these articular surfaces determine what kinds of movement are possible at the joint, joint surfaces seldom play a significant role in joint stability: Most joint sockets are just too shallow. Still, some joint surfaces have deep sockets or grooves that do provide stability. The best example is the ball and deep socket of the hip joint. Other examples are the elbow and ankle joints.

Ligaments

The capsules and ligaments of synovial joints help hold the bones together and prevent excessive or undesirable motions. As a rule, the more ligaments a joint has, the stronger it is. When other stabilizing factors are inadequate, however, undue tension is placed on the ligaments, and they fail. Once they are stretched, ligaments, like taffy, stay stretched. How-

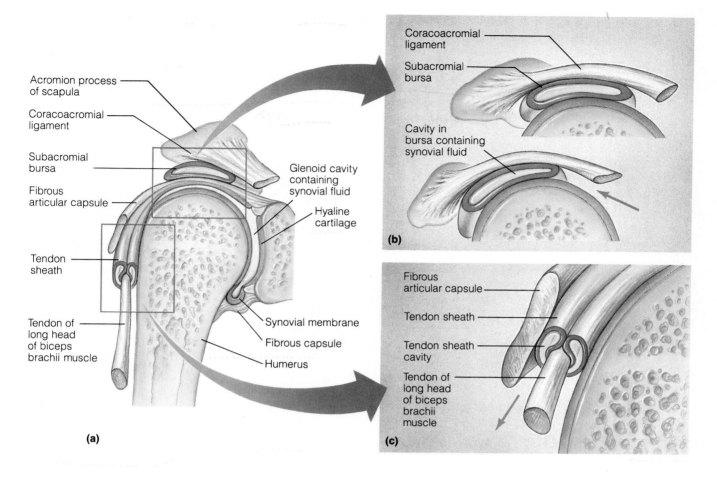

Figure 8.4   Bursae and tendon sheaths.   (**a**) Frontal section through the right shoulder joint. (**b**) Enlargement of (a), showing how a bursa eliminates friction where a ligament (or other structure) rubs against bone. (**c**) A tendon sheath wrapped around the tendon of the biceps brachii muscle.

ever, a ligament can stretch only about 6% beyond its normal length before it snaps.

Muscle Tone

Joints are stabilized by the muscle tendons that cross over them, just external to their joint capsules. For most joints, in fact, this is the most important stabilizing factor. These tendons are kept taut at all times by the tone of the muscles to which they attach. (**Muscle tone** is a constant, low level of contractile force generated by a muscle even when that muscle is not causing movement. Tone keeps the muscles healthy and ready to react to stimulation.) Muscle tone is very important in reinforcing the shoulder and knee joints. Likewise, tendons of the leg muscles support the joints in the arches of the foot.

People who are "double-jointed" can bend the thumb back to touch the forearm or place both heels behind the neck. Can you explain these abilities?

These people have looser and more stretchable joint ligaments and joint capsules than average. They do not have any "extra" joints (as the name "double-jointed" incorrectly implies). ■

Movements Allowed by Synovial Joints

As muscles contract, they cause bones to move at the synovial joints. The movements are of three basic types: (1) *gliding* of one bone surface across another;

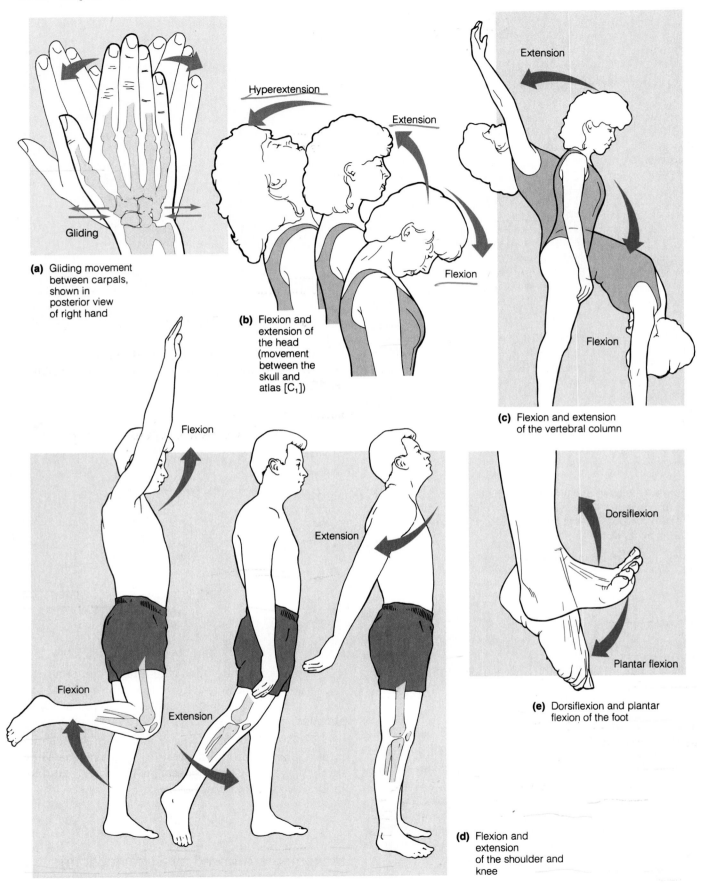

(a) Gliding movement between carpals, shown in posterior view of right hand

(b) Flexion and extension of the head (movement between the skull and atlas [C_1])

(c) Flexion and extension of the vertebral column

(d) Flexion and extension of the shoulder and knee

(e) Dorsiflexion and plantar flexion of the foot

Figure 8.5   Movements allowed by synovial joints.   (a) Gliding movements.
(b–g) Angular movements. **(h, i)** Rotation.

(2) *angular movements,* which change the angle between the two bones; and (3) *rotation* about a bone's long axis.

Gliding

Gliding (Figure 8.5a) is the simplest type of movement. In this movement, the nearly flat surfaces of two bones slip across each other. Gliding occurs at the joints between the carpals and tarsals and between the flat articular processes of the vertebrae.

Angular Movements

Angular movements (Figure 8.5b-g) increase or decrease the angle between two bones. These movements, which may occur along any plane of the body, include flexion, extension, abduction, adduction, and circumduction.

Flexion. **Flexion** is bending that *decreases the angle* of the joint and brings the two bones closer together (Figure 8.5b–d). This movement usually occurs in the sagittal plane of the body. Examples include bending the fingers, bending the forearm toward the arm at the elbow, and bending the trunk forward. As a less obvious example, the arm is flexed at the shoulder when the arm is lifted in an anterior direction (Figure 8.5d). Also, the thigh is flexed at the hip when the thigh is lifted anteriorly.

Extension. **Extension** is the reverse of flexion and occurs at the same joints (Figure 8.5b–d). It *increases the angle* between the joining bones and is a straightening action. Straightening the fingers after making a fist is an example of extension. Bending a joint back beyond its straight position is called **hyperextension** (literally, "superextension"). At the shoulder, extension carries the arm posteriorly (Figure 8.5d).

Dorsiflexion and Plantar Flexion of the Foot. The terms *extension* and *flexion* are not appropriate for describing the up-and-down movements of the foot at the ankle joint (Figure 8.5e). Because the foot normally joins the leg at a right angle, both the upward and the downward movements *decrease* this angle. Technically, both are flexion. Therefore, more specific terms are used: Lifting the foot so that its superior surface approaches the shin is called **dorsiflexion,** while depressing the foot (pointing the toes) is called **plantar flexion.**

Abduction. **Abduction,** from Latin words meaning "moving away," is movement of a limb *away* from the

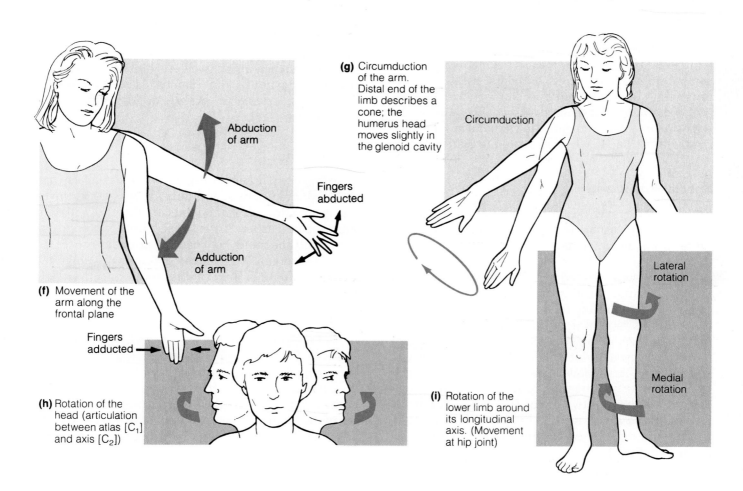

(g) Circumduction of the arm. Distal end of the limb describes a cone; the humerus head moves slightly in the glenoid cavity

Abduction of arm

Fingers abducted

Adduction of arm

(f) Movement of the arm along the frontal plane

Fingers adducted

(h) Rotation of the head (articulation between atlas [C_1] and axis [C_2])

Circumduction

Lateral rotation

Medial rotation

(i) Rotation of the lower limb around its longitudinal axis. (Movement at hip joint)

body midline. Raising the arm or thigh laterally is an example of abduction (Figure 8.5f). In reference to movement of the fingers or toes, *abduction* means spreading them apart. In this case, the "midline" is the longest digit: the third finger or the second toe. Note that bending the trunk away from the body midline to the right or left is called *lateral flexion* instead of abduction.

Adduction. Adduction ("moving toward") is the opposite of abduction. It is the movement of a limb *toward* the body midline (Figure 8.5f) or, in the case of the digits, toward the midline (longest digit) of the hand or foot.

Circumduction. Circumduction ("moving in a circle") is moving a limb or finger so that it describes a cone in space (Figure 8.5g). This is a complex movement that combines flexion, abduction, extension, and adduction. Circumduction combines almost all of the movements possible at the shoulder and hip joints. Therefore, circumducting your limbs is the best way for you to exercise many different limb muscles simultaneously.

Rotation

Rotation is the turning movement of a bone around its own long axis. Rotation is the only movement allowed between the first two cervical vertebrae (Figure 8.5h). It also occurs at the hip and shoulder joints (Figure 8.5i). Rotation may occur toward the median plane or away from it. For example, in **medial rotation** of the lower limb, the limb's anterior surface turns toward the median plane of the body. **Lateral rotation** is the opposite of that movement. The vertebral column also rotates, twisting the whole trunk to the right or left.

Special Movements

Certain movements do not fit into any of the above categories and occur at only a few joints. Some of these special movements are shown in Figure 8.6.

Supination and Pronation. The terms **supination** (soo″pĭ-na′shun; "turning backward") and **pronation** (pro-na′shun; "turning forward") refer to movements of the radius around the ulna (Figure 8.6a). Supination occurs when the forearm rotates laterally so that the palm faces anteriorly, whereas pronation occurs when the forearm rotates medially so that the palm faces posteriorly. Pronation brings the radius across the ulna so that the two bones form an X. A helpful memory trick: If you were to lift a cup of soup up to your mouth *on your palm,* you would be supinating ("soup"-inating").

Inversion and Eversion. Inversion and eversion are special movements of the foot (Figure 8.6b). To invert the foot, turn the sole medially; to evert the foot, turn the sole laterally.

Protraction and Retraction. Nonangular movements in the anterior and posterior directions are called **protraction** and **retraction,** respectively (Figure 8.6c). The mandible is protracted when you jut out your jaw and retracted when you bring it back. "Squaring" your shoulders in a military stance is another example of retraction.

Elevation and Depression. Elevation means lifting a body part superiorly (Figure 8.6d). For example, the scapulae are elevated when you shrug your shoulders. Moving the elevated part inferiorly is **depression.** During chewing, the mandible is elevated and depressed.

Opposition. In the palm of the hand, the saddle joint between metacarpal 1 and the carpals allows a movement called *opposition* of the thumb. This is the action by which you move your thumb to touch the tips of the other fingers on the same hand. This action is what makes the human hand such a fine tool for grasping and manipulating objects.

Types of Synovial Joints

The shapes of the articulating bone surfaces determine the movements allowed at a joint. Based on such shapes, our synovial joints can be classified as *plane, hinge, pivot, condyloid, saddle,* and *ball-and-socket joints* (Figure 8.7).

Plane Joints

In a **plane joint** (Figure 8.7a), the articular surfaces are essentially flat planes, and only short, gliding movements are allowed. Plane joints are the gliding joints introduced earlier (intertarsal joints, intercarpal joints, and the joints between the articular processes of the vertebrae). Their movements are **nonaxial;** that is, gliding does not involve rotation around any axis.

Hinge Joints

In a **hinge joint** (Figure 8.7b), the cylindrical end of one bone fits into a trough-shaped surface on another bone. Angular movement is allowed in just one direction, like a door on a hinge. Examples are the elbow joint, ankle joint, and the joints between the phalanges of the fingers. Hinge joints are classified as **uniaxial** (u″ne-aks′e-al; "one axis"): They allow movement around one axis only.

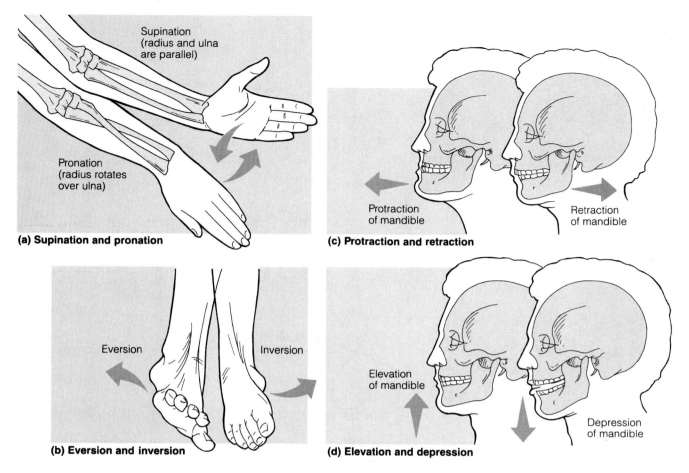

Figure 8.6   Some special body movements.

Pivot Joints

In a **pivot joint** (Figure 8.7c), the rounded end of one bone fits into a ring formed by another bone plus an encircling ligament. The rotating bone can turn only around its long axis. Thus, pivot joints are also uniaxial joints. One example is the proximal radioulnar joint, where the head of the radius rotates within a ring-like ligament secured to the ulna. Another example is the joint between the atlas and the dens of the axis (see Figure 7.17a on p. 161).

Condyloid Joints

In a **condyloid joint** (kon′dĭ-loid; "knuckle-like"), the egg-shaped articular surface of one bone fits into an oval concavity in another (Figure 8.7d). Both articular surfaces are oval. Condyloid joints allow the moving bone to travel (1) from side to side (abduction-adduction) and (2) back and forth (flexion-extension), but the bone cannot rotate around its long axis. Therefore,

movement occurs around two axes, so these joints are **biaxial** (*bi* = two). Examples are the knuckle joints (metacarpophalangeal) and the wrist joint.

Saddle Joints

In **saddle joints,** each articular surface has both convex and concave areas, just like a saddle. Nonetheless, these biaxial joints allow essentially the same movements as condyloid joints (Figure 8.7e). The best example of a saddle joint is the first carpometacarpal joint, in the ball of the thumb.

Ball-and-Socket Joints

In a **ball-and-socket joint** (Figure 8.7f), the spherical head of one bone fits into a round socket in another. Universal movement is allowed—in all axes, including rotation—so these are **multiaxial** ("many axes") joints. The shoulder and hip are ball-and-socket joints. *MOST MOVABLE*

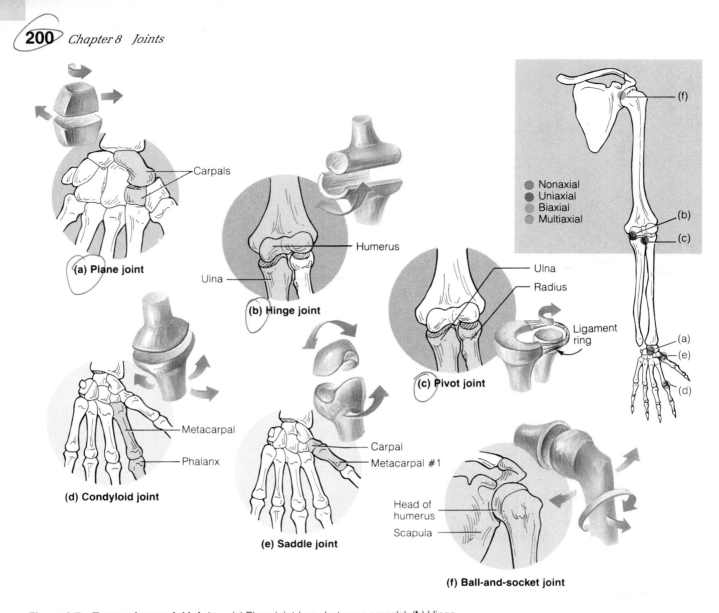

Figure 8.7   Types of synovial joints.   (a) Plane joint (e.g., between carpals). **(b)** Hinge joint (e.g., elbow joint). **(c)** Pivot joint (e.g., proximal radioulnar joint). **(d)** Condyloid joint (e.g., between metacarpals and phalanges). **(e)** Saddle joint (carpometacarpal joint of the thumb). **(f)** Ball-and-socket joint (e.g., shoulder joint).

Selected Synovial Joints

In this section, we consider five important synovial joints in detail: temporomandibular, shoulder, elbow, hip, and knee. While reading about these, keep in mind that they all contain articular cartilages, fibrous capsules, and synovial membranes.

Temporomandibular Joint

The **temporomandibular joint (TMJ),** or jaw joint, lies just anterior to the ear (Figure 8.8). At this joint, the head of the mandible articulates with the inferior surface of the squamous temporal bone. The head of the mandible is egg-shaped, while the articular surface on the temporal bone has a more complex shape: Posteriorly, it forms the concave **mandibular fossa.** Anteriorly, it forms a dense knob called the **articular tubercle.** A loose articular capsule encloses the joint, and the lateral aspect of this capsule thickens into a **lateral ligament.** Within the joint capsule is an articular disc, which divides the synovial cavity into superior and inferior compartments (Figure 8.8b). The two surfaces of the disc allow two distinct kinds of movement at the TMJ. First, the concave inferior surface receives the mandibular head and allows the familiar hinge-like movement of opening and closing the mandible. Second, the superior surface of the disc glides anteriorly with the mandibular head whenever the mouth is

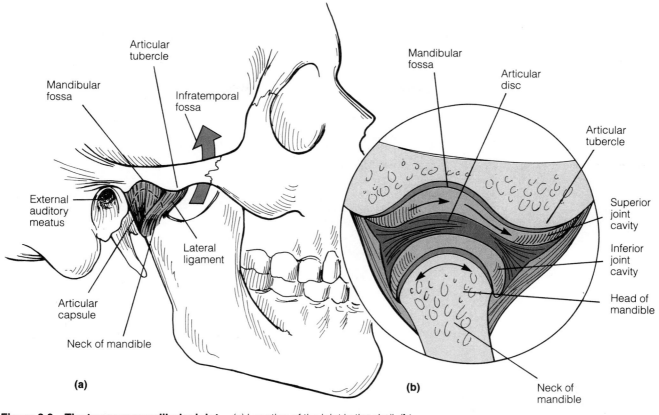

Figure 8.8   The temporomandibular joint.   **(a)** Location of the joint in the skull. **(b)** Enlargement of a sagittal section through the joint, showing the articular disc, the two compartments of the joint cavity, and the two main movements that occur (arrows).

opened wide. This anterior movement braces the mandibular head against the dense bone of the articular tubercle, so that the mandible is not forced superiorly through the thin roof of the mandibular fossa when one bites down on hard foods like nuts or hard candies. To demonstrate this anterior gliding of your mandible, place a finger on the mandibular head just anterior to your ear opening, then yawn.

Because of its shallow socket, the TMJ is the most easily dislocated joint in the body. Even a deep yawn can dislocate it. The joint almost always dislocates anteriorly; that is, the mandibular head glides anteriorly, ending up in a skull region called the infratemporal fossa (see Figure 8.8a). In such cases, the mouth remains wide open and cannot close. To realign a dislocated TMJ, the physician places his or her thumbs in the patient's mouth between the lower molars and the cheeks, then pushes the mandible inferiorly and posteriorly.

At least 5% of Americans suffer from painful conditions of the TMJ, called **temporomandibular disorders.** The most common symptoms are pain in the ear and face, tenderness of the jaw muscles, popping or clicking sounds during mouth opening, and stiffness of the TMJ. Usually caused by painful spasms of the chewing muscles, temporomandibular disorder affects people who respond to stress by grinding their teeth. However, it can also result from an injury to the TMJ or from poor occlusion of the teeth. Treatment usually focuses on getting the jaw muscles to relax (by massage, applying moist heat, or administering muscle-relaxant drugs), and patients often wear a bite plate at night to stop teeth grinding during sleep. In severe cases, surgery on the joint may be necessary. ■

Shoulder Joint

In the shoulder joint (Figures 8.4 and 8.9), stability has been sacrificed to provide the most freely moving joint of the body. This ball-and-socket joint is formed by the head of the humerus and the shallow glenoid cavity of the scapula. The glenoid cavity is slightly deepened by a rim of fibrocartilage called the **glenoid labrum** (*labrum* = lip) (Figure 8.9b). Still, this shallow cavity contributes little to joint stability. The articular capsule is remarkably thin and loose (qualities

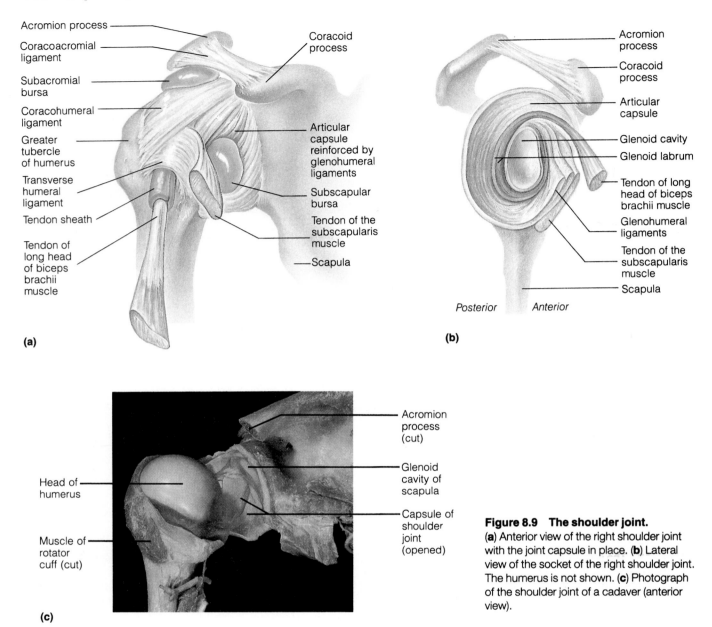

Acromion process

Coracoacromial ligament

Subacromial bursa

Coracohumeral ligament

Greater tubercle of humerus

Transverse humeral ligament

Tendon sheath

Tendon of long head of biceps brachii muscle

Coracoid process

Articular capsule reinforced by glenohumeral ligaments

Subscapular bursa

Tendon of the subscapularis muscle

Scapula

(a)

Acromion process

Coracoid process

Articular capsule

Glenoid cavity

Glenoid labrum

Tendon of long head of biceps brachii muscle

Glenohumeral ligaments

Tendon of the subscapularis muscle

Scapula

Posterior   Anterior

(b)

Head of humerus

Muscle of rotator cuff (cut)

Acromion process (cut)

Glenoid cavity of scapula

Capsule of shoulder joint (opened)

(c)

Figure 8.9   The shoulder joint.
(a) Anterior view of the right shoulder joint with the joint capsule in place. **(b)** Lateral view of the socket of the right shoulder joint. The humerus is not shown. **(c)** Photograph of the shoulder joint of a cadaver (anterior view).

that contribute to the joint's freedom of movement) and extends from the margin of the glenoid cavity to the anatomical neck of the humerus (Figure 8.9a). The only strong thickening of the capsule is the superior **coracohumeral ligament.** This ligament helps support the weight of the upper limb. The anterior part of the capsule thickens slightly into three **glenohumeral ligaments** (Figure 8.9b). These, however, are weak.

Muscle tendons that cross the shoulder joint contribute most to the joint's stability. The "superstabilizer" is the tendon of the long head of the biceps brachii muscle (Figure 8.9b). This tendon attaches to the superior margin of the glenoid labrum, travels within the joint cavity, and runs within the intertubercular groove of the humerus. The tendon secures the head

of the humerus tightly against the glenoid cavity. Four other tendons (and the associated muscles) make up the **rotator cuff,** or *musculotendinous cuff* (Figure 8.9c). This cuff encircles the shoulder joint and blends with the joint capsule. The rotator cuff muscles (Chapter 10, p. 264) include the subscapularis, supraspinatus, infraspinatus, and teres minor. The rotator cuff can be severely stretched when the arm is moved vigorously and is often injured in baseball pitchers who throw the ball too hard too often.

Shoulder dislocations are common injuries, often caused by falling on the shoulder. Since the joint's reinforcements are weakest inferiorly, the humerus tends to dislocate in that direction. ■

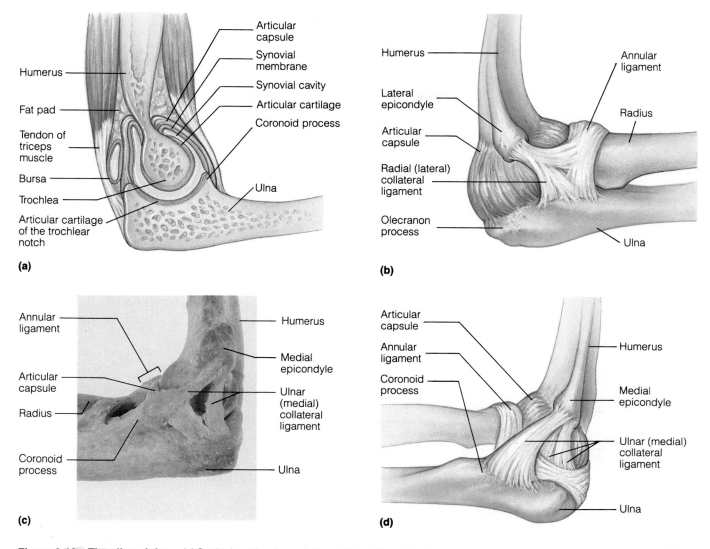

Figure 8.10   The elbow joint.   (**a**) Sagittal section through the middle of the right elbow joint (lateral view). (**b**) Lateral view of the right elbow joint. (**c**) and (**d**) Medial views of right elbow: cadaver photo (c) and diagram (d).

Elbow Joint

The elbow joint (Figure 8.10) is a hinge that allows only extension and flexion. In this joint, both the radius and ulna articulate with the condyles of the humerus. However, it is the close gripping of the humerus by the ulna's trochlear notch that forms the hinge and stabilizes the joint. The articular capsule attaches to the humerus and ulna (Figure 8.10b) and to the **annular ligament** (an'u-lar; "ring-like") of the radius, a ring around the head of the radius. Laterally and medially, the capsule thickens into strong ligaments that prevent lateral and medial movements: the **radial collateral ligament,** a triangular band on the lateral side (Figure 8.10b), and the **ulnar collateral ligament** on the medial side (Figure 8.10d). Tendons of several arm muscles (the biceps brachii, triceps brachii, and others) cross the elbow joint and provide stability (Figure 8.10a).

✚ The elbow joint is not easily dislocated, but the ulna can dislocate posteriorly if one falls on outstretched hands when the forearm is slightly flexed. ■

Hip Joint

The hip (coxal) joint, like the shoulder joint, has a ball-and-socket design (Figure 8.11). Its range of motion is wide but not nearly as wide as the shoulder's

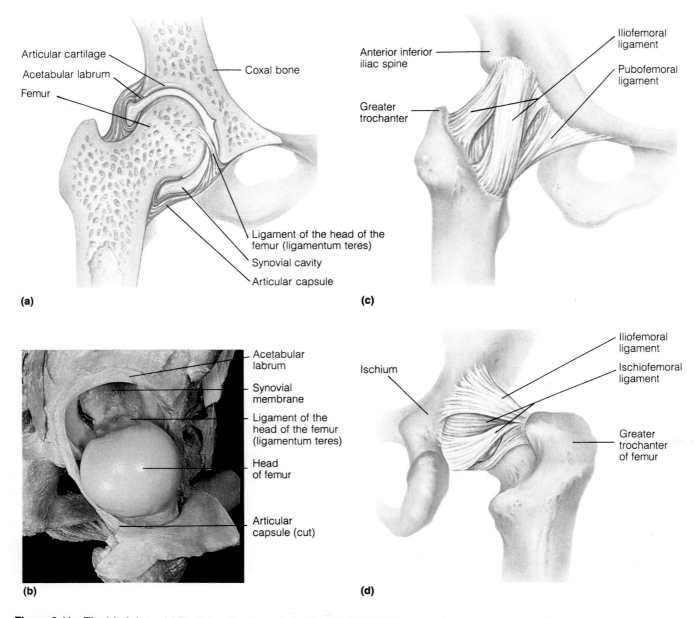

Figure 8.11 The hip joint. (a) Frontal section through the right hip joint. (b) Photograph of the interior of the hip joint, lateral view. (c) Anterior view of hip joint, with the capsule in place. (d) Posterior view.

range. Movements occur in all possible axes but are limited by the joint's ligaments and by its deep socket.

The hip joint is formed by the spherical head of the femur and the deeply cupped acetabulum of the hip bone. The depth of the acetabulum is enhanced by a circular rim of fibrocartilage called the **acetabular labrum** ("lip") (Figure 8.11c). The diameter of this labrum is less than that of the head of the femur. Thus, the femur cannot slip out of the socket, and hip dislocations are rare.

The joint capsule runs from the rim of the acetabulum to the neck of the femur (Figure 8.11c and d).

Three ligamentous thickenings of this capsule reinforce the joint: the **iliofemoral ligament,** a strong, V-shaped, anteriorly located ligament; the **pubofemoral ligament,** a triangular thickening of the capsule's inferior region; and the **ischiofemoral ligament,** a spiraling, posteriorly located ligament. These three ligaments are arranged in such a way that they "screw" the head of the femur into the acetabulum when a person stands erect, thereby increasing the stability of the joint.

The **ligament of the head of the femur** (Figure 8.11a) is a flat, intracapsular ligament that runs from the head of the femur to the inferior region of the acetabulum. This ligament remains slack during most hip movements, so it is not important in stabilizing the joint. Its mechanical function is unknown, but it does contain an artery that supplies the head of the femur. Damage to this artery leads to severe arthritis of the hip joint.

Muscle tendons that cross the hip joint contribute to its stability, as do the fleshy parts of many hip and thigh muscles that surround the joint. In this joint, however, stability comes chiefly from the cupped socket and the capsular ligaments.

✚ The femur can be dislocated posteriorly at the hip joint in car accidents when a flexed thigh hits the dashboard. ■

Knee Joint

The knee joint is the largest and most complex joint in the body (Figure 8.12). It primarily acts as a hinge but also permits some medial and lateral rotation when in the flexed position. In this joint, the wheel-shaped condyles of the femur roll along the flat-surfaced condyles of the tibia like tires rolling on a road. Sharing the cavity of the knee joint is an articulation between the patella and the inferior end of the femur (Figure 8.12a). This *femoropatellar joint* is a plane joint that allows the patella to glide across the distal femur as the knee bends.

The synovial cavity of the knee joint has a complex shape with several incomplete subdivisions and several extensions that lead to "blind alleys" (Figure 8.12a). At least a dozen bursae are associated with the knee joint, some of which are shown in the figure. The **subcutaneous prepatellar bursa** is often injured when the knee is bumped.

Two fibrocartilage menisci occur within the joint cavity, between the femoral and tibial condyles. These C-shaped **lateral** and **medial menisci** attach externally to the condyles of the tibia (Figure 8.12b). Besides deepening the shallow articular surfaces of the tibia, the menisci help prevent side-to-side rocking of the femur on the tibia. They also absorb shocks transmitted to the knee.

The capsule of the knee joint can be seen on the posterior and lateral aspects of the knee (Figure 8.12d), where it covers most parts of the femoral and tibial condyles. Anteriorly, however, the capsule is absent. Instead, this anterior area is covered by three broad ligaments that run inferiorly from the patella to the tibia (Figure 8.12c): the **patellar ligament,** flanked by the **medial** and **lateral patellar retinacula** (ret"ĭ-nak'u-la; "retainers"). The patellar ligament is actually a continuation of the tendon of the main muscles on the anterior thigh, the quadriceps femoris. Physicians tap the patellar ligament to test the knee-jerk reflex.

The joint capsule of the knee is reinforced by several capsular and extracapsular ligaments. All of these ligaments stretch taut when the knee is extended to prevent hyperextension of the leg at the knee. These ligaments are:

1. The **fibular** and **tibial collateral ligaments** (Figure 8.12c and e). The fibular collateral ligament descends from the lateral epicondyle of the femur to the head of the fibula. The tibial collateral ligament runs from the medial epicondyle of the femur to the medial condyle of the tibia. Besides checking leg extension, the collateral ligaments prevent the leg from moving laterally and medially at the knee.

2. The **oblique popliteal ligament** (pop"lĭ-te'al; "back of the knee") crosses the posterior aspect of the capsule (Figure 8.12d). It is actually a part of the tendon of the semimembranosus muscle, which stabilizes the joint.

3. The **arcuate popliteal ligament** arcs superiorly from the head of the fibula to the posterior aspect of the joint capsule (Figure 8.12d).

In addition, the knee joint has two important *intracapsular* ligaments (Figure 8.12b and e). These are called **cruciate ligaments** (kru'she-āt) because they cross each other (*crus* = cross). Each runs from the tibia superiorly to the femur. Each cruciate is named according to the location of its attachment to the tibia. The **anterior cruciate ligament** attaches to the *anterior* part of the tibia, in the intercondylar area. From there, it passes posteriorly to attach to the femur on the medial side of its lateral condyle. Functionally, the anterior cruciate ligament helps prevent anterior sliding of the tibia on the femur when the leg is flexed; it also prevents hyperextension of the leg at the knee (Figure 8.13). The **posterior cruciate ligament** arises from the *posterior* intercondylar area of the tibia. It passes anteriorly to attach to the femur on the lateral side of the medial condyle. This strong ligament prevents forward sliding of the femur or backward displacement of the tibia (Figure 8.13b). The posterior cruciate also acts with the anterior cruciate to lock the knee when one stands (see below).

The tendons of many muscles reinforce the joint capsule and act as the main stabilizers of the knee joint. Most important are the tendons of the quadriceps femoris and semimembranosus muscles. The greater the strength and tone of these muscles, the less the chance of knee injury. Athletes who are out of shape do suffer more knee injuries.

The knees have a built-in locking device that provides steady support for the body in the standing position. As one rises, the flexed leg begins to extend at the knee, and the femoral condyles roll like ball bearings on the tibial condyles. Then, as extension nears

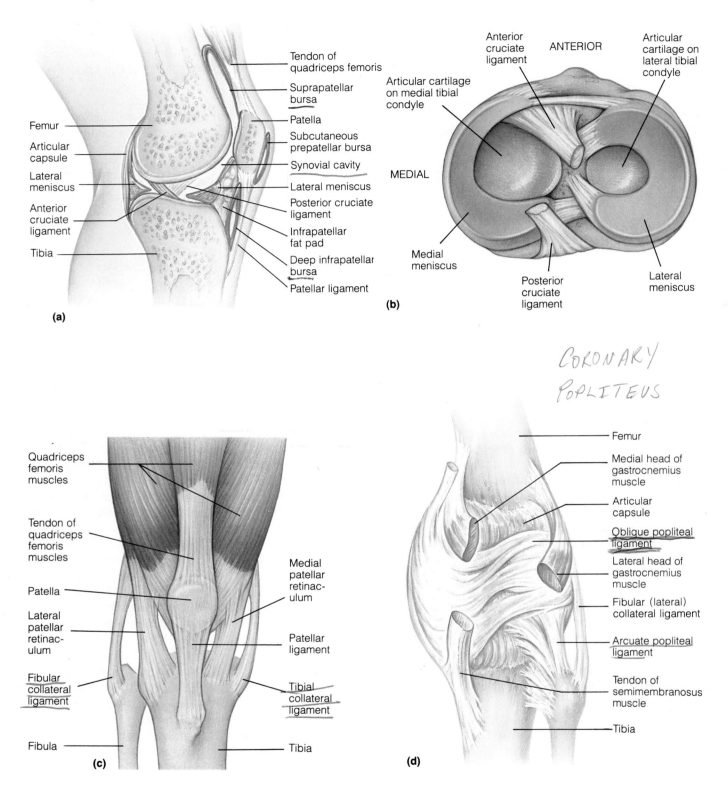

CORONARY
POPLITEUS

Figure 8.12   The knee joint.   (**a**) Sagittal section through the right knee joint. (**b**) Superior view of the right tibia in the knee joint, showing the menisci and cruciate ligaments. (**c**) Anterior view of knee. (**d**) Posterior view of the joint capsule, including ligaments. (**e**) Anterior view of flexed knee, showing the cruciate ligaments. Articular capsule has been removed, and the quadriceps tendon has been cut and reflected distally. (**f**) Photograph of an opened knee joint similar to (e).

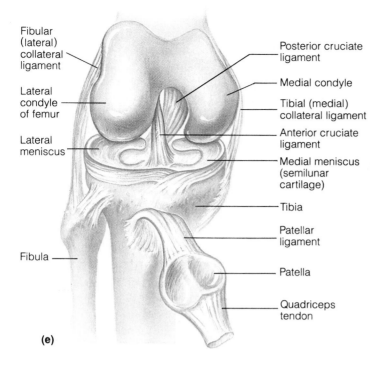

(e)

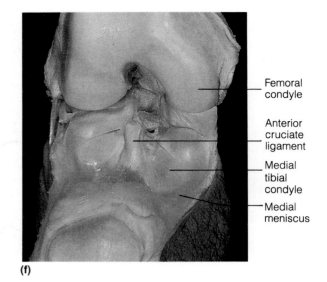

(f)

completion, the lateral femoral condyle stops rolling before the medial condyle stops. This causes the femur to rotate medially on the tibia until both cruciate and both collateral ligaments stretch tight and halt all movement. The tension in these ligaments locks the knee into a rigid structure that cannot be flexed again until it is unlocked. Unlocking is accomplished by a muscle called the popliteus (Table 10.15, p. 285), which rotates the femur laterally on the tibia.

Can you deduce why knee injuries occur so frequently in contact sports?

Knees are susceptible to sports injuries because their articular surfaces offer no stability. The nearly flat tibial surface provides no socket to secure the femoral condyles. Thus, the knee joint is very vulnerable to *horizontal* blows, such as tackling and body blocks in football. Most dangerous are *lateral* blows (Figure

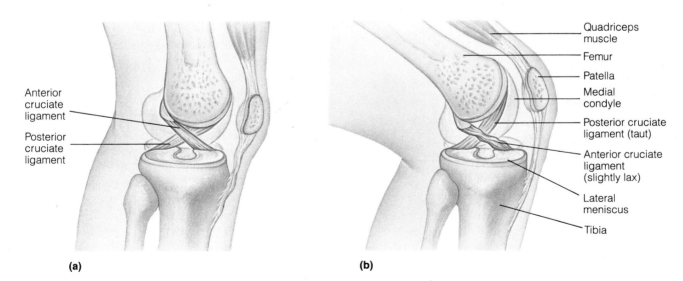

(a)

(b)

Figure 8.13   Knee joint movements.   The cruciate ligaments prevent undesirable movements at the knee joint. (**a**) When the knee is extended, the anterior cruciate becomes taut and prevents hyperextension at the joint. (**b**) When the knee is flexed, the posterior cruciate prevents posterior slipping movements of the tibia, while the anterior cruciate prevents anterior slipping movements.

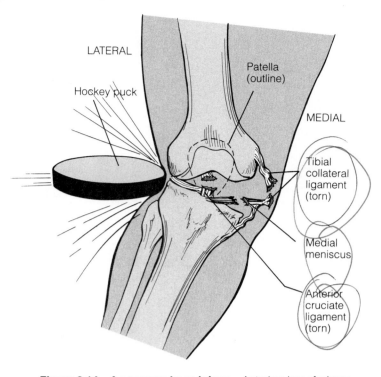

Figure 8.14 A common knee injury. Anterior view of a knee being hit by a hockey puck. In sports, most knee injuries result from blows to the lateral side. By separating the femur from the tibia medially, such blows tear both the tibial collateral ligament and the medial meniscus because the two are attached. The anterior cruciate ligament also tears.

8.14). These forces rip the tibial collateral ligament and the medial meniscus attached to it, plus the relatively weak anterior cruciate ligament. It is estimated that 50% of professional football players have serious knee injuries during their careers. ■

✚ Joint Injuries

Sprains

In a **sprain,** the ligaments reinforcing a joint are stretched or torn. Common sites of sprains are the lumbar region of the spine, the ankle, and the knee. Partly torn ligaments will repair themselves, but they heal slowly because ligaments are poorly vascularized. Sprains tend to be painful and immobilizing. Completely ruptured ligaments, by contrast, require prompt surgical repair because inflammation in the joint will break down the neighboring tissues and turn the injured ligament to "mush." Surgical repair can be

a difficult task: A ligament consists of hundreds of fibrous strands, and sewing one back together has been compared to sewing two hairbrushes together.

When important ligaments are too severely damaged to be repaired, they must be removed and replaced with substitute ligaments. For example, a piece of tendon from a muscle or collagen extracted from cattle hide and woven into bands can be stapled to the margins of the articulating bones. In another technique, carbon fibers are implanted in the patient's torn ligament to form a supporting mesh. The mesh is then invaded by fibroblasts, which lay down collagen fibers that rebuild the ligament.

Dislocations

A **dislocation (luxation)** occurs when the bones of a joint are forced out of alignment. This injury is usually accompanied by sprains, inflammation, pain, and difficulty in moving the joint. Dislocations may result from serious falls, and are common sports injuries. The jaw, shoulder, finger, and thumb joints are most commonly dislocated. Like fractures, dislocations must be reduced; that is, the bone ends must be returned to their proper positions by a physician. Attempts by an untrained person to "snap the bone back into its socket" are usually more harmful than helpful.

It is common for a person to experience repeated dislocations at the same joint. Can you imagine why?

The initial dislocation stretches the joint capsule and ligaments. Because the capsule is loose, the joint is now more likely to dislocate again. Injured ligaments eventually shorten to their original length, but this aspect of healing can take years. ■

Cartilage Injuries

Most injuries to joint cartilages involve torn menisci in the knee. Tearing often happens when a meniscus is simultaneously subjected to both high pressure and shear stress. (Recall from p. 120 that cartilage is weak in resisting shear.) For example, a tennis player lunging to return a ball sometimes twists the knee medially, ripping both the joint capsule and the medial meniscus attached to it. Because cartilage does not heal well, most sports physicians recommend that damaged knee cartilages be removed. This can be done by a remarkable procedure called **arthroscopic surgery** (ar-thro-skop'ik; "looking into joints"). The arthroscope, a small instrument bearing a tiny lens and light source, is attached to a television camera and enables the surgeon to view the joint through a small incision.

The surgeon removes the cartilage fragments and repairs the ligaments through the same tiny slit, minimizing tissue damage and scarring. Patients usually leave the hospital the same day. Removing parts or all of the menisci does not severely impair the mobility of the knee joint, but the joint is certainly less stable. ■

✚ Inflammatory and Degenerative Conditions

Inflammatory conditions that affect joints include inflammations of bursae and tendon sheaths and various forms of arthritis.

Bursitis and Tendonitis

Bursitis, inflammation of a bursa, usually results from a physical blow or friction, which causes the bursa to swell with fluid. Falling on one's knee can cause a painful bursitis of the subcutaneous prepatellar bursa (Figure 8.12a), known as **housemaid's knee.** Resting and rubbing one's elbows on a desk can lead to **student's elbow,** or **olecranon bursitis,** the swelling of a bursa just deep to the skin of the posterior elbow. Severe cases of bursitis may be treated by injecting inflammation-reducing drugs into the bursa. If the accumulation of fluid is excessive, the fluid may have to be removed by needle aspiration. **Tendonitis** is inflammation of tendon sheaths. Its causes, symptoms, and treatments mirror those of bursitis.

Bursitis and tendonitis may also result from arthritis or bacterial infection.

Arthritis

The term **arthritis** describes over 100 kinds of inflammatory or degenerative diseases that damage the joints. In all its forms, arthritis is the most widespread crippling disease in the United States. One in every seven Americans suffers its effects. All forms of arthritis have, to a greater or lesser degree, the same initial symptoms: pain, stiffness, and swelling of the joint.

Osteoarthritis (Degenerative Joint Disease)

Osteoarthritis (OA) accounts for about half of all arthritis cases. A chronic (long-term) degenerative condition, OA is often called "wear-and-tear arthritis." It is most common in the aged and is probably related to the normal aging process. OA affects women more

often than men, but 85% of all Americans develop this condition. OA affects the articular cartilages, causing them to soften, fray, crack, and erode.

The cause of OA is unknown. According to current theory, normal use causes joints to release enzymes that break down their cartilage, while the chondrocytes continually fix the damage by secreting more cartilage matrix. When destruction predominates, OA results. More specifically, OA may result from excessive strain on the joints, which causes too much of the cartilage-destroying enzyme to be released. Badly aligned or overworked joints are likely to develop OA.

As the disease progresses, bone spurs tend to grow around the margins of the damaged articular cartilages. These spurs encroach on the joint cavity and may restrict joint movement. Patients complain of stiffness upon waking in the morning, but this decreases within a half hour. The affected joints may make a crunching noise as they move and their roughened surfaces rub together.

OA is classified as a noninflammatory type of arthritis: Inflammation might be present, but it is not a primary symptom. The most commonly affected joints are the finger and knuckle joints and the weight-bearing joints of the lower limb (hip and knee). The nonsynovial joints between the vertebral bodies are also susceptible, especially in the cervical and lumbar regions of the spine.

The course of OA is slow and irreversible. Usually, its symptoms are controllable with a mild pain reliever like aspirin, along with moderate activity to keep the joints movable and with rest when the joint becomes very painful. OA usually does not cripple, but it can do so when the hip and knee joints are affected.

Rheumatoid Arthritis

Rheumatoid (roo'mah-toid) **arthritis (RA)** is a chronic inflammatory disorder. It is a complex disease: Along with pain and swelling of the joints, its symptoms include osteoporosis, muscle weakness, and problems with the heart and blood vessels. RA affects three times as many women as men. It usually develops gradually, or in spurts that are years apart. However, it can also appear quickly. There are tragic stories of women who develop severe RA within weeks and then suffer from it for the rest of their lives. Its onset is usually during a woman's childbearing years, but it may appear at any age. Although not as common as osteoarthritis, RA troubles millions, about 1% of all people. RA tends to affect many joints at the same time, particularly the small joints of the fingers, wrists, ankles, and feet. It usually affects the corresponding joints on both sides of the body (the right

A CLOSER LOOK   Joints: From Knights in Shining Armor to Bionic Man

There is a stark contrast between the time it took to develop joints for armor and the modern development of joint prostheses. The greatest challenge faced by visionary men of the Middle Ages and Renaissance was to design armor joints that allowed mobility while still protecting the human joints beneath. The ball-and-socket joints that they found so difficult to protect were the first to be engineered by modern-day "visionaries" for implantation into the body.

The history of joint prostheses (artificial joints) is less than 50 years old. It dates back to the 1940s and 1950s, when World War II and the Korean War produced large numbers of wounded who needed artificial limbs and joints. Today, about a quarter of a million arthritis patients receive total joint replacements each year, mostly because of the destructive effects of osteoarthritis or rheumatoid arthritis.

The human body attempts to reject any implanted foreign object or causes its corrosion. To produce strong, mobile, lasting joints, a substance was needed that was strong, nontoxic to the body, and resistant to the corrosive effects of organic acids present in blood. In 1963, Sir John Charnley, an English orthopedic surgeon, performed the first total hip replacement and revolutionized the therapy of arthritic hips. His device consisted of a Vitallium metal ball on a stem and a cup-shaped polyethylene plastic socket anchored to the pelvis by methyl methacrylate cement. This cement proved to be exceptionally strong and relatively problem free. It was not until ten years later that total knee joint replacements that operated smoothly and allowed all natural movements of the knee became available. Earlier devices locked abruptly when the knee was extended, causing the patient to fall or the joint to be wrenched free.

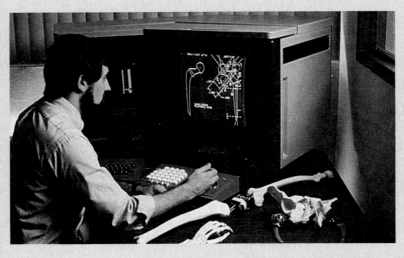

Replacement joints employing both metal and plastic are now available for a number of other joints, including the fingers, elbow, and shoulder. Current techniques have produced total hip and knee replacements that last about ten years in elderly patients who do not excessively stress the joint. Most such operations are done to reduce pain and restore about 80% of original joint function.

Replacement joints are not yet strong or durable enough for young, active people, but dramatic changes are occurring in the way joints are designed, manufactured, and installed. CAD/CAM (computer-aided design and computer-aided manufacturing) techniques are being used to design and create individualized artificial joints (see photo). The computer is fed the patient's X-rays along with information about the patient's problems. The computer draws from a database of hundreds of normal joints and generates possible designs and modifications for a joint prosthesis that can be reviewed in less than a minute. Once the design is selected, the computer produces a program that directs the machines that modify the implant or totally fabricate a new joint prosthesis.

The influence of CAD/CAM on the joint replacement business is enormous. Previously, designing and producing a customized joint took 12 weeks or more. The wait has been reduced to two weeks by this technique, and those "in the know" predict that a two-day wait will soon be the norm. Some prostheses, such as an artificial knee, can cost as much as $4500 to produce manually, but the CAD/CAM system can produce it for under $3000, and soon the cost may be even lower.

Joint replacement therapy is coming of age, but perhaps equally exciting is research into the capability of joint tissues to regenerate. We now know that if a joint with localized articular cartilage damage is slowly flexed and extended over a period of weeks, the joint cartilage sometimes regenerates itself. Furthermore, bone periosteum grafted onto joint surfaces may be able to produce new cartilage. These prospects of joint cartilage regeneration offer hope for younger patients, since they could stave off the need for a joint prosthesis for several years.

And so, through the centuries, the focus has shifted from jointed armor to artificial joints that can be put inside the body to restore lost joint function. Modern technology has in fact accomplished what the armor designers of the Middle Ages never even dreamed of.

and left elbows, for example). The course of RA is variable, marked by flare-ups and remissions (*rheumat* = susceptible to change and flux).

RA is an **autoimmune disease**—a disorder in which the body's immune system attacks its own tissues. The cause of this reaction is unknown, but RA might follow infection by certain bacteria and viruses. It has been postulated that these microorganisms produce molecules similar to natural molecules in the joints, and when the body develops the ability to destroy the foreign molecules, it can inappropriately destroy its own joint tissues as well.

RA begins with an inflammation of the synovial membrane. Capillaries in this membrane leak tissue fluid and white blood cells into the joint cavity. This excess synovial fluid then causes the joint to swell. With time, the inflamed synovial membrane thickens into a *pannus* ("rag"), a coat of granulation tissue that clings to the articular cartilages. The pannus erodes the cartilage and, often, the underlying bone. This erosion probably occurs because phagocytic white blood cells in the pannus leak digestive enzymes from their lysosomes. As the cartilage is destroyed, the pannus changes into a fibrous scar tissue that interconnects the bone ends. This scar tissue eventually ossifies, and the bone ends fuse together, immobilizing the joint. This condition, called **ankylosis** (ang″kĭ-lo′sis; "stiff condition"), often produces bent, deformed fingers (Figure 8.15).

Most drugs used to treat RA are either anti-inflammatory, or they suppress the immune system. None is highly successful, and there is no cure for this tragic disease. It is recommended that patients exercise to maintain as much mobility in the joints as possible. Joint prostheses are the last resort for severely crippled patients. Some RA sufferers have had over a dozen of their joints replaced by artificial joints! (See the box on the facing page)

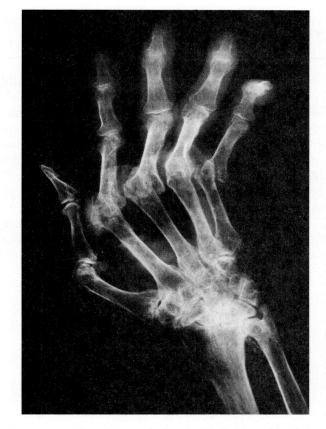

Figure 8.15 X-ray image of a hand with rheumatoid arthritis. Note the enlarged joints, a product of inflammation of the synovial membrane. Also note evidence of ankylosis (bone fusion) in the ring finger.

Gouty Arthritis (Gout)

Uric acid, a normal waste product of nucleic acid metabolism, is ordinarily eliminated in the urine without any problems. When uric acid rises to abnormal levels in the blood and body fluids, however, it precipitates as solid crystals of urate in the synovial membranes. An inflammatory response follows as the body tries to attack and digest the crystals. This produces an agonizingly painful attack of **gouty** (gow′te) **arthritis,** or gout. Initial attacks involve a single joint, usually in the lower limb, often at the base of the big toe. Gout is far more common in males than in females because men naturally have higher levels of uric acid in their blood.

Untreated gout can cause the ends of the articulating bones to fuse, immobilizing the joint. Fortunately, several drugs that effectively end or prevent gout attacks are available.

Lyme disease

Lyme disease is caused by bacteria called spirochetes and is transmitted by bites from ticks that live on deer. Arthritis is a major symptom, especially in the knee joint, along with a rash and flu-like symptoms. The symptoms may persist for years if the disease is untreated, although they eventually decline in most cases. Sometimes, however, they may proceed to neurological disorders and irregularities of the heartbeat. Since symptoms vary from person to person, the disease is difficult to diagnose. It is treatable with antibiotics, but it takes a long time to kill the infecting bacteria. The incidence of Lyme disease seems to be increasing in the United States.

The Joints Throughout Life

Synovial joints develop from mesenchyme that fills the spaces between the cartilaginous "bone models" in the late embryo. The outer region of this intervening mesenchyme condenses to become the fibrous joint capsule, while the inner region hollows to become the joint cavity. By week 8, these joints resemble adult joints in form and arrangement; that is, the synovial membranes are developed, and synovial fluid is being secreted into the joint cavities.

Excluding injuries, relatively few interferences with joint function occur until late middle age. At this time, the intervertebral discs are more likely to herniate, and osteoarthritis may develop.

Just as bones must be stressed to maintain their strength, joints must be exercised to keep their health. Exercise squeezes synovial fluid in and out of articular cartilages, providing the cartilage cells with the nourishment they require. And although exercise cannot prevent arthritis, it helps maintain joint integrity by strengthening the muscles that stabilize the joints. Exercise should not be overdone, however, because too much stress guarantees an early onset of osteoarthritis. Since the buoyancy of water relieves much of the stress on weight-bearing joints, people who exercise in pools often retain good joint function throughout life.

✚ Related Clinical Terms

Ankylosing spondylitis (ang"kĭ-lo'sing spon"dĭ-li'tis) (*ankyle* = stiff joint; *spondyl* = vertebra) An unusual kind of rheumatoid arthritis that mainly affects males. It usually begins in the sacroiliac joints and progresses superiorly along the spine. The vertebrae become interconnected by so much fibrous tissue that the spine becomes rigid ("poker back").

Arthrology (ar-throl'o-je) The study of joints.

Arthroplasty ("joint reforming") Replacing a diseased joint with an artificial joint (see the box on p. 210).

Chondromalacia patellae ("softening of cartilage by the patella") Damage and softening of the articular cartilages on the posterior surface of the patella and the anterior surface of the distal femur. This condition, seen most often in adolescent athletes, produces a sharp pain in the knee when the leg is extended (in climbing stairs, for example). Chondromalacia may result when the quadriceps femoris (the main group of muscles on the anterior thigh) pulls unevenly on the patella, persistently rubbing the patella against the femur in the knee joint. Chondromalacia can often be corrected by exercises that strengthen weakened parts of the quadriceps muscles.

Loose bodies Small, unattached pieces of bone or cartilage within a joint capsule. They can result from trauma to the joint. For example, blows to the knee can tear fragments of cartilage from knee menisci. Loose bodies may also follow erosion of the articular cartilage: Such erosion exposes the underlying bone, which then dies and allows fragmentation of bone particles into the joint cavities. Loose bodies can cause painful locking of a joint. Unless they are removed, they promote osteoarthritis.

Rheumatism A term commonly used by the layperson to indicate any disease involving pain in muscles or joints. It may apply to arthritis, bursitis, and so on.

Synovitis Inflammation of the synovial membrane of a joint. Synovitis can result from a blow, infection, or arthritis. In healthy joints, only small amounts of synovial fluid are present, but synovitis leads to the production of large amounts of fluid, causing the joints to swell. Such accumulation of fluid in the joint cavity is called *effusion.*

Tenosynovitis (ten'o-sin"o-vi"tis) Painful inflammation of a tendon or its sheath. Tenosynovitis occurs most often in the hands, wrists, feet, or ankles as a result of repeated, intense use and may be temporarily disabling. The condition is common in pianists and typists (see the discussion of carpal tunnel syndrome on p. 172). Tenosynovitis can also be caused by arthritis and bacterial infection of the tendon sheath. Therapy involves immobilization of the affected body region or surgery to drain an infected tendon sheath.

Chapter Summary

1. Joints (articulations) are sites where pieces of the skeleton meet. They hold bones together and allow various degrees of movement.

CLASSIFICATION OF JOINTS (p. 189)

1. Joints are classified structurally as fibrous, cartilaginous, or synovial. They are classified functionally as synarthrotic (no movement), amphiarthrotic (slight movement), or diarthrotic (free movement).

FIBROUS JOINTS (pp. 189–190)

1. In fibrous joints, the bones are connected by fibrous tissue. No joint cavity is present. Nearly all fibrous joints are synarthrotic.

2. The types of fibrous joints are sutures (between skull bones), syndesmoses ("ligament joints"), and gomphoses (articulation of the teeth with their sockets).

CARTILAGINOUS JOINTS (pp. 190–191)

1. In cartilaginous joints, the bones are united by cartilage. No joint cavity exists.

2. Synchondroses are immovable joints of hyaline cartilage (epiphyseal plates, for example). Symphyses are slightly movable fibrocartilage joints (intervertebral discs, pubic symphysis).

SYNOVIAL JOINTS (pp. 191–208)

1. Most joints in the body are synovial joints. All synovial joints are diarthrotic.

General Structure (pp. 191–194)

2. Synovial joints have a fluid-containing cavity and are covered by an articular capsule. The capsule has an outer fibrous region, often reinforced by ligaments, and an inner synovial membrane that produces synovial fluid. The articulating ends of bone are covered with cushions of articular cartilage. Nerves in the capsule provide a sense of "joint stretch." Some joints contain fibrocartilage discs (menisci), which allow two movements at one joint.

3. Synovial fluid is mainly a filtrate of the blood, but it also contains molecules that make it a friction-reducing lubricant.

How Do Synovial Joints Work? (p. 194)

4. Synovial joints reduce friction. The cartilage-covered bone ends glide on a slippery film of synovial fluid squeezed out of the articular cartilages. This mechanism is called weeping lubrication.

Bursae and Tendon Sheaths (p. 194)

5. A bursa is a fibrous sac lined with a synovial membrane and containing synovial fluid. Tendon sheaths, which are similar to bursae, wrap certain tendons. Bursae and tendon sheaths are lubricating devices that allow adjacent structures to move smoothly over one another.

Factors Influencing the Stability of Synovial Joints (pp. 194–195)

6. Joints are the weakest part of the skeleton. Factors that stabilize joints are the shapes of the articulating surfaces, ligaments, and the tone of muscles whose tendons cross the joint.

Movements Allowed by Synovial Joints (pp. 195–198)

7. Contracting muscles produce three common kinds of bone movements at synovial joints: (1) gliding, (2) angular movements (flexion, extension, abduction, adduction, and circumduction), and (3) rotation.

8. Special movements include supination and pronation, inversion and eversion, protraction and retraction, elevation and depression, and opposition.

Types of Synovial Joints (pp. 198–199)

9. The shapes of the articular surfaces reflect the kinds of movements allowed at a joint. Based on such shapes, joints are classified as plane (nonaxial), hinge or pivot (uniaxial), condyloid or saddle (biaxial), or ball-and-socket (multiaxial).

Selected Synovial Joints (pp. 200–208)

10. The temporomandibular joint is formed by (1) the head of the mandible and (2) the mandibular fossa and articular tubercle of the temporal bone. Because of its articular disc, this joint allows both a hinge-like opening of the mouth and an anterior gliding of the mandible. This joint often dislocates anteriorly and frequently experiences anxiety-induced temporomandibular disorders.

11. The shoulder joint is a ball-and-socket joint formed by the glenoid cavity and the head of the humerus. It is the most freely movable joint. Its socket is shallow, and its capsule is lax and reinforced by ligaments superiorly and anteriorly. The tendons of the biceps brachii and the rotator cuff (musculotendinous cuff) help stabilize the joint. The humerus often dislocates inferiorly at the shoulder joint.

12. The elbow is a hinge joint in which the ulna and radius articulate with the humerus. The elbow joint allows flexion and extension. Its articular surfaces are highly complementary and help stabilize it. Radial and ulnar collateral ligaments prevent side-to-side movement of the forearm.

13. The hip joint is a ball-and-socket joint formed by the acetabulum of the hip bone and the head of the femur. The hip joint is adapted for weight bearing. Its articular surfaces are deep and secure, and its capsule is heavy and strongly reinforced by three ligamentous thickenings.

14. The knee is a complex, shallow hinge joint formed by the articulation of the tibial and femoral condyles (and anteriorly by the patella with the femur). Extension, flexion, and some rotation are allowed. C-shaped menisci deepen the articular surfaces. The capsule, which is absent anteriorly, is reinforced by several capsular ligaments that help prevent hyperextension of the leg. The intracapsular cruciate ligaments help prevent displacement of the joint surfaces and lock the knee when one stands. The tone of the muscles crossing this joint is important for knee stability. In contact sports, lateral blows are responsible for many knee injuries.

JOINT INJURIES (pp. 208–209)

Sprains (p. 208)

1. Sprains are the stretching and tearing of joint ligaments. Since ligaments are poorly vascularized, healing is slow.

Dislocations (p. 208)

2. Joint dislocations move the surfaces of articulating bones out of alignment. They must be reduced.

Cartilage Injuries (pp. 208–209)

3. Sports injuries to the knee menisci are common and can result from twisting forces as well as from direct blows to the knee. Joint cartilages heal poorly, and may have to be removed.

INFLAMMATORY AND DEGENERATIVE CONDITIONS (pp. 209–211)

Bursitis and Tendonitis (p. 209)

1. Bursitis and tendonitis are inflammation of a bursa and a tendon sheath, respectively.

Arthritis (pp. 209–211)

2. Arthritis is the inflammation or degeneration of joints accompanied by pain, swelling, and stiffness. Arthritis includes many different diseases and affects millions of people.

3. Osteoarthritis is a degenerative condition that first involves the articular cartilages. It is common in the aged. Weight-bearing joints are most often affected.

4. Rheumatoid arthritis, the most crippling kind of arthritis, is an autoimmune disease involving severe inflammation of the joints (starting with the synovial membranes).

5. Gout is joint inflammation caused by a deposition of urate salts in the synovial membranes.

Lyme Disease (p. 211)

6. Arthritis is a symptom of Lyme disease, a condition caused by bacteria transmitted by tick bites.

THE JOINTS THROUGHOUT LIFE (p. 212)

1. Joints develop from mesenchyme between the bone models in the embryo.

2. Joints usually function well until late middle age, when osteoarthritis almost always appears.

Review Questions

Multiple Choice and Matching Questions

1. Match the terms in the key to the correct descriptions. (More than one choice might apply.)

Key: **(a)** fibrous joints **(b)** cartilaginous joints **(c)** synovial joints

_a,b_ **(1)** have no joint cavity
_a_ **(2)** types are sutures and syndesmoses
_a_ **(3)** dense connective tissue fills the space between the bones
_a_ **(4)** almost all joints of the skull
_b_ **(5)** types are synchondroses and symphyses
_c_ **(6)** all are diarthroses
_c_ **(7)** the most common type of joint in the body
_a,b_ **(8)** nearly all are synarthrotic
_c_ **(9)** shoulder, hip, knee, and elbow joints

2. Freely movable joints are (a) synarthroses, (b) diarthroses, (c) amphiarthroses, (d) synostoses, (e) all of these.

3. Characteristics of a synovial joint include (a) articular cartilage, (b) a joint cavity, (c) a lubricant, (d) articular capsule, (e) all of these.

4. In general, the most important factor determining the stability of synovial joints is (a) interlocking shapes of the articular surfaces, (b) reinforcing ligaments, (c) muscle tone, (d) the synovial fluid, which acts like glue, (e) the body's wrapping of skin, which holds the limbs together.

5. Characteristics of a symphysis include (a) presence of fibrocartilage, (b) ability to resist large compression and tension stresses, (c) presence of a joint cavity, (d) very high mobility, (e) both a and b.

6. Which specific joint does the following describe? "Articular surfaces are deep and secure; multiaxial; capsule heavily reinforced by ligaments; labrum helps prevent dislocation; the first joint to be built artificially; a very stable joint." (a) elbow, (b) hip, (c) knee, (d) shoulder.

7. Ankylosis means (a) twisting of the ankle, (b) tearing of a ligament, (c) displacement of a bone, (d) immobility of a joint due to fusion of its articular surfaces.

8. An autoimmune disorder in which joints are affected on both sides of the body and which involves pannus formation is (a) bursitis, (b) gout, (c) osteoarthritis, (d) rheumatoid arthritis, (e) Lyme disease.

9. Synovial joints are most richly innervated by nerve fibers that (a) monitor how much the capsule is stretched, (b) supply the articular cartilages, (c) cause the joint to move, (d) monitor pain when the capsule is injured.

10. Match the parts of a synovial joint listed in the key to their functions below. (More than one choice applies in some cases.)

Key: **(a)** articular cartilage **(b)** ligaments and fibrous capsule **(c)** synovial fluid **(d)** muscle tendons

_a_ **(1)** keeps bone ends from crushing when compressed; resilient
_b,d_ **(2)** resists tension placed on joints
_c_ **(3)** lubricant that minimizes friction and abrasion of joint surfaces
_c_ **(4)** keeps joints from overheating
_d_ **(5)** helps prevent dislocation

11. How does the femur move with respect to the tibia when the knee is locking during extension? (a) glides anteriorly, (b) flexes, (c) glides posteriorly, (d) rotates medially, (e) rotates laterally.

Short Answer and Essay Questions

12. Define joint.

13. Where does synovial fluid come from?

14. Explain weeping lubrication of the synovial joint surfaces.

15. Compare a bursa and a tendon sheath in terms of their structure, function, and locations.

16. Movements at joints may be nonaxial, uniaxial, biaxial, and multiaxial. Explain the meaning of each of these terms.

17. Compare and contrast the paired movements of flexion and extension with adduction and abduction.

18. How does rotation differ from circumduction?

19. Name two specific examples of each: hinge joint, plane joint, condyloid joint, ball-and-socket joint.

20. What are the functions of the menisci of the knee? Of the anterior and posterior cruciate ligaments?

21. What three specific ligaments and cartilages of the knee are often damaged in football and hockey injuries?

22. Why are sprains and injuries to joint cartilages particularly troublesome?

23. Name the most common direction in which each of the following joints tends to dislocate: (a) shoulder, (b) elbow, (c) hip.

24. Describe the two distinct types of movement that occur at the temporomandibular joint, and explain how both movements relate to the articular disc.

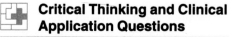

Critical Thinking and Clinical Application Questions

25. Harry was jogging down the road when he tripped in a pothole. As he fell, his ankle twisted violently to the side. The diagnosis was severe dislocation and spraining of the ankle. The surgeon stated that she would perform a reduction of the dislocation and then attempt to repair the injured ligament using ar-

throscopy. (a) What kinds of movements can normally occur at the ankle joint? (b) Was the doctor telling Harry that his bones were broken? (c) What is reduction of a joint? (d) Why was it necessary to repair ligaments surgically? (e) How will the use of arthroscopic surgery minimize Harry's recovery time and his suffering?

26. Dan Park, an exhausted anatomy student, was attending a lecture. After 30 minutes, he began to doze. He woke up a while later (the professor's voice was too loud to permit a good sleep) and let go with a tremendous yawn. To his great distress, he couldn't close his mouth—his lower jaw was stuck open. What do you think had happened?

27. Mrs. Estevez, who is 37 years old, visited her physician and complained of unbearable pain in several joints of both hands. The joints were very red, swollen, and warm to the touch. When asked if she had ever had such an episode in the past, she said she had had a similar attack 2 years earlier that had disappeared as suddenly as it had come. Her diagnosis was arthritis. (a) What type? (b) What is the precipitating cause of joint inflammation in this particular type of arthritis?

28. Sophie worked as a cleaning person for 30 years so that she could send her two children to college and provide them a better life. Several times she had been forced to call her employers to tell them she could not work that day because of a large swelling under the skin over one of her kneecaps. What was Sophie's condition, and what probably caused it?

9

Muscle Tissue

Student Objectives

Overview of Muscle Tissue (p. 217)

1. List four functional properties that distinguish muscle tissue from other tissues.

2. Compare and contrast skeletal, cardiac, and smooth muscle tissue.

Skeletal Muscle (pp. 217–228)

3. Name the layers of connective tissue that occur in and around a skeletal muscle, and briefly describe a muscle's blood and nerve supply.

4. Describe the various ways in which muscles attach to their origins and insertions.

5. Define muscle fascicles, and explain how they are arranged within muscles.

6. Describe the microscopic structure of a skeletal muscle cell (fiber) and the arrangement of its contractile filaments into sarcomeres and myofibrils.

7. Explain the sliding filament theory of muscle contraction.

8. Describe the sarcoplasmic reticulum and T tubules in muscle fibers.

9. Compare and contrast the three kinds of skeletal muscle fibers.

Cardiac Muscle (pp. 228–230)

10. Compare cardiac (heart) muscle to skeletal muscle.

11. Describe the structure and function of intercalated discs.

Smooth Muscle (pp. 230–231)

12. Compare smooth muscle fibers and skeletal muscle fibers in terms of their structure and functions.

13. Describe how smooth muscle cells form a sheet-like tissue.

Muscle Tissue Throughout Life (pp. 232–234)

14. Describe the embryonic development and capacity for regeneration of muscle tissue.

15. Explain the changes that occur in skeletal muscle with age.

Muscle means "little mouse" in Latin, a name that arose because flexing muscles look like mice scurrying beneath the skin. Indeed, the rippling muscles of weight lifters are often the first thing that comes to mind when one hears the word *muscle.* However, muscle is also the main tissue in the heart and in the walls of other hollow organs. In all its forms, muscle tissue makes up nearly half the body's mass.

Overview of Muscle Tissue

Functions

Muscle produces movement, maintains posture, stabilizes joints, and generates heat. Let us explore these properties one by one. Most of the musculature attaches to the skeleton and moves the body by moving the bones. The musculature in the walls of the visceral organs also produces movement by squeezing fluids and other substances through these hollow organs. Certain skeletal muscles contract periodically to maintain posture, stabilizing the joints so the body can stay in a standing or seated position without falling. The way in which muscles stabilize and strengthen joints was considered in Chapter 8 (p. 195). Finally, muscles generate heat as they contract, which allows the body to maintain a temperature of 37 °C (98.6 °F).

Muscle tissue has some special functional features that distinguish it from other tissues:

1. Contractility. Most significantly, muscle contracts forcefully. The long muscle cells shorten and generate a strong pulling force as they contract.

2. Excitability. Muscle cells are excited by nerve cells or other factors that cause an electrical impulse to travel across the muscle cell's plasma membrane. This impulse then signals the muscle cell to contract.

3. Extensibility. Muscle tissue can be stretched back to its original length after it contracts.

4. Elasticity. Muscle tissue can recoil and resume its resting length after being stretched.

Classification

There are three types of muscle tissue: *skeletal, cardiac,* and *smooth.* The properties of these tissues are summarized in Table 9.1.

Skeletal muscle tissue is packaged into **skeletal muscles,** discrete organs that attach to and move the skeleton, such as the biceps brachii muscle of the arm. Skeletal muscle tissue makes up a full 40% of body weight. Its cells show obvious stripes or striations, and its contraction is subject to voluntary control (you can "will" your arm to move).

Cardiac muscle tissue occurs only in the wall of the heart. Its cells are striated, like those of skeletal muscle, but its contraction is not voluntary. As a rule, we have no direct conscious control over how fast our hearts beat.

Most **smooth muscle tissue** occupies the walls of our hollow internal organs, such as the stomach, bladder, blood vessels, and respiratory passages. Its cells lack striations. Like cardiac muscle, smooth muscle is not subject to direct voluntary control.

Sometimes, cardiac and smooth muscle are together considered *visceral muscle,* a term reflecting the fact that both occur in the visceral organs.

Before exploring the differences between the three muscle types, we should point out some similarities they share. First, all muscle cells are elongated. For this reason, they are called **fibers.** *Again, muscle cells are called fibers.* Second, muscle contraction depends on **myofilaments** (*myo* = muscle), which fill most of the cytoplasm of each fiber. There are two kinds of myofilaments, one containing *actin* and the other containing *myosin.* Recall from Chapter 2 (p. 33) that these two proteins generate contractile force in every cell in the body and that this contractile property is most highly developed in muscle fibers. The final similarity among the three muscle tissues is one of terminology: The plasma membrane of muscle cells, instead of being called a plasmalemma, is called a **sarcolemma** (sar″ko-lem′ah) (*sarcos* = flesh or muscle; *lemma* = sheath), and the cytoplasm is called **sarcoplasm.** Nevertheless, the membranes and cytoplasm of muscle cells are not fundamentally different from those of other cell types.

Skeletal Muscle

The many skeletal muscles of the body vary widely in shape, ranging from spindle-shaped cylinders to triangles to broad sheets. Each muscle is a discrete organ made of several kinds of tissue. Although a muscle consists mostly of muscle tissue, it also contains connective tissue, blood in blood vessels, and nerves. In this section, we will study skeletal muscle anatomy from gross to microscopic levels (also see Table 9.2).

Table 9.1 Comparison of Skeletal, Cardiac, and Smooth Muscle

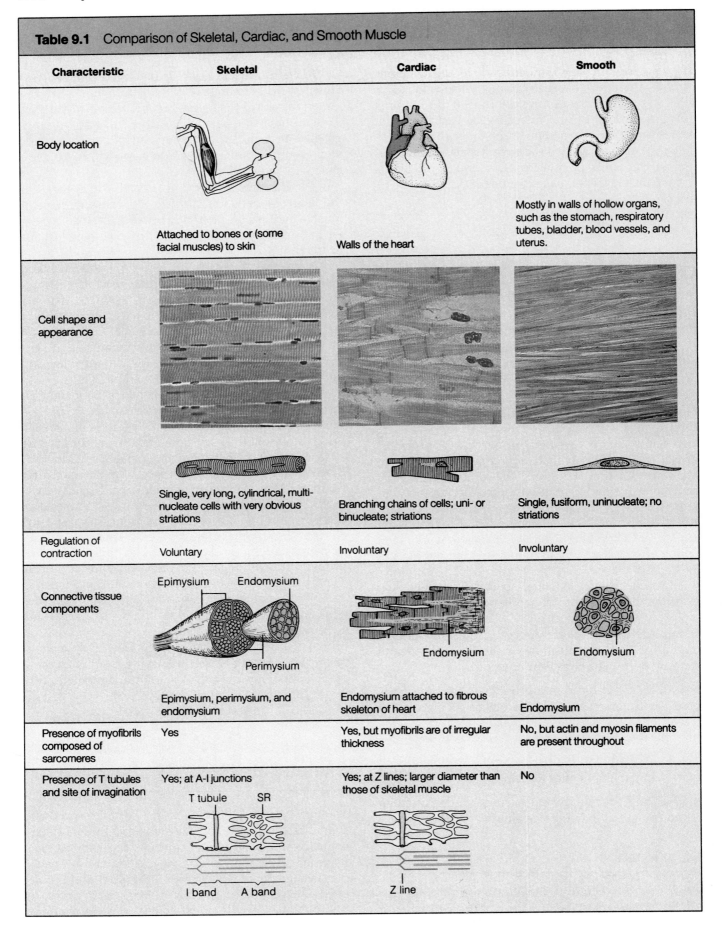

Characteristic	Skeletal	Cardiac	Smooth
Body location	Attached to bones or (some facial muscles) to skin	Walls of the heart	Mostly in walls of hollow organs, such as the stomach, respiratory tubes, bladder, blood vessels, and uterus.
Cell shape and appearance	Single, very long, cylindrical, multi-nucleate cells with very obvious striations	Branching chains of cells; uni- or binucleate; striations	Single, fusiform, uninucleate; no striations
Regulation of contraction	Voluntary	Involuntary	Involuntary
Connective tissue components	Epimysium, perimysium, and endomysium	Endomysium attached to fibrous skeleton of heart	Endomysium
Presence of myofibrils composed of sarcomeres	Yes	Yes, but myofibrils are of irregular thickness	No, but actin and myosin filaments are present throughout
Presence of T tubules and site of invagination	Yes; at A-I junctions	Yes; at Z lines; larger diameter than those of skeletal muscle	No

Table 9.1 (continued)

Characteristic	Skeletal	Cardiac	Smooth
Elaborate sarcoplasmic reticulum	Yes	Less than skeletal muscle; scant terminal cisternae	As well developed as in cardiac muscle.
Presence of gap junctions	No	Yes; at intercalated discs	Yes; in single-unit muscle
Fibers exhibit individual neuromuscular junctions	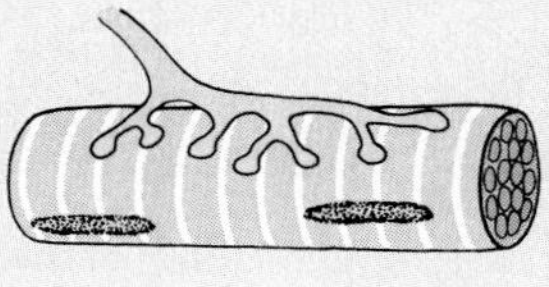 Yes	No	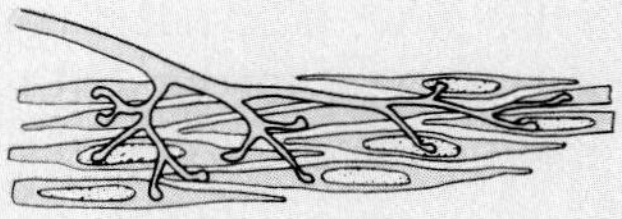 Not in single-unit muscle; yes in multiunit muscle
Source of Ca^{2+} for calcium pulse	Sarcoplasmic reticulum (SR)	SR and from extracellular fluid	SR and from extracellular fluid

Basic Features of a Skeletal Muscle

Connective Tissue and Fascicles

Several sheaths of connective tissue hold a skeletal muscle and its muscle fibers together. We will consider these sheaths from external to internal (Figure 9.1).

1. Epimysium. An "overcoat" of dense irregular connective tissue surrounds the whole muscle. This coat is the **epimysium** (ep″ĭ-mis′e-um), a name that means "outside the muscle." Sometimes, the epimysium blends with the deep fascia that lies between neighboring muscles.

2. Perimysium and Fascicles. Within each skeletal muscle, the muscle fibers are grouped into **fascicles** (fas′ĭ-klz; "bundles") that resemble bundles of sticks. Surrounding each fascicle is a layer of fibrous connective tissue called **perimysium** (per″ĭ-mis′e-um; "around the muscle [fascicles]").

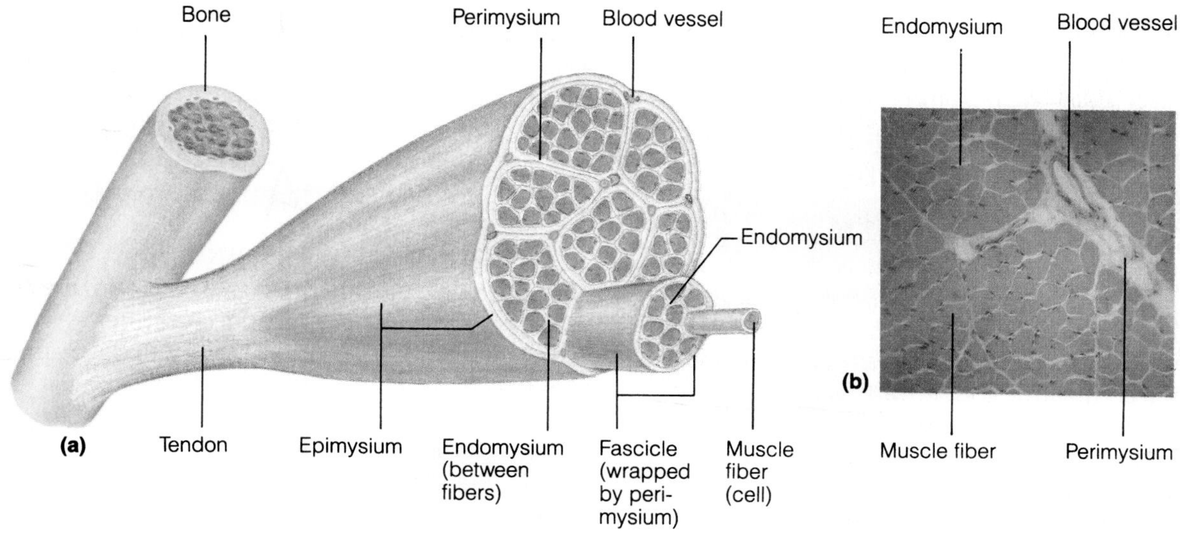

Figure 9.1 Connective tissue wrappings of skeletal muscle. (**a**) From external to internal, the connective tissue sheaths are epimysium, perimysium, and endomysium. (**b**) Photomicrograph of a cross section of a skeletal muscle (64 ×).

3. Endomysium. Within a fascicle, each muscle fiber is surrounded by a fine sheath of connective tissue consisting mostly of reticular fibers. This is the **endomysium** (en"do-mis'e-um; "within the muscle").

As shown in Figure 9.1, the epimysium, perimysium, and endomysium are continuous with the tendons that join muscles to bones. Therefore, when muscle fibers contract, they pull on the connective tissue sheaths, which transmit the force to the bone being moved. The connective tissue sheaths also provide muscle tissue with its natural elasticity and carry the blood vessels and nerves that serve the muscle.

Nerves and Blood Vessels

In general, each skeletal muscle is supplied by one nerve, one artery, and one or more veins—all of which enter or exit near the center of the muscle. The nerves and vessels branch repeatedly within the intramuscular connective tissue, with the smallest branches serving individual muscle fibers. Each muscle fiber in a skeletal muscle is contacted by one nerve ending, which signals the fiber to contract. Such a contact is called a *neuromuscular junction.* Further details of the innervation of muscle are considered in Chapter 13 (p. 367).

The rich blood supply to muscle reflects the high demand for nutrients and oxygen made by contracting muscle fibers. Capillaries in the endomysium are shown in Figure 9.2. These capillaries are long and winding when the muscle fibers are contracted, and they are stretched straight when the muscle is extended.

Muscle Attachments

Most skeletal muscles run from one bone to another, crossing at least one movable joint (Figure 9.3). When a muscle contracts, it causes one of these bones to move, while the other bone usually remains fixed. The less movable attachment of a muscle is called its **origin,** whereas the more movable attachment is the **insertion.** Thus, the insertion is pulled toward the origin. In the muscles of the limbs, the origin lies *proximal* to the insertion (Figure 9.3).

Muscles attach to their origins and insertions via strong fibrous connective tissue. In *direct,* or *fleshy, attachments,* the attaching strands of connective tissue are so short that the muscle fascicles themselves appear to reach the bone. In *indirect attachments,* by contrast, the fibrous connective tissue extends well beyond the muscle to form a rope-like **tendon** or a flat sheet called an *aponeurosis* (ap"o-nu-ro'sis). Indirect attachments are more common than direct attach-

Figure 9.2 Photomicrograph of the capillary network surrounding skeletal muscle fibers. The arterial supply was injected with purple gelatin to demonstrate the capillary bed. The long muscle fibers, which run horizontally across the photo, are stained blue. Note the wavy appearance of the purple capillaries (150 ×).

ments, and most muscles have tendons. Where tendons meet bones, raised bone markings (tubercles, trochanters, and crests, for example) often occur. Although most tendons and aponeuroses attach to bones, a few attach to skin, to cartilage, to sheets of fascia, or to a seam of fibrous tissue called a *raphe* (ra'fe; "seam").

Arrangement of the Fascicles

Muscle fascicles (bundles of fibers) are large enough to be seen with the unaided eye (Figure 9.4). One can learn a great deal about the action of a particular muscle from the arrangement of its fascicles. In different muscles, the fascicles are arranged in different patterns, including *parallel, pennate, convergent,* and *circular* patterns.

In a **parallel** arrangement, the long axes of the fascicles run parallel to the long axis of the muscle itself. Muscles with this arrangement are either strap-like, such as the sartorius muscle of the thigh, or spindle-shaped with an expanded belly, like the biceps brachii of the arm.

In a **pennate** (pen'āt) pattern, the fascicles are short and attach obliquely to a tendon that runs the whole length of the muscle. This pattern makes the muscle look like a feather (*penna* = feather). If the fascicles insert into only one side of the tendon, the muscle is **unipennate** (*uni* = one). The extensor dig-

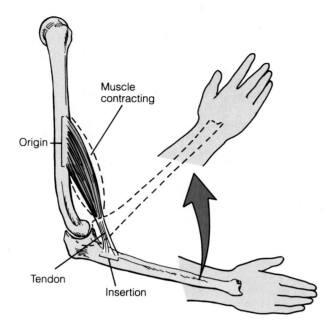

Figure 9.3 **Muscle attachments (origin and insertion).**
When a skeletal muscle contracts, its insertion moves toward its origin. The muscle illustrated here is the brachialis muscle of the upper limb. In this particular muscle, the origin is via a direct attachment to the humerus, and the insertion is via an indirect attachment (tendon) to the ulna.

itorum longus muscle on the anterior leg is a unipennate muscle. If the fascicles insert into the tendon from both sides, the arrangement is **bipennate** (*bi* = two). The rectus femoris muscle of the thigh is bipennate. A **multipennate** arrangement (not illustrated) looks like many feathers situated side by side, with all the quills inserting into one large tendon. The deltoid muscle, which forms the roundness of the shoulder, is multipennate.

In the **convergent** pattern, the origin of the muscle is broad, and the fascicles *converge* toward a tendon. Such a muscle can be triangular or fan-shaped. The pectoralis major muscle in the anterior thorax is an example.

Fascicles arranged in concentric rings form a **circular** pattern. Muscles with this arrangement surround external body openings, which they close by contracting. The general name for such a circular muscle is a *sphincter* (sfingk'ter; "squeezer"). Specific examples are the orbicularis oris muscle around the mouth and the orbicularis oculi around the eyes.

The arrangement of its fascicles influences both a muscle's range of motion and its power. Skeletal muscle fibers can shorten by about one-third of their resting length as they contract. The more parallel the fi-

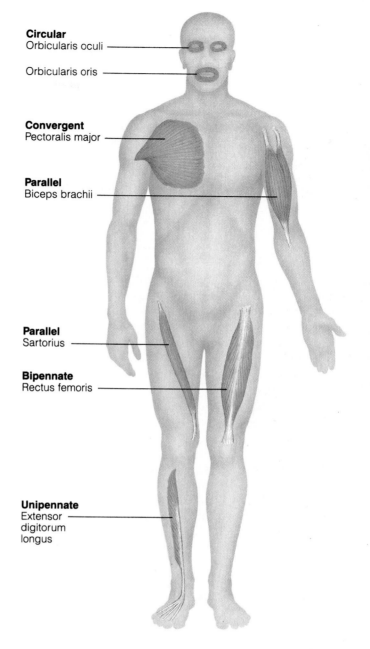

Figure 9.4 **The different arrangements of fascicles within various skeletal muscles.** The lines visible in the muscles are the fascicles.

bers are to the muscle's long axis, the more the whole muscle can shorten. Although muscles with the parallel arrangement of fascicles can shorten the most, they usually are not powerful. The power of a muscle depends more on the total number of fibers it contains. The stocky bipennate and multipennate muscles contain the most fibers; thus, they shorten very little but tend to be very powerful.

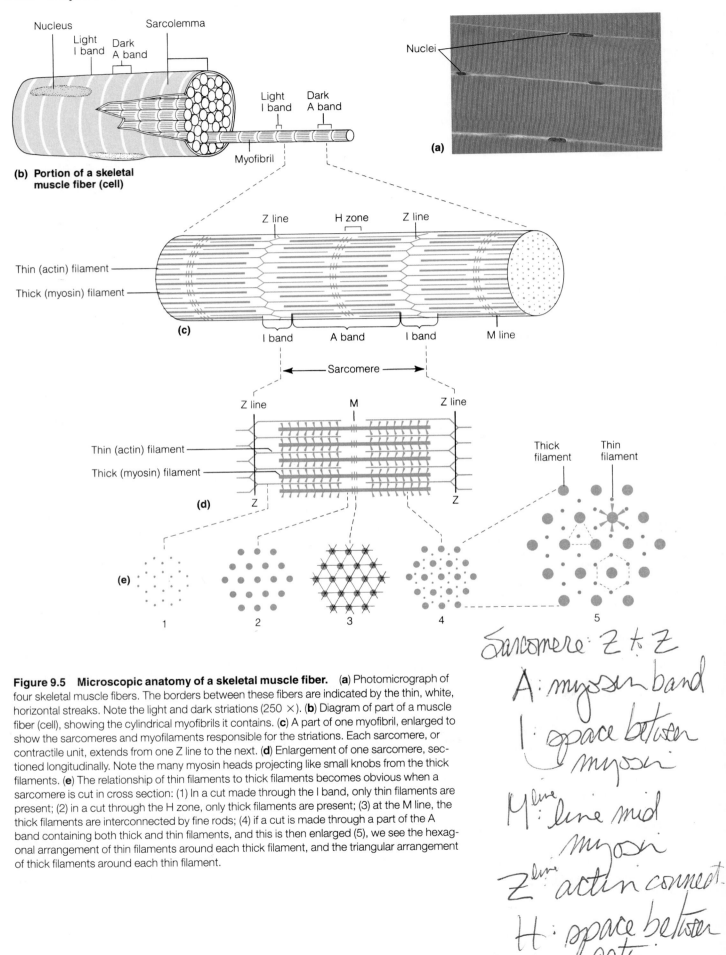

Figure 9.5 Microscopic anatomy of a skeletal muscle fiber. (a) Photomicrograph of four skeletal muscle fibers. The borders between these fibers are indicated by the thin, white, horizontal streaks. Note the light and dark striations (250 ×). (b) Diagram of part of a muscle fiber (cell), showing the cylindrical myofibrils it contains. (c) A part of one myofibril, enlarged to show the sarcomeres and myofilaments responsible for the striations. Each sarcomere, or contractile unit, extends from one Z line to the next. (d) Enlargement of one sarcomere, sectioned longitudinally. Note the many myosin heads projecting like small knobs from the thick filaments. (e) The relationship of thin filaments to thick filaments becomes obvious when a sarcomere is cut in cross section: (1) In a cut made through the I band, only thin filaments are present; (2) in a cut through the H zone, only thick filaments are present; (3) at the M line, the thick filaments are interconnected by fine rods; (4) if a cut is made through a part of the A band containing both thick and thin filaments, and this is then enlarged (5), we see the hexagonal arrangement of thin filaments around each thick filament, and the triangular arrangement of thick filaments around each thin filament.

Sarcomere: Z to Z
A: myosin band
I: space between myosin
M line: line mid myosin
Z line: actin connect.
H: space between actin

Microscopic and Functional Anatomy of Skeletal Muscle Tissue

The Fiber

The skeletal muscle fiber is long and cylindrical (Figure 9.5, and Table 9.2 on p. 227). These fibers are huge cells: Their diameter is 10 to 100 μm—up to ten times that of an average body cell—and their length is phenomenal, often many centimeters. It would be inaccurate, however, to call these fibers the biggest cells in the body, because each one is actually formed by the fusion of hundreds of embryonic cells. Because it forms this way, the skeletal muscle fiber contains many nuclei. These nuclei lie at the cell periphery, just deep to the sarcolemma (Figure 9.5b).

Myofibrils and Sarcomeres

Light and dark stripes called *striations* are visible in skeletal muscle fibers (Figure 9.5a and b). These striations are borne on long rods called **myofibrils** within the cytoplasm. (Note that these structures are *fibrils,* not *fibers* or *filaments.*) Myofibrils are unbranched cylinders that are present in large numbers, making up over 80% of the cytoplasm (Figure 9.6). You may think of the myofibrils as highly specialized cellular organelles unique to muscle fibers.

The different myofibrils in a fiber are separated by very thin streaks of cytoplasm (Figure 9.6). These streaks contain rows of mitochondria and glycogen granules that supply energy for muscle contraction. Unfortunately, distinguishing the individual myofibrils in histological sections is difficult (Figure 9.5a) because the striations of adjacent myofibrils line up almost perfectly.

Myofibrils consist of long rows of repeating segments called **sarcomeres** (sar′ko-mērz; "muscle segments") (Figure 9.5c and d). The sarcomere is the basic unit of contraction in skeletal muscle. The boundaries at the two ends of each sarcomere are called **Z lines.** In a three-dimensional view, these "lines" really are coin-shaped; thus, they are also called *Z discs.* Attached to each Z-disc and extending toward the center of the sarcomere are many fine myofilaments called **thin (actin) filaments.** Thin filaments consist primarily of the protein actin, although they contain other proteins as well. In the center of the sarcomere and overlapping the inner ends of the thin filaments is a cylindrical bundle of **thick (myosin) filaments.** These thick myofilaments consist largely of myosin molecules, but they also contain *ATPase enzymes* that split ATP (energy-storing molecules) to release the energy required for muscle contraction. Both

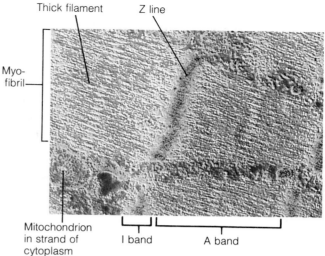

Figure 9.6 Skeletal muscle fiber: Electron micrograph of the cytoplasm (artificially colored). Several striped myofibrils are shown, separated from one another by strands of cytoplasm containing mitochondria. As in Figure 9.5, the various bands (A, I, Z) on the myofibrils are labeled. Magnified 4700 ×.

ends of a thick filament are studded with knobs called *myosin heads* (Figure 9.5d).

At this point, we can explain the pattern of striations in skeletal muscle fibers. The dark **A bands** correspond to the full length of the thick filaments. A bands also include the inner ends of the thin filaments, which overlap the thick filaments. The central part of each A band, where no thin filaments occur, is the **H zone.** An **M line** in the center of the H zone contains tiny rods that hold the adjacent thick filaments together (Figure 9.5d). Alternating with the A bands are the light **I bands,** the regions where only thin filaments occur. Notice that each I band lies within two different sarcomeres (Figure 9.5c) and is bisected by a Z line running through its center. Recall that the striations of adjacent myofibrils align perfectly, allowing us to see the various bands in whole muscle fibers (Figure 9.5a) as well as in individual myofibrils.

A longitudinal view of the myofilaments, like that pictured in Figure 9.5d, is somewhat misleading because it implies that each thick filament interdigitates (interlocks like the fingers of folded hands) with only four thin filaments. We can get a more accurate picture of the three-dimensional relationships between the filaments by viewing cross sections from different parts of a sarcomere (Figure 9.5e). In areas where the thick and thin filaments overlap, each thick filament is surrounded by a hexagonal arrangement of six thin filaments (part 5 in Figure 9.5e). Each thin filament, in turn, is bordered by a triangular arrangement of three thick filaments.

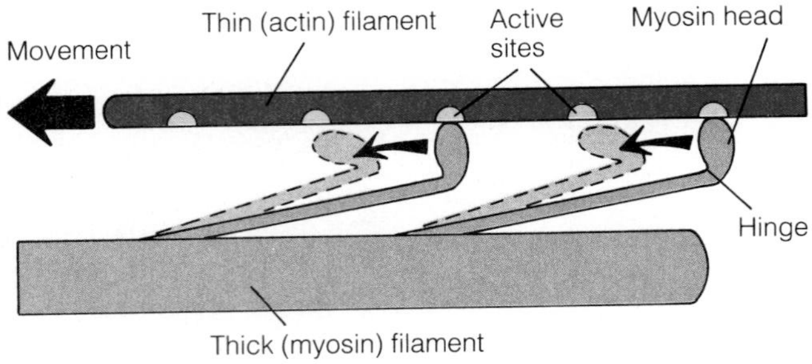

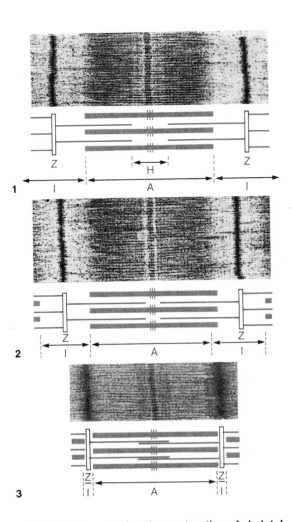

Figure 9.7 Sliding filament mechanism of contraction of a skeletal muscle. Myosin heads attach to actin in the thin filaments at "active sites," then pivot on "hinges" to pull the thin filaments inward toward the center of the sarcomere.

Figure 9.8 Striations during the contraction of skeletal muscle. A sarcomere is relaxed in (1), partly contracted in (2), and fully contracted in (3). For each step, an electron micrograph is shown above a diagram that interprets the micrograph. Note that the two Z lines are moving closer together and closer to the ends of the thick filaments (closer to the A band). This action causes the I band and H zone to shrink, but not the A band. The micrographs are magnified 20,000 ×.

Contraction

The mechanism of muscle contraction is explained by the **sliding filament theory** (Figure 9.7). Contraction results as the myosin heads of the thick filaments bind to the thin filaments at both ends of the sarcomere and pull them toward the center of the sarcomere by pivoting inward. After a given myosin head pivots, it lets go, "recocks," binds to the thin filament further along its length, and pivots again. This cycle is repeated many times during a single contraction. Thus, the thick filaments forcefully pull the two Z lines closer together, causing each sarcomere to shorten. It should be emphasized that the thick and thin filaments themselves do not shorten: They merely slide past one another. The sliding filament mechanism is initiated by the binding of calcium ions to the thin filaments and powered by ATP.

Figure 9.8 shows how contraction affects the striation pattern of skeletal muscle. As the Z lines move closer together, they cause the I bands to diminish in width, and the H zones disappear. The width of the A bands, however, stays the same (because the thick filaments do not change in length). When a muscle is stretched, on the other hand, the I bands and H zones widen as the Z lines move apart. Again, there is no change in the width of the A bands.

Muscle Extension

We have indicated that muscle tissue is extensible and that muscle fibers are stretched (extended) back to their original length after they contract. What is responsible for this stretching? Basically, a skeletal muscle is stretched by the movement *opposite* to that which it normally produces. For example, a muscle that normally *abducts* the arm at the shoulder is stretched when the arm is *adducted* at this joint. Similarly, the muscles that flex the vertebral column (as in bowing) are stretched and extended whenever we use other muscles to stand erect.

Stretching and the Force of Contraction

The ideal resting length for skeletal muscle fibers is the length at which they can generate the strongest pulling force. The highest force is generated when the fiber starts out just slightly stretched, so that its thin and thick filaments overlap to only a moderate extent (Figure 9.5d). Under these conditions, the myosin heads can move and pull along the whole length of the thin filaments. Under other conditions, contraction is suboptimal: If a muscle fiber is stretched so

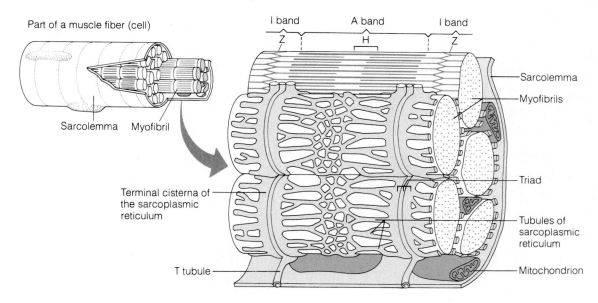

Figure 9.9 Sarcoplasmic reticulum and T tubules in the skeletal muscle fiber.
The sarcoplasmic reticulum and T tubules are in the cytoplasm of the muscle cell, between the myofibrils.

much that the thick and thin filaments do not overlap at all, the myosin heads have nothing to attach to, and no pulling force can be generated. Alternatively, if the sarcomeres are so compressed that the thick filaments are touching the Z lines, little further shortening can occur.

What is true for a muscle fiber is true for an entire muscle. Whole skeletal muscles have an optimal operational length of about 80% to 120% of their normal resting length. The muscle attachments allow the muscles to remain always near that optimal length; that is, the joints normally do not permit any bone to move so widely that its attached muscles could shorten or stretch beyond their optimal range.

Sarcoplasmic Reticulum and T Tubules

Skeletal muscle fibers (cells) contain two sets of intracellular tubules that participate in the regulation of muscle contraction: (1) the sarcoplasmic reticulum and (2) the T tubules (Figure 9.9). The **sarcoplasmic reticulum** is an elaborate smooth endoplasmic reticulum (p. 31) whose interconnecting tubules surround each myofibril like the sleeve of a loosely crocheted sweater surrounds your arm. Most of these tubules run longitudinally along the myofibril. However, the sarcoplasmic reticulum also forms larger, perpendicular cross channels over the A-I junctions called **terminal cisternae** ("end sacs"). These cisternae occur in pairs.

The sarcoplasmic reticulum stores large quantities of calcium ions (Ca^{2+}), which are released when the muscle is stimulated to contract. These calcium ions then diffuse through the cytoplasm to the thin filaments, where they trigger the sliding filament mechanism of contraction. After the contraction, the ions return to the sarcoplasmic reticulum for storage.

Muscular contraction is ultimately controlled by nerve-generated impulses that travel along the sarcolemma of the muscle cell. **T tubules** are deep indentations of this sarcolemma, running between each pair of terminal cisternae of the sarcoplasmic reticulum. Since T tubules are continuations of the sarcolemma, they can conduct impulses to the deepest regions of the muscle cell. Thus, they ensure that every myofibril in the muscle fiber contracts at the same time. The complex of a T tubule flanked by two terminal cisternae at each A-I junction is called a **triad** (tri′ad; "group of three").

Muscle Fiber Types

Not all skeletal muscle fibers are alike. These cells are divided into three classes based on the strength, speed, and endurance of their contraction. The classes of fibers are *red slow-twitch, white fast-twitch,* and *intermediate fast-twitch* (Figure 9.10). Most of the muscles in the body contain all three of these fiber types, but the proportions of each type differ among different muscles.

Red Slow-Twitch Fibers.   The red slow-twitch fibers are relatively thin. Their red color stems from their abundant content of *myoglobin* (mi″o-glob′in), an oxygen-binding pigment in their cytoplasm. Red fibers obtain their energy from aerobic (oxygen-requir-

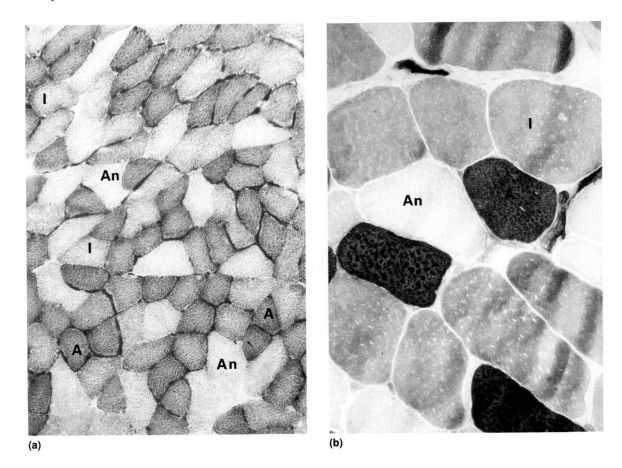

(a) (b)

Figure 9.10 The three types of fibers in each skeletal muscle. Fibers are shown in cross section. From smallest to largest, they are: red slow-twitch fibers (labeled A for aerobic), intermediate fast-twitch fibers (I), and white fast-twitch fibers (labeled An for anaerobic). The staining technique used in this preparation differentiates the fibers by the abundance of their mitochondria, mitochondrial enzymes, and other features. Magnifications: (**a**) 200 x; (**b**) 600 x.

ing) metabolic reactions. Thus, they have a relatively large number of mitochondria (where aerobic metabolism occurs) and a rich supply of capillaries. These fibers (1) contract slowly, (2) are extremely resistant to fatigue as long as sufficient oxygen is present, and (3) specialize in the delivery of prolonged contractions. There are many red slow-twitch fibers in the postural muscles of the lower back, muscles that must contract continuously to keep the spine straight and maintain posture. Because they are thin, red slow-twitch fibers do not generate much power.

White Fast-Twitch Fibers. White fast-twitch fibers are pale because they contain little myoglobin. About twice the diameter of red slow-twitch fibers, these fibers contain more myofilaments and generate much more power. These fibers depend on anaerobic pathways (no oxygen is used) to manufacture ATP. Consequently, white fibers contain few mitochondria or capillaries but abundant glycogen granules as a built-in fuel source. These fibers (1) contract rapidly, and (2) tire quickly (they are "fatigable fibers"). White fi-

bers are common in the muscles of the upper limbs, which often lift heavy objects for *brief* periods (in moving furniture across a room, for example).

Intermediate Fast-Twitch Fibers. The diameter of intermediate fast-twitch fibers falls between that of the other two fiber types. Like white fibers, intermediate fibers contract quickly; but like slow-twitch fibers, they are oxygen-dependent and have a high myoglobin content and a rich supply of capillaries. Since intermediate fast-twitch fibers depend largely on aerobic metabolism, they are fatigue-resistant, but less so than the red slow-twitch fibers. They are more powerful than the slow-twitch fibers, but less so than white fibers. Intermediate fast-twitch fibers are abundant in the muscles of the lower limbs, which must move the body for long periods during walking and jogging.

Because most muscles contain a mixture of these three fiber types, each muscle can perform different tasks at different times. A muscle in the calf of the leg, for example, sometimes propels the body in a short

Table 9.2   Structure and Organizational Levels of Skeletal Muscle

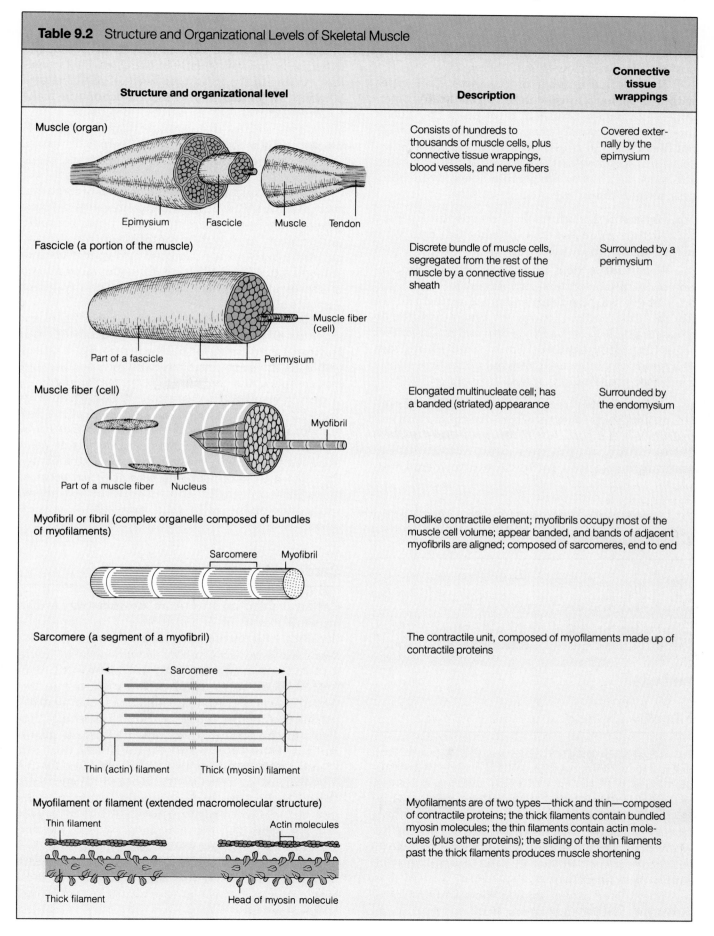

Structure and organizational level	Description	Connective tissue wrappings
Muscle (organ) Epimysium Fascicle Muscle Tendon	Consists of hundreds to thousands of muscle cells, plus connective tissue wrappings, blood vessels, and nerve fibers	Covered externally by the epimysium
Fascicle (a portion of the muscle) Part of a fascicle Muscle fiber (cell) Perimysium	Discrete bundle of muscle cells, segregated from the rest of the muscle by a connective tissue sheath	Surrounded by a perimysium
Muscle fiber (cell) Myofibril Part of a muscle fiber Nucleus	Elongated multinucleate cell; has a banded (striated) appearance	Surrounded by the endomysium
Myofibril or fibril (complex organelle composed of bundles of myofilaments) Sarcomere Myofibril	Rodlike contractile element; myofibrils occupy most of the muscle cell volume; appear banded, and bands of adjacent myofibrils are aligned; composed of sarcomeres, end to end	
Sarcomere (a segment of a myofibril) Sarcomere Thin (actin) filament Thick (myosin) filament	The contractile unit, composed of myofilaments made up of contractile proteins	
Myofilament or filament (extended macromolecular structure) Thin filament Actin molecules Thick filament Head of myosin molecule	Myofilaments are of two types—thick and thin—composed of contractile proteins; the thick filaments contain bundled myosin molecules; the thin filaments contain actin molecules (plus other proteins); the sliding of the thin filaments past the thick filaments produces muscle shortening	

sprint (using its white fibers), sometimes in long-distance races (using intermediate fast-twitch fibers), and sometimes just helps in maintaining a standing posture (using red slow-twitch fibers).

The relative number of the three fiber types within a person's musculature is genetically determined and remains constant throughout life. Therefore, exercise probably cannot transform the weaker red fibers into the more powerful white fibers. However, lifting heavy loads does increase the diameter and the strength of existing white fibers. By this process, weight lifters develop large muscles.

Based on this information, do you think weight lifting increases the number of *myofibrils* within one's white muscle fibers?

Weight lifting does lead to an increased production of contractile proteins, myofilaments, and myofibrils. In this way, the fibers enlarge. The fibers do not divide mitotically. However, when vigorously stressed, fibers do tear, and the torn and repaired fiber segments become new cells. Thus, weight lifting leads to hypertrophy (increased size) but not to hyperplasia (increased proliferation) of muscle fibers. ■

So far, we have compared the three kinds of skeletal muscle fibers in terms of speed of contraction and resistance to fatigue. But *all* of these skeletal fibers contract faster and tire more easily than do cardiac and smooth muscle fibers, which are considered next.

Cardiac Muscle

Cardiac muscle, the muscle tissue of the heart (Figure 9.11 and Table 9.1), comprises a thick layer of the heart wall called the myocardium. Its contractions pump blood through blood vessels of the circulatory system. Cardiac muscle is striated like skeletal muscle, and it contracts through the sliding filament mechanism.

Fibers

Cardiac muscle fibers are single cells (Figure 9.11b), not fused cell colonies like skeletal muscle fibers. Cardiac fibers average about 15 μm in width and 100 μm in length. Unlike all other muscle fibers, cardiac fibers branch. Each fiber contains one or two nuclei in its center—not at the periphery. The spaces between adjacent fibers are filled with an endomysium (delicate connective tissue), which contains the capillaries that supply these fibers.

The striated cardiac muscle fibers contain typical sarcomeres with A and I bands, H zones, and Z and M lines (Figure 9.12c). *Myofibrils,* however, are difficult to discern because they are not the simple cylinders present in skeletal muscle fibers. Instead, they branch extensively and vary greatly in diameter. This irregularity is due to the great abundance of mitochondria between the myofibrils. These mitochondria make large amounts of ATP by aerobic metabolism. Their abundance ensures that cardiac muscle is highly resistant to fatigue. If cardiac muscles were to tire easily, routine exercise would exhaust the heart and cause death.

As with skeletal muscle, the amount of force that cardiac muscle fibers can generate depends on their length. Significantly, cardiac muscle fibers normally remain just *under* their optimal length. Consequently, when they are stretched by a greater volume of blood returning to the heart, their contraction force increases and they can pump substantially larger amounts of blood.

Like skeletal muscle, cardiac muscle fibers are triggered to contract by Ca^{2+} entering the cytoplasm. However, the system for delivering the Ca^{2+} is less elaborate, the sarcoplasmic reticulum is smaller and less complex, and the T tubules are less abundant—occurring just *once* per sarcomere at the Z line (Figure 9.12c) instead of *twice* per sarcomere at the A–I junctions. These differences reflect the fact that a substantial amount of Ca^{2+} enters cardiac muscle cells from the tissue fluid outside the cell. Extracellular Ca^{2+} can diffuse rapidly into cardiac fibers because these fibers are smaller and therefore have a higher surface-to-volume ratio than do skeletal muscle fibers.

Cardiac Muscle Tissue

Unlike skeletal muscle fibers, neighboring cardiac muscle fibers attach to one another. This characteristic, along with the branching of the fibers, makes cardiac muscle tissue a complex network of cells (Figure 9.11b). The attachments occur at complex cell junctions called **intercalated discs** (in-ter′kah-la″ted; "inserted between"). At these junctions, the adjacent plasma membranes interlock through meshing "fingers" (Figure 9.12b). Intercalated discs are a combination of three types of cell junctions introduced in Chapter 4: (1) desmosomes, (2) long, desmosome-like junctions called *fascia adherens,* and (3) gap junctions. The first two of these junctions hold the cells together, while the gap junctions allow ions to pass between cells. The free movement of ions between cells allows the direct transmission of an electrical impulse through the entire network of cardiac muscle fibers. This impulse in turn signals all the muscle fibers in a heart chamber (atria or ventricles) to contract at the same time.

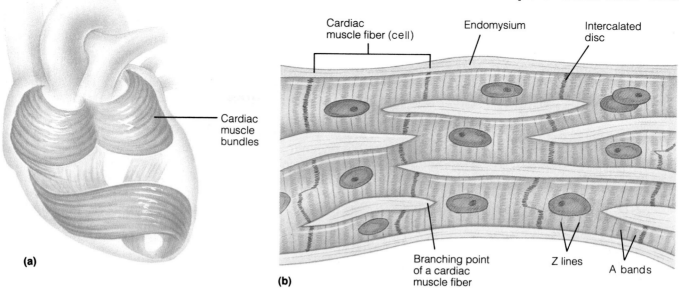

(a)

(b)

Figure 9.11   Location and basic structure of cardiac muscle.   **(a)** Arrangement of cardiac muscle in wall of the heart. **(b)** Cardiac muscle fibers are branching cells interconnected to form a network. Note the intercalated discs between the fibers.

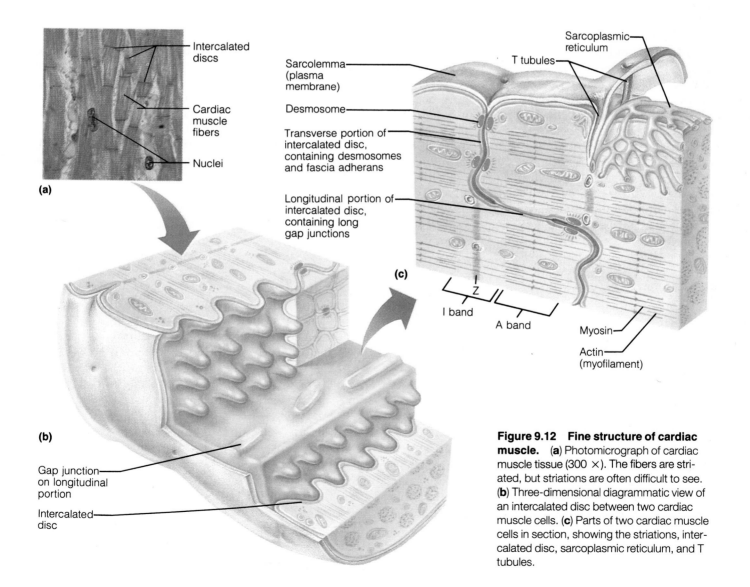

Figure 9.12   Fine structure of cardiac muscle.   **(a)** Photomicrograph of cardiac muscle tissue (300 ×). The fibers are striated, but striations are often difficult to see. **(b)** Three-dimensional diagrammatic view of an intercalated disc between two cardiac muscle cells. **(c)** Parts of two cardiac muscle cells in section, showing the striations, intercalated disc, sarcoplasmic reticulum, and T tubules.

Unlike skeletal muscle tissue, not all cells of cardiac muscle tissue are innervated. In fact, an isolated cardiac muscle fiber will contract in a rhythmic manner without any innervation at all. This inherent rhythmicity of cardiac muscle cells is the basis of the rhythmic heartbeat, as we will explain further in Chapter 17 (p. 470).

Smooth Muscle

Most smooth muscle of the body occurs in the walls of visceral organs like the intestines, bladder, and uterus (Table 9.1). Its fibers apparently contract by the sliding filament mechanism, but the details in this case are poorly understood.

Fibers

Smooth muscle fibers are spindle-shaped cells, each with one centrally located nucleus (Figure 9.13a). Typically, they have a diameter of 2–10 μm and a length of 100–500 μm. Skeletal muscle fibers are about 20 times wider and thousands of times longer.

No striations are visible in smooth muscle fibers viewed by light microscopy (the cytoplasm looks "smooth"), and electron microscopy confirms that there are no sarcomeres (Figure 9.13c). Interdigitating thick and thin filaments, however, are present, and fill almost the entire cytoplasm. These myofilaments lie nearly parallel to the long axis of the fiber and spiral down the fiber in a shallow helix like the stripes on a barber pole. The ratio of thick to thin filaments is far lower than that in skeletal muscle fibers (1:16, compared to 1:2). Even so, smooth muscle can generate as much pulling force (per cell volume) as skeletal muscle. It is not known how smooth muscle can generate so much pulling force with so few thick filaments. The explanation may lie in the structure of the thick filaments: They are very long and contain actin-gripping myosin heads along their whole length, not just at their ends.

As in skeletal and cardiac muscle, the entry of Ca^{2+} into the cytoplasm signals the smooth muscle fiber to contract. Some of these calcium ions enter from the extracellular fluid, whereas others are stored and released by an intracellular sarcoplasmic reticulum. Although this reticulum of tubules and sheets is difficult to demonstrate with standard techniques for electron microscopy, it is as large as in cardiac muscle fibers (1%–8% of the cell volume). T tubules and triads are absent from smooth muscle cells, as the sarcoplasmic reticulum directly touches the surface sarcolemma. Tiny spherical infoldings of this surface membrane called caveolae (Figure 9.13c), of unknown function, were previously thought to represent T tubules.

Like all cells in the body, smooth muscle fibers contain *intermediate filaments* (p. 33) that resist tension. The network of intermediate filaments forms a strong, cable-like cytoskeleton that harnesses the pull generated by the sliding of the myofilaments during contraction (Figure 9.13). The intermediate filaments attach to *dense bodies,* which scatter throughout the cytoplasm and occasionally anchor to the sarcolemma. These dense bodies also anchor groups of thin (actin) filaments and therefore correspond to the Z discs of skeletal muscle.

Smooth Muscle Tissue

Smooth muscle fibers are closely packed into a tissue (Table 9.1, p. 218). In most examples of smooth muscle tissue, the neighboring muscle fibers are interconnected by scattered gap junctions. Between these fibers is a thin endomysium containing capillaries and nerve fibers. The matrix of this connective tissue is secreted by the smooth muscle cells themselves. Smooth muscle lacks the coarser sheaths of connective tissue that associate with skeletal muscle and does not organize into a muscular organ.

Smooth muscle fibers usually organize into sheets. These sheets occur in the walls of arteries and veins and in the walls of hollow organs of the respiratory, digestive, urinary, and reproductive tracts (Figure 9.14). In most cases, at least two sheets of smooth muscle are present, with their muscle fibers oriented at right angles to each other. In one of these sheets, the **longitudinal layer,** the muscle fibers run parallel to the long axis of the organ. In the other, the **circular layer,** the fibers run around the circumference of the organ. The circular layer constricts the hollow organ, and the longitudinal layer shortens the organ's length. These muscle layers may generate alternate waves of contraction and relaxation that propel substances through the organ. This process is called *peristalsis* (per"ĭ-stal'sis; "around contraction"). For example, peristalsis moves food through the digestive tube.

We do not have direct voluntary control over the contraction of our smooth muscle, for it is innervated by a "nonconscious" part of the nervous system, the autonomic nervous system (Chapter 14). In general, only a few fibers in each muscle sheet are innervated, and the impulse spreads to the rest through gap junctions. The whole sheet then contracts as a single unit. This is called *single-unit innervation.* At other times, contractions occur without nervous input, stimulated by stretching of the muscle fibers or by hormones.

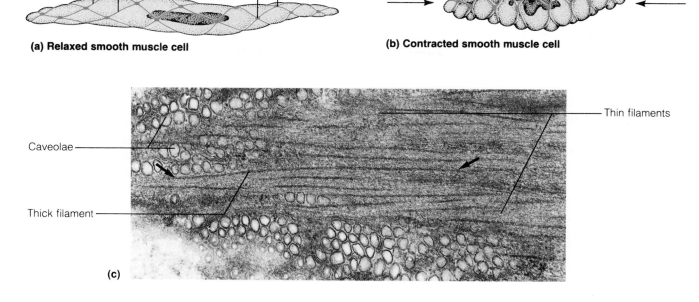

Intermediate filament bundles attached to dense bodies

(a) Relaxed smooth muscle cell

(b) Contracted smooth muscle cell

Caveolae

Thick filament

Thin filaments

(c)

Figure 9.13 A smooth muscle fiber.   (**a**) Relaxed smooth muscle fiber. (**b**) Contracted smooth muscle fiber. The intermediate filaments in the cytoplasm resist tension like wires and harness the pull generated during muscle contraction. (**c**) Electron micrograph of the cytoplasm (32,000 ×). Thick and thin filaments interdigitate, but no sarcomeres are present. The small white spheres are caveolae, visible where this section grazes the inside of the plasma membrane. The two arrows point to thick filaments.

There are exceptions to the above description: In the arrector pili muscles of the skin and the iris of the eye, every smooth muscle fiber is innervated individually. This is called *multiunit innervation.*

The contraction of smooth muscle is slow, sustained, and resistant to fatigue. Smooth muscle takes 30 times longer to contract and relax than does skeletal muscle. However, smooth muscle can maintain its contractile force for a long time without tiring (perhaps because the myofilaments lock together). This is a valuable feature, because the smooth muscle in the walls of small arteries and visceral organs must sustain a moderate degree of contraction without fatiguing, day in and day out. Since smooth muscle's energy requirements are low, mitochondria are not abundant in its fibers.

Smooth muscle can generate strong contractile forces even when it is vigorously stretched, perhaps because its thick filaments are so long and not readily pulled away from the thin filaments. This ability enables our hollow organs to contract even when they are greatly expanded with substances such as urine or feces.

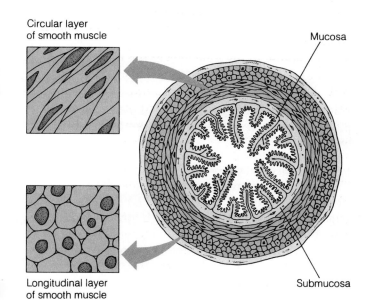

Circular layer
of smooth muscle

Mucosa

Longitudinal layer
of smooth muscle

Submucosa

Figure 9.14 Arrangement of smooth muscle sheets in walls of hollow organs.   As seen in this cross-sectional view of the intestine, two sheets of smooth muscle tissue are often present in the walls of the visceral organs. In one layer, the muscle fibers run circularly, and in the other the fibers run longitudinally. Note that in cross section (lower left), the round cell profiles resemble the stones on a cobblestone street.

Muscle Tissue Throughout Life

With rare exceptions, all muscle tissues develop from embryonic mesoderm cells called *myoblasts.* Myoblasts fuse to form skeletal muscle fibers, which then begin to manufacture thick and thin myofilaments and gain the ability to contract. Ordinarily, skeletal muscles are contracting by week 7, when the embryo is only about 2 cm long. Nerves grow into the muscle masses from the spinal cord, bringing the skeletal muscles under the control of the nervous system. At this time, the relative number of fast-twitch and slow-twitch fibers is determined.

In contrast, the myoblasts that produce cardiac and smooth muscle fibers do not fuse. However, the individual fibers of these tissues connect through gap junctions at a very early embryonic stage. Cardiac muscle begins pumping the blood just 3 weeks after fertilization.

Skeletal and cardiac muscle fibers undergo almost no division after birth. They do, however, lengthen and thicken during childhood and adolescence to keep up with the growing body. Furthermore, skeletal muscle fibers are surrounded throughout life by a few *satellite cells,* which resemble undifferentiated myoblasts. During youth, satellite cells fuse into the existing muscle fibers to help them grow, and following injury to a muscle, satellite cells fuse together to produce new skeletal muscle fibers. This capacity for regeneration is limited, however. Severely damaged skeletal muscle is replaced primarily by scar tissue.

Cardiac muscle has no satellite cells and no regenerative capacity whatsoever. This lack hinders recovery from heart attacks. Smooth muscle cells, by contrast, retain the capacity for division throughout life, and this tissue has a fair to good ability to regenerate.

Muscular dystrophy refers to a group of inherited muscle-destroying diseases that generally appear in childhood. The affected muscles enlarge because of a deposition of fat and connective tissue, but the muscle fibers themselves degenerate. The most common and serious form is **Duchenne muscular dystrophy.** This is inherited as a sex-linked recessive disease; that is, females carry and transmit the abnormal gene, which is expressed almost exclusively in males. This tragic disease is usually diagnosed when the boy is between 2 and 10 years old. Active, apparently normal children become clumsy and start to fall frequently as their muscles weaken. The disease progresses from pelvic muscles to shoulder muscles to the head and chest muscles. Victims rarely live past age 20 and usually die of respiratory failure and respiratory infections.

Recent research has pinned down the cause of Duchenne muscular dystrophy: The diseased muscle fibers lack a membrane protein called *dystrophin.* Without this protein, extracellular calcium ions leak into the muscle fibers and fatally disrupt muscle function.

Based on our previous discussion of normal myoblasts and satellite cells, can you guess how researchers hope one day to cure muscular dystrophy?

Scientists have successfully treated mice with the disease by injecting them with healthy myoblast cells that fuse with the unhealthy muscle fibers and induce them to produce dystrophin. A similar treatment may one day be possible for humans.

Our discussion of the muscular system during adulthood will begin with a commonly asked question. On average, the body strength of adult men is greater than that of adult women. Do you think this difference has a biological basis?

Evidence indicates that it does. Individuals vary, but on the average, women's skeletal muscles make up 36% of body mass, and men's skeletal muscles 42%. The greater muscular development of men is due mostly to the effects of androgen hormones (mainly testosterone) on skeletal muscle, not to effects of exercise. Since men are usually larger than women, the average difference in strength is even greater than the percentage difference in muscle mass would suggest (body strength *per unit muscle mass,* however, is the same in both sexes). Strenuous muscle exercise causes more enlargement of muscles in males than in females, again because of the influence of male sex hormone. Some athletes take large doses of synthetic male sex hormones ("steroids") to increase their muscle mass. This illegal and dangerous practice is discussed in the box on page 233. ∎

Because of its rich blood supply, skeletal muscle is amazingly resistant to infection throughout life. Given good nutrition and moderate exercise, skeletal muscle experiences relatively few problems.

As we age, the amount of connective tissue in our skeletal muscles naturally increases, the number of muscle fibers decreases, and the muscles become stringier, or more sinewy. Since skeletal muscles form so much of a person's body mass, body weight declines in many elderly people. Loss of muscle mass leads to a decrease in muscular strength, usually by 50% by age 80. Exercise helps offset the effects of aging on muscles, but muscles can also suffer indirectly: The aging of the heart and blood vessels affects nearly every organ in the body, and muscles are no exception. As atherosclerosis begins to diminish the flow of blood to the extremities, a circulatory condition called **intermittent claudication** (klaw′dĭ-ka′shun; "limping") may develop. In this condition,

A CLOSER LOOK   Athletes Looking Good—Doing Better With Anabolic Steroids?

Everyone loves a winner: Top athletes reap large social and monetary rewards. Thus, it is not surprising that some grasp at anything that increases their performance—including anabolic steroids. Anabolic steroids, variants of testosterone engineered by pharmaceutical companies, were introduced in the 1950s to treat victims of certain muscle-wasting diseases and anemia and to prevent muscle shrinkage in patients immobilized after surgery. Testosterone is responsible for the increase in muscle and bone mass and other physical changes that occur during puberty and convert boys into men. Testosterone stimulates the manufacture of muscle protein, thereby leading to an increase in the number of myofilaments and in the thickness of the skeletal muscle fibers. Convinced that megadoses of the steroids could produce enhanced masculinizing effects in grown men, many athletes were using the steroids by the early 1960s, and the practice is still going on today.

It has been difficult to determine the incidence of anabolic steroid use among athletes. The use of drugs has been banned by most international competitions, so users (and prescribing physicians or drug dealers) are naturally reluctant to talk about it. Nonetheless, there is little question that many professional bodybuilders and athletes competing in events that require great muscle strength (shot put, discus throwing, and weight lifting, for example) are heavy users. Other sports figures, such as football players, have also admitted to using steroids as an adjunct to training, diet, and psychological preparation for games. Advantages of anabolic steroids cited by athletes include an increase in muscle mass and strength, an increase in oxygen-carrying capability due to enhanced red blood cell volume, and an increase in aggressive behavior (the urge to "steamroller" the other guy.)

Typically, body builders who use steroids combine high doses (up to 200 mg/day) with heavy resistance training such as weight lifting. Intermittent use begins several months before an event, and both oral and injected steroid doses (a method called stacking) are increased gradually as the competition nears.

But do the drugs do all that is claimed? Research studies have reported increases in overall strength and body weight in steroid users. While these are results weight lifters dream about, there is a hot dispute over whether this also translates into athletic performance requiring fine muscle coordination and endurance. Users feel it does, but ex-users agree that the dangers of taking steroids outweigh any possible benefits.

The dangers of steroid use are substantial. Physicians maintain that steroids cause bloated faces (a sign of steroid excess); cause shriveled testes and infertility; damage the liver and promote liver cancer; and cause changes in blood cholesterol levels (which may predispose long-term users to coronary heart disease). Additionally, it now appears that the psychiatric hazards of anabolic steroids may be equally threatening: Recent studies have indicated that one-third of anabolic steroid users have serious mental problems. Manic behavior in which the users undergo Jekyll-Hyde personality swings and become extremely violent is common, as are depression and delusions.

Why some athletes use these drugs is an easy question to answer. Some admit to a willingness to do almost anything to win, short of killing themselves. Are they unwittingly doing this as well?

the restriction of blood supply to the legs leads to pain in the leg muscles during walking, forcing the person to stop and rest to get relief.

Smooth muscle tissue is remarkably trouble-free. The few clinical problems associated with this tissue tend to stem from external irritants. In the stomach and intestines, for example, irritation might result from ingestion of excess alcohol or spicy foods or from bacterial infections. Under such conditions, the mobility of the smooth muscle increases in an attempt to rid the body of the irritating agents, and diarrhea or vomiting occurs. The specific clinical problems of cardiac muscle are considered in Chapter 17.

✚ Related Clinical Terms

Fibromyositis (fi″bro-mi″o-si′tis) A group of common conditions involving inflammation and soreness of a muscle, its connective tissue coverings, and the capsules of nearby joints.

Lumbago (lum-ba′go) (*lumbus* = the loin) Acute pain in the lower back; backache. It may be due to a herniated disc but is usually due to muscle strain. (See below.) The injured muscles contract in spasms, causing rigidity in the lumbar region and painful movement. Backaches plague 80% of all Americans at some time in their lives.

Myalgia (mi-al′je-ah) (*algia* = pain) Muscle pain resulting from any muscle disorder.

Myopathy (mi-op′ah-the) (*path* = disease) Any disease of muscle.

RICE Acronym for **r**est, **i**ce, **c**ompression, and **e**levation, the standard treatment for a pulled muscle or excessively stretched tendons or ligaments.

Spasm A sudden, involuntary twitch of skeletal (or smooth) muscle, ranging in severity from merely irritating to very painful. Spasms may be due to chemical imbalances or to the injury of a muscle strain. Spasms of the eyelid or facial muscles, called *tics,* may result from psychological factors. Massaging the affected area may help to end the spasm. A *cramp* is a prolonged spasm that causes a muscle to become taut and painful.

Strain Tearing of a muscle, often due to a sudden movement that excessively stretches the muscle. Bleeding into the muscle and inflammation lead to pain.

Chapter Summary

OVERVIEW OF MUSCLE TISSUE (p. 217)

1. Our musculature produces movement, maintains posture, stabilizes joints, and generates body heat. Muscle tissue has the special properties of contractility, excitability, extensibility, and elasticity.

2. A muscle fiber is a muscle cell, the sarcoplasm of which is packed with myofilaments that generate contractile force.

3. Skeletal muscle attaches to the skeleton, has striated cells, and can be controlled voluntarily. Cardiac muscle occurs in the heart wall, has striated cells, and is controlled involuntarily. Smooth muscle occurs chiefly in the walls of hollow organs, has nonstriated cells, and is controlled involuntarily.

SKELETAL MUSCLE (pp. 217–228)

Basic Features of a Skeletal Muscle (pp. 219–222)

1. Skeletal muscles have various shapes (fusiform, triangular, and flat, for example). Each skeletal muscle is an organ. From large to small, the levels of organization in a muscle are as follows: whole muscle, fascicle, fiber, myofibril, sarcomere, and myofilament.

2. The connective tissue elements of a skeletal muscle are the epimysium (around the whole muscle), perimysium (around a fascicle), and endomysium (around fibers). A fascicle is a bundle of muscle fibers.

3. Every skeletal muscle fiber is stimulated to contract by a nerve cell. Skeletal muscle has a rich blood supply. Fine nerve fibers and capillaries occupy the endomysium.

4. Each muscle extends from an immovable (or less movable) attachment, called the origin, to a more movable attachment, called an insertion. Muscles attach to bones through tendons, aponeuroses, or direct (fleshy) attachments.

5. In different skeletal muscles, the fascicles are arranged in different patterns: parallel, pennate, convergent, and circular.

Microscopic and Functional Anatomy of Skeletal Muscle Tissue (pp. 223–228)

6. A skeletal muscle fiber is a long, striated cell formed from the fusion of many embryonic cells. It contains many peripherally located nuclei.

7. Myofibrils are striped cylinders that fill most of the cytoplasm of skeletal muscle cells. A myofibril consists of sarcomeres. A sarcomere extends from one Z line to the next. Thin (actin) filaments extend centrally from each Z line. Thick (myosin) filaments occupy the center of each sarcomere and overlap the inner ends of the thin filaments.

8. According to the sliding filament theory, the thin filaments are pulled toward the center of the sarcomere by a pivoting action of cross-bridges (myosin heads) on the thick filaments. Sliding of the filaments is triggered by a rise in cytoplasmic Ca^{2+} levels.

9. The myofilaments determine the striation pattern in skeletal (and cardiac) muscle fibers. There are A bands (where thick filaments are located), I bands (which contain only thin filaments and Z lines), plus M lines and H zones (where only thick filaments occur). During contraction, the Z lines move closer together, and the I bands and H zones narrow.

10. Between contractions, muscles are extended. They do not extend themselves but are stretched by the skeletal movements caused by opposing muscles.

11. The muscles attach to the skeleton in a way that keeps them at a near-optimal length for generating maximum contractile forces.

12. The sarcoplasmic reticulum is a specialized smooth endoplasmic reticulum in the muscle fiber. T tubules are deep invaginations of the sarcolemma. When a nerve cell stimulates a muscle fiber, it sets up an impulse in the sarcolemma that signals the sarcoplasmic reticulum to release Ca^{2+}, which then initiates the sliding of the myofilaments (muscle contraction).

13. There are three types of skeletal muscle fibers: (1) red slow-twitch (fatigue-resistant and best for maintaining posture), (2) white fast-twitch (for short bursts of power), and (3) intermediate fast-twitch (for long-term production of fairly strong contraction). Most muscles in the body contain a mixture of these fiber types.

CARDIAC MUSCLE (pp. 228–230)

Fibers (p. 228)

1. Cardiac muscle fibers are branching, striated cells with one or two central nuclei. They contain irregularly shaped myofibrils made of typical sarcomeres. Cardiac muscle contracts by the sliding filament mechanism.

2. Cardiac muscle cells have many mitochondria and depend on aerobic respiration to form ATP. They are very resistant to fatigue.

3. Contraction of cardiac muscle is triggered by Ca^{2+}, but some of this calcium comes from the extracellular fluid. The sarcoplasmic reticulum and T tubules are somewhat simpler than those of skeletal muscle.

Cardiac Muscle Tissue (pp. 228–230)

4. Adjacent cardiac cells are connected by intercalated discs. These specialized junctions contain desmosomes, fascia adherens, and gap junctions.

5. Networks of cardiac muscle cells contract together because they are electrically coupled by gap junctions. Cardiac muscle cells contract with their own inherent rhythm, the basis of the heartbeat.

SMOOTH MUSCLE (pp. 230–231)

Fibers (p. 230)

1. Smooth muscle fibers are small, spindle-shaped cells with one central nucleus. They have no striations or sarcomeres, but are filled with myofilaments that contract by the sliding filament mechanism.

2. The sarcoplasmic reticulum is as well developed as in cardiac muscle, although more difficult to demonstrate. T tubules are absent from smooth muscle cells.

Smooth Muscle Tissue (pp. 230–231)

3. Smooth muscle fibers are most often arranged in circular and longitudinal sheets.

4. Smooth muscle is innervated by involuntary nerve fibers, which usually innervate only a few muscle fibers per sheet. The impulse that signals contraction usually spreads from fiber to fiber through gap junctions.

5. Smooth muscle contracts for extended periods at low energy cost and without fatigue. Its fibers can generate substantial force even when they are highly stretched.

MUSCLE TISSUE THROUGHOUT LIFE (pp. 232–234)

1. Muscle tissue develops from embryonic mesoderm cells called myoblasts. Skeletal muscle fibers form by the fusion of many myoblasts. Smooth and cardiac muscle fibers develop from single myoblasts.

2. Mature skeletal muscle tissue has some ability to regenerate because of its satellite cells, but cardiac tissue cannot regenerate at all. Smooth muscle regenerates best.

3. On the average, men have more muscle mass than women have. This disparity is due to the effects of male sex hormones.

4. Skeletal muscles are richly vascularized and resistant to infection, but in old age they shrink, become fibrous, and lose strength.

Review Questions

Multiple Choice and Matching Questions

1. The connective tissue that encloses the sarcolemma of an individual muscle cell is called the (a) epimysium, (b) perimysium, (c) endomysium, (d) endosteum.

2. A fascicle is (a) a muscle, (b) a bundle of muscle cells enclosed by a connective tissue sheath, (c) a bundle of myofibrils, (d) a group of myofilaments.

3. A muscle in which the fascicles and fibers are arranged at an angle to a central longitudinal tendon has (a) a circular arrangement, (b) a longitudinal arrangement, (c) a parallel arrangement, (d) a pennate arrangement, (e) a convergent arrangement.

4. Thick and thin myofilaments have different properties. For each phrase below, indicate whether the filament described is thick or thin (write *thick* or *thin* in the blanks).

thin **(1)** contains actin
thick **(2)** contains myosin heads
thick **(3)** contains myosin
thin **(4)** does not lie in the H zone
thick **(5)** does not lie in the I band
thin **(6)** attaches to Z line

5. Skeletal muscle: levels of organization. Match each term given in the key with the correct description below.

Key: **(a)** a muscle **(b)** fascicle **(c)** fiber **(d)** myofibril **(e)** myofilament

_d_ **(1)** could be called an elaborate organelle; made of sarcomeres
_a_ **(2)** an organ
_b_ **(3)** a bundle of cells
_e_ **(4)** a group of large molecules
_c_ **(5)** a cell

6. Which of these bands or lines narrows when a skeletal or cardiac muscle fiber contracts? Write *yes* or *no* in each blank below.

_y_ **(1)** H band *thick*
_n_ **(2)** A band *thick*
_y_ **(3)** I band *thin + Z lines*
_n_ **(4)** M line *thick*

7. The function of T tubules in muscle contraction is (a) to make and store glycogen, (b) to release Ca^{2+} into the cell interior and then pick it up again, (c) to transmit an impulse deep into the muscle cell, (d) to make proteins.

8. Which fiber type would be the most useful in the leg muscles of a long-distance runner? (a) white fast-twitch, (b) white slow-twitch, (c) intermediate fast-twitch.

9. The ions that enter a muscle cell when an impulse passes over its sarcolemma and then trigger muscle contraction are (a) calcium, (b) chloride, (c) sodium, (d) potassium.

10. Fill in each blank with the correct single answer from the key.

Key: **(a)** skeletal muscle **(b)** cardiac muscle **(c)** smooth muscle

_b_ **(1)** striated and involuntary
_a_ **(2)** striated and voluntary
_c_ **(3)** not striated and involuntary
_b_ **(4)** its fibers form networks
_c_ **(5)** in wall of bladder
_b_ **(6)** has intercalated discs
_a_ **(7)** fiber is a giant, multinucleate cell
_c_ **(8)** best at regenerating when injured
_c_ **(9)** best at generating contractile force when highly stretched
_c_ **(10)** fibers in sheets in walls of most viscera

11. A skeletal muscle with fascicles in this pattern is likely to generate the most power: (a) circular arrangement, (b) longitudinal arrangement, (c) parallel arrangement, (d) pennate arrangement, (e) convergent arrangement.

Short Answer and Essay Questions

12. Name and explain the four special functional characteristics of muscle tissue.

13. (a) Distinguish a tendon from an aponeurosis and a fleshy attachment. (b) Define origin and insertion, and explain how they differ.

14. Explain the sliding filament theory of contraction by drawing and labeling a relaxed and a contracted sarcomere.

15. List the structural differences between the three distinct types of skeletal muscle fibers.

16. Which is more resistant to fatigue, cardiac muscle or skeletal muscle? What is the anatomical basis for this difference, and why is it important?

17. Tendons and aponeuroses are poorly vascularized (served by few blood vessels). Can the same be said of the skeletal muscles to which the tendons attach? Explain your answer.

18. Describe the structure and function of an intercalated disc.

19. Cindy Wong was a good anatomy student, but she realized she was mixing up the following "sound-alike" structures in skeletal muscle: myofilaments, myofibrils, fibers, and fascicles. Therefore, she wrote up a brief table to define and differentiate these four structures. Construct a table like hers.

Critical Thinking and Clinical Application Questions

20. Bernie Schwartz decided that his physique left much to be desired. He joined a health club and began to "pump iron" three times a week. After 3 months of training, during which he was able to lift increasingly heavy weights, his arm and chest muscles became much larger. Explain what happened to the fibers in these muscles.

21. Diego, who had not kept in shape, went out to play a game of touch football. As he was running, the calf region of his leg began to hurt. He went to a doctor the next day, who told him he had a strain. Diego kept insisting that no joints hurt. Clearly, he was confusing a *strain* with a *sprain.* Explain the difference.

22. Which of the following choices gives the best "reason" why the muscles that move the body and limbs are made of skeletal and not smooth muscle tissue? Explain your choice, and tell why each other answer is wrong. (a) Smooth muscle is much weaker than skeletal muscle tissue of the same size, so it could not move heavy limbs. (b) The smooth muscle cells do not contract by the sliding filament mechanism and therefore are inefficient. (c) Smooth muscle cells contract too slowly. If they moved the body, we could never move fast enough to survive dangerous situations. (d) Smooth muscle fatigues too easily.

23. Chickens carry out only brief bursts of flight, and their flying muscles consist of white fibers. The breast muscles of ducks, by contrast, consist of red and intermediate fibers. What can you deduce about the flying abilities of ducks?

10

Muscles of the Body

Student Objectives

Lever Systems: Bone-Muscle Relationships (pp. 238–240)

1. Explain the three types of lever systems in which muscles participate, and indicate the arrangement of elements (effort, fulcrum, and load) in each.

Interactions of Skeletal Muscles in the Body (p. 240)

2. Describe the functions of prime movers (agonists), antagonists, synergists, and fixators.

Naming the Skeletal Muscles (p. 241)

3. List the criteria used in naming muscles.

Development and Basic Organization of the Muscles (pp. 241–243)

4. Organize the body's muscles into four main groups based on their developmental origin.

Major Skeletal Muscles of the Body (pp. 243–292)

5. Name and identify the muscles described in Tables 10.1 to 10.17. State the origin, insertion, and action of each.

As we explained in the previous chapter, muscle tissue includes skeletal, cardiac, and smooth muscle. However, in the study of the muscular system, the **skeletal muscles** take center stage. They make up the flesh of our body, and also produce many types of movements: The blinking of an eye, standing on tiptoe, swallowing food, and wielding a sledgehammer are just a few examples. Before describing the individual muscles and their specific actions, we will explain the general principles of leverage, describe the ways in which muscles "play" with or against each other to bring about movements, and consider the criteria used in naming muscles.

Lever Systems: Bone-Muscle Relationships

The operation of most skeletal muscles involves the use of leverage. A **lever** is a rigid bar that moves on a fixed point, or **fulcrum,** when a force is applied to it. The applied force, or **effort,** is used to move a resis-

tance, or **load.** The bones of your skeleton act as levers, your joints as fulcrums. Muscle contraction provides the effort, applying this force where the muscle attaches to the bone. The load that is moved is the bone itself, along with the overlying tissues and anything else you are trying to move with that particular lever.

A lever allows a given effort to move a heavier load, or to move a load farther, than that effort otherwise could. If the load is close to the fulcrum and the effort is applied far from the fulcrum, a small effort exerted over a large distance can move a large load over a small distance (Figure 10.1a). We say that such a lever operates at a **mechanical advantage.** When a jack lifts a car, for example, the car moves up only a little with each large downward "push" of the jack handle, but not much muscular effort is needed. If, however, the load is farther from the fulcrum than is the effort, the effort applied must be greater than the load to be moved (Figure 10.1b). Such a lever operates at a **mechanical disadvantage.** Nonetheless, this disadvantage can be advantageous to us, for it lets the load be moved rapidly through a large distance. Wielding a shovel is an example. Shovel and jack are different kinds of levers, but both follow the same uni-

Figure 10.1 Lever systems operating at a mechanical advantage and a mechanical disadvantage. The equation at the top expresses the relationships among forces and distances for any lever system. (**a**) When the effort is farther from the fulcrum than is the load, a smaller effort (10 kg) can lift a larger load (1000 kg). Such a lever operates at a mechanical advantage. (**b**) By contrast, when the load is farther from the fulcrum than is the effort, a larger effort (100 kg) is needed to lift a smaller load (50 kg). Such a lever operates at a mechanical disadvantage.

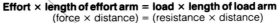

Effort × length of effort arm = load × length of load arm
(force × distance) = (resistance × distance)

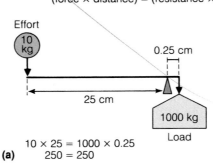

Effort
10 kg

0.25 cm

25 cm

1000 kg
Load

10 × 25 = 1000 × 0.25
(**a**)   250 = 250

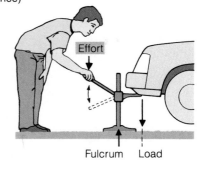

Effort

Effort
Fulcrum   Load

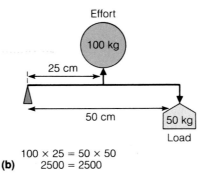

Effort
100 kg

25 cm

50 cm

50 kg
Load

100 × 25 = 50 × 50
(**b**)   2500 = 2500

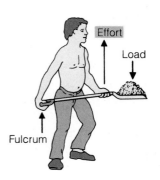

Effort
Load

Fulcrum

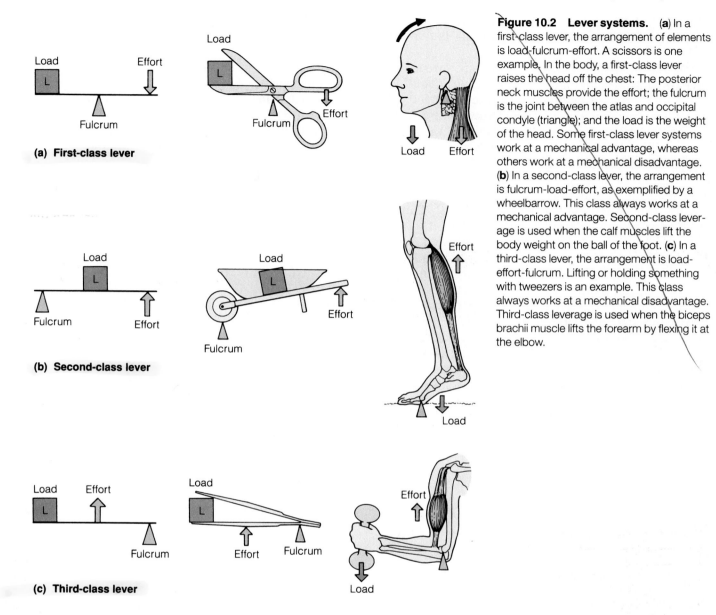

(a) First-class lever

(b) Second-class lever

(c) Third-class lever

Figure 10.2 Lever systems. (**a**) In a first-class lever, the arrangement of elements is load-fulcrum-effort. A scissors is one example. In the body, a first-class lever raises the head off the chest: The posterior neck muscles provide the effort; the fulcrum is the joint between the atlas and occipital condyle (triangle); and the load is the weight of the head. Some first-class lever systems work at a mechanical advantage, whereas others work at a mechanical disadvantage. (**b**) In a second-class lever, the arrangement is fulcrum-load-effort, as exemplified by a wheelbarrow. This class always works at a mechanical advantage. Second-class leverage is used when the calf muscles lift the body weight on the ball of the foot. (**c**) In a third-class lever, the arrangement is load-effort-fulcrum. Lifting or holding something with tweezers is an example. This class always works at a mechanical disadvantage. Third-class leverage is used when the biceps brachii muscle lifts the forearm by flexing it at the elbow.

versal **law of levers:** When the effort is farther from the fulcrum than is the load, the lever operates at a mechanical advantage; but when the effort is nearer than the load, the lever operates at a mechanical disadvantage.

Depending on the relative positions of the three elements—effort, fulcrum, and load—a lever belongs to one of three classes. In **first-class levers,** the effort is applied at one end, and the load is at the other, with the fulcrum somewhere between (Figure 10.2a). See-saws and scissors are first-class levers. First-class leverage also occurs when you lift your head off your chest. Some first-class levers operate at a mechanical

advantage (effort is farther from joint than load is), but others operate at a mechanical disadvantage (load is farther from joint than is effort).

In a **second-class lever,** the effort is applied at one end, and the fulcrum is at the other, with the load between (Figure 10.2b). A wheelbarrow is this kind of lever. Second-class leverage is not very common in the body. The best example is the act of standing on your toes. In this action, the joints in the ball of the foot are the fulcrum, the load is the whole weight of the body, and the calf muscles exert the effort, pulling the heel superiorly. Second-class levers all work at a mechanical advantage, because the muscle insertion (the effort) is farther from the fulcrum than is the load. Second-class levers are levers of strength, but speed and range of motion are sacrificed.

In **third-class levers,** effort is applied between the load and the fulcrum (Figure 10.2c). Lifting or holding something with tweezers uses third-class leverage. These levers work with speed but always at a mechanical disadvantage. Most skeletal muscles of the body are in third-class lever systems, as exemplified by the biceps brachii muscle of the arm. The fulcrum is the elbow joint; the force is exerted on the proximal region of the radius; and the load is the distal part of the forearm (plus anything carried in the hand). Third-class lever systems permit a muscle to be inserted very close to the joint across which movement occurs, allowing fast, extensive movements with relatively little shortening of the muscle. This in turn permits us to move our limbs quickly as we run and throw.

In summary, differences in the positioning of the three lever-elements modify the activity of muscles with respect to (1) speed of contraction, (2) range of movement, and (3) the weight of the load that can be lifted. In lever systems that operate at a mechanical disadvantage, force is lost, but a greater speed and range of movement is gained. Systems that operate at a mechanical advantage tend to be slower and more stable and are used where strength is a priority.

Interactions of Skeletal Muscles in the Body

It is easy to understand that different muscles often work together to bring about a single movement. What is not so obvious, however, is that muscles often work *against* one another as well. No single muscle can reverse the motion it produces (a muscle cannot "push" after pulling). Therefore, for whatever one muscle (or muscle group) can do, there must be other muscles that can "undo" the action. In general, groups of muscles that produce opposite movements lie on opposite sides of the same joint. Let us now further explore how muscles interact.

Muscles are classified into several *functional* types. The muscle that has the major responsibility for producing a specific movement is the **prime mover,** or **agonist** (ag'o-nist; "leader"), of that motion. For example, the biceps brachii muscle is a prime mover for flexing the forearm at the elbow (Figure 10.2c). Sometimes, two muscles contribute so heavily to the same movement that both are called agonists.

Muscles that oppose or reverse a particular movement act as **antagonists** (an-tag'o-nists; "against the leader"). When a prime mover is active, the antagonist muscles are being stretched and are often relaxed. Usually, however, the antagonists contract slightly during a movement to keep the movement from overshooting its mark or to slow the movement as it ends. Antagonists can also be prime movers in their own right; that is, an agonist for one movement serves as an antagonist for the opposite movement. For example, flexion of the forearm by the biceps brachii is antagonized by the triceps brachii muscle, the prime mover for extending the forearm.

In addition to prime movers and antagonists, most movements also involve one or more **synergists** (sin'er-jist; "together-worker"). Synergists help the prime movers, either (1) by adding a little extra force to the same movement or (2) by reducing undesirable extra movements that the prime mover may produce. This latter function deserves more explanation: Some prime movers span several joints and can cause movements at all the joints they span, but synergists act to cancel some of these movements. For example, the muscles that flex the fingers cross both the wrist and finger joints, but you can make a fist without flexing your wrist because synergists stabilize the wrist. Additionally, some prime movers can cause several kinds of movement *at the same joint,* and synergists cancel the particular movement that is inappropriate at the time.

Some synergists hold a bone firmly in place so that a prime mover has a stable base on which to move a body part. Such synergists are called **fixators.** An example is the muscles that fix the scapula when the arm moves. Muscles that maintain posture and stabilize joints also act as fixators.

In summary, although prime movers seem to "get all the credit" for causing movements, the actions of antagonistic and synergistic muscles are also important.

We have seen how important it is that muscles work together. When an inexperienced body-builder develops the muscles on one side of a limb more than on the other side he or she is said to be "muscle bound." Can you guess the symptoms of this condition?

The stronger muscle tone in the more developed muscles puts a constant tension on the joints, inhibiting the joints' flexibility of movement. For example, an overdeveloped biceps brachii keeps the forearm partly flexed at the elbow, making it difficult for the triceps brachii to perform its function of extending the forearm. Although muscle-bound athletes may have impressive physiques, their movements are awkward, and they perform poorly in sports. To avoid this problem, modern bodybuilding programs strive for an even development of opposing muscle groups. ■

Naming the Skeletal Muscles

Skeletal muscles are named by several criteria, each of which describes the muscle in some way. Paying attention to these cues can simplify the task of learning muscle names.

1. Location. Some names indicate where a muscle is located. For example, the brachialis muscle is in the arm (*brachium* = arm), and intercostal muscles lie between the ribs (*costal* = rib).

2. Shape. Some muscles are named for their shapes. For example, the deltoid is triangular (*delta* = triangle), and the right and left trapezius muscles together form a trapezoid.

3. Relative size of the muscle. The terms *maximus* (largest), *minimus* (smallest), *longus* (long), and *brevis* (short) are part of the names of some muscles—like the gluteus maximus and gluteus minimus muscles of the buttocks.

4. Direction of the fascicles and the muscle fibers. The names of some muscles tell the direction in which their fascicles (and muscle fibers) run. In muscles with the term *rectus* (straight) in their name, the fascicles run parallel to the body midline, whereas *transversus* and *oblique* mean that the fascicles run at a right angle and obliquely to the midline, respectively. Specific examples include the rectus abdominis, transversus abdominis, and external oblique muscles of the abdomen.

5. Location of attachments. The names of some muscles indicate their points of origin and insertion. Recall that the origin is the less movable attachment of a muscle, whereas the insertion is the more movable attachment. The origin is always named first. For instance, the brachioradialis muscle in the forearm originates on the bone of the brachium (arm), or humerus, and inserts on the radius.

6. Number of origins. When the term *biceps* ("two heads"), *triceps* ("three heads"), or *quadriceps* ("four heads") is in its name, a muscle has two, three, or four origins, respectively. For example, the biceps brachii has two origins.

7. Action. When muscles are named for their action, words such as *flexor, extensor, adductor,* or *abductor* appear in the name. For example, the adductor longus, on the medial thigh, adducts the thigh at the hip. The names of many forearm and leg muscles begin with *extensor* and *flexor,* indicating how they move the hand, foot, and digits.

Often, several different criteria are combined in the naming of a muscle. For instance, the name *extensor carpi radialis longus* tells us the muscle's action (extensor), the joint it acts on (*carpi* = wrist), and that it lies along the radius in the forearm (radialis). The name also indicates that the muscle is longer (longus) than some other wrist extensor muscles. While long names like this are difficult to pronounce, they are extremely informative.

Can you give some characteristics of the following muscles, based on their names: tibialis anterior; temporalis; erector spinae?

The tibialis anterior lies on the tibia in the anterior region of the leg; the temporalis is located in the temple on the temporal bone; and the erector spinae is a muscle in the back that extends the vertebral column (makes the spine erect). ■

Development and Basic Organization of the Muscles

We will cover the individual muscles of the body in a series of tables (Tables 10.1–10.17) on pages 246–292. Here, we provide a general overview of the muscles by classifying them into four groups based on their embryonic origin: (1) musculature of the visceral organs, (2) muscles of the trunk, (3) limb muscles, and (4) pharyngeal arch muscles. All of these muscles develop from the mesoderm germ layer (Chapter 3, p. 57), and Figure 10.3 reviews some parts of the mesoderm that are present during the second month of development. Notice the *myotomes, splanchnic mesoderm,* and *somatic mesoderm.*

1. Musculature of the visceral organs. Visceral musculature, the muscle in the walls of the visceral organs, develops from the splanchnic mesoderm around the early gut. Recall that visceral musculature includes both smooth and cardiac muscle. Since this musculature does not form skeletal muscles, it is not considered further in this chapter.

2. Trunk muscles. The trunk muscles are the skeletal muscles of the thorax, abdomen, and pelvis. This group also includes most muscles of the neck and a few in the head. These muscles arise from myotomes, the series of 30–40 muscle segments shown in Figure 10.3b. The *dorsal* regions of the myotomes become the deep muscles of the back (Table 10.4), whose main function is to extend the spine as we stand erect. The

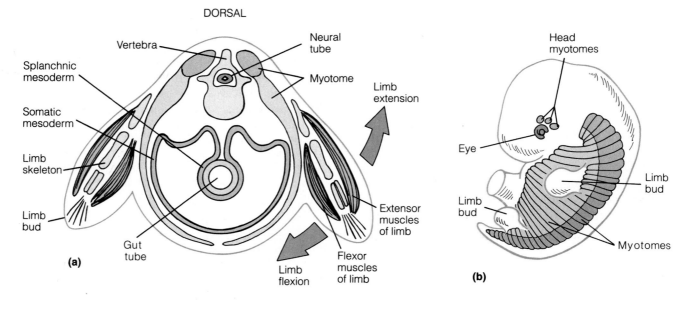

Figure 10.3 Development and basic organization of the muscles. (a) Cross section through the trunk and lower limbs of an embryo at 4 to 6 weeks (slightly older than the embryo in Figure 3.12 on p. 57) showing some divisions of the mesoderm and the formation of some basic muscle groups. (b) The myotomes in a 6-week embryo.

ventral regions of the myotomes, by contrast, become the muscles on the lateral and anterior parts of the trunk and neck. These ventrolateral muscles flex the vertebral column, among other functions. They include the muscles of the anterior neck (the suprahyoid and infrahyoid muscles in Table 10.3), the respiratory muscles of the thorax (Table 10.5), the muscles of the anterior abdominal wall (Table 10.6), and the muscles in the floor of the pelvis (Table 10.7).

A few myotomes also occur in the embryonic head (Figure 10.3b). These head myotomes give rise to the muscles that move the tongue (Table 10.2) and the muscles that move the eyes (see p. 417 in Chapter 15).

3. Limb muscles. Both the upper and lower limbs arise as limb buds filled with somatic mesoderm (p. 58). Some of this mesoderm becomes the limb bones, but most of it forms the skeletal muscles of the limbs. As shown in Figure 10.3a, the mesenchyme that lies *dorsal* to the limb bones becomes the *extensor* muscles of each limb, whereas the ventral mesenchyme becomes the limb *flexor* muscles.

Because they develop in the same way, the muscles of upper and lower limbs are similar and directly comparable. In the adult *upper* limb (Tables 10.10 to 10.13), the extensor muscles lie on the limb's poste-

rior side and extend the limb segments (forearm, hand, and fingers). The flexor muscles, by contrast, lie on the anterior side of the upper limb and flex the limb segments.

In the adult *lower* limbs (Tables 10.14 to 10.17), the extensor muscles occupy the anterior aspect of the limb. They extend the leg at the knee, dorsiflex the foot at the ankle, and extend the toes. The flexor muscles, by contrast, lie on the posterior aspect of the adult lower limb, and they flex the leg at the knee, plantar flex the foot at the ankle, and flex the toes.

The muscles that attach the limbs to their girdles, and the muscles that attach the girdles to the trunk, also belong to the limb muscle group (Tables 10.8, 10.9, and 10.14).

4. Pharyngeal arch muscles. The pharyngeal arch musculature surrounds the embryonic pharynx (the throat region of the digestive tube), and some of it migrates to form certain muscles in the head and neck of the adult. Unlike the visceral musculature around the rest of the digestive tube, this is skeletal muscle, not smooth muscle—presumably because the pharynx requires voluntary skeletal musculature to perform the voluntary act of swallowing. Indeed, the embryonic pharyngeal musculature becomes the adult *pharyngeal constrictor* muscles for swallowing (Table 10.3).

It also gives rise to the muscles of the larynx (voice box), the chewing muscles (Table 10.2), and the muscles of the face by which we express our emotions (Table 10.1). The pharyngeal arch muscles also include a major neck muscle called the sternocleidomastoid (Table 10.4) and a back muscle called the trapezius (Table 10.8).

Major Skeletal Muscles of the Body

There are over 600 muscles in the body, and learning them can be difficult. The first requirement is to make sure you understand all the body movements shown in Figures 8.5 and 8.6 (pp. 196 and 199). Some instructors require that their students learn only a few major groups of muscles, whereas others require a more detailed study. The information in Tables 10.1 to 10.17 should provide sufficient depth to satisfy either approach.

In these tables, the muscles are grouped by function and by location, roughly from head to foot. Every table starts by introducing a muscle group and then describes each muscle's shape, location, attachments, actions, and innervation. As you study each muscle, look at its attachments and the direction of its fascicles, and try to understand how these features determine its specific action. A good way to learn muscle actions is to act out the movements yourself and then feel the muscles contracting beneath your skin. Figure 10.4 summarizes all the superficial muscles in the body and will help you tie together the information in the tables. Few courses will ask students to memorize everything in the tables, but the details are provided for reference.

The following list summarizes the organization and sequence of the tables in this chapter:

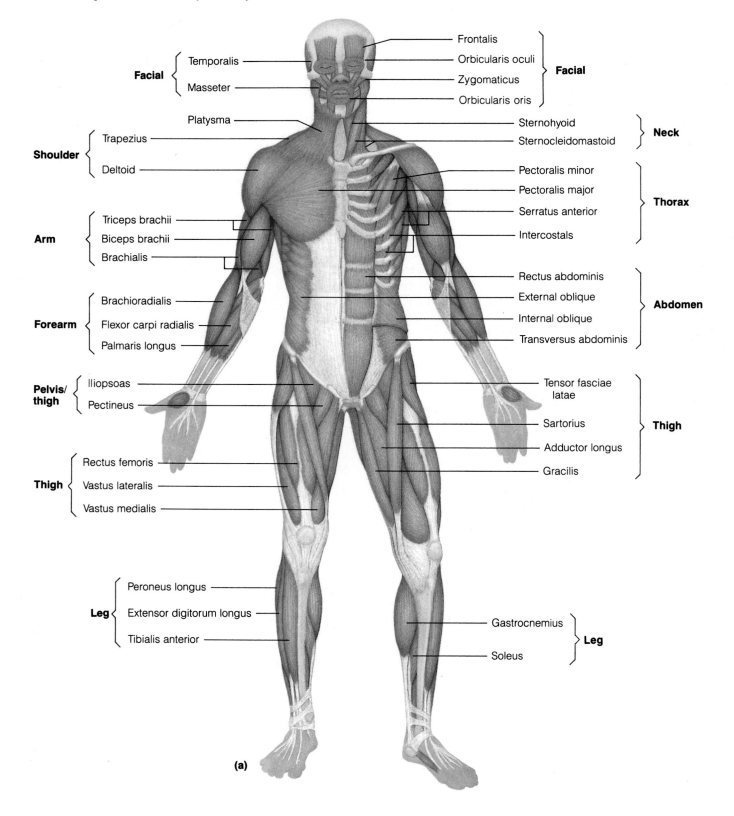

Facial
- Temporalis
- Masseter

Facial
- Frontalis
- Orbicularis oculi
- Zygomaticus
- Orbicularis oris

- Platysma

Neck
- Sternohyoid
- Sternocleidomastoid

Shoulder
- Trapezius
- Deltoid

Thorax
- Pectoralis minor
- Pectoralis major
- Serratus anterior
- Intercostals

Arm
- Triceps brachii
- Biceps brachii
- Brachialis

Abdomen
- Rectus abdominis
- External oblique
- Internal oblique
- Transversus abdominis

Forearm
- Brachioradialis
- Flexor carpi radialis
- Palmaris longus

Pelvis/thigh
- Iliopsoas
- Pectineus

Thigh
- Tensor fasciae latae
- Sartorius
- Adductor longus
- Gracilis

Thigh
- Rectus femoris
- Vastus lateralis
- Vastus medialis

Leg
- Peroneus longus
- Extensor digitorum longus
- Tibialis anterior

Leg
- Gastrocnemius
- Soleus

(a)

Figure 10.4   Superficial muscles of the body.   **(a)** Anterior view. The abdominal surface has been partly dissected on the left side of the body to show deeper muscles. **(b)** Posterior view.

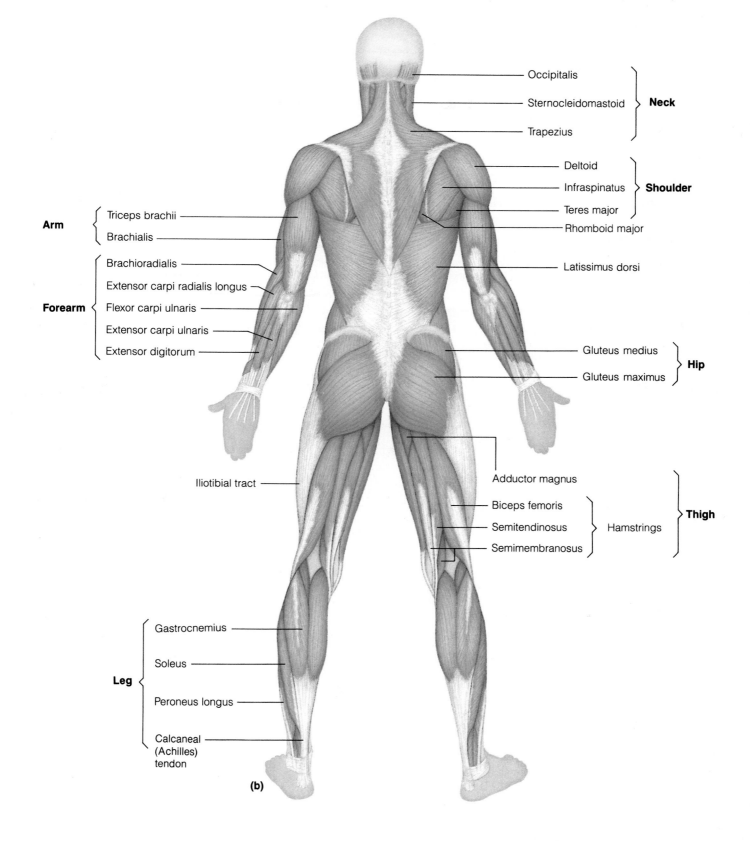

Occipitalis
Sternocleidomastoid } **Neck**
Trapezius

Deltoid
Infraspinatus } **Shoulder**
Teres major
Rhomboid major

Arm { Triceps brachii
Brachialis

Latissimus dorsi

Forearm { Brachioradialis
Extensor carpi radialis longus
Flexor carpi ulnaris
Extensor carpi ulnaris
Extensor digitorum

Gluteus medius
Gluteus maximus } **Hip**

Iliotibial tract

Adductor magnus

Biceps femoris
Semitendinosus } Hamstrings } **Thigh**
Semimembranosus

Leg { Gastrocnemius
Soleus
Peroneus longus
Calcaneal (Achilles) tendon

(b)

Table 10.1 Muscles of the Head, Part I: Facial Expression (Figure 10.5)

The muscles that promote facial expression lie in the face and scalp, just deep to the skin. They are thin, and adjacent muscles in this group tend to be fused. Unlike other skeletal muscles, facial muscles insert on the skin, not on bones. In the scalp, the main muscle is the **epicranius,** which has distinct anterior and posterior parts. In the face, the muscles clothing the facial bones lift the eyebrows and flare the nostrils, close the eyes and lips, and provide one of the best tools for influencing others—the smile. The great importance of the facial muscles in nonverbal communication becomes clear when they are paralyzed, as in some stroke victims. All muscles listed in this table are innervated by *cranial nerve VII, the facial nerve.* The muscles that move our eyeballs as we look in various directions are described in Chapter 15 (p. 417).

Muscle	Description	Origin (O) and insertion (I)	Action	Nerve supply
MUSCLES OF THE SCALP				
Epicranius (Occipitofrontalis) (ep"ĭ-kra'ne-us; ok-sip" ĭ-to-fron-ta'lis) (*epi* = over; *cran* = skull)	Bipartite muscle consisting of the frontalis and occipitalis muscles connected by a cranial aponeurosis, the galea aponeurotica; the alternate action of these two muscles pulls scalp forward and backward			
• **Frontalis** (fron-ta'lis) (*front* = forehead)	Covers forehead and dome of skull; no bony attachments	O—galea aponeurotica I—skin of eyebrows and root of nose	With aponeurosis fixed, raises the eyebrows (as in surprise); wrinkles forehead skin horizontally	Facial nerve (cranial VII)
• **Occipitalis** (ok-sip" ĭ-tal'is) (*occipito* = base of skull)	Overlies base of occiput; by pulling on the galea, fixes origin of frontalis	O—occipital bone I—galea aponeurotica	Fixes aponeurosis and pulls scalp posteriorly	Facial nerve
MUSCLES OF THE FACE				
Corrugator supercilii (kor'ah-ga-ter soo"per-sĭ'le-i) (*corrugo* = wrinkle; *supercilium* = eyebrow)	Small muscle; activity associated with that of orbicularis oculi	O—arch of frontal bone above nasal bone I—skin of eyebrow	Draws eyebrows together; wrinkles skin of forehead vertically (as in frowning)	Facial nerve
Orbicularis oculi (or-bik"u-lar-is ok'u-li) (*orb* = circular; *ocul* = eye)	Thin, flat sphincter muscle of eyelid; surrounds rim of the orbit; paralysis results in drooping of lower eyelid and spilling of tears	O—frontal and maxillary bones and ligaments around orbit I—tissue of eyelid	Protects eyes from intense light and injury; various parts can be activated individually; produces blinking, squinting, and draws eyebrows downward; strong closure of eyelids causes creasing of skin at lateral aspect of eye (these folds, called "crow's feet," become permanent with aging)	Facial nerve
Zygomaticus (zi-go-mat' ĭ-kus), major and minor (*zygomatic* = cheekbone)	Muscle pair extending diagonally from corner of mouth to cheekbone	O—zygomatic bone I—skin and muscle at corner of mouth	Raises lateral corners of mouth upward (smiling muscle)	Facial nerve
Risorius (ri-zor'e-us) (*risor* = laughter)	Slender muscle running beneath and laterally to zygomaticus	O—lateral fascia associated with masseter muscle I—skin at angle of mouth	Draws corner of lip laterally; tenses lips; synergist of zygomaticus	Facial nerve
Levator labii superioris (le-va'ter la'be-i soo-per"e-or'is) (*leva* = raise; *labi* = lip; *superior* = above, over)	Thin muscle between orbicularis oris and inferior eye margin	O—zygomatic bone and infraorbital margin of maxilla I—skin and muscle of upper lip	Opens lips; raises and furrows the upper lip; flares nostril (as in disgust)	Facial nerve
Depressor labii inferioris (de-pres'or la'be-i in-fer"e-or'is) (*depressor* = depresses; *infer* = below)	Small muscle running from lower lip to mandible	O—body of mandible lateral to its midline I—skin and muscle of lower lip	Draws lower lip downward (as in a pout)	Facial nerve

Muscle	Description	Origin (O) and insertion (I)	Action	Nerve supply
Depressor anguli oris (ang'gu-li or-is) (*angul* = angle, corner)	Small muscle lateral to depressor labii inferioris	O—body of mandible below incisors I—skin and muscle at angle of mouth below insertion of zygomaticus	Zygomaticus antagonist; draws corners of mouth downward and laterally (as in a "tragedy mask" grimace)	Facial nerve
Orbicularis oris (*or* = mouth)	Complicated, multilayered muscle of the lips with fibers that run in many different directions; most run circularly	O—arises indirectly from maxilla and mandible; fibers blended with fibers of other facial muscles associated with the lips I—encircles mouth; inserts into muscle and skin at angles of mouth	Closes lips; purses and protrudes lips (kissing muscle)	Facial nerve
Mentalis (men-ta'lis) (*ment* = chin)	One of the muscle pair forming a V-shaped muscle mass on chin	O—mandible below incisors I—skin of chin	Protrudes lower lip; wrinkles chin	Facial nerve
Buccinator (bu'sih-na"ter) (*bucc* = cheek or "trumpeter")	Thin, horizontal cheek muscle; principal muscle of cheek; deep to masseter (see also Figure 10.6)	O—molar region of maxilla and mandible I—orbicularis oris	Draws corner of mouth laterally; compresses cheek (as in whistling, blowing, and sucking); holds food between teeth during chewing; well developed in nursing infants	Facial nerve
Platysma (plah-tiz'mah) (*platy* = broad, flat)	Unpaired, thin, sheetlike superficial neck muscle; not strictly a head muscle, but plays a role in facial expression	O—fascia of chest (over pectoral muscles and deltoid) I—lower margin of mandible, and skin and muscle at corner of mouth	Helps depress mandible; pulls lower lip back and down, i.e., produces downward sag of mouth; tenses skin of neck (e.g., during shaving)	Facial nerve

Figure 10.5 Lateral view of muscles of the scalp, face, and neck.

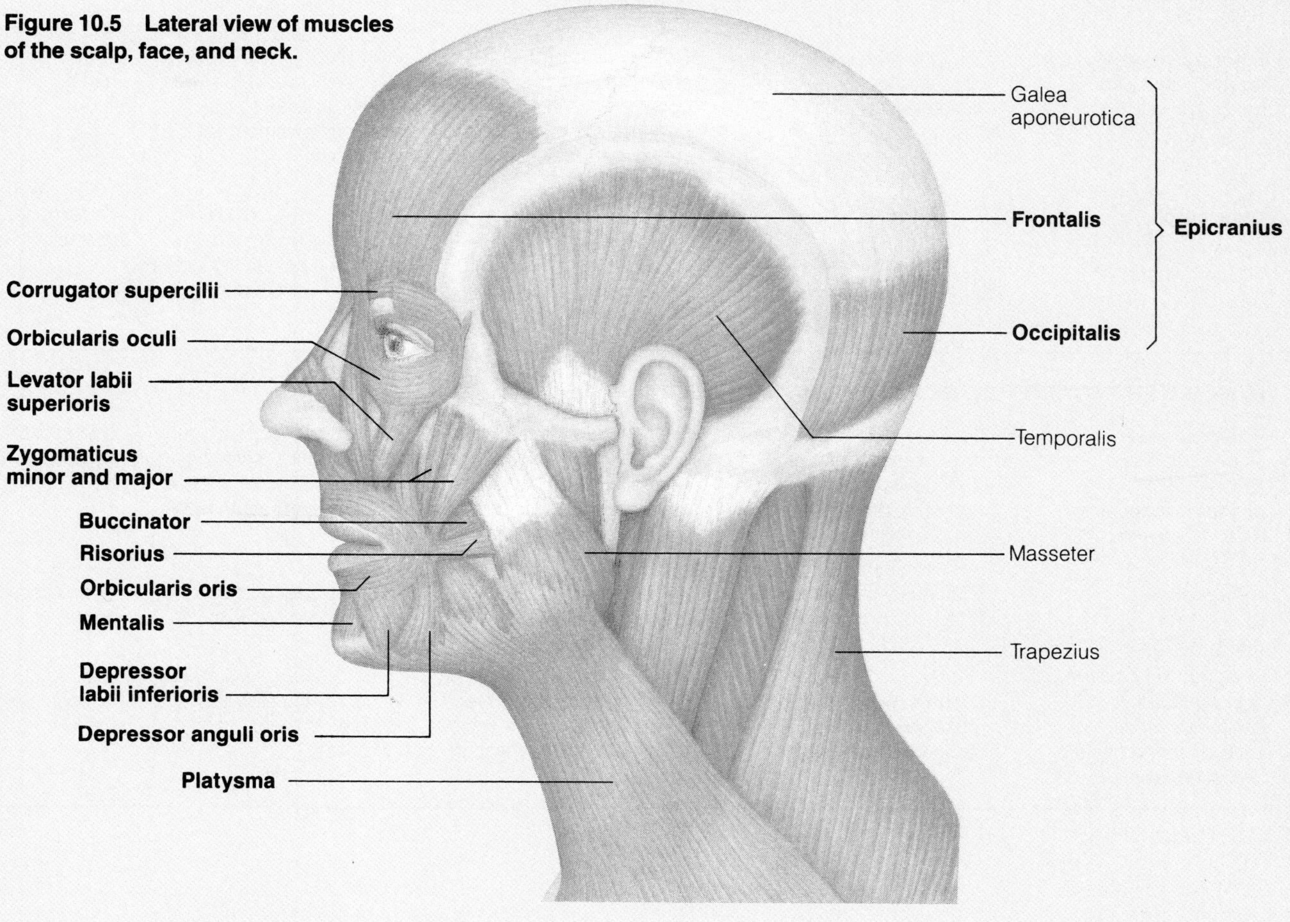

Table 10.2   Muscles of the Head, Part II: Mastication and Tongue Movement (Figure 10.6)

Four main pairs of muscles are involved in mastication (chewing). All are innervated by the mandibular division of *cranial nerve V, the trigeminal nerve.* The prime movers of jaw closure and biting are the powerful **masseter** and **temporalis** muscles. These can be felt bulging through the skin when the teeth are clenched. Grinding movements are brought about by the **pterygoid** muscles. The **buccinator** muscles in the cheeks (Table 10.1) also play a role in chewing. For lowering the mandible, gravity is usually sufficient, but if there is resistance to jaw opening, the jaw-opening muscles are activated (**digastric** and **mylohyoid** muscles: Table 10.3).

The tongue consists of muscle fibers that curl, squeeze, and fold the tongue during speaking and chewing. These **intrinsic tongue muscles,** which change the tongue's shape but do not really move it, are considered in Chapter 21 (p. 569) with the digestive system. Only the **extrinsic tongue muscles** are covered in this table. These move the tongue laterally, anteriorly, and posteriorly. The tongue muscles are all innervated by *cranial nerve XII, the hypoglossal nerve.*

Muscle	Description	Origin (O) and Insertion (I)	Action	Nerve supply
MUSCLES OF MASTICATION				
Masseter (mah-se′ter) (*maseter* = chewer)	Powerful muscle that covers lateral aspect of mandibular ramus	O—zygomatic arch I—angle and ramus of mandible	Prime mover of jaw closure; elevates mandible	Trigeminal nerve (cranial V)
Temporalis (tem″por-ă′lis) (*tempora* = time; pertaining to the temporal bone)	Fan-shaped muscle that covers parts of the temporal, frontal, and parietal bones	O—temporal fossa I—coronoid process of mandible via a tendon that passes beneath zygomatic arch	Closes jaw; elevates and retracts mandible; maintains position of the mandible at rest	Trigeminal nerve
Medial pterygoid (me′de-ul ter′ĭ-goid) (*medial* = toward median plane; *pterygoid* = winglike)	Deep two-headed muscle that runs along internal surface of mandible and is largely concealed by that bone	O—medial surface of lateral pterygoid plate of sphenoid bone, maxilla, and palatine bone I—medial surface of mandible near its angle	Synergist of temporalis and masseter muscles in elevation of the mandible; acts with the lateral pterygoid muscle to promote side-to-side (grinding) movements	Trigeminal nerve
Lateral pterygoid (*lateral* = away from median plane)	Deep two-headed muscle; lies superior to medial pterygoid muscle	O—greater wing and lateral pterygoid plate of sphenoid bone I—condyle of mandible and capsule of temporomandibular joint	Protrudes mandible (pulls it anteriorly); provides forward sliding and side-to-side grinding movements of the lower teeth	Trigeminal nerve
Buccinator	See Table 10.1	See Table 10.1	Trampoline-like action of buccinator muscles helps keep food between grinding surfaces of teeth during chewing	Facial nerve (cranial VII)
MUSCLES PROMOTING TONGUE MOVEMENTS (EXTRINSIC MUSCLES)				
Genioglossus (je″ne-o-glah′sus) (*geni* = chin; *glossus* = tongue)	Fan-shaped muscle; forms bulk of inferior part of tongue; its attachment to mandible prevents tongue from falling backward and obstructing respiration	O—internal surface of mandible near symphysis I—inferior aspect of the tongue and body of hyoid bone	Primarily protrudes tongue, but can depress or act in concert with other extrinsic muscles to retract tongue	Hypoglossal nerve (cranial XII)
Hyoglossus (hi′o-glos″us) (*hyo* = pertaining to hyoid bone)	Flat, quadrilateral muscle	O—body and greater horn of hyoid bone I—inferiolateral aspect of tongue	Depresses tongue and draws its sides downward	Hypoglossal nerve
Styloglossus (sti′-lo-glah′sus) (*stylo* = pertaining to styloid process)	Slender muscle running superiorly to and at right angles to hyoglossus	O—styloid process of temporal bone I—lateral inferior aspect of tongue	Retracts (and elevates) tongue	Hypoglossal nerve

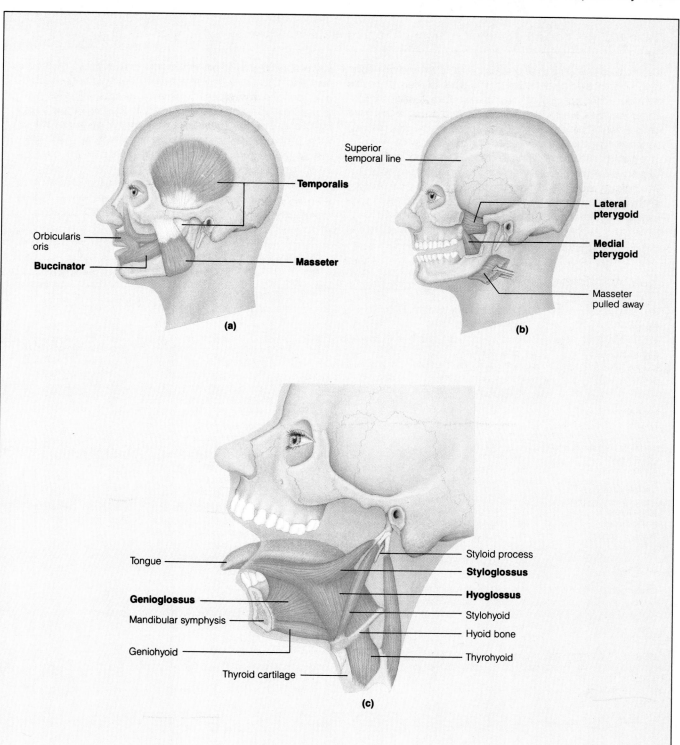

Figure 10.6   Muscles promoting mastication and tongue movements.   (**a**) Lateral view of the temporalis, masseter, and buccinator muscles. (**b**) Lateral view of the deep chewing muscles, the medial and lateral pterygoid muscles. (**c**) Extrinsic muscles of the tongue. Some suprahyoid muscles of the throat are also illustrated.

Table 10.3 Muscles of the Anterior Neck and Throat: Swallowing (Figure 10.7)

The neck is divided into two triangles (anterior and posterior) by the sternocleidomastoid muscle (Figure 10.7a). This table considers the muscles of the *anterior* triangle of the neck. These muscles are divided into **suprahyoid** and **infrahyoid** groups, which lie superior and inferior to the hyoid bone, respectively. Most of these muscles participate in swallowing.

Swallowing begins when the tongue and buccinator muscles squeeze food posteriorly along the roof of the mouth toward the pharynx. Then, many muscles contract in the pharynx and neck to complete the swallowing process: (1) The air passageway of the larynx closes so that food is not inhaled into the lungs. The suprahyoid muscles aid this action by lifting the larynx superiorly and anteriorly, under the cover of a protective flap (epiglottis). By pulling anteriorly on the hyoid bone, suprahyoid muscles also widen the pharynx to receive the food. (2) Then, the food is squeezed inferiorly into the esophagus by the muscles of the pharynx wall, the **pharyngeal constrictors.** (3) As swallowing ends, the hyoid bone and larynx are pulled inferiorly to their original position. This action is performed by the infrahyoid muscles.

Swallowing also includes mechanisms that prevent food from being squeezed superiorly into the nasal cavity, but these mechanisms will be considered later (Chapter 20, p. 538).

Muscle	Description	Origin (O) and insertion (I)	Action	Nerve supply
SUPRAHYOID MUSCLES (soo″prah-hi′oid)	Muscles that help form floor of oral cavity, anchor tongue, and move larynx superiorly during swallowing; lie superior to hyoid bone			
Digastric (di-gas′trik) (*di* = two; *gaster* = belly)	Consists of two bellies united by an intermediate tendon, forming a V shape under the chin	O—lower margin of mandible (anterior belly) and mastoid process of the temporal bone (posterior belly) I—by a connective tissue loop to hyoid bone	Acting in concert, the digastric muscles elevate hyoid bone and steady it during swallowing and speech; acting from behind, they open mouth (prime mover) and depress mandible	Trigeminal nerve (cranial V) for anterior belly; facial nerve (cranial VII) for posterior belly
Stylohyoid (sti″lo-hi′oid) (also see Figure 10.6)	Slender muscle below angle of jaw; parallels posterior belly of digastric muscle	O—styloid process of temporal bone I—hyoid bone	Elevates and retracts hyoid, thereby elongating floor of mouth during swallowing	Facial nerve
Mylohyoid (mi″lo-hi′oid) (*myle* = molar)	Flat, triangular muscle just deep to digastric muscle; this muscle pair forms a sling that forms the floor of the anterior mouth	O—medial surface of mandible I—hyoid bone and median raphe	Elevates hyoid bone and floor of mouth, enabling tongue to exert backward and upward pressure that forces food bolus into pharynx	Mandibular branch of trigeminal nerve
Geniohyoid (je′ne-o-hy″oid) (also see Figure 10.6) (*geni* = chin)	Narrow muscle in contact with its partner medially; runs from chin to hyoid bone	O—inner surface of mandibular symphysis I—hyoid bone	Pulls hyoid bone superiorly and anteriorly, shortening floor of mouth and widening pharynx for receiving food during swallowing	First cervical spinal nerve via hypoglossal nerve (cranial XII)
INFRAHYOID MUSCLES	Muscles causing depression of hyoid bone and larynx during swallowing and speaking; because of their ribbonlike appearance, often called *strap muscles*			
Sternohyoid (ster″no-hi′oid) (*sterno* = sternum)	Most medial muscle of the neck; thin; superficial except inferiorly, where covered by sternocleidomastoid	O—manubrium and medial end of clavicle I—lower margin of hyoid bone	Depresses larynx and hyoid bone if mandible is fixed; may also flex skull	C₁–C₃ through ansa cervicalis (slender nerve root in cervical plexus)
Sternothyroid (ster″no-thy′roid) (*thyro* = thyroid cartilage)	Lateral and deep to sternohyoid	O—posterior surface of manubrium of sternum I—thyroid cartilage	Pulls thyroid cartilage (plus larynx and hyoid bone) inferiorly	As for sternohyoid

Muscle	Description	Origin (O) and insertion (I)	Action	Nerve supply
Omohyoid (o"mo-hi'oid) (*omo* = shoulder)	Straplike muscle with two bellies united by an intermediate tendon; lateral to sternohyoid	O—superior surface of scapula I—hyoid bone, lower border	Depresses and retracts hyoid bone	As for sternohyoid
Thyrohyoid (thi"ro-hi'oid) (also see Figure 10.6)	Appears as a superior continuation of sternothyroid muscle	O—thyroid cartilage I—hyoid bone	Depresses hyoid bone and elevates thyroid cartilage	First cervical nerve via hypoglossal
Pharyngeal constrictor muscles—superior, middle, and inferior (far-rin'je-al)	Composite of three paired muscles whose fibers run circularly in pharynx wall; arranged so that the superior muscle is innermost and inferior one is outermost; substantial overlap	O—attached anteriorly to mandible and medial pterygoid plate (superior), hyoid bone (middle), and laryngeal cartilages (inferior), and run around to back of pharynx I—posterior median raphe of pharynx	Working as a group, all constrict pharynx during swallowing, which propels a food bolus to esophagus	Pharyngeal plexus [branches of vagus (X) and glossopharyngeal (IX) nerves]

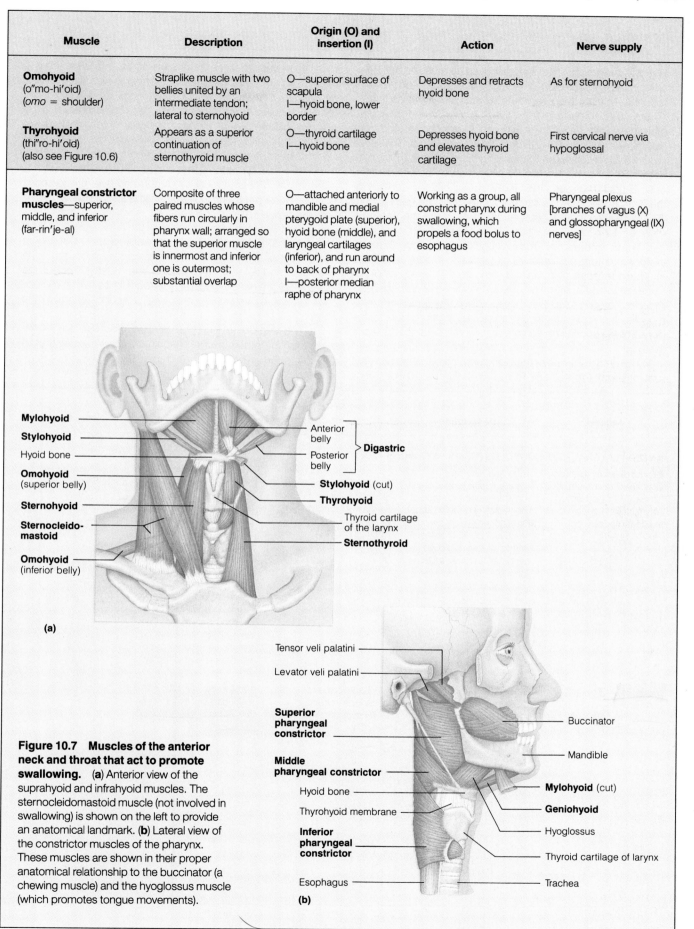

Figure 10.7 Muscles of the anterior neck and throat that act to promote swallowing. (a) Anterior view of the suprahyoid and infrahyoid muscles. The sternocleidomastoid muscle (not involved in swallowing) is shown on the left to provide an anatomical landmark. (b) Lateral view of the constrictor muscles of the pharynx. These muscles are shown in their proper anatomical relationship to the buccinator (a chewing muscle) and the hyoglossus muscle (which promotes tongue movements).

Table 10.4   Muscles of the Neck and Vertebral Column: Head Movements and Trunk Extension (Figure 10.8)

Head Movements. The head is moved by muscles that originate on the axial skeleton inferiorly. *Flexion* of the head is mainly brought about by the **sternocleidomastoid** muscles (Figure 10.8a), with some help from the suprahyoid and infrahyoid muscles (Table 10.3). The head can be *tilted* or *turned* from side to side as the neck is laterally flexed or rotated. These actions, which result when muscles contract on one side of the neck only, are performed by the sternocleidomastoids and the deeper neck muscles considered in this table. *Extension* of the head is aided by the trapezius muscle of the back (Figure 10.4a), but the main extensors of the head are the **splenius** muscles deep to the trapezius (Figure 10.8b).

Trunk Extension. Trunk extension is brought about by the *deep* or *intrinsic, muscles of the back* (Figure 10.8d and e). These muscles also maintain the normal curvatures of the spine, acting as postural muscles. As we consider these back muscles, keep in mind that they are deep; the superficial back muscles that cover them (Figure 10.4a) primarily run to the bones of the upper limb and are considered in Tables 10.8 and 10.9.

The deep muscles of the back form a broad, thick column running from the sacrum to the skull (Figure 10.8d). Many muscles of varying length contribute to this mass. It helps to regard each of these muscles as a string, which when pulled causes one or many vertebrae to extend or rotate on the vertebrae inferior to it. The largest of the deep back muscles is the **erector spinae** group. Many of the deep back muscles are long, so that large regions of the spine can be extended at once. When these muscles contract on just one side of the body, they help bend the spine laterally. Such lateral flexion is automatically accompanied by some rotation of the vertebral column. During vertebral movements, the articular facets of the vertebrae glide on each other.

In addition to the long muscles of the back are many short muscles that extend from one vertebra to the next (Figure 10.8e). These small muscles act primarily as synergists in the extension and rotation of the spine and as stabilizers of the spine. They are not described in this table, but you can deduce their actions by examining their origins and insertions in the figure.

The trunk muscles considered in this table are extensors. *Flexion* of the trunk is brought about by muscles that lie anterior to the vertebral column (Table 10.6).

Muscle	Description	Origin (O) and insertion (I)	Action	Nerve supply
ANTEROLATERAL NECK MUSCLES (Figure 10.8a and c)				
Sternocleidomastoid (ster"no-kli"do-mas'toid) (*sterno* = breastbone; *cleido* clavicle; *mastoid* = mastoid process)	Two-headed muscle located deep to platysma on anterolateral surface of neck; fleshy parts on either side of neck delineate limits of anterior and posterior triangles; key muscular landmark in neck; spasms of one of these muscles may cause torticollis (wryneck)	O—manubrium of sternum and medial portion of clavicle I—mastoid process of temporal bone	Prime mover of active head flexion; simultaneous contraction of both muscles causes neck flexion, generally against resistance as when one raises head when lying on back; (head flexion is ordinarily a result of combined effects of gravity and regulated relaxation of head extensors); acting alone, each muscle rotates head toward shoulder on opposite side and tilts or laterally flexes head to its own side	Accessory nerve (cranial nerve XI)
Scalenes (ska'lēnz)— anterior, middle, and posterior (*scalene* = uneven)	Located more laterally than anteriorly on neck; deep to platysma and sternocleidomastoid	O—transverse processes of cervical vertebrae I—anterolaterally on first two ribs	Elevate first two ribs (aid in inspiration); may be important in coughing; flex and rotate neck	Cervical nerves

Muscle	Description	Origin (O) and insertion (I)	Action	Nerve supply

INTRINSIC MUSCLES OF THE BACK (Figure 10.8b–e)

Splenius (sple′ne-us)— capitis and cervicis portions (kah-pit′us; ser-vĭs′us) (*splenion* = bandage; *caput* = head; *cervi* = neck) (Figure 10.8b)	Broad bipartite superficial muscle (capitis and cervicis parts) extending from upper thoracic vertebrae to skull; capitis portion known as "bandage muscle" because it covers and holds down deeper neck muscles	O—ligamentum nuchae,* spinous processes of vertebrae C_7–T_6 I—mastoid process of temporal bone and occipital bone (capitis); transverse processes of C_2–C_4 vertebrae (cervicis)	Act as a group to extend or hyperextend head; when splenius muscles on one side are activated, head is rotated and bent laterally toward same side	Cervical spinal nerves (dorsal rami)

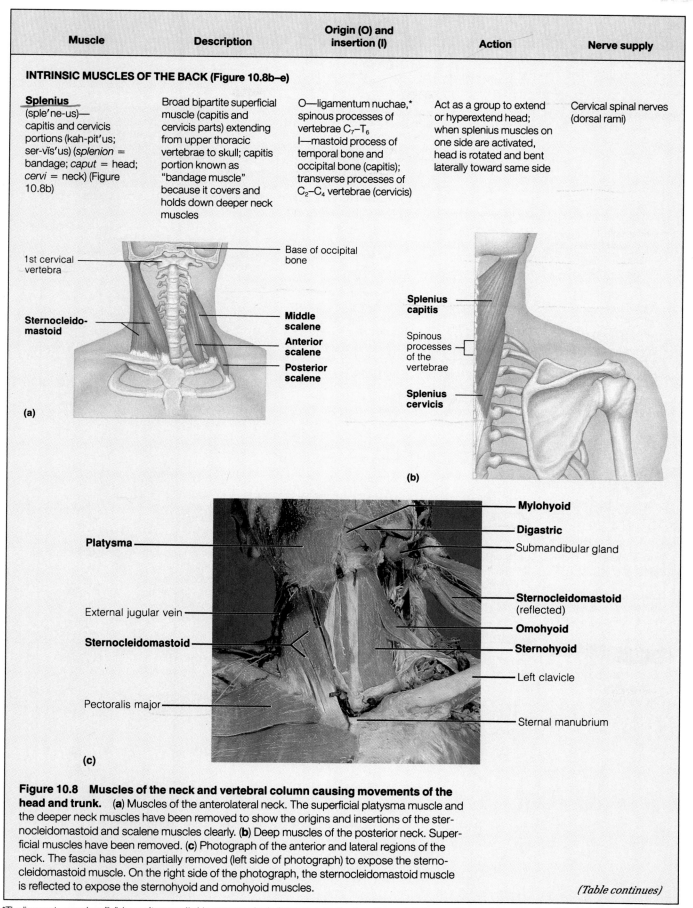

(a)

1st cervical vertebra

Base of occipital bone

Sternocleidomastoid

Middle scalene

Anterior scalene

Posterior scalene

(b)

Splenius capitis

Spinous processes of the vertebrae

Splenius cervicis

(c)

Platysma

External jugular vein

Sternocleidomastoid

Pectoralis major

Mylohyoid

Digastric

Submandibular gland

Sternocleidomastoid (reflected)

Omohyoid

Sternohyoid

Left clavicle

Sternal manubrium

Figure 10.8   Muscles of the neck and vertebral column causing movements of the head and trunk.   (**a**) Muscles of the anterolateral neck. The superficial platysma muscle and the deeper neck muscles have been removed to show the origins and insertions of the sternocleidomastoid and scalene muscles clearly. (**b**) Deep muscles of the posterior neck. Superficial muscles have been removed. (**c**) Photograph of the anterior and lateral regions of the neck. The fascia has been partially removed (left side of photograph) to expose the sternocleidomastoid muscle. On the right side of the photograph, the sternocleidomastoid muscle is reflected to expose the sternohyoid and omohyoid muscles.

(Table continues)

*The ligamentum nuchae (lig″ah-men′tum noo′ke) is a strong, elastic ligament extending from the occipital bone of the skull along the tips of the spinous processes of the cervical vertebrae. It binds the cervical vertebrae together and inhibits excessive head and neck flexion, thus preventing damage to the spinal cord in the vertebral canal.

Table 10.4 (continued)

Muscle	Description	Origin (O) and insertion (I)	Action	Nerve supply
Erector spinae (e-rek'tor spi'ne) (Figure 10.8d, left side)	Prime mover of back extension; erector spinae muscles, each consisting of three columns—the iliocostalis, longissimus, and spinalis muscles—form intermediate layer of intrinsic back muscles; erector spinae provide resistance that helps control action of bending forward at the waist and act as powerful extensors to promote return to erect position; during full flexion (i.e., when touching fingertips to floor), erector spinae are relaxed and strain is borne entirely by ligaments of back; on reversal of the movement, these muscles are initially inactive, and extension is initiated by hamstring muscles of thighs and gluteus maximus muscles of buttocks. As a result of this peculiarity, lifting a load or moving suddenly from a bent-over position is potentially dangerous (in terms of possible injury) to muscles and ligaments of back and intervertebral discs; erector spinae muscles readily go into painful spasms following injury to back structures			
• **Iliocostalis** (il″e-o-kos-tă′lis)— lumborum, thoracis, and cervicis portions (lum′bor-um; tho-ra′sis) (*ilio* = ilium; *cost* = rib)	Most lateral muscle group of erector spinae muscles; extend from pelvis to neck	O—iliac crests (lumborum); inferior 6 ribs (thoracis); ribs 3 to 6 (cervicis) I—angles of ribs (lumborum and thoracis); transverse processes of cervical vertebrae C_6–C_4 (cervicis)	Extend vertebral column, maintain erect posture; acting on one side, bend vertebral column to same side	Spinal nerves (dorsal rami)
• **Longissimus** (lon-jis′ ĭ-mus)— thoracis, cervicis, and capitis parts (*longissimus* = longest)	Intermediate tripartite muscle group of erector spinae; extend by many muscle slips from lumbar region to skull; mainly pass between transverse processes of the vertebrae	O—transverse processes of lumbar through cervical vertebrae I—transverse processes of thoracic or cervical vertebrae and to ribs superior to origin as indicated by name; capitis inserts into mastoid process of temporal bone	Thoracis and cervicis act together to extend vertebral column and acting on one side, bend it laterally; capitis extends head and turns the face toward same side	Spinal nerves (dorsal rami)
• **Spinalis** (spi-nă′lis)— thoracis and cervicis parts (*spin* = vertebral column, spine)	Most medial muscle column of erector spinae; cervicis usually rudimentary and poorly defined	O—spines of upper lumbar and lower thoracic vertebrae I—spines of upper thoracic and cervical vertebrae	Extends vertebral column	Spinal nerves (dorsal rami)
Semispinalis (sem′e-spĭ-nă′lis)— thoracis, cervicis, and capitis regions (*semi* = half; *thorac* = thorax) (Figure 10.8d, right side)	Composite muscle forming part of deep layer of intrinsic back muscles; extends from thoracic region to head	O—transverse processes of C_7–T_{12} I—occipital bone (capitis) and spinous processes of cervical (cervicis) and thoracic vertebrae T_1 to T_4 (thoracis)	Extends vertebral column and head and rotates them to opposite side; acts synergistically with sternocleidomastoid muscles of opposite side	Spinal nerves (dorsal rami)
Quadratus lumborum (kwod-ra′tus lum-bor′um) (*quad* = four-sided; *lumb* = lumbar region) (See also Figure 10.18a)	Fleshy muscle forming part of posterior abdominal wall	O—iliac crest and ilio-lumbar fascia I—transverse processes of upper lumbar vertebrae and lower margin of 12th rib	Flexes vertebral column laterally when acting separately; when pair acts jointly, lumbar spine is extended and 12th rib is fixed; maintains upright posture	T_{12} and upper lumbar spinal nerves (ventral rami)

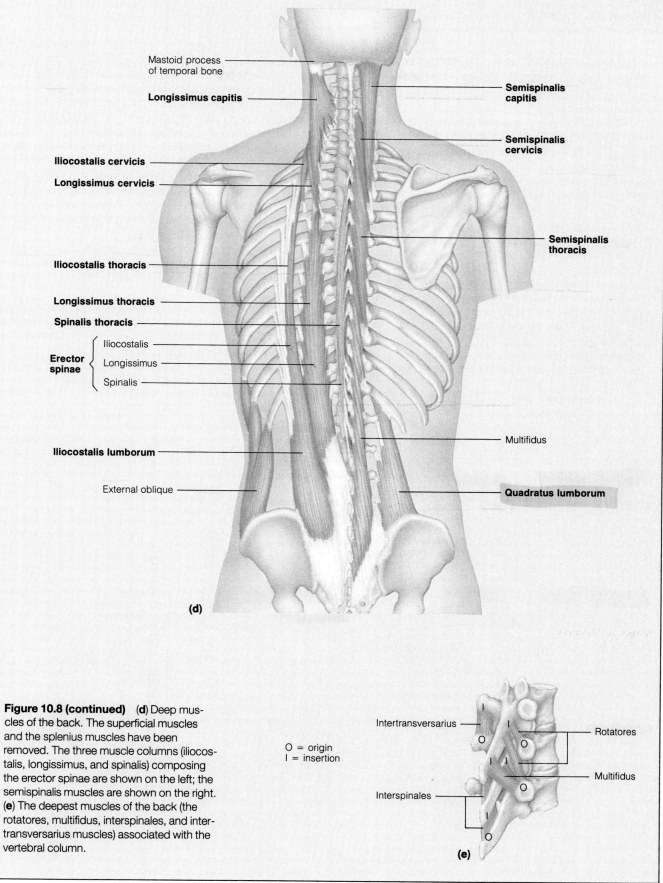

Mastoid process
of temporal bone

Longissimus capitis

Iliocostalis cervicis

Longissimus cervicis

Iliocostalis thoracis

Longissimus thoracis

Spinalis thoracis

**Erector
spinae** ⎰ Iliocostalis

Longissimus

Spinalis

Iliocostalis lumborum

External oblique

**Semispinalis
capitis**

**Semispinalis
cervicis**

**Semispinalis
thoracis**

Multifidus

Quadratus lumborum

(d)

Figure 10.8 (continued) **(d)** Deep muscles of the back. The superficial muscles and the splenius muscles have been removed. The three muscle columns (iliocostalis, longissimus, and spinalis) composing the erector spinae are shown on the left; the semispinalis muscles are shown on the right. **(e)** The deepest muscles of the back (the rotatores, multifidus, interspinales, and intertransversarius muscles) associated with the vertebral column.

O = origin
I = insertion

Intertransversarius

Interspinales

Rotatores

Multifidus

(e)

Go TO Pg 252

Table 10.5   Deep Muscles of the Thorax: Breathing (Figure 10.9)

An important function of the deep muscles of the thorax is to provide the movements necessary for ventilation, or breathing. Breathing has two phases—inspiration, or inhaling, and expiration, or exhaling—caused by cyclical changes in the volume of the thoracic cavity.

The thoracic muscles are very short: Most run only from one rib to the next. They form three layers in the wall of the thorax. The eleven **external intercostal muscles** form the most superficial layer. Their function is controversial, but they seem to lift the rib cage, which increases its anterior-posterior and lateral dimensions. This enlargement creates a vacuum that draws air into the lungs. Thus, the external intercostals seem to be inspiratory muscles. The eleven **internal intercostals** form the intermediate muscle layer. They may aid expiration during heavy breathing by depressing the rib cage, which decreases thoracic volume and helps expel air. The internal intercostals do not operate in normal quiet expiration, however. This is a passive phenomenon, resulting only from elastic recoil of the lungs. The third and deepest muscle layer of the thoracic wall attaches to the internal surfaces of the ribs. It has three discontinuous parts (from posterior to anterior): the *subcostals*, *innermost intercostals*, and *transversus thoracis*. These are small, and their function is unknown, so they are not listed in this table.

The **diaphragm,** the most important muscle of respiration (Figures 10.9b, 10.9c, and 1.4), forms a complete partition between the thoracic and abdominopelvic cavities. In the relaxed state, the diaphragm is dome-shaped, but it flattens as it contracts, increasing the volume of the thoracic cavity. Thus, the diaphragm is a powerful muscle of inspiration. It contracts rhythmically during respiration, but one can also contract it voluntarily to push down on the abdominal viscera and increase the pressure in the abdominopelvic cavity. This pressure helps to evacuate the contents of the pelvic organs (feces, urine, or a baby). It also helps in lifting heavy weights: When one takes a deep breath to fix the diaphragm, the abdomen becomes a firm pillar that will not buckle under the weight being lifted. The muscles of the anterior abdominal wall also help to increase the intra-abdominal pressure (Table 10.6).

With the exception of the diaphragm, which is innervated by *phrenic nerves* from the neck, all muscles described in this table are served by nerves running between the ribs, called *intercostal nerves*.

When breathing is forced and heavy, as during exercise, additional muscles become active in ventilation. For example, in forced inspiration, the scalene and sternocleidomastoid muscles of the neck help lift the ribs. Forced expiration is aided by muscles that pull inferiorly on the ribs (quadratus lumborum) or push the diaphragm superiorly by compressing the abdominal organs (abdominal wall muscles).

Muscle	Description	Origin (O) and insertion (I)	Action	Nerve supply
External intercostals (in″ter-kos′talz) (*external* = toward the outside; *inter* = between; *cost* = rib)	11 pairs lie between ribs; fibers run obliquely (downward and forward) from each rib to rib below; in lower intercostal spaces, fibers are continuous with external oblique muscle forming part of abdominal wall	O—inferior border of rib above I—superior border of rib below	With first ribs fixed by scalene muscles, pull ribs toward one another to elevate rib cage; aids in inspiration; synergists of diaphragm	Intercostal nerves
Internal intercostals (*internal* = toward the inside, deep)	11 pairs lie between ribs; fibers run deep to and at right angles to those of external intercostals (i.e., run downward and posteriorly); lower internal intercostal muscles are continuous with fibers of internal oblique muscle of abdominal wall	O—superior border of rib below I—inferior border (costal groove) of rib above	With 12th ribs fixed by quadratus lumborum, muscles of posterior abdominal wall and oblique muscles of the abdominal wall, draw ribs together and depress rib cage; aid in expiration; antagonistic to external intercostals	Intercostal nerves
Diaphragm (di′ah-fram) (*dia* = across; *phragm* = partition)	Broad muscle; forms floor of thoracic cavity; in relaxed state is dome-shaped; fibers converge from margins of thoracic cage toward a boomerang-shaped central tendon	O—inferior border of rib cage and sternum, costal cartilages of last six ribs and lumbar vertebrae I—central tendon	Prime mover of inspiration; flattens on contraction, increasing vertical dimensions of thorax; when strongly contracted, dramatically increases intra-abdominal pressure	Phrenic nerves

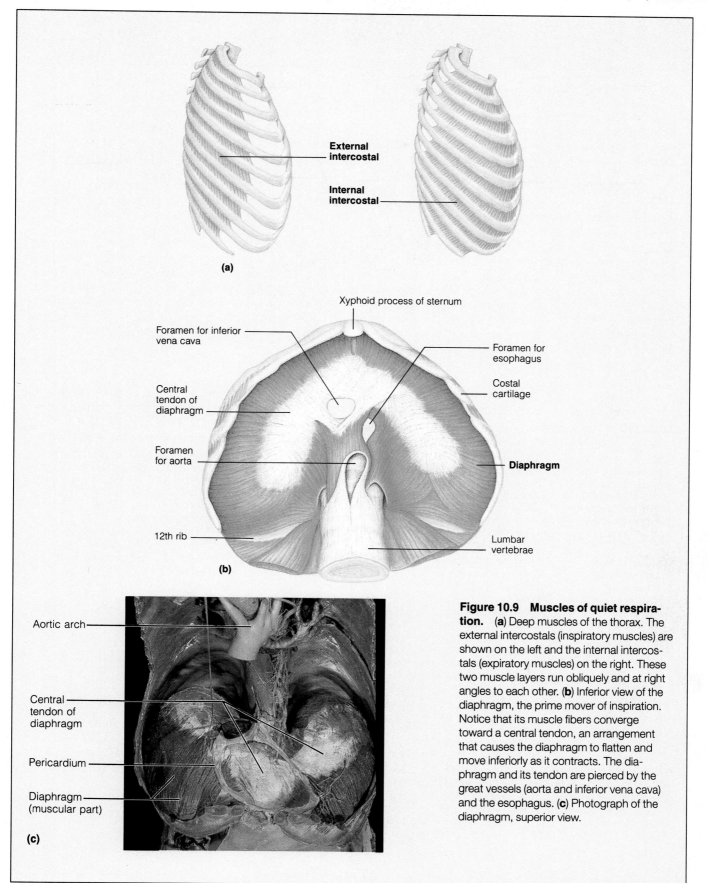

External intercostal

Internal intercostal

(a)

Xyphoid process of sternum

Foramen for inferior vena cava

Foramen for esophagus

Central tendon of diaphragm

Costal cartilage

Foramen for aorta

Diaphragm

12th rib

Lumbar vertebrae

(b)

Aortic arch

Central tendon of diaphragm

Pericardium

Diaphragm (muscular part)

(c)

Figure 10.9 Muscles of quiet respiration. **(a)** Deep muscles of the thorax. The external intercostals (inspiratory muscles) are shown on the left and the internal intercostals (expiratory muscles) on the right. These two muscle layers run obliquely and at right angles to each other. **(b)** Inferior view of the diaphragm, the prime mover of inspiration. Notice that its muscle fibers converge toward a central tendon, an arrangement that causes the diaphragm to flatten and move inferiorly as it contracts. The diaphragm and its tendon are pierced by the great vessels (aorta and inferior vena cava) and the esophagus. **(c)** Photograph of the diaphragm, superior view.

Table 10.6 Muscles of the Abdominal Wall: Trunk Movements and Compression of Abdominal Viscera (Figure 10.10)

The walls of the abdomen have no bony reinforcements (no ribs). Instead, the abdominal wall is a composite of sheet-like muscles. Forming the lateral part of the abdominal wall are three broad, flat muscle sheets, layered one on the next: the **external oblique, internal oblique,** and **transversus abdominis.** These are direct continuations of the three intercostal muscles in the thorax. The fascicles of the external oblique run inferomedially, at right angles to fascicles of the internal oblique muscle (which primarily run superomedially), whereas the fascicles of the deep transversus abdominis are strictly horizontal. These three muscles have different fascicle directions, an arrangement that lends strength to the abdominal wall. (Similarly, plywood is made of sheets with different grains and is stronger than regular wood.) These three muscles end anteriorly in broad, white aponeuroses (tendon-like sheets). The aponeuroses enclose a fourth muscle pair, the strap-like **rectus abdominis,** and extend medially to insert on the **linea alba** ("white line"), a tendinous raphe (seam) that runs vertically from the sternum to the pubic symphysis. The tight enclosure of the rectus abdominis within aponeuroses ensures that this vertical muscle cannot "bowstring" when it contracts to flex the trunk.

The four muscles of the abdominal wall perform many functions. They help contain the abdominal organs. These muscles also flex the trunk; therefore, performing sit-ups will cause the rectus abdominis to bulge beneath the skin. Other functions include lateral flexion and rotation of the trunk. In addition, these muscles help produce forced, heavy breathing: When they contract simultaneously, they pull the ribs inferiorly and squeeze the abdominal contents. This action in turn pushes the diaphragm superiorly, aiding forced expiration. When the abdominal muscles contract with the diaphragm, the increased intra-abdominal pressure helps promote micturition (voiding urine), defecation, vomiting, childbirth, sneezing, coughing, laughing, burping, screaming, and nose blowing. Next time you perform one of these activities, feel the abdominal wall muscles contract under your skin. These muscles also contract during heavy lifting—sometimes so forcefully that hernias result.

Muscle	Description	Origin (O) and insertion (I)	Action	Nerve supply
MUSCLES OF THE ANTERIOR AND LATERAL ABDOMINAL WALL				
	Four paired flat muscles; very important in supporting and protecting abdominal viscera and play an important role in promoting movement of vertebral column (flexion and lateral bending)			
Rectus abdominis (rek′tus ab-dom′ĭ-nis) (*rectus* = straight; *abdom* = abdomen)	Medial superficial muscle pair; extend from pubis to rib cage; ensheathed by aponeuroses of lateral muscles; segmented by three reinforcing tendinous intersections	O—pubic crest and symphysis I—xiphoid process and costal cartilages of ribs 5–7	Flex and rotate lumbar region of vertebral column; fix and depress ribs, stabilize pelvis during walking, increase intra-abdominal pressure	Intercostal nerves (T_6 or T_7–T_{12})
External oblique (o-blēk′) (*external* = toward outside; *oblique* = running at an angle)	Largest and most superficial of the three lateral muscles; fibers run downward and medially (same direction outstretched fingers take when hands put into pants pockets); aponeurosis turns under inferiorly forming inguinal ligament	O—by fleshy strips from outer surfaces of lower eight ribs I—most fibers into linea alba; some into pubic tubercle and iliac crest; majority of fibers insert anteriorly via a broad aponeurosis	When pair contracts simultaneously, aid rectus abdominis muscles in flexing vertebral column and in compressing abdominal wall and increasing intra-abdominal pressure; acting individually, aid muscles of back in trunk rotation and lateral flexion	Intercostal nerves (T_7–T_{12})
Internal oblique (*internal* = toward the inside; deep)	Fibers fan upward and forward and run at right angles to those of external oblique, which it underlies	O—lumbodorsal fascia, iliac crest, and inguinal ligament I—linea alba, pubic crest, last three ribs	As for external oblique	Intercostal nerves (T_7–T_{12}) and L_1
Transversus abdominis (trans-ver′sus) (*transverse* = running straight across or transversely)	Deepest (innermost) muscle of abdominal wall; fibers run horizontally	O—inguinal ligament, lumbodorsal fascia, cartilages of last six ribs; iliac crest I—linea alba, pubic crest	Compresses abdominal contents	Intercostal nerves (T_7–T_{12})

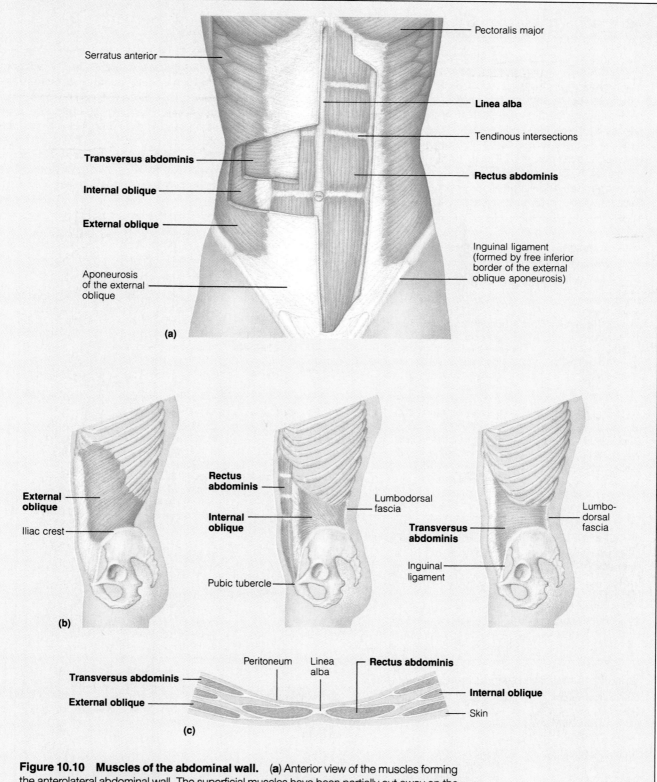

Figure 10.10 Muscles of the abdominal wall. (**a**) Anterior view of the muscles forming the anterolateral abdominal wall. The superficial muscles have been partially cut away on the left side of the diagram to reveal the deeper internal oblique and transversus abdominis muscles. (**b**) Lateral view of the trunk, illustrating the fiber direction and attachments of the external oblique, internal oblique, and transversus abdominis muscles. (**c**) Transverse section through the anterolateral abdominal wall (midregion), showing how the aponeuroses of the lateral abdominal muscles contribute to the rectus abdominis sheath.

Table 10.7   Muscles of the Pelvic Floor and Perineum: Support of Abdominopelvic Organs (Figure 10.11)

The pelvic floor is the **pelvic diaphragm** (Figure 10.11a), a sheet consisting of two muscles, the **levator ani** and the small **coccygeus.** This diaphragm supports the pelvic organs, seals the inferior opening of the bony pelvis, and lifts superiorly to help release feces during defecation. The pelvic diaphragm is pierced by the rectum and urethra (the tube for urine) and (in females) by the vagina. The body region inferior to the pelvic diaphragm is the *perineum* (Figure 10.11b and c). In the anterior half of the perineum is a triangular sheet of muscle called the **urogenital diaphragm.** This contains the **sphincter urethrae** muscle, which surrounds the urethra. One uses this

muscle voluntarily to prevent urination. Just inferior to the urogenital diaphragm is the superficial perineal space, which contains muscles **(bulbospongiosus, ischiocavernosus)** that help maintain erection of the penis and clitoris. In the posterior half of the perineum, circling the anus, lies the **external anal sphincter** (Figure 10.11b and p. 584). This muscle is used voluntarily to prevent defecation. Just anterior to this sphincter, at the exact midpoint of the perineum, is the **central tendon.** Many perineal muscles insert on this strong tendon and, in so doing, are able to support the heavy organs in the pelvis.

Muscle	Description	Origin (O) and insertion (I)	Action	Nerve supply
MUSCLES OF THE PELVIC DIAPHRAGM (Figure 10.11a)				
Levator ani (le-va′tor a′ne) (*levator* = raises; *ani* = anus)	Broad, thin, tripartite muscle (pubococcygeus, puborectalis, and iliococcygeus parts); its fibers extend inferomedially, forming a muscular "sling" around male prostate (or female vagina), urethra, and anorectal junction before meeting in the median plane	O—extensive linear origin inside pelvis from pubis to ischial spine. I—inner surface of coccyx, levator ani of opposite side, and (in part) into the structures that penetrate it	Supports and maintains position of pelvic viscera; resists downward thrusts that accompany rises in intrapelvic pressure during coughing, vomiting, and expulsive efforts of abdominal muscles; forms sphincters at anorectal junction and vagina	S_3, S_4, and inferior rectal nerve (branch of pudendal nerve)
Coccygeus (kok-sij′e-us) (*coccy* = coccyx)	Small triangular muscle lying posterior to levator ani; forms posterior part of pelvic diaphragm	O—spine of ischium I—sacrum and coccyx	Assists levator ani in supporting pelvic viscera; supports coccyx and pulls it forward after it has been reflected posteriorly by defecation and childbirth	S_4 and S_5
MUSCLES OF THE UROGENITAL DIAPHRAGM (Figure 10.11b)				
Deep transverse perineus (per″ĭ-ne′us) (*deep* = far from surface; *transverse* = across; *perine* = near anus)	Together the pair spans distance between ischial rami; in females, lies posterior to vagina	O—ischial rami I—midline central tendon of perineum; some fibers into vaginal wall in females	Supports pelvic organs; steadies central tendon	Pudendal nerve
Sphincter urethrae (*sphin* = squeeze)	Circular muscle surrounding membranous urethra and vagina (female)	O—ischiopubic rami I—midline raphe	Constricts urethra; helps support pelvic organs	Pudendal nerve
MUSCLES OF THE SUPERFICIAL SPACE (Figure 10.11c)				
Ischiocavernosus (is′ke-o-kav′ern-o′sus) (*ischi* = hip; *caverna* = hollow chamber)	Runs from pelvis to base of penis or clitoris	O—ischial tuberosities I—crus of corpus cavernosa of male penis or female clitoris	Retards venous drainage and maintains erection of penis or clitoris	Pudendal nerve
Bulbospongiosus (bul″bo-spun″je-o′sus) (*bulbon* = bulb; *spongio* = sponge)	Encloses base of penis (bulb) in males and lies deep to labia in females	O—central tendon of perineum and midline raphe of male penis I—anteriorly into corpus sponglosum of penis, or onto clitoris	Empties male urethra; assists in erection of penis in males and of clitoris in females	Pudendal nerve
Superficial transverse perineus (*superficial* = closer to surface)	Paired muscle bands posterior to urethral (and in females, vaginal) opening	O—ischial tuberosity I—central tendon of perineum	Stabilize and strengthen central tendon of perineum	Pudendal nerve

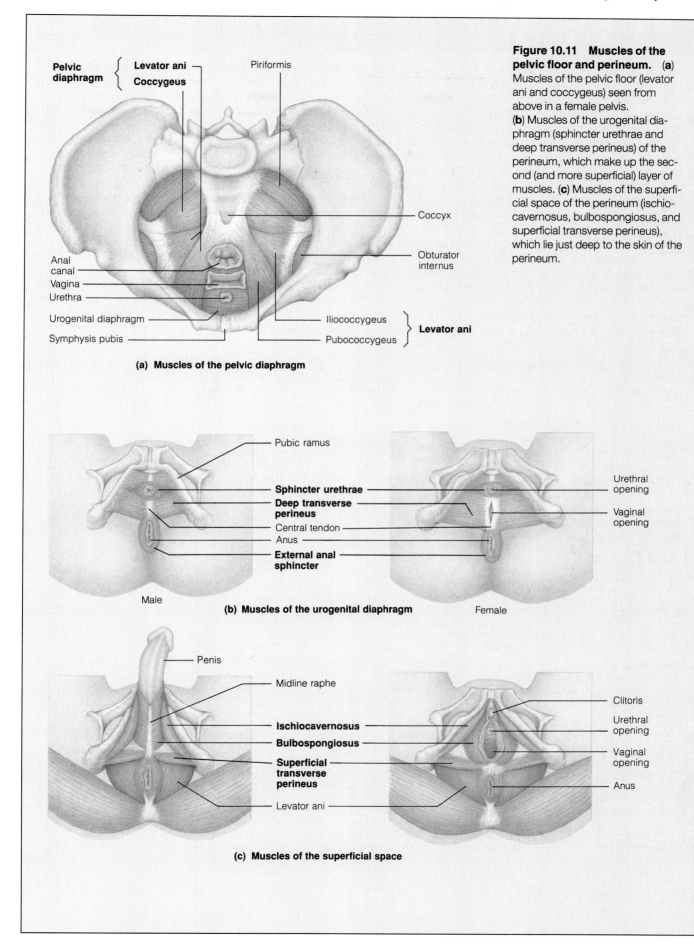

Figure 10.11 Muscles of the pelvic floor and perineum. (a) Muscles of the pelvic floor (levator ani and coccygeus) seen from above in a female pelvis. (b) Muscles of the urogenital diaphragm (sphincter urethrae and deep transverse perineus) of the perineum, which make up the second (and more superficial) layer of muscles. (c) Muscles of the superficial space of the perineum (ischiocavernosus, bulbospongiosus, and superficial transverse perineus), which lie just deep to the skin of the perineum.

(a) Muscles of the pelvic diaphragm

(b) Muscles of the urogenital diaphragm

(c) Muscles of the superficial space

Table 10.8   Superficial Muscles of the Anterior and Posterior Thorax: Movements of the Scapula (Figure 10.12)

Most superficial muscles on the thorax are *extrinsic shoulder muscles,* which run from the ribs and vertebral column to the shoulder girdle. They fix the scapula in place but also move the scapula to increase the arm's range of movements. The *anterior* muscles of this group are **pectoralis major, pectoralis minor, serratus anterior,** and **subclavius** (Figure 10.12a). The *posterior* muscles are **trapezius, latissimus dorsi, levator scapulae,** and the **rhomboids** (Figure 10.12b). The pectoralis major and latissimus dorsi move the scapula indirectly by moving the arm, so they are considered in Table 10.9.

The scapula undergoes a variety of movements on the posterior rib cage. In *elevation* (superior movement) of the scapula, the prime movers are the levator scapulae and the superior fascicles of the trapezius; the rhomboids are synergists. Note that the fascicles of all of these muscles run inferolaterally (Figure 10.12b); thus, they are ideal elevators of the scapula. *Depression* (inferior movement) of the scapula is mostly due to gravity (weight of the arm). When the scapula must be depressed against resistance, however, this action is done by the serratus anterior, pectoralis minor, the inferior part of the trapezius, and—

especially—the latissimus dorsi. The fascicles of all of these muscles run superiorly to insert on the scapula or upper humerus, so it is logical that they depress the scapula. *Protraction* or *abduction* of the scapula moves it laterally and anteriorly, as in punching. This action is mainly carried out by the serratus anterior, whose horizontal fibers pull the scapula anterolaterally. *Retraction* or *adduction* of the scapula moves it medially and posteriorly. This is brought about by the mid part of the trapezius and the superior part of the latissimus dorsi, whose fibers run horizontally from the vertebral column to pull the scapula and humerus medially. *Upward rotation* of the scapula (in which the scapula's inferior angle moves laterally) allows one to lift the arm above the head: The serratus anterior swings the inferior angle laterally, while the superior part of the trapezius, gripping the scapular spine, pulls the top of the scapula medially. In *downward rotation* of the scapula (where the inferior angle swings medially, as in paddling a canoe), the rhomboids pull the inferior part of the scapula medially, while the levator scapulae steadies the superior part.

Muscle	Description	Origin (O) and insertion (I)	Action	Nerve supply
MUSCLES OF THE ANTERIOR THORAX (Figure 10.12a)		CARTILAGES		
Pectoralis minor (pek"to-ra'lis mi'nor) (*pectus* = chest, breast; *minor* = lesser)	Flat, thin muscle directly beneath and obscured by pectoralis major	O—anterior surfaces of ribs 3–5 I—coracoid process of scapula	With ribs fixed, draws scapula forward and downward; with scapula fixed, draws rib cage superiorly	Medial pectoral nerve (C$_6$–C$_8$)
Serratus anterior (ser-a'tis) (*serratus* = saw)	Lies deep to scapula beneath and inferior to pectoral muscles on lateral rib cage; forms medial wall of axilla; origins have serrated, or sawtooth, appearance; paralysis results in "winging" of vertebral border of scapula away from chest wall, making arm elevation impossible	O—by a series of muscle slips from ribs 1–8 (or 9) I—entire anterior surface of vertebral border of scapula	Prime mover to protract and hold scapula against chest wall; rotates scapula so that its inferior angle moves laterally and upward; raises point of shoulder; important role in abduction and raising of arm and in horizontal arm movements (pushing, punching); called "boxer's muscle"	Long thoracic nerve (C$_5$–C$_7$)
Subclavius (sub-kla've-us) (*sub* = under, beneath; *clav* = clavicle)	Small cylindrical muscle extending from rib 1 to clavicle	O—costal cartilage of rib 1 I—groove on inferior surface of clavicle	Helps stabilize and depress pectoral girdle; paralysis produces no obvious effects	Nerve to subclavius (C$_5$ and C$_6$)
MUSCLES OF THE POSTERIOR THORAX (Figure 10.12b)				
Trapezius (trah-pe'ze-is) (*trapezion* = irregular four-sided figure)	Most superficial muscle of posterior thorax; flat, and triangular in shape; upper fibers run downward to scapula; middle fibers run horizontally to scapula; lower fibers run superiorly to scapula	O—occipital bone, ligamentum nuchae, and spines of C$_7$ and all thoracic vertebrae I—a continuous insertion along acromion and spine of scapula and lateral third of clavicle	Stabilizes, raises, retracts, and rotates scapula; middle fibers retract (adduct) scapula; superior fibers elevate scapula or can help extend head; inferior fibers depress scapula (and shoulder)	Accessory nerve (cranial nerve XI); C$_3$ and C$_4$

Muscle	Description	Origin (O) and insertion (I)	Action	Nerve supply
Levator scapulae (skap'u-le) (*levator* = raises)	Located at back and side of neck, deep to trapezius; thick, straplike muscle	O—transverse processes of C_1–C_4 I—superior vertebral border of scapula	Elevates/adducts scapula in concert with superior fibers of trapezius; tilts glenoid cavity downward; when scapula is fixed, flexes neck to same side	Cervical spinal nerves and dorsal scapular nerve (C_3–C_5)
Rhomboids (rom'boidz)— major and minor (*rhomboid* = diamond-shaped)	Two rectangular muscles lying deep to trapezius and inferior to levator scapulae; rhomboid minor is the more superior muscle; both muscles extend from vertebral column to scapula	O—spinous processes of C_7 and T_1 (minor) and spinous processes of T_2–T_5 (major) I—medial border of scapula	Act together (and with middle trapezius fibers) to retract scapula, thus "squaring shoulders"; rotate scapulae so that glenoid cavity rotates downward (as when arm is lowered against resistance; e.g., paddling a canoe); stabilize scapula	Dorsal scapular nerve (C_5)

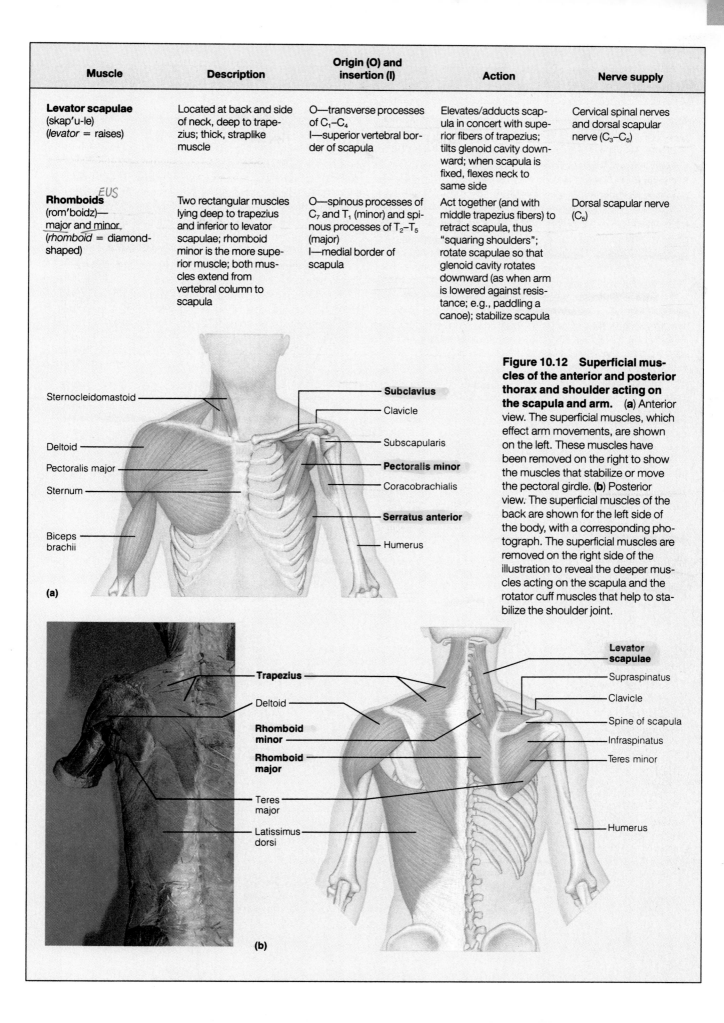

(a)

Sternocleidomastoid

Deltoid

Pectoralis major

Sternum

Biceps brachii

Subclavius

Clavicle

Subscapularis

Pectoralis minor

Coracobrachialis

Serratus anterior

Humerus

Figure 10.12 Superficial muscles of the anterior and posterior thorax and shoulder acting on the scapula and arm. (a) Anterior view. The superficial muscles, which effect arm movements, are shown on the left. These muscles have been removed on the right to show the muscles that stabilize or move the pectoral girdle. (b) Posterior view. The superficial muscles of the back are shown for the left side of the body, with a corresponding photograph. The superficial muscles are removed on the right side of the illustration to reveal the deeper muscles acting on the scapula and the rotator cuff muscles that help to stabilize the shoulder joint.

Trapezius

Deltoid

Rhomboid minor

Rhomboid major

Teres major

Latissimus dorsi

Levator scapulae

Supraspinatus

Clavicle

Spine of scapula

Infraspinatus

Teres minor

Humerus

(b)

Table 10.9   Muscles Crossing the Shoulder Joint: Movements of the Arm (Humerus) (Figure 10.13)

Recall that the shoulder joint is the most flexible joint in the body but pays the price of instability. Many muscles cross each shoulder joint to insert on the humerus. All muscles that act on the humerus originate from the pectoral girdle (although two of these muscles, the latissimus dorsi and pectoralis major, primarily originate from the axial skeleton).

As we consider the arm movements, remember that the term *arm* refers to the *upper* arm. Of the nine muscles covered here, the three largest ones are powerful prime movers: the **pectoralis major, latissimus dorsi,** and **deltoid** (Figure 10.13a and d). The other six are synergists and fixators. Four of these were introduced on p. 202 as muscles of the *rotator cuff:* **supraspinatus, infraspinatus, subscapularis,** and **teres minor.** This cuff reinforces the capsule of the shoulder joint to prevent dislocation of the humerus. The remaining two muscles are the **teres major** and **coracobrachialis.**

Generally speaking, muscles that originate *anterior* to the shoulder joint *flex* the arm (lift it anteriorly). These flexors include the pectoralis major, the anterior fibers of the deltoid, and the coracobrachialis. Muscles that originate *posterior* to the shoulder joint, by contrast, tend to *extend* the arm. These extensors include the latissimus dorsi, the posterior fibers of deltoid, and the teres major. The middle region of the deltoid muscle is the prime *abductor* of the arm and extends over the superior and lateral side of the humerus to pull it laterally. The main arm *adductors* are the pectoralis major anteriorly and the latissimus dorsi posteriorly. *Lateral* and *medial* rotation of the arm are primarily brought about by the small muscles.

The interactions among these nine muscles are complex, and each contributes to several movements. A summary of their actions is given on page 272 in Table 10.12 (Part I).

Muscle	Description	Origin (O) and insertion (I)	Action	Nerve supply
Pectoralis major (pek"to-ra'lis ma'jer) (*pect* = breast, chest; *major* = larger)	Large, fan-shaped muscle covering upper portion of chest; forms anterior axillary fold	O—clavicle, sternum, cartilage of ribs 1–6, and aponeurosis of external oblique muscle I—fibers converge to insert by a short tendon into greater tubercle of humerus	Prime mover of arm flexion; rotates arm medially; adducts arm against resistance; with scapula (and arm) fixed, pulls rib cage upward, thus can help in climbing, throwing, pushing, and in forced inspiration	Lateral and medial pectoral nerves
Latissimus dorsi (lah-tis'ĭ-mus dor'si) (*latissimus* = widest; *dorsi* = back)	Broad, flat, triangular muscle of lower back (lumbar region); extensive superficial origins; covered by trapezius superiorly; contributes to the posterior wall of axilla	O—indirect attachment via lumbodorsal fascia into spines of lower six thoracic vertebrae, lumbar vertebrae, lower 3 to 4 ribs, and iliac crest; also from scapula's inferior angle I—spirals around teres major to insert in floor of intertubercular groove of humerus	Prime mover of arm extension; powerful arm adductor; medially rotates arm at shoulder; depresses scapula; because of its power in these movements, it plays an important role in bringing the arm down in a power stroke, as in striking a blow, hammering, swimming, and rowing	Thoracodorsal nerve

Muscle	Description	Origin (O) and insertion (I)	Action	Nerve supply
Deltoid (del′toid) (*delta* = triangular)	Thick, multipennate muscle forming rounded shoulder muscle mass; responsible for roundness of shoulder; a site commonly used for intramuscular injection, particularly in males, where it tends to be quite fleshy	O—embraces insertion of the trapezius; lateral third of clavicle; acromion and spine of scapula I—deltoid tuberosity of humerus	Prime mover of arm abduction when all its fibers contract simultaneously; antagonist of pectoralis major and latissimus dorsi, which adduct the arm; if only anterior fibers are active, can act powerfully in flexion and medial rotation of humerus, therefore synergist of pectoralis major; if only posterior fibers are active, effects extension and lateral rotation of arm; active during rhythmic arm swinging movements during walking	Axillary nerve (C_5 and C_6)
Subscapularis (sub-scap″u-lar′is) (*sub* = under; *scapular* = scapula)	Forms part of posterior wall of axilla; tendon of insertion passes in front of shoulder joint; a rotator cuff muscle	O—subscapular fossa of scapula I—lesser tubercle of humerus	Chief medial rotator of humerus; assisted by pectoralis major; helps to hold head of humerus in glenoid cavity, thereby stabilizing shoulder joint	Subscapular nerves (C_5–C_7)
Supraspinatus (soo″prah-spi-nah′tus) (*supra* = above, over; *spin* = spine)	Named for its location on posterior aspect of scapula; deep to trapezius; a rotator cuff muscle	O—supraspinous fossa of scapula I—superior part of greater tubercle of humerus	Stabilizes shoulder joint; helps to prevent downward dislocation of humerus, as when carrying a heavy suitcase; assists in abduction	Suprascapular nerve
Infraspinatus (in″frah-spi-nah′tus) (*infra* = below)	Partially covered by deltoid and trapezius; named for its scapular location; a rotator cuff muscle	O—infraspinous fossa of scapula I—greater tubercle of humerus posterior to insertion of supraspinatus	Helps to hold head of humerus in glenoid cavity; rotates humerus laterally	Suprascapular nerve
Teres minor (te′rēez) (*teres* = round; *minor* = lesser)	Small, elongated muscle; lies inferior to infraspinatus and may be inseparable from that muscle; a rotator cuff muscle	O—lateral border of dorsal scapular surface I—greater tubercle of humerus inferior to infraspinatus insertion	Same action(s) as infraspinatus muscle	Axillary nerve
Teres major	Thick, rounded muscle; located inferior to teres minor; helps to form posterior wall of axilla (along with latissimus dorsi and subscapularis)	O—posterior surface of scapula at inferior angle I—crest of lesser tubercle on anterior humerus; insertion tendon fused with that of latissimus dorsi	Extends, medially rotates, and adducts humerus; synergist of latissimus dorsi	Lower subscapular nerve
Coracobrachialis (kor″ah-ko-bra″ke-al′is) (*coraco* = coracoid; *brachi* = arm)	Small, cylindrical muscle	O—coracoid process of scapula I—medial surface of humerus shaft	Flexion and adduction of the humerus; synergist of pectoralis major	Musculocutaneous nerve

(Table continues)

Table 10.9 (continued)

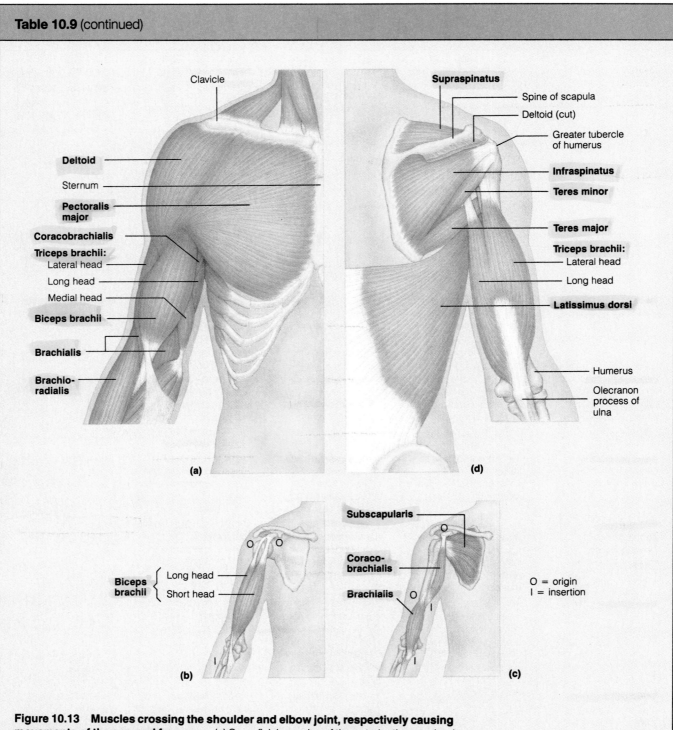

Figure 10.13   Muscles crossing the shoulder and elbow joint, respectively causing movements of the arm and forearm. (a) Superficial muscles of the anterior thorax, shoulder, and arm, anterior view. (b) The biceps brachii muscle of the anterior arm, shown in isolation. (c) The brachialis muscle arising from the humerus and the coracobrachialis and subscapularis muscles arising from the scapula, shown in isolation. (Note, however, that the coracobrachialis effects movements of the arm, not the forearm, and the subscapularis stabilizes the shoulder joint.) (d) The extent of the triceps brachii muscle of the posterior arm, shown in relation to the deep scapular muscles. The deltoid muscle of the shoulder has been removed.

Table 10.10   Muscles Crossing the Elbow Joint: Flexion and Extension of the Forearm (Figure 10.13)

This table focuses on muscles that lie in the arm but move the forearm. These muscles cross the elbow joint to insert on the forearm bones, which they flex and extend. Walls of fascia divide the arm into two muscle compartments: the *posterior extensors* and *anterior flexors*. The main *extensor* of the forearm is the **triceps brachii.**

The anterior muscles of the arm *flex* the forearm. In order of decreasing strength, these are the **brachialis, biceps brachii,** and **brachioradialis.** The biceps brachii and brachialis insert on the radius and ulna, respectively, and contract together. The brachioradialis (Figure 10.14a) is a weak flexor of the forearm and actually lies in the forearm more than in the arm.

The biceps brachii not only flexes the forearm but also *supinates* it by rotating the radius at the proximal radioulnar joint. This muscle cannot flex the forearm without also supinating it, so it is ineffective when one lifts a heavy object with a pronated hand that must stay pronated. (This is why doing chin-ups with palms facing anteriorly is harder than with palms facing posteriorly.)

The biceps and triceps also cross the shoulder joint, but they cause only weak movements at the shoulder.

The actions of the muscles in this table are summarized on page 272 in Table 10.12 (Part II).

Muscle	Description	Origin (O) and insertion (I)	Action	Nerve supply
POSTERIOR MUSCLES				
Triceps brachii (tri′seps bra′ke-i) (*triceps* = three heads; *brachi* = arm)	Large fleshy muscle; the only muscle of posterior compartment of arm; three-headed origin; long and lateral heads lie superficial to medial head	O—long head: infraglenoid tubercle of scapula; lateral head: posterior humerus; medial head: posterior humerus distal to radial groove I—by common tendon into olecranon process of ulna	Powerful forearm extensor (prime mover, particularly medial head); antagonist of forearm flexors; long head tendon may help stabilize shoulder joint and assist in arm adduction	Radial nerve
Anconeus (an-ko′ne-us) (*ancon* = elbow) (see Figure 10.15)	Short triangular muscle; closely associated (blended) with distal end of triceps on posterior humerus	O—lateral epicondyle of humerus I—lateral aspect of olecranon process of ulna	Abducts ulna during forearm pronation; synergist of triceps brachii in elbow extension	Radial nerve
ANTERIOR MUSCLES				
Biceps brachii (bi′seps) (*biceps* = two heads)	Two-headed fusiform muscle; bellies unite as insertion point is approached; tendon of long head helps stabilize shoulder joint	O—short head: coracoid process; long head: tubercle above glenoid cavity and lip of glenoid cavity; tendon of long head runs within capsule and descends into intertubercular groove of humerus I—by common tendon into radial tuberosity	Flexes elbow joint and supinates forearm; these actions usually occur at same time (e.g., when you open a bottle of wine, it turns the corkscrew and pulls the cork); weak flexor of arm at shoulder	Musculocutaneous nerve
Brachialis (bra′ke-al-is)	Strong muscle that is immediately deep to biceps brachii on distal humerus	O—front of distal half of humerus; embraces insertion of deltoid muscle I—coronoid process of ulna and capsule of elbow joint	A major forearm flexor (lifts ulna as biceps lifts the radius)	Musculocutaneous nerve
Brachioradialis (bra″ke-o-ra″de-a′lis) (*radi* = radius, ray) (also see Figure 10.14)	Superficial muscle of lateral forearm; forms lateral boundary of cubital fossa; extends from distal humerus to distal forearm	O—lateral supracondylar ridge at distal end of humerus I—base of styloid process of radius	Synergist in forearm flexion; acts to best advantage when forearm is partially flexed already	Radial nerve (an important exception: the radial nerve typically serves extensor muscles)

Handwritten annotation near Biceps brachii origin: SUPRAGLENOID TUBERCLE

Table 10.11   Muscles of the Forearm: Movements of the Wrist, Hand, and Fingers (Figures 10.14 and 10.15)

The many muscles in the forearm perform several basic functions: Some move the hand at the wrist, some move the fingers, and a few help supinate and pronate the forearm. Most of these muscles are fleshy proximally and have long tendons distally, almost all of which insert in the hand. At the wrist, these tendons are anchored firmly by band-like thickenings of deep fascia called **flexor** and **extensor retinacula** ("retainers"). These "wrist bands" keep the tendons from jumping outward when tensed. Crowded together in the wrist and palm, the muscle tendons are surrounded by slippery tendon sheaths that minimize friction as they slide against one another.

Many forearm muscles arise from the distal humerus; thus, they cross the elbow joint as well as the wrist and finger joints. However, their actions on the elbow are slight. At the *wrist* joint, the forearm muscles bring about flexion, extension, abduction, and adduction of the hand, but at the *finger* joints these muscles mostly just flex and extend the fingers. (The other movements of fingers are brought about by small muscles in the hand itself.)

Sheaths of fascia divide the forearm muscles into two main compartments, an *anterior flexor compartment* and a *posterior extensor compartment*. Both of these compartments, in turn, contain a superficial and a deep muscle layer. Most flexors in the anterior compartment originate from a com-

mon tendon on the medial epicondyle of the humerus. The flexors are innervated by two nerves that descend through the anterior forearm, the median and ulnar nerves—especially by the median nerve. Two of the anterior compartment muscles are not flexors but pronators, the **pronator teres** and **pronator quadratus.**

The posterior compartment muscles extend the hand and fingers. The only exception is the **supinator,** which helps supinate the forearm. Many of the posterior compartment extensors arise by a common tendon from the lateral epicondyle of the humerus. All are innervated by the radial nerve, the main nerve on the posterior aspect of the forearm.

Most muscles that move the palm and fingers are located in the forearm, not in the hand itself. The hand must perform so many different movements that it could not possibly contain all of its own muscles: If it did, it would be as large and clumsy as a baseball glove. Instead, the fingers are largely "operated" by tendons originating in the forearm, like a wooden puppet operated by strings. Still, small muscles do occur inside the hand, where they control our most delicate finger movements. These are described in Table 10.13.

The actions of the forearm muscles are summarized in Table 10.12 (Part III).

Muscle	Description	Origin (O) and insertion (I)	Action	Nerve supply
PART I: ANTERIOR MUSCLES (Figure 10.14)	These eight muscles of the anterior fascial compartment are listed from the lateral to the medial aspect. Most arise from a common flexor tendon attached to the medial epicondyle of the humerus and have additional origins as well. Most of the tendons of these flexors are held in place at the wrist by a thickening of deep fascia called the *flexor retinaculum*.			
SUPERFICIAL MUSCLES				
Pronator teres (pro′na′tor te′rēz) (*pronation* = turning palm posteriorly, or down; *teres* = round)	Two-headed muscle; seen in superficial view between proximal margins of brachio-radialis and flexor carpi radialis; forms medial boundary of cubital fossa	O—medial epicondyle of humerus; coronoid process of ulna I—by common tendon into lateral radius, midshaft	Pronates forearm; weak flexor of elbow	Median nerve
Flexor carpi radialis (flek′sor kar′pe ra″de-a′lis) (*flex* = decrease angle between two bones; *carpi* = wrist; *radi* = radius)	Runs diagonally across forearm; midway, its fleshy belly is replaced by a flat tendon that becomes cordlike at wrist	O—medial epicondyle of humerus I—base of second and third metacarpals; insertion tendon easily seen and provides guide to position of radial artery (used for pulse taking) at wrist	Powerful flexor of wrist; abducts hand; synergist of elbow flexion	Median nerve
Palmaris longus (pahl-ma′ris lon′gus) (*palma* = palm; *longus* = long)	Small fleshy muscle with a long insertion tendon; often absent; may be used as guide to find median nerve that lies lateral to it at wrist	O—medial epicondyle of humerus I—palmar aponeurosis	Weak wrist flexor; may tense fascia of palm (palmar aponeurosis) during hand movements	Median nerve

Muscle	Description	Origin (O) and insertion (I)	Action	Nerve supply
Flexor carpi ulnaris (ul-na′ris) (*ulnar* = ulna)	Most medial muscle of this group; two-headed; ulnar nerve lies lateral to its tendon	O—medial epicondyle of humerus, olecranon process, and posterior surface of ulna I—pisiform bone and base of fifth metacarpal	Powerful flexor of wrist; also adducts hand in concert with extensor carpi ulnaris (posterior muscle); stabilizes wrist during finger extension	Ulnar nerve
Flexor digitorum superficialis (di″ji-tor′um soo″per-fish″e-a′lis) (*digit* = finger, toe; *superficial* = close to surface)	Two-headed muscle; more deeply placed (therefore, actually forms an intermediate layer); overlain by muscles above but visible at distal end of forearm	O—medial epicondyle of humerus, coronoid process of ulna; shaft of radius I—by four tendons into middle phalanges of fingers 2–5	Flexes wrist and middle phalanges of fingers 2–5; the important finger flexor when speed and flexion against resistance are required	Median nerve
DEEP MUSCLES **Flexor pollicis longus** (pah′lĭ-kis) (*pollix* = thumb)	Partly covered by flexor digitorum superficialis; parallels flexor digitorum profundus laterally	O—anterior surface of radius and interosseous membrane I—distal phalanx of thumb	Flexes distal phalanx of thumb; weak flexor of wrist	Median nerve
Flexor digitorum profundus (pro-fun′dus) (*profund* = deep)	Extensive origin; overlain entirely by flexor digitorum superficialis	O—anteromedial surface of ulna and interosseous membrane I—by four tendons into distal phalanges of fingers 2–5	Slow-acting finger flexor; assists in flexing wrist; the only muscle that can flex distal interphalangeal joints	Medial half by ulnar nerve; lateral half by median nerve

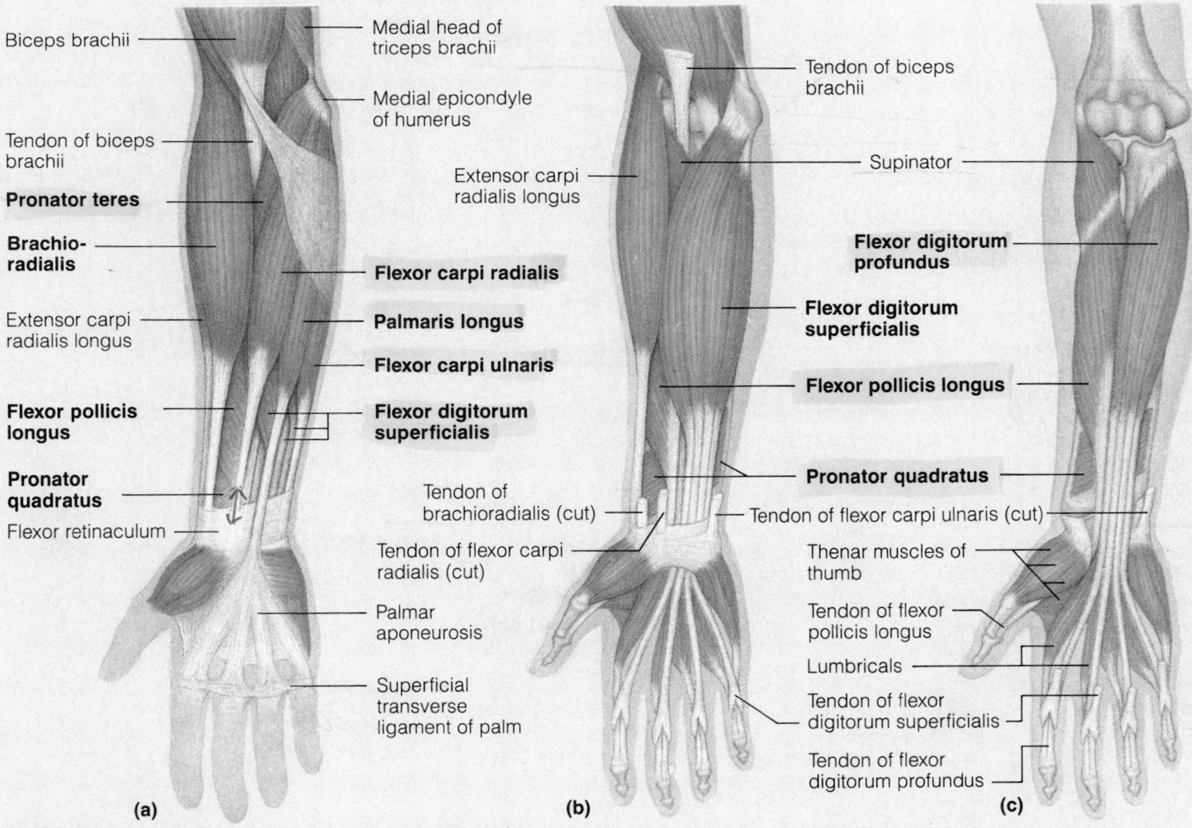

Figure 10.14 Muscles of the anterior fascial compartment of the forearm acting on the wrist and fingers. **(a)** Superficial view of the muscles of the right forearm and hand. **(b)** The brachioradialis, flexors carpi radialis and ulnaris, and palmaris longus muscles have been removed to reveal the position of the somewhat deeper flexor digitorum superficialis. **(c)** Deep muscles of the anterior compartment. Superficial muscles have been removed. The lumbricals and thenar muscles of the hand (intrinsic hand muscles) are also illustrated.

(Table continues)

Table 10.11 (continued)

Muscle	Description	Origin (O) and insertion (I)	Action	Nerve supply
Pronator quadratus (kwod-ra′tus) (*quad* = square, four-sided)	Deepest muscle of distal forearm; passes downward and laterally; only muscle that arises solely from ulna and inserts solely into radius	O—distal portion of anterior ulnar shaft I—distal surface of anterior radius	Pronates forearm; acts with pronator teres; also helps hold ulna and radius together	Median nerve
PART II: POSTERIOR MUSCLES (Figure 10.15)	These 11 muscles of the posterior fascial compartment are listed from the lateral to the medial aspect. They are all innervated by the radial nerve. More than half of the posterior compartment muscles arise from a common extensor origin tendon attached to the posterior surface of the lateral epicondyle of the humerus and adjacent fascia. The extensor tendons are held in place at the posterior aspect of the wrist by the *extensor retinaculum*, which prevents "bowstringing" of these tendons when the wrist is hyperextended. The extensor muscles of the fingers end in a broad hood over the dorsal side of the digits, the *extensor expansion*.			
SUPERFICIAL MUSCLES				
Extensor carpi radialis longus (ek-sten′sor) (*extend* = increase angle between two bones)	Parallels brachioradialis on lateral forearm, and may blend with it	O—lateral supracondylar ridge of humerus I—base of second metacarpal	Extends and abducts wrist	Radial nerve
Extensor carpi radialis brevis (breh′vis) (*brevis* = short)	Somewhat shorter than extensor carpi radialis longus and lies deep to it	O—lateral epicondyle of humerus I—base of third metacarpal	Extends and abducts wrist; acts synergistically with extensor carpi radialis longus to steady wrist during finger flexion	Deep branch of radial nerve
Extensor digitorum	Lies medial to extensor carpi radialis brevis; a detached portion of this muscle, called *extensor digiti minimi*, extends little finger	O—lateral epicondyle of humerus I—by four tendons into extensor expansions and distal phalanges of fingers 2–5	Prime mover of finger extension; extends wrist; can abduct (flare) fingers	Posterior interosseous nerve (branch of radial nerve)
Extensor carpi ulnaris	Most medial of superficial posterior muscles; long, slender muscle	O—lateral epicondyle of humerus and posterior border of ulna I—base of fifth metacarpal	Extends and adducts wrist (in conjunction with flexor carpi ulnaris)	Posterior interosseous nerve
DEEP MUSCLES **Supinator** (soo″pi-na′tor) (*supination* = turning palm anteriorly or upward)	Deep muscle at posterior aspect of elbow; largely concealed by superficial muscles	O—lateral epicondyle of humerus; proximal ulna I—proximal end of radius	Assists biceps brachii to supinate forearm; antagonist of pronator muscles	Radial nerve
Abductor pollicis longus (ab-duk′tor) (*abduct* = movement away from median plane)	Lateral and parallel to extensor pollicis longus; just distal to supinator	O—posterior surface of radius and ulna; interosseous membrane I—base of first metacarpal	Abducts and extends thumb	Posterior interosseous nerve (branch of radial nerve)
Extensor pollicis brevis and **longus**	Deep muscle pair with a common origin and action; overlain by extensor carpi ulnaris	O—dorsal shaft of radius and ulna; interosseous membrane I—base of proximal (brevis) and distal (longus) phalanx of thumb	Extends thumb	Posterior interosseous nerve
Extensor indicis (in′dĭ-kis) (*indicis* = index finger)	Tiny muscle arising close to wrist	O—posterior surface of distal ulna; interosseous membrane I—extensor expansion of index finger; joins tendon of extensor digitorum	Extends index finger	Posterior interosseous nerve

Handwritten annotations: NC— ; 1, 2, 3, 4 SEPARATE THIS MUSCLE, 5 ; NC— ; LATERAL ; SEPARATE ; + abduction

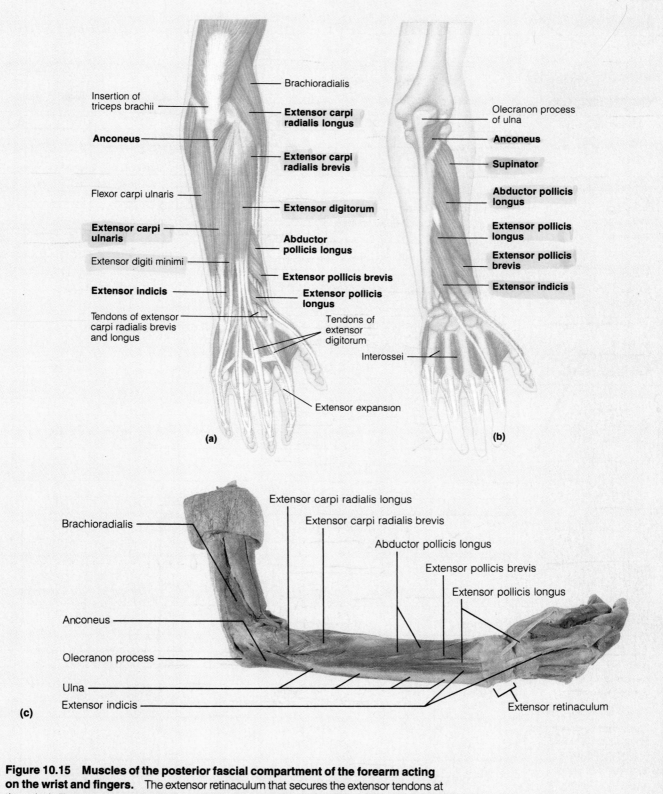

Figure 10.15   Muscles of the posterior fascial compartment of the forearm acting on the wrist and fingers.   The extensor retinaculum that secures the extensor tendons at the wrist is illustrated in part (c). (**a**) Superficial muscles of the right forearm, posterior view. (**b**) Deep posterior muscles of the right forearm; superficial muscles have been removed. The interossei, the deepest layer of intrinsic hand muscles, are also illustrated. (**c**) Photograph of the deep posterior muscles of the right forearm. Superficial muscles have been removed.

Table 10.12 Summary of Actions of Muscles Acting on the Arm, Forearm, and Hand (Figure 10.16)

PART I: MUSCLES ACTING ON THE ARM (HUMERUS) (PM = prime mover)	Actions at the shoulder					
	Flexion	**Extension**	**Abduction**	**Adduction**	**Medial rotation**	**Lateral rotation**
Pectoralis major	X (PM)			X (PM)	X	
Latissimus dorsi		X (PM)		X (PM)	X	
Deltoid	X (PM) (anterior fibers)	X (PM) (posterior fibers)	X (PM)		X (anterior fibers)	X (posterior fibers)
Subscapularis					X (PM)	
Supraspinatus			X			
Infraspinatus						X (PM)
Teres minor				X (weak)		X (PM)
Teres major		X		X	X	
Coracobrachialis	X			X		
Biceps brachii	X					
Triceps brachii				X		

PART II: MUSCLES ACTING ON THE FOREARM	Actions			
	Elbow flexion	**Elbow extension**	**Pronation**	**Supination**
Biceps brachii	X (PM)			X
Triceps brachii		X (PM)		
Anconeus		X		
Brachialis	X (PM)			
Brachioradialis	X			
Pronator teres			X	
Pronator quadratus			X	
Supinator				X

PART III: MUSCLES ACTING ON THE WRIST AND FINGERS	Actions on the wrist				Actions on the fingers	
	Flexion	**Extension**	**Abduction**	**Adduction**	**Flexion**	**Extension**
Anterior Compartment: Flexor carpi radialis	X (PM)		X			
Palmaris longus	X (weak)					
Flexor carpi ulnaris	X (PM)			X		
Flexor digitorum superficialis	X (PM)				X	
Flexor pollicis longus	X (weak)				X (thumb)	
Flexor digitorum profundus	X				X	
Posterior Compartment: Extensor carpi radialis longus and brevis		X	X			
Extensor digitorum		X (PM)				X (and abducts)
Extensor carpi ulnaris		X		X		
Abductor pollicis longus			X		(Abducts thumb)	
Extensor pollicis longus and brevis						X (thumb)
Extensor indicis						X (index finger)

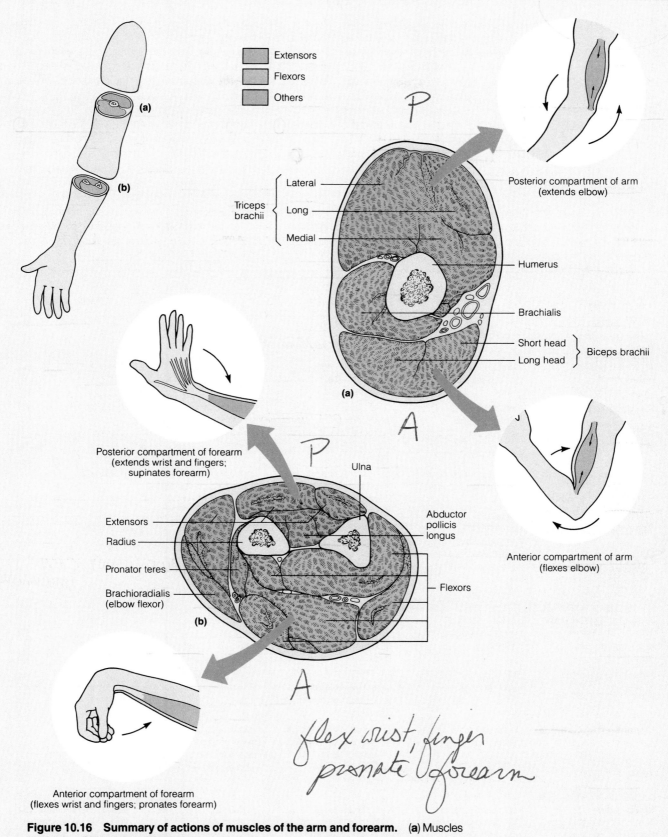

Extensors
Flexors
Others

(a)

(b)

Triceps brachii { Lateral, Long, Medial }

P

A

(a)

Posterior compartment of arm
(extends elbow)

Humerus

Brachialis

Short head
Long head } Biceps brachii

Anterior compartment of arm
(flexes elbow)

Posterior compartment of forearm
(extends wrist and fingers;
supinates forearm)

P

Ulna

Abductor
pollicis
longus

Extensors
Radius
Pronator teres
Brachioradialis
(elbow flexor)

Flexors

(b)

A

flex wrist, finger
pronate forearm

Anterior compartment of forearm
(flexes wrist and fingers; pronates forearm)

Figure 10.16   Summary of actions of muscles of the arm and forearm.   (a) Muscles
of the arm. (b) Muscles of the forearm.

This table considers the small muscles that lie entirely in the hand. All are in the palm, none on the hand's dorsal side. All move the metacarpals and fingers, but since they are small, they are weak. Therefore, they mostly control precise movements (such as threading a needle), leaving the powerful movements of the fingers ("power grip") to the forearm muscles. The intrinsic muscles include the main abductors and adductors of the fingers, as well as the muscles that produce the movement of opposition—moving the thumb toward the little finger—that enables one to grip objects in the palm (the handle of a hammer, for example). Many muscles in the palm are specialized in movement of the thumb, and surprisingly many move the little finger.

Movements of the thumb are defined differently from movements of other fingers, because the thumb lies at a right angle to the rest of the hand. The thumb flexes by bending medially along the palm, not by bending anteriorly, as do the other fingers. (Be sure to start with your hand in the anatomical position or this will not be clear!) The thumb straightens, or extends, by pointing laterally (as in hitchhiking), not posteriorly, as do the other fingers. To abduct the fingers is to splay them laterally, but to abduct the thumb is to point it anteriorly. Adduction of the thumb brings it back posteriorly.

The intrinsic muscles of the palm are divided into three groups: (1) those in the *thenar eminence* (ball of the thumb); (2) those in the *hypothenar eminence* (the ball of the little finger); and (3) muscles in the mid palm. The thenar and hypothenar muscles are almost mirror images of each other, each containing a small flexor, an abductor, and an opponens muscle. The midpalmar muscles are called **lumbricals** and **interossei**. We rely on these muscles to extend our fingers at the interphalangeal joints, a movement that the great finger extensors from the forearm cannot perform. The interossei are also the main abductors and adductors of the fingers.

Muscle	Description	Origin (O) and insertion (I)	Action	Nerve supply
THENAR MUSCLES IN BALL OF THUMB (the'nar) (*thenar* = palm)				
Abductor pollicis brevis (*pollex* = thumb)	Lateral muscle of thenar group; superficial	O—flexor retinaculum I—lateral base of thumb's proximal phalanx	Abducts thumb (at carpometacarpal joint)	Median nerve
Flexor pollicis brevis	Medial and deep muscle of thenar group	O—flexor retinaculum and nearby carpals I—lateral side of base of proximal phalanx of thumb	Flexes thumb (at carpometacarpal and metacarpophalangeal joints)	Median (or ulnar) nerve
Opponens pollicis (o-pōn'enz) (*opponens* = opposition)	Deep to abductor pollicis brevis, on metacarpal 1	O—flexor retinaculum and nearby carpals I—whole anterior side of metacarpal 1	Opposition: moves thumb to touch tip of little finger	Median nerve
Adductor pollicis	Fan-shaped with horizontal fibers; distal to other thenar muscles; oblique and transverse heads	O—bases of metacarpals 2-4 (oblique head); whole front of metacarpal 3 (transverse head) I—medial side of base of proximal phalanx of thumb	Adducts thumb	Ulnar nerve
HYPOTHENAR MUSCLES IN BALL OF LITTLE FINGER				
Abductor digiti minimi (dī'jī-ti min'ĭ-mi) (*digiti minimi* = little finger)	Medial muscle of hypothenar group; superficial	O—pisiform bone I—medial side of proximal phalanx of little finger	Abducts (and flexes) little finger at metacarpophalangeal joint	Ulnar nerve
Flexor digiti minimi brevis	Lateral deep muscle of hypothenar group	O—hamate bone and flexor retinaculum I—same as abductor digiti minimi	Flexes little finger at metacarpophalangeal joint	Ulnar nerve
Opponens digiti minimi	Deep to abductor digiti minimi	O—same as flexor digiti minimi brevis I—most of length of lateral side of metacarpal 5	Helps in opposition: brings metacarpal 5 toward thumb to cup the hand	Ulnar nerve
MIDPALMAR MUSCLES				
Lumbricals (lum'brĭ-k'lz) (*lumbric* = earthworm)	Four worm-shaped muscles in palm, one to each finger (except thumb); odd because they originate from the tendons of another muscle	O—lateral side of each tendon of flexor digitorum profundus in palm I—lateral edge of extensor expansion on back of first phalanx of fingers 2-5	By pulling on the extensor expansion over the first phalanx, they flex fingers at metacarpophalangeal joints but extend fingers at interphalangeal joints	Median nerve (lateral two) and ulnar nerve (medial two)

Muscle	Description	Origin (O) and Insertion (I)	Action	Nerve supply
Palmar interossei (in"ter-os'e-i) (*interossei* = between bones)	Four long, cone-shaped muscles in the spaces between the meta-carpals; lie ventral to the dorsal interossei	O—the side of each metacarpal that faces the mid axis of the hand (metacarpal 3); but absent from metacarpal 3 I—extensor expansion on first phalanx of each finger (except finger 3), on side facing mid axis of hand	Adductors of fingers: pull fingers in toward third digit; act with lumbricals to extend fingers at interphalangeal joints and flex them at metacarpo-phalangeal joints	Ulnar nerve
Dorsal interossei	Four bipennate muscles filling spaces between the metacarpals; deepest palm muscles, also visible on dorsal side of hand (Figure 10.15b)	O—sides of metacarpals I—extensor expansion over first phalanx of fingers 2–4 on side opposite mid axis of hand (finger 3), but on *both* sides of finger 3	Abduct (diverge) fingers; extend fingers at interphalangeal joints and flex them at metacarpo-phalangeal joints	Ulnar nerve

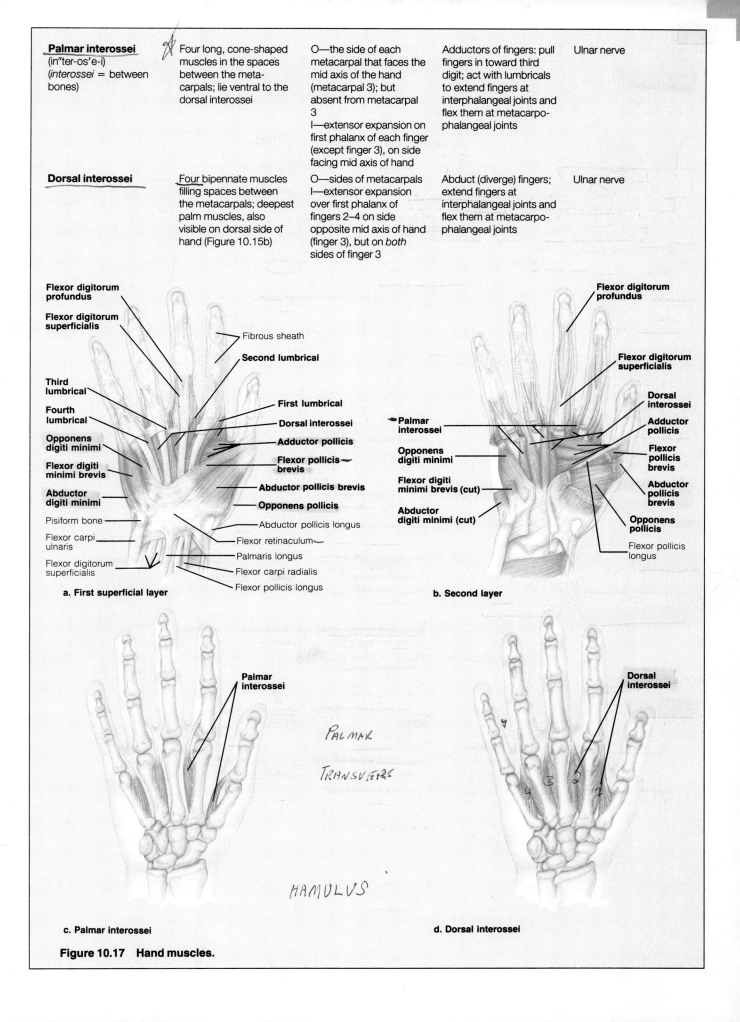

a. First superficial layer

b. Second layer

c. Palmar interossei

d. Dorsal interossei

Figure 10.17 Hand muscles.

Table 10.14 Muscles Crossing the Hip and Knee Joints: Movements of the Thigh and Leg (Figures 10.18 and 10.19)

In this table, we consider the movements at both the hip joint and the knee joint, because a number of muscles span both these joints.

The *anterior muscles* of the hip and thigh tend to flex the femur at the hip and to extend the leg at the knee—producing the foreswing phase of walking. The *posterior muscles* of the hip and thigh, by contrast, mostly extend the thigh and flex the leg—the backswing phase of walking. A third group of muscles, the *medial,* or *adductor, muscles* on the medial aspect of the thigh, move the thigh only, not the leg. In the thigh, the anterior, posterior, and adductor muscles are separated by walls of fascia into *anterior, posterior* and *medial compartments.* The deep fascia of the thigh (*fascia lata*) surrounds and encloses all three groups of muscles like a support stocking.

Movements at the hip joint are summarized first, then movements at the knee joint. The muscles that *flex* the thigh at the hip originate from the vertebral column and pelvis and pass anterior to the hip joint. These muscles include the **iliopsoas, tensor fasciae latae, rectus femoris,** and **pectineus** (Figure 10.18a). The thigh *extensors,* by contrast, arise posterior

to the hip joint and include the **gluteus maximus** and **hamstrings** (Figure 10.19a). As we mentioned above, the thigh **adductors** originate medial to the hip joint (Figure 10.18b). *Abduction* of the thigh is brought about mainly by the **gluteus medius** and **gluteus minimus,** buttocks muscles that lie lateral to the hip joint (Figure 10.19b). The adductors and abductors function during walking—not for moving the lower limb, but for shifting the trunk from side to side so that the body's center of gravity is always balanced directly over the limb that is on the ground. *Medial* and *lateral rotation* of the femur are accomplished by many different muscles.

At the knee joint, flexion and extension are the main movements. The only *extensor* of the leg at the knee is the four-part **quadriceps femoris** muscle from the anterior thigh (Figure 10.18a). The quadriceps is antagonized by the hamstrings of the posterior compartment (Figure 10.19a), which are the prime movers of knee *flexion.*

The actions of the muscles presented here are further summarized in Table 10.16 (Part I).

Muscle	Description	Origin (O) and insertion (I)	Action	Nerve supply
PART I: ANTERIOR AND MEDIAL MUSCLES (Figure 10.18)				
ORIGIN ON THE PELVIS				
Iliopsoas (il″e-o-so′us)	Iliopsoas is a composite of two closely related muscles (iliacus and psoas major) whose fibers pass under the inguinal ligament [see Table 10.6 (external oblique muscle) and Figure 10.10] to insert via a common tendon on the femur			
• **Iliacus** (il-e-ak′us) (*iliac* = ilium)	Large, fan-shaped, more lateral muscle	O—iliac fossa I—lesser trochanter of femur via iliopsoas tendon	Iliopsoas is the prime mover of hip flexion; flexes thigh on trunk when pelvis is fixed	Femoral nerve
• **Psoas** (so′us) **major** (*psoa* = loin muscle; *major* = larger)	Longer, thicker, more medial muscle of the pair. (Butchers refer to this muscle as the tenderloin)	O—by fleshy slips from transverse processes, bodies, and discs of lumbar vertebrae and T₁₂ I—lesser trochanter of femur via iliopsoas tendon	As above; also effects lateral flexion of vertebral column; important postural muscle	Ventral rami L₁–L₃
Sartorius (sar-tor′e-us) (*sartor* = tailor)	Straplike superficial muscle running obliquely across anterior surface of thigh to knee; longest muscle in body; crosses both hip and knee joints	O—anterior superior iliac spine I—winds around medial aspect of knee and inserts into medial aspect of proximal tibia	Flexes and laterally rotates thigh; flexes knee (weak); known as "tailor's muscle" because it helps effect cross-legged position in which tailors are often depicted	Femoral nerve

(Table continues)

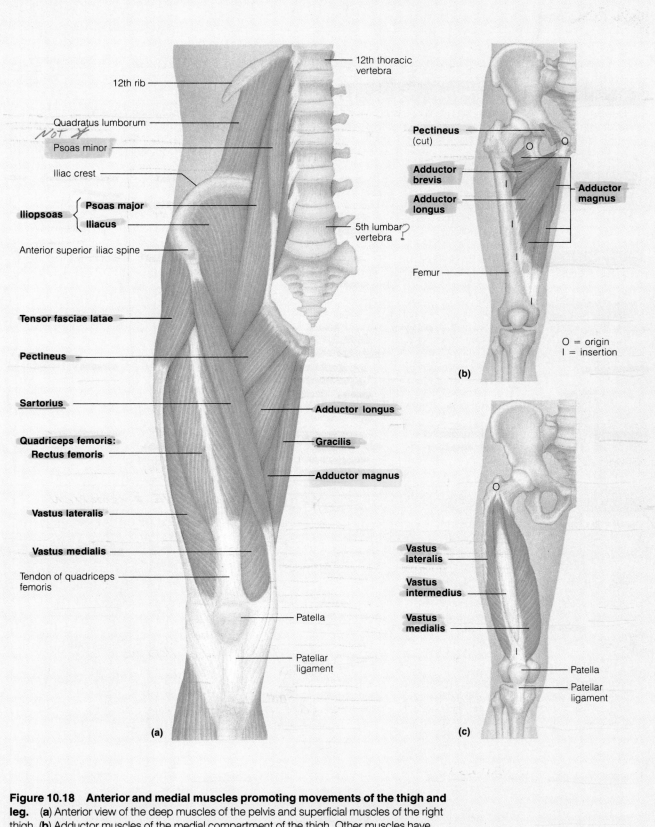

Figure 10.18   Anterior and medial muscles promoting movements of the thigh and leg.   (**a**) Anterior view of the deep muscles of the pelvis and superficial muscles of the right thigh. (**b**) Adductor muscles of the medial compartment of the thigh. Other muscles have been removed so that the origins and insertions of the adductor muscles can be seen. (**c**) The vastus muscles of the quadriceps group. The rectus femoris muscle of the quadriceps group and surrounding muscles have been removed to reveal the attachments and extent of the vastus muscles.

(Table continues)

10-2-MO

Table 10.14 (continued)

Muscle	Description	Origin (O) and insertion (I)	Action	Nerve supply

NC —

MUSCLES OF THE MEDIAL COMPARTMENT OF THE THIGH

Adductors
(ah-duk′torz)

Large muscle mass consisting of three muscles (magnus, longus, and brevis) forming medial aspect of thigh; arise from inferior part of pelvis and insert at various levels on femur; all used in movements that press thighs together, as when astride a horse; important in pelvic tilting movements that occur during walking and in fixing the hip when the knee is flexed; entire group innervated by obturator nerve. Strain or stretching of this muscle group is called a "pulled groin."

Muscle	Description	Origin (O) and insertion (I)	Action	Nerve supply
• **Adductor magnus** (mag′nus) (*adduct* = move toward midline; *magnus* = large)	A triangular muscle with a broad insertion; is a composite muscle that is part adductor and part hamstring in action	O—ischial and pubic rami and ischial tuberosity I—linea aspera and adductor tubercle of femur	Anterior part adducts and laterally rotates and flexes thigh; posterior part is a synergist of hamstrings in thigh extension	Obturator nerve and sciatic nerve
• **Adductor longus** (*longus* = long)	Overlies middle aspect of adductor magnus; most anterior of adductor muscles	O—pubis near pubic symphysis I—linea aspera	Adducts, flexes, and laterally rotates thigh	Obturator nerve
• **Adductor brevis** (*brevis* = short)	In contact with obturator externus muscle; largely concealed by adductor longus and pectineus	O—body and inferior ramus of pubis I—linea aspera above adductor longus	Adducts and laterally rotates thigh *DOES Ø FLEX*	Obturator nerve
Pectineus (pek-tin′e-us) (*pecten* = comb)	Short, flat muscle; overlies adductor brevis on proximal thigh; abuts adductor longus medially	O—pectineal line of pubis (on superior ramus) I—inferior to lesser trochanter of posterior *LATERAL TO ILLIACUS* aspect of femur	Adducts, flexes, and laterally rotates thigh	Femoral and ~~sometimes obturator~~ nerve
Gracilis (grah-si′lis) (*gracilis* = slender)	Long, thin, superficial muscle of medial thigh	O—inferior ramus and body of pubis I—medial surface of superior tibia shaft	Adducts thigh, flexes, and medially rotates leg, especially during walking *CROSSES 2 JOINTS; THEREFORE*	Obturator nerve

NC —

MUSCLES OF THE ANTERIOR COMPARTMENT OF THE THIGH   *DIFF ORIGINS, ONE INSERTION*

Quadriceps femoris
(kwod′ri-seps fem′o-ris)

Quadriceps femoris arises from four separate heads (*quadriceps* = four heads) that form the flesh of front and sides of thigh; these heads (rectus femoris, and lateral, medial, and intermediate vastus muscles) have a common insertion tendon, the quadriceps tendon, which inserts into the patella and then via the patellar ligament into tibial tuberosity. The quadriceps is a powerful knee extensor used in climbing, jumping, running, and rising from seated position; group is innervated by femoral nerve; the tone of quadriceps plays important role in strengthening knee joint

Muscle	Description	Origin (O) and insertion (I)	Action	Nerve supply
• **Rectus** (rek′tus) **femoris** (*rectus* = straight; *femoris* = femur)	Superficial muscle of anterior thigh; runs straight down thigh; longest head and only muscle of group to cross hip joint	O—anterior inferior iliac spine and superior margin of acetabulum I—patella and tibial tuberosity via patellar ligament	Extends knee and flexes thigh at hip *LEG* *CROSS 2 JOINTS, THEREFORE 2 ACTIONS*	Femoral nerve
• **Vastus lateralis** (vas′tus lat″er-a′lis) (*vastus* = large; *lateralis* = lateral)	Forms lateral aspect of thigh; a common intramuscular injection site, particularly in infants (who have poorly developed buttock and arm muscles)	O—greater trochanter, intertrochanteric line, linea aspera *(CREST IS POST)* I—as for rectus femoris	Extends knee *LEG* *ANTERIOR*	Femoral nerve
• **Vastus medialis** (me″de-a′lis) (*medialis* = medial)	Forms inferomedial aspect of thigh	O—linea aspera, intertrochanteric line I—as for rectus femoris	Extends knee; inferior fibers stabilize patella	Femoral nerve
• **Vastus intermedius** (in″ter-me′de-us) (*intermedius* = intermediate)	Obscured by rectus femoris; lies between vastus lateralis and vastus medialis on anterior thigh	O—anterior and lateral surfaces of proximal femur shaft I—as for rectus femoris	Extends knee	Femoral nerve

Muscle	Description	Origin (O) and insertion (I)	Action	Nerve supply
Tensor fasciae latae (ten'sor fă'she-e la'te) (*tensor* = to make tense; *fascia* = band; *lata* = wide)	Enclosed between fascia layers of anterolateral aspect of thigh; functionally associated with medial rotators and flexors of thigh	O—anterior aspect of iliac crest and anterior superior iliac spine I—iliotibial tract* *277*	Flexes and abducts thigh (thus a synergist of the iliopsoas and gluteus medius and minimus muscles); rotates thigh medially; steadies trunk on thigh by making iliotibial tract taut	Superior gluteal nerve

PART II: POSTERIOR MUSCLES (Figure 10.19)
GLUTEAL MUSCLES—ORIGIN ON PELVIS

Muscle	Description	Origin (O) and insertion (I)	Action	Nerve supply
Gluteus maximus (gloo'te-us mak'sĭ-mus) (*glutos* = buttock; *maximus* = largest)	Largest and most superficial of gluteus muscles; forms bulk of buttock mass; fibers are thick and coarse; important site of intramuscular injection (dorsal gluteal site); overlies large sciatic nerve; covers ischial tuberosity only when standing; when sitting, moves superiorly leaving ischial tuberosity exposed in the subcutaneous position	O—dorsal ilium, sacrum, and coccyx I—gluteal tuberosity of femur; iliotibial tract	Major extensor of thigh; complex, powerful, and most effective when thigh is flexed and force is necessary, as in rising from a forward flexed position and in thrusting the thigh posteriorly in climbing stairs and running; generally inactive during walking; laterally rotates thigh; antagonist of iliopsoas muscle	Inferior gluteal nerve *HELPS TO TENSE FASCIA*
Gluteus medius (me'de-us) (*medius* = middle)	Thick muscle largely covered by gluteus maximus; important site for intramuscular injections (ventral gluteal site); considered safer than dorsal gluteal site because there is less chance of injuring sciatic nerve	O—between anterior and posterior gluteal lines on lateral surface of ilium I—by short tendon into lateral aspect of greater trochanter of femur	*ANT. PORTION* Abducts and medially rotates thigh; steadies pelvis; its action is extremely important in walking; e.g., muscle of limb planted on ground tilts or holds pelvis in abduction so that pelvis on side of swinging limb does not sag; the foot of swinging limb can thus clear the ground	Superior gluteal nerve *POST. PORTION EXTENDS + LATERALLY ROTATES*
Gluteus minimus (mĭ'nĭ-mus) (*minimus* = smallest)	Smallest and deepest of gluteal muscles	O—between anterior and inferior gluteal lines on external surface of ilium I—anterior border of greater trochanter of femur	As for gluteus medius *(ANT PORTION ONLY)*	Superior gluteal nerve

LATERAL ROTATORS

Muscle	Description	Origin (O) and insertion (I)	Action	Nerve supply
Piriformis (pir'ĭ-form-is) (*piri* = pear; *forma* = shape)	Pyramidal muscle located on posterior aspect of hip joint; inferior to gluteus minimus; issues from pelvis via greater sciatic notch	O—anterolateral surface of sacrum (opposite greater sciatic notch) I—superior border of greater trochanter of femur	Rotates thigh laterally; since inserted above head of femur, can also assist in abduction of thigh when hip is flexed; stabilizes hip joint	S_1 and S_2, L_5
Obturator externus (ob"tu-ra'tor ek-ster'nus) (*obturator* = obturator foramen; *externus* = outside)	Flat, triangular muscle deep in upper medial aspect of thigh	O—outer surface of obturator membrane, external surface of pubis and ischium, and margins of obturator foramen I—by a tendon into trochanteric fossa of posterior femur	Rotates thigh laterally and stabilizes hip joint	Obturator nerve

(Table continues)

*The iliotibial tract is a thickened lateral portion of the *fascia lata* (the fascia that ensheaths all the muscles of the thigh). It extends as a tendinous band from the iliac crest to the knee.

10-2-mo

Table 10.14 (continued)

Muscle	Description	Origin (O) and insertion (I)	Action	Nerve supply
Obturator internus (in-ter′nus) (*internus* = inside)	Surrounds obturator foramen within pelvis; leaves pelvis via lesser sciatic notch and turns acutely forward to insert in femur	O—inner surface of obturator membrane, greater sciatic notch, and margins of obturator foramen I—greater trochanter in front of piriformis	As for obturator externus	L₅ and S₁
Gemellus (jĕ-meh′lis)— superior and inferior (*gemin* = twin, double; *superior* = above; *inferior* = below)	Two small muscles with common insertions and actions; considered extrapelvic portions of obturator internus	O—ischial spine (superior); ischial tuberosity (inferior) I—greater trochanter of femur	Rotate thigh laterally and stabilize hip joint	L₄, L₅, and S₁
Quadratus femoris (*quad* = four-sided square)	Short, thick muscle; most inferior of lateral rotator muscles; extends laterally from pelvis	O—ischial tuberosity I—greater trochanter of femur	Rotates thigh laterally and stabilizes hip joint	L₅ and S₁

MUSCLES OF THE POSTERIOR COMPARTMENT OF THE THIGH

Hamstrings	The hamstrings are fleshy muscles of the posterior thigh (biceps femoris, semitendinosus, and semimembranosus); they cross both the hip and knee joints and are prime movers of thigh extension and knee flexion; group has a common origin site and is innervated by sciatic nerve; ability of hamstrings to act on one of the two joints spanned depends on which joint is fixed; i.e., if knee is fixed (extended), they promote hip extension; if hip is extended, they promote knee flexion; however, when hamstrings are stretched, they tend to restrict full accomplishment of antagonistic movements; e.g., if knees are fully extended, it is difficult to flex the hip fully (and touch your toes), and when the thigh is fully flexed as in kicking a football, it is almost impossible to extend the knee fully at the same time (without considerable practice); name of this muscle group comes from old butchers' practice of using their tendons to hang hams for smoking; "pulled hamstrings" are common sports injuries in those who run very hard, e.g., football halfbacks			
• **Biceps femoris** (*biceps* = two heads)	Most lateral muscle of the group; arises from two heads	O—ischial tuberosity (long head), linea aspera and distal femur (short head) I—common tendon passes downward and laterally (forming lateral border of popliteal fossa) to insert into head of fibula and lateral condyle of tibia	Extends thigh and flexes knee; laterally rotates leg, especially when knee is flexed	Sciatic nerve
• **Semitendinosus** (sem″e-ten″di-no′sus) (*semi* = half; *tendinosus* = tendon)	Lies medial to biceps femoris; although its name suggests that this muscle is largely tendinous, it is quite fleshy; its slender tendon begins about two-thirds the way down thigh	O—ischial tuberosity in common with long head of biceps femoris I—medial aspect of upper tibial shaft	Extends thigh at hip; flexes knee; with semimembranosus, medially rotates leg	Sciatic nerve
• **Semimembranosus** (sem″e-mem″brah-no′sus) (*membranosus* = membrane)	Deep to semitendinosus	O—ischial tuberosity I—medial condyle of tibia	Extends thigh and flexes knee; medially rotates leg	Sciatic nerve

LONG HEAD CROSSES HIP

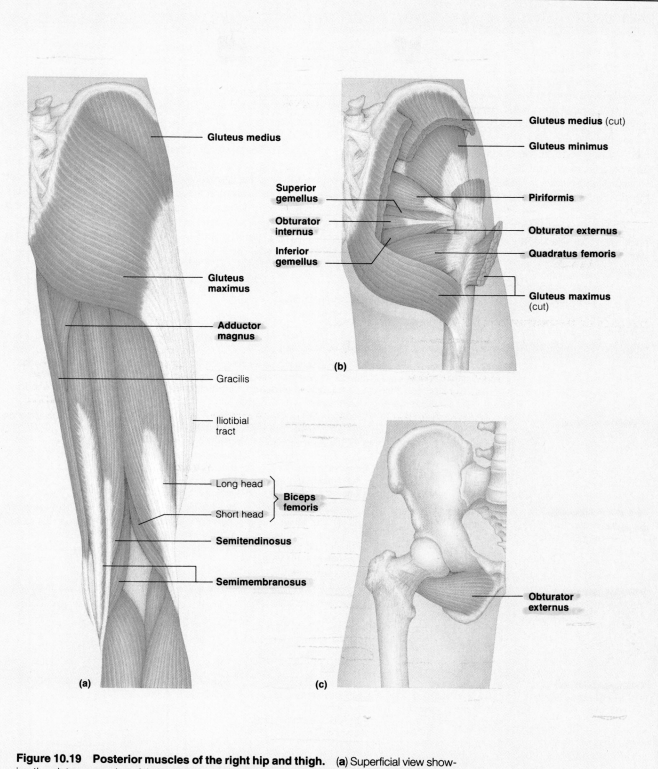

Figure 10.19 **Posterior muscles of the right hip and thigh.** **(a)** Superficial view showing the gluteus muscles of the buttock and hamstring muscles of the thigh. **(b)** Deep muscles of the gluteal region, which act primarily to rotate the thigh laterally. The superficial gluteus maximus and medius have been removed. **(c)** Anterior view of the isolated obturator externus muscle, showing its course as it travels from its origin on the anterior pelvis to the posterior aspect of the femur.

Table 10.15 Muscles of the Leg: Movements of the Ankle and Toes (Figures 10.20 to 10.22)

The deep fascia of the leg is continuous with the fascia lata that surrounds the thigh. Like a "knee sock" deep to the skin, this leg fascia surrounds the leg muscles and binds them tightly, preventing excess swelling of these muscles during exercise and also aiding venous return. Inward extensions from the leg fascia divide the leg muscles into *anterior, lateral,* and *posterior compartments* (Figure 10.23b), each with its own nerve and blood supply. Distally, the leg fascia thickens to form the **extensor, peroneal,** and **flexor retinacula,** "ankle bracelets" that hold the tendons in place where they run to the foot. As in the wrist and hand, the distal tendons are covered with slippery tendon sheaths.

The various muscles of the leg promote movements at the ankle joint (dorsiflexion and plantar flexion), at the intertarsal joints (inversion and eversion of the foot), or at the toes (flexion, extension). The muscles in the *anterior extensor compartment* of the leg (Figure 10.20) are directly comparable to the extensor muscle group of the forearm. Mostly, they extend the toes and dorsiflex the foot. Although dorsiflexion is not a powerful movement, it keeps the toes from dragging during walking. The *lateral compartment* muscles (Figure 10.21) are called *peroneal muscles* (*peron* = the fibula). They evert and plantar flex the foot and do not correspond to any muscles in the forearm. Muscles of the *posterior flexor compartment* (Figure 10.22) are comparable to the flexor muscle group of the forearm. They flex the toes and plantar flex the foot. Plantar flexion is the most powerful movement at the ankle: It lifts the weight of the entire body. Plantar flexion is necessary for standing on tiptoe and provides the forward thrust in walking and running.

The actions of the muscles in this table are summarized in Table 10.16 (Part II).

Muscle	Description	Origin (O) and insertion (I)	Action	Nerve supply
PART I: MUSCLES OF THE ANTERIOR COMPARTMENT (Figures 10.20 and 10.21)				
	All muscles of the anterior compartment are dorsiflexors of the ankle and have a common innervation, the deep peroneal nerve. Paralysis of the anterior muscle group causes *foot drop,* which requires that the leg be lifted unusually high during walking to prevent tripping over one's toes. "Shinsplints" is a painful inflammatory condition of the muscles of the anterior compartment.			
Tibialis anterior (tib"e-a'lis) (*tibial* = tibia; *anterior* = toward the front)	Superficial muscle of anterior leg; laterally parallels sharp anterior margin of tibia	O—lateral condyle and upper ⅔ of tibia; interosseous membrane I—by tendon into inferior surface of medial cuneiform and first metatarsal bone	Prime mover of dorsiflexion; inverts foot; assists in supporting medial longitudinal arch of foot *ADDUCTS*	Deep peroneal nerve
Extensor digitorum longus (*extensor* = increases angle at a joint; *digit* = finger or toe; *longus* = long)	On anterolateral surface of leg; lateral to tibialis anterior muscle	O—lateral condyle of tibia; proximal ¾ of fibula; interosseous membrane I—second and third phalanges of toes 2–5 via extensor expansion	Dorsiflexes foot; prime mover of toe extension (acts mainly at metatarsophalangeal joints)	Deep peroneal nerve
Peroneus tertius (per"o-ne'us ter'shus) (*perone* = fibula; *tertius* = third)	Small muscle; usually continuous and fused with distal part of extensor digitorum longus; not always present	O—distal anterior surface of fibula and interosseous membrane I—tendon passes anterior to lateral malleolus and inserts on dorsum of fifth metatarsal	Dorsiflexes and everts foot	Deep peroneal nerve
Extensor hallucis (hal'ü-sis) **longus** (*hallux* = great toe)	Deep to extensor digitorum longus and tibialis anterior; narrow origin	O—anteromedial fibula shaft and interosseous membrane I—tendon inserts on distal phalanx of great toe	Extends great toe; dorsiflexes foot	Deep peroneal nerve

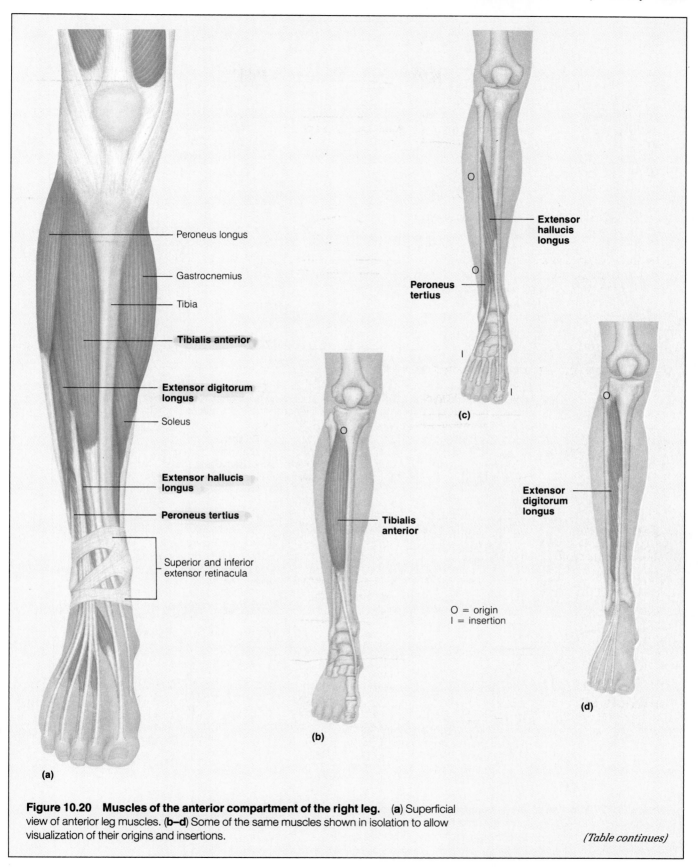

Figure 10.20   Muscles of the anterior compartment of the right leg.   **(a)** Superficial view of anterior leg muscles. **(b–d)** Some of the same muscles shown in isolation to allow visualization of their origins and insertions.

(Table continues)

Table 10.15 (continued)

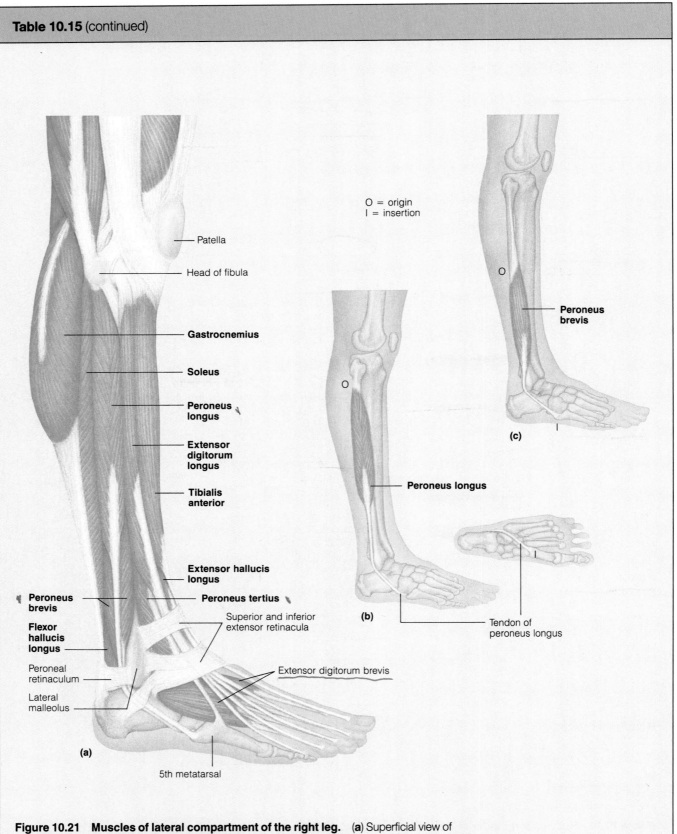

O = origin
I = insertion

Patella

Head of fibula

Gastrocnemius

Soleus

Peroneus longus

Extensor digitorum longus

Tibialis anterior

Extensor hallucis longus

Peroneus tertius

Peroneus brevis

Flexor hallucis longus

Superior and inferior extensor retinacula

Peroneal retinaculum

Lateral malleolus

Extensor digitorum brevis

(a)

5th metatarsal

(b)

Peroneus longus

Tendon of peroneus longus

Peroneus brevis

(c)

Figure 10.21   Muscles of lateral compartment of the right leg. (**a**) Superficial view of lateral aspect of the leg, illustrating the position of the lateral compartment muscles (peroneus longus and brevis) relative to anterior and posterior leg muscles. (**b**) View of peroneus longus in isolation; inset illustrates the insertion of the peroneus longus on the plantar surface of the foot. (**c**) Isolated view of the peroneus brevis muscle.

Muscle	Description	Origin (O) and insertion (I)	Action	Nerve supply

PART II: MUSCLES OF THE LATERAL COMPARTMENT (Figures 10.21 and 10.22)

These muscles have a common innervation, the superficial peroneal nerve. Besides plantar flexion and foot eversion, these muscles stabilize the lateral ankle and lateral longitudinal arch of the foot.

Muscle	Description	Origin (O) and insertion (I)	Action	Nerve supply
Peroneus longus (See also Figure 10.20) *[handwritten: HAS GROOVE THAT RUNS ACROSS]*	Superficial lateral muscle; overlies fibula	O—head and upper portion of fibula I—by long tendon that curves under foot to first *[handwritten: BASE (OF)]* metatarsal and medial cuneiform	Plantar flexes and everts foot; helps keep foot flat on ground *[handwritten: ABDUCTS]*	Superficial peroneal nerve
Peroneus brevis (*brevis* = short)	Smaller muscle; deep to peroneus longus; enclosed in a common sheath	O—distal fibula shaft I—by tendon running behind lateral malleolus to insert on proximal end of fifth metatarsal	Plantar flexes and everts foot *[handwritten: ABDUCTS]*	Superficial peroneal nerve

PART III: MUSCLES OF THE POSTERIOR COMPARTMENT (Figure 10.22)

The muscles of the posterior compartment have a common innervation, the tibial nerve. They act in concert to plantar flex the ankle.

SUPERFICIAL MUSCLES

Muscle	Description	Origin (O) and insertion (I)	Action	Nerve supply
Triceps surae (tri"seps sur'e) (See also Figure 10.21)	Refers to muscle pair (gastrocnemius and soleus) that shapes the posterior calf and inserts via a common tendon into the calcaneus of the heel; this calcaneal or Achilles tendon is the largest tendon in the body; prime movers of ankle plantar flexion			
• **Gastrocnemius** (gas"truk-ne'me-is) (*gaster* = belly; *kneme* = leg)	Superficial muscle of pair; two prominent bellies that form proximal curve of calf	O—by two heads from medial and lateral condyles of femur I—calcaneus via calcaneal tendon	Plantar flexes foot when knee is extended; since it also crosses knee joint, it can flex knee when foot is dorsiflexed	Tibial nerve
• **Soleus** (so'le-us) (*soleus* = fish) *[handwritten: Ø CROSS KNEE]*	Deep to gastrocnemius on posterior surface of calf	O—extensive cone-shaped origin from superior tibia, fibula, and interosseous membrane I—calcaneus via calcaneal tendon	Plantar flexes ankle; important locomotor and postural muscle during walking, running, and dancing	Tibial nerve
Plantaris (plan-tar'is) (*planta* = sole of foot)	Generally a small feeble muscle, but varies in size and extent; may be absent	O—posterior femur above lateral condyle I—via a long, thin tendon into calcaneus or calcaneal tendon	Assists in knee flexion and plantar flexion of foot	Tibial nerve

DEEP MUSCLES

Muscle	Description	Origin (O) and insertion (I)	Action	Nerve supply
Popliteus (pop-lit'e-us) (*poplit* = back of knee)	Thin, triangular muscle at posterior knee; passes downward and medially to tibial surface	O—lateral condyle of femur I—proximal tibia	Flexes and rotates leg medially to unlock knee from full extension when flexion begins	Tibial nerve
Flexor digitorum longus (*flexor* = decreases angle at a joint)	Long, narrow muscle; runs medial to and partially overlies tibialis posterior	O—posterior tibia I—tendon runs behind medial malleolus and splits into four parts to insert into distal phalanges of toes 2–5	Plantar flexes and inverts foot; flexes toes; helps foot "grip" ground	Tibial nerve
Flexor hallucis longus (See also Figure 10.21)	Bipennate muscle; lies lateral to inferior aspect of tibialis posterior	O—middle part of shaft of fibula; interosseous membrane I—tendon runs under foot to distal phalanx of great toe	Plantar flexes and inverts foot; flexes great toe at all joints; "push off" muscle during walking	Tibial nerve
Tibialis posterior (*posterior* = toward the back)	Thick, flat muscle deep to soleus; placed between posterior flexors	O—extensive origin from superior tibia and fibula and interosseous membrane I—tendon passes behind medial malleolus and under arch of foot; inserts into several tarsals and metatarsals 2–4	Prime mover of foot inversion; plantar flexes ankle; stabilizes medial longitudinal arch of foot (as during ice skating)	Tibial nerve

(Table continues)

Table 10.15 (continued)

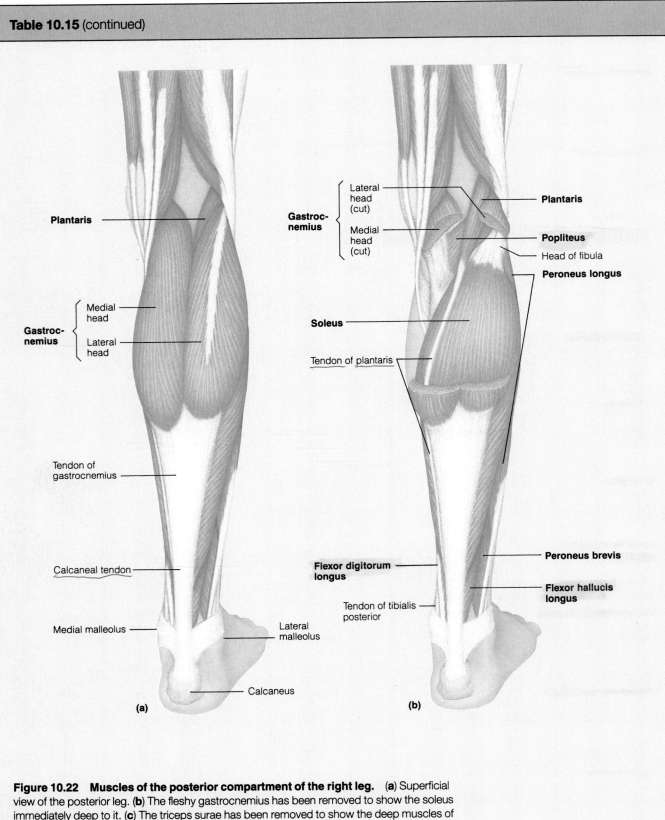

Figure 10.22   Muscles of the posterior compartment of the right leg.   **(a)** Superficial view of the posterior leg. **(b)** The fleshy gastrocnemius has been removed to show the soleus immediately deep to it. **(c)** The triceps surae has been removed to show the deep muscles of the posterior compartment. **(d–f)** Individual deep muscles are shown in isolation so that their origins and insertions may be visualized.

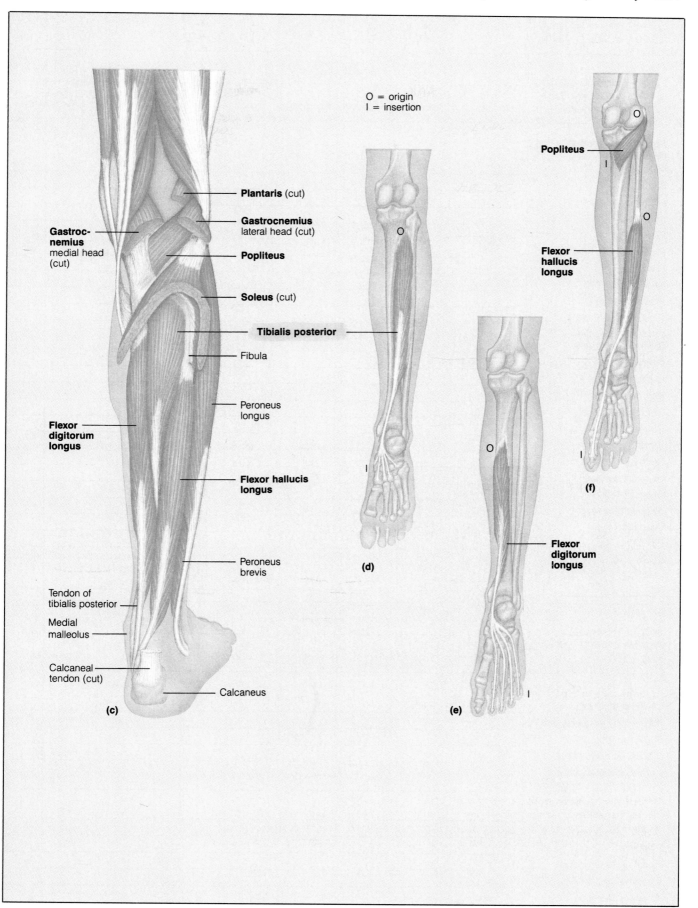

O = origin
I = insertion

Plantaris (cut)

Gastrocnemius
lateral head (cut)

Gastroc-
nemius
medial head
(cut)

Popliteus

Soleus (cut)

Tibialis posterior

Fibula

Peroneus
longus

Flexor
digitorum
longus

Flexor hallucis
longus

Peroneus
brevis

Tendon of
tibialis posterior

Medial
malleolus

Calcaneal
tendon (cut)

Calcaneus

(c)

(d)

Popliteus

Flexor
hallucis
longus

(f)

Flexor
digitorum
longus

(e)

Table 10.16 Summary of Actions of Muscles Acting on the Thigh, Leg, and Foot (Figure 10.23)

PART I: MUSCLES ACTING ON THE THIGH AND LEG (PM = prime mover)	Actions at the hip joint						Actions at the knee	
	Flexion	Extension	Abduction	Adduction	Medial rotation	Lateral rotation	Flexion	Extension
Anterior and medial muscles: Iliopsoas	X (PM)							
Sartorius	X					X	X	
Adductor magnus		X		X		X		
Adductor longus	X			X		X		
Adductor brevis	X			X		X		
Pectineus	X			X		X		
Gracilis				X			X	
Rectus femoris	X							X (PM)
Vastus muscles								X (PM)
Tensor fasciae latae	X		X		X			
Posterior muscles: Gluteus maximus		X (PM)				X		
Gluteus medius			X (PM)		X			
Gluteus minimus			X		X			
Piriformis			X			X		
Obturator internus						X		
Obturator externus						X		
Gemelli						X		
Quadratus femoris						X		
Biceps femoris		X (PM)					X (PM)	
Semitendinosus		X					X (PM)	
Semimembranosus		X					X (PM)	
Gastrocnemius							X	
Plantaris							X	
Popliteus							X (and rotates medially)	

PART II: MUSCLES ACTING ON THE ANKLE AND TOES	Action at the ankle joint				Action at the toes	
	Plantar flexion	Dorsiflexion	Inversion	Eversion	Flexion	Extension
Anterior compartment: Tibialis anterior		X (PM)	X			
Extensor digitorum longus		X				X (PM)
Peroneus tertius		X		X		
Extensor hallucis longus		X	X (weak)			X (great toe)
Lateral compartment: Peroneus longus and brevis	X			X		
Posterior compartment: Gastrocnemius	X (PM)					
Soleus	X (PM)					
Plantaris	X					
Flexor digitorum longus	X		X		X (PM)	
Flexor hallucis longus	X		X		X (great toe)	
Tibialis posterior	X		X (PM)			

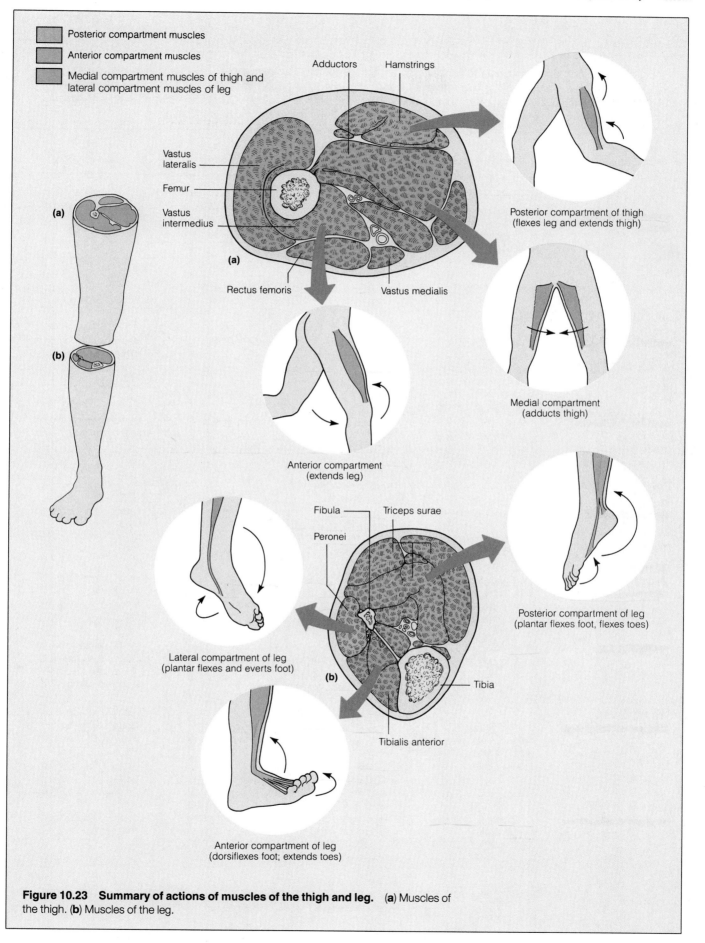

Posterior compartment muscles

Anterior compartment muscles

Medial compartment muscles of thigh and lateral compartment muscles of leg

(a)

(b)

Adductors   Hamstrings

Vastus lateralis

Femur

Vastus intermedius

(a)

Rectus femoris   Vastus medialis

Posterior compartment of thigh (flexes leg and extends thigh)

Medial compartment (adducts thigh)

Anterior compartment (extends leg)

Fibula   Triceps surae

Peronei

Posterior compartment of leg (plantar flexes foot, flexes toes)

Lateral compartment of leg (plantar flexes and everts foot)

(b)   Tibia

Tibialis anterior

Anterior compartment of leg (dorsiflexes foot; extends toes)

Figure 10.23   Summary of actions of muscles of the thigh and leg.   (a) Muscles of the thigh. (b) Muscles of the leg.

Table 10.17 Intrinsic Muscles of the Foot: Toe Movement and Foot Support (Figure 10.24)

The intrinsic muscles of the foot help to flex, extend, abduct, and adduct the toes. Furthermore, along with the tendons of some leg muscles that enter the sole, the foot muscles support the arches of the foot. There is a single muscle on the foot's dorsum (superior aspect), and many muscles on the plantar aspect (the sole). The plantar muscles occur in four layers, from superficial to deep. Overall, the foot muscles are remarkably similar to those in the palm of the hand.

Muscle	Description	Origin (O) and insertion (I)	Action	Nerve supply
MUSCLE ON DORSUM OF FOOT				
Extensor digitorum brevis (Figure 10.21a)	Small, four-part muscle on dorsum of foot; deep to the tendons of extensor digitorum longus; corresponds to the extensor indicis and extensor pollicis muscles of forearm	O—anterior part of calcaneus bone; extensor retinaculum I—base of proximal phalanx of big toe; extensor expansions on toes 2–4	Helps extend toes at metatarsophalangeal joints	Deep peroneal nerve
MUSCLES ON SOLE OF FOOT (Figure 10.24)				
FIRST LAYER				
Flexor digitorum brevis	Band-like muscle in middle of sole; corresponds to flexor digitorum superficialis of forearm and inserts into digits in the same way	O—tuber calcanei I—middle phalanx of toes 2–4	Helps flex toes	Medial plantar branch of tibial nerve
Abductor hallucis (hal′yu-kis) (*hallux* = the big toe)	Lies medial to flexor digitorum brevis; recall the similar thumb muscle, abductor pollicis brevis	O—tuber calcanei and flexor retinaculum I—proximal phalanx of big toe, medial side, via a sesamoid bone in the tendon of flexor hallucis brevis (see below)	Abducts big toe	Medial plantar branch of tibial nerve
Abductor digiti minimi	Most lateral of the three superficial sole muscles; recall the similar abductor muscle in palm	O—tuber calcanei I—lateral side of base of little toe's proximal phalanx	Abducts little toe	Lateral plantar branch of tibial nerve
SECOND LAYER				
Flexor accessorius (quadratus plantae)	Rectangular muscle just deep to flexor digitorum brevis in posterior half of sole; two heads	O—medial and lateral sides of calcaneus I—tendon of flexor digitorum longus in mid sole	Straightens out the oblique pull of flexor digitorum longus	Lateral plantar branch of tibial nerve
Lumbricals	Four little "worms," like lumbricals in hand	O—from each tendon of flexor digitorum longus I—extensor expansion on proximal phalanx of toes 2–5, medial side	Function same as lumbricals of hand	Medial plantar nerve (first lumbrical) and lateral plantar nerve (second to fourth lumbrical)
THIRD LAYER				
Flexor hallucis brevis	Covers metatarsal 1; splits into two bellies; recall the flexor pollicis brevis of thumb	O—mostly from cuboid bone I—via two tendons onto both sides of the base of the proximal phalanx of big toe; each tendon has a sesamoid bone in it	Flexes big toe's metatarsophalangeal joint	Medial plantar branch of tibial nerve
Adductor hallucis	Oblique and transverse heads; deep to lumbricals; recall adductor pollicis in thumb	O—from bases of metatarsals 2–4 and from peroneus longus tendon (oblique head); from a ligament across metatarsophalangeal joints (transverse head) I—base of proximal phalanx of big toe, lateral side	Helps maintain the transverse arch of foot; weak adductor of big toe	Lateral plantar branch of tibial nerve

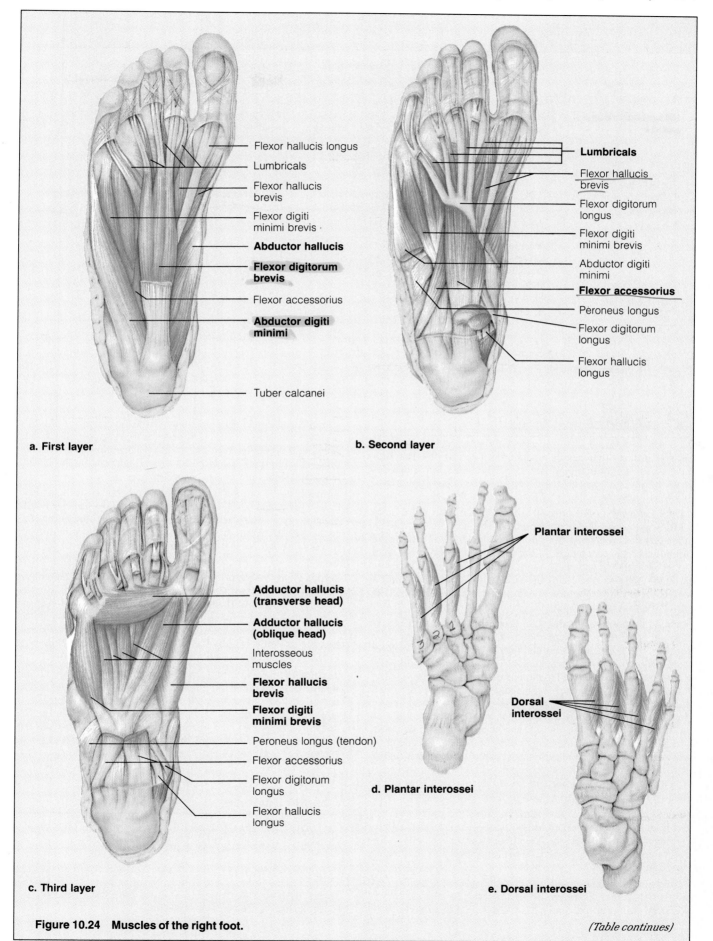

a. First layer

Flexor hallucis longus
Lumbricals
Flexor hallucis brevis
Flexor digiti minimi brevis
Abductor hallucis
Flexor digitorum brevis
Flexor accessorius
Abductor digiti minimi
Tuber calcanei

b. Second layer

Lumbricals
Flexor hallucis brevis
Flexor digitorum longus
Flexor digiti minimi brevis
Abductor digiti minimi
Flexor accessorius
Peroneus longus
Flexor digitorum longus
Flexor hallucis longus

c. Third layer

Adductor hallucis (transverse head)
Adductor hallucis (oblique head)
Interosseous muscles
Flexor hallucis brevis
Flexor digiti minimi brevis
Peroneus longus (tendon)
Flexor accessorius
Flexor digitorum longus
Flexor hallucis longus

d. Plantar interossei

Plantar interossei

e. Dorsal interossei

Dorsal interossei

Figure 10.24   Muscles of the right foot.

(Table continues)

Table 10.17 (continued)

Muscle	Description	Origin (O) and insertion (I)	Action	Nerve supply
Flexor digiti minimi brevis	Covers metatarsal 5; recall same muscle in hand	O—base of metatarsal 5 and tendon of peroneus longus I—base of proximal phalanx of toe 5	Flexes little toe at metatarsophalangeal joint	Lateral plantar branch of tibial nerve
FOURTH LAYER **Plantar and dorsal interossei**	Three plantar and four dorsal interossei; similar to the palmar and dorsal interossei of hand in locations, attachments, and actions; however, the long axis of foot around which these muscles orient is the second digit, not the third	See palmar and dorsal interossei (Table 10.13)	See palmar and dorsal interossei (Table 10.13)	Lateral plantar branch of tibial nerve

✚ Related Clinical Terms

Charley horse Tearing of a muscle, followed by bleeding into the tissues (hematoma) and severe pain; a common sports injury. Charley horse of the quadriceps femoris occurs frequently in football players.

Hallux valgus (hal'uks val'gus) (*valgus* = bent away from body midline) Permanent displacement of the big toe toward the other toes, often caused by tight shoes. In this condition, the sesamoid bones that underlie the head of the first metatarsal have been pushed laterally, away from the big toe. Thus, the abductor hallucis muscle of the sole, whose insertion tendon contains one of the sesamoids, cannot abduct the displaced toe back to its proper position.

Hernia (abdominal hernia) An abnormal protrusion of abdominal contents out of the abdominal cavity through a weak point in the muscles of the abdominal wall. The most commonly herniated elements are coils of small intestine and parts of the greater omentum (a large membrane, or mesentery, in the abdominal cavity). A hernia is usually caused by increased intra-abdominal pressure during lifting and straining. When an element herniates from the abdominal cavity, it does not pierce the skin, but usually lies deep to the skin or superficial fascia, where it can form a visible bulge on the body surface. The most common types of abdominal hernias occur in the groin region (inguinal hernia: p. 630), in the superior thigh (femoral hernia: p. 706), through the diaphragm (hiatal hernia), and in or near the navel (umbilical hernia: p. 696).

Mallet finger Extreme, persistent flexion of the distal phalanx of a finger, caused by forceful bending, as when a baseball hits the finger end-on. Part of the bony insertion of the extensor digitorum muscles is stripped off, so that the finger flexors flex the joint unopposed.

Rupture of calcaneus tendon The calcaneus (Achilles) tendon is the strongest tendon in the body, but its rupture is surprisingly common, particularly in older people as a result of stumbling and in sprinters who tense the tendon too hard while taking off for a race. The rupture leads to a gap just superior to the heel, and the calf bulges as the triceps surae muscles are released from their insertion. As a result, plantar flexion is very weak, but dorsiflexion is exaggerated.

Shinsplints Common term for pain in the anterior leg caused by swelling of the tibialis anterior muscle. This muscle swells after heavy exercise without adequate prior conditioning. Because it is tightly wrapped by fascia, the tibialis anterior cuts off its own circulation as it swells and presses painfully on its own nerves.

Tennis elbow Tenderness due to a degeneration of the tendon of origin of the forearm extensors at the lateral epicondyle of the humerus. The condition is caused and aggravated when these muscles contract forcefully to extend the hand at the wrist—as in executing a tennis backhand or lifting snow with a shovel. Despite its name, tennis elbow does not involve the elbow joint or the olecranon process.

Torticollis (tor"tĭ-kol'is; "turned neck") A condition in which the neck stays rotated to one side, keeping the head tilted in that direction ("wryneck"). Torticollis results from problems in a sternocleidomastoid muscle on one side of the neck only. The condition can be caused by spasms in the muscle or when the infant's muscle is stretched during birth. The damage causes the muscle to shrink in childhood.

Chapter Summary

LEVER SYSTEMS: BONE-MUSCLE RELATIONSHIPS (pp. 238–240)

1. A lever is a bar that is moved on a fulcrum. When an effort is applied to the lever, a load is moved. In the body, bones are the levers, joints are the fulcrums, and the effort is exerted by skeletal muscles pulling on their insertions.

2. When the effort is farther from the fulcrum than is the load, the lever works at a mechanical advantage (is slow and strong). When the effort is closer to the fulcrum than is the load, the lever works at a mechanical disadvantage (moves quickly and far but takes extra effort).

3. First-class levers (effort-fulcrum-load) may operate at a mechanical advantage or disadvantage. Second-class levers (fulcrum-load-effort) all work at a mechanical advantage. Third-class levers (fulcrum-effort-load) all work at a mechanical disadvantage. Most muscles of the body are in third-class lever systems to provide speed of movement.

INTERACTIONS OF SKELETAL MUSCLES IN THE BODY (p. 240)

1. Skeletal muscles are arranged in opposing groups across movable joints so that one group can reverse or modify the action of the other.

2. Prime movers (agonists) bear the main responsibility for a particular movement. Synergists help the prime movers, prevent undesirable movements, or stabilize joints (as fixators). Antagonists produce the opposite movement.

NAMING THE SKELETAL MUSCLES (p. 241)

1. Muscles are named for their location, shape, size, fascicle direction, location of attachments, number of origins, and action.

DEVELOPMENT AND BASIC ORGANIZATION OF THE MUSCLES (pp. 241–243)

1. Our musculature can be divided into four main groups: visceral musculature, trunk muscles, limb muscles, and pharyngeal arch muscles.

2. The trunk muscles derive from myotomes. They include dorsal muscles that extend the trunk and ventral muscles that flex the trunk. Limb muscles derive from the somatic mesoderm and exhibit the same arrangement in the upper and lower limbs: extensors on one side, flexors on the other. Pharyngeal arch muscles are skeletal muscles that surround the pharynx, but some migrate into the head and neck during development.

MAJOR SKELETAL MUSCLES OF THE BODY (pp. 243–292)

1. The muscles are divided into functional groups and presented in detail in Tables 10.1–10.17. The muscles of facial expression (Table 10.1) are thin and primarily insert on the facial skin. They close the eyes and lips, compress the cheeks, and allow us to smile and show our feelings.

2. Chewing muscles include the masseter and temporalis that elevate the mandible and the two pterygoids that grind food. Extrinsic muscles of the tongue anchor the tongue and control its movements (Table 10.2).

3. Deep muscles in the anterior neck help us swallow by lifting then depressing the larynx (suprahyoid and infrahyoid groups). Pharyngeal constrictors squeeze food into the esophagus (Table 10.3).

4. Muscles moving the head and muscles extending the spine are considered in Table 10.4. Flexion and rotation of the head and neck are brought about by the sternocleidomastoid and scalenes in the anterior and lateral neck. Extension of the spine and head is brought about by deep muscles of the back and the posterior neck (for example, the erector spinae and splenius muscles).

5. Movements of breathing are promoted by the diaphragm and by the intercostal muscles of the thorax (Table 10.5). Contraction of the diaphragm increases abdominal pressure during

straining, an action that is aided by the muscles of the abdominal wall.

6. The flat muscles of the abdominal wall (Table 10.6) are layered like plywood. This layering forms a strong wall that protects, supports, and compresses the abdominal contents. These muscles can also flex the trunk.

7. Muscles of the pelvic floor and perineum (Table 10.7) support our pelvic organs and inhibit urination and defecation.

8. The superficial muscles of the thorax attach to the pectoral girdle and fix or move the scapula during arm movements (Table 10.8).

9. All muscles that move the arm at the shoulder joint originate at least partly from the shoulder girdle (Table 10.9). Four muscles form the musculotendinous rotator cuff, which stabilizes the shoulder joint. As a rule, muscles located on the anterior thorax flex the arm (pectoralis major), whereas the posterior muscles extend it (latissimus dorsi). Abduction is mostly brought about by the deltoid. Adduction and rotation are brought about by many muscles on both the anterior and posterior sides.

10. Most muscles that lie within the arm move the forearm (Table 10.10). Anterior arm muscles are flexors of the forearm (biceps, brachialis), whereas posterior arm muscles are extensors of the forearm (triceps).

11. Forearm muscles primarily move the hand (Table 10.11). Anterior forearm muscles flex the hand and fingers (but include two pronators), whereas the posterior muscles are mostly extensors (but include the strong supinator).

12. The intrinsic muscles of the hand aid in fine movements of the fingers (Table 10.13) and in opposition, which helps us grip things in our palms. These small muscles are divided into thenar, hypothenar, and midpalmar groups.

13. Muscles that bridge the hip and knee joints move the thigh and the leg (Table 10.14). Anterior muscles mostly flex the thigh and extend the leg (quadriceps femoris), medial muscles adduct the thigh (adductors), and posterior muscles primarily abduct and extend the thigh and flex the leg (gluteals, hamstrings).

14. Muscles in the leg act on the ankle and toes (Table 10.15). Leg muscles in the anterior compartment are mostly dorsiflexors of the foot (extensor group). Muscles in the lateral compartment are everters and plantar flexors of the foot (peroneal group), and posterior leg muscles are strong plantar flexors (gastrocnemius and soleus, flexor group).

15. The intrinsic muscles of the foot (Table 10.17) support the foot arches and help move the toes. Most occur in the sole, arranged in four layers. They resemble the small muscles in the palm of the hand.

Review Questions

Multiple Choice and Matching Questions

1. A muscle that helps an agonist by causing a similar movement or by stabilizing a joint over which the agonist acts is (a) an antagonist, (b) a prime mover, (c) a synergist, (d) a fulcrum.

2. Match the muscles in column B to their embryonic origins in column A.

	Column A	Column B
_c_	(1) from head myotomes	(a) biceps brachii
_b_	(2) from trunk myotomes	(b) erector spinae
_e_	(3) limb extensor group	(c) tongue muscles
_a_	(4) limb flexor group	(d) chewing muscles
_d_	(5) pharyngeal arch muscles	(e) triceps brachii

3. The muscle that closes the eyes is (a) occipitalis, (b) zygomaticus, (c) corrugator supercilii, (d) orbicularis oris, (e) none of these.

4. The prime mover of inhalation (inspiration) in breathing is (a) the diaphragm, (b) internal intercostals, (c) external oblique, (d) smooth muscle inside the lungs.

5. The arm muscle that both flexes the forearm at the elbow and supinates it is the (a) brachialis, (b) brachioradialis, (c) biceps brachii, (d) triceps brachii.

6. The chewing muscles that protrude the mandible and grind the teeth from side to side are the (a) buccinators, (b) masseters, (c) temporalis, (d) pterygoids, (e) none of these.

7. Intrinsic muscles of the back that extend the spine (or head) include all of these except (a) splenius muscles, (b) semispinalis muscles, (c) scalene muscles, (d) erector spinae.

8. Which of the following is a large, deep muscle that protracts the scapula during punching? (a) serratus anterior, (b) rhomboids, (c) levator scapulae, (d) subscapularis.

9. Muscles that depress the hyoid bone and larynx include all but the (a) sternohyoid, (b) omohyoid, (c) geniohyoid, (d) sternothyroid.

10. The quadriceps femoris muscles include all but (a) vastus lateralis, (b) biceps femoris, (c) vastus intermedius, (d) rectus femoris.

11. A prime mover of thigh flexion at the hip is the (a) rectus femoris, (b) iliopsoas, (c) vastus intermedius, (d) gluteus maximus.

12. The prime mover of thigh extension at the hip in climbing stairs is the (a) gluteus maximus, (b) gluteus medius, (c) biceps femoris, (d) semimembranosus.

13. In walking, which two lower limb muscles keep the forward-swinging foot from dragging on the ground? (a) pronator teres and popliteus, (b) flexor digitorum longus and popliteus, (c) adductor longus and abductor digiti minimi in foot, (d) gluteus medius and tibialis anterior.

14. Someone who sticks out a thumb to hitch a ride is _____ the thumb. (a) extending, (b) abducting, (c) adducting, (d) opposing.

15. The major muscles used in doing a push-up are (a) biceps brachii and brachialis, (b) supraspinatus and subscapularis, (c) triceps brachii and pectoralis major, (d) coracobrachialis and latissimus dorsi.

16. The major muscles used in doing a chin-up are (a) triceps brachii and pectoralis major, (b) infraspinatus and biceps brachii, (c) serratus anterior and external oblique, (d) latissimus dorsi and brachialis.

Short Answer and Essay Questions

17. List four criteria used in naming muscles, and name a specific muscle that illustrates each criterion. (Do not use the same examples given in the text.)

18. Define and distinguish between first-, second-, and third-class levers. Which classes operate at a mechanical advantage?

19. What three muscles squeeze swallowed food through the pharynx to the esophagus?

20. (a) Name the four pairs of muscles that act together to compress the abdominal contents. (b) How do their fiber (fascicle) directions contribute to the strength of the abdominal wall? (c) Of these four muscles, which is the strongest flexor of the vertebral column?

21. List all six possible movements of the humerus that can occur at the shoulder joint, and name the prime mover(s) of each movement.

22. (a) Name two forearm muscles that are both powerful extensors and abductors of the hand at the wrist. (b) Name the only forearm muscle that can flex the distal interphalangeal joints. (c) Name the hand muscles responsible for opposition.

23. (a) Name the lateral rotators of the hip. (b) Explain the main function of each of the three gluteal muscles.

24. What is the functional reason why the muscle group on the dorsal leg (calf) is so much larger than the muscle group in the ventral region of the leg?

25. Name three muscles that help you keep your seat astride a horse.

26. Define (a) a fascia compartment in a limb and (b) a retinaculum in a limb.

Critical Thinking and Clinical Application Questions

27. Susan, a student nurse, was giving Mr. Graves a back rub. What two broad superficial muscles of his back were receiving most of her attention?

28. When Mrs. O'Brian returned to her doctor for a follow-up visit after childbirth, she complained that she was having problems controlling urine flow (was incontinent) when she sneezed. The physician gave Mrs. O'Brian instructions in performing exercises to strengthen the muscles of the pelvic floor. What are these specific muscles?

29. Mr. Ahmadi was advised by his physician to lose weight and start jogging. He began to jog daily. On the sixth day, he suddenly jumped out of the way of a speeding car. He heard a snapping sound that was immediately followed by pain in his right lower calf. A gap was visible between his swollen upper calf and his heel, and he was unable to plantar flex that foot. What do you think had happened?

30. Assume you are trying to lift a heavy weight off the ground with your right hand. Explain why it will be easier to flex your forearm at the elbow when your forearm is supinated than when it is pronated.

31. Trevor sent away for the *Build-Your-Body Mail-Order Course.* He performed only half of the types of exercises that were recommended, but he did these exercises with great enthusiasm for 8 months. That fall, he had a very impressive physique but did very poorly on the boxing team because he was slow and clumsy. His coach told him that he seemed to be muscle-bound. What had Trevor done wrong in his training program?

11

Fundamentals of the Nervous System and Nervous Tissue

Student Objectives

1. List the main functions of the nervous system.

Basic Divisions of the Nervous System (pp. 296–298)

2. Explain the structural and functional divisions of the nervous system.

Nervous Tissue (pp. 298–308)

3. Define neuron, describe its structural components, and relate each structure to its functional role.

4. Describe the fine structure of a synapse.

5. Classify neurons both structurally and functionally.

6. List the six types of supporting cells in nervous tissue, and distinguish them by shape and function.

7. Describe the structure of myelin sheaths.

8. Define nerve, and describe the structural components of nerves.

Function of Neurons and Impulse Conduction (pp. 308–310)

9. In simple terms, explain how action potentials are generated and propagated along axons.

10. Define reflex, and list the basic components of a reflex arc. Distinguish monosynaptic from polysynaptic reflexes.

Basic Neuronal Organization of the Nervous System (pp. 310–312)

11. Sketch a reflex arc consisting of a sensory neuron, interneuron, and motor neuron, and show how these neurons relate to the basic organization of the nervous system.

12. Distinguish gray matter from white matter in the central nervous system.

Development of the Nervous System and Nervous Tissue (pp. 312–313)

13. Describe the development of the nervous system in the embryo.

As you are driving down the freeway, a horn blares on your right. You swerve to your left. Charlie leaves a note on the kitchen table: "See you later—have the stuff ready at 6." You know that the "stuff" is chili with taco chips. You are dozing, and your infant son makes a soft cry. Instantly you awaken. What do these events have in common? All are everyday examples of the function of your nervous system, which has your body cells humming with activity nearly all of the time.

The nervous system is the master controlling and communicating system of the body. Every thought, action, instinct, and emotion reflects its activity. Cells of the nervous system communicate by means of electrical signals, which are rapid and specific, and usually cause almost immediate responses.

The nervous system has three overlapping functions: (1) It uses its millions of sensory receptors to monitor changes occurring both inside and outside the body. Each of these changes is called a *stimulus,* and the gathered information is called **sensory input.** (2) It processes and interprets the sensory input and makes decisions about what should be done at each moment—a process called **integration.** (3) It dictates a response by activating the *effector organs,* our muscles or glands; the response is called **motor output.** Some examples illustrate how these functions work together. When you are driving and see a red light ahead (sensory input), your nervous system integrates this information (red light means "stop"), and your foot goes for the brake (motor output). As another example, when you taste food your nervous system integrates this sensory information and signals your salivary glands to secrete more saliva into your mouth.

This chapter begins with an overview of the fundamental divisions of the nervous system. We then focus on the functional anatomy of nervous tissue, especially of the nerve cells, or *neurons,* which are the key to the efficient system of neural communication. Finally, we discuss how the arrangement of neurons determines the structural organization of the nervous system.

Basic Divisions of the Nervous System

We have only a single, highly integrated nervous system. However, for the sake of convenience, the nervous system can be divided into two principal parts (Figure 11.1). The **central nervous system (CNS)** consists of the *brain* and the *spinal cord,* which occupy the cranium and the vertebral canal, respectively. The CNS is the integrating and command center of the nervous system. It interprets incoming sensory information and dictates motor responses based on past experiences, reflexes, and current conditions. The **peripheral nervous system (PNS),** the part of the nervous system outside the CNS, consists mainly of the *nerves* that extend from the brain and spinal cord. *Cranial nerves* carry signals to and from the brain, while *spinal nerves* carry signals to and from the spinal cord. These peripheral nerves serve as communication lines that link all regions of the body to the central nervous system.

The PNS can be subdivided in several ways (Figure 11.2). Its **sensory,** or **afferent** (af'er-ent), **division** consists of nerve fibers that carry signals to the CNS from sensory receptors throughout the body (*afferent* means "carrying toward"). The **motor,** or **efferent** (ef'er-ent), **division** carries signals away from the CNS to the muscles and glands, causing these organs to contract or secrete (*efferent* means "carrying away"). Both the sensory and motor divisions are further divided according to the body region being served: The *somatic body region* consists of the structures external to the ventral body cavity (skin, skeletal musculature, bones), whereas the *visceral body region* contains the viscera within the ventral body cavity (digestive tube, lungs, heart, bladder, and so on). This scheme results in the following four subdivisions of the PNS: (1) *somatic sensory* (the sensory innervation of the outer part of the body); (2) *visceral sensory* (the sensory innervation of the viscera); (3) *somatic motor* (the motor innervation of most skeletal musculature); and (4) *visceral motor* (the motor innervation of muscle in the viscera and of glands). Because they are essential to an understanding of the nervous system, we will describe these four subdivisions more completely:

1. The **general somatic senses** are the senses whose receptors are spread widely through the outer part of the body (the term *general* means "widespread"). These include the many senses experienced on the skin and in the body wall (touch, pain, pressure, vibration, and temperature). Another group of general somatic senses is the **proprioceptive senses,** or **proprioception** (pro″pre-o-sep′shun). These senses, which are unfamiliar to many people, measure the amount of stretch in our muscles, tendons, and joint capsules. They inform us of the position and movement of our bodies in space, giving us a "body sense." (*Proprioception* literally means "sensing one's own body"). To demonstrate proprioception, flex and extend your fingers without looking at them—you will be able to feel exactly which joints are moving.

The **special somatic senses** are the somatic senses whose receptors are confined to relatively small areas rather than spread widely throughout the body (the term *special* means "localized"). Most special senses are confined to the head, including vision (receptors in the eye), smell (receptors in the nose), and hearing and balance (receptors in the ear). Note that the technical name for the sense of balance is **equilibrium**.

2. General visceral senses include stretch, pain, and temperature, which can be felt widely in the digestive and urinary tracts, reproductive organs, and other viscera. Nausea and hunger are also general visceral senses. Taste is a **special visceral sense**, because the taste receptors are localized on the tongue and on some surrounding structures.

3. The **general somatic motor** part of the PNS signals the contraction of all the skeletal muscles in the body (except the pharyngeal arch musculature: See below). We have voluntary control over the contraction of our skeletal muscles, so the somatic motor system is often called the *voluntary nervous system.* Because skeletal muscles are widely distributed throughout the body, there is no *special* somatic motor category.

4. The **general visceral motor** part of the PNS signals the contraction of the smooth muscle and cardiac muscle of the body and the secretion of glands. General visceral motor neurons make up the *autonomic*

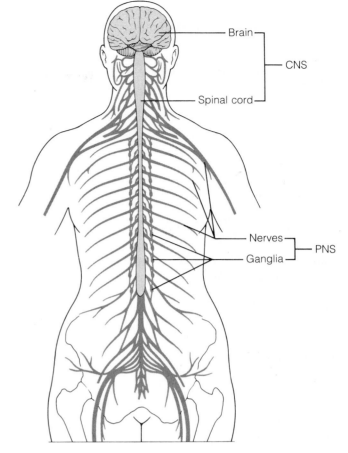

Figure 11.1   Basic divisions of the nervous system: The CNS and the PNS.

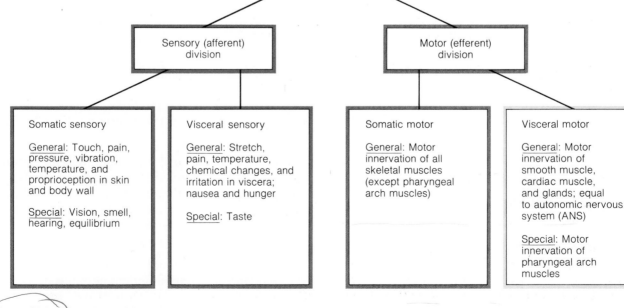

Figure 11.2   Flowchart of the nervous system.

(aw-to-nom'ik) *nervous system (ANS)*, which controls the function of the visceral organs (see Chapter 14). Because we generally have no voluntary control over such activities as pumping of the heart and movement of food through the digestive tract, the ANS is also called the *involuntary nervous system.*

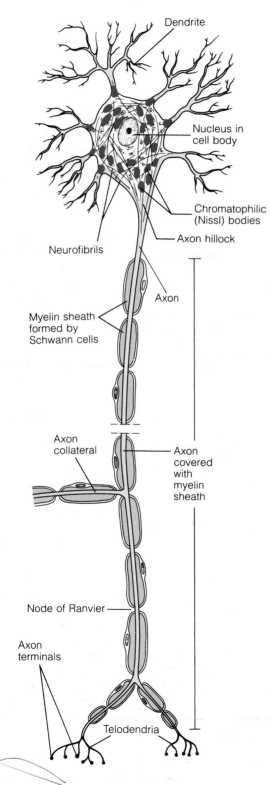

Figure 11.3   Structure of a typical large neuron (a motor neuron).

Visceral motor output also innervates the pharyngeal arch musculature. You will recall from page 242 that although this is skeletal musculature, it originates around the embryonic pharynx, a visceral organ. The motor innervation of this pharyngeal musculature, which is localized in the head and neck, is called **special visceral motor**.

We will return to these basic functional subdivisions of the nervous system throughout the next few chapters.

Nervous Tissue

The nervous system consists mostly of **nervous tissue**. Nervous tissue is highly cellular. In the CNS, for example, less than 20% of the tissue is extracellular space—the cells are densely packed and tightly intertwined. Although exceedingly complex, nervous tissue is made up of just two main types of cells: (1) *neurons,* the excitable nerve cells that transmit electrical signals, and (2) *supporting cells,* nonexcitable cells that surround and wrap the neurons. Both of these cell types develop from the same embryonic tissues, neural tube and neural crest (p. 55).

The Neuron

Overview

The human body contains many billions of **neurons**, or **nerve cells** (Figure 11.3), which are the basic structural units of the nervous system. Neurons are highly specialized cells that conduct electrical signals from one part of the body to another. These signals are transmitted along the plasma membrane, often taking the form of *nerve impulses,* or *action potentials.* Basically, an impulse is a reversal of electrical charge that travels along the neuronal membrane (see p. 308). Besides their ability to conduct electrical signals, neurons have other special characteristics:

1. They have *extreme longevity.* Given good nutrition, neurons can live and function optimally for a lifetime (over 100 years).

2. They *do not divide.* As the fetal neurons assume their roles as communicating links in the nervous system, they lose their ability to undergo mitosis. We pay a high price for this characteristic of neurons: Because they are unable to reproduce themselves, they cannot be replaced if they are destroyed.

3. They have an exceptionally *high metabolic rate,* requiring continuous and abundant supplies of oxygen and glucose. Neurons cannot survive for more than a few minutes without oxygen.

Neurons typically are large, complex cells. Although they vary in structure, they all have a *cell body* from which one or more *processes* project (Figures 11.3 and 11.4).

The Cell Body

The **cell body** is also called a *soma* (*soma* = body) or a *perikaryon* (per"ĭ-kar'e-on; "around the nucleus"). Neuron cell bodies vary widely in diameter, from 5 to 140 μm. Each consists of a single nucleus surrounded by cytoplasm. The large nucleus, said to resemble an owl's eye, is spherical and clear and contains an obvious nucleolus near its center (Figure 11.4). The cytoplasm contains all the usual cellular organelles as well as distinctive **chromatophilic (Nissl) bodies** (kro-mah'to-fil-ic; "color-loving; easily stained"). These bodies are large clusters of rough endoplasmic reticulum and free ribosomes that stain darkly with basic dyes. These cellular organelles continually renew the membranes of the cell and the protein part of the cytosol. **Neurofibrils** are bundles of intermediate filaments *(neurofilaments)* that run in a network between the chromatophilic bodies (Figure 11.3). Like all intermediate filaments (p. 33), neurofilaments keep the cell from being pulled apart when it is subjected to tensile forces. *Lipofuscin* (lip"o-fu'sin), a yellow-brown pigment, is a harmless by-product of lysosomal activity. Because lipofuscin is most abundant in neurons of elderly individuals, it may be related to aging. The cell body is the focal point for the outgrowth of the neuron processes during embryonic development. In most neurons, the plasma membrane of the cell body acts as a receptive surface that receives signals from other neurons.

Most neuron cell bodies are located within the CNS, where they are protected by the bones of the skull and vertebral column. Furthermore, clusters of cell bodies called **ganglia** lie along the nerves in the PNS (see Figure 11.1). The singular of *ganglia* is **ganglion** (gang'le-on), literally a "knot in a string."

Processes

Arm-like **processes** extend from the cell bodies of all neurons. These processes are of two types, *dendrites* and *axons,* which differ from each other both in structure and in the functional properties of their plasma membranes. The convention is to describe the processes by using a motor neuron as an example of a typical neuron (see Figure 11.3). Motor neurons do indeed resemble many neurons in the arrangement of their processes, but keep in mind that sensory neurons and many small neurons differ from the "typical" pattern presented here.

Dendrites.   Most neurons have numerous **dendrites**, processes that branch like the limbs on a tree (*dendro*

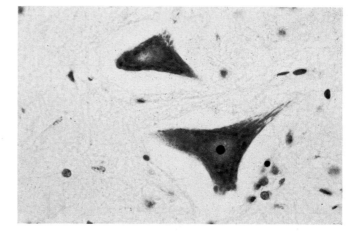

Figure 11.4   Neuron cell bodies in the CNS.   For this micrograph, motor neurons in the spinal cord were stained with a dye that reveals the nuclei and chromatophilic bodies but does not stain the cell processes beyond the bases of the dendrites (400 ×). Note the large nuclei and nucleoli of these neurons. The small nuclei between the neurons belong to neuroglia.

= tree). Virtually all organelles that occur in the cell body are also found within dendrites, and chromatophilic bodies extend into the basal part of each dendrite. Dendrites function as *receptive sites,* providing an enlarged surface area for signals from other neurons. Dendrites conduct electrical signals *toward* the cell body. It must be noted that dendrites do not conduct true impulses (action potentials) but another type of electrical signals, called *graded potentials* (see p. 308).

The Axon.   Each neuron has only one **axon** (ak'son; "axis, axle"). The axon arises from a cone-shaped region of the cell body called the **axon hillock** ("little hill") and then tapers to form a slender process that stays uniform in diameter for the rest of its length. Chromatophilic bodies and the Golgi apparatus are absent from the axon, as well as the axon hillock; otherwise, axons contain the same organelles found in the cell body. Neurofilaments, actin filaments, and microtubules are especially evident in axons, where they provide structural strength. These cytoskeletal elements also aid in the transport of substances to and from the cell body as the axonal cytoplasm is continually recycled and renewed. This movement of substances along axons is called **axonal transport.**

The axon of some neurons is short, but in others it can be extremely long. For example, the axons of the motor neurons that control muscles in the foot extend from the lumbar region of the spine to the sole, a distance of a meter or more (3–4 feet). Any long axon is called a **nerve fiber.**

Neurons with the longest axons have the largest cell bodies, whereas neurons with short axons tend to have small cell bodies. Can you think of an explanation for this relationship? (Answer on p. 300)

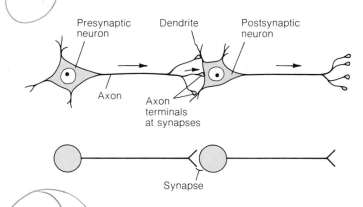

Figure 11.5 Two neurons communicating at synapses.
These synapses occur where the axon terminals of a presynaptic neuron contact the membrane of a postsynaptic neuron. Arrows indicate the direction of flow of electrical signals. The drawing below is a highly simplified version of the drawing above. Such simplified drawings are used to represent synapsing neurons in many figures in this text.

Recall that the cell body continuously maintains and renews the components of the axon. Therefore, a longer axon requires a larger cell body to house more cellular "machinery" for the axon-sustaining functions. ∎

Although axons branch far less frequently than dendrites, occasional branches do occur along their length. These branches, called **axon collaterals**, extend from the axon at more or less right angles (Figure 11.3). Whether an axon remains undivided or has collaterals, it usually branches profusely at its terminus (end): Ten thousand of these terminal branches, or **telodendria** (tel"o-den'dre-ah), per neuron is not unusual. The telodendria end in knobs that are variously called **axon terminals,** *end bulbs,* or *boutons* (bootonz'; "buttons"). These knobs contact other neurons to form specialized cell junctions called *synapses* (see below).

Functionally, axons are impulse generators and conductors that transmit nerve impulses *away* from the cell body. The nerve impulse is usually generated at the axon hillock and conducted along the axon to the axon terminals. There, it causes the release of chemicals called *neurotransmitters* into the extracellular space. The neurotransmitters excite or inhibit the neurons with which the axon is in close contact. Since each neuron both receives signals from and sends signals to scores of other neurons, it carries on "conversations" with many different neurons at the same time.

Axon diameter varies considerably among the different neurons of the body. Axons with larger diameters conduct impulses more rapidly than those with smaller diameters. This is because the resistance to the passage of an electrical current declines as the diameter of any "cable" increases—a basic law of physics.

Synapses

Neurons communicate with each other at synapses (Figures 11.5 to 11.7). A **synapse** (sin'aps; "union") is a cell junction that mediates the transfer of information from one neuron to the next. Because signals pass across most synapses in one direction only, synapses determine the direction of information flow through the nervous system.

The neuron that conducts signals toward a synapse is called the **presynaptic neuron** (Figure 11.5). Presynaptic neurons are information senders. The neuron that transmits electrical activity away from the synapse is called the **postsynaptic neuron**. Postsynaptic neurons are information recipients. As you might anticipate, most neurons function as both presynaptic and postsynaptic neurons, receiving information from some neurons and dispatching it to others.

Most synapses occur between the axon terminals of one neuron and the dendrites of another neuron. These are known as **axodendritic synapses** (Figure 11.6). However, many synapses also occur between axons and neuron cell bodies; these are the **axosomatic synapses**. Less common, and far less understood, are synapses between two axons *(axoaxonic)*, between two dendrites *(dendrodendritic)*, or between a dendrite and a cell body *(dendrosomatic)*.

Structurally, synapses are elaborate cell junctions. We will focus on the axodendritic synapse (Figure 11.7) because its structure is representative of most synapse types. In a synapse, the plasma membranes of the two neurons are separated by a narrow **synaptic cleft**. On the presynaptic side, the axon terminal contains **synaptic vesicles**. These are membrane-bound sacs filled with **neurotransmitters**, the molecules that transmit signals across the synapse. Mitochondria are abundant in the axon terminal, as the secretion of neurotransmitters requires a great deal of energy.

How does such a synapse work? When an impulse travels along the axon of the presynaptic neuron, it signals the synaptic vesicles to fuse with the presynaptic membrane (Figure 11.7b). The released neurotransmitter molecules diffuse across the synaptic cleft and bind to the postsynaptic membrane. This binding affects the membrane charge on the postsynaptic neuron, influencing the generation of a nerve impulse in that neuron (see p. 308 for more information).

Classification of Neurons

Neurons may be classified both by structure and by function.

Structural Classification. Neurons are grouped structurally according to the number of processes that

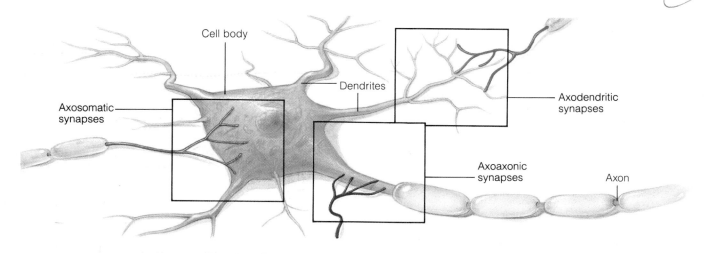

Figure 11.6   Some important types of synapses.   Several axons are synapsing on the large neuron in the center of this figure. Axodendritic, axosomatic, and axoaxonic synapses are shown. Axon terminals of presynaptic neurons are indicated in different colors in the different types of synapses.

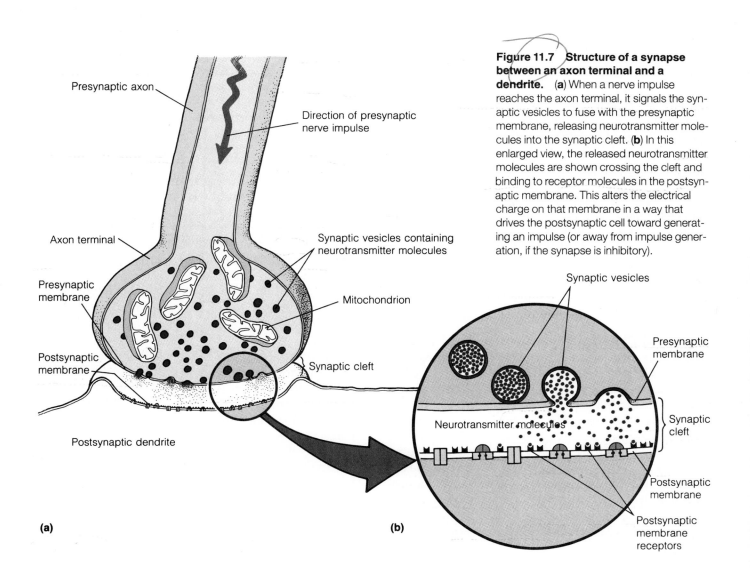

Figure 11.7   Structure of a synapse between an axon terminal and a dendrite.   (**a**) When a nerve impulse reaches the axon terminal, it signals the synaptic vesicles to fuse with the presynaptic membrane, releasing neurotransmitter molecules into the synaptic cleft. (**b**) In this enlarged view, the released neurotransmitter molecules are shown crossing the cleft and binding to receptor molecules in the postsynaptic membrane. This alters the electrical charge on that membrane in a way that drives the postsynaptic cell toward generating an impulse (or away from impulse generation, if the synapse is inhibitory).

(a)

(b)

Figure 11.8   The three types of neurons based on structural classification (number of processes projecting from the cell body). The "blue beads" that cover the axons in parts (a) and (c) are Schwann cells forming a myelin sheath (discussed later in the chapter).

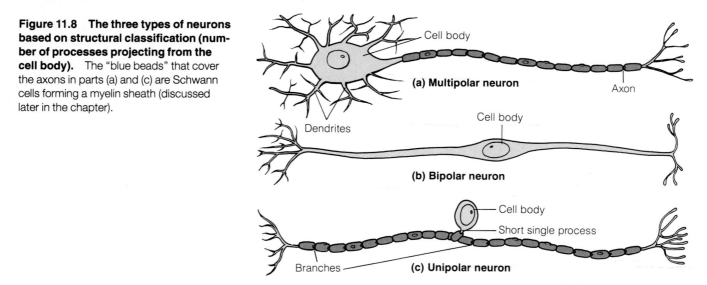

(a) Multipolar neuron

Cell body

Axon

Dendrites

Cell body

(b) Bipolar neuron

Cell body

Short single process

Branches

(c) Unipolar neuron

extend from the cell body (Figure 11.8). By this classification, neurons are *multipolar, bipolar,* or *unipolar* (*polar* = ends, poles).

Most neurons are **multipolar neurons;** that is, they have more than two processes. Usually, multipolar neurons have numerous dendrites and an axon. (However, some small multipolar neurons have no axon and rely strictly on their dendrites for conducting signals.) Well over 99% of neurons in the body belong to the multipolar class.

Bipolar neurons have two processes that extend from opposite sides of the cell body. These very rare neurons occur in some of the special sensory organs (inner ear, smell region of the nose, retina of eye), where they mostly serve as sensory neurons.

Unipolar neurons have a short, single process that emerges from the cell body and divides like a T into two long branches (Figure 11.8c). Most unipolar neurons start out as bipolar neurons whose two processes fuse together near the cell body during development. Therefore, they are more properly called *pseudounipolar neurons* (*pseudo* = false). Unipolar neurons make up the typical sensory neurons, which are discussed below.

Functional Classification. The functional classification scheme groups neurons according to the direction the nerve impulse travels relative to the CNS. Based on this criterion, there are *sensory neurons, motor neurons,* and *interneurons* (Figure 11.9).

Sensory neurons or *afferent neurons,* transmit impulses *toward* the CNS from sensory receptors in the PNS. Virtually all sensory neurons are unipolar, and their cell bodies are located in ganglia outside the CNS. As we mentioned above, the short, single process near the neuron cell body divides into two longer branches. One of these branches runs centrally into the CNS and is called the **central process,** while the

other branch extends peripherally to the receptors and is called the **peripheral process**. Both of these processes function as one, carrying impulses directly from the peripheral receptors to the CNS.

Are these processes of sensory neurons dendrites, or are they axons? The *central process* is clearly an axon because it (1) carries a nerve impulse and (2) carries that impulse away from the cell body—the two criteria that define an axon (p. 299). The *peripheral process,* by contrast, is ambiguous: It carries a nerve impulse, as does an axon; however, the signals it carries travel *toward* the cell body, a fundamental feature of dendrites. We choose to call the peripheral process an *axon* and a nerve fiber, because its fine structure is identical to that of true axons. For sensory neurons, we will use the term *dendrite* to mean only the small branches in the receptor region, at the very end of the peripheral process (Figure 11.9).

Motor neurons, or *efferent neurons,* carry impulses *away* from the CNS to the effector organs (muscles and glands). Motor neurons are multipolar, and (except for some neurons of the autonomic nervous system) their cell bodies are located within the CNS. Motor neurons form junctions with effector cells, signaling muscles to contract or glands to secrete.

Interneurons, or *association neurons,* lie between motor and sensory neurons and are confined entirely to the CNS (Figure 11.9). Interneurons are linked together into complex neural pathways and integrate information in complex ways. Interneurons make up 99.98% of the neurons of the body. This number reflects the vast amount of information processing that occurs in the human CNS. Almost all interneurons are multipolar neurons, but they show great diversity in size and in the branching patterns of their processes. The diverse neurons that are pictured in Figure 11.10 are all interneurons from the brain.

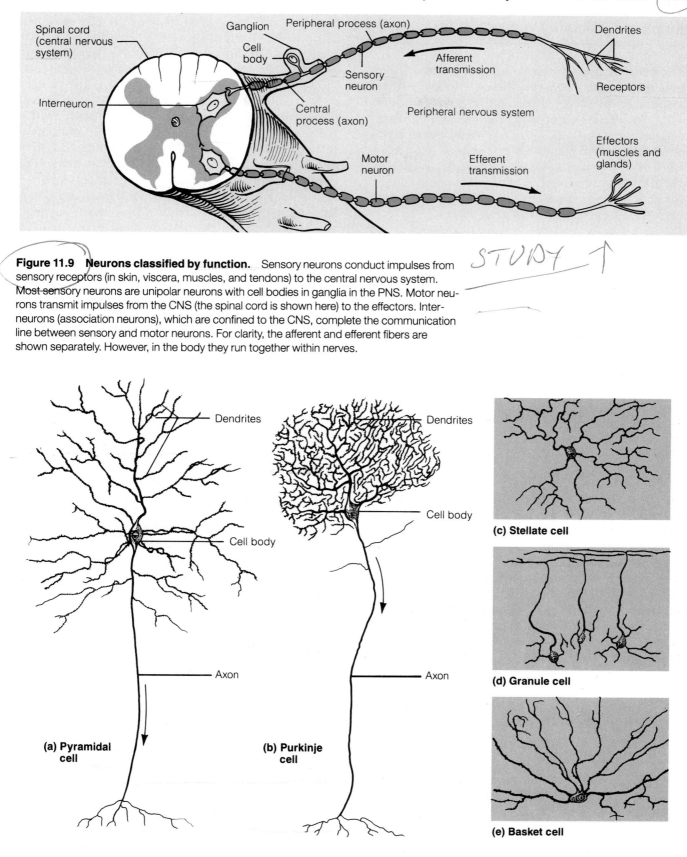

Figure 11.9   Neurons classified by function.   Sensory neurons conduct impulses from sensory receptors (in skin, viscera, muscles, and tendons) to the central nervous system. Most sensory neurons are unipolar neurons with cell bodies in ganglia in the PNS. Motor neurons transmit impulses from the CNS (the spinal cord is shown here) to the effectors. Interneurons (association neurons), which are confined to the CNS, complete the communication line between sensory and motor neurons. For clarity, the afferent and efferent fibers are shown separately. However, in the body they run together within nerves.

Figure 11.10   Variety of interneurons in the CNS (a–e).   Each neuron's distinct appearance depends partly on the length of its axon but mostly on the number and branching pattern of its dendrites. The neurons with more complex dendrites have more synaptic input; that is, more axon terminals synapse on such neurons.

Supporting Cells

All neurons associate closely with non-nervous **supporting cells**, of which there are six types. Four of these are in the CNS, and two are in the PNS (Figure 11.11). Each type of supporting cell has a unique function, but in general, these cells provide a supportive scaffolding for the neurons. Furthermore, supporting cells cover all parts of the neurons that are not involved in synapses. Such masking insulates the neurons and prevents the electrical activities of adjacent neurons from interfering with each other.

✚ The importance of supporting cells in insulating nerve fibers from one another is illustrated in the painful disorder called **tic douloureux** (tik doo″loo-roo′; "wincing in pain"). In this condition, the supporting cells around the sensory nerve fibers in the main nerve of the face (the trigeminal nerve) degenerate and are lost. As a result, impulses in nerve fibers that carry *touch* sensations proceed to influence and stimulate the uninsulated *pain* fibers in the same nerve, leading to a perception of pain by the brain. Because of this crossover, the softest touch to the face can produce agonizing pain. For more information on tic douloureux, see Table 13.2, page 371. ■

Supporting Cells in the CNS

The supporting cells in the CNS are collectively called the **neuroglia** (nu-rog′le-ah; "nerve glue") or **glial** (gle′al) **cells**. (Most authorities restrict the name *neuroglia* to the supporting cells in the CNS, but others consider all supporting cells neuroglia, including those in the PNS.) Like neurons, most glial cells have branching processes and a central cell body (Figure 11.11a–d). Neuroglia can be distinguished from neurons, however, by their much smaller size and by their darker-staining nuclei (Figure 11.4). Neuroglia outnumber neurons in the CNS by as much as 50 to 1, and they make up about half the mass of the brain. Unlike neurons, glial cells can divide throughout life.

✚ Can you deduce which of the two cell types in neural tissue—neurons or neuroglia—gives rise to more brain tumors?

The glial cells do. Since glial cells can divide regularly, they accumulate the "mistakes" in DNA replication that may transform them into neoplastic cells. This does not occur in neurons, which do not divide. Therefore, most tumors that originate in the brain (60%) are **gliomas** (tumors formed by uncontrolled proliferation of glial cells). ■

Star-shaped **astrocytes** (as′tro-sītz; "star cells") are the most abundant glial cells (Figure 11.11a). They have many radiating processes with bulbous ends.

Some of these bulbs cling to neurons, whereas others cling to capillaries. Because of these connections, some scientists believe that astrocytes transfer nutrients from the capillary blood to the neurons, thereby "nursing" the nerve cells. While their nutritive function is still disputed, most agree that astrocytes help control the ionic environment around neurons: The concentrations of various ions outside the axons must be kept within narrow limits for nerve impulses to be generated and conducted. Additionally, astrocytes recapture (and recycle) released neurotransmitters.

Microglia are the smallest and least abundant of the neuroglia (Figure 11.11b). They have elongated cell bodies and cell processes with many pointed projections, like a thorny bush. The microglia are phagocytes, the macrophages of the CNS. They engulf invading microorganisms and injured or dead neurons. The origin of microglia is controversial. Some authorities believe they originate, like the other macrophages of the body, from a type of blood cell called a monocyte. Others claim that microglia derive from the ectoderm of the embryonic neural tube, as do the other neuroglial cells.

You will recall from Chapter 3 (p. 55) that the CNS originates in the embryo as a hollow neural tube and retains a central cavity throughout life. **Ependymal cells** (ĕ-pen′dĭ-mal; "wrapping garment") form a simple epithelium that lines the central cavity of the spinal cord and brain (Figure 11.11c). Here, these cells provide a fairly permeable layer between the cerebrospinal fluid that fills this cavity and the tissue fluid that bathes the cells of the CNS. Ependymal cells bear cilia that help circulate the cerebrospinal fluid.

Oligodendrocytes (ol″ĭ-go-den′dro-sītz) (Figure 11.11d) have fewer branches than astrocytes. Indeed, their name means "few-branch cells." Oligodendrocytes line up in small groups along the thicker axons in the CNS. They wrap their cell processes around these axons, producing insulating coverings called *myelin sheaths* (discussed in detail below).

Supporting Cells in the PNS

The two kinds of supporting cells in the PNS are *satellite cells* and *Schwann cells*. These very similar cell types differ mainly in location. **Satellite cells** surround neuron cell bodies within ganglia (Figure 11.11e). Their name comes from a fancied resemblance to the moons (satellites) around a planet. **Schwann cells** (also called *neurolemmocytes*) surround all axons in the PNS and form myelin sheaths around many of these axons.

Myelin

Myelin (mi′ĕ-lin), a lipoprotein, is a fatty substance that surrounds the thicker axons of the body. It takes

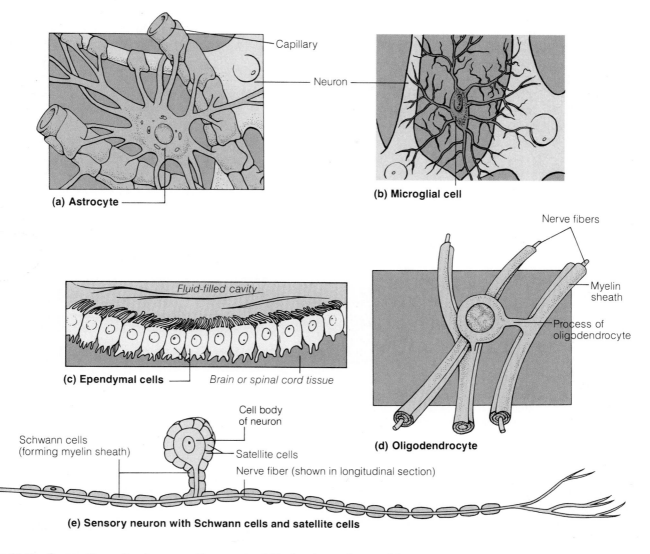

Figure 11.11   Supporting cells of nervous tissue.   (a–d) The four types of neuroglial cells in the CNS. In **(c)**, note that ependymal cells line the fluid-filled central cavity of the spinal cord and brain. **(e)** The two types of supporting cells in the PNS: Schwann cells and satellite cells around a sensory neuron.

the form of a segmented **myelin sheath** (Figures 11.12 and 11.13). This sheath consists of the plasma membrane of supporting cells—a membrane that is arranged in concentric rolls around the axon. By acting as an insulating layer that prevents the leakage of electrical current from the axon, the myelin sheath increases the speed of impulse conduction along the axon.

Myelin in the PNS.   As we stated above, the myelin sheaths in the PNS are formed by Schwann cells (Figure 11.12a). Myelin forms during the fetal period and the first year or so of postnatal life. To form myelin, the Schwann cells first indent to receive the axon and then wrap themselves around the axon repeatedly, in a jellyroll fashion (Figure 11.13a–d). Initially the wrapping is loose, but the cytoplasm of the Schwann cell gradually squeezes out from between the mem-

brane layers. When the wrapping process is finished, many concentric layers of Schwann cell plasma membrane enclose the axon. This coil of membranes is the true myelin sheath. The nucleus and most of the cytoplasm of the Schwann cell end up just external to the myelin layers. This external material is called the **neurilemma** ("neuron sheath") (Figure 11.13d).

Since the adjacent Schwann cells along a myelinated axon do not touch one another, there are gaps in the myelin sheath. These gaps, called **nodes of Ranvier**, or *neurofibral nodes,* occur at regular intervals about 1 mm apart (Figures 11.12 and 11.13e). As nerve impulses pass along the axon, they jump quickly from one node to the next without traveling along the myelin-covered neuronal membrane in between.

Only the thick, rapidly conducting axons are sheathed with myelin. Thin, slowly conducting axons that lack a myelin sheath are called **unmyelinated**

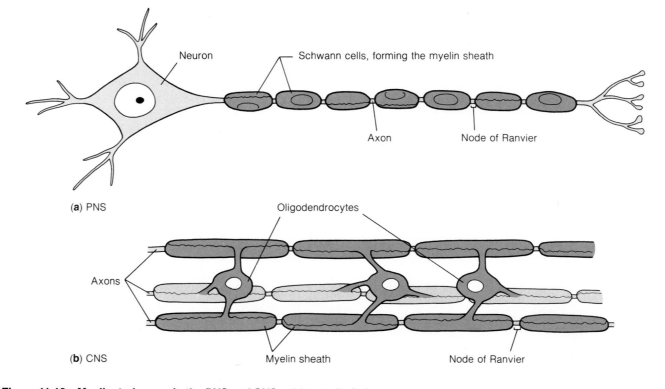

(a) PNS

(b) CNS

Figure 11.12 Myelinated axons in the PNS and CNS. **(a)** In the PNS, Schwann cells form the myelin sheath around nerve fibers. **(b)** In the CNS, oligodendrocytes form the myelin sheath. Note that each oligodendrocyte contributes myelin to several axons, whereas each Schwann cell is associated with just one axon.

axons (Figure 11.13f). Schwann cells do surround such axons but do not wrap them in concentric layers of membrane. A single Schwann cell can partly enclose 15 or more unmyelinated axons, each of which occupies a separate tubular recess in the surface of the Schwann cell.

Myelin in the CNS. Oligodendrocytes form the myelin sheaths in the brain and spinal cord (Figure 11.12b). In contrast to Schwann cells, each oligodendrocyte has multiple processes that coil around several different axons. Nodes of Ranvier are present, although they are more widely spaced than those in the PNS.

As in the PNS, the thinnest axons in the CNS are unmyelinated. These unmyelinated axons are covered by the long processes of glial cells that are so abundant in the CNS.

Can you predict the consequences of a widespread loss of myelin from the adult nervous system?

In **multiple sclerosis (MS),** which seems to be an autoimmune disease, the myelin sheaths gradually disappear, and the conduction of nerve impulses slows and then ceases. Affected people experience sensory disorders, their movements weaken, and they lose control of their muscles. Researchers hope to find a "cure" for this disease by blocking the ability of the immune system to attack myelin. ■

Nerves

A **nerve** is a cord-like organ in the peripheral nervous system (Figures 11.1 and 11.14). Each nerve consists of many axons (nerve fibers) arranged in parallel bundles and enclosed by successive wrappings of connective tissue. Almost all nerves contain both myelinated and unmyelinated sensory and motor fibers.

Within a nerve, each axon is surrounded by Schwann cells, then a delicate layer of loose connective tissue called **endoneurium** (en"do-nu're-um). Groups of axons are bound into bundles called **nerve fascicles** by a wrapping of connective tissue called the **perineurium**. Finally, the whole nerve is surrounded by a tough fibrous sheath, the **epineurium**. The layers of connective tissue in nerves correspond exactly to those in skeletal muscle (recall the endomysium, perimysium, and epimysium from p. 219). The connective tissue in a nerve contains the blood vessels that nourish the axons and Schwann cells.

In this chapter, we have defined *nerves, nerve fibers,* and *neurons.* Special effort must be made to avoid confusing these sound-alike terms. A neuron is a nerve cell, a nerve fiber is a long axon, and a nerve is a collection of nerve fibers in the PNS.

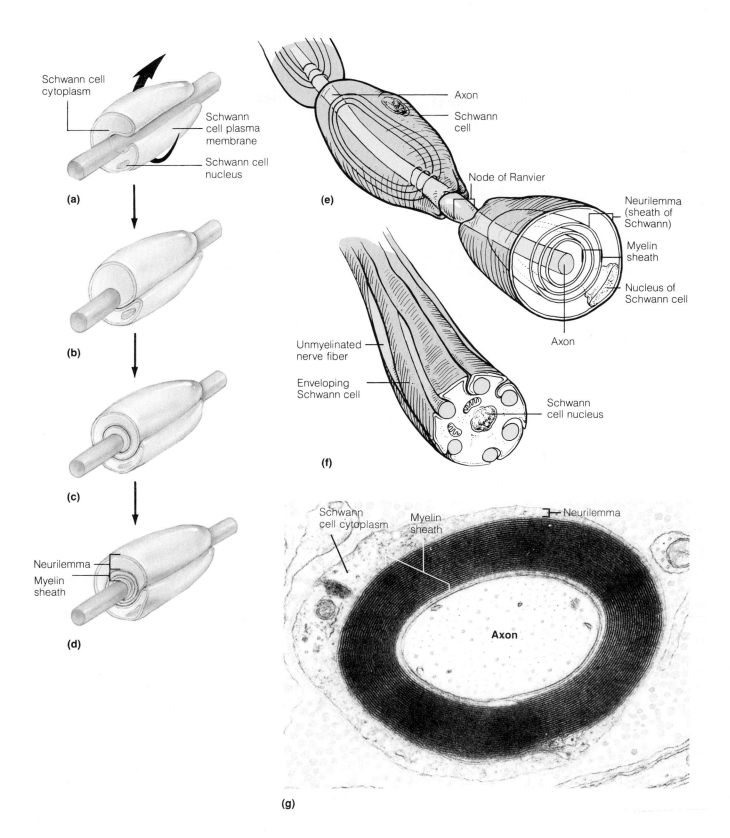

Figure 11.13   Relation of Schwann cells to axons in the PNS. **(a–d)** A Schwann cell forms a myelin sheath on a peripheral axon by coiling around it. **(e)** Three-dimensional "see-through" view of a myelinated axon in the PNS, showing adjacent Schwann cells and the bare node of Ranvier between them. **(f)** Unmyelinated axons in the PNS. These axons lie in tunnels in Schwann cells, but no coiling occurs. **(g)** Electron micrograph of a myelinated axon, cross-sectional view (20,200 ×).

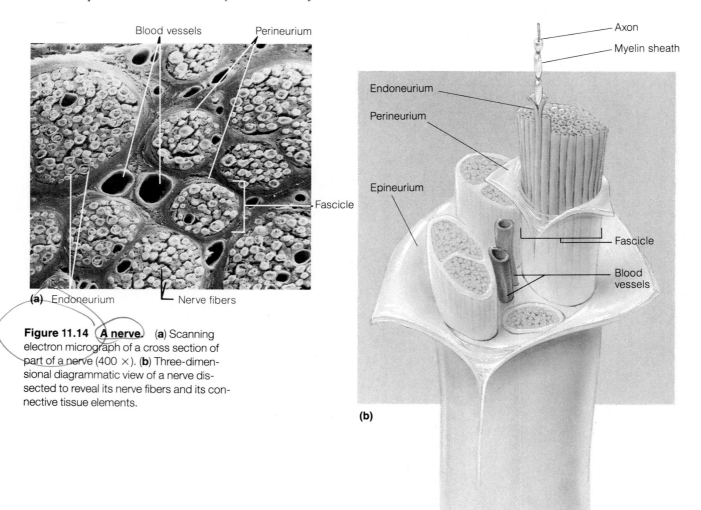

Figure 11.14 **A nerve.** **(a)** Scanning electron micrograph of a cross section of part of a nerve (400 ×). **(b)** Three-dimensional diagrammatic view of a nerve dissected to reveal its nerve fibers and its connective tissue elements.

Function of Neurons and Impulse Conduction

Nerve Impulses

We will now consider, in a simplified way, how neurons carry signals. In a resting (unstimulated) neuron, the plasma membrane is *polarized;* that is, its inner, cytoplasmic side is negatively charged with respect to its outer, extracellular side (Figure 11.15a). Furthermore, the concentration of sodium (Na+) ions is higher externally than internally, and the opposite is true for potassium (K+) ions. When an axon is stimulated experimentally (when it is pinched or shocked, for example), the permeability of the plasma membrane changes at the site of the stimulus, allowing positive Na+ ions to rush in (Figure 11.15b). As a result, the inner face of the axonal membrane becomes less negative, an event called *depolarization.*

If the depolarizing stimulus applied to the axon is strong enough, it will trigger a **nerve impulse**, or **action potential**, in which the membrane is not only depolarized, but its polarity is completely reversed (it

becomes negative externally and positive internally). Once begun, the nerve impulse travels rapidly and without decreasing in strength down the entire length of the axon (Figure 11.15c). The membrane then repolarizes itself (Figure 11.15d).

In the human body, impulse generation is not determined by direct application of the stimulus to an axon. Instead, the stimuli are applied to the dendrites and the cell body—the receptive zone of the neuron. The electrical response of this receptive zone determines whether the axon will generate an impulse. When the membrane of the receptive zone is stimulated, it does not undergo a polarity reversal. Instead, it undergoes a local depolarization in which the inner face of the membrane merely becomes less negative (but not positive). This local depolarization, called a **graded potential,** spreads from the receptive zone to the axon hillock (trigger zone), decreasing in strength as it travels. If this depolarizing signal is strong enough when it reaches the axon hillock, it acts as the trigger that initiates an action potential in the axon.

Many neurons do not receive stimuli directly from the environment but are "stimulated" only by signals received at synapses from other neurons. Synaptic input influences impulse generation in the fol-

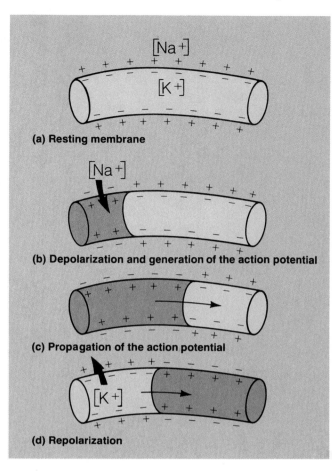

(a) Resting membrane

(b) Depolarization and generation of the action potential

(c) Propagation of the action potential

(d) Repolarization

Figure 11.15   A segment of an axon, illustrating the generation and propagation of an action potential (impulse).   See text for an explanation of stages (**a**) through (**d**).

defined as rapid, automatic motor responses to stimuli. Reflexes are unlearned, unpremeditated, and involuntary. Examples are jerking away your hand after it accidentally touches a hot stove or vomiting in response to spicy food that irritates your stomach. As you can see from these examples, reflexes are either *somatic reflexes* resulting in the contraction of skeletal muscles or *visceral reflexes* activating smooth muscle, cardiac muscle, or glands. Reflexes are mediated by chains of neurons called **reflex arcs** (Figure 11.16). Every reflex arc has five essential components, each of which activates the next (Figure 11.16a):

1. The *receptor:* This is the site where the stimulus acts—the dendritic endings of a sensory neuron, for example.

2. The *sensory neuron:* This transmits the afferent impulses to the CNS.

3. The *integration center:* In the simplest reflex arcs, the integration center is a single synapse between a sensory neuron and a motor neuron. In more complex reflexes, it involves multiple synapses that send signals through long chains of interneurons. The integration center lies in the CNS, exemplified by the spinal cord in Figure 11.16.

4. The *motor neuron:* This conducts efferent impulses from the integration center to an effector.

5. The *effector:* This is the muscle or gland cell that responds to the efferent impulses by contracting or secreting.

lowing way. First, neurotransmitters released by presynaptic neurons alter the permeability of the postsynaptic membrane to certain ions. This may lead to an inflow of positive ions, which, as we explained, depolarizes the postsynaptic membrane and drives the neuron toward impulse generation. Synapses that behave in this way are **excitatory synapses**. By contrast, **inhibitory synapses** have the opposite effect: They cause the external surface of the postsynaptic membrane to become even more positive, thereby reducing the ability of the postsynaptic neuron to generate an action potential. Thousands of excitatory and inhibitory synapses act on every neuron, competing to determine whether or not that neuron will generate an impulse.

Reflex Arcs

Having discussed how signals travel along and between neurons, we can now consider how neurons are linked together in chains that bring about simple behaviors. The simplest of all behaviors are **reflexes**,

The left side of Figure 11.16b illustrates the simplest of all reflexes, the **monosynaptic reflex** (mon′o-sĭ-nap″tik; "one synapse"). The example shown is the familiar "knee-jerk" reflex, in which a hammer hits the patellar ligament, thereby stretching the quadriceps muscles of the thigh. Stretching the muscle initiates an impulse in a sensory neuron that directly activates a motor neuron in the spinal cord, which then signals the quadriceps muscle to contract. This contraction counteracts the original stretching caused by the hammer.

Many skeletal muscles of the body participate in monosynaptic *stretch reflexes,* which help maintain equilibrium and upright posture. In these reflexes, the sensory neurons sense the stretching of muscles that occurs when the body starts to sway, then the motor neurons activate muscles that adjust the body's position to prevent the fall. Because they contain just one synapse, stretch reflexes are the fastest of all reflexes—for speed is essential to keep one from falling to the ground.

Far more common are **polysynaptic reflexes**, in which one or more interneurons are part of the reflex

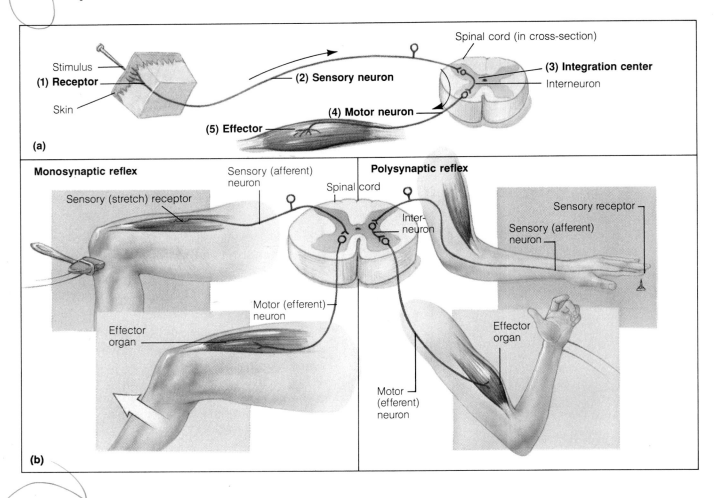

Figure 11.16 Simple reflex arcs. (a) The five basic components of all reflex arcs, as illustrated in a three-neuron reflex arc: At left, a pin stimulates a *receptor,* which sends an impulse through the *sensory neuron.* The signal then crosses the *integration center* (one or more synapses in the CNS) to activate the *motor neuron,* which activates the *effector.* (b) A monosynaptic stretch reflex (left) and a polysynaptic withdrawal reflex (right). A monosynaptic reflex arc has two neurons, whereas a polysynaptic reflex arc has more than two neurons (three, in this case).

pathway between the sensory and motor neurons. Most of the simple reflex arcs in the body contain a single interneuron and therefore have a total of three neurons. The *withdrawal reflexes,* by which we pull back from danger, are three-neuron reflexes. Such a reflex is shown on the right side of Figure 11.16b: Pricking a finger with a tack initiates an impulse in the sensory neuron, which activates the interneuron in the CNS. The interneuron then signals the motor neuron to contract the muscle that withdraws the hand.

The common three-neuron reflex arcs have a special importance in the science of neuroanatomy because they reveal the fundamental design of the entire nervous system. This is the topic of the next section.

Basic Neuronal Organization of the Nervous System

Figure 11.17 shows how three-neuron reflex arcs determine the basic structural plan of the nervous system. This figure emphasizes the reflex arcs that are associated with the spinal cord, although similar reflex arcs are also associated with the brain. In the lower half of the figure, note the locations of the sensory neurons (dorsally), motor neurons (ventrally), and interneurons (centrally). The cell bodies of the sensory neurons lie outside the CNS in **sensory ganglia,** and their central processes enter the *dorsal* aspect of the

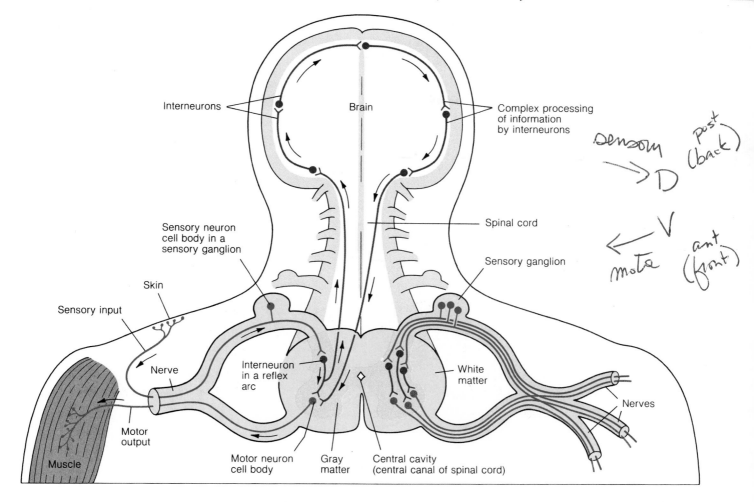

Figure 11.17   Simplified diagram of the human nervous system, based on the locations of sensory, motor, and interneurons.   In this posterior view, the spinal cord is cut in cross section, with the brain shown above. Most importantly, note the reflex arcs formed by the sensory neurons, interneurons, and motor neurons. Also note the arrangement of gray and white matter and the processing of information by circuits of interneurons in the spinal cord and brain. The arrangement of neurons in the spinal cord is realistically shown, but those in the brain are highly simplified. The exact arrangement of gray and white matter in the brain is considered in Chapter 12 (p. 319).

spinal cord to synapse with interneurons there. The cell bodies of the interneurons lie dorsal to those of the motor neurons in the CNS. The long axons of motor neurons exit the *ventral* aspect of the spinal cord and lie in the PNS. The *nerves* of the PNS consist of the motor axons plus the long peripheral processes of the sensory neurons. These motor and sensory nerve fibers extend throughout the body to reach the peripheral effectors and receptors.

While reflex arcs determine the basic organization of the CNS and PNS, the human nervous system is certainly more than just a series of simple reflex arcs. To appreciate its complexity, we must expand our consideration of interneurons. Interneurons include not only the intermediate neurons of the reflex

arcs but also all neurons that are confined entirely within the CNS. The complexity of the CNS arises from the vast number of interneurons in the spinal cord and brain. The interneurons are connected into complex neural circuits for processing information; that is, long chains of interneurons are interposed between each sensory and motor neuron (Figure 11.17). Although tremendously oversimplified, this is a helpful way to conceptualize the organization of neurons in the CNS.

The CNS has distinct regions of gray and white matter that reflect the arrangement of its neurons. The **gray matter** is a gray-colored zone that surrounds the hollow central cavity. It is H-shaped in the spinal cord, where its dorsal half contains cell bodies of in-

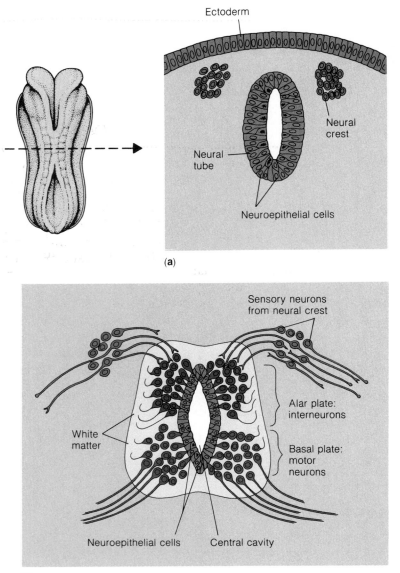

(a)

(b)

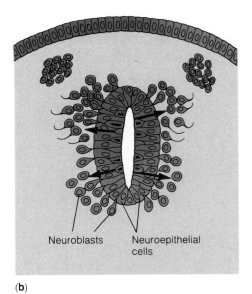

Figure 11.18 Development of the nervous system in weeks 4 and 5 of the embryonic period. (**a**) Cross section of the neural tube, the future spinal cord and brain. For orientation, a 22-day embryo is shown at upper left in dorsal view. (**b**) Neuroblasts (future neurons) arise through division of neuroepithelial cells and migrate externally. (**c**) Neuroblasts form the alar plate (future interneurons) and basal plate (future motor neurons). Neural crest cells form the sensory neurons.

(c)

terneurons and its ventral half contains the cell bodies of motor neurons (Figure 11.17). *Gray matter is the site where nerve cell bodies are clustered.* More specifically, gray matter consists of a mixture of neuron cell bodies, dendrites, and short, unmyelinated axons. External to the gray matter is **white matter**, which contains no neuron cell bodies but millions of axons. The white color of white matter comes from the myelin sheaths around many of the axons. Most of these axons ascend from the spinal cord to the brain or descend from the brain to the spinal cord, allowing these two regions of the CNS to communicate with each other. *White matter consists of axons running between different parts of the CNS.* Within the white matter, axons traveling to similar destinations form axon bundles called **tracts**.

In summary, the CNS has white matter external to gray matter, which surrounds the hollow central cavity. The anatomy of the CNS is described more completely in Chapter 12.

Development of the Nervous System and Nervous Tissue

Now that we have introduced the basic organization of the nervous system, we can consider how this organization develops in the embryo and fetus. Other developmental aspects of the nervous system are considered in later chapters.

Recall from Chapter 3 that the nervous system develops from the dorsal ectoderm, which invaginates to form the neural tube and the neural crest (Figure 11.18a and Figure 3.9 on p. 55). The neural tube, whose walls begin as a layer of **neuroepithelial cells**, becomes the CNS. These cells divide, migrate externally, and become **neuroblasts** (future neurons), which never again divide (Figure 11.18b). Just external to the neuroepithelium, the neuroblasts cluster into **alar** and **basal plates**, the future gray matter (Figure 11.18c). Ventrally, the neuroblasts of the basal plate become motor neurons and sprout axons that grow out to the effector organs. Dorsally, the neuroblasts of the alar plate become interneurons. Axons that sprout from the alar plate cells (and from some basal plate cells) form the white matter by growing outward along the length of the CNS. These events occur in both the spinal cord and brain. Most of the events described so far take place in the second month of development, but neurons continue to form until about the sixth month. Just before the neuroepithelial cells stop dividing, they give rise to the neuroglial cells. These neuroglia migrate throughout the white and gray matter during the second half of the prenatal period. When the neuroepithelium stops dividing, it becomes the ependymal cell layer.

Sensory neurons arise not from the neural tube but from the neural crest (Figure 11.18c). This fact explains why the cell bodies of sensory neurons lie *outside* the CNS. Like motor neurons and interneurons, sensory neurons stop dividing during the fetal period.

Programmed cell death is a normal part of the development of the nervous system. Of the neurons formed during the embryonic period, about two-thirds die before we are born. This production of excess numbers of neuroblasts ensures that all necessary neural connections will be made even if some neurons miss their targets.

From what you have learned about the development and growth of neurons, can you deduce why injuries to neural tissue during childhood and adulthood seldom lead to neuronal regeneration?

After the fetal period, there is no replacement of dead or damaged neurons during a person's lifetime. Therefore, neural injuries tend to cause permanent dysfunction. If the axons alone are cut, however, the cell bodies often survive, and the axons may regenerate. If a nerve is severed in the PNS, the sprouting axons can grow peripherally along solid bands formed by the surviving Schwann cells that surrounded the original axons. Thus, an eventual reinnervation of the target organ, with partial recovery of function, is sometimes possible in the PNS. In the CNS, however, the neuroglia never form bands to guide the regrowing axons and even hinder such axons by secreting growth-inhibiting chemicals. Therefore, there is no effective regeneration after injury to the spinal cord or brain (although the possibility is currently an area of intense medical research). ■

✚ Related Clinical Terms

Neuroblastoma (nu″ro-blas-to′mah) (*oma* = tumor) A malignant tumor in children arising from cells that have retained a neuroblast-like structure. Blastomas sometimes originate in the brain, but most are of neural crest origin in the PNS.

Neurologist (nu-rol′o-jist) A medical specialist in the study of the nervous system and its disorders.

Neuropathy (nu-rop′ah-the) Any disease of the nervous tissue, but particularly a degenerative disease of nerves.

Neurotoxin (nu″ro-tok′sin) Substance that is poisonous or destructive to nervous tissue; e.g., botulism and tetanus toxins.

Rabies (*rabies* = madness) A viral infection of the nervous system transferred to humans by the bites of infected mammals (dogs, bats, skunks, and so on). Once it enters the body, the virus is transported through peripheral nerve axons to the central nervous system, where it causes inflammation of the brain, delerium, and death. Because of extensive vaccination of dogs and careful medical treatment of animal bites, human rabies is now rare in the United States (usually fewer than five cases per year).

Shingles (Herpes zoster) A disease in which herpesviruses take advantage of axonal transport, the normal process by which substances are transported along the axons of neurons. Shingles is a viral infection of sensory neurons to the skin and is characterized by a rash of scaly, painful blisters usually confined to a narrow strip of skin, often on one side of the trunk. Shingles is caused by the varicella-zoster virus, which also produces chicken pox. The disease stems from a childhood infection of chicken pox, during which the viruses are transported from the skin lesions, through the peripheral processes (axons) of the sensory neurons, to the cell bodies in a sensory ganglion. There, in the neuronal nuclei, the viruses remain dormant for years, held in check by the immune system. When the immune system is weakened, the viruses multiply and travel back through the sensory axons to the skin, producing the rash of shingles. Often caused by stress, shingles attacks last for several weeks, separated by periods of healing and remission. Shingles is experienced by 1–2 of every 1000 people, mostly those over 50 years of age.

Chapter Summary

1. The nervous system controls most of the other organ systems in the body. Its chief functions are to monitor, integrate, and respond to information in the environment.

BASIC DIVISIONS OF THE NERVOUS SYSTEM (pp. 296–298)

1. The central nervous system consists of the spinal cord and brain. The peripheral nervous system is external to the CNS and consists mainly of nerves.

2. The PNS is divided into a sensory (afferent) division, which monitors the environment and conveys impulses toward the CNS, and a motor (efferent) division, which conveys impulses away from the CNS to activate the effectors (muscles and glands). The motor and sensory divisions are further subdi-

vided into somatic and visceral, special and general (see Figure 11.2).

3. Proprioception refers to a series of senses that monitor the degree of stretch in muscles, tendons, and joint capsules. Proprioception thus senses the position and movements of our body parts.

NERVOUS TISSUE (pp. 298–308)

The Neuron (pp. 298–303)

1. Neurons are long-lived, nondividing cells. Each has a cell body and cell processes. The processes are axons and dendrites.

2. The neuron cell body contains an "owl's eye" nucleus surrounded by cytoplasm that contains supportive neurofibrils and chromatophilic (Nissl) bodies (concentrations of rough endoplasmic reticulum and free ribosomes). As the chromatophilic bodies manufacture proteins and membranes for the neuron, the cell body continuously maintains and renews the contents of the cell, including the processes. Except for those found in ganglia, all neuron cell bodies are in the CNS.

3. Most neurons have a number of branched dendrites, receptive sites that conduct signals from other neurons toward the neuron cell body.

4. Most neurons have an axon, which generates and conducts nerve impulses away from the neuron cell body. Impulses begin at the cone-shaped axon hillock and end at the knob-like axon terminals, which participate in synapses and release neurotransmitter molecules.

5. A synapse is a functional junction between neurons. Synapses can be categorized as axodendritic, axosomatic, and axoaxonic (along with some lesser categories).

6. At synapses, information is transferred from a presynaptic neuron to a postsynaptic neuron. Synaptic vesicles in the presynaptic cell fuse with the presynaptic membrane and empty neurotransmitter molecules into the synaptic cleft. Taken up by the postsynaptic membrane, this neurotransmitter drives the postsynaptic neuron toward or away from firing.

7. Anatomically, neurons are classified by the number of processes issuing from their cell body as multipolar, bipolar, and unipolar.

8. Functionally, neurons are classified according to the direction in which they conduct impulses. Sensory neurons conduct impulses toward the CNS, motor neurons conduct away from the CNS, and interneurons lie in the CNS between sensory and motor neurons.

Supporting Cells (pp. 304–306)

9. Non-nervous supporting cells act to support, protect, nourish, and insulate neurons.

10. Neuroglial cells are the supporting cells in the CNS. They include star-shaped astrocytes, phagocytic microglia, ependymal cells that line the central cavity, and myelin-forming oligodendrocytes. Schwann cells (neurolemmocytes) and satellite cells are the supporting cells in the PNS.

11. Thick axons are myelinated. Myelin speeds impulse conduction along these axons.

12. The myelin sheath is a coat of supporting-cell membranes wrapped in layers around the axon. The myelin sheath is formed in the PNS by Schwann cells and in the CNS by oligodendrocytes. The sheath has gaps called nodes of Ranvier (neurofibral nodes). Unmyelinated axons are surrounded by supporting cells, but the membrane-wrapping process does not occur in such axons.

Nerves (p. 306)

13. A nerve is a bundle of axons in the PNS. Each axon is enclosed by an endoneurium; fascicles of axons are wrapped by a perineurium; and the whole nerve is surrounded by the epineurium.

FUNCTION OF NEURONS AND IMPULSE CONDUCTION (pp. 308–310)

Nerve Impulses (pp. 308–309)

1. In a resting (unstimulated) neuron, the plasma membrane is polarized, with a positive charge externally and a negative charge internally. A stimulus leads to depolarization. On the axon, a depolarization can trigger a complete charge reversal that travels quickly along the axon. This is an impulse, or action potential.

2. In the human body, stimuli applied to dendrites (or signals at synapses) produce minor depolarizations (graded potentials) that spread to the axon hillock and, if large enough, initiate an impulse there.

Reflex Arcs (pp. 309–310)

3. Reflexes are rapid, automatic responses to stimuli. Reflexes are classified as somatic or visceral.

4. Reflexes are mediated over chains of neurons called reflex arcs. The minimum number of elements in a reflex arc is five: receptor, sensory neuron, integration center, motor neuron, and effector.

5. A few fast reflexes for maintaining balance have only two neurons (sensory and motor). These are monosynaptic stretch reflexes.

6. Most reflexes in humans are polysynaptic. The simplest of these, such as withdrawal reflexes, have three neurons: a sensory neuron, interneuron, and motor neuron.

BASIC NEURONAL ORGANIZATION OF THE NERVOUS SYSTEM (pp. 310–312)

1. Simple three-neuron reflex arcs can be said to dictate the basic structural plan of the entire nervous system. Sensory neurons enter the spinal cord dorsally; motor axons exit it ventrally; and the intervening interneurons are confined to the CNS. The nerves in the PNS consist of the peripheral axons of the sensory and motor neurons. The cell bodies of motor neurons and interneurons make up the internal gray matter of the CNS, while the cell bodies of sensory neurons lie external to the CNS in sensory ganglia.

2. Throughout most of the CNS, the inner gray matter (in which neuron cell bodies are located) is surrounded by outer white matter (which consists of fiber tracts). The extreme center of the spinal cord and brain is a hollow central cavity.

DEVELOPMENT OF THE NERVOUS SYSTEM AND NERVOUS TISSUE (pp. 312–313)

1. The brain and spinal cord develop from the embryonic neural tube, which begins as a layer of dividing neuroepithelial cells. These cells migrate externally to become the neuroblasts of the alar plate (future interneurons) and basal plate (future motor neurons). Neuroblasts of the neural crest (external to the neural tube) become the sensory neurons. Neuroblasts sprout axons, which grow toward their targets.

2. After injury, effective axonal regeneration may occur in the PNS but not in the CNS.

Review Questions

Multiple Choice and Matching Questions

1. Which of the following structures is not part of the central nervous system? (a) the brain, (b) a nerve, (c) the spinal cord, (d) a tract.

2. Match the names of the supporting cells in column B with their correct descriptions in column A.

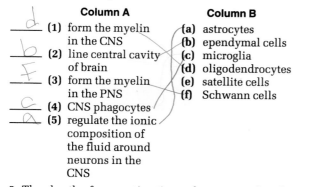

	Column A		Column B
_d_	**(1)** form the myelin in the CNS	**(a)**	astrocytes
_b_	**(2)** line central cavity of brain	**(b)**	ependymal cells
_f_	**(3)** form the myelin in the PNS	**(c)**	microglia
_c_	**(4)** CNS phagocytes	**(d)**	oligodendrocytes
_a_	**(5)** regulate the ionic composition of the fluid around neurons in the CNS	**(e)**	satellite cells
		(f)	Schwann cells

3. The sheath of connective tissue that surrounds a fascicle of nerve fibers is the (a) epineurium, (b) endoneurium, (c) perineurium, (d) neurilemma.

4. Circle the one structure that is in the somatic part of the human body: (a) bladder, (b) biceps muscle, (c) lung, (d) stomach.

5. Classify the following inputs and outputs as somatic sensory (SS), visceral sensory (VS), somatic motor (SM), or visceral motor (VM).

_SS_ **(1)** pain from skin
_SS_ **(2)** proprioception
_VM_ **(3)** efferent innervation of a gland
_SM_ **(4)** efferent innervation of the gluteus maximus
_VS_ **(5)** a stomach ache
_SS_ **(6)** a sound one hears
_VM_ **(7)** efferent innervation of the masseter (a pharyngeal arch muscle)

6. An example of an effector is (a) the eye, (b) a gland, (c) a sensory neuron, (d) a motor neuron.

7. Which of these parts of neurons occupies the gray matter in the spinal cord? (a) tracts of long axons, (b) motor neuron cell bodies, (c) sensory neuron cell bodies, (d) nerves.

8. A ganglion is a collection of (a) neuron cell bodies, (b) axons of motor neurons, (c) interneuron cell bodies, (d) axons of sensory neurons.

9. A synapse between an axon terminal and a neuron cell body is called (a) axodendritic, (b) axoaxonic, (c) axosomatic, (d) axoneuronic.

10. Myelin is most like which of the following cell parts introduced in Chapter 2? (a) cell nucleus, (b) smooth endoplasmic reticulum, (c) ribosomes, (d) the plasma membrane.

11. Match these parts of the adult central nervous system with the embryonic cells that give rise to them, as listed in the key.

Key: **(a)** alar plate cells, **(b)** basal plate cells, **(c)** neural crest cells

_c_ **(1)** sensory neurons
_b_ **(2)** motor neurons
_a_ **(3)** interneurons

Short Answer and Essay Questions

12. Define proprioception. *body sense*

13. Define interneuron.

14. Distinguish gray matter from white matter of the CNS in terms of location and composition.

15. How does the process of myelin formation differ in the CNS and PNS?

16. Describe the appearance of a cell nucleus of a neuron.

17. What are the composition and function of chromatophilic (Nissl) bodies?

18. Reexamine the different neurons shown in Figure 11.10, and indicate which one probably has the most axon terminals synapsing on it. Explain your reasoning.

19. List the five special senses.

20. Distinguish a nerve from a nerve fiber and a neuron.

21. What is the function of synaptic vesicles?

22. Explain why damage to peripheral nerve fibers is often reversible, whereas damage to CNS fibers rarely is.

Critical Thinking and Clinical Application Questions

23. Two anatomists were arguing about the sensory neuron. One said its peripheral process is an axon and gave two good reasons. But the other anatomist called the peripheral process a dendrite and gave one good reason. Cite all three reasons given, and state your own opinion.

24. A CT scan and other diagnostic tests indicated that Laressa had developed an oligodendroglioma. Can you deduce from its name what an oligodendroglioma is?

25. This is an event that received worldwide attention in 1962: A boy was playing in a train yard and fell under a train. His right arm was cut off cleanly by the train wheel. Surgeons reattached the arm, sewing nerves and vessels back together. The surgery proceeded very well. The arm immediately regained its blood supply, yet the boy could not move the limb or feel anything in it for months. Explain why it took longer to reestablish innervation than circulation.

26. Rochelle developed multiple sclerosis when she was 27. After 8 years she had lost a good portion of her ability to control her skeletal muscles. Why did this happen?

12

The Central Nervous System

Student Objectives

The Brain (pp. 317–345)

1. Describe the development of the five embryonic divisions of the brain.

2. Name the major regions of the adult brain.

3. Name and locate the ventricles of the brain.

4. List the major lobes, fissures, and functional areas of the cerebral cortex

5. Name three classes of fiber tracts in the white matter of the cerebrum.

6. Describe the form and function of the basal nuclei.

7. Name the divisions of the diencephalon and their functions.

8. Identify the three basic subdivisions of the brain stem, and note the major nuclei in each.

9. Describe the structure and function of the cerebellum.

10. Locate the reticular formation and the limbic system, and explain their functions.

11. Describe how meninges, cerebrospinal fluid, and the blood-brain barrier protect the CNS.

12. Explain the formation of cerebrospinal fluid and follow its circulation.

The Spinal Cord (pp. 345–351)

13. Describe the gross structure of the spinal cord, and the composition of its gray and white matter.

14. List the largest tracts in the spinal cord, and explain their place in the major neuronal pathways to and from the brain.

Injuries and Diseases of the Central Nervous System (pp. 351–356)

15. Describe the symptoms of concussions, contusions, cerebrovascular accidents, and Alzheimer's disease.

16. Explain the effects of severe injuries to the spinal cord.

The Central Nervous System Throughout Life (p. 356)

17. Describe these birth disorders: anencephaly, cerebral palsy, and spina bifida.

18. Explain the effects of aging on brain structure.

This chapter covers the brain and spinal cord, which make up the central nervous system (CNS). Historically, the CNS has been compared to the central switchboard of a telephone system that interconnects and directs a dizzying number of incoming and outgoing calls. Nowadays, many authorities envision the CNS as a kind of supercomputer. These analogies may explain some of the workings of the spinal cord, but neither really does justice to the fantastic complexity of the human brain. Whether we view it as an evolved biological organ, an impressive computer, or simply a miracle, the brain is certainly one of the most amazing things known and is the basis of each person's unique behavior.

In this chapter, we will use some directional terms that are unique to the CNS: The higher brain regions are said to lie **rostrally** (literally, "toward the snout"), whereas the inferior parts of the CNS are said to lie **caudally** ("toward the tail").

The Brain

The **brain** performs our most complex neural functions (intelligence, consciousness, instincts, and so on). Furthermore, through the cranial nerves that attach to it, the brain is involved in the sensory and motor innervation of the head. The unimpressive appearance of the brain (Figure 12.1) gives few hints of its many remarkable abilities. It is approximately two large handfuls of pinkish gray tissue, somewhat the consistency of cold oatmeal. The average man's brain weighs about 1600 g (3.5 pounds); that of the average woman, 1450 g. In terms of brain weight per body weight, however, males and females have equal brain sizes. Brain size does not seem to be the main determinant of brain power: The complexity of the neural "wiring" appears to be much more important. Albert Einstein's brain was only average in size, and he was a genius, whereas the brains of the prehistoric Neanderthal people (who produced, at most, a limited technology) were 15% larger than those of people living today.

Embryonic Development of the Brain

An examination of brain development will help us to understand the structures of the adult brain. As we mentioned previously, the brain arises as the rostral part of the neural tube in the 4-week embryo (Figure 12.2a). It immediately starts to expand, and constric-

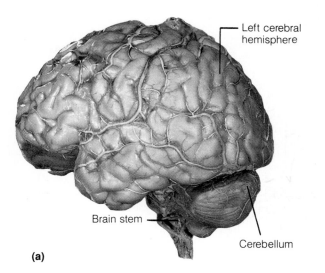

Left cerebral hemisphere

Brain stem

Cerebellum

(a)

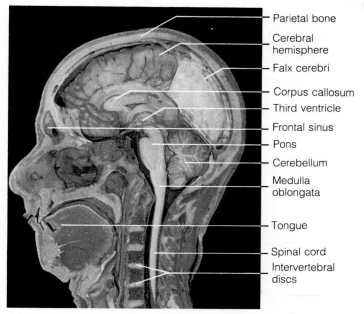

Parietal bone

Cerebral hemisphere

Falx cerebri

Corpus callosum

Third ventricle

Frontal sinus

Pons

Cerebellum

Medulla oblongata

Tongue

Spinal cord

Intervertebral discs

(b)

Figure 12.1   The human brain. (a) Complete brain in lateral view. **(b)** The head sectioned in the median plane to show the brain in place. Note how the brain fills the cranial cavity of the skull. (To provide a clearer view of the brain, the membrane called the falx cerebri has been partly removed.)

tions that define three **primary brain vesicles** appear (Figure 12.2b). These three vesicles are the **prosencephalon** (pros"en-sef'ah-lon), or **forebrain**; the **mesencephalon** (mes"en-sef'ah-lon), or **midbrain**; and the **rhombencephalon** (romb"en-sef'ah-lon), or **hindbrain**. (Note that the word stem *encephalos* means "brain.") The caudal portion of the neural tube becomes the spinal cord.

In week 5, the three primary vesicles give rise to five **secondary brain vesicles** (Figure 12.2c). The fore-

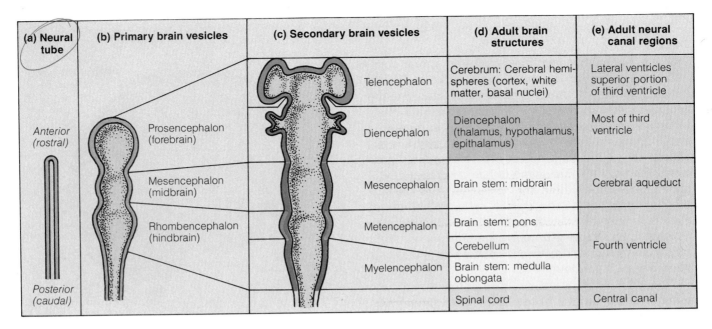

(a) Neural tube	(b) Primary brain vesicles	(c) Secondary brain vesicles	(d) Adult brain structures	(e) Adult neural canal regions
		Telencephalon	Cerebrum: Cerebral hemispheres (cortex, white matter, basal nuclei)	Lateral ventricles superior portion of third ventricle
	Prosencephalon (forebrain)	Diencephalon	Diencephalon (thalamus, hypothalamus, epithalamus)	Most of third ventricle
	Mesencephalon (midbrain)	Mesencephalon	Brain stem: midbrain	Cerebral aqueduct
	Rhombencephalon (hindbrain)	Metencephalon	Brain stem: pons	
			Cerebellum	Fourth ventricle
		Myelencephalon	Brain stem: medulla oblongata	
			Spinal cord	Central canal

Anterior (rostral) ... *Posterior (caudal)*

Figure 12.2   Embryonic development of the brain in simplified frontal sections. (**a**) The neural tube subdivides into (**b**) three primary brain vesicles (week 4), which divide into (**c**) five secondary brain vesicles (week 5), which differentiate into (**d**) adult brain structures. (**e**) Adult derivatives of the embryonic neural canal (the central cavity) in the brain and spinal cord.

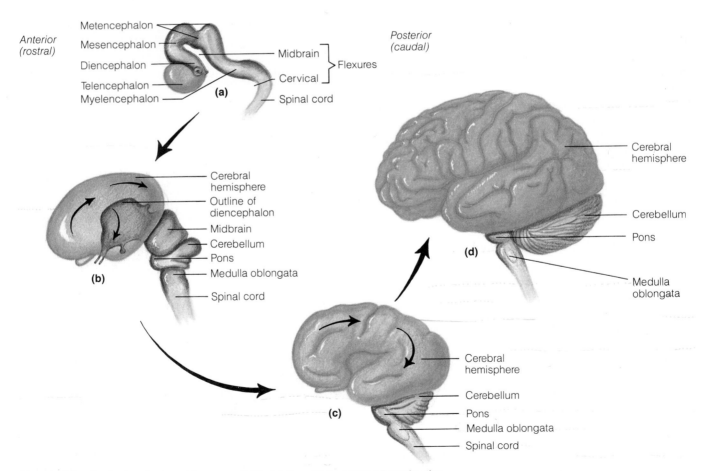

Figure 12.3   Brain development from week 5 to birth. (**a**) As early as 5 weeks, the growing brain is bent because of space restrictions in the skull. The brain at (**b**) 13 weeks, (**c**) 26 weeks, and (**d**) birth.

brain divides into the **telencephalon** ("end brain") and **diencephalon** ("inter-brain"). The mesencephalon remains undivided, but the hindbrain constricts to form the **metencephalon** ("after-brain") and the **myelencephalon** ("brain most like the spinal cord").

Each of the five secondary brain vesicles then develops rapidly to produce the major structures of the adult brain (Figure 12.2d). The greatest change occurs in the telencephalon, which has two lateral swellings that look much like Mickey Mouse's ears. These paired expansions become the large **cerebral hemispheres**, together called the **cerebrum** (ser-e'brum). The diencephalon develops three main divisions: thalamus, hypothalamus, and epithalamus. Farther caudally, the **cerebellum** (ser"e-bel'um) develops from the dorsal roof of the metencephalon, while the ventral part of the metencephalon becomes the **pons**. The myelencephalon is now called the **medulla oblongata** (mĕ-dul'ah ob"long-gah'tah). The midbrain, pons, and medulla form the **brain stem**. Finally, the central cavity of the neural tube enlarges in certain regions to form the hollow **ventricles** (ven'trĭ-klz; "little bellies") of the brain (Figure 12.2e).

During the late embryonic and the fetal periods, the brain continues to grow rapidly, and changes occur in the relative positions of its parts (Figure 12.3). Because the brain's growth is restricted by the membranous skull, two major flexures develop—a *midbrain flexure* and a *cervical flexure* (Figure 12.3a). The restricted space in the skull also forces the cerebral hemispheres to grow posteriorly over the rest of the brain (Figure 12.3b). Soon these hemispheres completely envelop the diencephalon and midbrain (Figure 12.3c). As each cerebral hemisphere grows, it bends into a horseshoe shape, as indicated by the two arrows in Figure 12.3c. By week 26, the continued growth of the cerebral hemispheres causes their surfaces to crease and fold (Figure 12.3d). This folding allows more neurons to fit within the limited space.

Regions and Organization of the Brain

We will discuss the brain in terms of the four regions shown in Figure 12.4. These regions are: (1) *cerebral hemispheres*, (2) *diencephalon*, (3) *brain stem* (midbrain, pons, and medulla), and (4) *cerebellum.* Most anatomists favor this scheme, but some classify the diencephalon as a part of the cerebrum or as part of the brain stem.

Recall from Chapter 11 (p. 311) that the CNS contains an inner region of gray matter, external to which lies white matter. The brain exhibits this basic design but also contains additional regions of gray matter that are not evident in the spinal cord (Figure 12.5). During brain development, certain groups of neurons migrate externally, forming collections of gray matter within the white matter. The most extreme examples

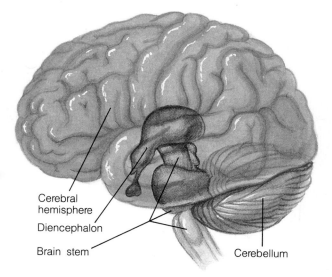

Cerebral hemisphere

Diencephalon

Brain stem

Cerebellum

Figure 12.4 Regions of the brain. The four main regions of the brain are the cerebral hemispheres, diencephalon, brain stem, and cerebellum.

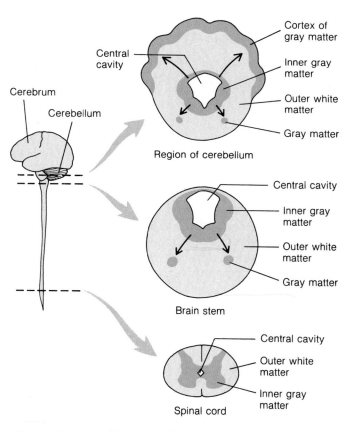

Cerebrum

Cerebellum

Central cavity

Cortex of gray matter

Inner gray matter

Outer white matter

Gray matter

Region of cerebellum

Central cavity

Inner gray matter

Outer white matter

Gray matter

Brain stem

Central cavity

Outer white matter

Inner gray matter

Spinal cord

Figure 12.5 Arrangement of gray and white matter in the brain (highly simplified). At center is a section through the brain stem, with a corresponding section through the region of the cerebellum shown above. The spinal cord is pictured below for reference. In each of these sections, the dorsal aspect is above. In general, white matter lies external to gray matter; however, collections of gray matter have migrated externally into the white matter in the developing brain (see arrows). The cerebrum resembles the cerebellum in its external cortex of gray matter.

Figure 12.6   The ventricles of the brain.
(a) Lateral view. (b) Anterior view. In (a) and (b), note that each lateral ventricle has three parts called the anterior, inferior, and posterior horns. The ventricles are drawn as if the brain were transparent.

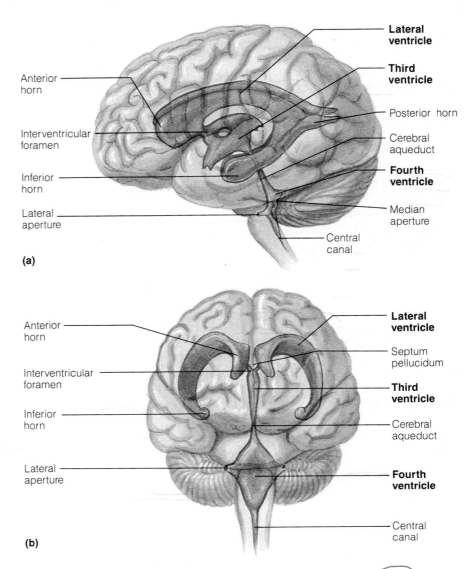

(a)

(b)

of this migration are in the cerebellum and cerebrum, where thick sheets of gray matter lie on the external brain surface. Each of these external sheets of gray matter is called a *cortex* ("bark on a tree"). All other gray matter in the brain is in the form of spherical or irregularly shaped clusters of neuron cell bodies called *brain nuclei* (not to be confused with cell nuclei). Functionally, the large amount of gray matter in the brain allows this part of the CNS to perform very complex neural functions—for gray matter contains many small interneurons for processing information.

To help you picture the spatial relationships among the basic brain regions, we will first consider the hollow ventricles that lie deep within the brain. Then we will explore the individual brain regions, from rostral to caudal. Finally, a summary of the brain regions is provided in Table 12.1 on page 341.

Ventricles of the Brain

The **ventricles** are central cavities filled with cerebrospinal fluid and lined by ependymal cells (Figure 12.6). They are continuous with each other and with the central canal of the spinal cord. The paired **lateral ventricles**, once called the *first and second ventricles,* lie in the cerebral hemispheres. Their C-shape reflects the bending of the cerebral hemispheres during development. Anteriorly, the two lateral ventricles lie close together, separated only by a thin median membrane called the *septum pellucidum* (sep′tum pě-lu′si-dum; "transparent wall") (Figure 12.13a, p. 331).

The **third ventricle** lies in the diencephalon. Anteriorly, it connects to each lateral ventricle through an **interventricular foramen** (Figure 12.6). In the mesencephalon, the central cavity remains thin and tube-like and is called the **cerebral aqueduct**. The **fourth ventricle** lies in the hindbrain, dorsal to the pons and superior half of the medulla. Caudally, it connects to the central canal in the inferior medulla and spinal cord. Three openings occur in the walls of the fourth ventricle: the paired **lateral apertures** in its side walls and the **median aperture** in its roof. These holes connect the ventricles with the *subarachnoid space,* a fluid-filled space that surrounds the whole CNS (p. 343). This continuity allows cerebrospinal fluid to fill both the ventricles and the subarachnoid space.

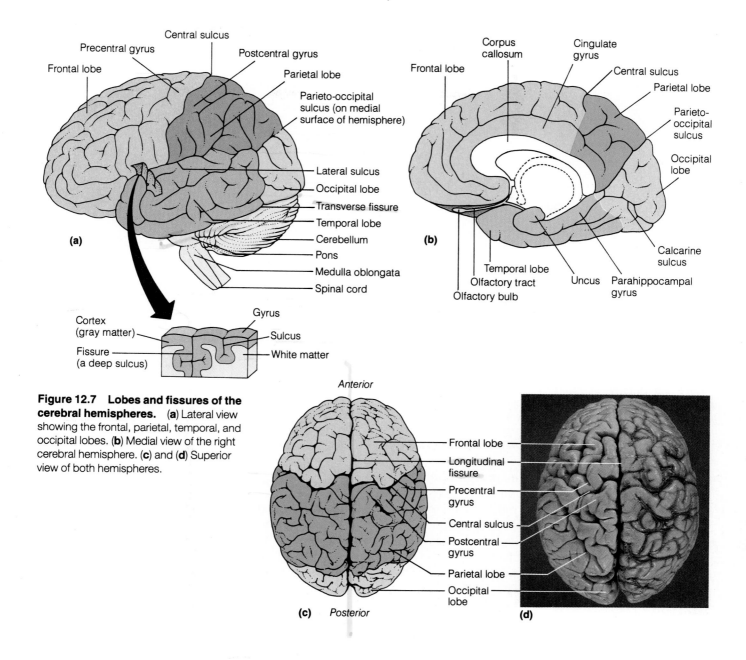

Figure 12.7 Lobes and fissures of the cerebral hemispheres. (**a**) Lateral view showing the frontal, parietal, temporal, and occipital lobes. (**b**) Medial view of the right cerebral hemisphere. (**c**) and (**d**) Superior view of both hemispheres.

The Cerebral Hemispheres

The **cerebral hemispheres** make up the most superior region of the brain (Figure 12.4). Together these hemispheres account for 83% of total brain mass. They so dominate the brain that many people mistakenly use the words *brain* and *cerebral hemispheres* interchangeably. Picture how a mushroom cap covers the top of its stalk, and you have a good analogy of how the cerebral hemispheres cover the diencephalon and the top of the brain stem.

Many distinctive grooves are evident on and around the cerebral hemispheres (Figure 12.7). The deepest of these grooves are called **fissures**. The **transverse fissure** separates the cerebral hemispheres from the cerebellum inferiorly (Figure 12.7a), while the

median **longitudinal fissure** separates the right and left cerebral hemispheres (Figure 12.7c). The many grooves on the surface of the cerebral hemispheres are called **sulci** (sul'ki). The singular of *sulci* is **sulcus**, Greek for "furrow." Between these sulci are twisted ridges of brain tissue called **gyri** (ji'ri; "twisters"). The more prominent gyri and sulci are similar in all people and are important anatomical landmarks.

Some deep sulci help to divide each cerebral hemisphere into five major lobes: the *frontal, parietal, occipital, temporal,* and *insula* lobes (Figure 12.7). Most of these lobes are named for the skull bones overlying them. The **frontal lobe** is separated from the **parietal lobe** by the **central sulcus**, which lies in the frontal plane. Bordering the central sulcus are two important gyri, the **precentral gyrus** anteriorly and the

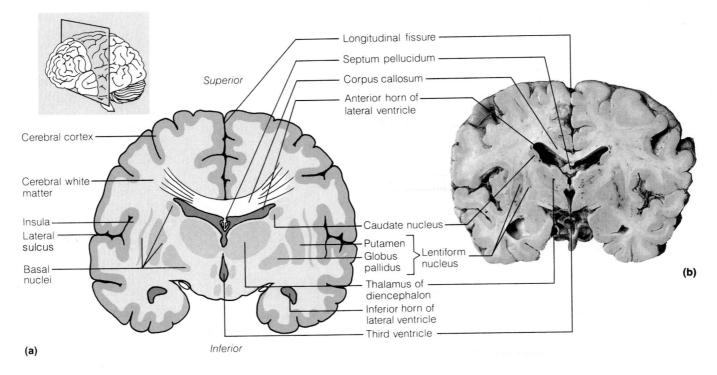

Superior

Longitudinal fissure
Septum pellucidum
Corpus callosum
Anterior horn of lateral ventricle

Cerebral cortex

Cerebral white matter

Insula

Lateral sulcus

Basal nuclei

Caudate nucleus
Putamen
Globus pallidus
Lentiform nucleus
Thalamus of diencephalon
Inferior horn of lateral ventricle
Third ventricle

(a)

Inferior

(b)

Figure 12.8   Internal structure of the forebrain.   The cerebrum is sectioned frontally to show the positions of the outer cerebral cortex, the intermediate white matter, and the deep basal nuclei. Since the cerebral hemispheres enclose the diencephalon, that region is also evident. Compare the diagram in (a) to the photo in (b).

postcentral gyrus posteriorly. The **occipital lobe** lies farthest posteriorly. It is separated from the parietal lobe by several landmarks, the most conspicuous of which is the **parieto-occipital sulcus**. This sulcus lies on the medial surface of the hemisphere (Figure 12.7b). On the lateral side of the hemisphere, the flap-like **temporal lobe** is separated from the overlying parietal and frontal lobes by the deep **lateral sulcus** (Figure 12.7a). This sulcus is so deep that, despite its name, it is actually a fissure. A fifth lobe of the cerebral hemisphere is the **insula** (in'su-lah; "island"). The insula is buried deep within the lateral sulcus and forms part of its floor (see Figure 12.8). The insula is covered by parts of the temporal, parietal, and frontal lobes.

The cerebral hemispheres fit snugly into the skull. The frontal lobes lie in the anterior cranial fossa, while the anterior parts of the temporal lobes fill the middle cranial fossa. The posterior cranial fossa houses the brain stem and cerebellum. The occipital lobes are located superior to the cerebellum and do not occupy a cranial fossa (see Figure 12.7a).

A frontal section through the forebrain shows the three largest regions within the cerebrum (Figure 12.8). These are a superficial **cerebral cortex** of gray matter, the **cerebral white matter** internal to this, and the **basal nuclei** deep in the white matter. These regions will now be described.

Cerebral Cortex

The cerebral cortex is the "executive suite" of the nervous system, the home of our "conscious mind." It enables us to be aware of ourselves and our sensations, to initiate and control voluntary movements, and to communicate, remember, and understand.

Since it is composed of gray matter, the cerebral cortex contains neuron cell bodies, dendrites, and very short unmyelinated axons, but no fiber tracts. It is only 2–4 mm thick (about 1/8 inch); however, its many convolutions triple its surface area, and it accounts for about 40% of the total mass of the brain. The cortex contains billions of neurons arranged in six layers.

In the late 1800s, anatomists studying the cerebral cortex were able to map subtle variations in the thickness of the six layers and in the structure of the contained neurons over the entire surface of the cerebral hemispheres. The most successful of these efforts was that of Korbinian Brodmann, who in 1909 divided the cortex into 52 different areas. Some of the most important **Brodmann areas** are numbered in Figure 12.9.

With a structural map emerging, early neurologists were anxious to localize *functional* regions of the cortex. In the 1800s, there were two opposing schools of thought concerning the localization of cerebral function. The *regional specialization theory* assumed that structurally distinct areas of the cortex carried out distinct functions. The *aggregate field view,* by contrast, proposed that mental functions were not localized but that the cortex acted as a whole. Thus, injury to a specific region would affect all higher mental functions equally. Today we know that both theories have merit. It has been proved that specific motor and sensory functions are indeed localized in discrete cortical regions, called *domains.* However, many higher mental functions, such as memory and language, appear to have overlapping domains and are more diffusely located.

Before discussing the functions of different cortical regions, we must consider how scientists determined these functions. Until recently, most information came from studying the mental defects of people and animals that had sustained localized brain injuries. Modern scientists study the activity of the brain directly using PET scans (p. 18).

Can you recall how the PET imaging technique is useful in determining cortical functions?

PET scans visualize areas of maximum blood flow to the cortex, which are areas of maximum mental activity as well. Therefore, the visual areas of the cortex "light up" on PET scans when a subject sees something, the auditory areas when a subject hears something, and so on. PET studies are revolutionizing our understanding of how the cerebral cortex works. ■

The cerebral cortex has been found to contain three kinds of functional areas: *motor areas* that control voluntary motor functions; *sensory areas* that provide for conscious awareness of sensation; and *association areas* that act mainly to integrate diverse information for purposeful action. As you read about these areas, do not confuse the sensory and motor *areas* of the cortex with sensory and motor *neurons;* all neurons in the cerebral cortex are interneurons (confined entirely to the CNS).

Motor Areas.   The cortical areas that control motor functions lie in the posterior part of the frontal lobe (Figure 12.9). They are the *primary motor cortex,* the *premotor cortex,* the *frontal eye field,* and *Broca's area.*

1. Primary motor cortex. The primary motor area, or *somatic motor area* is located along the precentral gyrus of the frontal lobe. It corresponds to area 4 of Brodmann. In this area are large neurons called *pyramidal cells* (Figure 11.10a, p. 303), which have long axons that project to the spinal cord. There, they signal the spinal motor neurons to bring about the skilled and voluntary movements of the limbs, trunk, and so on. The axons of pyramidal cells form the massive *pyramidal,* or *corticospinal, tract* (p. 351), which descends through the brain stem and through the spinal cord. The projection of these axons is **contralateral**; that is, they cross over to the opposite side of the brain and spinal cord (*contra* = opposite). Therefore, the

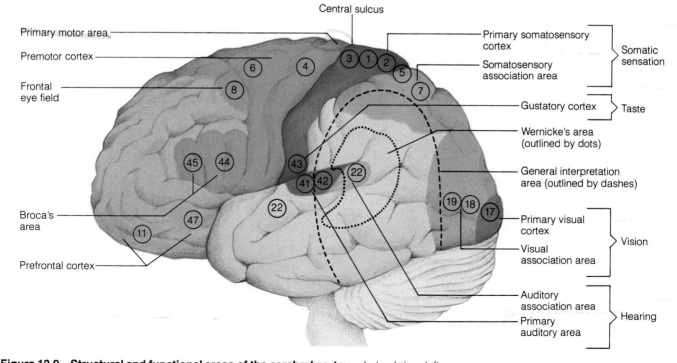

Figure 12.9   Structural and functional areas of the cerebral cortex.   Lateral view, left cerebral hemisphere. Numbers indicate Brodmann's structural areas, while colors define functional regions of the cortex.

left primary motor cortex controls muscles on the *right* side of the body, and vice versa.

The human body is represented spatially in the primary motor cortex of each hemisphere. In other words, the pyramidal cells that control foot movement are in one place, those that control hand movement are in another, and so on, as mapped on the left side of Figure 12.10. This map was first demonstrated by mildly stimulating parts of the precentral gyrus with an electrode and watching which part of the patient's body moved. Note that the body is represented upside down, with the head on the inferolateral part of the precentral gyrus and the toes at the superomedial end. Also note that the face and hand representations are disproportionately large: With so many pyramidal cells controlling them, the face and hand muscles can perform very delicate and skilled movements. The body map on the motor cortex is called a **motor homunculus** (ho-mung'ku-lus; "little man"). As we will see, there are many other parts of the CNS in which the body is represented spatially. This general principle is called **somatotopy** (so-mat'o-tōp'e), which literally means "body mapping."

Strokes, tumors, or wounds can destroy limited areas of the brain. Can you deduce some effects of localized lesions to the primary motor cortex?

Damage to an area of the primary motor cortex paralyzes the body muscles controlled by that area. Only *voluntary* control is lost, however, for the muscles can still contract reflexively. Because the projections are contralateral, lesions in the right cerebral hemisphere affect the left side of the body, and vice versa. ■

2. Premotor cortex. Just anterior to the precentral gyrus is the premotor cortex. It coincides with Brodmann area 6 (Figure 12.9). This region controls learned motor skills of a repetitious or patterned nature, like playing a musical instrument or typing. The premotor cortex signals large groups of muscles to contract in complex sequences. As far as we know, this region is not organized according to body region. You could think of the premotor cortex as a memory bank for skilled motor activities.

3. Frontal eye field. The frontal eye field lies anterior to the premotor cortex, in the inferior part of Brodmann area 8. It controls voluntary movements of the eyes. Damage to this area destroys the ability to look toward an object voluntarily.

4. Broca's (Bro'kahz) **area.** Broca's area lies anterior to the inferior region of the premotor cortex. It overlaps Brodmann areas 44 and 45. Until recently, Broca's area was thought to correspond to our *motor speech area,* a functional region that is known to occur in the frontal lobe of one hemisphere (usually the left) and to control the motor aspects of speech. New PET stud-

ies, however, show that Broca's area (and the corresponding area on the other hemisphere) becomes active just prior to *all* voluntary motor activities, not just speaking. The true function of Broca's area, therefore, may be to "preplan" all our voluntary movements, sending instructions to the primary motor and premotor cortex for execution.

Sensory Areas. The cortical areas involved with conscious awareness of sensation occur in the parietal, temporal, and occipital lobes. There is a distinct cortical area for each of the major senses:

1. Primary somatosensory cortex. The primary somatosensory cortex is located along the postcentral gyrus of the parietal lobe, just posterior to the primary motor cortex. It corresponds to Brodmann areas 1–3 and is involved with conscious awareness of the general somatic senses (skin senses and proprioception: Figure 11.1). This sensory information is picked up by sensory receptors in the periphery of the body and relayed through the spinal cord and brain stem to the somatosensory cortex. There, cortical neurons process the sensory information and identify the precise area of the body being stimulated. This ability to localize a stimulus precisely is called **spatial discrimination.** The projection is contralateral; that is, the right cerebral hemisphere receives its sensory input from the left half of the body. As in the primary motor cortex, somatotopy is exhibited: A **sensory homunculus** occurs on the postcentral gyrus (see the map in Figure 12.10). This map was demonstrated by stroking patients on various areas of their skin, then recording exactly where electrical activity resulted on the cerebral cortex. The amount of somatosensory cortex devoted to a body region is related to the sensitivity of that region (that is, to its number of sensory receptors). The lips and fingertips are our most sensitive body parts, and hence are the largest parts of the sensory homunculus.

Can you guess the effect of a stroke or injury that damages the primary somatosensory cortex?

Damage to the primary somatosensory cortex destroys the conscious ability to feel and localize touch, pressure, and vibrations on the skin. Most ability to feel pain and temperature is also lost, although these can still be felt in a vague, poorly localized way. ■

2. Somatosensory association area. The somatosensory association cortex lies posterior to the primary somatosensory cortex and corresponds roughly to areas 7 and 5 of Brodmann. With many connections to the primary somatosensory cortex, this association area integrates different sensory inputs (touch, pressure, and others) into a comprehensive evaluation of what is being felt. For example, when you reach into your pocket, the somatosensory association area

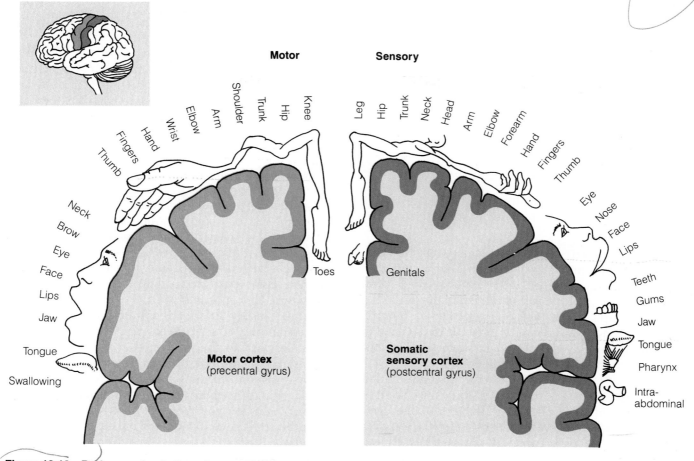

Motor

Sensory

Toes

Genitals

Motor cortex
(precentral gyrus)

Somatic sensory cortex
(postcentral gyrus)

Figure 12.10   Body mapping in the primary motor cortex and somatic sensory cortex of the cerebrum.   Frontal sections through the precentral motor gyrus (left) and the postcentral sensory gyrus (right). The amount of cortical tissue devoted to each body region is indicated by the size of that region on the "body maps."

draws upon stored memories of past sensory experiences and perceives the objects you feel as coins or keys. A person with damage to this area could not recognize these objects without looking at them.

3. Visual areas. The **primary visual cortex**, which corresponds to Brodmann area 17, is seen on the extreme posterior tip of the occipital lobe (Figure 12.9). Most of it, however, is located on the medial aspect of the occipital lobe, buried within the deep **calcarine sulcus** (kal'kar-in; "spur-shaped") (Figure 12.7b). The largest of all cortical sensory areas, the primary visual cortex receives visual information that originates on the retina of the eye. If the primary visual cortex is damaged, the person will have no conscious awareness of what is being viewed and will be functionally blind. There is a map of visual space on the primary visual cortex, analogous to the body map on the somatosensory cortex. The right half of visual space is represented on the left visual cortex, the left half on the right cortex.

The **visual association area** surrounds the primary visual area and covers much of the occipital

lobe. It coincides with Brodmann areas 18 and 19 (Figure 12.9). Communicating with the primary visual area, the visual association area interprets and evaluates visual stimuli in light of past visual experiences. People with damage to this area can see, but they do not understand what they are looking at.

Furthermore, recent experiments on monkeys indicate that complex visual processing extends well beyond the occipital lobe, involving the entire posterior half of the cerebral hemispheres.

4. Auditory areas. The **primary auditory cortex**, which functions in conscious awareness of sound, is on the superior edge of the temporal lobe, primarily inside the lateral sulcus. It corresponds to areas 41 and 42 of Brodmann (Figure 12.9). Sound waves excite the sound receptors of the inner ear and cause impulses to be transmitted to the primary auditory cortex. Here, this information is related to loudness, rhythm, and especially to pitch (high and low notes). There is a "pitch map" in the auditory cortex.

The **auditory association area** lies just posterior to the primary auditory area, in the posterior part of

Brodmann area 22. This area permits the evaluation of a sound—as a screech, thunder, music, and so on. Memories of past sounds seem to be stored here for reference. In one hemisphere (usually the left), the auditory association area lies in the center of **Wernicke's** (Ver'nĭ-kēz) **area**, a functional brain region involved in recognizing spoken words. Damage to Wernicke's area interferes with the ability to recognize and understand speech.

5. Gustatory (taste) cortex. The gustatory (gus'tah-to"re) cortex is involved in the conscious awareness of taste stimuli. It corresponds to area 43 of Brodmann and lies on the roof of the lateral sulcus. Logically, this taste area occurs on the tongue of the somatosensory homunculus.

6. Olfactory (smell) cortex. The primary olfactory cortex lies on the medial aspect of the cerebrum in a small region called the *piriform lobe* (pir'ĭ-form; "pear-shaped"), which is dominated by the hook-like **uncus** (Figure 12.7b). The olfactory nerves (cranial nerve I) from the nasal cavity transmit impulses that are ultimately relayed to the olfactory cortex. The outcome is conscious awareness of odors.

The olfactory cortex is part of a brain area called the *rhinencephalon* (ri"nen-sef'ah-lon; "nose brain"). The rhinencephalon includes all parts of the cerebrum that directly receive olfactory signals: the piriform lobe, **olfactory tract**, **olfactory bulb** (Figure 12.7b), and some minor structures. The rhinencephalon connects to the brain area that is involved in emotions, the *limbic system.* This relationship explains why smells often trigger emotions. The limbic system is discussed further on p. 339.

Association Areas. The association areas make up all the cortical regions other than the primary sensory areas and motor areas. Their name reflects the fact that some of these areas tie together, or make *associations* between, the different kinds of sensory information. They also seem to associate new sensory inputs with memories of past experiences. We have already explored the association areas that directly border the primary sensory regions (the somatosensory, visual, and auditory association areas). The remaining association areas are considered next.

1. Prefrontal cortex. The prefrontal cortex is the large region of the frontal lobe that lies anterior to the motor areas (Figure 12.9). The most complicated cortical region of all, it is involved with reasoning and complex learning abilities (called *cognition*) and personality. The prefrontal cortex is necessary for the production of abstract ideas, judgment, persistence, planning, social behavior, concern for others, and conscience. It also seems to be related to mood and has close links to the emotional (limbic) part of the forebrain. The tremendous elaboration of this prefrontal region distinguishes humans from other animals.

Can you predict some effects of injury to the prefrontal cortex?

Tumors or other lesions of the prefrontal cortex cause mental and personality disorders. Wide swings in mood may occur, and there may be a loss of attentiveness, of inhibitions, and of judgment. The affected person may be oblivious to social restraints, perhaps becoming careless about personal appearance or rashly attacking a 7-foot opponent rather than running. ■

2. General interpretation area. The general interpretation area, or *gnostic area* (nos'tik; "knowing"), is an ill-defined region that encompasses large parts of the temporal, parietal, and occipital lobes (Figure 12.9). The gnostic area occurs in one hemisphere only, usually the left. This region receives input from all the sensory association areas and appears to be a storage site for complex memory patterns associated with sensation. The gnostic area integrates all types of incoming sensory information into a comprehensive understanding of a situation, then sends its assessment forward to the prefrontal cortex, which adds emotional overtones to the assessment and decides how to respond to the experience. Suppose, for example, you drop a bottle of hydrochloric acid, and the acid splashes on you. You see the bottle shatter, you hear the crash, and you feel the burning on your skin. However, these individual perceptions do not dominate your consciousness. What does dominate is the overall message of "danger!"—by which time your leg muscles are already activated and are propelling you to the shower.

Damage to the gnostic area can produce quite profound effects. Can you guess what these might be?

Injury to the gnostic area causes one to become an imbecile—even if all the sensory association areas are unharmed—because one's ability to interpret experiences is lost. ■

Recent evidence suggests that the gnostic area and the prefrontal cortex work together to assemble our new experiences into logical constructs based on past experiences—or "stories." This would mean that our minds do not interpret the world objectively, but only according to what we already know and understand. Story telling may thus be "wired into" our brain, a possibility that would explain why stories have such great appeal in all human cultures.

3. Language comprehension areas. Traditionally, it has been thought that comprehension of the spoken word occurred in Wernicke's area, which is the (left) auditory association area plus some of the surrounding cortex. Recent PET studies, however, suggest that Wernicke's area is only involved in the superficial recognition of the words one hears, not in the deep com-

A CLOSER LOOK   "He-Brain" versus "She-Brain"

Women of this generation looking for equal treatment in the marketplace have been buffeted by the cultural notion that men are somehow more capable—the superior members of the human race. Yet, when all sexism is finally stripped away and differences in nature are disregarded, there *is* something different about males and females, and the difference is grounded in their biology.

As boys enter puberty, they tend to exhibit more aggressive behavior, and there is little doubt that the dramatic rise in male sex hormones surging through their blood is responsible for this change. The question now is how testosterone promotes or enables aggressive or violent behavior. For any hormone to influence behavior, it must first affect the brain. Radioisotope techniques have demonstrated that both male and female sex hormones (testosterone and estrogen, respectively) concentrate selectively in certain brain regions that play an important role in courtship, sex and mating behaviors, and aggression—behaviors in which the sexes differ most. There is also evidence that sexual differences in behavior exist long before the onset of puberty and its raging hormones. For example, newborn boys show more motor strength and muscle tone, while girls show greater sensitivity to touch, taste, and light and exhibit more reflex smiles. Comprehensive studies of schoolchildren have demonstrated that girls do better on verbal skills tests, whereas boys are more proficient in spatial-visual tasks. Further, although it was once presumed that damage to the left hemisphere would produce greater deficits in verbal tasks and that right hemispheric injury would impair spatial abilities, this has been borne out only for males. As a result of these studies, it has been hypothesized that females have more hemispheric overlap in verbal and spatial functions than do males, who demon-

strate earlier lateralization of cortical functioning.

Does greater lateralization for a given function imply superior performance for that function? Apparently not. Females show less lateralization of language functions, yet they tend to be superior to males in language skills. Also, greater lateralization carries a steep price tag, because if a highly lateralized functional area is damaged, the function is terminated. For example, loss of speech follows damage to the left hemisphere three times more often in males than in females.

How can these differences in behavior and skills be explained? The accounting may lie deep within the brain. In 1973, it was demonstrated for the first time that male and female brains differ structurally in some ways. The best example concerns a specific region (now called the *sexually dimorphic nucleus*) in the anterior part of the hypothalamus that has a distinctive synaptic pattern in each sex. Castration of male monkeys shortly after birth produced the female hypothalamic pattern, while injection of testosterone into females triggered the development of the male pattern. This discovery rocked the neuroscience community because it was the first evidence that there were structural brain differences in the sexes

and that sex hormones circulating before, at, or after birth could *change* the brain. Based on these and later observations, scientists concluded that the basic plan of the mammalian brain is female and stays that way unless "told to do otherwise" by masculinizing (male) hormones. The production of testosterone by male fetuses is the key to anatomical development as a male. It now appears that fetal testosterone is also responsible for development of the male brain pattern.

Since these discoveries, a number of other sex-related differences have been found in the central nervous system, including the following:

1. Females tend to have a longer left temporal lobe.

2. There are sex-related differences in the human corpus callosum. In females, its posterior portion is bulbous and wide, while in males the corpus callosum is generally cylindrical and fairly uniform in diameter. This may indicate that there are more communicating fibers between the hemispheres in females, which could support the hypothesis that the female brain is less lateralized.

3. Sex differences have been found in the spinal cord. Certain clusters of neurons that serve the external genitals are much larger in males than in females and they contain receptors for testosterone, but not estrogen. Testosterone exposure before birth produces the male pattern of these particular spinal cord neurons.

What should we make of these findings? Do these observed differences in the two sexes have little or nothing to do with behavioral subtleties, or do they underlie many (or most) manifestations of maleness or femaleness? It is too soon to know, and studies of sexual dimorphism of the brain are still "in diapers."

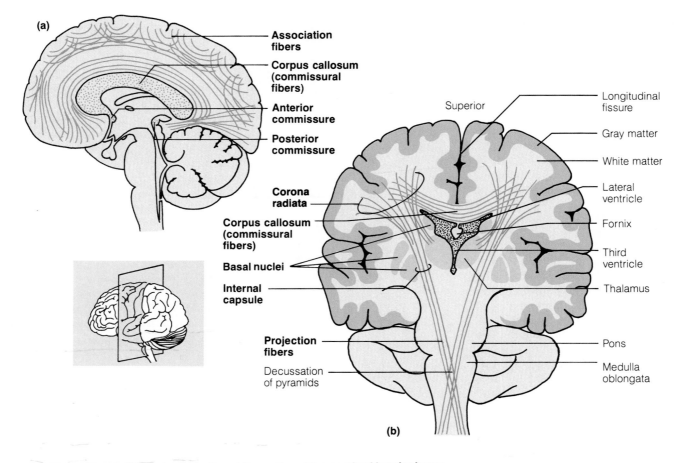

Figure 12.11 Types of fiber tracts in the white matter of the cerebral hemispheres.
(**a**) Midsagittal view of right hemisphere. Note the association fibers, and the corpus callosum, a large commissure that connects the right and left hemispheres. (**b**) Frontal section of the brain. Note the internal capsule, a bundle of projection fibers.

prehension of their meaning. Instead, speech comprehension seems to involve a part of the prefrontal cortex (usually the left one). This finding is controversial, but it is logical because it puts another of our highest thought processes in the prefrontal cortex. The prefrontal area of word comprehension has been located about halfway between the numbers 45 and 11 in Figure 12.9.

4. Visceral association area. The cortex of the insula may be involved in conscious perception of visceral sensations (upset stomach, full bladder, and so on).

Lateralization of Cortical Functioning. There is some division of labor between the right and left cerebral hemispheres. We have already explained, for example, that the two hemispheres control opposite halves of the body. Furthermore, the two hemispheres tend to specialize for different cognitive functions: In most people (90%), the left hemisphere has more control over language abilities, math abilities, and logic—whereas the right hemisphere is more involved with visual-spatial skills, intuition, emotion, and artistic and musical skills. In the remaining 10% of the pop-

ulation, either these roles of the hemispheres are reversed, or the two hemispheres share their cognitive functions equally. Functional differences between the hemispheres are greater in males than in females, as described in the box on page 327.

Despite their differences, the two cerebral hemispheres are nearly identical in structure and share most functions and memories. Such sharing is possible because the two hemispheres communicate through interconnecting fiber tracts called *commissures.* These are discussed in the next section.

Cerebral White Matter

Having considered the gray matter of the cerebral cortex, we will turn to the underlying **cerebral white matter** (Figure 12.11). Different areas of the cerebral cortex communicate extensively with one another and with the brain stem and spinal cord. This communication is accomplished by the many axons that form the cerebral white matter. Most of these axons are myelinated and bundled into large tracts. These axons and tracts are classified as (1) *commissural,* (2)

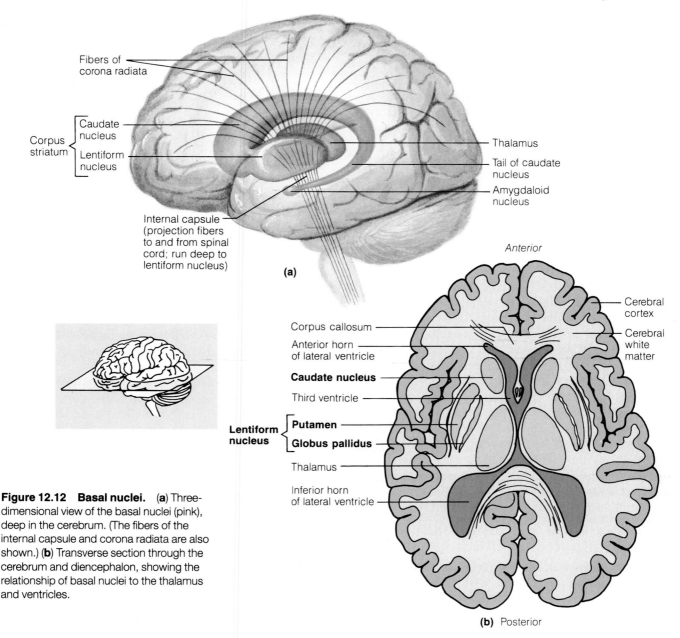

Figure 12.12   Basal nuclei. (a) Three-dimensional view of the basal nuclei (pink), deep in the cerebrum. (The fibers of the internal capsule and corona radiata are also shown.) (b) Transverse section through the cerebrum and diencephalon, showing the relationship of basal nuclei to the thalamus and ventricles.

association, and (3) *projection,* according to where they run.

1. Commissures, made of **commissural fibers,** run between the two hemispheres. Specifically, they interconnect the same gray areas of the right and left hemispheres, allowing the two hemispheres to function together as a coordinated whole. The largest commissure is the **corpus callosum** (kor′pus kah-lo′sum; "thickened body"). This broad band lies superior to the lateral ventricles, deep within the longitudinal fissure (Figure 12.11a). A less important commissure is the *anterior commissure.*

2. Association fibers connect different parts of the *same* hemisphere. Short association fibers connect

neighboring cortical areas, while long association fibers connect widely separated cortical lobes.

3. Projection fibers descend from the cerebral cortex to more caudal parts of the CNS or ascend to the cortex from lower centers. It is through projection fibers that sensory information reaches the cortex and motor instructions leave it. Projection fibers run vertically, whereas most commissural and association fibers run horizontally.

Deep to the cerebral white matter, the projection fibers form a compact band called the **internal capsule,** which passes between the thalamus and some of the basal nuclei (Figures 12.11b and 12.12a). Superior to the internal capsule, the projection fibers running to and from the cortex form a spreading fan, the **corona radiata** (kŏ-ro′nah ra-de-ah′tah; "radiating crown").

Basal Nuclei

Deep within the cerebral white matter lies a group of nuclei called the **basal nuclei** or **basal ganglia.*** They include the **caudate** ("tail-like"), the **lentiform** ("lens-shaped"), and the **amygdala**, or **amygdaloid** ("almond-shaped"), nuclei (Figure 12.12). The caudate nucleus arches superiorly over the thalamus and lies medial to the internal capsule. Together, the lentiform and caudate nuclei are called the **corpus striatum** (kor'pus stri-a'tum; "striated body") because some fibers of the internal capsule pass through and give them a striped appearance. The lentiform nucleus is divided into a medial part, the **globus pallidus** ("pale globe"), and a lateral part, the **putamen** (pu-ta'men; "pod") (Figures 12.12b and 12.8a).

The amygdala (ah-mig'dah-lah) sits on the tip of the tail of the caudate nucleus. Traditionally, it has been grouped with the basal nuclei, but functionally it belongs to the limbic system (p. 339).

Functionally, the basal nuclei can be viewed as complex neural calculators that cooperate with the cerebral cortex in controlling movements. Indeed, they communicate extensively with the cerebral cortex, receiving inputs from many cortical areas and sending almost all their output right back to the primary motor cortex (through a relay in the thalamus). The basal nuclei are important in *starting* and *stopping* the voluntary movements that are ordered and executed by the cortex. They also regulate the *intensity* of these movements, resembling the motor drive that regulates how fast a mechanical motor works.

 Can you deduce some disorders that result from dysfunctions of the basal nuclei?

Degenerative conditions of the basal nuclei produce dyskinesia (dis-ki-ne'ze-ah), literally "bad movements." For example, **Parkinson's disease** is characterized by slow movements, tremor of a hand, arm, or leg, and great difficulty in starting voluntary movements. In this condition, the basal nuclei are exerting *too little* motor drive. **Huntington's disease**, by contrast, involves *overstimulation* of the motor drive, so that the limbs jerk unstoppably in a dance-like manner. The symptoms of Huntington's disease are caused by degeneration of the corpus striatum, with the cerebral cortex eventually degenerating as well. Parkinson's disease results from degeneration of an extension of the basal nuclei in the midbrain called the substantia nigra (see p. 334). Huntington's disease is inherited, and Parkinson's disease may be caused by environmental toxins such as pesticides. ■

---

*Recall that a nucleus is a collection of neuron cell bodies within the CNS. Therefore, the term *basal nuclei* is correct. The more commonly used term *basal ganglia* is a misnomer because ganglia are PNS structures.

The Diencephalon

The **diencephalon** forms the central core of the forebrain and is surrounded by the cerebral hemispheres. It consists largely of three paired structures: the thalamus, hypothalamus, and epithalamus. These border the third ventricle and consist primarily of gray matter (Figure 12.13a).

The Thalamus

The egg-shaped **thalamus** (thal'ah-mus) makes up 80% of the diencephalon and forms the superolateral walls of the third ventricle. The right and left thalamus are joined by a small midline commissure, the *intermediate mass. Thalamus* is a Greek word meaning "inner room," which well describes this deep brain region.

The thalamus contains about a dozen major nuclei, each of which sends axons to the cerebral cortex (Figure 12.14). Some thalamic nuclei act as relay stations for the sensory information ascending to the primary sensory areas of the cortex. Afferent impulses from all the conscious senses converge on the thalamus and synapse in at least one of its nuclei. For example, the **ventral posterior lateral nucleus** receives general somatic sensory information. The **lateral** and **medial geniculate nuclei** (jĕ-nik'u-lāt; "knee shaped") receive visual and auditory information, respectively.

Sensory inputs are not the only type of information relayed through the thalamus. *Every part of the brain that communicates with the cerebral cortex must relay its signals through a nucleus of the thalamus.* The thalamus can therefore be thought of as the "gateway" to the cerebral cortex.

The thalamus not only relays information to the cortex but also actively processes the information as it passes through. The thalamic nuclei organize, amplify, or "tone down" the signals headed for the cerebral cortex. There are many other *relay nuclei* in the brain, all of which process and "edit" information before sending it along.

The division of labor among the different thalamic nuclei is remarkable. Each nucleus receives input from its own specific region of the CNS and in turn projects information to its own specific region of the cerebral cortex (Figure 12.14).

The Hypothalamus

The **hypothalamus** (hi"po-thal'ah-mus; "below the thalamus") is the inferior division of the diencephalon (Figures 12.13, 12.15, and 12.16). It forms the inferolateral walls of the third ventricle. On the underside of the brain, it lies between the **optic chiasma** (point of crossover of cranial nerves II, the optic nerves) and the posterior border of the **mammillary**

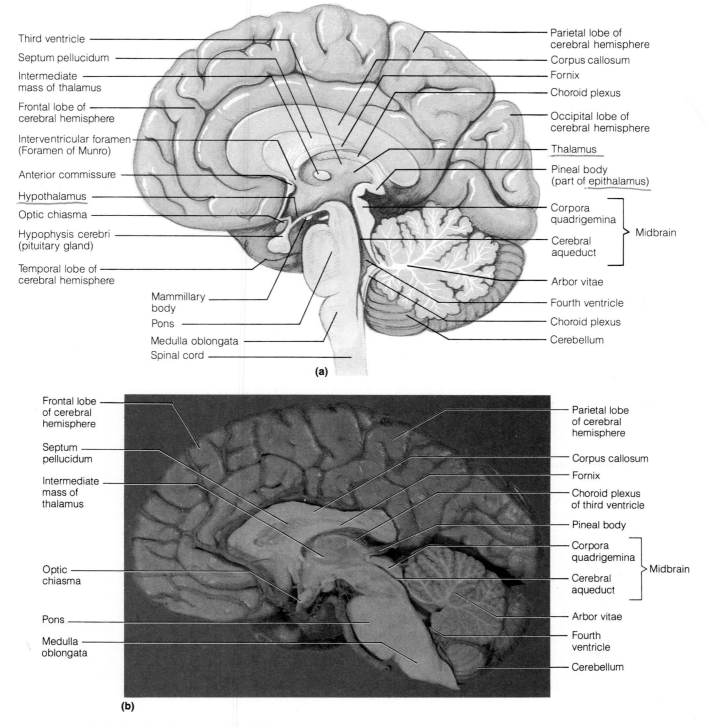

Figure 12.13   Brain sectioned along the midsagittal plane, showing the diencephalon and brain stem. (**a**) Diagram. (**b**) Photo.

bodies. The mammillary bodies are rounded bumps that bulge from the hypothalamic floor (*mammillary* = "little breast"). The *pituitary gland* also projects inferiorly from the hypothalamus. This gland secretes many hormones (Chapter 24, p. 663).

The hypothalamus, like the thalamus, contains about a dozen nuclei of gray matter (Figure 12.15). Functionally, the hypothalamus is the main visceral control center of the body, regulating many activities of the visceral organs. Its functions include the following:

1. Control of the autonomic nervous system. As you will recall, the autonomic nervous system is the system of peripheral motor neurons that regulates contraction of smooth and cardiac muscle and the secretion of glands (p. 297). The hypothalamus exerts control over the autonomic neurons. In doing so, it

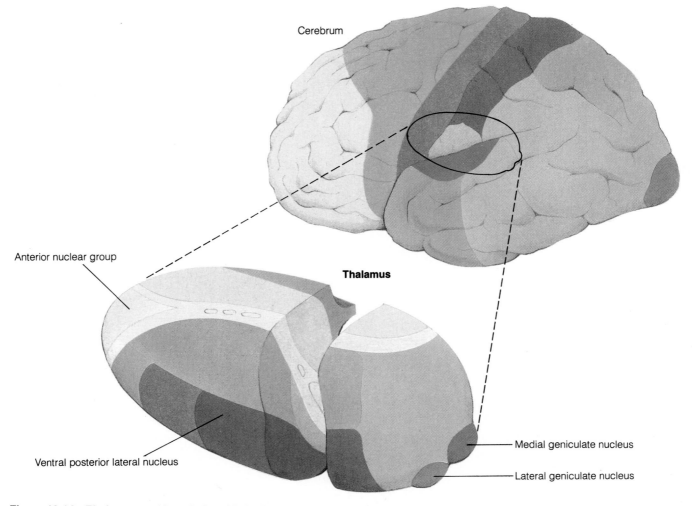

Cerebrum

Anterior nuclear group

Thalamus

Ventral posterior lateral nucleus

Medial geniculate nucleus

Lateral geniculate nucleus

Figure 12.14 Thalamus and its relationship to the cerebral cortex. Each nucleus of the thalamus sends its axons to its own region of the cerebral cortex, as shown by the color coding. Only the most important nuclei are named on this figure.

regulates heart rate and blood pressure, movement of the digestive tube, the secretion of sweat glands and salivary glands, and many other visceral activities.

2. Center for emotional response. The hypothalamus lies at the center of the emotional part of the brain, the limbic system. Regions involved in pleasure, rage, sex drive, and fear have been located in the hypothalamus.

3. Regulation of body temperature. The body's thermostat is in the hypothalamus. Some hypothalamic neurons sense blood temperature, then initiate the body's cooling or heating mechanisms as needed (sweating, shivering).

4. Hunger and thirst centers. By sensing the concentrations of nutrients and salts in the blood, certain hypothalamic neurons mediate feelings of hunger and thirst.

5. Sleep-wake cycles. Acting with other brain regions, the hypothalamus helps regulate the complex phe-

nomenon of sleep. The hypothalamus is responsible for the timing of the sleep cycle. The nuclei involved seem to be the **suprachiasmatic** (soo"prah-ki-az-mat'ik) **nucleus** above the optic chiasma and the **preoptic nucleus** anterior to that (Figure 12.15). The suprachiasmatic nucleus is the body's biological clock, regulating many daily rhythms *(circadian rhythms)*. It receives information on daylight-darkness cycles from the eye through the optic nerve, then sends signals to the preoptic nucleus. In response to such signals, the preoptic nucleus induces sleep.

6. Control of the endocrine system. The hypothalamus controls the secretion of hormones by the pituitary gland, which in turn regulates many functions of the visceral organs (Chapter 24).

Through experiments that stimulate or remove parts of the hypothalamus, it has become possible to localize each of the above functions to a general region of the hypothalamus. These regions are called

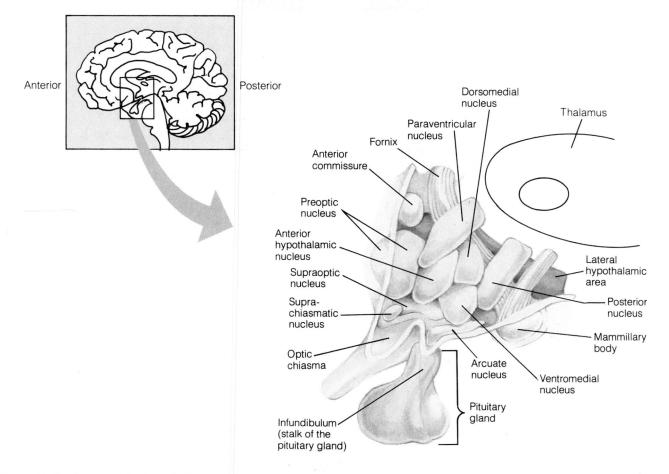

Figure 12.15   Nuclei of the hypothalamus.

functional centers. For the most part, the functional centers can be only roughly matched with specific structural nuclei.

Try to predict some of the functional disorders that result from injuring the hypothalamus.

Hypothalamic lesions cause a number of disorders in visceral functions and in emotions. These can include severe weight loss or obesity, sleep disturbances, dehydration, and a broad range of emotional disorders. ∎

The Epithalamus

The **epithalamus**, the most dorsal part of the diencephalon (Figure 12.13), forms part of the roof of the third ventricle. It contains one tiny group of nuclei and a small, unpaired knob called the **pineal gland** or **pineal body** (pin′e-al; "pinecone-shaped"). The pineal gland derives from the ependymal glial cells and is a hormone-secreting organ. Under the influence of the hypothalamus, the pineal gland secretes **melatonin** (mel″ah-to′nin). This hormone probably signals the body to prepare for the nighttime stage of the sleep-wake cycle.

The Brain Stem

From rostral to caudal, the three regions of the brain stem are the *midbrain, pons,* and *medulla oblongata* (Figures 12.13, 12.16, and 12.17). Each of these is roughly an inch long, and together they make up only 2.5% of total brain weight. The brain stem lies in the posterior cranial fossa of the skull, on the basiocciput. The brain stem has several general functions: (1) It produces rigidly programmed, automatic behaviors necessary for our survival; (2) it acts as a passageway for all the fiber tracts running between the cerebrum and the spinal cord; (3) 10 of the 12 pairs of cranial nerves attach to it, so it is heavily involved with the innervation of the face and head. The brain stem has the same structural plan as the spinal cord, with outer white matter surrounding an inner region of gray matter. However, there are also nuclei of gray matter located in the white matter (Figure 12.5).

The Midbrain

The midbrain lies between the diencephalon rostrally and the pons caudally (Figure 12.13). The hollow cerebral aqueduct runs through the midbrain, dividing it

Figure 12.16   Ventral view of the brain, showing the diencephalon and brain stem. Note that very little of the midbrain is visible—just its cerebral peduncles.

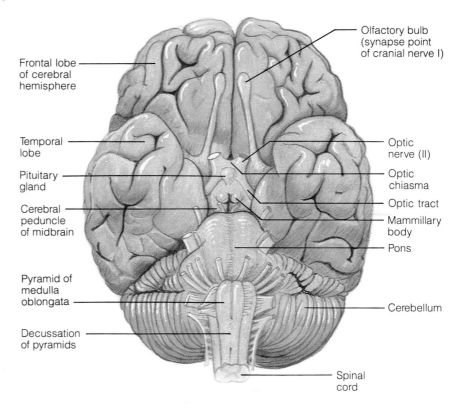

Olfactory bulb (synapse point of cranial nerve I)

Frontal lobe of cerebral hemisphere

Temporal lobe

Pituitary gland

Cerebral peduncle of midbrain

Pyramid of medulla oblongata

Decussation of pyramids

Optic nerve (II)

Optic chiasma

Optic tract

Mammillary body

Pons

Cerebellum

Spinal cord

into a *tectum* ("roof") dorsally and paired *cerebral peduncles* ventrally. On the ventral surface of the brain, the cerebral peduncles form vertical pillars (Figures 12.16 and 12.17a), which appear to hold up the forebrain, hence their name, meaning "the little feet of the cerebrum." These peduncles contain the large pyramidal (corticospinal) motor tracts descending toward the spinal cord. Dorsally, the midbrain exhibits another pair of bands, the **superior cerebellar peduncles** (Figure 12.17b). These connect the midbrain to the cerebellum.

A cross section through the midbrain reveals its internal structure (Figure 12.18a). Around the cerebral aqueduct is the **periaqueductal gray matter**. The function of this area is largely unknown, but stimulating it with electrodes produces analgesia (relief from pain) throughout the body.

The ventral part of the periaqueductal gray matter contains the cell bodies of motor neurons that contribute to two cranial nerves, the oculomotor (III) and the trochlear (IV). These nerves control most of the muscles that move the eyes (Chapter 13).

Many nuclei lie external to the periaqueductal gray matter, scattered within the surrounding white matter. The largest of these nuclei are the **corpora quadrigemina** (kwod″rĭ-jem′ĭ-nah; "quadruplets"), which make up the tectum. They are visible as four bumps on the dorsal surface of the midbrain and are more specifically named the superior and inferior colliculi (Figure 12.17b). The **superior colliculi** (kŏ-lik′u-li; "little hills") are visual nuclei that act in visual reflexes. For example, our eyes will track and fol-

low moving objects even if we are not consciously looking at the objects. The **inferior colliculi**, by contrast, belong to the auditory system. Among other functions, they act in reflexive responses to sound. Next time you are startled by a loud noise, notice how your head and eyes turn toward the sound. This "startle reflex" is mediated by the inferior colliculi.

Also embedded in the white matter of the midbrain are two pigmented nuclei: *substantia nigra* (*nigra* = black) and the *red nucleus* (Figure 12.18a). The band-like **substantia nigra** (sub-stan′she-ah ni′-grah) is in the cerebral peduncle, deep to the corticospinal tract. Its neuron cell bodies contain dark melanin pigment, a precursor of the neurotransmitter that is made and released by these neurons (dopamine). The substantia nigra is functionally linked to the basal nuclei (p. 330). The oval **red nucleus** lies deep to the substantia nigra. Its reddish hue is due to a rich blood supply and to the presence of iron pigment in the cell bodies of its neurons. The red nucleus has a minor motor function: It helps to bring about flexion movements of the limbs.

The midbrain also contains part of the *reticular formation*, a system of small nuclei that are scattered throughout the core of the brain stem. The reticular formation is discussed on pp. 339–341.

The Pons

The **pons** is a bulging region of the brain stem wedged between the midbrain and the medulla oblongata (Figure 12.13). Dorsally, it is separated from the cere-

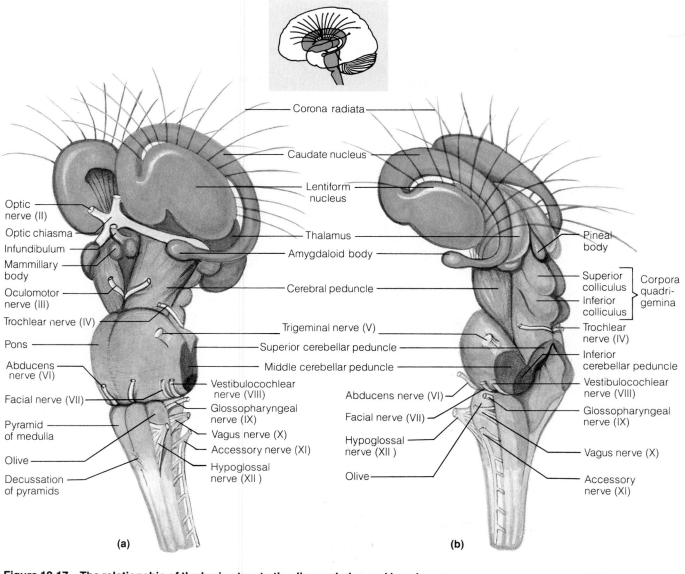

Figure 12.17   The relationship of the brain stem to the diencephalon and basal nuclei. The cerebral hemispheres have been removed. (**a**) Ventrolateral view. (**b**) Dorsolateral view.

bellum by the fourth ventricle. As its name suggests, the pons ("bridge") looks like a ventral bridge between the right and left halves of the cerebellum (Figure 12.16).

The following cranial nerves attach to the pons (Figure 12.17a): the trigeminal (V), which innervates the skin of the face and the chewing muscles; the abducens (VI), which innervates an eye-moving muscle; and the facial (VII), which supplies the muscles of facial expression, among other functions. These cranial nerves are discussed in more depth in Chapter 13.

Near the fourth ventricle, the pons contains the nuclei of the cranial nerves that attach to it (Figure 12.18c). **Cranial nerve nuclei** comprise two general groups: *motor nuclei* consisting of cell bodies of the motor neurons that contribute to cranial nerves; and

sensory nuclei consisting of interneurons that receive direct input from the sensory neurons in cranial nerves. The details of these complex nuclei are beyond the scope of this book and the nuclei are illustrated primarily for reference purposes.

Ventral to the cranial nerve nuclei, the pons contains a part of the reticular formation. Here, some *pons respiratory centers* help control our breathing movements—producing smooth transitions from inhalation to exhalation and vice versa.

Farthest ventrally, the pons contains the thick pyramidal motor tract descending from the cerebral cortex. Interspersed among the fibers of this tract are numerous *pontine nuclei*, relay nuclei in a path by which the cerebral motor cortex confers with the cerebellum about the coordination of voluntary move-

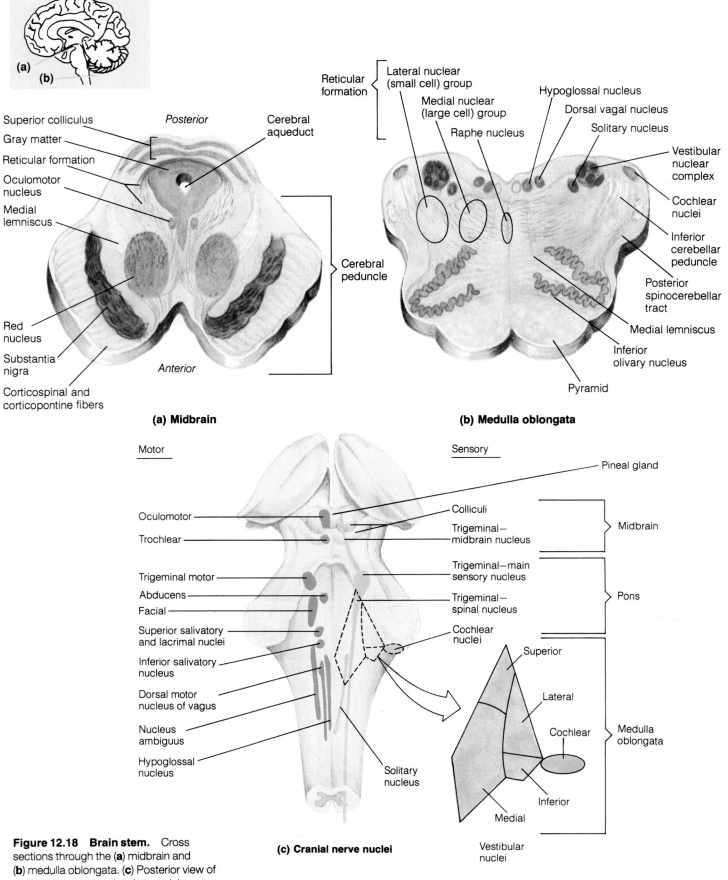

(a) Midbrain

Superior colliculus
Gray matter
Reticular formation
Oculomotor nucleus
Medial lemniscus
Red nucleus
Substantia nigra
Corticospinal and corticopontine fibers
Posterior
Cerebral aqueduct
Cerebral peduncle
Anterior

(b) Medulla oblongata

Reticular formation
Lateral nuclear (small cell) group
Medial nuclear (large cell) group
Raphe nucleus
Hypoglossal nucleus
Dorsal vagal nucleus
Solitary nucleus
Vestibular nuclear complex
Cochlear nuclei
Inferior cerebellar peduncle
Posterior spinocerebellar tract
Medial lemniscus
Inferior olivary nucleus
Pyramid

(c) Cranial nerve nuclei

Motor
Oculomotor
Trochlear
Trigeminal motor
Abducens
Facial
Superior salivatory and lacrimal nuclei
Inferior salivatory nucleus
Dorsal motor nucleus of vagus
Nucleus ambiguus
Hypoglossal nucleus

Sensory
Pineal gland
Colliculi
Trigeminal–midbrain nucleus
Trigeminal–main sensory nucleus
Trigeminal–spinal nucleus
Cochlear nuclei
Solitary nucleus

Midbrain
Pons
Medulla oblongata

Superior
Lateral
Cochlear
Inferior
Medial
Vestibular nuclei

Figure 12.18   Brain stem. Cross sections through the **(a)** midbrain and **(b)** medulla oblongata. **(c)** Posterior view of the brain stem, showing the cranial nerve nuclei.

ments. This interaction is described on page 338. The pontine nuclei send axons to the cerebellum in the thick **middle cerebellar peduncles** (Figure 12.17).

The Medulla Oblongata

The conical **medulla oblongata**, or simply **medulla**, is the most caudal part of the brain stem. It blends into the spinal cord at the level of the foramen magnum of the skull. As we mentioned earlier, a part of the fourth ventricle lies dorsal to the rostral half of the medulla. Here, the ventricle has a thin, sheet-like roof that contains a capillary-rich membrane called a *choroid plexus* (see Figure 12.13).

The medulla has several externally visible landmarks (Figures 12.16 and 12.17a). Flanking the ventral midline are two longitudinal ridges called **pyramids**. These are formed by the pyramidal tracts. In the caudal part of the medulla, most of these pyramidal fibers cross over to the opposite side of the brain. This crossover point is the **decussation of the pyramids** (de"kus-sa'shun; "a crossing"). As noted earlier, the result of this crossover is that each cerebral hemisphere controls the voluntary movements of the opposite side of the body.

Other external landmarks on the medulla are the inferior cerebellar peduncles and the olives. The **inferior cerebellar peduncles** (Figure 12.17b) are fiber tracts that connect the medulla to the cerebellum dorsally. Each **olive** (which does resemble an olive) lies just lateral to a pyramid and contains an **inferior olivary nucleus**, a large wavy fold of gray matter (Figure 12.18b). This nucleus is a relay station for sensory information traveling to the cerebellum, especially for proprioceptive information ascending from the spinal cord.

The five most inferior pairs of cranial nerves attach to the medulla (Figure 12.17). These nerves are the vestibulocochlear (VIII), which is the sensory nerve of hearing and equilibrium; the glossopharyngeal (IX), which innervates part of the tongue and pharynx; the vagus (X), which innervates many visceral organs in the thorax and abdomen; the accessory nerve (XI), which innervates some muscles of the neck; and the hypoglossal (XII), which innervates tongue muscles (Chapter 13). The sensory and motor nuclei of these cranial nerves lie deep in the medulla, near the fourth ventricle (Figure 12.18c and b). Only two of these nuclei are considered here: The **cochlear** (kok'le-ar) **nuclei** receive auditory information, and the **vestibular nuclei** receive information on equilibrium. The vestibular nuclei send axons to the spinal cord, where they help bring about subconscious movements by which the body maintains its equilibrium.

The most caudal part of the medulla contains some large nuclei in its dorsal roof, the **nucleus cuneatus** (ku-ne-a'tus) and the **nucleus gracilis** (gras'ĭ-

lis). These serve as relay nuclei in the pathway by which general somatic sensory information ascends through the spinal cord to the somatosensory cerebral cortex (see Figure 12.29a on p. 350).

The core of the medulla contains much of the reticular formation (Figure 12.18b), some nuclei of which perform autonomic (visceral motor) functions. The most important visceral centers in the medulla are the following:

1. The *cardiac center* adjusts the force and rate of the heartbeat.

2. The *vasomotor center* regulates blood pressure. It does so by stimulating or inhibiting the contraction of smooth muscle in the walls of blood vessels, thereby constricting or dilating the vessels. Constriction of arteries throughout the body causes blood pressure to rise, whereas dilation reduces blood pressure.

3. *Medullary respiratory centers* control the basic pattern and rate of breathing. Recall that the pons also contains some respiratory centers.

4. Additional centers regulate vomiting, hiccuping, swallowing, coughing, and sneezing.

Note that the above list contains many functions that we also attributed to the hypothalamus (pp. 331–332). The overlap is easily explained: The hypothalamus exerts its control over most visceral functions by relaying its instructions through the medulla's reticular centers, which carry them out.

The Cerebellum

Structure and Function

The cauliflower-like **cerebellum** makes up 11% of the mass of the brain and is exceeded in size only by the cerebrum. It is located dorsal to the pons and medulla, from which it is separated by the fourth ventricle (Figure 12.13). Functionally, the cerebellum smooths and coordinates body movements that are ordered by other brain regions and helps maintain posture and equilibrium.

The cerebellum consists of two expanded **cerebellar hemispheres** connected medially by the wormlike **vermis** (Figure 12.19). Its surface is folded into many plate-like ridges called **folia** (fo'le-ah; "leaves"), which are separated by grooves called **fissures**. Each hemisphere is subdivided into three lobes: the large **anterior** and **posterior lobes**, and the small **flocculonodular** (flok"u-lo-nod'u-lar) **lobe**. The flocculonodular lobes, shaped like an airplane propeller, are hidden ventral to the posterior lobe (Figure 12.19c). Functionally, the flocculonodular lobes adjust posture to maintain equilibrium, while the anterior and posterior lobes coordinate body movements.

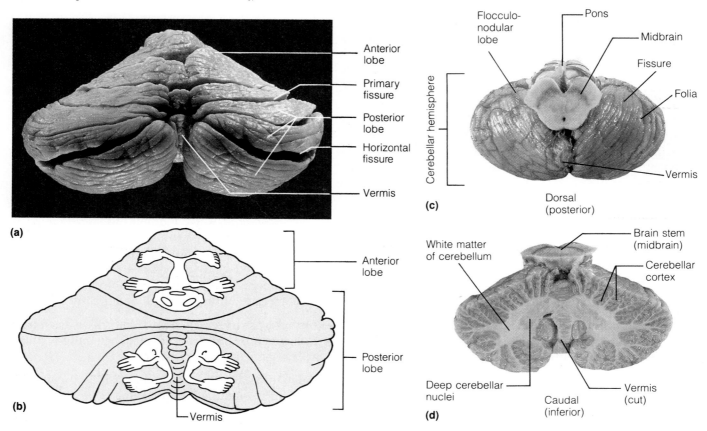

Figure 12.19   Cerebellum. (**a**) Posterior (dorsal) view of the cerebellum (superior is above). (**b**) A diagrammatic view of (a), showing the anterior and posterior lobes. The broad central region indicated by the "little men" coordinates ordinary, unskilled movements of the body, with each point on the cerebellar cortex influencing its own part of the body as indicated. The expanded lateral regions of the hemispheres, by contrast, control skilled and delicate movements. (**c**) Superior view of the cerebellum. Note the flocculonodular lobes ventrally. (**d**) The cerebellum sectioned to reveal its three layers. Like (a), this is a posterior view in which the midbrain is sectioned horizontally and the cerebellum is sectioned frontally.

Like the cerebrum, the cerebellum has three layers: an outer *cortex* of gray matter, internal *white matter,* and deeply situated masses of gray matter called *deep cerebellar nuclei* (Figure 12.19d). The cerebellar cortex determines how to smooth our body movements. The white matter consists of axons that carry information to and from the cortex. The deep cerebellar nuclei give rise to axons that relay the instructions from the cerebellar cortex to other parts of the brain.

To calculate how to coordinate body movements, the cerebellar cortex must continuously receive three types of information about how the body is moving: (1) *information on equilibrium,* relayed from receptors in the inner ear through the vestibular nuclei to the cerebellar cortex; (2) *information on the current movements of the limbs, neck, and trunk,* which travels from proprioceptors in muscles, tendons, and joints up the spinal cord to the cerebellar cortex; and (3) *information from the motor area of the cerebral cortex;* this information passes from the cerebral cortex through the pyramidal tract and pontine nuclei to the cerebellar cortex.

The main function of the cerebellum is to receive a preview of the intended movements being ordered by the motor cerebral cortex, to compare these intended movements to the body movements that are actually taking place, and then to send instructions to the cerebral cortex on how to resolve any differences between the intended and actual movements. Using this feedback from the cerebellum, the motor cerebral cortex can continuously readjust the motor commands it sends to the spinal cord, fine-tuning movements so that they are well coordinated.

Cerebellar Peduncles

The superior, middle, and inferior cerebellar peduncles are thick stalks of nerve fibers that connect the cerebellum to the brain stem (Figure 12.17b). Their fibers carry the information that travels from and to the cerebellum (as listed in the previous section). The **superior cerebellar peduncles**, connecting the cerebellum to the midbrain, carry instructions from the cerebellum toward the cerebral cortex. The **middle cerebellar peduncles**, connecting the pons to the cerebellum, carry the information from the motor cerebral cortex and the pontine nuclei. The **inferior cerebellar peduncles**, arising from the medulla, carry the information on equilibrium from the vestibular nuclei and the information on proprioception from the spinal cord.

Unlike the contralateral distribution of fibers to and from the cerebral cortex, virtually all fibers that enter and leave the cerebellum are **ipsilateral** (ip″si-lat′er-al; *ipsi* = same)—to and from the *same* side of the body.

Can you imagine some effects of damage to the cerebellum?

Damage to the anterior and posterior lobes leads to disorders of coordination. Sufferers may exhibit slow or jerky movements that are inaccurate and tend to overreach their targets. Such people are unable to touch their finger to their nose with their eyes closed. Speech may be slurred, slow, and singsong in pattern. Damage to the flocculonodular nodes, by contrast, leads to disorders of equilibrium. The affected individuals have a wide stance and an unsteady "drunken sailor" gait that predisposes them to falling. In all cases, damage in one half of the cerebellum (left or right) affects that same half of the body. ■

Functional Brain Systems

Functional brain systems are networks of neurons that function together but span large distances within the brain. The *limbic system,* spread widely throughout the forebrain, and the *reticular formation,* spanning the brain stem, are important functional systems.

The Limbic System

EMOTIONS MEMORY

The **limbic system** is a group of structures on the medial aspect of each cerebral hemisphere and diencephalon (Figure 12.20). In the cerebrum, the limbic structures form a broad ring (*limbus* = "headband"), which includes the *septal nuclei, cingulate gyrus, hippocampal formation,* and part of the amygdala. In the diencephalon, the main limbic structures are the hypothalamus and a nucleus of the thalamus called the *anterior nucleus.* The **fornix** ("arch") and other fiber tracts link the limbic system together. The limbic system overlaps the rhinencephalon ("smell brain": p. 326) in several places.

The limbic system is the "emotional brain." Of the cerebral limbic structures, two parts seem especially important in our emotions: the amygdala and the anterior part of the **cingulate gyrus** (sing′gu-lāt; "belt-shaped"). Lesions in the amygdala lead to personality changes, such as docile behavior, emotional instability, restlessness, and increased interest in fighting or eating. Damage to the anterior cingulate, by contrast, destroys the will: Patients with damage to this area have no desire to act.

The limbic system communicates with many other regions of the brain. Most output from the limbic system is relayed through the hypothalamus and the reticular formation, the brain regions that control our visceral responses. This fact explains why people under emotional stress get visceral illnesses such as high blood pressure. The limbic system also interacts with the prefrontal lobes of the cerebral cortex. Thus, there is a close interaction between our feelings (mediated by the emotional brain) and our thoughts (mediated by the thinking brain). As a result, we (a) react emotionally to the things we consciously understand to be happening and (b) are consciously aware of the emotional richness of our lives.

Some parts of the limbic system play a role in remembering information and facts. The key regions for this are the **hippocampus** (hip″o-kam′pus; "sea horse") and the amygdala, both in the medial aspect of the temporal lobe (Figure 12.20). People with bilateral damage to these regions cannot incorporate new facts into memories, though they retain their old memories from before the injury. These limbic structures help consolidate memories but do not store them. (Recall that memories are stored widely throughout the cortex.)

RETICULAR ACTIVATING SYS — AROUSAL

The Reticular Formation

The **reticular formation** runs through the central core of the medulla, pons, and midbrain (Figure 12.21). It consists of loosely clustered neurons in what is otherwise white matter. These reticular neurons form three columns that extend the length of the brain stem: (1) the midline **raphe** (ra′fe) **nuclei**, which are flanked laterally by (2) the **medial group** of nuclei and then (3) the **lateral group** (Figure 12.18b).

Because they have long, branching axons, single reticular neurons project to such widely separated regions as the thalamus, cerebellum, and spinal cord. Such widespread connections make reticular neurons ideal for governing the arousal of the brain as a whole. For example, certain reticular neurons send a continuous stream of impulses to the cerebrum (through relays in the thalamus), which maintains the cerebral cortex in an alert, conscious state (Figure 12.21). This arm of the reticular formation, which maintains consciousness and alertness, is called the *reticular activating system (RAS).*

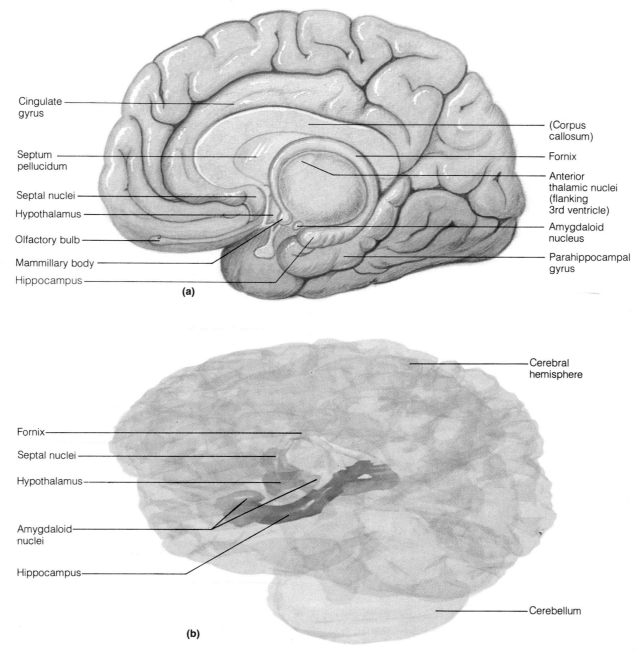

Cingulate gyrus

(Corpus callosum)

Fornix

Septum pellucidum

Anterior thalamic nuclei (flanking 3rd ventricle)

Septal nuclei

Hypothalamus

Amygdaloid nucleus

Olfactory bulb

Parahippocampal gyrus

Mammillary body

Hippocampus

(a)

Cerebral hemisphere

Fornix

Septal nuclei

Hypothalamus

Amygdaloid nuclei

Hippocampus

Cerebellum

(b)

Figure 12.20 The limbic system. (a) The limbic system (yellow) on the medial side of the brain. The brain stem has been removed. **(b)** Computer-generated image of the limbic structures.

Axons from all the great ascending sensory tracts synapse on RAS neurons, keeping these reticular neurons active and enhancing their arousing effect on the cerebrum. Visual, auditory, and touch stimuli keep us awake and mentally alert. For this reason, many students like to study in a crowded cafeteria, stimulated by the bustle of such an environment. The RAS also functions in sleep and in arousal from sleep.

Can you deduce the effects of drugs and injuries that depress the function of the RAS?

General anesthesia, alcohol, tranquilizers, and sleep-inducing drugs depress the RAS and lead to a loss of alertness or consciousness. Knockout punches produce the same effect because they twist the brain stem. Severe injury to the RAS is the cause of coma. ■

The reticular formation also has a *motor* arm, which sends axons to the spinal cord. Some of these axons control the skeletal muscles during coarse movements of the limbs (see *reticulospinal tract* in Table 12.3, p. 354), while others regulate visceral functions such as heart rate, blood pressure, and respiration (recall p. 337).

Protection of the Brain

Nervous tissue is soft and delicate, and the irreplaceable neurons can be injured by even slight pressure. However, the brain is protected from injury by a bony enclosure (the skull), by surrounding membranes called *meninges,* and by a watery cushion of *cerebrospinal fluid.* Furthermore, the brain is protected from harmful substances in the blood by the *blood-brain barrier.* We described the brain's bony enclosure, the cranium, in Chapter 7. Here we will consider the other protective devices.

Meninges

The **meninges** (mĕ-nin'jēz; "membranes") are three membranes of connective tissue that lie just external to the brain and spinal cord. Their functions are to (1) cover and protect the CNS, (2) enclose and protect the

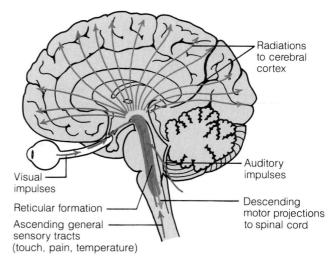

Figure 12.21   The reticular formation.   The reticular formation runs the length of the brain stem. A part of this formation, the reticular activating system (RAS) keeps the cerebral cortex alert and conscious. Ascending arrows represent sensory inputs to the RAS, which stimulate outputs from the RAS to the cerebrum. The reticular formation also controls some muscle activity throughout the body. This output is shown by the descending arrow at the bottom.

Table 12.1 Functions of Major Brain Regions	
Region	**Function**
Cerebral hemispheres (pp. 321–330)	Cortical gray matter localizes and interprets sensory inputs, controls voluntary and skilled skeletal muscle activity, and functions in intellectual and emotional processing; basal nuclei are subcortical motor centers important in initiation of skeletal movements
Diencephalon (pp. 330–333)	Thalamic nuclei are relay stations in conduction of (1) sensory impulses to cerebral cortex for interpretation, and (2) impulses from all other regions of the brain that communicate with the cerebral cortex; thalamus is also involved in memory processing
	Hypothalamus is chief integration center of autonomic (involuntary) nervous system; it functions in regulation of body temperature, food intake, water balance, thirst, and biological rhythms and drives; regulates hormonal output of anterior pituitary gland and is an endocrine organ in its own right (produces ADH and oxytocin); part of limbic system
Limbic system (p. 339)	Emotional brain; a functional system involving cerebral and diencephalon structures that mediates emotional response
Brain stem Midbrain (pp. 333–334)	Conduction pathway between higher and lower brain centers (e.g., cerebral peduncles contain the fibers of the pyramidal tracts); its superior and inferior colliculi are visual and auditory reflex centers; substantia nigra and red nuclei are subcortical motor centers; contains nuclei for cranial nerves III and IV
Pons (pp. 334–337)	Conduction pathway between higher and lower brain centers; pontine nuclei relay information from the cerebrum to the cerebellum; its respiratory nuclei cooperate with medullary respiratory centers to control respiratory rate and depth; houses nuclei of cranial nerves V–VII
Medulla oblongata (p. 337)	Conduction pathway between higher brain centers and spinal cord; site of decussation of the pyramidal tracts; houses nuclei of cranial nerves VIII–XII; contains nuclei cuneatus and gracilis (synapse points of ascending sensory pathways transmitting sensory impulses from skin and proprioceptors), and visceral nuclei controlling heart rate, blood vessel diameter, respiratory rate, vomiting, coughing, etc.; its inferior olivary nuclei provide the sensory relay to the cerebellum
Reticular formation (pp. 339–341)	A functional brain stem system that maintains cerebral cortical alertness (reticular activating system) and filters out repetitive stimuli; its motor nuclei help regulate skeletal and visceral muscle activity
Cerebellum (pp. 337–339)	Processes information received from cerebral motor cortex and from proprioceptors and visual and equilibrium pathways, and provides "instructions" to cerebral motor cortex and subcortical motor centers that result in proper balance and posture and smooth, coordinated skeletal muscle movements

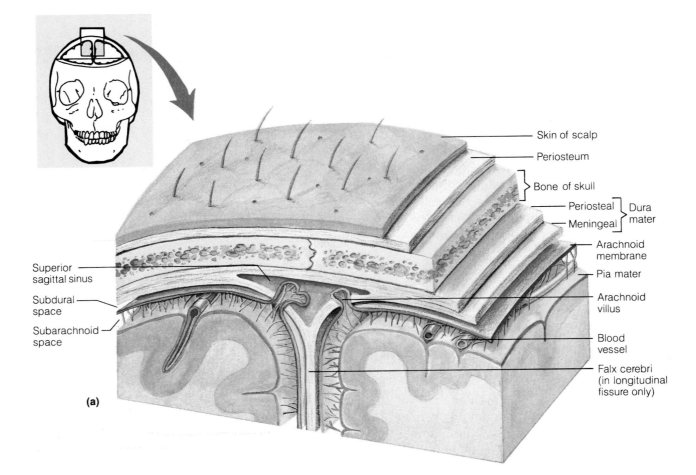

Skin of scalp
Periosteum
Bone of skull
Periosteal ⎱ Dura
Meningeal ⎰ mater
Arachnoid membrane
Pia mater
Arachnoid villus
Blood vessel
Falx cerebri (in longitudinal fissure only)

Superior sagittal sinus
Subdural space
Subarachnoid space

(a)

Figure 12.22   Meninges around the brain.   **(a)** A frontal section taken from the superior region of the head shows the meninges between the skull and brain (dura, arachnoid, and pia mater). In the midline, the dura mater forms the falx cerebri, a partition between the two cerebral hemispheres. A specialized vein, the superior sagittal sinus, lies in the dura mater superiorly. Arachnoid villi are also shown. **(b)** Posterior view of the brain in place, surrounded by the dura mater and dural sinuses.

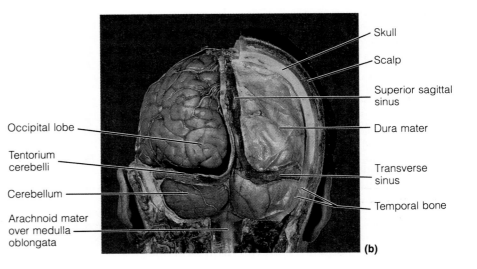

Skull
Scalp
Superior sagittal sinus
Dura mater
Transverse sinus
Temporal bone

Occipital lobe
Tentorium cerebelli
Cerebellum
Arachnoid mater over medulla oblongata

(b)

blood vessels that supply the CNS, and (3) contain the cerebrospinal fluid. From external to internal, the meninges are the *dura mater, arachnoid mater,* and *pia mater* (Figure 12.22). Cranial meninges are discussed here, meninges of the spinal cord on p. 345.

The Dura Mater.   The leathery **dura mater** (du'rah ma'ter) is by far the strongest of the meninges. Its name means "tough mother." Where it surrounds the brain, it is a two-layered sheet of fibrous connective

tissue. The more superficial **periosteal layer** attaches to the internal surface of the skull bones; that is, it is the bone periosteum. The deeper **meningeal layer** forms the true external covering of the brain. These two layers of dura mater are fused together, except where they separate to enclose the blood-filled *dural sinuses.* These sinuses act as veins, collecting blood from the brain and directing it to the large internal jugular veins of the neck. The largest dural sinus is the *superior sagittal sinus* in the superior midline.

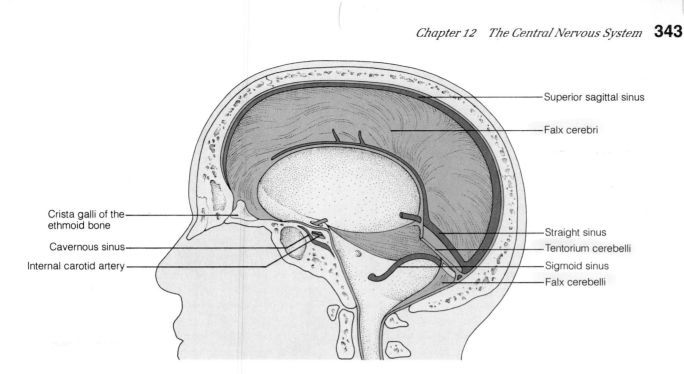

Superior sagittal sinus

Falx cerebri

Crista galli of the ethmoid bone

Cavernous sinus

Internal carotid artery

Straight sinus

Tentorium cerebelli

Sigmoid sinus

Falx cerebelli

Figure 12.23 Partitions of dura mater in the cranial cavity. Note the falx cerebri, falx cerebelli, and tentorium cerebelli. Some dural sinuses are also shown.

In several places, the meningeal dura extends inward to form flat partitions that subdivide the cranial cavity (Figure 12.23). These partitions, which limit movement of the brain within the cranium, include the following:

1. The **falx cerebri** (falks ser′ĕ-bri). This large vertical sheet lies in the median plane and dips into the longitudinal fissure between the cerebral hemispheres. It is sickle-shaped (*falx* = sickle). Anteriorly, it attaches to the crista galli of the ethmoid bone.

2. The **falx cerebelli** (ser″ĕ-bel′li). Continuing inferiorly from the posterior part of the falx cerebri, this vertical partition runs along the vermis of the cerebellum in the posterior cranial fossa.

3. The **tentorium** (ten-to′re-um) **cerebelli**. Resembling a tent over the cerebellum, this almost horizontal sheet lies in the transverse fissure between the cerebrum and cerebellum (*tentorium* means tent).

The Arachnoid Mater. The **arachnoid** (ah-rak′noid) **mater** lies just deep to the dura mater (Figure 12.22a). Between these two layers is a thin potential space called the **subdural space**. The dura and arachnoid surround the brain loosely, never dipping into the sulci on the cerebral surface. Deep to the arachnoid membrane is the wide **subarachnoid space**. Web-like threads span this space and hold the arachnoid mater to the underlying pia mater (this web is the basis of the name *arachnoid*, which means "spider-like"). The subarachnoid space is filled with cerebrospinal fluid and contains the largest blood vessels that supply the brain. Since the arachnoid is fine and elastic,

these important blood vessels are rather poorly protected.

Over the superior part of the brain, the arachnoid forms knob-like projections called **arachnoid villi** *(arachnoid granulations).* These villi project superiorly through the dura into the superior sagittal sinus (Figure 12.22a) and into some other dural sinuses as well. They function as valves that allow cerebrospinal fluid to pass from the subarachnoid space into the dural blood sinuses (see p. 344).

The Pia Mater. The **pia** (pi′ah) **mater** ("gentle mother") is a layer of delicate connective tissue richly vascularized with fine blood vessels. Unlike the other meninges, it clings tightly to the brain surface, following every convolution. As arteries enter the brain tissue, they carry ragged sheaths of pia mater internally for short distances.

Infections of the meninges are often extremely dangerous. Can you guess why?

Meningitis, or inflammation of the meninges, is caused by bacterial or viral infection. The infection can spread to the underlying nervous tissue, causing brain inflammation, or **encephalitis** (en″sef-ah-li′tis). Meningitis is usually diagnosed by taking a sample of cerebrospinal fluid from the subarachnoid space, and examining it for the presence of microbes. ◼

Cerebrospinal Fluid

Cerebrospinal fluid (CSF), located in and around the brain and spinal cord, forms a liquid cushion that gives buoyancy to the central nervous system. The

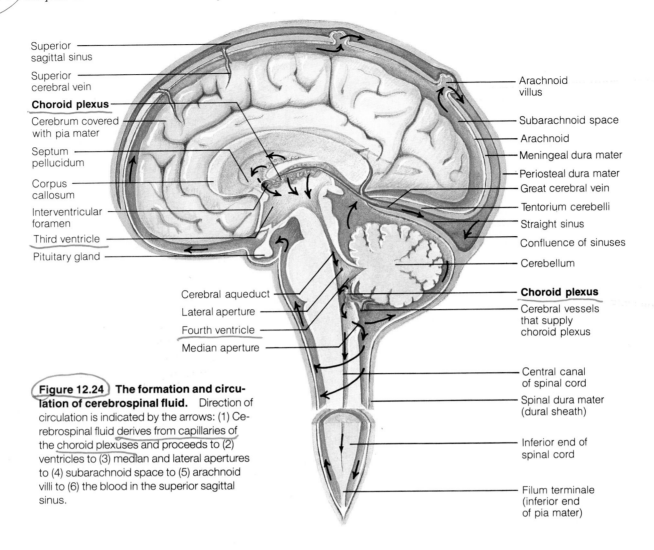

Superior sagittal sinus
Superior cerebral vein
Choroid plexus
Cerebrum covered with pia mater
Septum pellucidum
Corpus callosum
Interventricular foramen
Third ventricle
Pituitary gland

Arachnoid villus
Subarachnoid space
Arachnoid
Meningeal dura mater
Periosteal dura mater
Great cerebral vein
Tentorium cerebelli
Straight sinus
Confluence of sinuses
Cerebellum
Choroid plexus
Cerebral vessels that supply choroid plexus
Central canal of spinal cord
Spinal dura mater (dural sheath)
Inferior end of spinal cord
Filum terminale (inferior end of pia mater)

Cerebral aqueduct
Lateral aperture
Fourth ventricle
Median aperture

Figure 12.24 **The formation and circulation of cerebrospinal fluid.** Direction of circulation is indicated by the arrows: (1) Cerebrospinal fluid derives from capillaries of the choroid plexuses and proceeds to (2) ventricles to (3) median and lateral apertures to (4) subarachnoid space to (5) arachnoid villi to (6) the blood in the superior sagittal sinus.

brain actually floats in the CSF, which reduces its weight by 97% and prevents this delicate organ from crushing under its own weight. CSF also cushions the brain and spinal cord from blows and jolts. Additionally, although the brain does have a rich blood supply, CSF helps to nourish the brain and to remove waste products produced by neurons. Amazingly, very little CSF is needed to perform all these functions: Just 100–160 ml, about half a cup, is present at any time.

CSF is a watery broth that resembles the blood plasma from which it arises. However, its chemistry is somewhat different (more sodium and chloride ions and less protein).

Figure 12.24 diagrams the location and production of CSF. Most CSF is made in the **choroid plexuses**. These membranes attach to the roofs of the four brain ventricles and consist of a layer of ependymal cells covered externally by a capillary-rich layer of pia mater. CSF forms continuously from blood plasma by filtration from the capillaries of the pia mater and passes through the ependymal cells into the ventricles. After entering the ventricles, the CSF moves freely through these chambers (Figure 12.24). Some

CSF runs into the central canal of the spinal cord, but most enters the subarachnoid space through the lateral and median apertures in the walls of the fourth ventricle. In the subarachnoid space, CSF bathes the outer surfaces of the brain and spinal cord. From there, it enters the blood of the dural sinuses through the arachnoid villi. Cerebrospinal fluid arises from the blood and returns to it at a rate of about 500 ml a day.

Can you deduce what would happen if CSF were manufactured at an excessive rate or if its flow and drainage were physically blocked?

CSF would accumulate in the ventricles or subarachnoid space, exerting a crushing pressure on the brain. This condition, called **hydrocephalus** (hi"dro-sef'ah-lus; "water on the brain"), often results from a tumor or an inflammatory swelling that closes off the cerebral aqueduct or fourth ventricle or from meningitis that scars and closes the subarachnoid space and arachnoid villi. In some infants, hydrocephalus results from an overdeveloped choroid plexus that secretes excessive amounts of CSF. Hydrocephalus in a newborn causes enlargement of the head, which is possible because the young skull bones have not yet fused. In adults, however, hydrocephalus quickly causes brain damage because of the unyielding nature of the rigid cranium. Hydrocephalus is treated surgically by running a shunt (plastic tube) from the brain ventricles into a vein in the neck. ■

Blood-Brain Barrier

The brain has a rich supply of capillaries that provide its nervous tissue with nutrients, oxygen, and all other vital molecules. However, some blood-borne molecules that can cross other capillaries of the body cannot cross the brain capillaries. Blood-borne toxins, such as urea, mild toxins from food, and bacterial toxins, are prevented from entering brain tissue by the **blood-brain barrier**, which protects the neurons of the CNS.

The blood-brain barrier results primarily from special features of the endothelium (p. 68) that makes up the walls of the brain capillaries. The endothelial cells are joined together around their entire perimeters by tight junctions. Thus, brain capillaries are the least permeable capillaries in the body. Even so, the blood-brain barrier is not an absolute barrier. All nutrients and ions needed by the neurons pass through, some by special transport mechanisms in the plasma membranes of the endothelial cells. Furthermore, the barrier is ineffective against fat-soluble molecules, which easily diffuse through all cell membranes. Thus, the barrier allows blood-borne oxygen to reach brain neurons, as well as alcohol, nicotine, and anesthetics.

More information on the structure of the blood-brain barrier is provided on p. 486.

The Spinal Cord

External Structure and Protection

The **spinal cord** runs through the vertebral canal of the vertebral column, from the foramen magnum superiorly to the level of L_1 or L_2 inferiorly (Figure 12.25). Its function can be described in several ways: (1) Through the spinal nerves that attach to it, the spinal cord is involved in the sensory and motor innervation of the entire body inferior to the head. (2) The spinal cord provides a two-way conduction pathway for signals between the body and the brain. (3) The spinal cord is a major reflex center (recall Chapter 11, Figure 11.16).

Like the brain, the spinal cord is protected by bone, meninges, and cerebrospinal fluid (Figure 12.26). The tough dura mater, called the **spinal dural sheath**, differs somewhat from the dura around the brain: It does not attach to the surrounding bone and corresponds only to the meningeal layer of the brain's dura mater. Just external to the spinal dura is a rather large **epidural space** filled with fat and a network of veins. The fat forms a protective padding around the spinal cord. The subdural space, arachnoid mater, subarachnoid space, and pia mater resemble the corresponding elements around the brain. Inferiorly, the dura and arachnoid extend to the level of S_2, well beyond the end of the spinal cord (Figure 12.25a).

The spinal cord does not extend the full length of the vertebral canal but ends in the superior lumbar region. This is explained by fetal events. Until the third month of development, the spinal cord extends all the way to the coccyx, but thereafter it grows more slowly than the caudal part of the vertebral column. As the vertebral column grows caudally, the spinal cord assumes a progressively more rostral position. By the time of birth, the spinal cord ends at L_3. During childhood, the spinal cord reaches its adult position, terminating at the level of the intervertebral disc between L_1 and L_2. It is important to emphasize that this is merely the average level: The termination point varies among different people from the inferior margin of T_{12} to the superior margin of L_3.

Looking at Figure 12.25a, can you deduce the safest place along the vertebral canal to insert a needle to withdraw a sample of cerebrospinal fluid from the subarachnoid space?

Since the adult spinal cord ends superior to the midlumbar region, a needle can safely enter the subarachnoid space inferior to this region. This procedure is called a **lumbar puncture (spinal tap)**. The patient bends forward, and the needle is inserted between the spines and laminae of successive vertebrae, either between L_3 and L_4 or between L_4 and L_5. Lumbar punctures are used to sample the CSF for microbial or chemical analysis, to measure the pressure of the CSF, and to inject antibiotics, anesthesia, or X-ray contrast medium into the subarachnoid space. ■

At its inferior end, the spinal cord tapers into the **conus medullaris** (ko'nus med'u-lar"is; "the cone of the spinal cord") (Figure 12.25a and c). This cone in turn tapers into a long filament of connective tissue covered with pia mater, the **filum terminale** ("terminal filament"), which attaches inferiorly to the coccyx. This attachment helps to anchor the spinal cord in place so that it is not jostled by body movements. Furthermore, the spinal cord is anchored to the bony walls of the vertebral canal throughout its length by saw-toothed shelves of pia mater. These lateral shelves are the **denticulate ligaments** (den-tik'u-lāt; "toothed") (Figure 12.25d).

Thirty-one pairs of *spinal nerves* (PNS structures) arise from the spinal cord as nerve roots and

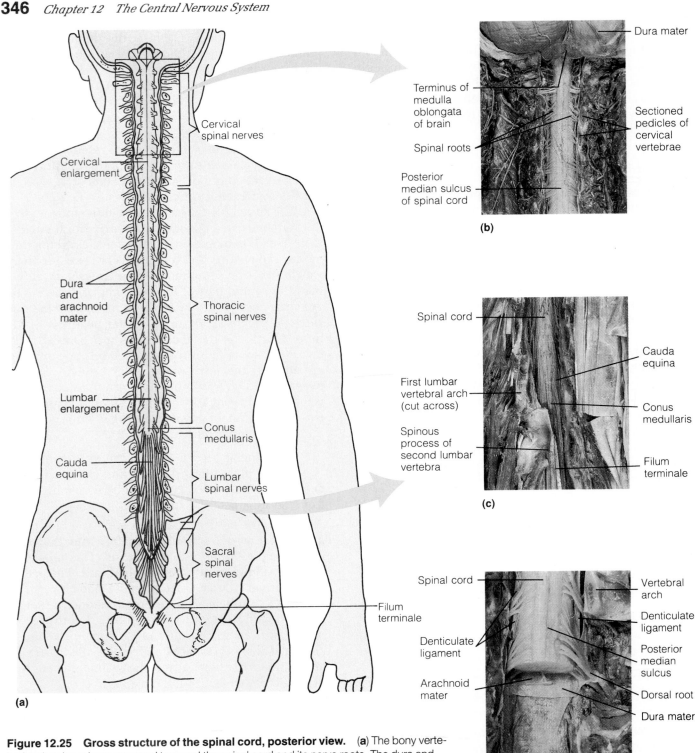

Figure 12.25   Gross structure of the spinal cord, posterior view. **(a)** The bony vertebral arches have been removed to reveal the spinal cord and its nerve roots. The dura and arachnoid mater are cut open and reflected laterally. **(b)** Photograph of the cervical part of the spinal cord. **(c)** Inferior end of the spinal cord, showing the conus medullaris, cauda equina, and filum terminale. **(d)** Enlargement of the spinal cord, showing the denticulate ligament.

exit through the intervertebral foramina to travel to the body regions they serve (Figure 12.25a). The spinal cord shows obvious enlargements in its cervical and lumbar regions, where the nerves serving the upper and lower limbs arise. These are the **cervical** and **lumbar enlargements**. Because the cord does not reach the inferior end of the vertebral column, the lumbar and sacral nerve roots must descend for some distance before reaching their intervertebral foramina. This collection of nerve roots at the inferior end of the vertebral canal is the **cauda equina** (kaw′dah e-kwi′nah; "horse's tail").

Just as the vertebral column is segmented, so too is the spinal cord. These segments of the spinal cord

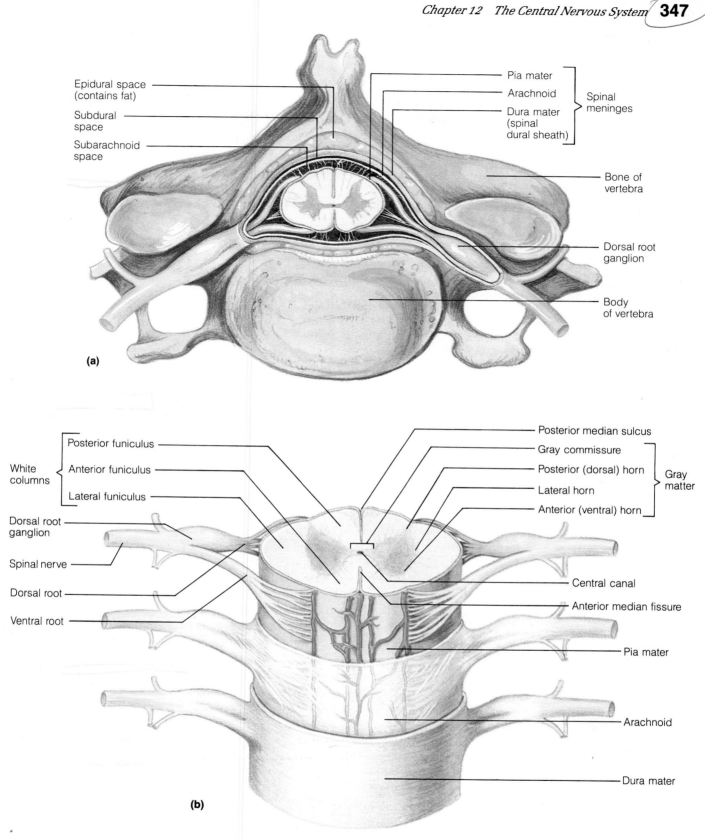

Epidural space (contains fat)

Subdural space

Subarachnoid space

Pia mater

Arachnoid

Dura mater (spinal dural sheath)

Spinal meninges

Bone of vertebra

Dorsal root ganglion

Body of vertebra

(a)

Posterior funiculus

White columns

Anterior funiculus

Lateral funiculus

Dorsal root ganglion

Spinal nerve

Dorsal root

Ventral root

Posterior median sulcus

Gray commissure

Posterior (dorsal) horn

Lateral horn

Anterior (ventral) horn

Gray matter

Central canal

Anterior median fissure

Pia mater

Arachnoid

Dura mater

(b)

Figure 12.26   The spinal cord.   (a) Cross section through the spinal cord in the cervical region, in relation to a surrounding cervical vertebra. **(b)** Three-dimensional ventral view of the spinal cord and its meningeal coverings.

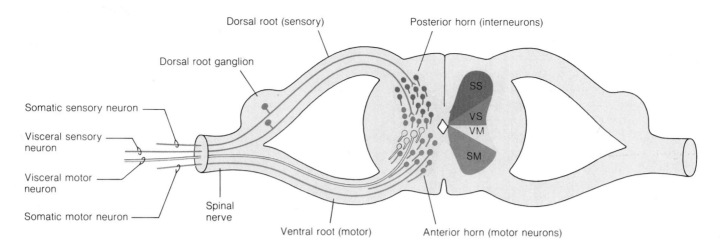

Dorsal root (sensory)

Posterior horn (interneurons)

Dorsal root ganglion

Somatic sensory neuron

Visceral sensory neuron

Visceral motor neuron

Somatic motor neuron

Spinal nerve

Ventral root (motor)

Anterior horn (motor neurons)

SS

VS

VM

SM

Figure 12.27 **Organization of the gray matter of the spinal cord.** The gray matter of the spinal cord can be divided into a sensory half dorsally and a motor half ventrally. SS: interneurons receiving input from somatic sensory neurons. VS: interneurons receiving input from visceral sensory neurons. VM: visceral motor cell bodies. SM: somatic motor cell bodies. Note that the dorsal and ventral roots are part of the PNS, not of the spinal cord.

are not visible externally but are defined by the numbering of the spinal nerves that issue from them (for example, sixth cervical, fifth thoracic, second lumbar, and third sacral). The segments of the spinal cord all lie *superior* to their corresponding vertebrae because of the rostral shift of the spinal cord during development. For example, spinal cord segment T_1 is at the level of vertebra C_7; cord segment T_{10} is at vertebra T_9; and cord segment L_3 is at vertebra T_{12}.

In a cross-sectional view, the spinal cord is wider laterally than anteroposteriorly (Figure 12.26b). Two median grooves mark its external surface: the **posterior median sulcus** and the **anterior median fissure**. These grooves run the length of the cord and partly divide it into right and left halves.

Gray Matter of the Spinal Cord and Spinal Roots

As mentioned earlier (p. 311), the spinal cord consists of an outer region of white matter and an inner region of gray matter (Figure 12.26b). As in other parts of the CNS, the gray matter of the cord consists of a mixture of neuron cell bodies, short unmyelinated axons and dendrites, and neuroglia. In cross section, the gray matter is shaped like the letter *H*. The crossbar of the H, the **gray commissure**, contains the narrow central cavity of the spinal cord, the **central canal**. The two posterior arms of the H are the **posterior horns** *(dorsal columns)*, while the two anterior arms are the **anterior horns** *(ventral columns)*. An additional pair of gray columns, the small **lateral horns**, is present in the thoracic and superior lumbar segments of the spinal cord.

Figure 12.27 shows the neuronal composition of the spinal gray matter (also see Figure 11.17). The posterior horns consist of interneurons. These interneurons receive information from sensory neurons whose cell bodies lie outside the spinal cord in *dorsal root ganglia* and whose axons reach the cord through *dorsal roots*. The anterior (and the lateral) horns contain cell bodies of motor neurons, which send their axons out of the cord through *ventral roots* to supply muscles (and glands). The size of the anterior horns varies along the length of the spinal cord, reflecting the amount of skeletal musculature innervated at each level. Thus, the anterior horns are largest in the cervical and lumbar regions of the cord, which innervate the upper and lower limbs, respectively.

The spinal gray matter can be divided further according to the innervation of the somatic and visceral regions of the body. The following four zones are evident within this gray matter (Figure 12.27): **somatic sensory (SS)**, **visceral sensory (VS)**, **visceral motor (VM)**, **somatic motor (SM)**.

White Matter of the Spinal Cord

The white matter of the spinal cord is composed of myelinated and unmyelinated axons. These fibers allow communication between different parts of the spinal cord and between the cord and brain. They run in three different directions:

1. **Ascending.** Most ascending fibers carry sensory information from the sensory neurons to the brain.

2. **Descending.** Most descending fibers carry motor instructions from the brain to the spinal cord.

3. **Commissural**. Some fibers cross from one side of the cord to the other.

Of these three types, the ascending and descending fibers make up most of the white matter.

The white matter on each side of the spinal cord is divided into three **white columns**, or **funiculi** (funik′u-li; "long ropes"), named according to their positions—the **posterior, anterior,** and **lateral funiculi** (Figure 12.26b). The anterior and lateral funiculi are continuous with one another and are divided by an imaginary line that extends out from the anterior gray horn. The posterior funiculus is also called the *dorsal white column.* As shown in Figure 12.28, the posterior funiculus is subdivided into a medial **fasciculus gracilis** ("slender bundle") and a lateral **fasciculus cuneatus** ("wedge-shaped bundle").

The three funiculi contain many fiber tracts, each made of axons with similar destinations and functions. The principal ascending and descending tracts of the spinal cord are shown in cross section in Figure 12.28. For the most part, these spinal tracts are named according to their origin and destination. We will now consider the major spinal tracts more closely.

Sensory and Motor Pathways

All major spinal tracts are parts of multineuron *pathways* that connect the brain to the body periphery, carrying sensory information to the brain and motor instructions to the effectors of the body (Figures 12.29 and 12.30). Before discussing them, we can make several generalizations about the great ascending and descending pathways:

1. Most of them cross over from one side of the CNS to the other (decussate) at some point along their course.

2. Most consist of a chain of two or three neurons, which contribute to successive tracts along the pathway.

3. Most pathways exhibit somatotopy. In this context, such "body mapping" means that axons in the tracts are arranged in a specific order, according to the body region they supply. For example, in one ascending tract, the axons transmitting impulses from the superior parts of the body lie lateral to those axons carrying impulses from the inferior body parts.

4. All pathways are paired (right and left), with one member of the pair present on each side of the spinal cord or brain.

Ascending (Sensory) Pathways

The ascending pathways (Figure 12.29) conduct sensory impulses superiorly through chains of two or three neurons (*first-, second-,* and *third-order neurons*) to various regions of the brain. These pathways

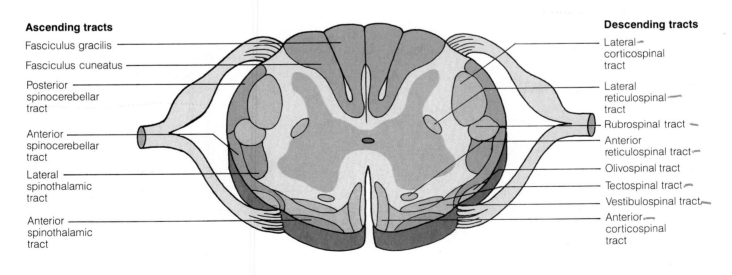

Ascending tracts

Fasciculus gracilis
Fasciculus cuneatus
Posterior spinocerebellar tract
Anterior spinocerebellar tract
Lateral spinothalamic tract
Anterior spinothalamic tract

Descending tracts

Lateral corticospinal tract
Lateral reticulospinal tract
Rubrospinal tract
Anterior reticulospinal tract
Olivospinal tract
Tectospinal tract
Vestibulospinal tract
Anterior corticospinal tract

Figure 12.28   Major fiber tracts in the white matter of the spinal cord.   The dorsal aspect is the superior part of this section. Ascending (sensory) tracts are blue, descending (motor) tracts are pink. These tracts are not as distinct as the figure indicates, and they overlap one another considerably. The white matter that lies directly ventral to the central canal is called the ventral white commissure. In this commissure, axons cross from one side of the cord to the other.

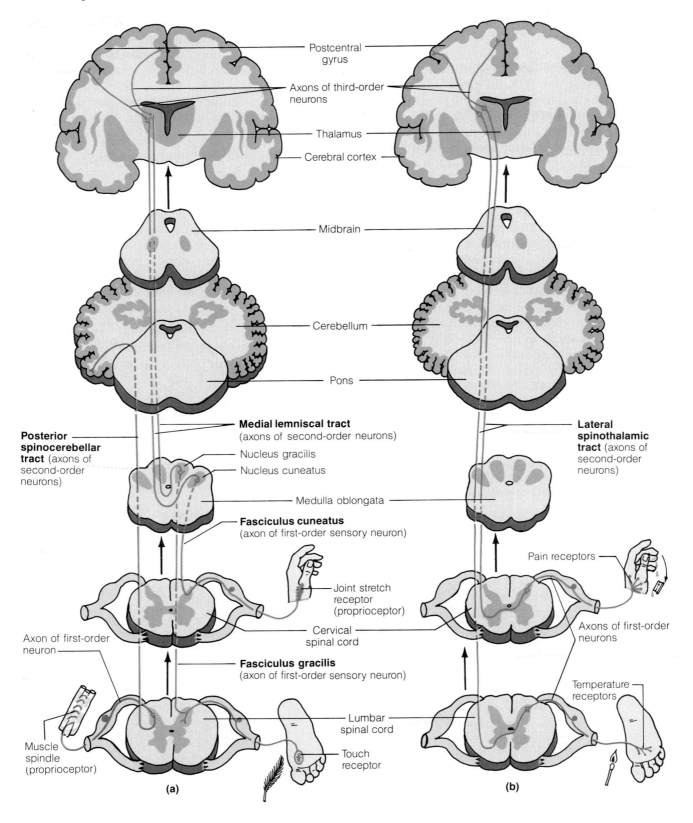

Postcentral gyrus

Axons of third-order neurons

Thalamus

Cerebral cortex

Midbrain

Cerebellum

Pons

Medial lemniscal tract (axons of second-order neurons)

Lateral spinothalamic tract (axons of second-order neurons)

Posterior spinocerebellar tract (axons of second-order neurons)

Nucleus gracilis

Nucleus cuneatus

Medulla oblongata

Fasciculus cuneatus (axon of first-order sensory neuron)

Joint stretch receptor (proprioceptor)

Pain receptors

Axons of first-order neurons

Axon of first-order neuron

Cervical spinal cord

Fasciculus gracilis (axon of first-order sensory neuron)

Muscle spindle (proprioceptor)

Temperature receptors

Lumbar spinal cord

Touch receptor

(a)

(b)

Figure 12.29   Ascending pathways for the general somatic senses.   These pathways relay sensory information from the body surface to the brain. Every pathway contains thousands of fibers, not just the few that are drawn. (**a**) Posterior spinocerebellar pathway (left) and dorsal column pathway (right). (**b**) Spinothalamic pathway.

carry general somatic sensory information. There are four main ascending pathways: The *dorsal column* and *spinothalamic* pathways transmit sensory impulses from the body to the primary somatosensory cortex for conscious interpretation. The *posterior* and *anterior spinocerebellar* pathways convey information on proprioception to the cerebellum, which uses this information to coordinate body movements. These pathways are summarized in Table 12.2 on p. 352, and each is described below.

The **dorsal column pathway** (Figure 12.29a, right side) carries information on fine touch, pressure, and proprioception. These are discriminative senses— senses we can localize very precisely on our body surface. In this pathway, the axons of the sensory neurons enter the spinal cord and send branches up the dorsal white column in the fasciculus gracilis or fasciculus cuneatus. These axons ascend to the medulla oblongata and synapse with second-order neurons in the nucleus gracilis or nucleus cuneatus. Axons from these nuclei form a tract called the **medial lemniscus** (lem-nis′kus; "ribbon"), which decussates and then ascends to the thalamus. From there, third-order neurons send axons to the primary somatosensory cortex on the postcentral gyrus. Here, the sensory information is processed and reaches the consciousness as precisely localized sensations.

The **spinothalamic pathway** (Figure 12.29b) carries information on pain, temperature, deep pressure, and the coarser aspects of touch—senses of which we are aware but cannot localize with exact precision on the body surface. In this pathway, the axons of sensory neurons enter the spinal cord, where they synapse on interneurons in the posterior gray horn. Axons of these second-order neurons decussate, enter the anterior and lateral funiculi as the **lateral** and **anterior spinothalamic tracts**, and ascend to the thalamus. From there, third-order axons project to the primary somatosensory cortex on the postcentral gyrus, where the information is processed into conscious sensations. Note that the brain usually interprets the sensory information carried by the spinothalamic pathway as unpleasant (pain, burns, cold, and so on).

The **posterior** and **anterior spinocerebellar pathways** carry information on proprioception to the cerebellum (Figure 12.29a, left side). Instead of crossing over once, these fibers either do not decussate at all or else cross twice, undoing the decussation.

Descending (Motor) Pathways

The descending tracts (Figure 12.30) that deliver motor instructions from the brain to the spinal cord can be divided into two groups: (1) pyramidal tracts and (2) all others. The following is an overview; more detailed information is provided in Table 12.3.

The **pyramidal**, or **corticospinal**, **tracts** (Figure 12.30a) control precise and skilled voluntary movements, such as writing and threading a needle. In these tracts, the axons of pyramidal neurons descend from the cerebral motor cortex to the spinal gray matter. There, the axons synapse with short interneurons that activate spinal motor neurons. In this way, the corticospinal tracts exert influence over the limb muscles, especially muscles that move the hand and fingers. The axons of the corticospinal tracts decussate along their course, most within the decussation of the pyramids (medulla) but some in the spinal cord.

The pyramidal tracts were considered so important by earlier anatomists that they named all other descending tracts "extrapyramidal tracts." These include the **tectospinal tract** from the superior colliculus (the tectum of the midbrain), the **vestibulospinal tract** from the vestibular nuclei, the **rubrospinal tract** from the red nucleus (*rubro* = red), and the **reticulospinal tract** from the reticular formation (Table 12.3). These tracts seem to bring about those body movements that are subconscious, coarse, or postural.

The cerebellum smooths and coordinates all movements dictated by the subcortical motor nuclei, just as it coordinates all skilled movements controlled by the pyramidal system. Therefore, it is not surprising that axons from the cerebellum project to the red nucleus, vestibular nuclei, and reticular nuclei.

It is now well known that pyramidal tract neurons project to and influence the activity of most of the "extrapyramidal" nuclei. Since *extrapyramidal* literally means "independent of the pyramidal (tracts)," modern anatomists have rejected that term, preferring instead to use the names of the individual motor pathways. (However, clinicians still classify many motor disorders as pyramidal or extrapyramidal in origin.)

Our discussion of sensory and motor pathways has focused on the innervation of body regions inferior to the head. In general, the pathways involving the head are similar to those for the trunk and limbs, except that the axons are located in cranial nerves and the brain stem, not in the spinal cord.

Injuries and Diseases of the Central Nervous System

Brain Damage

Brain dysfunctions are varied and extensive. We have mentioned some of these already. Here, we focus on traumatic brain injuries and degenerative brain diseases.

Table 12.2   Major Ascending (Sensory) Pathways and Spinal Cord Tracts

Spinal cord tract	Location (funiculus)	Origin	Termination	Function
Dorsal column pathways Fasciculus cuneatus and fasciculus gracilis	Posterior	Central axons of sensory (first-order) neurons enter dorsal root of the spinal cord and branch; branches enter posterior white column on same side without synapsing	By synapse with second-order neurons in nucleus cuneatus and nucleus gracilis in medulla; fibers of medullary neurons cross over and ascend in medial lemniscal tracts to thalamus, where they synapse with third-order neurons; thalamic neurons then transmit impulses to somatosensory cortex	Both tracts transmit sensory impulses from general sensory receptors of skin and proprioceptors, which are interpreted as discriminative touch, pressure, and "body sense" (limb and joint position) in opposite somatosensory cortex; cuneatus transmits afferent impulses from upper limbs, upper trunk, and neck; it is not present in spinal cord below level of T_6; gracilis carries impulses from lower limbs and inferior body trunk
Spinothalamic pathways Lateral spinothalamic	Lateral	Association (second-order) neurons of posterior horn; fibers cross to opposite side before ascending	By synapse with third-order neurons in thalamus; impulses then conveyed to somatosensory cortex by thalamic neurons	Transmits impulses concerned with pain and temperature to opposite side of brain; eventually interpreted in somatosensory cortex
Anterior spinothalamic	Anterior	Association neurons in posterior horns; fibers cross to opposite side before ascending	By synapse with third-order neurons in thalamus; impulses eventually conveyed to somatosensory cortex by thalamic neurons	Transmits impulses concerned with crude touch and pressure to opposite side of brain for interpretation by somatosensory cortex
Spinocerebellar pathways Posterior spinocerebellar*	Lateral (posterior part)	Association (second-order) neurons in posterior horn on same side of cord; fibers ascend without crossing	By synapse in cerebellum	Transmits impulses from trunk and lower limb proprioceptors on one side of body to same side of cerebellum; subconscious proprioception
Anterior spinocerebellar*	Lateral (anterior part)	Association (second-order) neurons of posterior horn; contains crossed fibers that cross back to the opposite side in the pons	By synapse in cerebellum	Transmits impulses from the trunk and lower limb on the same side of body to cerebellum; subconscious proprioception

*These spinocerebellar tracts carry information from the lower limb and trunk only. The corresponding tracts for the upper limb and neck (called *rostral spinocerebellar* and *cuneocerebellar tracts*) are beyond the scope of this book.

Traumatic Brain Injuries

Blows to the head, frequently suffered in traffic accidents, are the leading cause of accidental death in the United States. A **concussion** occurs when brain injury is slight and the symptoms are mild and transient. The victim may be dizzy, "see stars," or lose consciousness briefly, but no permanent neurological damage is sustained. The result of marked destruction of brain tissue, by contrast, is a **contusion**. A person suffering contusion of the cerebral cortex may stay conscious, but contusion of the brain stem usually produces coma.

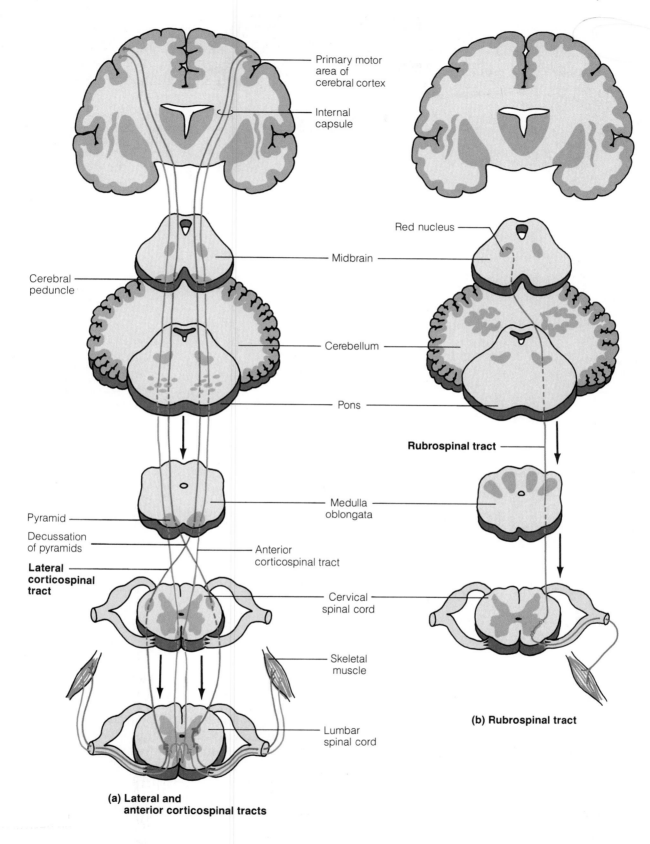

Figure 12.30   Descending motor pathways by which the brain influences movements.   (**a**) Pathways of the pyramidal (corticospinal) tracts. (**b**) Pathway of the rubrospinal tract (from red nucleus).

Table 12.3   Major Descending (Motor) Pathways and Spinal Cord Tracts

Spinal cord tract	Location (funiculus)	Origin	Termination	Function
Pyramidal				
Lateral corticospinal	Lateral	Pyramidal neurons of motor cortex of the cerebrum; decussate in pyramids of medulla	Anterior horn interneurons that influence motor neurons and occasionally directly to anterior horn motor neurons	Transmits motor impulses from cerebrum to spinal cord motor neurons (which activate skeletal muscles on opposite side of body); voluntary motor tract
Anterior corticospinal	Anterior	Pyramidal neurons of motor cortex; fibers cross over at the spinal cord level	Anterior horn (as above)	Same as lateral corticospinal tract
Other motor pathways (formerly called extrapyramidal)				
Tectospinal	Anterior	Superior colliculus of midbrain of brain stem (fibers cross to opposite side of cord)	Anterior horn (as above)	Turns neck so eyes can follow a moving object
Vestibulospinal	Anterior	Vestibular nuclei in medulla of brain stem (fibers descend without crossing)	Anterior horn (as above)	Transmits motor impulses that maintain muscle tone and activate limb and trunk extensors and muscles that move head; in this way, helps maintain balance during standing and moving
Rubrospinal	Lateral	Red nucleus of midbrain of brain stem (fibers cross to opposite side just inferior to the red nucleus)	Anterior horn (as above)	Transmits motor impulses concerned with muscle tone of distal limb muscles (mostly flexors) on opposite side of body
Reticulospinal (anterior, medial, and lateral)	Anterior and lateral	Reticular formation of brain stem (medial nuclear group of pons and medulla); both crossed and uncrossed fibers	Anterior horn (as above)	Transmits impulses concerned with muscle tone and many visceral motor functions; may control most unskilled movements

Following blows to the head, fatal consequences may result from subdural or subarachnoid hemorrhaging (bleeding from ruptured vessels into these spaces). People who initially are lucid after head trauma and then start to deteriorate neurologically are, in all probability, hemorrhaging intracranially. Accumulating blood within the skull increases intracranial pressure. If this pressure forces the brain stem through the foramen magnum, the control of heart rate, blood pressure, and respiration is lost. Surgeons treat intracranial hemorrhages by removing the blood mass and repairing the ruptured vessels.

Another consequence of serious head injury is swelling of the brain, which leads to brain-damaging pressure within the cranium. The swelling results both from the edema of inflammation and from water uptake by brain tissue (particularly by astrocytes). Anti-inflammatory drugs are routinely given to patients with head injuries in an attempt to keep cerebral edema from aggravating brain injury.

Degenerative Brain Diseases

The most important degenerative diseases of the brain are cerebrovascular accidents and Alzheimer's disease.

Cerebrovascular Accidents.   **Cerebrovascular accidents (CVA),** commonly called strokes, are the most common disorders of the nervous system and the third leading cause of death in the United States. Strokes occur when the flow of blood to a brain area is blocked and brain tissue dies from lack of oxygen. Such a deprivation of blood supply to a tissue is called **ischemia** (is-ke′me-ah; "to hold back blood").

The most common cause of CVA is blockage of a cerebral artery by a fixed or floating blood clot. Another common cause is progressive narrowing of brain vessels by atherosclerosis. Whatever the cause, the nature of the neurological deficits that follow depends on exactly which artery is affected and which area of the brain it supplies. People who survive a CVA may be paralyzed on one side of the body or have sensory deficits. If the language areas are damaged, difficulties in understanding or vocalizing speech develop.

Fewer than 35% of the people who survive an initial CVA are alive three years later, because those who suffer CVAs from blood clots are likely to have recurrent clotting problems—and more CVAs. Even so, the picture is not hopeless. Some patients recover some of their lost faculties, because undamaged neurons sprout new branches that spread into the damaged brain area and take over lost functions. However, neuronal spread is limited to only about 5 mm. Thus, large lesions always result in less than complete recovery.

Alzheimer's Disease.   **Alzheimer's** (Altz′hi-merz) **disease** is a progressive degenerative disease of the brain that ultimately results in dementia (mental deterioration). Between 5% and 15% of people over 65 develop this condition. Although usually confined to the elderly, Alzheimer's disease may begin in middle age. Victims exhibit widespread mental defects, including a loss of memory (particularly for recent events), shortened attention span, and disorientation. The disease worsens over a period of years. Ultimately, hallucinations occur.

Alzheimer's disease is associated with structural changes in the brain, particularly in the cerebral cortex and hippocampus, areas involved with thought functions and memory. The cerebral gyri shrink, and many neurons are lost. Additionally, PET scans show significant declines in the use of glucose in the affected regions (Figure 12.31). Microscopically, abnormal deposits of protein are present in the brain (and elsewhere in the body as well). These take the form of *amyloid plaques* scattered throughout the brain tissue and *neurofibrillar tangles* within neuron cell bodies. Despite intense research, the cause of Alzheimer's disease remains unknown. However, there is a preferential loss of the neurons that make a particular neurotransmitter called acetylcholine.

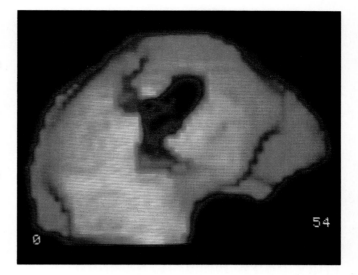

Figure 12.31   PET (positron emission tomography) scan of the cerebrum of a person with Alzheimer's disease.   The central blue area on this scan indicates a decline in the use of glucose and diminished neural activity in the cerebral cortex and hippocampus. PET scans can identify the early stages of Alzheimer's disease before any symptoms are evident.

Spinal Cord Damage

Any localized damage to the spinal cord or spinal roots leads to some functional loss, either *paralysis* (loss of motor function) or *paresthesia* (sensory loss: p. 389). Damage to the anterior horn or the ventral motor roots results in a complete or **flaccid paralysis** of the skeletal muscles served. No longer stimulated by neurons, the muscles shrink and waste away. Damage to the cerebral motor cortex results in **spastic paralysis**. In this condition, the spinal motor neurons and spinal reflexes remain intact. Thus, the muscles remain healthy, but their movements are no longer under voluntary control.

Spinal injuries can crush, tear, or cut through (transect) the spinal cord. Transection of the cord leads to a total loss of voluntary movements and conscious sensation in body regions inferior to the cut. If the damage occurs between the T_1 and L_2 segments of the cord, the lower limbs are affected. This condition is called **paraplegia** (par″ah-ple′je-ah). If the damage is in the cervical region, all four limbs are affected, a condition known as **quadriplegia** (*quadri* = four; *plegia* = a blow). Spinal transection superior to the mid neck destroys the ability to breathe because this is where the motor neurons to the diaphragm are located (C_3–C_5). Such injuries are always fatal unless the victim is kept alive by artificial resuscitation at the accident site and by a respirator thereafter. In evaluating spinal cord injuries, it is important to remember

GO TO CHP 15 →

that the spinal cord segments are slightly higher than the corresponding vertebral levels (p. 348).

 Over 10,000 Americans are paralyzed yearly in traffic and sports accidents. Can you name some specific types of accidents that might lead to spinal injuries?

Many spinal injuries result from whiplash in automobile accidents, from motorcycle accidents, from skiing accidents, from diving into shallow water, and from gunshot wounds. ■

The Central Nervous System Throughout Life

Established during the first month of development, the brain and spinal cord continue to grow and mature throughout the prenatal period. A number of congenital malformations originate during this period. The most serious of these are congenital hydrocephalus (see p. 344), anencephaly, and spina bifida.

In **anencephaly** (an″en-sef′ah-le; "without a brain") the cerebrum and part of the brain stem never develop. Presumably, anencephaly results from a failure of the rostral parts of the embryonic neural folds to fuse into a complete brain. The anencephalic newborn is totally vegetative, and mental life as we know it is absent. Death occurs soon after birth. About two in every thousand births are anencephalic.

Spina bifida (spi′nah bif′ĭ-dah; "forked spine") refers to a variety of birth defects involving incomplete formation of the bony vertebral arches (absence of vertebral laminae). The disorder most often affects the lumbosacral region. In **spina bifida cystica**, the most common variety, a sac-like cyst protrudes dorsally from the infant's spine. The cyst, called a **meningocele** (mĕ-ning′go-sēl), contains the meninges and cerebrospinal fluid. A large meningocele can also contain the spinal cord, in which case it is called a **myelomeningocele** (mi″ĕ-lo-mĕ-ning′go-sēl), a name meaning "spinal cord in a sac of meninges." The spinal cord starts out unharmed, but because the wall of the cyst is thin, physical damage and infection occur during and immediately after birth. This often leads to lifelong paralysis of the lower limbs and incontinence of the bowels and bladder. Furthermore, the movement of the spinal cord into the cyst pulls the medulla oblongata caudally through the tight foramen magnum of the skull. This movement constricts the brain ventricles, and hydrocephalus results. Immediate surgical intervention can relieve the hydrocephalus and sometimes prevent paralysis.

 Spina bifida babies should not be delivered vaginally, but by cesarean section. Try to say why.

Passage of the infant through the narrow birth canal is likely to compress the meningocele and damage the spinal cord. Planning for a cesarean delivery demands that the spina bifida be recognized prenatally—a fact emphasizing the importance of ultrasound exams for expectant mothers. ■

During difficult deliveries, the supply of oxygen to the infant's brain can be cut off temporarily. This—or any factors that damage the fetal brain in the later months of pregnancy—can lead to **cerebral palsy**, a neural disorder in which the voluntary muscles are poorly controlled or paralyzed. It reflects damage to the cerebral motor cortex or, less often, to the cerebellum or basal nuclei. In addition to spasticity, speech difficulties, and other motor impairments, about half of cerebral palsy victims have seizures, half are mentally retarded, and a third have some degree of deafness. Visual impairments are also common. Cerebral palsy is the largest single cause of crippling in children, affecting about 6 of every 1000 births.

After birth, different parts of the forebrain complete their development at different times. PET scans have revealed that the thalamus and the somatosensory cortex are active in a 5-day-old baby, but the visual cortex is not. This fact supports the observation that infants of this age respond to touch but have poor vision. By 11 weeks, more of the cortex is active, and the baby can reach for objects (a rattle, for example). By 8 months, the cortex is highly active, and the child can think about what she or he sees. Maturation of the nervous system continues through childhood and reflects the progressive myelination of its axons.

The brain reaches its maximum weight in the young adult. Over the next 60 years or so, neurons are damaged and die. The consequence of this slow loss of neurons is a steady decline in the weight and volume of the brain. However, the number of neurons lost is normally only a small percentage of the total, and the remaining neurons can change their synaptic connections, providing for continued learning throughout life.

Although a slow shrinkage of the brain is normal, in some people (boxers and alcoholics) the process is accelerated. Whether a boxer wins or loses the match, the likelihood of brain damage and atrophy increases with every blow that sends the brain rebounding within the skull. Everyone recognizes that alcohol has profound effects on the mind as well as on the body. Some of these effects are permanent. CT scans of chronic alcoholics reveal a reduction in the size and density of the brain—and this appears at a fairly early age. Like boxers, chronic alcoholics tend to show behavioral signs of senility unrelated to the aging process.

✚ Related Clinical Terms

Amyotrophic lateral sclerosis (ah-mi″o-trof′ik = muscle degeneration or muscle atrophy; skle-ro′sis = hardening) Chronic disease involving degeneration of the pyramidal neurons in the cerebral cortex, the motor neurons in the spinal cord, and a resultant wasting of the skeletal muscles. As the diseased pyramidal tracts deteriorate, astrocytes proliferate and form scar tissue that causes hardening (sclerosis) of the lateral aspects of the spinal cord. The ultimate cause is unknown, but viral infection is suspected. Initial symptoms include muscle weakness leading to awkwardness in using hands or difficulty in swallowing or speaking. The progress of ALS is relentless and ultimately fatal.

Cordotomy (kor-dot′o-me) A procedure in which a tract in the spinal cord is severed surgically, usually to relieve unremitting pain.

Encephalopathy (en-sef″ah-lop′ah-the) Any disease or disorder of the brain.

Hemiplegia (hem″ĭ-ple′je-ah) Paralysis of one side of the body (*hemi* = half). Hemiplegia often results from strokes that affect the cerebral motor cortex on one side of brain.

Meningioma (men-in″je-o′mah) A tumor that originates in the arachnoid mater and grows slowly. It may press on or enter the brain.

Microcephaly (mi″kro-sef′ah-le) Congenital condition involving the formation of a small brain and skull. May result from damage to the brain before birth or from premature fusion of the sutures of the skull. Most microcephalic children are severely retarded.

Monoplegia (*mono* = one) Paralysis of one limb.

Myelitis (mi″ĕ-li′tis) (*myel* = spinal cord) Inflammation of the spinal cord.

Myelogram (*gram* = recording) X ray of the spinal cord after injection of a contrast medium.

Chapter Summary

THE BRAIN (pp. 317–345)

1. The brain provides for voluntary movements, interpretation and integration of sensation, consciousness, and cognitive function. The brain is also involved with the innervation of the head through the cranial nerves. It weighs about 1500 grams.

Embryonic Development of the Brain (pp. 317–319)

2. The brain develops from the rostral part of the embryonic neural tube. By week 5, the early forebrain, midbrain, and hindbrain have become the telencephalon (cerebrum) and diencephalon, mesencephalon (midbrain), metencephalon (pons, cerebellum), and myelencephalon (medulla oblongata).

3. The cerebral hemispheres grow rapidly, enveloping the diencephalon and rostral brain stem. Their expanding surfaces fold into a complex pattern of sulci and gyri.

Regions and Organization of the Brain (pp. 319–320)

4. A widely used classification divides the brain into the cerebral hemispheres, diencephalon, brain stem, and cerebellum.

5. The brain stem consists of white matter external to central gray matter, whereas the cerebrum and cerebellum each have an additional, external cortex of gray matter.

Ventricles of the Brain (p. 320)

6. The two lateral ventricles are in the cerebral hemispheres; the third ventricle is in the diencephalon; the cerebral aqueduct is in the midbrain; and the fourth ventricle is in the hindbrain.

The Cerebral Hemispheres (pp. 321–330)

7. The large cerebral hemispheres exhibit gyri, sulci, and fissures. The five lobes of each hemisphere are the frontal, parietal, temporal, insula, and occipital.

8. The cerebral cortex is the site of conscious sensory perception, voluntary initiation of movements, and higher thought functions. Structural regions of the six-layered cerebral cortex were identified by Brodmann and others.

9. Functional areas of the cerebral cortex include the following: (a) motor areas in the frontal lobe: primary motor cortex, premotor cortex, frontal eye field, and Broca's area; (b) sensory areas in parietal, occipital, and temporal lobes: primary somatosensory cortex, primary visual cortex, primary auditory cortex, gustatory cortex, olfactory cortex; (c) association areas: various sensory association areas, prefrontal cortex for highest thought processes, general interpretation (gnostic) area, Wernicke's area for word recognition.

10. The cortex of each cerebral hemisphere receives sensory information from, and sends motor instructions to, the opposite (contralateral) side of the body. The body is mapped in an upside-down "homunculus" on the somatosensory and primary motor areas, examples of somatotopy.

11. The pyramidal, or corticospinal, tracts descend from the primary motor and premotor areas of the cortex and run through the brain stem to the spinal cord. There, they signal motor neurons to perform skilled voluntary movements.

12. In most people, the left cerebral hemisphere is specialized for language and math skills, and the right hemisphere is more concerned with visual-spatial and creative abilities.

13. In the white matter deep to the cerebral cortex, the types of fibers include commissural fibers, association fibers, and projection fibers.

14. The basal nuclei, embedded deep within the cerebral white matter, include the caudate, lentiform, and amygdala. Functionally, this group of nuclei works with the cerebral motor cortex to control (start and stop) complex movements.

The Diencephalon (pp. 330–333)

15. The diencephalon consists of the thalamus, hypothalamus, and epithalamus. It encloses the third ventricle.

16. The thalamus, an egg-shaped group of brain nuclei, is the gateway to the cerebral cortex. It is a major relay station for (1) sensory impulses ascending to the sensory cortex and (2) impulses from all brain regions that communicate with the cortex.

17. The hypothalamus, a series of brain nuclei, is the brain's most important visceral control center. It regulates sleep cycles, hunger, thirst, body temperature, secretion of the pituitary gland, the autonomic nervous system, and other activities.

18. The small epithalamus contains the pineal gland, which secretes a "nighttime" hormone called melatonin.

The Brain Stem (pp. 333–337)

19. The brain stem consists of the midbrain, pons, and medulla oblongata.

20. The midbrain is divided into a tectum and paired cerebral peduncles, with the latter containing the pyramidal motor tracts. In the tectum, the superior and inferior colliculi mediate visual and auditory reflexes. In the peduncles, the red nucleus and substantia nigra participate in motor functions. The periaqueductal gray matter relates to pain relief, but it also contains motor nuclei of cranial nerves III and IV (eye movers).

21. In the pons, nuclei of cranial nerves V–VII lie near the fourth ventricle. Centers in the reticular formation help regulate respiration. The ventral region of the pons contains the pyramidal tracts, plus the pontine nuclei that project to the cerebellum.

22. The medulla oblongata exhibits pyramids (formed by the pyramidal tracts), as well as the decussation of the pyramids. The olives contain relay nuclei to the cerebellum. Nuclei of cranial nerves VIII–XII lie near the fourth ventricle. Centers in the reticular formation regulate respiration, heart rate, blood pressure, and other visceral functions.

The Cerebellum (pp. 337–339)

23. The cerebellum smooths and coordinates body movements and helps maintain posture and equilibrium. Its gross parts (paired hemispheres and median vermis) are divided transversely into three lobes (anterior, posterior, and flocculonodular). Its surface is covered with folia (ridges) and fissures.

24. From superficial to deep, the three main "layers" of the cerebellum are the cortex, white matter, and the deep cerebellar nuclei.

25. The cerebellar cortex receives sensory information on current body movements, then compares this information to an image of how the body should be moving. In this way, the cerebellum carries out a dialogue with the motor cerebral cortex on how to coordinate precise voluntary movements.

26. The cerebellum connects to the brain stem by the superior, middle, and inferior cerebellar peduncles, thick fiber tracts that carry information to and from the cerebellum.

Functional Brain Systems (pp. 339–341)

27. The limbic system consists of many cerebral and diencephalic structures on the medial aspect of the forebrain. It is the "emotional-visceral" part of the brain. The hippocampus and amygdala help consolidate memories of facts.

28. The reticular formation includes diffuse nuclei that span the length of the brain stem. Its reticular activating system maintains the conscious state of the cerebral cortex, and the reticulospinal tracts signal many nonskilled body movements.

Protection of the Brain (pp. 341–345)

29. The brain is protected by skull bones, meninges, cerebrospinal fluid, and the blood-brain barrier.

30. The three meninges are the outer dura mater, the arachnoid, and the inner pia mater. They enclose both brain and spinal cord. Internal projections of the dura form septa that secure the brain in the skull (falx cerebri, falx cerebelli, tentorium cerebelli).

31. The cerebrospinal fluid, formed from blood plasma at the choroid plexuses in the roof of the ventricles, circulates through the ventricles and into the subarachnoid space. It returns to the blood through arachnoid villi at the superior sagittal sinus. Cerebrospinal fluid floats and cushions the CNS.

32. The blood-brain barrier reflects the relative impermeability of brain capillaries (complete tight junctions between endothelial cells). It lets water, oxygen, nutrients, and fat-soluble molecules enter the neural tissue but prevents entry of most harmful substances.

THE SPINAL CORD (pp. 345–351)

External Structure and Protection (pp. 345–348)

1. The spinal cord receives the spinal nerves that innervate the neck, limbs, and trunk. It also acts as a reflex center and contains axon tracts running to and from the brain. The spinal cord extends from the foramen magnum to the level of the first or second lumbar vertebra.

2. Thirty-one pairs of spinal nerve roots issue from the spinal cord. The most inferior bundle of roots resembles a horse's tail (cauda equina). The spinal cord is enlarged in its cervical and lumbar regions, reflecting the innervation of the limbs.

Gray Matter of the Spinal Cord and Spinal Roots (p. 348)

3. The H-shaped gray matter of the spinal cord has two anterior horns containing motor neurons and two posterior horns containing interneurons. The posterior horns can be subdivided into somatic and visceral sensory regions, the anterior horns into visceral and somatic motor regions.

4. Dorsal sensory roots and ventral motor roots are PNS structures that attach to the spinal cord.

White Matter of the Spinal Cord (pp. 348–349)

5. The white matter of the cord is divided into posterior, lateral, and anterior funiculi containing ascending and descending fiber tracts.

Sensory and Motor Pathways (pp. 349–351)

6. The ascending tracts in the spinal cord belong to sensory pathways that run from the body periphery to the brain. These include the spinocerebellar pathways (for proprioception) to the cerebellum and two pathways to the somatosensory cortex: the dorsal column pathway (for discriminative touch and proprioception) and the spinothalamic pathway (for pain, temperature, and coarse touch).

7. The descending tracts in the spinal cord belong to motor pathways that connect the brain to the body muscles. These include the corticospinal tracts that control skilled movements and other fiber tracts that control subconscious or coarse movements.

INJURIES AND DISEASES OF THE CENTRAL NERVOUS SYSTEM (pp. 351–356)

Brain Damage (pp. 351–355)

1. Blows to the head may cause concussions or contusions. These injuries can be aggravated by intracranial bleeding and swelling of the brain.

2. Cerebrovascular accidents (strokes) result when blood circulation to brain tissue is blocked and the tissue dies.

3. Alzheimer's disease is a degenerative brain disease affecting the cerebral cortex and hippocampus. Mental functions diminish progressively.

Spinal Cord Damage (pp. 355–356)

4. Transection or crushing of the spinal cord permanently eliminates voluntary movements and sensation inferior to the damage site.

THE CENTRAL NERVOUS SYSTEM THROUGHOUT LIFE (p. 356)

1. Birth defects and birth disorders that involve the brain include anencephaly, spina bifida, and cerebral palsy. Spina bifida is usually associated with myelomeningocele and hydrocephalus.

2. The brain stops growing in young adulthood. Neurons die throughout life and are not replaced. Thus, the weight and mass of the brain decline with age. Nonetheless, healthy people maintain nearly optimal intellectual function into old age.

Review Questions

Multiple Choice and Matching Questions

1. The primary motor cortex, Broca's area, and the premotor area are located in (a) the frontal lobe, (b) the parietal lobe, (c) the temporal lobe, (d) the occipital lobe.

2. Choose the proper response from the key for each of the following statements that describe various brain areas. Some letters will be used more than once.
Key: **(a)** cerebellum **(b)** corpora quadrigemina **(c)** corpus striatum **(d)** corpus callosum **(e)** hypothalamus **(f)** medulla **(g)** midbrain **(h)** pons **(i)** thalamus

_____ **(1)** basal nuclei involved in motor activities; related to Huntington's disease

_____ **(2)** region where there is a crossover of fibers of pyramidal tracts

_____ **(3)** control of temperature, autonomic nervous system, hunger, and water balance

_____ **(4)** houses the substantia nigra and red nucleus

_____ **(5)** mediate visual and auditory reflexes; found in midbrain

_____ **(6)** part of diencephalon with vital centers controlling heart rate, some aspects of emotion, and blood pressure

_____ **(7)** all inputs to cerebral cortex must first synapse in one of its nuclei

_____ **(8)** brain area that has folia and coordinates movements

_____ **(9)** brain region that contains the cerebral aqueduct

_____**(10)** associated with fourth ventricle and contains nuclei of cranial nerves V–VII

_____**(11)** thick tract between the two cerebral hemispheres

3. The innermost of the meninges, delicate and closely apposed to the brain surface, is the (a) pia mater, (b) alma mater, (c) arachnoid mater, (d) dura mater.

4. Cerebrospinal fluid is formed by the (a) arachnoid villi, (b) dural sinuses, (c) choroid plexuses, (d) all of these.

5. A patient suffered a cerebral hemorrhage that damaged the premotor cortex in her right cerebral hemisphere. As a result, she (a) cannot voluntarily move her left arm or leg, (b) feels no sensation on the left side of her body, (c) feels no sensation on the right side of her body, (d) can no longer play the violin.

6. Ascending pathways in the spinal cord convey (a) motor impulses, (b) sensory impulses, (c) pyramidal impulses, (d) commissural impulses.

7. Destruction of the anterior horn cells of the spinal cord results in loss of (a) intelligence, (b) all sensation, (c) pain, (d) motor control.

8. Fiber tracts allowing communication between neurons within the same cerebral hemisphere are (a) association tracts, (b) commissures, (c) projection tracts.

9. Some of the following brain structures consist of gray matter, some of white matter. Write *G* (for *gray)* or *W* (for *white)* as appropriate.

_____ **(1)** cortex of cerebellum
_____ **(2)** pyramids
_____ **(3)** internal capsule and corona radiata
_____ **(4)** red nucleus
_____ **(5)** medial lemniscus
_____ **(6)** cranial nerve nuclei
_____ **(7)** cerebellar peduncle

10. Which of these areas is most likely to store visual memories? (a) visual association area, (b) Wernicke's area, (c) premotor cortex, (d) primary somatosensory cortex.

11. A professor unexpectedly blew a loud horn in anatomy class, and all students looked up, startled. These reflexive movements of their necks and eye muscles were mediated by (a) cerebral cortex, (b) inferior olives, (c) raphe nuclei, (d) inferior colliculus, (e) nucleus gracilis.

Short Answer and Essay Questions

12. Make a diagram showing the five secondary brain vesicles of an embryo, and then list the basic adult brain regions derived from each.

13. (a) What is the advantage of having a cerebral cortex that is highly folded, as opposed to smooth and flat? (b) What is the name of the grooves between the gyri? (c) What deep groove divides the cerebrum into two hemispheres? (d) What groove divides the parietal lobe from the frontal lobe? (e) What groove divides the parietal from the temporal lobe?

14. (a) First, make a rough drawing of a lateral view of the left cerebral hemisphere. (b) You may be thinking, "But I just can't draw!" So, name which of the two hemispheres is involved in most people's ability to draw. (c) On your drawing, locate the following areas, and identify their major functions: primary motor cortex, premotor cortex, somatosensory association area, primary visual area, prefrontal cortex, Broca's area, gnostic area.

15. Unlike most senses, the projection of auditory information to the cerebral cortex is not just contralateral but ipsilateral as well. Explain what the terms *ipsilateral* and *contralateral* mean.

16. (a) What is the basic function of the basal nuclei? (b) Which basal nuclei form the lentiform nuclei? (c) Which one arches over the thalamus?

17. (a) Explain how the cerebellum attaches to the brain stem. (b) List some ways in which the cerebellum's structure is similar to that of the cerebrum.

18. Describe the role of the cerebellum in maintaining smooth, coordinated skeletal muscle activity.

19. (a) Where is the limbic system located? (b) What structures make up this system? (c) How is the limbic system important in behavior?

20. (a) Describe the location of the reticular formation in the brain. (b) What does RAS mean, and what is its function?

21. List four ways in which the CNS is protected.

22. How is cerebrospinal fluid formed and drained? Follow its pathway within and around the brain.

23. What constitutes the blood-brain barrier?

24. Differentiate a concussion from a contusion.

25. (a) What are the superior and inferior boundaries of the spinal cord in the vertebral canal? (b) A correctly executed lumbar spinal tap could poke the cauda equina or filum terminale, but not the conus medullaris. Explain.

26. (a) Which ascending tracts in the spinal cord carry sensory impulses concerned with discriminative touch and pressure? (b) Concerned with proprioception only?

27. (a) Is the thalamus mostly gray matter or white matter? (b) Describe the general function of the thalamus.

28. In the spinothalamic pathway, where are the cell bodies of the first-order neurons located? Of the third-order neurons?

29. Trace the descending pathway that is concerned with skilled voluntary movements.

30. As explained in this chapter's Related Clinical Terms, a cordotomy is sometimes done to relieve unending pain. Such a cut needs only to nick the spinal cord in a very specific place. Can you indicate that place on Figure 12.28?

31. Differentiate spastic paralysis and flaccid paralysis.

32. Define cerebrovascular accident, and describe its causes.

33. List a few changes in brain structure that occur with aging.

34. Define somatotopy, and give one example of it in the brain and one example in the spinal cord.

35. Nadia was a careful anatomy student, but she did not realize that a cerebral peduncle is different from a cerebellar peduncle. What is the difference?

Critical Thinking and Clinical Application Questions

36. Kimberly learned that the basic design of the CNS is gray matter on the inside and white matter on the outside. Then her friend Molly said that there is gray matter on the outside of the cerebrum of the brain. Kim became confused. Can you clear up her confusion?

37. Ralph had brain surgery to remove a small intracranial hematoma (blood mass). Ralph was allergic to general anesthesia, so the operation was done under local anesthesia. The surgeon removed a small part of the skull, and Ralph remained conscious. The operation went well, and Ralph asked the surgeon to stimulate his (unharmed) postcentral gyrus mildly with an electrode. The surgeon did so. What happened? Choose and explain: (a) Ralph was seized with uncontrollable rage. (b) He saw things that were not there. (c) He asked to see what was touching his hand, but nothing was. (d) He started to kick. (e) He heard his mother's voice from 30 years ago.

38. When their second child was born, Debbie and Trevor were told the baby had spina bifida and that he would be kept in intensive care for a week, after which time a "shunt would be put in." Also, immediately after the birth, an operation was performed on the infant's lower back. The parents were told that this operation went well but that their son would always be a "little weak in the ankles." Explain the statements in quotation marks in more informative and precise language.

39. Robert, a brilliant computer analyst, was hit on the forehead by a falling rock during mountain climbing. It was soon obvious to his coworkers that his behavior had changed dramatically. He had been a smart dresser but was now unkempt. One morning, he was seen defecating into the wastebasket. When he started revealing the most secret office gossip to everyone, his supervisor ordered Robert to report to the company doctor. What region of Robert's brain was affected by the cranial blow?

40. One war veteran was a quadriplegic, and another was a paraplegic. What can you say about the relative locations of their injuries?

41. Every time Spike went to a boxing match, he screamed for a knockout. Then his friend Rudy explained what happens when someone gets knocked out. What is the explanation?

13

The Peripheral Nervous System

The human brain, for all its sophistication, would be useless without its sensory and motor connections to the outside world. Our very sanity depends on a continual flow of sensory information from the environment. When blindfolded volunteers were suspended in a tank of warm water (a sensory-deprivation tank), they began to hallucinate: One saw pink and purple elephants, another heard a singing chorus, and others had taste hallucinations. Our sense of well-being also depends on our ability to carry out motor instructions sent from the CNS. For example, many victims of spinal cord injuries experience despair at being unable to move or take care of their own needs. The **peripheral nervous system (PNS)**—the nervous structures outside the brain and spinal cord—provides these vital links to the real world. Its nerves thread through virtually every part of the body, enabling the CNS to receive information and carry out its decisions.

The basic organization of the PNS is illustrated in Figure 13.1. Part (a) of the figure reviews the *functional* components of the PNS as presented in Chapter 11 (pp. 296–298). Recall that the sensory inputs and motor outputs carried by the PNS are categorized as *somatic* (outer body) or *visceral* (visceral organs) and as *general* (widespread) or *special* (localized). Part (b) of the figure shows the basic *structural* components of the PNS. These structural components are as follows:

1. The **sensory receptors.** Sensory receptors pick up stimuli (environmental changes) from inside and outside the body, then initiate impulses in sensory axons.

2. The **motor endings.** The motor endings are the endings of motor neurons that innervate the effectors (muscle fibers and glands).

3. The **nerves** and **ganglia.** As defined in Chapter 11, nerves are bundles of peripheral axons, and ganglia are clusters of peripheral cell bodies (such as the sensory cell bodies in Figure 13.1b). Almost all nerves contain both sensory and motor axons and are called *mixed nerves.* A few nerves, however, are purely sensory in function.

We begin this chapter describing the sensory receptors and the motor endings of the PNS. We then discuss the nerves of the body, starting with the *cranial nerves* (the nerves attached to the brain) and ending with the *spinal nerves* (the nerves attached to the spinal cord). Although the visceral portion of the PNS is mentioned, this chapter focuses on somatic functions. A more complete consideration of the visceral nervous system is deferred to Chapter 14.

Peripheral Sensory Receptors

Peripheral **receptors** are structures that pick up sensory stimuli and then set up signals in the sensory axons. Although there are many types of sensory receptors, most fit into two main categories: Many receptors, such as the one shown in Figure 13.1b, are modified *dendritic endings of sensory neurons.* Some receptors, however, are complete *receptor cells*—specialized epithelial cells or small neurons that transfer sensory information to the sensory neurons. Dendritic endings of sensory neurons monitor most types of general sensory information (such as touch, pain, pressure, temperature, proprioception), while the specialized receptor cells monitor most types of special sensory information (taste, vision, hearing, and equilibrium).

The sensory receptors may be classified in several different ways, according to: (a) their location in the body, (b) the type of stimulus they detect, and (c) their structure.

Classification by Location

The sensory receptors are divided into three classes based on their location or the location of the stimuli to which they respond. Receptors that are sensitive to stimuli arising outside the body are **exteroceptors** (ek"ster-o-sep'torz; *extero* = outside). As you might expect, most exteroceptors occur at or near the body surface. They include receptors for touch, pressure, pain, and temperature in the skin and most receptors of the special sense organs. **Interoceptors** (in"ter-o-sep'torz), also called *visceroceptors,* receive stimuli arising from the internal viscera, such as the digestive tube, bladder, and lungs. The different interoceptors monitor a variety of stimuli, including chemical changes, taste stimuli, stretching of tissues, and temperature. Their activation causes us to feel visceral pain, nausea, hunger, or fullness. **Proprioceptors** are located in our musculoskeletal organs; that is, in skeletal muscles, tendons, and joints (and also in ligaments, bone periosteum, and the fascia around muscles). As we explained earlier (p. 296), proprioceptors measure the degree of stretch of these locomotory organs, advising the CNS of our body movements.

Classification by Stimulus Detected

Another way to classify the sensory receptors is by the kinds of stimuli that most readily activate them. **Mechanoreceptors** respond to mechanical forces

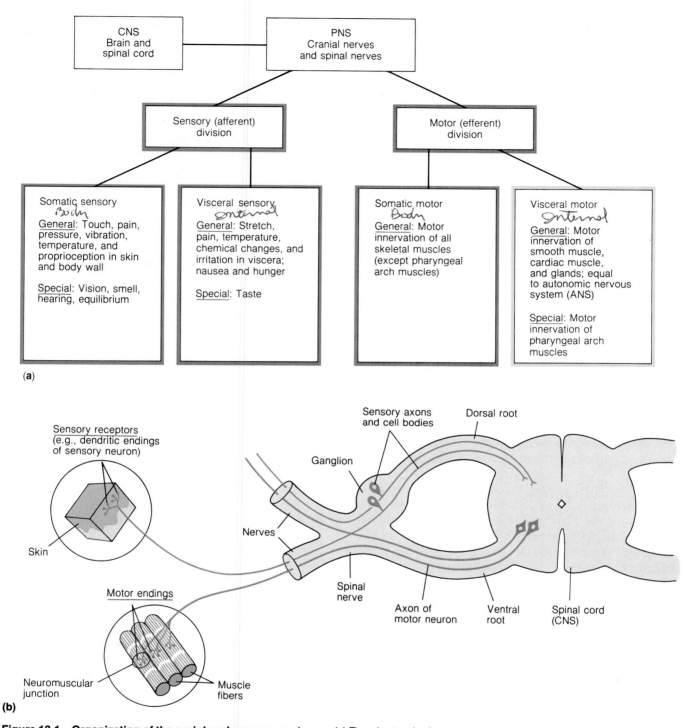

Figure 13.1 Organization of the peripheral nervous system. **(a)** Flowchart reviewing the types of sensory and motor information carried by nerve fibers of the peripheral nervous system (PNS); also see pp. 296–298. **(b)** Basic anatomical scheme of the PNS in the region of a spinal nerve. Note the sensory receptors and motor endings at left, the ganglion, and the sensory and motor axons that constitute nerves.

(touch, pressure, stretch, vibrations); **thermoreceptors** respond to temperature changes; **chemoreceptors** respond to chemicals in solution (molecules tasted or smelled, changes in blood chemistry); **photoreceptors** in the eye respond to light; and **nociceptors** (no″se-sep′torz) respond to harmful stimuli that result in pain (*noci* = harm).

Classification by Structure

Sensory receptors can also be classified by their structure. The special senses are discussed in Chapter 15; here we will consider only the **general sensory receptors.** These widely distributed receptors take the form of dendritic endings of sensory neurons and monitor touch, pressure, vibration, stretch, pain, temperature, and proprioception. These general receptors are divided into two broad structural groups: (1) *free (naked)* dendritic endings and (2) *encapsulated* dendritic endings surrounded by a capsule of connective tissue. The general sensory receptors are illustrated in Figure 13.2. Table 13.1 on p. 366 summarizes their locations and functions.

Before describing these receptors individually, we should point out that one receptor type can monitor several different kinds of stimuli, while different receptor types can respond to similar stimuli. The lack of a perfect "one receptor-one function" relationship should be kept in mind as you read about these receptors.

Free Dendritic Endings

Free dendritic endings *(naked dendritic endings)* of sensory fibers invade almost all tissues of the body but are particularly abundant in epithelia and in the connective tissue that underlies epithelia (Figure 13.2a). These receptors respond chiefly to pain and temperature (though some respond to tissue movements caused by pressure). One way to characterize the free endings is to say that they monitor the *affective* senses, those to which we have an emotional response—and we certainly respond emotionally to pain!

Certain free dendritic endings contribute to **Merkel discs** *(tactile menisci)*, which lie in the epidermis of the skin. Each consists of a disc-shaped epithelial cell innervated by a dendrite. Merkel discs seem to be receptors for light touch (also see p. 102 in Chapter 5). **Root hair plexuses** are free dendritic endings that wrap around hair follicles. These receptors for light touch monitor the bending of hairs. The tickle of a mosquito landing on your forearm is mediated by root hair plexuses.

Encapsulated Dendritic Endings

All **encapsulated dendritic endings** consist of one or more terminal fibers of sensory neurons enclosed in a capsule of connective tissue. All seem to be mechanoreceptors, and their capsules serve either to amplify the mechanical stimulus or to anchor these receptors to nearby moving tissues. The encapsulated receptors vary widely in shape, size, and distribution in the body.

Meissner's corpuscles. In **Meissner's corpuscles** *(tactile corpuscles)*, a few spiraling dendrites are surrounded by Schwann cells, which in turn are surrounded by an egg-shaped capsule of connective tissue. These corpuscles occur in the dermal papillae

beneath the epidermis of the skin. They are receptors for fine, discriminative touch and are especially numerous in sensitive and hairless areas of the skin, such as the fingertips, nipples, and lips. Apparently, Meissner's corpuscles perform the same "light touch" function in hairless skin that root hair plexuses perform in hairy skin.

End bulbs. **Krause's end bulbs**, once thought to be thermoreceptors for cold, are now known to be a type of Meissner's corpuscle for fine touch. Whereas true Meissner's corpuscles occupy the skin, however, Krause's end bulbs occur mostly in mucous membranes (on the surface of the eye and lining the mouth, for example).

Pacinian corpuscles. Pacinian corpuscles *(large lamellated corpuscles)* scatter throughout the deep connective tissues of the body. For example, they occur in the hypodermis layer deep to the skin. Although they are sensitive to deep pressure, they respond only to the initial application of the pressure before they tire and stop firing. Therefore, Pacinian corpuscles are *rapidly adapting* receptors that are best suited to monitor *vibration,* an on-off pressure stimulus. These large corpuscles, about 0.5–1 mm wide and 1–2 mm long, are easily visible to the naked eye. In section, a Pacinian corpuscle resembles a cut onion: Its single dendrite is surrounded by up to 60 layers of flattened Schwann cells, which in turn are covered by the capsule of connective tissue.

Ruffini's corpuscles. **Ruffini's corpuscles**, located in the dermis and elsewhere, contain a spray of dendritic endings enclosed in a thin, flattened capsule. Like Pacinian corpuscles, they respond to pressure and touch. However, they adapt slowly and thus can monitor *continuous* pressure placed on the skin.

Proprioceptors. Virtually all proprioceptors are encapsulated nerve endings. These stretch receptors in the locomotory organs include *muscle spindles, Golgi tendon organs,* and *joint kinesthetic receptors.*

Muscle spindles *(neuromuscular spindles)* are complex, fusiform structures that monitor the degree of stretch within skeletal muscles (Figure 13.2b). Each spindle contains several modified skeletal muscle fibers called **intrafusal muscle fibers** *(intra* = within; *fusal* = the spindle) surrounded by a connective tissue capsule. Intrafusal muscle fibers have fewer striations than do the ordinary muscle cells outside the spindles, which in this context are called **extrafusal muscle fibers** ("outside the spindle"). The intrafusal fibers are innervated by the dendrites of several sensory neurons. Some of these sensory dendrites twirl around the middle of the intrafusal fibers as **annulospiral** ("ring-spiral") **sensory endings**, while others

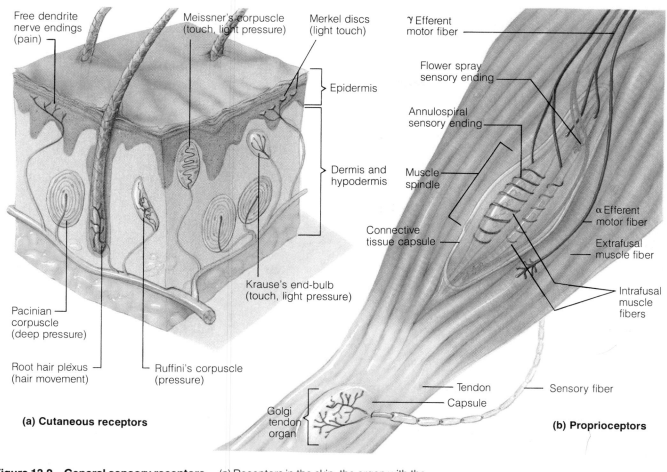

Figure 13.2   General sensory receptors.   (**a**) Receptors in the skin, the organ with the greatest variety of receptor types. Note the free dendritic nerve endings, Merkel discs, root hair plexus, Meissner's corpuscle, Krause's end bulb, Pacinian corpuscles, and Ruffini's corpuscles. (**b**) Proprioceptors: muscle spindle (above) and Golgi tendon organ (below).

supply the ends of the fibers in **flower spray sensory endings** that resemble tiny bouquets.

Recall that muscles are stretched by the contraction of antagonist muscles and also by the movements that occur when we tip off balance (p. 224). When a whole muscle is stretched, its intrafusal fibers are stretched also. This activates the sensory neurons that innervate the spindle, causing these neurons to signal the spinal cord and brain. The CNS then activates spinal motor neurons called α **(alpha) efferent neurons** that cause the entire muscle (extrafusal fibers) to generate contractile force and resist further stretching. This response can take the form of a monosynaptic spinal reflex that rapidly stops you from falling off balance on slippery ice, or the response can be controlled by the cerebellum. In the latter case, it is involved in the *regulation of muscle tone.* (Muscle tone is the steady force generated by noncontracting muscles to resist stretching.)

Motor axons also innervate the muscle spindle. These are the axons of spinal motor neurons called γ

(gamma) efferent axons, and they innervate the intrafusal fibers (Figure 13.2b). They let the brain preset the sensitivity of the spindle to stretch; that is, when the brain signals the gamma motor neurons to fire, the intrafusal muscle fibers contract and become tense so that only a little stretch is needed to stimulate the sensory dendrites. This quality makes the spindles highly sensitive to applied stretch. Gamma motor neurons are most active when balance reflexes must be razor sharp, as when a tightrope walker crosses a wire.

Golgi (gol'je) **tendon organs** are proprioceptors that monitor tension within tendons (Figure 13.2b). Each consists of an encapsulated bundle of tendon fibers (collagen fibers) in which sensory dendrites wind between and around the collagen fibers. Golgi tendon organs are stimulated when a contracting muscle pulls on its tendon. Their sensory neurons send this information to the cerebellum. They also induce a spinal reflex that both relaxes the contracting muscle and activates its antagonist. This relaxation

Table 13.1   General Sensory Receptors Classified by Structure and Function

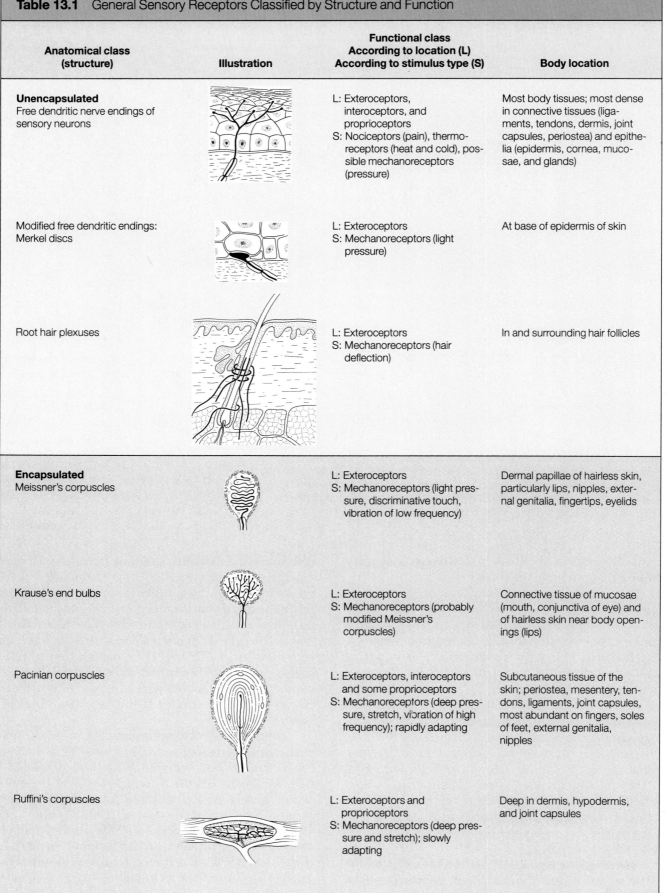

Anatomical class (structure)	Illustration	Functional class According to location (L) According to stimulus type (S)	Body location
Unencapsulated Free dendritic nerve endings of sensory neurons		L: Exteroceptors, interoceptors, and proprioceptors S: Nociceptors (pain), thermoreceptors (heat and cold), possible mechanoreceptors (pressure)	Most body tissues; most dense in connective tissues (ligaments, tendons, dermis, joint capsules, periostea) and epithelia (epidermis, cornea, mucosae, and glands)
Modified free dendritic endings: Merkel discs		L: Exteroceptors S: Mechanoreceptors (light pressure)	At base of epidermis of skin
Root hair plexuses		L: Exteroceptors S: Mechanoreceptors (hair deflection)	In and surrounding hair follicles
Encapsulated Meissner's corpuscles		L: Exteroceptors S: Mechanoreceptors (light pressure, discriminative touch, vibration of low frequency)	Dermal papillae of hairless skin, particularly lips, nipples, external genitalia, fingertips, eyelids
Krause's end bulbs		L: Exteroceptors S: Mechanoreceptors (probably modified Meissner's corpuscles)	Connective tissue of mucosae (mouth, conjunctiva of eye) and of hairless skin near body openings (lips)
Pacinian corpuscles		L: Exteroceptors, interoceptors and some proprioceptors S: Mechanoreceptors (deep pressure, stretch, vibration of high frequency); rapidly adapting	Subcutaneous tissue of the skin; periostea, mesentery, tendons, ligaments, joint capsules, most abundant on fingers, soles of feet, external genitalia, nipples
Ruffini's corpuscles		L: Exteroceptors and proprioceptors S: Mechanoreceptors (deep pressure and stretch); slowly adapting	Deep in dermis, hypodermis, and joint capsules

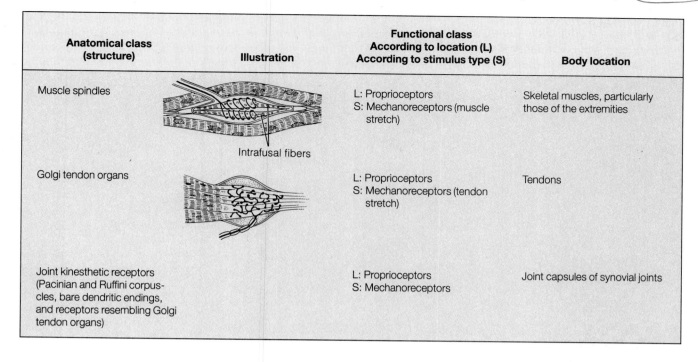

Anatomical class (structure)	Illustration	Functional class According to location (L) According to stimulus type (S)	Body location
Muscle spindles	Intrafusal fibers	L: Proprioceptors S: Mechanoreceptors (muscle stretch)	Skeletal muscles, particularly those of the extremities
Golgi tendon organs		L: Proprioceptors S: Mechanoreceptors (tendon stretch)	Tendons
Joint kinesthetic receptors (Pacinian and Ruffini corpuscles, bare dendritic endings, and receptors resembling Golgi tendon organs)		L: Proprioceptors S: Mechanoreceptors	Joint capsules of synovial joints

reflex is important in motor activities that involve rapid switching between flexion and extension, such as running.

Joint kinesthetic receptors (kin″es-thet′ik; "movement feeling") are proprioceptors that monitor stretch in the synovial joints. Specifically, they are sensory dendritic endings within the joint capsules. Four types of joint kinesthetic receptors are present within each joint capsule:

1. *Pacinian corpuscles* are rapidly adapting stretch receptors ideal for measuring acceleration and rapid movement of the joints.

2. *Ruffini's corpuscles* are slowly adapting stretch receptors ideal for measuring the positions of non-moving joints and the stretch of joints that undergo slow, sustained movements.

3. *Free dendritic endings* may be pain receptors.

4. Receptors resembling *Golgi tendon organs*, whose function in joints is not known.

Joint receptors, like the other classes of proprioceptors, send information on body movements to the cerebellum and cerebrum, as well as to spinal reflex arcs.

Peripheral Motor Endings

Having discussed the sensory nerve endings that receive stimuli, we now turn to the motor endings that activate the effectors (muscles and glands) of the body. We will describe the innervation of skeletal muscles first, then the innervation of visceral muscle and glands.

Innervation of Skeletal Muscle

Motor axons innervate skeletal muscle fibers at junctions called **neuromuscular junctions**, or **motor end plates** (Figures 13.1b and 13.3). A single neuromuscular junction is associated with each muscle fiber. These junctions are similar to the synapses between neurons. The neural part of the junction is a cluster of typical axon terminals (Figure 13.3a). These terminals are separated from the plasma membrane of the underlying muscle cell by a synaptic cleft. As in typical synapses, the axon terminals contain synaptic vesicles that release neurotransmitter when a nerve impulse reaches the terminals. The neurotransmitter chemical—acetylcholine—diffuses across the synaptic cleft and binds to receptor molecules on the muscle cell membrane (sarcolemma), setting up an impulse on the sarcolemma that signals the muscle cell to contract.

Although they resemble synapses, neuromuscular junctions have several unique features. Each axon terminal lies in a trough-like depression of the sarcolemma, which in turn shows groove-like invaginations (Figure 13.3b). The invaginations and the synaptic cleft are filled with a basal lamina (not illustrated), a structure never seen in the synapses between neurons. The basal lamina contains the enzyme acetylcholinesterase (as″ĕ-til-ko″lin-es′ter-ās). This enzyme breaks down acetylcholine immediately after the neurotransmitter signals a single contraction. This

Figure 13.3   The neuromuscular junction. (**a**) An axon of a motor neuron forms a neuromuscular junction with a skeletal muscle fiber. (**b**) Enlargement of a single axon terminal contacting a muscle cell.

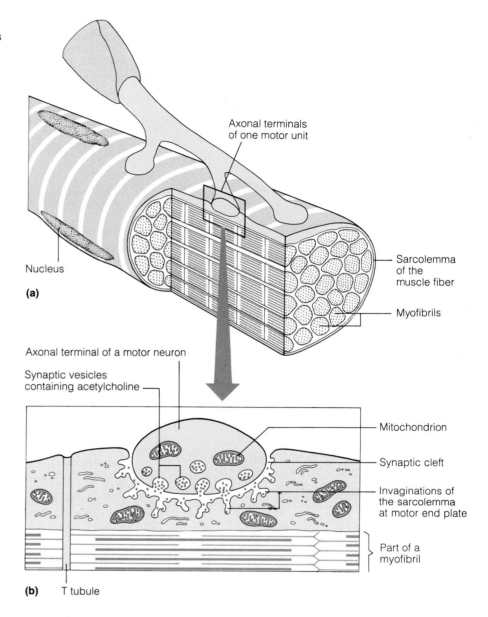

assures that each nerve impulse in the motor axon produces just one twitch of the muscle cell, preventing any undesirable additional twitches that would result if acetylcholine were to linger in the synaptic cleft.

Each motor axon branches to innervate a number of muscle fibers within a skeletal muscle (Figure 13.4). A motor neuron and all the muscle fibers it innervates are called a **motor unit**. When a motor neuron fires, all the skeletal muscle cells in the motor unit contract together. The average number of muscle fibers in a motor unit is 150, but it may be as high as several hundred or as low as 4. Muscles that require very fine control (such as the muscles moving the fingers and eyes) have small motor units, whereas bulky, weight-bearing muscles, whose movements are less precise (such as the hip muscles), have large motor units. The muscle fibers of a single motor unit are not clustered together but spread throughout the muscle. As a result, stimulation of a single motor unit causes a weak contraction of the entire muscle.

Innervation of Visceral Muscle and Glands

The contacts between visceral motor endings and the visceral effectors are much simpler than the elaborate neuromuscular junctions present on skeletal muscle. Near the smooth muscle or gland cells it innervates, a visceral motor axon swells into a row of knobs (varicosities) resembling beads on a necklace. These varicosities are the presynaptic terminals, which contain synaptic vesicles filled with neurotransmitter. Some of these axon terminals form shallow indentations on the membrane of the effector cell, but many remain a considerable distance from any cell. It takes time for

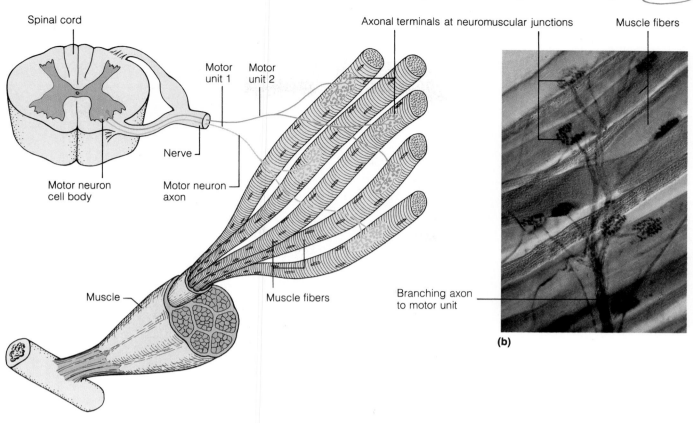

Figure 13.4   Motor units.   Each motor unit consists of one motor neuron and all the muscle fibers it innervates. (**a**) Drawing of two motor units. The muscle fibers of a motor unit are distributed throughout the entire skeletal muscle and are more widely scattered than is shown. (**b**) Photomicrograph of a part of a motor unit (115 X). Note the branching axon giving rise to clusters of axon terminals at neuromuscular junctions.

neurotransmitters to diffuse across these wide synaptic clefts, so visceral motor responses tend to be slower than somatic motor reflexes.

The motor innervation of cardiac muscle fibers in the heart resembles that of smooth muscle fibers and glands. However, the axonal endings maintain a uniform diameter and do not show varicosities at the sites where they release their neurotransmitters.

Cranial Nerves

The rest of this chapter is devoted to the specific nerves of the body, beginning with those in the head. Twelve pairs of **cranial nerves** attach to the brain and pass through various foramina in the skull (Figure 13.5a). These nerves are numbered from one to twelve (I–XII) in a rostral to caudal direction. The first two pairs attach to the forebrain, the rest to the brain stem. Other than the vagus nerve (X), which extends into the abdomen, the cranial nerves serve only head and neck structures.

The cranial nerves are described in detail in Table 13.2. The following is a general overview of the nerves and their basic functions:

   I. Olfactory. These are the sensory nerves of smell.

  II. Optic. This sensory nerve of vision is not a true nerve at all. Since it develops as an outgrowth of the brain, it is more correctly called a brain tract.

 III. Oculomotor. The name *oculomotor* means "eye mover." This nerve innervates four of the *extrinsic eye muscles*—muscles that move the eyeball in the orbit.

 IV. Trochlear. The name *trochlear* means "pulley." This nerve innervates an extrinsic eye muscle that hooks through a pulley-shaped ligament in the orbit.

  V. Trigeminal.   The name *trigeminal* means "threefold," which refers to this nerve's three

Figure 13.5 The twelve pairs of cranial nerves. (a) Note that almost all of the cranial nerves attach to the *ventral* side of the brain. (b) Summary of the cranial nerves by function. All cranial nerves that have a motor function also contain sensory fibers from proprioceptors in the muscles served.

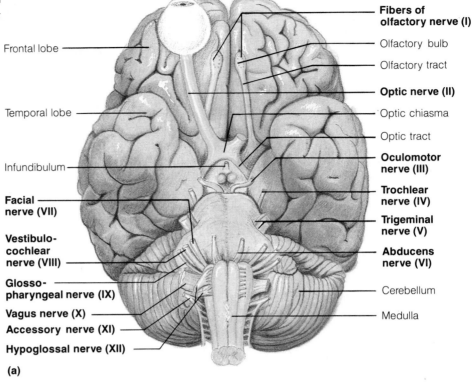

(a)

Cranial nerve	Name	Sensory function	Motor function	Parasympathetic fibers
I	Olfactory	Yes (smell)	No	No
II	Optic	Yes (vision)	No	No
III	Oculomotor	No	Yes	Yes
IV	Trochlear	No	Yes	No
V	Trigeminal	Yes (general sensation)	Yes	No
VI	Abducens	No	Yes	No
VII	Facial	Yes (taste)	Yes	Yes
VIII	Vestibulocochlear	Yes (hearing and balance)	No	No
IX	Glossopharyngeal	Yes (taste)	Yes	Yes
X	Vagus	Yes (taste)	Yes	Yes
XI	Accessory	No	Yes	No
XII	Hypoglossal	No	Yes	No

(b)

major branches. The trigeminal nerve provides sensory innervation to the face and motor innervation to the chewing muscles.

VI. Abducens. This nerve was so named because it innervates the muscle that *abducts* the eyeball (turns the eye laterally).

VII. Facial. This nerve innervates the muscles of *facial* expression and other structures.

VIII. Vestibulocochlear. This sensory nerve of hearing and equilibrium used to be called the *auditory nerve.*

IX. Glossopharyngeal. The name *glossopharyngeal* means "tongue and pharynx," structures that this nerve helps to innervate.

X. Vagus. The name *vagus* means "wanderer" or "vagabond." This nerve "wanders" beyond the head into the thorax and abdomen.

XI. Accessory. This nerve can be considered an *accessory* part of the vagus nerve, and we will treat it along with the vagus in this text discussion. This nerve used be called the *spinal accessory nerve.*

XII. Hypoglossal. The name *hypoglossal* means "below the tongue." This nerve runs inferior to the tongue and innervates the tongue muscles.

You may use the following saying to remember the first letters of the names of the 12 cranial nerves in their proper order: "**O**h, **O**h, **O**h, **t**o **t**ouch **a**nd **f**eel **v**ery **g**ood **v**elvet, **a**h!"

The cranial nerves contain the sensory and motor nerve fibers that innervate the head. The cell bodies of the *sensory* neurons lie either in peripheral receptor organs (the nose for smell, the eye for vision) or within **cranial sensory ganglia**, which lie along some cranial nerves (V, VII-X) just external to the brain. These sensory ganglia on the cranial nerves are directly comparable to the dorsal root ganglia on the spinal nerves (p. 348). The cell bodies of most *motor* neurons occur in cranial nerve nuclei in the ventral gray matter of the brain stem—just as cell bodies of spinal motor neurons occur in the ventral gray matter of the spinal cord.

Based on the types of sensory and motor fibers they contain, the twelve cranial nerves can be divided into three functional groups (Figure 13.5b). One group is purely sensory (I, II, VIII), containing *special somatic sensory* fibers for smell (I), vision (II), and hearing and equilibrium (VIII). A second group is primarily motor (III, IV, VI, XII), containing the *general somatic motor* fibers to skeletal muscles of the eye and tongue. A third group of cranial nerves (V, VII, IX, and X and XI) are more complex, containing both sensory and motor fibers. These nerves supply sensory innervation to the face (through *general somatic sensory* fibers) and to the mouth and viscera *(general visceral sensory)*, including the taste buds for the sense of taste *(special visceral sensory)*. These nerves also innervate all pharyngeal arch muscles *(special visceral motor)*, such as chewing muscles and the muscles of facial expression.

Additionally, several cranial nerves (III, VII, IX, X) contain *general visceral motor* fibers that stimulate visceral muscle and glands throughout much of the body. These motor fibers belong to a division of the autonomic nervous system (ANS) called the *parasympathetic division* (Figure 13.5b). Although the ANS is considered in Chapter 14, we must now mention that the ANS innervates body structures through chains of two motor neurons. The cell bodies of the second neurons occupy *autonomic motor ganglia* in the PNS.

Having read this overview, you can now "tackle" the individual cranial nerves in Table 13.2.

Table 13.2   Cranial Nerves

I   The olfactory (ol-fak′to-re) **nerves**

Origin and course: Olfactory nerve fibers arise from olfactory receptor cells located in olfactory epithelium of nasal cavity and pass through cribriform plate of ethmoid bone to synapse in olfactory bulb; fibers of olfactory bulb neurons extend posteriorly as olfactory tract, which runs beneath frontal lobe to enter cerebral hemispheres and terminates in primary olfactory cortex; see also Figure 15.3

Function: Purely sensory; carry afferent impulses for sense of smell

Fracture of ethmoid bone or lesions of olfactory fibers may result in partial or total loss of smell, a condition known as *anosmia* (an-oz′me-ah) ■

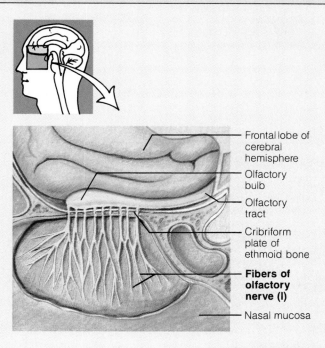

- Frontal lobe of cerebral hemisphere
- Olfactory bulb
- Olfactory tract
- Cribriform plate of ethmoid bone
- **Fibers of olfactory nerve (I)**
- Nasal mucosa

(table continues)

Table 13.2 (continued)

II   The optic nerves

Origin and course: Fibers arise from retina of eye to form optic nerve, which passes through optic foramen of orbit; the optic nerves converge to form the optic chiasma (ki-az'mah) where partial cross-over of fibers occurs, continue on as optic tracts, enter thalamus, and synapse there; thalamic fibers run (as the optic radiation) to occipital (visual) cortex, where visual interpretation occurs; see also Figure 15.15

Function: Purely sensory; carry afferent impulses for vision

Damage to optic nerve results in blindness in eye served by nerve; damage to visual pathway distal to optic chiasma results in partial visual losses; visual defects are called *anopsias* (an-op'se-as) ▪

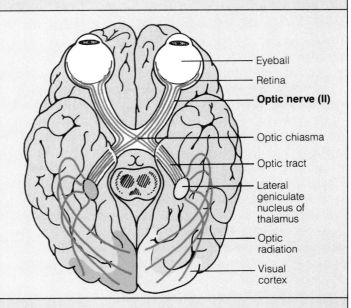

III   The oculomotor (ok"u-lo-mo'ter) nerves

Origin and course: Fibers extend from ventral midbrain (near its junction with pons) and pass through bony orbit, via *superior orbital fissure,* to eye

Function: Mixed nerves; although oculomotor nerves contain a few proprioceptive afferents, they are chiefly motor nerves (as implied by their name, meaning "motor to the eye"); each nerve includes:

- Somatic motor fibers to four of the six extrinsic eye muscles (inferior oblique and superior, inferior, and medial rectus muscles) that help direct eyeball, and to levator palpebrae superioris muscle, which raises upper eyelid
- Parasympathetic (autonomic) motor fibers to constrictor muscles of iris, which cause pupil to constrict, and to ciliary muscle, controlling lens shape for visual focusing
- Sensory (proprioceptor) afferents, which run from same four extrinsic eye muscles to midbrain

In oculomotor nerve paralysis, eye cannot be moved up, down, or inward, and at rest, eye rotates laterally (*external strabismus* [strah-biz'mus]) because the actions of the two extrinsic eye muscles not served by cranial nerve III are unopposed; upper eyelid droops *(ptosis)* and the person has double vision and trouble focusing on close objects ▪

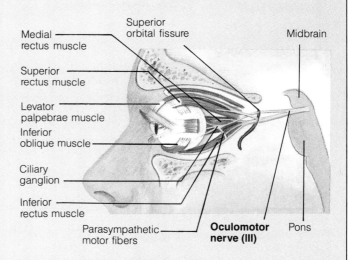

IV   The trochlear (trok'le-ar) nerves

Origin and course: Fibers emerge from dorsal midbrain and course ventrally around midbrain to enter orbits through *superior orbital fissures* along with oculomotor nerves

Function: Mixed nerves, primarily motor; supply somatic motor fibers to, and carry proprioceptor fibers from, one of the extrinsic eye muscles, the superior oblique muscle

Trauma to, or paralysis of, a trochlear nerve results in double vision and reduced ability to rotate eye inferolaterally ▪

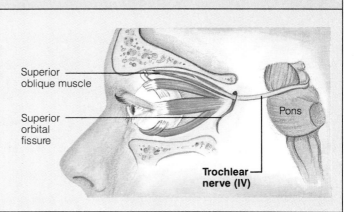

V   The trigeminal nerves

Largest of cranial nerves; fibers extend from pons to face, and form three divisions (*trigemina* = threefold): ophthalmic, maxillary, and mandibular divisions; major general sensory nerves of face; transmit afferent impulses from touch, temperature, and pain receptors; cell bodies of sensory neurons of all three divisions are located in large *trigeminal* (also called *semilunar* or *gasserian*) *ganglion*; the mandibular division also contains some motor fibers that innervate chewing muscles

Dentists desensitize upper and lower jaws by injecting local anesthetic (such as novocaine) into alveolar branches of maxillary and mandibular divisions, respectively; since this blocks pain-transmitting fibers of teeth, the surrounding tissues become numb

	Ophthalmic division (V₁)	**Maxillary division (V₂)**	**Mandibular division (V₃)**
Origin and course:	Fibers run from face to pons via superior orbital fissure	Fibers run from face to pons via foramen rotundum	Fibers pass through skull via foramen ovale
Function:	Conveys sensory impulses from skin of anterior scalp, upper eyelid, and nose, and from nasal cavity mucosa, cornea, and lacrimal gland	Conveys sensory impulses from nasal cavity mucosa, palate, upper teeth, skin of cheek, upper lip, lower eyelid	Conveys sensory impulses from anterior tongue (except taste buds), lower teeth, skin of chin, temporal region of scalp; supplies motor fibers to, and carries proprioceptor fibers from, muscles of mastication

✚ *Tic douloureux* (tik du″lu-ru′), or *trigeminal neuralgia* (nu-ral′je-ah), caused by inflammation of trigeminal nerve, is widely considered to produce most excruciating pain known; the stabbing pain lasts for a few seconds to a minute, but it can be relentless, occurring a hundred times a day; (the term *tic* refers to victim's wincing during pain); usually triggered by some sensory stimulus, such as brushing teeth or even a puff of air hitting the face, but may reflect pressure on trigeminal nerve root; analgesics only partially effective; in severe cases, nerve is cut proximal to trigeminal ganglion; this relieves the agony, but also results in loss of sensation on that side of face ■

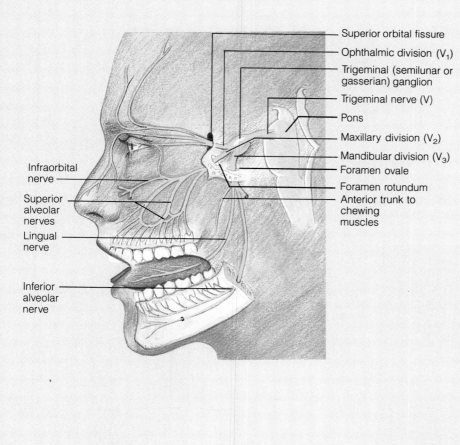

Distribution of sensory fibers of each division

(table continues)

Table 13.2 (continued)

VI The abducens (ab-du'senz) nerves

Origin and course: Fibers leave inferior pons and enter orbit via superior orbital fissure to run to eye

Function: Mixed nerve, primarily motor; supplies somatic motor fibers to lateral rectus muscle, an extrinsic muscle of the eye; convey proprioceptor impulses from same muscle to brain.

✚ In abducens nerve paralysis, eye cannot be moved laterally; at rest, affected eyeball rotates medially *(internal strabismus)* ■

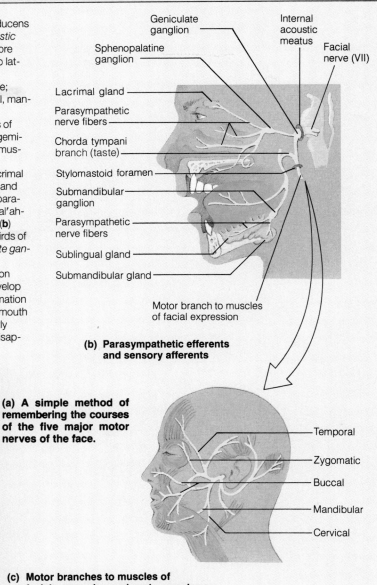

VII The facial nerves

Origin and course: Fibers issue from pons, just lateral to abducens nerves (see Figure 13.5), enter temporal bone via *internal acoustic meatus,* and run within bone (and through inner ear cavity) before emerging through *stylomastoid foramen*; nerve then courses to lateral aspect of face

Function: Mixed nerves that are the chief motor nerves of face; have five major branches on face: temporal, zygomatic, buccal, mandibular, and cervical see (**a**)

- Convey motor impulses to skeletal muscles of face (muscles of facial expression), except for chewing muscles served by trigeminal nerves, and transmit proprioceptor impulses from same muscles to pons; see (**c**)
- Transmit parasympathetic (autonomic) motor impulses to lacrimal (tear) glands, nasal and palatine glands, and submandibular and sublingual salivary glands. Some of the cell bodies of these parasympathetic motor neurons are in *sphenopalatine* (sfe"no-pal'ah-tīn) and *submandibular ganglia* on the trigeminal nerve; see (**b**)
- Convey sensory impulses from taste buds of anterior two-thirds of tongue; cell bodies of these sensory neurons are in *geniculate ganglion*; see (**b**)

✚ *Bell's palsy,* characterized by paralysis of facial muscles on affected side and partial loss of taste sensation, may develop rapidly (often overnight); cause is usually unknown, but inflammation of facial nerve is often suspect; lower eyelid droops, corner of mouth sags (making it difficult to eat or speak normally), eye constantly drips tears and cannot be completely closed; condition may disappear spontaneously without treatment ■

(b) Parasympathetic efferents and sensory afferents

(a) A simple method of remembering the courses of the five major motor nerves of the face.

(c) Motor branches to muscles of facial expression and scalp muscles

VIII   **The vestibulocochlear** (ves-tib"u-lo-kok'le-ar) **nerves**

Origin and course: Fibers arise from hearing and equilibrium apparatus located within inner ear of temporal bone and pass through internal acoustic meatus to enter brain stem at pons-medulla border; afferent fibers from hearing receptors (in cochlea) form *cochlear division;* those from equilibrium receptors (in semicircular canals and vestibule) form *vestibular division;* the two divisions merge to form vestibulocochlear nerve; see also Figure 15.22

Function: Purely sensory; vestibular branch transmits afferent impulses for sense of equilibrium, and sensory nerve cell bodies are located in *vestibular ganglia;* cochlear branch transmits afferent impulses for sense of hearing, and sensory nerve cell bodies are located in *spiral ganglion* within cochlea

Lesions of cochlear nerve or cochlear receptors result in *central* or *nerve deafness,* whereas damage to vestibular division produces dizziness, rapid involuntary eye movements, loss of balance, nausea and vomiting ■

Labels (figure): Semicircular canals — Vestibular ganglia — Internal acoustic meatus — Vestibular nerve — Vestibule — Cochlear nerve — Pons — Cochlea (containing spiral ganglion) — **Vestibulocochlear nerve (VIII)**

IX   **The glossopharyngeal** (glos"o-fah-rin'je-al) **nerves**

Origin and course: Fibers emerge from medulla and leave skull via *jugular foramen* to run to throat

Function: Mixed nerves that innervate part of tongue and pharynx; provide motor fibers to, and carry proprioceptor fibers from, superior pharyngeal muscles involved in swallowing and gag reflex; provide parasympathetic motor fibers to parotid salivary gland; (some of the nerve cell bodies of these parasympathetic motor neurons are located in *otic ganglion* on the trigeminal nerve)

Sensory fibers conduct taste and general sensory (touch, pressure, pain) impulses from pharynx and posterior tongue, from chemoreceptors in the carotid body (which monitor O_2 and CO_2 tension in the blood and help regulate respiratory rate and depth), and from pressure receptors of carotid sinus (which help to regulate blood pressure by providing feedback information); sensory neuron cell bodies are located in *superior* and *inferior ganglia*

Injury or inflammation of glossopharyngeal nerves impairs swallowing and taste, particularly for sour and bitter substances ■

Labels (figure): Parotid gland — Parasympathetic fibers — Pons — **Glossopharyngeal nerve (IX)** — Jugular foramen — Superior ganglion — Inferior ganglion — Otic ganglion — Carotid sinus — Pharyngeal muscles — Common carotid artery

(table continues)

Table 13.2 (continued)

X   The vagus (va'gus) nerves

Origin and course: The only cranial nerves to extend beyond head and neck region; fibers emerge from medulla, pass through skull via jugular foramen, and descend through neck region into thorax and abdomen; see also Figure 14.3

Function: Mixed nerves; nearly all motor fibers are parasympathetic efferents, except those serving skeletal muscles of pharynx and larynx (involved in swallowing); parasympathetic motor fibers supply heart, lungs, and abdominal viscera and are involved in regulation of heart rate, breathing, and digestive system activity; transmit sensory impulses from thoracic and abdominal viscera, from the carotid sinus (pressoreceptor for blood pressure) and the carotid and aortic bodies (chemoreceptors for respiration), and taste buds of posterior tongue and pharynx; carry proprioceptor fibers from muscles of larynx and pharynx

➕ Since nearly all muscles of the larynx ("voice box") are innervated by laryngeal branches of the vagus, vagal nerve paralysis can lead to hoarseness or loss of voice; other symptoms are difficult swallowing and impaired digestive system mobility. Total destruction of both vagus nerves is incompatible with life, because these parasympathetic nerves are crucial in maintaining normal state of visceral organ activity; without their influence, the activity of the sympathetic nerves, which mobilize and accelerate vital body processes (and shut down digestion), would be unopposed ■

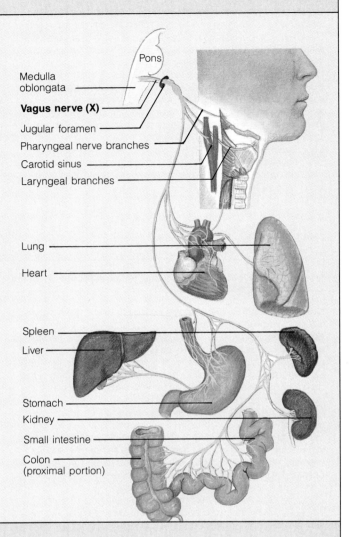

XI   The accessory nerves

Origin and course: Unique in that they are formed by union of a *cranial root* and a *spinal root;* cranial root emerges from lateral aspect of medulla of brain stem; spinal root arises from superior region (C_1–C_5) of spinal cord. Spinal portion passes upward along spinal cord, enters skull via foramen magnum, and temporarily joins cranial root; the resulting accessory nerve exits from skull through *jugular foramen;* then cranial and spinal fibers diverge; cranial root fibers join vagus, while spinal root runs to the largest neck muscles

Function: Mixed nerves, but primarily motor in function; cranial division joins with fibers of vagus nerve (X) to supply motor fibers to larynx, pharynx, and soft palate; spinal root supplies motor fibers to trapezius and sternocleidomastoid muscles, which together move head and neck, and conveys proprioceptor impulses from same muscles

➕ Injury to the spinal root of one accessory nerve causes head to turn toward injury side as result of sternocleidomastoid muscle paralysis; shrugging of that shoulder (role of trapezius muscle) becomes difficult ■

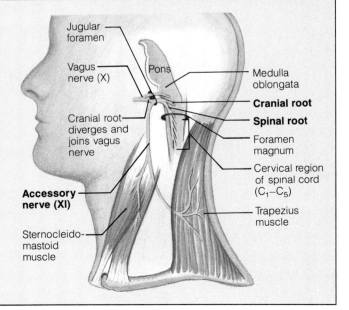

XII   The hypoglossal (hi"po-glos'al) **nerves**

Origin and course: As their name implies (*hypo* = beneath; *glossal* = tongue), hypoglossal nerves mainly serve the tongue; fibers arise by a series of roots from medulla and exit from skull via *hypoglossal canal* to travel to tongue; see also Figure 13.5

Function: Mixed nerves, but primarily motor in function; carry somatic motor fibers to intrinsic and extrinsic muscles of tongue, and proprioceptor fibers from same muscles to brain stem; hypoglossal nerve control allows not only food mixing and manipulation by tongue during chewing, but also tongue movements that contribute to swallowing and speech.

➕ Damage of hypoglossal nerves causes difficulties in speech and swallowing; if both nerves are impaired, the person cannot protrude tongue; if only one side is affected, tongue deviates (leans) toward affected side; eventually paralyzed side begins to atrophy ■

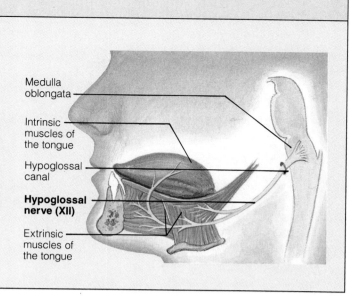

Medulla oblongata

Intrinsic muscles of the tongue

Hypoglossal canal

Hypoglossal nerve (XII)

Extrinsic muscles of the tongue

Spinal Nerves

General Features

Thirty-one pairs of **spinal nerves**, each containing thousands of nerve fibers, attach to the spinal cord (Figure 13.6). These short nerves have long branches that supply most of the body inferior to the head. The spinal nerves are named according to their point of issue from the vertebral column: There are 8 pairs of cervical spinal nerves (C_1–C_8), 12 pairs of thoracic nerves (T_1–T_{12}), 5 pairs of lumbar nerves (L_1–L_5), 5 pairs of sacral nerves (S_1–S_5), and 1 pair of coccygeal nerves (designated Co.). Notice that there are 8 pairs of cervical nerves but only 7 cervical vertebrae. This discrepancy is easily explained: The first cervical spinal nerve lies *superior* to the first vertebra, while the last cervical nerve exits *inferior* to the seventh cervical vertebra (leaving six nerves in between). Below the cervical region, every spinal nerve exits *inferior* to the verterbra of the same number.

As we mentioned in Chapter 12, each spinal nerve connects to the spinal cord by a **dorsal root** and a **ventral root** (Figures 13.7 and 12.26). Each root forms from a series of **rootlets** that attach along the whole length of the corresponding spinal cord segment (Figure 13.7a). The dorsal root contains *sensory* fibers arising from cell bodies in the dorsal root ganglion, while the ventral root contains *motor* fibers arising from the anterior gray column of the spinal cord. The spinal nerve lies at the lateral junction of the dorsal and ventral roots, just lateral to the dorsal root ganglion. The spinal nerves and dorsal root gan-

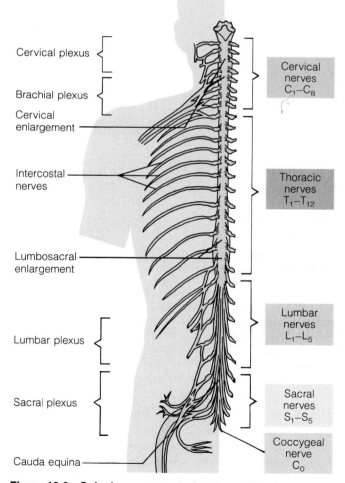

Cervical plexus

Brachial plexus

Cervical enlargement

Intercostal nerves

Lumbosacral enlargement

Lumbar plexus

Sacral plexus

Cauda equina

Cervical nerves C_1–C_8

Thoracic nerves T_1–T_{12}

Lumbar nerves L_1–L_5

Sacral nerves S_1–S_5

Coccygeal nerve C_0

Figure 13.6   Spinal nerves, posterior view. The short spinal nerves are shown on the right. Their ventral rami are shown on the left. Most ventral rami form nerve plexuses (cervical, brachial, lumbar, and sacral). The long, horizontal nerves in the region of the ribs are the intercostal nerves.

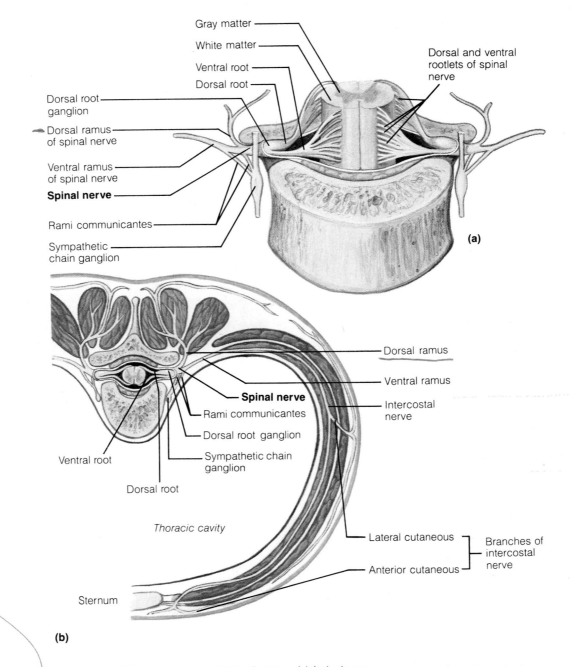

Figure 13.7 Roots and branches of the spinal nerves in the thorax. (a) A single segment of the spinal cord within one vertebra, with the associated nerves. The dorsal and ventral roots arise medially as rootlets and converge laterally to form the spinal nerve in the intervertebral foramen. (b) Cross section of thorax showing the main roots and branches of a spinal nerve. Note the dorsal and ventral roots and rami and the rami communicantes. In the thorax, each ventral ramus continues as an intercostal nerve.

glia both lie within the intervertebral foramina, against the bony pedicles of the vertebral arches.

Directly lateral to its intervertebral foramen, each spinal nerve branches into a **dorsal ramus** (ra′mus; "branch") and a **ventral ramus**. Connecting to the base of the ventral ramus are *rami communicantes* leading to *sympathetic chain ganglia*—visceral structures discussed in Chapter 14. Each of the branches of the spinal nerve, like the spinal nerve itself, contains both motor and sensory fibers.

The rest of this chapter focuses on the dorsal and ventral rami and their branches (Figure 13.7b). These rami supply the entire *somatic* region of the body (skeletal musculature and skin) from the neck inferiorly. The dorsal rami supply the dorsum of the body trunk (the "back"). The ventral rami are much thicker and supply a much larger area: the anterior and lateral regions of the trunk, and the limbs.

Before proceeding to the next section, you may want to review the difference between the *roots* and *rami*. The roots lie medial to the spinal nerves and are strictly sensory (dorsal root) or strictly motor (ventral root). The rami, by contrast, are lateral branches of the spinal nerves, and each contains both sensory and motor fibers.

Innervation of the Specific Regions of the Body

This section explores how the rami and their branches innervate the following regions of the body: the back, the thoracic and abdominal wall, the neck, the limbs, the joints, and the skin. Throughout the section, we will be mentioning the innervation of major groups of muscles. For more specific information on muscle innervation, see Tables 10.1–10.17 on pp. 246–292.

The Back

The pattern of innervation of the back by the dorsal rami follows a neat, segmented plan. Through several branches, each dorsal ramus innervates a horizontal strip of muscle and skin in line with its emergence point from the vertebral column (Figure 13.7b). This pattern is far simpler than that of the innervation of the rest of the body by the ventral rami.

Anterior Thoracic and Abdominal Wall

Only in the thorax are the ventral rami arranged in a simple and segmented pattern. The thoracic ventral rami run anteriorly, one deep to each rib, as the neatly segmented **intercostal nerves** (Figures 13.6 and 13.7b). These nerves supply the intercostal muscles, the skin of the anterior and lateral thorax, and most of the abdominal wall inferior to the rib cage. Along its course, each intercostal nerve gives off **lateral** and **anterior cutaneous branches** to the adjacent skin.

Two nerves in this thoracic series are unusual: The last (T_{12}) lies inferior to the twelfth rib, and is thus a *subcostal* ("below the ribs") *nerve* rather than an intercostal nerve. The first intercostal nerve (most superior) is exceptionally small because most fibers of T_1 enter the brachial plexus (next section).

Introduction to Nerve Plexuses

A feature of all spinal nerves except T_2–T_{12} is that their ventral rami branch and join one another lateral to the vertebral column, forming complex **nerve plexuses** ("networks") (Figure 13.6). Such interlacing networks occur in the cervical, brachial, lumbar, and sa-

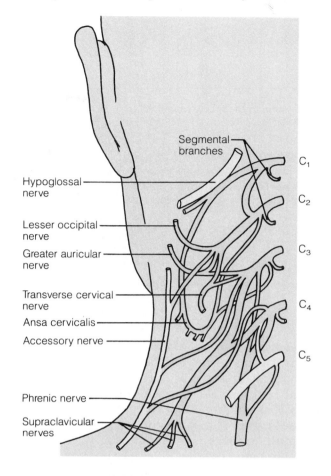

Figure 13.8 The cervical plexus. The nerves that are colored gray connect to the cervical plexus but do not belong to it. (See Table 13.3.)

cral regions and primarily serve the limbs. It must be emphasized that these plexuses are formed by *ventral* rami, not by dorsal rami. Within the plexuses, fibers from the different ventral rami crisscross each other and redistribute so that (1) each terminal branch of the plexus contains fibers from several different spinal nerves and (2) fibers from each ventral ramus travel to the body periphery via several different routes or branches. Thus, each muscle in a limb receives its nerve supply from more than one spinal nerve. This regrouping of fibers ensures that the destruction of a single spinal nerve cannot completely paralyze any limb muscle.

Cervical Plexus and the Neck

The **cervical plexus** is buried deep in the neck under the sternocleidomastoid muscle and is formed by the ventral rami of the first four cervical nerves (Figure 13.8). The plexus forms an irregular series of interconnecting loops from which branches arise. The functions of these branches are indicated in Table 13.3.

Table 13.3 Branches of the Cervical Plexus (See Figure 13.8)

Nerves	Ventral rami	Structures served
Cutaneous branches (superficial)		
Lesser occipital	C_2, C_3	Skin on posterolateral aspect of neck
Greater auricular	C_2, C_3	Skin of ear, skin over parotid gland
Transverse cervical	C_2, C_3	Skin on anterior and lateral aspect of neck
Supraclavicular (anterior, middle, and posterior)	C_3, C_4	Skin of shoulder and anterior aspect of chest
Motor branches (deep)		
Ansa cervicalis (superior and inferior roots)	C_1–C_3	Infrahyoid muscles of neck (omohyoid, sternohyoid, and sternothyroid)
Segmental and other muscular branches	C_1–C_5	Deep muscles of neck (geniohyoid and thyrohyoid) and portions of scalenes, levator scapulae, trapezius, and sternocleidomastoid muscles
Phrenic	C_3–C_5	Diaphragm (sole motor nerve supply)

Most are **cutaneous nerves** (nerves that supply only skin) that innervate the neck, the back of the head, and the most superior part of the shoulder. Other branches innervate muscles in the anterior neck region. The most important nerve from this plexus is the **phrenic** (fren'ik) **nerve** (which receives fibers from C_5 in addition to its main input from C_3 and C_4). The phrenic nerve courses inferiorly through the thorax and innervates the diaphragm (*phren* = the diaphragm). It provides both motor and sensory innervation to this most vital respiratory muscle.

Can you guess the common name for the condition in which the phrenic nerve induces abrupt, rhythmic contractions of the diaphragm? Can you deduce what would happen if the phrenic nerves were cut?

This first condition is hiccups, which often originate as a reflexive response to sensory irritation of the diaphragm or stomach. Swallowing spicy food or acidic soda pop can lead to hiccups in this way. If both phrenic nerves are cut or if the spinal cord is damaged superior to C_3–C_5, respiratory arrest will occur. ▪

Brachial Plexus and the Upper Limb

The **brachial plexus** (Figure 13.9) lies partly in the neck and partly in the axilla and gives rise to almost all of the nerves that supply the upper limb. The brachial plexus can sometimes be felt just superior to the clavicle at the lateral border of the sternocleidomastoid muscle.

The brachial plexus is formed by the intermixing of the ventral rami of the inferior four cervical nerves (C_5-C_8) and most of the T_1 ramus as well. Additionally, it may receive small branches from C_4 or T_2.

The brachial plexus is very complex (some consider it to be the anatomy student's nightmare). It is composed of four groups of stems and branches. From medial to lateral, these four components are the ventral rami (misleadingly called *roots**), which form *trunks*, which form *divisions*, which form *cords*. These elements are shown in Figure 13.9a. You can remember the sequence of roots, trunks, divisions, and cords by using the saying "**R**eally **t**ired? **D**rink **c**offee." We will now describe each of these components.

The five **roots** of the brachial plexus (ventral rami C_5-T_1) lie deep to the sternocleidomastoid. At the lateral border of this muscle, these rami unite to form the **upper**, **middle**, and **lower trunks**, each of which branches into an **anterior** and **posterior division**. The divisions give a general indication of which fibers ultimately serve the anterior part of the limb and which the posterior part. The divisions pass deep to the clavicle and enter the axilla, where they give rise to the **lateral**, **medial**, and **posterior cords**. Along the brachial plexus, small nerves branch off to supply muscles of the superior thorax and shoulder (and some skin of the shoulder as well).

Injuries to the brachial plexus are common and serious. Can you think of some causes and effects of such injuries?

Injury of the brachial plexus is usually due to stretching. This can occur when a football tackler yanks the arm of a halfback, for example. The plexus can also be stretched by blows to the top of the shoulder that force the humerus inferiorly. This injury re-

---

*Do not confuse these roots of the brachial plexus with the dorsal and ventral roots of spinal nerves.

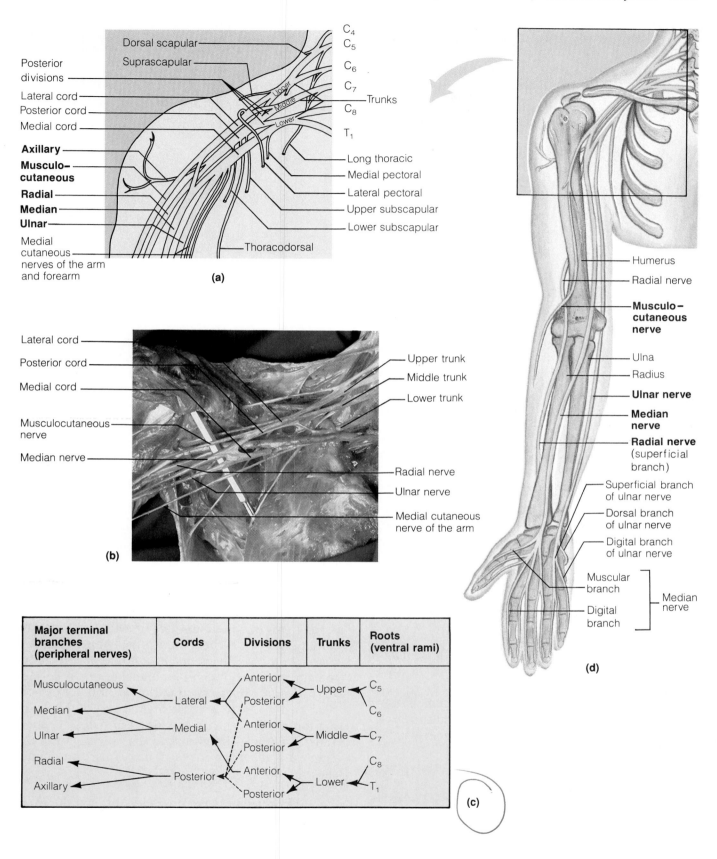

Figure 13.9 The brachial plexus. (a) Roots (rami C_5-T_1), trunks, divisions, and cords of the brachial plexus. (b) Photograph of the brachial plexus from a cadaver. (c) Flowchart summarizing the brachial plexus. (d) The major nerves of the upper limb. (See Table 13.4.)

Table 13.4 Branches of the Brachial Plexus (See Figure 13.9)

Nerves	Cord and spinal roots (ventral rami)	Structures served
Musculocutaneous	Lateral cord (C_5–C_7)	Muscular branches: flexor muscles in anterior arm (biceps brachii, brachialis, coracobrachialis) Cutaneous branches: skin on anterolateral forearm (extremely variable)
Median	By two branches, one from medial cord (C_8, T_1) and one from the lateral cord (C_5–C_7)	Muscular branches to flexor group of anterior forearm (palmaris longus, flexor carpi radialis, flexor digitorum superficialis, flexor pollicis longus, lateral half of flexor digitorum profundus, and pronator muscles); intrinsic muscles of lateral palm (thenar muscles and two lumbricals) Cutaneous branches: skin of lateral two-thirds of hand on palm side, and dorsum of fingers 2 and 3.
Ulnar	Medial cord (C_8 and T_1)	Muscular branches: flexor muscles in anterior forearm (flexor carpi ulnaris and medial half of flexor digitorum profundus); most intrinsic muscles of hand (hypothenar, all interossei, medial two lumbricals) Cutaneous branches: skin of medial third of hand, both anterior and posterior aspects
Radial	Posterior cord (C_5–C_8, T_1)	Muscular branches: posterior muscles of arm, forearm, and hand (triceps brachii, anconeus, supinator, brachioradialis, extensors carpi radialis longus and brevis, extensor carpi ulnaris, and several muscles that extend the fingers) Cutaneous branches: skin of posterolateral surface of entire limb (except dorsum of fingers 2 and 3)
Axillary	Posterior cord (C_5, C_6)	Muscular branches: deltoid and teres minor muscles Cutaneous branches: some skin of shoulder region
Dorsal scapular	Branches of C_5 rami	Rhomboid muscles and levator scapulae
Long thoracic	Branches of C_5–C_7 rami	Serratus anterior muscle
Subscapular	Posterior cord; branches of C_5 and C_6 rami	Teres major and subscapular muscles
Suprascapular	Upper trunk (C_5, C_6)	Shoulder joint; supraspinatus and infraspinatus muscles
Pectoral (lateral and medial)	Branches of lateral and medial cords (C_5–T_1)	Pectoralis major and minor muscles

sults, for example, when a cyclist is thrown from a bike and grinds the top of the shoulder into the pavement. Stretching or crushing the brachial plexus leads to paralysis or weakness of the upper limb. ■

The brachial plexus ends in the axilla, where its three cords give rise to the main nerves of the upper limb (Figure 13.9). Five of these nerves are especially important: the *axillary, musculocutaneous, median, ulnar,* and *radial nerves.* The distribution and targets of these nerves are described below. More detail is provided in Table 13.4.

The **axillary nerve** is a branch of the posterior cord. It runs posterior to the surgical neck of the humerus and innervates the deltoid and teres minor muscles. Its sensory fibers supply the capsule of the shoulder joint and a small region of skin on the inferior part of the shoulder and the superior part of the arm.

The **musculocutaneous nerve**, the main terminal branch of the lateral cord, courses within the anterior arm. Descending between the biceps brachii and brachialis muscles, it innervates these arm flexors. Distal to the elbow, it provides for skin sensation on the lateral forearm.

The **median nerve** innervates most muscles of the anterior forearm and the lateral palm. It descends through the arm without branching, lying medial and posterior to the biceps brachii muscle. Just distal to the elbow region, it gives off branches to all muscles in the flexor compartment of the forearm (except the flexor carpi ulnaris and the medial part of the flexor digitorum profundus). Reaching the hand, it innervates five intrinsic muscles in the lateral part of the palm, including the thenar muscles used to oppose the thumb.

The median nerve descends through the exact middle of the forearm and wrist, so it is often cut during wrist-slashing suicide attempts. Can you deduce some effects of severing this nerve near the wrist?

Destruction of the median nerve makes it difficult to oppose the thumb toward the little finger and thus to pick up small objects. ■

The **ulnar nerve** branches off the medial cord of the brachial plexus, then descends along the medial side of the arm. At the elbow, it passes posterior to the medial epicondyle of the humerus, then follows the ulna along the forearm. Here, it supplies the flexor carpi ulnaris and the medial (ulnar) part of the flexor digitorum profundus. It continues into the hand, where it innervates most of the intrinsic hand muscles and the skin on the medial side of the hand.

Because of its exposed position at the elbow, the ulnar nerve is vulnerable to injury. Striking the "funny bone"—the spot where the ulnar nerve rests against the medial epicondyle—causes tingling of the little finger. Even more commonly, the ulnar nerve is injured in the carpal region of the palm, another place where it is superficially located. People with damage to this nerve cannot adduct and abduct their fingers (because the interossei and medial lumbrical muscles are paralyzed), nor can they form a tight grip (because the hypothenar muscles are useless). Without the interossei muscles to extend the fingers at the interphalangeal joints, the fingers flex at these joints and the hand contorts into a **clawhand.** ■

The **radial nerve**, a continuation of the posterior cord, is the largest branch of the brachial plexus. It innervates almost the entire posterior side of the upper limb, including the limb extensor muscles. As it descends through the arm, the nerve wraps around the humerus in the radial groove, sending branches to the triceps brachii. The nerve then curves anteriorly around the lateral epicondyle at the elbow, where it divides into a superficial and deep branch. The superficial branch (Figure 13.9d) descends along the lateral edge of the radius to supply skin on the dorsal surface of the hand. The deep branch (not illustrated) runs posteriorly to supply most of the extensor muscles on the forearm.

Can you determine what disorders of movement may follow injury to the radial nerve?

Trauma to the radial nerve results in **wristdrop,** an inability to extend the hand at the wrist. If the lesion occurs far enough superiorly, the triceps muscle is paralyzed, so that the forearm cannot be actively extended at the elbow. Many fractures of the humerus follow the radial groove and harm the radial nerve there. This nerve can also be crushed in the axilla by improper use of a crutch or by "Saturday night paralysis," in which an intoxicated person falls asleep with an arm draped over the back of a chair. ■

Lumbar Plexus and the Lower Limb

The **lumbar plexus** (Figure 13.10) arises from the first four lumbar spinal nerves ($L_1–L_4$) and lies within the psoas major muscle in the posterior abdominal wall. Its smaller branches innervate parts of the abdominal wall and the psoas muscle itself, but the main branches descend to innervate the anterior thigh. The **femoral nerve,** the largest terminal branch of the lumbar plexus, runs deep to the inguinal ligament to enter the thigh. From there it divides into several large branches. Motor branches of the femoral nerve innervate the anterior thigh muscles, including the important quadriceps femoris. Cutaneous branches of the femoral nerve serve the skin of the anterior thigh and the medial surface of the leg from the knee to the foot. The **obturator nerve** passes through the large obturator foramen of the pelvis, enters the medial thigh, and innervates the adductor muscle group. These and other branches of the lumbar plexus are summarized in Table 13.5.

Compression of the spinal roots of the lumbar plexus by a herniated disc causes a major disturbance in gait. This is because the femoral nerve innervates muscles that flex the thigh at the hip (rectus femoris and iliacus), and muscles that extend the leg at the knee (the entire quadriceps femoris group). Other symptoms are pain or anesthesia of the anterior thigh and of the medial thigh if the obturator nerve is impaired. ■

Sacral Plexus and the Lower Limb

The **sacral plexus** arises from spinal nerves $L_4–S_4$ and lies immediately caudal to the lumbar plexus (Figure 13.11). Some fibers from the lumbar plexus contribute to the sacral plexus via the **lumbosacral trunk.** Therefore, the two plexuses are often referred to as the *lumbosacral plexus.* The sacral plexus has about a dozen named branches. About half of these branches serve the buttock and lower limb, while the others innervate parts of the pelvis and perineum. The most important branches are described below and summarized in Table 13.6.

The largest branch of the sacral plexus is the **sciatic** (si-at'ik) **nerve.** This is the thickest and longest nerve in the body. It supplies almost all of the lower limb, except the anterior and medial regions of the thigh. Actually, the sciatic is two nerves—the *tibial* and *common peroneal nerves*—wrapped in a common sheath. It leaves the pelvis by passing through the greater sciatic notch, then courses deep to the broad gluteus maximus muscle and enters the thigh just medial to the hip joint (*sciatic* literally means "of the hip"). From there the nerve descends through the posterior thigh deep to the hamstrings, which it innervates. Superior to the knee region, its two branches diverge.

The **tibial nerve** courses through the popliteal fossa (the region just posterior to the knee joint) then descends through the calf deep to the soleus muscle. At the ankle, it passes posterior to the medial malleolus and divides into the two main nerves of the sole of the foot, the medial and lateral **plantar nerves.** The tibial nerve and its branches supply almost all muscles in the posterior region of the lower limb. The nerve also innervates some skin on the lateral calf (through a **sural branch**) and on the sole.

The **common peroneal nerve** is really the "fibular nerve" (*perone* = fibula). It supplies most structures on the anterolateral aspect of the leg. It descends laterally from its point of origin in the popliteal fossa and enters the superior part of the leg, where it wraps around the neck of the fibula and divides into deep and superficial branches. The **superficial peroneal**

nerve supplies the peroneal muscles in the lateral compartment of the leg and most of the skin on the superior surface of the foot. The **deep peroneal nerve** serves the muscles of the anterior compartment of the leg—the extensors that dorsiflex the foot.

Except for the sciatic nerve, the largest branches of the sacral plexus are the **superior** and **inferior gluteal nerves.** These innervate the gluteal muscles. Other branches of the sacral plexus supply the thigh rotators and the muscles of the pelvic floor and perineum. The **pudendal nerve** (pu-den′dal; "shameful")

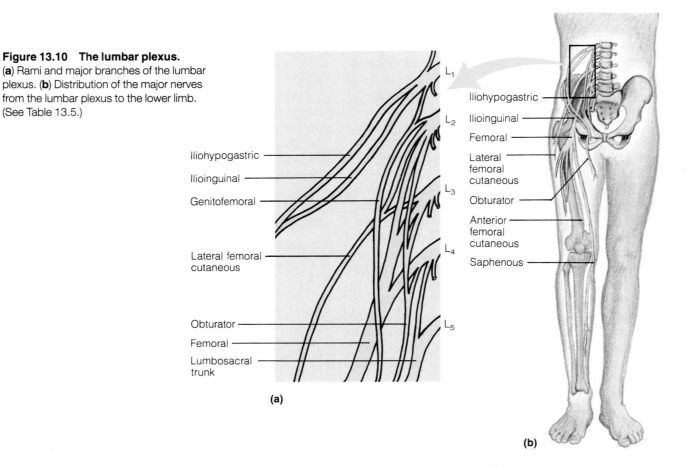

Figure 13.10   The lumbar plexus.
(**a**) Rami and major branches of the lumbar plexus. (**b**) Distribution of the major nerves from the lumbar plexus to the lower limb. (See Table 13.5.)

(a)

(b)

Table 13.5 Branches of the Lumbar Plexus (See Figure 13.10)		
Nerves	**Ventral rami**	**Structures served**
Femoral	L_2–L_4	Skin of anterior and medial thigh via *anterior femoral cutaneous* branch; skin of medial leg and foot, hip; and knee joints via *saphenous* branch; motor to anterior muscles (quadriceps and sartorius) of thigh; pectineus, iliacus
Obturator	L_2–L_4	Motor to adductor magnus (part), longus and brevis muscles, gracilis muscle of medial thigh, obturator externus; sensory for skin of medial thigh and for hip and knee joints
Lateral femoral cutaneous	L_2, L_3	Skin of lateral thigh; some sensory branches to peritoneum
Iliohypogastric	L_1	Skin of lower abdomen, lower back, and hip; muscles of anterolateral abdominal wall (obliques and transversus) and pubic region
Ilioinguinal	L_1	Skin of external genitalia and proximal medial aspect of the thigh; inferior abdominal muscles
Genitofemoral	L_1, L_2	Skin of scrotum in males, of labia majora in females, and of anterior thigh inferior to middle portion of inguinal region; cremaster muscle in males

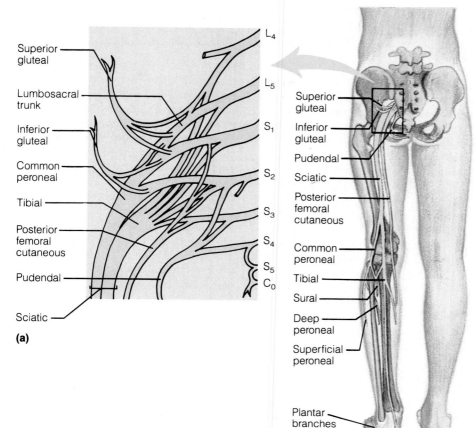

(a)

(b)

Figure 13.11   The sacral plexus.
(**a**) Rami and major branches of the sacral plexus. (**b**) Distribution of the major nerves from the sacral plexus to the lower limb. (See Table 13.6.)

Table 13.6	Branches of the Sacral Plexus (See Figure 13.11)	
Nerves	**Ventral rami**	**Structures served**
Sciatic nerve	$L_4, L_5, S_1–S_3$	Composed of two nerves (tibial and common peroneal) in a common sheath that diverge just proximal to the knee
• Tibial (including sural branch and medial and lateral plantar branches)	$L_4–S_3$	Cutaneous branches: to skin of posterior surface of leg and sole of foot Motor branches: to muscles of back of thigh, leg, and foot (to hamstrings except short head of biceps femoris, to posterior part of adductor magnus, to triceps surae, tibialis posterior, popliteus, flexor digitorum longus, flexor hallucis longus, and intrinsic muscles of foot)
• Common peroneal (superficial and deep branches)	$L_4–S_2$	Cutaneous branches: to skin of anterior surface of leg and dorsum of foot Motor branches: to short head of biceps femoris of thigh, peroneal muscles of lateral compartment of leg, tibialis anterior, and extensor muscles of toes (extensor hallucis longus, extensors digitorum longus and brevis)
Superior gluteal	L_4, L_5, S_1	Motor branches: to gluteus medius and minimus and tensor fasciae latae
Inferior gluteal	$L_5–S_2$	Motor branches: to gluteus maximus
Posterior femoral cutaneous	$S_1–S_3$	Skin of buttock, posterior thigh, and popliteal region; length variable; may also innervate part of skin of calf and heel
Pudendal	$S_2–S_4$	Supplies most of skin and muscles of perineum (region encompassing external genitalia and anus and including clitoris, labia, and vaginal mucosa in females and scrotum and penis in males); external anal sphincter

---

A CLOSER LOOK   Does Polio "Haunt" Overachievers?

Poliomyelitis, or infantile paralysis, was the most feared infectious disease in America during the 1940s and early 1950s. In 1949, a public service message in a nationwide magazine warned parents to take the following precautions during the annual summer polio epidemic: Make sure children avoid swimming in water not declared safe by their health departments, avoid crowds and "new contacts," and keep their hands and food scrupulously clean. Tragically, however, these precautions were often ineffective in preventing the spread of this debilitating disease.

Polio is caused by a virus that spreads through feces (in swimming pools, for example), on dirty hands, and by airborne droplets from coughs. The initial symptoms resemble those of all viral infections, including fever, headache, and stiffness of the neck and back. However, in about 10% of cases the disease progresses further: The virus enters and destroys some motor neurons of the spinal cord or brain, eliminating the motor functions of certain nerves and causing paralysis of muscles. Respiratory muscles may become paralyzed (note the iron lungs pictured in this box). Fortunately, only about 3% of those who contract polio are paralyzed permanently: The rest recover full or partial use of their muscles. Older children are usually affected more severely than younger children (as with many childhood diseases, younger children have milder symptoms).

During its peak years in the early 1950s, over 22 cases of polio were reported per every 100,000 Americans. Then, in 1955, the epidemic was stopped in its tracks by the development of the Salk and Sabin polio vaccines. In recent years, only a handful of polio cases have been reported in the United States.

While the devastating epidemics have faded into the past, many of the "recovered" survivors are now experiencing a "postpolio" deterioration. Vital people now in their 40s who fought their way back to health and active, vigorous lives have begun to notice a new, disturbing set of symptoms. Examples include extreme lethargy; sensitivity to cold; sharp, burning pains in their muscles and joints; and weakness and gradual loss of mass in the muscles that for so many years after their initial illness served them well.

Initially, these baffling symptoms were dismissed as the flu or as psychosomatic illness, or they were misdiagnosed as multiple sclerosis, arthritis, lupus, or amyotrophic lateral sclerosis, a devastating neurological disease that kills its victims swiftly. But as the number of similar complaints has increased, more attention has been given to this group of symptoms, which is now named *postpoliomyelitis muscular atrophy (PPMA) or postpolio syndrome.*

According to a 1983 study conducted by the Mayo Clinic, 25% of polio survivors are affected. It now appears that this estimate is extremely low. Moreover, certain survivors seem to be particularly vul-

---

innervates the muscles and skin of the perineum, mediates the act of erection, and is responsible for the voluntary control of defecation and urination (Table 10.7, p. 260). As it passes anteriorly to enter the perineum, this nerve lies just medial to the ischial tuberosity.

First, can you deduce some causes and effects of injury to the sciatic nerve? Then, can you think of a clinical procedure that would involve anesthetizing the pudendal nerves?

The sciatic nerve can be injured in many ways. **Sciatica** (si-at″ĭ-kah) is characterized by a stabbing pain over the course of this nerve. It often occurs when a herniated lumbar disc presses on the sacral dorsal roots within the vertebral canal. Wounds to the buttocks or posterior thigh can sever the sciatic nerve. When this occurs, the leg cannot be flexed at the knee (because the hamstrings are paralyzed), and the foot cannot be moved at the ankle. When only the tibial nerve is injured, the paralyzed muscles in the calf cannot plantar flex the foot, and a shuffling gait develops. When the common peroneal nerve is damaged, dorsiflexion is lost, and the foot drops into plantar flexion. This condition is called **footdrop**. The common peroneal nerve is susceptible to injury in the superolateral leg, where it is superficially located and easily crushed against the neck of the fibula.

The pudendal nerve may be injected with anesthetic to help block pain in the perineum during childbirth, or prior to surgery on the anal or genital regions. In such a **pudendal nerve block,** the injection needle is inserted into the nerve just medial to the ischial tuberosity. ■

Innervation of Joints of the Body

Since injuries to joints are common, health professionals should know which nerves serve which synovial joints. Memorizing so much information is almost impossible, however, because every joint capsule receives sensory branches from several different nerves (pp. 193–194). **Hilton's law** will help you deduce the nerves supplying any joint: By this simple rule, *any nerve that innervates a muscle producing movement at a joint also innervates the joint itself (and the skin*

nerable: those who contracted the disease after the age of ten, those who were severely affected, those who needed ventilatory support, and those in whom all four limbs were affected. The cause of PPMA is not known. The earlier belief that the polio virus was reactivated has now been discarded. A more likely explanation is that polio survivors, like all of us, continue to lose neurons throughout life. While unimpaired nervous systems can enlist nearby neurons to compensate for the losses, polio survivors who have already lost many neurons have drawn on that "pool" and may have few neurons left to take over. What is particularly ironic is that PPMA victims seem to be overachievers; that is, they are the ones who worked hardest to overcome their disease. The grueling hours of exercise that they assumed would make them strong may be the factor that ultimately caused their overworked motor neurons to "burn out."

Postpolio syndrome progresses slowly and is not life threatening. Currently, its victims are advised to conserve their energy by resting more and by using devices such as canes, walkers, and braces to sup-

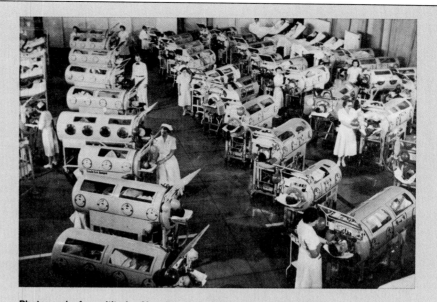

Photograph of a multitude of iron lungs in a March of Dimes Respiratory Center in the 1950s. The iron lungs provided mechanical breathing assistance for polio patients whose respiratory muscles were paralyzed.

port their wasting muscles. A major concern is determining how much more rest is required, so that the muscle atrophy caused by PPMA is not accentuated by disuse atrophy.

Many PPMA sufferers are emotionally devastated—surprised by the return of an old enemy, angry because no remedy currently exists, and understandably bitter because

all of their hard work has resulted in a new affliction. Their plight has finally begun to receive the attention it deserves from the medical and research community, and grants for research and postpolio assessment clinics have been established across the nation. But this is cold comfort to those experiencing the limbo of PPMA.

---

over the joint). Let us apply this rule to the knee joint: Recall that the knee is crossed by the quadriceps, hamstring, and gracilis muscles (Chapter 10). The nerves to these muscles are the femoral, the branches of the sciatic, and the obturator. Thus, all these nerves innervate the knee joint itself.

Now consider the elbow joint. It is crossed by the biceps brachii, the triceps brachii, and many flexor muscles on the forearm (including flexor carpi ulnaris). Can you deduce which four nerves send branches to the capsule of the elbow joint?

They are the musculocutaneous, radial, median, and ulnar nerves. ■

Innervation of the Skin: Dermatomes

The area of skin innervated by the cutaneous branches from a single spinal nerve is called a **dermatome,** literally a "skin segment." All spinal nerves except C_1 participate in dermatomes. The pattern of dermatomes is illustrated in Figure 13.12. The map was constructed by recording areas of numbness in patients who had experienced injuries to specific spi-

nal roots. In the *trunk* region, the dermatomes are almost horizontal, uniform in width, and in direct line with their spinal nerves. In the *limbs,* however, the pattern of dermatomes is less straightforward. In the upper limb, the skin is supplied by the nerves participating in the brachial plexus: the ventral rami of C_5 to T_1 (or T_2). In the lower limb, the lumbar nerves supply most of the anterior surface, while sacral nerves supply much of the posterior surface (this distribution basically reflects the areas supplied by the lumbar and sacral plexuses, respectively).

Adjacent dermatomes are not as cleanly separated from one another as a typical dermatome map indicates. On the trunk, neighboring dermatomes overlap each other by a full 50%, so that destruction of a single spinal nerve will not produce complete numbness anywhere. On the limbs, by contrast, the overlap is not complete: Some skin patches are innervated by just one spinal nerve.

Can you think of some clinical situations in which a knowledge of the dermatome map would prove useful?   (Answer on p. 388)

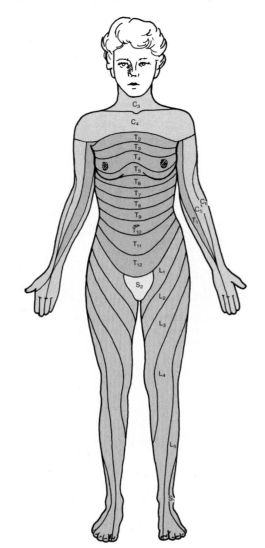

Figure 13.12   Map of dermatomes.   A dermatome is an area of skin innervated by the sensory fibers of a single spinal nerve (both ventral and dorsal rami). The pattern of dermatomes in the limbs is controversial, and studies using different mapping techniques have produced slightly different maps of limb dermatomes.

Clinicians use dermatomes to determine the level of spinal injuries. Numbness in all dermatomes below T_1, for example, indicates a spinal-cord injury at T_1. Furthermore, anesthetic may be injected into specific spinal nerves or roots to desensitize specific skin regions for surgery. Finally, the varicella-zoster virus that produces shingles (p. 313) resides in a single dorsal root ganglion, and a flare-up of this infection produces skin blisters along the entire course of that one dermatome. ■

The Peripheral Nervous System Throughout Life

The spinal nerves start to form late in the fourth week of development. Motor axons grow outward from the early spinal cord, while sensory peripheral axons grow out from the neural crest of the dorsal root ganglia (see Figure 11.18, p. 312). The bundles of these motor and sensory axons exit between successive vertebrae, where they converge to form the early spinal nerves. Each of the 31 spinal nerves sends motor fibers into a single myotome (muscle segment) and sends its sensory fibers to the overlying band of skin. (This development accounts for the segmented pattern of dermatomes.) Cranial nerves grow to innervate the head in a roughly comparable manner. Limb innervation is attained in a more complex way, as adjacent spinal nerves exchange fibers to form plexuses.

During week 5, nerves reach the organs they innervate. Shortly thereafter, some of the embryonic muscles migrate to new locations, and some skin dermatomes become displaced as they are pulled along the elongating limbs. Although they may migrate for considerable distances, these muscles and skin areas always retain their original nerve supply.

Aging of the nervous system has been discussed in previous chapters. In the PNS, sensory receptors atrophy to some degree with age, and tone in the muscles of the face and neck decreases. Reflexes slow a bit with age, but this change seems to reflect a slowdown in central processing more than changes in the peripheral nerve fibers.

✚ Related Clinical Terms

Myasthenia gravis (mi″es-the′ne-ah grǎ′vis) (*my* = muscle; *asthen* = weakness; *gravi* = severe, heavy)   A disorder of the neuromuscular junctions, this crippling disease is characterized by a progressive weakening of muscles. The muscles of the head, neck, and limbs are most seriously affected. This is an autoimmune disease in which the body's own defense molecules (antibodies) attach to and block the acetylcholine receptors in the sarcolemma of the motor end plates. (Once inactivated, these receptors cannot bind acetylcholine, so the muscle fibers cannot contract.)

Neuralgia (nu-ral′je-ah) (*algia* = pain)   Sharp, spasm-like pain along the course of one or more nerves, usually caused by inflammation or injury to the nerves.

Neuritis   Inflammation of a nerve. There are many different forms of neuritis with different effects (for example, increased or decreased nerve sensitivity, paralysis of the structure served, and pain).

Paresthesia (par"es-the'ze-ah; "faulty sensation") An abnormal sensation (numbness, burning, tingling) resulting from a disorder of a sensory nerve.

Chapter Summary

1. The PNS consists of sensory receptors, motor endings that innervate effectors, nerves, and ganglia. Most nerves contain both sensory and motor fibers (mixed nerves), but some are purely sensory.

PERIPHERAL SENSORY RECEPTORS (pp. 362–367)

1. Sensory receptors monitor stimuli (environmental changes) from both inside and outside the body. The widespread receptors for the general senses tend to be dendrites of sensory neurons, whereas the localized receptors for the special senses tend to be distinct receptor cells.

Classification by Location (p. 362)

2. Receptors are classified by location as exteroceptors (for external stimuli), interoceptors (for internal stimuli in the viscera), and proprioceptors (for measuring stretch in muscles and the skeleton).

Classification by Stimulus Detected (pp. 362–363)

3. Receptors are classified by stimuli detected as mechanoreceptors, thermoreceptors, chemoreceptors, photoreceptors, and nociceptors.

Classification by Structure (pp. 363–367)

4. Dendritic endings of sensory neurons monitor general sensory information. Structurally, they are either (1) free or (2) encapsulated dendritic endings. The free endings seem to be receptors for pain and temperature. The encapsulated endings are mechanoreceptors: They include Meissner's corpuscles and Krause's end bulbs (discriminative touch), Pacinian corpuscles and Ruffini's corpuscles (deep pressure), and most proprioceptors.

5. The proprioceptors include muscle spindles, Golgi tendon organs, and joint kinesthetic receptors. Muscle spindles are small bags of muscle fibers (intrafusal fibers) innervated by sensory dendrites within the skeletal muscles of the body. They monitor tension within muscles.

6. Golgi tendon organs are encapsulated dendritic endings embedded in muscle tendons. They monitor the tension generated in tendons during muscle contraction. Joint kinesthetic receptors occur in joint capsules. Some joint receptors measure tension placed on the joint (Pacinian corpuscles and Ruffini's corpuscles). Free nerve endings are receptors for joint pain.

PERIPHERAL MOTOR ENDINGS (pp. 367–369)

Innervation of Skeletal Muscle (pp. 367–368)

1. Motor axons innervate skeletal muscle fibers at synapse-like neuromuscular junctions (motor end plates). The motor axon terminal releases acetylcholine, which signals the muscle cell to contract. The synaptic cleft is a grooved gap filled with acetylcholinesterase, an enzyme that rapidly breaks down acetylcholine.

2. A motor unit consists of one motor neuron and all the skeletal muscle fibers it innervates. Motor units contain varying numbers of muscle fibers distributed widely within a muscle. All muscle fibers in a motor unit contract simultaneously.

Innervation of Visceral Muscle and Glands (pp. 368–369)

3. Visceral motor neurons do not form elaborate neuromuscular junctions with the visceral muscle and glands they innervate. Their axon terminals (varicosities) may even end some distance from the effector cells.

CRANIAL NERVES (pp. 369–377)

1. Twelve pairs of cranial nerves originate from the brain and issue through the skull to innervate the head and neck. Only the vagus nerves extend into the thoracic and abdominal cavities.

2. The cranial nerves can be grouped according to function. Some are purely sensory, containing special somatic sensory fibers (I, II, VIII); some are primarily motor, containing general somatic motor fibers (III, IV, VI, and XII); and some contain both sensory and motor fibers (V, VII, IX, X).

3. The following are the cranial nerves:

- The olfactory nerves (I): purely sensory; concerned with smell; attach to the olfactory bulb of the telencephalon.

- The optic nerves (II): purely sensory; transmit visual impulses from the retina to the thalamus.

- The oculomotor nerves (III): primarily motor; emerge from the midbrain and serve four extrinsic muscles that move the eye and some smooth muscle within the eye (Chapter 14); also carry proprioceptive impulses from the skeletal muscles served.

- The trochlear nerves (IV): primarily motor; emerge from the dorsal midbrain and carry motor and proprioceptive impulses to and from superior oblique muscles of the eyes.

- The trigeminal nerves (V): contain both sensory and motor fibers; emerge from the pons; the main sensory nerves of the face, nasal cavity, and mouth; three divisions—ophthalmic, maxillary, and mandibular; the mandibular branch contains motor fibers that innervate the chewing muscles.

- The abducens nerves (VI): primarily motor; emerge from the pons and supply motor and proprioceptive innervation to the lateral rectus muscles of the eyes.

- The facial nerves (VII): contain both sensory and motor fibers; emerge from the pons; motor nerves to muscles of facial expression and many glands in head; also carry sensory impulses from the taste buds of anterior two-thirds of the tongue.

- The vestibulocochlear nerves (VIII): purely sensory; enter brain at pons/medulla border; transmit impulses from the hearing and equilibrium receptors of the inner ears.

- The glossopharyngeal nerves (IX): contain both sensory and motor fibers; issue from the medulla; transmit sensory impulses from the posterior tongue (including taste) and the pharynx; innervate some superior pharyngeal muscles and the parotid glands.

- The vagus nerves (X): contain both sensory and motor fibers; arise from the medulla; most of its motor fibers are autonomic fibers; carries motor fibers to, and sensory fibers from, the pharynx, larynx, and visceral organs of the thorax and abdomen.

- The accessory nerves (XI): primarily motor; have a cranial root arising from the medulla and a spinal root arising from the cervical spinal cord; cranial root runs in vagus to supply muscles of larynx and pharynx; spinal root supplies trapezius and sternocleidomastoid.

- The hypoglossal nerves (XII): primarily motor; issue from the medulla; supply motor and proprioceptive innervation to the tongue muscles.

SPINAL NERVES (pp. 377–388)

General Features (pp. 377–379)

1. The 31 pairs of spinal nerves are numbered according to the region of the vertebral column from which they issue (cervical, thoracic, lumbar, sacral, coccygeal). There are eight cervical spinal nerves but only seven cervical vertebrae.

2. Each spinal nerve forms from the union of a dorsal sensory root and a ventral motor root. On each dorsal root is a dorsal

root ganglion (cell bodies of sensory neurons). The branches of each spinal nerve are the dorsal and ventral rami (somatic branches) and rami communicantes (visceral branches).

3. Dorsal rami serve the muscles and skin of the posterior trunk, while ventral rami serve the muscles and skin of the lateral and anterior trunk. Ventral rami also serve the limbs.

Innervation of the Specific Regions of the Body (pp. 379–388)

4. The back, from neck to sacrum, is innervated by the dorsal rami in a neatly segmented pattern.

5. The anterior and lateral wall of the thorax (and abdomen) are innervated by thoracic ventral rami—the segmented intercostal and subcostal nerves. Thoracic ventral rami do not form nerve plexuses.

6. The major nerve plexuses are networks of successive ventral rami that exchange fibers. They primarily innervate the limbs.

7. The cervical plexus (C_1–C_4) innervates the muscles and skin of the neck and shoulder. Its phrenic nerve (C_3–C_5) serves the diaphragm.

8. The brachial plexus serves the upper limb, the shoulder, and some thoracic muscles that move the arm. It arises primarily from C_5–T_1. From proximal to distal, the brachial plexus consists of "roots" (rami), trunks, divisions, and cords.

9. The main nerves arising from the brachial plexus are the axillary (to the deltoid muscle), musculocutaneous (to flexors on arm), median (to anterior forearm muscles and lateral palm), ulnar (to anteromedial muscles of the forearm and the medial hand), and radial (to the posterior part of the limb).

10. The lumbar plexus (L_1–L_4) innervates the anterior and medial muscles of the thigh, through the femoral and obturator nerves, respectively. The femoral nerve also innervates the skin on the anterior thigh and medial leg.

11. The sacral plexus (L_4–S_4) supplies the muscles and skin of the posterior thigh and almost all of the leg. Its main branch is the large sciatic nerve, which consists of the tibial nerve (to most hamstrings and all muscles of the calf and sole) and the common peroneal nerve (to muscles of the anterior and lateral leg and the overlying skin). Another branch of the sacral plexus is the pudendal nerve to the perineum.

12. The sensory nerves to a joint are branches from the nerves to the muscles that cross that joint (Hilton's law).

13. A dermatome is a segment of skin innervated by the sensory fibers of a single spinal nerve. Adjacent dermatomes overlap to some degree, more so on the trunk than on the limbs. Loss of sensation in specific dermatomes reveals sites of damage to spinal nerves or the spinal cord.

THE PERIPHERAL NERVOUS SYSTEM THROUGHOUT LIFE (p. 388)

1. In the middle of the embryonic period, each spinal nerve grows out between newly formed vertebrae to provide the motor innervation of an adjacent myotome (future trunk muscle) and the sensory innervation of the adjacent skin region (dermatome).

Review Questions

Multiple Choice and Matching Questions

1. Proprioceptors include all of the following except (a) muscle spindles, (b) Golgi tendon organs, (c) Merkel discs, (d) Pacinian corpuscles in joint capsules.

2. The large, onion-shaped pressure receptors in deep connective tissues are (a) Merkel discs, (b) Pacinian corpuscles, (c) free nerve endings, (d) Krause's end bulbs.

3. Match the receptor type from column B to the correct description in column A.

Column A
___ (1) pain and temperature receptors
___ (2) looks and functions like a Meissner's corpuscle but occurs in mucosae
___ (3) contains intrafusal fibers and flower spray endings
___ (4) discriminative touch receptor in hairless skin (fingertips)
___ (5) contains dendrites wrapped around thick collagen bundles in a tendon
___ (6) rapidly adapting deep pressure receptor
___ (7) slowly adapting deep pressure receptor

Column B
(a) Ruffini's corpuscle
(b) Golgi tendon organ
(c) muscle spindle
(d) Krause's end bulb
(e) free dendritic endings
(f) Pacinian corpuscle
(g) Meissner's corpuscle

4. Match the names of the cranial nerves in column B to the appropriate description in column A.

Column A
___ (1) moves four extrinsic eye muscles
___ (2) is the major sensory nerve of the face
___ (3) serves the sternocleidomastoid and trapezius
___ (4) are purely sensory (three nerves)
___ (5) serves the tongue muscles
___ (6) allows you to chew food
___ (7) is impaired in tic douloureux
___ (8) helps to regulate heart activity
___ (9) helps you to hear and keep your balance
___ (10) contain parasympathetic motor fibers (four nerves)
___ (11) serves muscles of facial expression

Column B
(a) abducens
(b) accessory
(c) facial
(d) glossopharyngeal
(e) hypoglossal
(f) oculomotor
(g) olfactory
(h) optic
(i) trigeminal
(j) vestibulocochlear
(k) vagus
(l) trochlear

5. For each of the following muscles or body regions, identify the plexus and the peripheral nerve (or branch of one) involved. Use one choice from Key A followed by one choice from Key B.

Structure innervated

b, 6 **(1)** the diaphragm
d, 8 **(2)** muscles of the posterior thigh and posterior leg
c, 2 **(3)** anterior thigh muscles
c, 5 **(4)** medial thigh muscles
a, 4 **(5)** anterior arm muscles that flex the forearm
a, 3 9 **(6)** muscles that flex the wrist and fingers (two nerves)
a, 7 **(7)** muscles that extend the wrist and fingers
a, 7 **(8)** skin and extensor muscles of the posterior arm
d, 1 **(9)** peroneal muscles, tibialis anterior, and toe extensors

Key A: Plexuses
(a) brachial
(b) cervical
(c) lumbar
(d) sacral

Key B: Nerves
(1) common peroneal
(2) femoral
(3) median
(4) musculocutaneous
(5) obturator
(6) phrenic
(7) radial
(8) tibial
(9) ulnar

6. Which one of the following contains only motor fibers? (a) dorsal root, (b) dorsal ramus, (c) ventral root, (d) ventral ramus.

7. The trigeminal nerve contains which class of nerve fibers? (a) somatic sensory only, (b) somatic motor and proprioceptor, (c) somatic sensory, visceral sensory, and visceral motor.

8. Which of the following is a chemoreceptor for a special sense and is also an interoceptor? (a) receptor in eye retina, (b) Pacinian corpuscle, (c) receptor cell in a taste bud, (d) Krause's end bulb.

Short Answer and Essay Questions

9. In the sensory receptors called "encapsulated dendritic endings," what is the "capsule" made of?

10. Define motor unit.

11. Name two ways in which the structure of a neuromuscular junction differs from that of a typical synapse between two neurons.

12. (a) Describe the roots to and the composition of a spinal nerve. (b) Name the branches of a spinal nerve (other than the rami communicantes), and explain what basic region of the body each branch supplies.

13. (a) Define nerve plexus. (b) List the spinal nerves of origin of the four major nerve plexuses, and name the general body regions served by each plexus.

14. In the brachial plexus, what specific rami make up each of the three trunks?

15. Use Hilton's law to deduce which nerves innervate the hip joint.

16. Dr. Omata noticed that her anatomy students often confused the glossopharyngeal (IX) and hypoglossal (XII) nerves because the two names sound alike. Thus, every year she made her class explain some basic differences between these two nerves. What would you tell her?

17. Adrian and Abdul, two anatomy students, were arguing about the facial nerve. Adrian said that it innervates all of the skin of the face, and that it is called the facial nerve for this reason. Abdul said the facial nerve does not innervate facial skin at all. Who was more correct?

18. Choose the correct answer, and explain why it is correct. Through which pair of intervertebral foramina do spinal nerves L_5 leave the vertebral canal? (a) through holes in the sacrum, (b) just superior to vertebra L_5, (c) just superior to S_1, (d) just inferior to S_4.

Critical Thinking and Clinical Application Questions

19. Harry fell to the ground from a tall ladder and suffered a fracture of the anterior cranial fossa of his skull. A watery, blood-tinged fluid dripped from his right nostril. Several days later, Harry complained that he could no longer smell. What nerve was damaged in the fall?

20. As Harry was falling off the ladder, he reached out and grabbed a tree branch with his right hand. Unfortunately, he had already fallen so far that he quickly lost his grip and kept falling. His right upper limb became weak and numb. What major nervous structure had he injured?

21. Frita, a woman in her early 70s, was having problems chewing. She was asked to stick out her tongue. It deviated to the right, and its right half was quite wasted. What nerve was injured?

22. Ted is a war veteran who was hit in the back with small pieces of a bomb. His skin is numb in the center of his buttocks and along the entire posterior side of a lower limb, but there is no motor problem. One of the following choices is the most likely site of his nerve injury. Choose the best answer, and explain your choice: (a) a few dorsal roots of the cauda equina, (b) spinal cord transection at C_6, (c) spinal cord transection at L_5, (d) femoral nerve transected in the lumbar region.

23. Jefferson, a football quarterback, suffered torn menisci in his right knee joint when he was tackled from the side. The same injury crushed his common peroneal nerve against the neck of his right fibula. What locomotory problems could Jefferson expect to experience from this nerve injury?

14

The Autonomic Nervous System and Visceral Sensory Neurons

Student Objectives

1. Define the autonomic nervous system (ANS), and explain its relationship to the peripheral nervous system as a whole.

Introduction to the Autonomic Nervous System (pp. 393–396)

2. Compare autonomic neurons to somatic motor neurons.

3. Describe the basic differences between the parasympathetic and sympathetic divisions of the autonomic nervous system.

The Parasympathetic Division (pp. 396–398)

4. Describe the anatomy of the parasympathetic division, and explain how it relates to the brain, cranial nerves, and sacral spinal cord.

The Sympathetic Division (pp. 398–404)

5. Describe the anatomy of the sympathetic division, and explain how it relates to the spinal cord and spinal nerves.

6. Explain the sympathetic function of the adrenal medulla.

Central Control of the Autonomic Nervous System (p. 405)

7. Explain how various regions of the CNS help to regulate the autonomic nervous system.

Visceral Sensory Neurons (pp. 405–406)

8. Describe the role and location of visceral sensory neurons relative to autonomic neurons.

9. Explain the concept of referred pain.

Visceral Reflexes (p. 406)

10. Explain how spinal and peripheral reflexes regulate some functions of visceral organs.

Pathology of the Autonomic Nervous System (pp. 406–407)

11. Briefly describe some diseases of the autonomic nervous system.

The Autonomic Nervous System Throughout Life (pp. 407–408)

12. Describe the embryonic development of the autonomic nervous system.

13. List some effects of aging on autonomic functions.

You wake up at night after having eaten at a restaurant where the food did not taste quite right. You find yourself waiting helplessly for your stomach to "decide" whether it can hold the food down. A few days later, you drive to work after drinking too much coffee and wish in vain that your full bladder would stop its uncomfortable contractions. Later that day, your bosses challenge a business decision you made, and you try not to let them see you sweat—but the sweat runs down your face anyway. All of these are examples of visceral functions that are not easily controlled by the conscious will and sometimes seem as if they had a "mind of their own." These functions are performed by the **autonomic nervous system (ANS),** a motor system that does indeed operate with a certain amount of functional independence (*autonomic* = "self-governing").

The ANS is the system of motor neurons that innervate the smooth muscle, cardiac muscle, and glands of the body. By controlling these effectors, the ANS regulates such visceral functions as heart rate, blood pressure, digestion, and urination. These functions are essential for maintaining the stability of the body's internal environment. The ANS is the *general visceral motor* division of the peripheral nervous system, according to the classification of nervous outputs

we have been using (see Figure 14.1). As shown in the figure, the ANS is distinct from the general *somatic* motor division (which innervates the somatic skeletal muscles) and the special visceral motor division (which innervates the pharyngeal arch muscles).

Although we will focus on autonomic (general visceral motor) functions, we will also consider the **general visceral senses.** The general visceral sensory system continuously monitors the activities of the visceral organs so that the autonomic motor neurons can make adjustments as necessary to ensure optimal performance of visceral functions.

Introduction to the Autonomic Nervous System

Comparison of the Autonomic and Somatic Motor Systems

Our previous discussions have focused largely on the *somatic* motor system, which innervates skeletal muscles. Recall that each somatic motor neuron runs from the central nervous system all the way to the

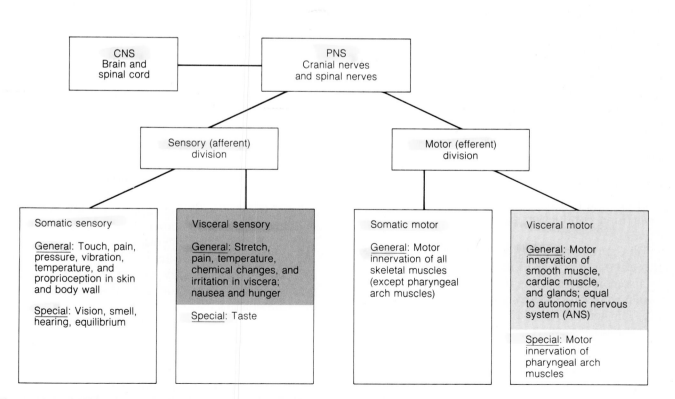

Figure 14.1   Position of the general visceral motor system (automonic nervous system), and the general visceral sensory system in the nervous system hierarchary.

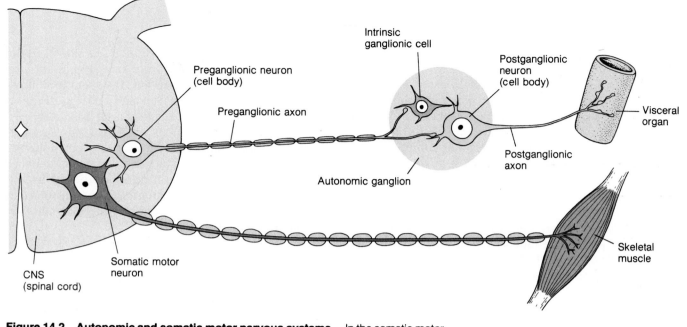

Figure 14.2   Autonomic and somatic motor nervous systems.   In the somatic motor system (red neurons), single neurons extend from the CNS to skeletal muscle cells. In the autonomic nervous system (yellow neurons), a chain of two motor neurons, preganglionic and postganglionic, runs between the CNS and the visceral effector cells.

muscle being innervated (Figure 14.2). Each motor unit consists of a single neuron, plus the skeletal muscle cells it innervates. Typical somatic motor axons are thick, heavily myelinated fibers that conduct nerve impulses rapidly.

By contrast, the comparable motor unit in the ANS includes a chain of *two* motor neurons (Figure 14.2). The first of these is called a **preganglionic neuron.** Its cell body lies within the spinal cord or brain. Its axon, the **preganglionic axon,** synapses with the second motor neuron, the **postganglionic neuron,** in a peripheral **autonomic ganglion.** The **postganglionic axon** then extends to the effector organ. The preganglionic axons are thin, lightly myelinated fibers, whereas the postganglionic axons are even thinner and are unmyelinated. Consequently, the conduction of impulses through the autonomic nervous system is slower than conduction through the somatic motor system.

The autonomic ganglia are *motor* ganglia containing the cell bodies of motor neurons. These ganglia also contain short neurons that are analogous to interneurons, called **intrinsic ganglionic cells** (Figure 14.2). The general function of these intrinsic neurons is to turn off the activity of a postganglionic neuron after it has fired.

We have established that two-neuron motor units are unique to the ANS. Can you deduce whether the visceral motor innervation of the *pharyngeal arch muscles* (chewing and swallowing muscles

and so on) is through two-neuron chains or by single motor neurons?

The innervation of the pharyngeal arch muscles is not autonomic (general visceral motor), but special visceral motor (Figure 14.1). Therefore, these special skeletal muscles are innervated by single motor neurons—just like the general skeletal muscles of the body. ■

Divisions of the Autonomic Nervous System

The ANS has two divisions, the *sympathetic* and *parasympathetic* (par"ah-sim"pah-thet'ik). These are shown issuing from the brain and spinal cord in Figure 14.3. Both divisions generally innervate the same visceral organs, but cause opposite effects: Where one division stimulates some smooth muscle to contract or a gland to secrete, the other division inhibits that action. Additionally, whereas the sympathetic division mobilizes the body during extreme situations (such as fear, exercise, or rage), the parasympathetic division enables us to unwind and rest and works to conserve body energy. In this way, the two divisions counterbalance each other to keep our body systems running smoothly. Let us now elaborate on these functional differences.

The **sympathetic division** is the "fight, fright, or flight" system. Its activity is evident during vigorous exercise, excitement, or emergency situations, such as

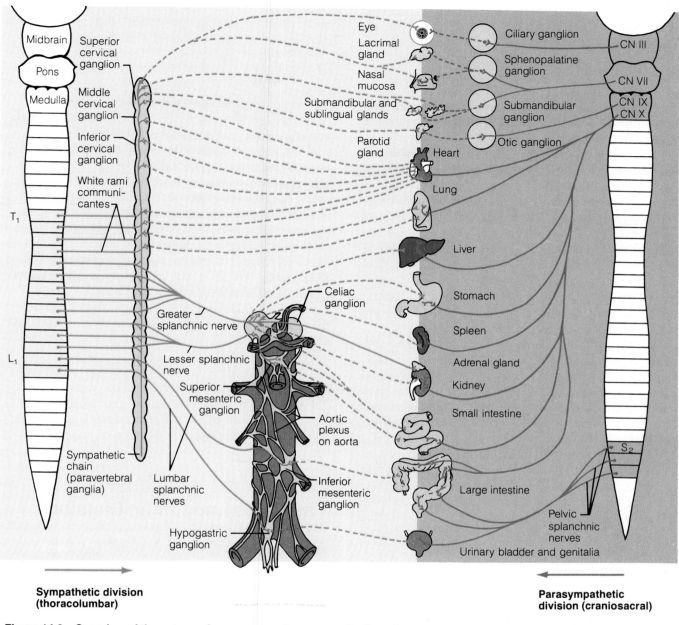

Figure 14.3   Overview of the autonomic nervous system, sympathetic and parasympathetic divisions, issuing from the brain and spinal cord.   Solid lines indicate preganglionic axons, while broken dotted lines indicate postganglionic axons. (Although there are four sympathetic lumbar splanchnic nerves, only two are illustrated for simplicity.) Note: CN = cranial nerve.

coming upon street toughs at night. A pounding heart, fast deep breathing, dilated eye pupils, and cold, sweaty skin are signs of mobilization of the sympathetic division. All of these symptoms help us survive danger: The increased heart rate and breathing rate deliver more blood and oxygen to the skeletal muscles used for fighting or running; widened pupils let in more light for clearer vision; and cold skin indicates that blood is being diverted from the integument toward more vital organs, such as the brain. Furthermore, the small air tubes in the lungs (bronchioles) dilate, increasing the uptake of oxygen; oxygen

consumption by the body's cells increases; and the liver releases more sugar into the blood, thereby providing for the increased energy needs of the cells. In this way, our "motors are revved" for vigorous activity. Temporarily nonessential functions, such as digestion and motility of the urinary tract, are inhibited: When you are running from a mugger, digesting lunch can wait.

The sympathetic division also innervates the blood vessels. Sympathetic input *constricts* most blood vessels by signaling the smooth muscle in their walls to contract, a mechanism called **vasoconstric-**

tion. In selected cases—for example, some blood vessels in the skeletal muscles—sympathetic input relaxes the muscular walls, so that the vessel *dilates* (widens). Sympathetic input constricts many more vessels than it dilates, so the overall effect is constriction, which causes the heart to work harder to pump blood around the vascular circuit. Therefore, sympathetic activity causes the blood pressure to rise during excitement and stress.

Unlike the sympathetic division, the **parasympathetic division** is most active when the body is at rest. This division is chiefly concerned with conserving body energy even as it directs vital "housekeeping" activities such as digestion and elimination of feces and urine. The "buzz-words" to remember are "resting and digesting." Parasympathetic function is best illustrated by a person who is relaxing after dinner and reading the newspaper. Heart rate and respiratory rates are at low-normal levels, and the gastrointestinal tract is digesting food. The pupils are constricted as the eye is focused for close vision.

As we explore the sympathetic and parasympathetic divisions in detail, we will refer to Table 14.1, p. 403, which lists their effects on individual visceral organs. The list seems long, but the effects are very easy to learn if you just remember "fight, fright, or flight" (sympathetic) versus "resting and digesting" (parasympathetic).

While it is easiest to think of the sympathetic and parasympathetic divisions as working independently of each other and in an all-or-none fashion, this is rarely the case. A dynamic antagonism exists between the two divisions, and they balance each other during times when we are neither highly excited nor completely at rest.

Along with these functional differences, there are anatomical and biochemical differences between the sympathetic and parasympathetic divisions. First, they issue from different regions of the CNS (Figure 14.3). The sympathetic is the **thoracolumbar division;** that is, its fibers come from the thoracic and superior lumbar parts of the spinal cord. The parasympathetic is the **craniosacral division:** Its axons emerge from the brain *(cranial part)* and the sacral spinal cord *(sacral part)*.

Another anatomical difference between the two divisions is shown in Figure 14.4: The sympathetic pathways have short preganglionic fibers and long postganglionic fibers, whereas the opposite is true for parasympathetic pathways. Therefore, all sympathetic ganglia lie near the spinal cord and vertebral column, whereas all parasympathetic ganglia lie far from the CNS, in or near the organs innervated.

Still another anatomical difference between the two divisions is that sympathetic axons branch profusely, while parasympathetic fibers do not (Figure 14.4). Such extensive branching allows each sympathetic neuron to influence a number of different visceral organs and enables many organs to mobilize simultaneously during the "fight, flight, or fright" response. Indeed, the literal translation of *sympathetic* ("experienced together") reflects the bodywide mobilization it provides. Parasympathetic effects, by contrast, are more localized and discrete.

There are biochemical differences between the two divisions of the ANS; that is, their postganglionic axons release different neurotransmitters (Figure 14.4). In the sympathetic division, the *post*ganglionic neurotransmitter is generally *norepinephrine* (also called *noradrenaline*), whereas in the parasympathetic division, the neurotransmitter is *acetylcholine* (ACh). At *pre*ganglionic axon terminals, the neurotransmitter is always acetylcholine—in both sympathetic and parasympathetic divisions. Neurons and nerve endings that produce acetylcholine are called **cholinergic** (ko″lin-er′jik). Those that produce noradrenaline are **adrenergic** (ad″ren-er′jik).

Not all postganglionic fibers in the sympathetic division are adrenergic: Those that innervate the sweat glands and the blood vessels in skeletal muscles are cholinergic (see Table 14.1).

The main differences between the parasympathetic and sympathetic divisions are summarized in Table 14.2 on p. 404. We will now explore the anatomical organization of each division, starting with the somewhat simpler parasympathetic division.

The Parasympathetic Division

The *cranial* part of the parasympathetic division innervates organs in the head, neck, thorax, and most of the abdomen (Figure 14.3). The *sacral* part supplies the rest of the abdominal organs and the pelvic organs.

Cranial Outflow

The cranial parasympathetic outflow is contained in several cranial nerves (Figure 14.3). More specifically, the preganglionic fibers run in the oculomotor, facial, glossopharyngeal, and vagus nerves. The preganglionic cell bodies occur in motor cranial nerve nuclei in the gray matter of the brain stem (shown in Figure 12.18c, p. 336). The precise locations of both the preganglionic and postganglionic neurons of the cranial parasympathetic pathways are described next.

Oculomotor Nerve (III)

The parasympathetic fibers of the oculomotor nerve innervate smooth muscles in the eye that cause the pupil to constrict and the lens of the eye to bulge—

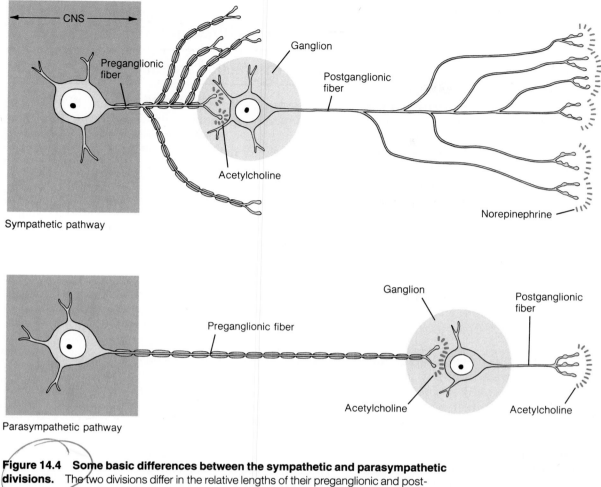

Figure 14.4 Some basic differences between the sympathetic and parasympathetic divisions. The two divisions differ in the relative lengths of their preganglionic and postganglionic axons, in the degree of axonal branching, and in the neurotransmitters released from the postganglionic axon terminals (acetylcholine and norepinephrine).

actions that allow focusing on close objects in the field of vision. The preganglionic axons found in the oculomotor nerve issue from cell bodies in the accessory *oculomotor nucleus* in the midbrain. The postganglionic cell bodies are in the **ciliary ganglion,** which lies in the posterior part of the orbit, just lateral to the optic nerve (Table 13.2, p. 372).

Facial Nerve (VII)

The parasympathetic fibers of the facial nerve stimulate the secretion of many glands in the head, including two salivary glands inferior to the mouth, called the submandibular and sublingual glands; the lacrimal (tear) gland above the eye; and mucus-secreting glands in the nasal cavity. In the pathway leading to the lacrimal and nasal glands, the preganglionic neurons originate in the *lacrimal nucleus* in the pons and synapse with postganglionic neurons in the **sphenopalatine ganglion,** just posterior to the maxilla (Table 13.2, p. 374). In the pathway leading to the submandibular and sublingual glands, the preganglionic neurons originate in the *superior salivatory nucleus* in

the pons and synapse with postganglionic neurons in the **submandibular ganglion,** deep to the mandibular angle.

Glossopharyngeal Nerve (IX)

The parasympathetic fibers of the glossopharyngeal nerve stimulate secretion of a large salivary gland called the parotid gland, which lies anterior to the ear. The preganglionic neurons originate in the *inferior salivatory nucleus* in the medulla and synapse with postganglionic neurons in the **otic ganglion** inferior to the foramen ovale of the skull (Table 13.2, p. 375).

The three cranial nerves considered so far (III, VII, IX) supply the entire parasympathetic innervation of the head. However, it is important to note that only *pre*ganglionic fibers run within these three nerves. Postganglionic fibers do not: This is because the distal ends of the preganglionic fibers "jump" over to branches of the *trigeminal* nerve (V), and then the postganglionic fibers travel in the trigeminal to their final destinations. This "hitchhiking" probably reflects the fact that the trigeminal nerve has the widest distribution within the face and thus is best suited for this "delivery" role.

Vagus Nerve (X)

Parasympathetic fibers from one additional cranial nerve, the vagus nerve, innervate the visceral organs of the thorax and most of the abdomen (Figure 14.3). The vagus nerve is an important part of the ANS: It contains nearly 90% of the preganglionic parasympathetic fibers in the body. Functionally, the parasympathetic fibers in the vagus nerve bring about typical "resting and digesting" activities. For example, they stimulate digestion, slow the heart rate, and constrict the bronchi in the lungs. The preganglionic cell bodies are mostly in the *dorsal motor nucleus* in the medulla, and the preganglionic axons run the entire length of the vagus nerve. It is interesting to note that most postganglionic neurons are confined within the walls of the organs being innervated. Their cell bodies form *intramural* (in"trah-mu'ral) *ganglia* (Latin for "within the walls").

Because the vagus nerve is essential to the functioning of so many organs, we will briefly follow its path through the body (Figure 14.5). As the vagus descends through the neck and trunk, it sends branches through many **autonomic nerve plexuses** to the organs being innervated. Specifically, it sends branches through the **cardiac plexus** to the heart, through the **pulmonary plexus** to the lungs, through the **esophageal plexus** to the esophagus, into the stomach wall, and through the **celiac plexus** and **superior mesenteric plexus** to the other abdominal organs (intestines, liver, pancreas, and so on).

Sacral Outflow

The sacral part of the parasympathetic outflow comes from the S$_2$–S$_4$ segments of the sacral spinal cord (Figure 14.3). It supplies the organs in the pelvis, including the distal half of the large intestine, the bladder, the reproductive organs (the uterus, for example), and the erectile tissues of the external genitalia. Parasympathetic effects on these organs include stimulation of defecation, voiding of urine, and erection (Table 14.1).

The preganglionic cell bodies of the sacral parasympathetics lie in the visceral motor region of the spinal gray matter (Figure 12.27, p. 348). The axons of these preganglionic neurons run in the ventral roots to the ventral rami, from which they branch to form **pelvic splanchnic nerves** (Figure 14.3). These nerves run through an autonomic plexus in the pelvic floor, the **inferior hypogastric plexus,** or *pelvic plexus* (Figure 14.5). Some of the preganglionic fibers synapse in ganglia in this plexus, but most synapse in intramural ganglia in the pelvic organs.

The Sympathetic Division

Basic Orientation and Structures

The sympathetic division issues from the thoracic and superior lumbar part of the spinal cord, from T$_1$ to L$_2$ (Figure 14.3). Its preganglionic cell bodies lie in the visceral motor region of the spinal gray matter, where they form the lateral gray horn (Figure 12.26b, p. 347).

The sympathetic division is more complex than the parasympathetic division, in part because it innervates more organs. It supplies not only all the visceral organs in the internal body cavities but also all visceral structures in the superficial regions of the body: the sweat glands, the hair-raising arrector pili muscles of the skin, and the smooth musculature in the walls of arteries and veins. It is easy to remember that the sympathetic system innervates these structures: Everyone knows we sweat under stress, our hair stands up in terror, and our blood pressure skyrockets (because of widespread vasoconstriction) when we get excited. The *parasympathetic* division does *not* innervate sweat glands, arrector pili, or (with minor exceptions) blood vessels.

The sympathetic division is also more complex because it has more ganglia. These fall into two classes: (1) *chain,* or *paravertebral, ganglia* and (2) *prevertebral,* or *collateral, ganglia.*

Chain Ganglia

The many **chain ganglia** line up vertically along both sides of the vertebral column, from the neck to the pelvis (Figures 14.5 and 14.6). Successive chain ganglia are interconnected by short nerves into long **sympathetic trunks** *(sympathetic chains).* Each sympathetic trunk resembles a strand of beads. Posteriorly, the chain ganglia attach to the ventral rami of the spinal nerves (Figure 14.6a). There is approximately one chain ganglion for each spinal nerve. However, the number of chain ganglia and spinal nerves is not identical, because some adjacent ganglia fuse during development. Such fusion is most evident in the neck region, where there are 8 spinal nerves but only 3 chain ganglia: the **superior, middle,** and **inferior cervical ganglia** (Figure 14.3). Furthermore, the inferior cervical ganglion usually fuses with the first thoracic ganglion to form the **stellate** ("star-shaped") **ganglion** in the superior thorax (Figure 14.5). Overall, there are 22–24 chain ganglia per side, and a typical person may have 3 cervical, 11 thoracic, 4 lumbar, 4 sacral, and 1 coccygeal ganglia.

The locations of the chain ganglia should be described more precisely for those students who must identify them in the laboratory: The cervical ganglia

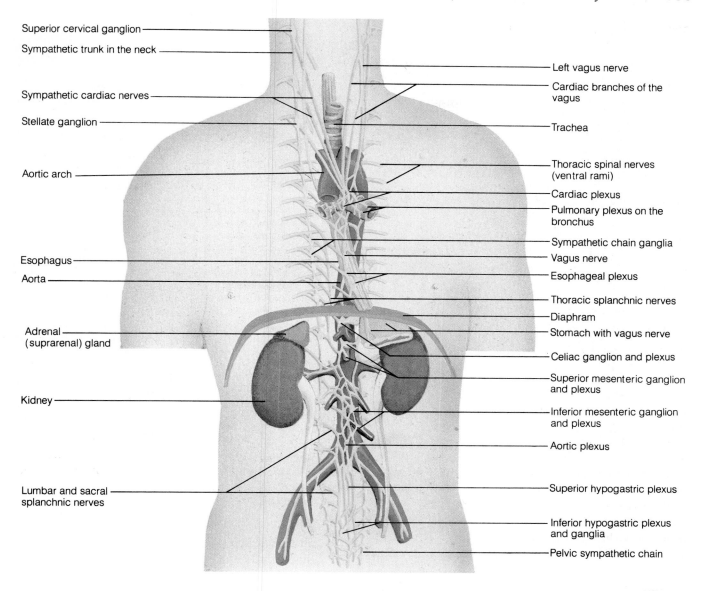

Superior cervical ganglion

Sympathetic trunk in the neck

Sympathetic cardiac nerves

Stellate ganglion

Aortic arch

Esophagus

Aorta

Adrenal (suprarenal) gland

Kidney

Lumbar and sacral splanchnic nerves

Left vagus nerve

Cardiac branches of the vagus

Trachea

Thoracic spinal nerves (ventral rami)

Cardiac plexus

Pulmonary plexus on the bronchus

Sympathetic chain ganglia

Vagus nerve

Esophageal plexus

Thoracic splanchnic nerves

Diaphram

Stomach with vagus nerve

Celiac ganglion and plexus

Superior mesenteric ganglion and plexus

Inferior mesenteric ganglion and plexus

Aortic plexus

Superior hypogastric plexus

Inferior hypogastric plexus and ganglia

Pelvic sympathetic chain

Figure 14.5   The vagus nerve, the autonomic nerve plexuses, and the autonomic ganglia throughout the body.   All autonomic plexuses are shared by both parasympathetic and sympathetic fibers, but the ganglia in these plexuses are almost exclusively sympathetic. Also note the sympathetic chain ganglia.

lie just anterior to the transverse processes of the cervical vertebrae; thoracic ganglia lie on the heads of the ribs; lumbar ganglia lie on the anterolateral sides of the vertebral bodies; and sacral ganglia lie medial to the sacral foramina.

Take care not to confuse the sympathetic chain ganglia with the *dorsal root ganglia.* Recall that dorsal root ganglia are sensory and lie along the dorsal roots in the intervertebral foramina, whereas the chain ganglia are motor and lie anterior to the ventral rami. Figure 14.7 shows the different locations of these two types of ganglia.

Prevertebral Ganglia

The **prevertebral ganglia** differ from the chain ganglia in several ways: (1) They are not paired and are not segmentally arranged; (2) they are confined to the abdomen and pelvis; and (3) they all lie anterior to the vertebral column (hence the name *prevertebral*), mostly on the abdominal aorta. The main prevertebral ganglia are the *celiac, superior mesenteric, inferior mesenteric,* and *inferior hypogastric ganglia* (Figures 14.3 and 14.5). These ganglia lie within the autonomic nerve plexuses of the same names.

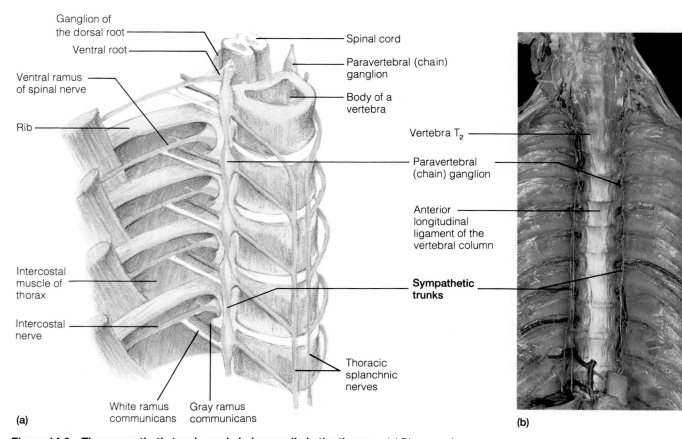

Figure 14.6   The sympathetic trunks and chain ganglia in the thorax.   (a) Diagram of the right sympathetic trunk in the posterior thorax, showing its relationship to the vertebral column. (b) In this cadaver photograph, the organs in the thorax have been removed to reveal the sympathetic trunks and chain ganglia on the posterior thoracic wall.

Sympathetic Pathways

Now that some basic parts of the sympathetic division have been introduced, we can examine how the preganglionic and postganglionic neurons fit into these parts (Figure 14.7). The preganglionic neurons, whose cell bodies lie in the spinal cord, send their motor axons out through the ventral root. These axons enter the spinal nerve and the first part of the ventral ramus. From there, they pass into the nearest chain ganglion by running through a **white ramus communicans** (plural: **rami communicantes** [kŏ-mu′nĭ-kahn-tās]). Once a preganglionic axon reaches a chain ganglion, it does one of several things:

1. It can synapse with a postganglionic neuron in the same chain ganglion (see neuron a in Figure 14.7).

2. It can pass through the chain ganglion without synapsing and run in a **splanchnic nerve** to synapse in a prevertebral ganglion (see neuron c in Figure 14.7).

3. The preganglionic fiber can ascend or descend in the sympathetic trunk and synapse in a chain gan-

glion or prevertebral ganglion at a different level of the body (see neuron b in Figure 14.7). Such ascending and descending in the sympathetic trunk allows sympathetic outputs, which come only from the thoracolumbar region, to supply the superior and inferior body regions, such as the head and pelvis. For example, the cervical ganglia are reached in this way.

Examples of each of these pathways will be given as we explore the innervation of specific body regions. Sympathetic pathways to the body periphery will be considered first, followed by the pathways to the head, thorax, abdomen, and pelvis.

Pathways to the Body Periphery

Sympathetic pathways to the body periphery innervate the sweat glands, arrector pili, and peripheral blood vessels. These pathways are represented by both (a) and (b) in Figure 14.7. In these pathways, the preganglionic fibers enter the chain ganglia and synapse with postganglionic cell bodies. From the chain ganglion, the postganglionic axons travel in a **gray**

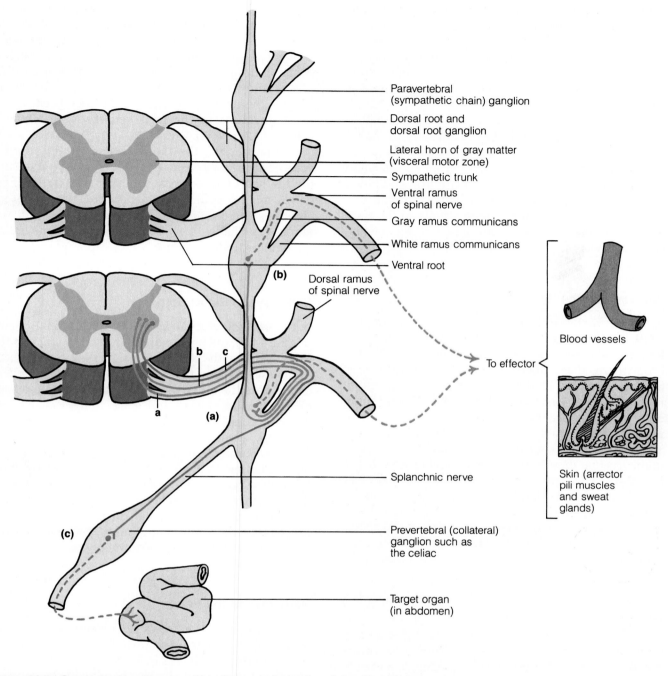

Figure 14.7   Sympathetic pathways.   This diagram indicates the relationship of the sympathetic neurons to the spinal cord, sympathetic trunks, and associated structures. Preganglionic neurons in the spinal cord, labeled a–c, send fibers to the chain ganglia. There, these fibers may synapse with a postganglionic neuron in a chain ganglion (neuron a) or may pass through to synapse in a prevertebral ganglion (neuron c). Also, a preganglionic fiber may ascend or descend in the sympathetic trunk (neuron b) before synapsing in a chain or prevertebral ganglion.

ramus communicans to the ventral ramus. The axons then follow branches of the ventral (and dorsal) ramus to the skin, where they supply the arrector pili and sweat glands. Anywhere along this path, the postganglionic axons can "jump onto" nearby blood vessels—then follow and innervate those vessels.

The Rami Communicantes

We have seen that the sympathetic chain ganglia connect to the ventral rami by two types of communicating rami: The *white rami communicantes* contain all the preganglionic fibers traveling to the chain ganglion, while the *gray rami communicantes* carry the postganglionic fibers headed for peripheral structures (Figure 14.7). White rami are white because preganglionic fibers are myelinated, and gray rami are gray because postganglionic fibers are unmyelinated. Note that gray rami lie *medial* to white rami, not lateral as one might expect.

Since the white rami carry the sympathetic outflow from the spinal cord, they occur only in the region of such outflow—on chain ganglia between T_1 and L_2. Gray rami, by contrast, occur on all the chain ganglia because sweat glands, arrector pili, and blood vessels must be innervated in *all* body segments.

Rami communicantes belong strictly to the sympathetic division: They are not part of the parasympathetic division.

Pathways to the Head

In the sympathetic innervation of the head (Figure 14.3), the preganglionic fibers originate in the first four thoracic segments of the spinal cord (T_1–T_4). From there, these fibers ascend in the sympathetic trunk to synapse in the superior cervical ganglion. From this ganglion, the postganglionic fibers "jump onto" large arteries that carry them to glands, smooth muscle, and vessels throughout the head. Functionally, these sympathetic fibers (1) stimulate the muscle in the iris that dilates the eye pupil and (2) inhibit the lacrimal, nasal, and salivary glands (the reason fear causes dry mouth). The sympathetic fibers also supply the *superior tarsal muscle,* a smooth muscle in the upper eyelid that prevents drooping of the eyelid whenever the eyes are open. This muscle also lifts the eyelid to open the eyes wide when one is frightened.

Can you describe how paralysis of the superior tarsal muscle would change a person's appearance?

Paralysis of this muscle leads to a permanent drooping of the upper eyelid, a condition called **ptosis** (to′sis; "falling"). Ptosis can reflect damage to the superior part of the sympathetic trunk as well as damage to the sympathetic nerves in the head. ■

Pathways to the Thorax

In the sympathetic innervation of thoracic organs, the preganglionic fibers originate at spinal levels T_1–T_6 (Figure 14.3). Some of these fibers synapse in the nearest chain ganglion, and the postganglionic fibers run directly to the organ being supplied. Fibers to the lungs and esophagus take this direct route, as do some fibers to the heart. Along the way, the postganglionic fibers contribute to the pulmonary, esophageal, and cardiac plexuses (Figure 14.5).

Many of the sympathetic fibers to the *heart,* however, take a less direct pathway. The preganglionic fibers ascend in the sympathetic trunk to synapse in the cervical chain ganglia (Figure 14.3). From there, the postganglionic fibers descend through the cardiac plexus and into the heart wall. Many nerves to the heart (a thoracic organ) arise in the neck because the heart develops in the neck region of the embryo (Chapter 17, p. 473).

Functionally, the thoracic sympathetic nerves speed heart rate and dilate the blood vessels that supply the heart wall (so that the heart muscle itself receives more blood). They also dilate the respiratory air tubes and inhibit the muscle and glands in the esophagus (Table 14.1). These effects are integral to the fight-or-flight response.

Pathways to the Abdomen

In the sympathetic innervation of abdominal organs (Figure 14.3), the preganglionic fibers originate in the inferior half of the thoracolumbar spinal cord (T_5–L_2). From there, these fibers pass through the adjacent chain ganglia and travel in **thoracic splanchnic nerves** (greater, lesser, and least) to synapse in prevertebral ganglia in the large plexuses on the abdominal aorta. These ganglia include the **celiac and superior mesenteric ganglia,** along with some smaller ones (Figures 14.3 and 14.5). Postganglionic fibers from these ganglia then follow the main branches of the aorta to the stomach, liver, kidney, spleen, and the intestines (not the distal half of the large intestine, however). For the most part, the sympathetic fibers *inhibit* the activity of these visceral organs (Table 14.1).

Pathways to the Pelvis

In the sympathetic innervation of pelvic organs, the preganglionic fibers originate in the most inferior part of the thoracolumbar spinal cord (T_{10}–L_2), then descend in the sympathetic trunk to the lumbar and sacral chain ganglia. Some fibers synapse there, and the postganglionic fibers run in *lumbar* and *sacral splanchnic nerves* to plexuses on the lower aorta and in the pelvis, namely, the **inferior mesenteric plexus, aortic plexus,** and the **hypogastric plexuses** (Figure 14.5). Other preganglionic fibers, by contrast, pass directly to these autonomic plexuses and synapse in prevertebral ganglia there—the **inferior mesenteric ganglia** and **inferior hypogastric ganglia.** Postganglionic fibers proceed from these plexuses to the pelvic organs, such as the bladder, reproductive organs, and the distal half of the large intestine. These sympathetic fibers inhibit urination and defecation and promote ejaculation (Table 14.1).

Table 14.1   Effects of the Parasympathetic and Sympathetic Divisions on Various Organs

Target organ/system	Sympathetic effects	Parasympathetic effects
Eye (iris)	Stimulates dilator muscles, dilates eye pupils	Stimulates constrictor muscles; constricts eye pupils
Eye (ciliary muscle)	No effect	Stimulates muscles, which results in bulging of the lens for accommodation and close vision
Glands (nasal, lacrimal, salivary, gastric, pancreas)	Inhibits secretory activity; causes vasoconstriction of blood vessels supplying the glands	Stimulates secretory activity
Sweat glands	Stimulates copious sweating (cholinergic fibers)	No effect
Adrenal medulla	Stimulates medulla cells to secrete epinephrine and norepinephrine	No effect
Arrector pili muscles attached to hair follicles	Stimulates to contract (erects hairs and produces goose bumps)	No effect
Heart muscle	Increases rate and force of heartbeat	Decreases rate; slows and steadies heart
Heart: coronary blood vessels	Causes vasodilation	Constricts coronary vessels
Bladder/urethra	Causes relaxation of smooth muscle of bladder wall; constricts urethral sphincter; inhibits voiding	Causes contraction of smooth muscle of bladder wall; relaxes urethral sphincter; promotes voiding
Lungs	Dilates bronchioles and mildly constricts blood vessels	Constricts bronchioles
Digestive tract organs	Decreases activity of glands and muscles of digestive system and constricts sphincters (e.g., anal sphincter)	Increases motility (peristalsis) and amount of secretion by digestive organs; relaxes sphincters to allow movement of foodstuffs along tract
Liver	Epinephrine stimulates liver to release glucose to blood	No effect
Gallbladder	Inhibits (gallbladder is relaxed)	Excites (gallbladder contracts to expel bile)
Kidney	Causes vasoconstriction; decreases urine output	No effect
Penis	Causes ejaculation	Causes erection (vasodilation)
Vagina/clitoris	Causes reverse peristalsis (contraction) of vagina	Causes erection (vasodilation) of clitoris
Blood vessels	Constricts most vessels and increases blood pressure; constricts vessels of abdominal viscera and skin to divert blood to muscles, brain, and heart when necessary; dilates vessels of the skeletal muscles (cholinergic fibers) during exercise	Little or no effect
Blood coagulation	Increases coagulation	No effect
Cellular metabolism	Increases metabolic rate	No effect
Adipose tissue	Stimulates lipolysis (fat breakdown)	No effect
Mental activity	Increases alertness	No effect

The Adrenal Medulla

An **adrenal (suprarenal) gland** lies on the superior part of each kidney (Figure 14.5). The internal portion of this gland, the **adrenal medulla,** is a major organ of the sympathetic nervous system. The adrenal medulla represents the largest and most specialized of all the sympathetic ganglia, a collection of modified postganglionic neurons that completely lack nerve processes (Figure 14.8). These cells do not innervate any single structure. Instead, they secrete great quantities of excitatory hormones into the blood of nearby capillaries during the fight-or-flight response. These hormones are norepinephrine (the chemical that is secreted by other postganglionic sympathetic neurons as a neurotransmitter) and a related excitatory molecule called **epinephrine (adrenaline).** Once released, these

Figure 14.8 The adrenal medulla.
Sympathetic innervation of the cells in the adrenal medulla. These cells secrete the excitatory hormones, epinephrine and nor-epinephrine, into the blood capillaries to produce excitatory effects throughout the body.

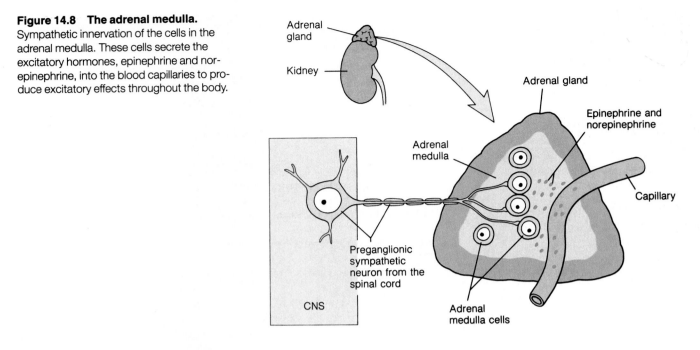

hormones travel throughout the body in the blood-stream, producing the widespread excitatory effects that we have all felt as a "surge of adrenaline."

The cells of the adrenal medulla are stimulated to secrete by the preganglionic sympathetic fibers that innervate them (Figure 14.8). These fibers arise from cell bodies in the T_6–T_{11} region of the spinal cord. From there, they run in the thoracic splanchnic nerves, pass through the celiac plexus, and reach the adrenal medulla (Figure 14.3). The adrenal medulla has a more concentrated sympathetic innervation than any other organ in the body.

Table 14.2 Anatomical and Physiological Differences Between the Parasympathetic and Sympathetic Divisions

Characteristic	Sympathetic	Parasympathetic
Origin	Thoracolumbar outflow: lateral horn of gray matter of spinal cord segments T_1–L_2	Craniosacral outflow: brain stem nuclei of cranial nerves III, VII, IX, and X; spinal cord segments S_2–S_4
Location of ganglia	Ganglia within a few cm of CNS: alongside vertebral column (paravertebral ganglia) and anterior to vertebral column (prevertebral ganglia)	Ganglia in or close to visceral organ served
Relative length of pre- and postganglionic fibers	Short preganglionic; long postganglionic	Long preganglionic; short postganglionic
Rami communicantes	Gray and white rami communicantes; white contain myelinated preganglionic fibers; gray contain unmyelinated postganglionic fibers	None
Degree of branching of preganglionic fibers	Extensive	Minimal
Functional goal	Prepares body to cope with emergencies and intense muscular activity	Maintenance functions; conserves and stores energy
Neurotransmitters	All preganglionic fibers release ACh; most postganglionic fibers release norepinephrine (adrenergic fibers); some postganglionic fibers (e.g., those serving sweat glands and blood vessels of skeletal muscles) release ACh; neurotransmitter activity augmented by release of adrenal medullary hormones (norepinephrine and epinephrine)	All fibers release ACh (cholinergic fibers)

Central Control of the Autonomic Nervous System

Although the ANS is not considered to be under direct voluntary control, its performance is nevertheless regulated by the central nervous system. Several levels of central control exist—by the brain stem and spinal cord, hypothalamus, and cerebral cortex.

Control by the Brain Stem and Spinal Cord

The reticular formation of the brain stem appears to exert the most direct influence over autonomic functions. Centers in the medulla oblongata (p. 337) regulate heart rate *(cardiac centers),* the diameter of blood vessels *(vasomotor center),* many types of digestive activities (*vomiting center,* for example), and respiration rate *(respiratory centers).* The pons also contains respiratory centers that interact with those in the medulla.

Control of autonomic functions at the level of the spinal cord involves the spinal visceral reflexes (p. 309). Note, however, that although the defecation and urination reflexes are integrated by the spinal cord, they are subject to conscious inhibition from the brain.

Control by the Hypothalamus

The integration center of the autonomic nervous system is the hypothalamus. In general, the medial and anterior parts of the hypothalamus direct parasympathetic functions, while lateral and posterior parts direct sympathetic functions. These hypothalamic centers exert their effects indirectly via relays through the reticular formation that, in turn, influence the preganglionic autonomic neurons in the spinal cord and brain. It is through the ANS that the hypothalamus controls heart activity, blood pressure, body temperature, and digestive functions. Recall from Chapter 12 (p. 339) that the emotional part of the cerebrum (limbic lobe) has strong links to the hypothalamus. In fact, the emotional response of the limbic lobe to danger and stress signals the hypothalamus to activate the sympathetic system in the fight-or-flight response.

Control by the Cerebral Cortex

It was once believed that the autonomic nervous system was not subject to voluntary control by the cerebral cortex. However, it now appears that people can learn to control some autonomic functions indirectly by developing extraordinary control over their emotions. For example, feelings of extreme calm (as in meditation) are associated with parasympathetic activation. Conversely, remembering a frightful experience is a way to voluntarily evoke sympathetic excitement.

Visceral Sensory Neurons

Overview of Visceral Sensory Neurons

The visceral division of the PNS contains sensory as well as motor (autonomic) neurons (Figure 14.1). General visceral sensory neurons monitor stretch, temperature, chemical changes, and irritation within the visceral organs. The brain interprets this visceral information as feelings of hunger, fullness, pain, nausea, or well-being. Almost all of the receptors for these visceral senses are free (unencapsulated) dendritic endings widely scattered throughout the visceral organs. Visceral sensations tend to be difficult to localize with precision: For example, we usually cannot distinguish whether gas pains originate in the stomach or in the intestine, or whether a pain in the lower abdomen originates from the uterus or the appendix.

Like somatic sensory neurons, the cell bodies of visceral sensory neurons are located in dorsal root ganglia (and in the sensory ganglia of cranial nerves), and their central processes reach the CNS through the dorsal roots. The long peripheral fibers of these sensory neurons accompany the autonomic motor fibers to the visceral organs. For example, many visceral sensory fibers accompany the *parasympathetic* fibers in the vagus nerve and monitor visceral sensations in the many organs that are served by this nerve. Other visceral sensory fibers accompany the *sympathetic* fibers along the following route: They run from the visceral organs into the autonomic plexuses and then through the splanchnic nerves, sympathetic trunk, rami communicantes, spinal nerves, and dorsal roots to enter the spinal cord. *All visceral pain fibers follow this sympathetic route to the CNS.*

The pathways by which visceral sensory information is relayed through the spinal cord to the cerebral cortex are just now being worked out. Basically, visceral inputs travel along the spinothalamic (and other) pathways to the thalamus. Neurons in the thalamus then relay this information to the cerebral cortex for conscious perception. Visceral sensory information also reaches the visceral control centers in the hypothalamus and medulla oblongata.

Visceral *pain* exhibits an unusual feature that is worth noting. Surprisingly, one often feels no pain when a visceral organ is cut or scraped. When pieces of mucous membrane are snipped from the uterus or

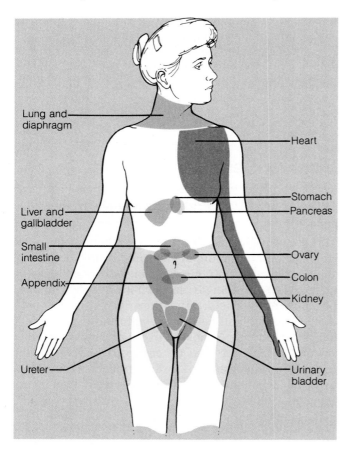

Lung and diaphragm

Heart

Liver and gallbladder

Small intestine

Appendix

Ureter

Stomach

Pancreas

Ovary

Colon

Kidney

Urinary bladder

Figure 14.9 Referred pain. A referred pain map showing the anterior cutaneous areas to which pain is referred from certain visceral organs.

Visceral Reflexes

Visceral sensory and autonomic neurons participate in **visceral reflex arcs.** Examples of visceral reflexes are the defecation reflex, in which the rectum is stretched by feces and the large intestine responds by contracting, and the micturition reflex, in which the bladder contracts after filling with urine. Many visceral reflexes are simple spinal reflexes in which sensory neurons activate spinal interneurons, which in turn activate preganglionic autonomic neurons. Other visceral reflex arcs, however, do not involve the central nervous system at all—they are strictly *peripheral reflexes.* In some of these peripheral reflexes, branches from visceral sensory fibers synapse with postganglionic motor neurons *within sympathetic ganglia.* Furthermore, complete three-neuron reflex arcs (with small sensory, motor, and intrinsic neurons) exist *entirely within the wall of the digestive tube.* These neurons make up the *enteric nervous system* (p. 568). The peripheral reflexes carry out localized autonomic responses involving small segments of an organ or just a few different visceral organs. These reflexes allow the peripheral part of the visceral nervous system to control some of its own activity, making it partly independent of the brain and spinal cord. This fact further illustrates the general concept that the autonomic nervous system operates partly on its own.

the intestines to be examined for cancer, for example, most patients report little discomfort. Visceral pain does result, however, from chemical irritation or inflammation of visceral organs, from spasms of the smooth muscle in these organs (cramping), and from excessive stretching of the organ. In these cases, visceral pain can be severe.

Referred Pain

People suffering from visceral pain often perceive this pain as somatic in origin—as if it originated from the skin or outer body wall. This phenomenon is called **referred pain** (Figure 14.9). For example, heart attacks can produce a sensation of pain in the superior thoracic wall and the medial aspect of the left arm. The cause of referred pain is not fully understood. However, it is known that both the affected organ and the region of the body wall to which the pain is referred are innervated by the same spinal segments. (For example, both the heart and the skin area to which heart pain projects are innervated by sensory neurons from T_1 to T_5.)

Pathology of the Autonomic Nervous System

Since the ANS is involved in nearly every important process that occurs within the body, it is not surprising that abnormalities of autonomic functioning can have far-reaching effects. Such abnormalities can impair blood delivery and elimination processes and can threaten life itself.

Raynaud's disease is characterized by intermittent attacks, during which the skin of the fingers and toes becomes pale, then blue and painful. Commonly provoked by exposure to cold, this disease is thought to be an exaggerated vasoconstriction response in the affected body regions. The severity of this disease ranges from mere discomfort to such extreme constriction of vessels that gangrene (tissue death) results. Raynaud's disease is treated with drugs that inhibit vasoconstriction. In severe cases, however, cutting the preganglionic sympathetic fibers serving the affected region is necessary. The involved vessels then dilate, reestablishing adequate blood delivery.

Can you determine why many people with high-pressure jobs and stressful life-styles develop high blood pressure?

Hypertension, or high blood pressure, can result from overactive sympathetic vasoconstriction promoted by continual stress. Hypertension is always serious because (1) it increases the work load on the heart, possibly precipitating heart disease and (2) it increases the wear and tear on the artery walls. Stress-induced hypertension is treated with drugs that prevent the muscle cells in the walls of blood vessels from binding with norepinephrine and epinephrine. ∎

The **mass reflex reaction,** a massive uncontrolled activation of both autonomic and somatic motor neurons, affects quadriplegics and paraplegics with spinal cord injuries above the level of T_6. The cord injury is followed by a temporary loss of all reflexes inferior to the level of the injury. When reflex activity returns, it is exaggerated because of the lack of inhibitory input from higher (brain) centers. Episodes of mass reflex begin, involving surges of motor output from large regions of the spinal cord. The usual trigger for such an episode is a strong stimulus to the skin or overfilling of a visceral organ, such as the bladder. During the mass reflex episode, the body goes into flexor spasms, the limbs move wildly, the colon and bladder empty, and profuse sweating begins. Most seriously, sympathetic activity raises the blood pressure to life-threatening levels. The precise mechanism of the mass reflex is unknown, but it has been called a type of epilepsy of the spinal cord.

Achalasia (ak″ah-la′ze-ah) **of the cardia** is a condition in which the esophagus loses its ability to propel swallowed food inferiorly. Additionally, the smooth muscle surrounding the inferior end of the esophagus (cardiac sphincter) contracts to block the passage of food into the stomach (*achalasia* means "failure to relax"). Accumulating food stretches the esophagus to enormous width, and meals cannot be kept down. This condition usually appears in young adults and is thought to result from an insufficient number of parasympathetic postganglionic neurons in the esophagus wall.

A similar condition involves immobility of the large intestine rather than the esophagus. Can you imagine the effects of such a condition?

Congenital megacolon, or **Hirschsprung's disease,** is a birth defect in which the parasympathetic innervation of the distal region of the large intestine fails to develop normally. Feces accumulate proximal to the immobile bowel segment, greatly distending this area (*megacolon* = enlarged large intestine). The condition is corrected surgically by removing the inactive part of the infant's intestine. ∎

The Autonomic Nervous System Throughout Life

Preganglionic neurons of the autonomic nervous system develop from the neural tube, as do somatic motor neurons (p. 313). Postganglionic neurons, however, develop from the neural crest—as do the visceral sensory neurons. Note that *all neurons with cell bodies in the PNS derive from neural crest, and all neurons with cell bodies in the CNS derive from neural tube.*

In the development of the sympathetic division (Figure 14.10), some cells migrate ventrally from the neural crest to form the chain ganglia. From there, other cells migrate ventrally to form both the prevertebral ganglia on the aorta and the adrenal medulla. Next, the chain and prevertebral ganglia receive axons from spinal preganglionic neurons, and they in turn send postganglionic axons to the visceral organs.

In the development of the parasympathetic division, the postganglionic neurons also appear to derive from the neural crest. However, some authorities propose that these neurons actually originate in the central nervous system. In either case, these postganglionic parasympathetic neurons reach the visceral organs by migrating along the growing preganglionic axons.

During youth, impairments of visceral nervous function are usually due to injuries to the spinal cord

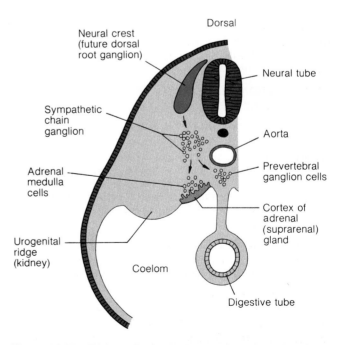

Figure 14.10   Embryonic development of some sympathetic structures, including the adrenal medulla.   *Transverse section through a 5-week embryo.*

or autonomic nerves. In old age, the efficiency of the ANS begins to decline. Elderly people are often constipated because the mobility of their gastrointestinal tract is reduced. Diminished tear formation leads to frequent eye infections, since tears contain bacteria-destroying enzymes. Whenever a young and healthy person rises to a standing position, the sympathetic division induces bodywide vasoconstriction. This response raises the blood pressure so that blood can be pumped to the head and brain. The response becomes sluggish with age, so elderly people may faint if they stand up too quickly. Although these age-related problems are distressing, they usually are not life-threatening and can be alleviated. For example, standing up more slowly gives the sympathetic nervous system time to adjust the blood pressure, eye drops are available for dry eyes, and drinking ample fluid helps alleviate constipation.

✚ Related Clinical Terms

Atonic bladder (a-tahn'ik; "without tone") A condition in which the bladder becomes flaccid and overfills, allowing urine to dribble out. Atonic bladder results from the temporary loss of the micturition reflex following injury to the spinal cord.

Horner's syndrome A condition that follows damage to the superior part of the sympathetic trunk on one side of the body. The following symptoms are seen in the head on the affected side only: drooping of the upper eyelid (ptosis), a constricted pupil, flushing of the face, and inability to sweat. Horner's syndrome usually arises from damage to the sympathetic trunk in the inferior region of the neck and can indicate a disease or infection in the neck.

Vagotomy (va-got'o-me) Cutting or severing of a vagus nerve, often done to decrease the secretion of stomach acid and other erosive digestive juices that aggravate ulcers.

Chapter Summary

1. The ANS is the system of motor neurons that innervates smooth muscle, cardiac muscle, and glands. It is the general visceral motor division of the PNS. The ANS largely operates below the level of consciousness.

INTRODUCTION TO THE AUTONOMIC NERVOUS SYSTEM (pp. 393–396)

Comparison of the Autonomic and Somatic Motor Systems (pp. 393–394)

1. In the somatic motor division of the nervous system, a single motor neuron forms the pathway from the CNS to the effectors. Autonomic motor pathways, by contrast, consist of chains of two neurons: the preganglionic neuron, whose cell body is in the CNS, and the postganglionic neuron, whose cell body is in an autonomic ganglion.

Divisions of the Autonomic Nervous System (pp. 394–396)

2. The ANS has parasympathetic and sympathetic divisions. Both innervate the same organs but produce opposite effects. The sympathetic division prepares us for "fight, fright, or flight," whereas the parasympathetic division is active during "resting and digesting." Details are listed in Table 14.1 (p. 403).

3. The parasympathetic division is craniosacral and has relatively long preganglionic axons. The sympathetic division is thoracolumbar and has relatively long postganglionic axons.

4. The two divisions differ in the neurotransmitters they release. Acetylcholine is released by parasympathetic postganglionic fibers, whereas norepinephrine is released by most sympathetic postganglionic fibers.

THE PARASYMPATHETIC DIVISION (pp. 396–398)

Cranial Outflow (pp. 396–398)

1. Cranial parasympathetic fibers arise in the brain stem nuclei of cranial nerves III, VII, IX, and X and synapse in ganglia in the head, thorax, and abdomen. Fibers in III serve smooth musculature in the eye via a relay in the ciliary ganglion. Fibers in VII serve the submandibular, sublingual, lacrimal, and nasal glands via relays in the submandibular and sphenopalatine ganglia. Fibers in IX serve the parotid gland via a relay in the otic ganglion.

2. Parasympathetic fibers in the vagus nerve (X) innervate organs in the thorax and most of the abdomen, including the heart, lungs, esophagus, stomach, liver, and most of the intestines. The fibers in the vagus are preganglionic. Almost all postganglionic neurons are located in intramural ganglia within the organ walls.

Sacral Outflow (p. 398)

3. Sacral parasympathetic preganglionic fibers issue from the visceral motor region of the gray matter of the spinal cord (S_2–S_4) and form the pelvic splanchnic nerves. Most of these fibers synapse in intramural ganglia in the pelvic viscera.

THE SYMPATHETIC DIVISION (pp. 398–404)

Basic Orientation and Structures (pp. 398–399)

1. The preganglionic sympathetic cell bodies are in the lateral horn of the spinal gray matter from the level of T_1 to L_2.

2. The sympathetic division supplies peripheral structures that the parasympathetic division does not (arrector pili, sweat glands, and the smooth muscle of blood vessels).

3. Sympathetic ganglia include (1) 22–24 pairs of chain ganglia (paravertebral ganglia), which run in sympathetic trunks on both sides of the vertebral column, and (2) unpaired prevertebral ganglia (collateral ganglia), most of which lie on the aorta in the abdomen.

4. Every preganglionic sympathetic fiber leaves the spinal cord through a ventral root and spinal nerve, then passes in a white ramus communicans to the nearest chain ganglion. There the fiber may (1) synapse with a postganglionic neuron, (2) pass through and travel in a splanchnic nerve to synapse in a prevertebral ganglion, or (3) ascend or descend in the sympathetic trunk to synapse in a ganglion at another body level.

Sympathetic Pathways (pp. 400–402)

5. White rami communicantes consist of myelinated preganglionic fibers that are headed for the chain ganglia, whereas gray rami communicantes consist of unmyelinated postganglionic fibers that are headed for the sweat glands, arrector pili, and blood vessels in the body periphery.

6. In the sympathetic pathway to the head, the preganglionic fibers synapse in the superior cervical ganglion. From there, most postganglionic fibers "jump onto" a large artery that distributes them to the glands and smooth musculature of the head.

7. In the sympathetic pathway to thoracic organs, most preganglionic fibers synapse in the nearest chain ganglion. The postganglionic fibers run directly to the organs (lungs, esophagus). Most postganglionic fibers to the heart, however, descend from the cervical chain ganglia in the neck.

8. In the sympathetic pathway to abdominal organs, the preganglionic fibers run in splanchnic nerves to synapse in prevertebral ganglia on the aorta. From these ganglia, the postganglionic fibers follow large arteries to the abdominal viscera (stomach, liver, kidney, most of intestine).

9. In the sympathetic pathway to pelvic organs, the preganglionic fibers synapse in chain ganglia or in prevertebral ganglia on the aorta, sacrum, and pelvic floor. Postganglionic fibers travel along plexuses to the pelvic organs.

The Adrenal Medulla (pp. 403–404)

10. The adrenal glands, one superior to each kidney, contain a medulla of modified postganglionic sympathetic neurons. These secrete epinephrine (adrenaline) and norepinephrine into the blood. This event is the "surge of adrenaline" felt during excitement.

11. The cells of the adrenal medulla are innervated by preganglionic sympathetic neurons, which signal them to secrete.

CENTRAL CONTROL OF THE AUTONOMIC NERVOUS SYSTEM (p. 405)

1. Visceral motor functions are regulated by the medulla oblongata, by the hypothalamus, or by the cerebral cortex acting through the hypothalamus. We can voluntarily regulate some autonomic activities by gaining extraordinary control over our emotions.

VISCERAL SENSORY NEURONS (pp. 405–406)

Overview of Visceral Sensory Neurons (pp. 405–406)

1. General visceral sensory neurons monitor temperature, pain, irritation, chemical changes, and stretch in the visceral organs.

2. Visceral sensory fibers run within the autonomic nerves, especially within the vagus and the sympathetic nerves. The sympathetic nerves carry all the pain fibers from the visceral organs of the body trunk.

3. A simplified description of the visceral sensory pathways to the brain is the following: sensory neurons to spinothalamic tract to thalamus to cerebral cortex.

4. Visceral pain is induced by stretching, infection, and cramping of internal organs but seldom by cutting or scraping these organs.

Referred Pain (p. 406)

5. Pain in visceral organs is referred to somatic regions of the body that receive innervation from the same spinal cord segments.

VISCERAL REFLEXES (p. 406)

1. Many visceral reflexes are spinal reflexes (for example, defecation and micturition reflexes). Some visceral reflexes, however, involve only peripheral neurons.

PATHOLOGY OF THE AUTONOMIC NERVOUS SYSTEM (pp. 406–407)

1. Most autonomic disorders reflect problems with the control of smooth muscle. Abnormalities in vascular control, which occur in Raynaud's disease, hypertension, and the mass reflex reaction, are most devastating.

THE AUTONOMIC NERVOUS SYSTEM THROUGHOUT LIFE (pp. 407–408)

1. Preganglionic neurons develop from the neural tube. Postganglionic neurons and visceral sensory neurons develop from the neural crest.

2. The efficiency of the ANS declines in old age: Gastrointestinal motility and production of tears decrease, and the vasomotor response to standing slows.

Review Questions

Multiple Choice and Matching Questions

1. All of the following characterize the ANS except (a) two-neuron motor chains, (b) preganglionic cell bodies in the CNS, (c) presence of neuron cell bodies in ganglia, (d) innervation of skeletal muscle.

2. Relate each of the following terms or phrases to either the sympathetic (S) or parasympathetic (P) division of the autonomic nervous system.

    _S_ **(1)** short preganglionic, long postganglionic fibers
    _P_ **(2)** intramural ganglia
    _P_ **(3)** craniosacral outflow
    _S_ **(4)** adrenergic fibers
    _S_ **(5)** cervical chain ganglia
    _P_ **(6)** otic and ciliary ganglia
    _S_ **(7)** more widespread response
    _S_ **(8)** increases heart rate, respiratory rate, and blood pressure
    _P_ **(9)** increases motility of stomach and secretion of lacrimal and salivary glands
    _S_ **(10)** innervates blood vessels
    _P_ **(11)** most active when you are swinging in a hammock
    _S_ **(12)** most active when you are competing in the Boston Marathon
    _S_ **(13)** gray rami communicantes
    _S_ **(14)** synapse in celiac ganglion

3. The thoracic splanchnic nerves contain which kind of fibers? (a) preganglionic parasympathetic, (b) postganglionic parasympathetic, (c) preganglionic sympathetic, (d) postganglionic sympathetic.

4. The prevertebral ganglia contain which kind of cell bodies? (a) preganglionic parasympathetic, (b) postganglionic parasympathetic, (c) preganglionic sympathetic, (d) postganglionic sympathetic.

5. Preganglionic sympathetic neurons develop from (a) neural crest, (b) neural tube, (c) alar plate, (d) endoderm.

6. Which is the best way to describe how the ANS is controlled? (a) completely under control of voluntary cerebral cortex, (b) entirely controls itself, (c) completely under control of brain stem, (d) little control by cerebrum, major control by hypothalamus, and major control by spinal and peripheral reflexes.

7. The white rami communicantes contain what kind of fibers? (a) preganglionic parasympathetic, (b) postganglionic parasympathetic, (c) preganglionic sympathetic, (d) postganglionic sympathetic.

8. Prevertebral sympathetic ganglia are involved with the innervation of the (a) abdominal organs, (b) thoracic organs, (c) head, (d) arrector pili, (e) all of these.

9. Orville said he had a heartache because he broke up with his girlfriend and put his hand over his heart on his anterior chest. Staci told him that if his heart really hurt, he could also be pointing somewhere else. Where? (a) his left arm, (b) his head, (c) his gluteal region, (d) his abdomen.

Short Answer and Essay Questions

10. Is the visceral sensory nervous system part of the autonomic nervous system? Explain your answer.

11. Describe the anatomical relationship of the white and gray rami communicantes to a spinal nerve and to the dorsal and ventral rami.

12. Why are gray rami communicantes gray?

13. Indicate the results of sympathetic activation of the following structures: sweat glands, eye pupils, adrenal medulla, heart, lungs, blood vessels of skeletal muscles, blood vessels of digestive viscera, salivary glands.

14. Which of the effects listed in the previous question would be reversed by parasympathetic activity?

15. Which ANS fibers release acetylcholine? Which release norepinephrine?

16. Imagine that a mad scientist invents a death-ray that destroys a person's ciliary, sphenopalatine, and submandibular ganglia (and nothing else). List all the symptoms the victim would show. Would the victim die, or would the scientist have to go back to the laboratory to invent a more effective death-ray?

17. Trace the sympathetic pathway (preganglionic and postganglionic axons) to the heart, and contrast this with the sympathetic pathway to the stomach.

18. A friend asks you how the parasympathetic division, which comes only from cranial and sacral regions, is able to innervate organs in the thorax and abdomen. How would you answer that question?

19. Describe the importance of the hypothalamus in controlling the autonomic nervous system. (You may want to reread the section on the hypothalamus in Chapter 12, pp. 330–333.)

20. What manifestations of decreased ANS efficiency are seen in elderly people?

Critical Thinking and Clinical Application Questions

21. One-year-old Jimmy has a swollen abdomen and is continually constipated. On examination, a mass is felt over the distal colon, and X-ray findings show the colon to be greatly distended in that region. What do you think is wrong with Jimmy?

22. Roweena Gibson, a high-powered marketing executive, develops a stomach ulcer. She complains of a deep abdominal pain that she cannot quite locate, plus a pain of her abdominal wall. Exactly where on the abdominal wall is the superficial pain most likely to be located?

23. Constipated people often have hardened feces that can tear the inner lining of the end of the large intestine and anus during defecation. Those with tears in the skin of the anus complain of sharp pain, whereas those with tears in the intestinal mucosa do not complain of such pain. How do you explain this difference?

15

The Special Senses

People are responsive creatures. The aroma of baking bread makes our mouths water. A flash of lightening makes us blink and a clap of thunder makes us recoil. Many such sensory stimuli greet us each day and are processed by our nervous systems.

We are usually told that we have five senses: touch, taste, smell, sight, and hearing. Actually, touch is a large group of general senses that we considered in Chapter 13. The other four traditional senses—*smell, taste, sight,* and *hearing*—are called special senses. Receptors for a fifth special sense, *equilibrium,* are housed in the ear, along with the organ of hearing.

In contrast to the widely distributed receptors for the general senses, the **special sensory receptors** are localized and confined to the head region. The receptors for the special senses are not dendritic endings of sensory neurons but distinct **receptor cells.** These receptors are neuron-like epithelial cells or small peripheral neurons that transfer sensory information to other neurons in afferent pathways to the brain. The special sensory receptors are either housed within complex sensory organs (eye and ear) or within distinctive epithelial structures (taste buds and olfactory epithelium).

This chapter explores the functional anatomy of the five special senses: the chemical senses of taste and smell, vision in the eye, and hearing and equilibrium in the ear.

The Chemical Senses: Taste and Smell

The receptors for taste (gustation) and smell (olfaction) are classified as **chemoreceptors** because they respond to chemical substances. The taste receptors are excited by food chemicals dissolved in saliva, the smell receptors by airborne chemicals that dissolve in fluids on the nasal membranes. The receptors for taste and smell complement each other and respond to many of the same stimuli.

Taste Buds

The taste receptors occur in **taste buds** in the mucosa of the mouth and pharynx. Of the 10,000 or so taste buds, most are on the surface of the tongue. A few others occur on the posterior region of the palate (roof of the mouth), the inner surface of the cheeks, and the epiglottis (a leaf-shaped flap behind the tongue).

Most taste buds occur in peg-like projections of the tongue mucosa called *papillae* (pah-pil'e) (Figure 15.1a). Specifically, they occur in small *fungiform papillae* that scatter over the entire surface of the tongue, and in large *circumvallate (vallate) papillae* that line up in an inverted V near the back of the tongue. The taste buds occur within the epithelium that covers the papillae, on the apical surface of the fungiform papillae, and in the side walls of the circumvallate papillae (Figure 15.1b).

Each taste bud is a globular collection of about 50 epithelial cells that resembles the bud of a tulip (Figure 15.1c). They contain three major cell types: **supporting cells,** the **gustatory cells** (receptor cells), and **basal cells.** Supporting cells are the most abundant. They insulate the gustatory receptor cells from each other and from the surrounding epithelium of the tongue. Sensory nerve fibers enter the taste buds and participate in synaptic junctions with the receptor cells.

Long microvilli project from the receptor and supporting cells and extend through a **taste pore** to the surface of the epithelium. There, these microvilli are bathed in saliva containing the dissolved molecules that stimulate taste. Such molecules bind to the plasma membrane of the microvilli. This binding alters the membrane charge of the receptor cells and causes them to generate impulses in the sensory nerve fibers that innervate them.

The cells in taste buds are replaced every 7–10 days by the division of the basal cells. Taste cells must be renewed rapidly and continuously because food scrapes the tongue during chewing, dislodging many cells from the taste buds at each meal.

Taste Sensations and the Gustatory Pathway

Taste sensations can be described in terms of four basic qualities: sweet, sour, salty, and bitter. Although there are no *structural* differences between the taste buds in different areas of the tongue, the tongue's anterior area is most sensitive to sweet and salty substances, the lateral area to sour, and the posterior area to bitter (Figure 15.2). These differences are not absolute, however, and most taste buds respond to several taste qualities.

Taste information reaches the brain stem and cerebral cortex through the *gustatory pathway.* Sensory fibers carrying taste information from the tongue occur primarily in two cranial nerves: The *facial nerve* transmits impulses from taste receptors in the anterior two-thirds of the tongue, whereas the *glossopharyngeal nerve* services the posterior third. Addi-

412

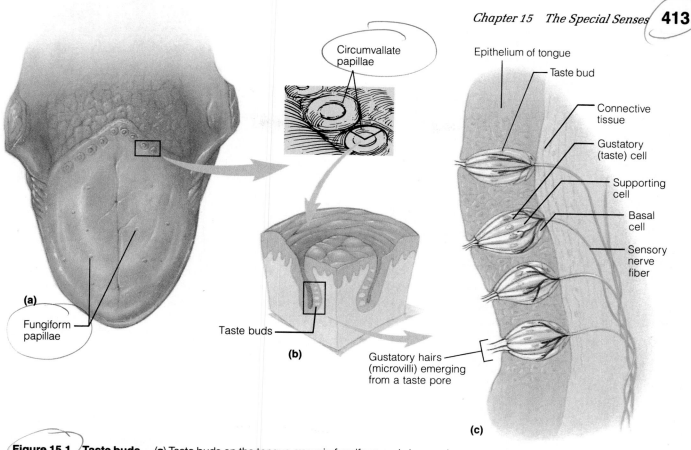

Figure 15.1   Taste buds.   (a) Taste buds on the tongue occur in fungiform and circumvallate (vallate) papillae, peg-like projections of the tongue mucosa. (b) A circumvallate papilla sectioned to show the taste buds in the epithelium of its lateral walls. (c) Enlarged view of four taste buds. Each bud contains gustatory cells (the taste receptor cells), supporting cells, and basal cells.

tionally, taste impulses from the few taste buds in the epiglottis and pharynx are conducted by the *vagus nerve.* All the sensory neurons that carry taste information synapse in a nucleus in the medulla called the *solitary nucleus.* From there, impulses are transmitted to the thalamus and ultimately to the gustatory area of the cerebral cortex in the parietal lobe.

By referring to the map of taste sensitivities in Figure 15.2, can you deduce which aspects of taste are diminished by injuries to the glossopharyngeal nerve?

Damage to the glossopharyngeal nerve reduces one's ability to taste bitter substances because this nerve supplies the tongue's posterior, bitter-sensitive region. Damage to the facial nerve, by contrast, diminishes one's ability to taste sweet, sour, and salty substances on the anterior surface of the tongue. ■

Olfactory Receptors

The receptors for smell lie in a yellow-tinged patch of epithelium on the roof of the nasal cavity (Figure 15.3a). Specifically, the **olfactory epithelium** covers the superior nasal concha and the superior part of the nasal septum. It lies out of the way of most inspired air that enters the nasal cavity. Sniffing draws air superiorly across the olfactory epithelium and thus intensifies the sense of smell.

The olfactory epithelium is a pseudostratified columnar epithelium (Figure 15.3b and c) that contains

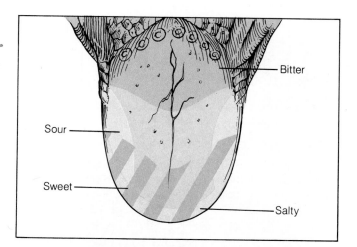

Figure 15.2   Patterns of maximum taste sensitivity on the tongue.   (Since the area that is most sensitive to sweet overlaps the sour and salty areas, the sweet area is indicated by pink stripes.)

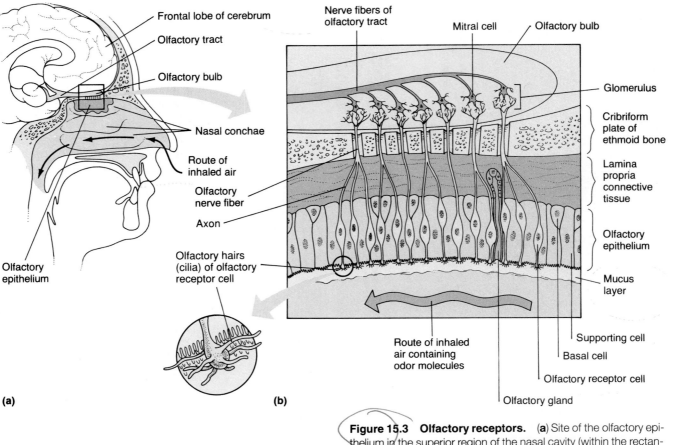

(a)

(b)

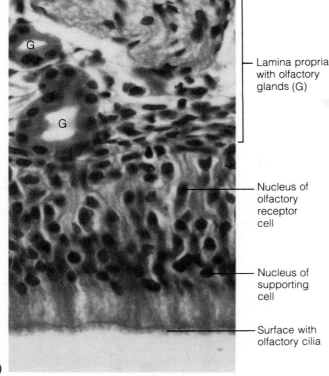

(c)

Figure 15.3 Olfactory receptors. (**a**) Site of the olfactory epithelium in the superior region of the nasal cavity (within the rectangle). (**b**) Enlarged view showing the cellular makeup of the olfactory epithelium and the course of olfactory nerve fibers to the overlying olfactory bulb. (**c**) Micrograph of the olfactory epithelium (720 ×). Note that the epithelium in (b) and (c) is oriented "upside down," with its apical surface below and its basal surface above.

millions of bowling-pin-shaped **olfactory receptor cells.** These are surrounded by columnar **supporting cells.** At the base of the epithelium lie short **basal cells.**

The olfactory receptor cells are unusual neurons, with cell bodies in the olfactory epithelium and axons that extend into the CNS (Figure 15.3b). Each receptor cell also has an apical dendrite that projects to the epithelial surface. There, the dendrite ends in a knob from which long **olfactory cilia** radiate. These cilia lie flat on the surface and act as the receptive structures for smell; that is, odor molecules bind to or dissolve in the plasma membrane of the cilia, thereby exciting the receptor cells. The surface of the olfactory epithelium is coated with a layer of mucus that is secreted by the supporting cells and by glands in the underlying connective tissue (the lamina propria). This mucus is a solvent that dissolves and "captures" the odor molecules from the air. It is renewed continuously, flushing away "old" odor molecules so that new odors always have access to the receptor cilia.

Unlike other cilia in the body, the olfactory cilia are largely immotile.

Olfactory receptor cells are the only neurons that undergo turnover throughout adult life. Their average life span is 60 days. They are replaced by the differentiation of basal cells in the olfactory epithelium.

The Olfactory Pathway

An axon extends from the base of each olfactory receptor cell through the connective tissue of the lamina propria (Figure 15.3b). There, the many axons gather into about 20 nerve bundles called the **fiber bundles of the olfactory nerve,** which penetrate the cribriform plate of the ethmoid bone and attach to the overlying *olfactory bulb* of the forebrain. Within this bulb, the olfactory nerve axons branch profusely and synapse with neurons called **mitral cells** (mi′tral; "cap-shaped") in complex synaptic clusters called **glomeruli** (glo-mer′u-li; "balls of yarn"). The mitral cells then relay the olfactory information to other parts of the brain. Specifically, mitral axons run through the *olfactory tract* to the primary olfactory area of the cerebral cortex (for conscious perception of smell) and to the limbic system (through which smells elicit emotions).

A great deal of olfactory information is processed locally in the olfactory bulb, long before it reaches the cerebral cortex or limbic system. Such processing, due to the action of inhibitory interneurons within the olfactory bulb, contributes to olfactory adaptation, the phenomenon by which we "get used to smells."

✚ Disorders of the Chemical Senses

It is possible to have disorders of either taste or olfaction, but olfactory disorders are more common. Absence of the sense of smell is called **anosmia** (an-oz′me-ah; "without smell"). Loss of smell can result from blows to the head that tear the olfactory nerves, from colds or allergies, or from physical blockage of the nasal cavity (by polyps, for example). The culprit in fully one-third of all cases of loss of chemical senses is zinc deficiency, and the cure is rapid once a zinc supplement is prescribed.

Brain disorders can distort the sense of smell. Some people have **uncinate** (un′sǐ-nāt) **fits,** olfactory hallucinations in which they perceive some imaginary odor, like that of gasoline or rotting meat. Uncinate fits are so called because the primary olfactory cortex is in the uncinate region, or uncus, of the cerebrum (p. 326). These fits may result from irritation of the olfactory pathway by brain surgery or head trauma. Olfactory auras—smells imagined by some epileptics just before they go into seizures—are brief uncinate fits. ■

The Eye

Vision is our dominant sense: It is estimated that 70% of all the sensory receptors in the body are in the eyes and that 40% of the cerebral cortex is involved in some aspect of processing of visual information. The visual receptor cells (photoreceptors) sense and encode the patterns of light that enter the eye, and the brain invests these signals with meaning, fashioning images of the world around us.

The Eye in the Orbit

The adult **eye,** or **eyeball,** is a sphere with a diameter of about 2.5 cm (1 inch). Only the anterior sixth of the eye's surface is visible. The rest of the eye lies in the cone-shaped *orbit,* where it is surrounded by a protective cushion of fat. Behind the eye, the posterior half of the orbit contains many structures that run to and from the eye. These include the optic nerve, the arteries and veins to the eye, and the extrinsic eye muscles.

We will first discuss the elements around the eye, then the eyeball itself.

Accessory Structures of the Eye

The accessory structures of the eye include the eyebrows, eyelids, conjunctiva, lacrimal apparatus, and extrinsic eye muscles (Figures 15.4 and 15.5).

Eyebrows

The **eyebrows** consist of coarse hairs in the skin on the superciliary arches (brow ridges of the skull). They shade the eyes from sunlight and prevent perspiration running down the forehead from reaching the eyes.

Eyelids

Anteriorly, the eyes are protected by the mobile **eyelids,** or **palpebrae** (pal′pĕ-bre). The upper and lower lids are separated by the **palpebral fissure** (eye slit) and meet each other at the medial and lateral **canthi** (eye corners). The medial canthus contains a fleshy elevation called the **caruncle** (kar′ung-k′l; "a bit of flesh"). Glands here produce the oily "eye-sand" that is left by the legendary Sandman at night. In most Asian people, a vertical fold of skin called the *epicanthic fold* occurs on both sides of the nose and sometimes covers the medial canthus.

The eyelids are thin, skin-covered folds supported internally by **tarsal plates** of connective tissue

Figure 15.4 Surface anatomy of the right eye and some accessory structures.

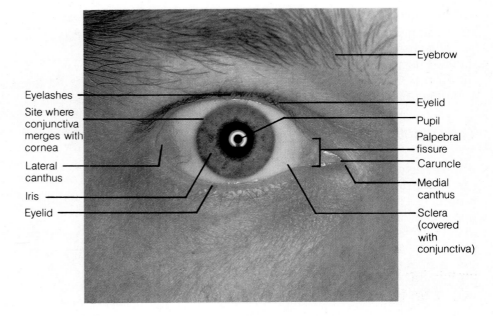

Eyelashes

Site where conjunctiva merges with cornea

Lateral canthus

Iris

Eyelid

Eyebrow

Eyelid

Pupil

Palpebral fissure

Caruncle

Medial canthus

Sclera (covered with conjunctiva)

(Figure 15.5a). These stiff plates give the eyelids their curved shape and serve as attachment sites for the eye-closing muscle, the *orbicularis oculi* (see p. 246).

The **levator palpebrae superioris** ("lifter of the upper eyelid") is the skeletal muscle that voluntarily opens the eye. It runs anteriorly from the posterior roof of the orbit, enters the upper eyelid, and inserts on the tarsal plate. The inferior part of the aponeurosis of this muscle contains slips of smooth muscle, called the *superior tarsal muscle,* an involuntary muscle that prevents the upper eyelid from drooping (see p. 402).

Projecting from the free margin of each eyelid are the **eyelashes.** The follicles of these hairs are richly innervated by nerve endings (root hair plexuses). Therefore, even a slight pressure on the eyelashes will trigger reflexive blinking.

Several types of glands occur within the eyelids. **Tarsal glands** are modified sebaceous glands embedded in the tarsal plates (Figure 15.5a). Their ducts open at the edge of the eyelids. These glands secrete an oil that may keep the two eyelids from sticking together. This oil also spreads over the surface of the eye and slows the evaporation of water. Other, more typical, sebaceous glands open into the hair follicles of the eyelashes, and modified sweat glands called *ciliary glands* lie between the hair follicles (*cilium* = eyelash).

Infection of a tarsal gland results in an unsightly cyst called a **chalazion** (kah-la′ze-on; "swelling"). Infection of the ciliary or sebaceous glands in the eyelid is called a **sty.** ∎

Conjunctiva

The **conjunctiva** (con″junk-ti′vah; "joined together") is a transparent mucous membrane. It covers the inner surfaces of the eyelids as the **palpebral conjunctiva** and folds back over the anterior surface of the eye as the **ocular,** or **bulbar, conjunctiva.** The latter covers only the white of the eye, not the cornea (the clear window over the iris and pupil). The ocular conjunctiva is a very thin membrane, and blood vessels are clearly visible beneath it. These vessels are responsible for "bloodshot" eyes. When an eye is closed, a slit-like space forms between the eye surface and the eyelids. This is the **conjunctival sac** (located where a contact lens would lie).

Microscopically, the conjunctiva consists of a stratified columnar epithelium underlain by a thin lamina propria of loose connective tissue. Its epithelium contains scattered goblet cells (mucous cells). The main function of the conjunctiva is to secrete a lubricating mucus that prevents the eyes from drying.

Inflammation of the conjunctiva, or **conjunctivitis,** is relatively common. Can you deduce how this condition affects one's appearance?

Conjunctivitis causes reddened, irritated eyes. **Pinkeye,** an infectious form of conjunctivitis caused by bacteria or viruses, is highly contagious.

Vitamin A is required to maintain the health of the epithelia throughout the body. Deficiency of this vitamin causes the conjunctiva to stop secreting mucus. The conjunctiva dries and becomes a scaly layer that severely impairs vision. ∎

Lacrimal Apparatus

The **lacrimal apparatus** (lak′rĭ-mal; "tear") keeps the surface of the eye moistened with lacrimal fluid (tears). This apparatus consists of a gland and ducts that drain the lacrimal fluid into the nasal cavity (Figure 15.5b). The **lacrimal gland** lies in the orbit superolateral to the eye. It produces the lacrimal fluid, which enters the superior part of the conjunctival sac through several small excretory ducts. Blinking of the eye then spreads this fluid inferiorly across the eyeball to the medial canthus, where it passes through tiny openings called **lacrimal puncta** ("points") into two small tubes called **lacrimal canals**. From these canals, the fluid drains into the **lacrimal sac** in the medial orbital wall. Finally, the fluid enters the **nasolacrimal duct,** which empties into the nasal cavity at the inferior nasal meatus. Since lacrimal fluid empties into the nasal cavity, we get the sniffles when we cry.

Lacrimal fluid contains mucus, antibodies, and **lysozyme,** an enzyme that destroys bacteria. When the eye surface is irritated by dust or by fumes (from an onion, for example), lacrimal secretion increases to wash away the irritant.

 People who have severe colds often have watery eyes as well. What is the connection?

The mucosa that lines the nasal cavity is continuous with the mucosa lining of the lacrimal apparatus. Therefore, a cold in the nasal passages can spread into the nasolacrimal duct and lacrimal sac, causing these passages to swell shut with inflammation. This blocks the drainage of lacrimal fluid from the eye surface, the fluid accumulates, and the eyes water. ■

Extrinsic Eye Muscles

The movement of each eyeball is controlled by six strap-like **extrinsic** (outer) **eye muscles** (Figures 15.6 and 15.7). These muscles originate from the walls of the orbit and insert onto the outer surface of the eyeball. They direct the gaze and hold the eyes in the orbits. Table 15.1 names the muscles and their innervation.

Four of the extrinsic eye muscles are *rectus* muscles (*rectus* = straight). These originate from a common tendinous ring, the **annular ring,** at the posterior point of the orbit. From there, the muscles run straight to their insertions on the anterior half of the eyeball. The **lateral rectus muscle** turns the eye laterally (outward), while the **medial rectus muscle** turns it medially (inward). The **superior** and **inferior rectus muscles** turn the eye superiorly and inferiorly, respectively. It is easy to deduce the actions of these muscles from their names and locations.

The actions of the two *oblique* muscles are not so easily deduced, because they take rather strange paths

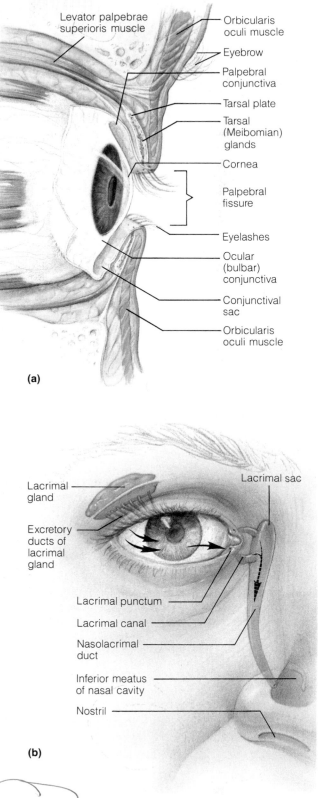

(a)

(b)

Figure 15.5   Accessory structures of the eye.   (a) Lateral view; some of the structures are shown in sagittal section. (b) The lacrimal (tear) apparatus. Arrows indicate the direction of flow of lacrimal fluid after its secretion by the lacrimal gland.

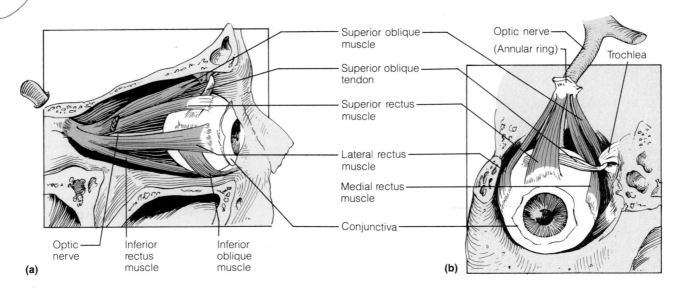

Figure 15.6   Extrinsic muscles of the eye.   (**a**) Lateral view of the right eye in the orbit.
(**b**) Right eye, superoanterior view.

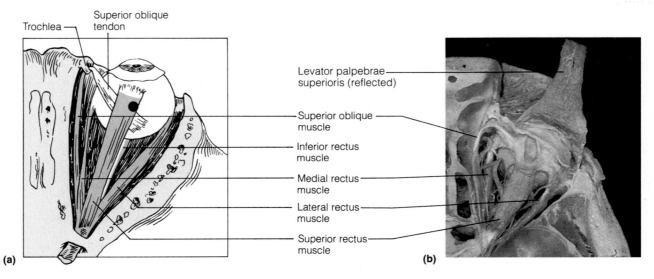

Figure 15.7   Superior view of the extrinsic muscles of the right eye.   (**a**) The dot indicates the exact center of the eye, the axis around which the eye turns. Note the superior
oblique and superior rectus muscles. From the angles at which these muscles approach the
eye, it is evident that the superior rectus turns the eye superiorly and medially and that the
superior oblique turns it inferiorly and laterally. On the inferior aspect of the eye, the inferior
rectus muscle turns the eye inferiorly and medially, and the internal oblique turns it superiorly
and laterally. (**b**) Same view from the orbit of a cadaver.

Table 15.1	Extrinsic Eye Muscles: Innervation and Action	
Name	**Controlling cranial nerve**	**Action**
Lateral rectus	VI (abducens)	Turns eye laterally
Medial rectus	III (oculomotor)	Turns eye medially
Superior rectus	III (oculomotor)	Elevates eye or rolls it superiorly
Inferior rectus	III (oculomotor)	Depresses eye or rolls it inferiorly
Inferior oblique	III (oculomotor)	Elevates eye and turns it laterally
Superior oblique	IV (trochlear)	Depresses eye and turns it laterally

through the orbit. The **superior oblique muscle** originates posteriorly near the annular ring, runs anteriorly along the medial orbit wall, then loops through a ligamentous sling, the **trochlea** ("pulley"), in the anteromedial part of the orbit roof. From there, its tendon runs posteriorly and inserts on the eye's posterolateral surface (Figure 15.7). Because its tendon approaches from an anterior direction, the superior oblique turns the eye inferiorly, and because its tendon also approaches from a medial direction, it turns the eye laterally. The **inferior oblique muscle** (Figure 15.6a) underlies the eye. It originates on the anteromedial part of the orbit floor and angles back to insert on the posterolateral part of the eye. Thus, the inferior oblique turns the eye superiorly and somewhat laterally.

People often ask why we have the two oblique muscles, when the four rectus muscles seem to provide all the eye movements we require—medial, lateral, superior, and inferior. The simplest way to answer this question is to point out that the superior and inferior recti cannot elevate or depress the eye *without also turning it medially.* These rectus muscles pull the eye medially because they approach the eye from a medial as well as from a posterior direction (Figure 15.7a). For an eye to be *directly* elevated or depressed, the lateral pull of the oblique muscles must cancel the medial pull of the superior and inferior recti.

The extrinsic muscles of both eyes are closely controlled by centers in the midbrain, so that the eyes move together in perfect unison when we direct them toward objects being viewed. Can you guess what happens to vision if this coordination is disrupted?

Double vision results, because the two eyes do not look at the same point in the visual field. This condition is called **strabismus** (strah-biz′mus), a term meaning "cross-eyed" or "squint-eyed." In a person with strabismus, the affected eye is turned either medially or laterally with respect to the normal eye. Strabismus results from weakness or paralysis of extrinsic eye muscles, from damage to the oculomotor nerve, or from other causes. ■

Structure of the Eyeball

The eye is a complex organ. Its various components not only protect and support the delicate photoreceptor cells but also gather, focus, and process light into precise images. Let us start with a brief overview of eye anatomy (Figure 15.8). Since the eyeball is shaped roughly like a globe of the earth, it is said to have poles: Its most anterior point is the **anterior pole,** its most posterior point, the **posterior pole.** The external wall of the eye consists of three layers, or *tunics,* and the internal cavity of the eye contains fluids called

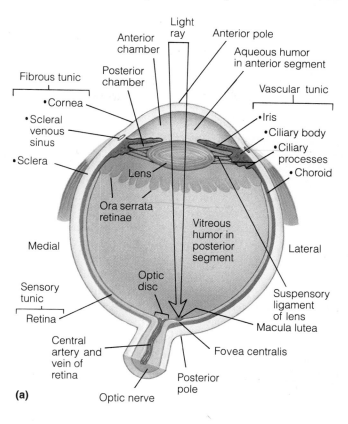

(a)

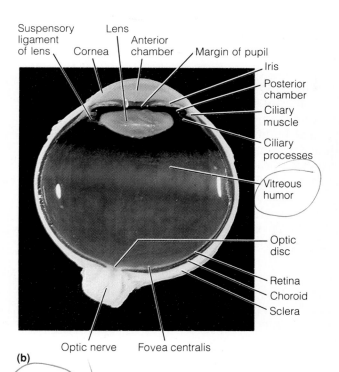

(b)

Figure 15.8   Superior view of a transverse section of the right eye.   (a) The three tunics that make up the wall of the eye are the fibrous tunic, the vascular tunic, and the sensory tunic. When an eye looks straight at an object, light rays from that object pass directly from the anterior pole to the posterior pole of the eye (see the arrow). (b) Photograph of a human eye in the same orientation as in part (a).

humors. The *lens,* a structure that helps to focus light, is supported vertically within the internal cavity and divides this cavity into anterior and posterior segments. The anterior segment is filled with the liquid *aqueous humor,* whereas the posterior segment is filled with the jelly-like *vitreous humor.*

The Fibrous Tunic

The three tunics that form the external wall of the eye are the *fibrous tunic, vascular tunic,* and *sensory tunic (retina).* The **fibrous tunic,** the most external of the three, is composed of dense connective tissue. It has two different regions: the sclera and the cornea (Figure 15.8). The white, tough **sclera** (skle'rah; "hard") forms the posterior five-sixths of the fibrous tunic. Seen anteriorly as the "white of the eye," the sclera protects and shapes the eyeball and provides a sturdy anchoring site for the extrinsic eye muscles. The sclera is continuous with—and corresponds to—the dura mater that covers the brain.

The anterior sixth of the fibrous tunic is the transparent **cornea,** through which light enters the eye. This round window bulges anteriorly from its junction with the sclera. The cornea consists of three layers of tissue: a thick layer of dense connective tissue sandwiched between a superficial *corneal epithelium* and a deep *corneal endothelium* (Figure 15.9). The connective tissue layer contains hundreds of sheets of collagen fibers stacked like the pages in a book. The clarity of the cornea is due to this regular alignment of collagen fibers. The cornea not only allows light into the eye interior but also serves as part of the light-bending apparatus of the eye (p. 426).

The superficial corneal epithelium is a stratified squamous layer that helps resist abrasion. The deep corneal endothelium is a simple squamous epithelium that continuously transports water out of the cornea into the aqueous humor. This action prevents the accumulation of tissue fluid between the collagen fibers, which would push these fibers out of alignment and reduce the clarity of the cornea.

The cornea is avascular—it receives nutrients and oxygen from the aqueous humor that lies posterior to it. The cornea also receives some oxygen from the air in front of it. Although the cornea lacks blood vessels, a single unusual vessel lies in the boundary zone between the sclera and the cornea, the **scleral venous sinus** (Figure 15.9). This sinus drains aqueous humor out of the eye.

The cornea is richly supplied with nerve endings, most of which are pain fibers. For this reason, some people can never adjust to wearing contact lenses. Touching the cornea causes reflexive blinking and an increased secretion of tears. Even with these protective responses, the cornea is vulnerable to damage by

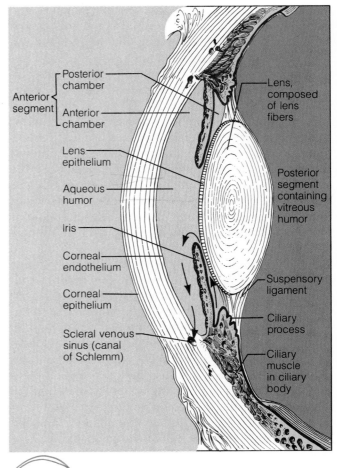

Figure 15.9 Structures in the anterior region of the eye (transverse section).

dust, slivers, and other objects. Luckily, its capacity for regeneration and healing is extraordinary.

A severely damaged cornea can be replaced surgically. The cornea is one of the few structures in the body that can be transplanted from one person to another without worry of rejection: Since it has no blood vessels, it is beyond the reach of the immune system that attacks other kinds of transplants. An "eye bank" is an institution that receives donor corneas for corneal transplants. ■

The Vascular Tunic

The **vascular tunic** is the middle coat of the eyeball. It is also called the *uvea* (u've-ah), a Latin word for "grape." Indeed, when the overlying sclera is removed, the dark vascular tunic makes the eye look like a large grape. The uvea has three distinct regions: the *choroid, ciliary body,* and *iris* (Figure 15.8).

The **choroid** (ko'roid; "membrane") is a highly vascular, darkly pigmented membrane that forms the

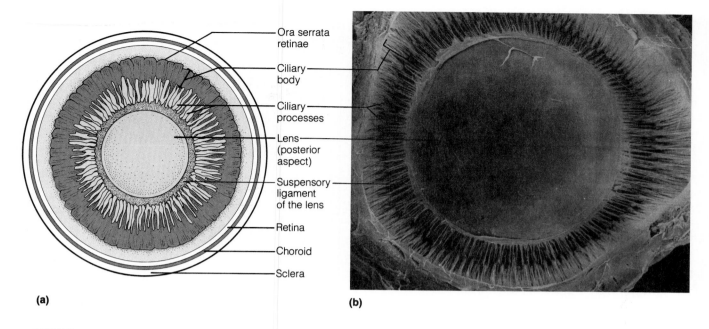

(a)

(b)

Figure 15.10   Posterior view of the anterior half of the eye.   (a) Note the ciliary processes on the ciliary body, the suspensory ligament of the lens, and the ora serrata retinae. (b) Scanning electron micrograph of the posterior aspect of the lens, suspensory ligament, and ciliary body. Note the fineness of the fibrils of the suspensory ligament of the lens, which holds the lens in the eye (28 ×).

posterior five-sixths of the vascular tunic. Its many blood vessels nourish the other tunics of the eye, including the part of the retina that lies just internal to the choroid. The brown pigment of the choroid is produced by melanocytes. This pigment helps absorb light, preventing it from scattering and reflecting within the eye and confusing vision. Overall, the choroid layer around the eye corresponds to the arachnoid and pia mater around the brain.

Anteriorly, the choroid is continuous with the **ciliary body,** a thickened ring of tissue that encircles the lens (Figures 15.8, 15.9, and 15.10). The ciliary body consists chiefly of smooth muscle called the **ciliary muscle** (Figure 15.9). As we will explain, this muscle acts to focus the lens. Nearest the lens, the posterior surface of the ciliary body is thrown into folds called **ciliary processes** (Figure 15.10). Fine fibrils extend from these processes and attach around the entire circumference of the lens. Collectively, this halo of fibrils is called the **suspensory ligament of the lens,** or the *zonule* (Figure 15.10b).

The **iris** is the visible, colored part of the eye (*iris* = rainbow). It lies between the cornea and lens, and its base attaches to the ciliary body (Figure 15.9). Its round central opening, the *pupil,* allows light to enter the eye. The iris contains smooth muscle fibers that are arranged in both circular and radiating patterns (the *sphincter* and *dilator pupillae* muscles). These muscles act to vary the size of the pupil. In bright light

and for close vision, the sphincter muscle contracts and constricts the pupil. In dim light and for distant vision, the dilator muscle contracts and widens the pupil, allowing more light to enter the eye. Constriction and dilation of the pupil are controlled by parasympathetic and sympathetic fibers, respectively (Chapter 14). The constriction of the pupils that occurs in intense (potentially damaging) light is a protective response called the **pupillary light reflex.**

Although irises come in many colors, they contain only brown pigment. All people (except albinos) have a layer of pigmented cells on the posterior surface of the iris, and brown-eyed people have many pigment cells in the body of the iris as well. Blue-eyed people, by contrast, have no pigment within the body of the iris. Their blue eye color represents the posterior pigmented layer as it appears through the colorless body of the iris. Hazel-eyed people have some pigment in the body of the iris but less pigment than do brown-eyed people. The eyes of newborn Caucasian children are gray or blue because the body of the iris is not yet pigmented.

The Sensory Tunic (Retina)

Composition of the Retina.   The deepest tunic of the eye is the delicate **retina** ("net") (Figure 15.11). It consists of two layers. The outer **pigmented layer,** which

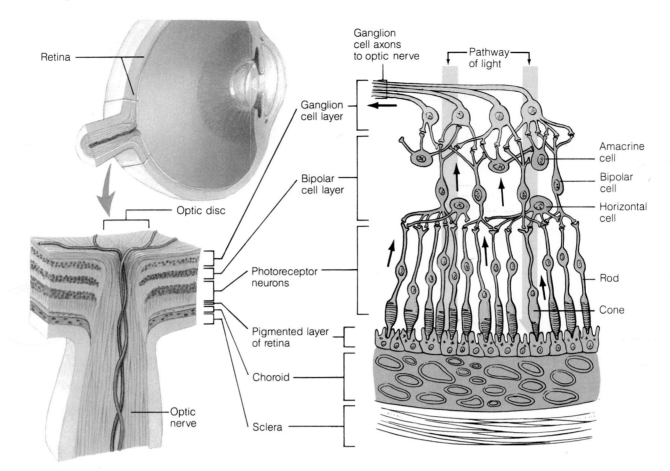

Figure 15.11 **The structure of the retina (enlargement of the posterior wall of the eye near the optic nerve).** The cellular structure of the retina is shown at right. Note that light rays must pass through the entire thickness of the neural retina to excite the photoreceptor cells. By contrast, neural signals flow through the retina in the opposite direction, as indicated by the black arrows.

lies against the choroid, is a single layer of flat-to-columnar epithelial cells. Like the pigmented cells in the choroid, these cells absorb light to prevent it from scattering within the eye. Far thicker than the pigmented layer is the inner **neural (nervous) layer,** a sheet of nervous tissue that contains the light-sensitive photoreceptor cells. The neural and pigmented layers of the retina are held together by a thin film of extracellular matrix, but they are not tightly fused. The retina is commonly called the *sensory tunic,* but only its neural layer plays a direct role in vision.

The neural layer of the retina contains three main types of neurons. From external to internal, these are the **photoreceptor cells, bipolar cells,** and **ganglion cells.** Light must pass through the transparent inner region of the retina to activate the photoreceptor cells, which abut the pigmented layer (Figure 15.11). (It would seem more logical to have the photoreceptors on the inner face of the retina where they could encounter light first, but that is not how the eye is structured.) When exposed to light, the photoreceptor cells

signal the bipolar cells, which in turn signal the ganglion cells to generate action potentials. Axons from the ganglion cells run along the internal surface of the retina and converge to form the *optic nerve,* which leaves the eye posteriorly. Along with its three main cell types, the retina also contains interneurons—amacrine cells, horizontal cells, and others—whose processes run laterally (parallel to the surface of the retina). It may surprise you that the retina contains interneurons, but this fact is easily explained: The retina develops as a part of the brain (p. 429). These retinal interneurons process and organize visual information before it is sent to higher brain centers for further processing. As a part of the brain, the retina also contains glial cells.

Photoreceptors. The photoreceptor cells are of two types: *rod cells* and *cone cells.* The more numerous **rod cells** are more sensitive to light and permit vision in dim light. Rod cells do not provide sharp images or color vision. Thus, things look gray and fuzzy when viewed in dim light. **Cone cells,** by contrast, operate best in bright light and provide for our high-acuity color vision.

The photoreceptors are considered neurons, but they also resemble tall epithelial cells turned upside down with their "tips" immersed in the pigmented layer (Figure 15.12). Both the rod and cone cells have an *outer segment* joined to an *inner segment* by a connecting stalk containing a cilium. The inner segment connects to the *cell body,* which is continuous with an *inner fiber* bearing synaptic endings. In each rod cell, the inner and outer segments together form a rod-shaped structure (hence the name *rod*), which connects to the cell body by an *outer fiber.* In each cone cell, by contrast, the inner and outer segments form a cone-shaped structure (hence the name *cone*), which joins to the cell body directly.

The **outer segments** are the receptor regions of the rod and cone cells. Each outer segment is a modified cilium whose plasma membrane has folded inward to form hundreds of membrane-covered discs. Light-absorbing *visual pigments* occur within the vast membrane of these discs, an association that greatly magnifies the surface area available for trapping light. When particles of light (photons) hit the visual pigment, the pigment splits into two molecules. Such splitting changes the membrane charge of the photoreceptors and causes these cells to signal the bipolar neurons with which they synapse. This initiates the flow of visual information to the brain.

The photoreceptors are highly vulnerable to damage by overly intense light or by heat. Like other neurons, these cells cannot regenerate if they are destroyed. However, they continually renew and replace their outer segments through the addition of new discs.

Regional Specializations of the Retina.   In certain parts of the eye, the retina differs from its "typical" structure described above. Here, we consider these specialized regions of the retina.

In the anterior part of the eye, the retina is incomplete. Its neural layer ends at the posterior margin of the ciliary body. This margin is the **ora serrata retinae** (o'rah sĕ-rah'tah), literally the "saw-toothed mouth of the retina" (Figure 15.10a). The pigmented layer of the retina, by contrast, extends anteriorly beyond the

Figure 15.12   Diagrammatic view of the photoreceptors (rod and cone cells) in the retina.   Note the various parts of these cells. The outer segments of the photoreceptors (below) indent the pigmented cells of the retina like pegs in sockets. At lower right, the tip of a rod's outer segment is sloughing off, to be digested by a lysosome within the pigment cell. As the tips of the rods are phagocytosed in this way, they are continuously replaced by new discs that are added to the other end of the stack of discs.

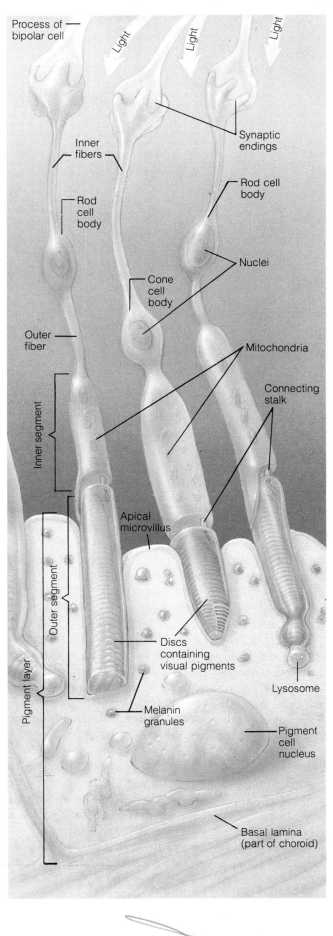

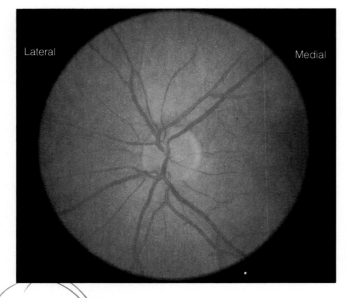

Lateral

Medial

Figure 15.13   The posteromedial wall of the retina as seen with an ophthalmoscope.   Note the optic disc, from which radiate the central vessels of the retina. The macula lutea is present but not visible, because it can be seen only in special red-free light.

ora serrata to cover the ciliary body and the posterior face of the iris.

In the posterior part of the eye, the retina exhibits several special areas (Figure 15.8). Lying precisely at the eye's posterior pole is the **macula lutea** (mak'u-lah lu'te-ah; "yellow spot"). In the middle of this spot is a tiny pit called the **fovea centralis** (fo've-ah sen-trah'lis; "central pit"). The fovea contains only cones; the macula contains mostly cones; and from the macula toward the periphery of the retina, the density of cones declines progressively. Because of its high concentration of cones, the fovea is the point on the retina that provides maximal visual acuity. Because the fovea lies directly in the anterior-posterior axis of the eye, we see things most clearly when we look straight at them—peripheral vision is not as sharp as central vision. A few millimeters medial to the fovea is the **optic disc** (*optic papilla*), a circular elevation. The axons of ganglion cells converge at the optic disc to exit the eye as the optic nerve. The optic disc is called the *blind spot* because it lacks photoreceptors, and light focused on it cannot be seen.

Blood Supply of the Retina.   The retina receives its blood from two sources. Its outer third, containing the photoreceptors, is supplied by capillaries in the choroid. Its inner two-thirds is supplied by the **central artery** and **vein of the retina.** These vessels enter and leave the eye by running through the center of the optic nerve (Figure 15.8). Radiating from the optic disc, they give rise to a rich network of tiny vessels that weave among the axons on the retina's inner face. This vascular network is clearly visible when one views the eye's interior with an ophthalmoscope, a

hand-held instrument that shines light through the pupil and illuminates the retina (Figure 15.13). Physicians observe the tiny retinal vessels for signs of hypertension, diabetes, and other diseases that damage our smallest blood vessels.

The structure and pattern of vascularization of the retina make it susceptible to **retinal detachment,** a condition that can cause blindness. In this condition, the loosely joined neural and pigmented layers of the retina separate ("detach") from one another. Retinal detachment inevitably begins with a tear in the retina, which may result from a small hemorrhage or a blow to the eye. This tear allows the jelly-like vitreous humor from the interior of the eye to seep through and wedge between the two retinal layers. With the neural layer detached, the photoreceptors are separated from their blood supply in the choroid. The photoreceptors are kept alive temporarily by nutrients from the inner retinal capillaries. However, these capillaries are too distant to supply them permanently, and the photoreceptors soon die. If the condition is diagnosed early, it is often possible to reattach the retina surgically or with a laser before blindness occurs. ■

Internal Chambers and Fluids

As we noted earlier, the lens and its halo-like suspensory ligament divide the eye into posterior and anterior segments (Figures 15.8a and 15.9). The **posterior segment** is filled with the clear **vitreous humor** (*vitreus* = glassy). This jelly-like substance contains fine fibrils of collagen and a ground substance that binds tremendous amounts of water. In fact, water makes up over 98% of the volume of the vitreous humor. The functions of this humor are to (1) transmit light, (2) support the posterior surface of the lens and hold the neural retina firmly against the pigmented layer, and (3) help maintain *intraocular pressure* (the normal pressure within the eye), thereby counteracting the pulling forces of the extrinsic eye muscles.

The **anterior segment** of the eye (Figure 15.9) is divided into an **anterior chamber** (between the cornea and iris) and a **posterior chamber** (between the iris and lens). The entire anterior segment is filled with **aqueous humor,** a clear fluid similar to blood plasma (*aqueous* = watery). Unlike the vitreous humor, which forms in the embryo and lasts a lifetime, aqueous humor is renewed continuously and is in constant motion (see the arrows in Figure 15.9). It forms as a filtrate of the blood from capillaries in the ciliary processes and enters the posterior chamber. From there, the aqueous humor flows through the pupil into the anterior chamber and drains into the scleral venous sinus, which returns it to the blood. Normally, the aqueous humor is produced and drained at the same rate. This equilibrium results in a constant intraocular pressure, which supports the eyeball internally. Furthermore, the aqueous humor supplies nutrients and oxygen to the avascular lens and cornea.

Can you deduce what would happen if aqueous humor drained more slowly than it formed?

This is the basis of **glaucoma** (glaw-ko'mah), a disease in which intraocular pressure increases to dangerous levels and damages the retina. Glaucoma results from a buildup of aqueous humor caused by blockage of the scleral venous sinus. The resulting destruction of the retina eventually causes blindness. (*Glaucoma* literally means vision "growing gray.") Vision can be saved if the condition is detected early. However, glaucoma often steals sight so slowly and painlessly that most people do not realize they have a problem until the damage is done. Late signs include blurred vision, seeing halos around lights, and headaches. The examination for glaucoma is simple: The examiner anesthetizes the sclera, pushes on it gently with an instrument called a tonometer, and measures the amount of deformation. This exam should be done yearly after age 40, for glaucoma affects fully 2% of people over that age. Early glaucoma is treated with eye drops that increase the rate of drainage of aqueous humor. ■

Lens

The **lens** is a thick, transparent disc that is biconvex (Figure 15.9). It changes shape to allow precise focusing of light on the retina. The lens is enclosed in a thin, elastic capsule and is held in place posterior to the iris by its suspensory ligament. Like the cornea, the lens lacks blood vessels, for vessels interfere with transparency.

The lens has two regions: the lens epithelium and the lens fibers. The **lens epithelium,** confined to the anterior surface of the lens, consists of cuboidal cells. Those epithelial cells around the edge of the lens continuously transform into the elongated **lens fibers** that form the bulk of the lens. The lens fibers, which are packed together like the layers in an onion, contain no nuclei and few organelles. They do, however, contain precisely folded proteins which make them transparent. New lens fibers are added continuously, so the lens enlarges throughout life. It grows denser, more convex, and less elastic with age. Thus, its ability to focus light is gradually impaired.

A **cataract** (kat'ah-rakt) is a clouding of the lens that causes the world to appear distorted, as if seen through frosted glass or through a waterfall (*cataract* = waterfall). Some cataracts are congenital, but most result from age-related hardening and thickening of the lens or a secondary consequence of diabetes mellitus. Recent evidence indicates that heavy smoking and excessive exposure to sunlight can also cause cataracts. No matter what the cause, cataracts

seem to result from an inadequate delivery of nutrients to the deeper lens fibers. Fortunately, the offending lens can be removed surgically and an artificial lens implanted to save the patient's sight. ■

The Eye as a Light-Gathering Device

In this section, we consider how light is focused on the retina. Any object that is being viewed can be said to consist of many small, individual points. Light radiates from each point in all directions, and some of the light rays enter the eye of the viewer (Figure 15.14). Rays from a distant point are parallel to one another as they reach the eye, while rays from a nearby point diverge markedly as they enter the eye. If we are to see clearly, the eye must be capable of

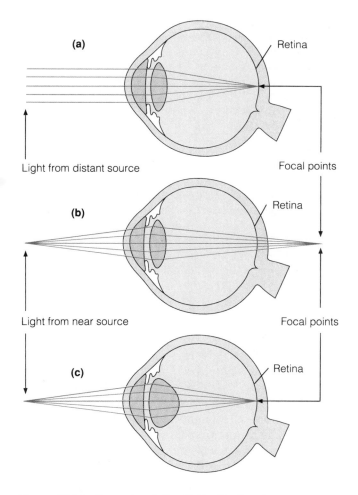

Figure 15.14   Accommodation (focusing for close vision). (a) The resting eye is set for distance vision—that is, parallel light rays from a distant point are focused directly on the retina. (b) The resting eye is not set for viewing nearby points—because divergent light rays are focused far behind the retina. (c) To focus on a nearby point, the lens rounds, and it bends the divergent rays more sharply, so that they converge on the retina. (For simplicity, the light-bending effect of the *cornea* is ignored in these diagrams. The cornea does most of the bending of the light at all times, even though the lens is responsible for accommodation.)

A CLOSER LOOK   If I Can't See Things Far Away, Am I Nearsighted or Farsighted?

It seems that whenever people who wear glasses or contact lenses get together and discuss their vision, one of them says something like, "Nearby objects appear blurry to me, but I can't remember if that means I'm nearsighted or farsighted." Or, someone else may say, "My glasses allow me to see faraway objects more clearly, so does that mean I am farsighted?" Here we will explain the meaning of nearsightedness and farsightedness as we explore the basis of eye-focusing disorders.

The eye that focuses images correctly on the retina is said to have **emmetropia** (em″ĕ-tro′pe-ah; literally "harmonious vision." Such an eye is shown in part (a) of the figure.

Nearsightedness is formally called **myopia** (mi″o′pe-ah; "short vision"). It occurs when the parallel light rays from distant objects are not focused on the retina, but in front of it (see part (b) in the figure). Therefore, *distant* objects appear blurry to myopic people. Nearby objects are in focus, however, because as an object nears the eye, its focal point naturally moves posteriorly and comes to lie on the retina. Myopia results from an eyeball that is too long, a lens that is too strong, or

a cornea that is too highly curved. Correction requires the use of *concave* corrective lenses that diverge the light rays before they enter the eye, so that the focal point is farther back. To answer the first question posed above, *near*sighted people see *near* objects clearly and need corrective lenses to focus on distant objects.

Farsightedness is formally called **hyperopia** (hi″per-o′pe-ah; "far vision"). It occurs when the parallel light rays from distant objects are focused *behind* the retina—at least in the resting eye, in which the lens is flat and the ciliary muscle is relaxed (see part (c) in the figure). Hyperopia usually results from an eyeball that is too short. People with hyperopia can see distant objects clearly because their ciliary muscles contract continuously to increase the light-bending power of the lens, which moves the focal point forward onto the retina. However, the diverging rays from *nearby* objects are focused so far behind the retina that even the full accommodative power of the lens cannot bring the focal point onto the retina. Therefore, nearby objects appear blurry. Furthermore, hyperopic individuals are subject to eyestrain as their end-

lessly contracting ciliary muscles tire from overwork. Correction of hyperopia requires *convex* corrective lenses that converge the light rays before they enter the eye. To answer the second question posed at the beginning of this essay, *far*sighted people can see *far*away objects clearly and require corrective lenses to focus on nearby objects.

Another eye-focusing disorder is **presbyopia** (pres″be-o′pe-ah), literally "old-person's vision." Developing between the ages of 45 and 65 in essentially all people, presbyopia is the loss of elasticity and accommodative power of the lens. A presbyopic eye that is otherwise normal can see distant objects clearly (recall that the normal eye of young and old alike is set for distance vision), but it cannot accommodate to focus on nearby objects. To allow near vision, corrective glasses with convex lenses are prescribed (think of this as supplementing the convex lens of the eye itself). However, many people with presbyopia do not have perfect *distance* vision either, so they wear bifocal glasses, with the upper part of the bifocals focused for distance vision, the lower part for closer vision.

Unequal curvatures in different

bending all these light rays so that they converge on the retina at a single *focal point.* The light-bending parts of the eye, called **refractory media,** are the cornea, the lens, and the humors. The cornea is responsible for most of the light-bending, the lens for some of it, and the humors a minimal amount.

Although the lens is not as powerful as the cornea in bending light, its curvature is adjustable. This adjustability allows the eye to focus on nearby objects—a process called **accommodation** (Figure 15.14). A resting eye, with its lens stretched flat by tension in the suspensory ligament, is "set" to focus the almost parallel rays from distant points. Therefore, distance vision is the natural state. The diverging rays from *nearby* points must be bent more sharply if they are to focus on the retina. This is accomplished by rounding of the lens: The ciliary muscle pulls the ciliary body both anteriorly and inward toward the pupil,

thereby releasing the tension on the suspensory ligament. No longer stretched, the lens rounds by its own elastic recoil. Accommodation is controlled by the parasympathetic fibers that signal the ciliary muscle to contract (p. 396).

Focusing on nearby objects is always accompanied by pupillary constriction, which prevents the most divergent light rays from entering the eye and passing through the extreme edges of the lens. Such rays would not be focused properly and would cause blurred vision.

For simplicity, our discussion of eye focusing has been confined to single-point images. We will not discuss how the eye focuses the "many-point" images of the large objects it views. However, it must be mentioned that the convex lens of the eye, just like the convex lens of a camera, produces images that are upside down and reversed from right to left. Therefore,

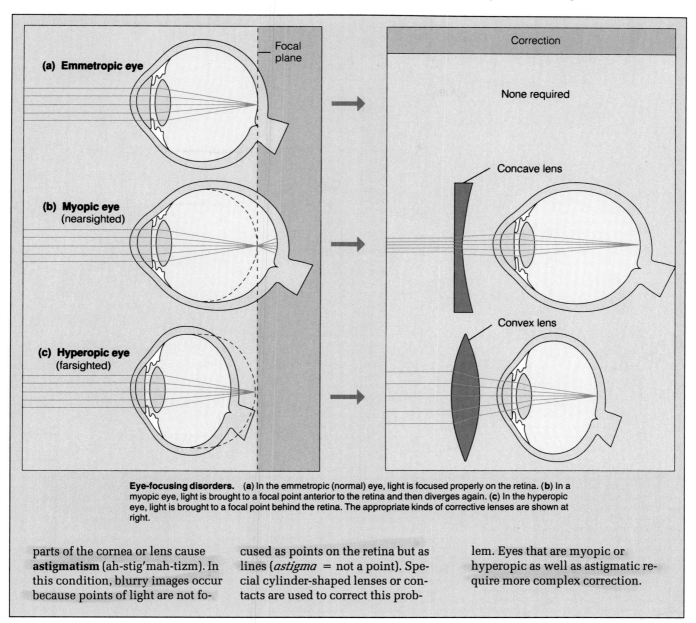

Eye-focusing disorders. (a) In the emmetropic (normal) eye, light is focused properly on the retina. (b) In a myopic eye, light is brought to a focal point anterior to the retina and then diverges again. (c) In the hyperopic eye, light is brought to a focal point behind the retina. The appropriate kinds of corrective lenses are shown at right.

parts of the cornea or lens cause **astigmatism** (ah-stig′mah-tizm). In this condition, blurry images occur because points of light are not fo- cused as points on the retina but as lines (*astigma* = not a point). Special cylinder-shaped lenses or contacts are used to correct this prob- lem. Eyes that are myopic or hyperopic as well as astigmatic require more complex correction.

an inverted and reversed image of the visual field blankets each retina. (The cerebral cortex then "flips" the image back, so that we see things as they are actually oriented.)

Common focusing disorders of the eye are discussed in the box above.

Visual Pathways

Visual Pathway to the Cerebral Cortex

Visual information travels to the cerebral cortex through the main **visual pathway.** This pathway begins in the retina as light activates the photoreceptors, which signal bipolar cells, which signal ganglion cells. Axons of the ganglion cells exit the eye in the **optic nerve** (Figure 15.15). At the X-shaped **optic chiasma** (ki-as′mah; "cross"), which lies anterior to the hypothalamus, the axons from the medial half of each eye decussate then continue in an **optic tract.** The paired optic tracts sweep posteriorly around the hypothalamus and send most of their axons to the **lateral geniculate nucleus (body) of the thalamus,** where they synapse. Axons of the thalamic neurons then project through the internal capsule to form the **optic radiation** of fibers in the cerebral white matter. These fibers project to the primary **visual cortex** in the occipital lobe, where the conscious perception of visual images (seeing) occurs.

The exchange of axons within the optic chiasma relates to depth perception (stereoscopic or three-dimensional vision). To explain this, we must divide the retina of each eye into a medial and lateral half (see the dashed line in Figure 15.15a). Because the lens system of each eye reverses all images, the medial half of each retina receives light rays from the lateral

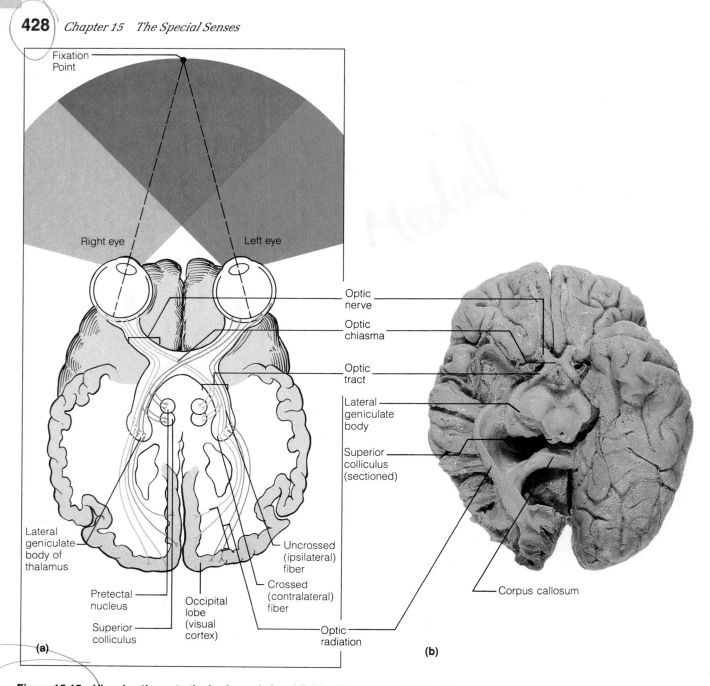

Figure 15.15   Visual pathway to the brain, and visual fields of the eyes.   (a) Visual pathway: In the visual pathway, ganglion cells from the retina form the optic nerves, optic chiasma, and optic tract to the thalamus. From the thalamus, the visual information is projected to the visual cortex in the occipital lobe of the cerebrum. **Visual fields**: Note that the right eye is on the left side of this figure, and vice versa. The visual fields of the two eyes overlap in the central, purple area (area of stereoscopic, or three-dimensional vision). **(b)** Photograph of the visual pathway.

(peripheral) part of the visual field (that is, from objects that lie far to the left or right rather than straight ahead). Correspondingly, the lateral half of each retina receives an image of the central part of the visual field. Only those axons from the *medial* halves of the two retinas cross over at the optic chiasma. The effect is that all information from the left half of visual space is directed through the right optic tract to be perceived by the right cerebral cortex. Likewise, the right half of visual space is perceived by the left visual cor-

tex. Each cerebral cortex receives an image of half the visual field, as viewed by the two different eyes from slightly different angles. The cortex then compares these two similar but different images and, in doing so, assembles a perception of depth.

The above relationships explain patterns of blindness that follow damage to different visual structures. Destruction of one eye or one optic nerve eliminates true depth perception and causes a loss of peripheral vision on the side of the damaged eye. For example, if the "left eye" in Figure 15.15a were lost, nothing could be seen in the visual area colored blue in that figure. On the other hand, if damage occurs beyond the optic chiasma—in an optic tract, the thalamus, or the visual cortex—then the entire opposite half of the visual field is lost. For example, a stroke affecting the left visual cortex leads to blindness (blackness) throughout the right half of the visual field.

A tumor in the hypothalamus or pituitary gland may compress the optic chiasma, destroying all the axons in this structure. Can you deduce how this affects vision?

Peripheral vision would be lost on both the right and left sides of the visual field. Central vision would remain intact, so the person could only see straight ahead (tunnel vision). ▪

Visual Pathways to Other Parts of the Brain

Some axons from the optic tracts send branches to the midbrain (Figure 15.15a). These branches end in the **superior colliculi,** reflex nuclei controlling the extrinsic eye muscles (p. 334), and in the **pretectal nuclei,** which mediate the pupillary light reflexes. Other branches from the optic tracts run to the *suprachiasmatic nucleus* of the hypothalamus (Figure 12.15). This nucleus is the timer that runs our daily biorhythms (p. 332), and it requires visual input to keep it in synchrony with the environmental cycle of daylight and darkness.

Embryonic Development of the Eye

The eyes develop as outpocketings of the brain (Figure 15.16). By week 4, paired lateral outgrowths called **optic vesicles** protrude from the diencephalon.

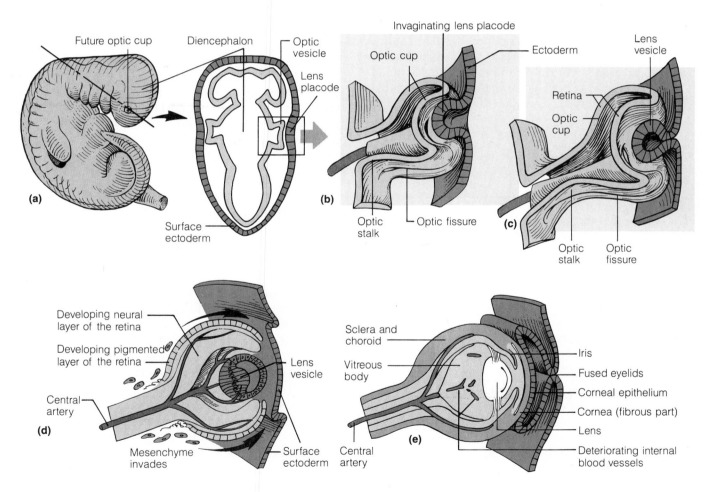

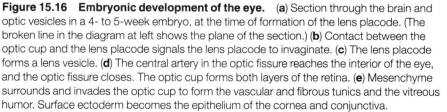

Figure 15.16   Embryonic development of the eye. (a) Section through the brain and optic vesicles in a 4- to 5-week embryo, at the time of formation of the lens placode. (The broken line in the diagram at left shows the plane of the section.) (b) Contact between the optic cup and the lens placode signals the lens placode to invaginate. (c) The lens placode forms a lens vesicle. (d) The central artery in the optic fissure reaches the interior of the eye, and the optic fissure closes. The optic cup forms both layers of the retina. (e) Mesenchyme surrounds and invades the optic cup to form the vascular and fibrous tunics and the vitreous humor. Surface ectoderm becomes the epithelium of the cornea and conjunctiva.

Soon, these hollow vesicles indent to form double-layered **optic cups.** The proximal parts of the outgrowths, called the **optic stalks,** form the basis of the optic nerves.

Once a growing optic vesicle reaches the overlying surface ectoderm, it signals the ectoderm to thicken and form a **lens placode** (plak′ōd; "plate"). By week 5, this placode has invaginated to form a **lens vesicle.** Shortly thereafter, the lens vesicle pinches off into the optic cup, where it becomes the lens.

The internal layer of the optic cup differentiates into the neural retina, whereas the external layer becomes the pigmented layer of the retina. The **optic fissure,** a groove on the underside of each optic stalk and cup, serves as a direct pathway for blood vessels to reach and supply the interior of the developing eye. When this fissure closes, the optic stalk becomes a tube through which the optic nerve fibers, originating in the retina, grow centrally to the diencephalon. The blood vessels that were originally within the optic fissure now lie in the center of the optic nerve.

The fibrous tunic, vascular tunic, and vitreous humor form from head mesenchyme that surrounds the early optic cup. The interior of the eyeball has a rich blood supply during development, but virtually all of these blood vessels degenerate. The vessels that remain are those in the vascular tunic and retina.

The Ear

The **ear** is the receptor organ for both hearing and equilibrium. Its three main regions are the *outer ear, middle ear,* and *inner ear* (Figure 15.17). The outer and middle ears participate in hearing only, whereas the inner ear functions in both hearing and equilibrium.

The Outer (External) Ear

The **outer (external) ear** consists of the auricle and the external auditory canal. The **auricle,** or **pinna,** is what most people call the ear—the shell-shaped projection that surrounds the opening of the external auditory canal. The auricle is composed of elastic cartilage covered with skin. Its rim is the **helix.** Its fleshy, dangling *lobule* ("earlobe") lacks supporting cartilage. The function of the auricle is to gather and funnel sound waves into the external auditory canal.

The **external auditory canal,** or external acoustic meatus (p. 146), is a short tube about 2.5 cm long. It runs medially from the auricle to the eardrum. Near the auricle, its wall consists of elastic cartilage, but its medial two-thirds is carved into the temporal bone.

The entire canal is lined with skin that contains hairs, sebaceous glands, and modified apocrine sweat glands called *ceruminous* (sĕ-roo′mĭ-nus) *glands.* The ceruminous and sebaceous glands secrete yellow-brown *cerumen,* or earwax (*cere* = wax). Earwax traps dust and repels insects, keeping these substances from entering the auditory canal.

Sound waves entering the external auditory canal hit the **tympanic membrane,** or eardrum (*tympanum* = drum). This thin, translucent membrane forms the boundary between the outer and middle ears. It is shaped like a flattened cone with its apex protruding medially into the middle ear cavity. Sound waves that travel through the air set the eardrum vibrating, and the eardrum in turn transfers the sound energy to tiny bones in the middle ear.

A torn tympanic membrane is called a **perforated eardrum.** Can you imagine some causes of this condition?

A perforated eardrum can be caused by a cotton swab or a sharp object placed in the external auditory canal. However, it usually results from a middle ear infection, in which pus accumulates medial to the eardrum. The resulting pressure causes this thin membrane to burst. Perforated eardrums heal well on their own, so they do not lead to permanent deafness. However, even a small amount of scarring can permanently diminish the acuity of hearing. ■

The Middle Ear

The **middle ear,** or *tympanic cavity,* is a small, air-filled space inside the petrous part of the temporal bone. The middle ear is lined by a thin mucous membrane and has the shape of a rounded box standing on end (Figure 15.17). Its *lateral boundary* is the tympanic membrane. Its *medial boundary* is a wall of bone that separates it from the inner ear. Two small holes penetrate this medial wall, a superior **oval window** *(vestibular window)* and an inferior **round window** *(cochlear window).* Superiorly, the tympanic cavity arches upward as the **epitympanic recess** (*epi* = over). Its *superior boundary* is the roof of the petrous bone. The bone here is so thin that middle ear infections can spread to the overlying brain. The *posterior wall* of the middle ear opens into the **mastoid antrum** (Figure 15.18), a canal leading to the *mastoid air cells.* Middle ear infections can spread through the mastoid antrum into these air cells in the mastoid process (p. 147). The *anterior wall* of the middle ear contains the opening of the auditory tube (Figures 15.17 and 15.18).

The **auditory tube** *(eustachian tube)* links the middle ear with the pharynx. About 4 cm long (1.5 inches), this tube runs medially, anteriorly, and inferi-

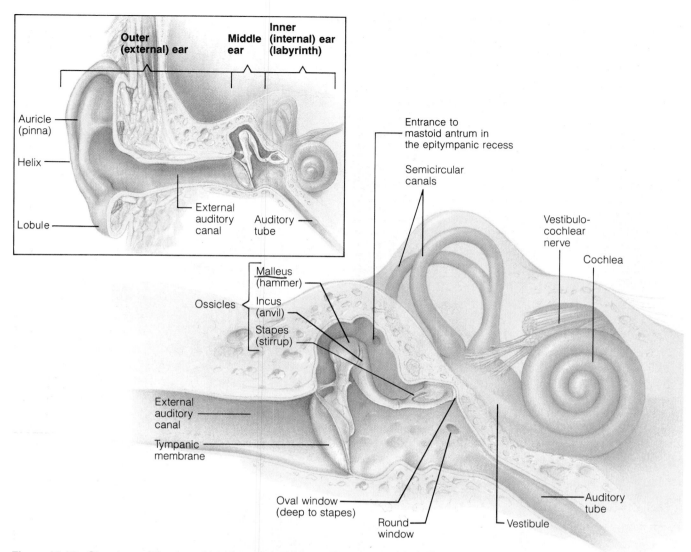

Figure 15.17   Structure of the ear.   Note the outer, middle, and inner ear, as labeled in the figure at upper left. In the inner ear, the bony labyrinth is shown, but the membranous labyrinth is not.

orly. Its lateral third consists of bone and occupies a groove on the inferior surface of the skull. The medial two-thirds of the tube is cartilage and opens into the superior part of the pharynx behind the nasal cavity. Normally, this tube is flattened and closed, but swallowing or yawning can open it briefly to equalize the air pressure in the middle ear with the pressure of the outside air. This is an important function because the eardrum does not vibrate freely unless the pressure on both its surfaces is the same. Differences in air pressure build up across the eardrum whenever we change altitude (as in an airplane), pushing the eardrum inward or outward. The next time your ears "pop" in such a situation, try yawning, for this is the best way to open your auditory tubes and equalize the air pressures.

Otitis media (o-ti′tis; "ear inflammation") is infection and inflammation of the middle ear. It usually starts as a throat infection that spreads to the middle ear through the auditory tube. Children are most susceptible because their auditory tubes are shorter than those of adults.

Children with otitis media sometimes have their eardrums lanced and have ear tubes inserted through the eardrum. Can you imagine the purpose of these procedures?

Fluid and pus can build up in the middle ear cavity and exert painful pressure within this enclosed space. A **myringotomy** (mir″ing-got′o-me), literally "lancing the eardrum," allows drainage and relieves the pressure. The tiny tube that is implanted in the eardrum during myringotomy permits the pus to continue draining into the external ear. This tube is left in the eardrum and falls out by itself within a year. These procedures are performed only in patients who have

10-16-MO

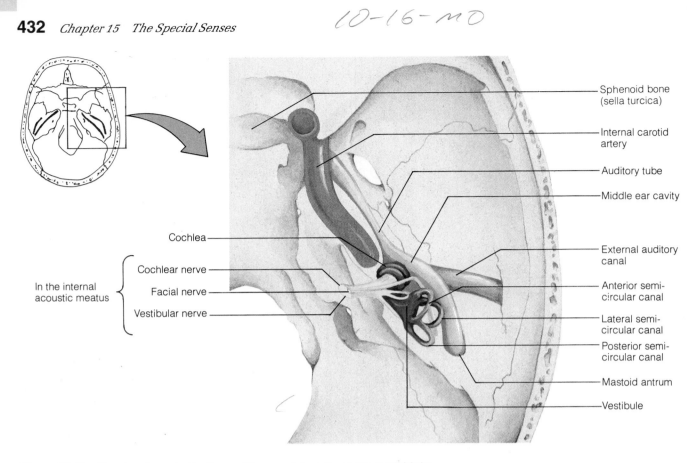

Figure 15.18   Ear structures in the floor of the cranial cavity of the skull (right ear). This is a superior view (structures are shown as though the surrounding temporal bone were transparent). For further orientation, see the inset at upper left and Figure 7.4c on p. 145.

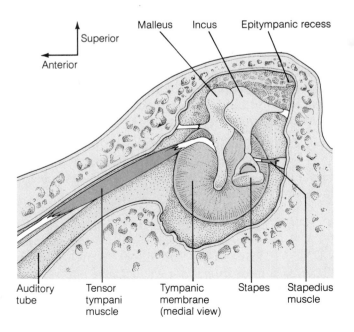

Figure 15.19   The three ossicles in the right middle ear, medial view.

persistent ear infections, as most cases of otitis media are cleared up by antibiotics. ∎

The tympanic cavity is spanned by the three smallest bones in the body, the **ossicles** (Figures 15.17 and 15.19). From lateral to medial, these are the *malleus, incus,* and *stapes.* The **malleus** (mal′e-us), or hammer, looks like a club with a knob on top. The **incus** (ing′kus), or anvil, resembles a tooth with two roots (Figure 15.19). The **stapes** (sta′pēz) looks exactly like the stirrup of a saddle (the part in which the rider's foot rests). The "handle" of the malleus attaches to the eardrum, and the base of the stapes fits into the *oval* window. Most people have trouble remembering whether the stapes fits into the oval window or into the round window inferior to it: To avoid this problem, remember that the footplate of a saddle stirrup is oval, not round.

Can you imagine the result of a condition in which the footplate of the stapes fuses to the oval window?

The stapes cannot move, and deafness results. This condition, called **otosclerosis** (o″to-sklĕ-ro′sis; "hardening of the ear"), is caused by an excessive growth of bone tissue in the walls of the middle ear cavity. A common age-related problem, otosclerosis affects 1 in every 200 people. It can be treated by a delicate and difficult surgery that removes the stapes and replaces it with a prosthetic (artificial copy). ■

Tiny ligaments suspend the ossicles in the middle ear, and tiny synovial joints link the ossicles into a chain. The chain of ossicles transmits the vibratory motion of the eardrum to the oval window, causing the stapes to set the fluid of the inner ear in motion. The ossicles amplify the pressure of the sound vibrations by concentrating the force from the large eardrum onto the much smaller oval window. Overall, the ossicles amplify sound pressure by about 20 times. Without them, we could hear only loud sounds.

Two tiny skeletal muscles occur in the middle ear cavity (Figure 15.19). The **tensor tympani** (ten′sor tim′pah-ni) originates on the auditory tube and inserts on the malleus. The **stapedius** (stah-pe′de-us) runs from the posterior wall of the middle ear to the stapes. When the ears are assaulted by very loud sounds, these muscles contract reflexively to limit the vibration of the ossicles and thus prevent damage to the hearing receptors.

The Inner (Internal) Ear

The **inner (internal) ear** is also called the **labyrinth** ("maze") because of its maze-like, highly complex shape (Figures 15.17, 15.18, and 15.20). It lies within the thick, protective walls of the petrous part of the temporal bone and consists of two main divisions: the bony labyrinth and the membranous labyrinth.

The **bony labyrinth** is a cavity carved in the petrous bone. Its system of twisting channels forms three divisions: From posterolateral to anteromedial, these are the *semicircular canals,* the *vestibule,* and the *cochlea* (Figure 15.17). Textbooks often illustrate the bony labyrinth as if it were a solid object; however, you should keep in mind that it is actually a *cavity.*

The **membranous labyrinth** is a continuous series of membrane-lined sacs and ducts that fit loosely within the bony labyrinth and (more or less) follow its contours (Figure 15.20). The main parts of the membranous labyrinth are (1) the *semicircular ducts* (one inside each semicircular canal), (2) the *utricle* and *saccule* (both in the vestibule), and (3) the *cochlear duct* (in the cochlea). The wall of the membranous labyrinth—its "membrane"—is a thin layer of connective tissue lined by a simple squamous epithelium. The parts of the bony and membranous labyrinths are summarized in Table 15.2 on p. 435.

The membranous labyrinth is filled with a clear fluid called **endolymph** (en′do-limf). External to the membranous labyrinth, the bony labyrinth is filled with another clear fluid called **perilymph** (per′ĭ-limf). It helps to point out that *endolymph* means "internal water" and *perilymph* means "surrounding water." Nowhere are the perilymph and endolymph continuous with one another.

The Vestibule

The **vestibule** is the central, egg-shaped cavity of the bony labyrinth (Figures 15.17 and 15.20). It lies just medial to the middle ear, and the oval window is in its lateral bony wall. Suspended within its perilymph are two egg-shaped parts of the membranous laby-

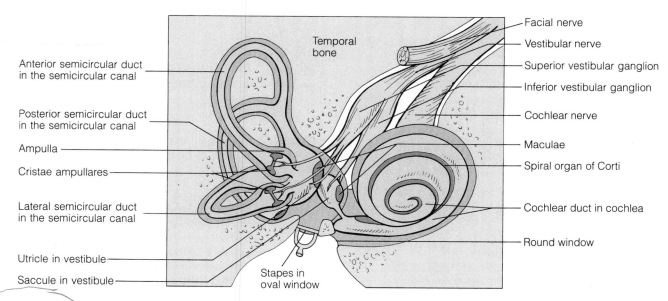

Figure 15.20   Membranous labyrinth of the inner ear, within the bony labyrinth.

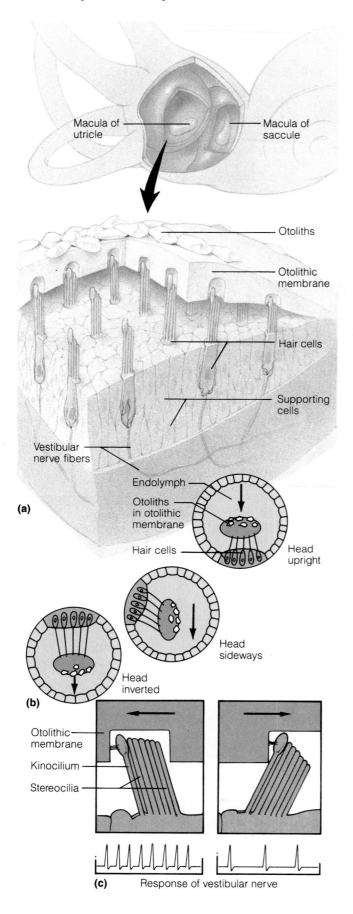

rinth, the **utricle** (u′trĭ-k′l) and the **saccule.** *Utricle* means "a leather bag," and *saccule* means "a little sac." The utricle is continuous with the semicircular ducts, the saccule with the cochlear duct.

The utricle and saccule each house a spot of sensory epithelium called a **macula** (mak′u-lah; "spot") (Figure 15.21a). Both the macula of the utricle and the macula of the saccule contain receptor cells that monitor the position of the head when the head is stationary. This aspect of the sense of balance is called *static equilibrium.* Furthermore, these receptor cells report on *linear acceleration;* that is, they monitor straight-line changes in the speed and the direction of head movements, but not rotational movements of the head.

Each macula is a patch of epithelium containing columnar **supporting cells** and scattered receptors called **hair cells** (Figure 15.21a). The hair cells synapse with sensory fibers of the vestibular part of the vestibulocochlear nerve, or the **vestibular nerve** for short. Named for the hairy look of its free surface, a hair cell has many stereocilia (long microvilli) and a single kinocilium (a true cilium) protruding from its apex. The tips of these stiff hairs are embedded in an overlying **otolithic** (o″to-lith′ik) **membrane.** This membrane is actually a jelly-like disc that contains heavy crystals of calcium carbonate called **otoliths** ("ear stones").

It is easy to understand how the otoliths contribute to the sense of equilibrium: Whenever the head is held still, gravity pulls on a heavy otolithic membrane and bends the hairs to one side (Figure 15.21b). Furthermore, whenever the body accelerates in a straight line, the heavy otolithic membrane lags behind, again causing the hairs to bend. In either case, bending of the hairs alters the membrane charge on the hair cells and changes the firing rate of the vestibular nerve (Figure 15.21c). This signal is interpreted by the brain as an indication of head position or head acceleration.

The macula in the utricle has a horizontal orientation, whereas the macula in the saccule is oriented vertically (see the uppermost diagram in Figure 15.21a). The arrangement of the two maculae at right

Figure 15.21 Anatomy and function of the maculae in the inner ear. (a) The upper diagram shows the locations of the maculae in the utricle and saccule, while the lower diagram is an enlargement showing that a macula is an epithelium containing hair cells. Note the overlying otolithic membrane. **(b)** Sections through the utricle, showing the effect of gravitational pull on the macula of the utricle when the head is held in various positions. In the macula of the utricle, the hairs bend when the head is held sideways. **(c)** When movement of the otolithic membrane (in the direction indicated by the arrow) bends the hairs of the hair cells in the direction of the kinocilium, the hair cells depolarize and signal the vestibular nerve to produce impulses. When the hairs are bent away from the kinocilium, the hair cells do not depolarize, causing the nerve fibers to slow their rate of impulse production.

Table 15.2	The Inner Ear: Basic Structures of the Bony and Membranous Labyrinths		
	Bony labyrinth	**Membranous labyrinth (within bony labyrinth)**	**Functions of the membranous labyrinth**
	1. Semicircular canals	Semicircular ducts	Equilibrium: rotational (angular) acceleration of the head
	2. Vestibule	Utricle and saccule	Equilibrium: static equilibrium and linear acceleration of the head
	3. Cochlea	Cochlear duct	Hearing

angles to one another ensures that every possible position of the head (and every direction of linear movement) stimulates at least one macula. The utricular macula responds best to acceleration in the *horizontal* plane and to holding the head *sideways,* because neither vertical (up and down) movements nor holding the head upright can displace its horizontal otolithic membrane. Conversely, the vertical saccular macula responds best to vertical movements and to holding the head upright.

The maculae are innervated by two branches of the vestibular nerve (Figure 15.20). The sensory neurons in this nerve are bipolar neurons, with cell bodies located in the *superior* and *inferior vestibular ganglia.* These ganglia lie in the internal acoustic meatus of the petrous temporal bone. The further path of vestibular nerve fibers within the brain is discussed shortly.

Having considered the receptors for static equilibrium and linear acceleration, we now turn to the receptors for *rotational* acceleration of the head.

The Semicircular Canals

The three **semicircular canals** of the bony labyrinth lie posterior and lateral to the vestibule (Figures 15.20 and 15.22). Each semicircular canal actually defines about two-thirds of a circle and has an expansion at one end called an **ampulla** ("flask"). Each canal lies in one of the three planes of space: The **anterior** and **posterior semicircular canals** lie in vertical planes at right angles to each other, whereas the **lateral semicircular canal** *(horizontal canal)* lies almost horizontally. The relationships of the three canals to one another are very clear from a superior view (Figure 15.18). Snaking through each semicircular canal is a membranous **semicircular duct.** Each duct has a swelling called a membranous **ampulla** within the corresponding bony ampulla (Figure 15.22b).

Each membranous ampulla houses a small crest called a **crista ampullaris** ("crest of the ampulla") (Figure 15.22b and c). The cristae contain the receptor cells that measure *rotational (angular) acceleration* of the head, as occurs when you twirl on the dance floor or do a somersault. Each crista is a patch of columnar epithelial cells raised on a long ridge. Like the maculae, it contains *supporting cells* and receptor *hair cells.* The "hairs" of these hair cells project into a tall, jelly-like mass that resembles a pointed cap, the **cupula** (ku'pu-lah; "little barrel"). The basal parts of the hair cells synapse with fibers of the vestibular nerve.

Since the three semicircular ducts lie in three different planes, each crista responds to head rotation in a different plane of space. When the head starts to rotate, the endolymph in the semicircular duct lags behind at first, pushing on the cupula and bending the hairs (Figure 15.22d). As their hairs bend, the hair cells depolarize and change the pattern of impulses carried by vestibular nerve fibers to the brain.

The Cochlea

The **cochlea** (kok'le-ah), from the Latin meaning "snail shell," is a spiraling chamber in the bony labyrinth (Figures 15.20 and 15.23). It is about half the size of a split pea. Its base (first part) attaches to the vestibule. From there, the cochlea coils for about 2½ turns around a pillar of bone called the **modiolus** (mo-di'o-lus) (Figure 15.23a). The modiolus is shaped like a screw whose tip lies at the apex of the cochlea, pointing anterolaterally. Just as screws have threads, the modiolus has a spiraling thread of bone called the **spiral lamina.** Running through the bony core of the modiolus is the cochlear division of the vestibulocochlear nerve, or the **cochlear nerve** for short.

The part of the membranous labyrinth within the cochlea is the **cochlear duct** (Figure 15.23b). It contains the receptors for hearing. This duct winds through the cochlea and ends blindly in the cochlear apex. The cochlear duct divides the cochlea into three separate chambers, or **scalas** (*scala* = a ladder): (1) The **scala vestibuli** is the part of the bony labyrinth shown superior to the cochlear duct in Figure 15.23b.

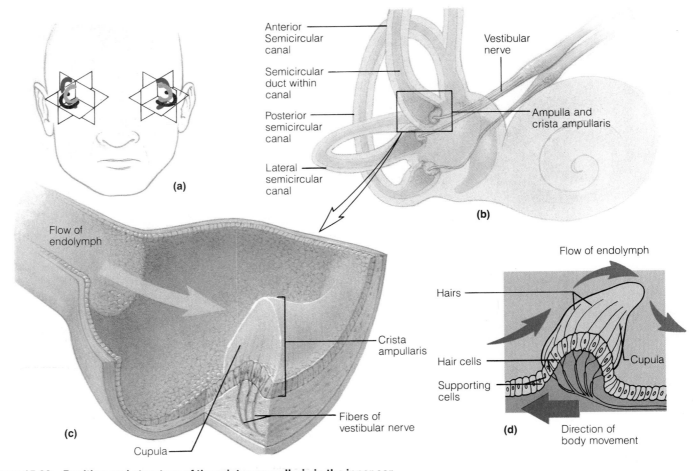

Figure 15.22 Position and structure of the cristae ampullaris in the inner ear.
(**a**) Positions of the three paired semicircular canals in the skull. The three canals lie at right angles to each other. (**b**) The semicircular ducts within the semicircular canals of the right ear. The cristae ampullaris are outlined in the three membranous ampullae. (**c**) Anatomy of a crista ampullaris. (**d**) During acceleratory rotation of the head, the endolymph in the semicircular duct lags behind. This fluid bends the cupula away from the direction of head movement, stimulating the hair cells.

It is continuous with the vestibule near the base of the cochlea and abuts the oval window. (2) The **scala media** is just another name for the cochlear duct itself. (3) The **scala tympani** is the part of the bony labyrinth shown inferior to the cochlear duct in Figure 15.23b. It ends at the round window at the base of the cochlea. The scala vestibuli and scala tympani are continuous with each other at the apex of the cochlea, in a region called the **helicotrema** (hel″ĭ-ko-tre′mah; "the hole in the spiral") (Figure 15.23a).

Since the cochlear duct is a part of the membranous labyrinth, it is filled with endolymph. The scala vestibuli and scala tympani, however, are divisions of the bony labyrinth and are therefore filled with perilymph.

The cochlear duct contains the receptors for hearing (Figure 15.23b). The "roof" of the cochlear duct, separating it from the scala vestibuli, is the **vestibular membrane.** The external wall of this duct is the *stria vascularis* ("vascularized streak"), an unusual epithelium that contains capillaries and secretes the en-

dolymph of the inner ear. The floor of the cochlear duct consists of the spiral lamina of bone plus an attached sheet of fibers called the **basilar membrane.** This membrane supports the **spiral organ of Corti,** the receptor epithelium for hearing. This tall epithelium (Figure 15.23c) is very similar to the receptor epithelia for equilibrium. It consists of columnar *supporting cells* and several long rows of cochlear *hair cells* (the receptor cells). The tips of the hairs (microvilli) are embedded in a gel-like **tectorial membrane** ("roofing membrane"). The hair cells synapse with sensory fibers of the cochlear nerve. These fibers belong to bipolar neurons, whose cell bodies occupy a **spiral ganglion** in the spiral lamina and modiolus (Figure 15.23a) and whose central fibers project to the brain.

How do sound waves stimulate the receptor (hair) cells in the spiral organ of Corti? A simplified explanation is presented here, and some additional information is provided in Figure 15.24. First, sound vibrations travel from the eardrum through the ossicles, causing the stapes to oscillate back and forth

10-16-MO

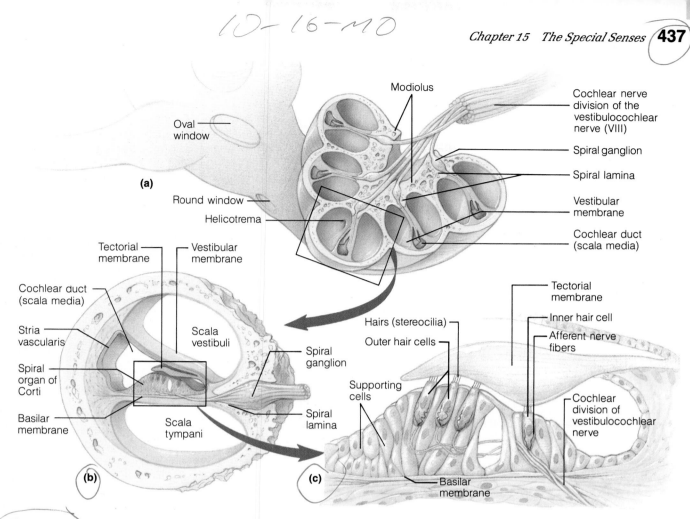

Figure 15.23   Anatomy of the cochlea.   (**a**) A section through the right cochlea (superior half removed). The area in the rectangle is the apex of the cochlea. (**b**) Magnified cross-sectional view of one turn of the cochlea, showing the three scalas: scala vestibuli, scala tympani, and cochlear duct (scala media). (**c**) Detailed structure of the spiral organ of Corti.

against the oval window. This oscillation sets up pressure waves in the perilymph of the scala vestibuli and in the endolymph of the cochlear duct. These waves cause the basilar membrane to vibrate up and down. The hair cells in the spiral organ move along with the basilar membrane (Figure 15.23c), but the overlying tectorial membrane (in which the hairs are anchored) does not move. Therefore, the movements of the hair cells cause their hairs to bend, and these cells release neurotransmitters each time such bending occurs. The neurotransmitters excite the cochlear nerve fibers, which carry the vibratory (sound) information to the brain.

Equilibrium and Auditory Pathways

As does all sensory information, information on equilibrium and hearing travels to the brain for processing and integration.

The **equilibrium pathway** transmits information on the position and movements of the head through the vestibular nerve to the brain stem. Equilibrium is the only special sense for which most information

goes to the *lower* brain centers—which are primarily reflex centers—rather than to the "thinking" cerebral cortex. This reflects the fact that our responses to body imbalance, such as stumbling, must be rapid and reflexive. (If we were to take time to think about correcting our fall, we would likely end up on the ground.) The vestibular nuclei in the medulla (p. 337) and the cerebellum (p. 337) are the major brain centers for processing information on equilibrium. Furthermore, a *minor* pathway to the cerebral cortex provides conscious awareness of the position and movements of the head. In this minor pathway, vestibular nerve fibers project to the vestibular nuclei, which project to the thalamus, which projects to the cerebrum.

The ascending **auditory pathway** transmits auditory information from the cochlear receptors to the cerebral cortex (Figure 15.25). First, impulses pass through the cochlear nerve to the **cochlear nuclei** in the medulla. From there, some neurons project to the **superior olivary nuclei,** which lie at the junction of the medulla and pons. Beyond this, the axons ascend in the **lateral lemniscus** (a fiber tract) to the **inferior colliculus** (the auditory reflex center in the midbrain),

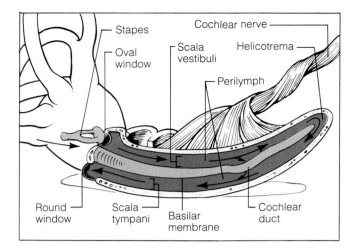

Figure 15.24   Role of the cochlea in hearing. The cochlea (in blue) is drawn as if uncoiled to make the events that occur in it easier to follow. Beating of the stapes on the oval window sets up vibrations in the perilymph of the scala vestibuli, which are transmitted to the endolymph of the cochlear duct. These vibrations cause the basilar membrane to vibrate, stimulating the cochlear nerve. The structure of the basilar membrane is such that it segregates sound according to frequency—its basal part (left) vibrates in response to low-pitched sounds, whereas its apical part (right) vibrates in response to high-pitched sounds. The vibrations of the basilar membrane set the perilymph vibrating in the underlying scala tympani (see the arrows in that scala). These vibrations then travel to the round window (at left), where they push on the membrane that covers that window, thereby dissipating their remaining energy into the air of the middle ear cavity. Without this release mechanism, echoes would reverberate within the rigid cochlear "box," disrupting sound reception.

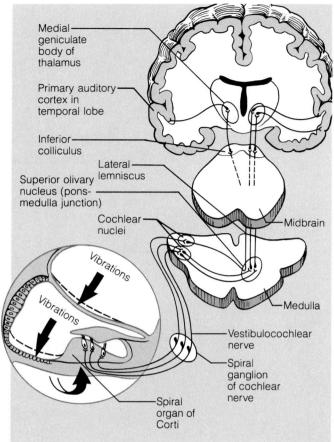

Figure 15.25   Simplified diagram of the auditory pathway, from the spiral organ of Corti to the primary auditory cortex in the temporal lobe. For clarity, only the pathway from the right ear is shown.

which projects to the **medial geniculate nucleus (body)** of the thalamus. Axons from the thalamic neurons then project to the primary auditory cortex, which provides conscious awareness of sound. The auditory pathway is unusual in that not all of its fibers cross over to the other side of the brain. Therefore, each primary auditory cortex receives impulses from both ears. Clinically, this phenomenon makes identifying damage to the primary auditory cortex on one side difficult, because such damage produces only a minimal amount of hearing loss.

Embryonic Development of the Ear

Development of the ear begins in the fourth week after conception (Figure 15.26). First the inner ear begins to form from a thickening of the surface ectoderm called the **otic placode,** which lies lateral to the hindbrain on each side of the head. The otic placode invaginates to form the **otic pit.** Then its edges fuse to form the **otic vesicle,** which detaches from the surface epithelium. The otic vesicle takes on a complex shape and becomes the membranous labyrinth. The mesenchyme tissue around the otic vesicle becomes the petrous temporal bone—that is, the walls of the bony labyrinth.

As the inner ear develops, the middle ear starts to form. Lateral outpocketings called *pharyngeal pouches* form from the endoderm-lined pharynx (p. 12). The middle ear cavity and the auditory tube develop from the first of the pharyngeal pouches. The ossicles, which bridge the middle ear cavity, develop from cartilage bars associated with the first and second pharyngeal pouches.

Turning to the external ear, the external auditory canal differentiates from the first **branchial groove** (brang'ke-al; "gill"), an indentation of the surface ectoderm. The auricle of the outer ear grows from a series of bulges around this branchial groove.

✚ Disorders of Equilibrium and Hearing

Motion Sickness

Motion sickness is a common disorder of equilibrium, in which motion such as riding in a car leads to nausea and vomiting. The cause of this condition has

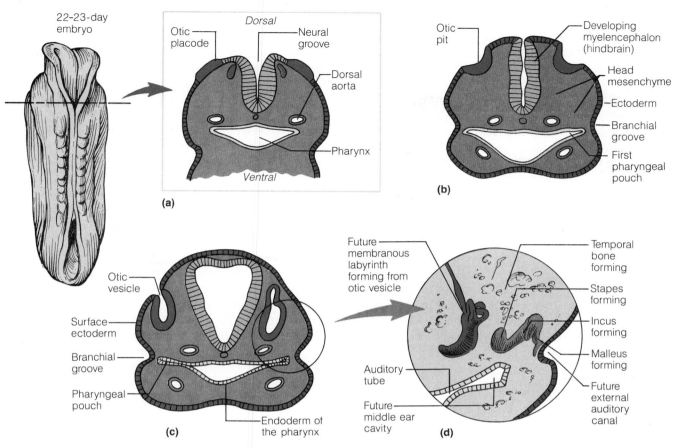

Figure 15.26   Embryonic development of the ear.   At left, a surface view of the embryo shows the plane of the sections. (**a**) By about day 22, otic placodes have formed as thickenings of the ectoderm. Endoderm (yellow) lines the pharynx. (**b**) The otic placodes invaginate to become otic pits, and branchial grooves start to push internally from the surface ectoderm. (**c**) At about 28 days, otic vesicles are forming from the otic pits. Also note the pharyngeal pouch. (**d**) Between weeks 5 and 8, the membranous labyrinth forms from the otic vesicle, while the first pharyngeal pouch gives rise to the auditory tube and middle ear cavity, and the branchial groove develops into the external auditory canal. Head mesenchyme forms the surrounding bony structures.

been difficult to determine. The most popular theory is that motion sickness is due to a mismatch of sensory inputs. For example, if you are in a rocking ship during a storm, visual inputs indicate that your body is fixed with reference to a stationary environment (your cabin), but your vestibular apparatus detects movement. The brain receives two different messages, and its "confusion" somehow leads to motion sickness. Another theory points out that the vestibular nuclei lie close to (and may project to) the centers in the medulla that control vomiting. Antimotion drugs can relieve the symptoms of motion sickness.

Meniere's Syndrome

Classic **Meniere's** (Men″ĕ-ārz′) **syndrome** is a disorder of the labyrinth that affects both the semicircular canals and the cochlea. The afflicted person has transient but repeated attacks of vertigo, nausea, and vomiting. Equilibrium is so disturbed that standing erect

is nearly impossible. A "howling" ringing of the ears is common, so hearing is impaired and ultimately lost. This syndrome probably results from a distortion of the membranous labyrinth by excessive amounts of endolymph. Less severe cases can be managed by antimotion drugs. For more debilitating attacks, diuretics (drugs that increase the output of urine) and a restriction of salt in the diet are used to decrease the body's volume of extracellular fluid, consequently decreasing the volume of the endolymph. Severe cases may require surgery to drain the excess endolymph from the inner ear. A last resort is removal of the entire labyrinth, which is usually deferred until hearing loss is complete.

Deafness

Any hearing loss, no matter how slight, is deafness of some sort. The two types of deafness are *conduction deafness* and *sensorineural deafness,* and they have different causes.

Conduction deafness occurs when sound vibrations cannot be conducted to the inner ear. It can be caused by earwax blocking the external auditory canal, a ruptured eardrum, otitis media, or otosclerosis.

Sensorineural deafness results from damage to the hair cells or to any part of the auditory pathway to the brain. Most often it results from the slow, normal loss of hearing receptor cells that proceeds throughout life. These hair cells can also be destroyed at an earlier age by one explosively loud noise or by repeated exposure to loud music, factory noise, or airport noise. Strokes and tumors that damage the auditory cortex can also cause sensorineural deafness. When deafness reflects damage to hair cells, hearing aids can help. Traditional hearing aids simply amplify sounds, an effective strategy if the loss of hair cells is not too great. For complete sensorineural deafness, **cochlear implants** are available. Placed in the temporal bone, these devices convert sound energy into electrical signals and deliver these signals directly to the cochlear nerve fibers. ■

The Special Senses Throughout Life

All special senses are functional, to a greater or lesser degree, at birth. Smell and taste are sharp in newborns, and infants relish food that adults consider bland (children have more taste buds than do adults). Most people experience no difficulties with their chemical senses throughout childhood and young adulthood, but starting in the fourth decade of life, the ability to taste and smell declines. This decline reflects a gradual loss of the chemoreceptors, which are replaced much more slowly than in younger years.

In the darkness of the uterus, the fetus cannot see. Nonetheless, even before the light-sensitive parts of the photoreceptors develop, the nerve connections of the visual pathway have been made and are functional. Visual experiences during the first 6 weeks of life help fine-tune these synaptic connections.

Congenital problems of the eyes are relatively rare, although maternal rubella may cause congenital blindness or cataracts. The eyeballs of newborns are foreshortened, so all babies are hyperopic. The newborn sees only gray tones. Eye movements are uncoordinated, and often only one eye is used at a time. By 5 months, infants can follow moving objects with their eyes, but their visual acuity is still not sharp. By the age of 5 years, depth perception is present, color vision is well developed, and the earlier hyperopia is usually gone.

As a person ages, the lens loses its clarity and becomes discolored. The dilator muscles of the iris become less efficient, so the pupils stay partly constricted. These two changes decrease the amount of light reaching the retina, and visual acuity is dramatically lower in people over 70. In addition, elderly persons are susceptible to certain conditions that cause blindness, such as glaucoma, retinal detachment, and diabetes mellitus. Age-related presbyopia is discussed in the box on page 426.

Newborn infants can hear, but their early responses to sounds are mostly reflexive—for example, crying and clenching the eyelids in response to a startling noise. By the third or fourth month, infants can localize sounds and will turn to the voices of family members. Critical listening begins in toddlers as they learn to speak.

Congenital abnormalities of the ears are fairly common. Examples include partly or fully missing pinnae and closed or absent external auditory canals. Maternal rubella during the first trimester of pregnancy can cause sensorineural deafness. Except for the common ear infections of childhood, few problems usually affect the ears until old age. By the 60s, however, deterioration of the spiral organ of Corti becomes noticeable. We are born with about 20,000 hair cells in each ear, but they are gradually lost and not replaced. The ability to hear high-pitched sounds fades first. This gradual loss of hearing with age is called **presbycusis** (pres"bĭ-ku'sis), literally "old hearing." Presbycusis is the most common type of sensorineural deafness. Although it is considered a disability of old age, it is becoming more common in younger people as the modern world grows noisier.

✚ Related Clinical Terms

Blepharitis (blef"ah-ri'tis) (*blephar* = eyelash; *itis* = inflammation of)   Inflammation of the margins of the eyelids.

Labyrinthitis   Inflammation of the labyrinth.

Ophthalmology (of"thal-mol'o-je) (*ophthalmos* = eye)   The science that studies the eye and eye diseases.

Ophthalmologist   A medical doctor whose specialty is treating eye disorders.

Optometrist   A licensed nonphysician who measures vision and prescribes corrective lens.

Otitis externa   Inflammation and infection of the external auditory canal, caused by bacteria or fungi that enter the canal from outside, especially when the canal is moist (e.g., after swimming).

Otorhinolaryngology (o"to-ri"no-lar"ing-gol'o-je)   The science that studies the ear, nose, and larynx and the diseases of these body regions.

Papilledema (pap"il-ĕ-de'mah) (*papilla* = nipple; *edema* = swelling)   A protrusion of the optic disc into the eyeball, caused by conditions that increase pressure inside the cranial cavity.

Scotoma (sko-to'mah) (*scoto* = darkness).   A blind spot in the visual field other than the normal blind spot caused by the optic disc. Scotoma often reflects the presence of a brain tumor pressing on nerve fibers along the visual pathway.

Trachoma (trah-ko'mah) (*trach* = rough)   A highly contagious infection of the conjunctiva and cornea, caused by bacteria called chlamydia. Common worldwide, trachoma blinds millions of people in third-world countries. It is effectively treated with eye ointments containing antibiotic drugs.

Weber's test   Hearing test in which a vibrating tuning fork is held to the forehead. In patients with normal hearing, the tone is heard equally well in both ears. It will be heard best in the "good" ear if sensorineural deafness is present, and best in the "bad" ear if conduction deafness is present.

Chapter Summary

THE CHEMICAL SENSES: TASTE AND SMELL (pp. 412–415)

Taste Buds (p. 412)

1. Most taste buds are on the tongue, in the epithelium of fungiform and circumvallate (vallate) papillae.

2. Taste buds contain gustatory receptor cells, supporting cells, and basal cells that give rise to the other two cell types. The receptor cells are excited when taste-stimulating chemicals bind to their microvilli.

Taste Sensations and the Gustatory Pathway (pp. 412–413)

3. The four basic qualities of taste—sweet, sour, salty, and bitter—are sensed best on different regions of the tongue.

4. The sense of taste is served by cranial nerves VII, IX, and X, which send impulses to the medulla. From there, impulses travel to the thalamus and the taste area of the cerebral cortex.

Olfactory Receptors (pp. 413–415)

5. The olfactory epithelium is located in the roof of the nasal cavity. This epithelium contains receptor, supporting, and basal cells.

6. The olfactory receptor cells are ciliated bipolar neurons. Odor molecules bind to the cilia, exciting the neurons. Axons of these receptor neurons form the filaments of the olfactory nerve (cranial nerve I).

The Olfactory Pathway (p. 415)

7. Olfactory nerve axons transmit impulses to the olfactory bulb. Here, these axons synapse with mitral cells that send olfactory information through the olfactory tract to the primary olfactory cortex and limbic system.

Disorders of the Chemical Senses (p. 415)

8. Disorders of smell include anosmia (inability to smell) and uncinate fits (smell hallucinations).

THE EYE (pp. 415–430)

The Eye in the Orbit (p. 415)

1. The eye is located in the bony orbit and is cushioned by fat. The cone-shaped orbit also contains nerves, vessels, and extrinsic muscles of the eye.

Accessory Structures of the Eye (pp. 415–419)

2. Eyebrows shade and protect the eye.

3. Eyelids protect and lubricate the eyes by reflexive blinking. Each eyelid contains a supporting tarsal plate, the roots of the eyelashes, and glands (tarsal, ciliary, sebaceous).

4. Muscles in the eyelids include the levator palpebrae superioris (opens the eye) and orbicularis oculi (closes the eye).

5. The conjunctiva is a mucosa that covers the inner surface of the eyelids (palpebral conjunctiva) and the white of the eye (ocular conjunctiva). Its mucus lubricates the eye surface.

6. The lacrimal gland secretes lacrimal fluid (tears), which is blinked medially across the eye surface and drained into the nasal cavity through the lacrimal canals, lacrimal sac, and nasolacrimal duct.

7. The six extrinsic eye muscles are the lateral and medial rectus (which turn the eye laterally and medially, respectively); superior and inferior rectus (which turn the eye superiorly and inferiorly, respectively, but also medially); and superior and inferior obliques (which turn the eye inferiorly and superiorly, respectively, but also laterally).

Structure of the Eyeball (pp. 419–425)

8. The wall of the eye has three tunics. The most external, fibrous tunic consists of the posterior sclera and the anterior cornea. The tough sclera protects the eye and gives it shape. The cornea is the avascular window through which light enters the eye.

9. The middle, pigmented vascular tunic (uvea) consists of the choroid, the ciliary body, and the iris. The choroid provides nutrients to the retina's photoreceptors and prevents the scattering of light within the eye. The ciliary body contains smooth ciliary muscles that control the shape of the lens and ciliary processes that secrete aqueous humor. The iris contains smooth muscle that changes the size of the pupil.

10. The sensory tunic, or retina, consists of an outer pigmented layer and an inner nervous layer. The nervous layer contains photoreceptors (rod and cone cells), other types of neurons, and neuroglia. Light influences the photoreceptors, which signal bipolar neurons, which signal ganglion neurons. The axons of ganglion neurons run along the inner retinal surface toward the optic disc.

11. The outer segments of the rods and cones contain light-absorbing pigment in membrane-covered discs. Light splits this pigment to initiate the flow of signals through the visual pathway.

12. Two important spots on the posterior retinal wall are (1) the macula lutea with its central fovea (area of highest visual acuity) and (2) the optic disc (blind spot) where axons of ganglion cells form the optic nerve.

13. The outer third of the retina (photoreceptors) is nourished by capillaries in the choroid, while the inner two-thirds is supplied by the central vessels of the retina.

14. The posterior segment of the eye, posterior to the lens, contains the gel-like vitreous humor. The anterior segment, anterior to the lens, is divided into anterior and posterior chambers by the iris. The anterior segment is filled with aqueous humor, which continually forms at the ciliary processes and drains into the scleral venous sinus.

15. The biconvex lens helps to focus light. It is suspended in the eye by the suspensory ligament (zonule) attached to the ciliary body. Tension in the zonule resists the lens's natural tendency to round up.

The Eye as a Light-Gathering Device (pp. 425–427)

16. As light enters the eye, it is bent by the cornea and the lens and focused on the retina. The cornea accounts for most of this refraction, but the lens allows focusing for different distances.

17. The resting eye is set for distance vision. Focusing on near objects requires accommodation (allowing the lens to round as ciliary muscles release tension on the suspensory ligament). The pupils also constrict. Both these actions are controlled by parasympathetic fibers in the oculomotor nerve.

18. Eye-focusing disorders include myopia (nearsightedness), hyperopia (farsightedness), presbyopia (loss of lens elasticity with age), and astigmatism.

Visual Pathways (pp. 427–429)

19. The visual pathway to the brain begins with some processing of visual information in the retina. From there, ganglion cell axons carry impulses via the optic nerve, optic chiasma, and optic tract to the lateral geniculate nucleus of the thalamus. Thalamic neurons project to the primary visual cortex.

20. At the optic chiasma, axons from the medial halves of the retinas decussate. This phenomenon provides each visual cortex with information on the opposite half of the visual field as seen by both eyes. The visual cortex compares the views from the two eyes and generates depth perception.

Embryonic Development of the Eye (pp. 429–430)

21. The eye starts as an optic vesicle, an outpocketing of the embryonic diencephalon. This vesicle then invaginates to form the optic cup, which becomes the retina. The overlying ectoderm folds to form the lens. The fibrous and vascular tunics derive almost entirely from mesenchyme around the optic cups.

THE EAR (pp. 430–440)
The Outer (External) Ear (p. 430)

1. The auricle and external auditory canal compose the outer ear, which acts to gather sound waves. The tympanic membrane (eardrum) transmits sound vibrations to the middle ear.

The Middle Ear (pp. 430–433)

2. The middle ear is a small chamber within the temporal bone. Its boundaries are the eardrum laterally, the bony wall of the inner ear medially, a bony roof superiorly, a posterior wall that opens into the mastoid antrum, and an anterior wall that opens into the auditory tube.

3. The auditory tube consists of bone and cartilage. It runs to the pharynx and equalizes air pressure across the eardrum.

4. The auditory ossicles (malleus, incus, stapes) span the middle ear cavity and transmit sound vibrations from the eardrum to the oval window. The ossicles amplify sounds. The tiny tensor tympani and stapedius muscles dampen the vibrations of very loud sounds.

The Inner (Internal) Ear (pp. 433–437)

5. The inner ear consists of the bony labyrinth (semicircular canals, vestibule, cochlea), which is a chamber that contains the membranous labyrinth (semicircular ducts, utricle and saccule, cochlear duct). The bony labyrinth contains perilymph, whereas the membranous labyrinth contains endolymph.

6. The saccule and utricle each contain a macula, a spot of receptor epithelium that monitors static equilibrium and linear acceleration. A macula contains hair cells, whose "hairs" are anchored in an overlying otolithic membrane. Forces on the otolithic membrane, caused by gravity and linear acceleration of the head, bend the hairs and initiate impulses in the vestibular nerve.

7. The semicircular ducts lie in three planes of space (anterior vertical, posterior vertical, and horizontal). Their cristae ampullaris contain hair cells that monitor rotatory acceleration. The "hairs" of these cells are anchored in an overlying cupula. Forces on the cupula, caused by rotational acceleration of the head, bend the hairs and initiate impulses in the vestibular nerve.

8. The coiled cochlea is divided into three parts (scalas). Running through its center is the cochlear duct (scala media), which contains the spiral organ of Corti. The latter is an epithelium that lies on the basilar membrane and contains the hair (receptor) cells for hearing. The other two parts of the cochlea are the scala vestibuli and the scala tympani.

9. In the mechanism of hearing, sound vibrations transmitted to the stapes vibrate the fluids in the cochlea. This vibrates the basilar membrane and spiral organ. This in turn bends the hairs of the receptor cells, whose tips are anchored in a nonmoving tectorial membrane. Bending of the hairs produces impulses in the cochlear nerve.

Equilibrium and Auditory Pathways (pp. 437–438)

10. Impulses generated by the equilibrium receptors travel along the vestibular nerve to the vestibular nuclei and the cerebellum. These brain centers initiate responses that maintain balance. There is also a minor equilibrium pathway to the cerebral cortex.

11. Impulses generated by the hearing receptors travel along the cochlear nerve to the cochlear nuclei in the medulla. From there, auditory information passes through several nuclei in the brain stem to the thalamus (medial geniculate nucleus) and auditory cortex.

Embryonic Development of the Ear (p. 438)

12. The membranous labyrinth develops from the otic placode, a thickening of ectoderm superficial to the hindbrain.

13. A pouch from the pharynx becomes the middle ear cavity and the auditory tube. The outer ear is formed by an external branchial groove (external auditory canal) and by swellings around this groove (auricle).

Disorders of Equilibrium and Hearing (pp. 438–440)

14. Motion sickness, caused by rocking movements, produces symptoms of nausea and vomiting. Meniere's syndrome is an overstimulation of the hearing and equilibrium receptors caused by an excess of endolymph in the membranous labyrinth.

15. Conduction deafness results from interference with the conduction of sound vibrations to the inner ear. Sensorineural deafness reflects damage to auditory receptor cells or neural structures.

THE SPECIAL SENSES THROUGHOUT
LIFE (p. 440)

1. The chemical senses are sharpest at birth and decline with age as the replacement of receptor cells becomes sluggish.

2. The eye is foreshortened (farsighted) at birth. Depth perception, coordination of eye movements, and color vision develop during early childhood.

3. With age, the lens loses elasticity and clarity, and visual acuity declines. Eye problems that may develop with age are presbyopia, cataracts, glaucoma, and retinal detachment.

4. Initially, infants respond to sound only in a reflexive manner. By 5 months, an infant can locate sound. Critical listening develops in toddlers.

5. Obvious age-related loss of hearing (presbycusis) occurs in the 60s and 70s.

Review Questions

Multiple Choice and Matching Questions

1. Damage to the olfactory tract would hurt your ability to (a) hear, (b) feel pain, (c) smell, (d) see.

2. Sensory impulses transmitted over the facial, glossopharyngeal, and vagus nerves are involved in the special sense of (a) taste, (b) vision, (c) equilibrium, (d) smell.

3. The part of the fibrous tunic that is white, tough, and opaque is the (a) choroid, (b) cornea, (c) retina, (d) sclera.

4. There are extrinsic eye muscles, so there must also be intrinsic (internal) eye muscles. Use logic to deduce which of the following are intrinsic eye muscles: (a) superior rectus, (b) orbicularis oculi, (c) smooth muscles of the iris and ciliary body, (d) levator palpebrae superioris.

5. The transmission of sound vibrations through the inner ear occurs chiefly through (a) nerve fibers, (b) air, (c) fluid, (d) bone.

6. Of the neurons in the retina, which form the optic nerve? (a) bipolar neurons, (b) ganglion neurons, (c) cone cells, (d) horizontal neurons.

7. Taste buds are on the (a) anterior part of the tongue, (b) posterior part of the tongue, (c) palate, (d) inner surface of the cheeks, (e) all of these, (f) a, b, and c.

8. Blocking the scleral venous sinus might result in (a) a sty, (b) glaucoma, (c) conjunctivitis, (d) Meniere's syndrome, (e) chalazion.

9. Conduction of sound from the middle ear to the inner ear occurs via vibration of the (a) malleus against the tympanic membrane, (b) stapes in the oval window, (c) incus in the round window, (d) tympanic membrane against the stapes.

10. Taste receptor cells are stimulated by (a) chemicals binding to their microvilli, (b) chemicals binding to the nerve fibers supplying them, (c) stretch on their microvilli, (d) impulses from the sensory nerve fibers supplying them, (e) none of these.

11. The structure that allows air pressure in the middle ear to be equalized with outside air pressure is (a) cochlear duct, (b) mastoid air cells, (c) endolymph, (d) tympanic membrane, (e) auditory tube.

12. Nearsightedness is more properly called (a) myopia, (b) hyperopia, (c) presbyopia, (d) emmetropia, (e) astigmatism.

13. Which lies closest to the exact posterior pole of the eye? (a) cornea, (b) optic disc, (c) optic nerve, (d) macula lutea, (e) the area where the central artery enters the retina, (f) optic foramen of the orbit.

14. The receptors for static equilibrium that report the position of the head in space relative to the pull of gravity are in the (a) organ of Corti, (b) maculae, (c) crista ampullaris, (d) cupula, (e) joint kinesthetic receptors.

15. Paralysis of a medial rectus muscle would affect (a) accommodation, (b) refraction, (c) depth perception, (d) pupil constriction.

16. Which series of reactions occurs when you look at a nearby object? (a) pupil constricts, suspensory ligament has less tension, lens thickens; (b) pupil dilates, suspensory ligament becomes taut, lens becomes less thick; (c) pupil dilates, suspensory ligament becomes taut, lens thickens; (d) pupil constricts, ciliary muscles relax, lens becomes less thick.

17. The order in which a light ray passes through the refractory media of the eye is (a) vitreous humor, lens, aqueous humor, cornea; (b) cornea, aqueous humor, lens, vitreous humor; (c) cornea, vitreous humor, lens, aqueous humor; (d) lens, aqueous humor, cornea, vitreous humor.

18. During embryonic development, the lens of the eye forms (a) as part of the choroid coat, (b) from the surface ectoderm, (c) as a part of the sclera, (d) as a part of the retina that breaks off.

19. The optic disc is the site where (a) more rods than cones occur, (b) the macula lutea is located, (c) only cones occur, (d) the optic nerve exits the eye.

20. Which of the following is not a cause of conduction deafness? (a) impacted cerumen, (b) otitis media, (c) cochlear nerve degeneration, (d) otosclerosis.

21. Ear stones (otoliths) are (a) a cause of deafness, (b) a type of hearing aid, (c) important in equilibrium, (d) the rock-hard petrous temporal bones.

Short Answer and Essay Questions

22. Maurice, an anatomy student, was arguing with his grandfather, Chester. Chester, who believed in folk wisdom, insisted that there are only five senses. Maurice, however, said that there are at least ten senses. Decide who was right, and then list all the senses you know. (Hint: Here is a chance to go back and review Figure 11.2.)

23. What is the precise location of the olfactory epithelium?

24. Why do you often have to blow your nose after crying?

25. What and where is the fovea centralis, and why is it important?

26. Name two special senses whose receptor cells are replaced throughout life and two special senses whose receptor cells are never replaced.

27. (a) Describe the embryonic derivation of the retina. (b) Explain how the middle ear cavity forms.

28. Describe some effects of aging on the eye and the ear.

29. Trace the auditory pathway to the cerebral cortex.

30. Distinguish the ciliary body from ciliary processes—in both location and function.

31. Compare and contrast the functions of the inferior oblique and superior rectus muscles.

32. What is the difference, if any, between a semicircular canal and semicircular duct? Between cochlea and cochlear duct?

Critical Thinking and Clinical Application Questions

33. Little Biff's uncle tells the physician that Biff, who is 3 years old, gets many "earaches," and that a neighbor says Biff needs "ear tubes put in." Upon questioning, the uncle reveals that Biff has not had a sore throat for a long time and that Biff is learning to swim. Does Biff have otitis media or otitis externa (Related Clinical Terms), and does he need ear tubes? Explain your reasoning.

34. Nine children attending the same day-care center developed red, inflamed eyes and eyelids. What is the most likely cause and name of this condition?

35. Dr. Nakvarati used an instrument to press on Mr. Cruz's eye during his annual physical examination on his sixtieth birthday. The eye deformed very little, indicating the intraocular pressure was too high. What was Mr. Cruz's probable condition?

36. Lionel suffered a ruptured artery in his middle cranial fossa, and a pool of blood compressed his left optic tract, destroying its axons. What part of the visual field was blinded?

37. Sylvia Marcus, aged 70, recently underwent surgery for otosclerosis. The operation was a failure and did not improve her condition. (a) What was the purpose of the surgery, and exactly what was it trying to accomplish? (b) Deduce why it is a difficult procedure with a relatively high failure rate.

16

Blood

Student Objectives

1. Distinguish the circulatory system from the cardiovascular system.

Overview: Composition of Blood (pp. 445–446)

2. Name the basic components of blood, and define hematocrit.

Blood Plasma (p. 446)

3. List some of the molecules in blood plasma.

Formed Elements (pp. 446–453)

4. Explain the technique for obtaining a blood smear.

5. Describe the special structural features and functions of erythrocytes.

6. List the five classes of leukocytes, along with the structural characteristics and functions of each.

7. Describe the structure of platelets and their role in clotting.

Blood Cell Formation (pp. 453–456)

8. Distinguish red from yellow bone marrow.

9. Describe the basic histologic structure of red marrow.

10. Explain the differentiation of the various types of blood cells.

The Blood Throughout Life (p. 456)

11. Describe the embryonic origin of blood cells. List four different organs that form blood cells in the fetus.

12. Name some blood disorders that become more common as the body ages.

The next four chapters discuss the **circulatory system.** This system can be subdivided into the **cardiovascular system** (the blood, the heart, and the blood vessels: Chapters 16–18) and the **lymphatic system** (vessels that carry a fluid called lymph: Chapter 19). We begin by considering **blood,** the fluid in the vessels of the cardiovascular system.

Blood is the river of life that surges within us, transporting nearly everything that must be carried from one place to another within the body. For many thousands of years, blood was considered magical, an elixir that held the mystical force of life—for when blood drained from the body, life departed as well. Today, blood retains an exalted position in the life-saving profession of medicine: Clinicians examine it more often than any other tissue when trying to determine the causes of diseases in their patients.

To get started, we need a brief overview of blood circulation, which is powered by the pumping action of the heart: Blood leaves the heart via the *arteries,* which branch repeatedly until they become tiny *capillaries.* By diffusing across the capillary walls, oxygen and nutrients leave the blood and enter the body tissues, and carbon dioxide and cellular wastes diffuse from the tissues to the bloodstream. From the capillaries, the oxygen-deficient blood flows into *veins,* which return it to the heart. Blood then flows to the lungs, where it picks up oxygen and releases carbon dioxide, and then returns to the heart to be pumped throughout the body once again.

Additional functions of the blood are to carry hormones (p. 662) from the endocrine glands to their target organs and to transport cells of the body's defense system to sites where they can fight infection. The blood also helps to regulate body temperature; that is, blood is diverted to or from the skin to control the amount of body heat lost across the body surface.

Blood accounts for about 8% of body weight. Its volume in adult males is 5 to 6 L (about 1.5 gallons) but is somewhat less than this in females (4 to 5 L).

Overview: Composition of Blood

Although blood appears to be a thick, homogeneous liquid, the microscope shows that it has both cellular and liquid components. Blood is a specialized type of connective tissue in which blood cells, called *formed elements,* are suspended in a nonliving fluid, called *plasma.*

When a sample of blood is spun in a centrifuge, the heavier formed elements are packed down by centrifugal force, and the less dense plasma remains at the top (Figure 16.1). The red mass at the bottom of the tube consists of *erythrocytes* (ĕ-rith'ro-sīts; "red cells"), the red blood cells that transport oxygen. A thin, whitish layer called the **buffy coat** is present at the junction between the erythrocytes and the plasma.

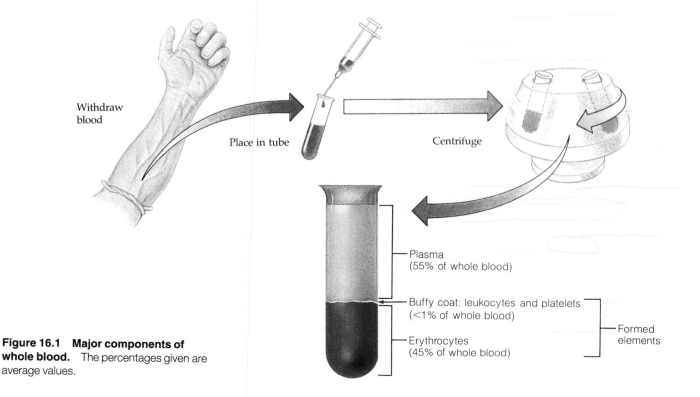

Withdraw blood

Place in tube

Centrifuge

Plasma
(55% of whole blood)

Buffy coat: leukocytes and platelets
(<1% of whole blood)

Erythrocytes
(45% of whole blood)

Formed elements

Figure 16.1 Major components of whole blood. The percentages given are average values.

This layer contains *leukocytes* (lu'ko-sīts; "white cells"), the white blood cells that act in various ways to protect the body. The buffy coat also contains *platelets (thrombocytes),* cell fragments that help stop bleeding. The percentage of the blood volume that is occupied by erythrocytes is known as the **hematocrit** (he-mat'o-krit; "blood fraction"). This averages about 45%. Leukocytes and platelets constitute less than 1% of the volume of blood. Plasma makes up the remaining 55% of whole blood.

Normal hematocrit values vary. In healthy males, the hematocrit is 47% ± 5%, whereas in healthy females it is 42% ± 5%. Values tend to be slightly higher in newborns (45% to 60%).

Blood Plasma

Blood plasma is a straw-colored, sticky fluid. Although it is about 90% water, it contains over 100 different kinds of molecules. These include ions (Na$^+$ and Cl$^-$, for example), nutrients (simple sugars, amino acids, lipids), wastes (urea, ammonia, carbon

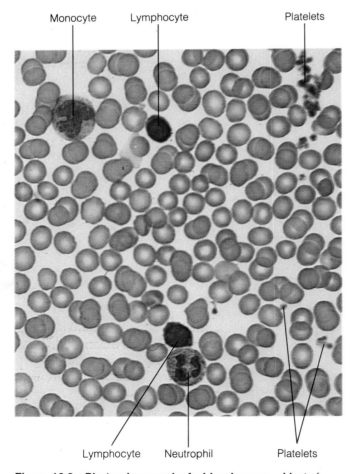

Monocyte   Lymphocyte   Platelets

Lymphocyte   Neutrophil   Platelets

Figure 16.2   Photomicrograph of a blood smear.   Most of the cells in the field of view are erythrocytes (red blood cells). Three kinds of leukocytes (white blood cells) are also present: monocytes, lymphocytes, and neutrophils. Also note the platelets. (750 ×).

dioxide), oxygen, hormones, and vitamins. Plasma also contains three main types of proteins: *albumin* (al-bu'min), *globulins* (glob'u-lins), and *fibrinogen* (fi-brin'o-jen). Albumin contributes to plasma osmotic pressure (the pressure that helps keep water in the bloodstream). The globulins include both antibodies and the blood proteins that transport lipids, iron, and copper. Fibrinogen is involved in blood clotting. If blood plasma is allowed to stand, it soon coagulates to form a clot and a clear fluid called **serum,** the blood plasma from which the clotting factors have been removed.

Formed Elements

The **formed elements of blood,** or **blood cells,** are *erythrocytes, leukocytes,* and *platelets.* These elements have some unusual features. (1) Two of the three are not even true cells: Erythrocytes have no nuclei or organelles, and platelets are just cell fragments. Only leukocytes are complete cells. (2) Most of the formed elements survive in the bloodstream for only a few days. (3) Most blood cells do not divide. Instead, they are continuously renewed by the division of cells in the bone marrow, where they originate. Short-lived and constantly replaced, blood cells may be compared to disposable products in our modern, throwaway society.

If you examine a smear of human blood under the light microscope (Figure 16.2), you will see disc-shaped erythrocytes, a variety of spherical and colorfully stained leukocytes, and tiny platelets that look like particles of debris. Erythrocytes vastly outnumber the other types of formed elements. Table 16.1 on page 452 summarizes the structures and functions of the different types of blood cells.

Before describing the individual kinds of blood cells, we will explain how blood smears are prepared for microscopic viewing. The blood technician puts a drop of fresh blood on a clean glass slide and then, using the edge of another slide, spreads the drop into a thin film. The film is then air dried, preserved in methanol (wood alcohol), and stained. Blood stain is a mixture of an acidic dye called *eosin* (e'o-sin), which is pink, and a basic dye called *methylene* (meth'ĭ-lēn) *blue,* which yields both blue and purple colors. There are several varieties of blood stain (Romanovsky's, Wright's, Giemsa's, Leishman's), but all are quite similar in composition.

Erythrocytes

Erythrocytes, or red blood cells (RBCs), are small cells about 8 μm in diameter (Figure 16.3). Shaped like biconcave discs—flattened discs with depressed cen-

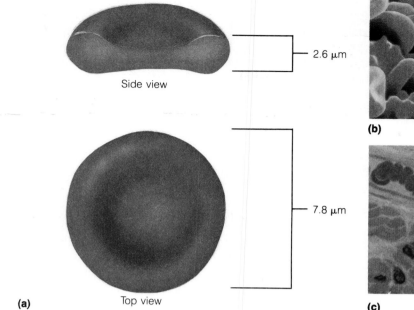

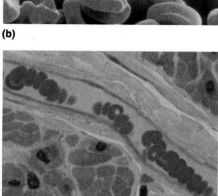

(b)

(c)

Figure 16.3   Structure of erythrocytes.   **(a)** An erythrocyte in cross-sectional view and in superior view. Note the biconcave shape. **(b)** Scanning electron micrograph of human erythrocytes. The small green structures are platelets (artificially colored; 2000 ×). **(c)** Photomicrograph of erythrocytes in a capillary (800 ×).

ters—their thin centers appear lighter in color than their edges (Figure 16.2). Erythrocytes are surrounded by a plasma membrane, but they possess no nuclei or organelles. Instead, their cytoplasm is packed with molecules of *hemoglobin,* the oxygen-carrying protein. The hemoglobin molecule consists of four chains of amino acids (four polypeptides), each of which bears an iron atom that is the binding site for oxygen molecules. The oxidation of iron atoms inside hemoglobin molecules gives blood its red color. Hemoglobin also attracts the eosin dye of blood stains, so erythrocytes stain a pink or orange-pink color in blood smears. Since normal red blood cells (7 to 8 μm in diameter) do not vary much in size, they are ideal measuring rods for determining the sizes of surrounding structures in histological sections. There are about 5 million erythrocytes in a cubic millimeter of blood (4.3 to 5.2 million in females, 5.1 to 5.8 million in males), or a total of 25 trillion erythrocytes in the bloodstream of a healthy adult.

Erythrocytes pick up oxygen at the lung capillaries and release it across other tissue capillaries throughout the body. Each of the erythrocyte's structural characteristics contributes to this respiratory function: (1) Their biconcave shape provides 30% more surface area than that of a comparable spherical cell, allowing rapid diffusion of oxygen into and out of the erythrocyte. (2) Discounting the water that is present in all cells, erythrocytes are over 97% hemoglobin. Without a nucleus or organelles, they are little more than bags of oxygen-carrying molecules. (3) Because erythrocytes lack mitochondria and generate

the energy they need by anaerobic mechanisms, they do not consume any of the oxygen they are transporting and are thus very efficient oxygen transporters.

Along with the oxygen it carries, the hemoglobin in erythrocytes also carries 20% of the carbon dioxide that is transported by the blood. Carbon dioxide binds to the amino-acid part of the hemoglobin molecule rather than to the iron. The transport of carbon dioxide does not interfere with oxygen transport in any way. For the details of gas transport by the blood, consult a physiology text.

The biconcave shape of the erythrocyte is maintained by a net of peripheral proteins on the inner surface of the plasma membrane. This net also resists the tearing forces that erythrocytes experience as they squeeze through the narrow capillaries. Furthermore, because the net is deformable, it gives erythrocytes enough flexibility to withstand moderate changes in shape—to twist, turn, and become cup-shaped as they travel through capillaries with diameters smaller than themselves and then to resume their biconcave shape.

Erythrocytes live for 100 to 120 days, much longer than most other types of blood cells. They originate from cells in red bone marrow, where they expel their nucleus and organelles before entering the bloodstream.

Anemia (ah-ne′me-ah; "lacking blood") is a condition is which the blood's capacity for carrying oxygen is abnormally low. There are many types of anemia. Some types result from low numbers of erythrocytes in the blood (due to blood loss, for ex-

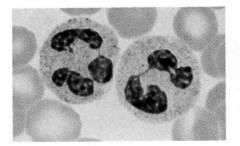
(a)

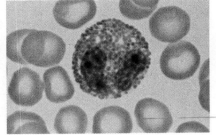
(b)

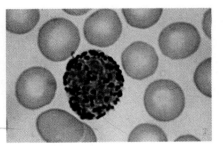
(c)

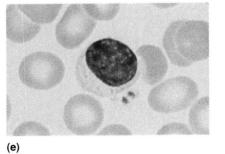

(d)

(e)

(f)

Figure 16.4 Leukocytes. (**a**) Neutrophil. (**b**) Eosinophil. (**c**) Basophil. (**d**) Small lymphocyte. (**e**) Medium-sized lymphocyte. (**f**) Monocyte. Orange-pink erythrocytes are seen throughout this figure (about 1,375 ×).

ample). Other types result from an abnormally small amount of hemoglobin in each erythrocyte (due, for example, to a deficiency of iron in the diet), and still others from a genetic defect in the structure of the hemoglobin molecules.

Can you deduce some symptoms that people with anemia exhibit?

Since their tissues are receiving inadequate amounts of oxygen, anemic individuals are constantly fatigued and often pale, short of breath, and chilly.

Polycythemia (pol″e-si-the′me-ah; "many blood cells") is an abnormal excess of erythrocytes in the blood. One variety, *polycythemia vera,* results from a cancer of the bone marrow that generates too many erythrocytes. Severe polycythemia causes an increase in the viscosity of the blood, which slows or blocks the flow of blood through the smallest vessels. ■

Leukocytes

Although **leukocytes,** or white blood cells (WBCs), are far less numerous than erythrocytes, they are crucial to the body's defense against disease. There are 4000 to 11,000 leukocytes per cubic millimeter of blood. They are the only formed elements that are complete cells, with nuclei and the usual organelles (Figure 16.4).

Leukocytes form a mobile army that continuously protects the body from infectious microorganisms such as bacteria, viruses, and parasites. As such, they

have some unusual functional characteristics. Unlike erythrocytes, which are confined to the bloodstream and carry out their functions in the blood, leukocytes function outside the capillaries in the loose connective tissues, where infections occur. Leukocytes leave the capillaries by actively squeezing between the endothelial cells that form the capillary walls, a process called **diapedesis** (di″ah-pĕ-de′sis; "leaping across"). Once outside the capillaries, leukocytes travel to the infection sites by amoeboid motion (by forming flowing cytoplasmic extensions that move them along).

Like other blood cells, leukocytes originate in the bone marrow and are released continuously into the blood. However, the bone marrow also *stores* leukocytes and releases large quantities of these blood cells under certain conditions. What conditions might invoke a massive release of stored leukocytes into the circulation?

Bone marrow releases stored leukocytes into the blood during serious infections to enhance the body's ability to fight these infections. Therefore, the clinician counts the leukocytes in a sample of a patient's blood when searching for evidence of infectious disease. When the leukocyte count exceeds 11,000 per cubic millimeter, the patient is said to have **leukocytosis.** ■

There are five distinct classes of leukocytes, all of which are roughly spherical in shape (Figure 16.4): *neutrophils* (nu′tro-filz), *eosinophils* (e″o-sin′o-filz), *basophils* (ba′so-filz), *lymphocytes* (lim′fo-sīts), and *monocytes* (mon'o-sīts). These cell types are divided into two main groups: **Granulocytes** contain many membrane-lined granules, whereas **agranulocytes** lack obvious granules. The relative abundances of these cells in healthy people are shown in Figure 16.5. When listing the five types of leukocytes from the most abundant to the least abundant, it helps to remember the phrase: "**N**ever **l**et **m**onkeys **e**at **b**ananas" (neutrophils, lymphocytes, monocytes, eosinophils, basophils).

Granulocytes

General Features.   The three types of **granulocytes** (also called *granular leukocytes*) are neutrophils, eosinophils, and basophils. They are larger and much shorter-lived than erythrocytes (Table 16.1). Along with their distinctive cytoplasmic granules, these cells have distorted, inactive nuclei. Although these nuclei are somewhat variable in shape, they all have purple-staining *lobes,* rounded masses joined by bridges of nuclear material. Because of the variability in the structure of their nuclei, granulocytes are also called *polymorphonuclear cells* ("many shapes of the nuclei").

Functionally, all granulocytes are phagocytic; that is, they engulf and digest foreign cells or molecules.

Neutrophils.   Bacteria-destroying **neutrophils** are the most abundant class of leukocyte, constituting about 60% of all white blood cells in healthy people (Figures 16.4a and 16.5). Their nucleus consists of two to six lobes interconnected by very thin threads of chromatin. Since the nuclei of neutrophils are more highly lobed and more distorted than nuclei of other granulocytes, many authorities reserve the name *polymorphonuclear cells* for neutrophils alone.

Neutrophils contain two kinds of cytoplasmic granules. Both kinds are so small that they can barely be seen with the light microscope. The more abundant *specific granules* stain a light pink, whereas *azurophilic* (azh"u-ro-fil'ik) *granules* stain reddish purple. Together, the two types of granules give the cytoplasm a light purple color. The name *neutrophil* means "neutral-loving" or "neutral-staining," indicating that the cytoplasm stains a neutral color that is intermediate between the red and blue staining of other blood cells.

The granules of neutrophils can be viewed as lysosomes that have been fortified with specific digestive enzymes that destroy the cell walls of bacteria. Indeed, the basic function of neutrophils is to phagocytize bacteria. As neutrophils digest the bacteria they have ingested, they also release powerful chemicals into the extracellular space. These chemicals kill bacteria in the tissue around the neutrophils but are so potent that they eventually destroy the neutrophils themselves. Dead neutrophils contribute to pus.

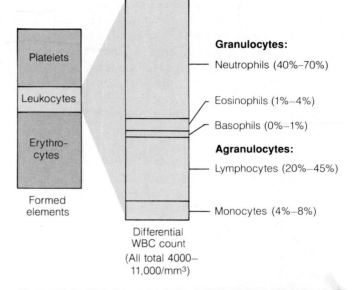

Figure 16.5   Relative percentages of the different types of leukocytes.   These values, including ranges for healthy individuals, are determined by counting the cells in blood smears (differential WBC count).

Eosinophils.   Eosinophils (Figure 16.4b) are relatively rare, accounting for 1% to 4% of all leukocytes. The nucleus usually has two lobes interconnected by a broad band and thus resembles an old-fashioned telephone receiver. The granules in the cytoplasm are large and stain red with the eosin dye of blood stains (*eosinophil* = eosin-loving). Like lysosomes, these granules contain a variety of digestive enzymes; unlike lysosomes, they lack enzymes that specifically digest bacteria.

Eosinophils perform several functions within the loose connective tissues: (1) They phagocytize antigen-antibody complexes (*antigens* are foreign molecules that enter the body, and *antibodies* are immune proteins that bind to antigens: p. 450). (2) Eosinophils somehow help to lessen the severity of allergies. (3) The most important role of eosinophils is to fight parasitic worms, such as platyhelminths (flatworms) and nematodes (roundworms). These worms are far too large for eosinophils to phagocytize, so eosinophils release the enzymes from their granules onto the parasite. These enzymes then digest and destroy the worm.

The most important enzyme within the eosinophil granules is called *major basic protein (MBP).* It forms large crystals that attract eosin dye, accounting for the red staining of eosinophils. MBP is the enzyme responsible for digesting parasites.

Basophils.   Basophils (Figure 16.4c) are the rarest white blood cells, averaging only 0.5% of all leukocytes, or 1 in 200. The nucleus usually has two lobes (Table 16.1), and it may be bent into the shape of a U or an S. The cytoplasm contains large granules that have an affinity for the basic dyes, staining a dark purple (*basophil* = base-loving). These granules contain histamine (p. 81) and other molecules that mediate inflammation. By secreting the contents of their granules at infection sites in the loose connective tissues, basophils increase the permeability of nearby capillaries and attract other leukocytes to the area (especially eosinophils). Basophils are weakly phagocytic, but what they phagocytize is not known.

The inflammation-mediating function of basophils is almost identical to that of *mast cells,* granulated cells in connective tissue that also secrete histamine (p. 81). Despite the similarities between these two cell types, their nuclei differ in shape (those of mast cells are oval, unlike the bilobed nuclei of basophils), and their granules differ in microscopic structure. Therefore, current evidence indicates that mast cells and basophils come from entirely different lines of blood cells.

Agranulocytes

The **agranulocytes** (which are also called *nongranular leukocytes*) are lymphocytes and monocytes. Although they resemble each other structurally, they are distinct and unrelated cell types.

Lymphocytes.   Lymphocytes (Figure 16.4d and e) are the most important cells of the immune system. Relatively common, they represent 20% to 45% of all leukocytes in the blood. The nucleus of a typical lymphocyte occupies most of the cell volume and is filled with condensed chromatin, which causes the nucleus to stain dark purple. The nucleus is usually spherical but may be slightly indented, and it is surrounded by a thin rim of pale blue cytoplasm. Lymphocytes are often classified according to size as small (5 to 8 μm), medium (10 to 12 μm), and large (14 to 17 μm). Most of the lymphocytes in the blood are small. Like other leukocytes, lymphocytes function not in the blood but in the connective tissues. In fact, most lymphocytes are firmly enmeshed in *lymphoid connective tissues,* where they play a crucial role in immunity (Chapter 19).

Lymphocytes are effective in fighting infectious organisms because each lymphocyte recognizes and acts against a *specific* foreign molecule. Any foreign molecule that induces a response from a lymphocyte is called an **antigen** (an'tĭ-jen; "induce against"). Most antigens are proteins or glycoproteins in the plasma membranes of foreign cells (or in the cell walls of bacteria) or else are proteins secreted by foreign cells

(bacterial toxins, for example). The two main classes of lymphocytes—**T cells** and **B cells**—attack antigens in different ways (Figure 16.6). A major type of T cell, the **cytotoxic** (killer) **T lymphocyte,** binds to an antigen-bearing cell and penetrates its membrane, thereby destroying that cell. B cells, by contrast, multiply to become **plasma cells** that secrete **antibodies.** Antibodies are proteins that bind to antigens and mark them for destruction by, for example, making them more recognizable to phagocytic cells. Despite their different actions, T and B lymphocytes cannot be distinguished from one another structurally, even under the electron microscope.

T and B lymphocytes tend to combat different classes of antigen-bearing cells. B lymphocytes and antibodies are best at fighting bacteria and bacterial toxins. T lymphocytes, by contrast, attack only eukaryotic cells (cells that are complex enough to have a nucleus and organelles). Thus, T cells attack invading fungi, cancer cells, and human cells that have been infected with viruses. T cells are primarily responsible for the rejection of transplanted organs.

Monocytes.   Monocytes, which make up 4% to 8% of white blood cells, are the largest leukocytes (Figure 16.4f). They resemble large lymphocytes, for both cell types exhibit a blue cytoplasm and a purple nucleus in blood smears. However, the nucleus of monocytes is often bowed into a distinctive kidney shape or horseshoe shape, and the nuclear chromatin is not as condensed (dark) as that in lymphocytes.

Like all leukocytes, monocytes use the bloodstream to reach the connective tissues. There, they transform into **macrophages,** oval cells with pseudopods (see Figure 4.14). Recall from Chapter 4 (p. 81) that macrophages are phagocytic cells with an appetite for a wide variety of foreign cells, molecules, and tiny particles of debris.

A **complete blood count (CBC)** is a common clinical procedure used to screen for diseases. A sample of blood is taken from the patient, and the following quantities are measured: the hematocrit, the hemoglobin content, and the concentrations of erythrocytes, leukocytes, and platelets (number per cubic millimeter). First, the living white and red cells are examined under a microscope for structural abnormalities. Then, a blood smear is prepared for a **differential WBC count,** in which the technician counts the percentages of each class of leukocytes.

Can you deduce the significance of an abnormally high concentration of neutrophils in a differential WBC count? Of a high concentration of eosinophils?

High numbers of neutrophils suggest the presence of a major bacterial infection somewhere in the body. High numbers of eosinophils can reflect infec-

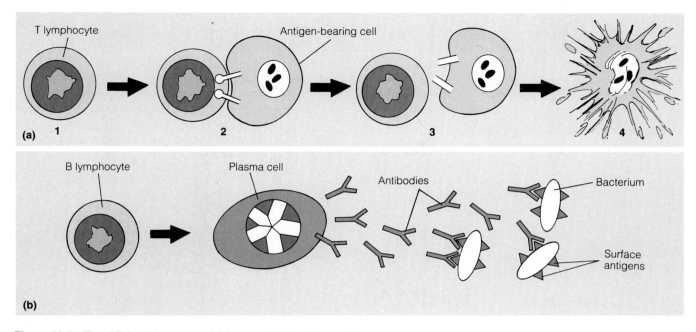

Figure 16.6 T and B lymphocytes. (**a**) A cytotoxic T lymphocyte binds to an antigen-bearing cell and destroys it: The T cell (step 1) adheres to the target cell (step 2) and secretes proteins that become inserted into the membrane of the target cell, forming open lesions. This is called the lethal hit. Next (step 3), the T cell detaches and the target cell later undergoes lysis (step 4). (**b**) A B lymphocyte gives rise to a plasma cell that secretes antibodies that bind to antigens on bacteria.

tion by parasitic worms or an allergy to ragweed during hay fever season. ■

Platelets

Platelets, also called *thrombocytes* (throm'bo-sīts; "clotting cells"), are not cells in the strict sense. They are disc-shaped fragments of cytoplasm enclosed by a plasma membrane that form by breaking off larger cells, called megakaryocytes (p. 456). In blood smears, each platelet exhibits a blue-staining outer region and an inner region that contains purple-staining secretory granules (Table 16.1). Platelets are only one-tenth to one-twentieth as abundant as erythrocytes.

Platelets plug small tears in damaged blood vessels and help initiate blood clotting. Immediately after a vessel is wounded, platelets adhere in large numbers to the edges of the gash then secrete several types of products. For example, some products from their secretory granules cause the bleeding vessel to constrict, thereby slowing the rate of bleeding. Furthermore, platelets secrete a molecule (thromboplastin) that helps initiate the clotting process. **Clotting** is a sequence of chemical reactions in blood plasma that ultimately generates a network of tough **fibrin** strands. This network forms among the accumulated platelets in the torn vessel, where it "glues" the platelets together and also traps any blood cells that enter it (Figure 16.7). This entire mass, called a **clot,** pro-

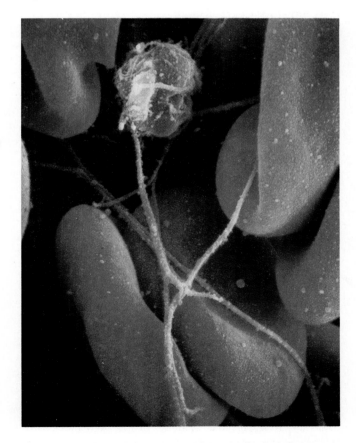

Figure 16.7 Scanning electron micrograph of erythrocytes trapped in a network of fibrin strands. The nearly spherical gray object at top center is a platelet (artificially colored; 15,000 ×).

Table 16.1   Summary of Formed Elements of the Blood

Cell type	Illustration	Description*	Number of cells/ mm³ (μl) of blood	Duration of development (D) and life span (LS)	Function
Erythrocytes (red blood cells; RBCs)		Biconcave, anucleate disc; salmon-colored; diameter 6–8 μm	4–6 million	D: 5–7 days LS: 100–120 days	Transport oxygen and carbon dioxide
Leukocytes (white blood cells, WBCs)		Spherical, nucleated cells	4000–11,000		
Granulocytes • Neutrophil		Nucleus multilobed; inconspicuous cytoplasmic granules; diameter 10–14 μm	3000–7000	D: 6–9 days LS: 6 hours to a few days	Phagocytize bacteria
• Eosinophil		Nucleus bi-lobed; red cytoplasmic granules; diameter 10–14 μm	100–400	D: 6–9 days LS: 8–12 days	Destroy antigen-antibody complexes; inactivate some inflammatory chemicals of allergy; kill parasitic worms
• Basophil		Nucleus lobed; large blue-purple cytoplasmic granules; diameter 10–12 μm	20–50	D: 3–7 days LS: ? (a few hours to a few days)	Release histamine and other mediators of inflammation
Agranulocytes • Lymphocyte		Nucleus spherical or indented; pale blue cytoplasm; diameter 5–17 μm	1500–3000	D: days to weeks LS: hours to years	Mount immune response by direct cell attack or via antibodies
• Monocyte		Nucleus U- or kidney-shaped; gray-blue cytoplasm; diameter 14–24 μm	100–700	D: 2–3 days LS: months	Phagocytosis; develop into macrophages in tissues
Platelets		Discoid cytoplasmic fragments containing granules; stain deep purple; diameter 2–4 μm	250,000 – 500,000	D: 4–5 days LS: 5–10 days	Seal small tears in blood vessels; instrumental in blood clotting

*Appearance when stained with Wright's stain.

vides a strong seal across the tear. Once the clot forms, the platelets within it contract, pulling the edges of the tear back together.

Platelets do not adhere to the insides of healthy vessels. However, they will stick to an untorn vessel if its lining is roughened by scarring, by inflammation, or by atherosclerosis (fat deposited on the inner walls of vessels; see p. 484). Undesirable clotting then occurs within that vessel. A clot that develops and persists in an unbroken blood vessel is called a **thrombus.**

 Thrombi can have dangerous consequences. Can you deduce what some of these are?

If the thrombus is large enough, it may block the flow of blood to the vessels beyond the occlusion and cause the death of the tissues supplied by those vessels. For example, if the blockage occurs in the coronary arteries that supply the heart (coronary thrombosis), the consequence may be the death of heart muscle and a fatal heart attack. If a thrombus breaks away from a vessel wall and floats freely in the bloodstream, it becomes an **embolus** (plural, **emboli**). An embolus poses no problem until it wedges in a vessel that is too narrow for it to pass through (*embolus* means "wedge"). For example, emboli that block lung vessels (pulmonary emboli) can dangerously impair the ability of the body to obtain oxygen, and emboli in the brain can cause strokes by blocking the blood supply to oxygen-sensitive brain cells. For more information on emboli, see the Related Clinical Terms on page 456. ■

Blood Cell Formation

The formation of blood cells is called **hematopoiesis** (hēm"ah-to-poi-e'sis) or **hemopoiesis** (*hemo, hemato* = blood; *poiesis* = to make). This process begins in the early embryo and continues throughout life. After birth, all blood cells originate in the bone marrow.

Bone Marrow: The Site of Blood Cell Formation

Bone marrow occupies the internal part of the bones (Figure 16.8). If all of the bone marrow in the skeleton were combined, it would form the largest organ in the human body except for the skin.

There are two types of bone marrow, red and yellow. Only **red marrow** actively generates blood cells. In fact, its red hue comes from the immature erythrocytes it contains. **Yellow marrow** is dormant: It manufactures blood cells only in emergencies that demand increased hematopoiesis. The color of yellow marrow comes from the many fat cells it contains. At birth, all marrow in the skeleton is red. In adults, red marrow remains throughout the axial skeleton and girdles and in the proximal epiphysis of the humerus and femur. Yellow marrow, however, occupies all other regions of the long bones of the limbs.

The histology of red bone marrow is shown in Figure 16.8b. Bone marrow contains many wide capillaries called *blood sinusoids.* Between these sinusoids is a reticular connective tissue (p. 82), in which reticular fibers form a complex network (see Figure 4.15d). The fibroblasts that cover and secrete this fiber network are called *reticular cells.* Within the network are fat cells and the immature blood cells in all stages of maturation. As the blood cells reach maturity, they migrate through the endothelial cells that form the wall of the sinusoids. In this way, new blood cells continuously enter the bloodstream.

The reticular tissue of the bone marrow also contains macrophages. These macrophages reach into the sinusoids to capture antigens from the blood. Such a "blood-cleaning" function is also performed by macrophages in the spleen and liver.

Stages of Blood Cell Formation

Figure 16.9 shows the pathways by which the blood cells in bone marrow mature. We will now consider these pathways.

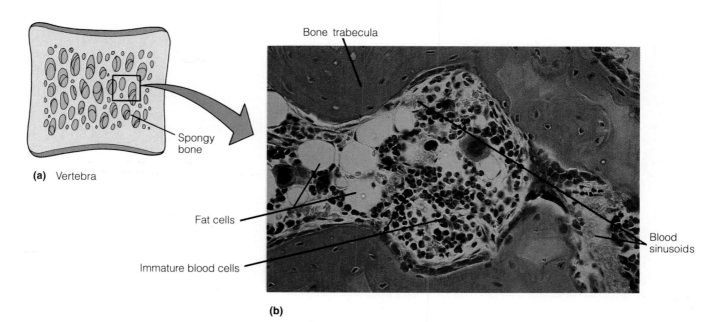

Figure 16.8   Bone marrow. (a) Red bone marrow in a vertebra. (b) Micrographic enlargement of (a) (300 ×). Note that fat cells are obvious components of all bone marrow, although they do not participate in hematopoiesis.

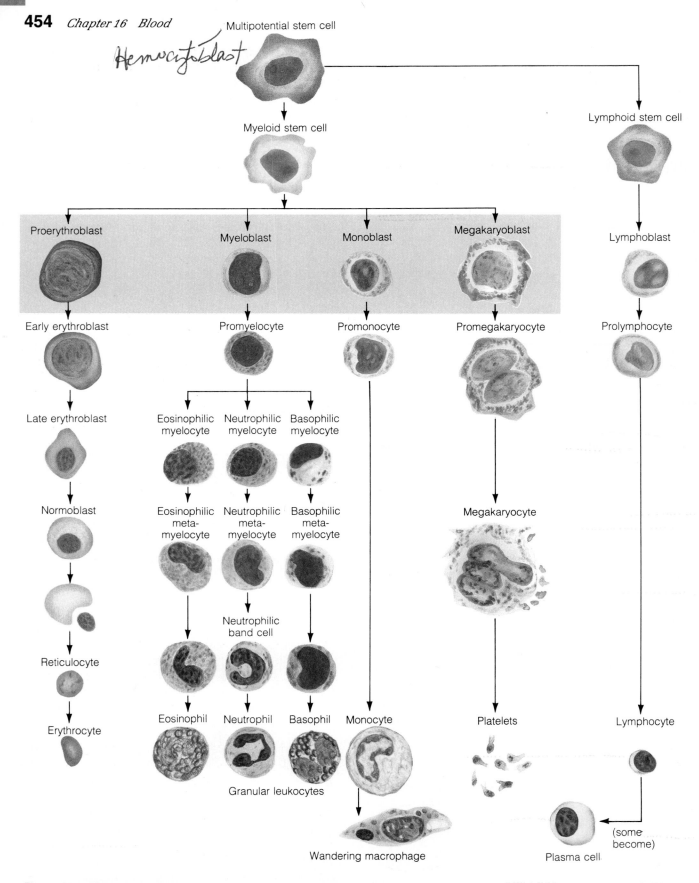

Hemocytoblast

Multipotential stem cell

Myeloid stem cell

Lymphoid stem cell

Proerythroblast

Myeloblast

Monoblast

Megakaryoblast

Lymphoblast

Early erythroblast

Promyelocyte

Promonocyte

Promegakaryocyte

Prolymphocyte

Late erythroblast

Eosinophilic myelocyte

Neutrophilic myelocyte

Basophilic myelocyte

Normoblast

Eosinophilic meta-myelocyte

Neutrophilic meta-myelocyte

Basophilic meta-myelocyte

Megakaryocyte

Neutrophilic band cell

Reticulocyte

Erythrocyte

Eosinophil

Neutrophil

Basophil

Monocyte

Platelets

Lymphocyte

Granular leukocytes

Wandering macrophage

(some become)

Plasma cell

Figure 16.9 Stages of differentiation of blood cells in the bone marrow. Multipotential hematopoietic stem cells (above) give rise to myeloid stem cells (which produce all blood cells but lymphocytes) and to lymphoid stem cells (which produce lymphocytes). The blast cells in the myeloid cell line (shown against the rectangular brown background) are committed cells. The durations of these developmental sequences are listed in Table 16.1.

Earliest Cell Stages

All blood cells arise from a common cell type, the **multipotential hematopoietic stem cell**, or **hemocytoblast.** These stem cells roughly resemble large lymphocytes (although their nucleus stains more lightly). They divide continuously, both to renew themselves and to produce cell lines that lead to the various blood cells. First, the cell line leading to lymphocytes—**lymphoid stem cells**—splits from the line leading to all other blood cells, the **myeloid** (mi′ĕ-loid) **stem cells.** Then, in the myeloid line, the stem cells become *committed cells;* that is, each can generate only a single type of blood cell. After commitment occurs, structural differentiation begins, and the cells divide as they differentiate. Let us now explore the specific structural changes that occur in each cell line.

Genesis of Erythrocytes

In the line that forms erythrocytes, the committed cells are **proerythroblasts** (pro″ĕ-rith′ro-blasts; "earliest red-formers"). These cells avidly accumulate iron for future use in manufacturing hemoglobin. Proerythroblasts give rise to early **erythroblasts** *(basophilic erythroblasts),* which act as ribosome-producing factories. Hemoglobin is manufactured on these ribosomes and accumulates during the next two stages: the **late erythroblast** and the **normoblast.** The "color" of the cytoplasm changes during these stages, as blue-staining ribosomes become masked by pink-staining hemoglobin. When the normoblast stage is reached, cell division stops. When the cytoplasm of the normoblast is almost filled with hemoglobin, the nucleus stops directing the cell's activities and shrinks. Then, the nucleus and almost all organelles are ejected, and the cell collapses and assumes its biconcave shape. The **reticulocyte** stage has now been reached. Reticulocytes contain a delicate network of blue-staining material (*reticulum* = network) representing clumps of ribosomes that remain behind when the other organelles are extruded. Reticulocytes enter the bloodstream and begin their task of transporting oxygen.

Erythrocytes remain in the reticulocyte stage for their first day or two in the circulation, after which their ribosomes are degraded by intracellular enzymes and lost. This fact has practical importance. Because they last for 1 to 2 days of the 100-day life span of erythrocytes, reticulocytes make up 1% to 2% of all erythrocytes in the blood of healthy people. Percentages outside this range indicate that a patient is producing erythrocytes at an accelerated or decreased rate. In screening for abnormal rates of erythrocyte production, a **reticulocyte count** is a routine part of many clinical blood tests.

What can you predict about the number of reticulocytes in the blood of a patient who has polycythemia vera (p. 448)? In a patient who has recently developed a degenerative disease of the bone marrow?

Since polycythemia vera accelerates the production of erythrocytes, more than 2% of the erythrocytes will be reticulocytes. Since degeneration of the bone marrow decreases the production of erythrocytes, the percentage of reticulocytes in the blood will be less than 1%. ■

Formation of Leukocytes and Platelets

The committed cells in each granulocyte line are called **myeloblasts** (mi′ĕ-lo-blasts). These cells accumulate lysosomes and become **promyelocytes** (pro-mi′ĕ-lo-sīts″). The distinctive granules of each granulocyte appear next, in the **myelocyte** stage. When this stage is reached, cell division ends. In the subsequent **metamyelocyte** stage, the nucleus stops functioning, and the first signs of nuclear distortion become obvious as the nuclei bend into thick "horseshoes." Neutrophils with such horseshoe nuclei are called **band cells** or *band forms.* The granulocytes then complete the maturation sequence and enter the bloodstream.

Although most neutrophils reach maturity before entering the circulation, some enter the blood while still in the band cell stage. Band cells normally make up 1% to 2% of the neutrophils in the blood. This percentage increases dramatically during acute bacterial infections, in which the marrow releases more immature neutrophils. Thus, band cells are counted in differential WBC counts. ■

The cell lines leading to monocytes and lymphocytes are also shown in Figure 16.9. In the line leading to monocytes, committed **monoblasts** enlarge and accumulate more lysosomes as they become **promonocytes** and then **monocytes.** In the cell line leading to lymphocytes, large **lymphoblasts** become **prolymphocytes** as their chromatin condenses. Much of this differentiation of lymphocytes takes place beyond the confines of the bone marrow, as we will explain in Chapter 19.

Leukemia, a form of cancer, is the uncontrolled proliferation of a leukocyte-forming cell line in the bone marrow. Leukemias are *acute* (quickly advancing) if they derive from blast cells, such as myeloblasts, lymphoblasts, or monoblasts. Leukemias are *chronic* (slowly advancing) if they involve the proliferation of later cell stages, such as myelocytes or mature lymphocytes. In all forms of leukemia, immature white cells flood into the bloodstream. More significantly, the nonfunctional cancer cells take over the bone marrow and crowd out the healthy blood cell lines. This process slows the production of normal blood cells, which has severe consequences to the patient.

Can you deduce what some of these consequences may be?

Because their functional blood cells are not being replaced, patients in the late stages of leukemia suffer from anemia, devastating infections, and from internal hemorrhaging (internal bleeding) due to clotting failure. Infections and hemorrhaging are the usual causes of death in people who succumb to leukemia. ∎

Other cells in the bone marrow become platelet-forming cells (Figure 16.9). In this line, immature **megakaryoblasts** (meg"ah-kar'e-o-blasts) undergo repeated mitoses, but no cytoplasmic division occurs. Furthermore, their nuclei never completely separate after mitosis. This process results in a giant cell with a large, multilobed nucleus that contains many times the normal number of chromosomes. This cell is the **megakaryocyte** (meg"ah-kar'e-o-sīt; "big nucleus cell"). The cytoplasm of megakaryocytes generates platelets in the following way: Megakaryocytes lie within the reticular connective tissue of bone marrow just external to the blood sinusoids. From there, cytoplasmic extensions of the megakaryocytes reach through the walls of the sinusoids into the bloodstream. These extensions then break apart to seed the blood with platelets.

The Blood Throughout Life

The first blood cells develop with the earliest blood vessels in the mesoderm around the yolk sac of the 3-week embryo. First, mesenchyme cells cluster into groups called *blood islands.* The outer cells in these clusters flatten and become the endothelial cells that form the walls of the earliest vessels. The inner cells of the blood islands become the earliest blood cells. Soon vessels form within the embryo itself, providing a route for blood cells to travel throughout the body.

Throughout the first two months of development, all blood cells form in the blood islands of the yolk sac. These islands not only form the multipotential stem cells that will last a lifetime but they also form primitive nucleated erythrocytes. Late in month 2, circulating stem cells from the yolk sac reach and lodge in the liver and spleen. These organs take over the blood-forming function and are the major hematopoietic organs until month 7. During their tenure, they produce the first leukocytes and the first platelet-forming cells, plus nucleated and non-nucleated erythrocytes. The bone marrow receives stem cells and begins low-level hematopoiesis during month 3. Bone marrow becomes the major hematopoietic organ in month 7 of development, and is the only hematopoietic organ from birth on. If there is a severe need for blood cell production, however, the liver and spleen may resume their blood cell-forming roles, even in the adult.

The most common diseases of the blood that appear with aging are chronic leukemias, anemias, and clotting disorders. However, these and most other age-related blood disorders are usually precipitated by disorders of the heart, blood vessels, or immune system. For example, the increased incidence of leukemias in old age is believed to result from the waning ability of the immune system to destroy cancer cells, and the formation of abnormal thrombi and emboli reflects the progress of atherosclerosis, which roughens the arterial walls.

✚ Related Clinical Terms

Bone marrow biopsy Obtaining a sample of bone marrow, usually with an aspiration needle that draws marrow from the sternum or the iliac bone. The marrow is examined to diagnose disorders of blood cell formation, leukemia, various infections of the marrow, and types of anemia resulting from damage to or failure of the marrow.

Bone marrow transplant The replacement of cancerous or defective bone marrow with normal marrow, which is usually obtained from a donor. Marrow aspirated from the donor is transfused into the veins of the recipient, where its cells travel in the circulation to repopulate the recipient's marrow. The lymphocytes produced by the transplanted marrow attack the body tissues of the recipient, so the patient must take drugs that suppress the activity of lymphocytes and the immune system (immunosuppressive drugs).

Embolus Any abnormal mass carried freely in the bloodstream. May be a blood clot, bubbles of air, fat masses, clumps of cells, or aggregates of debris. Blood clots are the most common kind of emboli (p. 452), but *fat emboli* are also common. Fat emboli enter the blood from the bone marrow following a bone fracture. *Bacterial emboli* (clusters of bacteria) can occur during blood-poisoning.

Hemorrhage (hem'ŏ-rij; "blood bursting forth") Any abnormal discharge of blood out of a vessel; bleeding.

Pus A thick, white-to-yellow fluid composed of dead, degenerating, and living leukocytes (mostly neutrophils and macrophages) and tissue debris that is liquefied by enzymes liberated from the protective leukocytes. Pus is produced in bacterial infections.

Sickle cell anemia An inherited disease, exhibited primarily in people of central African descent, which results from a defect in the hemoglobin molecule. When the concentration of oxygen in the blood is low, as during exercise or anxiety, the abnormal hemoglobin crystallizes, causing the circulating erythrocytes to distort into the shape of a crescent (thus "sickle cell"). The deformed erythrocytes are fragile and easily destroyed. Furthermore, these cells have difficulty passing through capillaries, so they may collect in small blood vessels and promote clotting. The disease has always been fatal in the past, but modern treatments are now allowing some patients to survive into adulthood. About 1 of every 700 African Americans has sickle cell anemia, but 1 in 12 carries the sickle cell trait.

Thalassemia (thal"ah-se'me-ah; "sea blood")   A group of inherited anemias characterized by an insufficient production of one polypeptide chain of hemoglobin. Seen most often in people of Mediterranean descent, such as Greeks and Italians. In the most common type, called beta-thalassemia, the erythrocytes are small, pale, and easily ruptured, so RBC counts are low. Symptoms include fatigue, enlargement of the spleen, and abnormal enlargement of the bone marrow and bones.

Thrombocytopenia (throm"bo-si"to-pe'ne-ah) (*penia* = lack)   Abnormally low concentration of platelets in the blood. Characterized by diminished clot formation and by internal bleeding from small vessels, thrombocytopenia may result from damage to the bone marrow, leukemia, autoimmune destruction of the platelets, or overactivity of the spleen (which removes and destroys platelets as well as other blood cells).

Chapter Summary

1. The circulatory system is subdivided into the cardiovascular system (heart, blood vessels, and blood) and the lymphatic system (lymph vessels and lymph).

OVERVIEW: COMPOSITION OF BLOOD (pp. 445–446)

1. Blood consists of plasma and formed elements (erythrocytes, leukocytes, and platelets). Erythrocytes make up about 45% of blood volume, plasma about 55%, leukocytes and platelets under 1%.

2. The volume percentage of erythrocytes is the hematocrit.

BLOOD PLASMA (p. 446)

1. Plasma is a fluid that is 90% water. The remaining 10% is solutes, including nutrients, respiratory gases, salts, hormones, and plasma proteins. Serum is plasma from which the clotting elements have been removed.

FORMED ELEMENTS (pp. 446–453)

1. Blood cells are short-lived and are replenished continuously by new cells from the bone marrow.

2. A blood smear is prepared by drawing a drop of blood into a film on a glass slide, then drying, preserving, and staining the film. Blood stains are mixtures of two dyes, eosin and methylene blue.

Erythrocytes (pp. 446–448)

3. Erythrocytes are the most abundant blood cells. They are anucleate, biconcave discs with a diameter of 7 to 8 μm. Essentially, they are bags of hemoglobin, the oxygen-transporting protein.

4. The main function of erythrocytes is to transport oxygen between the lungs and the body tissues.

5. Erythrocytes live about 4 months in the circulation.

Leukocytes (pp. 448–451)

6. Leukocytes fight disease in the loose connective tissues outside capillaries, using the bloodstream only as a transport system. They leave capillaries by diapedesis and crawl through the connective tissues to the infection sites.

7. There are only 4000 to 11,000 leukocytes, compared to about 5 million erythrocytes, in a cubic millimeter of blood.

8. Granulocytes (granular leukocytes) are neutrophils, eosinophils, and basophils. These short-lived phagocytes have distinctive cytoplasmic granules and distorted (lobed) nuclei.

9. Neutrophils are the most abundant leukocytes and have multilobed nuclei. Two types of small granules give their cytoplasm a light purple color in blood stains. The function of neutrophils is to destroy bacteria.

10. Eosinophils have bilobed nuclei and large, red-staining granules with crystals of major basic protein. Their main function is to destroy parasitic worms.

11. Basophils are rare leukocytes with bilobed nuclei and large, purple-staining granules full of chemical mediators of inflammation. They function like mast cells.

12. The agranulocytes (nongranular leukocytes) are lymphocytes and monocytes. Although similar to one another in structure, these cell types are unrelated.

13. Lymphocytes attack antigens in the specific immune response. These cells have a sparse, blue-staining cytoplasm and a dense, purple-staining, spherical nucleus. Cytotoxic T lymphocytes destroy foreign cells by causing their lysis. B lymphocytes produce plasma cells that secrete antibodies.

14. Monocytes, the largest leukocytes, resemble large lymphocytes but have a lighter-staining nucleus that may be C-shaped. They become macrophages in the connective tissues.

Platelets (pp. 451–453)

15. Platelets are disc-shaped, membrane-enclosed fragments of megakaryocyte cytoplasm. They contain several kinds of secretory granules. They plug tears in blood vessels, signal vasoconstriction, help initiate clotting, then retract the clot and close the tear.

BLOOD CELL FORMATION (pp. 453–456)

Bone Marrow: The Site of Blood Cell Formation (p. 453)

1. Collectively, bone marrow is the body's second largest organ. Red marrow (in the axial skeleton and girdles of adults) actively produces blood cells. Yellow marrow (in the limb bones of adults) is dormant.

2. Microscopically, bone marrow consists of wide capillaries (sinusoids) snaking through reticular connective tissue. The latter contains reticular fibers, fibroblasts, macrophages, fat cells, and immature blood cells in all stages of maturation. New blood cells enter the blood through the sinusoid walls.

Stages of Blood Cell Formation (pp. 453–456)

3. All blood cells continuously arise from multipotential hematopoietic stem cells (hemocytoblasts). As these cells divide, there is an early separation into lymphoid stem cells (future lymphocytes) and myeloid stem cells (precursors to all other blood cell classes). Then, stem cells commit to the specific blood cell lines, and structural differentiation begins.

4. Erythrocytes start as proerythroblasts and proceed through various stages, during which hemoglobin accumulates and the organelles and nucleus are extruded. The stage that enters the bloodstream is the reticulocyte, and this stage persists for 1 to 2 days in the circulation.

5. Granular leukocytes begin as myeloblasts, then proceed through various stages in which they gain specific granules and their nucleus distorts into a horseshoe shape. The cells then mature and enter the bloodstream. Some neutrophils enter the circulation as immature band cells.

6. Monocytes and lymphocytes look somewhat like stem cells. Structural changes are minimal in the developmental pathways of these cells.

7. In the platelet line, immature megakaryoblasts become giant megakaryocytes with multilobed nuclei. The cytoplasm of megakaryocytes breaks up to form platelets in the blood.

THE BLOOD THROUGHOUT LIFE (p. 456)

1. The blood stem cells develop in blood islands on the yolk sac in the 3-week embryo and then travel to the hematopoietic organs.

2. The hematopoietic organs in the fetus are the liver, spleen, and bone marrow.

3. Blood disorders that may develop with age include increased incidences of leukemia, anemia, and clotting disorders.

Review Questions

Multiple Choice and Matching Questions

1. The volume of blood in an adult averages about (a) 1 L, (b) 3 L, (c) 5 L, (d) 7 L.

2. All of the following are true of erythrocytes except (a) shaped like a biconcave disc, (b) life span of about 120 days, (c) contain hemoglobin, (d) contain lobed nuclei.

3. Rank these leukocytes in order of their relative abundance, from 1 (most abundant) to 5 (least abundant) in the blood of a normal, healthy person.

_____ **(a)** lymphocytes
_____ **(b)** basophils
_____ **(c)** neutrophils
_____ **(d)** eosinophils
_____ **(e)** monocytes

4. The white blood cell that releases histamine and other inflammatory chemicals is the (a) basophil, (b) neutrophil, (c) monocyte, (d) eosinophil.

5. Which of the following blood cells are phagocytic? (a) lymphocytes, (b) erythrocytes, (c) neutrophils, (d) a and c.

6. The blood cell that can attack a specific antigen is (a) a lymphocyte, (b) a monocyte, (c) a neutrophil, (d) a basophil, (e) an erythrocyte.

7. A normal leukocyte count (per cubic millimeter of blood) for adults is (a) 3 to 4 million, (b) 4.5 to 5 million, (c) 5, (d) 10,000.

8. Match the descriptions of the blood cells at left with their names at right. Some names will be used more than once.

Column A	Column B
_____ **(1)** destroys parasitic worms	**(a)** erythrocyte
_____ **(2)** has two types of granules	**(b)** neutrophil
_____ **(3)** does not use diapedesis	**(c)** eosinophil
_____ **(4)** secretes major basic protein	**(d)** basophil
_____ **(5)** the only cell that is not spherical	**(e)** lymphocyte
_____ **(6)** granulocyte that phagocytizes bacteria	**(f)** monocyte
_____ **(7)** the largest blood cell	
_____ **(8)** granulocyte with the smallest granules	
_____ **(9)** the most like a reticulocyte	
_____ **(10)** lives for about 4 months	
_____ **(11)** the most abundant	
_____ **(12)** T and B cells	

9. In monocytes stained with typical blood stain, the blue cytoplasm and the purple nuclei are colored by this dye: (a) eosin, (b) brilliant cresyl blue, (c) basin (as in basophilic), (d) methylene blue, (e) none of the above—you need two dyes for two colors.

Short Answer and Essay Questions

10. Which class or classes of formed elements form the buffy coat in a hematocrit tube?

11. (a) What is the basic functional difference between red and yellow bone marrow? (b) Where is yellow marrow located?

12. (a) Describe the steps of erythrocyte formation in the bone marrow. (b) What name is given to the immature type of red blood cell that is released into the circulation?

13. Describe the structure of platelets, and explain their functions.

14. What is the difference between a multipotential hematopoietic stem cell and a committed cell in the bone marrow?

15. What is the relationship between megakaryocytes and platelets?

16. Looking at a blood smear under the microscope, Tina became confused by the unusual nuclei of granulocytes. She kept asking why each eosinophil had two nuclei and why neutrophils had four or five nuclei. How would you answer her?

Critical Thinking and Clinical Application Questions

17. A young girl named Janie is diagnosed as having acute lymphocytic leukemia. Her parents cannot understand why infection is a major problem for Janie when her WBC count is so high. Could you provide an explanation for Janie's parents?

18. Freddy saw a neuron cell body on his slide that he could tell was as wide as ten erythrocytes. How wide was the neuron cell body (in micrometers)?

19. The Jones family let their dog, Rooter, lick their faces, and they kissed it on the mouth, even after it explored the neighborhood dump and trash cans. The same day that the veterinarian diagnosed Rooter as having a tapeworm, a blood test indicated that both the family's daughters had blood eosinophil levels of over 3000 per cubic millimeter. What is the connection?

20. A reticulocyte count indicated that 5% of Tyler's red blood cells were reticulocytes. His blood test also indicated he had polycythemia and a hematocrit of 65%. Explain the connection between these three facts.

21. Cancer patients being treated with chemotherapy drugs, which are designed to destroy rapidly mitotic cells, are monitored closely for changes in their RBC and WBC counts. Why?

17

The Heart

Student Objectives

1. Define the pulmonary and systemic circuits.

Orientation and Location (pp. 460–462)

2. Describe the orientation, location, and surface anatomy of the heart in the thorax.

Structure of the Heart (pp. 462–466)

3. Describe the layers of the pericardium and the tissue layers of the heart wall.

4. List the important structural features of each heart chamber.

Pathway of Blood Through the Heart (pp. 466–467)

5. Trace a drop of blood through the four chambers of the heart and the systemic and pulmonary circuits.

Heart Valves (pp. 467–470)

6. Name the heart valves, and describe their locations and functions. Indicate where on the chest wall the valves are heard.

Fibrous Skeleton (p. 470)

7. Locate the fibrous skeleton of the heart, and explain its functions.

Conducting System and Innervation (pp. 470–472)

8. Name the components of the conducting system of the heart, and trace the conduction pathway.

Blood Supply to the Heart (pp. 472–473)

9. Trace the coronary arteries and cardiac veins on the heart surface.

The Heart Throughout Life (pp. 473–477)

10. Explain how the heart develops, and describe some congenital heart defects.

11. List some effects of aging on the heart.

The ceaselessly beating heart in the thorax has intrigued people for thousands of years. The ancient Greeks believed the heart to be the seat of intelligence. Others thought it was the source of emotions. While these theories have proved false, we do know that the emotions affect heart rate. When your heart pounds or occasionally skips a beat, you become acutely aware of how much you depend on this dynamic organ for your very life.

The heart is a muscular double pump with two functions (Figure 17.1): (1) Its right side pumps blood to the lungs to pick up oxygen and dispel carbon dioxide; (2) its left side receives the oxygenated blood returning from the lungs and pumps this blood throughout the body to supply oxygen and nutrients to the body tissues. The blood vessels that carry blood to and from the lungs form the **pulmonary circuit** (*pulmonos* = lung), while the vessels that transport blood to and from all body tissues form the **systemic circuit.** To receive the blood returning from the pulmonary and systemic circuits, the heart has two receiving chambers called **atria** (*atrium* = entranceway). To pump blood around the two circuits, the heart has two main pumping chambers called **ventricles** ("hollow bellies"). Now let us look more closely at the structure and function of the heart.

Orientation and Location

The relative size and weight of the heart belie its incredible strength and fortitude. About the size of a fist, the hollow, cone-shaped heart looks enough like the popular valentine image to satisfy the sentimentalists among us (Figure 17.2). Typically, the heart weighs between 250 and 350 grams—less than a pound.

The heart lies in the thorax posterior to the sternum and costal cartilages and rests on the superior surface of the diaphragm. It is the largest organ in the mediastinum (the thick partition between the two lungs). The heart assumes an oblique position in the thorax, with its pointed **apex** lying to the left of the midline and anterior to the rest of the heart (Figure 17.2c). If you press your fingers between the fifth and sixth ribs just inferior to the left nipple, you may feel the beating of your heart where the apex contacts the thoracic wall. Cone-shaped objects have a base as well as an apex, and the heart's *base* is its broad posterior surface.

The heart is said to have four corners defined by four points projected onto the anterior thoracic wall

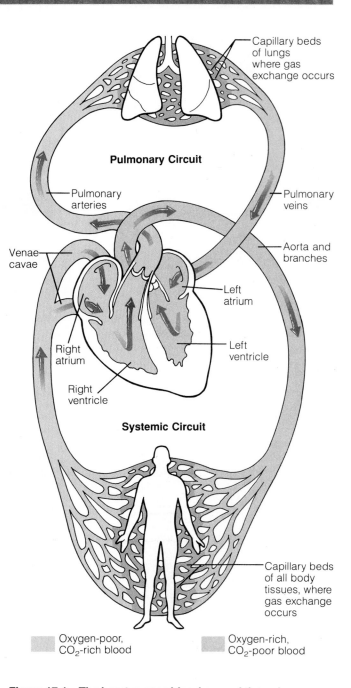

Figure 17.1   The heart pumps blood around the pulmonary and systemic circuits.   The right side of the heart is the pulmonary pump that propels oxygen-poor blood to the lungs. The left side of the heart is the systemic pump that pumps oxygen-rich blood to tissues throughout the body.

(Figure 17.2a). The *superior right* point lies where the costal cartilage of the third rib joins the sternum. The *superior left* point lies at the costal cartilage of the second rib, a finger's breadth lateral to the sternum. The *inferior right* point lies at the costal cartilage of the sixth rib, a finger's breadth lateral to the sternum.

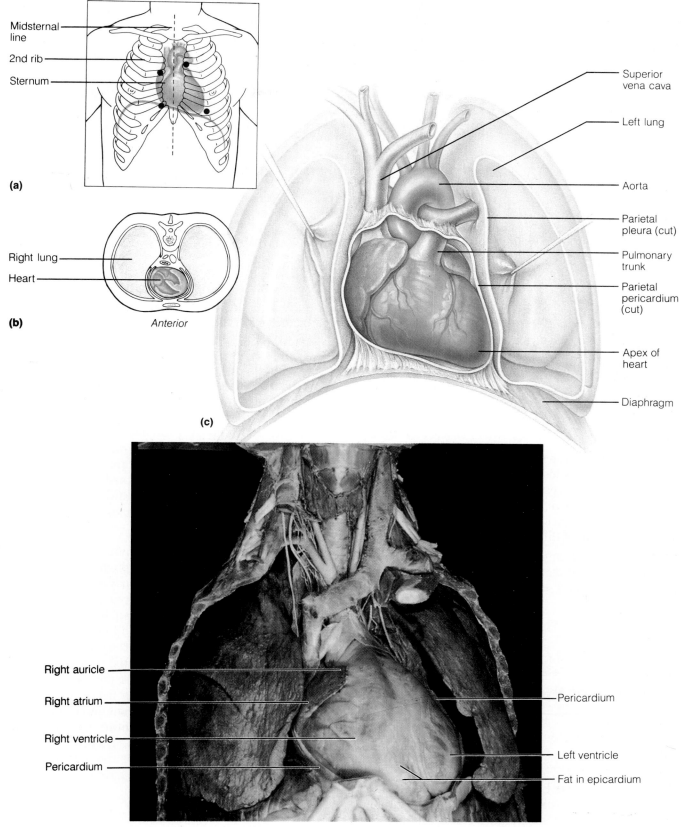

Figure 17.2   Location of the heart in the mediastinum of the thorax.   (**a**) Relation of the heart to the sternum and ribs in a person who is lying down (the heart lies slightly inferior to this position if the person is standing). Note the four corner points of the heart. (**b**) Cross section through the mid thorax showing the heart in the mediastinum between the lungs. (**c**) Relation of the heart and great vessels to the lungs. (**d**) Photograph of the heart in the mediastinum. Note the fat in the outer layer of the heart (epicardium).

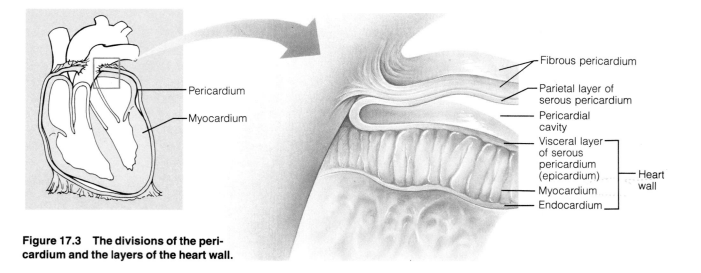

Figure 17.3   The divisions of the pericardium and the layers of the heart wall.

Finally, the *inferior left* point is the apex point. It lies in the fifth intercostal space in the midclavicular line (a line extending inferiorly from the midpoint of the left clavicle). By interconnecting these four corner points, one determines the normal size and location of the heart. The clinician must know these normal parameters, because an enlarged or displaced heart can indicate heart disease or other disease conditions.

Assume that the four corner points of a patient's heart, as revealed on an X-ray film, relate to the rib cage as follows: second right rib at sternum, tip of xiphoid process, seventh intercostal space lateral to the left nipple, and second left rib at sternum. Are the size and position of this patient's heart normal?

By sketching and interconnecting these four points on Figure 17.2a, you can see that this patient's heart is enlarged and perhaps displaced slightly to the left. ■

Structure of the Heart

Coverings

The heart is enclosed in a double membrane called the **pericardium** (per″ĭ-kar′de-um; "around the heart") or *pericardial sac* (Figure 17.3). The most superficial part of this sac is a layer of dense connective tissue called **fibrous pericardium.** This strong layer adheres to the diaphragm inferiorly. Superiorly, it is fused to the roots of the great vessels that leave and enter the heart. The fibrous pericardium acts as a

tough outer sac that holds the heart in place and prevents overfilling of the heart with blood.

Deep to the fibrous pericardium is the **serous pericardium,** a slippery serous membrane composed of two layers: The **parietal layer** adheres to the internal surface of the fibrous pericardium. Near the heart's superior margin, the parietal layer turns inferiorly and continues over the heart surface as the **visceral layer,** also called the *epicardium.* Between the parietal and visceral layers of serous pericardium is a slit-like space called the **pericardial cavity.** This space contains a film of serous fluid, a lubricant that reduces friction between the beating heart and the outer wall of the pericardial sac.

Infection and inflammation of the pericardium, or **pericarditis,** leads to a roughening of the serous lining of the pericardial cavity. As a consequence, the beating heart produces a rustling sound (pericardial friction rub) that can be heard with a stethoscope. Pericarditis may lead to the formation of painful adhesions of the heart to the outer pericardial wall. In severe cases of pericarditis, large amounts of inflammatory fluid exude into the pericardial cavity. This excess fluid compresses the heart, limiting the expansion of the heart between beats and diminishing its ability to pump blood. This condition is called **cardiac tamponade** (tam″po-nād′), literally "a heart plug." Physicians treat cardiac tamponade by inserting a syringe into the pericardial cavity and draining the excess fluid.

Cardiac tamponade also results if *blood* accumulates inside the pericardial cavity. Can you determine how blood could enter this cavity?

A penetrating wound to the heart wall, such as a stab wound, causes blood to leak out of the heart and overfill the pericardial cavity. ■

Layers of the Heart Wall

The wall of the heart has three layers: a superficial epicardium, a middle myocardium, and a deep endocardium (Figure 17.3). All three layers are richly supplied with blood vessels.

The **epicardium** ("upon the heart") is the visceral layer of the serous pericardium, as we explained above. This serous membrane is often infiltrated with fat, especially in older people (see Figure 17.2d). The **myocardium** ("muscle heart"), which forms the bulk of the heart, consists mainly of cardiac muscle and is the layer that actually contracts. The myocardium's elongated networks of cardiac muscle cells, called *bundles,* form circular and spiral patterns that serve to squeeze blood through the heart in the proper directions: inferiorly through the atria and superiorly through the ventricles (Figure 17.4). The **endocardium** ("inside the heart") is a glistening white sheet of endothelium resting on a thin layer of connective tissue. Located deep to the myocardium, it lines the heart chambers.

Heart Chambers

The four heart chambers are the *right* and *left atria* superiorly and the *right* and *left ventricles* inferiorly (Figure 17.5). Internally, the partition that divides the heart longitudinally is called either the **interatrial septum** or the **interventricular septum,** depending on which chambers it separates. Externally, two grooves on the heart's surface indicate the boundaries of the four chambers. The **atrioventricular groove,** or **coronary sulcus,** circles the boundary between the atria and ventricles like a crown (*corona* = crown). The **anterior interventricular sulcus** marks the anterior position of the interventricular septum that separates the right and left ventricles. This sulcus is continuous with the **posterior interventricular sulcus,** which separates the two ventricles on the heart's inferior surface.

Right Atrium

The **right atrium,** which forms the entire right border of the heart, is the receiving chamber for oxygen-poor blood returning from the systemic circuit. Blood enters the right atrium via three veins: (1) The *superior vena cava* returns blood from body regions superior to the diaphragm; (2) the *inferior vena cava* returns blood from body regions inferior to the diaphragm; and (3) the *coronary sinus* collects blood draining from the heart wall itself.

Externally, the right atrium exhibits no outstanding feature except the **right auricle,** a small appendage shaped like a dog's ear (*auricle* = little ear). This flap

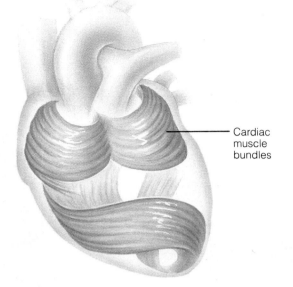
Cardiac muscle bundles

Figure 17.4   The circular and spiral arrangement of cardiac muscle bundles in the myocardium of the heart.

projects to the left from the superior corner of the atrium (see Figure 17.2d). Internally, the right atrium exhibits two basic divisions (Figure 17.5b): a smooth-walled posterior division and an anterior division lined by horizontal ridges called the **pectinate muscles** (*pectinate* = like the teeth of a comb). The two divisions are separated by a large, C-shaped ridge called the **crista terminalis** ("terminal crest"). Just posterior to the superior bend of the crista, the superior vena cava opens into the atrium. Just posterior to the inferior bend of the crista is the opening of the inferior vena cava. The coronary sinus opens into the atrium just anterior to the inferior end of the crista, while the **fossa ovalis** lies posterior to this end (Figures 17.5b and 17.6). The fossa ovalis is a depression in the interatrial septum that marks the spot where an opening existed in the fetal heart (the *foramen ovale:* p. 476).

Inferiorly and anteriorly, the right atrium opens into the right ventricle through the *tricuspid valve (right atrioventricular valve).*

Right Ventricle

The **right ventricle** (Figures 17.5 and 17.6) receives blood from the right atrium and pumps it into an artery called the **pulmonary trunk,** which carries the blood to the lungs for oxygenation. Externally, the right ventricle forms most of the anterior surface of the heart. Internally, its walls are marked by irregular ridges of muscle, called **trabeculae carneae** (trah-bek'u-le kar'ne-e; "crossbars of flesh") (Figure 17.5c). Cone-shaped **papillary muscles** project from the walls into the ventricular cavity (*papilla* = nipple).

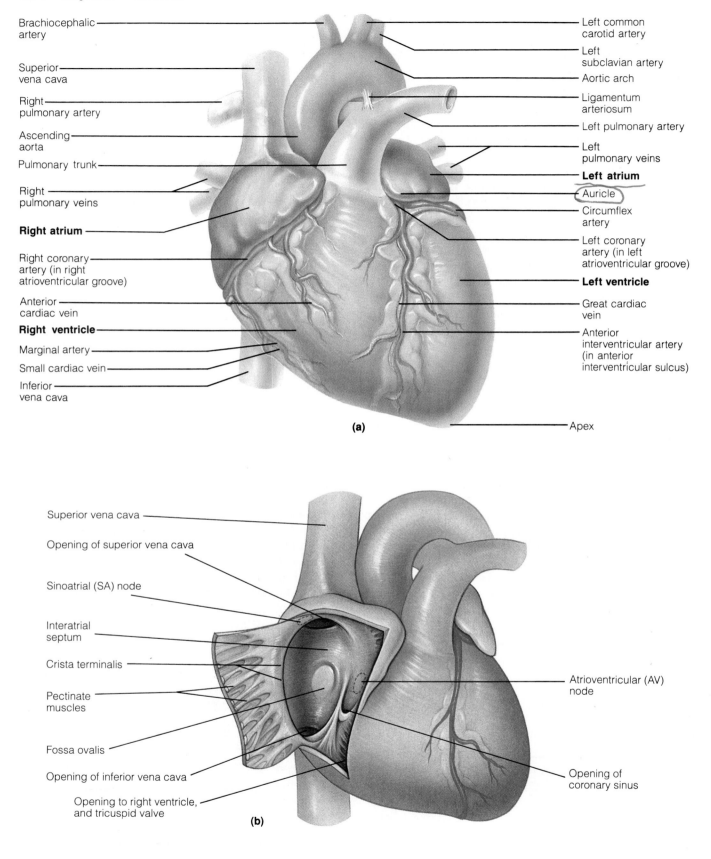

Brachiocephalic artery

Superior vena cava

Right pulmonary artery

Ascending aorta

Pulmonary trunk

Right pulmonary veins

Right atrium

Right coronary artery (in right atrioventricular groove)

Anterior cardiac vein

Right ventricle

Marginal artery

Small cardiac vein

Inferior vena cava

Left common carotid artery

Left subclavian artery

Aortic arch

Ligamentum arteriosum

Left pulmonary artery

Left pulmonary veins

Left atrium

Auricle

Circumflex artery

Left coronary artery (in left atrioventricular groove)

Left ventricle

Great cardiac vein

Anterior interventricular artery (in anterior interventricular sulcus)

Apex

(a)

Superior vena cava

Opening of superior vena cava

Sinoatrial (SA) node

Interatrial septum

Crista terminalis

Pectinate muscles

Fossa ovalis

Opening of inferior vena cava

Opening to right ventricle, and tricuspid valve

Atrioventricular (AV) node

Opening of coronary sinus

(b)

Figure 17.5 Gross anatomy of the heart. (**a**) Anterior view. (**b**) Anterior view emphasizing the right atrium, which is opened. The anterior wall of the atrium has been reflected to the side. (**c**) Frontal section showing the interior chambers and valves. (**d**) Inferior view. The surface shown rests on the diaphragm (dorsal aspect is at the top of this figure).

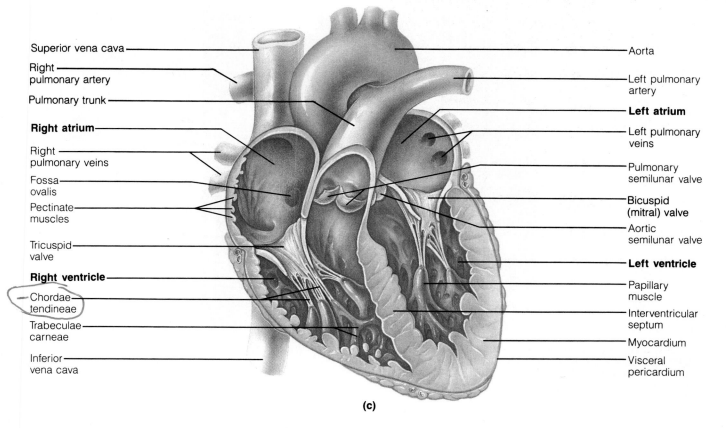

Superior vena cava

Right pulmonary artery

Pulmonary trunk

Right atrium

Right pulmonary veins

Fossa ovalis

Pectinate muscles

Tricuspid valve

Right ventricle

Chordae tendineae

Trabeculae carneae

Inferior vena cava

Aorta

Left pulmonary artery

Left atrium

Left pulmonary veins

Pulmonary semilunar valve

Bicuspid (mitral) valve

Aortic semilunar valve

Left ventricle

Papillary muscle

Interventricular septum

Myocardium

Visceral pericardium

(c)

Left pulmonary artery

Left pulmonary veins

Auricle

Left atrium

Great cardiac vein

Posterior vein of left ventricle

Left ventricle

Apex

Aorta

Superior vena cava

Right pulmonary artery

Right pulmonary veins

Base of heart

Right atrium

Inferior vena cava

Right coronary artery (in right atrioventricular groove)

Coronary sinus

Posterior interventricular artery (in posterior interventricular sulcus)

Middle cardiac vein

Right ventricle

(d)

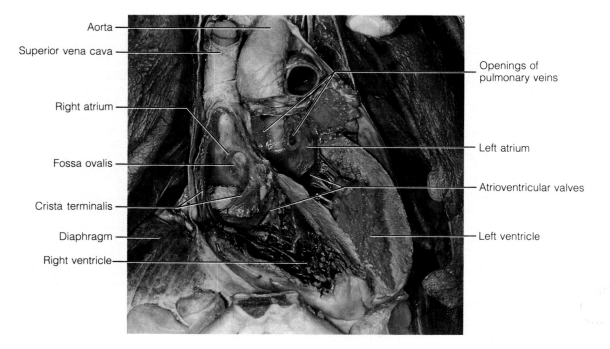

Aorta
Superior vena cava
Right atrium
Fossa ovalis
Crista terminalis
Diaphragm
Right ventricle

Openings of pulmonary veins
Left atrium
Atrioventricular valves
Left ventricle

Figure 17.6 Photograph of a heart from a cadaver, sectioned in the frontal plane. This is comparable to the diagram in Figure 17.5c, except that the bases of the aorta and pulmonary trunk have been removed.

Thin, strong bands called **chordae tendineae** (kor'de ten'dĭ-ne-e; "tendinous cords") project superiorly from the papillary muscles to the flaps (cusps) of the tricuspid valve. The popular expression "tugging on my heartstrings" is based on these bands. Superiorly, the opening between the right ventricle and the pulmonary trunk contains the *pulmonary semilunar valve.*

Left Atrium

The **left atrium** receives oxygen-rich blood returning from the lungs through four **pulmonary veins** (Figures 17.5c and 17.6). This atrium makes up most of the heart's posterior surface (base). The only part of the left atrium visible anteriorly is its triangular **left auricle** (Figure 17.5a). Internally, most of the atrial wall is smooth, with pectinate muscles lining the auricle only. Blood from the left atrium enters the left ventricle through the *bicuspid valve (left atrioventricular valve).*

Left Ventricle

The **left ventricle** receives oxygenated blood from the right atrium and pumps it into the **aorta,** the stem artery of the systemic circuit. This ventricle forms the apex of the heart and dominates the heart's inferior surface (Figure 17.5d). Like the right ventricle, it contains trabeculae carneae, papillary muscles, chordae tendineae, and the cusps of an atrioventricular valve

(bicuspid valve). Superiorly, the opening between the left ventricle and the aorta is covered by the *aortic semilunar valve.*

Pathway of Blood Through the Heart

We will summarize how blood passes through the four chambers of the heart by following the path of a single drop of blood around the pulmonary and systemic circuits (Figure 17.1). We begin with oxygen-poor systemic blood as it enters the right side of the heart through the superior and inferior venae cavae. This blood passes from the right atrium through the tricuspid valve to the right ventricle, aided by the contraction of the right atrium. Then, the right ventricle propels the blood through the pulmonary semilunar valve into the pulmonary trunk and around the pulmonary circuit for oxygenation. The freshly oxygenated blood returns via the pulmonary veins to the left atrium and passes through the bicuspid valve to the left ventricle, aided by the contraction of the left atrium. The left ventricle then propels the blood through the aortic semilunar valve into the aorta and its branches. After delivering oxygen and nutrients to the body tissues through the systemic capillaries, the oxygen-poor blood returns through the systemic veins to the right atrium—and the whole cycle repeats itself continuously.

Although each drop of blood passes through the heart chambers sequentially (one after another), this does not mean that the four chambers contract in that order. Rather, the two atria always contract *together,* followed by the simultaneous contraction of the two ventricles. This sequence constitutes a heartbeat, and the heart rate of a resting person averages 70 to 80 beats per minute. When a heart chamber is contracting, it is said to be in **systole** (sis′to-le; "contraction"). When a heart chamber is relaxing and filling with blood, it is in **diastole** (di-as′to-le; "expansion").

The muscular walls of the different heart chambers differ in thickness (Figure 17.5c). The walls of the atria are much thinner than those of the ventricles because the atria need exert little effort to propel blood inferiorly into the ventricles (much of the work is done by gravity). Furthermore, the wall of the left ventricle (the systemic pump) is at least three times as thick as that of the right ventricle (the pulmonary pump) (Figure 17.7). This reflects the fact that the systemic circuit is much longer than the pulmonary circuit and offers greater resistance to blood flow. Consequently, the left ventricle can generate much more pressure than the right and is a more powerful pump. The thick wall of the left ventricle gives this chamber a cylindrical shape, which flattens the cavity of the adjacent right ventricle into the shape of a crescent (Figure 17.7). (Think of a right hand grasping a clenched left fist.)

Heart Valves

Blood flows through the heart in one direction: from the atria to the ventricles and into the great arteries that leave the superior part of the heart. This one-way flow is enforced by the heart valves—the paired atrioventricular (AV) and semilunar valves—which open and close in response to blood pressure on their two sides. As illustrated in Figure 17.8, each heart valve consists of several *cusps* (flaps of endocardium reinforced by cores of dense connective tissue).

Structure

The two atrioventricular valves are located at the junctions of the atria and their respective ventricles. The **tricuspid,** or **right atrioventricular, valve** has three cusps. The **bicuspid,** or **left atrioventricular, valve** has only two cusps. This valve is also called the **mitral** (mi′tral) **valve** because its cusps resemble the two sides of a bishop's hat, or miter. The **aortic** and

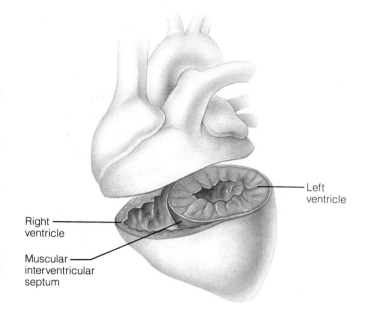

Right ventricle

Muscular interventricular septum

Left ventricle

Figure 17.7   Anatomical differences between the right and left ventricles.   The left ventricle has a thicker wall, and its cavity is circular. The cavity of the right ventricle, by contrast, is crescent-shaped.

pulmonary semilunar valves guard the bases of the large arteries issuing from the ventricles. Both of these semilunar valves have three pocket-like cusps, each shaped roughly like a crescent moon (*semilunar* = half-moon).

Functions

The two *atrioventricular valves* prevent the backflow of blood into the atria during contraction of the ventricles (Figure 17.9). While the ventricles are relaxed (in diastole), the cusps of the AV valves hang limply into the ventricular chambers. Blood flows into the atria and then through the open AV valves into the ventricles. When the ventricles start to contract, the pressure within them rises, forcing the blood superiorly against the valve cusps. This action forces the edges of the cusps together, closing the AV valves. The chordae tendineae and papillary muscles that attach to these valves serve as guy wires to anchor the cusps in their closed position. If the cusps were not anchored in this manner, they would be blown superiorly into the atria and would evert, in much the same way that an umbrella is blown inside out by a gusty wind.

Given the above information, you should be able to state the *precise* function of the papillary muscles. Can you do this?

The papillary muscles contract when the rest of the ventricle contracts (during ventricular systole), and they pull on the chordae tendineae to prevent the

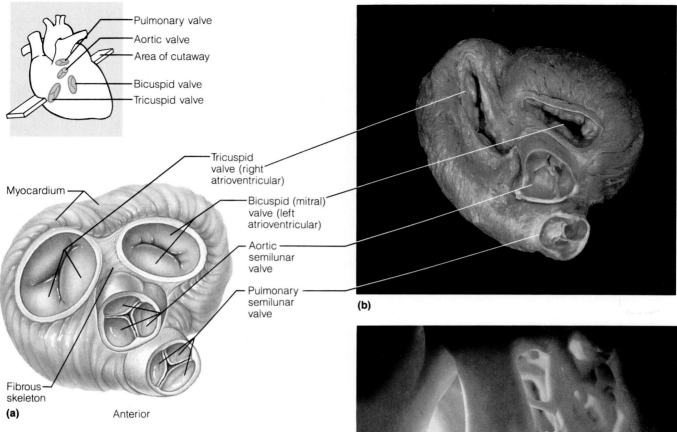

Figure 17.8   **Heart valves.**   (a) Superior view of the four heart valves (atria removed). The inset shows the plane of section through the heart. (b) Photograph similar to the view in part (a). (c) Photograph of the tricuspid valve. This inferior-to-superior view begins in the right ventricle and faces the right atrium.

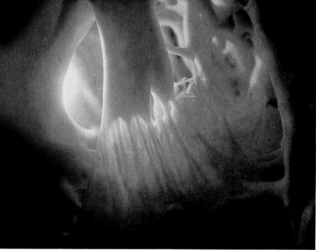

AV valves from everting. They do *not* act to open these valves. ■

The two *semilunar valves* prevent backflow from the great arteries into the ventricles (Figure 17.10). When the ventricles contract, the semilunar valves are forced open, and their cusps flatten against the arterial walls as the blood rushes past them. When the ventricles relax, the blood starts to flow backward toward the heart. This blood fills the cusps of the semilunar valves and closes the valves.

CAUSED BY VALVES

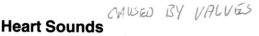

Heart Sounds

All four heart valves lie in roughly the same plane—the oblique plane of the coronary sulcus that extends between the heart's superior left and inferior right cor-

ners (Figures 17.8a and 17.11). The closing of the valves causes vibrations in the adjacent blood and heart walls. These vibrations account for the familiar "lub-dup" sounds of each heartbeat: The "lub" sound is produced by the closing of the AV valves at the start of ventricular systole, whereas "dup" is produced by the closing semilunar valves at the end of ventricular systole. Because the mitral valve closes slightly before the tricuspid closes, and the aortic valve generally closes just before the pulmonary valve closes, one can discern all four valve sounds when listening by stethoscope on the anterior chest wall. The clinician does not listen directly over the respective valves, for the sounds take oblique paths through the heart chambers to reach the chest wall. Handily, each valve is best heard near a different heart corner: the pulmonary valve near the superior left point, the aortic valve

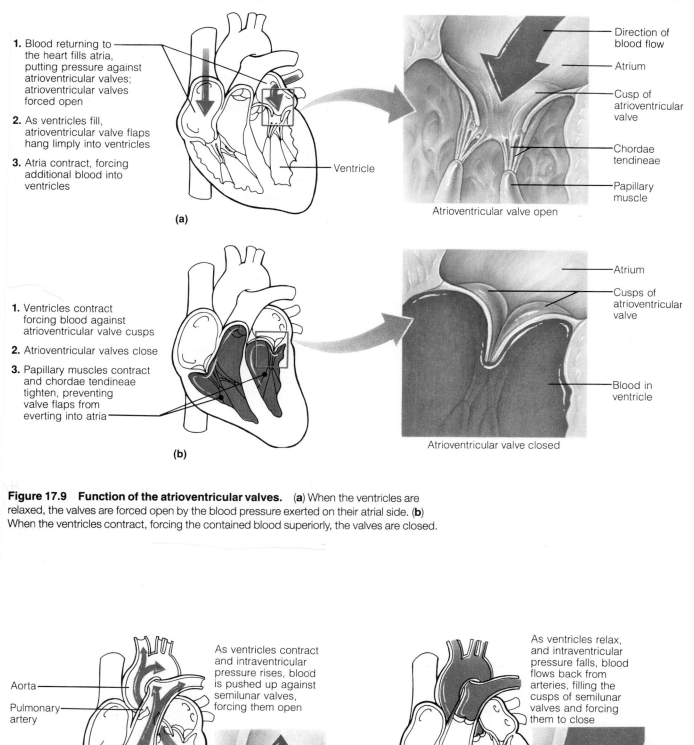

1. Blood returning to the heart fills atria, putting pressure against atrioventricular valves; atrioventricular valves forced open

2. As ventricles fill, atrioventricular valve flaps hang limply into ventricles

3. Atria contract, forcing additional blood into ventricles

Ventricle

(a)

Direction of blood flow

Atrium

Cusp of atrioventricular valve

Chordae tendineae

Papillary muscle

Atrioventricular valve open

1. Ventricles contract forcing blood against atrioventricular valve cusps

2. Atrioventricular valves close

3. Papillary muscles contract and chordae tendineae tighten, preventing valve flaps from everting into atria

(b)

Atrium

Cusps of atrioventricular valve

Blood in ventricle

Atrioventricular valve closed

Figure 17.9 Function of the atrioventricular valves. (**a**) When the ventricles are relaxed, the valves are forced open by the blood pressure exerted on their atrial side. (**b**) When the ventricles contract, forcing the contained blood superiorly, the valves are closed.

Aorta

Pulmonary artery

As ventricles contract and intraventricular pressure rises, blood is pushed up against semilunar valves, forcing them open

(a)

Semilunar valve open

As ventricles relax, and intraventricular pressure falls, blood flows back from arteries, filling the cusps of semilunar valves and forcing them to close

(b)

Semilunar valve closed

Figure 17.10 Function of the semilunar valves. (**a**) During ventricular contraction, the valves are open, and their flaps are flattened against the artery walls. (**b**) When the ventricles relax, the backflowing blood closes the valves.

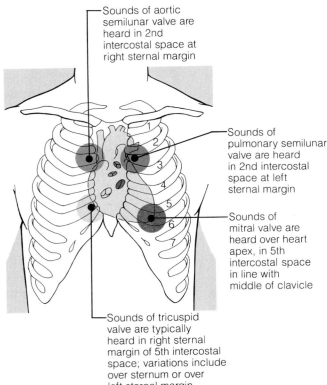

Sounds of aortic semilunar valve are heard in 2nd intercostal space at right sternal margin

Sounds of pulmonary semilunar valve are heard in 2nd intercostal space at left sternal margin

Sounds of mitral valve are heard over heart apex, in 5th intercostal space in line with middle of clavicle

Sounds of tricuspid valve are typically heard in right sternal margin of 5th intercostal space; variations include over sternum or over left sternal margin in 5th intercostal space

Figure 17.11 Points on the thorax surface where sounds of the heart valves are heard most clearly. These four points are almost identical to the four heart corners shown in Figure 17.2a, displaced only slightly so that one does not have to listen over a sound-blocking rib.

near the superior right point, the mitral valve at the apex point, and the tricuspid near the inferior right point. Figure 17.11 provides further details.

✚ Abnormal heart sounds, called **murmurs,** often arise from disorders of the heart valves. For example, if a valve is **incompetent** (does not close properly), leakage produces a swishing sound after the valve has closed. A **stenotic** valve is one in which fused or stiffened cusps have narrowed the opening (*stenosis* = narrowing). Stenosis of the aortic valve leads to a high-pitched sound during ventricular systole as blood passing through the constricted opening undergoes turbulence and vibration. Some common causes of valve disorders are an insufficient blood supply to the valves (caused by a heart attack), bacterial infection of the endocardium, and rheumatic fever. Both stenotic and incompetent valves increase the work load on the heart and decrease its pumping efficiency. Ultimately, the heart may weaken severely. A faulty valve may be replaced with a synthetic valve or with a valve taken from a pig's heart.

▦✚ Can you figure out which of the four heart valves is under the most strain and is therefore most often involved in valve disorders?

It is the mitral valve, which must resist the powerful contractions of the left ventricle. ▪

Fibrous Skeleton

The **fibrous skeleton** of the heart lies in the plane between the atria and the ventricles and surrounds all four heart valves like handcuffs (Figure 17.8a). Composed of dense connective tissue, the fibrous skeleton has four functions:

1. It anchors the valve cusps.

2. It prevents overdilation of the valve openings as blood pulses through them.

3. It is the point of insertion for the bundles of cardiac muscle in the atria and ventricles (Figure 17.4).

4. It blocks the direct spread of electrical impulses from atrial to ventricular muscle (see the next section).

Conducting System and Innervation

Conducting System

Cardiac muscle cells have an intrinsic ability to generate and conduct impulses, which in turn signal these cells to contract rhythmically. Again, these are properties of the heart muscle itself and do not depend on extrinsic nerve impulses. In fact, even if all nerve connections to the heart are severed, the heart continues to beat rhythmically, a fact amply demonstrated by transplanted hearts.

The **conducting system** of the heart (Figure 17.12) is a series of specialized cardiac muscle cells that carries impulses throughout the heart musculature, signaling the heart chambers to contract in the proper sequence. It also initiates each contraction sequence, thereby setting the basic rate of the heartbeat. The components of the conducting system are the *sinoatrial node,* the *atrioventricular node,* the *atrioventricular bundle,* the right and left *bundle branches,* and the *Purkinje fibers.*

The impulse that signals each heartbeat begins at the **sinoatrial (SA) node.** This crescent-shaped mass of muscle cells lies in the wall of the right atrium, just inferior to the entrance of the superior vena cava. The SA node sets the basic rate of the heartbeat by gener-

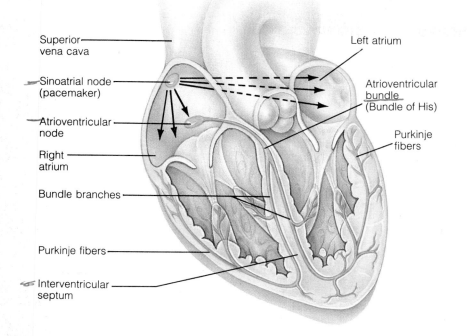

Superior vena cava

Sinoatrial node (pacemaker)

Atrioventricular node

Right atrium

Bundle branches

Purkinje fibers

Interventricular septum

Left atrium

Atrioventricular bundle (Bundle of His)

Purkinje fibers

ating 70 to 80 impulses per minute. Therefore, this node is the heart's pacemaker.

From the SA node, an impulse spreads quickly via gap junctions through the cardiac muscle fibers of both atria, signaling the atria to contract. The nonconducting tissue of the fibrous skeleton prevents this impulse from spreading directly to the walls of the ventricles. Instead, it reaches the ventricles by an indirect route: When the impulse reaches the **atrioventricular (AV) node** in the inferior part of the interatrial septum, it is delayed until the atria complete their contraction. After the delay, the impulse races through the **atrioventricular bundle,** which enters the interventricular septum and divides into right and left **bundle branches,** or *crura* ("legs"). About halfway down the septum, the crura become bundles of **Purkinje** (Pur-kin'je) **fibers,** also called *conduction myofibers.* These impulse-conducting fibers approach the apex of the heart, then turn superiorly into the ventricular walls. This arrangement ensures that the contraction of the ventricles begins at the apex of the heart and travels superiorly. This sequence ejects the contained blood superiorly into the great arteries that leave the ventricles.

The cells of the nodes and the AV bundle are small, but rather typical, cardiac muscle cells. The Purkinje fibers, by contrast, have a specialized morphology. They are large, barrel-shaped muscle cells that line up in long rows. They contain relatively few myofilaments, as these cells are specialized for conduction, not contraction. The large diameter of Purkinje fibers maximizes the speed of impulse conduction. Purkinje fibers occur in the deepest part of the endocardium of the ventricles, between the endocardium and myocardium layers.

The AV node is damaged by various forms of heart disease. Can you deduce the effect of damaging this node?

Because the atria and ventricles are separated by the electrically inert fibrous skeleton, the only route for impulse transmission from the atria to the ventricles is through the AV node. Therefore, damage to this node, called a **heart block,** interferes with the ability of the ventricles to receive the pacing impulses. Without these signals, the ventricles beat at an intrinsic rate that is slower than that of the atria—and too slow to maintain adequate circulation. In such cases, an artificial pacemaker set to discharge at the appropriate rate is usually implanted. ■

Innervation

Although the heart's inherent rate of contraction is set by the SA node, this rate can be altered by extrinsic neural controls. The nerves to the heart consist of *visceral sensory* fibers, heart-slowing *parasympathetic* fibers, and *sympathetic* fibers that increase the rate and force of heart contractions. The parasympathetic nerves arise as branches of the vagus nerve in the neck and thorax, whereas the sympathetic nerves travel to the heart from the cervical and upper thoracic chain ganglia (Figure 14.5, p. 399). All cardiac nerves pass through the cardiac plexus on the trachea before entering the heart. Although these autonomic nerve fi-

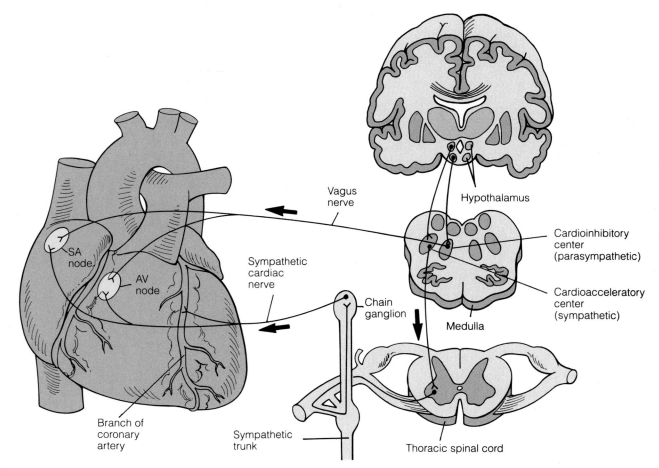

Figure 17.13   Autonomic innervation of the heart.

bers project to cardiac musculature throughout the heart, they project most heavily to the SA and AV nodes and the coronary arteries.

As we mentioned on page 337, the autonomic input to the heart is controlled by *cardiac centers* in the reticular formation of the medulla of the brain. In the medulla, the **cardioinhibitory center** influences parasympathetic neurons, whereas the **cardioacceleratory center** influences the sympathetic neurons.

The autonomic innervation of the heart is summarized in Figure 17.13.

Blood Supply to the Heart

Although the heart is filled with blood, this contained blood provides very little nourishment to the heart walls, which are too thick to make diffusion a practical means of providing nutrition. Instead, the functional blood supply of the heart comes from the right and left *coronary arteries* (Figure 17.14a). These arteries arise from the base of the aorta and encircle the heart in the coronary sulcus.

The **left coronary artery** arises from the left side of the aorta, passes posterior to the pulmonary trunk, then divides into two branches: the anterior interventricular and the circumflex arteries. The **anterior interventricular artery** descends in the anterior interventricular sulcus toward the apex of the heart. It sends branches deep into the interventricular septum and onto the anterior left walls of both ventricles. The **circumflex artery** follows the coronary sulcus posteriorly and supplies the left atrium and the posterior part of the left ventricle.

The **right coronary artery** emerges from the right side of the aorta and descends in the coronary sulcus on the anterior surface of the heart, between the right atrium and right ventricle. At the inferior border of the heart, the right coronary artery branches to form the **marginal artery.** Continuing into the posterior part of the coronary sulcus, the right coronary artery branches to form the **posterior interventricular artery** in the posterior interventricular sulcus. Overall, the branches of the right coronary artery supply the right atrium and almost all of the right ventricle.

The arrangement of the coronary arteries varies considerably. For example, in about 15% of people, the left coronary artery gives rise to *both* interventricular arteries. In other people (4%), a single coronary artery emerges from the aorta and supplies the entire heart.

Cardiac veins, which carry deoxygenated blood from the heart wall into the right atrium, also occupy the sulci on the heart surface (Figure 17.14b). The largest of these veins, the **coronary sinus,** occupies the posterior part of the coronary sulcus and returns almost all the venous blood from the heart to the right atrium. Draining into the coronary sinus are three large tributaries: the **great cardiac vein** in the anterior interventricular sulcus, the **middle cardiac vein** in the posterior interventricular sulcus, and the **small cardiac vein** running along the heart's inferior right margin. The anterior surface of the right ventricle contains several horizontal **anterior cardiac veins** that empty directly into the right atrium.

✚ Any blockage of the coronary arterial circulation can be serious—often it is fatal. **Angina pectoris** (an-ji'nah pek'tor-us), literally "choked chest," is thoracic pain caused by a fleeting deficiency in the delivery of blood to the heart musculature. Angina pectoris may result from stress-induced spasms of the coronary arteries or from a narrowing of these arteries due to atherosclerosis (fat deposited on the inner walls of arteries; see p. 484).

▦✚ Angina attacks occur most often during exercise. Can you guess why?

During exercise, the vigorously contracting heart may demand more oxygen than the narrowed coronary arteries can deliver.

In angina attacks, the myocardial cells weaken from a temporary lack of oxygen, but they do not die. Far more serious is prolonged blockage of a coronary artery resulting from an occluding blood clot, severe atherosclerosis, or severe vascular spasms. Under such conditions, the oxygen-starved cardiac muscle cells die. This is a **myocardial infarction,** or heart attack. A heart attack can kill the victim directly by weakening the heart, or it can kill indirectly by disrupting the conducting system and the heart rhythms. More information on heart attacks is provided in the box on page 474. ■

The Heart Throughout Life

Development of the Heart

An understanding of heart development is clinically important because congenital abnormalities of the heart account for nearly half of all deaths from birth defects. Congenital heart defects involve 1 in every 150 births.

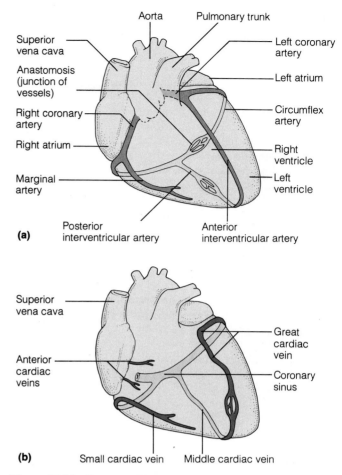

(a)

(b)

Figure 17.14 Coronary blood vessels. (a) The main coronary arteries. (b) The major cardiac veins.

As we explained in Chapter 16 (p. 456), all blood vessels begin as condensations of mesodermal mesenchyme called blood islands. The blood islands destined to become the heart form in the splanchnic mesoderm around the future head and neck of the embryonic disc (Figure 17.15). The heart folds neatly into the thorax region when the flat embryonic disc lifts up off the yolk sac to assume its three-dimensional body shape around day 20 to 21 (see pp. 56–57).

When the embryonic heart first reaches the thorax, it is a pair of tubes in the body midline (Figure 17.16a). These tubes fuse into a single tube about day 21. The heart starts pumping about day 22, by which time four bulges have developed along the heart tube (Figure 17.16b). These bulges are the earliest heart chambers and are unpaired. From tail to head, following the direction of blood flow, the four chambers are the *sinus venosus, atrium, ventricle,* and *bulbus cordis.*

1. Sinus venosus (ven-o'sus; "of the vein"). This chamber initially receives all blood from the veins of the embryo. It will become the smooth-walled part of

Heart Boosters, Retreads, and Replacements

The sharp, squeezing pain of angina pectoris typically lasts for no more than a few minutes and may not recur for months. However, its onset should be taken as a warning of a life-threatening condition that may eventually proceed to heart attack and cardiac arrest. Few experiences in life are more frightening than a heart attack: A sharp pain strikes with lightning speed through the chest (and perhaps the left arm and neck) and doesn't subside. Death occurs almost immediately in about one-third of all cases.

Between 1 and 2 million Americans will suffer a heart attack this year. Strange as it may seem, those who suffer angina or survive a heart attack should consider themselves lucky: They have been given a painful warning that, if heeded, allows time to make dietary changes, take heart-saving drugs, or have coronary bypass surgery (p. 485) to increase the blood supply to their oxygen-deprived heart. Not all people are so fortunate. For as many as 250,000 Americans yearly, a painless but fatal heart attack is the first and last symptom of heart disease. These people are victims of silent ischemia. They are at much greater risk of suffering a severe heart attack than are people who show obvious warning symptoms. The cause of silent heart disease is still a mystery. Some authorities suggest that these individuals have abnormal pain mechanisms. Another theory is that people with obvious atherosclerosis and a history of angina and ischemia attacks have gradually developed many interconnections between the vessels in the heart wall, which protect them from a fatal heart attack. People without symptoms, by contrast, have not built up these connections, so the first obstruction of a heart vessel is fatal.

The key to preventing heart attacks and cardiac death is to identify those at risk as early as possible by using sophisticated medical imaging techniques, such as CT scans, DSA, and MRI techniques (see pp. 17–20). Drug therapy can often help: Some

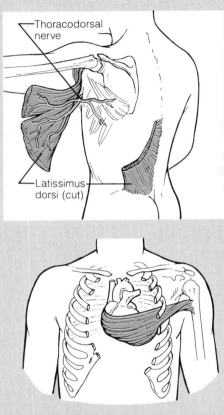

Schematic representation of the autograft heart wrap procedure.

drugs (beta blockers) lower the rate and force of heart contractions by blocking the sympathetic response of the heart to stress, thereby reducing the strain on the heart. Other drugs (calcium channel blockers) prevent the spasms in the coronary arteries that often trigger heart attacks.

But what are the options for someone whose heart is so devastated by disease that the drugs are too little, too late? Until recently, heart transplant surgery has been the only hope for a dying heart. But this procedure is riddled with problems. First, an acceptable tissue match must be found. Then, after complicated and traumatic surgery, the heart recipient must be dosed with drugs that suppress the T lymphocytes of the immune system—enough to thwart rejection of the donor heart, but not enough to be toxic to the recipient's kidneys. Although heart transplants have been

done for the past two decades, the survival rate still leaves much to be desired: about 80% the first year, minus 5% for each successive year. With these odds, not many heart recipients live to see their tenth post-surgical year.

The newest and most exciting technique is the autotransplant, in which the patient's own skeletal muscle is used to form a living patch in the heart wall or to augment the heart's pumping ability. The procedure, which was first done in February of 1985 on a woman with a large tumor in her heart, involves surgically removing the damaged region of the heart wall, suturing the severed edges together, and reinforcing the wall with borrowed latissimus dorsi muscle from the back (see illustration). The latissimus dorsi is freed from one of its attachments, wrapped around the heart area to be reinforced, and stitched into place. The muscle, however, remains attached to its insertion on the humerus and to its blood and nerve supply. A pacemaker is then attached to the patch of skeletal muscle to stimulate it intermittently, gradually working the rate of the pacemaker up to that of the heartbeat. Because the muscle tissue is "home grown," there is no problem of rejection. The major hurdle in autotransplants has been coaxing the skeletal muscle, which is built to work in short spurts, to perform continuously like cardiac muscle. But researchers are working on conditioning skeletal muscle with electrical shocks, thereby increasing its percentage of slow-twitch, fatigue-resistant red fibers. Also under investigation in animal studies are autotransplant "pouches," which are *complete* substitute hearts made of skeletal muscle bundles.

Although medical science is trying mightily to reproduce the human heart, our best course is still to practice healthy eating habits and have routine medical checkups to keep the heart from reaching the point of no return.

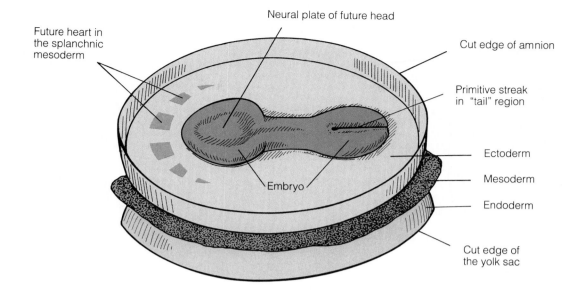

Future heart in
the splanchnic
mesoderm

Neural plate of future head

Cut edge of amnion

Primitive streak
in "tail" region

Embryo

Ectoderm

Mesoderm

Endoderm

Cut edge of
the yolk sac

Figure 17.15 Appearance of the heart in the embryonic disc (days 18–20). The amnion and the yolk sac have been removed to allow a clear view of the embryo. The heart originates around the future head and neck (left side of the figure).

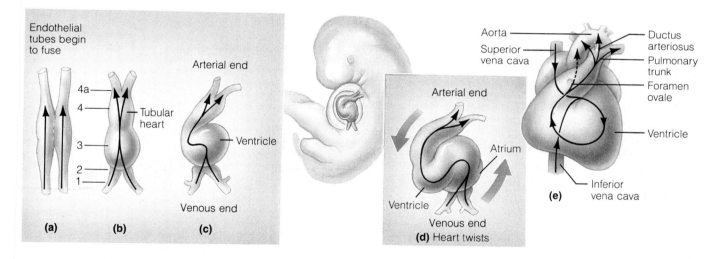

Endothelial
tubes begin
to fuse

4a
4 — Tubular
heart
3
2
1

(a) **(b)**

Arterial end

Ventricle

Venous end

(c)

Arterial end

Atrium

Ventricle

Venous end

(d) Heart twists

Aorta
Superior
vena cava

Ductus
arteriosus
Pulmonary
trunk
Foramen
ovale

Ventricle

Inferior
vena cava

(e)

Figure 17.16 Development of the human heart during week 4. Ventral view, with the cranial direction toward the top of the figures. **(a)** Around day 21. **(b)** Day 22. **(c)** Day 23. **(d)** Day 24. **(e)** Day 28. In part **(b)**, number 1 is the sinus venosus, 2 is the atrium, 3 is the ventricle, 4 is the bulbus cordis, and 4a is the truncus arteriosus.

the right atrium and the coronary sinus. The heartbeat originates in this embryonic chamber, so it should not surprise you that it gives rise to the sinoatrial node.

2. Atrium. This embryonic chamber eventually becomes the ridged parts of the right and left atria (the parts lined by pectinate muscles).

3. Ventricle. The strongest pumping chamber of the early heart, the ventricle gives rise to the *left* ventricle.

4. Bulbus cordis. This chamber, plus its most cranial extension, the *truncus arteriosus,* give rise to the pul-

monary trunk and first part of the aorta. It also gives rise to the *right* ventricle.

At the time these four chambers appear, the heart starts bending into an S shape (Figure 17.16c and d). The ventricle moves caudally and the atrium cranially, assuming their adult positions. This bending occurs because, whereas the ventricle and bulbus cordis grow quickly, the heart is unable to elongate within the confines of the pericardial sac.

During month 2 of development, the heart divides into its four definitive chambers. The division proceeds in a complex manner through a number of

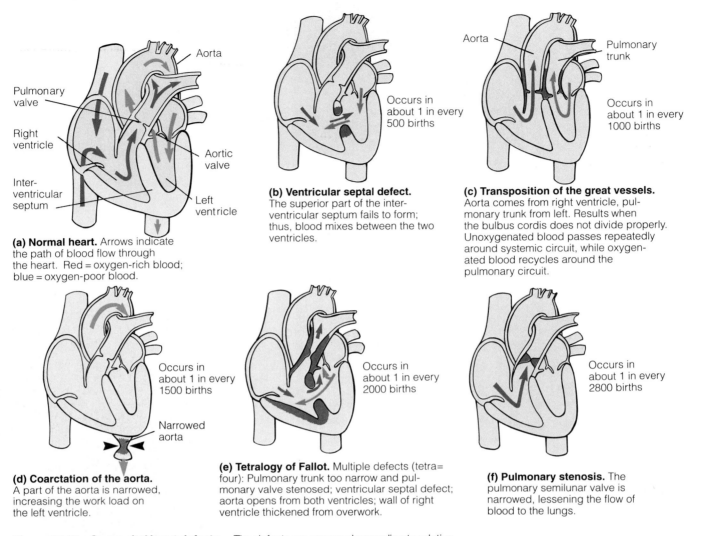

Aorta

Pulmonary
valve

Right
ventricle

Inter-
ventricular
septum

Aortic
valve

Left
ventricle

(a) Normal heart. Arrows indicate
the path of blood flow through
the heart. Red = oxygen-rich blood;
blue = oxygen-poor blood.

Occurs in
about 1 in every
500 births

(b) Ventricular septal defect.
The superior part of the interventricular septum fails to form;
thus, blood mixes between the two
ventricles.

Aorta

Pulmonary
trunk

Occurs in
about 1 in every
1000 births

(c) Transposition of the great vessels.
Aorta comes from right ventricle, pulmonary trunk from left. Results when
the bulbus cordis does not divide properly.
Unoxygenated blood passes repeatedly
around systemic circuit, while oxygenated blood recycles around the
pulmonary circuit.

Occurs in
about 1 in every
1500 births

Narrowed
aorta

(d) Coarctation of the aorta.
A part of the aorta is narrowed,
increasing the work load on
the left ventricle.

Occurs in
about 1 in every
2000 births

(e) Tetralogy of Fallot. Multiple defects (tetra=
four): Pulmonary trunk too narrow and pulmonary valve stenosed; ventricular septal defect;
aorta opens from both ventricles; wall of right
ventricle thickened from overwork.

Occurs in
about 1 in every
2800 births

(f) Pulmonary stenosis. The
pulmonary semilunar valve is
narrowed, lessening the flow of
blood to the lungs.

Figure 17.17   Congenital heart defects.   The defects are arranged according to relative
frequency of occurrence. Purple color is used to show the heart areas where the defects are
present.

stages, and congenital heart defects can result from
problems at any stage. Most of the interatrial septum
forms by growing caudally from the heart's roof. Most
of the interventricular septum forms by growing cranially from the heart's apex. Month 2 is also the time
when the bulbus cordis splits into the pulmonary
trunk and ascending aorta. The details of these events
are beyond the scope of this book, but two things must
be mentioned. First, the interatrial septum retains a
hole, the *foramen ovale,* which interconnects the two
atria until the time of birth, when it closes to become
the fossa ovalis (recall Figure 17.5c). Second, the superior (cranial) region of the interventricular septum
often fails to form. Its absence leaves a hole between
the two ventricles, which accounts for fully 30% of all
congenital heart defects.

Figure 17.17 illustrates and explains the common congenital heart defects, almost all of
which can be traced back to month 2 of development.

As you study the figure, note that the abnormalities
produce two basic kinds of disorders in the newborn:
They either (1) lead to a mixing of oxygen-poor systemic blood with the oxygenated pulmonary blood (so
that inadequately oxygenated blood reaches the body
tissues) or (2) involve narrowed valves and narrowed
vessels that greatly increase the work load on the
pumping ventricles. Modern surgical techniques can
usually correct these congenital defects. ■

The Heart in Adulthood and Old Age

In the absence of congenital heart problems, the resilient heart usually functions well throughout life. In a
person who exercises regularly and vigorously, the
heart gradually adapts to the increased demand by increasing in size. Researchers also have found that aerobic exercise helps clear fatty deposits from the walls
of the coronary vessels, thus retarding the process of

atherosclerosis. Barring some chronic illness, this beneficial response to exercise persists into old age.

Age-related changes that affect the heart include the following:

1. Hardening and thickening of the cusps of the heart valves. This occurs particularly where the stress of blood flow is greatest (mitral valve). Thus, heart murmurs are more common in elderly people.

2. Decline in cardiac reserve. Although the passing years seem to cause few changes in the resting heart rate, the aged heart is less able to increase its output in response to stressors that demand it pump more blood. Sympathetic control of the heart becomes less efficient, and heart rate gradually becomes more variable. There is a decline in the maximum heart rate, although this problem is much less severe in seniors who are physically active.

3. Fibrosis of cardiac muscle. Sometimes the nodes of the heart's conducting system become fibrosed (scarred) with age. Fibrosis hinders the initiation and transmission of contraction-signaling impulses, leading to abnormal heart rhythms and other conduction problems.

✚ Related Clinical Terms

Asystole (a-sis′to-le) (*a* = without) Failure of the heart to contract.

Cardiac catheterization Diagnostic procedure involving passage of a fine catheter (tubing) from a blood vessel at the body surface to the heart. Blood samples are withdrawn, and pressures within the heart can be measured. Findings help detect problems with the heart valves, heart deformities, and other heart malfunctions.

Cor pulmonale (*cor* = heart; *pulmonale* = lung) Enlargement or failure of the right ventricle as a result of elevated blood pressure in the pulmonary circuit. Cor pulmonale results from a blockage or constriction of the vessels in the lungs, which increases the resistance to blood flow and forces the right ventricle to pump harder. Acute cases may develop suddenly from an embolism that lodges in the pulmonary vessels. Chronic cases are usually associated with chronic lung diseases such as emphysema.

Endocarditis (en″do-kar-di′tis) Inflammation of the endocardium, usually confined to the endocardium of the heart valves. Endocarditis often results from infection by bacteria that have entered the bloodstream but may result from fungal infection or an autoimmune response. Drug addicts may develop endocarditis by injecting themselves with contaminated needles.

Mitral valve prolapse The most common disorder of heart valves, affecting up to 5% of the population. One or more cusps of the mitral valve "flop" up into the left atrium during ventricular systole, allowing a regurgitation of blood. A distinctive heart sound results (a click followed by a swish). Most cases are mild and harmless, but severe cases may lead to heart failure or a disruption of heart rhythm. Such cases are a common reason for valve replacement surgery.

Myocarditis (mi″o-kar-di′tis) Inflammation of the myocardium. It sometimes follows an untreated streptococcus infection in children. Myocarditis may be extremely serious because it can weaken the heart and impair its ability to pump.

Percussion Tapping the thorax or abdomen wall with the fingertips and using the nature of the resulting sounds to estimate location, density, and size of the underlying organs. Percussion of the thoracic wall can be used to estimate the size of a patient's heart.

Chapter Summary

1. The heart is a double pump whose right side pumps blood to and from the lungs for oxygenation and whose left side pumps blood throughout the body to nourish body tissues; that is, the right side is the pump of the pulmonary circuit, and the left side is the pump of the systemic circuit.

ORIENTATION AND LOCATION (pp. 460–462)

1. The cone-shaped human heart, about the size of a fist, lies obliquely within the mediastinum. Its apex points anteriorly and to the left. Its base is its posterior surface.

2. From an anterior view, the heart has a trapezoidal shape. The locations of its four corner points are shown in Figure 17.2a.

STRUCTURE OF THE HEART (pp. 462–466)

Coverings (p. 462)

1. The pericardium encloses the heart. It consists of a superficial layer (fibrous pericardium and parietal part of serous pericardium) and a deeper layer that covers the heart surface (visceral layer of serous pericardium, or epicardium). The pericardial cavity, between the two layers, contains lubricating serous fluid.

Layers of the Heart Wall (p. 463)

2. The layers of the heart wall, from external to internal, are epicardium, myocardium (which contains the cardiac muscle), and endocardium (endothelium and connective tissue).

Heart Chambers (pp. 463–466)

3. On the external surface of the heart, several sulci separate the four heart chambers (coronary sulcus, anterior and posterior interventricular sulci). Internally, the right and left sides of the heart are separated by the interatrial and interventricular septum.

4. The right atrium exhibits the following features: right auricle and pectinate muscles anteriorly; a smooth-walled posterior part; crista terminalis; openings of the coronary sinus and the superior and inferior venae cavae; SA and AV nodes; and fossa ovalis.

5. The right ventricle contains the following features: trabeculae carneae, papillary muscles and chordae tendineae, right atrioventricular (tricuspid) valve. The pulmonary semilunar valve lies at the base of the pulmonary trunk.

6. The left atrium has a large, smooth-walled, posterior region into which open the four pulmonary veins. Anteriorly, its auricle is lined by pectinate muscles.

7. The left ventricle, like the right ventricle, contains papillary muscles, chordae tendineae, trabeculae carneae, and an atrioventricular valve (bicuspid). The aortic semilunar valve lies at the base of the aorta.

PATHWAY OF BLOOD THROUGH THE HEART (pp. 466–467)

1. A drop of blood circulates in this order: right atrium to right ventricle to pulmonary circuit to left atrium to left ventricle to systemic circuit to right atrium.

2. In each heartbeat, both atria contract, followed by both ventricles.

HEART VALVES (pp. 467–470)

Structure (p. 467)

1. The four heart valves are the atrioventricular valves (tricuspid and bicuspid [mitral]) and the semilunar valves (aortic and pulmonary). All except the bicuspid have three cusps.

Functions (pp. 467–468)

2. The atrioventricular valves prevent backflow of blood into the atria during contraction of the ventricles. The pulmonary and aortic semilunar valves prevent backflow into the ventricles during relaxation of the ventricles.

Heart Sounds (pp. 468–470)

3. Using a stethoscope, physicians listen to the sounds produced by closing atrioventricular and semilunar valves. Each valve is best heard near a different heart corner on the anterior chest wall.

FIBROUS SKELETON (p. 470)

1. The fibrous skeleton of the heart surrounds the valves between the atria and ventricles. It anchors the valve cusps, is the point of insertion of the heart musculature, and blocks the direct spread of electrical impulses from atria to ventricles.

CONDUCTING SYSTEM AND INNERVATION (pp. 470–472)

Conducting System (pp. 470–471)

1. The conducting system of the heart is an interconnected series of cardiac muscle cells that initiates each heartbeat, sets the basic rate of the heartbeat, and coordinates the contraction of the heart chambers. The impulse that signals heart contraction travels from the SA node (the pacemaker) through the atrial myocardium to the AV node, to the atrioventricular bundle, bundle branches, Purkinje fibers, and the ventricular musculature.

Innervation (pp. 471–472)

2. Heart innervation consists of visceral sensory fibers, heart-slowing vagal parasympathetics, and sympathetics that accelerate heart rate.

BLOOD SUPPLY TO THE HEART (pp. 472–473)

1. The main vessels supplying the heart wall occupy the coronary sulcus and interventricular sulci. The right and left coronary arteries branch from the aorta to supply the heart wall. Venous blood, collected by the cardiac veins, empties into the coronary sinus and right atrium.

2. Constriction or blockage of the coronary arteries causes myocardial infarction (heart attack).

THE HEART THROUGHOUT LIFE (pp. 473–477)

Development of the Heart (pp. 473–476)

1. The heart develops from splanchnic mesoderm around the head and neck of the embryonic disc. When the heart folds into the thorax, it is a double tube that soon fuses into one and starts pumping blood (day 22). It soon bends into an S shape. The four earliest heart chambers are the sinus venosus, atrium, ventricle, and bulbus cordis.

2. The four final heart chambers are defined during month 2 through the formation of valves and dividing walls. Failures in this complex process account for most congenital defects of the heart (Figure 17.17).

The Heart in Adulthood and Old Age (pp. 476–477)

3. Age-related changes in the heart include hardening and thickening of the valve cusps, decline in reserve pumping capacity, fibrosis of cardiac muscle, and atherosclerosis.

Review Questions

Multiple Choice and Matching Questions

1. The most external part of the pericardium is (a) parietal serous, (b) fibrous, (c) visceral serous, (d) the pericardial cavity.

2. Which heart chamber forms most of the heart's inferior surface? (a) right atrium, (b) right ventricle, (c) left atrium, (d) left ventricle.

3. A part of the heart's conducting system that is located in the interventricular septum is the (a) atrioventricular node, (b) sinoatrial node, (c) atrioventricular bundle, (d) papillary muscles.

4. The chordae tendineae (a) open the atrioventricular valves, (b) prevent the atrioventricular valves from everting, (c) contract the papillary muscles, (d) open the semilunar valves.

5. How many cusps does the right atrioventricular valve have? (a) two, (b) three, (c) four.

6. Which artery lies in the posterior left part of the coronary sulcus of the heart? (a) posterior interventricular, (b) the cardiac artery, (c) circumflex artery, (d) marginal artery.

7. Freshly oxygenated blood is first received by the (a) right atrium, (b) left atrium, (c) right ventricle, (d) left ventricle.

8. The sequence of contraction of the heart chambers is (a) random, (b) left chambers followed by right chambers, (c) both atria followed by both ventricles, (d) right atrium, right ventricle, left atrium, left ventricle.

9. The middle cardiac vein runs with which artery? (a) marginal, (b) aorta, (c) coronary sinus, (d) anterior interventricular, (e) posterior interventricular.

10. The base of the heart (a) is its posterior surface, (b) lies on the diaphragm, (c) is the same as its apex, (d) is its superior border.

11. Circle the incorrect statement about the crista terminalis of the right atrium: (a) It separates the smooth-walled part from the part with pectinate muscles. (b) It is shaped like the letter C. (c) The coronary sinus and inferior vena cava open near its inferior part. (d) It lies mostly in the interatrial septum.

12. The aortic semilunar valve closes (a) at the same time the bicuspid valve closes, (b) just after the atria contract, (c) just before the ventricles contract, (d) just after the ventricles contract.

13. The ventricle of the embryonic heart gives rise to what adult structure(s)? (a) bulbus cordis, (b) both ventricles, (c) left ventricle, (d) the aorta, (e) none of these.

14. Which layer of the heart wall is the thickest? (a) endocardium, (b) myocardium, (c) epicardium, (d) endothelium.

15. Purkinje fibers in the heart are (a) neurons, (b) axons, (c) in the atria only, (d) sympathetic, (e) muscle.

16. The inferior left corner of the heart is located at the (a) second rib slightly lateral to sternum, (b) third rib at sternum, (c) sixth rib slightly lateral to sternum, (d) fifth intercostal space in midclavicular line.

17. The pump of the systemic circuit is the (a) left side of heart (left ventricle), (b) right side of heart (right atrium), (c) both sides of heart, (d) the respiratory pump.

18. The cardiac muscle of the heart wall receives its blood supply directly from (a) the coronary arteries, (b) the pulmonary arteries, (c) the blood inside the heart, (d) the pulmonary veins.

19. Use logic to deduce which of the congenital heart defects in Figure 17.17 is the most difficult to repair surgically: (a) tetralogy of Fallot, because it involves the most defects; (b) pulmonary stenosis, because the hard valves will break; (c) ventricular septal defect, because the hole is always so large; (d) transposition of the great vessels, because the aorta is so difficult to move.

Short Answer and Essay Questions

20. Ben Garber was annoyed when the teaching assistant who ran the discussion section of his anatomy class said that the atria pump blood to lungs, and the ventricles pump blood throughout the body. Can you correct this error?

21. Describe the location of the heart within the thorax.

22. Trace one drop of blood through all heart chambers and heart valves, and through the basic vascular circuits from the time it enters the left atrium until it enters the left atrium again.

23. (a) Name the elements of the heart's conducting system in order, beginning with the pacemaker. (b) What are the functions of this conducting system?

24. What is the difference, if any, between an auricle and an atrium of the heart?

Critical Thinking and Clinical Application Questions

25. After studying Figure 17.17 on congenital heart defects, classify the five defects presented according to whether they produce (1) mixing of oxygenated and unoxygenated blood, (2) increased work load for the ventricles, or (3) both of these problems.

26. Heather, a newborn baby, needed surgery because she was born with an aorta that arises from the right ventricle and a pulmonary trunk that arises from the left ventricle. What is this birth defect called, and what are its physiological effects?

27. Ms. Hamad, who is 73 years old, is admitted to the coronary care unit of a hospital with a diagnosis of left ventricular failure resulting from a myocardial infarction. Her heart rhythm is abnormal. Explain what a myocardial infarction is, how it is likely to have been caused, and why the heart rhythm is affected.

28. You have been called on to demonstrate where to listen for heart sounds. Explain where on the chest wall you would place a stethoscope to listen for (a) incompetence of the aortic semilunar valve and (b) stenosis of the mitral valve.

29. Mark was stabbed in the chest by his girlfriend's ex-lover. His face became blue, and he lost consciousness from lack of oxygenated blood to the brain. The diagnosis was cardiac tamponade rather than severe blood loss through internal bleeding. What is cardiac tamponade, how did it cause the observed symptoms, and how is it treated?

18

Blood Vessels

The blood vessels of the body form a closed delivery system that begins and ends at the heart. The idea that blood circulates in this way dates back to the 1620s and is based on the careful experiments of William Harvey, an English physician. Prior to that time, it was taught—as proposed by the ancient Greek physician Galen—that blood moved through the body like an ocean tide, first moving out from the heart, then ebbing back into the heart via the same vessels.

Blood vessels are not rigid tubes but dynamic structures that pulsate, constrict and relax, and even proliferate, as demanded by the changing needs of the body. In this chapter, we examine the structure and function of these important circulatory pathways through the body.

PART 1: GENERAL STRUCTURE OF BLOOD VESSELS

The three major types of blood vessels are *arteries, capillaries,* and *veins* (Figures 18.1 and 18.2). As the heart contracts, blood is forced into the large arteries that leave the ventricles. The blood then moves into successively smaller arteries, finally reaching their smallest branches, the *arterioles* (ar-te're-ōlz; "little arteries"), which feed into the capillaries of the body organs. Blood draining from the capillaries is collected by *venules,* small veins that merge to form larger veins that ultimately empty into the heart. This pattern of vessels applies to both the pulmonary and systemic circuits. Altogether, the blood vessels in an adult human body stretch for 100,000 km (60,000 miles), a distance of 2½ times around the world!

Notice that arteries are said to "branch," "diverge," or "fork" as they carry blood *away* from the heart. Veins, by contrast, are said to "join," "merge," "converge," or "serve as tributaries" as they carry blood *toward* the heart.

Structure of Blood Vessel Walls

The walls of all blood vessels, except the very smallest, are composed of three distinct layers, or *tunics* ("cloaks"). These tunics surround the central blood-filled space, the vessel **lumen** (Figure 18.1).

The innermost tunic of a vessel wall is the **tunica intima** (too'nĭ-kah in'tĭ-mah). You can think of this intima layer as being in *intimate* contact with the blood in the lumen. This tunic contains the *endothelium,* the simple squamous epithelium that lines the lumen of all vessels (Chapter 4, p. 68). The flat endothelial cells form a slick surface that minimizes the friction of blood moving across them. In vessels larger than about 1 mm in diameter, a thin layer of loose connective tissue, the **subendothelial layer,** lies just external to the endothelium.

The middle tunic, or **tunica media,** consists of circularly arranged smooth muscle cells and sheets of elastin. The activity of the smooth muscle is regulated by *vasomotor nerve fibers* of the sympathetic division of the autonomic nervous system. Contraction of these muscle cells decreases the size of the lumen, a process called *vasoconstriction.* Conversely, relaxation increases the diameter of the lumen, a process called *vasodilation.* Since small changes in vessel diameter greatly influence blood flow and blood pressure, the activities of the tunica media are critical in regulating circulatory dynamics. In arteries, which bear the greatest responsibility for maintaining blood pressure and continuous blood circulation, the tunica media is the bulkiest layer. Arteries are also highly elastic. These functional considerations are described shortly.

The outermost layer of the vessel wall is the **tunica adventitia** (ad"ven-tish'e-ah; "coming from abroad or from outside"). This tunic is a layer of connective tissue that protects the vessel, strengthens its wall, and anchors it to surrounding structures. The cells and fibers in this connective tissue run longitudinally. In the larger arteries and veins, the tunica adventitia itself contains tiny blood vessels. These vessels, the *vasa vasorum* (va'sah va-sor'um; "vessels of the vessels"), nourish the outer half of the vessel wall. The deeper half, by contrast, receives its nutrients from the blood in the lumen. Small vessels need no vasa vasorum because their walls are entirely supplied by luminal blood. The vasa vasorum to a large blood vessel may arise as tiny branches from that same vessel or as small branches from other, nearby blood vessels.

Since the heart can be viewed as an enlarged blood vessel, you should be able to match each layer of the heart wall (epicardium, endocardium, and myocardium: p. 463) with a corresponding vascular tunic. Can you do this?

The endocardium corresponds to the tunica intima, the myocardium to the tunica media, and the epicardium to the tunica adventitia. Furthermore, the coronary arteries and cardiac veins are the vasa vasorum of the heart. ■

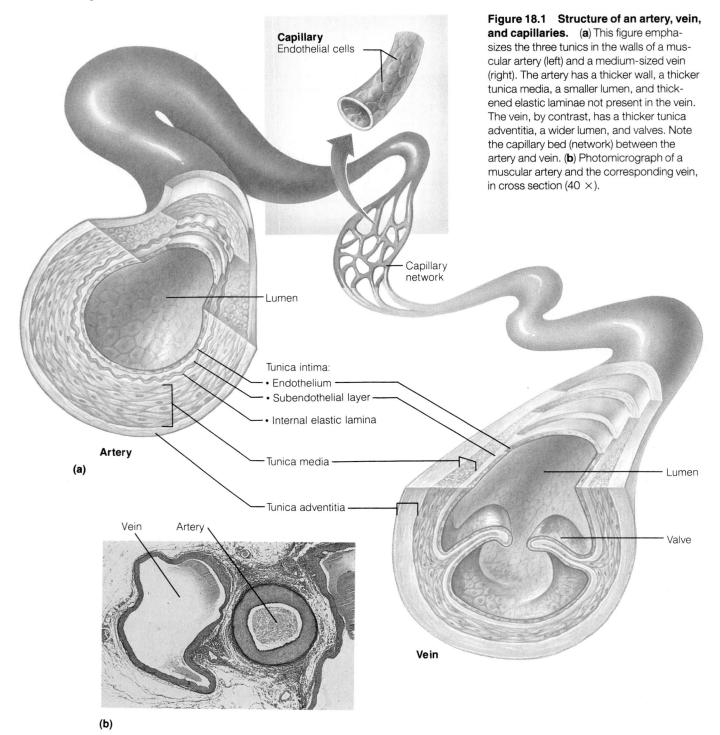

Capillary
Endothelial cells

Figure 18.1 Structure of an artery, vein, and capillaries. (**a**) This figure emphasizes the three tunics in the walls of a muscular artery (left) and a medium-sized vein (right). The artery has a thicker wall, a thicker tunica media, a smaller lumen, and thickened elastic laminae not present in the vein. The vein, by contrast, has a thicker tunica adventitia, a wider lumen, and valves. Note the capillary bed (network) between the artery and vein. (**b**) Photomicrograph of a muscular artery and the corresponding vein, in cross section (40 ×).

Capillary network

Lumen

Tunica intima:
• Endothelium
• Subendothelial layer
• Internal elastic lamina

Artery

(a)

Tunica media

Tunica adventitia

Lumen

Valve

Vein

Vein   Artery

(b)

Classes of Blood Vessels

Arteries

Arteries are vessels that carry blood away from the heart. Many people have the misconception that all arteries carry oxygen-rich blood, whereas all veins carry oxygen-poor blood. Although this fact is true of the systemic circuit, it is not true of the pulmonary circuit, whose arteries carry oxygen-poor blood to the lungs for oxygenation.

Our discussion of arteries will proceed from the largest to the smallest types.

Elastic Arteries

Elastic arteries are the large arteries near the heart—the aorta and its major branches. Their large lumen allows them to serve as low-resistance pathways for conducting blood between the heart and the medium-

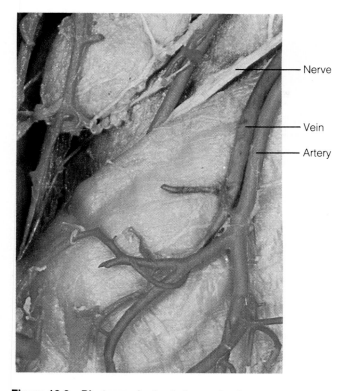

Nerve

Vein

Artery

Figure 18.2   Photograph of arteries and veins.   Vessels in lateral wall of the pelvis. Note that arteries and veins tend to run together, side by side.

sized arteries. For this reason, they are sometimes called *conducting arteries.* There is more elastin in the walls of elastic arteries than in any other type of vessel. This elastin is distributed throughout all three tunics, but it is most concentrated in the tunica media. There, it takes the form of concentric sheets interposed among layers of smooth muscle cells.

What is the function of the elastin in arteries? The heart pumps blood in a pulsatile manner (it contracts and relaxes) toward capillaries that are not strong enough to withstand pulsatile pressures. Elastic arteries smooth out these forceful pulsations by expanding as the heart forces blood into them and then recoiling to propel the blood onward when the heart relaxes. In this way, the arteries ensure that the pressure of arterial blood is steady by the time the blood reaches the capillaries.

Can you deduce some consequences of a loss of elasticity of the large arteries, as occurs in arteriosclerosis ("hardening of the arteries")?

Without the pressure-smoothing effect of the elastic arteries, the walls of arteries throughout the body experience higher maximum pressures. Battered by high pressures, the arteries may eventually rupture (causing strokes) or weaken and form balloon-like outpocketings (a condition called an *aneurysm*). Arteriosclerosis, a common condition, is described in detail in the box on pp. 484–485. ■

Muscular Arteries

Muscular (distributing) arteries lie distal to the elastic arteries. These "middle-sized" arteries range in diameter from about 1 cm to 0.3 mm. Constituting most of the named arteries seen in the anatomy laboratory, muscular arteries supply groups of organs, individual organs, and parts of organs. They are called muscular because their tunica media is thicker relative to the size of the lumen than that of any other vessel class. By actively altering the diameter of the artery, this muscular layer regulates the amount of blood flowing to the organ supplied according to the specific needs of that organ.

The tunica media of muscular arteries contains thin, scattered sheets of elastin. Additionally, a thickened sheet of elastin lies on each side of the tunica media: A wavy *internal elastic lamina* lies between the tunica media and intima, and an *external elastic lamina* lies between the tunica media and adventitia (only the internal elastic lamina is labeled in Figure 18.1a). The elastin in muscular arteries, like that in elastic arteries, helps smooth the pulsatile pressure produced by the heartbeat.

Arterioles

Arterioles are the smallest arteries. Their diameter ranges from about 0.3 mm to 10 μm. Their tunica media contains only one or two layers of smooth muscle cells. Larger arterioles exhibit all three tunics plus an internal elastic lamina. Smaller arterioles, which lead into the capillary beds, are little more than a single layer of smooth muscle cells spiraling around an underlying endothelium.

The diameter of each arteriole is regulated in two ways: (1) Local factors in the tissues signal the smooth musculature to contract or relax, thus regulating the amount of blood sent to each capillary bed, and (2) the sympathetic nervous system adjusts the diameter of arterioles throughout the body to regulate the systemic blood pressure. For example, a widespread sympathetic vasoconstriction raises blood pressure during fight-or-flight responses (Chapter 14, p. 396).

Capillaries

Capillaries are the smallest blood vessels. Their diameter is 8 to 10 μm, just large enough for erythrocytes to pass through in single file. Capillaries form networks called **capillary beds** that run throughout almost all body tissues (especially the loose connective tissues).

Capillary Function

Functionally, capillaries are the most important blood vessels. You will recall from Chapter 4 (p. 81) that all cells of the body are in contact with tissue fluid (inter-

A CLOSER LOOK   Atherosclerosis: Controlling the Silent Killer

Arterial disease is the primary cause of death in the Western world. **Arteriosclerosis** (ar-te″re-o-sklĕ-ro′sis), literally "hardening of the arteries," is a general term for the pathological thickening and loss of elasticity of arterial walls. It includes three distinct diseases: (1) the rare *Mönckeberg's arteriosclerosis,* characterized by scattered deposits of calcium in the tunica media of small arteries; (2) *medial arteriosclerosis,* in which the smooth muscle and elastin in the tunica media degenerate with age and are replaced by fibrous connective tissue; and (3) **atherosclerosis** (ath″er-o″sklĕ-ro′sis). Atherosclerosis is by far the most common and significant type of arterial disease.

In atherosclerosis, the arterial tunica intima thickens with deposits called **atheromas** ("growths that resemble oatmeal") or **atherosclerotic plaques.** These deposits narrow the arterial lumen (see the histological sections illustrated in this box). Once this happens, a roaming blood clot or an arterial spasm can block the vessel completely.

Atherosclerosis primarily affects the abdominal aorta, the coronary arteries to the heart wall, and the internal carotid arteries to the head. Narrowing and blockage of the cranial arteries lead to strokes, and blockage of atherosclerotic coronary arteries causes heart attacks (myocardial infarctions). Strokes and heart attacks account for almost half of all deaths in America—heart attacks alone account for one-third. More males than females are affected.

What triggers this scourge of the arteries? There are many theories.

According to the **response to injury** hypothesis, the initial event is injury to the tunica intima. This injury can be caused by the stress of persistently high blood pressure (hypertension), by a blow to the vessel, by viral infection, or by carbon monoxide in the blood (derived from cigarette smoke or car exhaust). Once a break occurs in the endothelium of the vessel, blood platelets cling to the injured site and initiate clotting to prevent blood loss. These platelets secrete platelet-derived growth factor, which signals smooth muscle cells in the tunica media to proliferate and invade the tunica intima. Many macrophages also invade this inner tunic. The smooth muscle cells, plus a connective-tissue matrix they secrete, form the core of the atherosclerotic plaque. Furthermore, the platelets secrete chemicals that

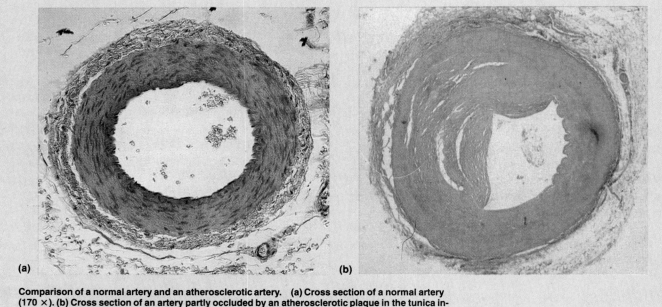

(a) **(b)**

Comparison of a normal artery and an atherosclerotic artery.   (a) Cross section of a normal artery (170 ×). (b) Cross section of an artery partly occluded by an atherosclerotic plaque in the tunica intima (30 ×).

stitial fluid) from which the cells receive nutrients and into which they deposit wastes. The function of the capillaries is to renew and refresh this tissue fluid. Because small molecules pass freely through the thin capillary walls, nutrients and oxygen from the blood

continuously enter the tissue fluid, and cellular wastes (carbon dioxide and nitrogenous wastes) are continuously removed.

Certain capillaries have additional functions. Oxygen enters the blood—and carbon dioxide leaves it—

increase the permeability of the endothelium to fats and cholesterol from the blood. These lipids then enter the smooth muscle cells, the macrophages, and the matrix, thereby thickening the plaque. Because the severity of plaque formation is increased by high levels of cholesterol in the blood, atherosclerosis is associated with fatty, high-cholesterol diets.

A more recent interpretation reverses the above hypothesis, proposing that blood-borne lipid invades the tunica intima very early, damaging that layer and leading to plaque formation.

Atherosclerosis grows more serious with age. As the plaques enlarge, they both weaken and harden the entire vessel wall. The muscle cells in the wall die, and the elastin sheets degenerate. These elements are slowly replaced by inelastic scar tissue and rigid calcium salts. As we discussed earlier (p. 483), inelastic arteries cannot expand to smooth out the pulsatile force of the heartbeat, so these events increase the risk of strokes, aneurysms, and other vascular problems. Advanced atherosclerotic plaques, called **complicated plaques,** have pits and ulcers on their surfaces. Large clots form on these rough surfaces and are increasingly likely to occlude the vessel or break off as emboli.

What can be done when the heart is at risk from atherosclerotic coronary arteries? Traditionally, the choice has been **coronary bypass surgery,** in which a vein or artery (saphenous vein of the leg, internal thoracic artery) is detached from another part of the body and implanted on the heart wall. This bypass vessel runs from the aorta to the distal, unclogged part of the damaged coronary artery. More recently, intravascular devices threaded into the obstructed arteries have become part of the ammunition of cardiovas-

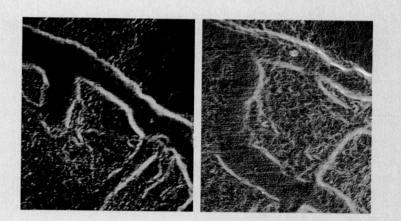

X-ray images of an occluded artery (left) and the same artery after being cleared by balloon angioplasty (right). These views were prepared by digital subtraction angiography, a computer imaging technique.

cular medicine. **Balloon angioplasty** uses a catheter with a balloon tightly packed into its tip (see photos). When the catheter reaches the obstruction, the balloon inflates, compressing the fatty mass against the vessel wall and widening the lumen. The *rotablator* uses a diamond-coated burr similar to a dental drill that literally "sands and polishes" the inside of a vessel. Another catheter device currently under development houses a fiberscope (containing hair-thin optical fibers that allow the surgeon to peer into an illuminated vessel), an inflatable cuff to stop the blood flow temporarily, and a power fiber that emits a laser beam to vaporize the arterial clogs. Although these intravascular devices are faster, cheaper, and much less risky than coronary bypass surgery, they share the same major shortcoming: They do nothing to stop the underlying disease, and new blockages often occur after a few months.

It was hoped that cholesterol-lowering drugs such as cholestyramine and lovastatin would act as a sort of cardiovascular Drāno and simply wash the fatty plaques off the arterial walls. These drugs, however,

are used only in patients whose low-cholesterol diets have failed to reduce blood cholesterol levels, and the side effects—nausea, bloating, and constipation—are so distressing that many people simply stop taking the drugs.

Other drugs are able to dissolve dangerous clots that block atherosclerotic arteries. These revolutionary drugs include **streptokinase,** a powerful enzyme extracted from bacteria, and **tissue plasminogen activator** (t-PA), a natural human product made by genetic engineering techniques. Injecting streptokinase or t-PA directly into the heart via a catheter restores blood flow quickly and puts an early end to many heart attacks in progress.

Factors contributing to atherosclerosis include a high-cholesterol diet, hypertension, emotional stress, smoking, obesity, and lack of exercise. Better management of these risk factors—especially of hypertension, smoking, and diet—has contributed to a 35% decrease in deaths from heart attacks in the United States since 1970. Much more must be done, however, before this "killer" is under control.

through the capillaries of the lungs. Capillaries of the small intestine receive digested nutrients. Capillaries in endocrine glands pick up hormones. Finally, capillaries in the kidneys are involved with the removal of nitrogenous wastes from the body.

Capillary Structure and Types

Structurally, a capillary is no more than a tube of endothelium surrounded by a basal lamina (Figure 18.3a). The endothelial cells are held together by tight

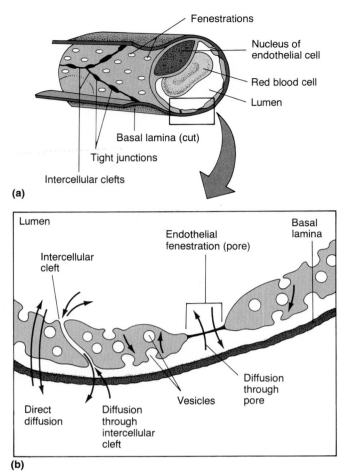

Figure 18.3   A capillary. (**a**) General structure of a fenestrated capillary. Continuous capillaries are similar, but they lack fenestrations (pores). The basal lamina fully surrounds the capillary. (**b**) Four pathways taken by molecules through the capillary wall. Most molecules apparently pass through the intercellular clefts.

junctions and occasional desmosomes. Tight junctions block the passage of small molecules (Chapter 4, p. 75), but such junctions do not surround the whole perimeter of the endothelial cells. Instead, they leave gaps of unjoined membrane, called **intercellular clefts,** through which molecules exit and enter the capillary.

Some capillaries are *fenestrated* (fen'is-tra-ted) and others are *continuous.* Fenestrated capillaries have pores, or **fenestrations** ("windows"), which pierce the endothelial cells (Figure 18.3a). These capillaries occur where there are exceptionally high rates of exchange of small molecules between the blood and the surrounding tissue fluid. For example, capillaries in the small intestine, which receive the digested nutrients from food, are fenestrated. So too are capillaries in the synovial membranes of joints, where large quantities of water molecules exit the blood to contribute to the synovial fluid. Unlike fenestrated capillaries, continuous capillaries have no endothelial pores. They occur in most organs of the body, such as skeletal muscles, skin, and the central nervous system.

Capillary Permeability

Basis of Capillary Permeability. There are four paths by which molecules pass into and out of capillaries (Figure 18.3b): (1) through the intercellular clefts, (2) through the pores of fenestrated capillaries, (3) by passing (diffusing) directly through the endothelial cell membranes, and (4) through cytoplasmic vesicles that invaginate from the plasma membrane and migrate across the cell. Most exchange of small molecules is thought to occur through the intercellular clefts. The pores of fenestrated capillaries may supplement these clefts, and the cytoplasmic vesicles apparently transmit a few larger molecules (such as small proteins). Carbon dioxide and oxygen seem to be the only important molecules that pass directly through the endothelial cells—these are uncharged molecules that easily diffuse through the lipid membranes of cells.

Low-Permeability Capillaries: The Blood Brain Barrier. The *blood-brain barrier* (introduced in Chapter 12) is a feature of brain capillaries that helps ensure that all but the most vital molecules remain in the blood and out of the brain tissue (see p. 345). We will now explain this barrier in more detail. Simply put, brain capillaries lack every structural feature that is associated with capillary permeability: Their tight junctions are complete, so intercellular clefts are absent. They are not fenestrated, and there are no endothelial vesicles. Instead, the vital molecules that must cross brain capillaries are "ushered through" by highly selective transport mechanisms in the plasma membranes of the endothelial cells.

Researchers are seeking ways to move drugs across the blood-brain barrier and into the brain. Chemotherapy drugs used to fight brain tumors and protein neurotransmitters used to treat certain mental disabilities are examples of beneficial drugs that are stopped by the barrier. Can you imagine at least one strategy for moving such drugs through the blood-brain barrier?

Researchers are taking three approaches to this problem: (1) injecting substances that temporarily open the tight junctions between endothelial cells so that the drugs can pass through, (2) linking the drugs to molecules that are normally transported across the endothelium by cellular transport mechanisms, and (3) surrounding the drug molecules with a coat of lipid (which will pass unhindered through the endothelium). ■

Capillary Beds

Figure 18.4 shows the structure of a typical capillary bed. A small arteriole leads to a **metarteriole** (a vessel structurally intermediate between an arteriole and a capillary), with true capillaries branching off along

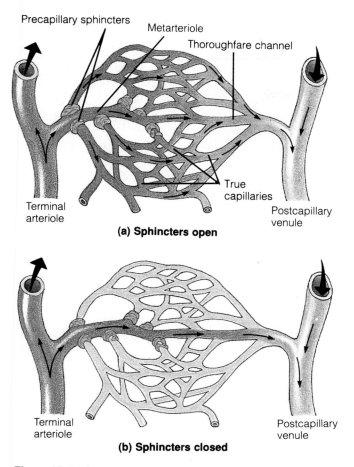

Figure 18.4 Anatomy of a capillary bed, as seen in a mesentery. (**a**) When precapillary sphincters relax, blood fills the true capillaries. (**b**) When these sphincters contract, they force most blood to flow straight from metarterioles to thoroughfare channels, bypassing the true capillaries.

the way. The metarteriole then continues into a **thoroughfare channel,** a vessel structurally intermediate between a capillary and a venule. The thoroughfare channel then joins a venule, receiving true capillaries along the way. A smooth muscle cell called a **precapillary sphincter** wraps around the root of each true capillary at the metarteriole. When these sphincters are relaxed, the true capillaries are wide open, and blood flows through them to supply the surrounding tissue. When the precapillary sphincters contract, by contrast, they close off the true capillaries and force blood to flow straight from the metarteriole into the thoroughfare channel and venule—thus bypassing the true capillaries. The sphincters open when the tissue needs blood and close when tissue needs are small (for example, when the nearby tissue cells already have adequate oxygen). In this way, the precapillary sphincters, metarterioles, and thoroughfare channels exercise precise control over the amount of blood supplying each tissue.

Although most tissues and organs have a rich capillary supply, there are exceptions. Tendons and ligaments are poorly vascularized. Epithelia and cartilage contain no capillaries but receive nutrients indirectly from vascularized connective tissues nearby. The lens and cornea of the eye have no capillary supply at all (they receive nutrients from the aqueous humor and other sources: See pp. 425 and 420).

Sinusoids

Some organs contain wide, leaky capillaries called **sinusoids** or **sinusoidal capillaries.** Each sinusoid follows a twisted path and exhibits both expanded and narrowed regions. Sinusoids are usually fenestrated, and their endothelial cells have fewer cell junctions than do ordinary capillaries. In some sinusoids, in fact, the intercellular clefts are wide open. Sinusoids occur wherever there is an extensive exchange of *large* materials, such as proteins or cells, between the blood and surrounding tissue. For example, sinusoids occur in the bone marrow and spleen, where many blood cells move through their walls. The large diameter and twisted course of sinusoids ensure that flowing blood slows when entering these vessels, allowing time for the many exchanges that occur across their walls.

Veins

Veins carry blood from the capillaries toward the heart. Blood loses most of its pressure while passing through the high-resistance arterioles and capillary beds, so the blood pressure is much lower in veins than in arteries. Veins in the systemic circuit carry blood that is relatively oxygen-poor, but the pulmonary veins carry oxygen-rich blood returning from the lungs.

Venules (8 to 100 μm in diameter) are the smallest veins, and **postcapillary venules** are the smallest venules. Postcapillary venules consist of an endothelium on which lie scattered, fibroblast-like cells, and they function like capillaries. In fact, more inflammatory fluid and leukocytes leave the circulation through postcapillary venules than through the capillaries themselves. The larger venules have one or two layers of smooth muscle cells (a tunica media) and a thin tunica adventitia as well.

Venules join to form **veins.** The walls of veins are thinner than those of arteries (Figure 18.1), but because the blood pressure within veins is low, there is no danger of their bursting. The lumens of veins are larger than those of arteries—in fact, the veins hold fully 65% of the body's blood at all times. In veins, the tunica adventitia is thicker than the tunica media (the opposite of the tunic makeup of arteries). In the body's largest veins—the venae cavae that return systemic blood to the heart—the adventitia layer is further thickened by longitudinal bands of smooth muscle. Veins have much less elastin in their walls than do arteries.

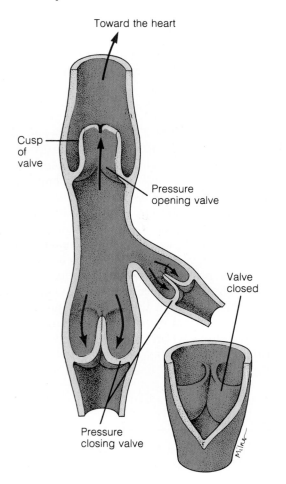

Toward the heart

Cusp
of
valve

Pressure
opening valve

Valve
closed

Pressure
closing valve

Figure 18.5 **The function of valves in a vein.** As the arrows indicate, the valves are opened by blood flowing toward the heart and are forced closed by backflow.

Because the pressure in veins is low, veins have valves to help direct blood toward the heart. Any backflow of blood forces these valves to close (Figure 18.5). **Venous valves** are infoldings of the tunica intima, each fold forming a cusp. Valves are most abundant in the veins of the limbs, where the superior flow of blood is opposed by gravity. A few valves are found in the veins of the head and neck, but there are none in the veins of the thoracic and abdominal cavities.

To demonstrate the effectiveness of your venous valves in preventing the backflow of blood, you may wish to try this simple experiment. Hang one hand by your side until the veins on its dorsal aspect become distended with blood. Next, place two fingertips against one of the distended veins, and, pressing firmly, move the superior finger proximally along the vein, and then release that finger. The vein remains flat and collapsed despite the pull of gravity. Finally, remove the distal fingertip, and watch the vein refill rapidly with blood.

Can you imagine what happens when the valves in veins weaken and fail?

This produces **varicose veins,** in which veins twist and swell with large amounts of pooled blood and the venous drainage is slowed considerably. Fully 15% of all adults suffer from varicose veins, usually in the lower limbs. Females are affected more often than males. Varicose veins can be hereditary but are also found in individuals whose jobs require prolonged standing in one position (store clerk, hairdresser, dentist, nurse). Obesity and pregnancy can cause or worsen the problem, because excessive weight constricts the leg-draining veins in the superior thigh. Another cause of varicose veins is elevated venous pressure. For example, straining to deliver a baby or to have a bowel movement raises the intra-abdominal pressure, preventing drainage of blood from the veins of the anal canal at the inferior end of the large intestine. The resulting varicosities in these anal veins are called **hemorrhoids.**

A severe case of varicose veins will so effectively slow the circulation through a body region that the tissues in that region die of oxygen starvation. To prevent this, physicians remove the affected veins or inject them with an irritating solution that causes them to scar and fuse shut. Alternate venous pathways then take over, enabling the body region to drain normally (see the discussion of venous anastomoses in the next section). ■

Vascular Anastomoses

Where vessels unite or interconnect, they form *vascular anastomoses* (ah-nas″to-mo′sēz; "coming together"). Most organs receive blood from more than one arterial branch, and nearby arteries often communicate with one another to form *arterial anastomoses.* Arterial anastomoses provide alternate pathways (*collateral channels*) for blood to reach a given body region. If one arterial branch is blocked or cut, the collateral channels can often provide the region with an adequate blood supply. Arterial anastomoses are abundant in the abdominal organs and around joints where active body movements may hinder blood flow through one channel. Because of the many anastomoses among the smaller branches of the coronary artery in the heart wall, a coronary artery can be 90% occluded by atherosclerosis before a myocardial infarction occurs. Anastomoses are poorly developed in the kidneys, in the spleen, in the parts of bone diaphyses nearest the epiphyses, and in the central artery of the retina. Thus, blockage of such arteries causes severe tissue damage.

Veins anastomose much more freely than arteries. You may be able to see *venous anastomoses* through the skin on the dorsum of your hand. Because of the abundance of venous anastomoses, occlusion of a vein rarely blocks blood flow or leads to tissue death.

PART 2: BLOOD VESSELS OF THE BODY

The complex system of blood vessels in the body is called the *vascular system.* Recall from Chapter 17 that this system has two basic circuits: The *pulmonary circuit* carries blood to and from the lungs for the uptake of oxygen and removal of carbon dioxide, while the *systemic circuit* carries oxygenated blood through the body and picks up carbon dioxide from body tissues (Figure 17.1, p. 460). Blood vessels in the systemic circuit also pick up nutrients from the digestive tract and deliver them to cells throughout the body, and receive nitrogenous wastes from the body cells and transport them to the kidneys for disposal in the urine.

As you read about the vessels, recall that arteries and veins tend to run together, side by side (Figure 18.2). In many places, these vessels also run with nerves. Also note that by convention, oxygen-rich blood is depicted as red, whereas oxygen-poor blood is depicted as blue, regardless of the vessel types.

Pulmonary Circulation

The **pulmonary trunk,** carrying oxygen-poor blood to the lungs, leaves the right ventricle of the heart, anterior to the aorta (Figure 18.6). It spirals to the aorta's left and ascends to reach the concavity of the aortic arch. There, the pulmonary trunk branches T-like into the **right** and **left pulmonary arteries.** Each pulmonary artery plunges into the medial surface of a lung and then divides into several **lobar arteries** (three in the right lung and two in the left lung). Within the lung, the lobar arteries branch along with the lung's air passageways (bronchi). As they decrease in size,

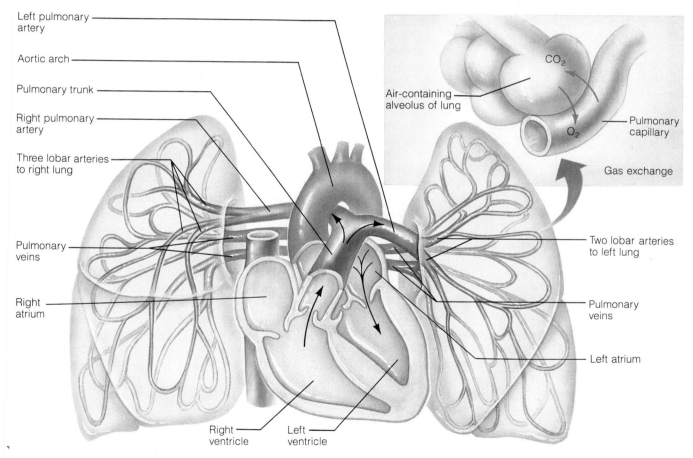

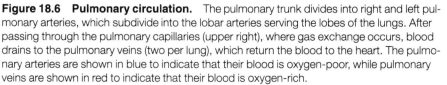

Figure 18.6   Pulmonary circulation.   The pulmonary trunk divides into right and left pulmonary arteries, which subdivide into the lobar arteries serving the lobes of the lungs. After passing through the pulmonary capillaries (upper right), where gas exchange occurs, blood drains to the pulmonary veins (two per lung), which return the blood to the heart. The pulmonary arteries are shown in blue to indicate that their blood is oxygen-poor, while pulmonary veins are shown in red to indicate that their blood is oxygen-rich.

Figure 18.7 Major arteries of the systemic circulation, anterior view.

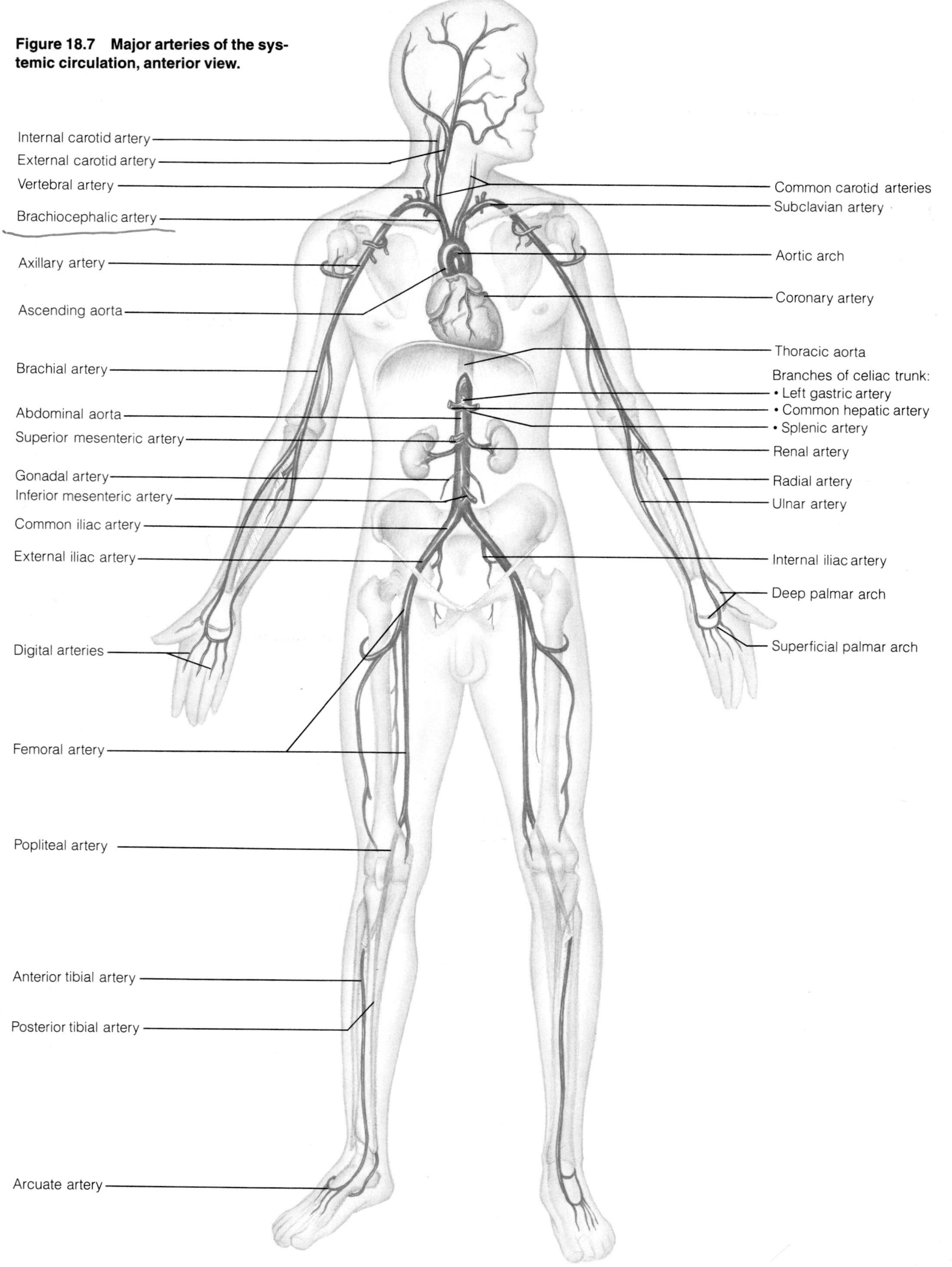

Internal carotid artery

External carotid artery

Vertebral artery

Brachiocephalic artery

Axillary artery

Ascending aorta

Brachial artery

Abdominal aorta

Superior mesenteric artery

Gonadal artery

Inferior mesenteric artery

Common iliac artery

External iliac artery

Digital arteries

Femoral artery

Popliteal artery

Anterior tibial artery

Posterior tibial artery

Arcuate artery

Common carotid arteries

Subclavian artery

Aortic arch

Coronary artery

Thoracic aorta

Branches of celiac trunk:
• Left gastric artery
• Common hepatic artery
• Splenic artery

Renal artery

Radial artery

Ulnar artery

Internal iliac artery

Deep palmar arch

Superficial palmar arch

the branching arteries become arterioles and finally the pulmonary capillaries that surround the delicate air sacs (lung alveoli). Gas exchange occurs across these capillaries, and the newly oxygenated blood enters venules and progressively larger veins. The largest venous tributaries form the two **pulmonary veins**—superior and inferior—that exit the medial aspect of each lung. In the mediastinum posterior to the heart, the four pulmonary veins run horizontally, just inferior to the pulmonary arteries, emptying into the left atrium.

The arteries and veins of the pulmonary circuit have thinner walls than do systemic vessels of comparable diameter—reflecting the fact that the maximum arterial pressure here is only one-sixth as high as that in the systemic circuit.

Systemic Circulation

Before tackling the systemic vessels, we should point out that they do not always match on the right and left sides of the body. Some of the large, deep vessels of the trunk region are asymmetrical. (Their initial symmetry is lost during embryonic development.) In the head and limbs, by contrast, almost all vessels are bilaterally symmetrical.

Systemic Arteries

As you read about the systemic arteries of the body, you may wish to refer to Table 18.1 on p. 500.

Aorta

The systemic arteries, which carry oxygenated blood to the capillaries of the body organs, begin with the **aorta** (Figures 18.7 and 18.8). The largest artery in the body, the aorta leaves the heart, arcs superiorly, then descends along the bodies of the vertebrae to the inferior part of the abdomen. The parts of the aorta are as follows:

Ascending Aorta. The ascending aorta (Figures 18.7 and 18.8) arises from the left ventricle of the heart and ascends for only about 5 cm. It begins posterior to the pulmonary trunk, passes to the right of that vessel, and then curves left to become the aortic arch. The only branches of the ascending aorta are the two *coronary arteries* that supply the wall of the heart (pp. 472–473).

Aortic Arch. Arching posteriorly and to the left, the aortic arch lies posterior to the manubrium of the sternum. The **ligamentum arteriosum,** a fibrous remnant of a fetal artery called the ductus arteriosus, intercon-

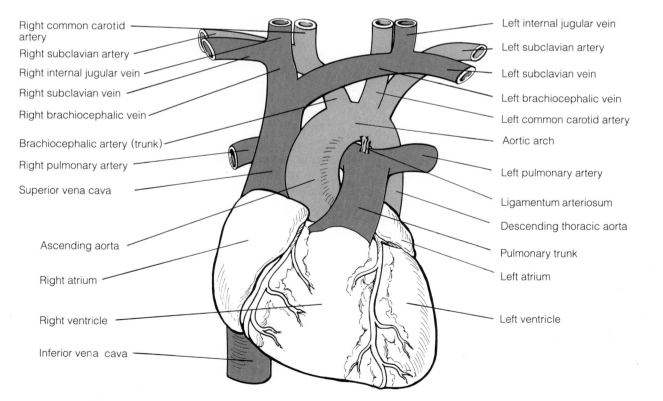

Figure 18.8 The great vessels that exit and enter the heart.

nects the aortic arch and the pulmonary trunk (Figure 18.8).

Three arteries branch from the aortic arch and run superiorly (Figures 18.7, 18.8, and 18.9). The first and largest branch is the **brachiocephalic artery** (bra″ke-o-sĕ-fal′ik; "arm-head"), or *innominate artery* (ĭ-nom′i-nāt; "no name"). This artery ascends to the base of the neck, then divides into the *right common carotid* and *right subclavian* arteries. The second and third branches of the aortic arch are the **left common carotid** and **left subclavian arteries,** respectively. As we will explain, these three arteries supply the head and neck, upper limbs, and the superior part of the thoracic wall. Note that the brachiocephalic artery on the right has no corresponding artery on the left—because the left common carotid and subclavian arteries arise directly from the aorta.

Descending Thoracic Aorta. Continuing from the aortic arch, the thoracic aorta descends along the bodies of the thoracic vertebrae (T_5–T_{12}) just to the left of the midline (Figures 18.7 and 18.8). The descending thoracic aorta sends many small branches to the thoracic organs and body wall.

Abdominal Aorta. The descending thoracic aorta pierces the diaphragm at the level of vertebra T_{12} and enters the abdominal cavity as the **abdominal aorta** (Figure 18.7). This part of the aorta lies on the lumbar vertebral bodies in the midline. The aorta ends at the level of vertebra L_4, where it divides into the right and left common iliac arteries, which supply the pelvis and lower limbs.

Arteries of the Head and Neck

Four pairs of arteries supply the head and neck (Figure 18.9). These are the *common carotid arteries* plus three branches from each subclavian artery: the *vertebral artery,* the *thyrocervical trunk,* and the *costocervical trunk.*

Common Carotid Arteries. Most parts of the head and neck receive their blood supply from the common carotid arteries. These arteries ascend through the anterior neck just lateral to the trachea. They are covered by the relatively thin sternocleidomastoid and infrahyoid muscles. At the superior border of the larynx—the level of the "Adam's apple"—each common carotid ends by dividing into an external and internal carotid artery.

The common carotid artery is more vulnerable to wounds than most other arteries in the body. Can you imagine why this is so?

The relatively superficial location of the common carotid artery makes it vulnerable to slashing wounds.

If this artery is cut, the victim can bleed to death in minutes. ■

The **external carotid arteries** supply most tissues of the head external to the brain and orbit. As each artery ascends, it sends a branch to the thyroid gland and larynx **(superior thyroid artery),** to the tongue **(lingual artery),** to the skin and muscles of the anterior face **(facial artery),** and to the posterior part of the scalp **(occipital artery).** Near the temporomandibular joint, the external carotid ends by splitting into the superficial temporal and maxillary arteries. The **superficial temporal artery** ascends just anterior to the ear and supplies most of the scalp. Branches of this vessel bleed profusely in scalp wounds. The **maxillary artery** runs anteriorly into the maxillary bone by passing through the chewing muscles. Along the way, it sends branches to the upper and lower teeth, the cheeks, nasal cavity, and muscles of mastication.

A clinically important branch of the maxillary artery is the *middle meningeal artery* (not illustrated). This branch enters the skull through the foramen spinosum and supplies the broad inner surfaces of the parietal bone and squamous region of the temporal bone, as well as the underlying dura mater.

Can you deduce the common cause and the effects of injuries to the middle meningeal artery?

Blows to the side of the head often tear this artery, producing an intracranial hematoma that can compress the cerebrum and disrupt brain function. ■

The **internal carotid arteries** supply the orbits and most of the cerebrum. Each internal carotid ascends through the superior neck directly lateral to the pharynx (Figure 18.9a) and enters the skull through the carotid canal in the temporal bone. From there, the internal carotid artery runs medially through the petrous region of the temporal bone, passes anteriorly along the body of the sphenoid bone, and bends superiorly to enter the sella turcica just posterior to the optic foramen (Figure 15.18, p. 432). Here, it gives off the **ophthalmic artery** to the eye and orbit and divides into the *anterior* and *middle cerebral arteries* (Figure 18.9a and c).

Each **anterior cerebral artery** supplies the medial surface of a cerebral hemisphere and anastomoses with its partner on the opposite side via a short **anterior communicating artery** (Figure 18.9c). Each **middle cerebral artery** runs through the lateral fissure of a cerebral hemisphere and supplies the lateral parts of the temporal and parietal lobes. Together, the anterior and middle cerebral arteries supply blood to over 80% of the cerebrum. The rest of the cerebrum is supplied by the posterior cerebral artery, which is described below.

Assume that a large embolus suddenly blocks an anterior cerebral artery, causing a cerebrovas-

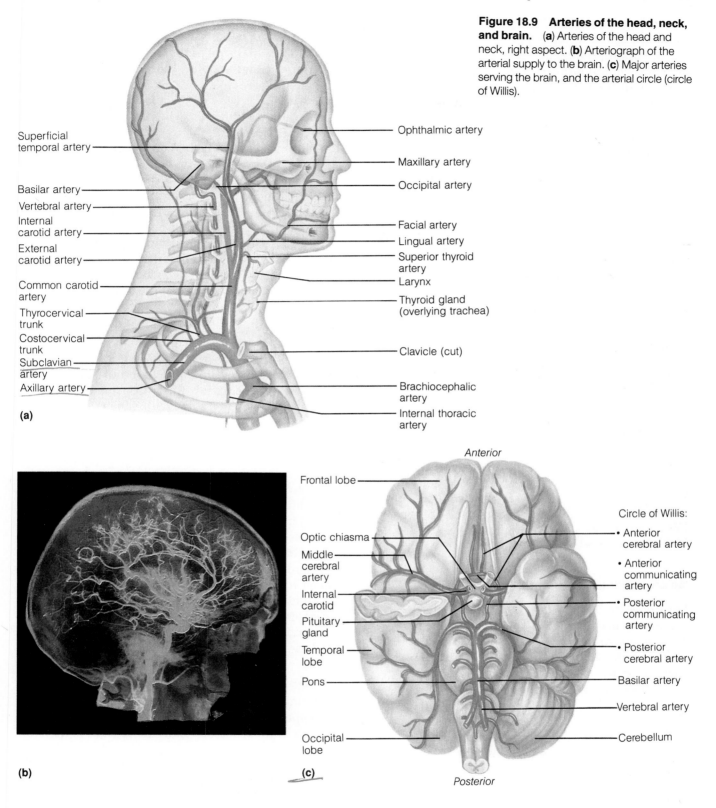

Figure 18.9   Arteries of the head, neck, and brain.   (a) Arteries of the head and neck, right aspect. (b) Arteriograph of the arterial supply to the brain. (c) Major arteries serving the brain, and the arterial circle (circle of Willis).

(a)

Superficial temporal artery

Basilar artery

Vertebral artery

Internal carotid artery

External carotid artery

Common carotid artery

Thyrocervical trunk

Costocervical trunk

Subclavian artery

Axillary artery

Ophthalmic artery

Maxillary artery

Occipital artery

Facial artery

Lingual artery

Superior thyroid artery

Larynx

Thyroid gland (overlying trachea)

Clavicle (cut)

Brachiocephalic artery

Internal thoracic artery

(b)

(c)

Anterior

Frontal lobe

Optic chiasma

Middle cerebral artery

Internal carotid

Pituitary gland

Temporal lobe

Pons

Occipital lobe

Circle of Willis:

• Anterior cerebral artery

• Anterior communicating artery

• Posterior communicating artery

• Posterior cerebral artery

Basilar artery

Vertebral artery

Cerebellum

Posterior

cular accident. What type of neural defects could result?

Since this artery supplies the medial (limbic) region of the cerebrum, emotional behavior could be affected. This artery also supplies the medial region of the frontal lobe, so cognitive functions could be diminished. ■

Vertebral Arteries.   The blood supply to the posterior brain comes from the right and left vertebral arteries. These spring from the subclavian arteries at the root of the neck (Figure 18.9a). They ascend through the foramina in the transverse processes of cervical vertebrae C_6 to C_1 and enter the skull through the foramen magnum. Along the way, they send branches to the cervical spinal cord. Within the cranium, the right and left vertebral arteries join to form the unpaired **basilar** (bas′ĭ-lar) **artery** (Figure 18.9c). This ascends along the ventral midline of the brain stem, sending branches to the cerebellum, pons, and inner ear. At the border of the pons and midbrain, the basilar artery divides into a pair of **posterior cerebral arteries.** These supply the occipital lobes and the inferior parts of the temporal lobes of the cerebral hemispheres.

Short **posterior communicating arteries** connect the posterior cerebral arteries to the middle cerebral arteries anteriorly. The two posterior communicating arteries and the single anterior communicating artery complete the formation of an arterial anastomosis called the **arterial circle (circle of Willis).** This circle forms a loop around the pituitary gland and optic chiasma, and it unites the brain's anterior and posterior blood supplies provided by the internal carotid and vertebral arteries. This anastomosis provides alternate routes for blood to reach the brain tissue in case of an occlusion of either a carotid or vertebral artery.

Thyrocervical and Costocervical Trunks. The rest of the neck receives its blood from two smaller branches of the subclavian arteries, the thyrocervical and costocervical trunks (Figure 18.9a). The **thyrocervical** (thi″ro-ser′vĭ-kal) **trunk,** which arises first, sends one branch anteriorly to the inferior part of the thyroid gland *(inferior thyroid artery)* and additional branches posteriorly over the scapula to help supply the scapular muscles. The **costocervical trunk** sends a branch superiorly into the deep muscles of the neck and a branch inferiorly to supply the two most superior intercostal spaces.

Arteries of the Upper Limb

The upper limb is supplied by arteries that arise from the subclavian artery (Figure 18.10). After giving off its branches to the neck, each subclavian artery runs laterally onto the first rib (Figure 7.20d, p. 166), lying deep to a clavicle. From here, the subclavian artery enters the axilla as the axillary artery.

Axillary Artery. The axillary artery descends through the axilla, giving off various branches. These include (a) **thoracoacromial** (tho″rah-ko-ah-kro′me-al) **trunk,** which arises just inferior to the clavicle and supplies much of the pectoralis and deltoid muscles; (b) **lateral thoracic artery,** which descends along the lateral edge of pectoralis minor and sends important branches to the breast; (c) **subscapular artery,** which serves the dorsal and ventral scapular regions and the latissimus dorsi muscle; and (d) **anterior** and **posterior circumflex humeral arteries,** which wrap around the surgical neck of the humerus and help supply the deltoid muscle and shoulder joint.

Brachial Artery. The axillary artery continues into the arm as the brachial artery. This artery descends along the medial side of the humerus deep to the biceps muscle and supplies the anterior arm muscles. One major branch, the **deep brachial artery** (also called *profunda brachii*), wraps around the posterior surface of the humerus with the radial nerve and serves the triceps muscle. As the brachial artery nears the elbow, it sends several small branches inferiorly. These anastomose with branches ascending from arteries in the forearm to supply the elbow joint. The brachial artery crosses the anterior aspect of the elbow joint in the midline of the arm, and the pulse of the brachial artery is easily felt here. This is also where one listens when measuring blood pressure. Immediately beyond the elbow joint, the brachial artery splits into the *radial* and *ulnar arteries,* which descend through the anterior aspect of the forearm.

Radial Artery. The radial artery descends along the medial margin of the brachioradialis muscle. It supplies muscles of the lateral anterior forearm, the lateral part of the wrist, and the thumb and index finger. At the root of the thumb, this artery lies very near the surface and provides a convenient site for taking the pulse (p. 701).

Ulnar Artery. The ulnar artery descends along the medial side of the anterior forearm. It lies between the superficial and deep flexor muscles and sends branches to the muscles that cover the ulna. Proximally, it gives off a major branch called the **common interosseous artery,** which splits immediately into *anterior* and *posterior interosseous arteries.* These descend along the respective surfaces of the interosseous membrane between the radius and the ulna. The anterior interosseous artery supplies the deep flexor muscles, while the posterior interosseous artery and its branches supply all the extensors on the posterior forearm.

Palmar Arches. In the palm, branches of the radial and ulnar arteries join to form two horizontal arches, the **superficial** and **deep palmar arches.** The superficial arch underlies the skin and fascia of the hand, while the deep arch lies against the metacarpal bones. The **digital arteries** that supply the fingers branch from these arches. (The radial and ulnar arteries also form a *carpal arch* on the dorsum of the wrist. Branches from this dorsal arch run distally along the metacarpal bones.)

Arteries of the Thorax

We will consider the thoracic wall first and then the deep thoracic viscera.

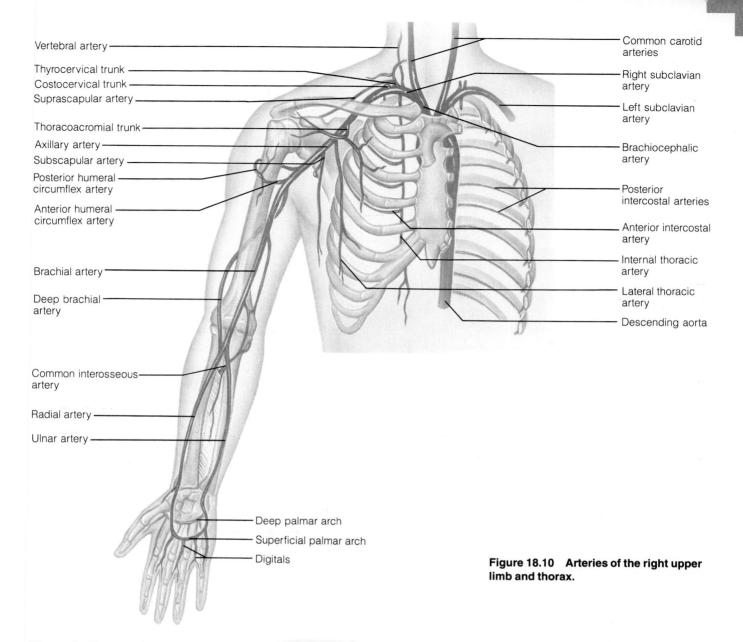

Vertebral artery

Thyrocervical trunk
Costocervical trunk
Suprascapular artery

Thoracoacromial trunk

Axillary artery
Subscapular artery
Posterior humeral circumflex artery
Anterior humeral circumflex artery

Brachial artery

Deep brachial artery

Common interosseous artery

Radial artery
Ulnar artery

Common carotid arteries
Right subclavian artery
Left subclavian artery
Brachiocephalic artery
Posterior intercostal arteries
Anterior intercostal artery
Internal thoracic artery
Lateral thoracic artery
Descending aorta

Deep palmar arch
Superficial palmar arch
Digitals

Figure 18.10   Arteries of the right upper limb and thorax.

Thoracic Wall.   The anterior thoracic wall receives its blood from the **internal thoracic artery** *(internal mammary artery)* (Figure 18.10). This vessel, which branches from the subclavian artery superiorly, descends just lateral to the sternum and just deep to the costal cartilages. **Anterior intercostal arteries** branch off at regular intervals and run horizontally to supply the ribs and the structures in the intercostal spaces. The internal thoracic artery also sends branches superficially to supply the mammary gland. It ends inferiorly at the costal margin, where it divides into a branch to the anterior abdominal wall and a branch to the anterior part of the diaphragm.

The internal thoracic artery can be used as the bypass vessel for coronary bypass surgery (see the box on pp. 484–485). The use of this artery has some advantages over the traditional use of the great saphenous vein of the leg (p. 509). First, as a thick-walled artery, the internal thoracic artery is less likely to be damaged by the pressure of the heartbeat and, therefore, is less likely to develop atherosclerosis itself. Second, the surgery is easier: The internal thoracic artery remains permanently attached to the subclavian artery superiorly, while its distal end is sutured onto the obstructed coronary artery in the heart wall. ■

The posterior thoracic wall receives its blood from the **posterior intercostal arteries.** The superior two pairs arise from the costocervical trunk, while the inferior nine pairs issue from the thoracic aorta. All of the posterior intercostal arteries run anteriorly in the costal grooves of their respective ribs. In the lateral thoracic wall, these arteries anastomose with the anterior intercostal arteries. Inferior to the twelfth rib, one pair of *subcostal arteries* branches from the thoracic aorta. Finally, a pair of *superior phrenic arteries*

495

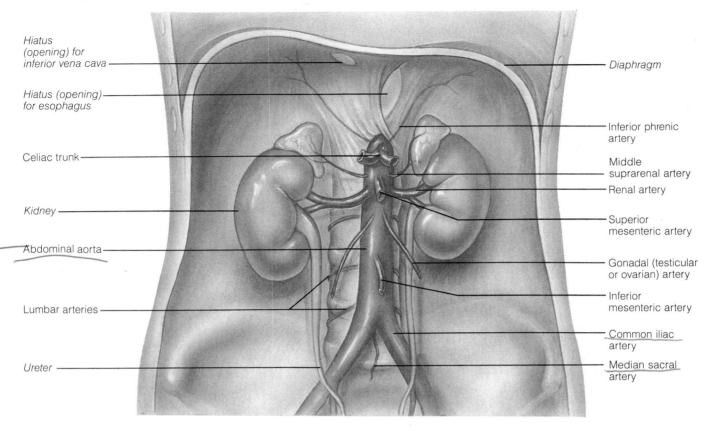

Hiatus
(opening) for
inferior vena cava

Hiatus (opening)
for esophagus

Celiac trunk

Kidney

Abdominal aorta

Lumbar arteries

Ureter

Diaphragm

Inferior phrenic
artery

Middle
suprarenal artery

Renal artery

Superior
mesenteric artery

Gonadal (testicular
or ovarian) artery

Inferior
mesenteric artery

Common iliac
artery

Median sacral
artery

Figure 18.11   Major branches of the abdominal aorta.

(not illustrated) leaves the most inferior part of the thoracic aorta. These small arteries supply the posterior, superior aspect of the diaphragm.

Thoracic Visceral Organs.   Many thoracic viscera receive their functional blood supply from small branches off the thoracic aorta. Since these vessels are so small, they are not illustrated. The *bronchial arteries* supply systemic (oxygenated) blood to the lung structures. Usually, two bronchial arteries serve the left lung, and one serves the right lung. In some individuals, they arise from posterior intercostal arteries instead of the aorta. Bronchial arteries enter the lung's medial surface along with the large pulmonary vessels.

The aorta also sends several small branches to the esophagus (directly anterior to it) and to the posterior part of the mediastinum and pericardium.

Arteries of the Abdomen

The arteries to the abdominal organs arise from the abdominal aorta (Figures 18.11, 18.12, and 18.13). In a person at rest, about half of the entire arterial flow is present in these vessels. We will present these arteries in their order of issue from the aorta, indicating the vertebral level at which each artery arises.

Inferior Phrenic Arteries.   The paired inferior phrenic arteries branch from the abdominal aorta at the level of T_{12}, just inferior to the aortic opening (hiatus) in the diaphragm (Figure 18.11). These arteries supply the inferior surface of the diaphragm.

Celiac Trunk.   The short, wide, unpaired celiac (se′le-ak) trunk (Figure 18.12) supplies the viscera in the superior part of the abdominal cavity (*coelia* = abdominal cavity). Specifically, the celiac trunk sends branches to the stomach, liver, pancreas, spleen, and a part of the small intestine (duodenum). It emerges from the aorta at the level of T_{12} and divides almost immediately into three branches: the *left gastric, splenic,* and *common hepatic arteries.*

The **left gastric artery** (*gaster* = stomach) runs superiorly and to the left, to the junction of the stomach with the esophagus. There it gives off several esophageal branches and descends along the right (lesser) curvature of the J-shaped stomach.

The **splenic artery** runs horizontally to the left, posterior to the stomach, to enter the spleen. Along the way, it lies superior to the pancreas, sending branches to this organ. Near the spleen, the splenic artery sends several branches superiorly to the stomach's dome (short gastric arteries) and sends a major branch along the stomach's left (greater) curvature.

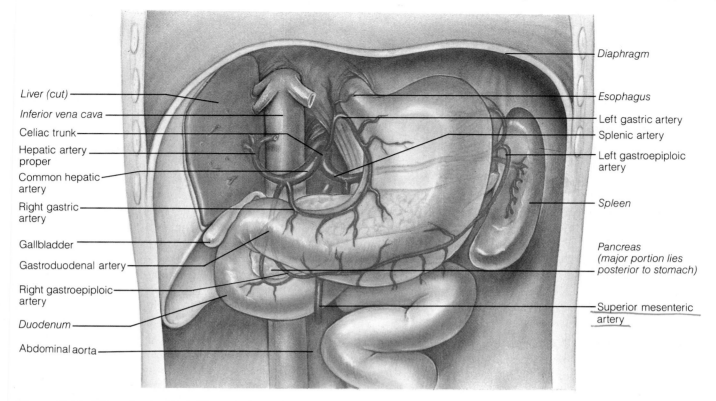

Liver (cut)
Inferior vena cava
Celiac trunk
Hepatic artery proper
Common hepatic artery
Right gastric artery
Gallbladder
Gastroduodenal artery
Right gastroepiploic artery
Duodenum
Abdominal aorta

Diaphragm
Esophagus
Left gastric artery
Splenic artery
Left gastroepiploic artery
Spleen
Pancreas (major portion lies posterior to stomach)
Superior mesenteric artery

Figure 18.12   The celiac trunk and its main branches.   The left half of the liver is not shown so that this artery can be seen.

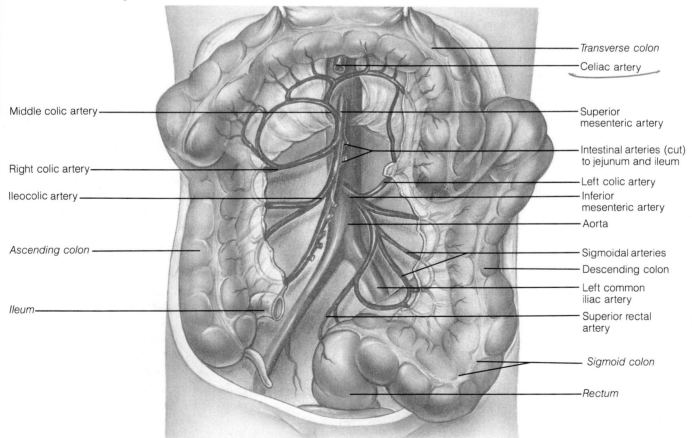

Middle colic artery
Right colic artery
Ileocolic artery
Ascending colon
Ileum

Transverse colon
Celiac artery
Superior mesenteric artery
Intestinal arteries (cut) to jejunum and ileum
Left colic artery
Inferior mesenteric artery
Aorta
Sigmoidal arteries
Descending colon
Left common iliac artery
Superior rectal artery
Sigmoid colon
Rectum

Figure 18.13   Distribution of the superior and inferior mesenteric arteries.   The transverse colon is pulled superiorly, and much of the mesentery (the yellow membrane) is not shown. In the body, the arteries in this figure are covered by the coils of the small intestine (jejunum and ileum, which are also not shown).

This latter branch is the **left gastroepiploic** (gas"tro-ep"ĭ-plo'ik) **artery.**

The **common hepatic artery** (*hepar, hepat* = liver) is the only branch of the celiac trunk that runs to the right. At the junction of the stomach with the small intestine (duodenum), this artery divides into an ascending branch, the *hepatic artery proper,* and a descending branch, the *gastroduodenal artery.* The **hepatic artery proper** divides into right and left branches just before entering the liver. The **right gastric artery,** which can arise either from the hepatic artery proper or from the common hepatic artery, runs along the stomach's lesser curvature from the right. The **gastroduodenal artery** (gas"tro-du"o-de'nal), the descending branch of the common hepatic artery, runs inferiorly on the head of the pancreas. It helps supply the pancreas, plus the nearby duodenum. Its major branch, the **right gastroepiploic artery,** runs along the stomach's greater curvature from the right.

Superior Mesenteric Artery. The large, unpaired superior mesenteric (mes"en-ter'ik) artery serves most of the intestine (Figures 18.11 and 18.13). It arises from the aorta, posterior to the pancreas, at the level of L_1. From there, this artery runs inferiorly and anteriorly to enter the mesentery, a drape-like membrane that supports the long, coiled parts of the small intestine (the jejunum and the ileum). The superior mesenteric artery angles gradually to the right as it descends through the mesentery. From its left side arise many **intestinal arteries** to the jejunum and ileum. From its right side emerge branches that supply much of the large intestine: the ascending colon, cecum, and appendix (via the **ileocolic artery**), and part of the transverse colon (via the **right** and **middle colic arteries**).

Suprarenal Arteries. The paired **middle suprarenal arteries** (Figure 18.11 and Table 18.1) emerge from the sides of the aorta, at L_1. They supply blood to the adrenal (suprarenal) glands on the superior poles of the kidneys. The adrenal glands also receive *superior suprarenal* branches from the nearby inferior phrenic arteries and *inferior suprarenal* branches (not illustrated) from the nearby renal arteries.

Renal Arteries. The paired renal arteries to the kidneys (*ren* = kidney) stem from the sides of the aorta, between vertebrae L_1 and L_2 (Figure 18.11). The kidneys remove nitrogenous waste molecules from the blood supplied by the renal arteries. As we mentioned earlier, the transportation of cellular wastes to the kidney is an important function of the vascular system, so the renal circulation is a major functional subdivision of the systemic circuit.

Gonadal Arteries. The paired arteries to the gonads are more specifically called **testicular arteries** in males and **ovarian arteries** in females (Figure 18.11). They branch from the aorta at L_2, the level where the gonads first develop in the embryo. The ovarian arteries extend inferiorly into the pelvis to serve the ovaries and part of the uterine tubes. The longer testicular arteries extend through the pelvis to the scrotum, where they serve the testes.

Inferior Mesenteric Artery. The unpaired inferior mesenteric artery (Figures 18.11 and 18.13) is the final major branch of the abdominal aorta, arising at the level of L_3. It serves the distal half of the large intestine—from the last part of the transverse colon to the middle part of the rectum. Its branches are the **left colic, sigmoidal,** and **superior rectal arteries.**

Lumbar Arteries. Four pairs of lumbar arteries arise from the posterolateral surface of the aorta in the lumbar region (Figure 18.11). These segmental arteries run horizontally to supply the posterior abdominal wall.

Median Sacral Artery. The median sacral artery issues from the most inferior part of the aorta. As it descends, this thin artery supplies the sacrum and coccyx along the midline.

Common Iliac Arteries. At the level of L_4, the aorta splits into the right and left common iliac arteries (Figure 18.14). These arteries supply the inferior part of the anterior abdominal wall, the pelvic organs, and the lower limbs (next section).

Arteries of the Pelvis and Lower Limbs

At the level of the sacroiliac joint on the pelvic brim, each common iliac artery forks into two branches, the *internal* and *external iliac arteries.*

Internal Iliac Arteries. The internal iliac arteries primarily supply the pelvic organs (Figure 18.15). More specifically, their branches carry blood to the pelvic walls, pelvic viscera, buttocks, medial thighs, and perineum. The following branches are the most important: Large **gluteal arteries** run posteriorly to the gluteal muscles. The **internal pudendal artery** leaves the pelvic cavity to supply the perineum and external genitalia. The **obturator artery** descends through the obturator foramen into the thigh adductor muscles. Other branches run to the bladder, rectum, uterus and vagina (in females), and pelvic reproductive glands (in males).

External Iliac Artery. The right and left external iliac arteries carry blood to the lower limbs (Figure 18.14). Originating from the common iliac arteries in the pelvis, each of these arteries descends along the arcuate line of the ilium bone. The artery sends some small

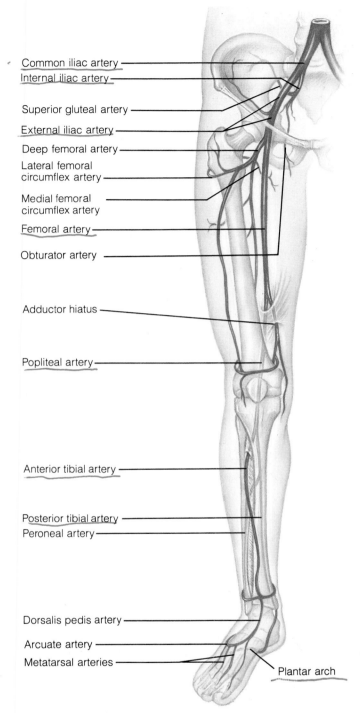

Common iliac artery

Internal iliac artery

Superior gluteal artery

External iliac artery

Deep femoral artery

Lateral femoral circumflex artery

Medial femoral circumflex artery

Femoral artery

Obturator artery

Adductor hiatus

Popliteal artery

Anterior tibial artery

Posterior tibial artery

Peroneal artery

Dorsalis pedis artery

Arcuate artery

Metatarsal arteries

Plantar arch

Figure 18.14   Arteries of the right pelvis and lower limb.

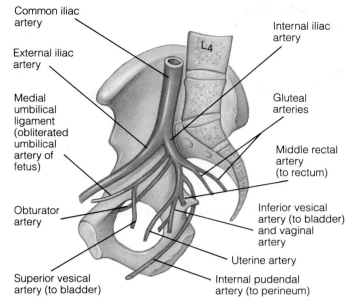

Common iliac artery

External iliac artery

Medial umbilical ligament (obliterated umbilical artery of fetus)

Obturator artery

Superior vesical artery (to bladder)

Internal iliac artery

L₄

Gluteal arteries

Middle rectal artery (to rectum)

Inferior vesical artery (to bladder) and vaginal artery

Uterine artery

Internal pudendal artery (to perineum)

Figure 18.15   Internal iliac artery.   The right half of the female pelvis, in medial view. The branches of the internal iliac artery are variable, arising in somewhat different places in different individuals.

tor magnus muscle (the adductor hiatus) and emerges posterior to the distal femur as the popliteal artery.

The superior part of the femoral artery is enclosed in a tube of dense fascia, but it is not protected by any overlying musculature. This lack of protection makes the proximal femoral artery a convenient place to take a pulse but also makes it susceptible to injury. For example, a butcher may slip while sawing meat and sever the femoral artery in the superior thigh. ■

Several arteries arise from the femoral artery in the thigh. The largest branch arises superiorly and is called the **deep femoral artery** (or *profunda femoris*). This vessel supplies the posterior thigh muscles. Proximal branches of the deep femoral artery are the **medial** and **lateral circumflex femoral arteries.** These circle the neck and upper shaft of the femur. The medial circumflex artery supplies the head of the femur, and if a fracture of the hip tears this artery, the bone tissue of the head of the femur dies. A long, descending branch of the lateral circumflex artery supplies the quadriceps femoris muscles.

Popliteal Artery.   The popliteal artery (pop″lĭ-te′al), the inferior continuation of the femoral artery, lies within the popliteal fossa (the region posterior to the knee), a deep location that offers protection from injury. The popliteal artery gives off several *genicular arteries* (jĕ-nik′u-lar; "knee") that circle the knee joint like horizontal hoops. Just inferior to the head of the fibula, the popliteal artery splits into the *anterior* and *posterior tibial arteries* of the leg.

branches to the anterior abdominal wall and enters the thigh by passing deep to the midpoint of the inguinal ligament. At this point, the external iliac artery is called the femoral artery.

Femoral Artery.   The femoral artery descends vertically through the thigh medial to the femur, along the anterior surface of the adductor muscles. Inferiorly, the femoral artery passes through a gap in the adduc-

Table 18.1   The Aorta and Its Branches

Aortic region	Major branches	Area served	Aortic region	Major branches	Area served
Ascending aorta	**Coronary arteries** (right and left)	Myocardium of the heart		**Left common carotid** and its branches	Same relative regions as right common carotid artery
Aortic arch	**Brachiocephalic** (innominate)	Branches provide arterial blood supply of head, neck, and upper extremity (right side)		**Left subclavian** and its branches	Same relative regions as right subclavian artery
	Right common carotid		Thoracic aorta	**Visceral branches**	Together serve viscera of thorax
	Right external carotid			Bronchial	Lungs and bronchi
	· Superior thyroid	Anterior neck and thyroid gland		Esophageal	Esophagus
	· Lingual	Tongue		**Parietal branches**	Thorax wall and diaphragm
	· Facial	Anterior facial structures		Posterior intercostals	Intercostal muscles, spinal cord, vertebrae, pleurae, and overlying skin
	· Occipital	Posterior scalp			
	· Maxillary	Deep facial structures, chewing muscles, teeth, nasal mucosae, dura mater		Superior phrenics	Posterior and superior surface of diaphragm
	· Superficial temporal	Lateral and superior scalp	Abdominal aorta	**Visceral branches**	Together supply abdominal and pelvic viscera
	Right internal carotid			Celiac trunk	
	· Ophthalmic	Orbit, eye, forehead, anterior scalp, nasal structures		*Common hepatic*	Liver, gallbladder, part of stomach
				Splenic	Spleen, part of pancreas, stomach
	· Anterior cerebral	Anterior medial surface of cerebral hemispheres		*Left gastric*	Stomach and inferior esophagus
	· Middle cerebral	Lateral and superior portion of cerebral hemispheres		Superior mesenteric	Small intestine, ascending colon, part of transverse colon
	Right subclavian			Middle suprarenals	Adrenal glands
	Vertebral	Deep structures (spinal cord and some muscles) of posterior neck		Renals	Kidneys
				Gonadals	Testes in males, ovaries in females
	· Basilar	Cerebellum, pons, inner ear, and via its terminal branches, the posterior cerebral arteries, the occipital and part of the temporal lobes of cerebral hemispheres		Inferior mesenteric	Distal colon (part of transverse, sigmoid, and descending portions)
				Parietal branches	
	Internal thoracic	Anterior thorax wall		Inferior phrenics	Inferior diaphragmatic surface
	Thyrocervical trunk	Inferior neck structures and some scapular muscles		Lumbars (several pairs)	Posterior abdominal wall
	Costocervical trunk	Deep neck muscles and superior intercostal muscles		Median sacral	Sacrum and coccyx
				Right and left common iliacs	
	Right axillary	Axilla		*Internal iliacs*	Lower abdominal wall, pelvic organs
	· Lateral thoracic	Wall of lateral thorax		*External iliacs*	
	· Subscapular	Scapula, dorsal thorax wall, latissimus dorsi		· Femoral	Thighs (and knee where femoral arteries become popliteal arteries)
	· Thoracoacromial	Superior shoulder, and pectoral region			
	Brachial	Arm		· Anterior tibials	Anterior leg muscles and dorsum of feet
	· Radial	Forearm and hand		· Posterior tibials	Flexor muscles of leg, plantar surfaces of feet, toes
	· Ulnar	Forearm and hand		· Peroneals	Lateral leg muscles

Anterior Tibial Artery. The anterior tibial artery runs through the anterior muscular compartment of the leg. More specifically, it descends along the interosseous membrane lateral to the tibia, sending branches to the extensor muscles along the way. At the ankle, it becomes the **dorsalis pedis artery,** literally the "artery of the dorsum of the foot." At the base of the metatarsal bones, the dorsalis pedis forms the **arcuate artery** that sends branches distally along the metatarsals.

✚ The dorsalis pedis artery is superficial and provides a place to feel the pulse (the "pedal pulse point"). The absence of this pulse indicates that the blood supply to the leg is inadequate. Checking the pedal pulse is routine in patients known to have impaired circulation to the legs and after surgery to the leg or groin. ■

Posterior Tibial Artery. The posterior tibial artery descends through the posteromedial part of the leg. It lies directly deep to the soleus muscle. Proximally, this artery gives off a large branch, the **peroneal artery,** which descends along the medial aspect of the fibula. Together, the posterior tibial and peroneal arteries supply the flexor muscles in the leg, and the peroneal arteries send branches to the peroneal muscles.

Inferiorly, the posterior tibial artery passes behind the medial malleolus of the tibia. On the medial side of the foot, it divides into *medial* and *lateral plantar* arteries. These arteries serve the sole, and the lateral plantar artery forms the **plantar arch.** Digital arteries to the toes arise from the plantar arch.

Systemic Veins

Having considered the arteries of the body, we now turn to the veins (Figure 18.16). Although most veins run with corresponding arteries, there are some important differences in the distributions of arteries and veins:

1. Whereas just one systemic artery leaves the heart (the aorta), three major veins enter the right atrium: the superior and inferior venae cavae and the coronary sinus.

2. All large and medium-sized arteries are located deeply for protection. By contrast, many *superficial veins* lie just beneath the skin, unattended by any arteries.

3. Several parallel veins often take the place of a single larger vein. Such multi-vein bundles, as well as networks of anastomosing veins, are called *venous plexuses.*

4. Two important body areas have unusual patterns of venous drainage. First, veins from the brain drain into *dural sinuses,* which are not typical veins but endothelium-lined channels supported by walls of dura mater. Second, venous blood draining from the digestive organs enters a special subcirculation, the *hepatic portal system,* and passes through capillaries in the liver before the blood reenters the general systemic circulation. The dural sinuses and hepatic portal system are considered below.

As you read about the veins of the body, you may wish to refer to the summary in Table 18.2 on p. 509.

Venae Cavae and Their Major Tributaries

The unpaired *superior* and *inferior venae cavae,* the two largest veins, empty directly into the right atrium of the heart (Figures 18.16 and 18.17). The name *vena cava* means "hollow, cave-like vein."

Superior Vena Cava. The superior vena cava receives the systemic blood from all body regions superior to the diaphragm. This vein arises from the union of the **left** and **right brachiocephalic veins** posterior to the manubrium and descends to enter the right atrium. Of the two brachiocephalic veins, the left is longer and nearly horizontal, while the right is vertical (Figure 18.8). Each brachiocephalic vein is formed by the union of an *internal jugular vein* and *subclavian vein.*

Inferior Vena Cava. The inferior vena cava, ascending through the posterior wall of the abdominal cavity, is the widest blood vessel in the body (Figures 18.16 and 18.17). It returns blood to the heart from all body regions inferior to the diaphragm. The inferior vena cava begins inferiorly on the body of vertebra L_5 and ascends on the right side of the vertebral bodies to the right of the abdominal aorta. Upon penetrating the diaphragm at T_8, the inferior vena cava empties into the right atrium.

Veins of the Head and Neck

Most blood draining from the head and neck enters three pairs of veins: (1) internal jugular veins from the dural sinuses, (2) external jugular veins, and (3) vertebral veins (Figure 18.18a). Although most of the extracranial veins have the same names as the extracranial arteries (facial, ophthalmic, occipital, and superficial temporal), their courses and interconnections differ substantially.

Dural Sinuses. Most veins of the brain drain into the cranial **dural sinuses** (Figure 18.18b). These sinuses form an interconnected series of channels in the

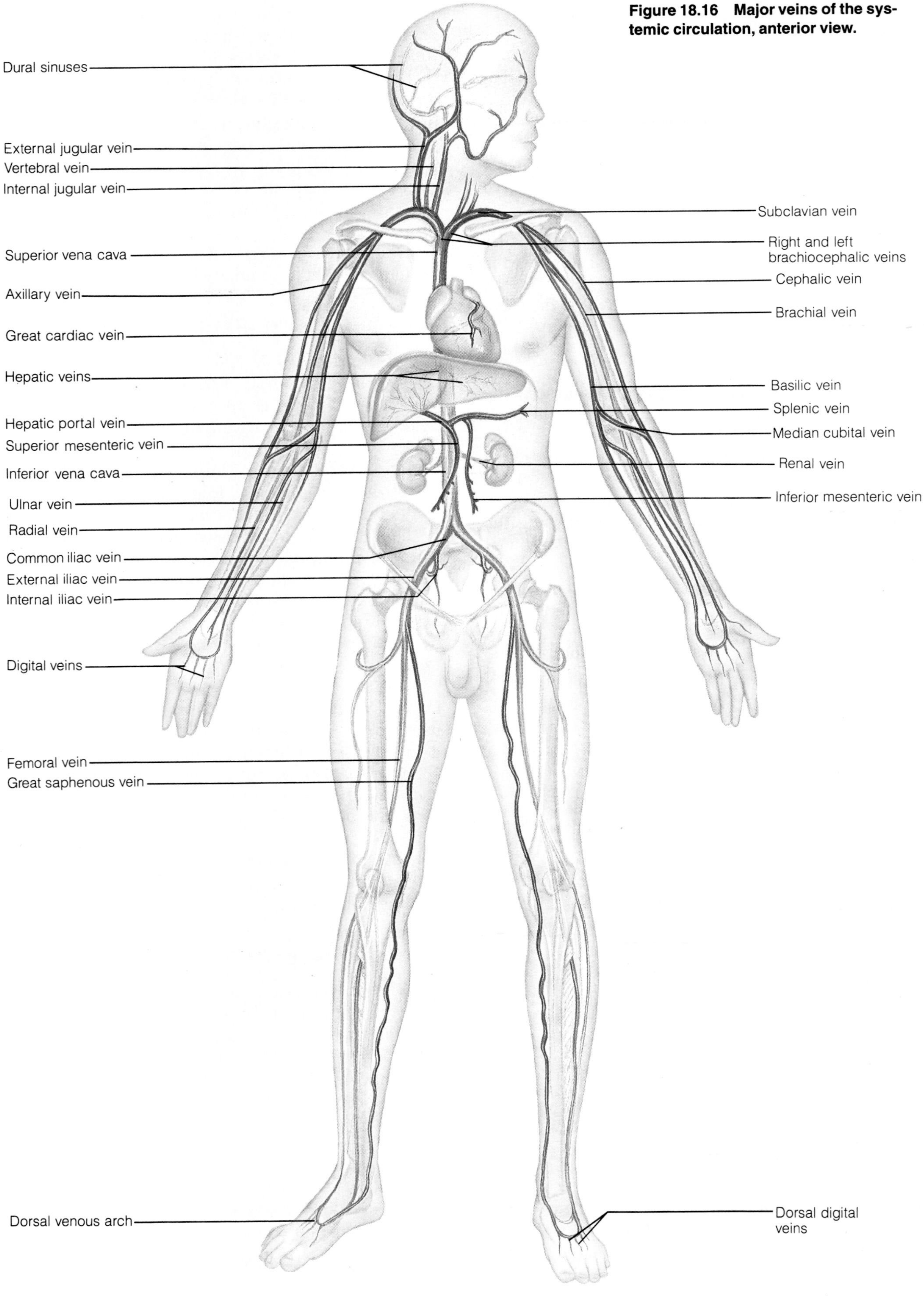

Figure 18.16   Major veins of the systemic circulation, anterior view.

Dural sinuses

External jugular vein
Vertebral vein
Internal jugular vein

Superior vena cava

Axillary vein

Great cardiac vein

Hepatic veins

Hepatic portal vein
Superior mesenteric vein

Inferior vena cava

Ulnar vein

Radial vein

Common iliac vein
External iliac vein
Internal iliac vein

Digital veins

Femoral vein
Great saphenous vein

Dorsal venous arch

Subclavian vein

Right and left
brachiocephalic veins

Cephalic vein

Brachial vein

Basilic vein

Splenic vein

Median cubital vein

Renal vein

Inferior mesenteric vein

Dorsal digital
veins

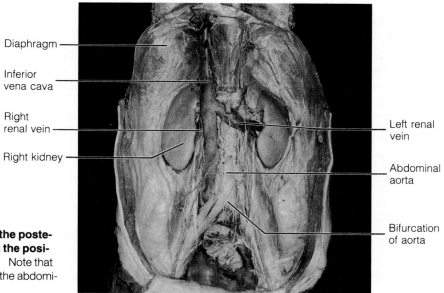

Figure 18.17   Photograph of the posterior abdominal wall, revealing the position of the inferior vena cava.   Note that this vessel lies just to the right of the abdominal aorta.

Diaphragm

Inferior vena cava

Right renal vein

Right kidney

Left renal vein

Abdominal aorta

Bifurcation of aorta

skull and lie between the two layers of cranial dura mater (p. 342). The **superior** and **inferior sagittal sinuses** lie in the falx cerebri between the cerebral hemispheres. They drain posteriorly into **transverse sinuses,** which run in shallow grooves on the internal surface of the occipital bone. Each transverse sinus in turn drains into an S-shaped (sigmoid) sinus, which becomes the **internal jugular vein** as it leaves the skull through the jugular foramen.

The paired **cavernous sinuses** flank the body of the sphenoid bone laterally, and each has an internal carotid artery *running within it.* The following cranial

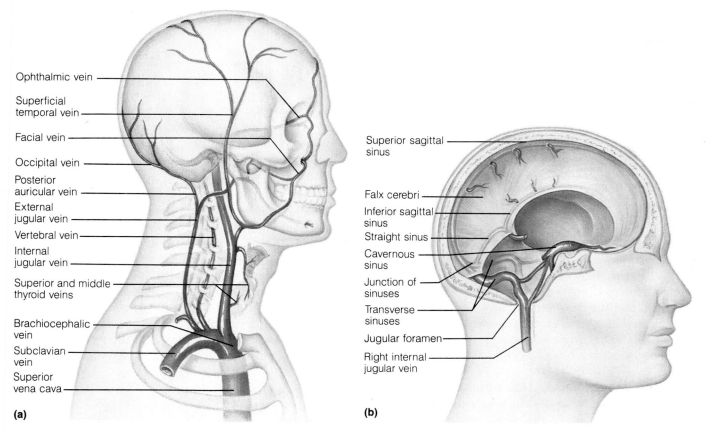

Ophthalmic vein

Superficial temporal vein

Facial vein

Occipital vein

Posterior auricular vein

External jugular vein

Vertebral vein

Internal jugular vein

Superior and middle thyroid veins

Brachiocephalic vein

Subclavian vein

Superior vena cava

(a)

Superior sagittal sinus

Falx cerebri

Inferior sagittal sinus

Straight sinus

Cavernous sinus

Junction of sinuses

Transverse sinuses

Jugular foramen

Right internal jugular vein

(b)

Figure 18.18   Venous drainage of the head, neck, and brain.   **(a)** Veins of the head and neck, right aspect. The external jugular vein has been pulled back slightly so that the vertebral and internal jugular veins can be seen. **(b)** Dural sinuses in the cranium, lateral view.

nerves also run within the cavernous sinus on their way to the orbit and the face: the oculomotor, trochlear, abducens, and the maxillary and ophthalmic divisions of the trigeminal nerve. The cavernous sinus communicates with the **ophthalmic vein** of the orbit (Figure 18.18a), which in turn communicates with the facial vein—the vein that drains the nose and upper lip. The other dural sinuses are small and relatively unimportant.

Squeezing pimples on the nose or upper lip can spread infection through the facial vein into the cavernous sinus and, from there, through the other dural sinuses in the skull. For this reason, the nose and upper lip are called the *danger triangle of the face.* ■

Blows to the head can rupture the internal carotid artery within the cavernous sinus. Leaked blood then accumulates within this sinus and exerts crushing pressure on the contained cranial nerves, disrupting their functions.

Such an injury to the cavernous sinus can lead to a loss of the ability to move the eyes. Can you tell why?

Recall that the oculomotor, trochlear, and abducens nerves, which run through this sinus, innervate the extrinsic eye muscles. When the functions of these nerves are disrupted, the eyes cannot move. ■

Internal Jugular Veins. The large internal jugular veins (jug'u-lar) drain almost all of the blood from the brain. From its origin at the base of the skull, each internal jugular vein descends through the neck. As it descends, it lies lateral first to the internal carotid artery and then to the common carotid artery. Along the way, the internal jugular vein receives blood from some deep veins of the face and neck—branches of the *facial* and *superficial temporal veins* (Figure 18.18a). At the base of the neck, the internal jugular vein joins the subclavian vein to form the brachiocephalic vein. The jugular veins are named for their end point, as *jugulum* means "the throat just above the clavicle."

Just as wounds to the neck can cut the carotid arteries, they can also sever the internal jugular veins. However, cut veins do not bleed as quickly, because blood pressure within veins is lower than the pressure within arteries. For this reason, the chances of surviving are higher if a neck wound affects the vein and not the artery. ■

External Jugular Veins. The external jugular vein is a superficial vein that descends vertically through the neck on the surface of the sternocleidomastoid muscle. Superiorly, its tributaries drain the posterior scalp, the lateral scalp, and some of the face (Figure 18.18a). The external jugular vein is not accompanied by any corresponding artery. Inferiorly, it empties into the subclavian vein.

Vertebral Veins. Unlike the vertebral arteries, the vertebral veins do not serve much of the brain. Instead, they drain only the cervical vertebrae, cervical spinal cord, and small muscles in the superior neck. Originating inferior to the occipital condyle, each vertebral vein descends through the transverse foramina of vertebrae C_1–C_6 in the form of a venous plexus. Emerging from C_6 as a single vein, the vertebral vein continues inferiorly to join the brachiocephalic vein in the root of the neck.

Veins of the Upper Limbs

Deep Veins. The deep veins of the upper limbs (Figure 18.19) follow the paths of their companion arteries and have the same names. All except the largest, however, are double veins that flank their artery on both sides. The **deep** and **superficial palmar venous arches** of the hand empty into the **radial** and **ulnar veins** of the forearm, which unite just inferior to the elbow joint to form the **brachial vein** of the arm. As the brachial vein enters the axilla, it empties into the **axillary vein,** which becomes the **subclavian vein** at the first rib.

Superficial Veins. The superficial veins of the upper limb are larger than the deep veins and are visible beneath the skin. These veins anastomose frequently along their course. The superficial veins begin with the *dorsal venous network* (not illustrated) on the dorsum of the hand. The **cephalic vein** starts at the lateral side of this network (Figure 18.19), then coils around the distal radius to enter the anterior forearm. From there, this vein ascends through the anterolateral side of the entire limb. The cephalic vein ends inferior to the clavicle, where it joins the axillary vein. The **basilic vein** arises from the medial aspect of the hand's dorsal network, then ascends along the posteromedial forearm and the anteromedial side of the arm. In the axilla, the basilic vein joins the brachial vein to become the axillary vein. On the anterior aspect of the elbow joint, in the region called the cubital fossa, the **median cubital vein** connects the basilic and cephalic veins. The median cubital vein is easy to find in most people and is used to obtain blood or to administer substances intravenously. The **median vein of the forearm** ascends in the center of the forearm. Its termination point at the elbow is highly variable.

Veins of the Thorax

Blood draining from the first few intercostal spaces enters the brachiocephalic veins. Most other thoracic structures, however, are drained by a group of veins called the *azygos system* (āz'ĭ-gos, or a-zi'gus). These

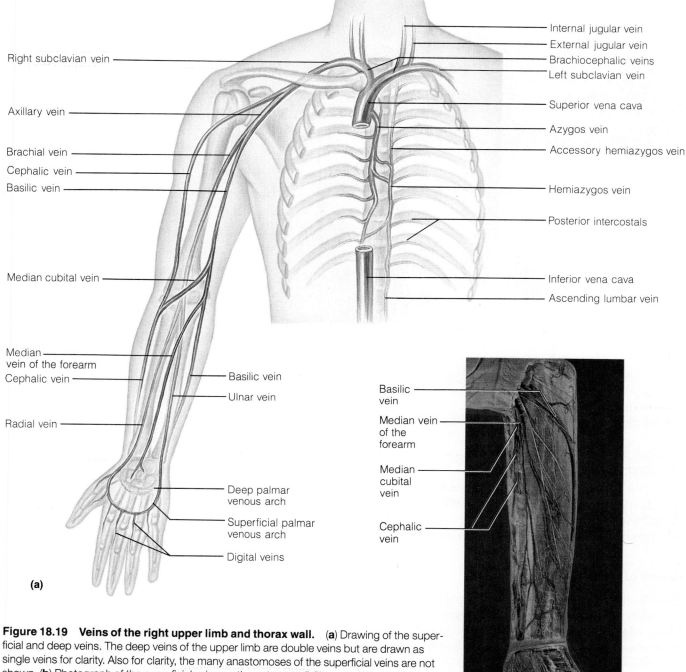

Internal jugular vein
External jugular vein
Brachiocephalic veins
Left subclavian vein
Superior vena cava
Azygos vein
Accessory hemiazygos vein
Hemiazygos vein
Posterior intercostals
Inferior vena cava
Ascending lumbar vein

Right subclavian vein
Axillary vein
Brachial vein
Cephalic vein
Basilic vein
Median cubital vein
Median vein of the forearm
Cephalic vein
Radial vein
Basilic vein
Ulnar vein
Deep palmar venous arch
Superficial palmar venous arch
Digital veins

(a)

Basilic vein
Median vein of the forearm
Median cubital vein
Cephalic vein

(b)

Figure 18.19 Veins of the right upper limb and thorax wall. (a) Drawing of the superficial and deep veins. The deep veins of the upper limb are double veins but are drawn as single veins for clarity. Also for clarity, the many anastomoses of the superficial veins are not shown. (b) Photograph of the superficial veins on the anteromedial aspect of the right forearm.

veins flank the vertebral column and ultimately empty into the superior vena cava (Figure 18.19a). The azygos system has the following vessels:

Azygos Vein. The azygos vein, whose name means "unpaired," ascends along the right or the center of the thoracic vertebral bodies. The azygos receives all of the right **posterior intercostal veins** (except the first), plus the subcostal vein. Superiorly, at the level of T$_4$, the azygos arches over the great vessels that run to the lung (the root of the right lung) and joins the superior vena cava.

Hemiazygos Vein. The hemiazygos vein (hem″ĭ-āz′ĭ-gos) ascends on the left side of the vertebral column (Figure 18.19). It corresponds to the inferior half of the azygos on the right (*hemiazygos* = half the azygos). This vein receives the ninth to eleventh left posterior intercostal veins and the subcostal vein. At about midthorax, the hemiazygos runs horizontally across the vertebrae and joins the azygos vein.

Accessory Hemiazygos. The accessory hemiazygos vein can be thought of as a superior continuation of the hemiazygos, receiving the fourth to eighth left

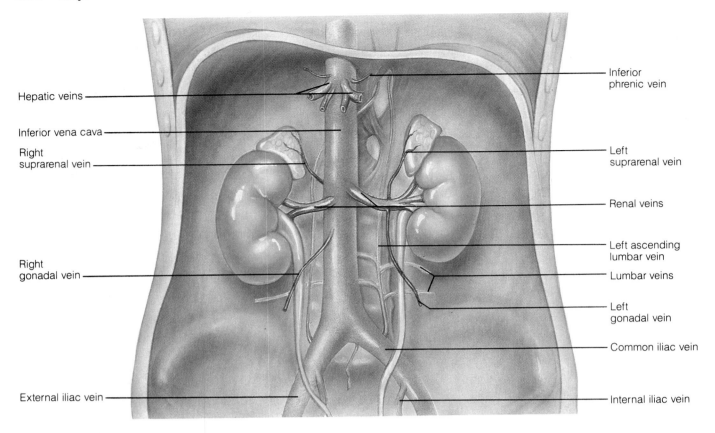

Hepatic veins

Inferior vena cava

Right
suprarenal vein

Right
gonadal vein

External iliac vein

Inferior
phrenic vein

Left
suprarenal vein

Renal veins

Left ascending
lumbar vein

Lumbar veins

Left
gonadal vein

Common iliac vein

Internal iliac vein

Figure 18.20   Tributaries of the inferior vena cava.

posterior intercostal veins. It also courses to the right to join the azygos.

The superior parts of the azygos and accessory hemiazygos veins also receive the small *bronchial veins* (not illustrated). These veins drain unoxygenated systemic blood from the bronchi in the lungs.

Veins of the Abdomen

Blood returning from the abdominopelvic viscera and the abdominal wall reaches the heart via the inferior vena cava (Figure 18.20). Most venous tributaries of this great vein share the names of the corresponding arteries.

Lumbar Veins.   Several pairs of lumbar veins drain the posterior abdominal wall. They run horizontally with the lumbar arteries.

Gonadal (Testicular or Ovarian) Veins.   The right and left gonadal veins ascend along the posterior abdominal wall with the gonadal arteries. The right vein drains into the anterior surface of the inferior vena cava at L_2. The left gonadal vein, by contrast, drains into the left renal vein.

Renal Veins.   The right and left renal veins drain the kidneys. Each lies just anterior to the corresponding renal artery.

Suprarenal Veins.   Although each adrenal gland has several main arteries, each has just one vein. The right suprarenal vein empties into the nearby inferior vena cava. The left suprarenal vein, by contrast, drains into the left renal vein.

Hepatic Veins.   The right and left hepatic veins exit the liver superiorly and empty into the most superior part of the inferior vena cava. These robust veins carry all the blood returning from the digestive organs in the abdominal and pelvic cavity. The hepatic veins are the end vessels of the hepatic portal system.

BEGINS +ENDS IN A CAPILLARY NETWORK

Hepatic Portal System.   The hepatic portal system is a specialized part of the vascular circuit that supplies the abdominal and pelvic digestive organs (Figures 18.21 and 18.22). This system of capillaries and veins serves a function unique to digestion: It picks up digested nutrients from the stomach and intestine and delivers these nutrients to the liver for processing and storage. As shown in Figure 18.21, capillaries in the stomach and intestine first receive the digested nutrients. These capillaries then drain into the tributaries of the **hepatic portal vein.** This vein delivers the nutrient-rich blood to capillaries in the liver (the *liver sinusoids*), through which nutrients reach the liver cells for processing. The liver cells also break down any toxins that entered the blood through the digestive tube. After passing through the liver sinusoids,

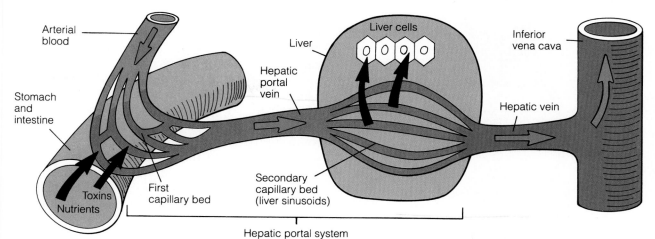

Figure 18.21   The basic scheme of the hepatic portal system and associated vessels. Nutrients and toxins are picked up from capillaries in the stomach and intestine and are transported to the liver cells for processing.

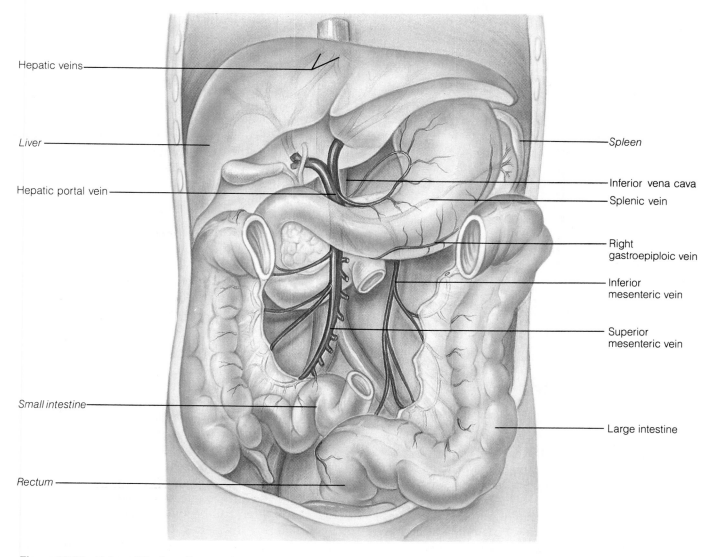

Figure 18.22   Veins of the hepatic portal system.

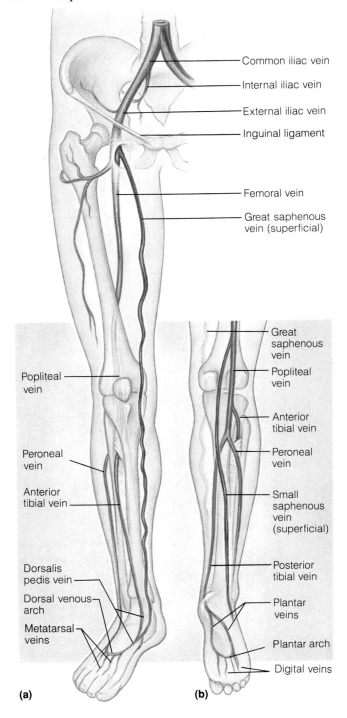

Common iliac vein

Internal iliac vein

External iliac vein

Inguinal ligament

Femoral vein

Great saphenous vein (superficial)

Popliteal vein

Peroneal vein

Anterior tibial vein

Dorsalis pedis vein

Dorsal venous arch

Metatarsal veins

(a)

Great saphenous vein

Popliteal vein

Anterior tibial vein

Peroneal vein

Small saphenous vein (superficial)

Posterior tibial vein

Plantar veins

Plantar arch

Digital veins

(b)

Figure 18.23   Veins of the right lower limb.   (a) Anterior view. (b) Posterior view of the leg and foot. The deep veins are actually double veins but are drawn as single veins for clarity.

the blood proceeds into the hepatic veins and inferior vena cava, thereby returning to the general systemic circulation. (For a more complete discussion of liver functions, see p. 586.)

The general definition of a portal system is a series of vessels in which *two* different capillary beds lie between the arterial supply and the final venous drainage. The hepatic portal system, as shown in Fig-

ure 18.21, neatly fits this definition. As you study the hepatic portal system, be careful not to confuse the *hepatic veins* with the *hepatic portal vein* (also called the *portal vein*).

The following veins of the hepatic portal system serve as tributaries of the hepatic portal vein (Figure 18.22):

Superior mesenteric vein. This large vein ascends just to the right of the superior mesenteric artery. It drains the entire small intestine, part of the large intestine (ascending and transverse colon), and some of the stomach. Its superior part lies posterior to the stomach and pancreas.

Splenic vein. This vessel runs horizontally, posterior to the stomach and pancreas. Its tributaries correspond to the branches of the splenic artery. Its right end joins the superior mesenteric vein to form the hepatic portal vein.

Inferior mesenteric vein. This vein ascends along the posterior abdominal wall, well to the left of the inferior mesenteric artery. Its tributaries drain the organs that are supplied by that artery (distal region of colon and superior rectum). The inferior mesenteric vein empties into the splenic vein posterior to the stomach and pancreas.

The hepatic portal vein itself is a short, vertical vessel that lies directly inferior to the liver and anterior to the inferior vena cava (Figure 18.22). Inferiorly, the portal vein begins posterior to the pancreas as the union of the superior mesenteric and splenic veins. Superiorly, it enters the underside of the liver. There the hepatic portal vein divides into right and left branches, whose smaller branches reach the liver sinusoids.

Along its way to the liver, the hepatic portal vein receives the right and left gastric veins from the stomach.

Veins of the Pelvis and Lower Limbs

Deep Veins.   Like those in the upper limb, most deep veins in the lower limbs share the names of the arteries they accompany, and all but the largest veins are double veins. Arising on the sole of the foot from the union of the medial and lateral **plantar veins,** the **posterior tibial vein** ascends deep within the calf muscles and receives the **peroneal vein** (Figure 18.23). The **anterior tibial vein** is the superior continuation of the dorsalis pedis vein of the foot. It ascends to the superior part of the leg, where it unites with the posterior tibial vein to form the **popliteal vein.** The popliteal vein ascends to become the **femoral vein,** which drains the thigh and becomes the **external iliac vein** upon entering the pelvis. In the pelvis, the external iliac vein unites with the **internal iliac vein** to form the **common iliac vein.**

Table 18.2   Major Systemic Veins Emptying into the Venae Cavae*

Vena cava	Major tributaries	Area drained
Superior vena cava	**Brachiocephalic**	Formed by union of internal jugular and subclavian on each side; the pair join to form the superior vena cava
	Internal jugular	Dural venous sinuses of the brain
	Vertebral	Posterior tissues of neck
	Subclavian	
	External jugular	Drains superficial tissues of head and neck
	Axillary	Continuation of subclavian; its tributaries together drain the shoulder, axilla, arm, and forearm. (Note that the median cubital vein connects the cephalic and basilic veins.)
	· Brachial	
	Radial	
	Ulnar	
	· Basilic†	
	· Cephalic†	
	Median cubital†	
	Median vein of the forearm†	
	Azygos	Vessels of the azygos system together drain much of the thorax wall
	Hemiazygos	
	Accessory hemiazygos	
Inferior vena cava	**Hepatic**	The hepatic portal system drains the digestive viscera via the superior and inferior mesenteric veins and splenic vein; that blood is shunted to the liver via the hepatic portal vein; the liver is drained by hepatic veins
	Liver sinusoids	
	Hepatic portal	
	Right suprarenal	Right adrenal gland
	Renal	Drain the kidneys; left renal vein also receives blood from left gonadal and left suprarenal veins
	Right gonadal	Right ovary or testis
	Lumbars (several pairs)	Posterior abdominal wall
	Common iliac	Together the common iliac veins receive the drainage of the pelvis and lower extremities
	Internal iliac	Pelvis; medial thigh
	External iliac	Lower limb
	Femoral	
	· Greater saphenous†	
	· Popliteal	
	Anterior tibial	
	Posterior tibial	
	Small saphenous†	

*Unless otherwise specified, each vein has right and left members.
†Indicates a superficial vein.

Superficial Veins.   Two large superficial veins, the *great* and *small saphenous veins* (sah-fe′nus; "obvious"), issue from the **dorsal venous arch** of the foot. The saphenous veins anastomose frequently with each other and with deep veins along their course. The **great saphenous vein** is the longest vein in the body. It ascends along the medial aspect of the entire limb to empty into the femoral vein just distal to the inguinal ligament. The **small saphenous vein** runs along the lateral side of the foot and then along the posterior calf. Posterior to the knee, it empties into the popliteal vein.

The great saphenous vein may be removed and sutured into the heart wall as a coronary bypass vessel. How must the surgeon proceed to accommodate this vein's numerous valves?

The surgeon must attach the vein to the heart *upside down,* so that the flow of the arterial blood opens, rather than closes, the valves.

The saphenous veins are more likely to become varicose than any other veins in the lower limb, because they are poorly supported by surrounding tissue. Furthermore, when valves begin to fail throughout the veins of a lower limb, the normal contractions of the leg muscles can squeeze blood out of the deep veins into the superficial veins through the anastomoses between these two groups of veins. This influx of blood engorges and weakens the saphenous veins even further. ■

✚ Portal-Systemic Anastomoses

About 10% of the adults in the United States (about 15 million people) are alcoholics. Advanced alcoholism leads to scarring and degeneration of the liver, called *cirrhosis,* which obstructs and slows the flow of blood through the liver sinusoids. This blockage raises the blood pressure throughout the hepatic portal system and results in **portal hypertension.** Fortunately, some veins of the portal system anastomose with veins that drain into the venae cavae, providing emergency pathways through which the "backed up" portal blood can return to the heart. These pathways are the *portal-systemic (portal-caval) anastomoses.* The main anastomoses are (1) veins in the inferior esophagus, (2) the hemorrhoidal veins in the wall of the anal canal, and (3) superficial veins in the anterior abdominal wall around the navel. These connecting veins are small, however, and they swell and burst when forced to carry large volumes of portal blood.

Can you imagine some symptoms exhibited by victims of cirrhosis as these anastomosing veins start to fail?

Alcoholics with cirrhosis of the liver may vomit blood (from torn esophageal veins), develop hemorrhoids (from swollen hemorrhoidal veins), and exhibit a snake-like network of distended veins through the skin around the navel. This network is called a *caput medusae* (kap'ut mĕ-du'se)—the Medusa head—after a monster in Greek mythology whose hair was made of writhing snakes. ■

The Blood Vessels Throughout Life

As we described in Chapter 16, the earliest blood vessels develop from blood islands around the yolk sac in the 2-week embryo. By the end of week 3, mesoderm *within* the embryo itself has begun to form blood vessels, which grow by sprouting extensions and joining together. At first, the vessels consist only of endothelium, but adjacent mesenchymal cells soon surround these tubes, forming the muscular and fibrous tunics of the vessel walls.

Fetal Circulation

By the start of the fetal period (month 3), the blood vessels are in place, and blood flows through each vessel in the same direction as it flows in adults. However, because the fetal cardiovascular system serves some functions that are unique to prenatal life, it contains some structures that are not present after birth (Figure 18.24). Most importantly, the fetus has large vessels that carry blood to and from the *placenta,* the organ that nourishes the developing fetus. The placenta resembles a thick pancake pushed against the lining of the mother's uterus. It is here that fetal blood picks up oxygen and nutrients and unloads its wastes through diffusional exchange with the mother's blood. However, fetal and maternal blood are not directly interconnected and do not mix.

Vessels To and From the Placenta

The fetal vessels that carry blood to and from the placenta are called *umbilical vessels* because they run within the umbilical cord (Figure 18.24a). The paired **umbilical arteries** branch from the internal iliac arteries in the pelvis and carry blood to the placenta for oxygenation. From the placenta, the unpaired **umbilical vein** returns this newly oxygenated blood to the fetus. The umbilical vein enters the umbilicus (navel) and ascends along the ventral abdominal wall to the liver. There, some of this returning placental blood empties into a branch of the hepatic portal vein, which carries it into the liver sinusoids to supply liver cells with nutrients obtained at the placenta. Most of the placental blood, by contrast, bypasses the liver sinusoids through a shunt called the **ductus venosus** ("venous duct"). In either case, all the placental blood proceeds to the inferior vena cava (via hepatic veins). In the inferior vena cava, this placental blood mixes with oxygen-poor blood returning from the lower parts of the body and proceeds directly into the heart's right atrium. The heart then pumps this blood throughout the systemic circuit, providing body tissues with oxygen and nutrients.

Most parts of the umbilical vessels are lost when the umbilical cord is cut after birth, but the parts that remain within the baby's body degenerate into fibrous bands called *ligaments* (Figure 18.24b). In children and adults, the remnant of the umbilical vein is the **ligamentum teres** ("round ligament"). The ductus venosus has become the **ligamentum venosum** on the liver's inferior surface. The umbilical arteries have become **medial umbilical ligaments** in the anterior abdominal wall inferior to the navel.

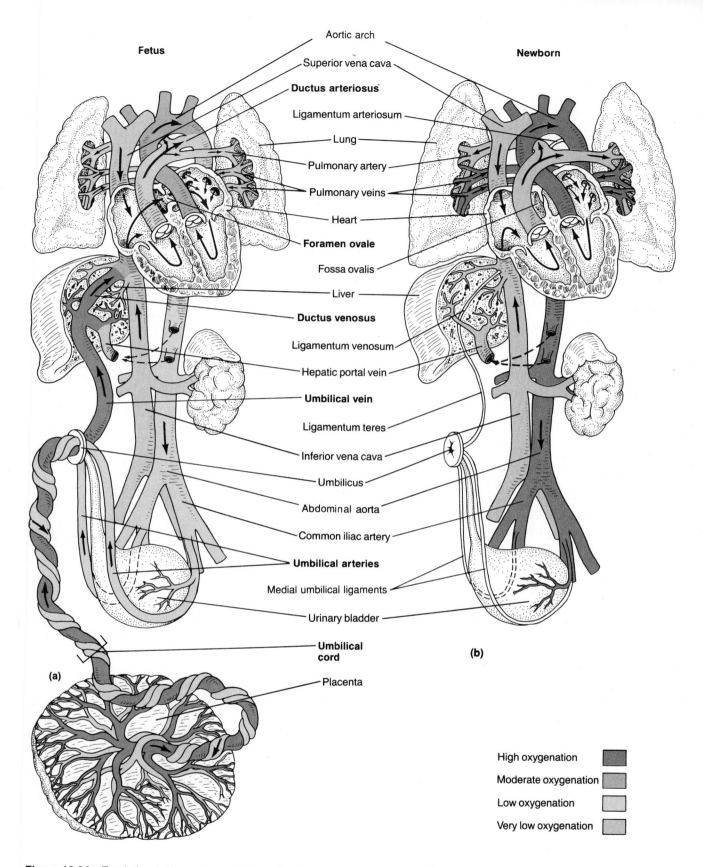

Fetus

Newborn

- Aortic arch
- Superior vena cava
- **Ductus arteriosus**
- Ligamentum arteriosum
- Lung
- Pulmonary artery
- Pulmonary veins
- Heart
- **Foramen ovale**
- Fossa ovalis
- Liver
- **Ductus venosus**
- Ligamentum venosum
- Hepatic portal vein
- **Umbilical vein**
- Ligamentum teres
- Inferior vena cava
- Umbilicus
- Abdominal aorta
- Common iliac artery
- **Umbilical arteries**
- Medial umbilical ligaments
- Urinary bladder
- **Umbilical cord**
- Placenta

(a)

(b)

High oxygenation

Moderate oxygenation

Low oxygenation

Very low oxygenation

Figure 18.24 Fetal circulation.   Arrows indicate the direction of blood flow. (**a**) Special adaptations for fetal life. The umbilical vein carries oxygen- and nutrient-rich blood from the placenta to the fetus. The umbilical arteries carry waste-laden, low-oxygen blood from the fetus to the placenta. The ductus arteriosus and foramen ovale shunt blood away from the nonfunctional lungs, and the ductus venosus enables much of the blood to bypass the liver circulation. The ductus venosus lies in a deep cleft between the two major lobes of the liver and is therefore a shunt *around* the liver. (**b**) Changes in the cardiovascular system at birth. The umbilical vessels close, as do the ductus venosus, ductus arteriosus, and foramen ovale.

511

Shunts Away From the Pulmonary Circuit

In the fetus, the respiratory organ is the placenta, not the lungs, which receive only a small trickle of blood necessary to support their growth. In both the fetus and the adult, however, the right atrium receives all blood returning from the systemic circuit, and the right ventricle pumps blood into the pulmonary trunk. This arrangement would seem to pose a problem to the fetus: Its right heart is "designed" to pump all systemic blood into the pulmonary circuit to the lungs—but its nonrespiratory lungs need little blood. Therefore, this blood is shunted to the left side of the heart and into the systemic circuit. Two shunts perform this function: the *foramen ovale* and the *ductus arteriosus.*

1. The foramen ovale. Recall from the previous chapter that there is an oval hole, the foramen ovale, in the interatrial septum of the fetal heart (Figure 18.24a). Each time the atria contract, the full right atrium squeezes some of its blood through this hole into the almost-empty left atrium. The foramen ovale is actually a valve with two flaps that prevent any blood from flowing in the other direction.

2. The ductus arteriosus. Most of the blood that enters the fetal right atrium continues into the right ventricle, which pumps it into the pulmonary trunk—seemingly on its way to overload the immature lungs. Handily, however, a wide arterial shunt connects the most cranial part of the pulmonary trunk to the adjacent aortic arch. This artery is the **ductus arteriosus.** Because of this shunt into the systemic circulation, only a small amount of blood proceeds to the pulmonary arteries and fetal lungs. Most of the blood continues into the aorta to nourish the tissues throughout the body, and some of this blood proceeds to the placenta to pick up more oxygen and nutrients.

Changes at Birth

At birth, the fetal circulatory pattern switches rapidly to the postnatal condition (Figure 18.24b). As the newborn takes its first breaths, the lungs receive more blood, and the ductus arteriosus constricts and closes. For the first time, oxygenated pulmonary blood begins pouring into the left atrium, raising the pressure within this chamber. The increased pressure pushes together the two valve flaps of the foramen ovale and closes the foramen. Both the foramen ovale and ductus arteriosus are now closed, and the postnatal circulatory plan is established.

Although the foramen ovale and ductus arteriosus close shortly after birth, they do not immediately fuse shut. It takes about 3 months for the ductus arteriosus to become the solid ligamentum arteriosum, and about a year for the flaps of the foramen ovale to fuse together.

Occasionally, either the ductus arteriosus or the foramen ovale stays patent (open) after birth. In such cases, the baby is blue from cyanosis. Patent ductus arteriosus and patent foramen ovale each characterize about 8% of all congenital heart defects, or 1 in every 1850 births. Both these conditions produce the same type of functional deficit. Do you know what this is?

They both lead to a mixing of oxygen-rich pulmonary blood with oxygen-poor systemic blood after birth. As a result, the blood reaching the tissues is not fully oxygenated. Still, the affected babies can survive, and surgeons usually wait until the child is a few years old (and can better withstand the dangers of thoracic surgery) before repairing the defects. ■

Blood Vessels in Adulthood

The most important aspect of the aging of the vascular system is the progression of atherosclerosis (p. 484). Because of the high fat content in the American diet, almost everyone in the United States develops atherosclerotic plaques in the major arteries before the age of 40. Although the degenerative process of atherosclerosis begins in youth, its consequences are rarely apparent until middle to old age, when it may lead to a myocardial infarction or stroke. Until puberty, the blood vessels of boys and girls look alike, but from puberty to about the age of 45 years, women have strikingly less atherosclerosis than men, probably because of protective effects of estrogen on a woman's blood vessels. This "gap between the sexes" closes between the ages of 45 and 65, and after age 65, the incidence of heart disease is the same in both sexes. Furthermore, when a woman does experience a heart attack, she is twice as likely to die from it as is a man. The reasons for this phenomenon are currently being investigated.

Related Clinical Terms

Aneurysm (an'u-rizm; "a widening") A balloon-like outpocketing of an artery (or vein) that places the vessel at risk for rupture. An aneurysm may reflect a congenital weakness of an artery wall or, more often, a gradual weakening of the vessel by hypertension or arteriosclerosis. The most common sites of aneurysms are the abdominal aorta and the arteries to the brain and kidneys.

Angiogram (an'je-o-gram") (*angio* = vessel; *gram* = writing) Diagnostic technique involving the infusion of a radiopaque substance into the bloodstream for X-ray examination of specific blood vessels.

Angiosarcoma Cancer originating from the endothelium of a blood vessel. Angiosarcoma may develop in liver vessels following exposure to chemical carcinogens.

Blue baby A baby with cyanosis (skin appears blue) due to relatively low levels of oxygen in the blood. This condition is caused by any congenital defect of the great vessels or heart that leads to low oxygenation of the systemic blood (ventricular septal defects, transposition of the great vessels, tetralogy of Fallot: p. 476). Patent foramen ovale and patent ductus arteriosus (p. 512) also produce this condition, as does failure of the lungs to inflate at birth.

Microangiopathic lesion An abnormal thickening of the basal lamina of capillaries, due to the deposit of glycoproteins. This results in thickened but leaky capillary walls. The condition is one of the hallmarks of long-standing diabetes mellitus.

Phlebitis (flĕ-bi'tis) (*phleb* = vein; *itis* = inflammation) Inflammation of a vein, accompanied by painful throbbing and redness of the skin over the inflamed vein. Phlebitis is most often caused by bacterial infection or local physical trauma.

Phlebotomy (flĕ-bot'o-me) (*tomy* = cut) An incision made in a vein for withdrawing blood or bloodletting.

Thrombophlebitis Undesirable intravascular clotting initiated by roughening of a venous lining. Thrombophlebitis often follows severe episodes of phlebitis. An ever-present danger is that the clot may detach and form an embolus.

Chapter Summary

■ PART 1: GENERAL STRUCTURE OF BLOOD VESSELS

1. The main types of blood vessels are arteries, capillaries, and veins. Arteries carry blood away from the heart and toward the capillaries. Veins carry blood away from the capillaries and toward the heart. Arteries "branch," whereas veins "serve as tributaries."

STRUCTURE OF BLOOD VESSEL WALLS (p. 481)

1. All but the smallest blood vessels have three tunics: tunica intima (with an endothelium), tunica media (mostly smooth muscle), and tunica adventitia (external connective tissue). The tunica media is thicker in arteries than in other classes of blood vessels.

2. Vasa vasorum are small arteries, capillaries, and veins that supply the outer wall of larger blood vessels.

CLASSES OF BLOOD VESSELS (pp. 482–488)

1. The three classes of arteries are elastic arteries, muscular arteries, and arterioles. Elastic (conducting) arteries are the largest arteries near the heart. They contain more elastin than does any other vessel type. The elastin in all arteries helps smooth the pressure pulses produced by the heartbeat.

2. Muscular (distributing) arteries carry blood to specific organs and organ regions. These arteries have a thick tunica media and are most active in vasoconstriction.

3. Arterioles, the smallest arteries, have one or two layers of smooth muscle cells in their tunica media. They regulate the flow of blood into capillary beds.

4. Capillaries, which have diameters just larger than erythrocytes, are endothelium-walled tubes arranged in networks called capillary beds. The most permeable capillaries have fenestrations (pores) in their endothelial cells. Other capillaries are continuous (no pores).

5. The basis of capillary permeability lies in (1) intercellular clefts between endothelial cells, (2) the pores of fenestrated capillaries, (3) direct diffusion of respiratory gases across the endothelium, and (4) endothelial vesicles.

6. Sometimes, blood is shunted straight through a capillary bed (from arteriole to metarteriole to thoroughfare channel to venule) and does not enter the true capillaries to the surrounding tissue. Precapillary sphincters determine how much blood enters the true capillaries.

7. Sinusoids are wide, twisted, leaky capillaries.

8. The smallest veins, postcapillary venules, function like capillaries.

9. Venous blood is under less pressure than arterial blood. Correspondingly, the walls of veins are thinner than those of arteries of comparable size. Veins also have a wider lumen, a thinner tunica media, and a thicker tunica adventitia.

10. Many veins have valves that prevent the backflow of blood. Varicose veins are veins whose valves have failed.

11. The joining together of arteries serving a common organ, called an arterial anastomosis, provides collateral channels for blood to reach the same organ. Anastomoses between veins are even more common than anastomoses between arteries.

■ PART 2: BLOOD VESSELS OF THE BODY

PULMONARY CIRCULATION (pp. 489–491)

1. The pulmonary trunk splits inferior to the aortic arch into the right and left pulmonary arteries. These arteries divide into lobar arteries, then branch repeatedly within the lung. Pulmonary venous tributaries empty newly oxygenated blood into the superior and inferior pulmonary veins of each lung. The four pulmonary veins extend from the lungs to the left atrium.

SYSTEMIC CIRCULATION (pp. 491–510)

1. Tables 18.1 and 18.2 summarize the main arteries and veins of the systemic circulation. The systemic circuit begins with the thoracic aorta (ascending, arch, descending) and ends with the two large venae cavae and the coronary sinus.

2. The hepatic portal system is a special subcirculation of veins that drain the digestive organs of the abdomen and pelvis. Figure 18.21 summarizes this system.

3. Cirrhosis of the liver leads to portal hypertension. Blood backs up through the portal-systemic anastomoses, overloading these delicate veins. This overload leads to esophageal bleeding, hemorrhoids, and caput medusae.

THE BLOOD VESSELS THROUGHOUT LIFE (pp. 510–512)

1. The first blood vessels develop from blood islands on the yolk sac in the 3-week embryo. Soon, vessels begin forming from mesoderm inside the embryo.

2. Blood vessels have the same positions in both the fetus and the newborn, and blood flows through the vessels in the same directions. However, the fetus also has vessels to and from the placenta (umbilical arteries, umbilical vein and ductus venosus) and shunts that bypass the almost-functionless pulmonary circuit (the foramen ovale in the interatrial septum and the ductus arteriosus between the pulmonary trunk and aortic arch).

3. Fetal shunts and vessels close shortly after birth. Failure of the foramen ovale or ductus arteriosus to close leads to a mixing of oxygenated and unoxygenated blood in the newborn.

4. The most important age-related vascular disorder is the progression of atherosclerosis.

Review Questions

Multiple Choice and Matching Questions

1. Which statement does not correctly describe veins? (a) They have less elastic tissue and smooth muscle than arteries. (b) They are subject to lower blood pressures than arteries. (c) They have larger lumens than arteries. (d) They always carry deoxygenated blood.

2. Which of the following tissues is mainly responsible for the vasoconstriction of arteries? (a) elastic tissue, (b) smooth muscle, (c) dense connective tissue, (d) adipose tissue.

3. Which of these layers of the artery wall thickens most in atherosclerosis? (a) tunica media, (b) tunica intima, (c) tunica adventitia, (d) tunica externa.

4. The structure of a capillary wall differs from that of an artery or vein in that (a) it has four tunics instead of three, (b) there is more smooth muscle, (c) it has only one layer, the endothelium, (d) it is much less leaky.

5. Blood flow through the capillaries is steady despite the rhythmic pumping action of the heart because of (a) the elasticity of the large arteries only, (b) the elasticity of all the arteries, (c) the ligamentum arteriosum, (d) the venous valves.

6. Fill in the blanks with the name of the appropriate vascular tunic (intima, media, or adventitia).

_____ **(1)** contains endothelium
_____ **(2)** tunic with the largest vasa vasorum
_____ **(3)** the thickest tunic in veins
_____ **(4)** the external tunic, mostly connective tissue
_____ **(5)** mostly smooth muscle
_____ **(6)** forms the valves of veins

7. Use logic to deduce the answer: Based on the vessels named *pulmonary trunk, thyrocervical trunk,* and *celiac trunk,* the name *trunk* must refer to (a) a vessel in the heart wall, (b) a vein, (c) a capillary, (d) a large artery from which other arteries branch.

8. Which of these vessels is bilaterally symmetrical (that is, an identical pair occurs on both sides of the body)? (a) internal carotid artery, (b) brachiocephalic artery, (c) azygos vein, (d) superior mesenteric vein.

9. Tracing the flow of arterial blood to the right hand, we find that blood leaves the heart and passes through the aorta, the right subclavian artery, the axillary and brachial arteries, and through either the radial or ulnar artery to a palmar arch. Which artery is missing from this sequence? (a) left coronary, (b) brachiocephalic, (c) cephalic, (d) right common carotid.

10. Which of the following do not drain directly into the inferior vena cava? (a) lumbar veins, (b) hepatic veins, (c) inferior mesenteric vein, (d) renal veins.

11. The profunda femoris (deep femoral) and profunda brachii (deep brachial) veins drain the (a) biceps brachii and hamstring muscles, (b) flexor muscles in the hand and leg, (c) triceps brachii and quadriceps femoris muscles, (d) intercostal spaces of the ribs.

12. The costocervical trunk (a) supplies some ribs and part of the neck, just as its name implies, (b) branches off the axillary artery, (c) supplies most of the muscles on the scapula, (d) gives off the internal carotid artery as its main branch.

13. A stroke that blocks a posterior cerebral artery will most likely affect (a) hearing, (b) seeing, (c) smell, (d) higher thought processes.

Short Answer and Essay Questions

14. What structural features seem to be responsible for the permeability of capillary walls?

15. Distinguish elastic arteries, muscular arteries, and arterioles, relative to their location, histology, and functions.

16. Name an organ containing fenestrated capillaries.

17. Physiologists often consider capillaries and postcapillary venules together as if they were identical. (a) What functions do these vessels share? (b) Structurally, how do they differ?

18. (a) Explain in your own words why varicose veins are more common in the lower limbs than elsewhere in the body. (b) Give a functional reason why valves are more abundant in the veins of the upper and lower limbs than in veins of the neck.

19. Are arteriosclerosis and atherosclerosis exactly the same? If not, explain the difference.

20. Jason told Megan that a sinusoid is nothing more than a cleft-like space in connective tissue. Is he right? Explain.

21. Sketch the arterial circle at the base of the brain, and label the important arteries that branch off this circle.

22. What is the name of the hole in the skull through which the internal jugular vein passes, and in what specific bone does it occur? (You may want to recheck Chapter 7.)

23. Name the vertebral level (T_1, L_3, and so on) at which each of the following arteries branches from the aorta: (a) inferior mesenteric, (b) renal, (c) celiac, (d) gonadal.

24. (a) Describe the common function of the foramen ovale and the ductus arteriosus in a fetus. (b) Contrast the location and function of the ductus venosus and ductus arteriosus.

25. (a) What are the two large tributaries that form the hepatic portal vein? (b) What is the function of the hepatic portal circulation? (c) Define a portal system.

26. List four different arteries to the stomach.

27. State the location of the azygos vein and the basic body regions drained by this vein.

Critical Thinking and Clinical Application Questions

28. In an eighth-grade health class, the teacher warned the students not to squeeze pimples or pluck hairs on their nose and upper lip. The students made fun of this warning, but they got the message when the teacher explained the danger triangle of the face. What is the danger of infections in this area?

29. Samantha received a small but deep puncture wound from broken glass in the exact midline of the anterior side of her distal forearm. She worried throughout the two-mile drive to the hospital that she would bleed to death because she had heard stories about people committing suicide by slashing their wrist. Look at Figure 18.10, and determine whether Sam's fear of bleeding to death is justified. Explain your reasoning.

30. Use logic to deduce which requires the more difficult and dangerous surgery on a child: closing a patent ductus arteriosus or closing a patent foramen ovale. Please explain your reasoning.

31. Your friend, who knows little about science, is reading a magazine article about a patient who had an "aneurysm at the base of his brain that suddenly grew much larger." The surgeons' first goal was to "keep it from rupturing," and the second goal was to "relieve the pressure on the brain stem and cranial nerves." The surgeons were able to "replace the aneurysm with a section of plastic tubing," so the patient recovered. Your friend asks you what all this means. Explain. (Hint: Start by checking this chapter's Related Clinical Terms.)

19

The Lymphatic System

Student Objectives

The Lymphatic Vessels (pp. 516–520)

1. Describe the structure and distribution of lymphatic vessels.

2. Explain the origin of lymph and the mechanisms of lymph transport.

3. List and explain the important functions of the lymphatic vessels.

Lymphoid Cells, Tissues, and Organs (pp. 520–529)

4. Describe the recirculation and activation of lymphocytes.

5. Relate the structure of lymphoid tissue to its infection-fighting function.

6. Describe the locations, histological structure, and functions of the lymph nodes, thymus, spleen, tonsils, Peyer's patches, and appendix.

The Lymphatic System Throughout Life (pp. 529–530)

7. Outline the development of the lymphatic vessels and lymphoid organs.

Having considered the cardiovascular system in the previous three chapters, we now turn to a closely related system, the lymphatic system. This system actually consists of two semi-independent parts: (1) the **lymphatic vessels**, which transport fluid that has escaped from the blood vessels back to the blood and (2) the **lymphoid tissues** and **organs**, which help provide the body's resistance and immunity to disease. Few people are familiar with the lymphatic system, yet it is of utmost importance to students entering the health professions: As we will see, the lymphatic vessels act as routes that disease organisms can use to travel throughout the body, while the lymphoid tissues and organs struggle to contain and destroy these organisms.

The Lymphatic Vessels

Basic Structure and Functions

An elaborate system of lymphatic vessels runs throughout the body. These vessels collect a fluid called **lymph** (*lympha* = clear water) from the loose connective tissue around the blood capillaries and carry this fluid to the great veins at the root of the neck (Figure 19.1). Since lymph flows only *toward* the heart, the lymphatic vessels form a one-way system rather than a full circuit.

There are several orders of lymphatic vessels (Figure 19.1b). The smallest vessels, those that first receive lymph, are the *lymph capillaries.* These drain into larger *lymphatic collecting vessels,* scattered along which are *lymph nodes.* The collecting vessels then drain into *lymph trunks,* which unite to form *lymph ducts,* which empty into the veins of the neck. We will consider each order of lymphatic vessel, but first let us explore the basic functions of the lymphatic system.

Recall from Chapter 4 that all blood capillaries are surrounded by a loose connective tissue that contains *tissue fluid* (see the "tissue space" in Figure 19.2a). This extravascular fluid arises from blood filtered through the capillary walls and consists of the small molecules of blood plasma (including water, ions, nutrient molecules, and respiratory gases). Tissue fluid is continuously leaving and re-entering the blood capillaries. For complex reasons, slightly *more* tissue fluid arises from the arteriole end ("upstream" end) of each capillary bed than reenters the blood at the venule end ("downstream" end). The lymphatic vessels function to collect this excess tissue fluid and

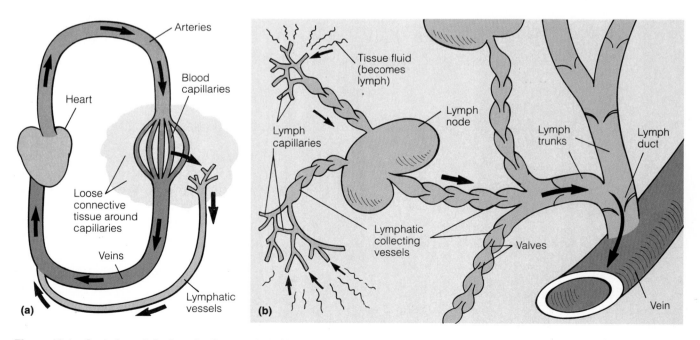

Figure 19.1 Overview of the lymphatic vessels. (**a**) Highly simplified scheme of the lymphatic vessels as they relate to the blood vessels of the cardiovascular circuit. (**b**) The types of lymphatic vessels and a lymph node. Lymph begins as tissue fluid in the loose connective tissue and ends up in the great veins at the root of the neck.

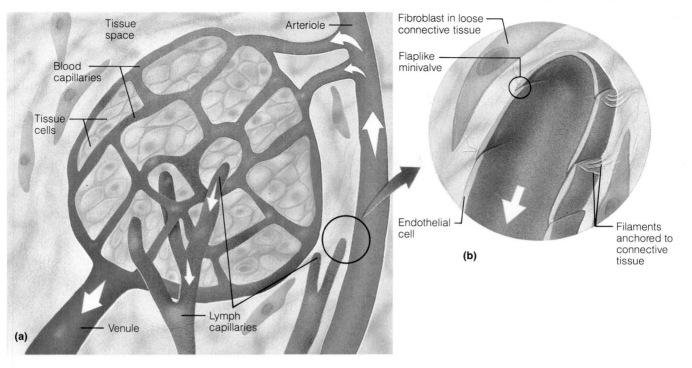

Figure 19.2   Distribution and structure of lymph capillaries.   (a) Structural relationship between blood capillaries and lymph capillaries. "Tissue space" is the matrix of loose connective tissue filled with tissue fluid, and "tissue cells" are any nearby cells nourished by the blood capillaries. (b) Lymphatic capillaries begin as blind-ended tubes. The endothelial cells of their walls loosely overlap one another, forming flap-like minivalves.

return it to the bloodstream. Indeed, any blockage of the lymphatic vessels causes the affected body region to swell with excess tissue fluid, a condition called *edema* (recall p. 91).

The lymphatic vessels also perform another function. Blood proteins leak slowly but steadily from blood capillaries into the surrounding tissue fluid, and *the lymph vessels return these leaked proteins to the bloodstream.* Recall that the proteins in blood generate osmotic forces that are essential for keeping water in the bloodstream (p. 446). If proteins were allowed to leak from the capillaries without compensation, a massive outflow of water would soon follow, and the entire cardiovascular system would collapse from lack of fluid in the blood.

Lymph Capillaries

Lymph capillaries, the permeable vessels that receive the tissue fluid, occur near blood capillaries in the loose connective tissue (Figure 19.2a). Like blood capillaries, they consist of a single layer of endothelial cells. Now known to be blind-ended vessels (Figure 19.2b), lymph capillaries are so remarkably permeable that they were once thought to be open at one end like a straw. We now know that their permeability re-

sults from the loose connections between the endothelial cells: These cells have few intercellular junctions, and the edges of adjacent cells overlap, forming easily opened valves. Bundles of fine filaments anchor the endothelial cells to the surrounding connective tissue. Therefore, any increase in the volume of the tissue fluid separates the valve flaps, exposing gaps in the wall and allowing the fluid to enter. Once this lymph enters the lymphatic capillaries, it cannot leak out, since backflow forces the valve flaps together.

The high permeability of lymph capillaries allows the uptake of large quantities of tissue fluid and large protein molecules. Unfortunately, any bacteria, viruses, or cancer cells in the loose connective tissue can also enter these permeable capillaries with ease. These pathogenic agents can then use the lymphatic vessels to travel throughout the body. However, this problem is partly solved by the lymph nodes, which destroy most pathogens in the lymph (see p. 524). ■

Lymph capillaries are widespread, occurring almost everywhere blood capillaries occur. However, lymph capillaries are absent from bone and teeth, bone marrow, and the entire central nervous system. Lymph vessels are not needed in the CNS, where excess tissue fluid drains through the nervous tissue into the cerebrospinal fluid. The cerebrospinal fluid then returns this tissue fluid to the blood at the superior sagittal sinus.

Highly specialized lymph capillaries called *lacteals* (lak′te-alz) are present in the finger-like villi of the intestinal mucosa. Lacteals receive digested fats from the intestine, which causes the lymph draining from the digestive viscera to become milky white (*lacte* means "milk"). This creamy lymph is called *chyle* (kīl; "juice"), and, like all lymph, it is carried to the bloodstream.

Lymphatic Collecting Vessels

From the lymph capillaries, lymph enters the **lymphatic collecting vessels**. These lymph vessels run along with corresponding blood vessels: In general, the superficial lymphatic vessels in the skin travel with superficial *veins,* whereas the deep lymphatic vessels of the trunk and digestive viscera travel with the deep *arteries.*

Lymphatic collecting vessels are narrow and delicate, so they usually are not seen in the dissecting laboratory. They exhibit the same tunics as blood vessels (tunica intima, tunica media, and tunica adventitia), but their walls are always much thinner. This thinness reflects the fact that lymph flows under very low pressure, since lymphatic vessels are not connected to the heart. To direct the flow of lymph, lymphatic collecting vessels contain more valves than do veins (Figure 19.1b). At the base of each valve, the vessel bulges, forming a pocket in which lymph collects to force the valve shut. Because of these bulges, the lymph vessel resembles a string of beads. This distinctive appearance allows physicians to recognize lymph vessels in X-ray films taken after these vessels are injected with radiopaque dye. This radiographic procedure is called **lymphangiography** (lim′fan″je-og′rah-fe; "lymph vessel picturing").

Unaided by the force of the heartbeat, lymph is propelled through lymph vessels by a series of weaker mechanisms. Both the bulging of contracting skeletal muscles and the pulsations of nearby arteries push on the lymph vessels, squeezing lymph through them. Additionally, the muscular tunica media of the lymph vessels contracts to help propel the lymph, and the normal movements of the limbs and trunk keep the lymph flowing. Despite these propulsion mechanisms, the transport of lymph is sporadic and slow. Many people whose jobs demand prolonged standing develop severe edema around the ankles by the end of the workday. The edema usually disappears if the legs are exercised (by walking home, for example). We all know people who bounce and wiggle their legs while sitting, a seemingly useless nervous habit that actually performs the important function of moving lymph up the legs.

Try to imagine the appearance of a body region (a limb, for example) whose lymph vessels have been blocked or removed. What would such a region look like?

The affected region would be swollen and puffy with edema. For example, in the tropical disease **elephantiasis** (el″ĕ-fan-ti′ah·sis), parasitic worms enter between the toes and move superiorly to block the lymph vessels that drain the lower limb or external genitalia. The resulting edema, along with a massive proliferation of fibrous (scar) tissue, produces a grossly enlarged lower limb or, in males, a swollen scrotum. During a radical mastectomy (removal of a cancerous breast), the lymph vessels and nodes that drain a woman's upper limb are removed at the axilla. Such women often suffer severe edema in the affected arm for several months, until the lymph vessels grow back. (Lymph vessels regenerate quite well and can even grow through scar tissue.) ■

Lymph Trunks

The largest lymphatic collecting vessels converge to form **lymph trunks** (Figures 19.3 and 19.4a). Lymph trunks drain large areas of the body and are large enough to be found by a skilled dissector. There are five major lymph trunks:

1. **Lumbar trunks.** These paired trunks lie along the sides of the aorta in the inferior abdomen. They receive all lymph draining from the lower limbs, the pelvic organs, and from some of the anterior abdominal wall.

2. **Intestinal trunk.** This unpaired trunk lies near the posterior abdominal wall in the midline. It receives fatty lymph (chyle) from the stomach, intestines, and other digestive organs.

3. **Bronchomediastinal** (brong″ko-me″de-ah-sti′-nal) **trunks.** Ascending near the sides of the trachea, these paired trunks collect lymph from the thoracic viscera and thoracic wall.

4. **Subclavian trunks.** Located near the base of the neck, these paired trunks receive lymph from the upper limb. They also drain the inferior neck and the superior thoracic wall.

5. **Jugular trunks.** Located at the base of each internal jugular vein, these paired trunks drain lymph from the head and neck.

Lymph Ducts

The lymph trunks drain into the largest lymphatic vessels, the **lymph ducts** (Figures 19.3 and 19.4). Some people have two lymph ducts, while others have just one.

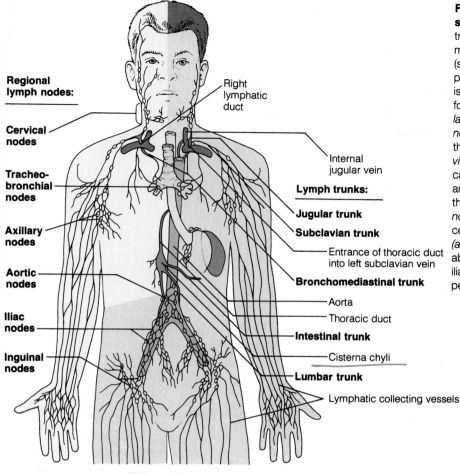

Regional lymph nodes:

Cervical nodes

Tracheo-bronchial nodes

Axillary nodes

Aortic nodes

Iliac nodes

Inguinal nodes

Right lymphatic duct

Internal jugular vein

Lymph trunks:

Jugular trunk

Subclavian trunk

Entrance of thoracic duct into left subclavian vein

Bronchomediastinal trunk

Aorta

Thoracic duct

Intestinal trunk

Cisterna chyli

Lumbar trunk

Lymphatic collecting vessels

Figure 19.3   The major lymphatic vessels and lymph nodes.   Note the lymph trunks and ducts in the thorax and abdomen. The superior right quarter of the body (shaded green) is drained by the right lymphatic duct, while the remainder of the body is drained by the thoracic duct. Also note the following lymph nodes: The superficial *axillary nodes* in the armpit and the *inguinal nodes* in the superior thigh filter lymph from the upper and lower limbs, respectively. *Cervical nodes* along the jugular veins and carotid arteries receive lymph from the head and neck. Nodes in the mediastinum of the thorax, such as the deep *tracheobronchial nodes,* receive lymph from the thoracic viscera. Deep nodes along the abdominal aorta *(aortic nodes)* drain lymph from the posterior abdominal wall, and deep nodes along the iliac arteries *(iliac nodes)* drain lymph from pelvic organs and the lower limb.

Right jugular trunk

Right lymphatic duct

Right subclavian trunk

Right subclavian vein

Right bronchomediastinal trunk

Brachiocephalic veins

Superior vena cava

Azygos vein

Cisterna chyli

Right lumbar trunk

(a)

Left jugular trunk

Internal jugular veins

Left subclavian trunk

Left subclavian vein

Left broncho-mediastinal trunk

Entrance of thoracic duct into left subclavian vein

Esophagus

Trachea

Ribs

Thoracic duct

Hemiazygos vein

Left lumbar trunk

Intestinal trunk

Inferior vena cava

Azygos vein on vertebral bodies

Thoracic duct

Aorta

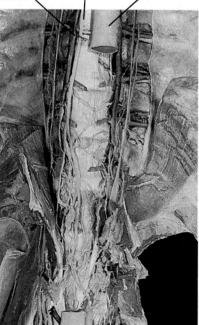

(b)

Figure 19.4   Lymphatic trunks and ducts.   (a) Enlarged anterior view of the posterior thoracic and abdominal walls, showing the major lymph trunks and ducts. In this specimen, the thoracic duct drains into the left subclavian vein, but in other specimens the duct drains into the left brachiocephalic or left internal jugular vein. **(b)** Photograph of the thoracic duct on the vertebral column.

1. **Thoracic duct.** The thoracic duct occurs in everyone. Its most inferior part, at the union of the lumbar and intestinal trunks, is the **cisterna chyli** (sis-ter'nah ki'li; "sac of chyle"). The cisterna chyli lies on the bodies of vertebrae L_1 and L_2. From there, the thoracic duct ascends along the vertebral bodies. In the superior thorax, it turns to the left and empties into the venous circulation at the junction of the left internal jugular and left subclavian veins (Figure 19.4a). The thoracic duct is often joined near its end by the left jugular, subclavian, and/or bronchomediastinal trunks. Alternately, any or all of these three lymph trunks can empty separately into the nearby veins.

2. **Right lymphatic duct.** The *right* jugular, subclavian, and bronchomediastinal trunks empty into the neck veins at or near the junction of the right internal jugular and subclavian veins. In some people, each of these trunks enters the veins separately, but in others, the three right lymph trunks join to form the **right lymphatic duct**, which alone empties into the neck veins.

Overall, the thoracic duct may drain lymph from fully three-fourths of the body. The right lymphatic duct, if present, drains the superior right quarter of the body (Figure 19.3).

In summary, the lymphatic vessels (1) return excess tissue fluid to the blood, (2) return leaked proteins to the blood, and (3) carry absorbed fat from the intestine to the blood (through lacteals). Additionally, the lymph nodes along the lymph vessels fight disease. This function is discussed further in the next section.

Lymphoid Cells, Tissues, and Organs

The *lymphoid organs*—the lymph nodes, the spleen, the thymus, the tonsils, the aggregated lymphoid follicles in the small intestine (Peyer's patches), and the appendix—are important components of the immune system (Figure 19.5). These organs consist largely of *lymphoid tissue,* which houses and helps to produce *lymphocytes,* the main cells of the immune system. Lymphocytes are essential in the body's defense against infectious microorganisms and cancer, and they also provide long-term immunity to disease. Although all lymphoid organs have roles in protecting the body, only the lymph nodes filter lymph.

We will investigate the component parts of the lymphoid organs—the lymphocytes and lymphoid tissue—before considering the lymphoid organs themselves. This plan will also give us a chance to explore some basic concepts of immunity.

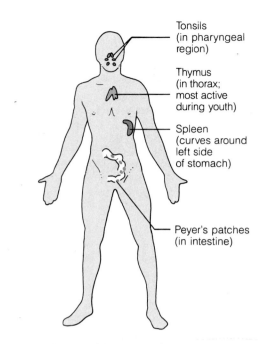

Figure 19.5 Lymphoid organs. Location of the tonsils, spleen, thymus, and Peyer's patches.

Lymphocytes

Function and Recirculation of Lymphocytes

Infectious microorganisms, such as bacteria and viruses, often enter the body by penetrating the epithelial surfaces and then proliferate in the underlying loose connective tissues. There, these invaders are fought by the inflammatory response; by nonspecific phagocytes, such as macrophages (Figure 19.6); and finally by **lymphocytes,** the most effective of our defense cells. Recall from Chapter 16 (p. 450) that lymphocytes are white blood cells and that each lymphocyte recognizes its own type of foreign molecule, called an *antigen.* Also recall that *B lymphocytes* multiply to become plasma cells that secrete antibodies, while *cytotoxic T lymphocytes* destroy antigen-bearing cells by penetrating their membranes with a "lethal hit" (Figure 16.6, p. 451). These B and T cells travel in the blood and lymph streams to reach and protect loose connective tissues throughout the body. They continuously enter and exit these connective tissues (including the often-infected lymphoid connective tissue) by squeezing through the walls of capillaries and venules. This repeated movement of lymphocytes between the circulatory vessels and the connective tissues, called **recirculation**, ensures that lymphocytes reach infection sites quickly.

Activation of Lymphocytes

Immature lymphocytes must pass through several stages before they are ready to attack antigens (Figure 19.7). Most lymphocytes pass through these activation stages during infancy and childhood, but many are activated in adulthood as well. The stages are:

Origin. Like all blood cells, lymphocytes originate from hematopoietic stem cells in bone marrow.

Gaining immunocompetence. From the bone marrow, some immature lymphocytes (lymphoid stem cells) enter the bloodstream and travel to the thymus. These cells, which will become T lymphocytes (*T* stands for *thymus*), stay in the thymus for 2 to 3 days, dividing rapidly. As they divide, each T lymphocyte gains the ability to recognize its specific antigen type—even though it has not yet encountered that antigen. This ability to recognize a unique antigen is called *immunocompetence.* B lymphocytes also develop immunocompetence, but the site where this development occurs was long a mystery. In birds, it occurs in a pouch off the intestine called the bursa of Fabricius, and immunologists long sought a "bursa equivalent" in humans (the B in *B cell* stands for *bursa*). Techniques for labeling B lymphocytes in their earliest stages have finally shown that human B cells gain immunocompetence while they are still in the bone mar-

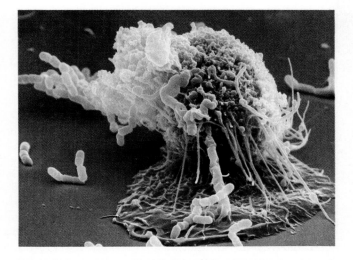

Figure 19.6   A macrophage phagocytizing bacteria. Recall that macrophages are the general phagocytic cells of the body, engulfing and digesting a wide variety of foreign molecules and cells. In this scanning electron micrograph (4300 ×), the macrophage is pulling sausage-shaped *Escherichia coli* bacteria toward it with its long cytoplasmic extensions. Macrophages digest the phagocytized organisms into antigen molecules, which they then present to T lymphocytes, an important step in lymphocyte activation (see p. 522).

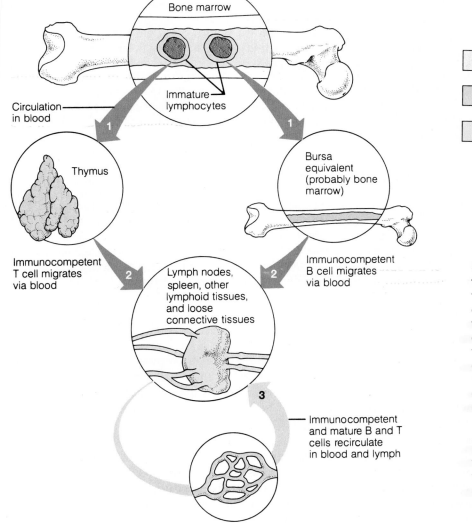

Figure 19.7   Lymphocyte activation. Immature lymphocytes originate in the bone marrow (top). (1) Immature lymphocytes destined to become T cells migrate to the thymus, where they develop immunocompetence. B cells, by contrast, remain in the bone marrow while developing immunocompetence. (2) After leaving the thymus or bone marrow, immunocompetent B and T cells "seed" the body's loose connective tissues, especially the lymphoid tissues in lymphoid organs such as the lymph nodes and spleen. Antigen challenge occurs in this connective tissue, and lymphocytes become fully activated (mature). (3) Immunocompetent and mature lymphocytes recirculate continuously in the bloodstream and lymph and throughout the lymphoid tissues of the body.

In October 1347, several ships made port in Sicily, and within days all of the sailors they carried were dead of bubonic plague. By the end of the fourteenth century, approximately 70% of the population of Europe had been wiped out by this "Black Death." In January 1987, the U.S. Secretary of Health and Human Services warned that *acquired immune deficiency syndrome (AIDS)* might be the plague of the twentieth century. These are strong words, but are they true?

Although AIDS was first identified in the United States in 1981 among homosexual men and intravenous drug users of both sexes, it had begun to afflict the heterosexual population of Africa several years earlier. In fact, the earliest proven AIDS death has now been traced back to 1959. Symptomatic AIDS has two forms: (1) AIDS-related complex (ARC) is characterized by severe weight loss, night sweats, and swollen lymph nodes. (2) In full-blown AIDS, these same symptoms are accompanied by increasingly frequent opportunistic infections—including a rare type of pneumonia called *Pneumocystis carinii pneumonia* and the bizarre malignancy *Kaposi's sarcoma*—in people with no previous history of immune deficiency disease. (Kaposi's sarcoma is a cancer or cancer-like condition of the endothelial cells of blood vessels, evidenced by purple lesions of the skin.) Some AIDS victims develop slurred speech and severe dementia. The course of AIDS is inexorable, ending in complete debilitation and death from cancer or overwhelming infection.

AIDS is caused by the human immunodeficiency virus (HIV), which is transmitted solely through body secretions—blood, semen, and possibly vaginal secretions. Most commonly, HIV enters the body via blood transfusions or blood-contaminated needles and during intimate sexual contact in which the mucosa is torn (and bleeds) or where open lesions induced by sexually transmitted diseases allow the virus access to the blood. Although HIV has also been detected in saliva and tears, it is not believed to be spread via those secretions.

HIV destroys helper T cells, resulting in depression of cell-mediated immunity. (Helper T cells are the cornerstone of the immune system, for without them lymphocytes cannot be activated.) Although antibody levels and the activation of cytotoxic T cells initially rise in response to exposure to the virus, in time a profound deficit of normal antibodies develops, and the cytotoxic T cells become unresponsive to viral cues. The whole immune system is turned topsy-turvy. Although the helper T cells are the main targets of HIV, macrophages, neutrophils, and brain neurons are also at risk for HIV infection. The virus invades the brain, perhaps via infected macrophages.

This invasion accounts for the neurological symptoms (dementia) of some AIDS patients. Although there are exceptions, most AIDS victims die from within a few months to 8 years after diagnosis.

There are probably 100 asymptomatic carriers of the virus for every diagnosed case of AIDS: The virus may lurk undetectable in T cells and macrophages, and the disease has a long incubation period (from a few months to 10 years) between the initial exposure and the appearance of clinically recognizable symptoms. Not only has the number of identified cases jumped exponentially in the at-risk populations, but the "face of AIDS" is also changing. Victims now include people who do not belong to the original high-risk groups. Before reliable testing of donated blood became available, some people contracted the virus from blood transfusions. The virus can also be transmitted from an infected mother to her fetus. Although homosexual men still account for the bulk of cases in which the virus is transmitted by sexual contact, more and more heterosexuals are contracting this disease. Particularly disturbing is a rapid increase in diagnosed cases among teenagers. The number of cases in inner-city ghettos, the corridors of poverty where intravenous drug use is the chief means of AIDS transmission, is alarming. The drug community accounts for about 25% of all

row. Thus, our bursa-equivalent is probably the bone marrow.

Antigen challenge. B and T lymphocytes become fully activated only after entering the bloodstream, traveling to an infected loose connective tissue, and encountering their antigen there. This initial encounter between lymphocyte and antigen is the *antigen challenge.* In this encounter, the antigen must bind to specific receptor molecules on the plasma membrane of the lymphocyte. For a T lymphocyte to be activated, the antigen must be presented by an *antigen-presenting cell,* usually a macrophage that has recently phagocytized microorganisms containing that antigen (Figure 19.6). As the macrophage presents the antigen, it secretes a chemical that signals the T cell to divide rapidly and gain full functional activity. Another, distinct type of lymphocyte, called a **helper T cell**, participates in this stage of lymphocyte activation: Helper T cells secrete another chemical that greatly stimulates the proliferation of B and cytotoxic T lymphocytes.

AIDS cases, and the bulk of AIDS cases in newborns occur where drug abuse abounds. This shocking revelation prompted the National Research Council to recommend that the government provide sterile needles to users of intravenous drugs in an attempt to suppress that means of transmission. This suggestion has proved controversial, however.

Though the disease is not spread by casual contact, some extremists are demanding that AIDS-infected children be kept out of the schools. Some employers are demanding AIDS testing before hiring, and insurance companies are denying insurance to those who will not submit to an AIDS test.

Although the diagnostic tests to identify carriers of the AIDS virus have become increasingly sophisticated and home screening tests are being developed, no cure has yet been found. However, over 100 drugs are now in the FDA pipeline, and vaccines to protect against AIDS infection now seem possible by the turn of the century. AZT (zidovudine), a drug fully approved for AIDS treatment, prolongs life by inhibiting viral replication. Several new and still experimental drugs appear to be valuable anti-AIDS agents. Transfer factor, one of the lymphokines (immune-activating chemicals) released by cytotoxic T cells, has shown promise in ARC patients. Synthetic chemicals that bind to the HIV virus and block its ability to attach to the body's cells, and drugs that inactivate the enzymes needed for HIV replication, also look promising.

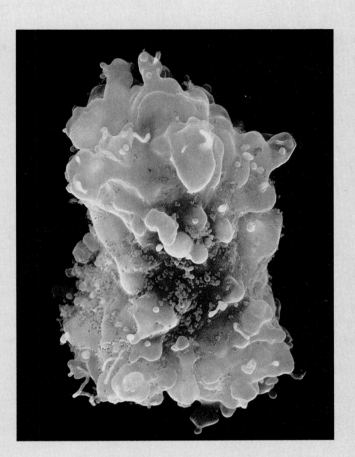

AIDS viruses attack a T cell: The HIV viruses, artificially colored blue in this scanning electron micrograph, surround a T lymphocyte. The virus invades helper T cells, killing them and seriously weakening the immune system of the infected person (14,000 ×).

Besides the drugs undergoing clinical tests, a number of nonapproved drugs are being prescribed or independently sought out and purchased in other countries by activist groups of AIDS patients. Some of the nonapproved techniques will prove useless, such as the controversial hyperthermia technique in which an AIDS patient's blood is run through a heater (raising the temperature to 115°F) and then back into the body. Other techniques will have harmful effects, but when life itself is at stake, this may be a small gamble. Perhaps, as urged in the media, the best defense is to practice "safe sex" by using condoms and knowing one's sexual partner. The alternative is abstinence.

Can you imagine what would happen if a virus were selectively to destroy helper T cells throughout the body?

This is the *human immunodeficiency virus,* which causes acquired immune deficiency syndrome (AIDS). With the destruction of the helper T cells, lymphocyte production declines and the immune system is greatly weakened. The patient becomes vulnerable to infectious organisms that seldom affect people with healthy immune systems. AIDS patients typically die from these infections within a few years of the first appearance of symptoms. AIDS is considered in more detail in the box above. ■

As a newly activated T or B cell divides within the loose connective tissue, it produces many copies of itself, all of which recognize the same antigen as did the original cell. This cell division produces two types of cells: effector and memory lymphocytes. The short-lived *effector lymphocytes* mount an attack against the pathogen and then die. For example, effector B lymphocytes become plasma cells that secrete antibodies and die in a few weeks. *Memory lymphocytes,* by contrast, remain inactive during the current infection and wait until the body encounters their specific antigen again—maybe decades later. When a memory lymphocyte finally encounters its antigen, its proliferative response and its attack are most vigorous and rapid. Memory lymphocytes are the basis of acquired immunity; that is, they prevent us from getting measles, mumps, and chickenpox (and many other diseases) more than once. Most of the recirculating lymphocytes that patrol the body are memory lymphocytes. Memory lymphocytes come in both B and T varieties.

Lymphoid Tissue

Lymphoid tissue is a reticular connective tissue. It is the type of loose connective tissue where (1) most infections occur, (2) most infectious microorganisms are destroyed, (3) most recirculating lymphocytes gather, (4) most antigen challenges occur, and (5) most effector and memory lymphocytes are produced. Lymphoid tissue dominates all the lymphoid organs except the thymus, and it also occurs in localized patches in the walls of the digestive and respiratory tubes. The structural features of lymphoid tissue serve its infection-fighting role. Its basic tissue framework is a network of reticular fibers secreted by reticular cells (fibroblasts). Within the spaces of this network reside vast numbers of T and B lymphocytes that arrive continuously from the capillaries coursing through this tissue. Macrophages on the fiber network kill invading microorganisms by phagocytosis and activate nearby lymphocytes by presenting them with antigens.

Lymphoid tissue contains scattered, spherical clusters of densely packed lymphocytes, called lymphatic (lymphoid) **follicles** or *nodules* (see Figure 19.8b and c). These follicles often exhibit lighter-staining centers, called **germinal centers**. Each follicle derives from the activation of a single memory B cell, whose rapid division generates the thousands of lymphocytes in the follicle. Newly produced B cells migrate away from the follicle to become plasma cells.

Lymphoid Organs

Lymph Nodes

Lymph nodes (Figure 19.8) are bean-shaped organs situated along the lymphatic collecting vessels. Non-scientists call them "lymph glands"—but this term is not correct. There are about 500 lymph nodes in the human body, ranging from 1 to 25 mm in diameter (up to 1 inch). Large clusters of lymph nodes occur near

the body surface in the inguinal, axillary, and cervical regions. These superficial nodes, plus some important groups of deep nodes, are pictured and described in Figure 19.3.

Lymph nodes have one basic function: They destroy infectious microorganisms and cancer cells in the lymph, effectively preventing them from using lymph vessels to spread throughout the body. Let us now examine the histological structure of lymph nodes to understand how they perform this function.

Each lymph node is surrounded by a fibrous **capsule** of dense connective tissue (Figure 19.8b). Fibrous strands called **trabeculae** (trah-bek'u-le; "beams") extend inward from the capsule to divide the node into compartments. Lymph enters the convex aspect of the node through several **afferent lymphatic vessels**. It exits the indented region on the other side, the **hilus** (hi'lus), through **efferent lymphatic vessels**. Within the node, lymph percolates through **lymph sinuses.** These sinuses are large lymph capillaries spanned internally by a network of reticular fibers lined by endothelial cells. Many macrophages live on this fiber network, phagocytizing the disease organisms and foreign particles in the lymph that flows through the sinuses. In general, all lymph filters through several nodes, so the lymph is usually free of pathogens by the time it reaches the great veins of the neck.

Lymph nodes are more than just lymph filters. The regions between the lymph sinuses are filled with typical lymphoid tissue, which constitutes most of the mass of the node. As the lymph percolates through the lymph sinuses, some of the contained antigens leak into this lymphoid tissue. There, the antigens activate lymphocytes and add to the body's valuable supply of memory lymphocytes.

Lymph nodes have two histologically distinct regions, an external **cortex** ("outer bark") and a **medulla** ("middle") near the hilus. All follicles and most B cells occupy the lymphoid tissue of the most superficial part of the cortex. The rest of the cortical lymphocytes are primarily T cells. Thin, inward extensions from the cortical lymphoid tissue, called **medullary cords,** help define the medulla. The medullary cords contain both T and B lymphocytes, plus plasma cells.

Sometimes, lymph nodes are overwhelmed by the very agents they are trying to destroy. For example, when large numbers of bacteria or viruses are trapped in the nodes, the nodes become enlarged, inflamed, and very tender to the touch. Such an infected lymph node is called a *bubo* (bu'bo). Buboes are the most obvious symptom of bubonic plague, the "Black Death" that killed over 70% of Europe's population in the 1300s. Furthermore, metastasizing cancer cells that enter lymph vessels and are trapped in the local lymph nodes continue to multiply there. The fact that

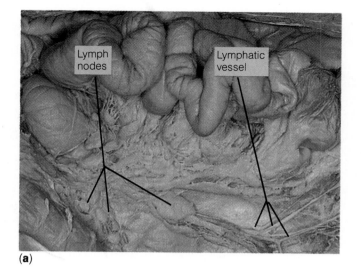

(a)

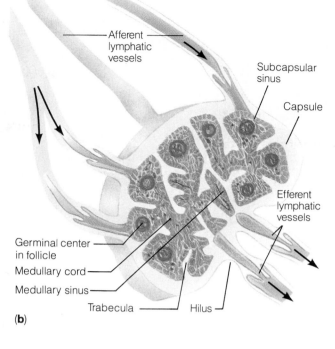

(b)

Figure 19.8 Structure of a lymph node. (a) Photograph of lymph nodes and associated lymphatic vessels that drain the intestines. (b) Diagram of a lymph node (longitudinal section) and the associated lymphatic vessels. The lymphatic vessels and lymph sinuses are colored green, and the lymphoid tissue is blue. Lymphatic follicles occur in the outer cortex region of the node. The medulla is the internal region that contains the medullary cords and medullary sinuses. Arrows indicate the direction of lymph flow into and out of the node. (c) Photomicrograph of a part of a lymph node (28 ×).

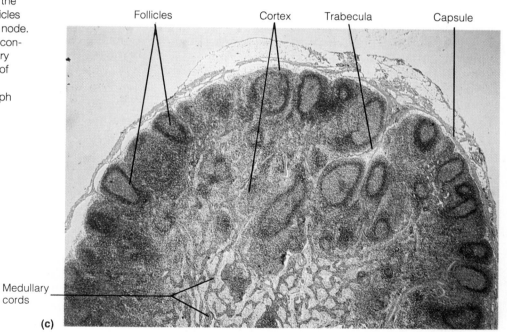

(c)

cancer-infiltrated lymph nodes are swollen but not painful helps to distinguish cancerous nodes from those infected by microorganisms. (Pain results from inflammation, and cancer does not induce the inflammatory response.) Potentially cancerous lymph nodes can be located by palpation. For example, a physician examining a patient for breast cancer feels for swollen axillary lymph nodes. Physicians can also locate enlarged, cancerous lymph nodes by using CT scans and lymphangiograms. ■

Spleen

The soft, blood-rich **spleen** (Figures 19.9 and 19.10) is the largest lymphoid organ. Its size varies greatly among different people, but on average the spleen is the size of a fist. This unpaired organ is located in the superior left part of the abdominal cavity just posterior to the stomach (Figure 19.5). The spleen is roughly egg-shaped, although its anterior surface is concave rather than convex. The large *splenic vessels* enter and exit the anterior surface along a line called

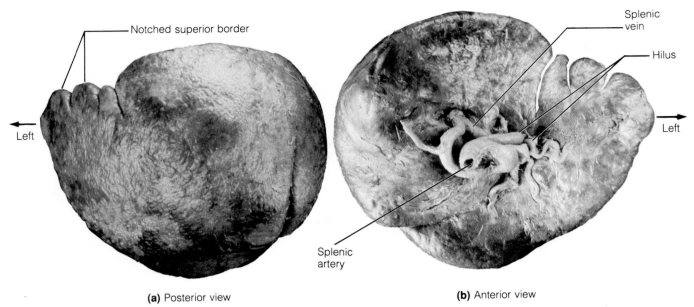

(a) Posterior view

(b) Anterior view

Figure 19.9   Photograph of the spleen.   Note that the long axis of the spleen, with the hilus, is usually oriented horizontally, not vertically.

the **hilus.** (*Hilus* literally means the point where a seed attaches to its stalk, a good analogy.)

The spleen has two main blood-cleansing functions: (1) It removes blood-borne antigens (this is its immune function), and (2) it removes and destroys aged or defective blood cells. Additionally, the spleen is a site of hematopoiesis in the fetus and stores blood platelets throughout life.

The spleen, like the lymph nodes, is surrounded by a fibrous capsule from which trabeculae extend inward (Figure 19.10b). The larger branches of the splenic artery run in the capsule and trabeculae, and send smaller arterial branches into the substance of the spleen. These branches are called **central arteries** because they are enclosed by (lie near the center of) thick sleeves of lymphoid tissue. Collectively, all the sleeves of lymphoid tissue make up the **white pulp** of the spleen. Blood-borne antigens enter this lymphoid tissue to be destroyed as they activate the immune response. Surrounding the white pulp is **red pulp**. This contains venous sinuses (blood sinusoids) and a reticular connective tissue that is exceptionally rich in macrophages **(the splenic cords)**. Whole blood leaks from the sinuses into this connective tissue, whose macrophages then phagocytize any defective blood cells. Hence, red pulp is responsible for the spleen's ability to dispose of worn-out blood cells, whereas white pulp serves the immune function of the spleen.

In histological sections (Figure 19.10c), the white pulp forms what appear to be islands in a sea of red pulp. The naming of the pulp regions reflects their appearances in fresh spleen tissue rather than in stained histological sections. With many stains, the white pulp actually appears darker than the red pulp.

Splenomegaly (sple″no-meg′ah-le) (*mega* = big), or enlargement of the spleen, usually results from blood diseases such as mononucleosis, malaria, leukemia, and polycythemia vera. A simple way to identify an enlarged spleen is to feel for the spleen's notched superior border through the skin of the abdominal wall just anterior to the costal margin. A healthy spleen will never reach this far anteriorly.

Because the capsule of the spleen is relatively thin, physical injury or a serious infection may cause the spleen to rupture. Can you imagine some effects of a ruptured spleen?

Rupture of the spleen leads to severe internal bleeding, so the spleen must be removed quickly and its artery tied off, a surgical procedure called a **splenectomy**. A person can live a relatively healthy life without a spleen, because macrophages in the bone marrow and liver can assume most of the spleen's functions. Such an individual will be more susceptible to infections, however, so surgeons performing splenectomies now leave some of the spleen in place if possible. Alternatively, healthy fragments of the spleen are reimplanted immediately after the operation. In such cases, the spleen can sometimes regenerate. ∎

Thymus

The two-lobed **thymus** lies in the superior thorax, just posterior to the sternum (Figure 19.11a). As we discussed earlier, the thymus programs immature lymphocytes into T lymphocytes. More specifically, it secretes the hormones thymosin and thymopoietin,

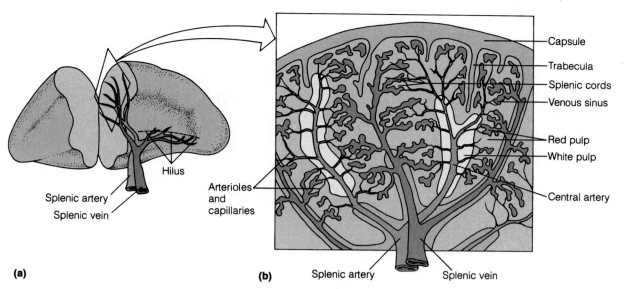

- Capsule
- Trabecula
- Splenic cords
- Venous sinus
- Red pulp
- White pulp
- Central artery

Hilus

Splenic artery

Splenic vein

Arterioles and capillaries

Splenic artery

Splenic vein

(a)

(b)

Figure 19.10   Structure of the spleen. (a) Gross structure, anterior view. (b) Diagram of the histological structure of the spleen. (c) Photomicrograph of spleen tissue (85 ×). The white pulp, a lymphoid tissue with many lymphocytes, is surrounded by red pulp containing abundant erythrocytes.

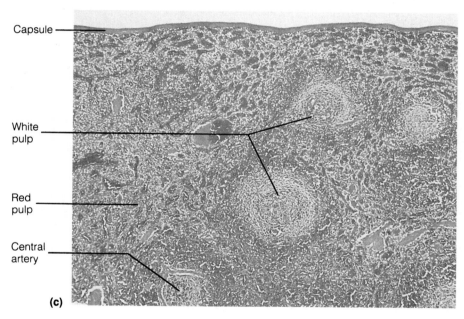

Capsule

White pulp

Red pulp

Central artery

(c)

which cause T lymphocytes to gain immunocompetence. Prominent in newborns, the thymus continues to increase in size during childhood, when it is most active. During adolescence, it starts to atrophy gradually. By old age, it is almost entirely replaced by fibrous and fatty tissue and is difficult to distinguish from the surrounding connective tissue. Although it atrophies with age, the thymus continues to produce immunocompetent T cells (but at a reduced rate) throughout adulthood.

The thymus contains numerous **lobules** arranged somewhat like the fronds in a head of cauliflower. Each lobule in turn contains an outer cortex and an inner medulla (Figure 19.11b). The **cortex** stains dark because it is packed with rapidly dividing T lympho-

cytes gaining immunocompetence. The **medulla** of the lobules contains fewer lymphocytes and appears lighter. The function of the thymic medulla is uncertain, but this region does contain distinctive structures called **thymic (Hassall's) corpuscles** (Figure 19.11c). Thymic corpuscles seem to be collections of degenerating epithelial cells, and their number and size increase with age.

The thymus differs from the other lymphoid organs in two basic ways: (1) It functions strictly in lymphocyte maturation and is thus the only lymphoid organ that does not directly fight antigens. In fact, the *blood-thymus barrier,* analogous to the blood-brain barrier, keeps blood-borne antigens from leaking out of thymic capillaries and prematurely activating the

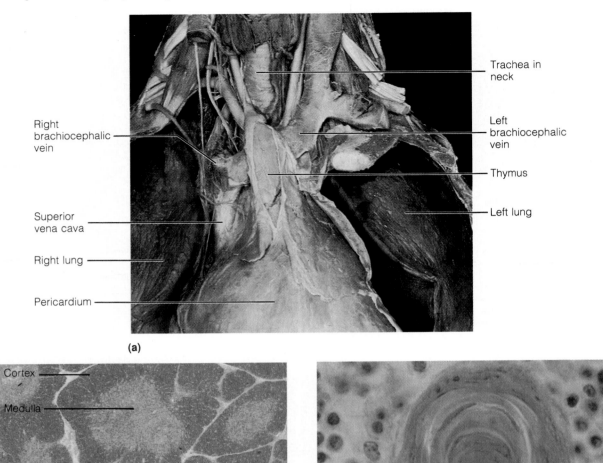

(a)

(b)

(c)

Figure 19.11   Structure of the thymus.   **(a)** Photograph of the thymus in the thorax of an adult. Here the thymus lies in the superior thorax, but it may extend superior or inferior to this. **(b)** Photomicrograph of a portion of the thymus, showing several lobules with cortex and medulla regions (25 ×). **(c)** Photomicrograph of a thymic (Hassall's) corpuscle in the thymic medulla (950 ×). The corpuscle appears as the whorled structure in the center of the photomicrograph.

immature thymic lymphocytes. (2) The tissue framework of the thymus is not a true lymphoid connective tissue. The thymus arises like a gland from the epithelium lining the embryonic pharynx, so its basic tissue framework consists of star-shaped epithelial cells rather than reticular fibers. These *epithelial reticular cells (thymocytes)* secrete the hormones that stimulate T cells to become immunocompetent.

Tonsils

The **tonsils** (Figure 19.12), perhaps the simplest lymphoid organs, are mere swellings of the mucosa lining the pharynx. There are four groups of tonsils, whose precise locations are indicated in Figure 20.3 on page 536. The *palatine tonsils* lie on the lateral sides of the pharyngeal wall, just inferior to the posterior edge of the palate. These are the largest tonsils and the ones most often infected and removed during childhood (in a surgical procedure called *tonsillectomy*). The *lingual tonsils* lie on the posterior surface of the tongue, the *pharyngeal tonsil* (adenoids) lies on the pharyngeal roof, and the *tubal tonsils* surround the openings of the auditory tubes into the pharynx. The tonsils gather and remove many of the pathogens that enter the pharynx in inspired air and swallowed food.

The histological structure of tonsils is shown in Figure 19.12. Recall that all mucosae consist of an ep-

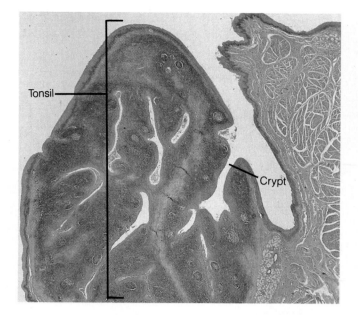

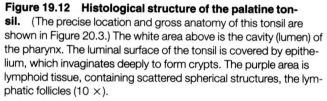

Figure 19.12   Histological structure of the palatine tonsil. (The precise location and gross anatomy of this tonsil are shown in Figure 20.3.) The white area above is the cavity (lumen) of the pharynx. The luminal surface of the tonsil is covered by epithelium, which invaginates deeply to form crypts. The purple area is lymphoid tissue, containing scattered spherical structures, the lymphatic follicles (10 ×).

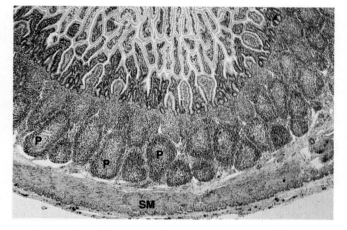

Figure 19.13   Histological structure of aggregated lymphoid follicles (Peyer's patches) in the wall of the ileum of the small intestine (20 ×). P = Peyer's patches. SM = smooth muscle in the external part of the intestinal wall.

ithelium underlain by a lamina propria connective tissue (Chapter 4, p. 90). In the tonsils, the lamina propria consists of lymphoid tissue packed with lymphocytes and scattered lymphatic follicles. The overlying epithelium invaginates deep into the interior of the tonsil, forming blind-ended **crypts.** These crypts trap bacteria and particulate matter, and the trapped bacteria work their way through the epithelium into the lymphoid tissue, where they cause the activation of lymphocytes. Although this "strategy" of trapping bacteria in crypts leads to tonsil infections during childhood, it generates a wide variety of memory lymphocytes to protect us throughout life.

Aggregated Lymphoid Follicles and the Appendix

Aggregated lymphoid follicles, or **Peyer's** (Pi'erz) **patches,** are clusters of large lymphatic follicles in the wall of the distal part (ileum) of the small intestine (Figures 19.5 and 19.13). Similarly, lymphoid tissue is heavily concentrated in the wall of the **appendix,** a tubular offshoot of the first part (cecum) of the large intestine. Peyer's patches and the appendix are in an ideal position to destroy bacteria (which are present in large numbers in the intestine), thereby preventing these pathogens from breaching the intestinal wall. Furthermore, these two intestinal lymphoid organs generate many memory lymphocytes for long-term immunity. Peyer's patches, the appendix, and the tonsils are part of a collection of lymphoid tissue that

scatters throughout the wall of the entire digestive tract and is called **gut-associated lymphoid tissue (GALT).** Similarly, lymphoid tissue occupies the walls of the respiratory tubes, where it is called **bronchus-associated lymphoid tissue (BALT).** Collectively, GALT and BALT protect the respiratory and digestive tracts from the never-ending onslaught of foreign matter that enters these cavities.

The Lymphatic System Throughout Life

The lymphatic system develops from a number of sources. Both lymphatic vessels and the main clusters of lymph nodes grow from *lymphatic sacs,* which are projections from the large veins of the embryo. The thymus originates as an outgrowth of the endoderm lining the embryonic pharynx (see Figure 24.13), detaches from the pharynx, and migrates caudally into the thorax. In the fetal period, the thymus first receives immature lymphocytes that will become T cells. All the other lymphoid organs and tissues (spleen, lymph nodes, GALT) arise from mesodermal mesenchyme. Except for the spleen, these organs are poorly developed before birth. However, shortly after birth, they become heavily populated by circulating lymphocytes and start to gain their functional properties. Evidence indicates that the young thymus secretes hormones that signal the development of the other lymphoid organs.

The lymphoid organs and immune system normally serve us well until late in life, when their efficiency begins to wane and the ability to fight infection declines. Old age is accompanied by an increased susceptibility to disease. The greater incidence of cancer in the elderly is assumed to be another example of the declining ability of the immune system to destroy harmful cells. Just why this decline occurs is not really known, but "genetic aging" and its consequences are probably at fault.

✚ Related Clinical Terms

Hodgkin's disease   A malignancy of the lymph nodes characterized by swollen, nonpainful nodes, fatigue, and, often, persistent fever and night sweats. Hodgkin's disease is treated with radiation therapy, and the cure rate is high relative to other cancers.

Lymphadenopathy (lim-fad″ĕ-nop′ah-the) (*adeno* = gland; *pathy* = disease)   Any disease of the lymph nodes.

Lymphangitis (lim″fan-ji′tis) (*angi* = vessel)   Inflammation of a lymph vessel. Like large blood vessels, the lymph vessels are supplied with blood by vasa vasorum. When lymph vessels are infected and inflamed, their vasa vasorum become congested with blood. The superficial lymphatic vessels then become visible through the skin as red lines that are tender to the touch.

Lymphedema   Swelling due to accumulation of lymphatic fluid in the loose connective tissues.

Lymphoma   Any neoplasm (tumor) of the lymphoid tissue, whether benign or malignant.

Mononucleosis   A viral disease common in adolescents and young adults. Symptoms include fatigue, fever, sore throat, and swollen lymph nodes. Mononucleosis is caused by the Epstein-Barr virus, which specifically attacks B lymphocytes. This attack leads to a massive activation of T lymphocytes, which in turn attack the virus-infected B cells. Large numbers of oversized T lymphocytes circulate in the bloodstream. (These lymphocytes were originally misidentified as monocytes: *mononucleosis* = a condition of monocytes.) Mononucleosis is transmitted in saliva ("kissing disease") and usually lasts 4 to 6 weeks.

Tonsillitis   Congestion of the tonsils, typically with infecting bacteria, which causes them to become red, swollen, and sore.

Chapter Summary

1. The lymphatic system consists of (1) lymphatic circulatory vessels that carry lymph and (2) the lymphoid tissues and lymphoid organs of the immune system.

THE LYMPHATIC VESSELS (pp. 516–520)

Basic Structure and Functions (pp. 516–517)

1. The vessels of the lymphatic system, from smallest to largest, are lymph capillaries, collecting vessels (with lymph nodes), trunks, and ducts.

2. Lymph is excess tissue fluid, which originates because slightly more fluid leaves blood capillaries than returns there. Lymph vessels pick up this excess fluid and return it to the great veins at the root of the neck.

3. Lymph vessels also retrieve blood proteins that leak from capillaries and return these proteins to the bloodstream.

Lymph Capillaries (pp. 517–518)

4. Lymph capillaries weave through the loose connective tis-

sues of the body. These blind-ended tubes are highly permeable because their endothelial cells are loosely joined. The permeable lymph capillaries may also pick up disease-causing microorganisms and cancer cells, which thus threaten to spread widely in the lymph vessels.

5. Lymph capillaries called lacteals absorb digested fat from the small intestine.

Lymphatic Collecting Vessels (p. 518)

6. Lymphatic collecting vessels run alongside arteries and veins but have thinner walls and many more valves and resemble a string of beads.

7. Lymph flows very slowly through lymphatic collecting vessels. Flow is maintained by normal body movements, contractions of skeletal muscles, arterial pulsations, and contraction of smooth muscle in the wall of the lymph vessel. Lymphatic valves prevent backflow.

Lymph Trunks (p. 518)

8. The lymph trunks (lumbar, intestinal, bronchomediastinal, subclavian, and jugular) each drain a large body region. All except the intestinal trunk are paired.

Lymph Ducts (pp. 518–520)

9. The right lymphatic duct drains lymph from the superior right quarter of the body. The thoracic duct receives lymph from the rest of the body. These ducts empty into the junction of the internal jugular and subclavian veins. The thoracic duct starts at the cisterna chyli at L_1–L_2 and ascends along the thoracic vertebral bodies.

LYMPHOID CELLS, TISSUES, AND ORGANS (pp. 520–529)

1. Lymphoid organs and lymphoid tissues house millions of lymphocytes, important cells of the immune system.

Lymphocytes (pp. 520–524)

2. B and T lymphocytes fight infectious microorganisms in the loose and lymphoid connective tissues of the body—B cells by producing antibody-secreting plasma cells, and cytotoxic T cells by directly killing antigen-bearing cells.

3. Lymphocytes patrol connective tissues throughout the body by passing in and out of the circulatory vessels (recirculation).

4. Lymphocytes arise from stem cells in the bone marrow. T cells develop immunocompetence in the thymus, while B cells develop immunocompetence in the bone marrow. Immunocompetent lymphocytes then circulate to the loose and lymphoid connective tissues, where antigen binding (the antigen challenge) leads to lymphocyte activation.

5. The antigen challenge involves a complex interaction among the activating lymphocyte, an antigen-presenting cell (for activation of T cells only), and a helper T lymphocyte. A newly activated T or B cell divides quickly to produce many short-lived effector lymphocytes and some long-lived memory lymphocytes. Recirculating memory lymphocytes provide long-term immunity.

Lymphoid Tissue (p. 524)

6. Lymphoid tissue is a loose reticular connective tissue. Often infected, it is packed with B and T lymphocytes from the blood. The functions of lymphoid tissue are antigen destruction, lymphocyte activation, and the production of memory cells.

7. Lymphoid tissue contains follicles (nodules) with germinal centers. Each follicle contains thousands of B lymphocytes, all derived from one activated B cell.

Lymphoid Organs (pp. 524–529)

8. Clustered along the lymphatic collecting vessels, bean-shaped lymph nodes cleanse the lymph stream of infectious agents and cancer cells. Figure 19.3 shows the major groups of lymph nodes in the body.

9. Each lymph node has a fibrous capsule, cortex, medulla, hilus, and lymph sinuses. Lymph enters the node via afferent lymphatic vessels and exits via efferent vessels at the hilus. Within the node, lymph percolates through the sinuses where macrophages cleanse the lymph.

10. Within a lymph node, lymphoid tissue lies between the sinuses. This lymphoid tissue receives some of the antigens that pass through the node, leading to lymphocyte activation and memory cell production. These lymphocytes break free and circulate.

11. The spleen lies in the superior left part of the abdominal cavity. The splenic vessels enter and exit the hilus on the anterior surface.

12. The spleen has two main functions: (1) cleansing antigens from the blood and (2) destroying worn-out blood cells. The first function is performed by the white pulp, the second by the red pulp. White pulp consists of sleeves of lymphoid tissue, each surrounding a central artery. Red pulp consists of venous sinuses and a blood-filled reticular connective tissue (splenic cords) whose macrophages remove worn-out blood cells.

13. The thymus in the superior, anterior thorax is most active during youth. Its hormones, secreted by epithelial reticular cells, signal the contained T lymphocytes to gain immunocompetence.

14. The thymus exhibits lobules, each with an outer cortex packed with maturing T cells and an inner medulla containing degenerative thymic (Hassall's) corpuscles.

15. The thymus neither directly fights antigens nor contains true lymphoid tissue.

16. The tonsils in the pharynx, aggregated lymphoid follicles (Peyer's patches) in the small intestine, and the wall of the appendix are parts of GALT (gut-associated lymphoid tissue). GALT prevents pathogens from breaching the mucous membrane of the digestive tract. BALT (bronchus-associated lymphoid tissue) is the corresponding tissue in the respiratory tubes.

THE LYMPHATIC SYSTEM THROUGHOUT LIFE (pp. 529–530)

1. Lymphatic vessels develop from lymphatic sacs attached to the embryonic veins. The thymus develops from endoderm, and the other lymphoid organs derive from mesenchyme.

2. The thymus is the first lymphoid organ to appear in the embryo. It signals the development of most other lymphoid organs.

3. Lymphoid organs become populated by lymphocytes, which arise from hematopoietic tissue.

4. With aging, the immune system becomes less responsive. Thus, the elderly suffer more often from infections and cancer.

Review Questions

Multiple Choice and Matching Questions

1. Lymphatic vessels (a) carry both blood and lymph, (b) cannot be distinguished from blood vessels from their external appearance, (c) transport leaked plasma proteins and fluids to the cardiovascular system, (d) have no valves, (e) a and c.

2. The sac-like initial portion of the thoracic duct is the (a) lacteal, (b) right lymphatic duct, (c) lymphatic sac, (d) cisterna chyli.

3. Lymph capillaries (a) are open-ended like drinking straws, (b) have continuous tight junctions like those of the capillaries of the blood-brain barrier, (c) contain endothelial cells separated by flap-like valves that open wide, (d) have special barriers that stop cancer cells from entering, (e) all of the above.

4. The basic structural framework of most lymphoid organs consists of (a) areolar connective tissue, (b) hematopoietic tissue, (c) reticular connective tissue, (d) adipose tissue.

5. Lymph nodes cluster in all the following body areas except (a) the brain, (b) the axillae, (c) the groin, (d) the neck.

6. The germinal centers in lymph nodes are sites of (a) the lymph sinuses, (b) proliferating B lymphocytes, (c) T lymphocytes, (d) a and c, (e) all of the above.

7. The red pulp of the spleen (a) contains venous sinuses and a macrophage-rich connective tissue, (b) is another name for the fibrous capsule, (c) is lymphoid tissue, (d) is the part of the spleen that destroys worn-out erythrocytes, (e) a and d.

8. The lymphoid organ that functions primarily during youth and then atrophies most markedly in adulthood is the (a) spleen, (b) thymus, (c) palatine tonsils, (d) bone marrow.

9. Collections of lymphoid tissue that guard mucosa surfaces include all of the following except (a) GALT, (b) BALT, (c) the tonsils, (d) Peyer's patches, (e) the thymus.

10. All of these cell types contribute to the destruction of infectious foreign cells, but only one is phagocytic. Which one? (a) macrophage, (b) B lymphocyte, (c) cytotoxic T lymphocyte, (d) helper T lymphocyte.

11. Lymphocytes that develop immunocompetence in the thymus are (a) B lymphocytes, (b) T lymphocytes.

12. Which of the following lymphoid organs have a cortex and a medulla? More than one choice is correct. (a) lymph nodes, (b) spleen, (c) thymus, (d) Peyer's patches, (e) tonsils.

13. Which one of the following lymphoid organs does not contain lymph follicles or germinal centers? (a) lymph nodes, (b) spleen, (c) thymus, (d) Peyer's patches, (e) tonsils.

14. Which of the following is important for the activation of a B lymphocyte during the antigen challenge? (a) the antigen, (b) helper T cell, (c) chemicals that stimulate the lymphocytes to divide, (d) all of these.

15. Developmentally, the embryonic lymphatic vessels are most closely associated with the (a) veins, (b) arteries, (c) nerves, (d) thymus.

16. It sometimes is difficult to distinguish the different lymphoid organs from one another in histological sections. How would you tell the thymus from a lymph node? (a) Only the thymus has a cortex and medulla; (b) lymphocytes are far less densely packed in the thymus than in the lymph node; (c) the thymus contains no blood vessels; (d) only the thymus has distinct lobules and thymic (Hassall's) corpuscles.

17. This chapter did not mention where helper T cells become immunocompetent. Based on their name, however, you can deduce that this happens in the (a) lymphoid tissue, (b) bone marrow, (c) bursa, (d) thymus, (e) trabeculae of the spleen and lymph nodes.

Short Answer and Essay Questions

18. What are some of the basic differences between blood, tissue fluid, and lymph?

19. Compare the basic functions of a lymph node to those of the spleen.

20. If you saw a blood vessel and a lymphatic collecting vessel running side by side, how could you tell them apart?

21. What is the difference between splenomegaly and splenectomy?

22. Trace the entire course of the thoracic duct.

23. List and briefly explain three important functions of the lymphatic vessels.

24. Describe the three steps of lymphocyte activation, starting with the most immature lymphocyte precursors (lymphoid stem cells).

25. Explain why no lymph vessels are necessary in the central nervous system.

26. What is lymphocyte recirculation? What role does it play?

27. Which three of the six groups of lymph nodes shown in Figure 19.3 are easily felt (palpated) through the skin during a physical examination?

Critical Thinking and Clinical Application Questions

28. A friend tells you that she has tender, swollen "glands" along the left side of the front of her neck. You notice that she has a bandage on her left cheek that is not fully hiding a large infected cut there. Exactly what are her swollen "glands," and how did they become swollen?

29. When young Joe Chang went sledding, the runner of a friend's sled hit him in the left side and ruptured his spleen. Joe almost died before he got to a hospital. What is the immediate danger of a ruptured spleen?

30. The man in the hospital bed next to Joe is an alcoholic with cirrhosis of the liver and portal hypertension. His spleen is seriously enlarged. Based on what you remember from Chapter 18, how could portal hypertension lead to splenomegaly?

31. Mrs. Roselli has undergone a left radical mastectomy. Her left arm is severely swollen and painful, and she is unable to raise it higher than her shoulder. (a) Explain the origin of her signs and symptoms. (b) Is she likely to have relief from these symptoms in time? Explain.

32. Traci arrives at the clinic complaining of pain and redness of her right ring finger. The finger and the dorsum of her hand have edema, and red streaks are apparent on her right forearm. Antibiotics are prescribed, and the nurse applies a sling to the affected arm. Why is it important that Traci not move the affected arm excessively?

20

The Respiratory System

GAS EXCHANGE = MAIN FX

Student Objectives

Functional Anatomy of the Respiratory System (pp. 534–550)

1. Identify the respiratory tubes and passageways in descending order, from the nose to the alveoli in the lungs. Distinguish the structures of the conducting zone from those of the respiratory zone.

2. List and describe several protective mechanisms of the respiratory system.

3. Describe the structure and functions of the larynx.

4. List the histological changes in the walls of the respiratory passages, from the upper to the lower parts of the respiratory tree.

5. Describe the structure of a lung alveolus and of the respiratory membrane.

6. Describe the gross structure of the lungs and the pleurae.

Ventilation (pp. 550–554)

7. Explain the relative roles of the respiratory muscles and lung elasticity in the act of ventilation.

8. Define surfactant, and explain its function in ventilation.

9. Explain how the brain and peripheral chemoreceptors control the breathing rate.

Diseases of the Respiratory System (pp. 554–556)

10. Compare the causes and consequences of chronic bronchitis, emphysema, and lung cancer.

The Respiratory System Throughout Life (pp. 556–557)

11. Trace the development of the respiratory system in the embryo and fetus.

12. Describe the normal changes that occur in the respiratory system from infancy to old age.

In a popular rock song of the early 1970s, the singer says that all he needs is the air that he breathes and to love his lover. The admission in a love song that respiration comes before romance is a powerful testimony to the fact that breathing is our most urgent need. The trillions of cells in the body require a continuous supply of oxygen to carry out their vital functions. We cannot live without oxygen for even a few minutes, as we can without food or water. As our cells use oxygen, furthermore, they give off carbon dioxide, a waste product the body must get rid of. The major function of the **respiratory system** is to fulfill these needs, that is, to supply the body with oxygen and dispose of carbon dioxide. To accomplish this, at least four distinct processes, collectively called **respiration**, must occur:

1. **Pulmonary ventilation.** Air must be moved in and out of the lungs so that the gases in the air sacs (alveoli) of the lungs are continuously changed and refreshed. This movement is commonly called **ventilation** or breathing.

2. **External respiration.** Gas exchange (oxygen loading and carbon dioxide unloading) occurs between the blood and air at the lung alveoli.

3. **Transport of respiratory gases.** Oxygen and carbon dioxide must be transported between the lungs and the cells of the body. This is accomplished by the cardiovascular system, which uses blood as the transporting fluid.

4. **Internal respiration.** At the systemic capillaries, gases are exchanged between the blood and the tissue cells.*

In this chapter, we focus on pulmonary ventilation and external respiration because they alone are the special responsibility of the respiratory system. However, unless gas transport and internal respiration also occur, the respiratory system cannot accomplish its primary goal of obtaining oxygen and eliminating carbon dioxide. Thus, the respiratory and cardiovascular systems are closely coupled, and if either system fails, the body's cells begin to die from oxygen starvation.

Besides its role in gas exchange, the respiratory system is also intimately involved with the sense of smell and the vocalizations of speech.

---

*The use of oxygen and the production of carbon dioxide by tissue cells, called *cellular respiration,* is the cornerstone of all energy-producing chemical reactions in the body. For a discussion of cellular respiration, consult a physiology or cell biology text.

Functional Anatomy of the Respiratory System

The organs of the respiratory system include the *nose* and *nasal cavity,* the *pharynx,* the *larynx,* the *trachea,* the *bronchi* and their smaller branches, and the *lungs,* which contain the terminal air sacs, or *alveoli* (Figure 20.1). Functionally, these respiratory structures are divided into respiratory and conducting zones. The *respiratory zone,* the actual site of gas exchange in the lungs, is composed of tiny structures called respiratory bronchioles, alveolar ducts, and alveoli. The *conducting zone* includes all other respiratory passageways, which serve as fairly rigid conduits by which air reaches the sites of gas exchange. The structures of the conducting zone also filter, humidify, and warm the incoming air. Thus, the air reaching the lungs contains much less dust than it did when it entered the nose and is warm and damp, like the air of the tropics. The functions of the major organs of the respiratory system are summarized in Table 20.1 on page 549.

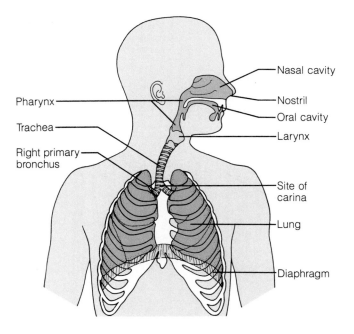

Figure 20.1 Organs of the respiratory system. The structures shown are all in the conducting zone. (Structures of the respiratory zone are small and lie deep within the lung.) Organs superior to the larynx are called *upper respiratory structures,* while the larynx and all structures inferior to it are the *lower respiratory structures.*

The Nose and the Paranasal Sinuses

The Nose

The **nose** is the only externally visible part of the respiratory system. Unlike the often poetic references to other facial features, such as the eyes or lips, the nose is usually the target of irreverence: We are urged to keep our nose to the grindstone and to keep it out of someone else's business. However, considering its important functions, the nose deserves more esteem. The nose (1) provides an airway for respiration, (2) moistens and warms entering air, (3) filters inhaled air and cleanses it of foreign particles, (4) serves as a resonating chamber for speech, and (5) houses the olfactory (smell) receptors.

The structures of the nose are divided into the *external nose* and the internal *nasal cavity.* The skeletal framework of the **external nose** is shown in Figure 20.2. This framework consists of the frontal and nasal bones superiorly (forming the root and bridge, respectively), the maxillary bones laterally, and flexible plates of hyaline cartilage inferiorly (the lateral, septal, and alar cartilages). Noses vary a great deal in size and shape, largely because of differences in the nasal cartilages. The skin covering the nose's anterior and lateral aspects is thin and contains many sebaceous glands that open into some of the largest skin pores found anywhere on the face.

The **nasal cavity** (Figures 20.1 and 20.3) lies in and posterior to the external nose. During breathing, air enters this cavity by passing through the **external nares** (na' rēz), or *nostrils.* The nasal cavity is divided into right and left halves by the *nasal septum* in the midline (see Figure 7.10 on p. 153). Posteriorly, the nasal cavity is continuous with the nasal part of the pharynx (nasopharynx) through the **internal nares**, also called the *posterior nares* or *choanae* (ko-a'ne; "funnels").

Let us review the bony boundaries of the nasal cavity (covered in detail in Chapter 7, p. 152): The roof of the nasal cavity is formed by the ethmoid and sphenoid bones of the skull. Its floor is formed by the *palate* (pal'at), which separates the nasal cavity from the mouth inferiorly and keeps food out of the airways. Anteriorly, where the palate contains the horizontal processes of the palatine bones and the palatine process of the maxillary bone, it is called the **hard palate**. The posterior part is the muscular **soft palate**.

The portion of the nasal cavity that lies just superior to the nostrils, within the flared wings of the external nose, is the **vestibule** ("porch, entranceway"). The vestibule is lined with skin containing sebaceous and sweat glands and numerous hair follicles. The nose hairs, or *vibrissae* (vi-bris'e) (*vibro* = to quiver), filter large particles, such as insects and lint, from the inspired air. The remainder of the nasal cavity is lined with two types of mucous membrane: (1) the small patch of *olfactory mucosa* near the roof of the nasal cavity, which houses the receptors for smell (pp. 413–414), and (2) the *respiratory mucosa,* which lines the vast majority of the nasal cavity (Figure 20.4). The respiratory mucosa consists of (a) a pseudostratified ciliated columnar epithelium containing scattered goblet cells and (b) the underlying lamina propria connective tissue. This lamina propria is richly supplied with compound tubuloalveolar glands (p. 77) that contain mucous cells and serous cells. By definition, *mucous cells* secrete mucus, and *serous cells* secrete a watery fluid containing digestive enzymes. Each day, the nasal glands and the epithelial goblet cells secrete about a quart of mucus containing lysozyme, an enzyme that digests and destroys bacteria. This sticky mucus forms a sheet that covers the surface of the mucosa and traps inhaled dust, bacteria, pollen, viruses, and other debris from the air.

Thus, an important function of the respiratory mucosa is to filter the inhaled air. The ciliated cells in its epithelial lining create a gentle current that moves the sheet of contaminated mucus posteriorly to the pharynx, where it is swallowed. Thus, particles filtered from the air are ultimately destroyed by digestive juices in the stomach of the digestive system. We

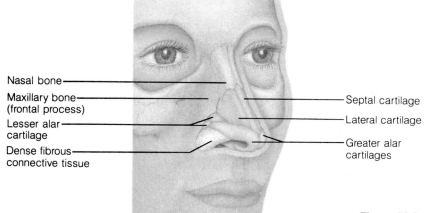

Nasal bone
Maxillary bone (frontal process)
Lesser alar cartilage
Dense fibrous connective tissue
Septal cartilage
Lateral cartilage
Greater alar cartilages

Figure 20.2   Skeletal framework of the external nose.

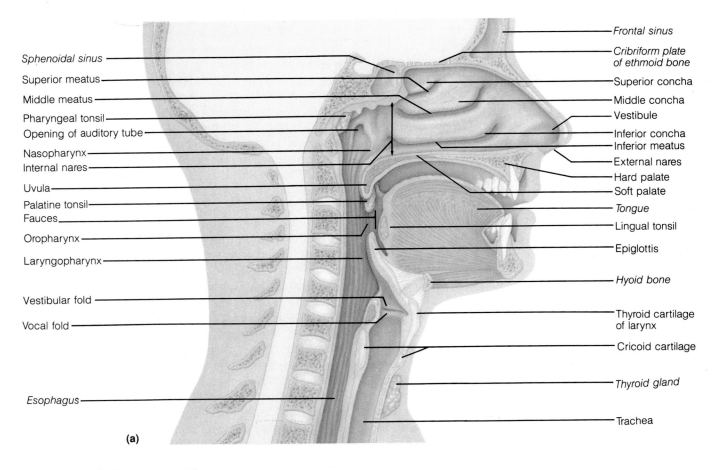

Sphenoidal sinus
Superior meatus
Middle meatus
Pharyngeal tonsil
Opening of auditory tube
Nasopharynx
Internal nares
Uvula
Palatine tonsil
Fauces
Oropharynx
Laryngopharynx
Vestibular fold
Vocal fold
Esophagus

Frontal sinus
Cribriform plate of ethmoid bone
Superior concha
Middle concha
Vestibule
Inferior concha
Inferior meatus
External nares
Hard palate
Soft palate
Tongue
Lingual tonsil
Epiglottis
Hyoid bone
Thyroid cartilage of larynx
Cricoid cartilage
Thyroid gland
Trachea

(a)

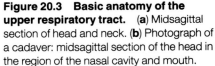

Figure 20.3   Basic anatomy of the upper respiratory tract.   (a) Midsagittal section of head and neck. (b) Photograph of a cadaver: midsagittal section of the head in the region of the nasal cavity and mouth.

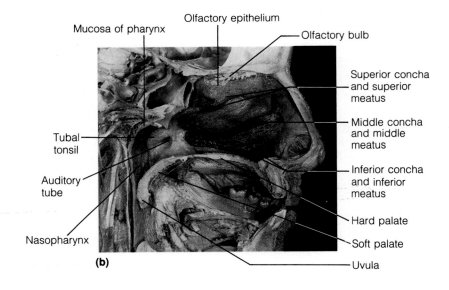

Mucosa of pharynx
Olfactory epithelium
Olfactory bulb
Superior concha and superior meatus
Middle concha and middle meatus
Inferior concha and inferior meatus
Hard palate
Soft palate
Tubal tonsil
Auditory tube
Nasopharynx
Uvula

(b)

usually are unaware of the action of our nasal cilia, but when they are exposed to cold temperatures, these cilia become sluggish, allowing mucus to accumulate in the nasal cavity and then dribble out the nostrils. This is the reason you might have a "runny nose" on a crisp, wintry day.

A rich plexus of thin-walled veins occupies the lamina propria of the nasal mucosa and warms the incoming air that flows across the mucosal surface.

When the temperature of inhaled air drops, the vascular plexus responds by engorging with warm blood, thereby intensifying the air-heating process. Because of the abundance and the superficial location of these vessels, nosebleeds are common and often profuse.

Projecting medially from each lateral wall of the nasal cavity are three mucosa-covered, scroll-like structures, the *superior* and *middle conchae* of the ethmoid bone and the *inferior concha*, which is a sep-

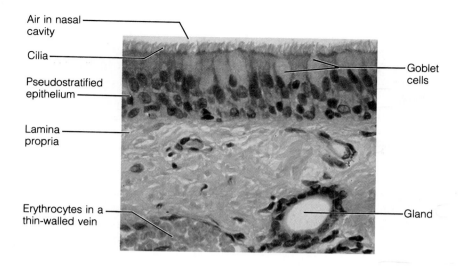

Air in nasal cavity

Cilia

Pseudostratified epithelium

Lamina propria

Erythrocytes in a thin-walled vein

Goblet cells

Gland

Figure 20.4   Photomicrograph of the respiratory mucosa that lines the nasal cavity and paranasal sinuses (400 ✕).   For a surface view of the ciliated epithelium, see Figure 1.11c on p. 16.

arate bone (Figure 20.3). The groove inferior to each concha is a *meatus.* As inhaled air rushes over the curved conchae, the resulting turbulence greatly increases the amount of contact between the nasal mucosa and this inspired air. The gaseous part of the inhaled air swirls through the twists and turns of the conchae, but the heavier, nongaseous particles are deflected onto the mucus-coated surfaces, where they become trapped. As a result, few particles larger than 4 μm get past the nasal cavities.

The nasal mucosa is richly supplied with sensory nerve endings. Can you propose a function for the sneezing that follows a tickle in one's nose?

The sneeze reflex is stimulated when irritating particles (dust, pollen, and so on) contact the sensitive mucosa of the nasal cavity. The sneeze forces air outward in a violent burst, as a crude way of expelling the irritant from the nose. ■

The Paranasal Sinuses

The nasal cavity is surrounded by a ring of air-filled cavities called *paranasal sinuses* located in the frontal, sphenoid, ethmoid, and maxillary bones (Figure 7.11, p. 153). Recall from Chapter 7 that these sinuses open into the nasal cavity, are lined by the same mucosa, and perform the same air-processing functions. Their mucus drains into the nasal cavities, and the suctioning effect caused by nose blowing helps to drain them.

Cold viruses, streptococcal bacteria, and various allergens can cause **rhinitis** (ri-ni′tis; "nose inflammation"). This is inflammation of the nasal mucosa, accompanied by an excessive production of mucus. The mucous secretion results in nasal congestion, a runny nose, and postnasal drip.

Have you ever noticed that colds start in your nasal cavity and may proceed to your sinuses, your throat, and finally your lungs? Can you explain this nose-to-lungs progression?

The infectious agent (the cold virus) enters your nose in the inhaled air and then spreads progressively along the mucosa lining the rest of the respiratory tract until your immune system finally stops the infection.

Sinusitis—infection and inflammation of the paranasal sinuses—is difficult to treat and can result in marked changes in voice quality. When the passages that connect the paranasal sinuses to the nasal cavity become blocked with mucus or infectious material, the air in the sinus cavities is absorbed. The result is a partial vacuum and a *sinus headache* localized over the inflamed areas. ■

The Pharynx

The **pharynx** (far′ingks) is the funnel-shaped passageway that connects the nasal cavity and mouth superiorly to the larynx and esophagus inferiorly (Figures 20.1 and 20.3a). The pharynx descends from the base of the skull to the level of the sixth cervical vertebra and serves as a common passageway for both food and air. Commonly called the *throat,* the pharynx vaguely resembles a short length of garden hose.

On the basis of location and function, the pharynx is divided into three regions. From superior to inferior, they are the *nasopharynx, oropharynx,* and *laryngopharynx* (lah-ring″go-far′ingks). The muscular wall of the pharynx is composed of skeletal muscle throughout its length (see Table 10.3, pp. 250–251), but the composition of its mucosal lining varies among the three pharyngeal regions.

The Nasopharynx

The **nasopharynx** lies directly posterior to the nasal cavity, inferior to the sphenoid bone and superior to the level of the soft palate (Figure 20.3a). Because it lies superior to the point where food enters the body, the nasopharynx serves only as an air passageway.

During swallowing, the soft palate and its pendulous **uvula** (u'vu-lah; "little grape") reflect superiorly, an action that effectively closes off the nasopharynx and prevents food from entering the nasal cavity. When we giggle, this sealing action fails, and swallowed fluids can spray from the nose.

The nasopharynx is continuous with the nasal cavity through the internal nares, and its ciliated pseudostratified epithelium takes over the mucus-propelling job where the nasal mucosa leaves off. The mucosa high on its posterior wall contains masses of lymphoid tissue, the **pharyngeal tonsil** (also called the *adenoids*), which destroy pathogens entering the nasopharynx in the air. (The function of tonsils is described in detail in Chapter 19, pp. 528–529.)

Given the location of the adenoids, try to imagine how inflammation of this tonsil may affect breathing.

Infected and enlarged adenoids obstruct the flow of air through the nasopharynx. Mouth breathing then becomes necessary, so the air is not properly moistened, warmed, or filtered before reaching the lungs.

An *auditory tube,* which drains the middle ear, opens into each lateral wall of the nasopharynx. A thick ridge of pharyngeal mucosa arcs over this opening, forming the **tubal tonsil** (Figure 20.3b). Because of its location, the tubal tonsil offers some protection against middle ear infections likely to spread from the pharynx.

The Oropharynx

The **oropharynx** lies posterior to the oral cavity (mouth) and is continuous with it through an archway called the **fauces** (faw'sēz; "throat") (Figure 20.3a). The oropharynx extends inferiorly from the soft palate to the epiglottis (a flap posterior to the tongue). Both swallowed food and inhaled air pass through the oropharynx.

As the nasopharynx blends into the oropharynx, the lining epithelium changes from pseudostratified columnar epithelium to a thick, protective *stratified squamous epithelium.* This structural adaptation reflects the increased friction and greater chemical trauma accompanying the passage of swallowed food. Two pairs of tonsils are embedded in the mucosa of the oropharynx: The **palatine tonsils** lie in the lateral walls of the fauces, and the **lingual tonsils** cover the posterior surface of the tongue.

The Laryngopharynx

Like the oropharynx superior to it, the **laryngopharynx** serves as a common passageway for food and air and is lined with a stratified squamous epithelium.

The laryngopharynx lies directly posterior to the larynx and opens into both the esophagus and the larynx: The esophagus conducts food and fluids to the stomach, and the larynx conducts air to the respiratory tract. *STOP* /

The Larynx

Basic Anatomy

The **larynx** (lar'ingks), or voice box, extends from the level of the fourth to the sixth cervical vertebra. Superiorly, it attaches to the hyoid bone and opens into the laryngopharynx (Figure 20.3a). Inferiorly, the larynx is continuous with the trachea (windpipe).

The larynx has three important functions. Its two main tasks are to provide an open airway and to act as a switching mechanism to route food and air into the proper channels. For these purposes, the inlet (superior opening) to the larynx is closed during swallowing and open during breathing. Because the larynx houses the vocal cords, its third function is voice production.

The framework of the larynx is an intricate arrangement of nine cartilages connected by membranes and ligaments (Figure 20.5). The large, shield-shaped **thyroid cartilage** is formed by the fusion of two cartilage plates. It resembles an upright open book, with the book's "spine" lying in the anterior midline of the neck. This "book spine" is the ridge-like **laryngeal** (lah-rin'je-al) **prominence**, which is obvious externally as the **Adam's apple.** The thyroid cartilage is larger in males than in females because male sex hormones stimulate its growth during puberty. Inferior to the thyroid cartilage is the **cricoid** (kri'koid) **cartilage**, which is shaped like a signet ring and is perched on top of the trachea.

As shown in Figure 20.5b, three pairs of small cartilages lie just superior to the cricoid cartilage in the posterior part of the larynx: the **arytenoid cartilages** (ar″ĭ-te'noid; "ladle-like"), the **corniculate cartilages** (kor-nik'u-lāt; "little horn"), and the **cuneiform cartilages** (ku-ne'ĭ-form; "wedge-shaped"). The most important of these are the pyramid-shaped arytenoids, which anchor the vocal cords.

The ninth cartilage of the larynx, the spoon-shaped **epiglottis** (ep″ĭ-glot'is), is composed of elastic cartilage and is almost entirely covered by a mucosa. The stalk of the epiglottis attaches anteriorly to the internal aspect of the angle of the thyroid cartilage. From there, the epiglottis extends superoposteriorly and attaches to the posterior aspect of the tongue. (*Epiglottis* literally means "upon the back of the tongue"). During swallowing, the whole larynx is pulled superiorly, and the epiglottis tips inferiorly to cover and seal the laryngeal inlet. Because this action keeps food out of the lower respiratory tubes, the epi-

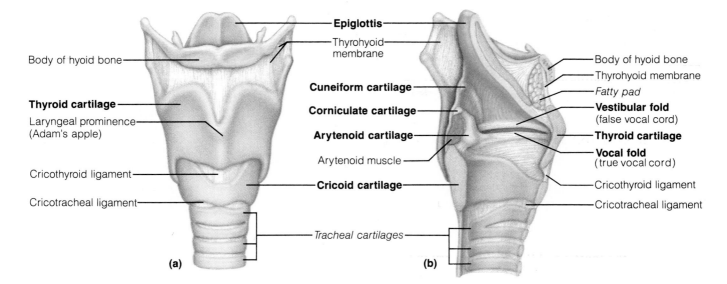

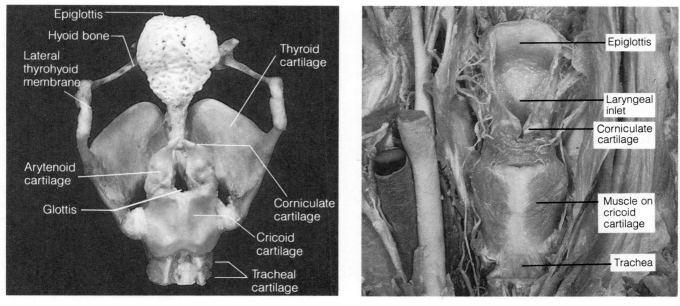

Figure 20.5 Anatomy of the larynx. (a) Anterior view of the skeleton of the larynx. (b) Sagittal section of the larynx. The anterior aspect is to the right. (c) Photograph of the cartilages of the larynx, posterior view. (d) Posterior view of the larynx in the neck; this view is from within the pharynx.

glottis has been called the "guardian of the airways." If anything (other than air) enters the larynx, it initiates the cough reflex, which expels the substance and prevents it from continuing into the lungs. Because this protective reflex does not work when we are unconscious, it is never a good idea to administer fluids when attempting to revive someone who has lost consciousness.

One might argue that the larynx is poorly adapted for our survival. It lies so far inferiorly in the neck and must ascend so far during swallowing that it sometimes cannot reach the protective cover of the epiglottal lid before food enters the laryngeal inlet. Humans

are the only animals that routinely choke to death, yet the low position of the larynx is essential to our ability to talk: Its inferior location allows for an exceptionally long pharynx, which acts as a resonating chamber for the sounds of speech.

Within the larynx, *vocal ligaments* run anteriorly from the arytenoid cartilages to the thyroid cartilage. These ligaments, composed largely of elastic fibers, form the core of a pair of mucosal folds called the **vocal folds** or **(true) vocal cords** (Figures 20.5b and 20.6). Because the mucosa in this region is avascular, the vocal folds appear pearly white. As air from the lungs moves superiorly across the vocal folds, it

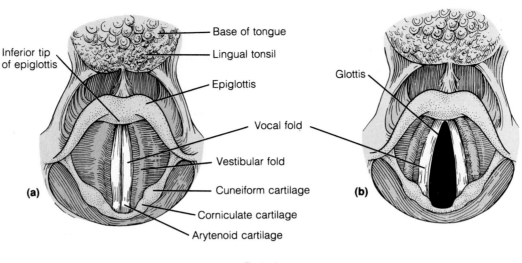

Figure 20.6 Superior view of the larynx. (a) Closed glottis (vocal folds together). (b) Open glottis (vocal folds apart).

makes them vibrate, producing the basic sounds of speech. The medial opening between the vocal folds through which air passes is called the **glottis**, or *rima glottidis* ("fissure of the glottis"). Another pair of horizontal mucosal folds lies directly superior to the vocal folds. These are the **vestibular folds, or false vocal cords**. They play no part in sound production.

The epithelium lining the superior part of the larynx, an area subject to food contact, is stratified squamous, but inferior to the vocal folds, the epithelium is pseudostratified ciliated columnar. In this epithelium, the power stroke of the cilia is directed toward the pharynx so that dust-filtering mucus is continuously moved superiorly from the lungs. Sometimes, we help move this mucus up out of the larynx by "clearing the throat."

Voice Production

Speech involves the intermittent release of exhaled air and the opening and closing of the glottis. The length of the vocal folds and the size of the glottis are altered by the action of the intrinsic laryngeal muscles, most of which move the arytenoid cartilages. As the length and tension of the vocal folds change, the pitch of the sound changes. Generally, the tenser the vocal folds, the faster the exhaled air causes them to vibrate, and the higher the pitch. As a boy's larynx enlarges during puberty, his laryngeal prominence grows anteriorly into a large Adam's apple, lengthening the vocal folds. Because longer vocal folds vibrate more slowly than short folds do, the voice becomes deeper. For this reason, most men have lower voices than females or young boys.

The voices of adolescent boys frequently "crack," alternating between high-pitched and low-pitched sounds. Can you guess why this occurs?

Voice cracking occurs because the boys have not yet learned to control the action of their longer vocal folds. ■

Loudness of the voice depends on the force with which air rushes across the vocal folds. The greater the force, the stronger the vibrations, and the louder the sound. Our vocal folds do not move at all when we whisper, but they vibrate vigorously when we yell.

Although the vocal folds produce the sounds of speech, the quality of the voice depends on many other structures. As we mentioned above, the entire height of the pharynx acts as a resonating chamber to amplify the quality of sound. The oral cavity, nasal cavity, and paranasal sinuses also contribute to vocal resonance. In addition, normal speech and good enunciation depend on the "shaping" of sounds into recognizable consonants and vowels by the pharynx, tongue, soft palate, and lips.

You probably know someone who has lost the ability to speak for a day or two while suffering from a bad cold. What do you think is the cause of this sudden loss of speech?

Inflammation of the larynx, or **laryngitis**, causes swelling of the vocal folds, which interferes with their vibration. This produces a change in the tone of the voice, hoarseness, or (in severe cases) an inability to speak above a whisper. Overuse of the voice, infections, and inhalation of irritating chemicals can all cause laryngitis. ■

Sphincter Functions of the Larynx

The vocal folds act as a sphincter under certain conditions, such as straining to defecate or urinate. During such abdominal straining, the glottis closes, confining inhaled air in the lower respiratory tract. The abdominal muscles contract, and the intra-abdominal pressure rises. These events, collectively known as **Valsalva's maneuver,** help to evacuate the rectum or bladder and can also stabilize the trunk region of the body when one lifts a heavy load.

Innervation

The larynx receives its sensory and motor innervation through several branches of the vagus nerves in the superior neck and from the *recurrent laryngeal nerves.* The recurrent laryngeal nerves branch off the vagus in the superior thorax and loop superiorly to reascend through the neck. More specifically, the left recurrent laryngeal nerve loops under the aortic arch, and the right recurrent laryngeal nerve loops under the right subclavian artery. These backtracking nerves are so unusual that the ancient Greeks mistook them for slings supporting the great arteries.

When surgeons perform surgery on the anterior neck, they must be careful not to cut the recurrent laryngeal nerves. Can you deduce the effect of damaging these nerves?

Damage to these nerves disrupts speech. Transection of one recurrent laryngeal nerve immobilizes one vocal fold, producing a degree of hoarseness. However, the other vocal fold can compensate, and speech remains almost normal. If both nerves are transected, speech (except for whispering) is lost entirely.

Because each recurrent laryngeal nerve is related to the apex of a lung in the thorax, cancers in the apical region of the lung can compress this nerve and cause hoarseness. ∎

The Trachea

The **trachea** (tra′ke-ah), or windpipe, descends from the larynx through the neck and into the mediastinum; it ends by dividing into the two primary bronchi in the mid thorax (Figures 20.1 and 20.7). Unlike most other organs in the neck, the trachea is very mobile: It stretches and descends during inhalation, recoils during exhalation, and can be moved from side to side by probing fingers. Early anatomists mistook the trachea for a rough-walled artery (*trachea* = rough).

The tracheal wall contains 16 to 20 C-shaped rings of hyaline cartilage (Figure 20.7a) joined to one another by intervening membranes of fibroelastic connective tissue. Consequently, the trachea is flexible enough to permit twisting and elongation, but the cartilage rings prevent it from collapsing and keep the airway patent despite the pressure changes that occur during breathing. The open posterior parts of the cartilage rings, which abut the esophagus, contain smooth muscle fibers of the **trachealis muscle** and soft connective tissue. Because the posterior wall of the trachea is not rigid, the esophagus can expand anteriorly as swallowed food passes through it. Contraction of the trachealis muscle decreases the diameter of

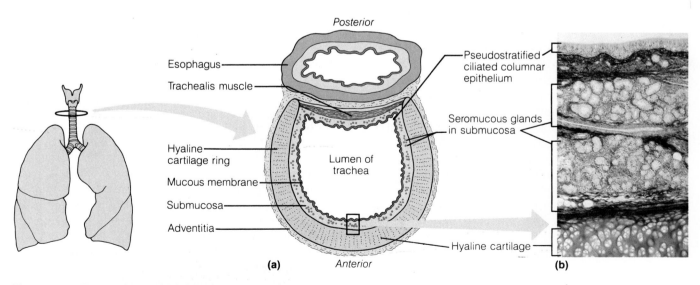

Esophagus

Trachealis muscle

Hyaline cartilage ring

Mucous membrane

Submucosa

Adventitia

Posterior

Lumen of trachea

Anterior

Pseudostratified ciliated columnar epithelium

Seromucous glands in submucosa

Hyaline cartilage

(a)   **(b)**

Figure 20.7   The trachea.   (a) Gross view of the trachea at left, and cross-sectional view at center. The cross section shows the relationship of the trachea to the esophagus, the position of the cartilage rings in the tracheal wall, and the trachealis muscle connecting the free ends of the cartilage rings. **(b)** Photomicrograph of a portion of the tracheal wall (50 ×). The lamina propria is the dark blue layer between the epithelium and the gland-filled submucosa.

the trachea: During coughing and sneezing, this action helps to expel mucus from the trachea by accelerating the exhaled air to a speed of 100 mph! A ridge on the internal aspect of the last tracheal cartilage, called the **carina** (kah-ri′nah; "keel"), marks the division of the trachea into the primary bronchi (Figure 20.1). The mucosa that lines the carina is highly sensitive to irritants, and the cough reflex often originates here.

Figure 20.7b shows the microscopic structure of the trachea. The trachea wall consists of several layers that are common to many tubular organs of the body. From internal to external, these layers are the *mucous membrane, submucosa,* and *adventitia.* The **mucous membrane,** as usual, consists of an inner epithelium and a lamina propria. The epithelium is the same air-filtering pseudostratified epithelium that occurs throughout most of the respiratory tract. Its cilia continuously propel mucous sheets loaded with filtered dust superiorly toward the pharynx. The lamina propria contains a good number of elastic fibers and is separated from the submucosa by a distinct sheet of elastin (not illustrated). This elastin, which occurs in all smaller air tubes as well, stretches during inhalation and recoils during exhalation. The **submucosa** ("below the mucosa") is another layer of connective tissue. It contains glands with both serous and mucous cells (seromucous glands), which help produce the mucous sheets within the trachea. The **adventitia,** the most external layer, is a connective tissue that contains the tracheal cartilages.

The Bronchi and Subdivisions: The Bronchial Tree

The Conducting Zone

The right and left **primary bronchi,** also called the *principal bronchi* (brong′ki), are formed by the division of the trachea (Figure 20.8a). This division occurs at the level of the sternal angle (T_4) in the cadavers studied in the anatomy laboratory, but in living, standing people, the division typically occurs at T_7. Each primary bronchus runs obliquely in the mediastinum before plunging into the medial depression (hilus) of a lung. The primary bronchi lie directly posterior to the large pulmonary vessels that run to and from the lungs (Figure 20.8b). The right primary bronchus is wider, shorter, and more vertical than the left.

When a child inhales a small object, such as a button or marble, in which primary bronchus is the object likely to lodge?

The right one. Because of its greater width and more vertical orientation, the right bronchus acts more like a direct continuation of the trachea than does the left bronchus. ■

As they approach and enter the lungs, the primary bronchi divide into **secondary (lobar) bronchi**—three on the right and two on the left, each of which supplies one lung lobe. The secondary bronchi branch into third-order **tertiary (segmental) bronchi,** which in turn divide repeatedly into smaller bronchi (fourth-order, fifth-order, and so on). Overall, there are about 23 orders of air tubes in the lungs, the tiniest almost too small to be seen without a microscope. Tubes smaller than 1 mm in diameter are called **bronchioles** ("little bronchi"), and the most minute of these, the **terminal bronchioles,** are less than 0.5 mm in diameter. Because of its complex branching pattern, the conducting network in the lungs is often called the **bronchial tree** or *respiratory tree.* A resin cast of the bronchial tree and its accompanying pulmonary arterial supply is pictured in Figure 20.8b.

The tissue composition of the walls of the primary bronchi mimics that of the trachea, but as the conducting tubes become smaller, a number of changes occur:

1. **The cartilage changes.** The cartilage rings are replaced by irregular *plates* of cartilage as the primary bronchi enter the lungs (Figure 20.8c). By the level of the bronchioles, supportive cartilage is no longer present in the tube walls. However, elastic fibers are present in the tube walls throughout the bronchial tree.

2. **The epithelium changes.** The mucosal epithelium thins as it changes from pseudostratified columnar to simple columnar and then to simple cuboidal in the smallest bronchioles. Neither cilia nor mucus-producing cells are present in the smallest bronchioles, where the sheets of air-filtering mucus end. Any airborne dust particles that travel beyond the bronchioles are not trapped in mucus but are removed by macrophages in the alveoli.

3. **Smooth muscle becomes important.** A complete layer of smooth muscle first appears in the walls of the large bronchi and is present throughout the smaller bronchi and bronchioles. The function of this musculature is to regulate the amount of air entering the lung alveoli. The musculature relaxes to widen the air tubes during sympathetic excitement when respiratory needs are great, and it constricts the air tubes under parasympathetic direction when respiratory needs are low (Chapter 14, p. 403).

Abnormally strong, spasmodic contractions of the bronchial smooth musculature, called bronchoconstriction, is the dominant feature of **bronchial asthma,** a respiratory condition that affects about 5% of adults and 10% of children. Can you imagine the effects of this condition?

(Answer on page 543)

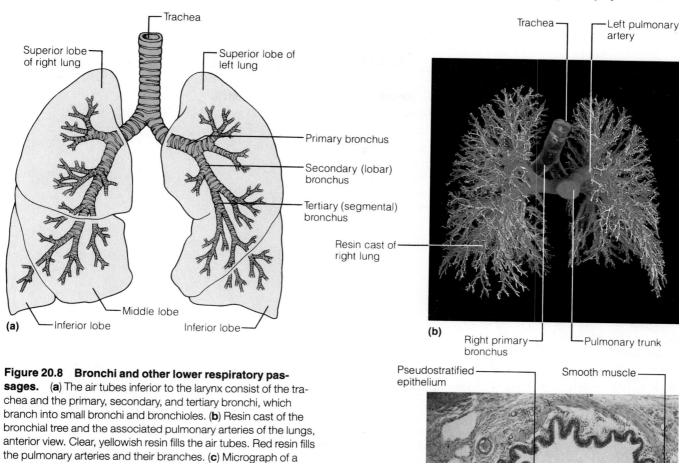

Figure 20.8   Bronchi and other lower respiratory passages.   (**a**) The air tubes inferior to the larynx consist of the trachea and the primary, secondary, and tertiary bronchi, which branch into small bronchi and bronchioles. (**b**) Resin cast of the bronchial tree and the associated pulmonary arteries of the lungs, anterior view. Clear, yellowish resin fills the air tubes. Red resin fills the pulmonary arteries and their branches. (**c**) Micrograph of a bronchus in the lung (100 ×). The layers in the bronchus wall are essentially the same as those in the trachea.

During asthma attacks, narrowing of the bronchi leads to shortness of breath, wheezing, and coughing. The bronchi not only constrict, but also become obstructed as their mucosa swells with inflammation and their epithelia secrete large quantities of sticky mucus. The ultimate cause of asthma is not known, but people with asthma are known to be hypersensitive to irritants in the air or to stress. Attacks may be triggered by inhalation of substances to which the sufferer is allergic, by inhalation of dust or smoke, by respiratory infections, by emotional upset, or by the mild shock of breathing cold air. Some of these stressors produce their effects by stimulating mast cells in the lung to release inflammatory chemicals. Other stressors seem to induce reflexes that increase parasympathetic impulses to the bronchi. ■

The Respiratory Zone

The respiratory zone begins as the terminal bronchioles feed into **respiratory bronchioles** within the lung (Figure 20.9). Protruding from these smallest bronchioles are scattered **alveoli** (*alveol* = small cavity). The respiratory bronchioles lead into **alveolar ducts**, straight ducts whose walls consist almost entirely of alveoli. The alveolar ducts then lead into terminal clusters of alveoli called **alveolar sacs**. Many people mistakenly equate alveoli with alveolar sacs, but they are not the same thing: The alveolar sac is analogous to a bunch of grapes, in which the individual grapes are the alveoli. The opening from an alveolar duct into an alveolar sac is called an **atrium** (meaning an "entrance chamber").

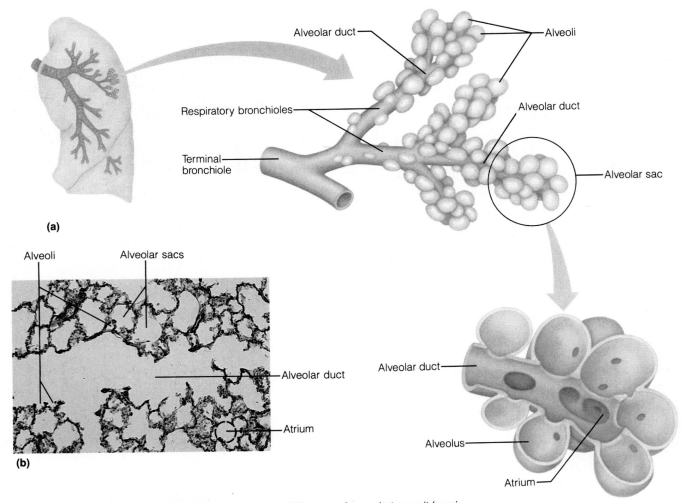

Figure 20.9 Structures of the respiratory zone. (a) Diagram of a respiratory unit (respiratory bronchiole, alveolar ducts, alveolar sacs, and alveoli). (b) Photomicrograph of a part of the lung (66 ×).

About 300 million air-filled alveoli crowd together within the lungs (Figure 20.9b), accounting for most of lung volume and providing a tremendous surface area for gas exchange. The total area of all alveoli in an average pair of lungs is 140 square meters, or 1500 square feet. This is 40 times greater than the surface area of the skin!

The wall of each alveolus consists of a single layer of squamous epithelial cells called **type I cells** underlain by a delicate basal lamina. The extreme thinness of this wall (0.5 μm) is hard to imagine—a sheet of tissue paper is much thicker. The external surfaces of the alveoli are densely covered with a "cobweb" of pulmonary capillaries (Figure 20.10a), with each capillary surrounded by a thin sleeve of the finest areolar connective tissue. Together, the alveolar and capillary walls and their fused basal laminas form the **respiratory membrane** (also called the *air-blood barrier*). Gas is present on one side of the membrane, and blood flows past on the other (Figure 20.10b and

c). Gases pass easily through this thin membrane: The oxygen diffuses from the alveolus into the blood, and carbon dioxide exits the blood to enter the gas-filled alveolus.

Scattered among the type I squamous cells in the alveolar walls are cuboidal epithelial cells called **type II cells** (Figure 20.10b). The type II cells secrete a fluid that coats the internal alveolar surfaces. This fluid contains a detergent-like substance called *surfactant,* without which the alveolar walls would stick together during exhalation (p. 552).

The lung alveoli have three other significant features: (1) They are surrounded by fine elastic fibers of the same type that surround the entire respiratory tree. (2) Open pores are present between adjacent alveoli (Figure 20.10b). These **alveolar pores** allow equalization of air pressure throughout the lung and provide alternate air routes to any alveoli whose bronchi have collapsed through disease. (3) **Alveolar macrophages** crawl freely along the internal alveolar sur-

(a)

Figure 20.10   Anatomy of alveoli and the respiratory membrane.   **(a)** Scanning electron micrograph of casts of several alveoli and the surrounding pulmonary capillaries (255 ×). **(b)** and **(c)** Detailed anatomy of the alveoli and the respiratory membrane. This membrane is composed of a squamous alveolar type I cell, a capillary endothelial cell, and the fused basal laminas of these cells. Also note the alveolar type II cells, macrophages, and alveolar pores. Oxygen (O_2) diffuses from the alveolar air into the pulmonary capillary blood, while carbon dioxide (CO_2) diffuses from the pulmonary blood into the alveolus.

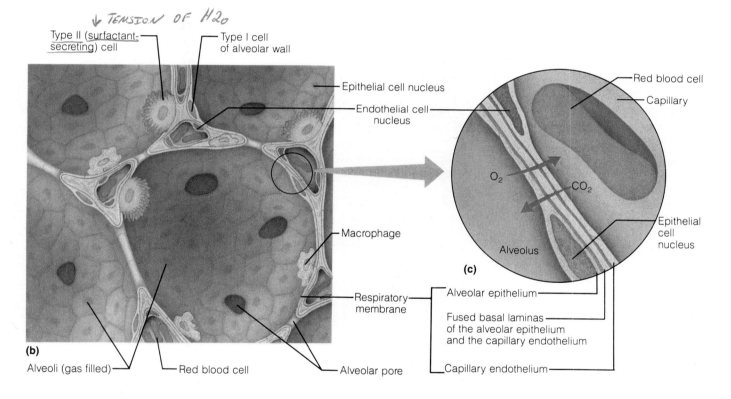

faces. Also called *dust cells,* these macrophages remove the tiniest airborne dust particles that have passed beyond the mucous filters of the respiratory tree. Since the alveoli are "dead ends," there must be some way to keep aged macrophages from accumulating within the alveoli. Most macrophages are moved superiorly into the bronchioles, where ciliary action carries them to the pharynx. In this manner, we each clear and swallow over 2 million dust cells per hour!

The Lungs and Pleural Cavities

Gross Anatomy of the Lungs

The paired **lungs** occupy all the thoracic cavity except the mediastinum, which houses the heart, great blood vessels, trachea, primary bronchi, esophagus, and other organs (Figure 20.11). Each lung is roughly cone-shaped. The anterior, lateral, and posterior sur-

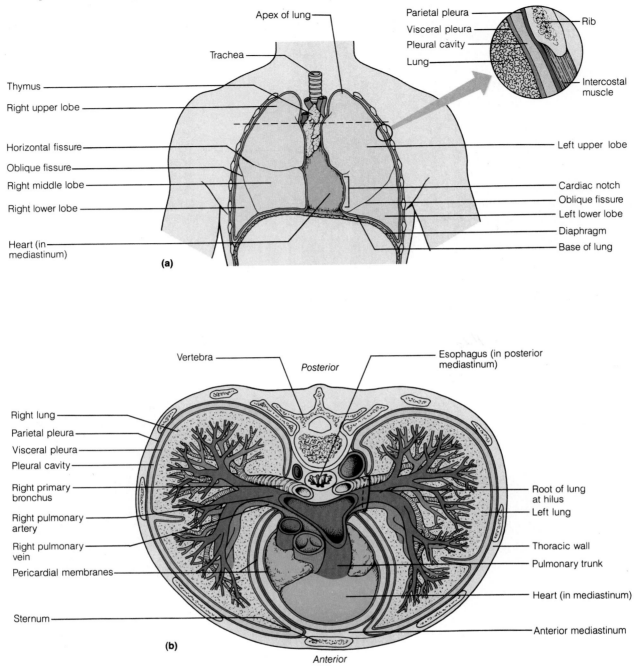

Figure 20.11 Anatomical relationships of organs in the thoracic cavity. (a) Anterior view of the thoracic organs. The lungs flank the central mediastinum. The inset at upper right depicts the pleura and the pleural cavity. (b) Transverse section through the superior part of the thorax, showing the lungs and the main organs in the mediastinum. The plane of section is shown by the dotted line in part (a).

faces of a lung lie in close contact with the ribs and form a continuously curving surface called the *costal surface.* Just deep to the clavicle is the **apex,** the rounded, superior tip of the lung. The concave inferior surface that rests on the diaphragm is the **base.** On the *medial (mediastinal) surface* of each lung is an indentation, the **hilus,** through which blood vessels, bronchi, lymph vessels, and nerves enter and exit the lung. Collectively, these structures attach the lung to

the mediastinum and are called the **root** of the lung. The largest components of this root are the pulmonary artery and veins and the primary bronchus. Figure 20.12 (parts b and c) shows the precise arrangement of the root structures at the hilus of each lung.

Because the heart is tilted slightly to the left of the median plane of the thorax, the two lungs differ slightly in shape and size. The left lung is somewhat smaller than the right, and the **cardiac notch**—a con-

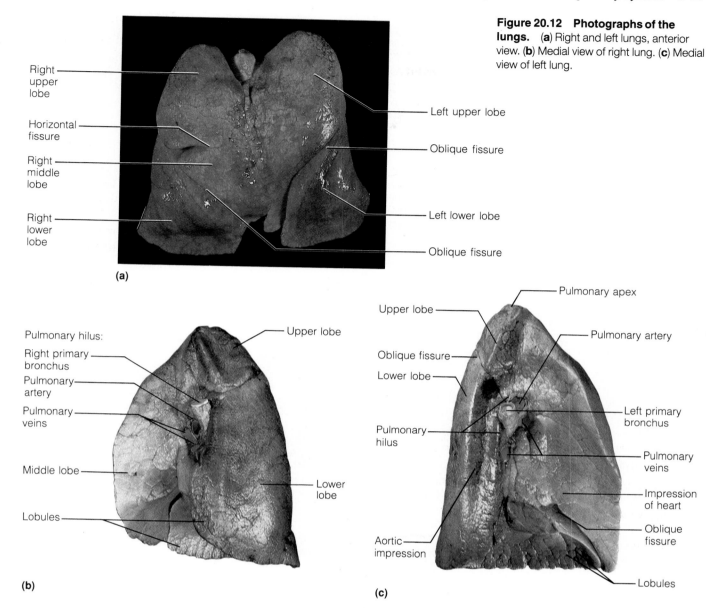

Figure 20.12   Photographs of the lungs. (**a**) Right and left lungs, anterior view. (**b**) Medial view of right lung. (**c**) Medial view of left lung.

Labels in figure:
(a) Right upper lobe; Horizontal fissure; Right middle lobe; Right lower lobe; Left upper lobe; Oblique fissure; Left lower lobe; Oblique fissure

(b) Pulmonary hilus: Right primary bronchus; Pulmonary artery; Pulmonary veins; Middle lobe; Lobules; Upper lobe; Lower lobe

(c) Upper lobe; Oblique fissure; Lower lobe; Pulmonary hilus; Aortic impression; Pulmonary apex; Pulmonary artery; Left primary bronchus; Pulmonary veins; Impression of heart; Oblique fissure; Lobules

cavity in the left lung's medial aspect—is molded to and accommodates the heart (Figure 20.11a). Several deep fissures divide the two lungs into different patterns of **lobes**. The left lung is divided into two lobes, the **upper lobe** and the **lower lobe**, by the **oblique fissure**. The right lung is partitioned into three lobes, the **upper, middle,** and **lower lobes,** by the **oblique** and **horizontal fissures**. The upper lobes are also called superior lobes, the lower lobes, inferior lobes. As we mentioned earlier, each lobe is served by a secondary bronchus and its branches.

Each of the lobes, in turn, contains a number of pyramid-shaped **bronchopulmonary segments** (Figure 20.13). These segments are separated from one another by thin partitions of dense connective tissue. Each bronchopulmonary segment receives air from an individual tertiary (third-order) bronchus. Each lung contains ten bronchopulmonary segments arranged in similar, but not identical, patterns in the two lungs.

The bronchopulmonary segments have clinical importance: They limit the spread of some diseases within the lung, for infections do not easily cross the partitions between them. Furthermore, because of the partitions, surgeons can neatly remove one or several segments without damaging any others or cutting any major blood vessels. ■

The smallest subdivision of the lung that can be seen with the naked eye is the **lobule**. The lobules appear on the lung surface as hexagons ranging from the size of a pencil eraser to the size of a penny (Figure 20.12b and c). Each lobule is served by a large bronchiole and its branches. In most city dwellers and in smokers, the connective tissue that separates the individual lobules is blackened with carbon.

As we mentioned earlier, the lungs consist largely of air tubes and spaces. The balance of the lung tissue, its *stroma* (stro'mah; "supporting mattress"), is a

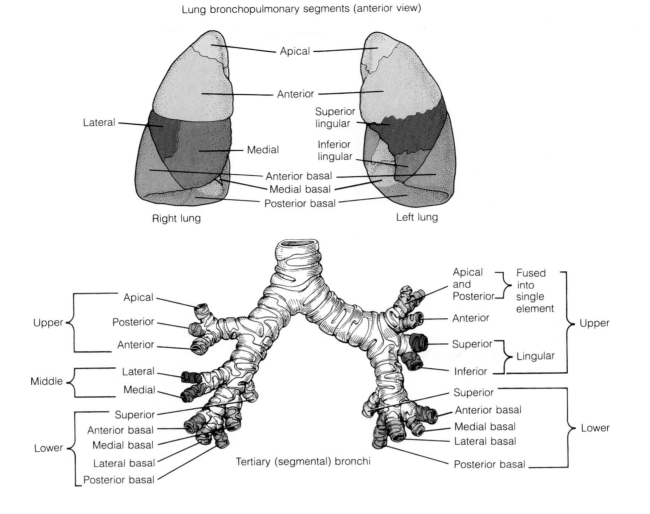

Lung bronchopulmonary segments (anterior view)

Apical

Anterior

Lateral

Superior lingular

Inferior lingular

Medial

Anterior basal

Medial basal

Posterior basal

Right lung

Left lung

Apical — Upper
Posterior
Anterior

Apical and Posterior — Fused into single element
Anterior
Superior — Lingular
Inferior
Upper

Lateral — Middle
Medial

Superior — Lower
Anterior basal
Medial basal
Lateral basal
Posterior basal

Tertiary (segmental) bronchi

Superior
Anterior basal
Medial basal
Lateral basal
Posterior basal
Lower

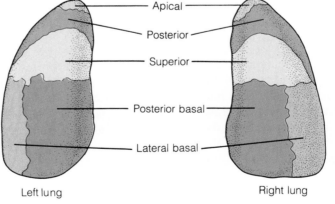

Lung bronchopulmonary segments (posterior view)

Apical

Posterior

Superior

Posterior basal

Lateral basal

Left lung

Right lung

Figure 20.13 Bronchopulmonary segments. Also shown are the tertiary (third-order) bronchi that supply each bronchopulmonary segment. Most introductory anatomy courses do not require that their students learn the individual bronchopulmonary segments, but we provide this figure for reference.

Table 20.1   Principal Organs of the Respiratory System

Structure	Description, general and distinctive features	Function
Nose	Jutting external portion supported by bone and cartilage; internal nasal cavity divided in half by midline nasal septum and lined with mucosa	Produces mucus; filters, warms, and moistens incoming air; resonance chamber for speech
	Roof of nasal cavity contains olfactory epithelium	Receptors for sense of smell
	Nasal cavity surrounded by paranasal sinuses	Same as nasal cavity; also lighten skull
Pharynx	Passageway connecting nasal cavity to larynx and oral cavity to esophagus; three subdivisions: nasopharynx, oropharynx, and laryngopharynx	Passageway for air and food
	Houses tonsils	Facilitates exposure of immune system to inhaled antigens
Larynx	Connects pharynx to trachea; framework of cartilage and dense connective tissue; opening (glottis) can be closed by epiglottis or vocal folds	Air passageway; prevents food from entering lower respiratory tract
	Houses true vocal cords	Voice production
Trachea	Flexible tube running from larynx and dividing inferiorly into two primary bronchi; walls contain C-shaped cartilages that are incomplete posteriorly where trachealis muscle occurs	Air passageway; filters, warms, and moistens incoming air
Bronchial tree	Consists of right and left primary bronchi, which subdivide within the lungs to form secondary and tertiary bronchi and bronchioles; bronchiolar walls contain complete layer of smooth muscle; constriction of this muscle impedes expiration	Air passageways connecting trachea with alveoli; warms and moistens incoming air
Alveoli	Microscopic chambers at termini of bronchial tree; walls of simple squamous epithelium underlain by thin basement membrane; external surfaces intimately associated with pulmonary capillaries	Main sites of gas exchange
	Special alveolar cells produce surfactant	Reduces surface tension; helps prevent lung collapse
Lungs	Paired composite organs located within pleural cavities of thorax; composed primarily of alveoli and respiratory passageways; stroma is fibrous elastic connective tissue, allowing lungs to recoil passively during expiration	House passageways smaller than primary bronchi
Pleurae	Serous membranes; parietal pleura lines thoracic cavity; visceral pleura covers external lung surfaces	Produce lubricating fluid and compartmentalize lungs

framework of connective tissue containing many elastic fibers. As a result, the lungs are light, soft, spongy, elastic organs that each weigh only about 0.6 kg (1.25 pounds). The elasticity of healthy lungs helps to reduce the effort of breathing, as described shortly.

Blood Supply and Innervation of the Lungs

The **pulmonary arteries** deliver oxygen-poor blood to the lungs for oxygenation (Figures 20.8b and 20.11b). In the lung, these arteries branch along with the bronchi, generally lying *posterior* to the corresponding bronchi. The smallest arteries feed into the **pulmonary capillary networks** around the alveoli (Figure 20.10a). Oxygenated blood is carried from the alveoli of the lungs to the heart by the **pulmonary veins**. Most of the tributaries to these veins run *anterior* to the corresponding bronchi within the lungs. However, some important venous tributaries run in the partitions of connective tissue between the lung lobules and the bronchopulmonary segments.

The *bronchial arteries* and *veins* provide and drain systemic blood to and from the lung tissues. These small vessels enter and exit the lungs at the hilus, and within the lung they run along the branching bronchi.

The lungs are innervated by sympathetic, parasympathetic, and visceral sensory fibers. These nerve fibers enter each lung through the *pulmonary plexus* on the lung root and run along the bronchial tubes and blood vessels within the lungs. As we mentioned earlier, the parasympathetic fibers constrict the air tubes, and the sympathetic fibers dilate them.

The Pleura

The **pleura** (ploo′rah; "the side") is a thin, double-layered serosa in the thoracic cavity (Figures 20.11a and 20.14). The **parietal pleura** lines the internal surface of the thoracic wall, the superior surface of the diaphragm, and the lateral surfaces of the mediastinum. From the mediastinum, the parietal pleura reflects lat-

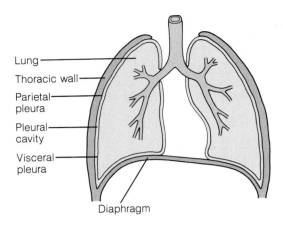

Lung
Thoracic wall
Parietal pleura
Pleural cavity
Visceral pleura
Diaphragm

Figure 20.14 Diagram of the pleurae and pleural cavities.

erally to enclose the root of the lung. At the hilus of the lung, the parietal pleura becomes the **visceral pleura** (also called the *pulmonary pleura*), which covers the external lung surface. The visceral pleura also dips into and lines the lung fissures.

The **pleural cavity** is the space between the parietal and visceral pleurae. This "cavity" is actually a slit-like potential space filled with a layer of *pleural fluid.* Secreted by the pleura, this lubricating fluid allows the lungs to glide without friction over the thoracic wall during breathing movements. The fluid also holds the parietal and visceral pleurae together, just as an intervening film of oil or water would hold together two glass plates. The two pleurae can slide easily from side to side across each other, but their separation is strongly resisted. Consequently, the lungs cling tightly to the thoracic wall and are forced to expand and recoil as the size (volume) of the thoracic cavity increases and decreases during breathing (see next section).

The pleurae also help divide the thoracic cavity into three separate compartments—the central mediastinum and two lateral pleural compartments, each containing a lung. This compartmentalization helps to prevent one moving organ (lung or heart) from interfering with another. Compartmentalization also limits the spread of local infections and the extent of traumatic injury.

Pleurisy (pleuritis)—inflammation of the pleura—often results from lung infections such as pneumonia. Can you predict some symptoms of pleurisy?

The rubbing of the two inflamed pleural membranes against one another produces a stabbing chest pain with each breath. Actually, this pain originates only from the parietal pleura because the visceral pleura, as a visceral structure, is relatively insensitive (see the discussion of visceral pain on pp. 405–406).

If pleurisy is prolonged, the inflamed pleurae may secrete an excessive amount of pleural fluid, which then overfills the pleural cavity and exerts pressure on the lungs. This type of pleurisy hinders breathing movements, but it is much less painful than the rubbing type. ■

Ventilation

The Mechanism of Ventilation

Breathing, or **pulmonary ventilation**, consists of two phases: *inspiration* (inhalation), the period when air flows into the lungs, and *expiration* (exhalation), the period when gases exit the lungs. The mechanical factors that promote these gas flows are the topic of this section (see Figure 20.15). For more information on the *muscles* of ventilation, see Table 10.5 on pp. 256–257.

↑ TO ↓ , PASSIVE

Inspiration

The process of **inspiration** is easy to understand if you visualize the thoracic cavity as a box with a single entrance at the top: the tube-like trachea. The volume of this box is changeable and can be increased by enlarging all of its diameters, thereby decreasing the gas pressure within it. The decrease in internal gas pressure in turn causes air to enter the box from the atmosphere, since gases always flow along their pressure gradients. The inrushing air enters the lungs.

The above conditions exist during normal quiet inspiration, when the inspiratory muscles—the diaphragm and external intercostal muscles—are activated. Here is how quiet inspiration works:

1. Action of the diaphragm. When the dome-shaped diaphragm contracts, it moves inferiorly and flattens (this is called "superoinferior expansion" in Figure 20.15a). As a result, the vertical dimension (height) of the thoracic cavity increases.

2. Action of the intercostal muscles. The external intercostal muscles contract to raise the ribs. Since the ribs normally extend anteroinferiorly from the vertebral column, lifting them enlarges both the right-to-left dimension of the thoracic cavity ("lateral expansion" in Figure 20.15b) and the anterior-to-posterior diameter of the thorax ("anteroposterior expansion" in Figure 20.15c).

Although these actions expand the thoracic dimensions by only a few millimeters along each plane, this is sufficient to increase the volume of the thoracic cavity by almost half a liter (a pint)—the usual vol-

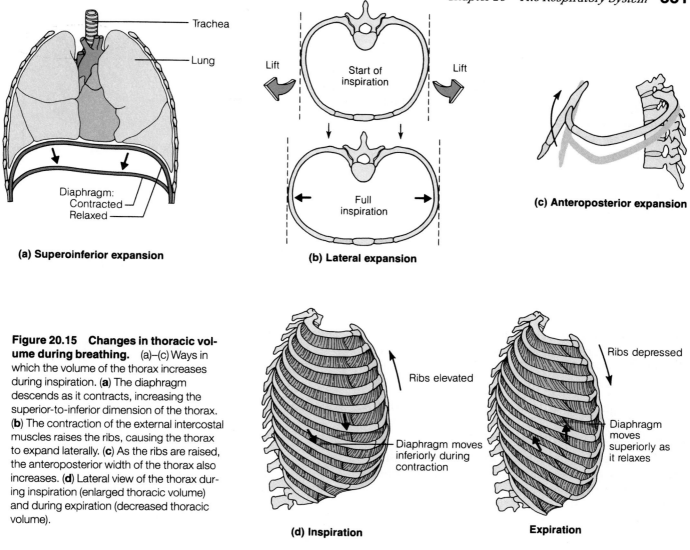

Figure 20.15   Changes in thoracic volume during breathing. (a)–(c) Ways in which the volume of the thorax increases during inspiration. (a) The diaphragm descends as it contracts, increasing the superior-to-inferior dimension of the thorax. (b) The contraction of the external intercostal muscles raises the ribs, causing the thorax to expand laterally. (c) As the ribs are raised, the anteroposterior width of the thorax also increases. (d) Lateral view of the thorax during inspiration (enlarged thoracic volume) and during expiration (decreased thoracic volume).

ume of air that enters the lungs during a normal quiet inspiration.

During deep or forced inspiration, the thoracic volume is increased by the contraction of additional muscles. The rib cage is elevated by the scalenes and sternocleidomastoid muscle of the neck and by the pectoral muscles of the chest. Additionally, the back extends as the thoracic curvature is straightened by the erector spinae muscles.

Expiration

Quiet **expiration** in healthy individuals is chiefly a passive process. As the inspiratory muscles relax, the rib cage drops under the force of gravity, and the relaxing diaphragm moves superiorly. At the same time, the many elastic fibers within the lungs recoil. Thus, the volumes of the thorax and the lungs decrease simultaneously. The decrease in volume raises the pressure within the lungs, and air moves along its pressure gradient out of the lungs.

By contrast, *forced* expiration is an active process produced by contraction of muscles in the ab-dominal wall, primarily the oblique and transversus abdominis muscles. These contractions (1) increase the intra-abdominal pressure, which forces the diaphragm superiorly, and (2) sharply depress the rib cage and so decrease thoracic volume. The internal intercostal muscles, quadratus lumborum, and the latissimus dorsi also help to depress the rib cage.

You may have heard of someone suffering a collapsed lung. Can you imagine some factors that would cause a lung to collapse?

A lung collapses if air enters the pleural cavity, a phenomenon called **pneumothorax** (nu"mo-tho'raks; "air thorax"). This breaks the seal of pleural fluid that holds the lung to the thoracic wall, allowing the elastic lung to collapse like a deflating balloon. Pneumothorax may result from a penetrating wound to the thorax or from lung disease that erodes a hole through the external surface of the lung. A pneumothorax is reversed surgically by closing the "hole" through which air enters the pleural cavity, then gradually withdrawing the air from this cavity with chest tubes.

This treatment allows the lung to reinflate and resume its normal function.

Obstruction of a bronchus may also cause the lung to collapse as the air beyond the point of blockage is gradually reabsorbed into the pulmonary capillaries. Bronchi may be obstructed by a plug of mucus, an inhaled object, a tumor, or enlarged lymph nodes.

Note that since the lungs are in completely separate pleural cavities, one lung can collapse without affecting the function of the other. ■

In a healthy lung, the alveoli remain open at all times and do not collapse during exhalation. At first glance this phenomenon seems to contradict the laws of physics, for a watery film coats the internal surfaces of the alveoli, and water molecules have a high attraction for one another (called *surface tension*) that should collapse the alveoli after each breath. The reason this does not occur is that the alveolar film also contains **surfactant** (ser-fak′tant), detergent-like molecules secreted by the type II alveolar cells. Surfactant interferes with the cohesiveness of water molecules, thereby greatly reducing surface tension and enabling the alveoli to remain open.

Pulmonary surfactant is not produced until the end of fetal life. What might its absence mean for an infant born prematurely?

In such premature infants, the alveoli do collapse during exhalation and must be completely reinflated during each inspiration, an effort that requires tremendous expenditure of energy. This can lead to exhaustion and respiratory failure. The condition, called **infant respiratory distress syndrome (IRDS)**, is responsible for one-third of all infant deaths. It is treated by using positive-pressure respirators to force air into the alveoli, keeping them inflated between breaths. ■

Neural Control of Ventilation

Although our tide-like breathing seems beautifully simple, its control is fairly complex. The basic pattern of breathing is generated by neurons in the reticular formation of the medulla and pons, but other parts of the brain can modify this pattern.

Respiratory Centers in the Medulla

The pace-setting nucleus in the medulla oblongata is called the **inspiratory center**, or, more precisely, the **dorsal respiratory group (DRG)** (Figure 20.16). Neurons of this center fire in a spontaneous and rhythmic way, producing the oscillating pattern of breathing.

When the inspiratory neurons fire, nerve impulses travel along the phrenic and intercostal nerves to stimulate contraction of the diaphragm and exter-

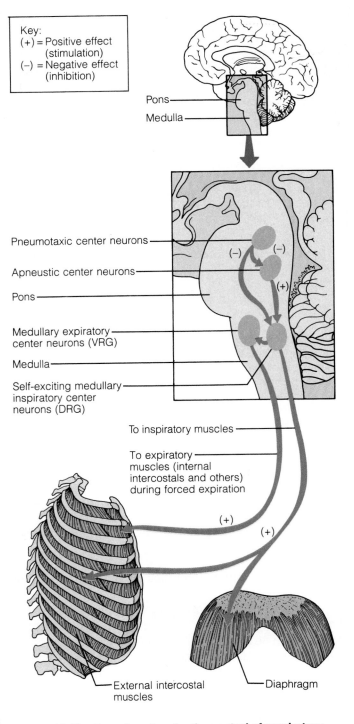

Key:
(+) = Positive effect (stimulation)
(−) = Negative effect (inhibition)

Figure 20.16 Neural centers for the control of respiratory rhythm. Respiratory centers in the medulla and pons are shown as ovals in the enlarged view.

nal intercostal muscles, respectively. The inspiratory center then becomes dormant, and expiration occurs passively as the inspiratory muscles relax and the lungs recoil. This cyclic on-off activity of the inspiratory neurons repeats continuously and produces our baseline respiratory rate of 12–15 breaths per minute.

The function of the second medullary center, the **expiratory center,** or **ventral respiratory group (VRG),** is not as well understood. It seems to control *forced* expiration (and perhaps also forced inspiration). It does not promote expiration during normal quiet breathing.

Respiratory Centers in the Pons

Although the medullary inspiratory center generates the basic respiratory rhythm, breathing becomes abnormal if connections between the pons and medulla are cut. Breathing is still rhythmic, but it occurs in gasps. The centers in the pons influence those in the medulla, producing smooth transitions from inspiration to expiration, and vice versa.

The **pneumotaxic** (nu"mo-tak'sik) **center,** the more superior center (Figure 20.16), continuously sends inhibitory impulses to the inspiratory center of the medulla. When its signals are particularly strong, the duration of inspiration shortens, and breathing quickens. The quick breaths are shallow. Shortening the inspiratory phase prevents overinflation of the lungs.

The **apneustic** (ap-nu'stik) **center** provides inspiratory drive by continuously stimulating the medullary inspiratory center. Its effect is to prolong inspiration or to cause breath holding in the inspiratory phase. Breathing becomes very deep and slow when the fibers running from the pneumotaxic center to the apneustic center are severed, a fact indicating that the apneustic center is normally inhibited by the pneumotaxic center.

Influence of Chemical Factors

Although the medullary inspiratory center sets a baseline ventilatory rate, this is modified by input from receptors that sense the chemistry of blood. These chemoreceptors respond to falling concentrations of oxygen, rising levels of carbon dioxide, or increased acidity of the blood by signaling the respiratory center to increase the depth and rate of breathing. This action brings the blood gases back to their normal concentrations. The chemoreceptors are of two types: The *central chemoreceptors* are neurons in the medulla. They lie near the inspiratory center but are distinct from it. The *peripheral chemoreceptors* are the **aortic bodies** on the aortic arch and the **carotid bodies** in the fork of the common carotid artery (Figure 20.17). The aortic bodies send their sensory information to the medulla through the vagus nerve, while the carotid bodies send theirs through the glossopharyngeal nerve.

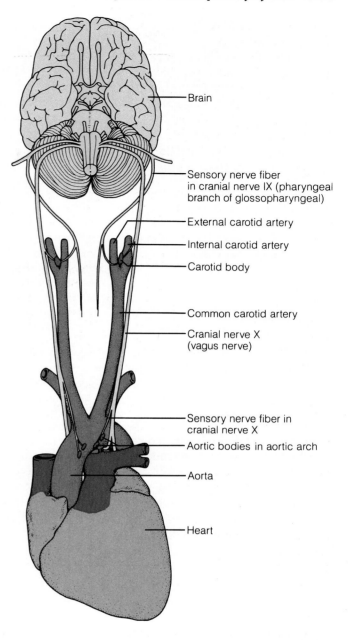

- Brain
- Sensory nerve fiber in cranial nerve IX (pharyngeal branch of glossopharyngeal)
- External carotid artery
- Internal carotid artery
- Carotid body
- Common carotid artery
- Cranial nerve X (vagus nerve)
- Sensory nerve fiber in cranial nerve X
- Aortic bodies in aortic arch
- Aorta
- Heart

Figure 20.17   Location of the peripheral chemoreceptors in the carotid and aortic bodies.   Also shown are the sensory pathways from these receptors through cranial nerves IX and X to the respiratory centers in the medulla.

Influence of Higher Brain Centers

Higher centers of the brain can modify the basic ventilatory rate and rhythm.

Hypothalamic Centers.   Strong emotions and pain acting through the limbic system activate sympathetic centers in the hypothalamus, which can then influence ventilation by signaling the respiratory centers in the brain stem. For example, have you ever touched

---

A CLOSER LOOK Lung Cancer: The Facts Behind the Smoke Screen

Lung cancer accounts for fully one-third of all cancer deaths in the United States, and its incidence is increasing daily. In both sexes, lung cancer is the most prevalent type of malignancy. It has a notoriously low cure rate: The overall 5-year survival rate is just 7%, and the average person survives only about 9 months after diagnosis. The survival rate is low because lung cancer is tremendously aggressive and metastasizes rapidly and widely and because most cases are not diagnosed until they are well advanced.

For years, America was remarkably unaware of the link between lung cancer and cigarette smoking, even though over 90% of lung cancer patients were smokers. As late as the 1950s, professional athletes were being used to advertise cigarettes, and cigarettes were promoted as a harmless way to keep one's weight down ("reach for a cigarette rather than a sweet"). As late as the 1960s, smoking was widely held to be socially desirable, even romantic,

despite the fact that it causes bad breath, yellow teeth, and leaves an unpleasant residue on the smoker's fingers, as well as on the clothes and hair of all those exposed to the smoke. It is impossible to consider smoking glamorous in light of recent scientific studies that strongly link the smoking habit to premature wrinkling of the skin of the face. Smoking a single cigarette increases one's heart rate, constricts the peripheral blood vessels throughout the body, disrupts the flow of air in the lungs, and affects one's brain and mood. Furthermore, secondhand smoke can harm the nonsmoker. Not until 4 to 5 hours after a nonsmoker leaves a smoky room is the carbon monoxide cleared from the body. "Passive smokers" who are frequently exposed to tobacco smoke experience respiratory irritations, and most studies indicate that they are at increased risk for lung cancer as well.

Ordinarily, nasal hairs, sticky mucus, and the action of cilia do a

fine job of protecting the lungs from chemical and biological irritants, but when one smokes, these cleansing devices are overwhelmed and eventually become nonfunctional. Continuous irritation prompts the production of more mucus, but smoking paralyzes the cilia that clear this mucus and also depresses the activity of lung macrophages. The result is a pooling of mucus in the lower respiratory tree and an increased frequency of pulmonary infections, including pneumonia and COPD. However, it is the irritant effects of the tobacco tars—the 15 or so carcinogens in tobacco smoke—that eventually translate into lung cancer. These carcinogens eventually cause the epithelial cells lining the bronchial tree to proliferate wildly and lose their characteristic histological structure.

The three most common types of lung cancer are (1) **squamous cell carcinoma** (20% to 40% of cases), which arises in the epithelium of the larger bronchi and tends to form

---

something cold and clammy and then gasped? That response was mediated through the hypothalamus.

Cortical Controls. Although breathing is normally regulated involuntarily by the respiratory centers in the brain stem, we can also exert conscious control over the rate and depth of our breathing—choosing to hold our breath or voluntarily taking a deep breath, for example. This voluntary control involves direct signals from the motor cerebral cortex to the motor neurons that stimulate the respiratory muscles, bypassing the medullary centers. Our ability to voluntarily hold our breath is limited, however, for the respiratory centers of the brain stem automatically reinitiate breathing when the concentration of carbon dioxide in the blood reaches critical levels. That is the reason why drowning victims are always found to have inhaled water into their lungs.

Diseases of the Respiratory System

The respiratory system is particularly vulnerable to infectious diseases because it is open to airborne pathogens. We have already discussed many of these respiratory infections, such as rhinitis and laryngitis, but here we concentrate on the most disabling respiratory disorders: the group of diseases collectively called **chronic obstructive pulmonary disease (COPD)**. The most lethal lung disease, **lung cancer**, is given special consideration in the box above. These disorders are "living proof" of the devastating effects of cigarette smoking on the lungs.

The chronic obstructive pulmonary diseases, exemplified by *obstructive emphysema* and *chronic*

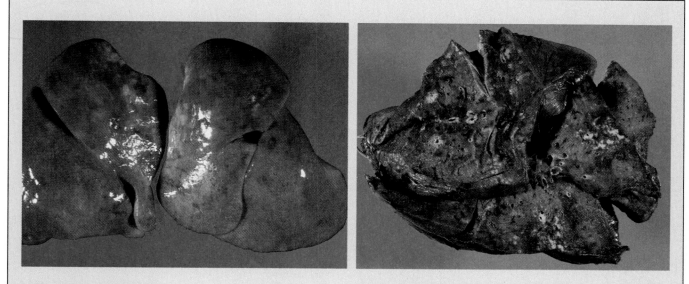

Healthy lungs of a nonsmoker.

Lung of a smoker. Smoker's lungs always contain patches of carbon. Cancer may develop in the air tubes.

masses that cavitate and bleed; (2) **adenocarcinoma** (25% to 35%), which originates in the peripheral areas of the lung as solitary nodules that develop from bronchial mucous glands and alveolar epithelial cells; and (3) **small cell carcinoma** (10% to 20%), also called **oat cell carcinoma**, which consists of lymphocyte-like epithelial cells that originate in the primary bronchi and grow aggressively in cords or small grape-like clusters within the mediastinum.

The most effective treatment for lung cancer is complete removal of the diseased lung because this procedure has the best potential for halting metastases and prolonging life. However, removal is a choice that is open to very few lung cancer patients—for most patients, the chances of surviving lung cancer are too poor to justify a traumatic and useless surgery. In most cases, radiation therapy and chemotherapy are the only options.

It is estimated that smoking contributes to one-seventh of all deaths in the United States. For smokers, the best chance of avoiding these statistics is to quit smoking. The incidence of lung cancer is about 20 times greater in smokers than in nonsmokers, but studies suggest this ratio drops to 2 to 1 for ex-smokers who have not smoked in 15 years.

bronchitis, are a major cause of death and disability in the United States and are becoming more common. These diseases share certain features: (1) The patient almost invariably has a history of smoking; (2) **dyspnea** (disp-ne′ah; "bad breathing"), difficult or labored breathing, occurs; (3) coughing and pulmonary infections occur frequently; and (4) most COPD victims ultimately develop respiratory failure.

Obstructive emphysema (em″fĭ-se′mah; "to inflate") is distinguished by a permanent enlargement of the alveoli accompanied by a deterioration of the alveolar walls (Figure 20.18). The disease is associated with chronic inflammation of the lungs and increased activity of lung macrophages, whose lysosomal enzymes seem responsible for destroying the alveolar walls. Chronic inflammation also leads to fibrosis (formation of scar tissue), and the lungs become progressively less elastic, making expiration difficult and exhausting. For complex reasons, the bronchioles

open during inspiration but collapse shut during expiration, trapping huge volumes of air in the alveoli. This retention of air leads to the development of a permanently expanded "barrel chest." Widespread damage to the pulmonary capillaries increases the resistance in the pulmonary vascular circuit, forcing the heart's right ventricle to enlarge through overwork. Cigarette smoking is the most common cause of emphysema, but hereditary factors seem important in some patients.

In **chronic bronchitis**, inhaled irritants lead to a prolonged secretion of excess mucus by the mucosa of the lower respiratory passages and to inflammation and fibrosis of this mucosa. These responses obstruct the airways, severely impairing ventilation and gas exchange. Infections occur frequently because bacteria and viruses thrive in the stagnant pools of mucus. Patients with chronic bronchitis are sometimes called "blue bloaters" because lowered blood oxygenation

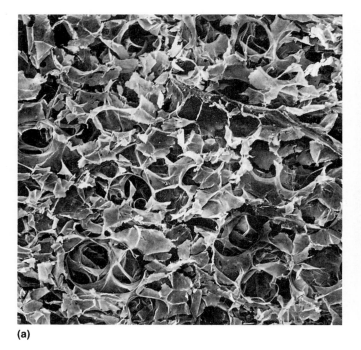

(a)

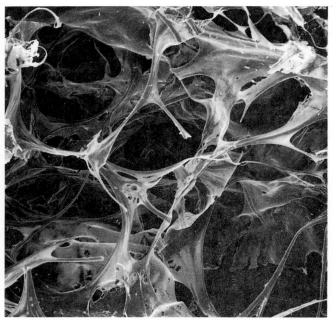

(b)

Figure 20.18 Scanning electron micrograph of the alveoli (a) in a normal lung and (b) in a lung with emphysema. Note that the alveoli are much larger in the emphysematic lung (magnification: 25 ×).

occurs early in the disease, often resulting in cyanosis. However, the degree of dyspnea is usually moderate when compared to that of emphysema sufferers. So significant is cigarette smoking as a causative factor of chronic bronchitis that this disease would be an insignificant health problem if cigarettes were unavailable. Air pollution is another causative factor, but a minor one.

The Respiratory System Throughout Life

Since embryos develop in a craniocaudal (head-to-tail) direction, the upper respiratory structures appear before the lower ones. By week 4, a thickened plate of ectoderm called the **olfactory placode** (plak′ōd; "plate") has appeared on each side of the future face (Figure 20.19a and b). These placodes quickly invaginate to form *olfactory pits* that form the nasal cavity (including the olfactory epithelium in its roof). The nasal cavity then connects with the future pharynx of the developing foregut, which forms at the same time.

The lower respiratory organs develop from a tubular outpocketing off the foregut called the **laryngotracheal bud** (Figure 20.19c). The proximal part of this bud forms the trachea, and its distal part branches repeatedly to form the bronchi and their subdivisions, including (eventually) the lung alveoli. The respiratory tubes, like the gut tube from which they arise, are lined by endoderm and covered by splanchnic mesoderm. The endoderm becomes the lining epithelium (and glands) of the trachea, bronchial tree, and alveoli. The splanchnic mesoderm, by contrast, gives rise to all other layers of the tracheal and bronchial walls (including cartilage, smooth muscle, and lamina propria) and to the stroma of the lungs.

The respiratory system reaches functional maturity relatively late in development. No alveoli appear until the start of month 7 (25 weeks), and the alveolar type I cells do not attain their extreme thinness until the time of birth. It is not until 26–30 weeks that a prematurely born baby can survive and breathe on its own. Infants born before this time are most severely threatened by infant respiratory distress syndrome resulting from inadequate production of surfactant (p. 552).

During fetal life, the lungs are filled with fluid, and all respiratory exchanges occur across the placenta. At birth, the first breaths bring air to the lungs, and the alveoli inflate and begin to function in gas ex-

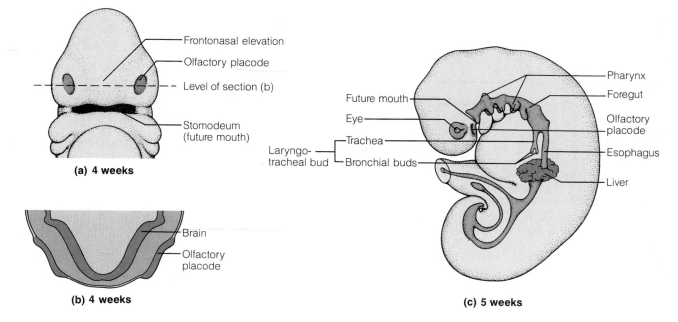

Figure 20.19   Embryonic development of the respiratory system.   (a) Ventral view of an embryo's face and head, showing the ectodermal olfactory placodes. **(b)** Transverse section through the region of the dashed line in part (a). Ventral aspect is below. **(c)** Development of the lower respiratory structures. The laryngotracheal bud is shown as an outpocketing from the foregut.

change. However, it takes nearly 2 weeks for the lungs to become fully inflated. During this time, the fluid that originally filled the lungs is absorbed into the alveolar capillaries.

Important congenital defects of the respiratory system include cleft palate (p. 183) and **cystic fibrosis (CF).** CF is an inherited disease in which the functions of exocrine glands are disrupted throughout the body. CF affects the respiratory system by causing an oversecretion of a viscous mucus by the bronchial glands. This mucus clogs the respiratory passageways and predisposes the child to early death through repeated respiratory infections. CF accounts for about 5% of childhood deaths. Modern antibiotics allow two-thirds of all CF sufferers to survive at least into their 20s. Researchers recently isolated the gene that, when defective, is responsible for CF. Eventually, it may be possible to deliver nondefective copies of this gene to the mucus-secreting cells in the respiratory ducts of children with CF, thereby providing a cure.

At birth, only one-sixth of the final number of lung alveoli are present. The lungs continue to mature throughout childhood, and more alveoli are formed until young adulthood. Research has revealed that when smoking begins in the early teens, the lungs never completely mature, and those additional alveoli never form.

Under normal conditions, the respiratory system works so efficiently that we are not even aware of it. Most problems that do occur are the result of external factors—for example, viral or bacterial infections, irritants that trigger asthma in susceptible individuals, or obstruction of the trachea by a piece of food. For many years, tuberculosis and bacterial pneumonia were the worst killers in the United States and Europe. Antibiotics have decreased their threat to a large extent, but they are still dangerous diseases. Influenza, a respiratory flu virus, killed at least 20 million people worldwide in 1918, and government health agencies still keep a watchful eye on the spread of new flu outbreaks each year. By far the most troublesome respiratory disorders, however, are COPD and lung cancer.

As we age, the thoracic wall becomes more rigid, and the lungs gradually lose their elasticity. Both of these factors result in a slowly decreasing ability to ventilate the lungs. The levels of oxygen in the blood may fall slightly, while levels of carbon dioxide rise. Additionally, just as the overall efficiency of the immune system declines with age, many of the respiratory system's protective mechanisms become less effective. The activity of the cilia in its epithelial lining decreases, and the macrophages in the lungs become sluggish. The net result is that the elderly are more at risk for respiratory infections, particularly pneumonia and influenza.

➕ Related Clinical Terms

Adenoidectomy Surgical removal of infected pharyngeal tonsil (adenoids).

Bronchoscopy (*scopy* = viewing) Use of a viewing tube to examine the internal surface of the main bronchi in the lung. The tube is inserted through the nose or mouth and guided inferiorly through the larynx and trachea. Forceps may be attached to the tip of the tube to remove trapped objects, take biopsy samples, or retrieve samples of mucus for examination.

Croup Disease in children in which viral infection causes the air passageways to narrow through inflammation. Coup is characterized by coughing that sounds like the bark of a dog, hoarseness, and wheezing or grunting sounds during inspiration. Most cases pass after a few days, but severe cases may require a tracheotomy (see below) to bypass the obstructed upper respiratory tubes.

Deviated septum Condition in which the nasal septum takes a more lateral course than normal and may obstruct breathing. Deviated septum is often manifest in old age but may also be congenital or result from a blow to the nose.

Epistaxis (ep"ĭ-stak'sis) (*epistazo* = to bleed at the nose) Nosebleed; nasal hemorrhage. Epistaxis commonly follows trauma to the nose or excessive nose blowing. Most nasal bleeding is from the highly vascularized anterior part of the nasal septum and can be stopped by pinching the nostrils together or packing them with cotton.

Nasal polyps Mushroom-like growths (benign neoplasms) of the nasal mucosa. Polyps may occur in response to continual nasal irritation and may block airflow.

Pneumonia Infectious inflammation of the lungs, in which fluid accumulates in the alveoli. Over 50 different varieties are known: Most are viral or bacterial (but the type of pneumonia often associated with AIDS, *Pneumocystis carinii* pneumonia, is caused by a fungus). Pneumonia is a leading cause of death in the United States (the sixth most common cause), because almost any severely ill person can develop it.

Sudden infant death syndrome (SIDS) Unexpected death of an apparently healthy infant during sleep. Commonly called crib death, SIDS is one of the most frequent causes of death in infants under 1 year old. The cause is unknown, but it may reflect immaturity of the brain's respiratory control centers.

Tracheotomy A surgical technique for opening an airway when the upper respiratory tract is obstructed. In a tracheotomy, the surgeon makes a vertical incision between the second and third tracheal rings in the anterior neck. A tube is then placed in the opening to keep this airway open.

Tuberculosis (TB) An infectious disease caused by the bacterium *Mycobacterium tuberculosis,* which primarily enters the body in inhaled air. TB typically affects the lungs but can spread through the lymph vessels to other organs as well. A massive inflammatory and immune response contains the primary infection within fibrous or calcified nodules in the lungs (tubercles), but the bacteria often survive, break out, and cause repeated infections. Symptoms of TB are coughing, fever, and chest pain. Antibiotics effectively treat TB. People with AIDS are at increased risk for contracting tuberculosis.

Chapter Summary

1. Respiration involves four processes: ventilation, external respiration, transport of respiratory gases in the blood, and internal respiration. Both the respiratory system and the cardiovascular system participate in respiration.

FUNCTIONAL ANATOMY OF THE RESPIRATORY SYSTEM (pp. 534–550)

1. The organs of the respiratory system include a conducting zone (nose to terminal bronchioles), which warms, moistens, and filters the inhaled air, and a respiratory zone (respiratory bronchioles to alveoli), where gas exchange occurs.

The Nose and the Paranasal Sinuses (pp. 535–537)

2. The nose and nasal cavity provide an airway for respiration and house the olfactory receptors.

3. The external nose is shaped by bone and by cartilage plates. The nasal cavity begins at the external nares and ends posteriorly at the internal nares. Air-swirling conchae occupy its lateral walls. The paranasal sinuses drain into the nasal cavity.

4. The mucosa that lines the nasal cavity and paranasal sinuses processes inspired air. Its epithelium (a pseudostratified ciliated columnar epithelium with goblet cells) is covered with a mucous sheet that filters dust particles. Its lamina propria contains glands that contribute to the mucous sheet and blood vessels that warm and moisten the air.

The Pharynx (pp. 537–538)

5. The pharynx has three regions. The nasopharynx (behind the nasal cavity) is an air passageway. The oropharynx (behind the mouth) and the laryngopharynx (behind the larynx) are passageways for both food and air.

6. The soft palate flips superiorly to seal off the nasopharynx during swallowing. The tubal and pharyngeal tonsils occupy the mucosa that lines the nasopharynx. The oropharynx contains the lingual and palatine tonsils. The laryngopharynx opens into the laryngeal inlet anteriorly and into the esophagus inferiorly.

The Larynx (pp. 538–541)

7. The larynx, or voice box, is the entrance gate to the trachea and lower respiratory tubes. The larynx has a skeleton of nine cartilages: the thyroid, epiglottis, cricoid, paired arytenoids, and some minor cartilages.

8. The epiglottis acts as a lid that prevents food or liquids from entering the lower respiratory channels. The larynx moves superiorly under this protective flap during swallowing.

9. The larynx contains the vocal cords (vocal folds). Exhaled air causes these folds to vibrate, producing the sounds of speech. The vocal folds extend anteriorly from the arytenoids to the thyroid cartilage, and muscles that move the arytenoids change the tension on the folds to vary the pitch of the voice. The pharynx, nasal cavity, lips, and tongue also aid in articulation.

10. The main sensory and motor nerves of the larynx (and of speech) are the recurrent laryngeal branches of the vagus.

The Trachea (pp. 541–542)

11. The trachea extends from the larynx in the neck to the primary bronchi in the thorax. This tube is reinforced by C-shaped cartilage rings, which keep it open. The trachealis muscle narrows the trachea, increasing the speed of airflow during coughing.

12. The wall of the trachea contains several layers: a mucous membrane, submucosa, and an outer adventitia. The mucous membrane consists of an air-filtering pseudostratified ciliated columnar epithelium, and a lamina propria rich in elastic fibers.

The Bronchi and Subdivisions:
The Bronchial Tree (pp. 542–545)

13. The right and left primary bronchi supply the lungs. Within the lungs, they subdivide into secondary bronchi, tertiary bronchi, and so on, finally to bronchioles and terminal bronchioles.

14. As the respiratory tubes become smaller, the cartilage in their walls is reduced and finally lost (bronchioles); the epithelium thins and loses its air-filtering function (smallest bronchioles); smooth muscle becomes increasingly important; and elastic fibers continue to surround all of the tubes.

15. The terminal bronchioles lead into the respiratory zone: the respiratory bronchioles, alveolar ducts, and alveolar sacs. All of these contain tiny chambers called alveoli. Gas exchange occurs in the alveoli, across the thin respiratory membrane. This air-blood barrier consists of alveolar type I cells, fused basal laminae, and capillary endothelial cells.

16. Alveoli also contain alveolar type II cells (which secrete surfactant) and freely wandering alveolar macrophages (dust cells). The macrophages remove dust particles that reach the alveoli.

The Lungs and Pleural Coverings (pp. 545–550)

17. The lungs flank the mediastinum in the thoracic cavity. Each lung is suspended in a pleural cavity by its root (the vessels supplying it) and has a base, an apex, and medial and costal surfaces. The root structures enter the lung hilus.

18. The right lung has three lobes (upper, middle, and lower lobes, as defined by oblique and horizontal fissures), whereas the left lung has two lobes (upper and lower lobes, as defined by the oblique fissure). The lobes are divided into bronchopulmonary segments (supplied by third-order bronchi).

19. The lungs consist primarily of air tubes and alveoli, but they also contain a stroma of elastic connective tissue.

20. The pulmonary arteries carry deoxygenated blood and generally lie posterior to their corresponding bronchi within the lungs. The pulmonary veins, which carry oxygenated blood, tend to lie anterior to their corresponding bronchi. The lungs are innervated by the pulmonary plexus.

21. Pleurae are serous membranes. The parietal pleura lines the thoracic wall, diaphragm, and mediastinum. The visceral pleura covers the external surfaces of the lungs. The slit-like pleural cavity between the parietal and visceral pleura is filled with a serous fluid that holds the lungs to the thorax wall and reduces friction during breathing movements.

VENTILATION (pp. 550–554)

The Mechanism of Ventilation (pp. 550–552)

1. Ventilation consists of inspiration and expiration. Inspiration occurs when the diaphragm and intercostal muscles contract, increasing the volume of the thorax. As the intrathoracic pressure drops, air rushes into the lungs.

2. Expiration is largely passive, occurring as the inspiratory muscles relax. The volume of the thorax decreases, and the lungs recoil elastically, raising pressure and forcing air out of the lungs.

3. Surface tension of the alveolar fluid threatens to collapse the alveoli after each breath. This tendency is resisted by surfactant secreted by alveolar type II cells.

Neural Control of Ventilation (pp. 552–554)

4. The respiratory centers in the medulla oblongata are the inspiratory center (dorsal respiratory group) and the expiratory center (ventral respiratory group). The inspiratory center is responsible for the rhythmicity of breathing, the expiratory center for forced expiration.

5. Respiratory centers in the pons, the pneumotaxic and apneustic centers, influence the activity of the medullary inspiratory center.

6. Chemoreceptors monitor concentrations of respiratory gases and acid in the blood. The central chemoreceptors are neurons in the medulla, and the peripheral chemoreceptors are the carotid and aortic bodies.

7. Emotions, pain, and other stressors can alter ventilation by acting through the hypothalamus. Ventilation can also be controlled voluntarily by the cerebral cortex for short periods of time.

DISEASES OF THE RESPIRATORY SYSTEM (pp. 554–556)

1. The major respiratory disorders are COPD (emphysema and chronic bronchitis) and lung cancer. A significant cause is cigarette smoking.

2. Emphysema is characterized by permanent enlargement and destruction of alveoli. The lungs lose elasticity, and expiration becomes an active, exhaustive process.

3. Chronic bronchitis is characterized by an excessive production of mucus in the lower respiratory passages, which impairs ventilation and leads to infection. Patients may become cyanotic as a result of low oxygen levels in the blood.

4. Lung cancer is extremely aggressive and metastasizes rapidly. The 5-year survival rate is less than 10%.

THE RESPIRATORY SYSTEM THROUGHOUT LIFE (pp. 556–557)

1. The nasal cavity develops from the olfactory placode; the pharynx forms as part of the foregut; and the lower respiratory tubes grow from an outpocketing of the embryonic pharynx (laryngotracheal bud). The epithelium lining the lower respiratory tubes derives from endoderm. Mesoderm forms all other parts of these tubes and the lung stroma as well.

2. The respiratory system completes its development very late in the prenatal period: No alveoli or surfactant appear until the start of month 7, and only one-sixth of the final number of alveoli are present at birth. Respiratory immaturity, especially a lack of surfactant, is the main cause of death of premature infants.

3. With age, the thorax becomes more rigid, the lungs become less elastic, and ventilation capacity declines. Respiratory infections become more common in old age.

Review Questions

Multiple Choice and Matching Questions

1. The molecule that keeps the alveoli from collapsing between breaths by reducing surface tension is called (a) water, (b) mucus, (c) surfactant, (d) lysozyme.

2. When the inspiratory muscles contract, (a) only the lateral diameter of the thoracic cavity increases, (b) only the anteroposterior diameter of the thoracic cavity increases, (c) the volume of the thoracic cavity decreases, (d) both the lateral and the anteroposterior dimensions of the thoracic cavity increase, (e) the diaphragm bulges superiorly.

3. Damage to which of the following would cause breathing to stop? (a) the pneumotaxic center, (b) the reticular formation in the medulla, (c) the apneustic center, (d) the primary motor cerebral cortex, (e) the hypothalamus, (f) a and c.

4. The skeleton of the external nose consists of (a) cartilage and bone, (b) bone only, (c) hyaline cartilage only, (d) elastic cartilage only.

5. The mucous sheets that cover the inner surfaces of the nasal cavity and bronchi are secreted by which cells? (a) serous cells in tubuloacinar glands, (b) mucous cells in tubuloacinar glands and epithelial goblet cells (c) ciliated cells, (d) alveolar type II cells.

6. Circle the single false statement about the vocal cords: (a) They are the same as the vocal folds. (b) They attach to the arytenoid cartilages. (c) Exhaled air flowing through the glottis vibrates them to produce sound. (d) They are also called the vestibular folds.

7. In both lungs, the surface that is the largest is (a) costal, (b) mediastinal, (c) inferior (base), (d) superior (apex).

8. Match the proper type of lining epithelium from the key with each respiratory structure listed below.

Key: (a) stratified squamous, (b) pseudostratified columnar, (c) simple squamous, (d) stratified columnar

_____ (1) nasal cavity
_____ (2) nasopharynx
_____ (3) laryngopharynx
_____ (4) trachea and bronchi
_____ (5) alveoli (type I cells)

9. Match the proper type of bronchus or bronchiole (at right) with the lung region supplied by that air tube and its branches (at left).

Lung region	Air tube
_____ (1) bronchopulmonary segment	(a) primary bronchus
_____ (2) lobule	(b) secondary bronchus
_____ (3) alveolar ducts and sacs	(c) tertiary bronchus
_____ (4) whole lung	(d) large bronchiole
_____ (5) lobe	(e) respiratory bronchiole

10. An alveolar sac (a) is an alveolus, (b) relates to an alveolus as a bunch of grapes relates to one grape, (c) is a huge, sac-like alveolus in an emphysema patient, (d) is the same as an alveolar duct, (e) all of the above.

11. The respiratory membrane (air-blood barrier) consists of (a) alveolar type I cell, basal laminae, endothelial cell; (b) air, connective tissue, lung; (c) type II cell, dust cell, type I cell; (d) pseudostratified epithelium, lamina propria, capillaries.

12. Which of the following is not in the conducting zone of the respiratory system? (a) pharynx, (b) alveolar duct, (c) trachea, (d) terminal bronchioles, (e) larynx.

13. The trachealis muscle and the smooth muscle around the bronchi develop from the following embryonic layer: (a) ectoderm, (b) endoderm, (c) mesoderm, (d) neural crest.

14. A serous cell of a gland secretes (a) the slippery serous fluid in the body cavities, (b) mucus, (c) a watery fluid containing enzymes, (d) tissue fluid.

15. The function of alveolar type I cells is (a) to produce surfactant, (b) to propel mucous sheets, (c) to phagocytize dust particles, (d) to allow rapid diffusion of respiratory gases.

Short Answer and Essay Questions

16. Trace the route of exhaled air from an alveolus to the external nares. Name all the structures through which the air passes.

17. (a) Why is it important that the trachea be reinforced with cartilage rings? (b) Of what advantage is it that the rings are incomplete posteriorly?

18. The cilia lining the upper respiratory passages (superior to the larynx) beat inferiorly while the cilia lining the lower respiratory passages (larynx and inferior) beat superiorly. What is the functional "reason" for this difference?

19. What is the function of the abundant elastin fibers that occur in the stroma of the lung and around all respiratory tubes from the trachea through the respiratory tree?

20. Describe the functional relationships between changes in thoracic volume and gas flow into and out of the lungs.

21. (a) Which tonsils lie in the fauces of the oropharynx? (b) Do the lingual tonsils lie near the superior part of the epiglottis? Explain.

22. It is easy to confuse the hilus of the lung with the root of the lung. Define both structures, and contrast hilus and root.

23. Three terms that are easily confused are *choanae, conchae,* and *carina.* Define each of these, clarifying the differences.

24. Briefly explain the anatomical "reason" why most men have deeper voices than boys or women.

25. Define the parietal pleura and the visceral pleura.

26. Sketch a picture of the right and left lungs in anterior view, showing all the fissures and lung lobes, as well as the cardiac notch.

Critical Thinking and Clinical Application Questions

27. (a) Two girls in a high school cafeteria were giggling over lunch, and both accidentally sprayed milk out their nostrils at the same time. Explain in anatomical terms why swallowed fluids can sometimes come out the nose. (b) A boy in the same cafeteria then stood on his head and showed he could drink milk upside down without any of it entering his nasal cavity or nose. What prevented the milk from flowing downward into his nose?

28. A surgeon had to remove three adjacent bronchopulmonary segments from the left lung of a patient with tuberculosis. Almost half of the lung was removed, yet there was no severe bleeding, and relatively few blood vessels had to be cauterized (closed off). Why was the surgery so easy to perform?

29. While diapering his 1-year-old boy (who puts almost everything in his mouth), Mr. Gregoire failed to find one of the small safety pins previously used. Two days later, his son developed a cough and became feverish. What probably had happened to the safety pin, and where (anatomically) would you expect to find it?

30. Mr. and Ms. Rao took their sick 5-year-old daughter to the doctor. The girl was breathing entirely through her mouth, her voice sounded odd and whiny, and a pus-like fluid was dripping from her nose. Which one of the four sets of tonsils was most likely infected in this child?

31. A taxi driver was carried into an emergency room after being knifed once in the left side of the thorax. The diagnosis was pneumothorax and a collapsed lung. (a) Explain exactly why the lung collapsed. (b) Explain why only one lung (not both) collapsed.

32. Mr. Jackson was driving to work when his car was hit broadside by a car that ran a red light. When freed from the wreckage, Mr. Jackson was deeply cyanotic, and his breathing had stopped. His heart was still beating, but his pulse was fast and thready. His head was cocked at a peculiar angle, and the emergency personnel said he seemed to have a fracture at the level of vertebra C_2. (a) How might these findings account for the cessation of breathing? (b) Why was Mr. Jackson cyanotic? Define cyanosis (see Chapter 5).

21

The Digestive System

Student Objectives

Overview of the Digestive System (pp. 562–564)

1. Describe the overall function of the digestive system, and differentiate the alimentary canal from the accessory digestive organs.

2. List the major processes that occur during digestion.

3. Draw the major subdivisions of the anterior abdominal wall.

Anatomy of the Digestive System (pp. 564–594)

4. Explain the location and function of the peritoneum and peritoneal cavity. Define mesentery.

5. Describe the four layers of the wall of the alimentary canal.

6. Describe the location, gross and microscopic anatomy, and basic functions of the mouth and teeth, pharynx, esophagus, stomach, small intestine, and large intestine.

7. Describe the gross and microscopic anatomy of the liver, gallbladder, and pancreas.

8. Name the mesenteries associated with the abdominal digestive organs.

The Digestive System Throughout Life (pp. 594–595)

9. Explain how the digestive organs develop in the embryo, and define the foregut, midgut, and hindgut.

10. Describe some abnormalities of the digestive organs at different stages of life.

Children typically feel a special fascination for the workings of the digestive system: They seem to relish crunching a potato chip, swallowing carbonated sodas, and making "mustaches" with milk on their lips. They delight in listening to their stomach "growl," and their earliest questions reveal a tremendous curiosity about bodily wastes. As adults, we know that a healthy digestive system is essential to the maintenance of life, for it converts food into the raw materials that build and fuel the body's cells. Specifically, the digestive system takes in food, breaks it into nutrient molecules, absorbs these molecules into the bloodstream, and then rids the body of the indigestible remains.

Overview of the Digestive System

Organization

The organs of the digestive system (Figure 21.1) can be divided into two main groups: (1) the *alimentary canal* (*aliment* = nourishment) and (2) the *accessory digestive organs.*

The **alimentary canal,** also called the *gastrointestinal (GI) tract,* is the muscular digestive tube that winds through the body. The organs of the alimentary canal are the *mouth, pharynx, esophagus, stomach,*

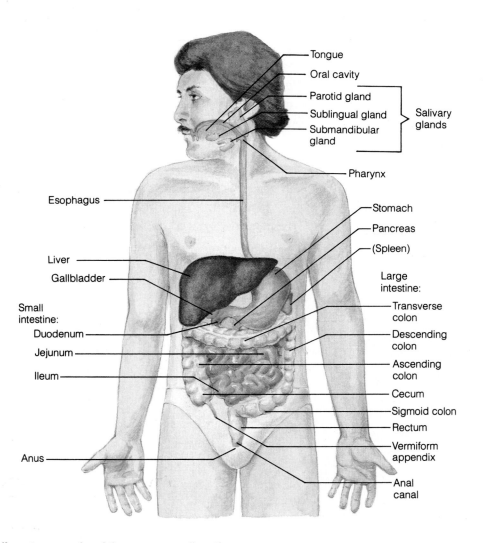

Figure 21.1   The alimentary canal and the accessory digestive organs.

small intestine (small bowel), and *large intestine* (large bowel). The large intestine leads to the terminal opening, or *anus.* In a cadaver, the alimentary canal is about 9 m (30 feet) long, but in a living person, it is considerably shorter because of its relatively constant muscle tone. Food material in the alimentary canal is technically considered to be *outside* the body because the canal is open to the external environment at both ends.

The **accessory digestive organs** are the *teeth, tongue, gallbladder,* and a number of large digestive glands—the *salivary glands, liver,* and *pancreas.* The teeth and tongue are in the mouth, and the digestive glands and gallbladder lie external to the alimentary canal and connect to it by ducts. The accessory digestive glands produce saliva, bile, and digestive enzymes—secretions that contribute to the breakdown of foodstuffs.

Digestive Processes

The processing of food by the digestive system involves six essential activities: ingestion, propulsion, mechanical digestion, chemical digestion, absorption, and defecation. Figure 21.2 summarizes the roles of specific organs in these processes.

1. **Ingestion** is simply the process of taking food into the mouth.

2. **Propulsion** is the process of moving food through the alimentary canal. It includes swallowing, which is initiated voluntarily, and peristalsis, an involuntary process. *Peristalsis,* the major means of propulsion, involves alternate waves of contraction and relaxation of musculature in the organ walls (Figure 21.3a). The net effect is the squeezing of food from one organ to the next, but some mixing occurs as well.

3. **Mechanical digestion** physically prepares food for chemical digestion by enzymes. Mechanical processes include chewing, churning food in the stomach, and **segmentation,** or rhythmic local constrictions of the intestine (Figure 21.3b). Segmentation mixes food with digestive juices and increases the rate of nutrient absorption by moving different parts of the food mass repeatedly over the intestinal wall.

4. **Chemical digestion** is a series of steps in which complex food molecules (carbohydrates, proteins, and lipids) are broken down to their chemical building blocks (simple sugars, amino acids, fatty acids, and glycerol). Chemical digestion is carried out by enzymes that are secreted into the lumen (central cavity) of the alimentary canal.

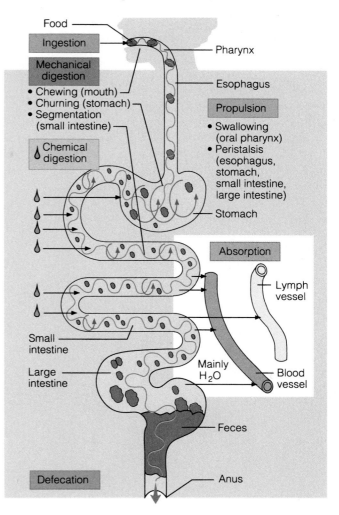

Figure 21.2 Schematic summary of digestive processes.
These processes include ingestion, mechanical digestion, propulsion, chemical (enzymatic) digestion, absorption, and defecation. The specific sites of these activities are indicated. The mucosa of almost the entire alimentary canal secretes mucus, which protects and lubricates.

ALL ENZYMES = PROTEINS

Ø ALL PROTEINS = ENZYMES

5. **Absorption** is the transport of digested end products from the lumen of the alimentary canal into the blood and lymph capillaries in its wall.

6. **Defecation** is the elimination of indigestible substances from the body as feces.

Abdominal Regions and Quadrants

Most digestive organs are contained in the abdominopelvic cavity, the largest division of the ventral body cavity. To map the positions of these abdominopelvic organs, clinicians divide the anterior abdomi-

CARBOHYDRATES ⟶ MONOSACCHARIDES

PROTEINS ⟶ AMINO ACIDS

FATS ⟶ FATTY ACIDS + GLYCEROL

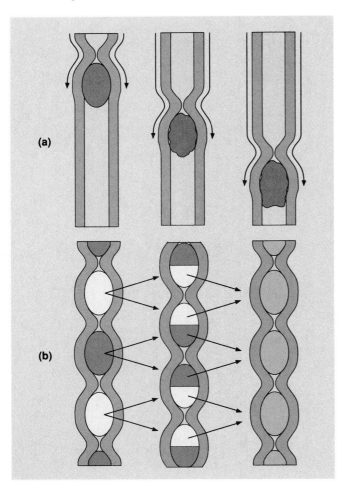

(a)

(b)

Figure 21.3 Peristalsis and segmentation. (a) In peristalsis, adjacent segments of the alimentary canal alternately contract and relax, moving food distally along the canal. (b) In segmentation, nonadjacent segments of the intestine alternately contract and relax. Because the active segments are separated by inactive regions, the food is moved onward and then backward. This movement results in mixing of the food rather than in forward propulsion.

nal wall into a number of regions. In one scheme, four lines forming a pattern similar to a tic-tac-toe grid divide the abdominal wall into nine areas (Figure 21.4a and b; also recall Chapter 1, pp. 15). The two vertical lines of the grid extend inferiorly from the midpoint of each clavicle. The superior horizontal line is in the *subcostal* ("below ribs") *plane,* and connects the inferior points of the costal margins. The inferior horizontal line is in the *transtubercular plane* and connects the tubercles (widest points) of the iliac crests. All of these bony landmarks can be felt on the body's surface. The superior three of the nine regions are the **right** and **left hypochondriac regions** (hi"po-kon'dre-ak; "deep to the ribs") and the central **epigastric region** (ep"ĭ-gas'trik; "superficial to the stomach"). The middle three regions are the **right** and **left lumbar regions** (or *lateral regions*), and the central **umbilical**

region. The inferior three regions are the **right** and **left iliac regions,** or **inguinals,** and the central **hypogastric** ("below the stomach") **region.** The hypogastric region is also called the **pubic region.** Because many abdominal organs move, their positions on the abdominal grid are only approximate.

In a simpler scheme, a vertical and a horizontal line intersecting at the navel (Figure 21.4c) define four regions, the **right** and **left upper** and **lower quadrants.**

Anatomy of the Digestive System

The Peritoneal Cavity and Peritoneum

Recall from Chapter 1 that all divisions of the ventral body cavity contain slippery *serous membranes.* The **peritoneum** (per"ĭ-to-ne'um) of the abdominopelvic cavity is the most extensive of these membranes (Figure 21.5). The **visceral peritoneum** covers the external surfaces of most digestive organs and is continuous with the **parietal peritoneum,** which lines the walls of the abdominopelvic cavity. Between the visceral and parietal peritoneum is the **peritoneal cavity,** a slit-like potential space containing serous fluid. This lubricating fluid, which is secreted by the peritoneum, allows the organs to glide easily along one another as they move during digestion.

Peritonitis is inflammation and infection of the peritoneum. Can you guess what may cause this condition?

Peritonitis can arise from a piercing wound to the abdomen, from a perforating ulcer that leaks stomach juices into the peritoneal cavity, or from poor sterile technique during abdominal surgery. Most commonly, however, peritonitis results from a burst appendix that leaks feces into the peritoneal cavity. Peritonitis is a dangerous condition that can lead to death if it is *generalized* (widespread within the peritoneal cavity). Treatment involves cleaning out the peritoneal cavity and administering large doses of antibiotics. ■

A **mesentery** (mes'en-ter"e) is a double layer of peritoneum—a sheet of two serous membranes fused back to back—that extends to the digestive organs from the body wall (Figure 21.5a). Mesenteries hold the organs in place, store fat, and provide a route by which circulatory vessels and nerves reach the organs in the peritoneal cavity. Mesenteries are described more completely later in the chapter.

Not all digestive organs are suspended within the peritoneal cavity by a mesentery. Some parts of the in-

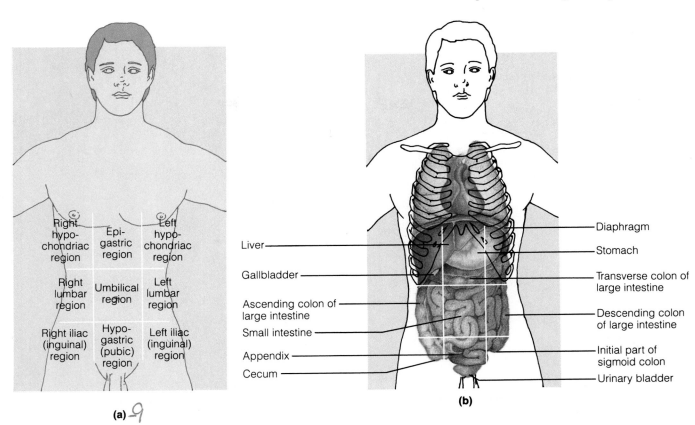

(a)

(b)

Liver

Gallbladder

Ascending colon of large intestine

Small intestine

Appendix

Cecum

Diaphragm

Stomach

Transverse colon of large intestine

Descending colon of large intestine

Initial part of sigmoid colon

Urinary bladder

Right hypo-chondriac region

Epi-gastric region

Left hypo-chondriac region

Right lumbar region

Umbilical region

Left lumbar region

Right iliac (inguinal) region

Hypo-gastric (pubic) region

Left iliac (inguinal) region

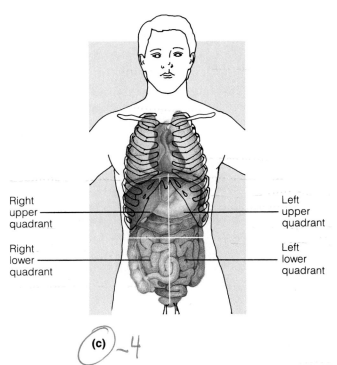

(c)

Right upper quadrant

Left upper quadrant

Right lower quadrant

Left lower quadrant

Figure 21.4   Divisions of the anterior abdominal wall for mapping the digestive organs in the abdominopelvic cavity.   (a) The nine surface regions of the anterior abdominal wall. (b) The abdominal viscera as they relate to the nine surface regions. (c) A second, simpler scheme of four quadrants centered at the navel.

testine, for example, originate in the cavity but then fuse to the dorsal abdominal wall during development (Figure 21.5b). In so doing, these organs lose their mesentery and lodge behind the peritoneum. Such organs are called **retroperitoneal** (*retro* = behind). By contrast, the digestive organs that keep their mesentery and remain within the peritoneal cavity are called **intraperitoneal** or **peritoneal** organs. The stomach is an example of such an organ. (The specific intraperitoneal and retroperitoneal organs are listed in Table 21.1 on page 592.)

It is not completely understood why some digestive organs are located retroperitoneally while others are suspended freely. One explanation is that the an-

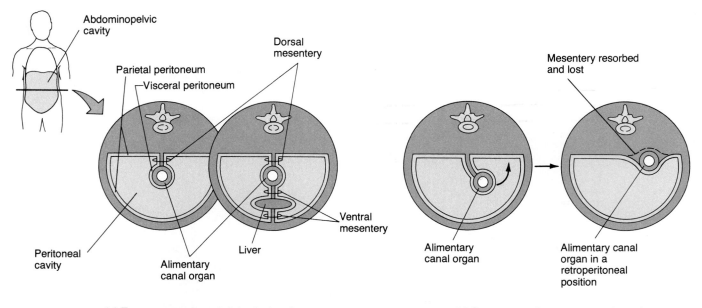

(a) Transverse section of abdominal cavity

(b) Some organs become retroperitoneal

Figure 21.5   The peritoneum and peritoneal cavity. **(a)** Simplified cross sections through the abdominopelvic cavity. The peritoneum is shown as a distinct line around the peritoneal cavity. Most regions of the abdominopelvic cavity contain a dorsal mesentery (cross section at left), while the superior part of the abdominal cavity also contains a ventral mesentery (cross section at right). Note that the slit-like peritoneal cavity is actually much thinner than indicated here, because it is nearly "filled" by the organs within it. **(b)** Some parts of the alimentary canal lose their mesentery during development and become retroperitoneal.

choring of various segments of the alimentary canal to the posterior abdominal wall minimizes the chances of knotting or kinking of the alimentary canal during peristaltic movements.

Histology of the Alimentary Canal Wall

The walls of every organ of the alimentary canal, from the esophagus to the anal canal, are made up of the same four tissue layers (Figure 21.6). In fact, similar layers occur in the hollow organs of the urinary and reproductive systems as well. From the lumen outward, these layers are the *mucosa, submucosa, muscularis externa,* and *serosa.*

WHAT MAKE UP THE MUCOSA?   3 SUBLAYERS

The Mucosa

The innermost layer is the **mucosa,** or *mucous membrane.* More complex than other mucous membranes in the body, the typical digestive mucosa contains three sublayers: (1) a lining epithelium, (2) a lamina propria, and (3) a muscularis mucosae.

The lining **epithelium** borders the lumen of the alimentary canal and performs many functions related to digestion (absorption and mucus production, for example). This epithelium is continuous with the

ducts and secretory cells of the digestive glands. Most of these glands lie fully within the wall and are called *intrinsic glands.* Larger glands, such as the liver and pancreas, extend beyond the walls of the alimentary canal and are called *extrinsic (accessory) glands.*

The **lamina propria** is a loose areolar connective tissue. Its capillaries nourish the lining epithelium and absorb digested nutrients. The lamina propria includes the majority of the gut-associated lymphoid tissue (GALT), which serves as a line of defense against bacteria and other microorganisms in the alimentary canal (p. 529).

External to the lamina propria is the **muscularis mucosae,** a thin layer of smooth muscle that produces local movements of the mucosa. For example, the twitching of this muscle layer dislodges sharp food particles that become embedded in the mucosa.

The Submucosa

Just external to the mucosa is the **submucosa,** a layer of connective tissue containing blood vessels, lymph vessels, and nerve fibers. Its rich vascular network sends branches to all other layers of the wall. Its connective tissue is a type intermediate between loose areolar and dense irregular, a "moderately dense" connective tissue. The many elastic fibers within the

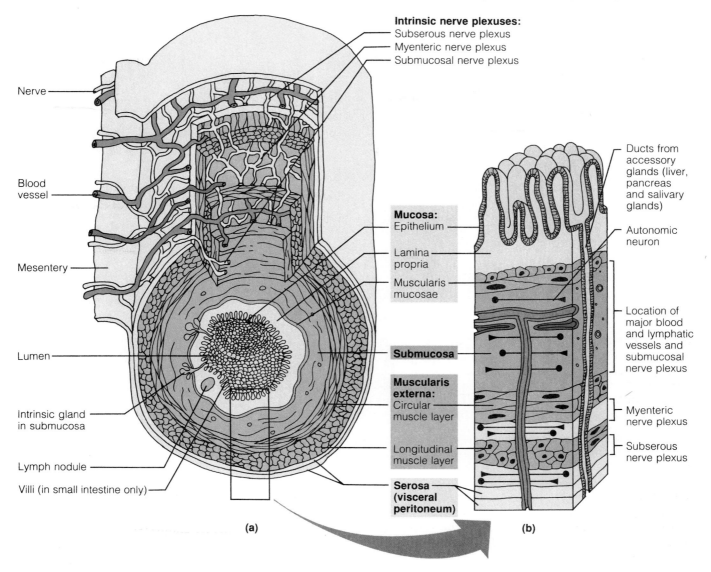

Intrinsic nerve plexuses:
Subserous nerve plexus
Myenteric nerve plexus
Submucosal nerve plexus

Nerve

Blood vessel

Mesentery

Lumen

Intrinsic gland in submucosa

Lymph nodule

Villi (in small intestine only)

Mucosa:
Epithelium

Lamina propria

Muscularis mucosae

Submucosa

Muscularis externa:
Circular muscle layer

Longitudinal muscle layer

Serosa (visceral peritoneum)

Ducts from accessory glands (liver, pancreas and salivary glands)

Autonomic neuron

Location of major blood and lymphatic vessels and submucosal nerve plexus

Myenteric nerve plexus

Subserous nerve plexus

(a)

(b)

Figure 21.6   Histology of the alimentary canal (from esophagus through large intestine). (**a**) Cross-sectional view, with some of the wall removed to show the intrinsic nerve plexuses. (**b**) Enlargement of a portion of part (a).

submucosa enable the stomach to return to shape after holding a large meal.

The Muscularis Externa

External to the submucosa is the **muscularis externa,** also simply called the *muscularis.* Throughout most of the alimentary canal, this tunic consists of two layers of smooth muscle, an inner *circular layer* whose fibers orient around the canal, and an outer *longitudinal layer* whose fibers orient along the length of the canal. Contractions of the muscularis externa are responsible for peristalsis and segmentation. In some places, the circular layer thickens to form sphincters that act as valves to prevent backflow of food from one organ to the next.

The Serosa

The **serosa,** or visceral peritoneum, is the outermost layer of the intraperitoneal organs of the alimentary canal. Like all serous membranes, it is formed of a simple squamous epithelium (mesothelium) underlain by a thin layer of areolar connective tissue.

The esophagus, which lies within the thoracic mediastinum and is not suspended in the peritoneal cavity, has no serosa. Instead, its external layer is an *adventitia,* an ordinary fibrous connective tissue that binds the esophagus to surrounding structures. Retroperitoneal organs have both a serosa and an adventitia—a serosa on the anterior side facing the peritoneal cavity and an adventitia on the posterior side embedded in the posterior abdominal wall.

FIBRO SEROUS LAYER (BETTER NAME)

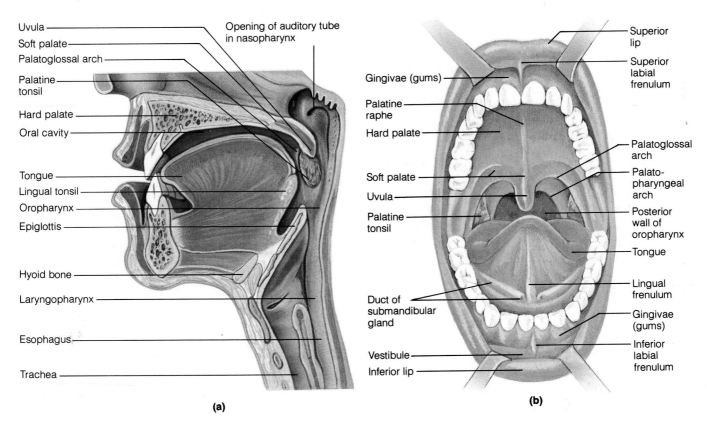

Figure 21.7 Anatomy of the mouth. (a) Sagittal section of the oral cavity and pharynx. (b) Anterior view of mouth.

Nerve Plexuses

Visceral nerve plexuses occur within the wall of the alimentary canal (Figure 21.6). The **myenteric nerve plexus** (mi-en-ter'ik; "intestinal muscle") lies in the muscularis externa between the circular and longitudinal layers, and the **submucosal nerve plexus** occupies the submucosa. The myenteric plexus innervates the muscularis externa and controls peristalsis and segmentation. The submucosal plexus, by contrast, primarily innervates the mucosa: It signals the glands there to secrete and the muscularis mucosae to contract. Both plexuses contain parasympathetic and sympathetic motor components and visceral sensory fibers, all of which link the alimentary canal to the brain and bring digestion under the influence of the central nervous system.

Despite this external influence from the brain, digestive activity is largely automatic, controlled by an internal nervous system of *enteric neurons* (*enteric* = gut) in both the myenteric and submucosal plexuses. Within the wall of the alimentary canal, enteric neurons form independent reflex arcs of sensory, intrinsic, and motor neurons. Thus, enteric neurons are able to bring about independently the normal movements of peristalsis and segmentation. The classical autonomic nervous system (parasympathetic and sympathetic) merely speeds or slows this inherent activity.

The parasympathetic and sympathetic inputs to the digestive organs are postganglionic sympathetic fibers, preganglionic parasympathetic fibers, and postganglionic parasympathetic neurons (Chapter 14). Recall that parasympathetic input stimulates digestive functions whereas sympathetic input inhibits digestion.

The wall of the alimentary canal has a third nerve plexus: the **subserous nerve plexus** within the serosa (Figure 21.6b). This minor plexus consists of parasympathetic and sympathetic fibers that extend internally to supply the two main plexuses.

The Mouth and Associated Organs

The Mouth

Food enters the alimentary canal through the mouth, where it is chewed, manipulated by the tongue, and moistened with saliva. The **mouth,** or <u>oral cavity</u> (Figure 21.7), is a mucosa-lined cavity whose boundaries are the lips anteriorly, the cheeks laterally, the palate superiorly, and the tongue inferiorly. Its anterior opening is the **oral orifice.** Posteriorly, the mouth borders the fauces of the oropharynx.

The mouth is divided into the vestibule and the oral cavity proper. The **vestibule** (ves′tĭ-būl; "porch") is the slit between the teeth and the cheeks (or lips). When you brush the outer surface of your teeth, your toothbrush is in the vestibule. The **oral cavity proper** is the region of the mouth that lies internal to the teeth.

Histology of the Mouth. The walls of the oral cavity consist of only a few layers of tissue: an internal mucosa (an epithelium and lamina propria only), a thin submucosa in some areas, and an external layer of muscle or bone. The lining of the mouth, a thick stratified squamous epithelium, protects it from abrasion by sharp pieces of food during chewing. The tongue, palate, lips, and gums may show slight keratinization of their lining epithelium, which provides extra protection against abrasion.

The Lips and Cheeks. The **lips,** or **labia,** and the **cheeks** have a core of skeletal muscle covered by skin. The orbicularis oris muscle, which encircles the oral orifice, forms the bulk of the lips. The cheeks are formed largely by the buccinator muscles. The lips and cheeks help keep food inside the mouth during chewing.

The lips are thick flaps extending from the inferior boundary of the nose to the superior boundary of the chin. The region of the lip where one applies lipstick or lands a kiss is called the *red margin.* This is a transition zone where the highly keratinized skin meets the oral mucosa. The red margin is poorly keratinized and translucent, revealing the color of blood in the underlying capillaries. Because there are no sweat or sebaceous glands in the red margin, it must be moistened with saliva periodically to prevent drying and cracking. The **labial frenulum** (fren′u-lum; "little bridle of the lip") is a median fold that connects the internal aspect of each lip to the gum (Figure 21.7b).

The Palate. The **palate,** which forms the roof of the mouth, has two distinct parts: the *hard palate* anteriorly and the *soft palate* posteriorly (Figure 21.7). (Recall Chapter 20, p. 535.) The bony hard palate forms a rigid surface against which the tongue forces food during chewing. The muscular soft palate is a mobile flap. Dipping inferiorly from its free edge is the fingerlike uvula. The soft palate rises to close off the nasopharynx during swallowing. (To demonstrate this action, try to breathe and swallow at the same time.) Laterally, the soft palate is anchored to the tongue by the **palatoglossal arches** and to the wall of the oropharynx by the **palatopharyngeal arches** (Figures 21.7 and 21.9). These two folds form the boundaries of the fauces, the arched area of the oropharynx that contains the palatine tonsils.

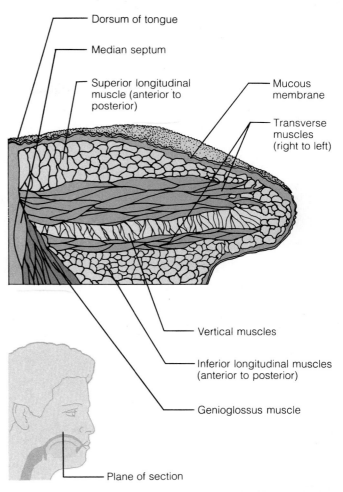

Figure 21.8 labels:
- Dorsum of tongue
- Median septum
- Superior longitudinal muscle (anterior to posterior)
- Mucous membrane
- Transverse muscles (right to left)
- Vertical muscles
- Inferior longitudinal muscles (anterior to posterior)
- Genioglossus muscle
- Plane of section

Figure 21.8 Frontal section through the left half of the tongue, showing its intrinsic skeletal muscles and the genioglossus, an extrinsic muscle.

The Tongue

The **tongue** occupies the floor of the mouth and fills most of the oral cavity when the mouth is closed (Figure 21.7a). The tongue is predominantly a muscle constructed of interlacing fascicles of skeletal muscle fibers. During chewing, the tongue grips food and constantly repositions it between the teeth. Tongue movements also mix the food with saliva and form it into a compact mass called a *bolus* (bo′lus; "lump"). Then, during swallowing, the tongue moves posteriorly to push the bolus into the pharynx. In speech, the tongue helps to form some consonants (k, d, t, and l, for example). Finally, the tongue houses most of the taste buds.

The tongue has both intrinsic and extrinsic muscle fibers. The *intrinsic muscles* are confined within the tongue and are not attached to bone. Their fibers run in several different planes (Figure 21.8). These muscles change the shape of the tongue but do not change its position. The *extrinsic muscles,* including the genioglossus muscle shown in Figure 21.8, extend

Know Bones That Make Up Hard Palate! Pg 535

Soft Palate Contains Connective Tissue, Covered By Epithelium. Lymphatic Tissue.

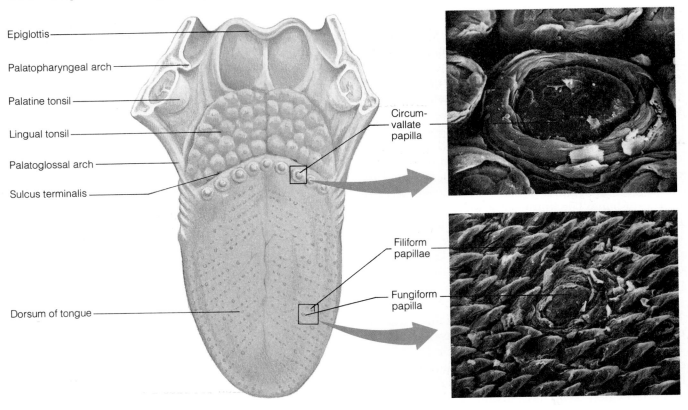

Epiglottis

Palatopharyngeal arch

Palatine tonsil

Lingual tonsil

Palatoglossal arch

Sulcus terminalis

Dorsum of tongue

Circum-vallate papilla

Filiform papillae

Fungiform papilla

Figure 21.9   The superior surface of the tongue.   Also shown are the locations and detailed structures of the circumvallate, fungiform, and filiform papillae. (Top right, 300 ×; bottom right, 140 ×).

to the tongue from bones of the skull or the soft palate (see Tables 10.2 and 10.3 on pp. 248–251). These extrinsic muscles alter the position of the tongue: They protrude it, retract it, and move it laterally. The tongue is divided by a median septum of connective tissue, and each half contains identical groups of muscles.

A fold of mucosa on the undersurface of the tongue, the **lingual frenulum,** secures the tongue to the floor of the mouth (Figure 21.7b). It also limits posterior movements of the tongue.

In some people, the lingual frenulum is abnormally short or extends exceptionally far anteriorly. Can you deduce the effect of this condition?

People with such a frenulum are often referred to as "tongue-tied" because of the speech distortions that result when movement of the tongue is restricted. This congenital condition, called **ankyloglossia** (ang″kĭ-lo-glos′e-ah; "fused tongue"), is corrected surgically by snipping the frenulum. ■

The superior surface of the tongue bears *papillae,* peg-like projections of the mucosa. The three major types of papillae—filiform, fungiform, and vallate (circumvallate)—are shown in Figure 21.9. The coni-

cal and pointed **filiform papillae** (fil′ĭ-form; "thread-shaped") roughen the tongue so that it can grip and manipulate food during chewing. These are the smallest and most numerous papillae and line up in parallel rows. Keratinization stiffens them and gives the tongue surface its whitish appearance.

Fungiform papillae (fun′jĭ-form) resemble tiny mushrooms (*fungi* = mushroom). While less abundant than filiform papillae, they are scattered widely over the tongue surface. Each fungiform papilla has a vascular core that gives it a red color.

Ten to twelve large **vallate** or **circumvallate papillae** (ser″cum-val′āt) line up in a V-shaped row, two-thirds of the way posteriorly on the tongue surface. This row lies directly anterior to a groove, the **sulcus terminalis,** which marks the border between the mouth and pharynx. Each circumvallate papilla is surrounded by a circular ridge (*circumvallate* means "surrounding wall"), from which it is separated by a deep furrow. Both the circumvallate and fungiform papillae house taste buds (p. 412).

The posterior third of the tongue lies in the oropharynx, not in the mouth. It is covered not with papillae but with *lingual tonsils,* which form large bumps (see Figure 21.9 and p. 528).

SALIVARY AMYLASE (ENZYME)
AMYL - REFERS TO CARBOHYDRATE

The Salivary Glands

Saliva is a complex mixture of water, ions, mucus, and enzymes that performs many functions: It moistens the mouth, dissolves food chemicals so that they can be tasted, wets food, and binds the food together into a bolus. Enzymes contained in the saliva begin the digestion of starches. Many bacteria inhabit the mouth and produce acids that dissolve the enamel of the teeth, thereby initiating tooth decay; saliva contains a bicarbonate buffer that neutralizes this acid. Finally, saliva contains bactericidal enzymes, antibodies, and a cyanide compound, all of which keep the mouth clean by killing oral bacteria.

All **salivary glands** are compound tubuloalveolar glands. Small *intrinsic salivary glands* are scattered within the mucosa of the tongue, palate, lips, and cheeks. Saliva from these glands keeps the mouth moist at all times. Large *extrinsic salivary glands,* by contrast (Figure 21.10), secrete saliva only as we eat. These are the *parotid, submandibular,* and *sublingual glands.* They lie external to the mouth but are connected to the mouth through their ducts.

The largest extrinsic gland is the **parotid** (pah-rot'id) **gland.** True to its name (*par* = near; *otid* = the ear), it lies anterior to the ear, between the masseter muscle and the skin (Figure 21.10a). The duct of this gland, called the **parotid duct,** runs parallel to the zygomatic arch, penetrates the muscle of the cheek, and opens into the mouth lateral to the second upper molar. The branches of the facial nerve run through the parotid gland on their way to the muscles of facial expression. For this reason, surgery on this gland can lead to facial paralysis.

Mumps is caused by a virus that spreads from one person to another in the saliva. Its dominant symptom is inflammation and swelling of the parotid gland. Why do you think people with mumps say it hurts to open their mouth or chew?

Movements of the mandible pull and push on the irritated parotid glands, causing pain. ■

The **submandibular gland** (*submaxillary gland*) is about the size of a walnut. It lies along the medial surface of the mandibular body, just anterior to the angle of the mandible. The duct of this gland raises the mucosa of the floor of the mouth and opens directly lateral to the tongue's frenulum (Figure 21.7b). The **sublingual gland** lies in the floor of the oral cavity, inferior to the tongue. Its 10 to 12 ducts open into the mouth, directly superior to the gland (Figure 21.10a).

The histology of the salivary glands is shown in Figure 21.10b. The secretory cells of these glands are *serous cells,* which produce a watery secretion containing the enzymes and ions of saliva, and *mucous cells,* which produce mucus. The parotid glands contain only serous cells; the submandibular and intrinsic glands contain approximately equal numbers of serous and mucous cells; and the sublingual glands

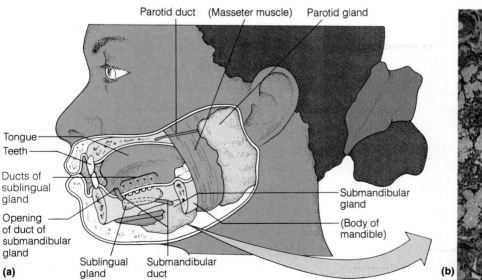

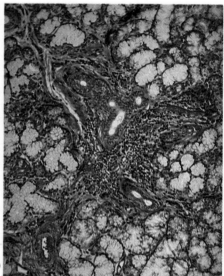

Figure 21.10   The major salivary glands.   (a) Locations of the parotid, submandibular, and sublingual glands. **(b)** Photomicrograph of the sublingual gland (75 ×). This gland consists of a series of compound tubuloalveolar glands exactly like the one shown in Figure 4.11h on page 78. The sublingual gland contains mostly mucous cells (white), with a few serous cells (pink). The red, Y-shaped structure at center is a duct.

(Figure 21.10b) contain mostly mucous cells. The smallest *ducts* of all salivary glands are formed by a simple cuboidal epithelium.

The Teeth

The **teeth** lie in sockets (alveoli) in the gum-covered margins of the mandible and maxilla. We masticate, or chew, by raising and lowering the mandible and by moving it from side to side while continually using the tongue to position food between the teeth. In the process, the teeth tear and grind the food, breaking it into smaller fragments.

Dentition and the Dental Formula.

By age 21, two sets of teeth, the primary and permanent dentitions, have formed (Figure 21.11).

The primary dentition consists of the **deciduous teeth** (de-sid'u-us; "falling off"), also called **milk teeth** or *baby teeth.* The first teeth to appear, about 6 months after birth, are the lower central incisors. Additional pairs of teeth erupt at 1- to 2-month intervals until all 20 milk teeth have emerged, by about 2 years of age. As the deeper, **permanent teeth** enlarge and develop, the roots of the milk teeth are resorbed, and the milk teeth loosen. The milk teeth typically fall out between the ages of 6 and 12 years. Generally, all permanent teeth except the third molars have erupted by the end of adolescence. The third molars, also called *wisdom teeth,* emerge between the ages of 17 and 25 years. There are 32 permanent teeth in a full set, but the wisdom teeth often fail to erupt, and in some people they are completely absent.

Many adults have their wisdom teeth removed by a dentist or oral surgeon. Can you guess why this is done?

Instead of emerging normally, a tooth may remain embedded deep in the jawbone and push on the roots of the other teeth. Such an *impacted tooth* causes pressure and pain and must be removed. Wisdom teeth are the most commonly impacted teeth. ■

Teeth are classified according to their shape and function as incisors, canines, premolars, and molars (Figure 21.11). The chisel-shaped **incisors** are adapted for nipping off pieces of food. The cone-shaped **canines** (eyeteeth) tear and pierce. The **premolars** (bicuspids) and **molars** have broad crowns with rounded *cusps* (surface bumps) and grind food. The molars (literally "millstones"), which have four or five cusps, are the best grinders. During chewing, the upper and lower molars lock repeatedly together, the cusps of the uppers fitting into valleys in the lowers, and vice versa. This action generates tremendous crushing forces.

The *dental formula* is a shorthand way of indicating the numbers and relative positions of the dif-

ferent classes of teeth in the mouth. This formula is written as a ratio, uppers over lowers, for just half of the mouth (because right and left halves are the same). The total number of teeth is calculated by multiplying the dental formula by two. The dental formula for the permanent dentition (two incisors, one canine, two premolars, and three molars) is written as:

$$2\text{-}1\text{-}2\text{-}3 \times 4 \qquad \frac{2\text{I, 1C, 2P, 3M}}{2\text{I, 1C, 2P, 3M}} \times 2 \ (32 \text{ teeth})$$

This is the dental formula for the deciduous teeth:

$$2\text{-}1\text{-}2 \times 4 \qquad \frac{2\text{I, 1C, 2M}}{2\text{I, 1C, 2M}} \times 2 \ (20 \text{ teeth})$$

The teeth are supplied by the following nerves and arteries (also see Table 13.1 on p. 366): The upper teeth are served by the superior alveolar nerves, which are branches of the maxillary division of the trigeminal nerve. The lower teeth are served by the inferior alveolar nerves, which are branches of the mandibular division of the trigeminal nerve. The arterial supply is from the superior and inferior alveolar arteries, branches of the maxillary artery from the external carotid.

Tooth Structure.

Each tooth has two main regions: the exposed **crown,** and the **root(s)** in the socket (Figure 21.12). These regions are connected by a **neck** near the gum line. The crown is covered with a layer of **enamel,** about 2.5 mm thick. Enamel, which directly bears the forces of chewing, is the hardest substance in the body. Calcium salts make up 99% of its mass, and it contains no cells or vessels. Its densely packed hydroxyapatite (mineral) crystals are arranged in force-resisting rods or prisms that orient perpendicular to the tooth's surface.

Dentin (den'tin; "tooth") underlies the enamel cap and forms the bulk of the tooth. Dentin is bone-like in its mineral and collagen components but is harder than bone and lacks internal blood vessels. Dentin contains unique radial striations called *dentinal tubules* (Figure 21.12). Each tubule is filled with an elongated cellular process of an *odontoblast* (o-don'to-blast; "tooth former"), the cell type that secretes and maintains the dentin. The spherical cell bodies of odontoblasts lie just deep to the dentin, within the underlying pulp cavity.

The **pulp cavity,** in the center of the tooth, is filled with **pulp,** a loose connective tissue containing the tooth's vessels and nerves. Pulp supplies nutrients for the tooth's hard tissues and provides for tooth sensation. The part of the pulp cavity in the root is the **root canal.** At the tip of each root is the **apical foramen,** the opening into the root canal.

CUSPID = CANINE (1 ROOT)

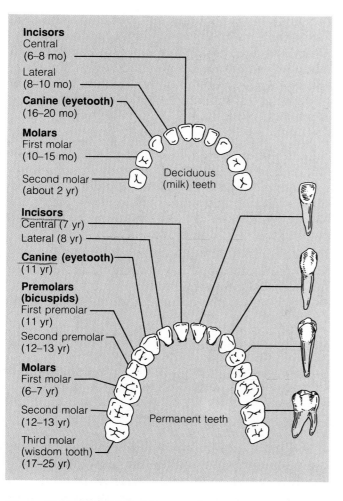

Incisors
Central
(6–8 mo)
Lateral
(8–10 mo)
Canine (eyetooth)
(16–20 mo)
Molars
First molar
(10–15 mo)
Second molar
(about 2 yr)

Deciduous
(milk) teeth

Incisors
Central (7 yr)
Lateral (8 yr)
Canine (eyetooth)
(11 yr)
**Premolars
(bicuspids)**
First premolar
(11 yr)
Second premolar
(12–13 yr)
Molars
First molar
(6–7 yr)
Second molar
(12–13 yr)
Third molar
(wisdom tooth)
(17–25 yr)

Permanent teeth

Figure 21.11   Human deciduous and permanent teeth.
Approximate ages at which the teeth erupt are shown in parentheses at left. Only the lower jaw is shown (the upper teeth are similar). Shapes of the individual teeth are shown at right. Note that all incisors, canines, and premolars have one root and that the lower molars have two roots. The upper molars have three roots.

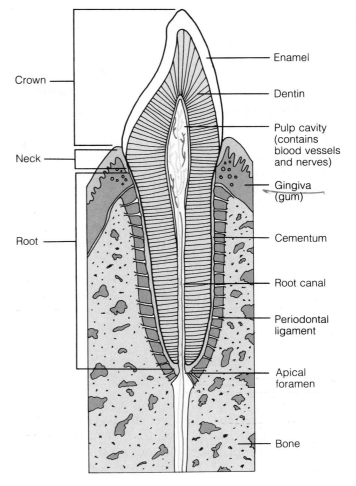

Crown
Neck
Root

Enamel
Dentin
Pulp cavity
(contains
blood vessels
and nerves)
Gingiva
(gum)
Cementum
Root canal
Periodontal
ligament
Apical
foramen
Bone

Figure 21.12   Longitudinal section of a canine tooth within its bony alveolus (socket).   In the dentin, the many parallel lines represent dentinal tubules.

When a tooth is damaged by a blow or by a deep cavity, the pulp may die and become infected. In such cases, **root canal therapy** must be performed. In this procedure, all of the pulp is drilled out, and the pulp cavity is sterilized and filled with an artificial, inert material. The tooth is then capped. ■

The external surface of the tooth root is covered by a calcified connective tissue called **cementum**. Essentially a bone layer, cementum attaches the tooth to the **periodontal ligament** (per″i-o-don′tal; "around the tooth"). This ligament anchors the tooth in the bony socket of the jaw. The periodontal ligament is continuous with the gum, or **gingiva** (jin-ji′vah), at the neck of the tooth.

Dental **cavities,** or **caries** (kar′ēz; "rottenness"), result from a gradual demineralization of the enamel and dentin by bacterial action. The decay process begins with the accumulation of **dental plaque,** a film of sugar, bacteria, and other debris that adheres to the teeth. Metabolism of the trapped sugars by the bacteria produces acids, which dissolve the calcium salts from the teeth. Once the salts have leached out, the remaining organic matrix is broken down by protein-digesting enzymes from the bacteria. Frequent brushing helps prevent tooth decay by removing plaque.

Even more serious than tooth decay is the effect of plaque on the *gums.* As plaque accumulates around the necks of the teeth, the contained bacteria release toxins that irritate the gums and cause them to pull away from the teeth. The plaque calcifies into a layer called **calculus** (kal′ku-lus; "stone"), on which more plaque accumulates, further inflaming the gums. This condition is **gingivitis** (jin″jĭ-vi′tis). Gingivitis can be reversed if the calculus is removed, but if it is neglected, the bacteria invade the periodontal tissues, forming pockets of infection that destroy the periodontal ligament and dissolve away the bone around the tooth. This condition, which often results in the loss of teeth, is called **periodontitis.** It begins around age 35 and eventually affects 95% of people. (This is the reason why so many people wear dentures.) Still, tooth loss is not inevitable. Even advanced cases of periodontitis can be treated by cleaning the infected pockets around the roots and cutting the gums to shrink the pockets.

Given the above considerations, can you deduce why dentists recommend that people floss between their teeth every day, beginning at a young age?

Flossing removes plaque and will minimize one's chances of developing periodontitis and losing one's teeth later in life. ■

The Pharynx

From the mouth, swallowed food passes posteriorly into the **oropharynx** and then the **laryngopharynx** (see Figure 21.7a and p. 537). Both of these are passageways for food, fluids, and inhaled air.

The histology of the pharyngeal wall resembles that of the mouth. The oropharynx and laryngopharynx are lined by a stratified squamous epithelium, which protects them against abrasion. (Swallowed food often contains rough particles, even after mastication.) The external muscle layer consists primarily of three *pharyngeal constrictor muscles:* superior, middle, and inferior (Figure 10.7b, p. 251). These muscles of swallowing encircle the pharynx and partially overlap one another. Like three stacked, clutching fists, they contract in sequence, from superior to inferior, to squeeze the bolus into the esophagus. The pharyngeal constrictors are *skeletal* muscles, as swallowing is a voluntary action.

The Esophagus

Gross Anatomy

The **esophagus** (literally, "carry food") is a muscular tube that propels swallowed food to the stomach. Its lumen is collapsed when the esophagus is empty. The esophagus begins as a continuation of the pharynx in the mid neck and descends through the thorax on the anterior surface of the vertebral column (Figure 21.1). Inferiorly, it pierces the diaphragm through the *esophageal hiatus* (hi-a′tus; "gap, opening") to enter the ab-

domen. Its abdominal part, which is only about 2 cm long, joins the stomach at the **cardiac orifice** (see Figure 21.14). Here, a *cardiac sphincter* acts to close off the lumen and prevent regurgitation of acidic stomach juices into the esophagus. (The only anatomical evidence of this sphincter is a minimal thickening of smooth muscle in the wall.) The edges of the esophageal hiatus also help prevent regurgitation.

In a **hiatal hernia,** the superior part of the stomach pushes through an enlarged esophageal hiatus into the thorax. This condition follows a weakening of the diaphragmatic muscle fibers around the hiatus. It often leads to an erosion of the inside of the esophagus. Can you guess why?

Since the diaphragm no longer reinforces the action of the cardiac sphincter, the acidic stomach juices are persistently regurgitated, eroding the wall of the esophagus and causing a burning pain. ■

Microscopic Anatomy

Unlike the mouth and pharynx, the esophagus wall contains all four layers of the alimentary canal described on page 566. These layers are shown in Figure 21.13. The following features are of interest:

1. The lining epithelium is a nonkeratinized stratified squamous epithelium. At the junction of the esophagus and stomach, this abrasion-resistant layer changes abruptly to the simple columnar epithelium of the stomach, which is specialized for secretion (Figure 21.13a).

2. When the esophagus is empty, its mucosa and submucosa are thrown into longitudinal folds, but during passage of a bolus, these folds flatten out.

3. The wall of the esophagus contains mucous glands. These primarily are compound tubuloalveolar glands, which extend from the submucosa to the lumen. As a bolus passes, it compresses these glands, causing them to secrete a lubricating mucus, which smoothes the further passage of the bolus through the esophagus.

4. The muscularis externa consists of skeletal muscle in the superior third of the esophagus, a mixture of skeletal and smooth muscle in the middle third, and smooth muscle in the inferior third. This arrangement is easy to understand if the esophagus is viewed as the zone where the skeletal muscle of the mouth and pharynx gives way to the smooth muscle of the stomach and intestines.

5. The most external esophageal layer is an adventitia, not a serosa. There is no serosa, because the thoracic segment of the esophagus is not suspended in the peritoneal cavity.

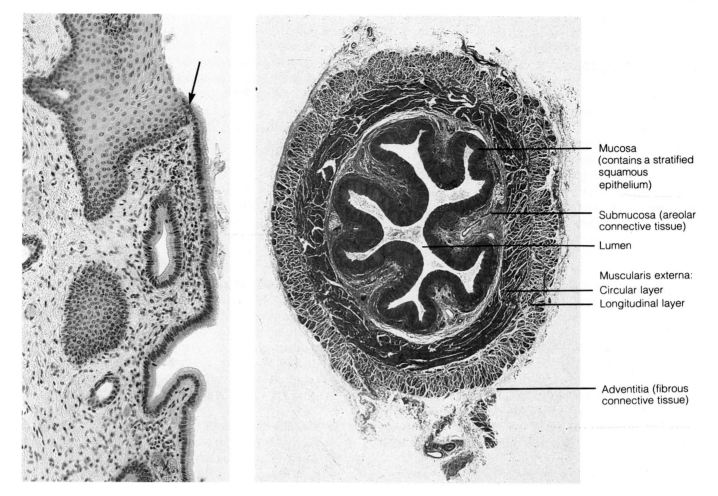

Mucosa
(contains a stratified squamous epithelium)

Submucosa (areolar connective tissue)

Lumen

Muscularis externa:
Circular layer
Longitudinal layer

Adventitia (fibrous connective tissue)

(a)   **(b)**

Figure 21.13 Microscopic structure of the esophagus. At right (part b), the esophagus is shown in cross section (5 ×). Folds of the mucosa and submucosa give the central lumen an irregular shape. The micrograph at left (part a) shows how the epithelium lining the esophagus (stratified squamous) changes abruptly into the epithelium lining the stomach (simple columnar) at the junction of the esophagus and stomach (132 ×).

The Stomach

The J-shaped **stomach** (Figure 21.14) is the widest part of the alimentary canal and extends from the esophagus to the small intestine. It is a temporary storage tank in which food is churned and turned into a paste called **chyme** (kīm; "juice"). The stomach also starts the breakdown of food proteins by secreting **pepsin,** a protein-digesting enzyme. Pepsin can function only under acidic conditions, and the stomach wall secretes hydrochloric acid. Stomach juices are acidic enough to dissolve steel, and this acidity destroys many harmful bacteria in the food. Although most nutrients are absorbed in the small intestine, some substances are absorbed through the stomach.

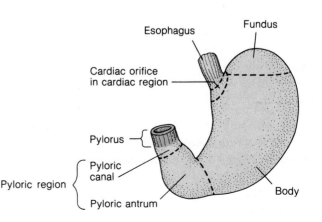

Esophagus

Fundus

Cardiac orifice in cardiac region

Pylorus

Pyloric canal

Pyloric region

Pyloric antrum

Body

Figure 21.14 The stomach: Shape and basic regions.

These substances include water, electrolytes, and some drugs (aspirin, alcohol). Food remains in the stomach for roughly 4 hours.

Gross Anatomy

The stomach lies in the superior left portion of the peritoneal cavity (Figure 21.1). Specifically, it lies in the left hypochondriac, epigastric, and umbilical regions of the abdomen. The stomach is directly inferior to the diaphragm and is partly hidden posterior to the left side of the liver. The spleen and pancreas lie posterior to the stomach. Although the stomach is anchored at both ends by esophageal and intestinal attachments, it is quite mobile in between. It tends to lie high and run horizontally in short, stout people (steerhorn stomach) and is elongated vertically in many tall, thin people (J-shaped stomach). A full, J-shaped stomach may extend low enough to reach the pelvis!

The stomach is very distensible: It easily holds 1.5 L of food and has a maximum holding capacity of about 4 L (1 gallon). Internally, the empty stomach contains longitudinal folds of the mucosa called **rugae** (roo′ge; "wrinkles") (Figure 21.15a and b). These folds flatten as the stomach fills, providing a large surface area to accommodate food and fluids. The convex left surface of the stomach is its **greater curvature**, and the concave right margin is the **lesser curvature.**

The main regions of the stomach are shown in Figure 21.14. The **cardiac region,** or **cardia** ("near the heart"), is a ring-shaped zone encircling the cardiac orifice at the junction with the esophagus. The **fundus,** the stomach's dome, is tucked under the diaphragm. The **body** is the large midportion of the stomach. It terminates at the funnel-shaped *pyloric region,* whose wider first part is the **pyloric antrum** ("cave") and whose narrower, second part is the **pyloric canal.** The pyloric region ends at the terminus of the stomach, the **pylorus.** The pylorus contains the **pyloric sphincter** (Figure 21.15a), which controls the entry of chyme into the intestine (*pylorus* means "gatekeeper").

The nerves and vessels of the stomach are as follows: The sympathetic fibers derive from the thoracic splanchnic nerves via the celiac plexus, whereas the parasympathetic fibers are from the vagus (Chapter 14). The arteries to the stomach arise from the celiac trunk and include the right and left gastrics, short gastrics, and right and left gastroepiploics. The corresponding veins drain into the portal, splenic, and superior mesenteric veins (Chapter 18).

Microscopic Anatomy

The wall of the stomach exhibits the typical layers of the alimentary canal, but its muscularis externa and mucosa show special features. Along with circular and longitudinal layers of smooth muscle, the muscularis externa also has another (the innermost) layer that runs *obliquely* (Figure 21.15a and c). The stomach musculature produces complex waves of peristalsis that churn and pummel the food into smaller fragments.

The epithelium that lines the stomach is a simple columnar epithelium consisting entirely of goblet cells ("surface epithelium" in Figure 21.15c). These cells secrete a protective coat of mucus, thereby shielding the stomach wall from the destructive effects of acid and pepsin in the lumen. The stomach mucosa is exposed to some of the harshest conditions in the entire alimentary canal.

The surface of the stomach mucosa is dotted with millions of cup-shaped **gastric pits,** which open into tubular **gastric glands** (Figure 21.15c and d). Goblet cells invariably line the gastric pits, but the cells lining the gastric glands vary among the different regions of the stomach. In the pyloric and cardiac regions, the cells of the glands are primarily mucous cells. In the fundus and body, by contrast, the gastric glands contain the following cell types, which produce the majority of the stomach secretions (Figure 21.15c):

1. **Mucous neck cells** occur in the upper ends, or necks, of the gastric glands. They secrete a different type of mucus from that secreted by the surface goblet cells. The specific function of this mucus is not known.

2. **Parietal (oxyntic) cells** occur mainly in the middle regions of the glands. They produce the stomach's hydrochloric acid (HCl), by pumping H^+ and Cl^- ions into the lumen of the gland. Although parietal cells appear spherical when viewed by light microscopy, they actually have three thick prongs like those of a pitchfork. Many long microvilli cover each prong, providing a large surface area so that H^+ and Cl^- can move quickly out of the cells. The cytoplasm contains many mitochondria that supply the large amount of energy expended in pumping ions. Parietal cells also secrete **gastric intrinsic factor,** a protein necessary for the absorption of Vitamin B_{12} by the small intestine. The body uses this vitamin in the manufacture of red blood cells.

3. **Chief (zymogenic) cells** occur mainly in the basal regions of the glands. They make and secrete **pepsinogen** (pep-sin′o-jen), the precursor form of pepsin. Pepsinogen is activated to pepsin when it encounters acid in the apical region of the gland. Pepsinogen, like all enzymes, is itself a protein. Therefore, chief cells have the features of protein-secreting cells: a well-developed rough endoplasmic reticulum (rough ER) and Golgi apparatus, plus secretory granules in the apical cytoplasm.

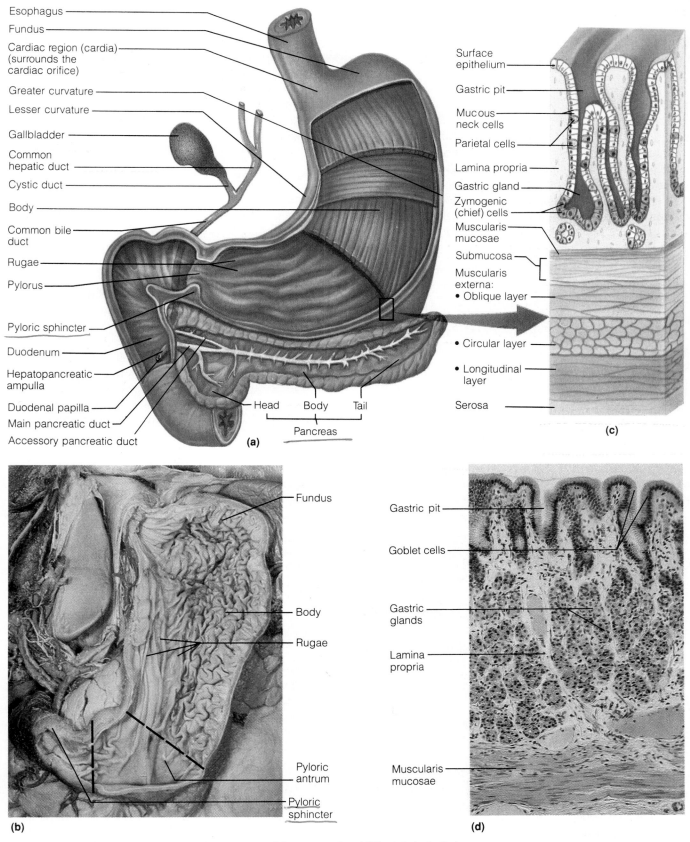

Figure 21.15 Gross and microscopic anatomy of the stomach.   (a) The inferior half of
this stomach has been opened to reveal its internal surface. In the superior half, the muscularis externa is dissected to show the fiber directions of the three muscle layers. The pancreatic, cystic, and hepatic ducts are also shown. (b) Photograph of the internal surface of the
stomach. Note the longitudinal folds, called rugae. (c) Enlarged view of the stomach wall,
showing gastric pits and glands in section. (d) Micrograph of the stomach mucosa, a view
similar to part (c) (132 ×).

PEPSIN
HYDROCHLORIC ACID

Other epithelial cell types occur in the gastric glands, but also extend beyond these glands. Such cells include:

1. *Enteroendocrine cells* (en"ter-o-en'do-krin; "gut endocrine") are hormone-secreting cells scattered throughout the lining epithelium and glands of the alimentary canal. These cells (not illustrated) release their hormones into the capillaries of the underlying lamina propria. One of these hormones, **gastrin,** signals parietal cells to secrete HCl when food enters the stomach. Most enteroendocrine cells that produce gastrin are in the stomach's pyloric region.

2. *Undifferentiated stem cells* (not illustrated) are located throughout the stomach, at the junction of the gastric glands and the gastric pits. These cells divide continuously, replacing the entire lining epithelium of goblet cells every 3 to 7 days. Such rapid replacement is vital because individual goblet cells can survive for only a few days in the harsh environment of the stomach.

The Small Intestine

The **small intestine** is the longest part of the alimentary canal (Figure 21.1). Most enzymatic digestion and virtually all absorption occurs here. Most digestive enzymes that operate within the small intestine are not secreted by the intestine; rather, they enter from the pancreas. During digestion, the small intestine undergoes active segmentation movements, shuffling the chyme back and forth and maximizing its contact with the nutrient-absorbing mucosa. Peristalsis propels chyme through the small intestine in about 3 to 6 hours.

Gross Anatomy

The small intestine is a convoluted tube that runs from the pyloric sphincter, in the epigastric region of the abdomen, to the first part of the large intestine, in the lower right quadrant. Although the small intestine may be 6 m long in cadavers, its length averages only 2 m (6 feet) during life.

The small intestine has three subdivisions (Figure 21.1): the **duodenum** (du"o-de'num; "twelve finger-widths long"), **jejunum** (jě-joo'num; "empty"), and **ileum** (il'e-um; "twisted intestine"). These segments contribute approximately 5%, 40%, and 60% of the length of the small intestine, respectively. Most of the C-shaped duodenum lies retroperitoneally. The jejunum and ileum form sausage-like coils that are suspended in the peritoneal cavity and framed by the large intestine. The jejunum makes up the superior left part of this coiled intestinal mass, whereas the ileum makes up the inferior right part.

Although the *duodenum* (Figure 21.16) is the shortest subdivision of the small intestine, it has the

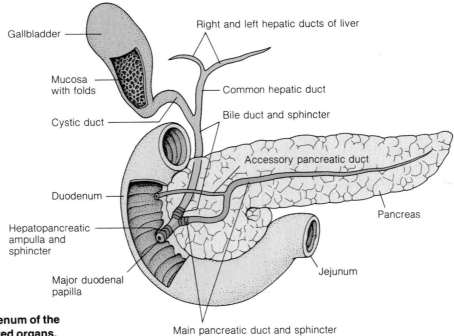

Gallbladder

Mucosa with folds

Cystic duct

Duodenum

Hepatopancreatic ampulla and sphincter

Major duodenal papilla

Right and left hepatic ducts of liver

Common hepatic duct

Bile duct and sphincter

Accessory pancreatic duct

Pancreas

Jejunum

Main pancreatic duct and sphincter

Figure 21.16 The duodenum of the small intestine, and related organs. Opening into the duodenum are various ducts from the pancreas, gallbladder, and liver. This is a more complete illustration of the structures inferior to the stomach shown in Figure 21.15a.

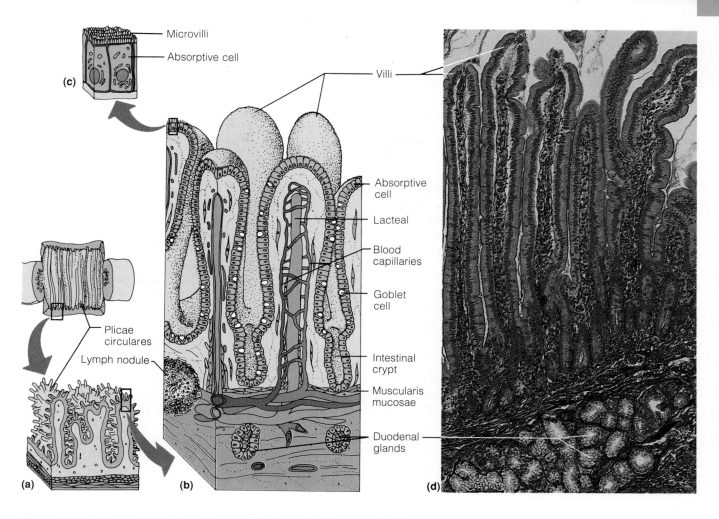

Figure 21.17   The small intestine: Structural features that increase its internal surface area for the absorption of nutrients.   The portion shown here is from the duodenum. (**a**) Intestine cut open to show plicae circulares, which in turn are covered with finger-like villi. (**b**) Villi and the inner layers of the wall of the small intestine. (**c**) Two absorptive cells with many microvilli on their free (luminal) surfaces. (**d**) Photomicrograph of the wall of the small intestine, showing villi. Also note the duodenal glands in the submucosa (150 ×).

most features of interest. It receives digestive enzymes from the pancreas and bile from the liver and gallbladder. The *bile duct* and the *main pancreatic duct* enter the wall of the duodenum, where they form a bulb called the **hepatopancreatic ampulla** (hep″ah-to-pan″kre-ah′tik am-pul′ah; "flask from the liver and pancreas"). This ampulla opens into the duodenum via a volcano-shaped mound called the **major duodenal papilla.** Entry of bile and pancreatic juice into the duodenum is controlled by sphincters of smooth muscle that surround the hepatopancreatic ampulla and the ends of the pancreatic and bile ducts.

The nerves and vessels of the small intestine are as follows: Nerve fibers include parasympathetics from the vagus and sympathetics from the thoracic splanchnic nerves, both relayed through the superior mesenteric (and celiac) plexus. The arterial supply is primarily from the superior mesenteric artery (p. 498). The veins run parallel to the arteries and typically drain into the superior mesenteric vein. From there,

the nutrient-rich venous blood from the small intestine drains into the hepatic portal vein, which carries it to the liver.

Microscopic Anatomy

Modifications for Absorption.   As we've already noted, nearly all nutrient absorption occurs in the small intestine, an organ highly adapted for this function. Its length alone provides a huge surface area for absorption, and its wall has three structural modifications that amplify this surface enormously: *plicae circulares, villi,* and extensive *microvilli* (Figure 21.17). Most absorption occurs in the proximal region of the small intestine, so these specializations decrease in number toward the distal end.

The *plicae circulares* (pli′ke), or *circular folds,* are permanent, transverse ridges of the mucosa and submucosa. They are nearly 1 cm tall. Besides increasing the absorptive surface area, these folds force

579

A CLOSER LOOK Peptic Ulcers: "Something Is Eating at My Stomach"

Archie, a 53-year-old factory worker, began to experience a burning pain in his upper abdomen an hour or two after each meal. At first, he blamed the quality of his home cooking, but he experienced the same symptoms after eating at the factory cafeteria or at restaurants. Archie always responded to stress by drinking and smoking heavily, and his abdominal pain became markedly worse during a hectic week when he worked 15 hours overtime on the assembly line. Finally, after 2 months of increasingly severe pain, Archie consulted a physician and was told that he had a peptic ulcer.

Peptic ulcers affect 1 of every 8 Americans. A peptic ulcer is a crater-like erosion in the mucosa of any part of the alimentary canal that is exposed to the secretions of the stomach. Gastric acid and pepsin cause this damage—people whose stomachs fail to secrete these substances never develop peptic ulcers. A few peptic ulcers occur in the lower esophagus, following the regurgitation of stomach contents, but most (98%) occur in the pyloric part of the stomach (gastric ulcers) or the

first part of the duodenum (duodenal ulcers). Duodenal ulcers are about three times more common than gastric ulcers. In the past, more men than women developed peptic ulcers, but the incidence is now nearly equal for both sexes. Peptic ulcers may appear at any age, but they develop most frequently between the ages of 50 and 70 years. After developing, they tend to recur—healing, then flaring up periodically—for the rest of one's life.

Gastric and duodenal ulcers may produce a gnawing or burning pain in the epigastric region of the abdomen. This pain often appears 1 to 3 hours after a meal (or causes one to awaken at night) and is relieved by eating. Other potential symptoms include loss of appetite, burping, nausea, and vomiting. Not all people with ulcers experience the above symptoms, however, and many exhibit no symptoms at all.

Despite years of intensive study, the cause of peptic ulcers remains incompletely understood. For over a century, it has been "common knowledge" that stress causes ulcers, and the stereotypical ulcer patient has been the overworked busi-

ness executive. Recent studies have not been able to demonstrate such a link between stress and ulcers, however, and many researchers now doubt the validity of this claim. Nonetheless, a stressful life-style does seem to aggravate existing ulcers. A new theory proposes ulcers are associated with a strain of acid-resistant bacteria that inhabit the stomach of 30% of all people. The tendency to develop ulcers is known to be inherited to a certain extent.

Duodenal and gastric ulcers seem to be separate and distinct conditions, each of which has multiple causes. *Duodenal* ulcers seem to result when acidic secretions from the stomach overwhelm the defenses that guard the mucosa of the duodenum from these secretions. Many duodenal ulcers are caused by an oversecretion of acid and pepsin by the stomach or by a too-frequent emptying of the stomach into the duodenum. Less often, a duodenal ulcer reflects a failure of the duodenum to secrete enough protective mucus or bicarbonate to buffer the stomach acids. In *gastric* ulcers, the stomach secretes acid at a normal rate, but the defenses in the stomach

---

chyme to spiral through the intestinal lumen, slowing the movement of chyme and allowing time for full nutrient absorption.

Villi are finger-like projections of the mucosa that give it a velvety texture, much like the soft nap of a Turkish towel. Over 1 mm tall, villi are large enough to be seen with the naked eye. They are covered by a simple columnar epithelium made up primarily of **absorptive cells** specialized for absorbing digested nutrients (Figure 21.17b and c). Within the lamina-propria core of each villus is a network of blood capillaries and a wide lymph capillary called a **lacteal.** With the exception of the absorbed fats, which

enter the lacteals, all end products of nutrient digestion enter the blood capillaries.

The villi are able to move during digestion because each contains a slip of smooth muscle in its core. (This muscle extends into the villus from the muscularis mucosae.) Such movements increase the amount of contact between the villi and the nutrients in the intestinal lumen, increasing absorption efficiency. The movements also squeeze lymph through the lacteals.

Microvilli occur on most epithelial surfaces in the body, but those in the small intestine are exceptionally long and densely packed (Figure 21.17c). Be-

mucosa are inadequate. Gastric ulcers are usually associated with gastritis; that is, they occur within larger areas of inflamed stomach mucosa. Often, the gastritis (and the ulcer) seem to be caused by a persistent reflux of digestive juices from the duodenum into the stomach.

The anatomy of a peptic ulcer is shown in the figure in this box. It is a round, sharply defined crater in the mucosa. Typical ulcers are 1 to 4 cm in diameter and have eroded to the depth of the muscularis mucosae. The base of the ulcer contains necrotic tissue, granulation tissue, and scar tissue. Eroded blood vessels may sometimes be seen there as well.

Peptic ulcers can produce serious complications. In about 20% of cases, eroded blood vessels bleed into the alimentary canal, causing vomiting of blood and blood in the feces. In such cases, anemia may result from a severe loss of blood. In 5% to 10% of ulcer patients, scarring within the stomach obstructs the pyloric opening, blocking digestion (see "Pyloric stenosis" in this chapter's Related Clinical Terms). About 5% of peptic ulcers *perforate,* allowing the contents of the stomach or duodenum to leak into the peritoneal cavity. This can cause either

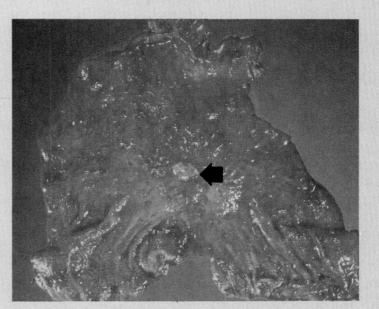

A peptic ulcer in the mucosa of the stomach is visible at the arrow.

peritonitis or the digestion and destruction of the nearby pancreas. A perforated ulcer is a life-threatening condition.

In spite of these potential complications, most peptic ulcers heal readily and respond well to treatment. The first steps in treatment are to avoid smoking, alcohol, and aspirin, all of which aggravate ulcers. Antacid drugs are often administered to neutralize the stomach acids. Physicians can also prescribe drugs that coat the stomach mucosa

with a protective layer or drugs that slow the secretion of acid by the stomach's parietal cells. These simple measures usually allow the ulcer to heal in a few months, and only one-third of patients require prolonged drug therapy. In rare, persistent cases, surgery to cut the vagus nerves to the stomach or to remove part of the stomach may be necessary. Both of these treatments cure the ulcer by decreasing the production of acid and pepsin.

---

sides amplifying the absorptive surface, the plasma membranes of these microvilli contain enzymes that complete the final stages of breakdown of nutrient molecules.

The amount of absorptive surface in the small intestine is remarkable. Together, the plicae circulares, villi, and microvilli increase the intestinal surface area to about 200 square meters. That matches the floor area of an average two-story house!

Histology of the Wall.   All typical layers of the alimentary canal occur in the small intestine. The lining

epithelium contains not only absorptive cells but also scattered goblet cells and enteroendocrine cells.

1. *Absorptive cells* (Figure 21.17c). The uptake of digested nutrients by absorptive cells is an energy-demanding process, and these cells contain many mitochondria. An abundant endoplasmic reticulum assembles the newly absorbed lipid molecules into lipid-protein complexes called *chylomicrons* (ki″lo-mi′kronz), which enter the lacteal capillaries. It is in this form that absorbed fat enters the circulation.

2. *Goblet cells* (Figure 21.17b). These cells secrete a

coat of lubricating mucus onto the internal surface of the intestine. This mucus forms a protective barrier that prevents enzymatic digestion of the intestinal wall.

3. *Enteroendocrine cells.* The enteroendocrine cells of the duodenum secrete several hormones. These include *cholecystokinin* (ko"le-sis"to-ki'nin; "gallbladder activator"), which signals the gallbladder to release stored bile, and *secretin,* which signals the pancreas to secrete pancreatic juice.

Between the villi, the mucosa contains tubes called **intestinal crypts,** *intestinal glands,* or *crypts of Lieberkühn* (Lēb'er-kun) (Figure 21.17b). The epithelial cells that line these crypts secrete *intestinal juice,* a watery liquid that carries the nutrients absorbed from the intestinal lumen. The intestinal crypts also renew the mucosal epithelium: Undifferentiated epithelial cells in them divide rapidly and move continuously onto the villi. These are among the most quickly dividing cells of the body, replacing the inner epithelium of the small intestine every 3 to 6 days. Such rapid replacement is necessary because individual epithelial cells cannot long withstand the destructive effects of the digestive enzymes in the intestinal lumen.

Treatments for cancer, such as radiation therapy and chemotherapy, stop the replication of cellular DNA throughout the body and thus preferentially destroy the most quickly dividing cells. Can you explain why nausea, diarrhea, and vomiting usually follow cancer treatments?

These treatments usually obliterate the rapidly dividing cells of the intestinal epithelium, along with the targeted cancer cells. Such destruction of the lining of the alimentary canal disrupts digestion and causes the symptoms listed above. ∎

Although the intestinal crypts contain mostly undifferentiated cells, they also contain mature *Paneth cells* (not illustrated). These cells secrete enzymes that destroy certain bacteria, thereby helping to determine which bacteria types live in the intestinal lumen. Bacteria are indeed permanent residents of the intestinal lumen, and are called the *intestinal flora.* These bacteria manufacture some essential vitamins, which the intestines absorb. For example, vitamin B_{12} is manufactured by the intestinal bacteria.

The lamina propria and submucosa of the small intestine contain many areas of lymphoid tissue, including *aggregated follicles (Peyer's patches)* in the ileum (p. 529).

The submucosa is a typical connective tissue. In the duodenum only, it contains a set of compound tubular **duodenal glands** (also called *Brunner's glands),* whose ducts open into the intestinal crypts (Figure 21.17b). These glands secrete an alkaline mucus that helps neutralize the acidic chyme from the stomach and contributes to the protective layer of mucus on the inner surface of the small intestine.

The outer layers of the small intestine (muscularis externa and serosa) show no unusual features.

The Large Intestine

The **large intestine** is the last major subdivision of the alimentary canal (Figure 21.18). The digested material that reaches it contains few nutrients, but the residue remains there for 12 to 24 hours. With the exception of a small amount of digestion of this residue by the many bacteria in the large intestine, no further food breakdown occurs. The primary function of the large intestine is to absorb electrolytes and water from the digested mass, resulting in semisolid feces. Propulsion through the large intestine is sluggish and weak, except for **mass peristaltic movements,** which pass over the colon a few times a day to force the feces powerfully toward the rectum.

Gross Anatomy

General Features. The large intestine frames the small intestine on 3½ sides, resembling an open rectangle (Figure 21.1). Its diameter is greater than that of the small intestine, but its length (1.5 m) is less than that of the small intestine (2 m). The large intestine has the following subdivisions: *cecum, appendix, colon, rectum,* and *anal canal* (Figure 21.18a).

Over most of its length, the large intestine exhibits three special features: teniae coli, haustra, and epiploic appendages. **Teniae coli** (te'ne-e ko'li; "ribbons of the colon") are three longitudinal strips spaced at equal intervals around the cecum and colon. They represent the longitudinal layer of the muscularis externa, which is very thin except at these teniae. Since the teniae maintain muscle tone, they cause the large intestine to pucker into sacs, or **haustra** (haw'strah; "to draw up"). **Epiploic appendages** (ep"ĭ-plo'ik; "membrane-covered") are fat-filled pouches of visceral peritoneum that hang from the intestine. Their significance is unknown.

The Cecum and Appendix. The large intestine begins with the sac-like **cecum** (se'kum; "blind pouch") in the right iliac fossa. The ileum of the small intestine opens into the cecum's medial wall. Internally, this opening is surrounded by the **ileocecal valve** (Figure 21.18b), which is formed by two raised edges of the mucosa. The action of this valve is quite simple: A sphincter in the distal ileum keeps the valve closed until there is food in the stomach, at which time the sphincter reflexively relaxes, opening the valve. As the cecum fills, its walls stretch, pulling the valve

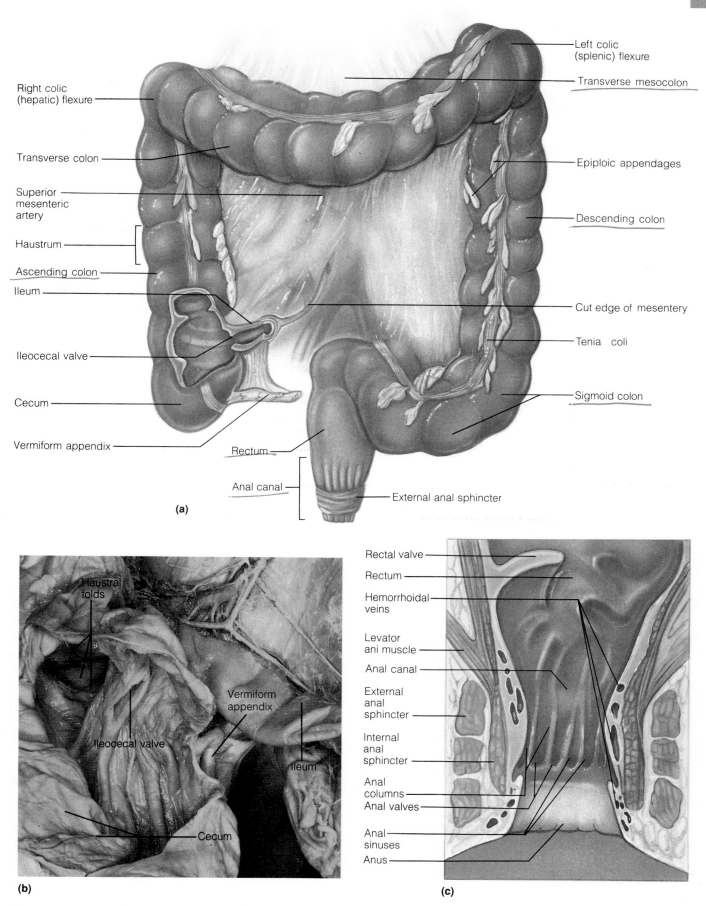

Figure 21.18 Gross anatomy of the large intestine. (**a**) Diagram. (**b**) The cecum: Photograph from a cadaver showing a cecum that has been opened by cutting through its anterior wall. Note the ileocecal valve internally and the vermiform appendix externally. (**c**) The inferior rectum and anal canal.

583

edges together and closing the ileocecal valve. This action prevents reflux of feces from the cecum into the ileum.

The **vermiform appendix** (ver′mĭ-form, "worm-shaped") is a blind tube that opens into the postero-medial wall of the cecum. It is almost always illustrated hanging inferiorly, but it typically lies "tucked up" posterior to the cecum in the right iliac fossa. The appendix has large masses of lymphoid tissue in its wall and probably functions like the tonsils, gathering antigens and generating memory lymphocytes for long-term immunity. This function predisposes the appendix, like the tonsils, to serious infections.

Appendicitis, acute inflammation of the appendix, results from a blockage (often by stool) that traps infectious bacteria within its lumen. Appendicitis often leads to ischemia and gangrene (death and decay) of the appendix. If the appendix ruptures, bacteria and feces spray over the abdominal organs, causing peritonitis. Because the symptoms of appendicitis vary greatly, this condition is notoriously difficult to diagnose. Often, however, the first symptom is pain in the umbilical region, followed by loss of appetite, fever, nausea, vomiting, and relocalization of pain to the lower right quadrant of the abdominal surface. Immediate surgical removal of the appendix, called **appendectomy** (ap″en-dek′to-me), is the usual treatment. Appendicitis occurs most often between the ages of 15 and 25, when the entrance to the appendix is widest. ■

The Colon. The **colon** (ko′lon) has several distinct segments (Figure 21.18a). From the cecum, the **ascending colon** ascends along the right side of the posterior abdominal wall and reaches the level of the right kidney. Here it makes a right-angle turn, the **right colic flexure.** This is also called the **hepatic flexure** because the liver lies directly superior to it. From this flexure, the **transverse colon** extends to the left across the peritoneal cavity. Directly anterior to the spleen, it bends acutely at the **left colic (splenic) flexure** and descends along the left side of the posterior abdominal wall as the **descending colon.** Inferiorly, the colon enters the true pelvis as the S-shaped **sigmoid colon** *(sigma* = Greek letter corresponding to the letter *s).*

The sigmoid colon is the most frequent site of a condition called **diverticulosis.** When the diet lacks bulk, the contractions of the circular muscle in the colon become more powerful, increasing the pressure on its walls. This pressure promotes the formation of sacs called *diverticula* (di″ver-tik′u-lah), small outward herniations of the mucosa through the colon wall. Diverticulosis generally leads to nothing more than dull pain, although it may rupture an artery in the colon and produce bleeding from the anus. More serious is **diverticulitis,** a condition in which the diverticula become infected and may perforate, leaking feces into the peritoneal cavity. Diverticulosis, the milder condition, occurs in 30% to 40% of all Americans over age 50. Its symptoms are generally relieved by a high-fiber diet. ■

Rectum. In the pelvis, the sigmoid colon joins the **rectum** (Figure 21.18a), which descends along the inferior half of the sacrum. The rectum has no teniae coli. Instead, its longitudinal muscle layer is complete and well-developed, so that it can generate strong contractions for defecation. The word *rectum* means "straight," but the rectum actually has several tight bends. Internally, these bends are represented as three transverse folds, the **rectal valves,** or *rectal folds* (Figure 21.18c). These valves separate feces from flatus; that is, they stop feces from being passed along with gas.

Anal Canal. The last segment of the large intestine is the **anal canal.** About 3 cm long, it begins where the rectum passes through the levator ani, the muscle that forms the pelvic floor. Thus, the anal canal lies entirely external to the abdominopelvic cavity.

Internally, the superior half of the anal canal exhibits longitudinal folds of mucosa, the **anal columns** (Figure 21.18c). Neighboring anal columns join each other inferiorly at crescent-shaped transverse folds called **anal valves.** The pockets just superior to these valves are **anal sinuses.** These sinuses release mucus when they are compressed by feces, providing lubrication that eases fecal passage during defecation.

The horizontal line along which the anal valves lie is called the *pectinate* ("comb-shaped") *line.* The mucosa superior to this line is innervated by visceral sensory fibers, so it is relatively insensitive to pain. Inferior to the pectinate line, however, the mucosa is sensitive to pain because it is innervated by somatic nerves.

Hemorrhoids (recall p. 488) are varicosities of the *hemorrhoidal veins* in the anal canal (Figure 21.18c). From this description, can you deduce some symptoms of hemorrhoids?

Because they are stretched and inflamed, these swollen veins throb, itch, bleed, and bulge into the lumen of the anal canal. Severe hemorrhoids are treated by burning or cutting them away or, more commonly, by tying them at their base with rubber bands, whereupon they wither, die, and fall away. ■

The wall of the anal canal contains two sphincter muscles: an **internal anal sphincter** of smooth muscle and an **external anal sphincter** of skeletal muscle. The former is a thickening of the circular layer of the

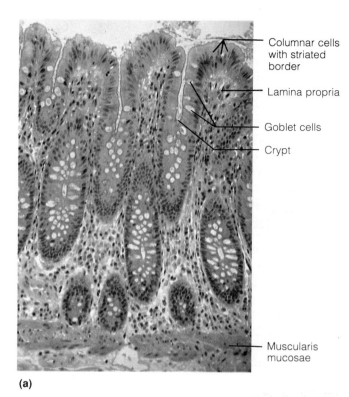

Columnar cells with striated border

Lamina propria

Goblet cells

Crypt

Muscularis mucosae

(a)

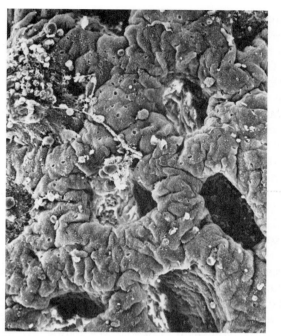

(b)

Figure 21.19   The mucosa of the large intestine.   (a) Photomicrograph. Note the abundance of goblet cells (120 ×). (b) Scanning electron micrograph of the mucosal surface of the colon, showing the entrances to the intestinal crypts (525 ×).

muscularis, whereas the latter is a distinct muscle (also see Figure 10.11b on p. 261). The external sphincter contracts voluntarily to inhibit defecation, whereas the internal sphincter contracts involuntarily to prevent feces from leaking from the anus between defecations. It also inhibits defecation during emotional stress. During toilet training, children learn to control the external anal sphincter.

Defecation.   The rectum is usually empty, but when feces are forced into it by mass peristaltic movements, the stretching of the rectal wall initiates the defecation reflex. Mediated by the spinal cord, this parasympathetic reflex signals the walls of the sigmoid colon and rectum to contract and the anal sphincters to relax. If one decides to delay defecation, the reflexive contractions end, and the rectum relaxes. Another mass movement occurs a few minutes later, initiating the defecation reflex again—and so on, until one chooses to defecate or the urge to defecate becomes unavoidable.

During defecation, the musculature of the rectum contracts to expel the feces. We aid this process voluntarily by contracting the diaphragm and the abdominal wall muscles to increase intra-abdominal pressure. We also contract the levator ani muscle (p. 260), which lifts the anal canal superiorly. This lifting ac-

tion leaves the feces inferior to the anus and thus outside the body.

Vessels and Nerves.   The first half of the large intestine—to a point two-thirds of the way along the transverse colon—is supplied by the superior mesenteric vessels. Its sympathetic innervation is from the superior mesenteric and celiac ganglia and plexuses, and its parasympathetic innervation is from the vagus.

The distal half of the large intestine is supplied by the inferior mesenteric vessels, although the lower rectum and the anal canal are served by rectal branches of the internal iliac vessels. The sympathetic innervation of the distal half of the large intestine is via the hypogastric and inferior mesenteric plexuses, and the parasympathetic innervation is from the pelvic splanchnic nerves. The final portion of the anal canal is innervated by somatic nerves (the pudendal nerve, for example) rather than by visceral nerves.

Microscopic Anatomy

The wall of the large intestine (Figure 21.19) differs in several ways from that of the small intestine. Villi are absent, which reflects the fact that very little nutrient absorption occurs in the large intestine. Intestinal crypts are present, however, as simple tubular glands.

The internal surface of the colon—and the intestinal crypts—are lined by a simple columnar epithelium containing the same cell types found in the small intestine. Goblet cells are more abundant here, however, and they secrete large amounts of lubricating mucus, which eases the passage of feces toward the end of the alimentary canal. The *absorptive cells* (labelled "Columnar cells" in Figure 21.19a) take in water and electrolytes. Finally, undifferentiated stem cells occur at the bases of the intestinal crypts, and full epithelial replacement occurs every week or so.

The other layers of the wall are rather typical. The lamina propria and submucosa contain more lymphoid tissue than they contain elsewhere in the alimentary canal, but considering the extensive bacterial flora of the large intestine, this is not surprising. The specializations of the muscularis externa and serosa are the teniae coli and epiploic appendages.

The *anal canal* is a zone of epithelial transition, where the simple columnar epithelium of the intestine abruptly changes to stratified squamous epithelium. This transition occurs near the level of the pectinate line. At the extreme inferior end of the anal canal, the mucosa merges with the true skin that surrounds the anus.

The Liver

The ruddy **liver,** the largest gland in the body, weighs about 1.4 kg (3 pounds) in an average adult. Amazingly versatile, it performs over 500 functions! Its digestive function is to produce **bile,** a green, alkaline liquid that is stored in the gallbladder and secreted into the duodenum. Bile salts emulsify fats in the small intestine; that is, they break up fatty nutrients into tiny particles that are more accessible to digestive enzymes from the pancreas. The liver also performs many metabolic functions. For example, it picks up glucose from nutrient-rich blood returning from the alimentary canal and stores this as glycogen for subsequent use by the body. The liver also processes fats and amino acids and stores certain vitamins. It detoxifies many poisons and drugs in the blood. It also manufactures the blood proteins. Almost all of these functions are carried out by a type of cell called a **hepatocyte** (hep'ah-to-sīt"), or simply a *liver cell.*

Gross Anatomy

The liver lies inferior to the diaphragm, in the right superior part of the abdominal cavity (Figures 21.1 and 21.20c). It fills much of the right hypochondriac and epigastric regions and extends into the left hypochondriac region. The liver lies almost entirely within the rib cage, which protects this highly vascular organ from blows that could cause it to rupture. The liver is shaped like a triangular wedge, the base of which faces right and the apex of which lies just inferior to the level of the left nipple. The liver has two surfaces, **diaphragmatic** and **visceral** (Figures 21.20 and 21.21). The diaphragmatic surface faces anteriorly and superiorly, the visceral surface posteroinferiorly. Most of the liver is covered with a layer of visceral peritoneum. The superior part of the liver, however, is fused to the diaphragm and is therefore devoid of peritoneum. This is the liver's **bare area.**

A deep, vertical **fissure** divides the liver into a large **right lobe** and a smaller **left lobe.** Two other lobes, the **quadrate lobe** and the **caudate lobe,** are visible on the visceral surface just to the right of the fissure (Figures 21.20b and 21.21). These lobes are anatomically part of the right lobe, but functionally they belong to the left lobe, with which they share nerves and vessels.

An important area near the center of the visceral surface is the **porta hepatis** (por'tah hep-ah'tis; "gateway to the liver"). Most of the major vessels and nerves that enter and leave the liver do so here (Figure 21.21). The right and left branches of the *hepatic portal vein,* which carry nutrient-rich blood from the stomach and intestines, enter the porta hepatis, as do the right and left branches of the *hepatic artery,* carrying oxygen-rich blood to the liver. The **right** and **left hepatic ducts,** which carry bile from the respective liver lobes, exit from the porta hepatis and fuse to form the **common hepatic duct,** which extends inferiorly toward the duodenum. Autonomic nerves reach the liver from the celiac plexus and consist of both sympathetic and parasympathetic (vagal) fibers.

Other important structures on the liver's visceral surface are the *gallbladder* and the *inferior vena cava,* which lie to the right of the caudate and quadrate lobes, respectively (Figure 21.21). The inferior vena cava receives the *hepatic veins* carrying blood from the liver.

Many structures relate to the liver's fissure. The *falciform ligament,* a mesentery that binds the liver to the anterior abdominal wall, attaches to the anterior aspect of the fissure (Figure 21.20a). The inferior margin of the falciform ligament contains the **round ligament,** or **ligamentum teres** (*teres* = round). Extending to the liver from the navel, the cord-like ligamentum teres is the remnant of the umbilical vein of the fetus (see p. 510). Posteriorly, the superior half of the fissure contains the **ligamentum venosum** (Figure 21.21), a cord-like remnant of the ductus venosus of the fetus (see p. 510).

Microscopic Anatomy

The liver contains large numbers of **classical liver lobules** (Figure 21.22), each about the size of a sesame seed. A lobule is shaped like a hexagonal (six-sided)

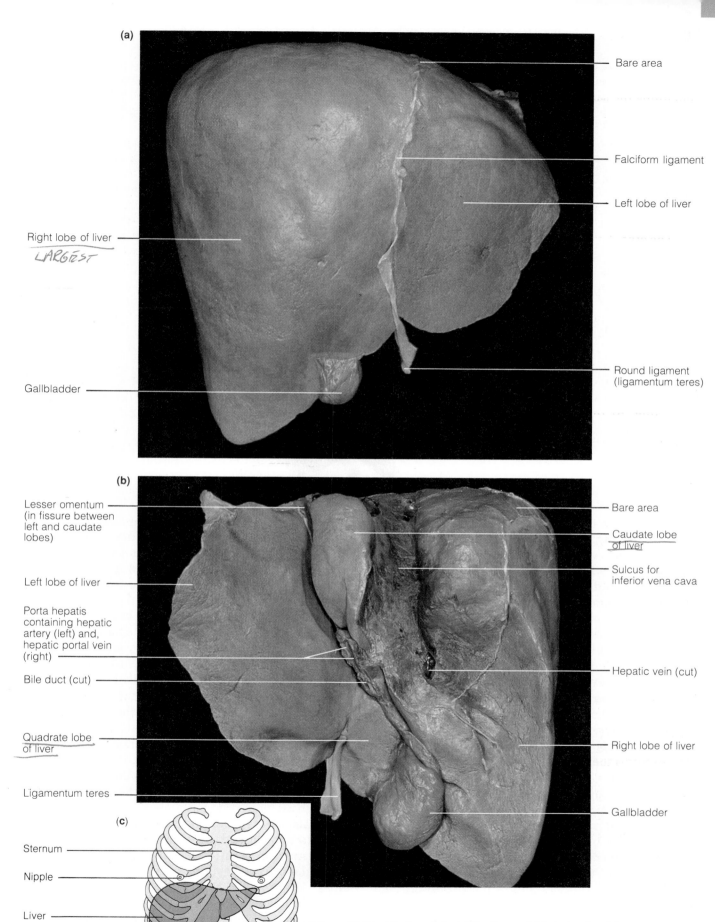

(a)

Bare area

Falciform ligament

Left lobe of liver

Right lobe of liver

LARGEST

Round ligament
(ligamentum teres)

Gallbladder

(b)

Lesser omentum
(in fissure between
left and caudate
lobes)

Bare area

Caudate lobe
of liver

Sulcus for
inferior vena cava

Left lobe of liver

Porta hepatis
containing hepatic
artery (left) and,
hepatic portal vein
(right)

Bile duct (cut)

Hepatic vein (cut)

Quadrate lobe
of liver

Right lobe of liver

Ligamentum teres

Gallbladder

(c)

Sternum

Nipple

Liver

Figure 21.20 Photographs of the liver. (a) Anterior view (diaphragmatic surface). (b) Posteroinferior view (visceral surface).
(c) Location of the liver with respect to the rib cage.

587

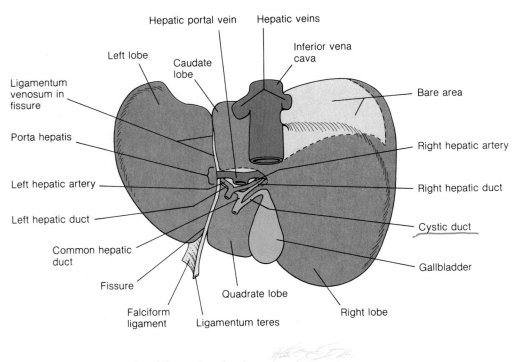

Left lobe

Hepatic portal vein   Hepatic veins

Caudate lobe

Inferior vena cava

Ligamentum venosum in fissure

Bare area

Porta hepatis

Right hepatic artery

Left hepatic artery

Right hepatic duct

Left hepatic duct

Cystic duct

Common hepatic duct

Gallbladder

Fissure

Quadrate lobe

Right lobe

Falciform ligament   Ligamentum teres

Figure 21.21   The liver, posteroinferior view (visceral surface).

solid, and consists of plates of liver cells, or hepatocytes, radiating out from a **central vein** (Figure 21.22c). If you were to take a thick paperback book and open it so wide that its two covers touched each other, you would have a rough model of the liver lobule, with the spreading pages representing the plates of hepatocytes, and the hollow cylinder formed by the rolled spine representing the central vein. The hepatocytes in the plates are organized like bricks in a wall.

At almost every corner of the lobule is a **portal triad** (tri′ad; "three"), or **portal tract** or *portal canal*. The portal triad contains three main vessels: a branch of the *hepatic artery,* a branch of the hepatic *portal vein,* and a **bile duct** (which carries bile away from the liver lobules). Note that the blood vessels bring both arterial and venous blood to the lobule. The arterial blood supplies the hepatocytes with oxygen, and blood from the portal vein supplies the hepatocytes with nutrients from the intestine.

Between the plates of hepatocytes are large capillaries, the **liver sinusoids.** Blood from both the portal vein and the hepatic artery percolates inward from the triad regions through these sinusoids to the central vein (Figure 21.22d). From there, the central veins form tributaries that ultimately lead to the hepatic veins and then to the inferior vena cava outside the liver.

Inside the sinusoids are star-shaped **Kupffer cells,** or *hepatic macrophages.* These cells destroy bacteria from the blood as it flows past. Thus, even though microorganisms in the intestine may enter the intestinal capillaries, few of them make it past the liver. Besides cleansing the blood of microorganisms, Kupffer cells also destroy worn-out blood cells—as do macrophages of the spleen and bone marrow.

The liver sinusoids are lined by an exceptionally leaky, fenestrated endothelium. Vast quantities of blood plasma pour out of the sinusoids, bathing the hepatocytes with fluid. Hepatocytes require a large blood supply because so many of their functions depend on interactions with the blood.

Hepatocytes possess large numbers of many different organelles that enable them to carry out their many functions. For example, the abundant rough ER manufactures the blood proteins; the well-developed smooth ER helps produce bile salts and detoxifies poisons from the blood; and the large Golgi apparatus packages the abundant secretory products of the endoplasmic reticulum. Energy for these functions is provided by an exceptional abundance of mitochondria. Additionally, hepatocytes are loaded with glycogen granules, a feature that reflects their role in storing sugar to regulate blood glucose levels. Finally, hepatocytes have a great capacity for cell division and regeneration: Judging from experiments on laboratory

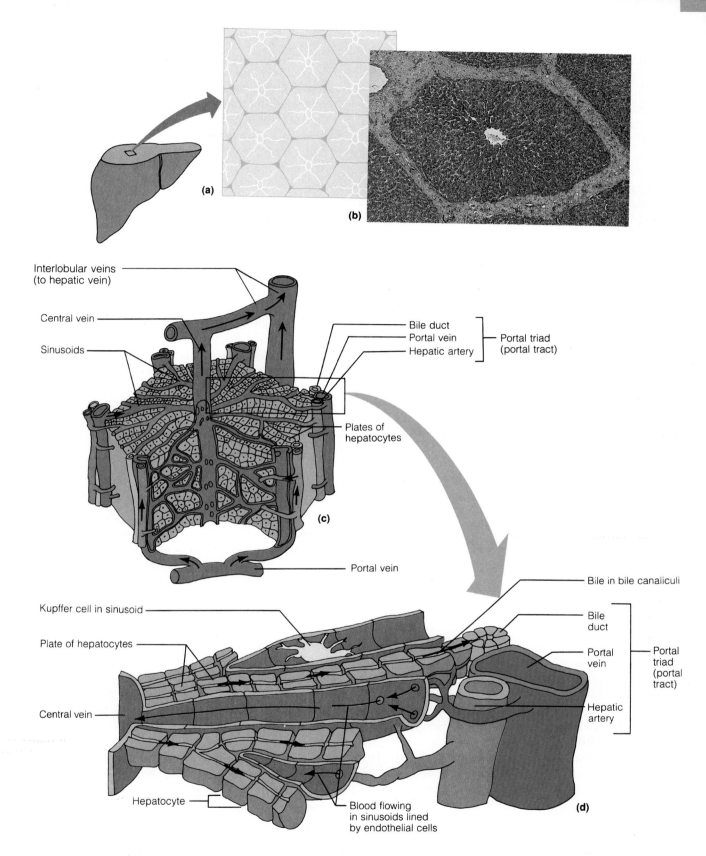

Figure 21.22 Microscopic anatomy of the liver. (a) Schematic view of the cut surface of the liver, showing the hexagonal nature of its lobules. (b) Photomicrograph of one liver lobule, in section (50 ×). (c) Schematic three-dimensional representation of a liver lobule. Blood flows inward from the corners to the central vein. (d) Enlarged view of a small part of a liver lobule, illustrating the components of the portal triad (portal tract) region, the positions of the bile canaliculi, and Kupffer cells in the sinusoids. The directions of blood and bile flow are also indicated.

animals, if half of a person's liver were removed, it would all regenerate in a few weeks!

Collectively, hepatocytes produce about 500 to 1000 ml of bile each day. The secreted bile enters tiny intercellular channels, called **bile canaliculi** ("little canals"), which lie between adjacent hepatocytes (Figure 21.22d). These canaliculi carry bile outward through each lobule, emptying into the bile ducts in the portal triads. From there, the bile flows through progressively larger ducts, exiting the liver through the hepatic ducts at the porta hepatis. The additional bile-carrying ducts, which lead to the duodenum, are shown in Figure 21.16 on page 578.

Hepatitis (hep"ah-ti'tis) is inflammation of the liver. It usually results from viral infection following ingestion of contaminated water or transmission of the virus through blood (by means of blood transfusions or contaminated needles). Hepatitis is also transmitted by the fecal-mouth route. For this reason, food-service workers are instructed to scrub their hands after using a washroom.

Cirrhosis (sǐ-ro'sis; "orange-colored") is a progressive inflammation of the liver that usually results from chronic alcoholism. The alcohol-poisoned hepatocytes are continuously replaced, but the liver's connective tissue regenerates faster. Therefore, the liver becomes fibrous, and its function declines. The scar tissue impedes the flow of blood through the liver, causing portal hypertension (p. 510). The patient may grow confused or comatose as toxins accumulate in the blood and depress brain functions. ■

The Gallbladder

The **gallbladder** is a muscular sac that rests in a shallow depression on the visceral surface of the liver (Figure 21.21). Its function is to store and concentrate bile produced by the liver. Its rounded head, or *fundus,* protrudes from the inferior margin of the liver (Figure 21.20a). Internally, its mucosa is folded into a honeycomb pattern (Figure 21.16). Like the rugae of the stomach, these folds allow the mucosa to expand as the gallbladder fills. They also maximize the area across which water from the bile can be absorbed.

The gallbladder's duct is the **cystic duct** (*cyst* = bladder). As shown in Figure 21.16, this duct joins the common hepatic duct from the liver to form the **bile duct** *(common bile duct),* which empties into the duodenum. The liver secretes bile continuously, but the sphincters at the end of the bile duct and hepatopancreatic ampulla are closed when bile is not needed for digestion. At these times, bile backs up into the gallbladder for storage. Later, when fatty chyme from a meal enters the duodenum, the gallbladder's muscular wall contracts, and the sphincters at the end of the duct system relax, expelling bile into the duodenum.

Histologically, the wall of the gallbladder has fewer layers than the wall of the alimentary canal: (1) a mucosa consisting of a simple columnar epithelium and a lamina propria; (2) one layer of smooth muscle, and (3) a thick, outer layer of connective tissue covered by a serosa. The cells of the lining epithelium concentrate the bile by absorbing some of its water and ions.

Bile is the normal vehicle in which cholesterol is excreted from the body, and bile salts keep the cholesterol dissolved within bile. Too much cholesterol (or too few bile salts) can lead to the crystallization of cholesterol in the gallbladder. Can you guess the clinical name and symptoms of this condition?

The crystallized cholesterol forms **gallstones** that can plug the cystic duct, producing agonizing pain when the gallbladder or its duct contracts. Treatments for gallstones include administering drugs that might dissolve the stones, pulverizing the stones with ultrasound vibrations, and surgically removing the gallbladder. Gallstones are easy to diagnose, because they show up well with ultrasound imaging (p. 19). ■

The Pancreas

The **pancreas** ("all meat") is both an exocrine and an endocrine gland. Its exocrine function is to produce most of the enzymes that digest foodstuffs in the small intestine, and its main endocrine function is to produce hormones that regulate the levels of sugar in the blood. The pancreas lies retroperitoneally in the epigastric and left hypochondriac regions of the abdomen (Figure 21.23). It is shaped like a tadpole, with head, body, and tail regions (Figure 21.15a). Its head lies in the C-shaped curvature of the duodenum, and its tail extends to the left to touch the spleen.

The pancreas receives blood through branches of the hepatic, splenic, and superior mesenteric vessels. Its autonomic nerves are from the celiac plexus. The sympathetic input derives from the thoracic splanchnic nerves, and the parasympathetic input is from the vagus.

The **main pancreatic duct** extends through the length of the pancreas (Figure 21.16). As we've mentioned, this duct joins the bile duct to form the hepatopancreatic ampulla and empties into the duodenum at the major duodenal papilla. An **accessory pancreatic duct** lies in the head of the pancreas and drains into the main duct.

Microscopically, the pancreas contains many compound acinar glands, which open into the two large ducts like clusters of grapes attached to a main vine. The acini of these glands consist of serous **acinar cells** (Figure 21.24a). These cells make, store, and secrete the pancreatic enzymes, a variety of enzymes capable of digesting all categories of foodstuffs. The

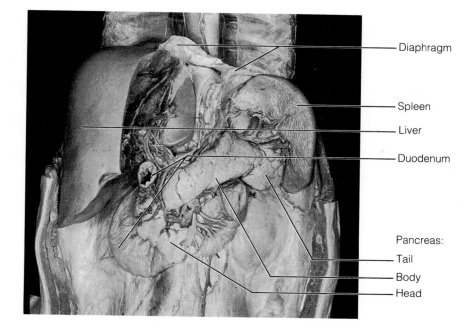

Diaphragm

Spleen
Liver
Duodenum

Pancreas:
Tail
Body
Head

Figure 21.23 Photograph of the pancreas and some adjacent organs in the posterior abdominal wall. The stomach and much of the liver have been removed.

enzymes are stored in inactive form in intracellular secretory granules called **zymogen granules** (zi'mo-jen; "fermenting"). The acinar cells also contain an elaborate rough ER and Golgi apparatus, typical features of cells that secrete proteins. From each acinus, the secreted product travels through the pancreatic duct system to the duodenum, where the enzymes are activated. Furthermore, the epithelial cells that line the smallest pancreatic ducts secrete a bicarbonate-rich fluid that helps neutralize acidic chyme in the duodenum.

The pancreas is also an endocrine gland. It secretes two major hormones, insulin and glucagon, which lower and raise blood sugar levels, respectively. The hormone-secreting cells are clustered into spherical bodies called **pancreatic islets (islets of Langerhans),** which are scattered among the exocrine acini (Figure 21.25). The endocrine pancreas is discussed in detail on page 674.

Mesenteries *Ø A LOT OF TIME*

We now consider the mesenteries that support the abdominal digestive organs (Figure 21.26). Recall that mesenteries are double-layered sheets of peritoneum that connect the peritoneal organs to the dorsal and ventral body wall (Figure 21.5a). Most mesenteries are *dorsal* mesenteries, extending dorsally from the alimentary canal to the posterior abdominal wall. In the superior abdomen, however, a *ventral* mesentery extends ventrally from the stomach and liver to the anterior abdominal wall. As you read about the different

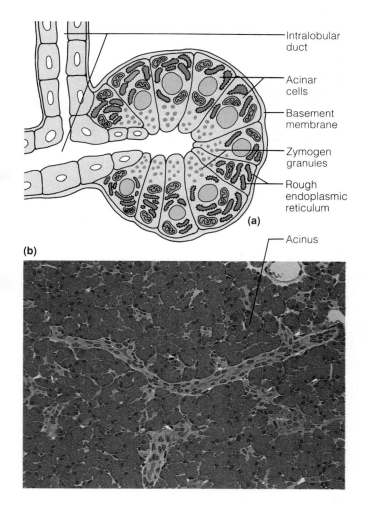

Intralobular duct

Acinar cells

Basement membrane

Zymogen granules

Rough endoplasmic reticulum

(a)

(b)

Acinus

Figure 21.24 The pancreas: Histology. (a) One acinus (a secretory unit). Note the individual acinar cells, which secrete the pancreatic digestive enzymes. (b) Photomicrograph of pancreatic acinar tissue (200 ×). A single acinus is pointed out.

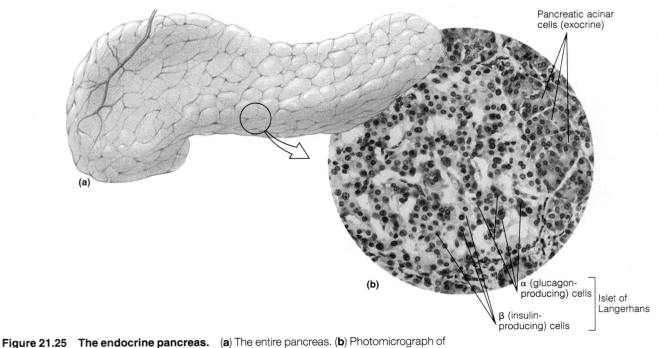

(a)

Pancreatic acinar
cells (exocrine)

(b)

α (glucagon-
producing) cells

Islet of
Langerhans

β (insulin-
producing) cells

Figure 21.25   The endocrine pancreas.   (a) The entire pancreas. **(b)** Photomicrograph of
pancreas tissue (300 ×) showing a lighter-staining pancreatic islet (islet of Langerhans) surrounded by pancreatic acini.

Table 21.1 Summary of Intraperitoneal and Retroperitoneal Digestive Organs in the Abdomen and Pelvis	
Intraperitoneal organs (and their mesenteries)	**Retroperitoneal organs**
Liver (falciform ligament and lesser omentum)	Duodenum (almost all of it)
Stomach (greater and lesser omentum)	Ascending colon
Ileum and jejunum (mesentery proper)	Descending colon
Transverse colon (transverse mesocolon)	Rectum
Sigmoid colon (sigmoid mesocolon)	Pancreas

parts of the dorsal and ventral mesenteries, note that some mesenteries are called *ligaments,* even though these peritoneal sheets are not the same as the fibrous ligaments that interconnect bones. For a summary of the mesenteries described below, see Table 21.1.

The two ventral mesenteries are the falciform ligament and the lesser omentum. The **falciform ligament** (fal′sĭ-form; "sickle-shaped") binds the anterior aspect of the liver to the anterior abdominal wall and diaphragm (Figure 21.26a). The **lesser omentum** (o-men′tum; "fatty skin") runs from the fissure of the liver and the porta hepatis to the lesser curvature of the stomach (Figure 21.26b).

All remaining mesenteries are dorsal mesenteries. The **greater omentum** (Figure 21.26a and d) connects the greater curvature of the stomach to the posterior abdominal wall, but in a very roundabout way:

It is tremendously elongated and extends inferiorly to cover the transverse colon and coils of small intestine like a butterfly net. The greater omentum contains a great deal of fat. This mesentery has a remarkable ability to limit the spread of infections within the peritoneal cavity. For example, it can wrap and enclose an inflamed appendix.

In the small intestine, the duodenum is retroperitoneal, but the long coils of the jejunum and ileum are supported by *the* **mesentery,** or *mesentery proper* (Figures 21.26c). This sheet fans inferiorly from the posterior abdominal wall like long, pleated curtains.

Most parts of the large intestine are retroperitoneal and lack a mesentery. The transverse and sigmoid colons, however, are exceptions: The transverse colon is held to the posterior abdominal wall by the **transverse mesocolon** (mez″o-ko′lon; "mesentery of

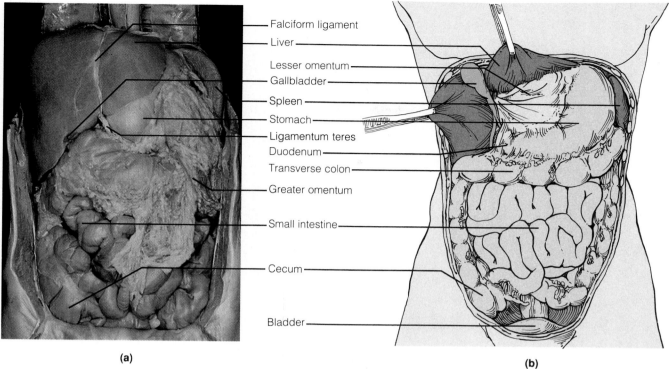

(a)

(b)

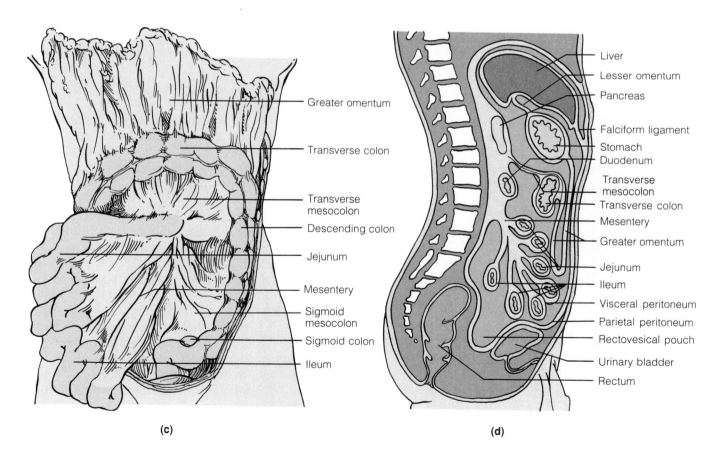

(c)

(d)

Figure 21.26 The mesenteries. (a) Superficial view of the abdominal organs, with only the anterior abdominal wall removed. The greater omentum is in its normal position covering the abdominal viscera. (b) The lesser omentum attaches the lesser curvature of the stomach to the posterior half of the fissure of the liver (the liver is lifted out of the way, and the greater omentum has been removed). (c) The greater omentum and transverse colon have been reflected superiorly to reveal the mesentery proper ("Mesentary"), the transverse mesocolon, and the sigmoid mesocolon. (d) Sagittal section through the abdominopelvic cavity, showing the attachments of the various mesenteries to the posterior and anterior body wall.

Labels for (a): Falciform ligament, Liver, Lesser omentum, Gallbladder, Spleen, Stomach, Ligamentum teres, Duodenum, Transverse colon, Greater omentum, Small intestine, Cecum, Bladder

Labels for (c): Greater omentum, Transverse colon, Transverse mesocolon, Descending colon, Jejunum, Mesentery, Sigmoid mesocolon, Sigmoid colon, Ileum

Labels for (d): Liver, Lesser omentum, Pancreas, Falciform ligament, Stomach, Duodenum, Transverse mesocolon, Transverse colon, Mesentery, Greater omentum, Jejunum, Ileum, Visceral peritoneum, Parietal peritoneum, Rectovesical pouch, Urinary bladder, Rectum

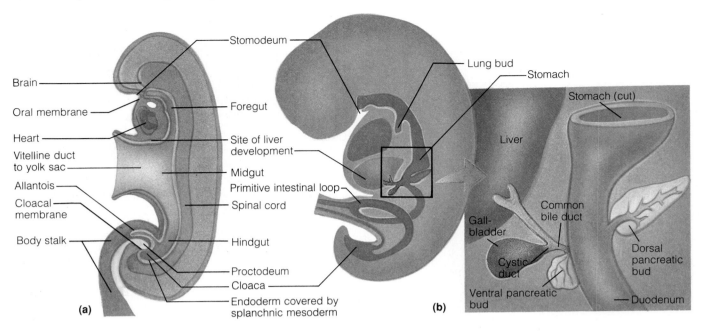

Figure 21.27 Embryonic development of the digestive system. (a) Embryo at 3 weeks (23 days). The primitive gut has formed (foregut, midgut, hindgut). The midgut is still open and continuous with the yolk sac. The oral and cloacal membranes will break through in a few weeks to form the oral and anal openings. (b) By 5 weeks of development, the liver and pancreas are budding off of the distal foregut, as shown in the enlargement. The pancreas forms from two pancreatic buds (ventral and dorsal) that later join.

the colon"), a nearly horizontal sheet. This is fused to the underside of the greater omentum, so it can be viewed only inferiorly (Figure 21.26c). The **sigmoid mesocolon** is the mesentery that connects the sigmoid colon to the posterior pelvic wall.

The Digestive System Throughout Life

Embryonic Development

Recall from Chapter 3 (p. 56) that the alimentary canal originates when the flat embryo folds into the shape of a cylinder, enclosing a tubular part of the yolk sac within its body (Figure 21.27a). This is the *primitive gut,* a tube of endoderm that is covered by splanchnic mesoderm. The endoderm gives rise to the lining epithelium of the alimentary canal and the secretory cells of the digestive glands. The splanchnic mesoderm gives rise to all other layers in the wall of the alimentary canal.

Initially, the middle region of the primitive gut is open to the yolk sac through the **vitelline duct** (vi-tel'in; "yolk"). This duct is a key landmark that di-

vides the gut into three basic regions: foregut, midgut, and hindgut (Figure 21.27a). The **foregut** becomes the first segment of the digestive system, from the pharynx to the point in the duodenum where the bile duct enters. The **midgut** continues from the duodenum to a point two-thirds of the way along the transverse colon. The **hindgut** forms the rest of the large intestine. The abdominal foregut, midgut, and hindgut are supplied by the celiac, superior mesenteric, and inferior mesenteric arteries, respectively.

The caudal portion of the early hindgut joins a tube-like outpocketing called the **allantois** (ah-lan'to-is; "sausage"). The expanded junction between the hindgut and the allantois is the **cloaca** (klo-a'kah; "sewer"). The cloaca gives rise to the rectum and most of the anal canal, among other structures.

In the mouth region of the embryo, the endoderm-lined gut touches the surface ectoderm to form an **oral membrane,** which lies in a depression called the **stomodeum** (sto"mo-de'um; "on the way to becoming the mouth"). Similarly, at the end of the hindgut, endoderm meets ectoderm to form the **cloacal membrane** in a pit called the **proctodeum** (prok"to-de'um; "on the way to becoming the anus"). The oral membrane lies at the future mouth/pharynx boundary (the fauces), and the cloacal membrane lies in the future anal canal, roughly where the pectinate line will occur. The oral and cloacal membranes are reabsorbed during month 2, thereby opening the alimentary canal to the outside.

During weeks 4 and 5, the embryonic gut starts to elongate, bend, and form outpocketings (Figure 21.27b). Salivary glands arise as outpocketings from the mouth; the pharynx develops four pairs of lateral *pharyngeal pouches* (Figure 1.6b, p. 11); and the future lungs and trachea bud off the distal pharynx. A spindle-shaped enlargement of the abdominal foregut is the first sign of the stomach. The liver and pancreas arise as buds from the last part of the foregut. The midgut elongates into a loop, the **primitive intestinal loop.** In months 2 and 3, this loop rotates and elongates to bring the intestines into their final positions.

The Digestive System in Later Life

Unless abnormal interferences occur, the digestive system operates through childhood and adulthood with relatively few problems. However, contaminated or spicy foods sometimes cause an inflammation of the alimentary canal called **gastroenteritis** (gas″tro-en″tĕ-ri′tis). As you might guess, the symptoms of gastroenteritis include nausea, vomiting, cramps, loss of appetite, or diarrhea. As we've mentioned, appendicitis is common in teenagers and young adults. Gallstones and ulcers are problems of middle age.

During old age, the activity of the digestive organs declines. Fewer digestive juices are produced, the absorption of nutrients becomes less efficient, and peristalsis slows. So much water is reabsorbed from the slow-moving fecal mass in the large intestine that the feces become hard and compacted. The result is a decrease in the frequency of bowel movements and, often, constipation.

Diverticulosis and cancer of the digestive organs are other common problems of the aged. Cancer of the stomach, colon, liver, or pancreas rarely exhibits early signs, and the cancer often progresses to an inoperable stage before the person seeks medical attention. These forms of cancer are deadly—colon cancer is the second leading cause of cancer deaths in the United States. However, if detected early, cancers of the digestive viscera are sometimes treatable. Perhaps the best advice is to have regular medical checkups. Half of all rectal cancers can be felt digitally during rectal exams, and nearly 80% of colon cancers can be seen and removed during a colonoscopy (see "Endoscopy" in the Related Clinical Terms). Evidence suggests that diets high in plant fiber decrease the incidence of colon cancer.

Related Clinical Terms

Ascites (ah-si′tēz) (*asci* = bag, bladder) Abnormal accumulation of serous fluid within the peritoneal cavity. The fluid oozes from capillaries in the walls of the intestines. Excessive ascites causes visible bloating of the abdomen. Ascites may be caused by portal hypertension following liver cirrhosis or by heart or kidney disease.

Colitis (ko-li′tis) Inflammation of the large intestine, or colon.

Endoscopy (en-dos′ko-pe) (*endo* = inside; *scopy* = viewing) The viewing of the lining of a ventral body cavity or tubular organ with a flexible, tube-like device called an endoscope. The endoscope contains a lens and a light radiating from its tip. Endoscopes are used to view the internal surfaces of various parts of the alimentary canal, including the stomach (gastroscopy), colon (colonoscopy), and sigmoid colon (sigmoidoscopy).

Enteritis (*enteron* = intestine) Inflammation of the intestine, especially the small intestine.

Intestinal obstruction Total or partial blockage of the large or small intestine. It most often results from a paralysis of the ileum (in which trauma to the body or viscera inhibits the parasympathetic stimulation of peristalsis). Obstruction may also be caused by a twisting of the intestine or by intestinal tumors. Common symptoms include cramps, vomiting, and failure to pass gas and feces.

Liver biopsy (bi′op-se) (*bio* = living; *opsis* = vision, viewing) Removal of a small piece of living tissue from the liver, which is then examined for signs of disease. The puncturing needle is inserted through the seventh, eighth, or ninth intercostal space, in the right midaxillary line (straight inferiorly from the axilla) after the patient has exhaled as much air as possible. (Exhalation minimizes the chances of the needle piercing the lung.)

Pancreatitis (pan″kre-ah-ti′tis) A rare but serious inflammation of the pancreas. It most often results when the pancreatic enzymes are activated in the pancreatic duct rather than in the small intestine, causing pancreatic tissue to be digested. This painful condition, which may arise from blockage of the pancreatic duct by gallstones, can lead to nutritional deficiencies, destruction of pancreatic function, and death.

Pyloric stenosis (*stenosis* = narrowing, constriction) Congenital condition in which the pyloric sphincter of the stomach is abnormally constricted. The characteristic symptom of projectile vomiting usually does not appear until the baby begins to eat solid food. The stenosis usually can be repaired surgically. This condition occurs in about 1 in 400 newborns and can also occur in adults through scarring caused by an ulcer or by a tumor that blocks the pyloric opening.

Rectocele (rek′to-sēl) (*recto* = rectum; *cele* = sac) Condition in females in which the rectum pushes on the vagina and bulges into the posterior vaginal wall. Rectocele usually results from tearing of the supportive muscles of the pelvic floor during childbirth, which then allows the unsupported pelvic viscera to sink inferiorly.

Tracheoesophageal fistula (*fistula* = pipe, connection) Abnormal passage between the esophagus and trachea. In the embryo, the future trachea forms from, and separates from, the pharynx and esophagus of the foregut. If the separation does not proceed correctly, the esophagus may empty directly into the trachea. Thus, when the newborn baby drinks, the milk passes into the lungs, and the baby chokes. This defect, which involves about 1 in every 3000 births, is usually corrected surgically.

Chapter Summary

OVERVIEW OF THE DIGESTIVE SYSTEM (pp. 562–564)

Organization (pp. 562–563)

1. The digestive system includes the alimentary canal (mouth, pharynx, esophagus, stomach, and small and large intestines) and accessory digestive organs (teeth, tongue, salivary glands, liver, gallbladder, and pancreas).

Digestive Processes (p. 563)

2. The digestive system carries out the processes of ingestion, propulsion, mechanical digestion, chemical digestion, absorption, and defecation.

Abdominal Regions and Quadrants (pp. 563–564)

3. The nine regions of the anterior abdominal wall are defined by two horizontal planes (subcostal and transtubercular) and by the vertical midclavicular planes. These regions are two hypochondriacs and an epigastric, two lumbars (laterals) and an umbilical, and two iliacs (inguinals) and a hypogastric (pelvic). The anterior abdominal wall can also be divided into four quadrants.

ANATOMY OF THE DIGESTIVE SYSTEM (pp. 564–594)

The Peritoneal Cavity and Peritoneum (pp. 564–566)

1. The serous membrane in the abdominopelvic cavity is the peritoneum, which has a parietal layer (on the internal surface of the body wall) and a visceral layer (on the viscera). The slit between the visceral and parietal peritoneum is the peritoneal cavity. It holds serous fluid, which decreases friction as the organs move.

2. A mesentery is a double layer of peritoneum that tethers the movable digestive organs to the body wall. Mesenteries also store fat and carry blood vessels and nerves. Visceral organs that lack a mesentery and are fused to the posterior body wall are called retroperitoneal organs.

Histology of the Alimentary Canal Wall (pp. 566–568)

3. The esophagus, stomach, and intestine share the same tissue layers: an inner mucosa, a fibrous submucosa, a muscularis externa, and an outer serosa (visceral peritoneum) or adventitia. The mucosa consists of a lining epithelium, lamina propria, and muscularis mucosae.

4. Visceral nerve plexuses (myenteric, submucosal, subserous) occur in the wall of the alimentary canal. These plexuses contain parasympathetic, sympathetic, and visceral sensory fibers, as well as enteric neurons (neurons confined in the wall of the alimentary canal).

The Mouth and Associated Organs (pp. 568–574)

5. Food enters the alimentary canal through the mouth (oral cavity), which consists of an external vestibule and an internal oral cavity proper.

6. The mouth is lined by a stratified squamous epithelium, which resists abrasion by food fragments.

7. The lips and cheeks keep food inside the mouth during chewing. The red margin is the red part of the lips that borders the oral orifice.

8. The tongue is mucosa-covered skeletal muscle. Its intrinsic muscles change its shape, and its extrinsic muscles change its position. There are three classes of papillae on the tongue's superior surface: filiform papillae, which grip food during chewing; and fungiform and circumvallate papillae, which contain taste buds.

9. Saliva is produced by intrinsic salivary glands in the oral mucosa and by three pairs of large extrinsic salivary glands—parotid, submandibular, and sublingual. The salivary glands are compound tubuloalveolar glands containing varying amounts of serous and mucous cells.

10. Teeth tear and grind food in the chewing process (mastication). The 20 deciduous teeth begin to fall out at age 6 and are gradually replaced during childhood and youth by the 32 permanent teeth.

11. Teeth are classified as incisors, canines, premolars, and molars. Each tooth has an enamel-covered crown and a cementum-covered root. The bulk of the tooth is dentin, which surrounds the central pulp cavity. A periodontal ligament secures the tooth to the bony alveolus.

The Pharynx (p. 574)

12. During swallowing, food passes from the mouth through the oropharynx and laryngopharynx, which are lined by a stratified squamous epithelium. The pharyngeal constrictor muscles squeeze food into the esophagus during swallowing.

The Esophagus (p. 574)

13. The esophagus descends from the pharynx through the posterior mediastinum and into the abdomen. There it joins the stomach at the cardiac orifice, where a functional sphincter prevents superior regurgitation of stomach contents.

14. The esophageal mucosa contains a stratified squamous epithelium. The muscularis consists of skeletal muscle superiorly and smooth muscle inferiorly. The esophagus has an adventitia rather than a serosa.

The Stomach (pp. 575–578)

15. The J-shaped stomach churns food into chyme and secretes HCl and pepsin. It lies in the superior left part of the abdomen. Its major regions are the cardia, fundus, body, pyloric region, and pylorus (containing the pyloric sphincter). Its right and left borders are the lesser and greater curvatures. When the stomach is empty, its internal surface exhibits rugae.

16. The internal surface of the stomach is lined by goblet cells. This surface is dotted with gastric pits that lead into tubular gastric glands. Secretory cells in the gastric glands include pepsinogen-producing chief cells, parietal cells that secrete HCl, mucous neck cells, and enteroendocrine cells that secrete hormones (including gastrin).

17. The stomach wall is protected against self-digestion and acid by an internal coat of mucus and by the rapid regeneration of its lining epithelium.

The Small Intestine (pp. 578–582)

18. The small intestine is the main site of digestion and nutrient absorption. Its segments are the duodenum, jejunum, and ileum.

19. The duodenum lies retroperitoneally in the superior abdomen. The bile duct and pancreatic duct join to form the hepatopancreatic ampulla and empty into the duodenum through the major duodenal papilla.

20. The small intestine has four features that increase its surface area and allow rapid absorption of nutrients: its great length, plicae circulares, villi, and abundant microvilli on the absorptive cells. Absorbed nutrients enter capillaries in the core of the villi.

21. The epithelium lining the internal intestinal surface contains absorptive, goblet, and enteroendocrine cells. Between the villi are the intestinal crypts, which secrete intestinal juice and continuously renew the lining epithelium.

22. Other special features of the small intestine are (1) aggregated lymphoid follicles in the ileum and (2) the duodenal mucous glands.

The Large Intestine (pp. 582–586)

23. The large intestine compacts feces by absorbing water. It forms an open rectangle around the small intestine. Its subdivisions are the cecum and appendix, colon (ascending, transverse, descending, and sigmoid), rectum, and anal canal. Special features on the external surface of the colon and cecum are teniae coli, haustra, and epiploic appendages.

24. The cecum lies in the right iliac fossa and contains the ileocecal valve. Attached to the cecum is the vermiform appendix, which contains abundant lymphoid tissue.

25. Near the end of the large intestine, the sigmoid colon enters the pelvis and joins the rectum. The rectum pierces the pelvic floor and joins the anal canal, which ends at the anus. Internally, the rectum contains the transverse rectal folds, and the anal canal contains anal columns, anal valves, and anal sinuses.

26. The defecation reflex produces a contraction of the rectal walls when feces enter the rectum. During defecation, a person strains to increase the intra-abdominal pressure and lifts the anal canal. Defecation can be inhibited by the anal sphincters.

27. Like the small intestine, the large intestine is lined by a simple columnar epithelium with absorptive and goblet cells and contains intestinal crypts. It lacks villi and secretes more mucus and contains more lymphoid tissue than the small intestine.

The Liver (pp. 586–590)

28. The liver performs many functions, including processing of absorbed nutrients and secreting fat-emulsifying bile.

29. The liver lies in the superior right region of the abdomen. It has diaphragmatic and visceral surfaces and four lobes (right, left, quadrate, and caudate).

30. Most vessels enter or leave the liver through the porta hepatis. Other structures on the visceral surface are the inferior vena cava, hepatic veins, gallbladder, and ligamentum venosum.

31. Several structures attach to the liver's fissure, including the falciform and round ligaments. The liver's bare area lies against the diaphragm.

32. Classical liver lobules are hexagonal structures. They consist of plates of hepatocytes that are separated by blood sinusoids and converge on a central vein. Portal triads (portal vein, hepatic artery, bile duct) occur at most corners of the lobule.

33. Blood flows through the liver in the following sequence: branches of the hepatic artery and portal vein, through the sinusoids to supply the hepatocytes, to the central veins, hepatic veins, inferior vena cava, and heart. In the sinusoids, Kupffer cells (macrophages) remove debris from the blood.

34. Hepatocytes perform almost all liver functions. These cells secrete bile into the bile canaliculi, which proceeds to the bile ducts, to the hepatic ducts, and through the common hepatic duct to the gallbladder and duodenum.

The Gallbladder (p. 590)

35. The gallbladder, a green muscular sac, stores and concentrates bile. Its duct is the cystic duct. When fatty chyme enters the small intestine, the gallbladder squeezes bile into the bile duct and duodenum.

The Pancreas (pp. 590–591)

36. The pancreas runs horizontally across the posterior abdominal wall, between the duodenum and the spleen.

37. The pancreas is an exocrine gland containing many compound acinar glands that empty into the main and accessory pancreatic ducts. The serous acinar cells secrete digestive enzymes, and the smallest duct cells secrete an alkaline fluid. Both products are emptied to the duodenum.

38. The pancreas is also an endocrine gland, with hormone-secreting cells contained in pancreatic islets.

Mesenteries (pp. 591–594)

39. The ventral mesenteries are the falciform ligament and lesser omentum. The dorsal mesenteries are the greater omentum, mesentery proper, transverse mesocolon, and sigmoid mesocolon.

THE DIGESTIVE SYSTEM THROUGHOUT LIFE (pp. 594–595)

Embryonic Development (pp. 594–595)

1. As the 3-week-old embryo lifts off the yolk sac to assume its cylindrical body shape, it encloses the primitive gut, a tube of endoderm covered by splanchnic mesoderm. The endoderm becomes the lining epithelium (and gland cells), and the splanchnic mesoderm gives rise to all other layers of the wall of the alimentary canal.

2. The embryonic gut is divided into a foregut, midgut, and hindgut, which form distinct regions of the adult digestive system.

3. The primitive embryonic gut tube, straight at first, soon grows outpocketings (liver, pancreas, pharyngeal pouches), shows swellings (stomach, cloaca), and lengthens into a primitive intestinal loop. This loop rotates and elongates to bring the intestines into their final positions.

The Digestive System in Later Life (p. 595)

4. Various diseases may plague the digestive organs throughout life. Appendicitis is common in young adults; gastroenteritis and food poisoning can occur at any time; and ulcers and gallbladder problems increase in middle age.

5. The efficiency of digestive processes declines in the elderly, and constipation becomes common. Diverticulosis and cancers (such as colon cancer and stomach cancer) appear with increasing frequency in an aging population.

Review Questions

Multiple Choice and Matching Questions

1. Which of these digestive organs is not a part of the alimentary canal? (a) stomach, (b) liver, (c) small intestine, (d) large intestine, (e) pharynx.

2. Which of these organs is retroperitoneal? (a) pharynx, (b) stomach, (c) ascending colon, (d) ileum.

3. Which statement is true of the peritoneal cavity? (a) It is the same thing as the abdominopelvic cavity. (b) It is filled with air. (c) Like the pleural and pericardial cavities, it is a potential space containing serous fluid. (d) It contains the pancreas and all of the duodenum.

4. The submucosal nerve plexus (a) innervates the mucosa layer, (b) lies in the mucosa layer, (c) controls peristalsis, (d) contains only motor neurons.

5. Mesenteries. Match each of the descriptions at right with the appropriate mesentery at left.

Descriptions	Mesenteries
_____ **(1)** connects the ileum and jejunum to the posterior abdominal wall	**(a)** greater omentum
_____ **(2)** connects fissure of liver to the anterior abdominal wall	**(b)** lesser omentum
_____ **(3)** connects the large intestine to the pelvic wall	**(c)** falciform ligament
_____ **(4)** attaches to greater curvature of stomach; has the most fat	**(d)** mesentery proper
_____ **(5)** runs from stomach's lesser curvature to fissure of the liver	**(e)** transverse mesocolon
_____ **(6)** a mesentery of the large intestine that is fused to the underside of the greater omentum	**(f)** sigmoid mesocolon

6. Match each of the organs at left with the type of inner epithelium that lines its lumen, from the column at right.

Organ	Epithelium types
_____ **(1)** oral cavity	**(a)** simple squamous
_____ **(2)** oropharynx	**(b)** simple cuboidal
_____ **(3)** esophagus	**(c)** simple columnar
_____ **(4)** stomach	**(d)** stratified squamous
_____ **(5)** small intestine	**(e)** stratified cuboidal
_____ **(6)** colon	

7. This pointed type of tongue papilla is not involved with taste reception: (a) filiform, (b) fungiform, (c) circumvallate, (d) dermal.

8. Circle the one false statement about the gallbladder: (a) It makes bile. (b) Its duct is called the cystic duct. (c) It has a fundus lying inferior to the liver. (d) It has mucosal folds similar to the stomach rugae.

9. The duct of the submandibular gland opens into the mouth (a) in the vestibule, (b) near the tongue's frenulum, (c) at the mandibular angle, (d) through the upper second molars.

10. The products of protein digestion are called amino acids. These nutrients are absorbed into the bloodstream largely through cells lining the (a) stomach, (b) small intestine, (c) large intestine, (d) bile duct, (e) lacteals.

11. The hepatopancreatic ampulla lies in the wall of the (a) liver, (b) pancreas, (c) duodenum, (d) stomach.

12. Which two structures produce alkaline secretions that neutralize the acidic stomach chyme as it enters the duodenum? (a) intestinal flora and pancreatic acinar cells, (b) gastric glands and gastric pits, (c) lining epithelium of intestine and Paneth cells, (d) pancreatic ducts and duodenal glands.

13. Which correctly describes the flow of blood through the classical liver lobule and beyond? (More than one choice is correct.) (a) portal vein branch to sinusoids to central vein to hepatic vein to inferior vena cava, (b) porta hepatis to hepatic vein to portal vein, (c) portal vein branch to central vein to hepatic vein to sinusoids, (d) hepatic artery branch to sinusoids to central vein to hepatic vein.

14. The exocrine glands in the pancreas are of this type: (a) simple tubular, (b) simple acinar, (c) compound tubular, (d) compound acinar, (e) compound tubuloalveolar.

15. Which of these embryonic structures gives rise to the rectum and most of the anal canal? (a) proctodeum, (b) stomodeum, (c) cloaca, (d) midgut.

16. A 3-year old girl was rewarded with a hug because she was now completely toilet trained. Which muscle had she learned to control? (a) levator ani, (b) internal and external obliques of her abdominal wall, (c) external anal sphincter, (d) internal anal sphincter.

17. Which cell type fits this description? It occurs in the stomach mucosa, has three prongs, contains many mitochondria and many microvilli, and pumps hydrogen ions. (a) absorptive cell, (b) parietal cell, (c) goblet cell, (d) muscularis externa cell, (e) mucous neck cell.

18. The only feature in this list that is shared by both the small and large intestines is (a) intestinal crypts, (b) Peyer's patches (aggregated follicles), (c) teniae coli, (d) haustra, (e) plicae circulares, (f) intestinal villi.

19. The only part of the digestive tube whose wall does not secrete any mucus is the (a) mouth (with its intrinsic salivary glands), (b) esophagus, (c) stomach, (d) small intestine, (e) large intestine, (f) none of the above.

Short Answer and Essay Questions

20. Make a simple drawing of the organs of the alimentary canal and label each organ. Then add three labels to your drawing— *salivary glands, liver,* and *pancreas*—and use arrows to show where each of these glandular organs empties its secretion into the alimentary canal.

21. Name the layers of the wall of the alimentary canal. Note the tissue composition and the major function of each layer.

22. Define mesentery.

23. (a) Write the dental formulas for both the deciduous and the permanent teeth. (b) What and where is pulp?

24. What are the effects of aging on the activity of the digestive system?

25. Bianca went on a trip to the Bahamas during spring vacation and did not study for her anatomy test scheduled early the following week. On the test, she mixed up the following pairs of structures: (a) the rectal valves and the anal valves, (b) pyloric region and pylorus of the stomach, (c) anal canal and anus, (d) villi and microvilli (in the small intestine), (e) hepatic vein and hepatic portal vein, (f) gastric pits and gastric glands. Can you help her by defining and differentiating all of these sound-alike structures?

26. Make a rough sketch of the visceral surface of the liver, and label the following five structures: fissure, porta hepatis, inferior vena cava, gallbladder, caudate lobe.

27. Name four structural features that increase the absorptive surface area of the small intestine.

28. List the three major vessels of the portal triad, and describe the location of this triad with respect to the liver lobule.

29. What is the function of lacteals in digestion?

30. Name three organelles that are abundant within hepatocytes, and explain how each of these organelles contributes to liver functions.

**Critical Thinking and
Clinical Application Questions**

31. In this chapter, we divided the anterior abdominal wall into nine regions and also into four quadrants. This question tests your understanding by asking you to combine both schemes: List which of the nine regions lie within (or overlap) (a) the upper right quadrant and (b) the lower left quadrant.

32. A 21-year-old man with severe appendicitis did not seek treatment in time and died a week after his abdominal pain and fever began. Explain why appendicitis can quickly lead to death.

33. Duncan, an inquisitive 8-year-old, saw his grandfather's dentures soaking overnight in a glass. He asked his grandfather how his teeth had fallen out. Reconstruct the kind of story the man is likely to have told.

34. Eva, a middle-aged attorney, complains of a burning pain in the "pit of her stomach," usually beginning about 2 hours after eating and lessening after she drinks a glass of milk. When asked to indicate the site of pain, she points to her epigastric region. Her GI tract is examined by X-ray fluoroscopy. A gastric ulcer is visualized. What are the possible consequences of nontreatment?

35. A doctor used an endoscope and located some polyps (precancerous tumors) in the wall of the large intestine of an important government official in his seventies. What is an endoscope?

36. The janitor who cleaned the anatomy lab had a protruding abdomen that looked like the biggest "beer-belly" the students had ever seen. However, the rest of his body did not look fat at all. Another janitor on the floor told some students that the man was a reformed alcoholic and that he had over "100 pounds of fluid in his belly." What was the man's probable condition? (Hint: See this chapter's Related Clinical Terms.)

22

The Urinary System

Student Objectives

Kidneys (pp. 601–611)

1. Describe the location, coverings, and external gross anatomy of the kidney.
2. Describe the internal gross anatomy and the main blood vessels of the kidney.
3. Identify the segments of the uriniferous tubule, and explain their specific roles in forming urine.
4. Describe the structure and functions of the capillaries and arterioles in the kidney.

Ureters (pp. 611–613)

5. Describe the location, histology, and function of the ureters.

Urinary Bladder (pp. 613–614)

6. Describe the shape, location, histology, and function of the bladder.

Urethra (pp. 614–616)

7. For both sexes, describe the structure and function of the urethra.

Micturition (pp. 616–617)

8. Define micturition, and describe the micturition reflex.

The Urinary Organs Throughout Life (pp. 617–619)

9. Describe the embryonic development of the urinary organs.
10. List several effects of aging on the structure and function of the urinary system.

The kidneys (Figure 22.1) maintain the purity and constancy of the blood and the other extracellular body fluids. Much like a water purification plant that keeps a city's water drinkable and disposes of its wastes, the kidneys are usually unappreciated until they malfunction and the body fluids become contaminated. Every day, the kidneys filter many liters of fluid from the blood, allowing toxins, metabolic wastes, and excess ions to leave the body in urine, while returning needed substances from the filtrate to the blood. The main waste products excreted in urine are three nitrogen compounds: (1) *urea*, derived from the breakdown of amino acids during normal recycling of the body's proteins; (2) *uric acid*, which results from the turnover of nucleic acids; and (3) *creatinine* (kre-at′ĭ-nin), formed by the breakdown of creatine phosphate, a molecule in muscle that stores energy for the manufacture of adenosine triphosphate (ATP). Although the lungs, liver, and skin also participate in excretion, the kidneys are the major excretory organs.

Disposing of wastes and excess ions is only one aspect of the work of the kidneys. As they perform these excretory functions, they also regulate the volume and chemical makeup of the blood, maintaining the proper balance between water and salts and between acids and bases. Frankly, this would be tricky work for a chemical engineer, but the kidneys do it efficiently most of the time.

Besides the urine-forming kidneys, the other organs of the urinary system are the paired *ureters* (u-re′terz; "pertaining to urine"), the *urinary bladder,* and the *urethra* (u-re′thrah) (Figure 22.2). The ureters are tubes that carry urine from the kidney to the bladder; the bladder is a temporary storage reservoir for urine; and the urethra is a tube that carries urine to the body exterior.

Kidneys

Gross Anatomy

Location and External Anatomy

The red-brown, bean-shaped **kidneys** lie in the superior lumbar region of the posterior abdominal wall (Figures 22.1, 22.2, and 22.3). They extend from the level of the eleventh or twelfth thoracic vertebra superiorly to the third lumbar vertebra inferiorly and thus receive some protection from the inferior two ribs (Figure 22.2b). The right kidney is crowded by the liver and lies slightly inferior to the left kidney. An average adult kidney is about 12 cm tall, 6 cm wide, and 3 cm thick (the size of a large bar of soap). The lateral surface of each kidney is convex. The medial surface is concave and has a vertical cleft called the **renal hilus.** It is at the hilus that vessels and nerves enter and leave the kidney. On the superior portion of each kidney lies an adrenal (suprarenal) gland, an endocrine gland that is functionally unrelated to the kidney.

Several layers of supportive tissue surround each kidney (Figure 22.3a). The **renal capsule** (*ren* = kidney) adheres directly to the kidney surface. This thin but tough layer of dense connective tissue maintains the kidney's shape and forms a barrier that can inhibit the spread of infection from surrounding regions into the kidneys. Just external to the renal capsule is an **adipose capsule** consisting of *perirenal fat* (per′e-re″nal; "around the kidney"), and external to that is an envelope of **renal fascia.** Finally, another layer of fat lies posterior to the renal fascia, the *pararenal fat* (par′ah-re″nal; "near the kidney"). The perirenal and pararenal fat layers cushion the kidney against blows.

The fatty and fascia encasements hold the kidneys in place within the body. A loss of this fat (as with rapid weight loss) can lead to damage to the kidneys. Can you guess how such damage could develop?

Unsupported, the kidneys drop to a lower position, kinking the ureter and blocking the drainage of

(text continues on p. 604)

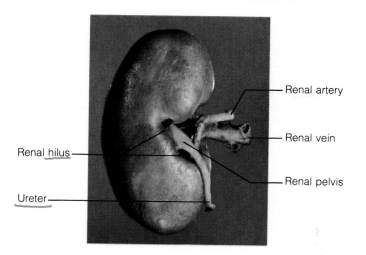

Figure 22.1 Photograph of the left kidney from a cadaver (posterior view).

Renal artery

Renal vein

Renal hilus

Renal pelvis

Ureter

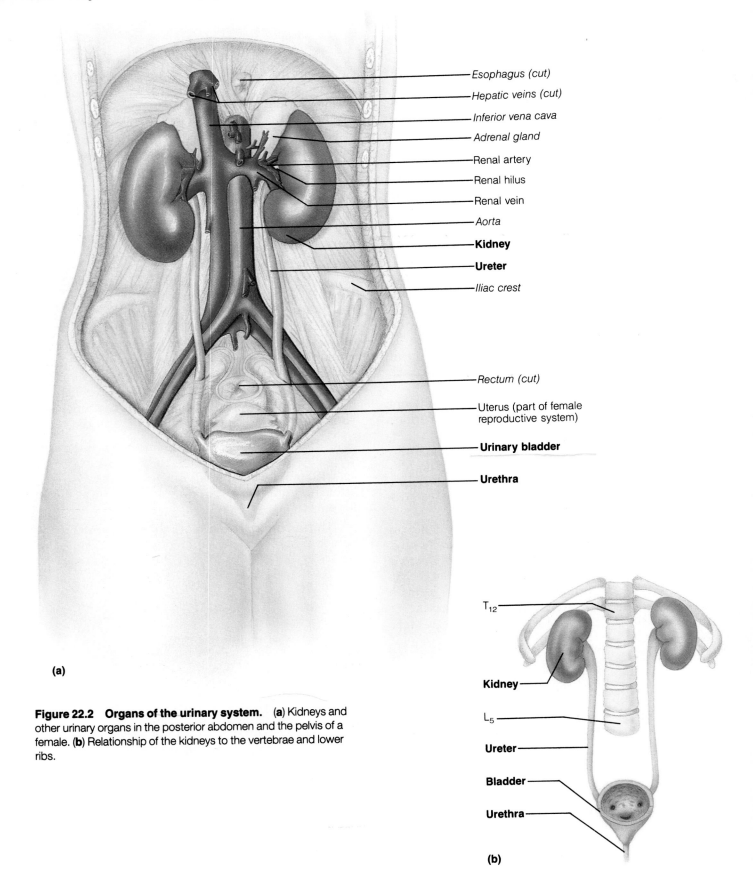

(a)

Esophagus (cut)
Hepatic veins (cut)
Inferior vena cava
Adrenal gland
Renal artery
Renal hilus
Renal vein
Aorta
Kidney
Ureter
Iliac crest

Rectum (cut)
Uterus (part of female reproductive system)
Urinary bladder
Urethra

T₁₂
Kidney
L₅
Ureter
Bladder
Urethra

(b)

Figure 22.2 Organs of the urinary system. (a) Kidneys and other urinary organs in the posterior abdomen and the pelvis of a female. (b) Relationship of the kidneys to the vertebrae and lower ribs.

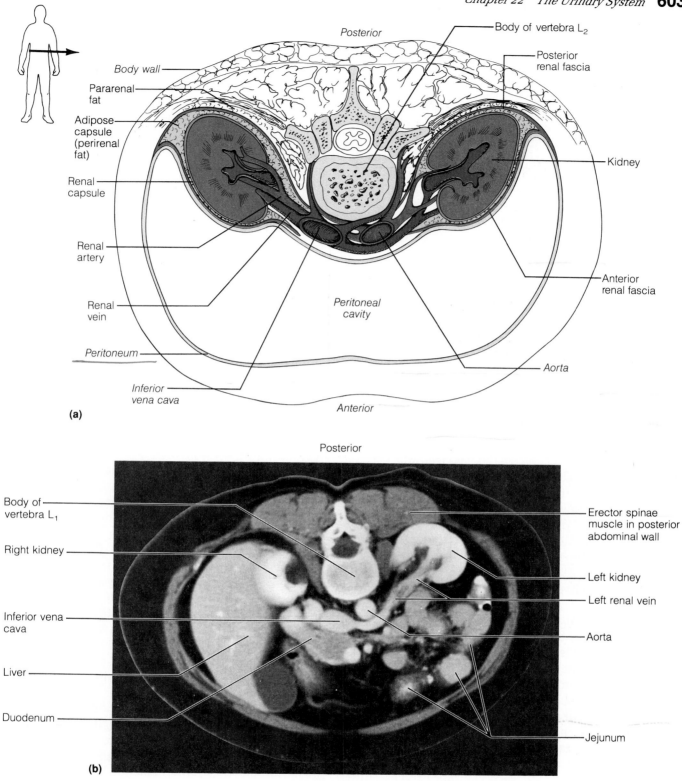

Figure 22.3 Position of the kidneys within the posterior abdominal wall (transverse section). (**a**) The kidneys are retroperitoneal and surrounded by layers of fascia and fat. These layers are the renal capsule, adipose capsule of perirenal fat, renal fascia, and the pararenal fat posteriorly. (**b**) CT scan through the abdomen, showing the relationship of the kidneys to the peritoneal organs of the digestive system.

urine. Urine accumulates in the kidney and exerts pressure on its tissue. This is **hydronephrosis** (hi″dro-ně-fro′sis; "water in the kidney"), and it may lead to necrosis of the kidney and renal failure. ■

 Surgery is performed on kidneys for a variety of reasons (tumor removal and removal of a kidney for a kidney transplant, for example). Surgeons usually approach the kidney by cutting through the posterolateral abdominal wall, where the kidney lies closest to the body surface (Figure 22.3a). However, the incision must be made inferior to the level of T_{12} to avoid puncturing the pleural cavity, which lies posterior to the superior third of each kidney. (Recall that puncturing the pleural cavity leads to pneumothorax and collapse of the lung: p. 551.) ■

Internal Anatomy

A frontal section through a kidney reveals three distinct regions (Figure 22.4): the *cortex,* the *medulla,* and the medial *sinus.* The most superficial region, the **renal cortex,** is light in color and has a granular appearance. Deep to the cortex is the darker **renal medulla,** which consists of cone-shaped masses called **medullary pyramids** or **renal pyramids.** The broad base of each pyramid faces the cortex, whereas the apex, or **papilla,** points internally. The medullary pyramids exhibit striations, for they contain roughly parallel bundles of tiny urine-collecting tubules. The **renal columns,** inward extensions of the renal cortex, separate adjacent pyramids from one another.

The human kidney is said to have *lobes.* A kidney lobe is defined as a single medullary pyramid along with the cortical tissue that surrounds that pyramid. There are five to eleven lobes (and pyramids) in each kidney.

The *renal sinus* is a large space within the medial part of the kidney opening to the exterior through the hilus. This sinus is actually a "filled space" containing the renal vessels and nerves, as well as some fat. The largest elements within the renal sinus, however, are urine-carrying tubes called the renal pelvis and calyces. The **renal pelvis** (*pelvis* = basin), a flat, funnel-shaped tube, is simply the expanded superior part of the ureter. Branching extensions of the renal pelvis form two or three **major calyces** (ca′lih-sēs), each of which subdivides to form several **minor calyces,** cup-shaped tubes that enclose the papillae of the pyramids (*calyx* = cup). The calyces collect urine draining from the papillae and empty it into the renal pelvis. The urine then flows through the renal pelvis and into the ureter, which transports it to the bladder for storage.

 Infection of the renal pelvis and calyces produces **pyelitis** (pi″ě-li′tis), literally "inflamma-

tion of the renal pelvis." When the infection involves the rest of the kidney as well, it produces **pyelonephritis** (pi″ě-lo-ně-fri′tis; *nephros* = kidney). These infections are usually caused by the spread of fecal bacteria *Escherichia coli* from the anal region superiorly through the urinary tract. Less often, pyelonephritis results from blood-borne bacteria that lodge in the kidneys and proliferate there. In severe cases of pyelonephritis, the kidney swells, abscesses form, and the renal pelvis fills with pus. If untreated, the infected kidneys may be severely damaged. However, timely treatment with antibiotics usually achieves total remission. ■

Gross Vasculature and Nerve Supply

Since the kidneys continuously cleanse the blood and adjust its composition, it is not surprising that they have a rich blood supply (Figure 22.4b). Under normal resting conditions, the large **renal arteries** deliver about one-fourth of the heart's systemic output to the kidneys. The renal arteries issue at right angles from the abdominal aorta, between the first and second lumbar vertebrae (see Figure 22.2a and Figure 18.11 on p. 496). Because the aorta lies slightly to the left of the body midline, the right renal artery is longer than the left. As each renal artery approaches a kidney, it divides into five **segmental arteries** that enter the hilus (Figure 22.4b). Within the renal sinus, each segmental artery branches into several **lobar arteries.** These then divide to form **interlobar arteries,** which lie in the renal columns between the medullary pyramids.

At the medulla-cortex junction, the interlobar arteries branch into **arcuate arteries** (ar″ku-āt; "shaped like a bow"), which arch over the bases of the medullary pyramids. Small **interlobular arteries** radiate from the arcuate arteries to supply the cortical tissue. As their name implies, the interlobular arteries divide the cortical tissue into *lobules.* (Note that *interlobular* and *interlobar* arteries are not the same.) More than 90% of the blood entering the kidney perfuses the cortex.

The *veins* of the kidney essentially trace the pathway of the arteries in reverse: Blood leaving the renal cortex drains sequentially into the **interlobular, arcuate, interlobar,** and **renal veins** (there are no lobar or segmental veins). The renal vein issues from the kidney at the hilus and empties into the inferior vena cava (Figure 22.2a). Since the inferior vena cava lies on the right side of the vertebral column, the left renal vein is about twice as long as the right. Along its course, each renal vein lies anterior to the renal artery, and both blood vessels lie anterior to the renal pelvis at the hilus of the kidney (Figures 22.1 and 22.4b).

The nerve supply of the kidney is provided by the *renal plexus,* a network of autonomic fibers and au-

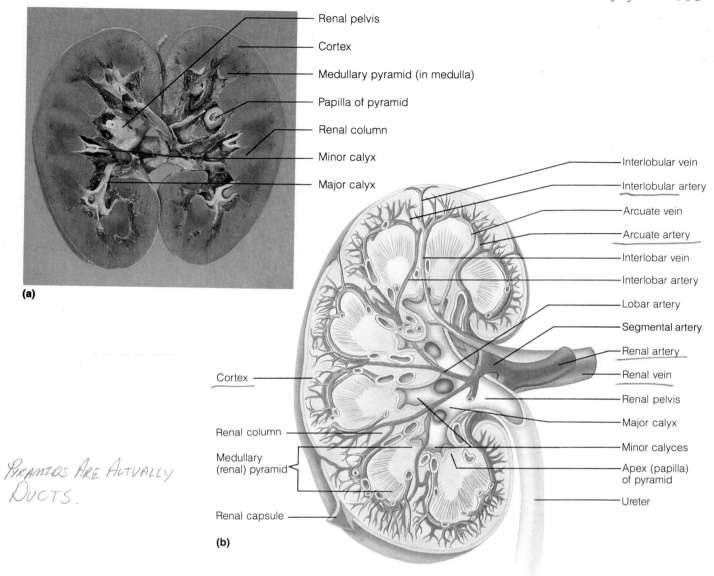

(a)

Renal pelvis

Cortex

Medullary pyramid (in medulla)

Papilla of pyramid

Renal column

Minor calyx

Major calyx

Interlobular vein

Interlobular artery

Arcuate vein

Arcuate artery

Interlobar vein

Interlobar artery

Lobar artery

Segmental artery

Renal artery

Renal vein

Renal pelvis

Major calyx

Minor calyces

Apex (papilla) of pyramid

Ureter

Cortex

Renal column

Medullary (renal) pyramid

Renal capsule

(b)

PYRAMIDS ARE ACTUALLY DUCTS.

Figure 22.4   Internal anatomy of the kidney. (**a**) Photograph of a frontally sectioned kidney. (**b**) Diagram of a frontally sectioned kidney, emphasizing major blood vessels. (This is a posterior view of the kidney's anterior half.) The renal sinus is the kidney's medial region and contains the renal pelvis, calyces, lobar vessels, nerves, and fat.

tonomic ganglia on the renal arteries. (The renal plexus is a branch of the celiac plexus: Figure 14.5, p. 399). The renal plexus is supplied by sympathetic fibers from the most inferior thoracic splanchnic nerve, the first lumbar splanchnic nerve, and other sources. These nerve fibers mainly control the diameters of the kidney arteries.

Microscopic Anatomy

The main structural and functional unit of the kidney is the **uriniferous tubule** (u″rĭ-nif′er-us; "urine-carry-

ing") (Figure 22.5). More than a million of these epithelium-lined tubes crowd together within each kidney. Each uriniferous tubule has two basic parts: (1) a *nephron* and (2) a *collecting tubule.* As we will see, the nephrons form the urine, and the collecting tubules help concentrate urine (remove some water from it) as it travels toward the calyx system.

The Nephron

Basic Functions. The **nephron** produces urine through three interacting mechanisms: filtration,

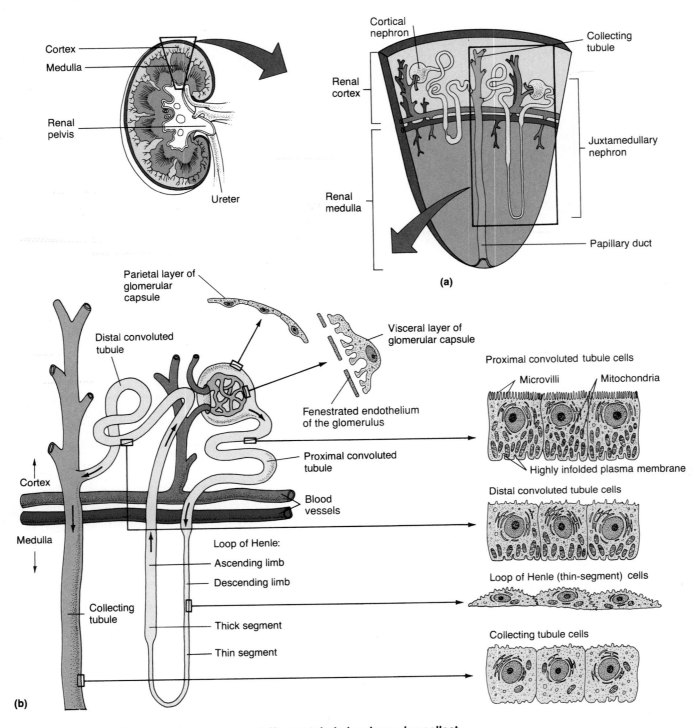

Figure 22.5   Location and structure of the uriniferous tubule (nephron plus collecting tubule). (**a**) A wedge-shaped section (lobe) of kidney tissue, showing the location of nephrons and collecting tubules in the kidney. Although thousands of collecting tubules and nephrons occupy each lobe of the kidney, only a few are shown here. (**b**) Schematic view of a uriniferous tubule, depicting the ultrastructure of the epithelial cells that form its various regions.

reabsorption, and secretion (Figure 22.6). In **filtration,** a filtrate of the blood leaves the kidney capillaries and enters the nephron (see arrow [a] in the figure). This filtrate resembles tissue fluid, containing all the small molecules of blood plasma. As it proceeds through the nephron, the filtrate is processed into urine by the mechanisms of reabsorption and secretion. During **reabsorption,** most of the nutrients, water, and essen-

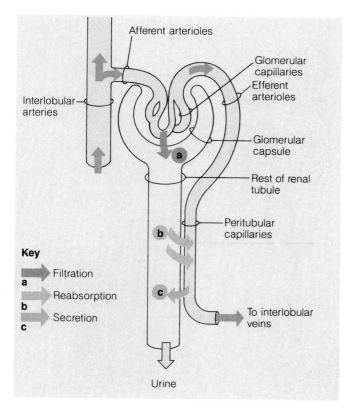

Afferent arterioles

Glomerular capillaries

Efferent arterioles

Glomerular capsule

Interlobular arteries

Rest of renal tubule

Peritubular capillaries

Key

→ Filtration
a

→ Reabsorption
b

→ Secretion
c

To interlobular veins

Urine

Figure 22.6 Basic kidney functions: The urine-forming part of the kidney, depicted as a single, large nephron. Millions of these nephrons act together in a kidney. The three major mechanisms by which the kidneys form urine are (**a**) glomerular filtration, (**b**) tubular reabsorption, and (**c**) tubular secretion.

tial ions are reclaimed from the filtrate, and returned to the blood of surrounding capillaries (arrow [b]). In fact, 99% of the volume of the renal filtrate is reabsorbed in this manner. As the essential molecules are reclaimed from the filtrate, the remaining wastes and unneeded substances contribute to the urine that ultimately leaves the body. Supplementing this passive method of waste disposal is the active process of **secretion,** which moves additional undesirable molecules into the nephron from the blood of surrounding capillaries (arrow [c]). Now let us explore the basic divisions of the nephron and see how each contributes to the processes of filtration, reabsorption, and secretion.

Renal Corpuscle and Filtration. The first part of the nephron is the spherical **renal corpuscle** (Figures 22.7a and 22.8). Renal corpuscles occur strictly in the cortex of the kidney. Each renal corpuscle consists of a tuft of capillaries called a **glomerulus** (glo-mer′u-

lus; "ball of yarn") surrounded by a cup-shaped, hollow **glomerular capsule** (also called *Bowman's capsule*). The glomerulus lies within its glomerular capsule like a fist pushed deeply into an inflated balloon (Figure 22.7a). This capillary tuft is supplied by an afferent arteriole and drained by an efferent arteriole (see p. 611). The endothelium of the glomerulus is fenestrated. Thus, these capillaries are exceptionally porous: They allow large quantities of fluid and small molecules to pass from the capillary blood into the hollow interior of the glomerular capsule, the **capsular space.** This fluid is the filtrate that is ultimately processed into urine.

The external layer, or **parietal layer,** of the glomerular capsule is a simple squamous epithelium. This layer simply contributes to the structure of the capsule and plays no part in the formation of filtrate. The capsule's **visceral layer,** which clings to the glomerulus, consists of highly modified, branching epithelial cells called **podocytes** (pod′o-sītz; "foot cells"). The branches of the octopus-like podocytes terminate in **foot processes,** or **pedicels** ("little feet"), which interdigitate with one another as they surround the glomerular capillaries. Thin clefts between the foot processes, called **filtration slits** or *slit pores* (Figure 22.7a), allow the filtrate to pass into the capsular space.

The **filtration membrane (filtration barrier)** is the actual filter that lies between the blood in the glomerulus and the capsular space. It consists of three layers (Figure 22.7c): (1) the fenestrated endothelium of the capillary, (2) the filtration slits between the foot processes of podocytes, and (3) an intervening basement membrane consisting of the fused basal laminae of the endothelium and the podocyte epithelium. The capillary pores (fenestrations) restrict the passage of blood cells, but the basement membrane restricts all but the smallest plasma proteins. This membrane permits free passage only of small molecules such as water, ions, glucose, amino acids, and urea. Therefore, the basement membrane seems to be the main molecular filter of the filtration membrane. Current evidence indicates that the filtration slits between the podocytes play almost no part in holding back molecules. Thus, the name *filtration slits* is probably inaccurate.

Other Parts of the Nephron. After forming in the renal corpuscle, the filtrate proceeds into the long tubular section of the nephron (Figure 22.5b). This section begins as the elaborately coiled *proximal convoluted tubule,* makes a hairpin loop called the *loop of Henle,* winds and twists again as the *distal convoluted tubule,* and ends by joining a collecting tubule. This meandering structure of the nephron increases

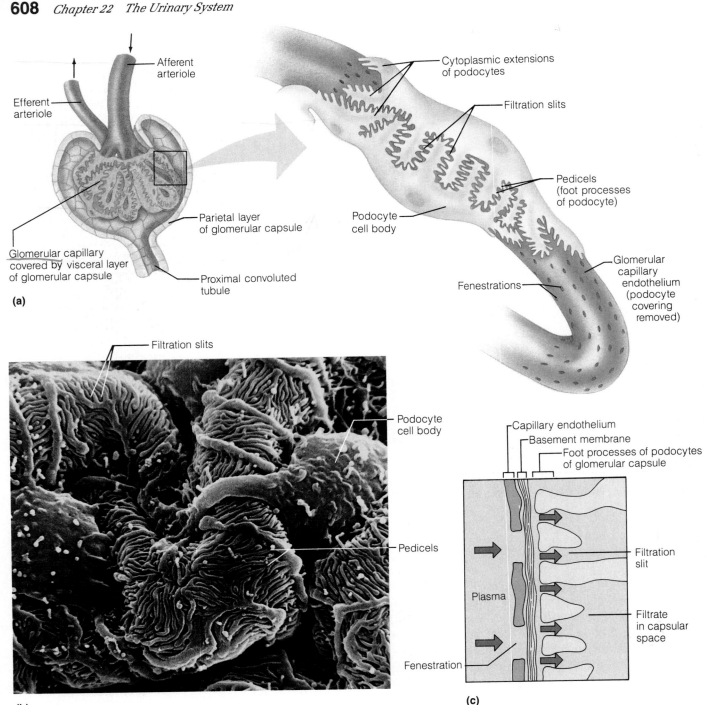

Figure 22.7 Renal corpuscle and the filtration membrane. (a) The renal corpuscle, at left, consists of a glomerulus (a tuft of glomerular capillaries) surrounded by a glomerular capsule, like a fist pushed deep within a balloon. The enlargement at right shows the relationship of the visceral (inner) layer of the glomerular capsule to the glomerular capillaries. The visceral epithelium (cells called podocytes) is pictured incompletely to show the fenestrations in the underlying capillary wall. (b) Scanning electron micrograph of podocytes clinging to the glomerular capillaries (artificially colored; 3900 ×). This is the same view as in the right half of part (a). (c) Diagram of a section through the filtration membrane, showing its three layers. The interior of the glomerular capillary is at left, the glomerular capsular space is at right.

Figure 22.8   Photomicrograph of the renal cortex.   Note the renal corpuscles and the sections through the various tubules. In the proximal convoluted tubules, the lumens appear "fuzzy" because they are filled with the long microvilli of the epithelial cells that form the tubule walls. In the distal tubules, by contrast, the lumens are clear (magnification: 400 ×).

Renal corpuscle:

Glomerular capsular space

Glomerulus

Artery

Distal convoluted tubules

Proximal convoluted tubules

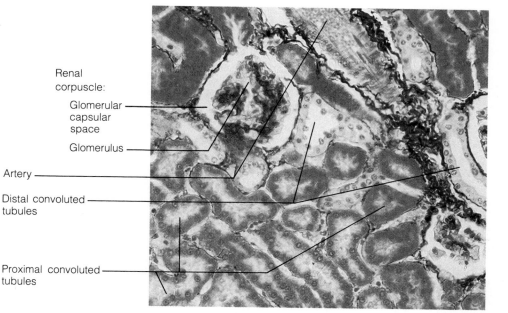

its length and enhances its capabilities in processing the filtrate that flows through it.

Each part of the nephron has a unique cellular anatomy that reflects its filtrate-processing function. The **proximal convoluted tubule,** confined entirely to the renal cortex, is most active in the processes of reabsorption and secretion. Its walls are formed by cuboidal epithelial cells whose luminal (exposed) surfaces have long microvilli that nearly fill the tubule lumen (Figures 22.5b and 22.8). These microvilli increase the surface area of these cells tremendously, maximizing their capacity for reabsorbing water, ions, and solutes from the filtrate. Furthermore, the cells of the proximal tubule contain many mitochondria (which provide the energy for reabsorption) and a highly infolded plasmalemma on their basal and lateral cell surface. This expanded basolateral membrane contains many ion-pumping enzymes that are responsible for reabsorbing molecules from the filtrate.

The U-shaped **loop of Henle** (or *loop of the nephron*) consists of a descending and an ascending limb. The first part of the **descending limb** is continuous with the proximal tubule and has a similar structure. The rest of the descending limb is the **thin segment.** The most narrow part of the nephron, the thin segment is lined by a permeable simple squamous epithelium. The thin segment may continue into the ascending limb and inevitably joins the **thick segment** of the ascending limb. The cell structure of this thick segment resembles that of the distal convoluted tubule.

The **distal convoluted tubule,** like the proximal convoluted tubule, is confined to the renal cortex, is lined by a simple cuboidal epithelium, and is specialized for the selective secretion and reabsorption of

ions. It is less active in reabsorption than the proximal tubule, however, and its cells do not exhibit an abundance of absorptive microvilli (Figure 22.5b). Still, cells of the distal tubule do have many mitochondria and infoldings of the basolateral membrane, features that are typical of most ion-pumping cells in the body.

Cortical nephrons represent 85% of nephrons within the kidneys (see Figure 22.5a). Cortical nephrons are located almost completely in the cortex: Their loops of Henle dip only a short way into the medulla. The remaining nephrons (15%), which lie close to the cortex-medulla junction, are **juxtamedullary** ("near the medulla") **nephrons.** Their loops of Henle deeply invade the medulla, and their thin segments are much longer than those of cortical nephrons. These long loops of Henle (in conjunction with nearby collecting tubules) contribute to the kidney's ability to produce a concentrated urine.

Collecting Tubules

Urine passes from the distal tubules of the nephrons into the **collecting tubules.** Each collecting tubule receives urine from several nephrons and extends straight through the cortex into the deep medulla (Figure 22.5a). There it joins with adjacent collecting tubules to form larger **papillary ducts,** which empty into the minor calyces through the renal papillae. Histologically, the collecting tubules consist of a simple cuboidal epithelium (Figure 22.5b), which thickens to become a simple columnar epithelium in the papillary ducts. The collecting tubules conserve body fluids: More specifically, when the body must conserve water, the posterior part of the pituitary gland secretes antidiuretic hormone (Chapter 24, p. 667), which increases the permeability of the collecting ducts. This

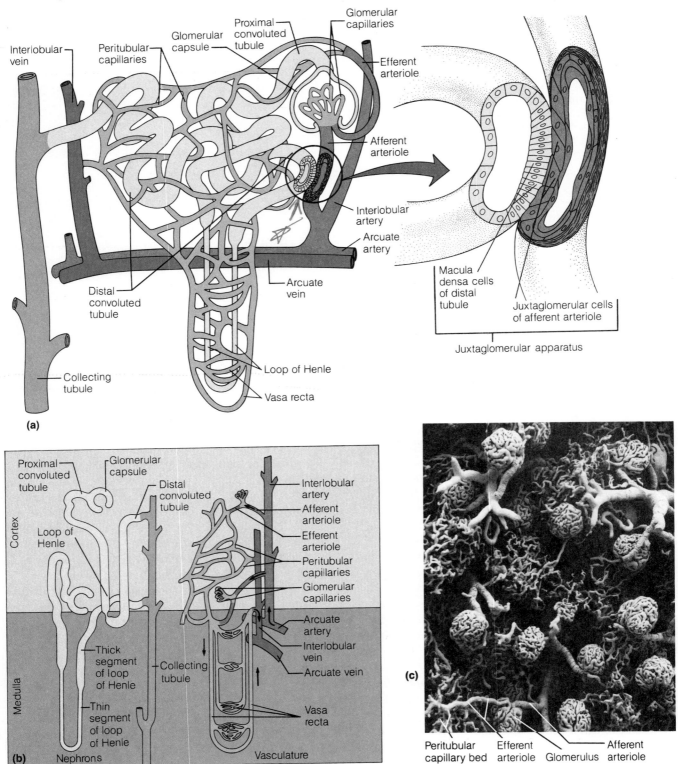

(a)

(b) Nephrons   Vasculature

(c)

Peritubular   Efferent   Afferent
capillary bed   arteriole   Glomerulus   arteriole

Figure 22.9   Anatomy of the vasculature around nephrons.   (a) Structure of a juxtamedullary nephron and its associated capillaries. The juxtaglomerular apparatus is enlarged
at right. (b) Comparison of the tubular and vascular anatomy of cortical nephrons and juxtamedullary nephrons. (c) Scanning electron micrograph of a cast of the blood vessels that are
associated with nephrons (65 ×).

increases the reabsorption of water from the urine into the surrounding blood vessels and decreases the total amount of urine produced.

Microscopic Blood Vessels Associated with Uriniferous Tubules

Every nephron is closely associated with two capillary beds: the *glomerulus* and the *peritubular capillaries* (Figure 22.9).

As we discussed above, the capillaries of the glomerulus produce the filtrate that ultimately becomes urine. The glomerulus differs from all other capillary beds in the body: It is both fed and drained by arterioles—an **afferent arteriole** and an **efferent arteriole,** respectively. The *afferent arterioles* arise from the interlobular arteries that run through the renal cortex (also see Figure 22.4b). Because arterioles are high-resistance vessels and the efferent arteriole is narrower than the afferent arteriole, the blood pressure in the glomerulus is extraordinarily high for a capillary bed. This pressure easily forces the filtrate out of the blood into the glomerular capsule. The kidneys generate 1 L (about 1 quart) of this filtrate every 8 minutes, but, as we've mentioned, only 1% ends up as urine; the other 99% is reabsorbed by the uriniferous tubule and returned to the blood in the peritubular capillary beds.

The **peritubular capillaries** *(intertubular capillaries)* arise from the efferent arterioles draining the glomeruli (Figure 22.9). These capillaries cling closely to the nephrons and empty into nearby venules of the renal venous system. The peritubular capillaries are adapted for absorption: They are low-pressure, porous capillaries that readily absorb solutes and water from the tubule cells as these substances are reabsorbed from the filtrate. Furthermore, all molecules that are *secreted* by the nephrons are derived from the blood of the peritubular capillaries.

In the deepest part of the renal cortex, efferent arterioles from the juxtamedullary glomeruli continue into thin-walled looping vessels called **vasa recta** (va'sah rek'tah; "straight vessels"). These hairpin loops extend into the medulla, running alongside the loops of Henle. The vasa recta are part of the kidney's urine-concentrating mechanism.

Figure 22.10 provides a complete summary of the microscopic and gross blood vessels of the kidney.

Juxtaglomerular Apparatus

The **juxtaglomerular apparatus** (juks"tah-glo-mer'u-lar), pictured on the right side of Figure 22.9a, represents a specialized contact between the first part of the distal convoluted tubule and the afferent arteriole near the glomerulus (*juxta* = near; *glomerular* = the glomerulus). At their point of contact, both structures are modified.

The wall of the afferent arteriole contains **juxtaglomerular cells,** modified smooth muscle cells with secretory granules that contain a hormone called **renin** (re'nin; "kidney hormone"). The juxtaglomerular cells seem to be mechanoreceptors that secrete their renin in response to falling blood pressure in the afferent arteriole. The **macula densa** (mak'u-lah děn'sah; "dense spot") is the region of the distal tubule adjacent to the juxtaglomerular cells. The tall, closely packed epithelial cells of the macula densa seem to be chemoreceptors that signal the juxtaglomerular cells to secrete renin whenever the solute concentration in the renal filtrate falls below a certain level. Renin signals a number of physiological responses, but the ultimate effect of this hormone is to increase blood pressure, blood solute concentration, and blood volume whenever these parameters have fallen toward unacceptable levels.

Ureters

Gross Anatomy

The **ureters** are slender tubes that convey urine from the kidneys to the bladder (Figures 22.2 and 22.11). Each ureter begins superiorly (at the level of L_2) as a continuation of the renal pelvis. From there, it descends retroperitoneally through the abdomen, crosses the pelvic brim at the sacroiliac joint, and enters the true pelvis. The ureter then enters the posterolateral corner of the bladder and runs medially within the posterior bladder wall before opening into the bladder's interior. This oblique pathway prevents backflow of urine from the bladder into the ureters—

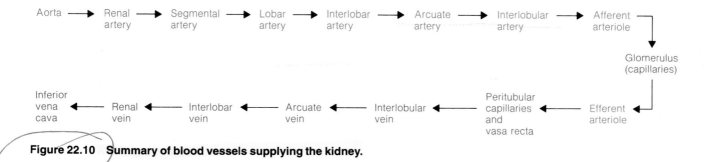

Figure 22.10   Summary of blood vessels supplying the kidney.

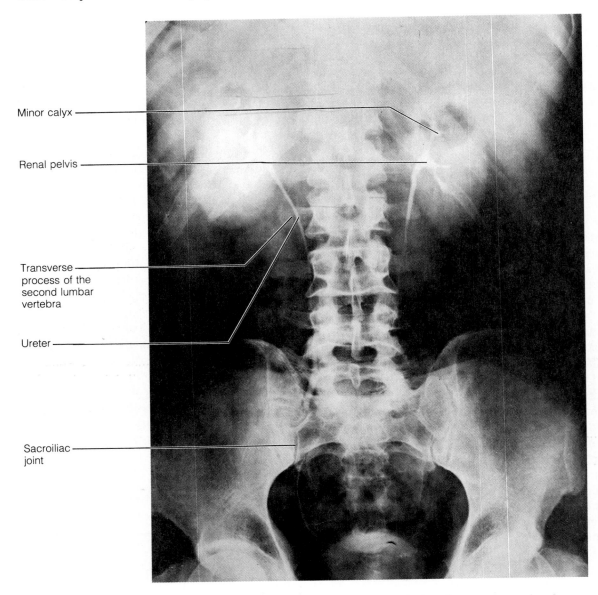

Minor calyx

Renal pelvis

Transverse process of the second lumbar vertebra

Ureter

Sacroiliac joint

Figure 22.11 X-ray image of the ureters (pyelogram) in the posterior abdominal wall and the pelvis. Also note the calyces and the renal pelvis (above).

any increase of pressure within the bladder compresses the bladder wall, thereby closing the distal ends of the ureters.

The radiographic procedure for examining the ureters and renal calyces is called **pyelography** (pi"ĕ-log'rah-fe; "recording the renal pelvis"). The clinician identifies the ureters on X-ray films by tracing their descent along a line defined by the transverse processes of the lumbar vertebrae (Figure 22.11).

Can you deduce the path by which radiologists introduce X-ray contrast medium into the ureters and calyces for pyelography?

The contrast medium can be injected into the ureters through a catheter in the bladder (retrograde pyelography). Alternatively, the medium can be injected into a vein so that it reaches the ureters when excreted by the kidney (intravenous pyelography). ■

Microscopic Anatomy

The histological structure of the ureters is the same as that of the renal calyces and renal pelvis. The walls of these tubes have three basic layers: a mucosa, a muscularis, and an adventitia (Figure 22.12). The lining **mucosa** is composed of a lamina propria and a transitional epithelium that stretches when the ureters fill with urine. The lamina propria is a stretchy, fibroelastic connective tissue containing scattered patches of lymphoid tissue. The middle **muscularis** layer con-

Figure 22.12   Structure of the ureter (cross-sectional view, 10 ×).   Note the basic layers. The mucosa folds into longitudinal ridges, giving the lumen an irregular, branched appearance. These mucosal folds stretch and flatten to accommodate large pulses of urine.

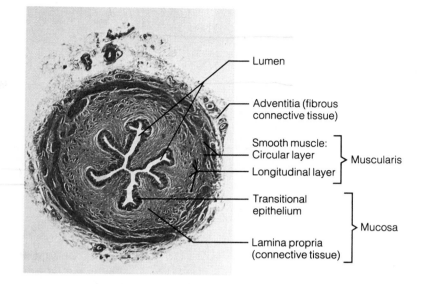

- Lumen
- Adventitia (fibrous connective tissue)
- Smooth muscle: Circular layer } Muscularis
- Longitudinal layer
- Transitional epithelium } Mucosa
- Lamina propria (connective tissue)

sists of two layers: an inner longitudinal layer and an outer circular layer of smooth muscle. A third layer of muscularis, an external longitudinal layer, appears in the inferior third of the ureter. The external **adventitia** of the ureter wall is a typical connective tissue.

The ureters play an active role in transporting urine. Distension of the ureter by entering urine stimulates its muscularis to contract, setting up peristaltic waves that propel urine to the bladder. (Urine does *not* reach the bladder through gravity alone.) Although the ureters are innervated by both sympathetic and parasympathetic nerve fibers, neural control of their peristalsis appears to be insignificant compared to the local stretch response of ureteric smooth muscle.

On occasion, calcium, magnesium, or uric acid salts in urine may crystallize and precipitate in the renal pelvis, forming kidney stones, or **renal calculi** (kal'ku-li; "little stones"). Most calculi are under 5 mm in diameter and therefore can pass through the urinary tract without causing serious problems. Larger calculi, however, cause excruciating pain when they obstruct a ureter, blocking the drainage of urine and increasing the pressure within the kidney. Pain also results when the contracting walls of the ureter close in on the sharp calculi during periodic waves of peristaltic constriction. Pain due to kidney stones radiates to the posterior abdominal wall on the same side of the body. Calculi tend to lodge in three especially narrow regions of the ureters: (1) at the level of L_2, where the renal pelvis first narrows into the ureter; (2) at the sacroiliac joint, where the ureter enters the true pelvis; and (3) where the ureters enter the bladder. The clinician should be aware of these three points when searching for kidney stones on X-ray films.

Predisposing conditions for kidney stones are frequent bacterial infections of the urinary tract, urinary retention, high concentrations of calcium in the blood, and alkaline urine. Surgical removal has been the traditional treatment for kidney stones. Renal calculi often recur, however, so clinicians have long sought alternatives to the trauma of repeated surgeries. In one technique, an ultrasound probe inserted into the ureter shatters the stones with sound waves. In a related technique, ultrasonic shock waves from outside the patient's body break up the calculi.

Can you deduce why people with a history of kidney stones are encouraged to drink large quantities of water?

When the urine is dilute, the salts it contains cannot precipitate out of solution to form calculi. ■

Urinary Bladder

The **urinary bladder** is a collapsible, muscular sac that stores and expels urine. It lies inferior to the peritoneal cavity on the pelvic floor, just posterior to the pubic symphysis (Figure 22.13). In males, the bladder lies anterior to the rectum, and the prostate gland (a male reproductive gland) surrounds the urethra directly inferior to the bladder. In females, the bladder lies just anterior to the vagina and uterus.

A full bladder is roughly spherical and expands superiorly into the abdominal cavity (Figure 22.13). An empty bladder, by contrast, lies entirely within the pelvis and has the shape of an upside-down pyramid (Figure 22.14). This shape gives the bladder four triangular surfaces and four corners, or angles. The two *posterolateral angles* receive the ureters, and the *inferior angle (neck)* drains into the urethra. At the bladder's *anterior angle* is a fibrous band called the

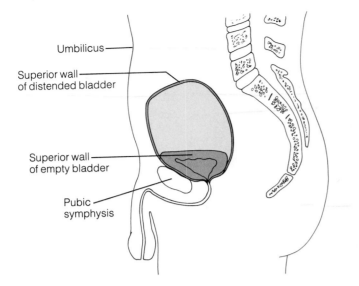

Figure 22.13 The urinary bladder: Positions and shapes of a full and an empty bladder.

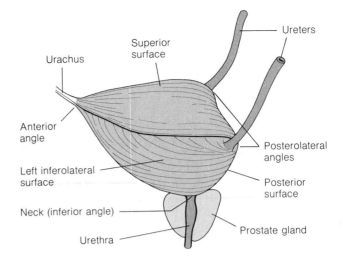

Figure 22.14 The empty bladder: An inverted pyramid with four surfaces and four angles. This is a male bladder, oriented like the bladder in Figure 22.13. Its four surfaces are the superior, posterior, and left and right inferolateral surfaces. Note that a tube or band (ureter, urethra, urachus) attaches to each angle of the bladder.

urachus (u'rah-kus; "urinary canal of the fetus"). The urachus is the closed remnant of an embryonic tube called the allantois (p. 594).

The interior of the bladder has openings for both ureters and the urethra (Figure 22.15). These three openings define a triangular region on the posterior wall, called the **trigone** (tri'gōn; "triangle"). The trigone is of special clinical importance because infections tend to persist in this region.

The bladder is supplied by the following vessels and nerves: The arteries are branches of the internal iliac arteries, primarily the *superior* and *inferior vesicle arteries* (*vesicle* = sac = the bladder) (see Figure 18.15 on p. 499). Veins draining the bladder form a plexus on the bladder's inferior and posterior surfaces, which empties into the internal iliac veins. Nerves extend to the bladder from the inferior hypogastric plexus. They consist of parasympathetic fibers (ultimately from the pelvic splanchnic nerves), a few sympathetic fibers (ultimately from the lower thoracic and upper lumbar splanchnic nerves), and visceral sensory fibers.

The wall of the bladder has three layers (Figure 22.16): (1) a mucosa with a distensible transitional epithelium and a lamina propria, (2) a thick muscular layer, and (3) a fibrous adventitia (except on the superior surface of the bladder, which is covered by parietal peritoneum). The muscular layer, called the **detrusor muscle** (de-tru'sor; "to thrust out"), consists of highly intermingled smooth muscle fibers arranged in inner and outer longitudinal layers and a middle circular layer. This muscle squeezes urine from the bladder during urination.

The bladder is very distensible and uniquely suited for its function of storing urine. When there is

little urine in it, the bladder collapses into its basic pyramidal shape. Its walls are thick, and its mucosa is thrown into folds, or rugae (Figure 22.15). But as urine accumulates, the rugae flatten, and the wall of the bladder thins as it stretches. These changes allow the bladder to store larger amounts of urine without a significant rise in internal pressure (at least until 300 ml has accumulated). A moderately full adult bladder holds about 500 ml (1 pint) of urine. However, the bladder can hold more than twice that amount if necessary.

Urethra

The **urethra** is a thin-walled tube that drains urine from the bladder and conveys it out of the body (Figure 22.15). In both sexes, this tube consists of smooth muscle and an inner mucosa. (In males, however, the muscle layer becomes very thin toward the distal end of the urethra.) The lining epithelium changes from (1) a transitional epithelium near the bladder to (2) a stratified and pseudostratified columnar epithelium in mid urethra (sparse in females), and then to (3) a stratified squamous epithelium near the end of the urethra.

At the bladder-urethra junction, a thickening of the detrusor muscle forms the **internal urethral sphincter.** This is an involuntary sphincter (smooth muscle) that keeps the urethra closed when urine is not being passed. It prevents dribbling of urine between voidings. A second sphincter, the **external urethral sphincter** *(sphincter urethrae),* surrounds the

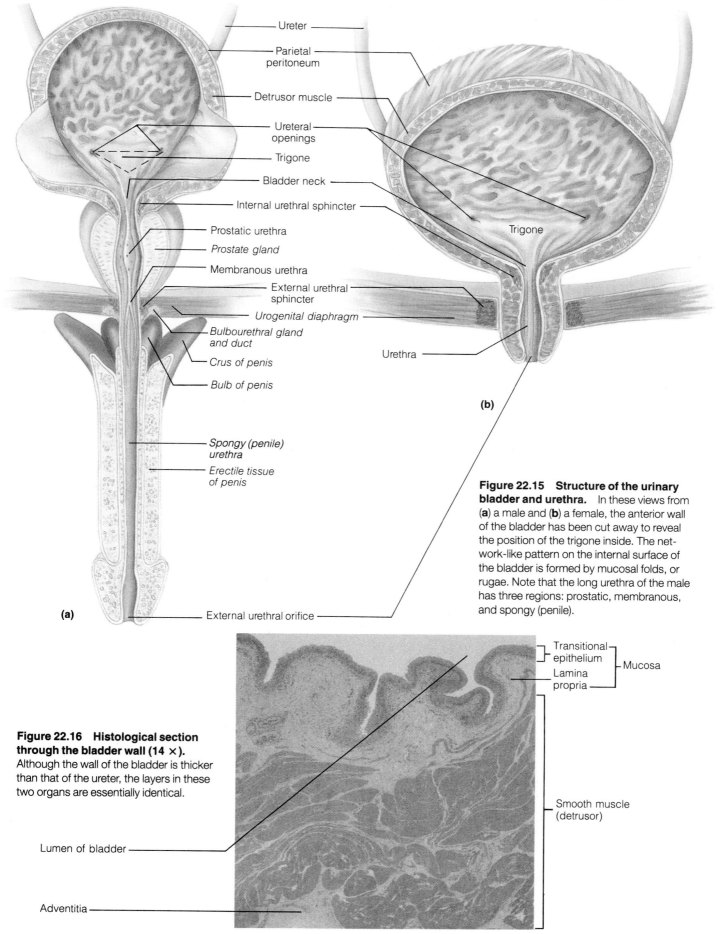

Ureter

Parietal peritoneum

Detrusor muscle

Ureteral openings

Trigone

Bladder neck

Internal urethral sphincter

Prostatic urethra

Prostate gland

Membranous urethra

External urethral sphincter

Urogenital diaphragm

Bulbourethral gland and duct

Crus of penis

Bulb of penis

Spongy (penile) urethra

Erectile tissue of penis

(a)

External urethral orifice

Trigone

Urethra

(b)

Figure 22.15   Structure of the urinary bladder and urethra.   In these views from (**a**) a male and (**b**) a female, the anterior wall of the bladder has been cut away to reveal the position of the trigone inside. The network-like pattern on the internal surface of the bladder is formed by mucosal folds, or rugae. Note that the long urethra of the male has three regions: prostatic, membranous, and spongy (penile).

Transitional epithelium

Lamina propria

Mucosa

Smooth muscle (detrusor)

Figure 22.16   Histological section through the bladder wall (14 ×). Although the wall of the bladder is thicker than that of the ureter, the layers in these two organs are essentially identical.

Lumen of bladder

Adventitia

urethra within the sheet of muscle called the *urogenital diaphragm.* This external sphincter is a skeletal muscle that we use to willfully inhibit urination until the proper time. The levator ani muscle of the pelvic floor also serves as a voluntary constrictor of the urethra. (For more information on the sphincter urethrae and levator ani muscles, see Table 10.7 on p. 260.)

The length and functions of the urethra differ in the two sexes. In females, the urethra is just 3 or 4 cm long (1.5 inches) and is bound to the anterior wall of the vagina by connective tissue. It opens to the outside at the **external urethral orifice** *(external urethral meatus).* This urethral opening lies anterior to the vaginal opening and posterior to the clitoris (Figure 23.20, p. 644). The orifice is small and often difficult to locate.

In males, the urethra is about 20 cm long (8 inches) and has three named regions. The **prostatic urethra,** about 2.5 cm long, runs within the prostate gland. The **membranous urethra** runs for about 2.5 cm through the membrane-like urogenital diaphragm. The **spongy,** or **penile, urethra,** about 15 cm long, passes through the entire penis and opens at the tip of the penis via the **external urethral orifice.** The male urethra carries semen as well as urine (although never simultaneously) from the body. The reproductive function of the male urethra is covered in Chapter 23.

From the above facts about the male and female urethra, can you deduce why urinary tract infections are more common in females than in males?

Because the female urethra is short and its external orifice lies close to the anus, fecal bacteria can easily enter the urethra. (To minimize this risk, females should always wipe from front to back after defecation.) An infection of the urethra, **urethritis,** can spread superiorly to cause a bladder infection, **cystitis,** and even kidney infections (pyelitis or pyelonephritis). Symptoms of urinary tract infection include a burning sensation during urination, increased urgency and frequency of urination, fever, and sometimes cloudy or blood-tinged urine. When the kidneys are involved, back pain and a severe headache often occur as well. ■

Micturition

Micturition (mik″tu-rish′un), also called voiding or urination, is the act of emptying the bladder (Figure 22.17). Ordinarily, when about 200 ml of urine has accumulated, the distension of the bladder wall activates stretch receptors there, triggering a visceral reflex arc. Sensory impulses are transmitted to the sacral region of the spinal cord, and motor impulses return to the bladder via the pelvic splanchnic nerves

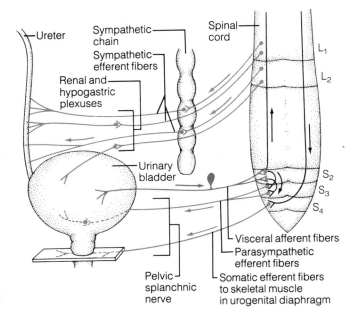

Figure 22.17 Micturition reflex arcs and the innervation of the bladder. Stretching of the bladder wall by accumulating urine sends visceral afferent impulses to the sacral spinal cord. Efferent impulses then travel to the detrusor muscle of the bladder via parasympathetic fibers in the pelvic splanchnic nerves. These splanchnic nerves also carry the somatic motor fibers to the voluntary external urethral sphincter in the urogenital diaphragm. Sympathetic fibers to the bladder (above) seem to relax the detrusor muscle and stimulate contraction of the internal urethral sphincter.

of the parasympathetic division. This causes the detrusor muscle to contract and the internal urethral sphincter to relax. At this point, one can voluntarily delay micturition by closing the external urethral sphincter. After a few minutes, however, the bladder contracts again. If urination is delayed again, the reflex is damped once more. However, the urge to void eventually becomes irresistible.

During micturition, the squeezing action of the bladder's detrusor muscle is assisted by the muscles of the abdominal wall, which contract to raise the intra-abdominal pressure.

Incontinence, the inability to control micturition, is normal in babies who have not learned to use their external urethral sphincter. Reflexive voiding occurs each time a baby's bladder fills enough to activate its stretch receptors, but the internal sphincter prevents dribbling of urine between voidings just as it does in adults. After the age of toilet training, incontinence is usually a result of emotional problems or nerve damage (a stroke or a spinal lesion). Incontinence can also occur during pregnancy as the heavy uterus presses down on the woman's bladder and stretches the sphincter in her urogenital diaphragm. In **stress incontinence,** a sudden increase in intra-abdominal pressure (during laughing, sneezing, or coughing) forces urine through the external sphincter.

In **urinary retention,** the bladder is unable to expel its contained urine. Urinary retention is common in patients recovering from surgeries in which general anesthesia was given. (Following anesthetization, it seems to take a little time for the detrusor muscle to resume its activity.) When urinary retention is prolonged, a slender rubber drainage tube called a **catheter** (kath'ĕ-ter; "thrust in") must be inserted through the urethra to drain the urine and prevent overfilling of the bladder.

Which do you think is easier to catheterize—the male urethra or the female urethra?

The straight, short urethra of females is much easier to catheterize than the long, curved urethra of males (see Figures 22.13 and 22.15). ■

The Urinary Organs Throughout Life

The embryo develops not just one pair of kidneys, but three, one after another. These different kidneys develop from the *urogenital ridges,* paired elevations of intermediate mesoderm on the dorsal abdominal wall (Figure 22.18). Of the three kidneys that form, only the last one persists to become the adult kidney. Initially, during week 4, the first kidney, or **pronephros** (pro-nef'ros), forms as a set of nephrons then quickly degenerates. Although the pronephros is never functional and is gone by week 6, it sends a **pronephric**

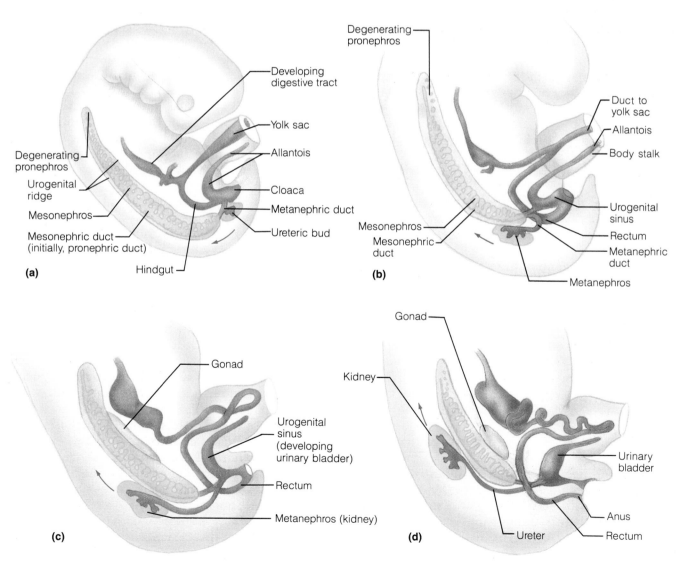

Figure 22.18 Development of the urinary organs in the embryo. (a) Week 5. (b) Week 6. (c) Week 7. (d) Week 8.

duct to the cloaca, and this duct is used by the kidneys that develop later. As the second nephron system, the **mesonephros** ("middle kidney"), claims the pronephric duct, this duct becomes the **mesonephric duct.** The nephrons of the mesonephros, in turn, degenerate once the third kidney, the **metanephros** (met"ah-nef'ros), becomes functional.

The metanephros is the definitive kidney (*metanephros* = "ultimate kidney"). It develops in the pelvic region in the following way: Starting in week 5, a hollow **ureteric bud** grows from the mesonephric duct into the urogenital ridge, inducing the mesoderm there to form the nephrons. The ureteric bud, in turn, develops into the renal pelvis, calyces, and collecting tubules. Its unexpanded proximal part, now called the metanephric duct, becomes the ureter.

After forming in the pelvis, the metanephric kidneys ascend to their final position in the abdomen, receiving their blood supply from successively higher sources as they ascend. Usually, the lower renal vessels degenerate as the upper ones appear. However, these early vessels often fail to degenerate. In fact, 30% of people have multiple renal arteries.

The metanephric kidneys are actively forming urine by month 3 of fetal life. It may surprise you that the kidneys operate in the fetus, since the *placenta* performs all excretory functions before birth (p. 649). Urine production in the fetus has an important function, however: The urine is voided into the amniotic sac, where it helps maintain the proper volume of amniotic fluid.

As the metanephric kidneys develop, the *cloaca* (p. 594) divides into two parts: (1) the future rectum and anal canal and (2) the **urogenital sinus,** into which the urinary and genital ducts empty (Figure 22.18b). The urogenital sinus becomes the urinary bladder and the urethra. The *allantois* (p. 594), an extension of the urogenital sinus into the umbilical cord, becomes the urachus of the bladder.

As the embryonic kidneys ascend through the narrow pelvic brim, they face a tight squeeze. The right and left kidneys come close together, and in about 1 out of every 600 people, they fuse into one U-shaped **horseshoe kidney.** This condition is usually harmless, but it can be associated with obstructed drainage and hydronephrosis. In **pelvic kidney,** one of the two kidneys stays in the bony pelvis throughout life. Since a pelvic kidney often blocks the birth canal, a woman with such a kidney may have difficulty in giving birth.

Congenital polycystic kidney is a rare, inherited condition in which the infant's kidneys contain many urine-filled sacs (cysts). These cysts are dilations along the collecting ducts in the renal cortex and medulla. Because the cysts impede the drainage of urine and crush the surrounding renal tissue, most children with this disease die of kidney failure early in life.

However, dialysis machines (which cleanse the blood) and kidney transplants are improving their chance of survival.

Adult polycystic kidney disease, another inherited condition, is distinct from congenital polycystic kidney. The cysts, which represent dilations along any part of the nephron, develop so gradually that they produce no symptoms until about 40 years of age. Then, both kidneys begin to enlarge with many blister-like cysts containing blood, mucus, or urine. The damage caused by these cysts progresses slowly, and many victims live without problems until their 60s. Ultimately, however, the kidneys become "knobby" and grossly enlarged, reaching weights of up to 14 kg (30 pounds) each. Most victims die from kidney failure or hypertension. Adult polycystic disease is relatively common, affecting 1 in 500 people.

Because an infant's bladder is small and its kidneys are unable to concentrate urine for the first 2 months, a newborn baby voids 5 to 40 times a day, depending on the amount of fluid intake. As the child grows, she or he voids less often but produces progressively more urine. By adolescence, the adult rate of urine output (averaging 1500 ml/day) is achieved.

Control of the voluntary urethral sphincter is attained as the nervous system matures. By 15 months, most toddlers are aware of having voided. By 18 months, most children can hold urine in the bladder for about 2 hours; this is the first sign that toilet training for micturition can begin. Daytime control usually is achieved well before nighttime control. As a rule, it is unrealistic to expect complete nighttime control before the age of 4 years.

From childhood through late middle age, most problems that affect the urinary system are infections. *Escherichia coli* are normal residents of the lower digestive tract and seldom cause problems there, but these bacteria produce 80% of all urinary tract infections. Sexually transmitted diseases, which are chiefly infections of the reproductive tract, can also inflame the urinary tract, clogging some urinary ducts. Childhood streptococcal infections, such as a severe strep throat or scarlet fever, can lead to long-term inflammatory damage of the kidneys.

Only about 3% of elderly people have histologically normal kidneys, and kidney function declines with advancing age. The kidneys shrink, the nephrons decrease in size and number, and the tubules become less efficient at secretion and reabsorption. By age 70, the rate of filtration is only about half that of middle-aged adults. This slowing is believed to result from narrowing of the renal arteries by atherosclerosis. Diabetics are particularly at risk for renal disease, and over 50% of those who have had diabetes mellitus for 20 years (regardless of their age) are in renal failure owing to the vascular ravages of this disease.

The bladder of an aged person is shrunken, and the desire to urinate is often delayed. Loss of muscle tone in the bladder causes an annoying frequency of urination. Many seniors eventually experience incontinence, and with it a tremendous blow to their self-esteem. Urinary retention is a common problem in older men with hypertrophy of the prostate gland, a condition that constricts the urethra (p. 632). However, an inability to void may also be associated with a cold or the flu, certain drugs, and urinary tract obstruction. A regular regimen of exercise that keeps the body in optimal condition and promotes alertness to voiding signals may sometimes prevent incontinence and retention.

✚ Related Clinical Terms

Acute glomerulonephritis (GN) (glo-mer″u-lo-nĕ-fri′tis) Inflammation of the glomeruli, often arising from an immune response. In some cases, circulating immune complexes (antibodies bound to foreign substances, such as streptococcal bacteria) become trapped in the basement membrane of the glomerulus. In other cases, an autoimmune attack mounted against one's own kidney tissue leads to glomerular damage. In either case, the subsequent inflammatory response damages the renal filtration membrane, increasing its permeability. Blood proteins and even blood cells start to enter the renal filtrate and appear in the urine. The loss of plasma proteins diminishes the osmotic pressure of the blood, so large amounts of fluid seep from the bloodstream into the interstitial spaces, causing edema throughout the body. The kidneys may shut down temporarily, but if the patient is maintained on a dialysis machine, normal renal function usually returns in a few months. If the glomerular damage is permanent, however, chronic GN and, ultimately, renal failure result.

Bladder cancer This form of cancer, which accounts for about 3% of cancer deaths, usually involves neoplasms of the bladder's lining epithelium. Bladder cancer may be induced by organic carcinogens in the urine. (These carcinogens are first absorbed from the environment and appear in the urine after being processed by the kidneys.) Smoking (tobacco tars), exposure to certain industrial chemicals, and certain artificial sweeteners have been linked to bladder cancer. Blood in the urine is a common warning sign.

Cystocele (sis′to-sēl) (*cyst* = the bladder; *cele* = sac, hernia) Condition in which the urinary bladder pushes on the vagina and bulges into the anterior superior vaginal wall. Cystocele may result from tearing of the supportive muscles of the pelvic floor during childbirth, which then causes the unsupported pelvic viscera to sink inferiorly.

Cystoscopy (sis-tos′ko-pe) The threading of a thin viewing tube through the urethra into the bladder to examine the inner surface of the bladder mucosa.

Kidney transplant The transplant of a functioning kidney from a donor to a recipient whose kidneys are failing. Kidney transplants are far more common than transplants of any other major organ. They have a high rate of success (80% to 90%) and are comparatively easy to perform (kidneys are easy to detach because just a few vessels need to be cut). A single kidney is transplanted, because one is sufficient to carry out excretory functions. As with other organ transplants, the recipient must take drugs that suppress rejection by the immune system. Interestingly, the kidney is usually transplanted into the iliac fossa of the pelvis, not into the lumbar region.

Nephrotoxin A substance that is toxic to the kidney (heavy metal, organic solvent, or bacterial toxin).

Renal infarct Area of dead, or necrotic, renal tissue. It may result from infection, hydronephrosis, or blockage of the blood supply to the kidney. A common cause of localized renal infarct is blockage of an interlobar artery. Because interlobar arteries are end arteries (that is, they do not anastomose), their obstruction leads to ischemic necrosis of the regions they supply.

Urologist (u-rol′o-jist) Physician who specializes in diseases of the urinary structures in both sexes and in diseases of the *reproductive* tract of males.

Chapter Summary

1. In forming urine, the kidneys cleanse the blood of nitrogenous wastes, toxins, excess ions and water, and other unnecessary or undesirable substances. Kidneys also maintain the proper chemistry of the blood and other body fluids. The organs for transporting and storing urine are the ureters, bladder, and urethra.

KIDNEYS (pp. 601–611)

Gross Anatomy (pp. 601–605)

1. The bean-shaped kidneys lie retroperitoneally in the posterior abdominal wall, extending from the level of T_{11} or T_{12} to L_3.

2. The supportive and protective layers around the kidney are the renal capsule, the renal fascia, and several layers of cushioning fat.

3. A kidney has an external cortex, a deeper medulla consisting of medullary pyramids, and a medial renal sinus. The renal sinus contains the main vessels and nerves of the kidney as well as the renal pelvis. Extensions of the renal pelvis, the major and minor calyces, collect urine draining from the papillae of the medullary pyramids. The medial, slit-like opening of the renal sinus is the hilus.

4. The vascular pathway through a kidney is as follows: renal artery → segmental arteries → lobar arteries → interlobar arteries → arcuate arteries → interlobular arteries → afferent arterioles → glomeruli → efferent arterioles → peritubular capillary beds (or vasa recta) → interlobular veins → arcuate veins → interlobar veins → renal veins.

5. The nerve supply of the kidneys, consisting largely of sympathetic fibers to the arteries, is from the renal plexus.

Microscopic Anatomy (pp. 605–611)

6. The main structural and functional unit of the kidney is the uriniferous tubule, which consists of a nephron and collecting tubule. The nephron, which forms the urine, consists of a renal corpuscle, a proximal convoluted tubule, a loop of Henle, and a distal convoluted tubule.

7. The production of urine by the nephron involves three processes: filtration (producing a filtrate of the blood, which will be modified into urine); reabsorption (reclaiming the desirable molecules from the filtrate and returning these to the blood); and secretion (pumping undesirable molecules from the peritubular capillaries into the nephron).

8. Filtration occurs at the renal corpuscles in the renal cortex. Each renal corpuscle consists of a glomerulus (a capillary tuft) surrounded by a glomerular capsule. The visceral (inner) layer of this capsule consists of podocytes, which surround the capillaries of the glomerulus.

9. In passing from the glomerulus to the capsular space, the filtrate passes through the filtration membrane. This membrane consists of the fenestrated glomerular endothelium, the basement membrane, and the filtration slits between podocytes. The basement membrane is the main filter, holding back molecules that are larger than the size of small proteins.

10. From the glomerular capsule, the filtrate enters the proximal convoluted tubule (in the renal cortex). The simple cuboidal epithelial cells of the proximal tubule show extreme specializations for reabsorption and secretion: long, dense microvilli; many mitochondria; and infoldings of the basolateral plasmalemma.

11. The loop of Henle consists of descending and ascending limbs. The first part of the descending limb, which resembles the proximal tubule, narrows to join the thin segment, whose walls consist of a simple squamous epithelium. The thin segment joins the thick segment of the ascending limb, which is structurally similar to the distal convoluted tubule.

12. The distal convoluted tubule (in the renal cortex) is active in reabsorption and secretion, but less so than the proximal tubule. Its simple cuboidal epithelial cells exhibit abundant mitochondria and infoldings of the basolateral membrane, but no elaboration of the surface microvilli.

13. There are two kinds of nephrons: cortical and juxtamedullary. The loops of Henle of juxtamedullary nephrons project deeply into the medulla and contribute to the mechanism for concentrating urine.

14. Each nephron is associated with two capillary beds: (1) the glomerulus (for filtration); and (2) the peritubular capillaries (involved in reabsorption and secretion). Glomeruli are supplied by afferent arterioles and drained by efferent arterioles, which in turn are continuous with the peritubular capillaries. Efferent arterioles near the medulla are continuous with vasa recta around the loops of Henle.

15. The juxtaglomerular apparatus occurs at the point of contact between the afferent arteriole and the distal convoluted tubule. It consists of juxtaglomerular cells and a macula densa. Functionally, it relates to the hormonal control of blood pressure and blood volume.

URETERS (pp. 611–613)

Gross Anatomy (pp. 611–612)

1. The ureters are slender tubes descending retroperitoneally from the kidneys to the bladder. Each ureter enters the true pelvis by passing over the sacroiliac joint.

Microscopic Anatomy (pp. 612–613)

2. From external to internal, the layers of the ureter wall are (1) an adventitia of connective tissue, (2) a muscularis of smooth muscle, and (3) a highly distensible mucosa consisting of a lamina propria and a transitional epithelium. The muscularis squeezes urine inferiorly through the ureters.

URINARY BLADDER (pp. 613–614)

1. The urinary bladder, a distensible muscular sac for storing urine, lies in the pelvis just posterior to the pubic symphysis.

2. Whereas a full bladder can balloon superiorly into the abdominal cavity, an empty bladder is an inverted pyramid with four angles (neck with urethra, anterior angle with urachus, and two posterolateral angles where ureters enter). The trigone, an elevated triangle on the inner surface of the posterior wall, is bordered by the ureteric and urethral openings.

3. The bladder wall, from internal to external, consists of a mucosa with a transitional epithelium, the detrusor muscle, and an adventitia (a serosa on the superior surface of the bladder).

URETHRA (pp. 614–616)

1. The urethra is the tube that conveys urine from the bladder out of the body. Various types of stratified epithelium line its lumen.

2. Where the urethra leaves the bladder, it is surrounded by the internal urethral sphincter, an involuntary sphincter of smooth muscle. Where the urethra passes through the urogenital diaphragm, it is surrounded by the external urethral sphincter, a voluntary sphincter of skeletal muscle.

3. In females, the urethra is 3 to 4 cm long and conducts only urine. In males, the urethra is about 20 cm long and conducts both urine and semen.

MICTURITION (pp. 616–617)

1. Micturition is emptying the bladder.

2. As accumulating urine stretches the bladder, it initiates the micturition reflex, a spinal reflex that causes the detrusor muscle to contract and the internal urethral sphincter to relax (open).

3. Because the external urethral sphincter is controlled voluntarily, micturition can be delayed temporarily.

THE URINARY ORGANS THROUGHOUT LIFE (pp. 617–619)

1. Three sets of kidneys (pronephric, mesonephric, and metanephric) develop in sequence from the intermediate mesoderm of the embryo. The metanephros—the definitive kidney—forms in the pelvis as the ureteric bud signals nephrons to form in the intermediate mesoderm. The metanephric kidney moves cranially from the pelvis to the lumbar region.

2. The embryonic cloaca forms the urogenital sinus, which becomes both the bladder and the urethra.

3. The kidneys of newborns cannot concentrate urine. Their bladders are small, and voiding is frequent. Neuromuscular maturation generally allows toilet training to begin by 18 months of age.

4. From youth to middle age, the most common urinary problems are bacterial infections.

5. With age, nephrons are lost, the filtration rate decreases, and the kidneys become less efficient at concentrating urine. The capacity and tone of the bladder decrease with age, leading to frequent voiding and (often) incontinence. Urinary retention is a common problem of elderly men.

Review Questions

Multiple Choice and Matching Questions

1. The inferior border of the right kidney is at the level of which vertebra? (a) T_{12}, (b) L_1, (c) L_2, (d) L_3.

2. The capillaries of the glomerulus differ from other capillary networks in the body in which way? (a) They form an anastomosing network. (b) They drain into arterioles instead of ven-

ules. (c) They contain no endothelium. (d) They are sites where a fluid leaves the blood. (e) They contain portal veins.

3. Which one of these structures occurs exclusively in the renal medulla? (a) renal corpuscles, (b) distal convoluted tubules, (c) vasa recta, (d) proximal convoluted tubules.

4. Which one of the following is not always part of (or in) the renal cortex? (a) glomerulus, (b) distal convoluted tubules, (c) loops of Henle, (d) proximal convoluted tubules, (e) renal columns.

5. A commonly used surgical approach to the kidney is (a) posterolaterally, (b) through the vertebral canal, (c) inferiorly through the bladder and ureters, (d) through the navel.

6. The arrangement of the major blood vessels and urine-carrying vessels at the hilus of the kidney is as follows (from anterior to posterior): (a) renal artery branches, renal pelvis, renal vein; (b) renal vein, renal pelvis, renal artery branches; (c) renal pelvis, renal vein, renal artery branches; (d) renal vein, renal artery branches, renal pelvis.

7. The part of the uriniferous tubule whose epithelial cells contain the longest microvilli and the most mitochondria is the (a) glomerular capsule (podocytes), (b) proximal tubule, (c) thin segment, (d) distal tubule.

8. Parts of the uriniferous tubule that do not contain a simple cuboidal epithelium are (a) proximal and distal convoluted tubules, (b) parietal layer of glomerular capsule and thin segment, (c) collecting tubule and thick ascending limb, (d) thick part of the descending limb and proximal convoluted tubule.

9. Which of the following is the layer of fat that lies closest to the kidney? (a) renal capsule, (b) adipose capsule of perirenal fat, (c) renal fascia, (d) pararenal fat.

10. Assign each of these urine-carrying ducts a number from 1 to 5, according to the path taken by the urine (the urine first passes through #1, then through #2, and so on).

_____ **(a)** renal pelvis
_____ **(b)** papillary duct
_____ **(c)** ureter
_____ **(d)** major calyx
_____ **(e)** minor calyx

11. Only one part of the uriniferous tubule originates from the embryonic ureter (ureteric bud) and is not part of the nephron. Which? (a) glomerular capsule, (b) proximal convoluted tubule, (c) loop of Henle, (d) distal convoluted tubule, (e) collecting tubule.

12. The main function of the transitional epithelium in the ureter is (a) not the same as that of the transitional epithelium in the bladder, (b) protection against kidney stones, (c) secretion of mucus, (d) reabsorption, (e) stretching.

13. Jim was standing at a urinal in a crowded public rest room, and a long line was forming behind him. He became anxious (a sympathetic response) and found he could not micturate no matter how hard he tried. Use logic to deduce Jim's problem. (a) His internal urethral sphincter was constricted and would not relax. (b) His external urethral sphincter was constricted and would not relax. (c) His detrusor muscle was contracting too hard. (d) He almost certainly had a burst bladder.

14. A major function of the collecting tubules is (a) secretion, (b) filtration, (c) concentrating urine, (d) lubrication with mucus, (e) stretching.

15. What is the glomerulus? (a) It is the same as the renal corpuscle. (b) It is the same as the uriniferous tubule. (c) It is the same as the nephron. (d) It is capillaries.

16. Urine passes through the ureters by which mechanism? (a) ciliary action, (b) peristalsis and gravity, (c) gravity alone, (d) as the bladder expands with urine, it exerts suction on the ureters and renal pelvis.

17. The part of the filtration membrane that prevents the passage of substances that are smaller than blood cells but larger than small proteins is (a) podocytes, (b) capillary endothelial pores, (c) basement membrane, (d) filtration slits.

Short Answer and Essay Questions

18. What is the functional importance of the fat layers around the kidneys?

19. Urinary structures that are often confused are the ureters and the urethra; the perirenal and pararenal fat; the interlobar and interlobular arteries; and the renal sinus and renal pelvis. For each pair, differentiate one structure from the other.

20. Trace the path taken by the renal filtrate (and urine) from the glomerulus to the urethra. Name every microscopic and gross tube and structure that it passes through on its journey.

21. Explain the location and function of the peritubular capillaries.

22. Describe the basic process and purpose of tubular reabsorption.

23. List all layers of the filtration membrane in a renal corpuscle, and describe the function of each layer.

24. Name (a) the four angles of the empty bladder and (b) the tube or band that attaches to each angle.

25. Define micturition, and describe the micturition reflex.

26. Describe the changes that occur in the anatomy and function of the kidney with age.

27. In adult polycystic kidney disease, what are the cysts?

28. How does the particular course of the ureters through the bladder wall minimize the chance of hydronephrosis?

29. What three openings define the trigone of the bladder?

Critical Thinking and Clinical Application Questions

30. While repairing a frayed utility wire, Herbert, an electrician, falls from a telephone pole to the ground. Medical examination reveals a fracture of his lumbar spine and transection of the lumbar region of the spinal cord. How will Herbert's micturition be controlled from this time on? Will his detrusor muscles still function? Explain the reasoning behind your responses.

31. What is cystitis? Why do women suffer more often from cystitis than do men?

32. Hattie, aged 55, is awakened by excruciating pain that radiates from her right abdomen to the loin and groin regions on the same side. The pain is not continuous but recurs at intervals of 3 to 4 minutes. Diagnose her problem, and cite factors that might favor its occurrence. Also, explain why Hattie's pain comes in "waves."

33. Felicia, a medical student, arrived late to a surgery where she was to observe the removal of an extensive abscess from the fat around a patient's kidney. Felicia was startled to find the surgical team working to reinflate the patient's lung and to remove air from the pleural cavity. How could renal surgery lead to such events?

34. Len, a radiologist, was looking for renal calculi in a patient's right ureter in a pyelogram. Which three regions of the ureter did he scrutinize first?

35. Why should parents teach their young daughters to wipe from front to back (after defecation)?

23

The Reproductive System

Student Objectives

The Male Reproductive System (pp. 623–633)

1. Describe the location, structure, and function of the testes.
2. Describe the histology of the testes, and outline the events of spermatogenesis.
3. Describe the location, structure, and functions of the accessory organs of the male reproductive system.
4. Describe the structure of the penis, and explain the mechanism of erection.

The Female Reproductive System (pp. 634–646)

5. Describe the location, structure, and function of the ovaries.
6. Explain the phases of the ovarian cycle and the stages of oogenesis.
7. Describe the anatomy of the uterine tubes in terms of their function.
8. Explain the location, regions, supportive structures, and layers of the uterus.
9. Outline the three phases of the uterine cycle.
10. Explain the structure of the vagina in terms of its functions.
11. Describe the anatomy of the female external genitalia.
12. Describe the anatomy of the mammary glands, and explain their clinical importance.

Pregnancy and Childbirth (pp. 646–653)

13. Describe the processes of implantation and placenta formation.
14. Define a placental (chorionic) villus, and explain its functions.
15. Describe some changes that occur in a woman's body when she is pregnant.
16. List the three phases of labor.

The Reproductive System Throughout Life (pp. 653–657)

17. Compare and contrast the prenatal development of the male and female sex organs.
18. Note the anatomical changes that occur during puberty and menopause.

Most organ systems of the body function almost continuously to maintain the well-being of the individual. The reproductive system, however, appears to "slumber" until puberty, after which it plays an important role in our adult lives. Although male and female reproductive organs are quite different, their common purpose is to produce offspring.

In both females and males, the sex organs, or **genitalia** (jen″ĭ-ta′le-ah), are divided into *primary* and *accessory* types. The **primary sex organs**, or **gonads** (go′nadz; "seeds"), are the male testes and the female ovaries. The gonads produce the sex cells or **gametes** (gam′ēts)—the male sperm and the female ovum (egg)—that fuse to form a fertilized egg. All other genitalia in both sexes are **accessory sex organs**. These include the internal glands and ducts that nourish the gametes and transport them toward the outside of the body, as well as the external genitalia.

Besides producing sex cells, the gonads also secrete *sex hormones* and therefore function as endocrine glands. Recall from Chapter 4 (p. 77) that hormones are messenger molecules that travel through the bloodstream to signal various physiological responses in the cells that take them up. Sex hormones play vital roles in the development, maintenance, and function of all sex organs. Other hormones, secreted by the pituitary gland at the base of the brain, also influence reproductive functions.

We begin by considering the male reproductive system (because it is somewhat simpler) and progress to the more complex female reproductive system.

The Male Reproductive System

Figure 23.1 presents an overview of the male reproductive organs. The male gonads are the sperm-producing *testes*, which lie within the scrotum. From the testes, the sperm travel to the outside of the body through a system of ducts, including (in order): the duct of the *epididymis*, the *ductus deferens*, the *ejaculatory duct*, and finally the *urethra*, which opens at the tip of the *penis*. The accessory sex glands, which empty their secretions into the ducts during ejaculation, are the *seminal vesicles, prostate gland*, and *bulbourethral glands.*
COWPERS

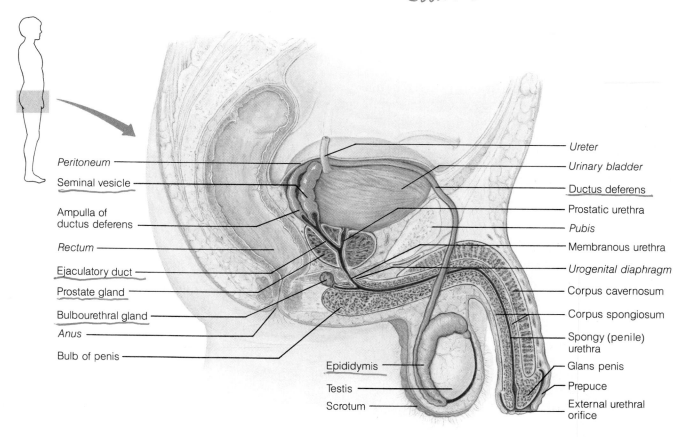

Figure 23.1 Reproductive organs of the male, sagittal view. A portion of the pubic bone (pubis) is in three-dimensional view to indicate how the ductus deferens enters the pelvic cavity.

Peritoneum
Seminal vesicle
Ampulla of ductus deferens
Rectum
Ejaculatory duct
Prostate gland
Bulbourethral gland
Anus
Bulb of penis
Epididymis
Testis
Scrotum

Ureter
Urinary bladder
Ductus deferens
Prostatic urethra
Pubis
Membranous urethra
Urogenital diaphragm
Corpus cavernosum
Corpus spongiosum
Spongy (penile) urethra
Glans penis
Prepuce
External urethral orifice

623

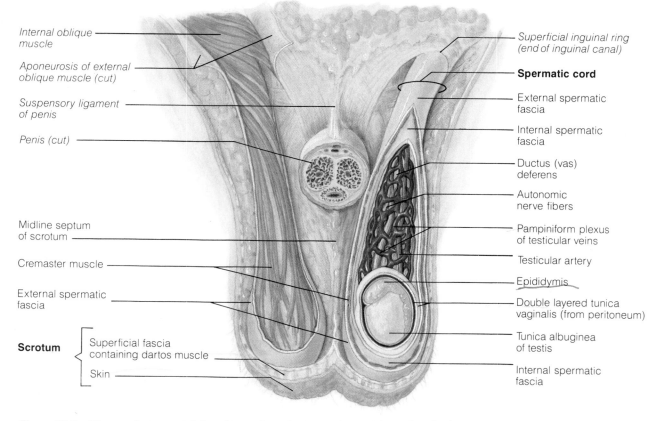

Internal oblique muscle

Aponeurosis of external oblique muscle (cut)

Suspensory ligament of penis

Penis (cut)

Midline septum of scrotum

Cremaster muscle

External spermatic fascia

Scrotum {
Superficial fascia containing dartos muscle

Skin

Superficial inguinal ring (end of inguinal canal)

Spermatic cord

External spermatic fascia

Internal spermatic fascia

Ductus (vas) deferens

Autonomic nerve fibers

Pampiniform plexus of testicular veins

Testicular artery

Epididymis

Double layered tunica vaginalis (from peritoneum)

Tunica albuginea of testis

Internal spermatic fascia

Figure 23.2   The scrotum, containing the testis and spermatic cord (anterior view). The scrotum has been opened and its anterior wall removed.

The Scrotum

The **scrotum** (skro′tum; "pouch") is a sac of skin and superficial fascia that hangs inferiorly external to the abdominopelvic cavity at the root of the penis (Figure 23.2). The scrotum is covered with sparse hairs and skin that is more heavily pigmented than any found elsewhere on the body. The paired, oval **testes** (tes′tēz; "witnesses"), or testicles, lie suspended in the scrotum. A septum in the midline divides the scrotum into right and left halves, providing one compartment for each testis. Their superficial location seems to put the testes—which contain a male's entire genetic heritage—in a rather vulnerable position. However, viable sperm cannot be produced at core body temperature (37 °C); thus, the scrotum, whose superficial position provides a temperature about 3 °C cooler, is an essential adaptation. Furthermore, the scrotum responds to changes in external temperature. For example, when it is cold, the testes are pulled up toward the warm body wall, and the scrotal skin wrinkles to increase its thickness against heat loss. When it is warm, the scrotal skin is flaccid and loose, and the testes hang low to increase the skin surface available for cooling (sweating). These actions are performed by two muscles in the scrotum: The **dartos muscle** (dar′tos; "skinned"), a layer of smooth muscle in the superficial fascia, is responsible for wrinkling the

scrotal skin. The **cremaster muscles** (kre-mas′ter; "a suspender"), consisting of bands of skeletal muscle that extend inferiorly from the internal oblique muscles of the trunk, are responsible for elevating the testes.

Doctors often recommend that men do not wear tight underwear or jeans, especially if they want to father children. Can you imagine why?

Tight clothing holds the testes close to the warmth of the body, inhibiting sperm production. ■

The Testes

Gross Anatomy

Each testis (Figure 23.3) averages about 2.5 cm (1 inch) in width and 4 cm in height. Within the scrotum, each testis is partially enclosed by a serous sac called the **tunica vaginalis** (vaj-in-al′is; "ensheathing coat") (Figure 23.3a). Derived from an outpocketing of the peritoneal cavity in the abdomen, the tunica vaginalis consists of a superficial parietal layer, an intermediate cavity containing serous fluid, and a deeper visceral layer that hugs the surface of the testis. Just deep to the visceral layer lies the **tunica albuginea** (al″bu-jin′e-ah; "white coat"), the fibrous capsule of the testis. Septal extensions of the tunica

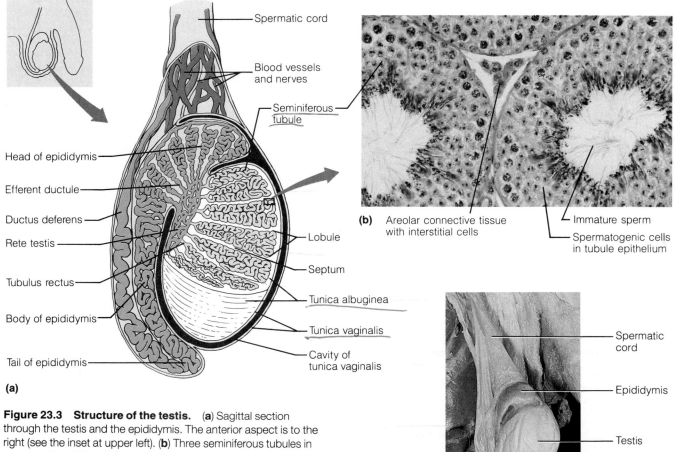

(b) Areolar connective tissue with interstitial cells

Immature sperm

Spermatogenic cells in tubule epithelium

Spermatic cord

Epididymis

Testis

(c)

Spermatic cord

Blood vessels and nerves

Seminiferous tubule

Head of epididymis

Efferent ductule

Ductus deferens

Rete testis

Tubulus rectus

Body of epididymis

Tail of epididymis

Lobule

Septum

Tunica albuginea

Tunica vaginalis

Cavity of tunica vaginalis

(a)

Figure 23.3   Structure of the testis. **(a)** Sagittal section through the testis and the epididymis. The anterior aspect is to the right (see the inset at upper left). **(b)** Three seminiferous tubules in cross section (250 ×). Note the spermatogenic (sperm-forming) cells in the tubule epithelium and the interstitial cells in the connective tissue between the tubules. **(c)** External view of the testis from a cadaver; same orientation as in part (a).

albuginea run internally to divide the testis into 250 to 300 wedge-shaped compartments called **lobules**. Each lobule contains one to four coiled **seminiferous tubules** (sem″ĭ-nif′er-us; "sperm-carrying"), the actual "sperm factories." Most seminiferous tubules are looped like hairpins.

Posteriorly, the seminiferous tubules of each lobule converge to form a **tubulus rectus,** a straight tubule that conveys sperm into the **rete testis** (re′te; "network of the testis"), a complex network of tiny branching tubes. The rete testis lies in the *mediastinum testis,* a region of dense connective tissue in the posterior part of the testis. From the rete testis, sperm leave the testis through about a dozen *efferent ductules* and enter the *epididymis,* a comma-shaped structure that hugs the outer surface of the testis (also see Figures 23.3c and 23.1).

The testes receive their arterial blood from the long testicular arteries, which branch from the aorta in the superior abdomen (see Figure 18.11, p. 496). The testicular veins, which roughly parallel the testicular arteries in the posterior abdominal wall, arise from a venous network in the scrotum called the **pam-**

piniform plexus (pam-pin′ĭ-form; "tendril-shaped") (Figure 23.2). The veins of this plexus surround the testicular arteries like climbing vines. These veins absorb heat from the arterial blood, cooling it before it enters the testes and thereby keeping the testes cool. The testes are innervated by both divisions of the autonomic nervous system. Associated visceral sensory nerves transmit impulses that result in agonizing pain and nausea when the testes are hit forcefully.

It is worth emphasizing that the testicular vessels and nerves descend to the testis from a superior location—generally from the level of L_2 on the posterior abdominal wall (Figures 18.11 and 18.20). This level is the initial location of the testes in the embryo.

The Seminiferous Tubules and Spermatogenesis

A section through a lobule of the mature testis (Figure 23.3b) reveals several *seminiferous tubules* separated from each other by an areolar connective tissue. These sperm-forming tubules consist of a thick stratified epithelium surrounding a hollow central lumen. This epithelium consists of spherical *spermatogenic*

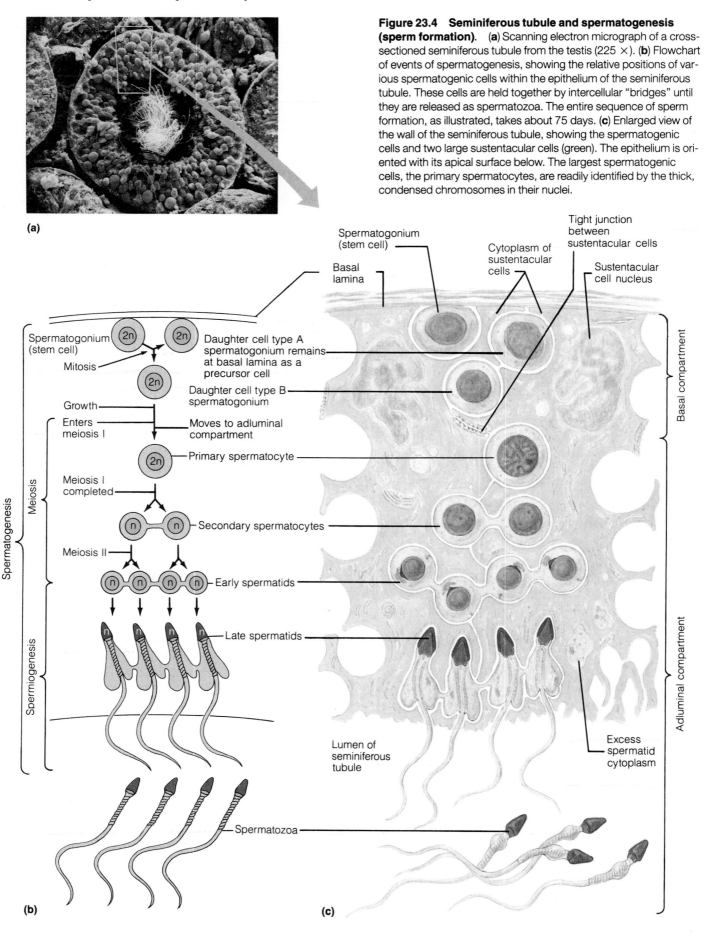

Figure 23.4 Seminiferous tubule and spermatogenesis (sperm formation). (a) Scanning electron micrograph of a cross-sectioned seminiferous tubule from the testis (225 ×). (b) Flowchart of events of spermatogenesis, showing the relative positions of various spermatogenic cells within the epithelium of the seminiferous tubule. These cells are held together by intercellular "bridges" until they are released as spermatozoa. The entire sequence of sperm formation, as illustrated, takes about 75 days. (c) Enlarged view of the wall of the seminiferous tubule, showing the spermatogenic cells and two large sustentacular cells (green). The epithelium is oriented with its apical surface below. The largest spermatogenic cells, the primary spermatocytes, are readily identified by the thick, condensed chromosomes in their nuclei.

(a)

Spermatogonium (stem cell)

Basal lamina

Cytoplasm of sustentacular cells

Tight junction between sustentacular cells

Sustentacular cell nucleus

Spermatogonium (stem cell)

Mitosis

Daughter cell type A spermatogonium remains at basal lamina as a precursor cell

Daughter cell type B spermatogonium

Growth

Enters meiosis I

Moves to adluminal compartment

Primary spermatocyte

Meiosis I completed

Secondary spermatocytes

Meiosis II

Early spermatids

Late spermatids

Spermatogenesis

Meiosis

Spermiogenesis

Lumen of seminiferous tubule

Excess spermatid cytoplasm

Spermatozoa

Basal compartment

Adluminal compartment

(b)

(c)

("sperm-forming") *cells* embedded in columnar *sustentacular* ("supporting") *cells* (Figure 23.4).

Spermatogenic Cells.   The spermatogenic cells are stages in the process of sperm formation or **spermatogenesis** (sper″mah-to-jen′ĕ-sis). This process, which begins at puberty, forms about 400 million sperm per day in the adult male. The least differentiated cells in the sequence are **spermatogonia** (sper″mah-to-go′ne-ah; "sperm seed"), which lie peripherally on the epithelial basal lamina. As these cells differentiate, they move inward toward the lumen to produce *primary* and *secondary spermatocytes, spermatids,* and *spermatozoa.* Spermatogenesis may be divided into three successive stages (Figure 23.4b and c):

Stage 1: Formation of spermatocytes. The spermatogonia divide vigorously and continuously by mitosis, with each division producing two distinctive daughter cells. A **type A** daughter cell remains at the basal lamina to maintain the germ cell line, while a **type B** cell moves toward the lumen to become a **primary spermatocyte** (sper″mah′to-sīt).

Stage 2: Meiosis. The spermatocytes undergo **meiosis** (mi-o′sis; "a lessening"), two successive divisions in which the number of chromosomes is reduced to half that of typical body cells. That is, these divisions reduce the chromosome number from 46, or 23 pairs (the 2*n* or *diploid* number), to 23 single chromosomes (the *n,* or *haploid* number). Meiosis is an essential part of gamete formation in both sexes: It ensures that the diploid number of chromosomes is reestablished after fertilization (when two haploid gametes join to make a diploid zygote), and it prevents the chromosomes from doubling in number with each new generation. The details of meiosis can be found in any biology textbook. We need only note that the two nuclear divisions of meiosis are called meiosis I and II (Figure 23.4b). By definition, the cells undergoing meiosis I are primary spermatocytes, which produce two **secondary spermatocytes,** which undergo meiosis II. As each secondary spermatocyte completes meiosis II, it produces two small cells called **spermatids.** Thus, four spermatids result from the meiotic division of each original spermatocyte.

Stage 3: Spermiogenesis. The third and final stage of spermatogenesis is **spermiogenesis** (sper″me-o-jen′ĕ-sis), in which spermatids differentiate into sperm (Figure 23.5). Each spermatid undergoes a streamlining process as it fashions a tail and sheds its superfluous cytoplasmic baggage. The resulting sperm cell, or **spermatozoon** (sper″mah-to-zo′on; "animal seed"), has two main regions: a *head* and a *tail.* A thickened *midpiece* represents the first part of the tail. The **head** of the sperm consists almost entirely of its flattened nucleus, which contains highly condensed chromatin. Adhering to the top of the nucleus is a helmet-like **acrosome** (ak′ro-sōm; "tip piece"). The lysosome-like acrosome is manufactured by the Golgi apparatus and contains digestive enzymes that enable a sperm to penetrate and enter an egg. The **midpiece** of the sperm contains mitochondria spiraled tightly around the core of the tail. The long **tail** is an elaborate flagellum whose motile cytoskeleton has grown out from a centriole near the nucleus. The mitochondria provide the energy needed for the whip-like movements of the tail that will propel the sperm through the female reproductive tract.

The newly formed sperm detach from the epithelium of the seminiferous tubule and enter the lumen of the tubule. These young sperm look mature, but they are nearly immotile and do not gain the ability to swim until they have left the testis (p. 629).

Overall, the process of spermatogenesis is controlled by the stimulating action of two hormones: **follicle-stimulating hormone (FSH)** from the anterior part of the pituitary gland and **testosterone,** the primary male sex hormone from the testes.

Sustentacular Cells.   The spermatogenic cells are surrounded by and actually embedded in the plasma membrane of **sustentacular cells,** or *Sertoli* (Ser-to′le) *cells,* which extend from the basal lamina to the lumen of the seminiferous tubule (Figure 23.4c). The sustentacular cells, bound to each other by tight junctions on their lateral membranes, divide the seminiferous tubule into two compartments: (1) The **basal compartment** extends from the basal lamina to the tight junctions and contains the spermatogonia and earliest primary spermatocytes. (2) The **adluminal** ("near the lumen") **compartment** lies internal to the tight junctions. It includes the more advanced spermatogenic cells and the lumen of the tubule.

The tight junctions between the sustentacular cells form the **blood-testis barrier** (Figure 23.6). This barrier prevents the membrane antigens (mostly proteins) of differentiating sperm cells from escaping through the basal lamina into the bloodstream. Since sperm do not form until puberty, they are absent when the immune system develops tolerance to the body's own tissues early in life. Moreover, all spermatogenic cells that form after meiosis I are genetically different from other body cells. Consequently, if the blood-testis barrier did not exist, antigens on the sperm cell membranes would enter the blood and provoke an autoimmune response to one's own sperm. The resulting destruction of sperm would cause sterility. Early spermatogenic cells must pass through this barrier as they move toward the lumen. More specifically, the tight junctions between the sustentacular cells open to let primary spermatocytes pass into the adluminal com-

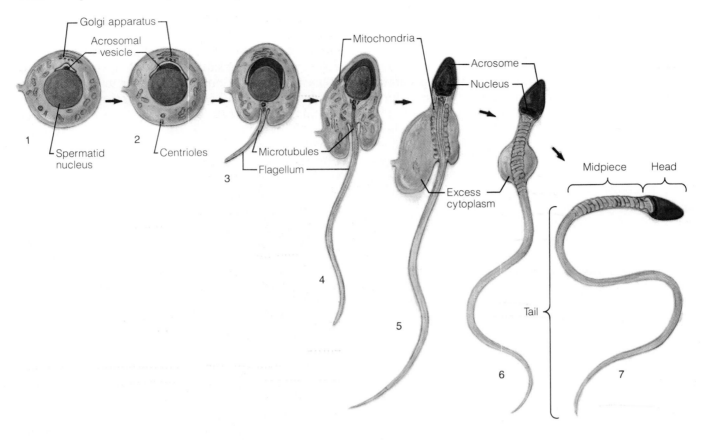

Figure 23.5 Spermiogenesis: Transformation of a spermatid into a sperm. The steps in spermiogenesis are as follows: (1) The Golgi apparatus produces enzyme-filled vesicles that fuse together to form the acrosome. (2) The acrosome positions itself at the anterior end of the nucleus, and the centrioles move to the opposite end of the nucleus. (3) Microtubules assemble from a centriole and grow to form the flagellum of the sperm tail. (4) Mitochondria multiply. (5) The mitochondria position themselves around the proximal core of the flagellum, the chromatin condenses in the nucleus, the nucleus flattens, and excess cytoplasm is shed from the cell. (6) Structure of an immature sperm that has just been released into the tubule lumen from a sustentacular cell. (7) A structurally mature sperm with a streamlined shape that allows active swimming.

partment—much as locks in a canal are opened to let a boat pass.

Sustentacular cells assist sperm production in many ways. They convey nutrients to the spermatogenic cells, actively move these cells toward the tubule lumen, and phagocytize the cytoplasm that is shed as spermatids become sperm. Sustentacular cells also secrete **testicular fluid** into the tubule lumen. The pressure of this fluid pushes the sperm out of the testes. Furthermore, sustentacular cells secrete **androgen-binding protein,** a protein that concentrates male sex hormone (testosterone) in the vicinity of the spermatogenic cells. Testosterone in turn stimulates spermatogenesis. Finally, sustentacular cells secrete the hormone **inhibin,** which slows the rate of sperm production (when necessary) by affecting the pituitary gland.

Interstitial Cells. The loose connective tissue between the seminiferous tubules contains clusters of **interstitial cells,** or *Leydig* (Li′dig) *cells* (Figure 23.3b). Interstitial cells are spherical or polygon-shaped, and they stain pink with the dye eosin. They manufacture the male sex hormones, or **androgens,** the main example of which is testosterone. After it is secreted into the nearby blood capillaries and lymph capillaries, testosterone circulates throughout the body and maintains all male sex characteristics and sex organs. In fact, all male genitalia atrophy if the testes (and testosterone) are removed.

Secretion of testosterone by the interstitial cells is controlled by a hormone from the anterior part of the pituitary gland, called **luteinizing** (lŏo′te-in-īz″ing) **hormone (LH).** In stimulating testosterone secretion, LH controls testosterone's effects on the entire male reproductive system.

Testicular cancers affect about 1 of every 50,000 males and develop most often in young men (ages 15 to 35). Knowing that cancers often derive

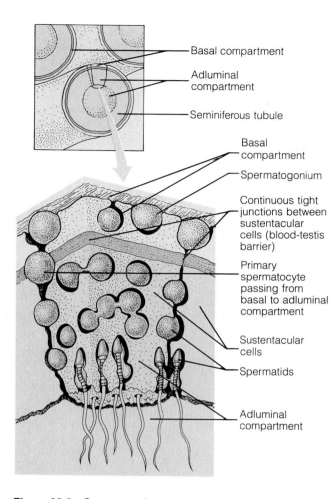

Basal compartment
Adluminal compartment
Seminiferous tubule

Basal compartment
Spermatogonium
Continuous tight junctions between sustentacular cells (blood-testis barrier)
Primary spermatocyte passing from basal to adluminal compartment
Sustentacular cells
Spermatids
Adluminal compartment

Figure 23.6 Sustentacular cells, various spermatogenic cells, and the blood-testis barrier. Continuous tight junctions interconnect sustentacular cells laterally at their bases, thus separating a basal compartment containing spermatogonia (gray) from an adluminal compartment containing more advanced spermatogenic cells (blue). The tight junctions form the blood-testis barrier.

from the body's most rapidly dividing cells, can you deduce which cells in the testes give rise to most testicular cancers?

Early-stage spermatogenic cells produce most testicular cancers. The tumors produced by these cells are firm, usually painless lumps in the testis. Males are advised to examine their testicles regularly for such lumps, which can be felt through the skin of the scrotum. If detected early, most testicular cancers are curable. Note, however, that most masses that form lumps in the scrotum are not tumors but relatively harmless pockets of fluid called varicoceles and hydroceles (see the Related Clinical Terms on p. 657). ■

The Male Duct System

As we've discussed, sperm leaving the seminiferous tubules travel through the *tubuli recti* and *rete testis*

and then leave the testis through the *efferent ductules* (Figure 23.3a). These small tubes are lined by simple cuboidal and columnar epithelia, some cells of which bear cilia or flagella that help move the sperm along. The main force in moving sperm out of the testes, however, seems to be the flushing action of the testicular fluid (recall that sperm cannot yet swim at this stage). From the testis, the sperm enter the epididymis.

The Epididymis

The **epididymis** (ep″ĭ-did′ĭ-mis; "beside the testis") is a comma-shaped organ that arches over the posterior and lateral side of the testis (Figure 23.3a and c). It can be divided into three regions: a head, body, and tail. The *head* contains the efferent ductules, which empty into the **duct of the epididymis.** This highly coiled duct completes the head and forms all of the *body* and *tail.* With an uncoiled length of over 6 m (20 feet), the duct of the epididymis is longer than the entire intestine!

Histologically, the duct of the epididymis is dominated by a tall, pseudostratified columnar epithelium (Figure 23.7a). The luminal surface of this epithelium bears many long microvilli called **stereocilia** (ster″e-o-sil′e-ah). These are not cilia, and they do not move. Instead, they provide the tall epithelial cells with a vast surface area for reabsorbing testicular fluid and for transferring nutrients and secretions to the many sperm that are stored in the central lumen. External to the epithelium lies a layer of smooth muscle.

The immature, nearly immotile sperm that leave the testes are moved slowly through the duct of the epididymis. During this journey, which takes about 20 days, the sperm gain the ability to swim. This maturation is stimulated by a protein secreted by the epididymis epithelium. Sperm are ejaculated from the epididymis, *not* directly from the testes. At the beginning of ejaculation, the smooth muscle in the walls of the epididymis contracts, expelling sperm from the tail of the epididymis into the next segment of the duct system, the ductus deferens.

Sperm can be stored in the epididymis for several months. If held longer, they are eventually phagocytized by the epithelial cells of the epididymis.

The Ductus Deferens

The **ductus deferens** (def′er-ens; "carrying away"), or *vas deferens,* is about 45 cm (18 inches) long. This duct runs superiorly through the scrotum, pierces the anterior abdominal wall, and enters the pelvic cavity (Figure 23.1). From there, it runs posteriorly along the lateral wall of the true pelvis, arches medially over the ureter, and descends along the posterior wall of the bladder. Its distal end expands as the **ampulla** ("flask") **of the ductus deferens** and then joins with

Figure 23.7   The epididymis and ductus deferens in histological cross sections. (a) The duct of the epididymis. (There appear to be several ducts, but these are in fact multiple coils of a single duct in section). Note the thick epithelium and the stereocilia (75 ×). (b) Ductus deferens. Note the thick layer of smooth muscle (25 ×).

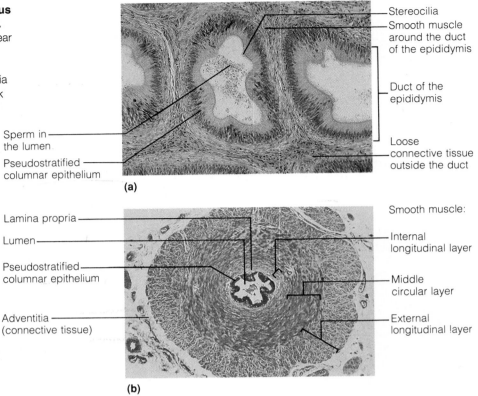

(a)

(b)

the duct of the seminal vesicle (a gland) to form the short **ejaculatory duct.** Each ejaculatory duct runs within the prostate gland, where it empties into the urethra.

The histology of the ductus deferens is shown in Figure 23.7b. Its wall consists of an inner mucosa (with the same pseudostratified epithelium as in the epididymis, plus a lamina propria), an extremely thick muscularis, and an outer adventitia. During ejaculation, the smooth muscle in the muscularis creates strong peristaltic waves that rapidly propel sperm through the ductus deferens to the urethra.

Some men take responsibility for birth control by having a **vasectomy** (vah-sek′to-me; "cutting the vas"). In this minor surgery, the physician makes a small incision into the side of the scrotum and then transects the ductus and ligates (ties off) the cut ends. Although sperm continue to be produced, they can no longer exit the body and are phagocytized in the epididymis. The operation is irreversible in about half of cases, probably because lymphocytes enter the cut end of the ductus deferens, leading to an autoimmune destruction of the sperm-forming cells in the testes. (The blood-testis barrier has thus been circumvented.)

While performing a vasectomy, a physician must be able to distinguish the ductus deferens from the testicular vessels that surround it (p. 625). Can you guess the "trick" that physicians use to identify the ductus deferens during surgery?

Its muscular layer is so thick that the ductus deferens feels like a hard wire when squeezed between one's fingertips. ∎

The Spermatic Cord

The ductus deferens is the largest component of the **spermatic cord,** a tube of fascia that also contains the testicular vessels and nerves (Figures 23.2 and 23.8). The medial part of the spermatic cord lies in the scrotum, and its lateral part runs through the **inguinal canal,** an obliquely oriented trough in the anterior abdominal wall (Figure 23.8). The inguinal canal is formed by the inguinal ligament, which is the free inferior margin of the aponeurosis of the external oblique muscle (p. 258). This canal begins medially at the **superficial inguinal ring** (a V-shaped opening in the external oblique aponeurosis) and runs laterally to the **deep inguinal ring** (a weak point in the fascia of the anterior body wall where the ductus deferens and testicular vessels enter the pelvic cavity).

Because the deep inguinal ring—and an area of fascia just medial to it called the inguinal triangle—are weak areas in the abdominal wall, coils of intestine can push anteriorly through them into the inguinal canal. This condition is an **inguinal hernia,** and the herniated elements may follow the inguinal canal all the way into the scrotum (Figure 23.8). Such hernias are often caused by lifting or straining that raises intra-abdominal pressure and forces the intes-

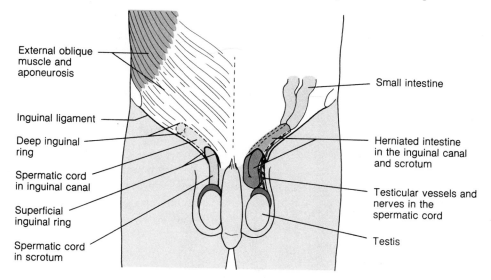

External oblique muscle and aponeurosis

Inguinal ligament

Deep inguinal ring

Spermatic cord in inguinal canal

Superficial inguinal ring

Spermatic cord in scrotum

Small intestine

Herniated intestine in the inguinal canal and scrotum

Testicular vessels and nerves in the spermatic cord

Testis

Figure 23.8   Spermatic cord, inguinal canal, and hernias.   Frontal view of the inguinal region, showing the position of the spermatic cord (right side of the body) and an inguinal hernia (left side of the body). The inguinal canal lies just deep to the aponeurosis of the external oblique muscle. Note the herniated intestine passing through the inguinal canal toward the scrotum. (The greater omentum may also herniate from the abdominal cavity by this path.) An inguinal hernia forms a bulge in the groin region.

tine through the body wall. At least 2% of adult males develop inguinal hernias. Surgical repair involves pushing the herniated intestine back into the abdominal cavity and closing off the enlarged opening with surgical thread or a patch. Females, like males, have two inguinal rings and an inguinal canal. Their inguinal canals are quite narrow, however, because they contain only a thin band called the round ligament of the uterus (p. 641). For this reason, inguinal hernias are rare in females. ■

The Urethra

The male urethra carries the ejaculated sperm from the ejaculatory ducts to the outside of the body (recall Chapter 22, p. 616). Figure 23.9a reviews its three parts: the *prostatic urethra* in the prostate gland, the *membranous urethra* in the urogenital diaphragm, and the *spongy (penile) urethra* in the penis. The mucosa of the distal part of the urethra contains outpocketings called *urethral glands* (not shown). These minor glands secrete a mucus that helps lubricate the urethra just before ejaculation.

Accessory Glands

The male accessory glands include the paired *seminal vesicles,* the single *prostate gland,* and paired *bulbourethral glands* (Figure 23.1). These glands pro-

duce the bulk of the **semen** (sperm plus secretions of accessory glands and ducts).

The Seminal Vesicles

The **seminal vesicles** *(seminal glands)* lie on the posterior surface of the bladder (Figure 23.1). Each of these hollow glands is about the shape and length of a finger (5 to 7 cm). Each seminal vesicle is pouched, coiled, and folded back on itself, however, so its true (uncoiled) length is about 15 cm. Internally, the mucosa of the seminal vesicle is folded into a honeycomb pattern of crypts and blind chambers, and the lining epithelium is a secretory pseudostratified columnar epithelium. The external wall is a thick layer of smooth muscle underlying a fibrous capsule of connective tissue. The muscle contracts during ejaculation to empty the gland.

The seminal vesicles produce about 60% of the volume of semen. Their secretion is a viscous fluid that contains a sugar called fructose (which nourishes the sperm) and other chemical substances. As we noted earlier, the duct of each seminal vesicle joins the ductus deferens on the same side of the body to form an ejaculatory duct within the prostate gland. Sperm and seminal fluid mix in the ejaculatory duct and enter the prostatic urethra together during ejaculation.

The fluid secreted by the seminal vesicles contains a yellow pigment that makes semen fluoresce

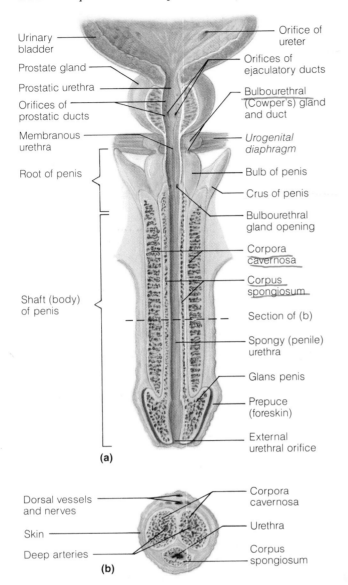

Urinary bladder

Prostate gland

Prostatic urethra

Orifices of prostatic ducts

Membranous urethra

Root of penis

Shaft (body) of penis

Orifice of ureter

Orifices of ejaculatory ducts

Bulbourethral (Cowper's) gland and duct

Urogenital diaphragm

Bulb of penis

Crus of penis

Bulbourethral gland opening

Corpora cavernosa

Corpus spongiosum

Section of (b)

Spongy (penile) urethra

Glans penis

Prepuce (foreskin)

External urethral orifice

(a)

Dorsal vessels and nerves

Skin

Deep arteries

Corpora cavernosa

Urethra

Corpus spongiosum

(b)

Figure 23.9   Male urethra and penis.   (a) Longitudinal (frontal) section through the penis. **(b)** Cross section through the penis, with dorsal aspect above (the dorsal aspect is the posterior aspect of the erect penis).

under ultraviolet light. In this way, semen residues are identified in criminal investigations of sexual assault.

The Prostate Gland

The **prostate** (pros'tāt) **gland** is the size and shape of a chestnut. It encircles the first part of the urethra just inferior to the bladder (Figure 23.10a). Enclosed by a capsule of connective tissue, the prostate gland consists of 20 to 30 compound tubuloalveolar glands embedded in a mass of dense connective tissue and smooth muscle called the **fibromuscular stroma** (Figure 23.10b). The muscle of this stroma contracts dur-

ing ejaculation to squeeze the prostatic secretion into the urethra.

The prostatic secretion accounts for about one-third of the volume of semen. This secretion is a milky fluid containing various substances that (1) enhance sperm motility and (2) clot then liquefy the semen. (Ejaculated semen first clots like blood and then liquefies, enabling the sperm to swim out.)

The prostatic secretion also includes various enzymes of unknown function. One of these enzymes, *acid phosphatase,* has clinical importance: An elevated level of acid phosphatase in a man's blood is a sure sign that prostatic cancer (discussed shortly) has spread to the bone marrow.

Almost all males develop prostatic tumors during middle or old age. Can you imagine what symptom leads to the detection of most tumors of the prostate?

The enlarging prostate compresses the prostatic urethra, causing difficulty in urination. Most prostatic tumors are detected when men seek medical treatment for this problem. To help confirm that the prostate is enlarged, the physician performs a **rectal exam**: Since the prostate lies just anterior to the rectum (Figure 23.10a), a finger in the anal canal can easily feel this gland through the anterior rectal wall. A tumorous prostate gland is removed by a surgical procedure called **prostatectomy** ("cutting out the prostate").

Although most prostatic tumors are benign, malignancies are common. In fact, fully 6% of men die from prostate cancer. Figure 23.10b shows the glandular structure of the prostate. Cells of the *main glands* and *submucosal glands* give rise to malignancies, whereas cells of the *mucosal glands* give rise to the benign tumors.

Along with its susceptibility to tumors, the prostate is also subject to infections (in sexually transmitted diseases, for example). **Prostatitis** (pros"tah-ti'tis), inflammation of the prostate, is the single most common reason for a man to consult a urologist. ■

The Bulbourethral Glands

The **bulbourethral** (bul"bo-u-re'thral) **glands** are pea-sized glands situated inferior to the prostate gland. More specifically, they lie within the urogenital diaphragm (Figures 23.1 and 23.9). These compound tubuloalveolar glands produce a mucus that enters the spongy urethra. Some of this secretion is released when a male becomes sexually excited (that is, *prior to* ejaculation). It is believed that this mucus neutralizes traces of acidic urine in the urethra and lubricates the urethra to smooth the passage of semen during ejaculation.

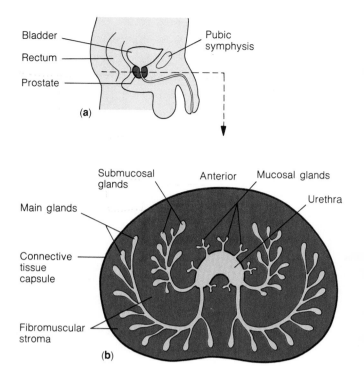

Figure 23.10   The prostate gland. (**a**) The location of the prostate gland in the pelvis. (**b**) Diagram of a transverse section through the prostate gland. Note the three classes of glands (main, submucosal, mucosal).

The Penis

The **penis** (Latin for "tail"), the male organ of sexual intercourse, delivers sperm into the female reproductive tract (Figure 23.9). The penis and scrotum make up the external reproductive structures, or external genitalia, of the male. The penis consists of an attached **root** and a free **shaft** or **body** that ends in an enlarged tip called the **glans penis**. The skin covering the penis is loose and slides distally to form a cuff around the glans, called the **prepuce** (pre'pūs), or **foreskin**. Frequently, the foreskin is removed shortly after birth in a surgical procedure called **circumcision** (ser"kum-sizh'un; "cutting around").

Internally, the penis contains the spongy urethra and three long cylindrical bodies (corpora) of erectile tissue. Each of these three **erectile bodies** is a thick tube covered by a sheath of dense connective tissue and filled with a network of fibrous partitions. This sponge-like network, in turn, is riddled with vascular spaces. The midventral erectile body surrounding the spongy urethra is the **corpus spongiosum** (kor'pus spun"je-o'sum; "spongy body"). It is enlarged *distally*, where it forms the glans penis, and *proximally*, where it forms a part of the root called the **bulb of the penis**.

The bulb is secured to the urogenital diaphragm and is covered externally by the sheet-like bulbospongiosus muscle (see Figure 10.11, p. 261). The paired, dorsal erectile bodies, the **corpora cavernosa** (kav"er-no'sah; "cavernous bodies"), make up most of the mass of the penis. Their proximal ends in the root are the **crura** ("legs") **of the penis** (singular, *crus*). Each crus is anchored to the pubic arch of the bony pelvis and is surrounded by an ischiocavernosus muscle (Figure 10.11).

The main vessels and nerves of the penis are shown in Figure 23.9b. Most of these lie near the dorsal midline. The sensory *dorsal nerves* are branches of the pudendal nerve from the sacral plexus, and the *dorsal arteries* are branches of the internal pudendal arteries from the internal iliac arteries. Two *dorsal veins* (superficial and deep) lie precisely in the dorsal midline and drain all blood from the penis. Finally, a *deep artery* runs within each corpus cavernosum. The autonomic nerves to the penis, which follow the arteries and supply the erectile bodies, arise from the inferior hypogastric plexus in the pelvis.

The chief phases of the male sexual response are (1) erection of the penis, which allows it to penetrate into the female vagina, and (2) ejaculation, which expels semen into the vagina. *Erection* results from engorgement of the erectile bodies with blood. During sexual excitement, through the action of parasympathetic nerve fibers, the arteries supplying the erectile bodies dilate, increasing the flow of blood to the vascular spaces within. As the erectile bodies begin to swell with blood, they press on the small veins that normally drain them, slowing venous drainage and maintaining engorgement.

Whereas erection is under parasympathetic control, *ejaculation* is under sympathetic control. Ejaculation begins with a strong, sympathetically induced contraction of the smooth musculature throughout the reproductive ducts and glands. This action squeezes the semen toward and into the urethra. Simultaneously, the bulbospongiosus muscle of the penis undergoes rapid contraction, squeezing the semen onward through the penile urethra and out of the body.

The Male Perineum

The male **perineum** (per"ĭ-ne'um; "around the anus") contains the scrotum, the root of the penis, and the anus (Figure 25.14, p. 697). More specifically, the perineum is defined as the diamond-shaped area between the pubic symphysis anteriorly, the coccyx posteriorly, and the ischial tuberosities laterally. The floor of the perineum is formed by muscles described in Chapter 10 (pp. 260–261).

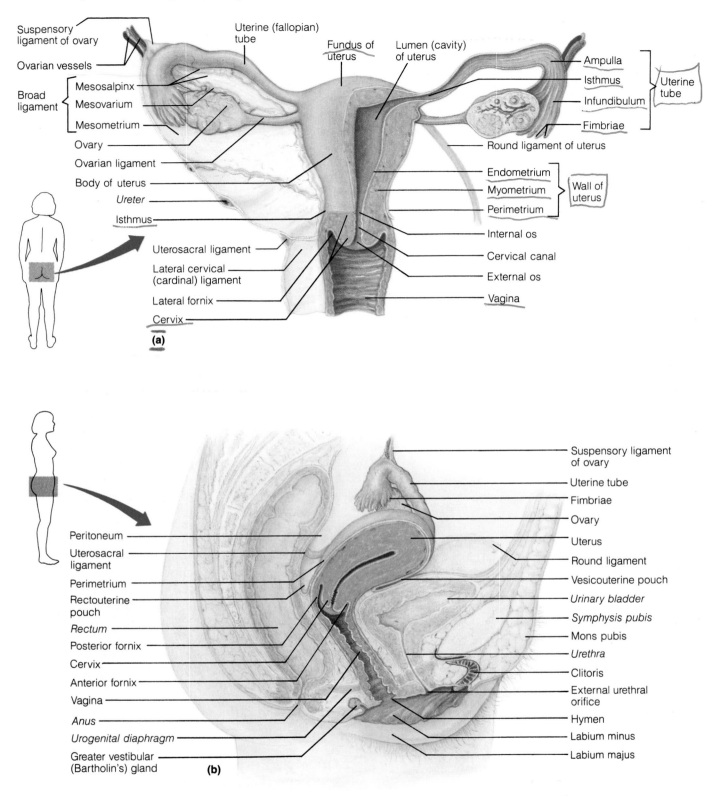

Suspensory ligament of ovary

Ovarian vessels

Broad ligament
Mesosalpinx
Mesovarium
Mesometrium

Ovary

Ovarian ligament

Body of uterus

Ureter

Isthmus

Uterosacral ligament

Lateral cervical (cardinal) ligament

Lateral fornix

Cervix

Uterine (fallopian) tube

Fundus of uterus

Lumen (cavity) of uterus

Ampulla
Isthmus
Infundibulum
Fimbriae
Uterine tube

Round ligament of uterus

Endometrium
Myometrium
Perimetrium
Wall of uterus

Internal os

Cervical canal

External os

Vagina

(a)

Peritoneum

Uterosacral ligament

Perimetrium

Rectouterine pouch

Rectum

Posterior fornix

Cervix

Anterior fornix

Vagina

Anus

Urogenital diaphragm

Greater vestibular (Bartholin's) gland

Suspensory ligament of ovary

Uterine tube

Fimbriae

Ovary

Uterus

Round ligament

Vesicouterine pouch

Urinary bladder

Symphysis pubis

Mons pubis

Urethra

Clitoris

External urethral orifice

Hymen

Labium minus

Labium majus

(b)

Figure 23.11 Female internal reproductive organs. (a) Posterior view of the reproductive organs in the pelvis. On the right side, the posterior walls of the various organs and the broad ligament have been removed. (b) Midsagittal section of the female pelvis.

The Female Reproductive System

The female reproductive organs differ from those of males in several important ways. They not only produce gametes (ova, or "eggs"), but also prepare to support a developing embryo during pregnancy. Furthermore, the female reproductive organs undergo changes according to a reproductive cycle, the **menstrual cycle,** which averages 28 days in length.

Figure 23.11 provides an overview of the female's internal reproductive organs. The female's gonads are the ovum-producing *ovaries.* The female's accessory ducts include the *uterine tubes,* where fertilization takes place; the *uterus,* where the embryo develops; and the *vagina,* which acts as a birth canal and receives the penis during sexual intercourse. The female's external genitalia are referred to as the *vulva.* The milk-producing *mammary glands* are actually part of the integumentary system rather than the reproductive system. Since they nourish the infant, however, the discussion of the mammary glands is included here.

The Ovaries

Basic Anatomy

The paired, almond-shaped **ovaries** flank the uterus on each side. They measure about 3 cm by 1.5 cm by 1 cm. Each ovary lies within the peritoneal cavity, against the bony lateral wall of the true pelvis. The surfaces of the ovaries are smooth in young girls but after puberty become scarred and pitted from the monthly release of ova.

The ovaries are held in place by mesenteries and ligaments. The mesentery that suspends the ovary in the peritoneal cavity is the horizontal **mesovarium** (mez"o-va're-um; "mesentery of the ovary"). This is part of the **broad ligament,** a large fold of peritoneum that drapes over the uterus and the uterine tubes like a tent (Figure 23.11a). The **suspensory ligament of the ovary** is a lateral continuation of the broad ligament, and it attaches the ovary to the lateral pelvic wall (Figure 23.11a). Finally, the ovary is anchored to the uterus medially by the **ovarian ligament,** a distinct fibrous band enclosed within the broad ligament.

The ovaries are served by the *ovarian arteries* (branches from the abdominal aorta) and by the *ovarian branch of the uterine arteries* (Figure 23.12). The ovarian artery, veins, and nerves reach the ovary by traveling within the suspensory ligament and then through the mesovarium. The ovaries are innervated by both divisions of the autonomic nervous system.

Figure 23.13 shows the internal structure of the ovary. The ovary is surrounded by a fibrous capsule

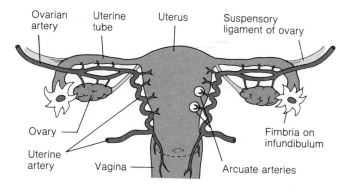

Figure 23.12   Arteries of the internal female genitalia (posterior view).   Corresponding veins run with these arteries.

called the **tunica albuginea,** which is much thinner than the tunica albuginea of the testis. Externally, the tunica albuginea is covered by a simple cuboidal epithelium called the **germinal epithelium** (a misnomer, since this is simply a mesothelium and does *not* germinate the ova). Internally, the main substance of the ovary is divided into an outer *cortex* and an inner *medulla.* The **ovarian cortex** houses the gametes, which are called **oocytes** (o'o-sīts; "egg cells") at this stage. All oocytes occur within sac-like, multicellular structures called **follicles** ("little bags"), which increase markedly in size as they mature. The cortical tissue between the follicles is a cell-rich connective tissue. The deep **ovarian medulla** is a loose connective tissue containing the largest blood vessels, nerves, and lymph vessels of the ovary. These vessels enter the ovary through the *hilus,* a horizontal slit in the anterior ovarian surface where the mesovarium attaches.

Go toward then come back to these

The Ovarian Cycle

The monthly **ovarian cycle** is the menstrual cycle as it relates to the ovary. While reading about this cycle, refer to Figures 23.14 and 23.19.

Follicular Phase (First Half of the Ovarian Cycle).
From before birth to the end of a female's reproductive years, the ovarian cortex contains many thousands of follicles, most of which are **primordial follicles** (Figure 23.14, follicle 1). Each of these immature follicles, from which all later follicle stages arise, consists of an oocyte surrounded by one layer of flat supportive cells called **follicular cells.** At the start of each ovarian cycle, six to twelve primordial follicles start to grow. Follicular growth continues through the first 2 weeks of each cycle, which make up the **follicular phase** (follicles 1–6 in Figure 23.14). During this phase, follicular growth is stimulated by *follicle-stimulating hormone (FSH)* and *luteinizing hormone (LH)* from the anterior part of the pituitary gland. Together, FSH and LH are called **gonadotropins** (go-nad"o-

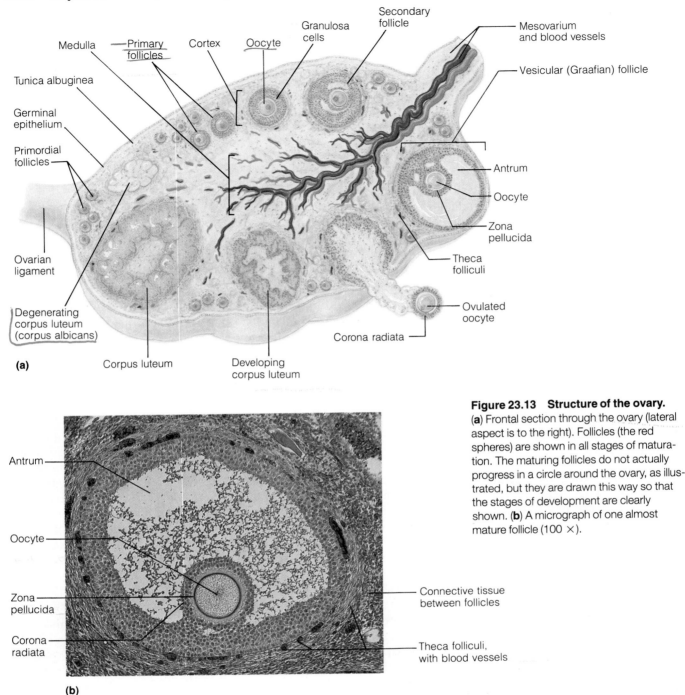

(a)

Medulla, Primary follicles, Cortex, Oocyte, Granulosa cells, Secondary follicle, Mesovarium and blood vessels, Tunica albuginea, Germinal epithelium, Primordial follicles, Ovarian ligament, Degenerating corpus luteum (corpus albicans), Corpus luteum, Developing corpus luteum, Corona radiata, Ovulated oocyte, Theca folliculi, Zona pellucida, Oocyte, Antrum, Vesicular (Graafian) follicle

(b)

Antrum, Oocyte, Zona pellucida, Corona radiata, Connective tissue between follicles, Theca folliculi, with blood vessels

Figure 23.13 Structure of the ovary.
(a) Frontal section through the ovary (lateral aspect is to the right). Follicles (the red spheres) are shown in all stages of maturation. The maturing follicles do not actually progress in a circle around the ovary, as illustrated, but they are drawn this way so that the stages of development are clearly shown. (b) A micrograph of one almost mature follicle (100 ×).

trōp′inz), which literally means "gonad-nourisher hormones." Of the many follicles that start growing, most degenerate along the way, leaving only one follicle each month to expel its oocyte from the ovary for potential fertilization. Such expulsion, called *ovulation,* occurs on about day 14, the midpoint of the ovarian cycle.

We will now describe in detail the structural stages through which follicles pass during the follicular phase. First, the flat follicular cells of primordial follicles become cuboidal, and the oocyte grows in diameter (follicle 1 becomes follicle 2 in Figure 23.14). Now the follicle is a **primary follicle**. Next, the follic-

ular cells multiply to form a stratified epithelium around the oocyte (follicle 3 in the figure). From this point on, the follicular cells are called **granulosa cells.**

In the next stage (follicle 4), a glycoprotein coat called the **zona pellucida** (zo′nah pĕ-lu′sid-ah; "transparent belt") appears around the oocyte. This coat acts as a protective shell that sperm ultimately must penetrate to fertilize the oocyte. As the granulosa cells continue to divide, a layer of connective tissue condenses around the exterior of the primary follicle. This layer is the **theca folliculi** (the′kah fŏ-lik′u-li; "box around the follicle"). Structurally, its cells are

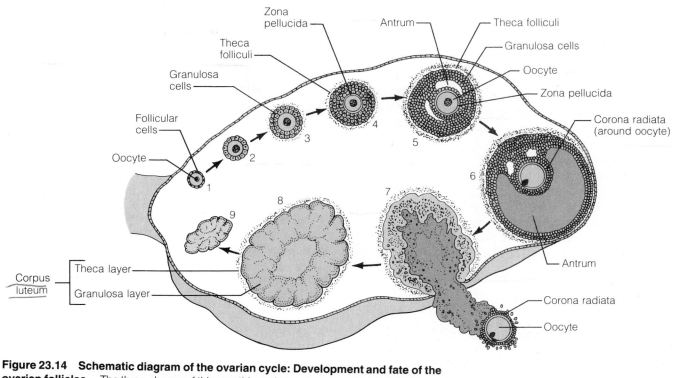

Figure 23.14   Schematic diagram of the ovarian cycle: Development and fate of the ovarian follicles.   The three phases of this monthly cycle are the follicular phase (follicles (1–6), ovulatory phase (follicle 7), and luteal phase (follicle 8). (1) Primordial follicle, an oocyte surrounded by flat follicular cells. (2) An early primary follicle, as evidenced by the cuboidal nature of its follicular cells. (3 and 4) The primary follicle grows as its follicular cells divide to become several layers of granulosa cells. The zona pellucida and theca folliculi appear. (5) Secondary follicle, as evidenced by the appearance of an antrum filled with fluid. (6) The mature vesicular (Graafian) follicle has a huge antrum and is ready to be ovulated at mid cycle. (7) Ovulation: the ruptured follicle and the ovulated oocyte surrounded by its corona radiata of granulosa cells. (8) The corpus luteum formed from the ruptured follicle consists of theca and granulosa layers. (9) The corpus albicans scar.

spindle-shaped, like fibroblasts. Functionally, however, the internal cells of the theca secrete hormones, and the external cells resemble smooth muscle fibers. The internal theca cells, with help from the granulosa cells, produce the female sex hormones, or **estrogens** (es′tro-jens; "generators of sexual activity"). This production of estrogen is stimulated by the pituitary gonadotropins (LH and FSH) during the follicular phase. Estrogen stimulates the growth and activity of all female sex organs and signals the mucosa of the uterus to repair itself after each menstrual period (p. 643).

In the next stage of follicle development (follicle 5), a clear liquid gathers between the granulosa cells, coalescing to form a fluid-filled cavity called the **antrum** ("cave"). The presence of an antrum distinguishes the new **secondary follicle** from the primary follicle. The antrum expands with fluid until it isolates the oocyte, along with a surrounding coat of granulosa cells called a **corona radiatia** (kŏ-ro′nah ra-de-ah′tah; "radiating crown"), on a stalk at one end of the follicle (Figure 23.13b). Finally, one follicle attains full size (follicle 6), reaching a diameter of 2 cm

(almost 1 inch!). This mature follicle, the **vesicular (Graafian) follicle**, is ready to be ovulated.

Ovulation.   Ovulation occurs about halfway through each ovarian cycle (follicle 7 in Figure 23.14). In this process, an oocyte exits from one of the woman's two ovaries, enters the peritoneal cavity, and is swept into a uterine tube. The signal for ovulation is the sudden release of a large quantity of luteinizing hormone (LH) from the pituitary gland, just before day 14. The mechanics of ovulation are as follows: The ovarian wall over the follicle bulges, thins, and oozes fluid. It then ruptures, and the oocyte exits, surrounded by its corona radiata. The forces responsible for this process are not fully understood, but they probably involve an enzymatic breakdown of the follicle wall, with this initial rupture being followed by a violent contraction of the muscle-like cells of the external theca layer.

The Corpus Luteum and the Luteal Phase (Second Half of the Ovarian Cycle).   After ovulation, the part of the follicle that remains in the ovary collapses, and its wall is thrown into wavy folds (follicle 8 in Figure

23.14). This structure, now called the **corpus luteum** ("yellow body"), consists of the remaining granulosa and theca layers.

The corpus luteum is not a degenerative structure but an endocrine gland that persists through the second half, or **luteal phase,** of each ovarian cycle. It secretes a hormone called **progesterone** (pro-jes'tě-rōn; "hormone that prepares for pregnancy"), as well as estrogen. Progesterone acts on the mucosa of the uterus, signaling it to prepare for implantation of an embryo. If there is no implantation, however, the corpus luteum dies after 2 weeks. In this case, all that ultimately remains is a scar called a **corpus albicans** (al'bĭ-kans; "white body") (follicle 9 in Figure 23.14). The corpus albicans remains in the ovary for several months, shrinking until it is finally removed by macrophages.

Oogenesis

Just as sperm form by spermatogenesis, ova form in the ovary by **oogenesis** (o"o-jen'ě-sis; "egg generation") (Figure 23.15). Oogenesis, like spermatogenesis, includes the chromosome-reduction divisions of meiosis (2*n* to *n:* p. 627). Oogenesis is a process that takes many years to complete. First, in the fetal period, the stem germ cells, **oogonia,** give rise to the female's lifelong supply of oocytes, which are immediately arrested in an early stage of the first meiotic division. Now called **primary oocytes,** they remain "stalled" in meiosis 1 for decades until they are ovulated from their follicle. Only at ovulation, under the influence of the LH surge that signals ovulation, does a primary oocyte finish meiosis 1 and enter meiosis 2 as a **secondary oocyte**—but then it arrests again, not finishing meiosis 2 until a sperm penetrates its membrane. Only after the completion of meiosis 2 is the egg technically called an **ovum.**

Figure 23.15 not only summarizes the events of oogenesis but also relates these events to the development of the follicles in the ovary. As you study this figure, notice that since meiosis is a *pair* of cell divisions, it necessarily produces four daughter cells: One of these cells is the large ovum, and the other three are small cells called **polar bodies.** The polar bodies degenerate quickly, without being fertilized or contributing to the developing embryo.

The Uterine Tubes

The **uterine** (u'ter-in) **tubes,** also called **fallopian tubes** and **oviducts,** form the first part of the female duct system (Figure 23.11a). Functionally, they receive the ovulated oocyte and provide a site for fertilization. Each uterine tube begins laterally near an ovary and ends medially where it empties into the superior part of the uterus. The uterine tube has several named regions. Its lateral region is an open funnel called the **infundibulum** (in"fun-dib'u-lum; "funnel"). The margin of the infundibulum is surrounded by ciliated, finger-like projections called **fimbriae** (fim'bre-e; "fibers, a fringe") that drape over the ovary. Medial to the infundibulum is the expanded **ampulla** ("flask"). This region, which forms half the length of the uterine tube, is the site where fertilization usually occurs. Finally, the medial third of the uterine tube is the **isthmus** (is'mus; "a narrow passage").

Unlike the reproductive ducts of males, which are directly continuous with the tubules of the testes, the uterine tubes have little or no direct contact with the ovaries. An ovulated oocyte is cast into the peritoneal cavity of the pelvis, where many oocytes are lost. However, the uterine tube performs a complex series of movements to capture the oocyte. More specifically, the infundibulum bends to cover the ovary while the fimbriae stiffen and sweep the ovarian surface. The beating cilia on the fimbriae then produce currents in the peritoneal fluid that carry the oocyte into the uterine tube, where it begins its journey toward the uterus.

This journey is aided by the structural features of the uterine tube. Sheets of smooth muscle in the wall generate waves of peristalsis. Furthermore, the highly folded mucosa of the uterine tube has a simple columnar epithelium that contains ciliated cells. Together, the muscular peristalsis and the beating of the cilia propel the oocyte toward the uterus. Alternating with ciliated cells in the lining epithelium are tall, nonciliated cells that secrete substances that nourish the oocyte.

Externally, the uterine tube is covered by visceral peritoneum and is supported by a short mesentery called the **mesosalpinx** (mez"o-sal'pinks), a part of the broad ligament (Figure 23.11a). *Mesosalpinx* literally means "mesentery of the trumpet," a reference to the trumpet shape of the uterine tube. It helps to associate the uterine tube with the term *salpinx* because most clinical terms refer to it as such. For example, a **salpingectomy** is surgical removal of a uterine tube, **salpingitis** is an inflammation of the uterine tube, and **pyosalpinx** is pus *(pyon)* filling the uterine tube.

✚ Infections often spread from the vagina and uterus to the uterine tubes and even to the ovaries and pelvic peritoneum. The microorganisms of gonorrhea and chlamydia (common sexually transmitted diseases) can follow this route, causing a severe inflammation. Such widespread infection of the internal female organs is called **pelvic inflammatory disease (PID).**

PID can lead to sterility. Can you explain precisely why?

PID can scar and block the uterine tubes, so sperm cannot reach the ova.

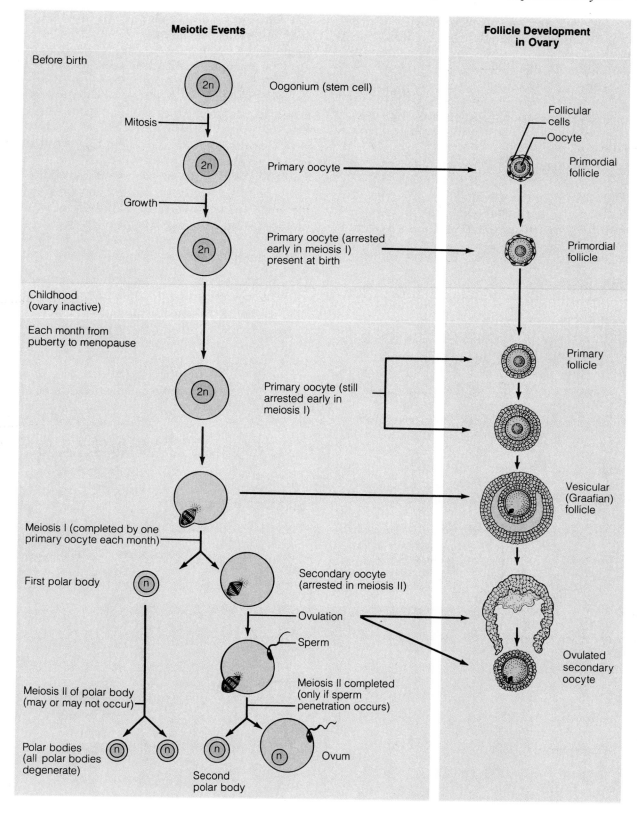

Figure 23.15 Oogenesis. Left, flowchart of the events of meiosis. Right, correlation with follicle development and ovulation in the ovary.

You may know someone who had her uterine tubes "tied" surgically as a form of birth control. Such **tubal ligation** usually involves threading a cutting instrument and a viewing tube (endoscope) through a small incision in the anterior abdominal wall and then tying and/or cutting the uterine tubes. Tubal ligation usually is irreversible, although in some cases delicate microsurgery can successfully reopen or reattach the uterine tubes.

The uterine tube is the most common site of *ectopic pregnancies.* As explained in Chapter 3 (p. 62), ectopic pregnancy is the implantation of an embryo outside the uterus. Implantations in the uterine tube often result in hemorrhaging that threatens a woman's life. ■

The Uterus

Gross Anatomy

The **uterus** (womb) lies in the pelvic cavity, anterior to the rectum and posterosuperior to the bladder (Figure 23.11b). It is a hollow, thick-walled organ whose functions are to receive, retain, and nourish a fertilized egg throughout pregnancy. In a woman who has never been pregnant, the uterus is about the size and shape of a small, inverted pear, but it is somewhat larger in women who have had children. Normally, the uterus is tilted anteriorly, or **anteverted,** at the superior part of the vagina. However, the uterus is frequently inclined posteriorly, or **retroverted,** in older women.

The major portion of the uterus is called the **body** (Figures 23.11a and 23.16). The rounded region superior to the entrance of the uterine tubes is the **fundus,** and the slightly narrowed region between the body and the cervix is the **isthmus.** The narrow neck of the uterus is the **cervix,** the inferior tip of which projects into the vagina. The hollow central lumen of the uterus is quite small (except during pregnancy). It is divided into the *cavity of the body* and the **cervical canal.** The cervical canal communicates with the vagina inferiorly via the **external os** (*os* = mouth) and with the cavity of the body superiorly via the **internal os.** The mucosal lining of the cervical canal contains *cervical glands* that secrete a mucus that fills the cervical canal and covers the external os, presumably to block the spread of bacteria from the vagina into the uterus. Cervical mucus also blocks the entry of sperm, except at mid cycle, when it becomes less viscous and allows the sperm to pass through.

Cancer of the cervix is common among women between the ages of 30 and 50. Risk factors include frequent cervical inflammations, sexually transmitted diseases, multiple pregnancies, and—according to recent evidence—sexual intercourse with men who have penile warts. The cancer cells arise from the epithelium covering the cervical tip. In a Papanicolaou (Pap) smear, or cervical smear test, some of these epithelial cells are scraped away and then examined for abnormalities. A Pap smear is the most effective way to detect this slow-growing cancer, and women are advised to have one every 1 to 3 years. ■

Supports of the Uterus

Several ligaments and mesenteries help hold the uterus in place (Figure 23.11a). The uterus is anchored to the lateral pelvic walls by its **mesometrium**

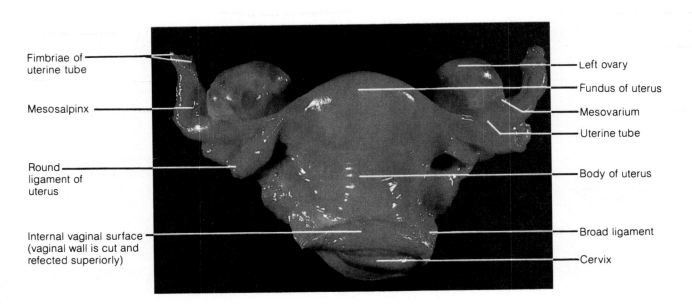

Fimbriae of
uterine tube

Mesosalpinx

Round
ligament of
uterus

Internal vaginal surface
(vaginal wall is cut and
refected superiorly)

Left ovary

Fundus of uterus

Mesovarium

Uterine tube

Body of uterus

Broad ligament

Cervix

Figure 23.16 Anterior view of the internal reproductive organs from a female cadaver. Compare this to the posterior view of the same organs in Figure 23.11a.

Figure 23.17   The female pelvic organs in anterior view.   Note the uterus and the vesicouterine and rectouterine pouches of the peritoneal cavity. The rectouterine pouch is the most inferior point of the peritoneal cavity. (*Males* have a pouch equivalent to the rectouterine pouch, but since males have no uterus, their corresponding pouch is called the rectovesical pouch [*vesical* = bladder].)

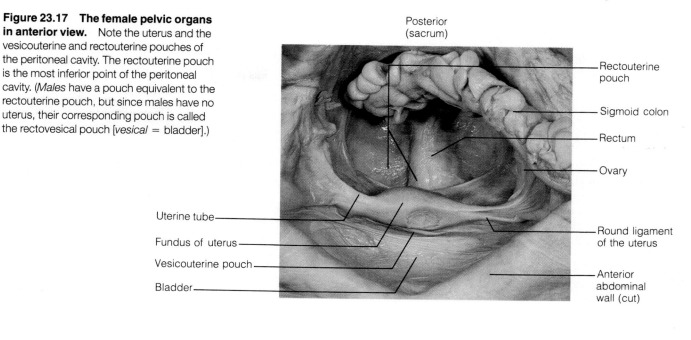

(mez″o-me′tre-um; "mesentery of the uterus"), which is the largest division of the broad ligament. Inferiorly, the **lateral cervical (cardinal) ligaments** run horizontally from the uterine cervix and superior vagina to the lateral pelvic walls. These "ligaments" are thickenings of the fascia of the pelvis. The uterus is bound to the anterior body wall by the paired **round ligaments of the uterus**. Each round ligament starts as a continuation of the ovarian ligament on the superolateral aspect of the uterus, descends through the mesometrium and inguinal canal, and anchors in one of the labia majora of the external genitalia.

Despite the above ligaments, most support of the uterus is provided by the muscles of the pelvic floor, namely the muscles of the urogenital and pelvic diaphragms (Table 10.7, pp. 260–261). These muscles are sometimes torn during childbirth. Subsequently, the unsupported uterus may sink inferiorly, until the tip of the cervix protrudes through the external vaginal opening. This condition is called **prolapse of the uterus.** ■

The undulating course of the peritoneum around and over the various pelvic organs produces several cul-de-sacs, or blind-ended peritoneal pouches (Figure 23.17). For example, the *vesicouterine* (ves″ĭ-ko-u′ter-in) *pouch* lies between the uterus and bladder (*vesico* = bladder), and the *rectouterine pouch* lies between the uterus and rectum. The rectouterine pouch forms the most inferior part of the peritoneal cavity. Therefore, pus from intraperitoneal infections and blood from internal wounds accumulate in this pouch.

The Uterine Wall

The wall of the uterus is composed of three basic layers (Figure 23.11a): an outer *perimetrium* (per″ĭ-me′tre-um; "around the uterus"), a middle *myometrium* (mi″o-me′tre-um; "muscle of the uterus"), and an inner *endometrium* ("within the uterus"). The **perimetrium,** the outer serous membrane, is the visceral peritoneum. The **myometrium,** the bulky middle layer, consists of interlacing bundles of smooth muscle. The myometrium contracts during childbirth to expel the baby from the mother's body. The **endometrium** is the mucosal lining of the uterine cavity (Figure 23.18). It consists of a simple columnar epithelium, with secretory and ciliated cells, underlain by a connective tissue stroma (lamina propria). If fertilization occurs, the embryo burrows into the endometrium and resides there for the rest of its development. The endometrium has two chief divisions: *stratum functionalis* and *stratum basalis.* The thick, inner **stratum functionalis,** or **functional layer,** undergoes cyclic changes in response to varying levels of ovarian hormones in the blood and is shed during menstruation (about every 28 days). The thin **stratum basalis,** or **basal layer,** does not shed and is responsible for forming a new functional layer after menstruation ends. The endometrium contains straight tubular **uterine glands** that change in length as the endometrium thickens and thins.

To understand the changes in the endometrium during the menstrual cycle, it is essential to understand the blood supply of the uterus (Figures 23.12 and 23.18b). The **uterine arteries** arise from the inter-

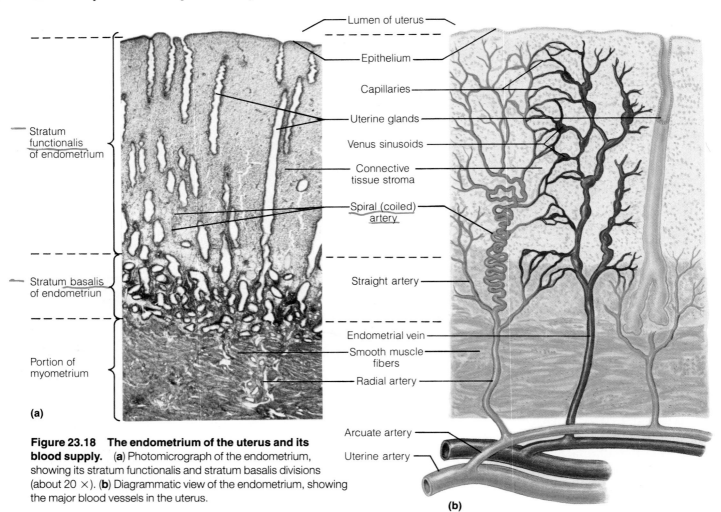

(a)

Figure 23.18 The endometrium of the uterus and its blood supply. **(a)** Photomicrograph of the endometrium, showing its stratum functionalis and stratum basalis divisions (about 20 ×). **(b)** Diagrammatic view of the endometrium, showing the major blood vessels in the uterus.

(b)

nal iliac arteries in the pelvis, ascend along the sides of the uterine body, and send branches into the uterine wall. These divide into **arcuate arteries** that course through the myometrium. **Radial arteries** then reach the endometrium, where they give off **straight arteries** *(basal arteries)* to the stratum basalis and **spiral (coiled) arteries** to stratum functionalis. The spiral arteries undergo repeated degeneration and regeneration during successive menstrual cycles, and they undergo spasms that cause the functionalis layer to shed during menstruation. Veins in the endometrium are thin-walled and form an extensive network with occasional sinusoidal enlargements.

The Uterine Cycle

The **uterine cycle** is the menstrual cycle as it involves the endometrium. Specifically, it is a series of cyclic phases that the endometrium experiences month after month as it responds to changing levels of ovarian hormones in the blood. These endometrial phases are closely coordinated with the phases of the ovarian cycle, which in turn are dictated by the gonadotropin

hormones from the pituitary gland. The three phases of the uterine cycle are the (1) **menstrual phase** (days 1–5), in which the stratum functionalis is shed; (2) **proliferative phase** (days 6–14), in which the functionalis rebuilds; and (3) **secretory phase** (days 15–28), in which the endometrium prepares for implantation of an embryo. Figure 23.19 describes these phases in detail and explains their relation to the ovarian cycle.

The Vagina

The **vagina** ("sheath") is a thin-walled tube that lies inferior to the uterus, anterior to the rectum, and posterior to the urethra and bladder (Figure 23.11b). Its external opening is the **vaginal orifice.** The vagina is often called the *birth canal,* as it provides a passageway for delivery of an infant and for menstrual flow. It also receives the penis (and semen) during sexual intercourse.

The highly distensible wall of the vagina consists of three coats: an outer *adventitia* of fibrous connec-

Figure 23.19   The female menstrual cycle: Structural and hormonal changes. The time bar at the bottom of the figure, reading Days 1 to 28, applies to all four parts of this figure. (**a**) The fluctuating levels of pituitary gonadotropins in the blood (FSH = follicle-stimulating hormone; LH = luteinizing hormone). These hormones regulate the events of the ovarian cycle. (**b**) The fluctuating blood levels of ovarian hormones (estrogen and progesterone) that signal the endometrial changes of the uterine cycle. (**c**) Structural changes in the ovarian follicles during the ovarian cycle. (**d**) Changes in the endometrium of the uterus during the uterine cycle.

The phases of the ovarian cycle, depicted in (c), are as follows:

1. Days 1–14: Follicular phase. The ovarian follicles grow under the influence of FSH and LH from the pituitary gland; see (a). The follicles secrete increasing amounts of estrogen into the blood; see (b).

2. Day 14: Ovulatory phase. Ovulation is signaled by a surge of LH from the pituitary gland, and then the corpus luteum forms.

3. Days 14–28: Luteal phase. The corpus luteum secretes progesterone and estrogen into the blood. High levels of progesterone inhibit the secretion of LH and FSH by the pituitary; see (a). Without LH to sustain it, the corpus luteum dies at the end of the luteal phase, and a new ovarian cycle begins (unless an embryo implants in the uterus).

The phases of the uterine cycle, depicted in (d), are as follows:

1. Days 1–5: Menstrual phase. In this phase, the uterus sheds all but the deepest part of its endometrium: The thick functional layer of the endometrium detaches from the uterine wall, a process that is accompanied by bleeding for 3 to 5 days. The detached tissue and blood pass out through the vagina as the menstrual flow. By day 5, the growing ovarian follicles are starting to produce more estrogen; see (b).

2. Days 6–14: Proliferative phase. In this phase, the endometrium rebuilds itself: Under the influence of rising blood levels of estrogen, the basal layer of the endometrium generates a new functional layer. This new layer thickens, its glands enlarge, and its spiral arteries increase in number (also see Figure 23.18). Consequently, the endometrium once again becomes velvety, thick, and well vascularized. All the changes in the endometrium in the proliferative phase are signaled by estrogen. Ovulation occurs in the ovary at the end of this phase (day 14), forming the corpus luteum; see (c).

3. Days 15–28: Secretory phase. In this phase, the endometrium prepares for implantation of an embryo. Rising levels of progesterone from the corpus luteum act on the estrogen-primed endometrium, causing the spiral arteries to elaborate and coil more tightly and converting the functional layer to a secretory mucosa. The uterine glands enlarge, coil, and begin secreting nutritious glycoproteins into the uterine cavity. These nutrients sustain the embryo until it has implanted in the blood-rich endometrial lining. All the events of the secretory phase are signaled by progesterone.

If fertilization and implantation occur, the endometrium remains in this secretory phase throughout pregnancy. If fertilization does not occur, however, the degenerating corpus luteum stops secreting progesterone. Blood progesterone levels fall, depriving the endometrium of hormonal support, and the spiral arteries kink and constrict spasmodically. Denied oxygen and nutrients, the endometrial cells begin to die from ischemia, and as their lysosomes rupture, the functional layer begins to "self-digest," setting the stage for menstruation to begin on day 28. The spiral arteries constrict one final time and then suddenly open wide. As blood gushes into the weakened capillary beds, they fragment, causing the functional layer to slough off. The menstrual cycle starts over again on this first day of menstrual flow.

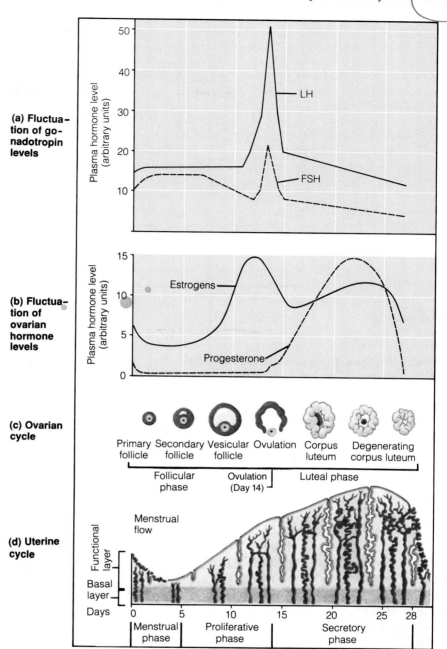

(**a**) Fluctuation of gonadotropin levels

(**b**) Fluctuation of ovarian hormone levels

(**c**) Ovarian cycle

Primary follicle · Secondary follicle · Vesicular follicle · Ovulation · Corpus luteum · Degenerating corpus luteum

Follicular phase · Ovulation (Day 14) · Luteal phase

(**d**) Uterine cycle

Menstrual phase · Proliferative phase · Secretory phase

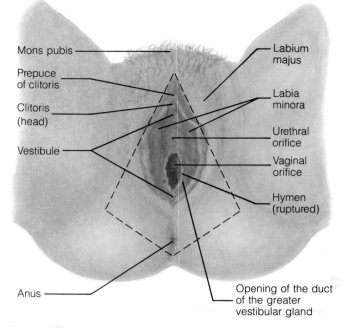

Mons pubis

Prepuce of clitoris

Clitoris (head)

Vestibule

Anus

Labium majus

Labia minora

Urethral orifice

Vaginal orifice

Hymen (ruptured)

Opening of the duct of the greater vestibular gland

Figure 23.20   The external genitalia (vulva) of the female. The diamond-shaped region enclosed by the dotted lines is the perineum.

tive tissue, a *muscularis* of smooth muscle, and an inner *mucosa.* The mucosa is marked by transverse folds (rugae) (Figure 23.11a). These ridges stimulate the penis during intercourse, and they flatten as the vagina expands during childbirth. The mucosa consists of a lamina propria and an inner epithelium. The lamina propria contains elastic fibers that help to return the vagina to its original shape after it is expanded. Internally, the stratified squamous epithelium of the vagina is a protective layer that can withstand the rubbing friction of intercourse and bacterial attack. The vaginal lumen is quite acidic (pH 3.5 to 4). This acidity helps keep the vagina healthy and free of infection, but it is also hostile to sperm.

Near the vaginal orifice, the mucosa elaborates to form an incomplete diaphragm called the **hymen** (hi'men; "membrane") (Figure 23.20). The hymen is very vascular and tends to bleed when ruptured during the first sexual intercourse. However, its durability varies: In some females, it ruptures during a sports activity, during insertion of a tampon, or in a pelvic examination. Occasionally, the hymen is so tough that it must be breached surgically if intercourse is to occur.

The superior end of the vagina widens to surround the tip of the cervix (Figure 23.11a). The recess that encircles the cervical tip is called the vaginal **fornix.** The posterior part of this recess, the **posterior fornix,** is much deeper than the lateral and anterior fornices (Figure 23.11b).

The posterior fornix of the vagina lies very near the peritoneal cavity (rectouterine pouch) (Figure 23.11b). Therefore, attempts by untrained persons to enter the uterus with a sharp instrument may puncture the vaginal wall at the posterior fornix and cause hemorrhaging and subsequent peritonitis. This often happened in the past during unskilled attempts to induce abortion ("scrape the uterus") with a coat hanger, wire, or other unsterilized sharp object. The legalization of abortions has all but eliminated this type of problem.

Recall that any excess fluid in the peritoneal cavity (blood from an internal injury or pus from an infection, for example) accumulates in the rectouterine pouch. By looking at Figure 23.11b, can you deduce how physicians can remove such fluid in a way that avoids the trauma of surgically opening the abdominal wall?

To drain excess fluid from the peritoneal cavity, the physician inserts a syringe needle superiorly through the vagina and posterior fornix into the rectouterine pouch. ■

The External Genitalia and Female Perineum

The female reproductive structures that lie external to the vagina are the **external genitalia, vulva** ("covering"), or **pudendum** ("shameful") (Figure 23.20). These structures include the mons pubis, the labia, the clitoris, and structures associated with the vestibule.

The **mons pubis** (mons pu'bis; "mountain on the pubis") is a fatty, rounded pad overlying the pubic symphysis. After puberty, pubic hair covers the skin of this area. Extending posteriorly from the mons are two elongated, hair-covered, fatty skin folds, the **labia majora** (la'be-ah mah-jor'ah; "larger lips"). These labia are the female counterpart, or *homologue,* of the male scrotum (that is, they derive from the same embryonic structure). The labia majora enclose the **labia minora** (mi-nor'ah; "smaller lips"), two thin, hairless skin folds. The labia minora enclose a recess called the **vestibule** ("entrance hall"), which contains the external openings of the urethra and vagina. In the vestibule, the vaginal orifice lies posterior to the urethral orifice. Flanking the vaginal orifice are the paired *greater vestibular glands* (not illustrated), which lie deep to the posterior part of the labia on each side. These pea-sized glands secrete lubricating mucus into the vaginal orifice on sexual arousal, facilitating entry of the penis. At the extreme posterior point of the vestibule, the right and left labia minora come together to form a ridge called the *frenulum of the labia* ("little bridle of the lips") or the *fourchette* (foor-shet').

Just anterior to the vestibule is the **clitoris** (klĭ′to-ris, or klit′o-ris; "hill"), a small protruding structure composed largely of erectile tissue. It is sensitive to touch and swells with blood during sexual stimulation. The clitoris is homologous to the penis, having both a glans and a body (although there is no urethra within it). It is hooded by a fold of skin, the **prepuce of the clitoris,** formed by the anterior junction of the two labia minora. Like the penis, the clitoris contains paired *corpora cavernosa,* which continue posteriorly into *crura* that extend along the bony pubic arch. Unlike the penis, the clitoris contains no corpus spongiosum. The homologous *bulbs of the vestibule,* however, lie along each side of the vaginal orifice. During sexual stimulation, these vestibular bulbs engorge with blood and may help grip the penis within the vagina.

The female **perineum** is a diamond-shaped region between the pubic arch anteriorly, the coccyx posteriorly, and the ischial tuberosities laterally (Figure 23.20). In the exact center of the perineum, just posterior to the fourchette, is the *central tendon (perineal body)* (see Figure 10.11b on p. 261). This knob is the insertion tendon of most muscles that support the pelvic floor (Table 10.7).

Because of its location directly posterior to the vagina, the central tendon is sometimes torn by an infant's head during childbirth. Once torn, the central tendon heals poorly, allowing the pelvic organs to sink inferiorly and the uterus to prolapse. To prevent such problems, an **episiotomy** (e-piz″e-ot′o-me; "cutting of the vulva") is performed in over 50% of deliveries in the United States. In this procedure, the vaginal orifice is widened by a posterior cut through the fourchette at the time the baby's head appears at the vestibule. The posterior incision either passes straight through the central tendon or just lateral to it. After the birth, this clean incision is stitched together and generally heals without difficulty. ■

The Mammary Glands

Mammary glands (breasts) are present in both sexes but function only in females (Figure 23.21), in whom they produce milk to nourish an infant.

Embryonically, the mammary glands form as part of the skin. They are modified sweat glands that arise along a line that runs between the axilla and groin on the lateral trunk of the embryo. This embryonic *milk line* persists only in the mid thorax, where the definitive breasts form. About 1 in 500 people develops extra (accessory) nipples or breasts, which can occur anywhere along the milk line.

The base of the cone-shaped female breast extends from the second rib superiorly to the sixth rib inferiorly. Its medial border is the sternum, and its lat-

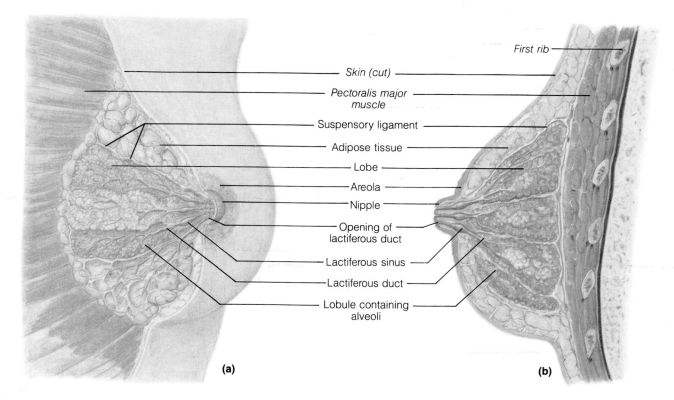

First rib

Skin (cut)

Pectoralis major muscle

Suspensory ligament

Adipose tissue

Lobe

Areola

Nipple

Opening of lactiferous duct

Lactiferous sinus

Lactiferous duct

Lobule containing alveoli

(a)   (b)

Figure 23.21 Structure of a lactating (milk-secreting) mammary gland.
(a) Anterior view of a partly dissected breast. (b) Sagittal section of a breast.

eral boundary is the midaxillary line (a line dropped straight inferiorly from the middle of the axilla). The muscles deep to the breast are the pectoralis major and minor and parts of the serratus anterior and external oblique. The arteries to the breast are the lateral thoracic artery and cutaneous branches of the internal thoracic and posterior intercostals (Chapter 18, p. 495). The lymph vessels draining the breast, which can spread cancerous cells throughout the body during breast cancer, drain into *parasternal lymph nodes* (nodes along the internal thoracic arteries) and, most importantly, into the *axillary lymph nodes* in the axilla.

The **nipple,** the central protruding area from which an infant sucks milk, is surrounded by a ring of pigmented skin, the **areola** (ah-re′o-lah; "a small area"). Large sebaceous glands in the areola produce an oily sebum during nursing that reduces chapping and cracking of the skin of the nipple.

Internally, the mammary gland consists of 15 to 25 **lobes** (Figure 23.21), each of which is a distinct compound alveolar gland opening at the nipple. The lobes are separated from one another by fibrous connective tissue and a large amount of adipose tissue. The interlobar connective tissue forms **suspensory ligaments** that extend from the underlying skeletal muscles to the overlying dermis. As their name implies, the suspensory ligaments provide natural support for the breasts. In advanced stages of breast cancer, these ligaments contract and dimple the skin like an orange peel. The lobes of the breast consist of smaller units called **lobules.** The lobules in turn are clusters of tiny acini or alveoli (a bunch of grapes is a good analogy for a lobule). The alveoli are lined by a simple cuboidal epithelium of milk-secreting cells. From the alveoli, milk passes through progressively larger ducts until it reaches the **lactiferous ducts,** which lie within and deep to the nipple (*lactiferous* = milk-carrying). Just deep to the areola, each lactiferous duct has a dilated region called a **lactiferous sinus.** Milk accumulates in the lactiferous sinuses during nursing.

The glandular structure of the breast is undeveloped in nonpregnant women. During childhood, a girl's breasts consist only of rudimentary ducts (as do the *male* breasts throughout life). When a girl reaches puberty, these ducts grow and branch but still contain no lobules or acini (breast enlargement at puberty is due to fat deposition). The glandular acini finally appear about halfway through pregnancy, and milk production starts a few days after childbirth.

✚ After lung cancer, breast cancer is the leading cause of death in American women: One in ten will develop this condition, and 1 in 20 will die from it. Breast cancer usually arises from the epithelial cells of the ducts, not from the acini. A small cluster of cancer cells grows to form a lump in the breast, from which cells eventually metastasize.

✚ Can you imagine how most breast cancers are detected?

Over 90% of all breast cancers are discovered by women during monthly self-examinations in which the breasts are felt for lumps or irregularities. Therefore, this simple examination should be a health maintenance priority in every woman's life. ■

Breast cancer is treated by radiation and chemotherapy and usually by **mastectomy** (mas-tek′to-me; "breast cutting"), the removal of all or some of the breast. Until the 1970s, the standard treatment was **radical mastectomy,** removal of the entire affected breast plus all underlying muscles, fascia, and associated lymph nodes. Medical records indicate that this painful and disfiguring treatment is no more effective at halting the cancer than is less extensive surgery. Therefore, most physicians now recommend **lumpectomy,** in which only the cancerous part (lump) of the breast is excised, or **simple mastectomy,** in which the breast alone is removed, with perhaps some of the axillary lymph nodes.

✚ Do you think *males* can develop breast cancer?

Since male breasts also contain lactiferous ducts, males can indeed develop breast cancer. Male breast cancer is extremely rare but usually fatal. ■

Pregnancy and Childbirth

In this section, we consider the events that occur in the female reproductive tract during pregnancy. We focus on the mother rather than on her embryo and fetus (embryology is discussed in Chapter 3).

Pregnancy

Fertilization

Before fertilization can occur, sperm deposited in the vagina must swim through the uterus to reach the ovulated oocyte in the uterine tube. Once a sperm approaches the immediate vicinity of the oocyte, it undergoes the **acrosomal reaction,** the release of digestive enzymes from its acrosome (Figure 23.22). Recall that the oocyte is surrounded by a corona radiata of granulosa cells and, deep to that, by the zona pellucida, both of which must be breached before the oocyte itself can be penetrated. Hundreds of acrosomes must rupture to break down the "cement" that holds the granulosa cells together and to digest holes

GOTO P 50
CHP 3

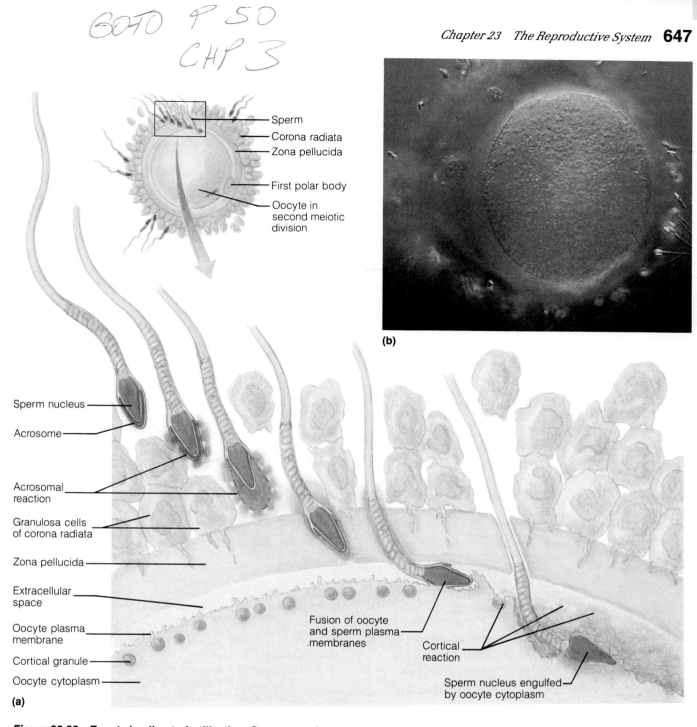

Sperm
Corona radiata
Zona pellucida

First polar body
Oocyte in second meiotic division

(b)

Sperm nucleus
Acrosome
Acrosomal reaction
Granulosa cells of corona radiata
Zona pellucida
Extracellular space
Oocyte plasma membrane
Cortical granule
Oocyte cytoplasm

(a)

Fusion of oocyte and sperm plasma membranes

Cortical reaction

Sperm nucleus engulfed by oocyte cytoplasm

Figure 23.22 Events leading to fertilization: Sperm entering an oocyte and the cortical reaction. (a) The steps by which a sperm penetrates an oocyte are depicted from left to right. Penetration of the corona radiata and zona pellucida around the oocyte is accomplished by the release of acrosomal enzymes from many different sperm. Fusion of the plasma membrane of a single sperm with the oocyte membrane induces the cortical reaction. In this reaction, cortical granules in the oocyte release their contents into the extracellular space, preventing entry of more than one sperm. (b) Photomicrograph of live human sperm surrounding an oocyte (750 ×).

through the zona pellucida. But once a path has been cleared and a single sperm makes contact with the oocyte membrane, its nucleus is pulled into the oocyte cytoplasm. The *tail* of the sperm, however, does not enter the oocyte.

As soon as one sperm succeeds in attaching to the oocyte membrane, this contact triggers a series of responses that prohibit other sperm from entering the oocyte. Most dramatically, it induces the **cortical reaction** (Figure 23.22a), wherein granules in the oocyte

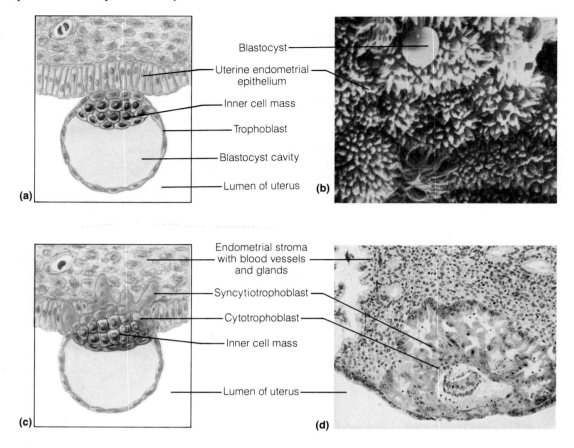

Figure 23.23   Implantation of the blastocyst.   (**a**) Diagram of a blastocyst that has just adhered to the uterine endometrium (day 6). (**b**) Scanning electron micrograph of a blastocyst as it begins to implant in the endometrium. (**c**) Implanting embryo at a slightly later stage (day 7), showing the cytotrophoblast and syncytiotrophoblast. (**d**) Light micrograph of an implanting embryo (day 10).

cytoplasm spill their contents into the extracellular space beneath the zona pellucida. This spilled material binds water, and as it swells, it detaches all sperm still in contact with the oocyte membrane.

<u>Fertilization</u> occurs at the moment the male and female <u>chromosomes combine within the ovum</u>. The fertilized egg (zygote) then initiates cleavage (cell division). By the fourth day after fertilization, the embryo is in the multicellular *blastocyst* stage and has entered the uterus (see p. 51).

Implantation

When the blastocyst reaches the uterus, it floats freely in the uterine cavity for about 2 days, receiving nourishment from uterine secretions. Then, about 6 days after fertilization, the blastocyst begins **implantation,** the act of burrowing into the endometrium (Figure 23.23a). At this time, the blastocyst consists of an *inner cell mass* (the future embryo) and an outer **trophoblast.** *Trophoblast* means "nourishment generator": This layer will soon provide the embryo with nourishment from the mother's uterus. In the first step of implantation, trophoblast cells proliferate and form

two distinct layers (Figure 23.23c). The inner layer is the **cytotrophoblast** (si"to-trof'o-blast; "cellular part of the trophoblast"), and its cells retain their cell boundaries. In the outer layer, by contrast, the cells lose their plasma membranes and fuse into a multinuclear mass of cytoplasm called the **syncytiotrophoblast** (sin-sit"e-o-trof'o-blast) (*syncytio* = fused cells), which projects invasively into the endometrium and digests the uterine cells it contacts. As the endometrium is eroded, the blastocyst becomes deeply embedded within this thick, velvety lining (Figure 23.24a). Cleft-like spaces called **lacunae** then open within the syncytiotrophoblast and quickly fill with maternal blood leaked from degraded endometrial blood vessels (Figure 23.24b and c). Embryo-derived tissues have now made their first contact with maternal blood, their ultimate source of nourishment.

Formation of the Placenta

Events now proceed toward the formation of the placenta, the organ that nourishes the developing fetus. Both embryonic (trophoblastic) and maternal (endometrial) tissues contribute to this organ. First, the proliferating trophoblast gives rise to a layer of **extraembryonic mesoderm** on its internal surface (Figure 23.24b and c). The trophoblast and extraembryonic mesoderm together are now called the **chorion** (ko're-on; "membrane"). The chorion then folds into finger-like **chorionic villi,** which contact the lacunae containing maternal blood (Figure 23.24c). At this same stage, the embryo's body connects to the chorion outside of it through a **body stalk** *(connecting stalk)* made of extraembryonic mesoderm. This stalk forms the core of the future umbilical cord as the umbilical arteries and umbilical vein grow into it from the embryo's body. Next, the mesodermal cores of the chorionic villi develop capillaries that connect to the umbilical blood vessels growing from the embryo (Figure 23.24d). With the establishment of this vascular connection, the embryo's own blood now courses within the chorionic villi, very close to (but not mixing with) the maternal blood just external to these villi. Nutrients, wastes, and other substances begin exchanging across the villi, between the two separate bloodstreams. This exchange is occurring by the end of the first month of pregnancy.

Formation of the placenta is completed during months 2 and 3 (Figure 23.24d and e). The chorionic villi that lie nearest the umbilical cord grow in complexity, whereas the villi around the rest of the embryo regress and ultimately disappear. The part of the mother's endometrium adjacent to the complex villi and umbilical cord is called the **decidua basalis** (de-sid'u-ah), whereas the endometrium on the luminal side of the implanted embryo is the **decidua capsularis**. The decidua capsularis expands to accommodate the growing fetus, which completely fills the uterine lumen by the start of month 4. At this time, the decidua basalis and chorionic villi together make up a thick, pancake-shaped disc at the end of the umbilical cord—the **placenta** ("cake") (Figure 23.24e). The placenta continues to nourish the fetus for 6 more months, until birth. The placenta detaches and sloughs off after birth, so the name of its maternal portion—*decidua* ("that which falls off")—is very appropriate.

It is worth reiterating that the placenta is not called the placenta until it achieves its pancake shape at the start of month 4 (week 13), even though its chorionic villi are established and functioning long before this time (Figure 23.25).

Anatomy of the Placenta

The exchanges that occur across the chorionic villi between maternal and fetal blood are absolutely essential to prenatal life. These exchanges provide the fetus with nutrients and oxygen, dispose of its wastes, and allow it to send hormonal signals to the mother. To travel between maternal and fetal blood, substances must pass through the *placental barrier.* This

consists of all three layers of each chorionic villus (syncytiotrophoblast, cytotrophoblast, extraembryonic mesoderm) plus the endothelium of the fetal capillaries in the villi (Figure 23.24f). The cytotrophoblast and mesoderm layers are soon lost from the villi, thereby thinning the barrier and allowing a very efficient exchange of molecules.

Many kinds of molecules pass across the chorionic villi of the placenta, by a number of mechanisms. Sugars, fats, and oxygen diffuse from mother to fetus, whereas urea and carbon dioxide diffuse from fetus to mother. Maternal antibodies are actively transported across the placenta and give the fetus some resistance to disease. The placenta is fairly effective at blocking the passage of bacteria from mother to fetus, but many viruses pass through (poliovirus, German measles virus, AIDS virus), as do many drugs and toxins (alcohol, mercury, heroin). Because the placenta is not an effective barrier to drugs, the mother must minimize her intake of harmful substances during pregnancy to minimize the chances of birth defects (Chapter 3).

Finally, many *hormones* cross the placenta from fetus to mother. The syncytiotrophoblast is the hormone-manufacturing layer of the placenta, and it secretes its hormones directly into the maternal blood. Placental hormones include estrogen, progesterone, and **human chorionic gonadotropin (HCG)**. The major function of these hormones is to maintain the uterus in its pregnant state and prevent spontaneous abortion. HCG appears in maternal blood shortly after the syncytiotrophoblast forms, a mere 1 week after fertilization. The presence of this hormone in the urine of a pregnant woman is the basis for most home pregnancy tests.

Anatomical Changes in the Mother During Pregnancy

As pregnancy progresses, the female reproductive organs become increasingly vascularized and engorged with blood, and the vagina develops a purple hue (Chadwick's sign). This enhanced vascularity causes a marked increase in vaginal sensitivity and sexual intensity. The breasts, too, become engorged with blood, and—prodded by rising levels of estrogen and progesterone—they enlarge and their areolae darken. Some women develop a yellowish brown pigmentation of the skin on the nose and cheeks, a condition called **chloasma** (klo-az'mah; "to be green") or the "mask of pregnancy."

The degree of uterine enlargement during pregnancy is nothing less than remarkable (Figure 23.26). Overall, the mass of the uterus increases by a factor of 24, primarily because of growth and division of the myometrial muscle cells. Initially smaller than a fist, the uterus fills most of the pelvic cavity by 16 weeks

(text continues on p. 652)

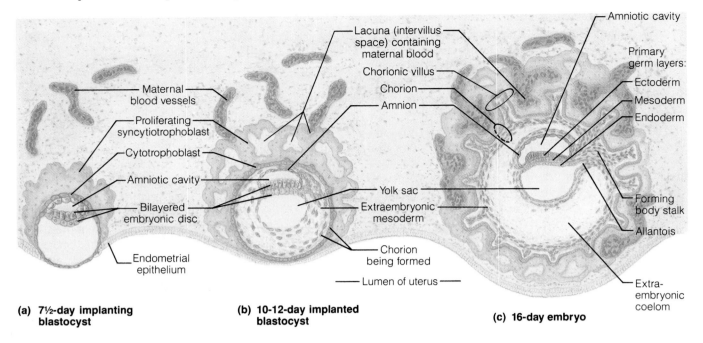

(a) 7½-day implanting blastocyst

(b) 10-12-day implanted blastocyst

(c) 16-day embryo

Figure 23.24 Stages of placenta formation. (a) Implanting blastocyst at 7 1/2 days after fertilization. The syncytiotrophoblast is eroding into the endometrium. (b) Implantation is completed by day 12, and extraembryonic mesoderm appears deep to the cytotrophoblast. Spaces called lacunae appear in the syncytiotrophoblast and will soon fill with maternal blood. (c) By 16 days, the chorionic villi are elaborating, and the body stalk is present (future umbilical cord). (d) At 4 1/2 weeks, the embryo's umbilical vessels are in place, and exchanges between maternal and embryonic blood are occurring across the chorionic villi. (e) At 13 weeks, the placenta is fully formed from the chorionic villi and the endometrium (decidua basalis) as a "pancake" around the end of the umbilical cord. (f) Detailed anatomy of the vascular relationships in the placenta, centering around the chorionic villi. Nutrients and other substances pass across the walls of the villi from maternal blood (in lacunae) to fetal blood (in capillaries in the villi).

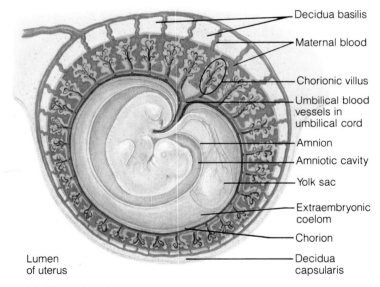

(d) 4½-week embryo

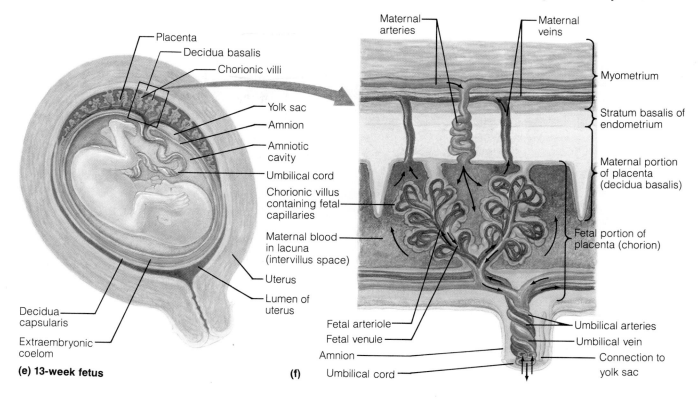

Maternal arteries

Maternal veins

Placenta

Decidua basalis

Chorionic villi

Yolk sac

Amnion

Amniotic cavity

Umbilical cord

Chorionic villus containing fetal capillaries

Maternal blood in lacuna (intervillus space)

Uterus

Lumen of uterus

Decidua capsularis

Extraembryonic coelom

(e) 13-week fetus

Myometrium

Stratum basalis of endometrium

Maternal portion of placenta (decidua basalis)

Fetal portion of placenta (chorion)

Fetal arteriole

Fetal venule

Amnion

(f)   Umbilical cord

Umbilical arteries

Umbilical vein

Connection to yolk sac

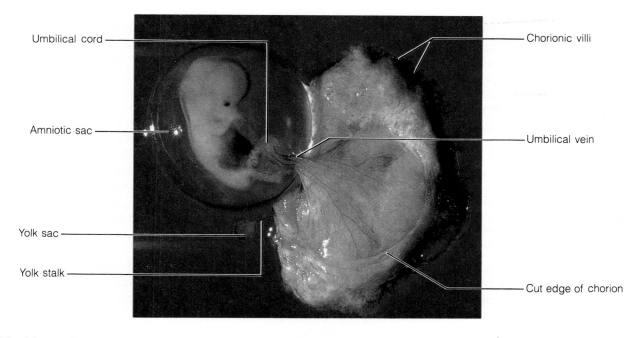

Umbilical cord

Amniotic sac

Yolk sac

Yolk stalk

Chorionic villi

Umbilical vein

Cut edge of chorion

Figure 23.25   A late embryo at about 50 days after conception and the embryonic membranes that will contribute to the placenta (at right).

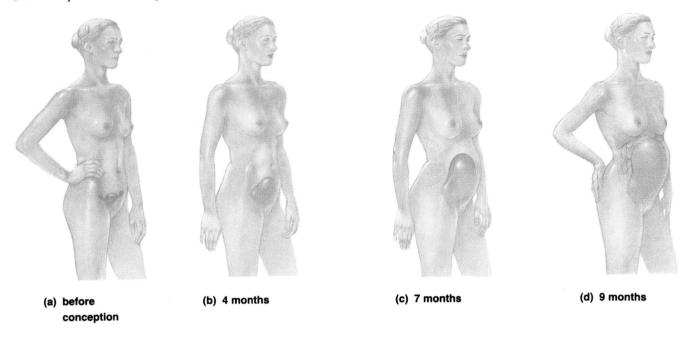

(a) before conception   **(b) 4 months**   **(c) 7 months**   **(d) 9 months**

Figure 23.26   Enlargement of the uterus during pregnancy.   (**a**) Before conception. (**b**) At 4 months, the fundus of the uterus is halfway between the pubic symphysis and the umbilicus. (**c**) At 7 months, the fundus is well above the umbilicus. (**d**) At 9 months, the fundus reaches the xiphoid process.

(Figure 23.26b). As pregnancy continues, the uterus pushes higher and higher into the abdominal cavity (Figure 23.26c). As birth nears, the uterus reaches the level of the xiphoid process of the sternum and occupies most of the abdominal cavity (Figure 23.26d).

Normal pregnancy is responsible for a considerable weight gain. Because some women are overweight or underweight before pregnancy, it is almost impossible to state the ideal or desirable weight gain. However, a gain of about 13 kg (29 pounds) usually occurs.

Childbirth

Parturition (par″tu-rish′un; "bring forth young") is the culmination of pregnancy—giving birth to the baby. It occurs an average of 266 days after fertilization and 280 days after the last menstrual period. The events that expel the infant from the uterus are referred to as **labor** (Figure 23.27). The initiation of labor is very complex: It involves stimulation of uterine-muscular contraction by a hormone called *oxytocin* (which is secreted from the posterior part of the mother's pituitary gland) and by *prostaglandins* (pros″tah-glan′dinz) from the placenta. Labor involves three successive stages: the dilation, expulsion, and placental stages.

The **dilation (first) stage** (Figure 23.27a and b) begins with the first regular contractions of the uterus. As the head is forced against the cervix, the cervix softens, thins, and dilates. This stage ends when the cervix is fully dilated by the baby's head (about 10 cm in diameter). The dilation stage is the longest part of labor, lasting 6 to 12 hours (or more).

The **expulsion (second) stage** (Figure 23.27c) lasts from full dilation to delivery, or actual childbirth. The uterine contractions become stronger during this stage, and the mother has an increasing urge to bear down with her abdominal muscles. Although this phase can take up to 2 hours, it typically last 50 minutes in a first birth and 20 minutes in subsequent births.

The **placental (third) stage** (Figure 23.27d), or the delivery of the placenta, is accomplished within 15 minutes after birth of the infant. Forceful uterine contractions continue after birth and compress the uterine blood vessels. This compression both limits bleeding and causes the placenta to detach from the uterine wall. The placenta (afterbirth) is easily removed by a slight tug on the umbilical cord. The blood vessels in the umbilical cord are then counted to verify that there are two umbilical arteries and one umbilical vein (see p. 510). This is done because the absence of one umbilical artery is often associated with cardiovascular disorders in the infant.

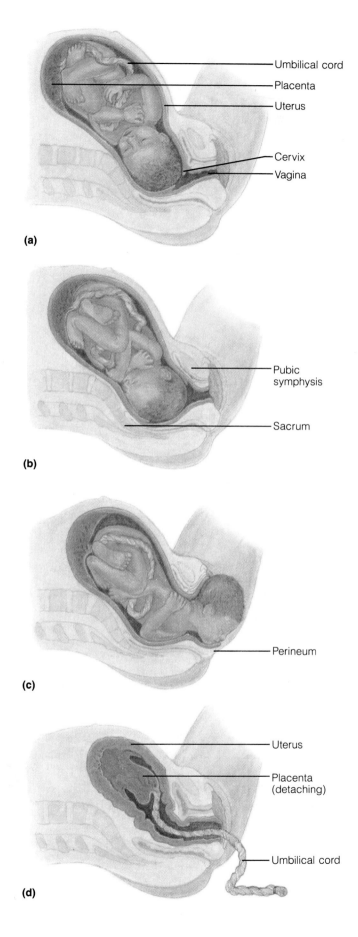

(a)
- Umbilical cord
- Placenta
- Uterus
- Cervix
- Vagina

(b)
- Pubic symphysis
- Sacrum

(c)
- Perineum

(d)
- Uterus
- Placenta (detaching)
- Umbilical cord

The Reproductive System Throughout Life

Embryonic Development of the Sex Organs

The gonads of both males and females begin their development during week 5 as masses of intermediate mesoderm called the **gonadal ridges** (Figure 23.28). The gonadal ridges form bulges on the dorsal abdominal wall in the lumbar region, just medial to the mesonephros (transient kidney: p. 618). The **mesonephric ducts** (future male ducts) develop medial to the **paramesonephric ducts** (future female ducts), and both sets of ducts empty into the cloaca (future bladder and urethra). At this time, the embryo is said to be in the **sexually indifferent stage** because the gonadal ridge and ducts are structurally identical in both sexes.

In the week after the gonadal ridges appear, *primordial germ cells* migrate to them from the yolk sac and seed the developing gonads with germ cells destined to become male spermatogonia or female oogonia. Once these germ cells are in place, they signal the gonadal ridges to differentiate into testes or ovaries, according to the genetic makeup of the embryo. In *male* embryos, this differentiation begins in week 7. Testes cords (the future seminiferous tubules) grow from the gonadal surface into the inner part of the gonad, where they connect to the mesonephric duct via the efferent ductules. With further development, the mesonephric duct becomes the major male sex ducts (the epididymis, ductus deferens, and ejaculatory ducts). The paramesonephric ducts play no part in male development and soon degenerate.

In female embryos, the process begins slightly later, in about week 8. The outer, or cortical, part of the immature ovaries forms the ovarian follicles. Shortly thereafter, the paramesonephric ducts differentiate into most structures of the female duct system (uterine tubes, uterus, the superior part of the vagina). The mesonephric ducts, by contrast, degenerate.

Figure 23.27 Stages of labor. (**a**) Dilation stage (early). The baby's head has entered the true pelvis. The widest dimension of the head is along the left-right axis. (**b**) Late dilation stage. The baby's head rotates so that its greatest dimension is in the anteroposterior axis as it moves through the pelvic outlet. Dilation of the cervix is nearly complete. (**c**) Expulsion stage. The baby's head extends at the neck as it reaches the perineum and is delivered. The rest of the body is narrower than the head and follows more easily. (**d**) Placental stage. After the baby has been delivered, the placenta detaches and is itself delivered.

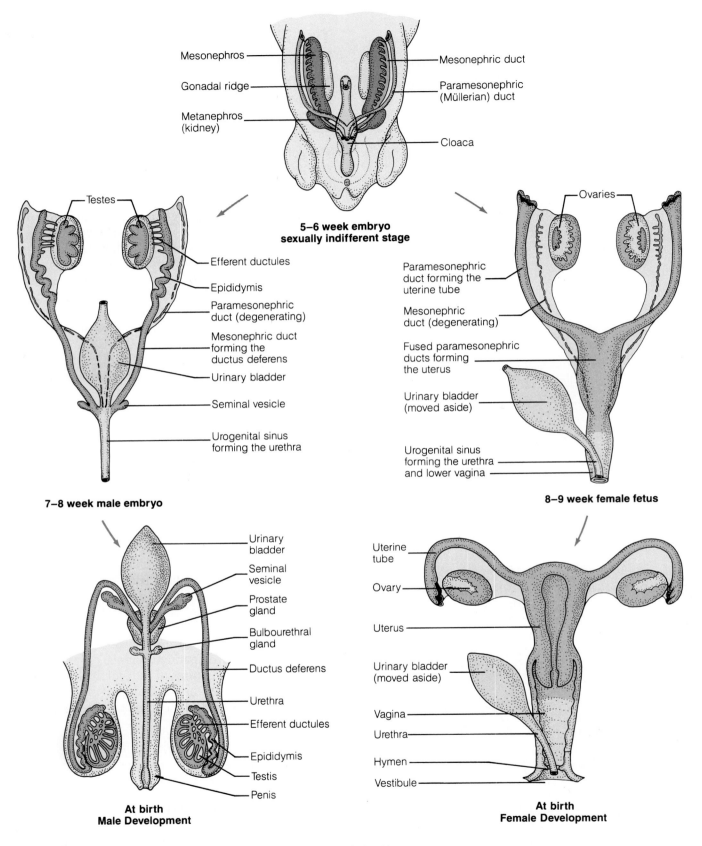

Mesonephros

Gonadal ridge

Metanephros (kidney)

Mesonephric duct

Paramesonephric (Müllerian) duct

Cloaca

5–6 week embryo sexually indifferent stage

Testes

Efferent ductules

Epididymis

Paramesonephric duct (degenerating)

Mesonephric duct forming the ductus deferens

Urinary bladder

Seminal vesicle

Urogenital sinus forming the urethra

7–8 week male embryo

Ovaries

Paramesonephric duct forming the uterine tube

Mesonephric duct (degenerating)

Fused paramesonephric ducts forming the uterus

Urinary bladder (moved aside)

Urogenital sinus forming the urethra and lower vagina

8–9 week female fetus

Urinary bladder

Seminal vesicle

Prostate gland

Bulbourethral gland

Ductus deferens

Urethra

Efferent ductules

Epididymis

Testis

Penis

At birth Male Development

Uterine tube

Ovary

Uterus

Urinary bladder (moved aside)

Vagina

Urethra

Hymen

Vestibule

At birth Female Development

Figure 23.28   Development of the internal reproductive organs in both sexes.

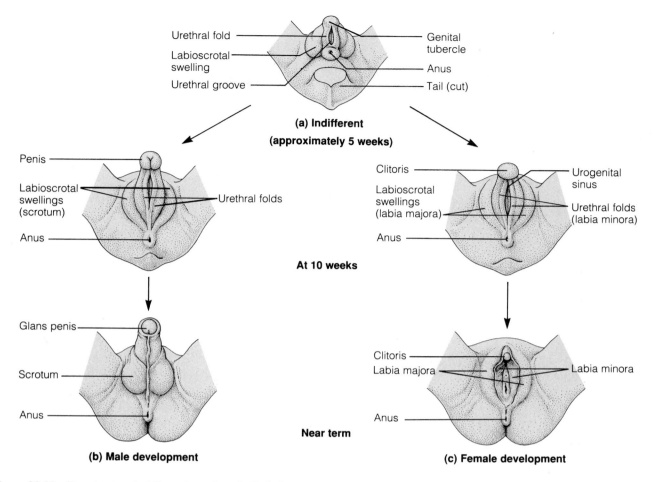

Urethral fold
Labioscrotal swelling
Urethral groove
Genital tubercle
Anus
Tail (cut)

(a) Indifferent
(approximately 5 weeks)

Penis
Labioscrotal swellings (scrotum)
Urethral folds
Anus

Clitoris
Labioscrotal swellings (labia majora)
Urogenital sinus
Urethral folds (labia minora)
Anus

At 10 weeks

Glans penis
Scrotum
Anus

Clitoris
Labia majora
Labia minora
Anus

Near term

(b) Male development

(c) Female development

Figure 23.29   Development of the external genitalia in both sexes.   The two pictures at the bottom of the figure show the fully developed perineal region.

Like the gonads, the *external* genitalia develop from identical structures in both sexes (Figure 23.29). During the sexually indifferent stage, both male and female embryos have a small projection called the **genital tubercle** on their external pelvic surface. The urogenital sinus of the cloaca (future urethra and bladder) lies deep to the tubercle, and the **urethral groove,** which serves as the external opening of the urogenital sinus, lies between the genital tubercle and the anus. The urethral groove is flanked laterally by the **urethral folds** *(genital folds)* and **labioscrotal swellings**.

During week 8, the external genitalia begin to develop rapidly. In males, the genital tubercle enlarges, forming most of the penis (the glans and the erectile bodies). The urethral folds fuse in the midline to form the penile urethra. The two labioscrotal swellings also join in the midline to become the scrotum. In females, the genital tubercle gives rise to the clitoris, the unfused urethral folds become the labia minora, and the unfused labioscrotal folds become the labia majora. The urethral groove persists in females as the vestibule.

Hypospadias (hi″po-spa′de-as), which occurs in male infants only, is the most common congenital abnormality of the urethra. It results from a failure of the two urethral folds to fuse completely, producing urethral openings on the undersurface of the penis. Hypospadias affects 1 in 300 male births. This problem is corrected surgically when the boy is about 12 months old.

Can you deduce how hypospadias affects urination?

More urine exits from the undersurface of the penis than from its tip. ■

Descent of the Gonads

In both sexes, the gonads originate in the lumbar region and then descend caudally during the fetal period. In the male fetus, the testes descend toward the scrotum, followed by their blood vessels and nerves (Figure 23.30a). Recall that the testes require such a cool, superficial location for sperm formation. The

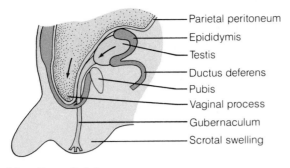

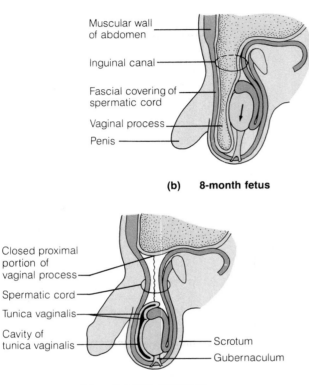

(a) 7-month fetus

- Parietal peritoneum
- Epididymis
- Testis
- Ductus deferens
- Pubis
- Vaginal process
- Gubernaculum
- Scrotal swelling

(b) 8-month fetus

- Muscular wall of abdomen
- Inguinal canal
- Fascial covering of spermatic cord
- Vaginal process
- Penis

(c) 1-month-old infant

- Closed proximal portion of vaginal process
- Spermatic cord
- Tunica vaginalis
- Cavity of tunica vaginalis
- Scrotum
- Gubernaculum

Figure 23.30 Descent of the testes. Sagittal sections through the pelvis of the male fetus. **(a)** The testes descend dorsal to the parietal peritoneum, reaching the inguinal region in the 3-month fetus. At this time, the vaginal process (an extension of the peritoneal cavity) pushes caudally into the future scrotum. **(b)** In month 7, the testes descend rapidly into the scrotum via the inguinal canal. **(c)** Finally, the caudal part of the vaginal process becomes the tunica vaginalis ventral to the testis, whereas the proximal part of the vaginal process is obliterated.

testes reach the pelvis near the inguinal region in month 3, when a finger-like outpocketing of the peritoneal cavity, the **vaginal process,** pushes through the muscles of the anterior abdominal wall to form the inguinal canal. The testes proceed no further until month 7, when they suddenly and quickly pass through the inguinal canal and enter the scrotum (Figure 23.30b). After the testes have reached the scrotum, the proximal part of each vaginal process closes off, whereas its caudal part becomes the sac-like tunica

vaginalis (Figure 23.30c). In some boys, the vaginal process does not close at all.

> Can you guess the clinical consequence of a vaginal process that remains open?

This open path to the scrotum forms a "built-in" route for an inguinal hernia. Indeed, such **congenital inguinal hernias** are the most common hernias. Since herniated intestinal coils are easily trapped and constricted within the vaginal process, congenital inguinal hernias are usually repaired surgically before the boy reaches 5 years of age. ■

The mechanism causing the descent of the testis is unknown, but descent may result from a rapid growth of the caudal body (future pelvic region) that pulls the testis caudally with it. The final descent through the inguinal canal is known to be stimulated by testosterone. Overall, the path of testicular descent seems to be guided by a fibrous cord called the **gubernaculum** (gu″ber-nak′u-lum; "governor"), which extends caudally from the testis to the floor of the scrotal sac (Figure 23.30).

Like the testes, the female ovaries descend during fetal development—but only into the pelvis, where the tent-like broad ligament blocks their further descent. Each ovary is guided in its descent by a gubernaculum that anchors in the labium majus and is homologous to the gubernaculum in the male (Figure 23.30a). The female's gubernaculum becomes both the round ligament of the uterus (its caudal portion) and the ovarian ligament (its cranial portion).

Puberty and Menopause

Puberty is the period of life, generally between age 10 and 15, when the reproductive organs grow to their adult size and reproduction becomes possible. These changes occur under the influence of rising levels of gonadal hormones (testosterone in males and estrogen in females).

The events of puberty occur in the same sequence in all individuals, but the age at which they occur varies widely. In males, the event that signals puberty's onset is enlargement of the testes and scrotum around the age of 13 years, followed by the appearance of pubic, axillary, and facial hair. The latter three features are **secondary sex characteristics** (features induced in the *nonreproductive* organs by sex hormones). Other secondary sex characteristics include the enlargement of the larynx that deepens the male voice and an increased oiliness of the skin that can lead to acne. The musculoskeletal system also increases in mass at puberty, accompanied by a growth spurt that lasts up to 6 years. Sexual maturation is evidenced by the presence of mature sperm in the semen.

In females, the first sign of puberty is budding breasts, often apparent by age 11. Menarche (mĕ-nar'ke; "month beginning"), the first menstruation, usually occurs 1 to 2 years later. Dependable ovulation and fertility await the maturation of pituitary hormonal controls, which takes nearly 2 more years. In the meantime, the estrogen-induced secondary sexual characteristics appear at around age 13, including (1) an increase in subcutaneous fat, especially in the hips and breasts, and (2) widening and lightening of the bones of the pelvic girdle (adaptations to facilitate childbirth). Furthermore, the ovaries secrete some androgens that signal the development of pubic and axillary hair, apocrine sweat glands, and oiliness of the skin. The estrogen-induced female growth spurt lasts from about age 12 to age 15 to 17.

Most women reach the peak of their reproductive abilities in their late 20s. Later, when few follicles remain in the ovaries, ovulation and menstruation cease. This event, called *menopause,* normally occurs between 46 and 54 years.

Although low-level estrogen production continues for a while after menopause, the ovaries finally become nonfunctional as hormone-secreting organs. Deprived of the stimulatory effects of estrogen, the reproductive organs and breasts start to atrophy. As the vagina becomes dry, intercourse may become painful, and vaginal infections become more common. The skin thins gradually, and bone mass declines slowly, sometimes resulting in osteoporosis. Slowly rising blood cholesterol levels place postmenopausal women at risk for cardiovascular disorders.

There is no equivalent of menopause in men. Although aging men do exhibit a steady decline in testosterone and a longer latent period for regaining erections between orgasms, a male's reproductive capability seems unending. Healthy men can father offspring into their 80s.

✚ Related Clinical Terms

Cryptorchidism (krip-tor'kĭ-dizm) (*crypt* = hidden; *orcho* = testis)   Congenital condition in which one or both testes fails to descend into the scrotum. The testis may be in the inguinal canal (usually) or in the pelvis. Cryptorchidism occurs in about 2% of full-term male births, but in most such cases the testis completes its descent, entirely on its own, before the first birthday. A testis that remains undescended is sterile because of the heat of the body. However, the interstitial cells secrete testosterone in normal quantities. Cryptorchidism may be corrected surgically.

Endometriosis (en"do-me"tre-o'sis)   A condition in which fragments of the uterine endometrium migrate out through the uterine tubes into the pelvic cavity. These fragments implant in the peritoneal coverings of various pelvic organs. The pain associated with menstruation is extreme because the excess endometrial tissue undergoes menstrual bleeding. (Furthermore, menstrual blood trapped in the pelvic cavity forms enlarging cysts that add to the pain.) Endometriosis is treated with drugs that suppress menstruation or, in severe cases, by removal of the uterus.

Gynecology (gi"nĕ-kol'o-je) (*gyneco* = woman; *logy* = study of)   Specialized branch of medicine that deals with the diagnosis and treatment of female reproductive disorders.

Hydrocele (hi'dro-sēl) (*hydro* = water; *cele* = sac)   A swelling in the scrotum caused by an excessive accumulation of fluid in the cavity of the tunica vaginalis. Hydrocele may result from an infection or injury to the testis that causes the layers of the tunica vaginalis to secrete excess serous fluid. Most hydroceles are small and do not require treatment. Large hydroceles can be treated by aspirating the fluid with a needle inserted in the tunica vaginalis and then removing the serous lining of the tunica vaginalis.

Hysterectomy (his"tĕ-rek'to-me) (*hyster* = uterus; *ectomy* = cut out)   Surgical removal of the uterus. It is a very common operation, performed most often to treat tumors and cancers of the uterus, which are themselves common.

Laparoscopy (lap"ah-ros'ko-pe) (*lapar* = the flank; *scopy* = observation)   Examination of the abdominopelvic cavity with a laparoscope, a viewing device at the end of a thin tube inserted through the anterior abdominal wall. Laparoscopy is often used to assess the condition of the pelvic reproductive organs of females.

Mammography (mah-mog'rah-fe) (*mamma* = breast; *graphy* = writing)   X-ray examination of the breasts for signs of breast cancer.

Oophorectomy (o"of-o-reck'to-me) (*oophor* = ovary)   Surgical removal of the ovary, usually to remove ovarian cysts or ovarian cancer.

Orchitis (or-ki'tis) (*orcho* = testis)   Inflammation of the testes, sometimes caused by the mumps virus.

Ovarian cancer   Malignancy of the ovary. The fourth most common cause of cancer death in women, it mostly affects women over 40. Women who have never had children are at highest risk.

Placenta previa (pre've-ah; "appearing in front of")   Condition in which the embryo implants in the inferior part—rather than in the normal superior part—of the uterine wall. The placenta may cover over the internal os of the cervix. Placenta previa is usually associated with bleeding in the last 3 months of pregnancy as some of the placenta pulls away from the uterus. To limit the bleeding, bed rest is prescribed. Placenta previa affects about 1 in 200 pregnancies.

Varicocele (var'ĭ-ko-sēl")   Varicose veins in the pampiniform plexus of the spermatic cord. About 10% to 15% of men have this condition, which is usually harmless. Sometimes, however, the drainage of the pampiniform veins is so greatly inhibited that these veins cannot perform their testis-cooling function (p. 625). The testes heat up, leading to low sperm count or sterility. In fact, varicocele is the most common cause of fertility problems in males.

Chapter Summary

1. The function of the reproductive system is to produce off-spring. The gonads produce gametes (sperm or ova) and sex hormones. All other reproductive organs are accessory organs.

THE MALE REPRODUCTIVE SYSTEM (pp. 623–633)

The Scrotum (p. 624)

1. The scrotum is the sac of skin and subcutaneous tissue that contains the testes. It provides a temperature slightly lower than that of the body, as is required for viable sperm production. The dartos muscle, cremaster muscle, and pampiniform plexus help regulate the temperature of the testes in the scrotum.

The Testes (pp. 624–629)

2. Each testis lies in a serous sac called the tunica vaginalis. It is divided into many lobules that contain sperm-producing seminiferous tubules separated by areolar connective tissue.

3. The thick epithelium of the seminiferous tubule consists of spermatogenic cells (spermatogonia, primary and secondary spermatocytes, and spermatids), embedded in columnar sustentacular cells. Spermatogenic cells move toward the tubule lumen as they differentiate into sperm by a process called spermatogenesis.

4. The three stages of spermatogenesis are (1) formation of spermatocytes, (2) meiosis, and (3) spermiogenesis.

5. In spermiogenesis, spermatids become tadpole-shaped sperm cells by forming an acrosome, growing a flagellum, flattening the nucleus, and then shedding their superfluous cytoplasm.

6. Sustentacular (Sertoli) cells form the blood-testis barrier, nourish the spermatogenic cells, and move them toward the lumen. They also secrete testicular fluid for sperm transport, androgen-binding protein, and the hormone inhibin.

7. Interstitial (Leydig) cells are oval cells in the connective tissue between seminiferous tubules. They secrete androgens (mostly testosterone) under the influence of luteinizing hormone from the pituitary gland.

The Male Duct System (pp. 629–631)

8. From the seminiferous tubules, sperm travel through the tubuli recti and rete testes, then out of the testes through the efferent ductules into the epididymis.

9. The comma-shaped epididymis hugs the posterolateral surface of the testis. The duct of the epididymis, lined with a pseudostratified columnar epithelium with stereocilia, is the site where testicular fluid is absorbed and sperm gain the ability to swim. Ejaculation begins with contraction of smooth muscle in the duct of the epididymis.

10. The ductus (vas) deferens extends from the epididymis in the scrotum to the ejaculatory duct in the pelvic cavity. The thick layers of smooth muscle in its wall propel sperm into the urethra by peristalsis during ejaculation.

11. The spermatic cord contains the ductus deferens and the testicular vessels. This cord runs through the anterior abdominal wall from the scrotum to the superficial inguinal ring and then continues through the inguinal canal to the deep inguinal ring.

12. The urethra runs from the bladder to the tip of the penis. It conducts semen and urine to the body exterior.

Accessory Glands (pp. 631–633)

13. The seminal vesicles, long, pouched glands posterior to the bladder, secrete a sugar-rich fluid that contributes 60% of the ejaculate.

14. The prostate gland, surrounding the prostatic urethra, is a group of compound glands embedded in a fibromuscular stroma. Added to semen, its secretion enhances sperm motility (among other functions).

15. The bulbourethral glands in the urogenital diaphragm secrete mucus into the urethra before ejaculation to lubricate this tube for passage of the semen.

The Penis (p. 633)

16. The penis has a root, a shaft, and a glans covered by foreskin. Its main nerves and vessels lie dorsally. Most of its mass consists of vascular erectile bodies (a corpus spongiosum and two corpora cavernosa). Engorgement of these bodies with blood causes erection.

The Male Perineum (p. 633)

17. The diamond-shaped male perineum contains the scrotum, the root of the penis, and the anus.

THE FEMALE REPRODUCTIVE SYSTEM (pp. 634–646)

1. The female reproductive system produces ova and sex hormones and houses a developing infant until birth. It undergoes a menstrual cycle of about 28 days.

The Ovaries (pp. 635–638)

2. The almond-shaped ovaries lie on the lateral walls of the pelvic cavity and are held in the peritoneal cavity by various mesenteries and ligaments.

3. The mesovarium (mesentery of the ovary) is part of a fold of peritoneum called the broad ligament. Other parts of the broad ligament are the mesosalpinx, the mesometrium, and the suspensory ligament of the ovary.

4. Each ovary is divided into an outer cortex (follicles separated by connective tissue) and an inner medulla (connective tissue containing the main ovarian vessels and nerves).

5. A newborn female has many thousands of primordial follicles in her ovaries. A primordial follicle consists of an oocyte surrounded by a layer of flat follicular cells.

6. During the follicular phase of each monthly ovarian cycle (days 1–14), six to twelve follicles start maturing under the influence of gonadotropin hormones (follicle-stimulating hormone, luteinizing hormone) from the anterior part of the pituitary gland. The follicles progress from primordial follicles to early primary follicles, to growing primary follicles, to secondary follicles, to mature vesicular follicles ready for ovulation. Generally, only one follicle per month completes the maturation process.

7. The theca folliculi around a growing follicle cooperates with the follicle itself to produce estrogens under the stimulatory influence of LH and FSH.

8. Upon stimulation by LH, ovulation occurs at mid cycle (day 14), releasing the oocyte from the ovary into the pelvic peritoneal cavity. Ovulation seems to involve a weakening and rupture of the follicle wall followed by violent muscular contraction of the external theca cells.

9. In the luteal phase of the ovarian cycle (days 15–28), the ruptured follicle becomes a corpus luteum that secretes progesterone and estrogen. If fertilization does not occur, the corpus luteum degenerates in about 2 weeks into a corpus albicans scar.

10. Oogenesis, or the production of female gametes, takes decades to complete. The stem cells (oogonia) appear in the ovarian follicles of the fetus. Primary oocytes stay in meiosis 1 until ovulation occurs years later, and each secondary oocyte then stays in meiosis 2 until it is penetrated by a sperm.

The Uterine Tubes (pp. 638–640)

11. Each uterine tube extends from an ovary to the uterus. Its

regions, from lateral to medial, are the infundibulum, ampulla, and isthmus. The fimbriated distal end creates currents that help draw an ovulated oocyte into the uterine tube.

12. The wall of the uterine tube contains a muscularis and a highly folded inner mucosa with a simple columnar epithelium. Both the smooth muscle and ciliated epithelial cells help propel the oocyte toward the uterus.

The Uterus (pp. 640–642)

13. The hollow uterus is shaped like an inverted pear. It has a fundus, body, and cervix. Cervical glands fill the cervical canal with a bacteria-blocking mucus.

14. The uterus is supported by the broad, lateral cervical, and round ligaments. Most support, however, comes from the muscles of the pelvic floor.

15. The uterine wall consists of the outer perimetrium, the muscular myometrium, and a thick mucosa called the endometrium. The endometrium consists of a stratum functionalis, which sloughs off monthly (except in pregnancy) and an underlying stratum basalis, which replenishes the functionalis.

16. The uterine cycle has three phases: menstrual, proliferative, and secretory. The first two phases are a shedding and then a rebuilding of the endometrium in the 2 weeks before ovulation. The third phase prepares the endometrium to receive an embryo in the 2 weeks after ovulation.

The Vagina (pp. 642–644)

17. The vagina is a highly distensible tube that runs from the uterus to the body's exterior. It receives the penis and semen during intercourse and acts as the birth canal.

18. The layers of the vagina wall are an outer adventitia, a middle muscularis, and an inner mucosa consisting of an elastic lamina propria and a stratified squamous epithelium. The lumen is acidic.

19. The vaginal fornix is a ring-like recess around the tip of the cervix in the superior vagina.

The External Genitalia and Female Perineum (pp. 644–645)

20. The female external genitalia (vulva) include the mons pubis, the labia majora and minora, the clitoris with its erectile bodies, and the vestibule containing the vaginal and urethral orifices. The mucus-secreting greater vestibular glands and the bulbs of the vestibule (erectile bodies) lie just deep to the labia.

21. The central tendon of the perineum lies just posterior to the vaginal orifice and the fourchette.

The Mammary Glands (pp. 645–646)

22. The mammary glands develop from the skin of the embryonic milk lines. Internally, each breast consists of 15 to 25 lobes (compound alveolar glands) that secrete milk. The lobes are separated by adipose tissue and by supportive suspensory ligaments. The full glandular structure of the breast does not develop until the second half of pregnancy.

PREGNANCY AND CHILDBIRTH (pp. 646–652)

Pregnancy (pp. 646–652)

1. Fertilization usually occurs in the ampulla of a uterine tube. Many sperm must release enzymes from their acrosomes to penetrate the oocyte's corona radiata and zona pellucida. Entry of one sperm into the oocyte prevents the entry of others.

2. Once fertilized, the zygote divides, travels to the uterus, and implants in the endometrium. As it implants, its trophoblast divides into a cytotrophoblast and a syncytiotrophoblast. Lacunae open in the syncytiotrophoblast and fill with blood from endometrial vessels, forming the first contact between maternal blood and embryonic tissue.

3. The placenta acts as the respiratory, nutritive, and excretory organ of the fetus and produces the hormones of pregnancy. It is formed both from embryonic tissue (chorionic villi) and from maternal tissue (endometrial decidua).

4. The placental exchange barrier consists of the layers of the chorionic villi plus the endothelium of the fetal capillaries inside those villi. Of these layers, the syncytiotrophoblast secretes the placental hormones. No direct mixing of maternal and fetal blood occurs across a healthy placenta.

5. During pregnancy, the maternal reproductive organs become increasingly vascularized, and the breasts enlarge. The uterus expands superiorly and eventually grows to occupy most of the abdominal cavity.

Childbirth (p. 652)

6. The three stages of labor are dilation, expulsion of the infant, and the placental stage.

THE REPRODUCTIVE SYSTEM THROUGHOUT LIFE (pp. 653–657)

Embryonic Development of the Sex Organs (pp. 653–655)

1. The gonads of both sexes develop in the dorsal lumbar region from the gonadal ridges of intermediate mesoderm. The primordial germ cells migrate to the gonads from the endoderm of the yolk sac. The mesonephric ducts form most of the male ducts and glands. The paramesonephric ducts form the uterine tubes and uterus of the female.

2. The external genitalia arise from the genital tubercle (male penis and female clitoris), urethral folds (penile urethra and labia minora), and labioscrotal swellings (scrotum and labia majora).

Descent of the Gonads (pp. 655–656)

3. The testes form in the dorsal abdomen and descend into the scrotum. The ovaries also descend, but only into the pelvis.

Puberty and Menopause (pp. 656–657)

4. Puberty is the interval when reproductive organs mature and become functional. It starts in males with penile and scrotal growth and in females with breast development.

5. During menopause, ovarian function declines, and ovulation and menstruation cease.

Review Questions

Multiple Choice and Matching Questions

1. The structures that draw an ovulated oocyte into the female duct system are (a) microvilli, (b) stereocilia, (c) amoebas, (d) fimbriae.

2. The male homologue of the female clitoris is the (a) penis, (b) scrotum, (c) penile urethra, (d) testis.

3. Which of the following is correct about the female vestibule? (a) The vaginal orifice is the most anterior of the two orifices in the vestibule. (b) The urethra is between the vaginal orifice and the anus. (c) The anus is between the vaginal orifice and the urethra. (d) The urethral orifice is anterior to the vaginal orifice.

4. Match each hollow organ listed below with its lining epithelium. (Some choices may be used more than once.)

Organ	Lining epithelium
_____ **(1)** duct of epididymis	**(a)** simple squamous
_____ **(2)** uterine tube	**(b)** simple cuboidal
_____ **(3)** ductus deferens	**(c)** simple columnar
_____ **(4)** uterus	**(d)** pseudostratified columnar
_____ **(5)** vagina	**(e)** stratified squamous

5. Which of the following produces the male sex hormones? (a) seminal vesicles, (b) seminiferous tubules, (c) developing follicles of the testes, (d) interstitial cells.

6. The uterine cycle can be divided into three continuous phases. Starting from the first day of the cycle, their order is (a) menstrual, proliferative, secretory; (b) menstrual, secretory, proliferative; (c) secretory, menstrual, proliferative; (d) proliferative, menstrual, secretory; (e) proliferative, secretory, menstrual.

7. All of the following are true of the gonadotropins except that they are (a) secreted by the pituitary gland, (b) LH and FSH, (c) hormones with important functions in both males and females, (d) the sex hormones secreted by the gonads.

8. The hormone-secreting layer of the placental barrier is (a) the extraembryonic mesoderm, (b) the lacunae, (c) endothelium, (d) cytotrophoblast, (e) none of the above.

9. Ovulation occurs during what part of the ovarian cycle? (a) beginning, (b) middle, (c) end, (d) random (differs every month).

10. Which of the following cannot pass through the placental barrier? (a) blood cells, (b) sugars, (c) alcohol, (d) oxygen, (e) antibodies.

11. A function of the stereocilia in the duct of the epididymis and ductus deferens is (a) absorption and secretion, (b) beating forward to move sperm through, (c) initiating ejaculation, (d) beating backward to keep the sperm in these ducts for 20 days.

12. The broad ligament of the uterus in the female pelvis (a) is really a mesentery, (b) consists of the mesovarium and mesosalpinx but not the mesometrium, (c) is the same as the round ligament, (d) is the same as the lateral cervical ligament, (e) b and c.

13. The embryonic paramesonephric duct gives rise to the (a) uterus and uterine tubes, (b) ovary, (c) epididymis and vas deferens, (d) urethra.

14. Specifically, which glands within the prostate give rise to benign tumors (as opposed to malignant cancer)? (a) submucosal, (b) main, (c) mucosal, (d) fibromuscular stroma.

15. Circle the false statement about the uterine cervix: (a) It dilates at the end of the first stage of labor. (b) All of it projects into and lies within the vagina. (c) The cervical glands secrete mucus. (d) It contains the cervical canal.

16. If the uterine tube is a trumpet ("salpinx"), what part of it represents the wide, open end of the trumpet? (a) isthmus, (b) ampulla, (c) infundibulum, (d) flagellum.

17. The myometrium is the muscular layer of the uterus, and the endometrium is the _____ layer. (a) serosa, (b) adventitia, (c) submucosa, (d) mucosa.

Short Answer and Essay Questions

18. Describe the major structural and functional regions of a sperm.

19. List three secondary sex characteristics of females.

20. Is the vas deferens the same as the ductus deferens? If they differ, tell how.

21. Trevor had difficulty in distinguishing spermatogenesis from spermiogenesis. Can you help him by defining these terms?

22. (a) When asked to describe the path of the ductus deferens, Ryan said it runs from the scrotum directly up to the penile urethra. Ryan was told that he had made a common error and that the path of this duct is longer and more complex than he had thought. Trace the entire path of the ductus deferens. (b) To expand on part (a), trace the pathway of an ejaculated sperm from the male epididymis all the way into the uterine tube of a female.

23. In menstruation, the stratum functionalis is shed from the endometrium. Explain the hormonal and physical factors responsible for this shedding. (Hint: See Figure 23.19.)

24. The epithelium of the vagina fights bacteria, and the cervical glands of the uterus prevent the spread of vaginal bacteria. Explain how each of these mechanisms works.

25. Outline the structural changes experienced by a maturing follicle in the follicular stage of the ovarian cycle.

26. Name three sets of arteries that must be cauterized during a simple mastectomy.

27. Some anatomy students were saying that the bulbourethral glands (and the urethral glands) of males act like city workers who come around and clear parked cars from the street before a parade. What did they mean by this analogy?

28. A man swam in a cold lake for an hour and then noticed that his scrotum was shrunken and wrinkled. His first thought was that he had lost his testicles. What had really happened?

29. Determine whether this statement is true or false, and then explain your answer: Complete continuation and a perfect mixing of fetal and maternal blood occur across the placenta.

30. What is the blood-testis barrier, and why is it necessary?

 ### Critical Thinking and Clinical Application Questions

31. Circle the correct answer, and explain your choice: A person who has had this type of mastectomy will have great difficulty adducting the arms, as in doing push-ups. (a) radical mastectomy, (b) simple mastectomy, (c) lumpectomy. (Hint: Try to remember the main muscle used in doing push-ups.)

32. Based on what you know about the sources of semen, deduce whether a vasectomy diminishes the volume of the ejaculate very much.

33. Adolf, a 68-year-old man, has trouble urinating and is given a rectal exam. What is Adolf's most probable condition, and what is the purpose of the rectal exam?

34. Gina gave birth to six children while living in a commune in the mountains. Now 44 years old, she visits a doctor in town, complaining of general discomfort in her perineal region and incontinence for urination. The doctor examines her and sees her cervix protruding from the vaginal orifice. What is Gina's condition, and what has caused it?

35. What is a Pap smear, and what is its purpose?

36. A woman in substantial pain called her doctor and explained (between sobs) that she was about to have her baby "right now." The doctor calmed her and asked how she had come to that conclusion. She said that her water had broken and that her husband could see the full width of the baby's head. (a) Was she right? (b) What stage of labor was she in? (c) Do you think she had time to make it to the hospital 60 miles away? Why or why not?

37. Lucy had both her left ovary and her right uterine tube removed surgically at age 17 because of a cyst and a tumor in these organs. Now, at age 32, she remains healthy and is expecting her second child. How could Lucy conceive a child with just one ovary and one uterine tube, widely separated on opposite sides of the pelvis like this?

24

The Endocrine System

The endocrine system is a series of ductless glands that secrete messenger molecules called **hormones** into the circulation. The circulating hormones travel to distant body cells and signal characteristic physiological responses in these cells. Through its hormonal signals, the endocrine system controls and integrates the functions of other organ systems in the body. In playing a general integrative role, the endocrine system resembles the nervous system, with which it closely interacts. Hormones travel more slowly than nerve impulses, however, so the endocrine system tends to regulate slow processes, such as growth and metabolism, rather than processes that demand rapid responses, such as the contraction of skeletal muscle. Some major processes controlled by the endocrine system are growth of the body and of the reproductive organs; mobilization of body defenses against stress; maintenance of the proper blood chemistry; and control of the rate of oxygen use by the body's cells. The scientific study of hormones and the endocrine glands is called **endocrinology**.

Endocrine Organs and Hormones: An Overview

Endocrine Organs

The major endocrine organs are shown in Figure 24.1. Compared to most other organs of the body, the endocrine organs are small and unimpressive. Indeed, to collect 1 kg of hormone-producing tissue (2.2 pounds), you would need to collect *all* of the endocrine tissue from nine adults! In addition, the anatomical continuity typical of most organ systems does not exist among the endocrine organs. Instead, endocrine organs are scattered about in widely separated parts of the body.

The endocrine cells of the body are partly contained within "pure" endocrine organs and partly within organs of other body systems. The purely endocrine organs are the *pituitary gland* at the base of the brain; the *pineal gland* in the roof of the diencephalon; the *thyroid* and *parathyroid glands* in the neck; and the *adrenal glands* on the kidneys (each adrenal gland is actually two glands, an *adrenal cortex* and an *adrenal medulla*). Organs within other organ systems that contain endocrine cells include the *pancreas, thymus, gonads,* and others. The *hypothalamus* of the brain also falls in this latter category: Along with its typical nervous functions, the hypothalamus produces hormones and is therefore considered a *neuroendocrine* organ.

Most endocrine cells—like most gland cells in the body—are of *epithelial* origin. However, the endocrine system is so diverse that it also includes hormone-secreting neurons, muscle cells, and fibroblast-like cells.

The endocrine glands are richly supplied with blood and lymph vessels that receive the hormonal secretions of the endocrine cells. Typically, endocrine cells are arranged in small clusters, cords, or branching networks, in a manner that maximizes their contact with large numbers of capillaries.

Hormones

Classes of Hormones

The body produces many different kinds of hormones, all with distinct chemical structures. Most hormones, however, belong to two broad molecular categories: (1) amino acid-based molecules, and (2) steroid molecules. **Amino acid-based hormones** include modified amino acids (or *amines*), peptides (short chains of amino acids), and proteins (long chains of amino acids). **Steroids**, by contrast, are lipid molecules derived from cholesterol.

Basic Hormone Action

All major hormones circulate throughout the entire body, exit the bloodstream at the capillaries, and encounter virtually all tissues. Nevertheless, a given hormone influences the activity of only specific tissue cells, called its **target cells**. The ability of a target cell to respond to a hormone depends on the presence of specific receptor molecules on the cell's plasma membrane or in its interior, to which that particular hormone can bind. Once binding has occurred, the target cell reacts in an automatic, or preprogrammed, way.

Each kind of hormone produces its own characteristic effects within the body. These effects, however, depend on the preprogrammed responses of its target cells, not on any information contained in the hormone molecule itself. To illustrate this point, hormones with similar molecular structures often have very dissimilar functions. Hormones are just molecular triggers—they do not carry information.

Control of Hormone Secretion

Endocrine cells release their hormones into the surrounding extracellular space, and from there the hormones immediately enter the adjacent capillaries.

The various endocrine cells of the body are stimulated to manufacture and secrete their hormones by three major types of stimuli (Figure 24.2): *humoral, neural,* and *hormonal.*

Humoral Stimuli. Some endocrine glands secrete their hormones in direct response to changing levels of ions or nutrients in the blood. Such glands are said to be controlled by **humoral stimuli.** (The word *humoral* refers to the blood and other body fluids.) This is the simplest of all endocrine control mechanisms. For example, the cells of the parathyroid gland (Figure 24.2a) directly monitor the concentration of calcium ions (Ca^{2+}) in the blood, detect any decline in this concentration, and then respond by secreting a hormone that acts to reverse the decline.

Neural Stimuli. The secretion of a few endocrine glands is controlled by nervous input (Figure 24.2b). For example, sympathetic nerve fibers stimulate the cells in the adrenal medulla to release epinephrine and norepinephrine during "fight, flight, or fright" situations (also see p. 403 in Chapter 14).

Hormonal Stimuli. Finally, many endocrine glands secrete their hormones in response to hormones received from other endocrine glands; that is, certain hormones have the sole purpose of signaling the secretion of other hormones (Figure 24.2c). For example, the hypothalamus of the brain secretes some hormones that stimulate the anterior part of the pituitary gland to secrete its hormones. To carry this sequence even further, the pituitary hormones then proceed to stimulate hormonal secretion by the body's largest endocrine glands: the thyroid, adrenal cortex, and gonads. This particular interaction (hypothalamus → pituitary → other endocrine glands) lies at the very core of the science of endocrinology and will be considered again later in the chapter.

No matter how it is stimulated, hormone secretion is always limited by *negative feedback* control: As soon as a hormone rises beyond a certain concentration in the blood, its rate of secretion is forced to drop. This ensures that hormone concentrations stay within a narrow, "desirable" range in the blood.

The Major Endocrine Organs

The Pituitary Gland

Introduction

The **pituitary gland**, or **hypophysis** (hi-pof'ĭ-sis; "undergrowth [from the brain]") is an important endocrine organ that secretes at least nine major hormones

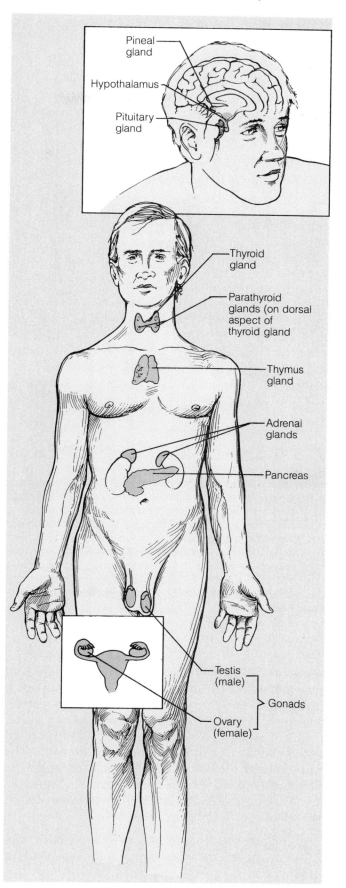

Figure 24.1 Location of the major endocrine organs of the body. Endocrine cells also occur in the heart, alimentary canal, kidney, skin, and placenta.

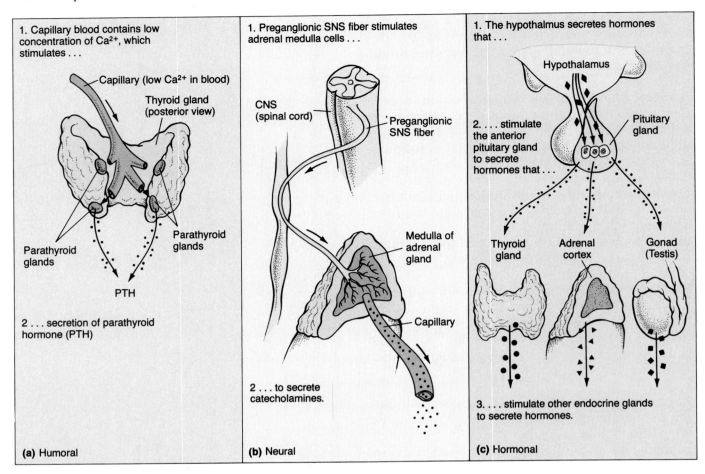

1. Capillary blood contains low concentration of Ca²⁺, which stimulates . . .

Capillary (low Ca²⁺ in blood)

Thyroid gland (posterior view)

Parathyroid glands

Parathyroid glands

PTH

2 . . . secretion of parathyroid hormone (PTH)

(a) Humoral

1. Preganglionic SNS fiber stimulates adrenal medulla cells . . .

CNS (spinal cord)

Preganglionic SNS fiber

Medulla of adrenal gland

Capillary

2 . . . to secrete catecholamines.

(b) Neural

1. The hypothalmus secretes hormones that . . .

Hypothalamus

Pituitary gland

2. . . . stimulate the anterior pituitary gland to secrete hormones that . . .

Thyroid gland

Adrenal cortex

Gonad (Testis)

3. . . . stimulate other endocrine glands to secrete hormones.

(c) Hormonal

Figure 24.2   Control of hormone release: Three different mechanisms.
(a) Control by humoral stimuli: Cells in the parathyroid gland secrete their hormone in response to falling concentrations of calcium in the blood. **(b)** Control by neural stimuli: Sympathetic neurons (preganglionic SNS fibers) stimulate cells of the adrenal medulla to release their hormones into the bloodstream. **(c)** Control by hormonal stimuli: The hypothalamus of the brain releases hormones that stimulate the anterior pituitary gland to secrete its own hormones. Some of the pituitary hormones then proceed to signal the release of hormones by the thyroid gland, the adrenal cortex, and the gonads. In this way, the hypothalamus regulates hormone secretion by the body's major endocrine organs and thereby controls much of the endocrine system.

(Figure 24.3). It sits in the sella turcica of the sphenoid bone in the floor of the skull, just inferior to the brain. Usually said to be the size and shape of a pea, the pituitary gland is more accurately described as a pea on a stalk. The stalk, which connects superiorly to the hypothalamus, is the **infundibulum** ("funnel"). The infundibulum projects inferiorly from a part of the hypothalamus called the tuber cinereum, which lies between the optic chiasma anteriorly and the mammillary bodies posteriorly.

Judging from Figure 24.3, can you deduce why tumors of the pituitary gland can lead to blindness?

As it enlarges, a tumorous pituitary gland can expand superiorly to compress the visual axons in the optic chiasma. Additionally, as the tumor continues to enlarge, it can compress the overlying hypothalamus, leading to emotional disturbances, such as uncontrolled rage. ■

The pituitary gland has two basic divisions, an anterior **adenohypophysis** (ad″ĕ-no-hi-pof′ĭ-sis; "glandular hypophysis") and a posterior **neurohypophysis** (nu″ro-hi-pof′ĭ-sis; "neural hypophysis"). The adenohypophysis is composed of glandular tissue, whereas the neurohypophysis is a part of the brain. Each of these divisions has three subdivisions (Figure 24.3 and Table 24.1). In the adenohypophysis, the largest subdivision is the anteriormost **pars distalis (anterior lobe)**. Just posterior to this lies the **pars intermedia**, and just superior to it lies the **pars tuberalis**, which wraps around the infundibulum like a tube. In the neurohypophysis, the three subdivisions are (from

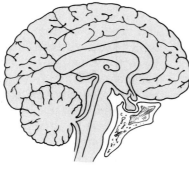

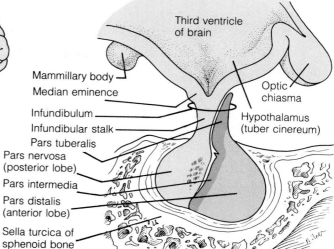

Figure 24.3   The pituitary gland (hypophysis).   The basic regions of the pituitary gland. The anterior aspect is to the right. The adenohypophysis is indicated in pink and the neurohypophysis is colored yellow. The view at left shows the location of the pituitary gland at the base of the brain.

inferior to superior) the **pars nervosa (posterior lobe)**, the **infundibular stalk**, and the cone-shaped **median eminence** of the hypothalamus.

Arterial blood reaches the hypophysis through two branches of the internal carotid artery (Figure 24.4). The **superior hypophyseal artery** supplies the entire adenohypophysis and the infundibulum, while the **inferior hypophyseal artery** supplies the pars nervosa. The veins from the extensive capillary beds in the pituitary gland drain into the cavernous sinus (p. 503) and other nearby dural sinuses.

The Adenohypophysis

As we explore the adenohypophysis, we will concentrate on its largest division, the *pars distalis.* The pars distalis makes and secretes at least seven major hormones and contains five different types of endocrine cells (see Table 24.2 on p. 668). These cells secrete protein hormones and therefore exhibit the cytoplasmic features of protein-secreting cells (secretory granules and a well-developed rough endoplasmic reticulum [rough ER] and Golgi apparatus). The five cell classes are as follows:

1. Somatotropic cells. These cells secrete **growth hormone (GH)**, which is also called **somatotropin (SH)** (so"mah-to-tro'pin; "body changer"). This hormone stimulates growth of the entire body by signaling body cells to increase their production of proteins and by stimulating growth of the epiphyseal plates of the skeleton (p. 130). Somatotropic cells are the most abundant type of endocrine cells in the pars distalis.

2. Lactotropic cells (*lact* = milk). These cells secrete **prolactin (PRL)** (pro-lak'tin), a hormone that stimulates manufacture of milk by the breasts.

3. Thyrotropic cells. This class of cells secretes **thyroid-stimulating hormone (TSH)**, the hormone

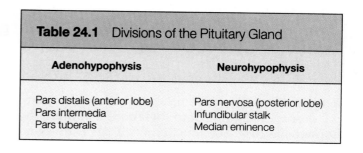

Table 24.1 Divisions of the Pituitary Gland	
Adenohypophysis	**Neurohypophysis**
Pars distalis (anterior lobe)	Pars nervosa (posterior lobe)
Pars intermedia	Infundibular stalk
Pars tuberalis	Median eminence

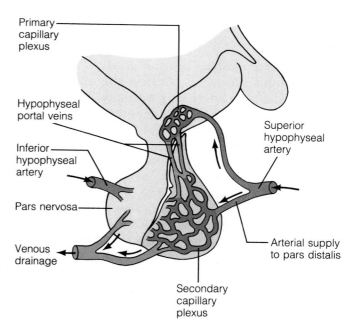

Figure 24.4   The major vessels of the pituitary gland.

that signals the thyroid gland to secrete its own hormone (thyroid hormone).

4. Corticotropic cells. These cells secrete at least two hormones, which are split from a common parent molecule. These hormones are **adrenocorticotropic**

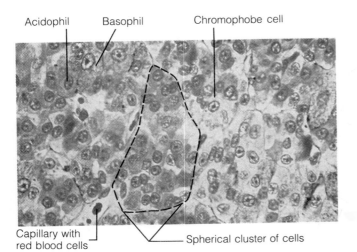

Acidophil   Basophil   Chromophobe cell

Capillary with red blood cells — Spherical cluster of cells

Figure 24.5   Histology of the pars distalis of the adenohypophysis (350 ×). Note the acidophil, basophil, and chromophobe cells.

hormone **(ACTH)** (ad-re″no-kor″tĭ-ko-trōp′ik; "adrenal cortex-changing") and **melanocyte-stimulating hormone (MSH)**. The function of ACTH is to signal the adrenal cortex to secrete its hormones (glucocorticoids), while MSH apparently darkens skin pigmentation by stimulating the melanocytes in the epidermis.

5. Gonadotropic cells. These cells secrete both of the gonadotropin hormones, **follicle-stimulating hormone (FSH)** and **luteinizing hormone (LH)**. These hormones act on the gonads, stimulating maturation of the sex cells and inducing the secretion of sex hormones (Chapter 23). In females, FSH and LH stimulate the maturation of the egg-containing ovarian follicles and the secretion of estrogen and progesterone from theca and granulosa cells in the ovary (p. 637). Furthermore, a large amount of LH is secreted in the middle of the female menstrual cycle to induce ovulation (p. 643). In males, LH signals the secretion of androgens (mainly testosterone) by interstitial cells in the testes (p. 628), and FSH stimulates the maturation of sperm cells and the production of androgen-binding protein by sustentacular cells (pp. 627–628).

Notice that four of the seven hormones secreted by the pars distalis (TSH, ACTH, FSH, LH) regulate the hormonal secretion of other endocrine glands. Such hormones are called **tropic hormones** (tro′pik; "changing"). The remaining three adenohypophyseal hormones (GH, PRL, and MSH) exert their major effects on nonendocrine target tissues.

The endocrine cells of the pars distalis are clustered into spheres and branching cords separated by

capillaries (Figure 24.5). When tissue of the pars distalis is stained with typical histological dyes, its five cell types group into three categories: (1) **acidophils**, which stain with acidic stains (somatotropic cells and lactotropic cells); (2) **basophils**, which stain with basic stains (thyrotropic, corticotropic, and gonadotropic cells); and (3) **chromophobes** ("color avoiders"), which stain poorly. Chromophobes are either immature cells or cells whose supply of hormone has been depleted.

The pars distalis is by far the most important division of the adenohypophysis. The other divisions, the *pars intermedia* and *pars tuberalis,* are poorly understood. They secrete some of the same hormones that are secreted by the pars distalis: Cells in the pars intermedia secrete MSH and ACTH, and the pars tuberalis may contain gonadotropic cells.

Control of Hormone Secretion by the Adenohypophysis

The secretion of hormones by the adenohypophysis is controlled by the brain (Figure 24.6). More specifically, the hypothalamus of the brain secretes peptide hormones called **releasing hormones (releasing factors)**, which then signal the cells in the adenohypophysis to release their hormones. The hypothalamus also secretes **inhibiting hormones**, which *turn off* the secretion of hormones by the adenohypophysis when necessary. There are distinct releasing and inhibiting hormones for almost every adenohypophyseal hormone; these include *growth hormone-releasing hormone, prolactin-inhibiting hormone, gonadotropin-releasing hormone,* and so on.

Here is the precise pathway taken by the releasing and inhibiting hormones (Figure 24.6): They are manufactured within neuron cell bodies in the hypothalamus and then secreted like neurotransmitters from the axon terminals of these neurons. (Note that *neurons* are serving as endocrine cells.) The releasing hormones enter a **primary capillary plexus** in the median eminence and travel inferiorly in **hypophyseal portal veins** to a **secondary capillary plexus** in the pars distalis. Leaving this plexus, the releasing hormones attach to the adenohypophyseal cells and stimulate these cells to secrete hormones (GH, LH, TSH, PRL, and so on) back into the secondary plexus. From there, the newly secreted adenohypophyseal hormones proceed into the general circulation and travel to their target organs throughout the body.

The primary and secondary capillary plexuses in the pituitary gland, plus the intervening hypophyseal portal veins, make up the *hypophyseal portal system.* This system fits the standard definition of a portal system (p. 508): It consists of two capillary beds separated by veins, with the first bed receiving molecules that are destined for tissues supplied by the second capillary bed.

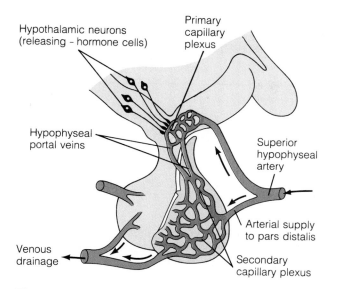

Figure 24.6   Structural and functional relationships between the adenohypophysis and the hypothalamus. The hypothalamic neurons (upper left) secrete releasing hormones, which travel through the hypophyseal portal veins to the pars distalis. These releasing hormones signal the adenohypophyseal cells to release their hormones.

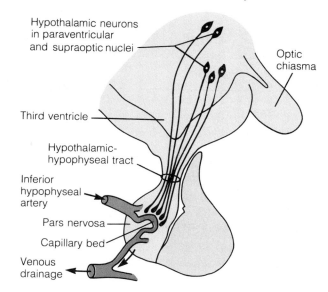

Figure 24.7   The neurohypophysis. Hypothalamic neurons, whose cell bodies lie in the paraventricular and supraoptic nuclei, send axons through the neurohypophysis and secrete the neurohypophyseal hormones from their axon terminals in the pars nervosa.

In summary, we have established that the hypothalamus controls the secretion of hormones by the adenohypophysis, which in turn controls the secretion of hormones by the thyroid gland, adrenal cortex, and gonads. In this way, the brain controls the body's largest and most important endocrine glands (Figure 24.2c).

The Neurohypophysis

The neurohypophysis, which secretes two hormones, is structurally a part of the brain (Figure 24.7). It consists of nervous tissue that contains unmyelinated axons and neuroglial cells. Its axons make up the **hypothalamic-hypophyseal tract**. This tract arises from neuron cell bodies in the supraoptic and paraventricular nuclei in the hypothalamus and ends in axon terminals in the pars nervosa. The neurohypophyseal hormones are synthesized in the neuron cell bodies, transported along the axons, and stored in the axon terminals. When the neurons fire, they release the stored hormones into a capillary bed in the pars nervosa for distribution throughout the body. Therefore, the neurohypophysis does not manufacture hormones but only stores and releases hormones produced in the hypothalamus.

The two neurohypophyseal hormones are (1) **antidiuretic hormone (ADH)** (an″tĭ-di″u-ret′ik; "inhibiting urination"), also called *vasopressin;* and (2) **oxytocin** (ok″sĭ-to′sin; "childbirth hormone"). ADH is manufactured in the neurons of the supraoptic nucleus, oxytocin in the paraventricular nucleus (see Table 24.2). Both of these hormones are peptides.

ADH targets the collecting tubules in the kidney, which respond by reabsorbing more water from the urine and returning it to the bloodstream. In this way, ADH helps the body to retain as much fluid as possible during times of dehydration (thirst) or fluid loss (severe bleeding, for example). Oxytocin induces contraction of the smooth musculature of both male and female reproductive organs. Most importantly, it signals the myometrium of the uterus to contract, expelling the infant during childbirth. Oxytocin also induces contraction of muscle-like cells (myoepithelial cells) around the secretory alveoli in the breast to eject milk during breast-feeding.

Sometimes, such as when a pregnancy has lasted far beyond the due date, it is necessary for a physician to induce labor. Can you guess how labor is induced?

There are several ways, but the most effective way involves injecting the mother with natural or synthetic oxytocin, which initiates uterine contractions. Less often, such oxytocics are used to stop bleeding after the delivery (by causing constriction of the ruptured blood vessels at the site of placental attachment) and to stimulate the ejection of milk. ∎

The Thyroid Gland

The butterfly-shaped **thyroid gland** is located in the anterior neck, on the trachea just inferior to the larynx. Its two lateral **lobes** (the butterfly "wings") are connected by a median bridge called the **isthmus**. The

Table 24.2 Cells and Hormones of the Pituitary Gland

Cell type	Hormone	Target/effects
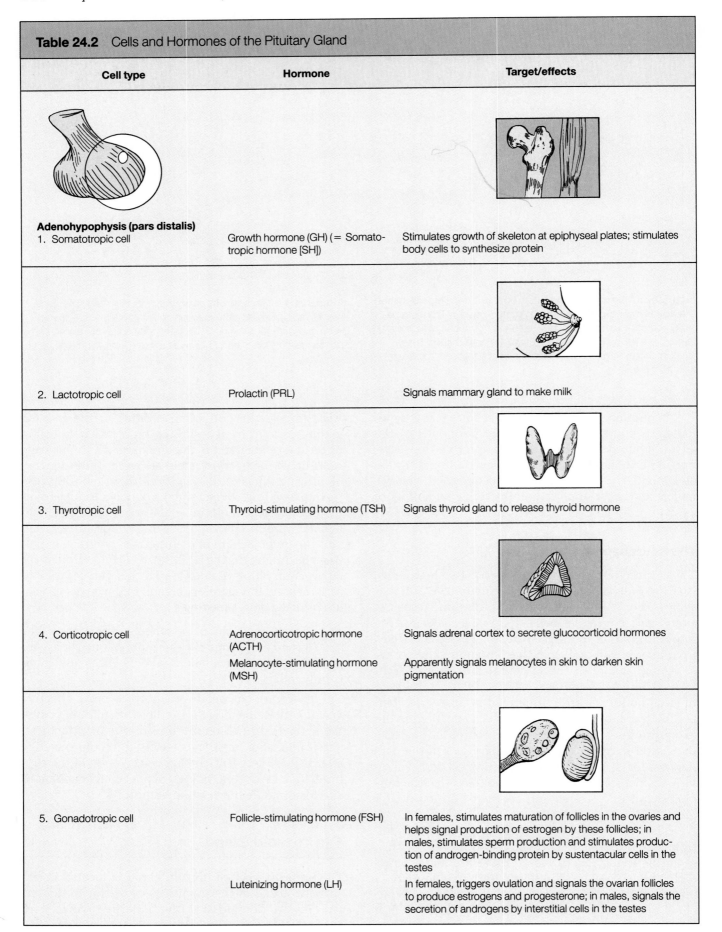 **Adenohypophysis (pars distalis)** 1. Somatotropic cell	Growth hormone (GH) (= Somatotropic hormone [SH])	Stimulates growth of skeleton at epiphyseal plates; stimulates body cells to synthesize protein
2. Lactotropic cell	Prolactin (PRL)	Signals mammary gland to make milk
3. Thyrotropic cell	Thyroid-stimulating hormone (TSH)	Signals thyroid gland to release thyroid hormone
4. Corticotropic cell	Adrenocorticotropic hormone (ACTH)	Signals adrenal cortex to secrete glucocorticoid hormones
	Melanocyte-stimulating hormone (MSH)	Apparently signals melanocytes in skin to darken skin pigmentation
5. Gonadotropic cell	Follicle-stimulating hormone (FSH)	In females, stimulates maturation of follicles in the ovaries and helps signal production of estrogen by these follicles; in males, stimulates sperm production and stimulates production of androgen-binding protein by sustentacular cells in the testes
	Luteinizing hormone (LH)	In females, triggers ovulation and signals the ovarian follicles to produce estrogens and progesterone; in males, signals the secretion of androgens by interstitial cells in the testes

Table 24.2   (continued)

Cell type	Hormone	Target/effects
Neurohypophysis		
1. Neurons from supraoptic nucleus of hypothalamus	Antidiuretic hormone (ADH) (= vasopressin)	Stimulates kidneys (collecting tubules) to reclaim more water from the urine
2. Neurons from paraventricular nucleus of hypothalamus	Oxytocin	Signals contraction of smooth muscle of male and female reproductive tracts; initiates labor, initiates ejection of milk from breast

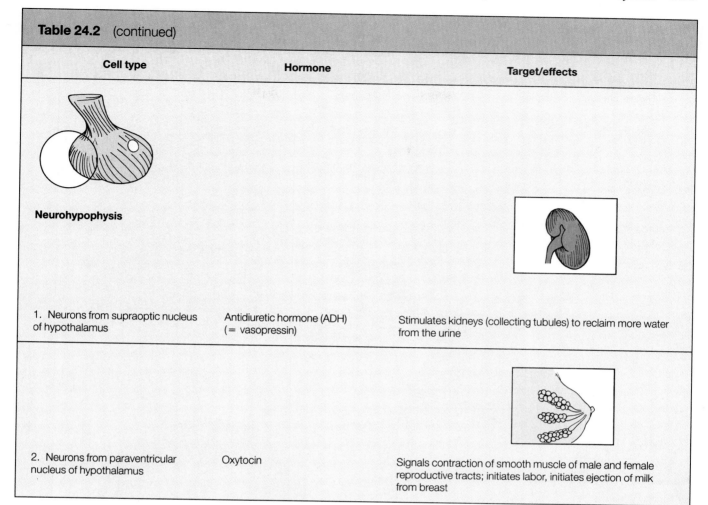

thyroid gland is the largest pure endocrine gland in the body and has a prodigious blood supply from the superior and inferior thyroid arteries (Figure 24.8a).

Internally, the thyroid gland is composed of hollow, spherical **follicles**, separated from one another by an areolar connective tissue rich in capillaries (Figure 24.8b). The walls of each follicle are formed by cuboidal or squamous epithelial cells called **follicle (follicular) cells**, and the central lumen is filled with a jelly-like substance called **colloid** (kol'oid; "glue-like"). This colloid consists of **thyroglobulin** (thi″ro-glob'u-lin), a protein from which thyroid hormone is ultimately derived.

The follicle cells of the thyroid gland secrete **thyroid hormone (TH)**. This name actually applies to two similar hormones called thyroxine (thi-rok'sin), or T_4, and triiodothyronine (tri″i-o″do-thi′ro-nēn), or T_3. Each of these hormone molecules is constructed from a pair of amino acids and contains the element iodine, which is essential for the function of the hormone. Thyroid hormone affects many target cells throughout the body. Its main function is to increase the basal metabolic rate (the rate at which the body uses oxygen to transform nutrients into energy).

Can you deduce the symptoms exhibited by people whose thyroid gland secretes excessive amounts of TH?

A person who secretes an excess of TH is "hyperactive," nervous, and continually feels warm, whereas a person who does not produce enough TH feels sluggish and cold. This and other disorders of the endocrine glands are covered in more detail in the box on pp. 676–677. ■

TH is manufactured and secreted by the thyroid follicles in the following way. The follicle cells continuously synthesize the precursor protein thyroglobulin and secrete it into the center of the follicle for iodination and storage. The thyroid gland is the only endocrine gland that stores its hormone extracellularly and in large quantities—its supply of TH is enough to last several months. To initiate secretion of the stored TH into the blood, the *pituitary* gland releases thyroid-stimulating hormone (TSH, *not* TH), which signals the follicle cells to reclaim their thyroglobulin by endocytosis. Next, TH is cleaved off the thyroglobulin molecules by lysosomal enzymes in the cytoplasm of the follicle cell. Then, this TH diffuses

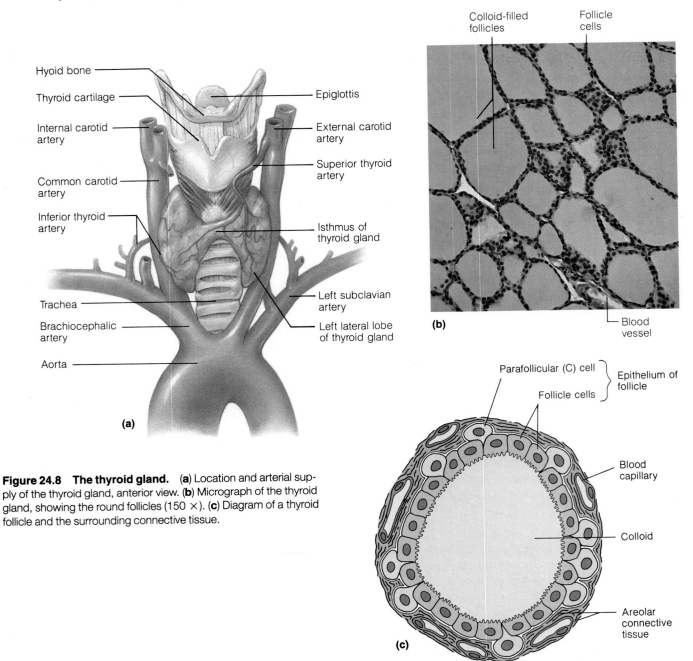

Figure 24.8 The thyroid gland. (a) Location and arterial supply of the thyroid gland, anterior view. (b) Micrograph of the thyroid gland, showing the round follicles (150 ×). (c) Diagram of a thyroid follicle and the surrounding connective tissue.

out of the follicle cells into the capillaries around the follicle, entering the circulation.

Another population of endocrine cells in the thyroid gland secretes an entirely different hormone. These **parafollicular**, or **C**, **cells** lie within the follicular epithelium but appear to project into the surrounding connective tissue (Figure 24.8c). Parafollicular cells secrete the protein hormone **calcitonin** (kal″sĭ-to′nin), which depresses excessive levels of Ca^{2+} in the blood by slowing the calcium-releasing activity of osteoclasts in bone. Calcitonin has no demonstrable effect in adults—it seems to act only in childhood, when the skeleton grows quickly and osteoclast activity is high.

The Parathyroid Glands

The small, yellow-brown **parathyroid glands** lie on the posterior surface of the thyroid gland (Figure 24.9). They are sometimes embedded in the substance of that gland, but they always remain distinct organs,

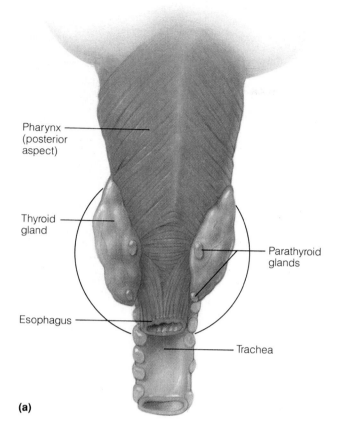

(a)

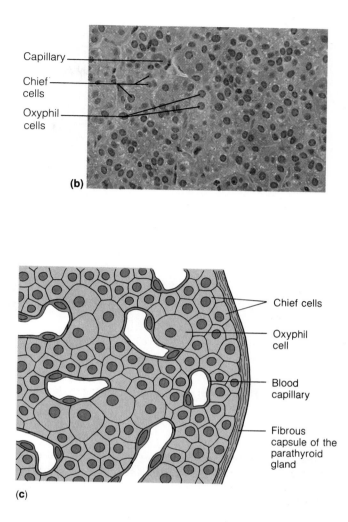

Capillary

Chief cells

Oxyphil cells

(b)

Chief cells

Oxyphil cell

Blood capillary

Fibrous capsule of the parathyroid gland

(c)

Figure 24.9   The parathyroid glands.   (a) Location of the parathyroid glands on the posterior aspect of the thyroid gland. This is a posterior view of the pharynx and trachea. (b) Micrograph of a section through a parathyroid gland (540 ×). (c) The histology of the parathyroid gland.

defined by their own capsules of connective tissue. Most people have two pairs of parathyroid glands, but the precise number varies. As many as eight glands have been reported, and some may be located in other regions of the neck or even in the thorax.

Histologically, the endocrine cells of the parathyroid gland are arranged in thick, branching cords (Figure 24.9b and c). The two types of cells in these cords are (1) small and abundant **chief cells** and (2) rare, larger **oxyphil cells** (ok'sĭ-fil; "acid-loving" ["acid-staining"]). The chief cells secrete the parathyroid hormone, which is a small protein molecule. Therefore, chief cells show the typical features of protein-secreting cells (abundant rough ER and secretory granules). The function of oxyphil cells, by contrast, is unknown.

Parathyroid hormone, also called **PTH** and **parathormone**, increases the concentration of Ca^{2+} in the blood whenever this concentration falls too low. This hormone is essential to life because low Ca^{2+} leads to

lethal neuromuscular disorders. Parathyroid hormone raises blood calcium levels by various mechanisms: (1) It stimulates osteoclasts to release more Ca^{2+} from bone; (2) it increases the reabsorption of Ca^{2+} by the distal tubules of the kidney; and (3) it activates vitamin D, which stimulates uptake of Ca^{2+} by the intestine.

Notice that parathyroid hormone and calcitonin have *opposite*, or antagonistic, effects: PTH raises blood calcium, whereas calcitonin lowers it.

Years ago, it was observed that while some patients recovered uneventfully after partial (or even total) removal of the *thyroid* gland, others exhibited uncontrolled muscle spasms and severe pain and rapidly deteriorated to their death. Can you explain why thyroid surgery sometimes proved lethal?

The lethal neuromuscular disorders resulted when the surgeon unknowingly removed the parathyroid glands along with the thyroid. ∎

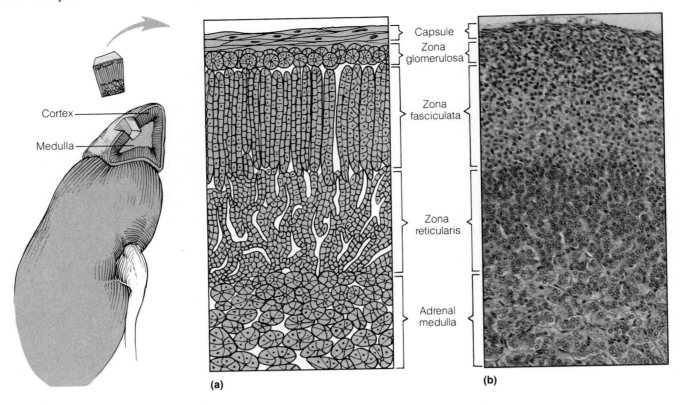

Figure 24.10   The adrenal gland.   Histology of the adrenal cortex and adrenal medulla: (a) Diagram. (b) Photomicrograph (50 ×).

The Adrenal (Suprarenal) Glands

The paired **adrenal (suprarenal) glands** are pyramidal or crescent-shaped organs perched on the superior surface of the kidneys (Figure 24.10). Although the name of this gland was officially changed from *adrenal* ("near the kidney") to *suprarenal* ("above the kidney") in the 1960s, the new name has not caught on, and most scientists still use *adrenal.*

Each adrenal gland is supplied by up to 60 small *suprarenal arteries,* which form three groups: the superior suprarenal arteries from the inferior phrenic artery, the middle suprarenal arteries from the aorta (p. 498), and the inferior suprarenal arteries from the renal artery. The single *suprarenal vein* drains into the renal vein (left) or the inferior vena cava (right) (see p. 506). The nerve supply, consisting almost exclusively of sympathetic fibers to the adrenal medulla, is richer than the nerve supply of any other visceral organ (p. 404).

Each adrenal gland is two endocrine glands in one. The internal *adrenal medulla,* more like a "knot" of nervous tissue than a gland, derives from the neural crest and acts as part of the sympathetic nervous system. The external *adrenal cortex,* surrounding the medulla and forming the bulk of the gland, derives from embryonic mesoderm. The cortex and medulla secrete hormones of entirely different chemical types, but all adrenal hormones help us cope with "extreme situations"—situations associated with danger, terror, or stress.

The Adrenal Medulla

The centrally located **adrenal medulla** is discussed in Chapter 14 as a part of the autonomic nervous system (pp. 403–404), so it is covered only briefly here. Its spherical *chromaffin* (kro-maf'in) *cells* are modified postganglionic sympathetic neurons that secrete catecholamines (the amine hormones epinephrine and norepinephrine) into the blood to enhance the fight, fright, and flight response. These cells store their catecholamines in cytoplasmic vesicles that can be stained with salts containing chromium metal (*chromaffin cells* = cells with an affinity for chromium). Within the adrenal medulla, the chromaffin cells are arranged in spherical clusters (with some branching cords).

The Adrenal Cortex

The thick **adrenal cortex** secretes a variety of hormones, all of which are *steroids.* Microscopically, the adrenal cortex exhibits three distinct layers, or zones (Figure 24.10b). From external to internal these are as follows:

1. Zona glomerulosa (glo-mer-u-lōs′ah). The cells in this zone are arranged in spherical clusters (*glomerulus* = ball of yarn).

2. Zona fasciculata (fah-sik″u-lah′tah). The cells in this zone are arranged in parallel cords (*fascicle* = bundle of parallel sticks) and contain an abundance of lipid droplets.

3. Zona reticularis (rĕ-tik′u-lar″is). The cells in this inner layer are arranged in a branching network (*reticulum* = network) and stain intensely with the pink dye eosin.

The hormones secreted by the adrenal cortex are called **corticosteroids** (kor″tĭ-ko-ste′roids). Along with the sex hormones from the testes and ovaries, corticosteroids are the major steroid hormones produced by the human body. There are two main classes of corticosteroids: *mineralocorticoids* (min″er-al-o-kor′tĭ-koids) and *glucocorticoids* (gloo″ko-kor′tĭ-koids). The **mineralocorticoids** are primarily produced in the zona glomerulosa. The main mineralocorticoid, **aldosterone** (al-dos′ter-ōn), is secreted in response to a decline in blood volume or in blood pressure. To compensate for this decline, aldosterone signals the kidneys (distal tubules) to reabsorb more water and sodium into the blood. The **glucocorticoids**, of which **cortisol** is the main type, are primarily produced in the zona fasciculata and zona reticularis. Glucocorticoids are released during stressful situations (fasting, anxiety, the stress of a crowded city, and so on), and help the body cope with stress. In essence, glucocorticoids keep blood glucose levels high enough to nourish the brain's activities while forcing most other body cells to switch to fats and amino acids as food sources. When present in excess, gluco-

corticoids depress the inflammatory response and inhibit the action of the immune system.

The zona reticularis also secretes some androgens (male sex hormones) and tiny amounts of estrogens (female sex hormones).

Based on the information above, can you answer the following clinical questions?

Which type of hormone is used as a drug to decrease the severity of inflammation in patients with inflammatory disorders?

Glucocorticoids are used because of their anti-inflammatory effects. For example, such hormones help suppress the inflammation of rheumatoid arthritis and severe allergies.

Tumors of the zona reticularis can produce marked effects on *females.* Can you deduce what these effects are?

The proliferating tumor cells secrete large amounts of androgens, leading to masculinization of females. The voice deepens, hair grows on the female's face, and so forth. ■

Structure of Steroid-Secreting Cells

Cells that secrete steroid hormones have many distinctive ultrastructural features (Figure 24.11). Whereas protein-secreting gland cells have an elaborate rough ER and abundant secretory granules, steroid-secreting cells have an abundant *smooth* endoplasmic reticulum (smooth ER) and no secretory granules at all. The abundance of smooth ER is readily explained: This organelle carries out most stages in the manufacture of lipid-based steroid molecules. The lack of secretory granules is also readily explained:

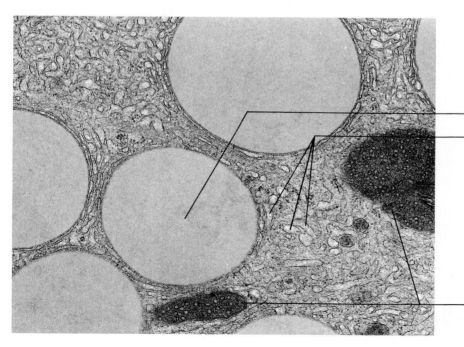

Figure 24.11 Ultrastructure of a cell that secretes steroid hormones (interstitial cell in the testis) Electron micrograph of a representative part of the cytoplasm (55,000 ×). The special features are labeled.

— Lipid droplet

— Smooth endoplasmic reticulum

— Mitochondria with tubular cristae

Steroids are secreted not by exocytosis (fusion of secretory granules to the plasma membrane), but through a direct and continuous outward diffusion of these hormones across the plasma membrane. The mitochondria of steroid cells have unusual swollen cristae (shaped like tubes instead of shelves) that carry out some of the steps of steroid synthesis. Finally, lipid droplets are abundant in the cytoplasm of steroid-secreting cells, because lipid provides the raw material from which steroids are made. These features characterize not only the cells of the adrenal cortex but also the cells in the testis and ovary that secrete the steroid sex hormones: the interstitial cells of Leydig, theca folliculi cells, and cells of the corpus luteum (Chapter 23).

The Pineal Gland

The **pineal gland** is a small, pine cone-shaped body at the end of a short stalk on the roof of the diencephalon (Figure 24.1 and Figure 12.13 on p. 331). Its endocrine cells, *pinealocytes,* are arranged in both spherical clusters and branching cords. Pinealocytes are star-shaped cells with long, branching cell processes. Within the adult pineal gland, dense particles of calcium lie between the cell clusters, forming the "pineal sand." Pinealocytes secrete the hormone **melatonin**, which helps regulate our circadian rhythms (Chapter 12, p. 333). Melatonin might also prevent premature maturation of the gonads.

Radiologists can easily locate the pineal gland on radiographs, and they use this gland as a landmark to determine the orientation of the brain on X-ray images. Can you guess what makes the pineal gland so radiopaque?

X-rays cannot penetrate the dense calcium mineral in the pineal sand. ∎

The Pancreas

Located in the posterior wall of the abdominal cavity, the tadpole-shaped *pancreas* is a mixed gland composed of both exocrine and endocrine cells. The exocrine *acinar cells,* forming the bulk of the gland, secrete digestive enzymes that are ducted into the small intestine during the digestion of food. (Recall Chapter 21, p. 590).

The endocrine cells of the pancreas are contained in spherical bodies called **pancreatic islets (islets of Langerhans)** (Figure 24.12). About a million of these islets are scattered among the exocrine acini. In each islet, the endocrine cells are arranged in twisted, branching cords, separated by capillaries. The main cell types in the islets are alpha and beta cells. The **alpha**, or **A, cells** secrete **glucagon** (gloo'kah-gon).

This hormone signals *liver* cells to release glucose from their glycogen stores, thus raising blood sugar levels whenever these levels have fallen too low. **Beta, or B, cells** secrete **insulin** ("hormone from the islets"). This hormone signals most cells of the body to take up glucose from the blood, thus lowering excessive blood sugar levels (after we digest a sugary snack, for example). At other times, the beta cells secrete *amylin,* a hormone that counteracts the action of insulin on cells of the body. Most alpha cells lie at the periphery of the pancreatic islets, whereas the more abundant beta cells distribute uniformly throughout the islets.

The pancreatic islets also contain two rare cell types (not illustrated). *Delta,* or *D, cells* secrete **somatostatin** (so"mah-to-stat'in), a hormone that inhibits the secretion of glucagon and insulin by the nearby alpha and beta cells. Finally, **F cells** secrete **pancreatic polypeptide**, a hormone that may act to inhibit the exocrine activity of the pancreas.

The pancreatic hormones are small proteins (insulin, amylin, glucagon, pancreatic polypeptide) and peptides (somatostatin). Correspondingly, the islet cells contain the typical features of cells engaged in protein secretion.

The Thymus

Located deep to the sternum in the thorax is the lobulated *thymus.* Recall that the thymus is an important organ of the immune system: T lymphocytes arise here from lymphocyte-precursor cells (p. 521). This transformation seems to be stimulated by **thymic hormones**, which are secreted by the structural cells of the thymus, the epithelial reticular cells. Thymic hormones are a family of peptide molecules, including thymopoietin (thi"mo-poi'ě-tin) and thymosin (thi'-mo-sin). The thymus is considered in detail in Chapter 19 (pp. 527–528).

The Gonads

The gonads (testes and ovaries) are the main source of the steroid sex hormones. In the male *testes,* interstitial cells between the sperm-forming tubules secrete **androgens** (mainly **testosterone**). These sex hormones maintain the reproductive organs and the secondary sex characteristics of males and help promote the formation of sperm. In the female *ovaries,* the theca folliculi and corpus luteum secrete **estrogens** and **progesterone**. Estrogens maintain the reproductive organs and secondary sex characteristics of females, and progesterone signals the uterus to prepare for pregnancy. Further details are presented in Chapter 23 (pp. 627–629, 635–638, and p. 643.).

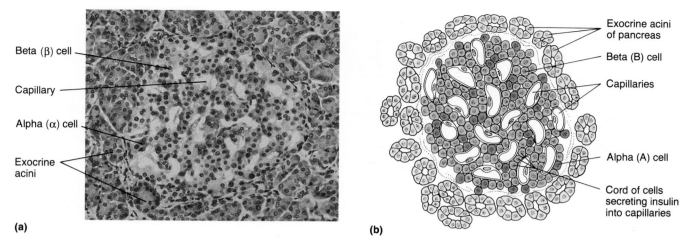

Figure 24.12   A pancreatic islet. (**a**) Micrograph (250 ×). In the spherical islet, the alpha and beta cells are stained differentially. (**b**) Diagram of an islet and the surrounding acini, similar to that shown in part (a).

Other Endocrine Structures

Other endocrine cells occur within various organs of the body, including the following:

1. The heart. The atria of the heart contain some specialized cardiac muscle cells that secrete **atrial natriuretic hormone (ANH)** (na″tre-u-ret′ik; "producing salty urine"). This peptide hormone decreases blood volume, blood pressure, and blood sodium concentration when these parameters become elevated. ANH accomplishes this task by signaling the kidney to increase its production of salty urine and by other mechanisms.

Pharmaceutical companies are exploring the possibility of manufacturing ANH to treat vascular hypertension. Can you explain their rationale?

Recall that hypertension, or high blood pressure, may lead to life-threatening vascular damage (strokes, heart failure, aneurysms). As a molecule that naturally lowers blood pressure, ANH may offer great clinical benefits. ∎

2. The gastrointestinal tract and its derivatives. *Enteroendocrine cells* are hormone-secreting cells scattered within the epithelial lining of the alimentary canal (recall Chapter 21, p. 578). Similar endocrine cells also occur within organs that derive from the embryonic gut, such as the respiratory tubes, pancreas, prostate, and thyroid gland. Collectively, these scattered epithelial cells make up the **diffuse neuroendocrine system (DNES)**. There are over 35 kinds of DNES cells, whose amine and peptide hormones perform such functions as regulating digestion, controlling aspects of blood chemistry, and adjusting local blood flow. Many of the hormones secreted by DNES cells are chemically identical to neurotransmitter molecules, and some of these hormones signal nearby target cells without first entering the bloodstream. Because such characteristics are neuron-like, the DNES epithelial cells are called neuroendocrine cells.

3. The placenta. Besides sustaining the fetus during pregnancy, the placenta secretes several steroid and protein hormones that influence the course of pregnancy. Placental hormones include estrogens, progesterone, human chorionic gonadotropin, and others (p. 649).

4. The kidney. The kidneys secrete several hormones. For example, the muscle cells of the juxtaglomerular apparatus secrete the protein hormone *renin* (p. 611), which indirectly signals the adrenal cortex to secrete aldosterone. Unidentified cells in the kidney secrete *erythropoietin* (ĕ-rith″ro-poi′ĕ-tin; "red-maker"), a protein hormone that signals the bone marrow to increase the production of red blood cells.

5. The skin. The epidermis of the skin produces vitamin D, a steroid hormone essential for calcium metabolism. When modified cholesterol molecules in epidermal cells are exposed to ultraviolet radiation, they are converted to a form of vitamin D. This compound then enters the capillaries of the dermis, travels throughout the bloodstream, undergoes chemical modification in the liver, and becomes fully activated in the kidney. The activated form of vitamin D signals the intestine to absorb Ca^{+2} from the diet. Without this vitamin, the bones weaken from insufficient calcium content.

A CLOSER LOOK   Hormones Out of Balance: Disorders of the Endocrine Glands

Most disorders of the endocrine glands reflect either a hypersecretion (oversecretion) or a hyposecretion (undersecretion) of specific hormones. Hypersecretion often results from a tumor in an endocrine gland as rapidly proliferating tumor cells secrete their hormones at an uncontrolled rate. Hyposecretion, by contrast, can result from factors that damage the endocrine glands (infections, autoimmune attack, physical trauma). A sampling of several noteworthy endocrine disorders is presented in this box.

Pituitary Disorders

Some disorders of the adenohypophysis affect the secretion of growth hormone (GH). For example, a tumor that leads to hypersecretion of GH in children causes **gigantism**, in which the body grows exceptionally rapidly and the person becomes abnormally tall (often reaching a height of 8 feet). If excessive amounts of GH are secreted after adult height has been attained and the epiphyseal plates have closed,

acromegaly (ak″ro-meg′ah-le) results. This is characterized by the enlargement of bony areas still responsive to GH, namely, the bones of the hands, feet, and face (*acromegaly* means "enlargement of the extremities"). The effects of acromegaly can be seen in the photographs below.

Hyposecretion of GH in children produces **pituitary dwarfs**, whose bodies are proportioned normally but who seldom reach 4 feet in height. If treated with injections of GH before they stop growing, many children with this condition can reach a near-normal height.

Diabetes insipidus (di″ah-be′tēz in-sĭ′pĭ-dus; "passing of dilute [urine]") is a relatively rare condition in which the pars nervosa of the pituitary does not secrete sufficient amounts of antidiuretic hormone. Consequently, the patient produces large quantities of urine and compensates by drinking large amounts of water. Diabetes insipidus can be caused by a blow to the head that damages the posterior pituitary or by a nearby tumor that compresses the pars nervosa.

Disorders of the Thyroid Gland

In **Graves' disease**, the thyroid gland secretes excessive amounts of thyroid hormone (TH). This seems to be an autoimmune disease, in which the body makes abnormal antibodies that mimic TSH and stimulate the secretion of TH by follicle cells of the thyroid. Typical symptoms of Graves' disease include an elevated metabolic rate, sweating, nervousness, and weight loss despite adequate food intake. The eyeballs may protrude, perhaps because of edema in the orbital tissue behind the eyes, or perhaps because the abnormal antibodies affect the extrinsic eye muscles. Graves' disease develops most often in women of middle age.

Insufficient secretion of TH produces different effects at different stages of life. In adults, hyposecretion of TH results in **myxedema** (mik″sĕ-de′mah; "mucous swelling"). Symptoms include a low metabolic rate, lethargy, constant chilliness, puffy eyes, edema, and mental sluggishness (but not mental retardation). When myxedema results

A person with acromegaly. This condition results from hypersecretion of growth hormone in the adult. Note the enlarged jaw, nose, and hands. From left to right, the same woman is shown at age 16, age 33, and age 52.

from an insufficient amount of iodine in the diet, the thyroid gland enlarges to raise a large, visible lump in the anterior neck, an **endemic** (en-dem′ik) **goiter**. In this condition, the cells of the thyroid follicles produce colloid but are unable to iodinate it or make functional hormones. The pituitary gland then secretes increasing amounts of TSH in a futile attempt to stimulate the thyroid to produce TH, but this action only causes the follicles to accumulate more and more colloid that cannot be processed. This excessive accumulation of colloid is the main reason that the thyroid gland swells. Before the marketing of iodized salt, parts of the midwestern United States were called the "goiter belts." Because these areas had iodine-poor soil and no access to iodine-rich shellfish, goiters were common there.

In children, an insufficient production of thyroid hormone leads to **cretinism** (kre′tĭ-nizm). The affected child has a short, disproportionate body and a thick tongue and neck and is mentally retarded.

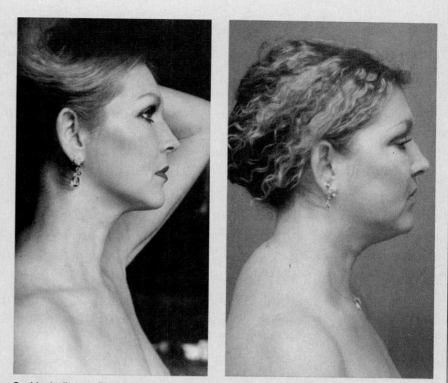

Cushing's disease. Photographs of a professional model before and after she developed Cushing's disease. In the photo at right, notice the swollen face and the "hump" on the posterior neck.

Disorders of the Adrenal Cortex

An excessive secretion of glucocorticoid hormones leads to **Cushing's disease** or **syndrome**. This can be caused by an ACTH-secreting tumor of the pituitary gland or by a tumor of the adrenal cortex. Mild cases can also result from large doses of glucocorticoid drugs, which are used to suppress inflammation. Cushing's syndrome is characterized by high concentrations of glucose in the blood and the loss of protein from muscles and bones. The so-called cushingoid signs include a swollen "moon" face, redistribution of fat to the abdomen and posterior neck (causing a "buffalo hump"), and depression of the immune and inflammatory responses (see the photographs above).

Addison's disease, the major hyposecretory disorder of the adrenal cortex, usually involves deficits in both glucocorticoids and mineralocorticoids. Concentrations of glucose and sodium in the blood drop,

and severe dehydration and low blood pressure are common.

A Disorder of the Pancreas: Diabetes Mellitus

Diabetes mellitus affects about 2% of Americans and has a strong hereditary component. It results from an insufficient secretion of insulin or from resistance of the body's cells to the effects of insulin. When insulin activity is absent or deficient, glucose cannot enter most cells, so blood sugar levels remain high, and glucose appears in the urine. (*Diabetes mellitus* literally means "passing [urine] that contains honey.") When sugars cannot be used as cellular fuel, the body's cells use fats for this purpose, and the acidic breakdown products of fat metabolism accumulate in the blood. If the disease is untreated, this acidity depresses almost all physiologic functions and leads to death.

There are two types of diabetes mellitus. **Type I**, or **insulin-dependent, diabetes** develops suddenly,

usually before the age of 15. It results from an autoimmune destruction of insulin-secreting beta cells in the pancreas. Insulin must be administered to type I diabetics several times each day to control blood glucose levels. After 20 or 30 years with this disease, most type I diabetics develop life-threatening complications: the lipid in their blood predisposes them to atherosclerosis (p. 484), and the excessive sugar in their body fluids disrupts capillary function (see *microangiopathic lesions* on p. 513).

Type II, or **non-insulin-dependent, diabetes** develops slowly, usually appears after age 40, and accounts for over 90% of all cases of diabetes. Most type II diabetics produce insulin, but their body cells have a lowered sensitivity to the effects of insulin. Type II diabetes is usually less serious and more easily managed than type I. It can usually be controlled by dietary means, such as losing weight, avoiding sugar-rich food, and eating small, frequent meals.

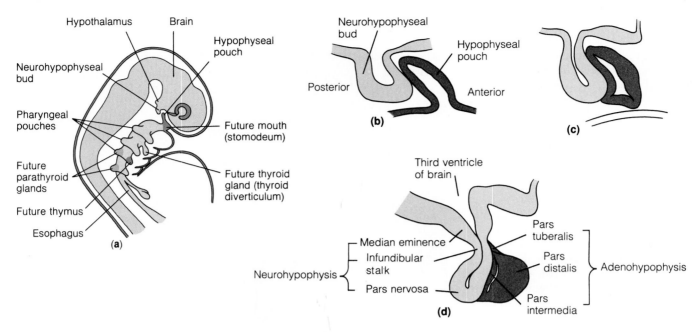

Figure 24.13 Embryonic development of some major endocrine organs. (a) An embryo in week 5 or 6. Four pharyngeal pouches are present on each side of the pharynx, and the hypophyseal pouch is pushing superiorly toward the brain from the roof of the mouth. Also note the future thyroid gland (thyroid diverticulum) on the pharyngeal floor. (b)–(d) Stages in the formation of the pituitary gland during weeks 6 to 8. The hypophyseal pouch (adenohypophysis) meets the neurohypophyseal bud (neurohypophysis) that is growing inferiorly from the floor of the brain.

The Endocrine System Throughout Life

Since the endocrine organs are so diverse and so widely distributed throughout the body, it is no surprise that their embryonic origins are diverse: These organs arise from all three germ layers.

The pituitary gland has a dual origin (Figure 24.13a–d). The adenohypophysis arises as a pouch of ectoderm—the **hypophyseal pouch**, or *Rathke's pouch*—from the roof of the mouth. This pouch contacts the future neurohypophysis, which grows inferiorly from the floor of the brain as the **neurohypophyseal bud**.

The thyroid gland forms from a thickening of endoderm on the floor of the pharynx (Figure 24.13a). This thickening first appears on the posterior part of the future tongue and then migrates caudally into the neck. Both the parathyroid glands and the thymus arise from the endoderm lining the pharyngeal pouches and then migrate to their final positions in the neck and thorax. The pineal gland arises from ependymal cells that cover the roof of the embryonic diencephalon.

The adrenal gland has a dual embryonic origin. Its medulla originates from the neural crest cells of nearby sympathetic chain ganglia (Chapter 14). The adrenal cortex, by contrast, develops from mesoderm lining the coelom on the dorsal abdominal wall.

Barring disease of the endocrine glands (hypersecretory and hyposecretory disorders), most endocrine organs operate smoothly throughout life until old age. The aging process may bring about changes in the rates of hormone secretion or in the sensitivity of target cell receptors to their hormones. Research on normal endocrine functioning in the elderly is difficult, however, because functioning is often altered by the chronic illnesses common in that age group.

The adenohypophysis changes with age. The amounts of connective tissue and aging pigment (lipofuscin) in the gland increase, vascularization decreases, and the number of hormone-secreting cells declines. These changes may or may not affect hormone production. For example, blood levels of ACTH remain constant, whereas levels of TSH and gonadotropins actually *increase* with age.

The adrenal cortex also shows structural changes with age, but normal rates of glucocorticoid secretion appear to persist as long as the person is healthy. No age-related changes in the release of catecholamines by the adrenal medulla have been found.

The synthesis and release of thyroid hormones diminish somewhat with age. Typically, the thyroid follicles are loaded with colloid in the elderly, and fibrosis of the gland occurs.

The parathyroid glands change little with age, and the concentrations of PTH in the blood remain at fairly normal values. However, the decline in estrogen secretion after menopause may sensitize women to the bone-demineralizing effects of PTH.

✚ Related Clinical Terms

Hypophysectomy Surgical removal of the pituitary gland.

Pheochromocytoma (fe″o-kro″mo-si″to′mah; "dusky color tumor") Tumor of the chromaffin cells of the adrenal medulla. It results in excessive secretion of catecholamine hormones, which produces the symptoms of a prolonged sympathetic response (especially hypertension).

Prolactinoma (*oma* = tumor) The most common type of pituitary gland tumor (30% to 40% or more), evidenced by a hypersecretion of prolactin, an excessive secretion of milk, and menstrual disturbances in women.

Psychosocial dwarfism Dwarfism (and failure to thrive) resulting from stress and emotional disorders that suppress the hypothalamic release of growth hormone-releasing hormone and thus adenohypophyseal secretion of GH.

Thyroid storm (thyroid crisis) A sudden and dangerous increase in the effects of thyroid hormone, due to excessive amounts of TH in the circulation. Metabolic rate is greatly increased, as indicated by fever, a rapid heart rate, high blood pressure, nervousness, and tremors. Precipitating factors include stressful situations, excessive intake of TH supplements, and trauma to the thyroid gland.

Chapter Summary

1. Endocrine organs are ductless glands that release hormones into the blood or lymph.

2. Hormones are messenger molecules that signal physiological changes in target cells.

3. Hormonally regulated processes include reproduction, growth, mobilization of body defenses against stressors, maintenance of the proper chemistry of the blood and body fluids, and regulation of cellular metabolism.

ENDOCRINE ORGANS AND HORMONES: AN OVERVIEW (pp. 662–663)

Endocrine Organs (p. 662)

1. Endocrine organs are small and widely separated from one another within the body. The pure endocrine organs are the pituitary, thyroid, parathyroid, adrenal, and pineal glands. Other organs that contain endocrine cells are the gonads, pancreas, kidney, alimentary canal, heart, thymus, and skin. The hypothalamus of the brain is a neuroendocrine organ.

2. Endocrine organs are richly vascularized.

3. Although most endocrine cells are modified epithelial cells, others are neurons, muscle cells, or fibroblast-like cells.

Hormones (p. 662)

4. Most hormones are either amino-acid derivatives (amines, peptides, proteins) or steroids (lipid-based molecules derived from cholesterol).

5. Hormones produce their effects by leaving the capillaries and binding to specific receptor molecules on or in their target cells. Such binding triggers a preprogrammed response in the target cell.

6. Endocrine organs are stimulated to release their hormones by humoral, neural, or hormonal stimuli.

7. The hypothalamus of the brain regulates many functions of the endocrine system through the hormones it secretes.

THE MAJOR ENDOCRINE ORGANS (pp. 663–674)

The Pituitary Gland (pp. 663–667)

1. The pituitary gland is suspended from the diencephalon of the brain by a stalk (infundibulum) and lies in the sella turcica of the sphenoid bone. It consists of an anterior adenohypophysis and a posterior neurohypophysis.

2. The adenohypophysis has three parts: pars distalis (anterior lobe), pars intermedia, and pars tuberalis. The neurohypophysis also has three parts: pars nervosa (posterior lobe), infundibular stalk, and median eminence.

3. The pituitary gland receives its rich blood supply from the superior and inferior hypophyseal arteries.

4. The largest part of the adenohypophysis is the pars distalis. It secretes seven protein hormones from five types of cells: (1) somatotropic cells (growth hormone: GH); (2) lactotropic cells (prolactin: PRL); (3) thyrotropic cells (thyroid-stimulating hormone: TSH); (4) corticotropic cells (adrenocorticotropic hormone and melanocyte-stimulating hormone: ACTH and MSH); and (5) gonadotropic cells (follicle-stimulating hormone and luteinizing hormone: FSH and LH). Cells in the pars distalis cluster into spheres (and some branching cords).

5. The basic functions of each adenohypophyseal hormone are these: GH stimulates growth of the body and skeleton; PRL signals milk production; TSH signals the thyroid gland to secrete thyroid hormone; ACTH signals the adrenal cortex to secrete glucocorticoids; MSH causes the skin to darken; FSH and LH signal the maturation of sex cells and the secretion of sex hormones. Four of these seven hormones stimulate other endocrine glands to secrete and are called tropic hormones: FSH, LH, ACTH, TSH.

6. The hypothalamus of the brain controls the secretion of hormones from the adenohypophysis. Certain hypothalamic neurons make releasing hormones and inhibiting hormones, which they send through the hypophyseal portal system to the pars distalis, signaling cells there to secrete their hormones. The adenohypophyseal hormones then enter a secondary capillary plexus and travel to their target cells throughout the body.

7. The neurohypophysis, which consists of nervous tissue, contains the hypothalamic-hypophyseal axon tract. The cell bodies of the neurons that form this tract are located in the paraventricular and supraoptic nuclei of the hypothalamus. These neu-

rons synthesize oxytocin and ADH, respectively, and store them in their axon terminals. As these neurons fire, they release their stored hormones into capillaries in the pars nervosa.

8. The neurohypophyseal hormones have the following functions: ADH increases reabsorption of water from the urine, and oxytocin induces labor and ejection of milk from the breasts.

The Thyroid Gland (pp. 667–670)

9. The thyroid gland lies on the superior trachea. Internally, it consists of spherical follicles covered by epithelial follicle cells and separated by a capillary-rich connective tissue. The follicles are filled with a colloid of thyroglobulin (a storage protein containing thyroid hormone).

10. Thyroid hormone (TH) contains iodine and increases basal metabolic rate. It is synthesized continuously by follicle cells and stored within the follicles until TSH from the pituitary gland signals the follicle cells to reclaim the TH and secrete it into the extrafollicular capillaries.

11. Parafollicular cells protrude from the thyroid follicles. These cells secrete the hormone calcitonin, which can lower blood calcium concentrations in children.

The Parathyroid Glands (pp. 670–671)

12. Several pairs of parathyroid glands lie on the dorsal aspect of the thyroid gland. Their chief cells are arranged in thick, branching cords and secrete parathyroid hormone. This hormone raises low blood calcium levels.

The Adrenal (Suprarenal) Glands (pp. 672–674)

13. The paired adrenal glands lie on the superior surface of each kidney. Each adrenal gland has two distinct parts, an outer cortex and an inner medulla.

14. The adrenal medulla consists of spherical clusters (and some branching cords) of chromaffin cells. These cells secrete excitatory catecholamines into the blood—the surge of adrenaline that is experienced during fight, fright, or flight situations.

15. The adrenal cortex has three layers: outer zona glomerulosa, middle zona fasciculata, and inner zona reticularis. The name of each zone describes its histological structure.

16. The steroid hormones secreted by the adrenal cortex (corticosteroids) include mineralocorticoids (mostly from the zona glomerulosa), glucocorticoids (mostly from the zona fasciculata and reticularis), and some sex hormones (from zona reticularis). Mineralocorticoids (mainly aldosterone) conserve water and sodium by increasing reabsorption of these substances by the kidney. Glucocorticoids (mainly cortisol) help the body cope with stress by stabilizing blood glucose levels. They also inhibit inflammation and actions of the immune system.

17. Steroid-secreting cells, including the cells in the gonads that secrete sex hormones, have an abundant smooth ER, tubular cristae in their mitochondria, abundant lipid droplets, and no secretory granules. These features relate to steroid synthesis.

The Pineal Gland (p. 674)

18. The pineal gland, on the roof of the diencephalon, contains pinealocytes, which cluster into spherical clumps and cords separated by dense particles of calcium called pineal sand.

19. Pinealocytes secrete the hormone melatonin, which helps regulate circadian rhythms.

The Pancreas (p. 674)

20. The endocrine structures in the pancreas are the spherical pancreatic islets. These islets consist of alpha (A), beta (B), delta (D) and F cells arranged in twisting cords.

21. Alpha cells secrete glucagon, which raises blood sugar levels, whereas beta cells secrete insulin, which lowers blood sugar levels.

The Thymus (p. 674)

22. The thymus, an important organ of the immune system, secretes thymic hormones that are essential for the production of T lymphocytes.

The Gonads (p. 674)

23. Various cells in the ovaries and testes secrete the steroid sex hormones, estrogens and androgens.

OTHER ENDOCRINE STRUCTURES (p. 675)

1. The heart acts as an endocrine organ: Some muscle cells in its atria secrete atrial natriuretic hormone, which stimulates loss of body fluids through the production of a sodium-rich urine.

2. Endocrine cells are scattered within the epithelium of the digestive tube and other gut-derived organs (respiratory tubes, and so on). These epithelial cells, which have some neuron-like properties, make up the diffuse neuroendocrine system (DNES). There are many classes of DNES cells, some of which secrete hormones that regulate digestion.

3. The following organs also have endocrine functions: placenta (hormones of pregnancy); kidney (renin and erythropoietin); and skin (vitamin D).

THE ENDOCRINE SYSTEM THROUGHOUT LIFE (pp. 678–679)

1. The endocrine glands have diverse developmental origins from all three germ layers. The adenohypophysis arises from ectoderm on the roof of the mouth, and the neurohypophysis arises from the floor of the brain. The endocrine organs in the neck (thyroid, parathyroids, thymus) derive from the endoderm of the pharynx. The pineal forms from the ependyma of the roof of the brain. The adrenal medulla develops from sympathetic chain ganglia, and the adrenal cortex arises from the mesoderm of the dorsal abdominal wall.

2. The efficiency of some endocrine organs gradually decreases as the body ages.

Review Questions

Multiple Choice and Matching Questions

1. The major stimulus for the release of parathyroid hormone is (a) hormonal, (b) humoral, (c) nervous.

2. The major stimulus for the release of estrogens is (a) hormonal, (b) humoral, (c) nervous.

3. Choose from the following key to identify the hormones described below.

Key: **(a)** melatonin   **(b)** antidiuretic hormone
      (c) somatotropin   **(d)** luteinizing hormone
      (e) TH   **(f)** TSH
      (g) prolactin   **(h)** oxytocin
      (i) cortisol   **(j)** parathyroid hormone

_____ **(1)** stimulates cell division in the epiphyseal plates of growing bones

_____ **(2)** involved in water balance; causes the kidneys to conserve water

_____ **(3)** stimulates milk production

_____ **(4)** stimulates milk ejection

_____ **(5)** tropic hormone that signals the gonads to secrete sex hormones

_____ **(6)** increases basal metabolic rate

_____ **(7)** tropic hormone that stimulates the thyroid gland to secrete thyroid hormone

_____ **(8)** adjusts blood sugar levels and helps the body cope with stress

———— **(9)** secreted by the pineal gland

———— **(10)** increases blood calcium levels

———— **(11)** secreted by the neurohypophysis (two possible choices)

———— **(12)** the only steroid hormone in the list

4. The pars distalis of the adenohypophysis does not secrete (a) antidiuretic hormone, (b) growth hormone, (c) gonadotropins, (d) TSH.

5. If adenohypophyseal secretion is deficient in a growing child, the child will (a) develop acromegaly, (b) become a dwarf, but have normal body proportions, (c) mature sexually at an age earlier than normal, (d) have an endemic goiter.

6. Which is true of hormones in general? (a) Exocrine glands produce them. (b) They travel throughout the body in blood. (c) They affect only non-hormone-producing organs. (d) All steroid hormones (sex hormones, glucocorticoids, mineralocorticoids) produce very similar physiological effects in the body.

7. The pure endocrine organs of the body (a) tend to be very large organs, (b) are closely connected with each other, (c) all contribute directly to the same function (digestion), (d) tend to lie near the midline of the body.

8. Structure of various endocrine cells. Write (a) or (b) in each blank as appropriate.

Key: **(a)** Secretes protein hormones: This cell has a well-developed rough ER and secretory granules.

(b) Secretes steroid hormones: This cell has a well-developed smooth ER, no secretory granules, unusual mitochondria, and lipid droplets.

———— **(1)** any endocrine cell in the pars distalis

———— **(2)** interstitial (Leydig) cell in the testis

———— **(3)** chief cell in the parathyroid gland

———— **(4)** zona fasciculata cell

———— **(5)** theca cell in the ovary that secretes sex hormones

———— **(6)** parafollicular cells in the thyroid gland

9. Histology of the pure endocrine glands. Match each gland below with the best approximation of its histological structure from the key.

Key: **(a)** spherical clusters of cells

(b) parallel cords of cells

(c) branching cords of cells

(d) follicles

(e) nervous tissue

———— **(1)** pars nervosa of pituitary gland

———— **(2)** zona glomerulosa of adrenal gland

———— **(3)** pars distalis of pituitary gland

———— **(4)** thyroid gland

———— **(5)** zona fasciculata of adrenal gland

10. What are the divisions of the neurohypophysis? (a) anterior lobe and posterior lobe, (b) pars emphasis, metropolis, and hypothesis, (c) pars distalis, tuberalis, and intermedia, (d) pars nervosa, infundibular stalk, and median eminence, (e) pars glomerulosa, fasciculata, and reticularis.

11. Which type of cell secretes releasing hormones? (a) a neuron, (b) a chromaffin cell, (c) a cell in the pars distalis, (d) a parafollicular cell.

12. Which of these endocrine organs is shaped like a butterfly? (a) pituitary gland, (b) pineal gland, (c) thyroid gland, (d) parathyroid gland, (e) adrenal gland.

13. Of the following endocrine structures, which develops from the ectoderm of the roof of the mouth? (a) neurohypophysis, (b) adenohypophysis, (c) thyroid, (d) pineal, (e) hypothalamus.

14. Chromaffin cells occur in the (a) parathyroid gland, (b) pars distalis, (c) pituitary gland, (d) adrenal gland, (e) pineal gland.

15. The hormone secreted by the heart, atrial natriuretic hormone, has exactly the opposite function of this hormone secreted by the zona glomerulosa: (a) antidiuretic hormone, (b) epinephrine, (c) cortisol, (d) aldosterone, (e) androgens.

16. The anterior lobe of the pituitary gland is the same as the (a) neurohypophysis, (b) pars nervosa, (c) pars distalis, (d) hypothalamus.

17. There is plenty of iodine in seawater and seafood. Therefore, in the days before iodine was added to commercial salt, endemic goiters were most common in this part of the United States: (a) East Coast, (b) West Coast, (c) Gulf Coast, (d) Midwest.

Short Answer and Essay Questions

18. Define hormone.

19. (a) Describe the body location of each of the following endocrine organs: anterior and posterior lobe of the pituitary gland; pineal gland; thyroid gland; parathyroid glands; and adrenal glands. (b) List the hormones secreted by each of these organs.

20. The adenohypophysis secretes so many hormones that it is often called the master endocrine organ, but it too has a "master." What controls the release of anterior pituitary hormones?

21. Define tropic hormone, and name a specific tropic hormone.

22. As Brandon was reading about the endocrine system, he realized that he had been confusing the thyroid gland, the thymus, and the thyroid cartilage with each other. Explain the difference among these three structures in terms of basic location and function. (Figures 19.11a and 24.8 will be of some help.)

23. Joshua explained to his classmate Jennifer that the thyroid gland contains parathyroid cells in its follicles and that the parathyroid cells secrete parathyroid hormone and calcitonin. Jennifer told him he was all mixed up again. Can you correct Josh's mistakes?

24. (a) Explain precisely where parafollicular cells are located in relation to the thyroid follicles. (b) Although it was not mentioned explicitly in this chapter, both the parafollicular cells and pancreatic islet cells are part of DNES. What do the letters DNES stand for?

25. Name a hormone secreted by a muscle cell and a hormone secreted by a neuron.

26. Are TH and TSH the same thing? If not, define both, and explain the difference.

Critical Thinking and Clinical Application Questions

27. The brain senses when we are in a stressful situation, and the hypothalamus responds to this by secreting a releasing hormone called corticotropin-releasing hormone. Through a long sequence of events, this hormone helps the body to deal with the stressful situation. Outline this entire sequence, starting with corticotropin-releasing hormone and ending with the release of cortisol. (Be sure to trace the hormones through the hypophyseal portal system and out of the pituitary gland.)

28. Jeremy, a 5-year-old boy, has been growing by leaps and bounds. His height is 70% above normal for his age group. A CT scan reveals a pituitary tumor. (a) What hormone is being secreted in excess? (b) What name is given to the condition that Jeremy will exhibit if corrective measures are not taken?

29. An accident victim who had not been wearing a seat belt received trauma to his forehead when he was thrown against the windshield. The physicians in the emergency room worried that his brain stem may have been driven inferiorly through the foramen magnum. To help assess this, they quickly took a standard X-ray film of his head and searched for the position of the pineal gland. How could anyone expect to find this tiny, boneless gland in an X-ray film?

30. Mrs. Giardino had an abnormally high concentration of calcium in her blood, and her physicians were certain she had a tumor of the parathyroid gland. Yet when surgery was performed on her neck, the surgeon could not find the parathyroid glands at all. Where should the surgeon look next to find the tumorous parathyroid gland?

31. A carnival sideshow came to a small town. Many health professionals in the town felt that such "freak" shows were cruel and exploitative, and asked the town council to enforce truth-in-advertising laws. The council agreed. They required that the fat man be billed as a "person with hypothyroidism," the dwarf and giant be called "people with pituitary disorders," the bearded lady be called a "woman with a tumor of the adrenal cortex," and the woman who could protrude her eyes from the orbits be called "a person with Graves' disease." Explain how each of these endocrine disorders produced the characteristic features of these five people.

25

Surface Anatomy

Student Objectives

1. Define surface anatomy, and explain why it is an important field of study. Define palpation.

The Head (pp. 684–688)

2. Describe and palpate the major surface features of the cranium and face.

The Neck (pp. 688–691)

3. Describe and define the important surface features of the neck, including the anterior and posterior triangles.

The Trunk (pp. 691–697)

4. Describe the easily palpated bony and muscular landmarks of the back. Locate the vertebral spines on the living body.

5. List the bony surface landmarks of the thoracic cage, and explain how they relate to the major soft organs of the thorax. Explain how to find any rib (second to eleventh).

6. Name and palpate the important surface features on the anterior abdominal wall.

7. Identify and explain the following: linea alba, umbilical hernia, examination for an inguinal hernia, linea semilunaris, and McBurney's point.

8. Explain how to palpate a full bladder and the bony boundaries of the perineum.

Upper Limb (pp. 698–703)

9. Locate and palpate the main surface features of the upper limb.

10. Define and explain the significance of the cubital fossa, pulse points in the distal forearm, and the anatomical snuff box.

Lower Limb (pp. 703–707)

11. Describe and palpate the surface landmarks of the lower limb.

12. Explain exactly where to administer an injection in the gluteal region and in the other major sites of intramuscular injection.

There is an old joke about a student who flunked anatomy because he was caught counting his ribs during an exam. Actually, that student was practicing the study of **surface anatomy**, an extremely valuable branch of anatomical and medical science. Surface anatomy does indeed study the external surface of the body. But more importantly, it also studies *internal* organs as they relate to external surface landmarks and as they are seen and felt through the skin. Feeling internal structures through the skin with the fingers is called **palpation** (pal-pa'shun; "touching").

Surface anatomy is living anatomy, better studied in live people than in cadavers. It can provide a great deal of information about the living skeleton (almost all bones can be palpated) and about the muscles and vessels that lie near the body surface. Furthermore, a skilled examiner can learn much about the heart, lungs, and other deep organs by performing a surface assessment. Thus, surface anatomy serves as the basis of the standard physical examination. If you are planning a career in the health sciences or physical education, your study of surface anatomy will show you where to take pulses, where to insert tubes and needles, where to make surgical incisions, where to locate broken bones and inflamed muscles, and where to listen for the sounds of the lungs, heart, and intestines.

The study of surface anatomy is influenced by the thickness of the subcutaneous layer deep to the skin. This fatty layer tends to be thicker in small children and women than in men and can be very thick in obese people. Obviously, surface anatomy is more difficult to study on those with a thick layer of subcutaneous fat. The examiner can often compensate, however, by pressing harder than usual to find the bony and muscular landmarks being sought. This is called *deep palpation.*

In this chapter, we take a regional approach to surface anatomy, exploring the head first and proceeding to the trunk and the limbs. We ask you to observe and palpate your own body as you work through the chapter, because your body is the best learning tool of all. To aid your exploration of living anatomy, Figures 25.1, 25.2, and 25.3 review the bones and muscles you will encounter, and Figure 25.4 previews various points where the arterial pulse can be palpated.

The Head

The head (Figures 25.5 and 25.6) is divided into the cranium and the face.

Cranium

Run your fingers over the superior surface of your head. Notice that the underlying cranial bones lie very near the surface. Proceed to your forehead and palpate the *superciliary arches* ("brow ridges") directly superior to your orbits (Figure 25.5a). Then move your hand to the posterior surface of the skull, where you can feel the knob-like *external occipital protuberance.* By running your finger directly laterally from this protuberance, feel the ridge-like *superior nuchal line* on the occipital bone. This line marks the superior extent of the muscles of the posterior neck and serves as the boundary between the head and the neck. Now feel the prominent *mastoid process* on each side of the cranium just posterior to your ear.

Place a hand on your temple, and clench your teeth together in a biting action. You should be able to feel the *temporalis muscle* bulge as it contracts. Next, raise your eyebrows, and feel your forehead wrinkle. This wrinkling is produced by the subcutaneous *frontalis muscle,* a muscle of facial expression. The frontalis inserts superiorly onto a broad aponeurosis called the *galea aponeurotica* (Table 10.1, p. 246), which covers the superior surface of the cranium. This aponeurosis binds tightly to the overlying subcutaneous tissue and skin to form the true **scalp**. Push on your scalp, and confirm that it slides freely over the underlying cranial bones. Because the scalp is only loosely bound to the skull, people can easily be "scalped" (in industrial accidents, for example). The scalp is richly vascularized by a large number of arteries running through its subcutaneous tissue. Most arteries of the body constrict and close after they are cut or torn, but those in the scalp are unable to do so because they are held open by the dense connective tissue surrounding them.

 What do these facts suggest about the amount of bleeding that accompanies scalp wounds?

Scalp wounds bleed profusely. However, because the scalp is so well vascularized, these wounds heal quickly. ■

Face

The surface of the face is divided into many different regions, including the orbital, nasal, oral (mouth), and auricular (ear) areas. In the region of the eyes (Figure 25.6), trace a finger around the entire bony margin of an orbit. The *lacrimal fossa* in the lacrimal and max-

(text continues on p. 688)

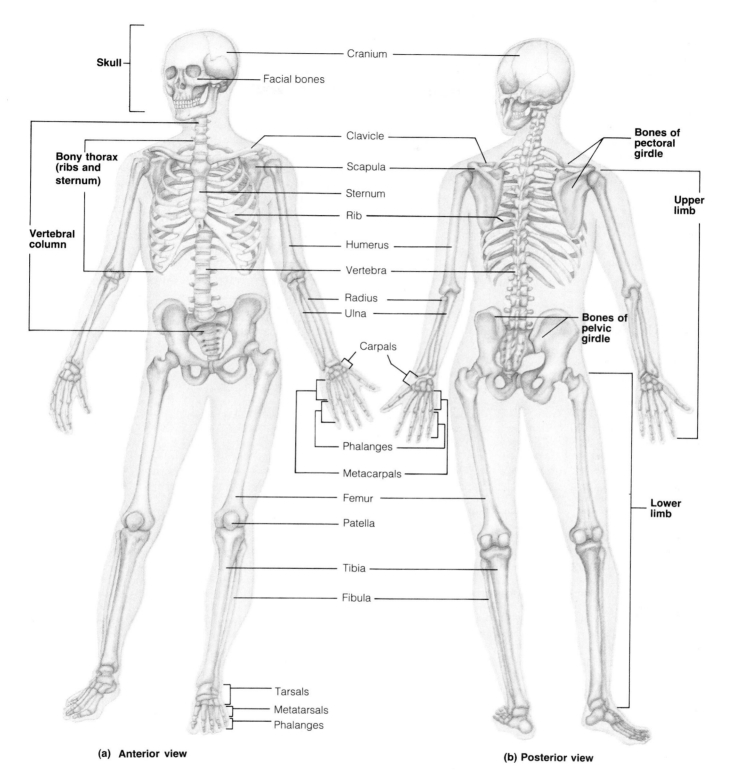

Skull

Cranium

Facial bones

Bony thorax (ribs and sternum)

Clavicle

Scapula

Sternum

Rib

Humerus

Vertebra

Vertebral column

Radius

Ulna

Carpals

Phalanges

Metacarpals

Femur

Patella

Tibia

Fibula

Tarsals

Metatarsals

Phalanges

Bones of pectoral girdle

Upper limb

Bones of pelvic girdle

Lower limb

(a) Anterior view

(b) Posterior view

Figure 25.1   Bones of the human skeleton.

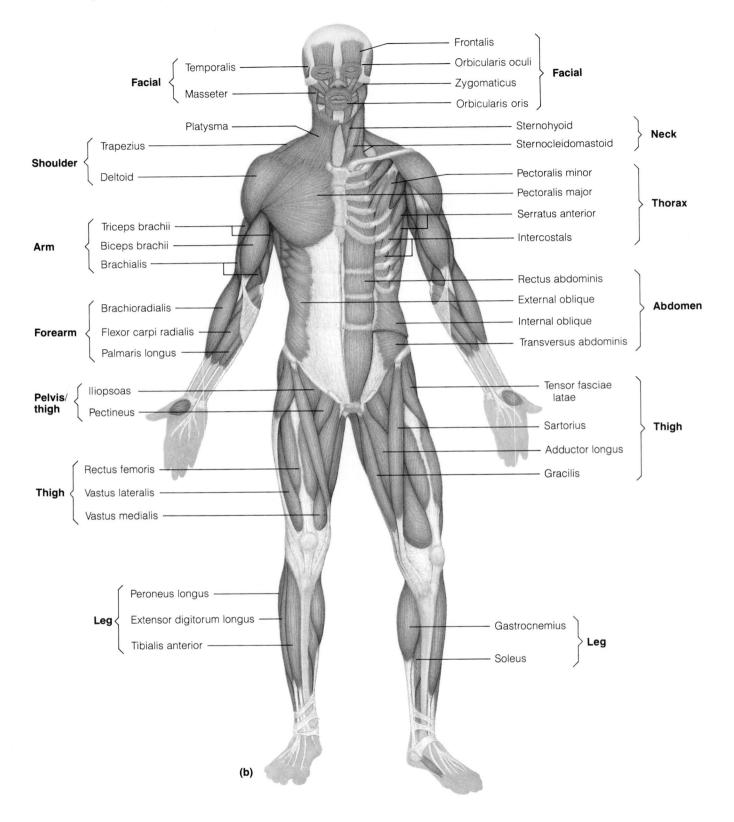

Facial
- Temporalis
- Masseter

Facial
- Frontalis
- Orbicularis oculi
- Zygomaticus
- Orbicularis oris

Platysma

Shoulder
- Trapezius
- Deltoid

Neck
- Sternohyoid
- Sternocleidomastoid

Thorax
- Pectoralis minor
- Pectoralis major
- Serratus anterior
- Intercostals

Arm
- Triceps brachii
- Biceps brachii
- Brachialis

Abdomen
- Rectus abdominis
- External oblique
- Internal oblique
- Transversus abdominis

Forearm
- Brachioradialis
- Flexor carpi radialis
- Palmaris longus

Pelvis/ thigh
- Iliopsoas
- Pectineus

Thigh
- Tensor fasciae latae
- Sartorius
- Adductor longus
- Gracilis

Thigh
- Rectus femoris
- Vastus lateralis
- Vastus medialis

Leg
- Peroneus longus
- Extensor digitorum longus
- Tibialis anterior

Leg
- Gastrocnemius
- Soleus

(b)

Figure 25.2   Anterior view of the superficial muscles of the body.   (On the left side of the trunk, some deeper muscles are shown.)

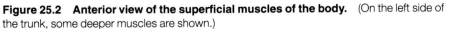

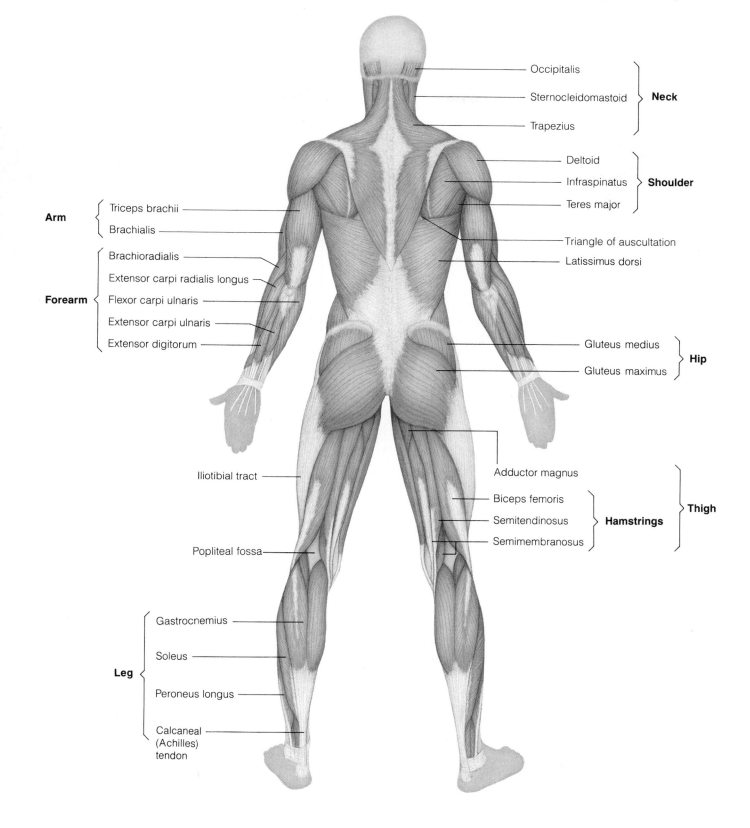

Figure 25.3 Posterior view of the superficial muscles of the body.

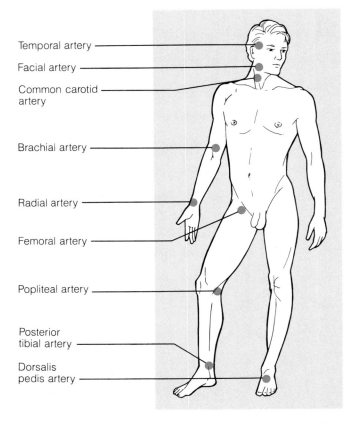

Figure 25.4 Points on the body surface where the pulse is most easily felt. The arteries indicated are discussed throughout the chapter, as are some additional pulse points that are not pictured here.

ymeter, a device that measures blood oxygen levels across the surface of the skin, may be clamped to the ear lobe. The auricle has several other regions (see Figure 25.5b), but they are not important to our discussion.

Next, place a finger on your temple just anterior to the auricle (Figure 25.5a). There, you will be able to feel the pulsations of the *superficial temporal artery,* which ascends to supply the scalp.

Run your hand anteriorly from the ear toward the orbit, and feel the *zygomatic arch* just deep to the skin. This "cheek bone" can easily be broken by blows to the face. Next, place your fingers on the skin of your face, and feel it bunch and stretch as you contort your face into a series of smiles, frowns, and grimaces. You are now monitoring the action of the subcutaneous *muscles of facial expression* (Table 10.1, p. 246).

In the region of the lower jaw, palpate the parts of the bony *mandible:* its anterior *body* and its posterior ascending *ramus.* Press on the skin over the mandibular ramus, and feel the *masseter muscle* bulge when you bite down. Palpate the anterior border of the masseter, and trace it to the mandible's inferior margin. At this point, you will be able to detect the pulse of your *facial artery* (Figure 25.5a). Finally, to feel the *temporomandibular joint,* place a finger directly anterior to the tragus of your ear, and open and close your mouth several times. The bony structure you feel moving is the *head of the mandible.*

The Neck

Bony Landmarks

Begin by running your fingers inferiorly along the back of your neck, in the posterior midline. There you can feel the *spinous processes* of the cervical vertebrae (the spine of C$_7$ is especially prominent). Now, place a finger on your chin, and run it inferiorly along the neck's anterior midline (Figure 25.7). The first hard structure you encounter will be the U-shaped *hyoid bone,* which lies in the angle between the floor of the mouth and the vertical part of the neck (Figure 25.8). Directly inferior to this, you will feel the *laryngeal prominence* (Adam's apple) of the thyroid cartilage. This prominence begins with a V-shaped notch superiorly and continues inferiorly as a sharp vertical ridge. Just inferior to the prominence, your finger will sink into a soft depression (formed by the *cricothyroid ligament*) and then will proceed onto the rounded surface of the *cricoid cartilage.* Now swallow several times, and feel the whole larynx move up and down. Continuing inferiorly onto the upper trachea, you

illary bones, which contains the tear-gathering lacrimal sac, may be felt on the medial side of the eye socket. Next, you may wish to review the surface anatomy of the eye and eyelids, using Figure 15.4 on page 416. Then, touch the most superior part of your nose, its *root* between the eyebrows (Figure 25.6). Just inferior to this, between your eyes, is the *bridge* of the nose formed by the nasal bones. Continue your finger inferiorly along the nose's anterior margin, the **dorsum nasi,** to its tip, the **apex.** Place one finger in a *nostril* and another finger on the flared wing, the **ala,** that defines the nostril's lateral border. Then feel the **philtrum** (fil′trum; "love"), the shallow vertical groove on your upper lip below the nose.

Grasp your *auricle,* the shell-like part of the external ear that surrounds the opening of the *external auditory canal* (Figure 25.5b). Just posterior to this canal, you can feel the deep ear well, the **concha.** Directly anterior to the canal is a stiff projection called the **tragus** (tra′gus; "goat"). The tragus helps protect the opening to the canal. Now trace the ear's outer rim, or *helix,* to the *lobule* (ear lobe) inferiorly. The lobule is easily pierced, and since it is not highly sensitive to pain, it provides a convenient place to obtain a drop of blood for clinical blood analysis. A *pulse ox-*

The labels in the figure read:
Temporal artery
Facial artery
Common carotid artery
Brachial artery
Radial artery
Femoral artery
Popliteal artery
Posterior tibial artery
Dorsalis pedis artery

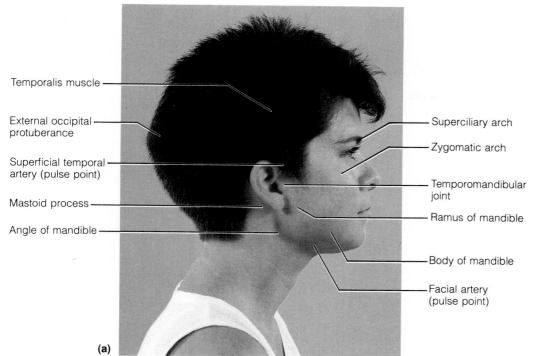

Temporalis muscle

External occipital protuberance

Superficial temporal artery (pulse point)

Mastoid process

Angle of mandible

Superciliary arch

Zygomatic arch

Temporomandibular joint

Ramus of mandible

Body of mandible

Facial artery (pulse point)

(a)

Figure 25.5   Surface anatomy of the lateral aspect of the head. (**a**) Overview. (**b**) Close-up of an auricle.

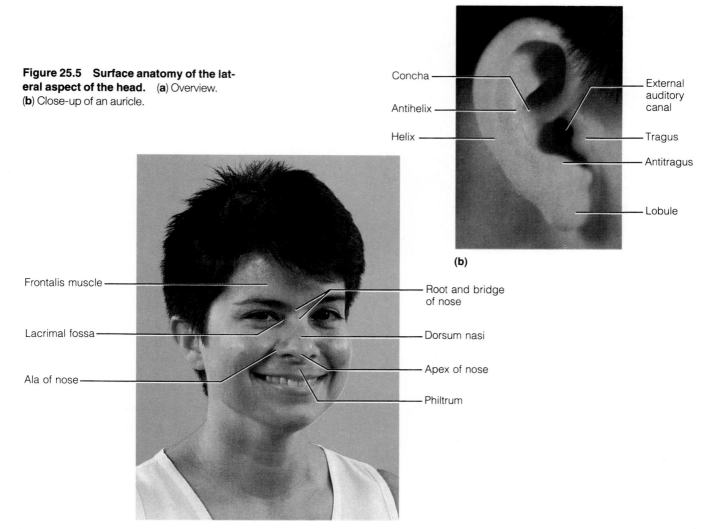

Concha

Antihelix

Helix

External auditory canal

Tragus

Antitragus

Lobule

(b)

Frontalis muscle

Lacrimal fossa

Ala of nose

Root and bridge of nose

Dorsum nasi

Apex of nose

Philtrum

Figure 25.6   Surface structures of the face.

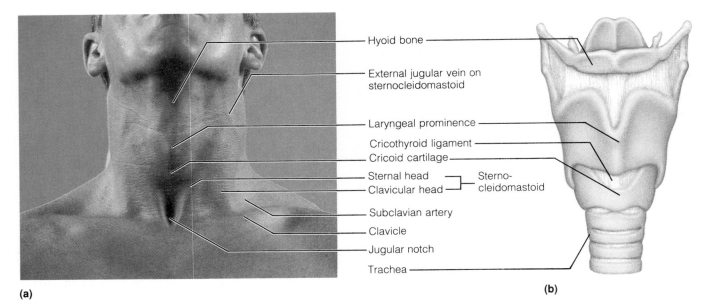

(a)

(b)

Figure 25.7   Anterior surface of the neck. (a) Photograph. (b) Diagram of the underlying skeleton of the larynx.

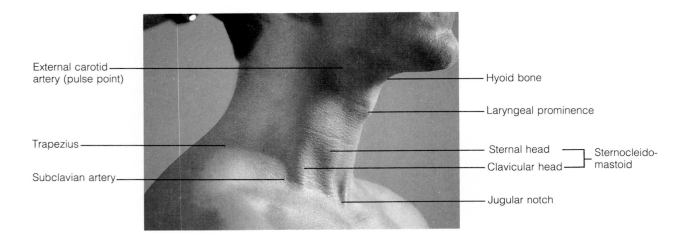

Figure 25.8   Lateral surface of the neck.

might be able to feel the *isthmus of the thyroid gland* as a spongy cushion over the second to fourth tracheal rings (also see Figure 24.8a on p. 670). Then, try to palpate the two soft lateral *lobes* of your thyroid gland along the sides of the trachea. Next, move your finger all the way inferiorly to the root of the neck, and rest it in the *jugular notch.* This is the depression in the superior part of the sternum between the two clavicles. By pushing deeply at this point, you can feel the cartilage rings of the trachea.

Muscles

The *sternocleidomastoid* is the most prominent muscle in the neck and the neck's most important surface landmark. You can best see and feel it when your head is turned to the side. If you stand in front of a mirror and turn your face sharply from right to left several times, you will be able to see both heads of this muscle, the *sternal head* medially and the *clavicular head* laterally (Figures 25.7 and 25.8). Physicians palpate the sternocleidomastoid as they search for any swelling of the *cervical lymph nodes* that lie both superficial and deep to this muscle. (Swollen cervical nodes provide evidence of infections or cancer of the head and neck.) The *common carotid artery* and *internal jugular vein* lie just deep to the sternocleidomastoid, a relatively superficial location that exposes these vessels to danger in slashing wounds to the neck. Just lateral to the inferior part of the sternocleidomastoid lies the large *subclavian artery* on its way to supply the upper limb. By pushing on the subclavian artery at this point, one can stop the bleeding from a wound anywhere in this limb. Just anterior to the sternocleidomastoid, superior to the level of your larynx, you can feel a carotid pulse—the pulsations of the *external carotid artery* (Figure 25.8). The *external jugular vein* descends vertically, just superficial to the sternocleidomastoid and deep to the skin (Figure 25.7). To make this vein "appear" on your neck, stand before a mirror, and gently compress the skin superior to your clavicle with your fingers.

Another large muscle in the neck, on the posterior aspect, is the *trapezius* (Figure 25.8). You can feel this muscle contract just deep to the skin as you shrug your shoulders.

Triangles of the Neck

The sternocleidomastoid muscles divide each side of the neck into the posterior and anterior triangles (Figure 25.9a). The **posterior triangle** is defined by the sternocleidomastoid anteriorly, the trapezius posteriorly, and the clavicle inferiorly. The **anterior triangle** is defined by the inferior margin of the mandible su-

periorly, the midline of the neck anteriorly, and the sternocleidomastoid posteriorly.

The contents of these two triangles are shown in Figure 25.9b. The posterior triangle contains many important nerves and blood vessels, including the *accessory nerve* (cranial nerve XI), most of the *cervical plexus,* and the *phrenic nerve.* In the inferior part of the triangle are the external jugular vein, the trunks of the *brachial plexus,* and the subclavian artery. These structures are relatively superficial and are easily cut by lacerations or slashing wounds to the neck. In the neck's anterior triangle, the important structures include the *submandibular gland,* the *suprahyoid* and *infrahyoid muscles,* and the parts of the carotid arteries and jugular veins that lie superior to the sternocleidomastoid.

A wound to the posterior triangle of the neck can lead to long-term loss of sensation in the skin of the neck and shoulder, as well as a partial paralysis of the sternocleidomastoid and trapezius muscles. Can you explain these effects?

They result from damage to the cervical plexus and accessory nerve, which supply these skin regions and muscles. ■

The Trunk

The trunk of the body consists of the thorax, abdomen, pelvis, and perineum. The *back* includes parts of all of these regions, but for convenience it is treated separately.

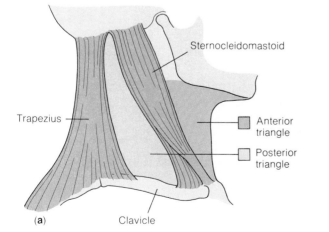

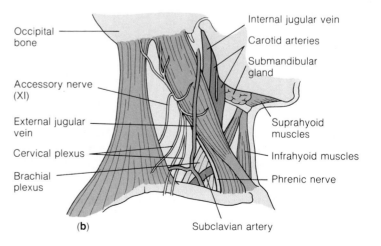

Figure 25.9 Anterior and posterior triangles of the neck. (a) Boundaries of the triangles. (b) Some contents of the triangles.

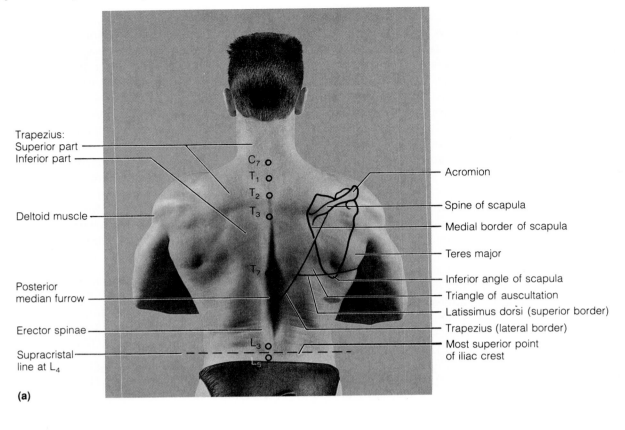

Trapezius:
Superior part
Inferior part

C$_7$ o
T$_1$ o
T$_2$ o
T$_3$ o

Deltoid muscle

T$_7$

Posterior
median furrow

Erector spinae

Supracristal
line at L$_4$

L$_3$ o
L$_5$ o

Acromion

Spine of scapula

Medial border of scapula

Teres major

Inferior angle of scapula

Triangle of auscultation

Latissimus dorsi (superior border)

Trapezius (lateral border)

Most superior point
of iliac crest

(a)

Trapezius

Biceps brachii

Triceps brachii

Superior border of
latissimus dorsi

Deltoid:
Anterior part

Intermediate part

Posterior part

Teres major

Inferior angle
of scapula

Erector spinae

(b)

Figure 25.10   Surface anatomy of the back, two different poses, (a) and (b).

The Back

Bones

The vertical groove in the center of the back is called the **posterior median furrow** (Figure 25.10a). The *spinous processes* of the vertebrae are visible in this furrow when the spinal column is flexed. Palpate a few of these processes on your back (C_7 and T_1 are the most prominent and the easiest to find). Although locating structures on your own back is difficult, you should be able to palpate the posterior parts of some ribs, as well as the prominent *spine of the scapula* and the scapula's long *medial border.* The scapula lies superficial to ribs 2 to 7, its *inferior angle* is at the level of the spinous process of vertebra T_7, and the medial end of the scapular spine lies opposite the spinous process of T_3.

Now feel the *iliac crests* (superior margins of the iliac bones) in your lower back. You can find these crests effortlessly by resting your hands on your hips. Locate the most superior point of each crest, a point that lies roughly halfway between the posterior median furrow and the lateral side of the body (Figure 25.10a). A horizontal line through the two superior points, the **supracristal line**, intersects L_4, providing a simple way to locate that vertebra. The ability to locate L_4 is essential for performing a *lumbar puncture,* a procedure in which the clinician inserts a needle into the vertebral canal of the spinal column directly superior or inferior to L_4. (Lumbar punctures for withdrawing cerebrospinal fluid are described on p. 345.)

The *sacrum* is easy to palpate just superior to the cleft in the buttocks, and you can feel the *coccyx* in the extreme inferior part of that cleft, just posterior to the anus.

Muscles

The largest superficial muscles of the back are the *trapezius* superiorly and *latissimus dorsi* inferiorly (Figures 25.3 and 25.10b). Furthermore, the deeper *erector spinae* muscles are very evident in the lower back, flanking the vertebral column like thick vertical cords. Feel your erector spinae muscles contract and bulge as you extend your spine.

The superficial muscles of the back fail to cover a small area of the rib cage called the **triangle of auscultation** (aw″skul-ta′shun) (Figure 25.10a). This triangle lies just medial to the inferior part of the scapula, and its three boundaries are formed by the trapezius medially, the latissimus dorsi inferiorly, and the scapula laterally (also see Figure 25.3). The physician places a stethoscope over the skin of this triangle to listen for lung sounds (*auscultation* = listening). To hear the lungs clearly, the doctor first asks the patient to fold the arms together in front of the chest and then flex the trunk.

Can you deduce the precise reason for having the patient take this action?

This action draws the scapula anteriorly and enlarges the triangle of auscultation as much as possible. ∎

The Thorax

Bones

Start exploring the anterior surface of your bony *thoracic cage* (Figure 25.11) by defining the extent of the *sternum* (Figure 25.1a). Use a finger to trace your sternum's triangular *manubrium,* the flat sternal *body,* and the tongue-shaped *xiphoid process.* Now palpate the ridge-like *sternal angle,* where the manubrium meets the body of the sternum. Locating the sternal angle is important because it directs you to the second ribs (which attach to it), and once you find the second rib, you can count down to identify every other rib in the thorax (except the first and sometimes the twelfth rib, which lie too deeply to be palpated). The sternal angle is a highly reliable landmark—it is easy to locate, even in overweight patients.

By locating your individual ribs, you attain a series of horizontal lines of "latitude" by which to map and locate the underlying visceral organs of the thoracic cavity. Such mapping also requires lines of "longitude," so let us construct some vertical lines on the wall of the trunk. Lift an arm straight up in the air, and extend a line inferiorly from the center of the axilla onto the lateral thoracic wall. This is the **midaxillary line** (Figure 25.12a). Now estimate the midpoint of your *clavicle,* and run a vertical line inferiorly from that point to the groin. This is the **midclavicular line**, and it will pass about 1 cm medial to the nipple.

Next, feel along the V-shaped inferior edge of your rib cage, the *costal margin.* At the superior angle of the costal margin, the **infrasternal angle**, lies the *xiphisternal joint.* Deep to the xiphisternal joint, the heart lies on the diaphragm.

The thoracic cage provides many valuable landmarks for locating the vital organs of the thoracic and abdominal cavities. On the anterior thoracic wall, ribs 2–6 define the superior-to-inferior extent of the female breast, and the fourth intercostal space defines the location of the *nipple* in men, children, and small-breasted women. The right costal margin runs across the anterior surface of the liver and gallbladder (see Figure 21.4 on p. 565). Surgeons must be aware of the inferior margin of the *pleural cavities* (Figure 25.12): If they accidentally cut into one of these cavities, a lung collapses (p. 551). The inferior pleural margin lies adjacent to vertebra T_{12} near the posterior midline (Figure 25.12b) and runs horizontally across

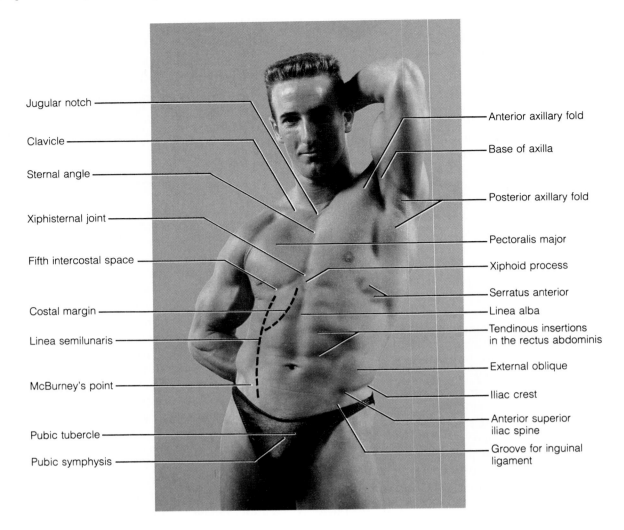

Jugular notch

Clavicle

Sternal angle

Xiphisternal joint

Fifth intercostal space

Costal margin

Linea semilunaris

McBurney's point

Pubic tubercle

Pubic symphysis

Anterior axillary fold

Base of axilla

Posterior axillary fold

Pectoralis major

Xiphoid process

Serratus anterior

Linea alba

Tendinous insertions in the rectus abdominis

External oblique

Iliac crest

Anterior superior iliac spine

Groove for inguinal ligament

Figure 25.11   The anterior thorax and abdomen.

the back to reach rib 10 at the midaxillary line. From there, the pleural margin ascends to rib 8 in the midclavicular line (Figure 25.12a) and to the level of the xiphisternal joint near the anterior midline. The *lungs* do not fill the inferior region of the pleural cavity. Instead, their inferior borders run at a level that is two ribs superior to the pleural margin, until they meet that margin near the xiphisternal joint.

The relation of the *heart* to the thoracic cage is considered in Chapter 17 (p. 460) and illustrated in Figure 17.2a. In essence, the superior right corner of the heart lies at the junction of the third rib and the sternum; the superior left corner lies at the second rib, near the sternum; the inferior left corner lies in the fifth intercostal space in the midclavicular line; and the inferior right corner lies at the sixth rib, near the sternum. You may wish to outline the heart on your chest by connecting the four corner points with a washable marker.

Can you recall the clinical importance of the four corners of the heart, with respect to the heart valves?

Clinicians listen for the sounds of the four heart valves near each of these heart corners (Figure 17.11 and p. 468). ■

Muscles

The main superficial muscles of the anterior thoracic wall are the *pectoralis major* and the anterior slips of the *serratus anterior*. Using Figure 25.11 as a guide, try to palpate these two muscles on your chest. They both contract during push-ups, and you can confirm this by pushing yourself up from your desk with one arm while palpating the muscles with your opposite hand.

The Abdomen

Bony Landmarks

The anterior abdominal wall (Figure 25.11) extends inferiorly from the costal margin to an inferior boundary that is defined by the following landmarks:

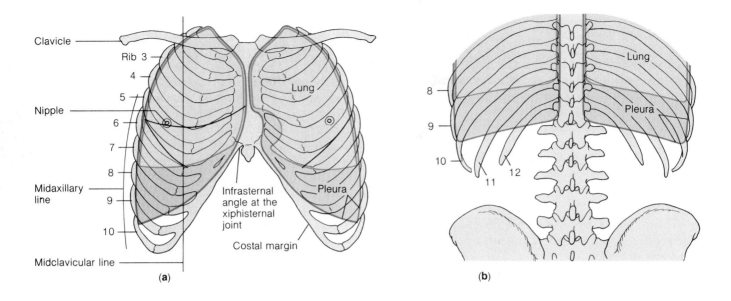

Figure 25.12 The bony rib cage as it relates to the underlying lungs and pleural cavity. Both the pleural cavities (blue) and the lungs (red) are outlined. (**a**) Anterior view. (**b**) Posterior view.

1. Iliac crest. Recall that the iliac crests are the superior margins of the iliac bones, and you can locate them by resting your hands on your hips.

2. Anterior superior iliac spine. Representing the most anterior point of the iliac crest, this spine is a prominent landmark. It can be palpated in everyone, even those who are overweight.

3. Inguinal ligament. The inguinal ligament, indicated by a groove on the skin of the groin, runs medially from the anterior superior iliac spine to the pubic tubercle of the pubic bone.

4. Pubic crest. You will have to press deeply to feel this crest on the pubic bone near the median *pubic symphysis.* The *pubic tubercle,* the most lateral point of the pubic crest, is easier to palpate, but you will still have to push deeply.

Inguinal hernias lie directly superior to the inguinal ligament and may exit from a medial opening called the *superficial inguinal ring* (Chapter 23, p. 630). To locate this ring, palpate the pubic tubercle (Figure 25.13). To test whether a male has an inguinal hernia, the physician pushes on the skin that overlies the pubic tubercle, forces a finger into the superficial inguinal ring, and asks the patient to cough. If the patient has an inguinal hernia (or an open vaginal process that will lead to such a hernia: p. 656), the cough will push an intestinal coil through the inguinal canal to touch the physician's fingertip. ■

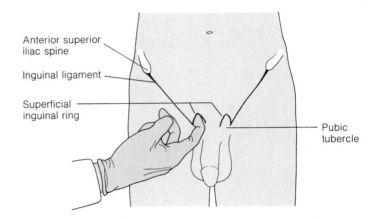

Figure 25.13 Clinical examination for an inguinal hernia in a male. The examiner palpates the patient's pubic tubercle, pushes superiorly to invaginate the scrotal skin into the superficial inguinal ring, and asks the patient to cough. If an inguinal hernia exists, it will push inferiorly, and the examiner can feel it touch his or her fingertip.

Muscles and Other Surface Features

The central landmark of the anterior abdominal wall is the *umbilicus* (navel). Running superiorly and inferiorly from the umbilicus is the *linea alba* ("white line"), represented in the skin of lean people by a vertical groove (Figure 25.11). The linea alba is a tendinous line that extends from the xiphoid process to the pubic symphysis, just medial to the rectus abdominis muscles (Table 10.6, p. 258). The linea alba is a favored site for surgical entry into the abdominal cavity because the surgeon can make a long cut through this line with no muscle damage and a minimum of bleeding.

Several kinds of hernias involve the umbilicus and the linea alba. In an **acquired umbilical hernia**, the linea alba weakens until intestinal coils push through it just superior to the navel. The herniated coils form a bulge just deep to the skin.

Given the fact that acquired umbilical hernias follow a weakening of the anterior abdominal wall, can you deduce which people are most likely to develop such hernias?

These hernias are most common in people over 40 years of age whose abdomens have been stretched by obesity or pregnancy.

Another type of umbilical hernia is a **congenital umbilical hernia**, present in some infants: The umbilicus opens into the umbilical cord in the fetus, but it generally closes a few days after birth through the constriction of a ring of fascia that surrounds the umbilicus. In some infants, however, this ring does not constrict, and intestinal loops protrude through the opening. The umbilical hernia is seen as a cherry-sized bulge deep to the skin of the navel that enlarges whenever the baby cries. Congenital umbilical hernias are usually harmless, and most correct themselves automatically before the child's second birthday. ■

McBurney's point is the spot on the anterior abdominal skin that lies directly superficial to the base of the appendix (Figure 25.11). It is located one-third of the way along a line between the right anterior superior iliac spine and the umbilicus. Try to find it on your body.

Can you guess the clinical importance of McBurney's point?

McBurney's point is the most common site of incision in appendectomies, and it is often the place where the pain of appendicitis is experienced most acutely. Palpation of McBurney's point that causes strong pain after the pressure is removed (rebound tenderness) can indicate appendicitis. This is not a precise method of diagnosis, however—contact a physician if you suspect someone has appendicitis. ■

Flanking the linea alba are the vertical, strap-like *rectus abdominis* muscles (Figures 25.11 and 25.2). Feel these muscles contract just deep to your skin as you do a bent-knee sit-up (or as you bend forward after leaning back in your chair). In the skin of lean people, the lateral margin of each rectus muscle makes a groove known as the **linea semilunaris** (lin'e-ah sem''ĭ-lu-nar'is; "half-moon line"). On your right side, estimate where your linea semilunaris crosses

the costal margin of the rib cage. The fundus of the *gallbladder* lies just deep to this spot, so this is the standard point of incision for gallbladder surgery. In muscular people, three horizontal grooves can be seen in the skin covering the rectus abdominis. These grooves represent the *tendinous insertions* (or *intersections* or *inscriptions*), fibrous bands that subdivide the rectus muscle. Because of these subdivisions, each rectus abdominis muscle presents four distinct bulges.

The only other major muscles that can be seen or felt through the anterior abdominal wall are the lateral *external obliques*. Feel these muscles contract as you cough, strain, or raise your intra-abdominal pressure in some other way.

Recall that the anterior abdominal wall can be divided into four quadrants (Figure 21.4c, p. 565). A clinician who needs to listen to a patient's **bowel sounds** places the stethoscope over all four abdominal quadrants, one after another. Normal bowel sounds, which result as peristalsis moves air and fluid through the intestine, are high-pitched gurgles that occur every 5 to 15 seconds.

Can you deduce why clinicians find it valuable to listen to bowel sounds?

Abnormal bowel sounds can indicate intestinal disorders. An absence or decrease in bowel sounds indicates a halt in intestinal activity, which follows a long-term obstruction of the intestine, surgical handling of the intestine, peritonitis, or other conditions. Loud tinkling or splashing sounds, by contrast, indicate an increase in intestinal activity. Such loud sounds may accompany gastroenteritis (inflammation and upset of the GI tract), a partly obstructed intestine, or the early stages of complete intestinal obstruction. ■

The Pelvis and Perineum

Most of the bony surface features of the *pelvis* are best studied with the abdomen or the lower limb, so they are not considered here. Most *internal* pelvic organs are not palpable through the skin of the body surface. A full *bladder,* however, becomes firm and can be felt through the abdominal wall just superior to the pubic symphysis. A bladder that can be palpated more than a few centimeters above this symphysis is dangerously full and retaining urine, and it should be drained by catheterization. The surface anatomy of the *perineum,* which contains the anus and external genitalia, is reviewed in Figure 25.14. Four palpable bony structures define the corners of the diamond-shaped perineum: the pubic symphysis, the ischial tuberosities, and the coccyx.

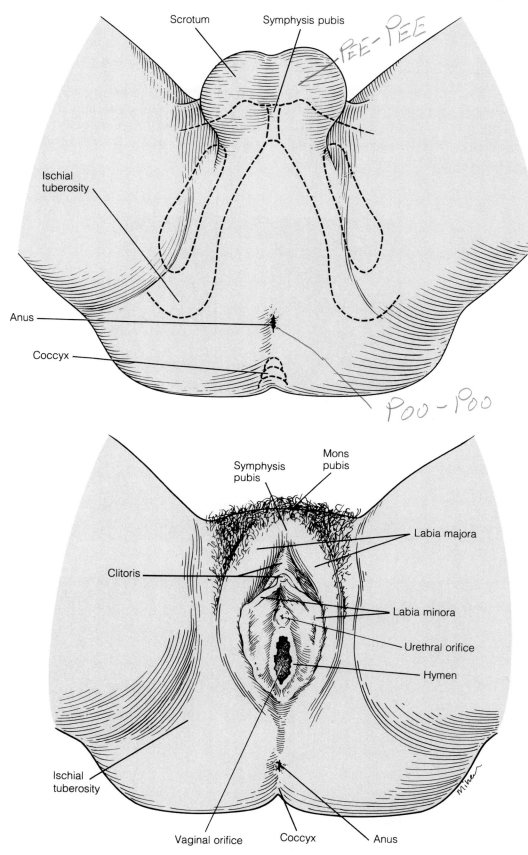

Figure 25.14   Surface features of the perineum   In both sexes, the bony boundaries of the diamond-shaped perineum are the pubic symphysis anteriorly, the two ischial tuberosities laterally, and the coccyx posteriorly.

Upper Limb

Axilla

The **base of the axilla** is the groove in which the underarm hair grows (Figure 25.11). Deep to this base lie the *axillary lymph nodes* (which swell and can be palpated in breast cancer), the large *axillary vessels* serving the upper limb, and much of the brachial plexus. The base of the axilla forms a "valley" between two thick, rounded ridges, the **axillary folds**. Just anterior to the base, clutch your **anterior axillary fold**. This fold is formed by the pectoralis major muscle. Then grasp your **posterior axillary fold**. This fold is formed by the latissimus dorsi and teres major muscles of the back as they course toward their insertions on the humerus.

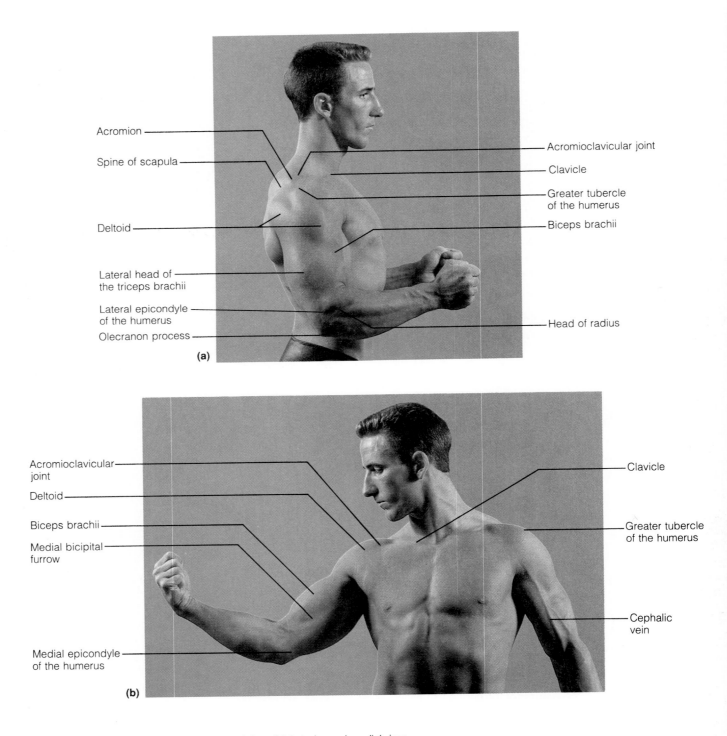

Acromion

Spine of scapula

Deltoid

Lateral head of the triceps brachii

Lateral epicondyle of the humerus

Olecranon process

Acromioclavicular joint

Clavicle

Greater tubercle of the humerus

Biceps brachii

Head of radius

(a)

Acromioclavicular joint

Deltoid

Biceps brachii

Medial bicipital furrow

Medial epicondyle of the humerus

Clavicle

Greater tubercle of the humerus

Cephalic vein

(b)

Figure 25.15 Shoulder and arm. (a) Lateral view. (b) Anterior and medial view.

Shoulder

Re-locate the prominent spine of your scapula posteriorly (Figure 25.15). Follow the spine to its lateral end, the flattened *acromion* on the shoulder's summit. Then, palpate your *clavicle* anteriorly, tracing this bone from the sternum to the shoulder. Notice the clavicle's curved shape. Now locate the junction between the clavicle and the acromion on the superolateral surface of your shoulder, at the *acromioclavicular joint.* To find this joint, thrust your arm anteriorly repeatedly until you can palpate the precise point of pivoting action. Next, place your fingers on the *greater tubercle* of the humerus. This is the most lateral bony landmark on the superior surface of the shoulder. It is covered by the thick *deltoid muscle,* which forms the rounded superior part of the shoulder. Intramuscular injections are often given into the deltoid, about 5 cm (2 inches) inferior to the greater tubercle (Figure 25.16a).

Arm

Recall that according to anatomists, the arm runs only from the shoulder to the elbow, and not beyond. In the arm, the *humerus* can be palpated along its entire length, especially along its medial and lateral sides. Feel the *biceps brachii* muscle contract on your anterior arm when you flex your forearm against resistance. The medial boundary of the biceps is represented by the **medial bicipital furrow** (Figure 25.15b). This groove contains the large *brachial artery,* and by pressing on it with your fingertips you can feel your *brachial pulse.* By pressing harder on this artery, one

can stop bleeding from a hemorrhage in more distal parts of the limb. The brachial artery is the artery used in measuring blood pressure with a sphygmomanometer (sfig″mo-mah-nom′ĕ-ter), a device whose cuff is wrapped around the arm superior to the elbow. Next, extend your forearm against resistance, and feel your *triceps brachii* muscle bulge in the posterior arm. All three heads of the triceps (lateral, long, and medial) are visible through the skin of a muscular person (Figure 25.17).

Elbow Region

In the distal part of your arm, near the elbow, palpate the two projections of the humerus, the *lateral* and *medial epicondyles* (Figure 25.15). Midway between the epicondyles, on the posterior side, feel the *olecranon process* of the ulna, which forms the point of the elbow. Confirm that the two epicondyles and the olecranon all lie in the same horizontal line when the elbow is extended. If these three bony processes do not line up, the elbow is dislocated. Now feel the posterior surface of the medial epicondyle: You are palpating your ulnar nerve.

Can you recall the effect of accidentally hitting the ulnar nerve here, posterior to the medial epicondyle?

As we explained in Chapter 13 (p. 383), banging the ulnar nerve at the "funny bone" sends a sharp twinge of pain along the medial side of the forearm and hand. ■

On the anterior surface of the elbow is a triangular depression called the **cubital fossa** or *antecubital*

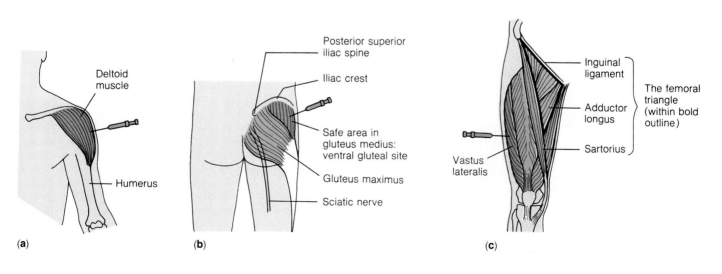

(a) (b) (c)

Figure 25.16 Three major sites of intramuscular injections. **(a)** Deltoid muscle on arm. **(b)** Ventral gluteal site (gluteus medius). **(c)** Vastus lateralis in lateral thigh. Part (c) also shows the femoral triangle.

Figure 25.17   The three heads of the triceps brachii muscle insert on a large tendon.

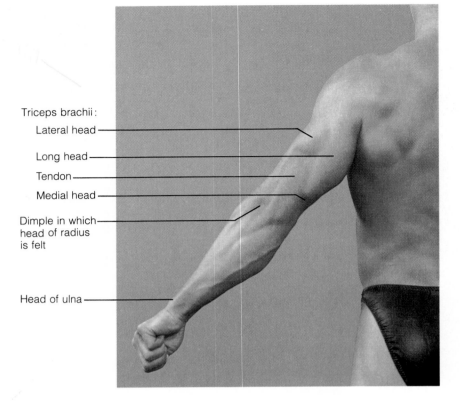

Triceps brachii:

Lateral head

Long head

Tendon

Medial head

Dimple in which head of radius is felt

Head of ulna

---

fossa (*ante* = in front of; *cubit* = elbow) (Figure 25.18). The triangle's superior *base* is formed by a horizontal line between the humeral epicondyles, and its two inferior sides are defined by the *brachioradialis* and *pronator teres* muscles (Figure 25.18b). Try to define these boundaries on your own limb: To find the brachioradialis muscle, flex your forearm against resistance, and watch this muscle bulge through the

skin of your lateral forearm. To feel your pronator teres contract, palpate the cubital fossa as you pronate your forearm against resistance.

The cubital fossa contains the superficial *median cubital vein* (Figure 25.18a). Clinicians often draw blood from this superficial vein and insert intravenous (IV) catheters into it to administer medications, transfused blood, and nutrient fluids. The clinician

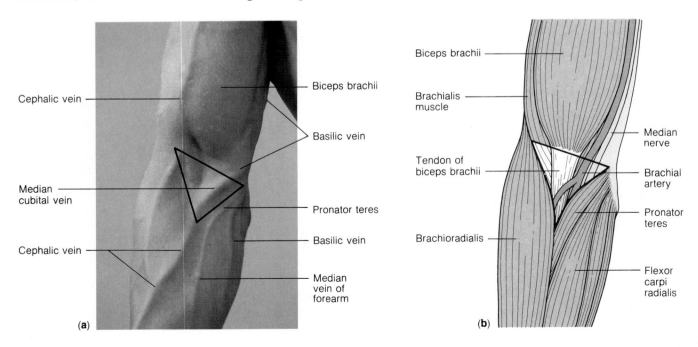

Cephalic vein

Median cubital vein

Cephalic vein

Biceps brachii

Basilic vein

Pronator teres

Basilic vein

Median vein of forearm

(a)

Biceps brachii

Brachialis muscle

Tendon of biceps brachii

Brachioradialis

Median nerve

Brachial artery

Pronator teres

Flexor carpi radialis

(b)

Figure 25.18   The cubital fossa on the anterior surface of the right elbow (outlined by the triangle).   (a) Photograph. (b) Diagram of deeper structures in the fossa.

must be aware that the large *brachial artery* lies just deep to the median cubital vein (Figure 25.18b), so a needle must be inserted into the vein from a shallow angle (almost parallel to the skin) to avoid puncturing the artery. Other structures that lie deep in the cubital fossa are also shown in Figure 25.18b.

The median cubital vein interconnects the larger cephalic and basilic veins of the upper limb. Recall from Chapter 18 (p. 504) that the *cephalic vein* ascends along the lateral side of the forearm and arm, whereas the *basilic vein* ascends through the limb's medial side. These veins are visible through the skin of lean people (Figure 25.18a).

Forearm and Hand

The two parallel bones of the forearm are the medial *ulna* and the lateral *radius*. You can feel the ulna along its entire length as a sharp ridge on the posterior forearm (confirm that this ridge runs inferiorly from the olecranon process). As for the radius, you can feel its distal half, but most of its proximal half is covered by muscle. You can, however, feel the rotating *head* of the radius. To do this, extend your forearm, and note that a dimple forms on the posterior lateral surface of the elbow region (Figure 25.17). Press three fingers into this dimple, and rotate your free hand as if you were turning a doorknob. You will feel the head of the radius rotate as you perform this action.

Both the radius and ulna have a knob-like *styloid process* at their distal end. Figure 25.19 shows a way to locate these processes. Do not confuse the ulna's styloid process with the conspicuous *head of the ulna,* from which the styloid process stems. Confirm that the styloid process of the radius lies about 1 cm (0.4 inch) distal to that of the ulna.

✚ **Colles' fracture** of the wrist is an impacted fracture in which the distal end of the radius is pushed proximally into the shaft of the radius. This usually occurs when someone falls on outstretched hands, and it most often happens to elderly women with osteoporosis. Colles' fracture bends the wrist into curves that resemble those on a fork.

▦ Can you deduce how physicians use palpation to diagnose a Colles' fracture?

In this fracture, the physician can feel that the styloid process of the radius has moved proximally from its normal position. ▪

Next, feel the major groups of muscles within your forearm. Flex your hand and fingers against resistance, and feel the anterior *flexor muscles* contract. Then extend your hand at the wrist, and feel the tightening of the posterior *extensor muscles.*

Near the wrist, the anterior surface of the forearm reveals many significant features (Figure 25.20). Flex

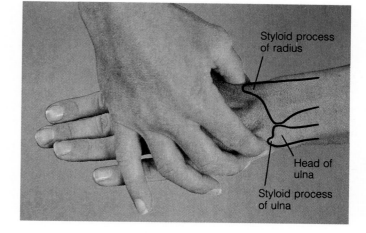

Figure 25.19 A way to locate the styloid processes of the ulna and radius.   The right hand is palpating the left hand in this picture. Note that the head of the ulna is not the same as its styloid process. The styloid process of the radius lies about 1 cm distal to the styloid process of the ulna.

your fist against resistance, and the tendons of the main wrist flexors will bulge the skin of the distal forearm. Most obvious will be the tendons of the *flexor carpi radialis* and *palmaris longus* muscles. (The palmaris longus, however, is absent from at least one arm in 30% of all people, so your forearm may exhibit just one prominent tendon instead of two.) The *radial artery* lies just lateral to (on the thumb side of) the flexor carpi radialis tendon, where the pulse is easily detected (Figure 25.20b). Feel your radial pulse here. The *median nerve* (which innervates the thumb) lies deep to the palmaris longus tendon. Finally, the *ulnar artery* lies on the medial side of the forearm, just lateral to the tendon of the *flexor carpi ulnaris.* Using Figure 25.20b as a guide, locate and feel your ulnar arterial pulse.

By extending your thumb and pointing it posteriorly, you will form a triangular depression in the base of the thumb on the back of your hand. This is the **anatomical snuff box** (Figure 25.21). Its two elevated borders are defined by the tendons of the thumb extensor muscles, *extensor pollicis brevis* and *extensor pollicis longus.* The radial artery runs within the snuff box, so this is another site for taking a radial pulse. The main bone on the floor of the snuff box is the scaphoid bone of the wrist, but the styloid process of the radius is also present here. (If displaced by a bone fracture, the radial styloid process will be felt outside of the snuff box rather than within it.) The "snuff box" took its name from the fact that people once put snuff (tobacco for sniffing) in this hollow before lifting it up to the nose.

On the dorsum of your hand, observe the superficial veins just deep to the skin. This is the *dorsal ve-*

Figure 25.20 The anterior surface of the forearm and fist.
(a) The entire forearm. (b) Enlarged view of the distal forearm and hand: The tendons of the flexor muscles guide the clinician to several sites for taking pulses.

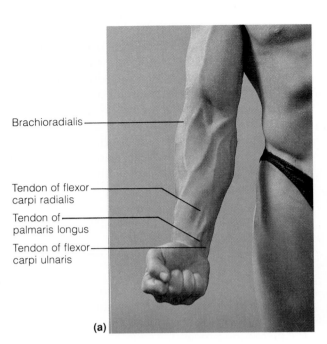

Brachioradialis

Tendon of flexor carpi radialis

Tendon of palmaris longus

Tendon of flexor carpi ulnaris

(a)

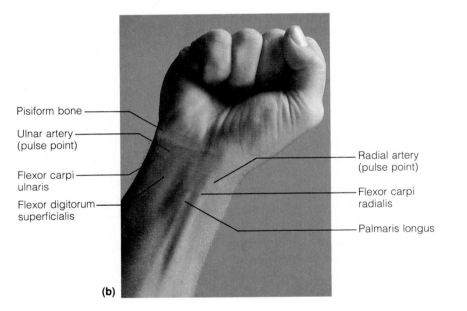

Pisiform bone

Ulnar artery (pulse point)

Flexor carpi ulnaris

Flexor digitorum superficialis

Radial artery (pulse point)

Flexor carpi radialis

Palmaris longus

(b)

Figure 25.21 The dorsum of the hand. Note especially the anatomical snuff box and the dorsal venous network.

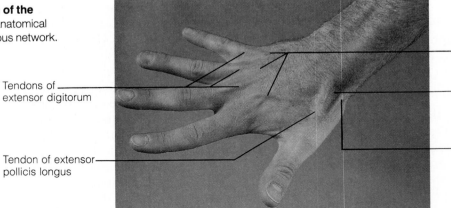

Tendons of extensor digitorum

Tendon of extensor pollicis longus

Dorsal venous network

Anatomical snuff box

Tendon of extensor pollicis brevis

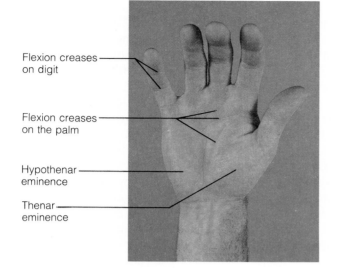

Figure 25.22 The palmar surface of the hand.

Labels (top to bottom):
- Flexion creases on digit
- Flexion creases on the palm
- Hypothenar eminence
- Thenar eminence

nous network, which drains superiorly into the cephalic vein. This venous network provides a site for drawing blood and inserting intravenous catheters and is preferred over the median cubital vein for these purposes. Next, extend your hand and fingers, and observe the tendons of the *extensor digitorum.*

The anterior surface of the hand also contains some features of interest (Figure 25.22). These features include the *epidermal ridges* ("fingerprints") and many *flexion creases* in the skin. Grasp your *thenar eminence* (the bulge on the palm that contains the thumb muscles) and your *hypothenar eminence* (the bulge on the medial palm that contains muscles of the little finger).

Lower Limb

Gluteal Region

Locate your iliac crests again, and trace each to its most posterior point, the small but sharp *posterior superior iliac spine* (Figure 25.23). You may have difficulty feeling this spine, but it is indicated by a distinct dimple in the skin that is easy to find. This dimple lies two to three fingers' breadths lateral to the midline of the back. The dimple also indicates the position of the *sacroiliac joint*, where the hip bone attaches to the sacrum of the spinal column. In *bone marrow biopsies*, the physician inserts a needle into the iliac bone 1 cm (0.4 inch) inferolateral to this dimple. A sample of marrow is then aspirated and examined for evidence of blood disorders or leukemia.

Dominating the gluteal region are the two *prominences* ("cheeks") of the buttocks. These are formed by subcutaneous fat and by the thick *gluteus maximus* muscles. The midline groove between the two prominences is called the **natal cleft** (na'tal; "rump") or *gluteal cleft*. The inferior margin of each prominence is the horizontal **gluteal fold**, which roughly corresponds to the inferior margin of the gluteus maximus. Try to palpate the *ischial tuberosity* just above the medial side of each gluteal fold (it will be easier to feel if you sit down or flex your thigh first). The ischial tuberosities are the robust inferior parts of the ischial bones, and they support the body's weight during sitting. Next, palpate the *greater trochanter* of the femur

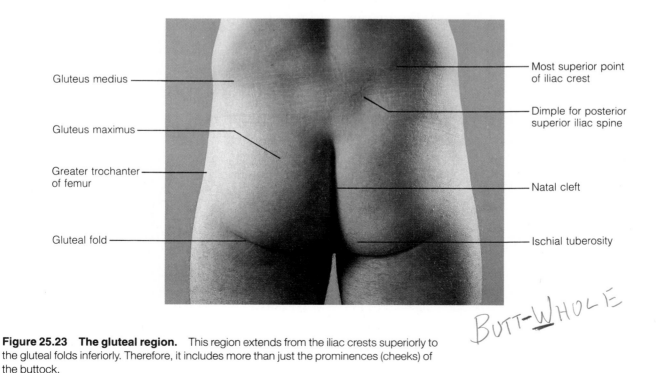

Labels (left, top to bottom):
- Gluteus medius
- Gluteus maximus
- Greater trochanter of femur
- Gluteal fold

Labels (right, top to bottom):
- Most superior point of iliac crest
- Dimple for posterior superior iliac spine
- Natal cleft
- Ischial tuberosity

Figure 25.23 The gluteal region. This region extends from the iliac crests superiorly to the gluteal folds inferiorly. Therefore, it includes more than just the prominences (cheeks) of the buttock.

on the lateral side of your hip (Figures 25.23 and 25.24). This trochanter lies just anterior to a hollow and about 10 cm (one hand's breadth) inferior to the iliac crest. Here is how to confirm that you have found the greater trochanter: Because this trochanter is the most superior point on the lateral femur, you can feel it move with the femur as you alternately flex and extend your thigh.

The gluteal region is a major site for administering intramuscular injections. When inserting the needle to apply such injections, extreme care must be taken to avoid piercing a major nerve that lies just deep to the gluteus maximus muscle. Can you guess what nerve this is?

It is the thick *sciatic nerve,* which innervates much of the lower limb. Furthermore, the needle must avoid the gluteal nerves and gluteal blood vessels, which also lie deep to the gluteus maximus.

To avoid harming these structures, the injections are applied to the gluteus *medius* (not maximus)

muscle superior to the cheeks of the buttocks, in a safe area called the **ventral gluteal site** (Figure 25.16b). To locate this site, it once was recommended that each buttock be divided into four quadrants and the injection be made into the superior lateral quadrant, but that did not provide a wide enough margin of safety for avoiding the sciatic nerve. A better way of finding the safe area is to draw a line laterally from the posterior superior iliac spine (dimple) to the greater trochanter and then proceed 5 cm (2 inches) superiorly from the midpoint of that line. Another safe way to locate the ventral gluteal site is to approach the lateral side of the patient's left hip with your extended right hand (or the right hip with your left hand), then place your thumb on the anterior superior iliac spine and your index finger as far posteriorly on the iliac crest as it can reach. The heel of your hand comes to lie on the greater trochanter, and the needle is then inserted in the angle of the V formed between your thumb and index finger. Note that the needle must always enter at least 4 cm (1.5 inches) inferior to the iliac crest, or else you will pierce bone instead of muscle.

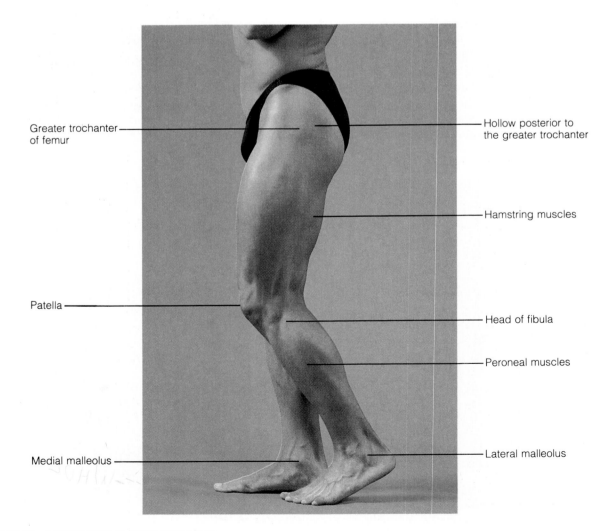

Figure 25.24   Lateral surface of the lower limb.

Gluteal injections are not given to small children because their "safe area" is too small to locate with certainty and because the gluteal muscles are thin at this age. Instead, infants and toddlers receive shots in the prominent vastus lateralis muscle of the thigh. ■

Thigh

The thigh is pictured in Figures 25.24, 25.25, and 25.26. Much of the femur is clothed by thick muscles, so the thigh has few palpable bony landmarks. Distally, you can feel the *medial* and *lateral condyles of the femur* and the *patella* anterior to the condyles (see Figure 25.25c and a). Next, palpate your three groups of thigh muscles: the *quadriceps femoris muscles* anteriorly, the *adductor muscles* medially, and the *hamstrings* posteriorly. The *vastus lateralis,* the lateral muscle of the quadriceps group, is a site for in-

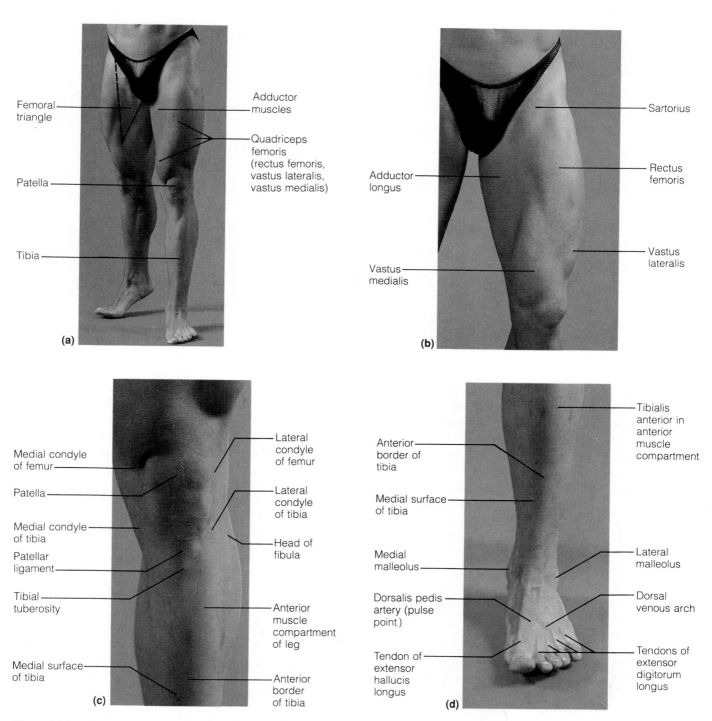

Figure 25.25 Anterior surface of the lower limb. (a) Both limbs, with the right limb revealing its medial surface. The femoral triangle is outlined on the right limb. (b) Enlargement of the left thigh. (c) The left knee region. (d) The dorsum of the foot.

Figure 25.26 Posterior surface of the lower limb. Notice the diamond-shaped popliteal fossa posterior to the knee.

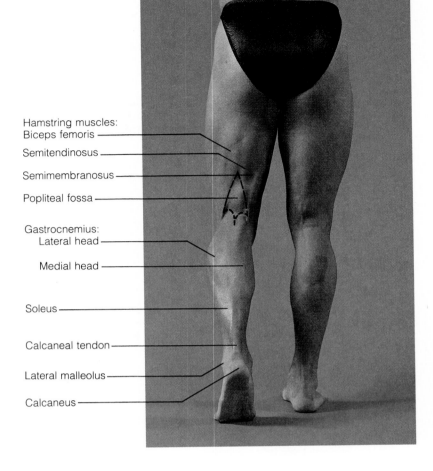

Hamstring muscles:
Biceps femoris
Semitendinosus
Semimembranosus
Popliteal fossa
Gastrocnemius:
Lateral head
Medial head
Soleus
Calcaneal tendon
Lateral malleolus
Calcaneus

tramuscular injections. Such injections are administered about halfway down the length of this muscle (Figure 25.16c).

The anterosuperior surface of the thigh exhibits a three-sided depression called the **femoral triangle** (Figure 25.25a). As shown in Figure 25.16c, the superior border of this triangle is formed by the inguinal ligament, and its two inferior borders are defined by the *sartorius* and *adductor longus* muscles. The large *femoral artery* and *vein* (the main vessels to the lower limb) descend vertically through the center of the femoral triangle. To feel the pulse of your femoral artery, press inward just inferior to your midinguinal point (halfway between the anterior superior iliac spine and the pubic tubercle). Be sure to push hard, because the artery lies somewhat deep. By pressing very hard on this point, one can stop the bleeding from a hemorrhage in the distal limb. The femoral triangle also contains most of the *inguinal lymph nodes* (which are easily palpated if swollen), and it is the site of **femoral hernias.** In a femoral hernia, coils of the intestine push inferiorly from the abdominal cavity, descend deep to the inguinal ligament, and enter the superior thigh. There, they form a distinct bulge deep to the skin of the femoral triangle. (Note that a femoral

hernia, which lies inferior to the inguinal ligament, is entirely different from an *inguinal* hernia, which lies superior to the inguinal ligament.)

On the posterior of the knee is a diamond-shaped hollow called the **popliteal fossa** (Figure 25.26). Large muscles define the four borders of this fossa: The *biceps femoris* forms the superolateral border, the *semitendinosus* and *semimembranosus* define the superomedial border, and the two heads of the *gastrocnemius* form the two inferior borders. The *popliteal artery* and *vein* (main vessels to the leg) lie deep within this fossa. You may be able to feel a popliteal pulse if you flex your leg at the knee and push your fingers firmly into the popliteal fossa. If a physician is unable to feel a patient's popliteal pulse, the femoral artery may be narrowed by atherosclerosis.

Superficially, the popliteal fossa is covered by a roof of strong fascia that prevents any expansion of this space. Therefore, anything that causes the popliteal fossa to swell is accompanied by pressure and pain. Such swelling can be caused by an infected lymph node in the popliteal fossa, varicosity of the popliteal vein, or a ballooning aneurysm in the popliteal artery. ■

Leg and Foot

Locate your patella again, then follow the thick *patellar ligament* inferiorly from the patella to its insertion on the superior tibia (Figure 25.25c). Here you can feel a rough projection, the *tibial tuberosity.* Continue running your fingers inferiorly along the tibia's sharp *anterior border* and its flat *medial surface*—bony landmarks that lie very near the surface throughout their length.

Can you deduce what superficial structure is bruised when someone receives a "bang on the shin"?

It is the exposed medial surface of the shaft of the tibia. ∎

Now, return to the superior part of your leg, and palpate the expanded *lateral* and *medial condyles of the tibia* just inferior to the knee. (You can distinguish the tibial condyles from the *femoral* condyles because you can feel the tibial condyles move with the tibia during knee flexion.) Feel the bulbous *head of the fibula* in the superolateral region of the leg (Figures 25.24 and 25.25c). Try to feel the *common peroneal nerve* (nerve to the anterior leg and foot) where it wraps around the fibula's *neck* just inferior to its head. This nerve is often bumped against the bone here and damaged.

In the most distal part of the leg, you can feel the *lateral malleolus* of the fibula as the lateral prominence of the ankle (Figure 25.25d). Notice that this lies slightly inferior to the *medial malleolus* of the tibia, which forms the ankle's medial prominence. Place your finger just posterior to the medial malleolus to feel the pulse of your *posterior tibial artery.*

Next, palpate the main muscle groups of your leg, starting with the calf muscles posteriorly (Figure 25.26). Standing on tiptoes will help you feel the *lateral* and *medial heads of the gastrocnemius* and, inferior to these, the broad *soleus* muscle. Also feel the tension in your *calcaneal (Achilles) tendon* and the point of insertion of this tendon onto the calcaneus

bone of the foot. Return to the anterior surface of the leg, and palpate the *anterior muscle compartment* (Figure 25.25c and d) while alternately dorsiflexing then plantar flexing your foot. You will feel the tibialis anterior and extensor digitorum muscles contracting then relaxing. Then, palpate the *peroneal muscles* that cover most of the fibula laterally (Figure 25.24). The tendons of these muscles pass posterior to the lateral malleolus and can be felt at a point posterior and slightly superior to that malleolus.

Now, observe the dorsum (superior surface) of your foot. You may see the superficial *dorsal venous arch* overlying the proximal part of the metatarsal bones (Figure 25.25d). This arch gives rise to both saphenous veins (the main superficial veins of the lower limb). Visible in lean people, the *great saphenous vein* ascends along the medial side of the entire limb, and the *small saphenous vein* ascends through the center of the calf (see Figure 18.23 on p. 508).

As you extend your toes, observe the tendons of the *extensor digitorum longus* and *extensor hallucis longus* muscles on the dorsum of the foot. Finally, place a finger on the extreme proximal part of the space between the first and second metatarsal bones. Here you should be able to feel the pulse of the *dorsalis pedis artery.*

We hope that this chapter has reinforced some of the factual information you learned in earlier chapters and has helped you appreciate the human body as a living structure. A study of one's own surface anatomy can be a profound and humbling experience—it forces us to admit that we are physical as well as mental beings and that the objects we see in anatomy books and in the anatomy laboratory really do exist inside our own bodies. If you are planning a career in the health-related fields or in physical education, your study of living anatomy will help ease the transition from working with dead specimens in your anatomy course to working with the bodies of active, responsive, sensitive people in a clinical situation.

Chapter Summary

1. Surface anatomy is the study of the external features of the body and of those internal features that can be observed or palpated through the body surface. Surface anatomy reveals more about the skeleton and muscles than about other organs, but it also provides information on many of our deep viscera.

2. Surface anatomy provides the simplest and least invasive method of assessing the health of the living body, so it forms the basis of the routine physical examination. Surface landmarks

also show the clinician where to take pulses, make injections, listen to internal organs, make surgical incisions, feel for broken bones, and more.

THE HEAD (pp. 684–688)

Cranium (p. 684)

1. In the cranium, the bony and muscular structures that can be felt through the skin include the superciliary arches, external occipital protuberance, mastoid process, and temporalis muscles.

2. The scalp covers the superior surface of the cranium and consists of skin, a highly vascular subcutaneous layer, and the galea aponeurotica.

Face (pp. 684–688)

3. The parts of the nose include its root, bridge, dorsum, apex, and alae. Parts of the auricle include the concha, tragus, helix, and lobule.

4. Palpable features of the face include the zygomatic arch, mandibular body and ramus, masseter muscle, temporomandibular joint, and the pulse points of the superficial temporal and facial arteries.

THE NECK (pp. 688–691)

Bony Landmarks (pp. 688–690)

1. By running a finger inferiorly along the anterior midline of the neck, one can palpate the hyoid bone, laryngeal prominence, cricoid cartilage, thyroid gland, trachea, and jugular notch.

Muscles (pp. 690–691)

2. The sternocleidomastoid muscle is an important surface landmark of the neck. Near or on this muscle, one can locate swollen cervical lymph nodes, the external carotid pulse, the subclavian artery, and the external jugular vein.

Triangles of the Neck (p. 691)

3. The posterior triangle of the neck is defined by the sternocleidomastoid anteriorly, the trapezius posteriorly, and the clavicle inferiorly. The anterior triangle is defined by the inferior margin of the mandible superiorly, the midline of the neck anteriorly, and the sternocleidomastoid posteriorly.

4. Many nerves and arteries are located relatively close to the surface in the posterior triangle of the neck: the accessory nerve, the brachial and cervical plexuses, the subclavian artery, and the external jugular vein.

THE TRUNK (pp. 691–697)

The Back (pp. 692–693)

1. In the posterior median furrow of the back, one can count and identify the vertebral spinous processes. The most easily located vertebral spines are these: C_7 and T_1 (the most prominent spines), T_3 (at the level of the medial root of the spine of the scapula), T_7 (level of the inferior scapular angle), and L_4 (level of the supracristal line).

2. Superficial muscles of the back include the trapezius and latissimus dorsi. These two muscles, along with the medial border of the scapula, define the triangle of auscultation. Physicians place a stethoscope on this triangle to listen for lung sounds.

The Thorax (pp. 693–694)

3. The sternal angle is an important surface landmark on the anterior thorax. By finding the second rib at the sternal angle, one can count inferiorly to identify the other ribs. The midaxillary and midclavicular lines are imaginary vertical lines that extend inferiorly from the center of the axilla and clavicle, respectively. The apex of the V-shaped costal margin (infrasternal angle) is at the xiphisternal joint.

4. The thoracic cage provides many reliable bony landmarks for locating the soft organs in the thoracic cavity. By orienting to specific ribs, one can map the location of the pleural cavity, the lungs, and the heart.

The Abdomen (pp. 694–696)

5. Surface landmarks on the anterior abdominal wall include the iliac crest, anterior superior iliac spine, groove for the in-

guinal ligament, pubic tubercle, pubic crest, pubic symphysis, umbilicus, groove for linea alba, rectus abdominis muscle, and linea semilunaris. A physician's finger placed in the superficial inguinal ring can detect an inguinal hernia.

6. The umbilicus and linea alba are sites of umbilical hernias, and the linea alba is a common incision site for surgical entry into the abdominal cavity.

7. McBurney's point, two-thirds of the way between the umbilicus and the right anterior superior iliac spine, is usually the site of maximum rebound tenderness during appendicitis and is the most common site of incision for appendectomy.

8. The fundus of the gallbladder underlies the right costal margin, where the linea semilunaris intersects this margin.

9. Physicians listen for bowel sounds over all four quadrants of the anterior abdominal wall.

The Pelvis and Perineum (pp. 696–697)

10. A full urinary bladder can be palpated through the anterior abdominal wall superior to the pubic symphysis. Four palpable structures define the corners of the diamond-shaped perineum: pubic symphysis, coccyx, and the two ischial tuberosities.

UPPER LIMB (pp. 698–703)

Axilla (p. 698)

1. Important superficial structures associated with the axilla include the axillary base, axillary lymph nodes, and axillary vessels. The anterior and posterior walls of the axilla are called axillary folds.

Shoulder (p. 699)

2. Skeletal landmarks that can be palpated in the shoulder region include the acromion of the scapula, the acromioclavicular joint, and the greater tubercle of the humerus.

3. Intramuscular injections are often given into the deltoid muscle, about 5 cm (2 inches) inferior to the greater tubercle.

Arm (p. 699)

4. On the arm, one can palpate the humerus and the biceps and triceps brachii muscles. The medial bicipital furrow, on the medial side of the arm, indicates the position of the brachial artery.

Elbow Region (pp. 699–701)

5. Bony landmarks in the elbow region include the two epicondyles of the humerus and the olecranon process of the ulna, all lying in the same horizontal plane.

6. The cubital fossa (antecubital fossa) is the triangular depression on the anterior surface of the elbow. The median cubital vein runs through it. Clinicians often insert needles into this superficial vein to draw blood or to administer IV medications and fluids.

Forearm and Hand (pp. 701–703)

7. In the forearm, the palpable bony structures include the entire posterior margin of the ulna and the distal half and head of the radius. At the wrist, the styloid process of the radius lies 1 cm distal to the styloid process of the ulna.

8. One can palpate the flexor and extensor muscle groups on the anterior and posterior forearm, respectively.

9. On the anterior forearm near the wrist, one sees the tendons of the flexor carpi radialis, palmaris longus (absent in some people), and flexor carpi ulnaris. A radial arterial pulse is felt just lateral to the flexor carpi radialis tendon, an ulnar arterial pulse just lateral to the flexor carpi ulnaris tendon.

10. The tendons of the thumb extensor muscles define the anatomical snuff box on the posterior base of the thumb. A radial pulse and the styloid process of the radius are felt here.

11. The dorsal venous network on the posterior surface of the hand is an important site for drawing blood and administering IV medications. On the palmar surface of the hand, features include flexion creases and the thenar and hypothenar eminences.

LOWER LIMB (pp. 703–707)

Gluteal Region (pp. 703–705)

1. The posterior point of the iliac crest is the posterior superior iliac spine, represented by a dimple in the skin. Other bony landmarks in the gluteal region are the ischial tuberosities and the greater trochanter of the femur.

2. The two gluteal prominences, separated by the natal cleft, consist of gluteus maximus muscles and subcutaneous fatty tissue.

3. The gluteal region is a major site for administering intramuscular injections. For safety, the needle is inserted into the gluteus medius muscle at the ventral gluteal site. Gluteal injections are not administered to small children.

Thigh (pp. 705–706)

4. Bony landmarks that can be palpated in the thigh include the patella and the condyles of the femur. The main groups of thigh muscles are the quadriceps, adductors, and hamstrings. The vastus lateralis is a common site for intramuscular injections.

5. The femoral triangle on the anterior superior thigh contains the femoral artery and vein and most of the inguinal lymph nodes. A femoral hernia bulges the skin of the femoral triangle. One feels a femoral pulse just inferior to the midinguinal point.

6. The popliteal fossa is a diamond-shaped hollow on the posterior aspect of the knee, defined by four muscular boundaries. It contains the deeply located popliteal artery and vein and is covered superficially by a layer of dense fascia.

Leg and Foot (p. 707)

7. Bony landmarks that can be palpated in the leg include the tibial tuberosity and tibial condyles, the anterior border and medial surface of the tibia, the head of the fibula, and the lateral and medial malleolus. The pulse of the posterior tibial artery is felt just posterior to the medial malleolus.

8. The main muscles of the leg, which can be palpated as they contract, include the gastrocnemius and soleus on the calf, the muscles in the anterior compartment, and the peroneal muscles on the fibula.

9. On the dorsum of the foot, one can observe the dorsal venous arch and feel the pulse of the dorsalis pedis artery.

Review Questions

Multiple Choice and Matching Questions

1. A blow to the cheek is most likely to break what superficial bone or bone part? (a) superciliary arches, (b) the philtrum, (c) zygomatic arch, (d) the tragus.

2. Rebound tenderness (a) occurs in appendicitis, (b) is whiplash of the neck, (c) is a sore foot from playing basketball, (d) occurs when the larynx falls back into place after swallowing.

3. A pressure point is a place where one presses on an artery through the body surface to stop bleeding farther distally. Which of the following is not a pressure point for any major artery? (a) neck just superior to the clavicle and lateral to the sternocleidomastoid, (b) the sciatic artery in middle of gluteus maximus, (c) medial bicipital furrow on arm, (d) midinguinal point in femoral triangle.

4. The anatomical snuff box (a) is in the nose, (b) contains the styloid process of the radius, (c) is defined by tendons of the flexor carpi radialis and palmaris longus, (d) cannot really hold snuff.

5. Some landmarks on the body surface can be seen or felt, but others are abstractions that you must construct by drawing imaginary lines. Which of the following pairs of structures is abstract and invisible? (a) umbilicus and costal margin, (b) anterior superior iliac spine and natal cleft, (c) linea alba and linea semilunaris, (d) McBurney's point and midaxillary line, (e) philtrum and sternocleidomastoid.

6. Many pelvic organs can be palpated by placing a finger in the rectum or vagina, but only one pelvic organ is readily palpated through the skin. This is the (a) nonpregnant uterus, (b) prostate gland, (c) full bladder, (d) ovaries, (e) rectum.

7. A muscle that contributes to the posterior axillary fold is (a) pectoralis major, (b) latissimus dorsi, (c) trapezius, (d) infraspinatus, (e) pectoralis minor, (f) a and e.

8. Which of the following is not a pulse point? (a) anatomical snuff box, (b) inferior margin of mandible anterior to masseter muscle, (c) center of distal forearm at palmaris longus tendon, (d) medial bicipital furrow on arm, (e) dorsum of foot between the first two metatarsals.

9. Which pair of ribs inserts on the sternum at the sternal angle? (a) first, (b) second, (c) third, (d) fourth, (e) fifth.

10. The inferior angle of the scapula is at the same level as the spinous process of this vertebra: (a) C_5, (b) C_7, (c) T_3, (d) T_7, (e) L_4.

11. The supracristal line on the back lies at the same level as the spinous process of this vertebra: (a) C_5, (b) C_7, (c) T_3, (d) T_7, (e) L_4.

12. An important bony landmark that can be recognized by a distinct dimple in the skin is (a) posterior superior iliac spine, (b) styloid process of the ulna, (c) shaft of the radius, (d) acromion.

13. A nurse missed a patient's median cubital vein while trying to withdraw blood and then inserted the needle far too deeply into the cubital fossa. This error could cause any of the following problems, except this one (Hint: See Figure 25.18): (a) paralysis of the ulnar nerve, (b) paralysis of the median nerve, (c) bruising the insertion tendon of the biceps brachii muscle, (d) blood spurting from the brachial artery.

14. Which of these organs is almost impossible to study with the techniques of surface anatomy? (a) heart, (b) lungs, (c) brain, (d) nose.

15. A preferred site for inserting an intravenous medication line into a blood vessel is (a) medial bicipital furrow on arm, (b) external carotid artery, (c) dorsal venous arch of hand, (d) popliteal fossa.

16. One listens for bowel sounds with a stethoscope that is placed (a) on the four quadrants of the abdominal wall; (b) in the triangle of auscultation; (c) in the right and left midaxillary line, just superior to the iliac crests; (d) inside the patient's bowels (intestines), on the tip of an endoscope.

Short Answer and Essay Questions

17. Define palpation.

18. Why is a knowledge of surface anatomy valuable in clinical situations and for evaluating athletic injuries?

19. Explain how one locates the proper site for intramuscular injections into (a) the ventral gluteal site and (b) the deltoid muscle.

20. Ashley, a pre-physical therapy student, was trying to locate the vertebral spinous processes on the flexed back of her friend Amber, but she kept losing count. Amber told her to check her count against several reliable "guideposts" along the way: the spinous processes of C_7, T_3, T_7, and L_4. Can you describe how to find each of these four particular vertebrae without having to count any vertebrae?

21. Describe three different places on or near the wrist where you can take someone's pulse, and tell exactly how to find each of those places.

22. During a physical examination, how would a physician distinguish between an inguinal hernia and a femoral hernia?

23. Do you have a palmaris longus muscle in your own right arm, or not? Explain how you know this.

24. Where is a standard site for inserting the needle into the iliac bone for a bone marrow biopsy?

25. If necessary, review the surface anatomy of the heart from Chapter 17 to answer this question: Where is the apex point of the heart on the thoracic wall?

26. How does one find the midinguinal point?

27. Locate the standard points of surgical incision for reaching both the appendix and the gallbladder.

28. Gregory hit his funny bone. What nerve was hit against what bony process?

29. Refer to the picture of the pulse points in Figure 25.4, and explain in your own words where the three most inferior of these points are located.

Critical Thinking and Clinical Applications Questions

30. Walking to her car after her sixty-fifth birthday party, Mrs. Schultz tripped on ice and fell forward on her outstretched palms. When she arrived at the emergency room, her right wrist and hand were bent like a fork handle. Dr. Jefferson felt that the styloid process of her radius was outside of the anatomical snuff box and slightly proximal to the styloid process of the ulna. When he checked her elbow, he found that the olecranon process lay 2 cm proximal to the two epicondyles of the right humerus. Explain all these observations, and describe what had happened to Mrs. Schultz's limb.

31. One of a group of rabbit hunters was accidentally sprayed with buckshot in both of his gluteal prominences. When his companions saw that he would survive, they laughed and joked about where he had been shot. They were horrified and ashamed a week later when it was announced that their friend would be permanently paralyzed and without sensation in both legs from the knee down, as well as on the back of his thighs. What had happened?

32. Four close friends, who were taking an anatomy course, spent their weekends working in a meat packing plant. One day, the supervisor was pressuring them to work faster, and one of the friends was cut deeply by a sharp butcher knife that severed all the arteries in the anterior part of his proximal right forearm. His buddies rushed him to the hospital, and as they wheeled him in the door, the emergency room doctor noticed that all three friends were pressing their hands on the injured man in different places. One was pressing on his neck above the right clavicle and lateral to the sternocleidomastoid, another was pressing on the medial side of his right arm, and another was pressing on the center of his cubital fossa. Exactly what were they doing, and why did the physician say that by doing this they had saved their friend's life?

33. An athletic trainer was helping a college basketball player find the site of a pulled muscle. The trainer asked the athlete to extend her thigh at the hip forcefully, but she found this action too painful to perform. Then the trainer palpated her posterior thigh and felt some swelling of the muscles there. In simplest terms, which basic muscle group was injured?

34. The drug fight was over, and George felt exhilarated because his gang had won. The only wound he had received was a relatively shallow knife slash across the inferior left side of his neck. Then George's nightmare began: He suddenly realized his left upper limb was paralyzed; parts of the skin on his neck were numb; a considerable amount of blood was flowing (but not spurting) from the wound; and he could not turn his head to the right or shrug his right shoulder. Worst of all, he was gasping for breath. Based on your knowledge of surface anatomy, explain as many of George's symptoms as you can.

35. Two-year-old Daisy fell against the edge of a table and opened a 7-cm gash in her scalp. A doctor stitched her up, and the wound healed. For the next year, her parents told their friends that they had been frightened by how much the wound bled and were amazed that it had healed in only 2 weeks. Are these observations unusual for a scalp wound? Explain.

36. Gabrielle, a talented college athlete, told her physician that she was experiencing great pain in the hollow behind her left knee. Clinical tests indicated that the pain was caused by a popliteal lymph node swollen with cancer cells that had spread from a bone cancer in the tibia. Gabrielle was told that her prognosis was good because the cancer had been detected very early and that she would not need to have her leg amputated. The following semester, in her anatomy course, Gabrielle learned that cancerous lymph nodes do not hurt (Chapter 19, p. 525). If that is so, why had her swollen popliteal node caused so much pain?

The Metric System

MEASURE-MENT	UNIT AND ABBREVIATION	METRIC EQUIVALENT	METRIC TO ENGLISH CONVERSION FACTOR	ENGLISH TO METRIC CONVERSION FACTOR
Length	1 kilometer (km)	= 1000 (10^3) meters	1 km = 0.62 mile	1 mile = 1.61 km
	1 meter (m)	= 100 (10^2) centimeters = 1000 millimeters	1 m = 1.09 yards 1 m = 3.28 feet 1 m = 39.37 inches	1 yard = 0.914 m 1 foot = 0.305 m
	1 centimeter (cm)	= 0.01 (10^{-2}) meter	1 cm = 0.394 inch	1 foot = 30.5 cm 1 inch = 2.54 cm
	1 millimeter (mm)	= 0.001 (10^{-3}) meter	1 mm = 0.039 inch	
	1 micrometer (μm) [formerly micron (μ)]	= 0.000001 (10^{-6}) meter		
	1 nanometer (nm) [formerly millimicron (mμ)]	= 0.000000001 (10^{-9}) meter		
	1 angstrom (Å)	= 0.0000000001 (10^{-10}) meter		
Area	1 square meter (m^2)	= 10,000 square centimeters	1 m^2 = 1.1960 square yards 1 m^2 = 10.764 square feet	1 square yard = 0.8361 m^2 1 square foot = 0.0929 m^2
	1 square centimeter (cm^2)	= 100 square millimeters	1 cm^2 = 0.155 square inch	1 square inch = 6.4516 cm^2
Mass	1 metric ton (t)	= 1000 kilograms	1 t = 1.103 ton	1 ton = 0.907t
	1 kilogram (kg)	= 1000 grams	1 kg = 2.205 pounds	1 pound = 0.4536 kg
	1 gram (g)	= 1000 milligrams	1 g = 0.0353 ounce 1 g = 15.432 grains	1 ounce = 28.35 g
	1 milligram (mg)	= 0.001 gram	1 mg = approx. 0.015 grain	
	1 microgram (μg)	= 0.000001 gram		
Volume (solids)	1 cubic meter (m^3)	= 1,000,000 cubic centimeters	1 m^3 = 1.3080 cubic yards 1 m^3 = 35.315 cubic feet	1 cubic yard = 0.7646 m^3 1 cubic foot = 0.0283 m^3
	1 cubic centimeter (cm^3 or cc)	= 0.000001 cubic meter = 1 milliliter	1 cm^3 = 0.0610 cubic inch	1 cubic inch = 16.387 cm^3
	1 cubic millimeter (mm^3)	= 0.000000001 cubic meter		
Volume (liquids and gases)	1 kiloliter (kl or kL)	= 1000 liters	1 kL = 264.17 gallons	1 gallon = 3.785 L
	1 liter (l or L)	= 1000 milliliters	1 L = 0.264 gallons 1 L = 1.057 quarts	1 quart = 0.946 L
	1 milliliter (ml or mL)	= 0.001 liter = 1 cubic centimeter	1 ml = 0.034 fluid ounce 1 ml = approx. $\frac{1}{4}$ teaspoon 1 ml = approx. 15–16 drops (gtt.)	1 quart = 946 ml 1 pint = 473 ml 1 fluid ounce = 29.57 ml 1 teaspoon = approx. 5 ml
	1 microliter (μl or μL)	= 0.000001 liter		
Time	1 second (s)	= $\frac{1}{60}$ minute		
	1 millisecond (ms)	= 0.001 second		
Temperature	Degrees Celsius (°C)		$°F = \frac{9}{5}°C + 32$	$°C = \frac{5}{9}(°F - 32)$

Answers to Multiple Choice and Matching Questions

Chapter 1
1. c
2. (1) d; (2) b; (3) a
3. (a) wrist; (b) hipbone; (c) nose; (d) toes; (e) scalp
4. b
5. (a) D; (b) V; (c) D; (d) V; (e) V
6. (a) 2; (b) 3; (c) 1; (d) 4
7. b
8. d
9. c
10. b
11. c
12. e

Chapter 2
1. a
2. a
3. b
4. a
5. (a) Golgi apparatus (b) rough ER (c) condensed (d) smooth (e) nucleosome (f) microtubules (g) mitochondria
6. d
7. b
8. a
9. (a) metaphase (b) prophase (c) telophase (d) prophase (e) anaphase
10. (a) microtubules (b) intermediate filaments (c) microtubules (d) actin microfilaments (e) intermediate filaments
11. (1) b; (2) g; (3) c or b; (4) d; (5) e; (6) f; (7) a
12. b
13. c
14. d

Chapter 3
1. (1) a; (2) b; (3) b; (4) a
2. b
3. b
4. c
5. (1) e; (2) g; (3) a; (4) f; (5) i; (6) b; (7) h; (8) d; (9) c
6. (1) f; (2) g; (3) a; (4) c; (5) d; (6) e; (7) b
7. c
8. a
9. c
10. d
11. c
12. a, d
13. c
14. e
15. a

Chapter 4
1. (1) a; (2) c; (3) d; (4) b; (5) b.
2. c and e
3. (1) b; (2) f; (3) a; (4) d; (5) g.
4. (1) d; (2) a; (3) c; (4) b; (5) e.
5. b
6. c
7. a
8. (1) cilia; (2) microvilli; (3) desmosome; (4) tight junction; (5) basement membrane.
9. b

Chapter 5
1. a
2. c
3. (1) e; (2) b or d; (3) a; (4) c; (5) d
4. d
5. d
6. b
7. b
8. a
9. c
10. c
11. b
12. a
13. d
14. b
15. d

Chapter 6
1. e
2. a
3. e
4. (1) T; (2) F; (3) T; (4) F; (5) F; (6) T; (7) T
5. b
6. c
7. d
8. b
9. c
10. (a) 3; (b) 2; (c) 4; (d) 1; (e) 5; (f) 6
11. c
12. b
13. e
14. c
15. b
16. b

Chapter 7
1. (1) b, g; (2) h; (3) d; (4) d, f; (5) e; (6) c; (7) a, b, d, h; (8) i.
2. (1) g; (2) f; (3) b; (4) a; (5) b; (6) c; (7) d; (8) e.
3. (1) b; (2) c; (3) e; (4) a; (5) h; (6) e; (7) f.
4. b
5. c

Chapter 8
1. (1) a, b; (2) a; (3) a; (4) a; (5) b; (6) c; (7) c; (8) a, b; (9) c
2. b
3. e
4. c
5. e
6. b
7. d
8. d
9. a
10. (1) a; (2) b, d; (3) c; (4) c; (5) b, d
11. d

Chapter 9
1. c
2. b
3. d
4. (1) thin; (2) thick; (3) thick; (4) thin; (5) thick; (6) thin
5. (1) d; (2) a; (3) b; (4) e; (5) c
6. (1) yes; (2) no; (3) yes; (4) no

7. c
8. c
9. a
10. (1) b; (2) a; (3) c; (4) b; (5) c; (6) b; (7) a; (8) c; (9) c; (10) c
11. d

Chapter 10
1. c
2. (1) c; (2) b; (3) e; (4) a; (5) d
3. e
4. a
5. c
6. d
7. c
8. a
9. c
10. b
11. b
12. a
13. d
14. a
15. c
16. d

Chapter 11
1. b
2. (1) d; (2) b; (3) f; (4) c; (5) a
3. c
4. b
5. (1) SS; (2) SS; (3) VM; (4) SM; (5) VS; (6) SS; (7) VM
6. b
7. b
8. a
9. c
10. d
11. (1) c; (2) b; (3) a

Chapter 12
1. a
2. (1) c; (2) f; (3) e; (4) g; (5) b; (6) e; (7) i; (8) a; (9) g; (10) h; (11) d
3. a
4. c
5. d
6. b
7. d
8. a
9. (1) G; (2) W; (3) W; (4) G; (5) W; (6) G; (7) W
10. a
11. d

Chapter 13
1. c
2. b
3. (1) e; (2) d; (3) c; (4) g; (5) b; (6) f; (7) a
4. (1) f; (2) i; (3) b; (4) g,h,j; (5) e; (6) i; (7) i; (8) k; (9) j; (10) c,d,f,k; (11) c
5. (1) b,6; (2) d,8; (3) c,2; (4) c,5; (5) a,4; (6) a,3&9; (7) a,7; (8) a,7; (9) d,1
6. c
7. c
8. c

Chapter 14
1. d
2. (1) S; (2) P; (3) P; (4) S; (5) S; (6) P; (7) S; (8) S; (9) P; (10) S; (11) P; (12) S; (13) S; (14) S
3. c
4. d
5. b
6. d
7. c
8. a
9. a

Chapter 15
1. c
2. a
3. d
4. c
5. c
6. b
7. f
8. b
9. b
10. a
11. e
12. a
13. d
14. b
15. c
16. a
17. b
18. b
19. d
20. c
21. c

Chapter 16
1. c
2. d
3. (a) 2; (b) 5; (c) 1; (d) 4; (e) 3
4. a
5. c
6. a
7. d
8. (1) c; (2) b; (3) a; (4) c; (5) a; (6) b; (7) f; (8) b; (9) a; (10) a; (11) a; (12) e
9. d

Chapter 17
1. b
2. d
3. c
4. b
5. b
6. c
7. b
8. c
9. e
10. a
11. d
12. d
13. c
14. b
15. e
16. d
17. a
18. a
19. a

Chapter 18
1. d
2. b
3. b

4. c
5. b
6. (1) intima; (2) adventitia; (3) adventitia; (4) adventitia; (5) media; (6) intima
7. d
8. a
9. b
10. c
11. c
12. a
13. b

Chapter 19
1. c
2. d
3. c
4. c
5. a
6. b
7. e
8. b
9. e
10. a
11. b
12. a and c
13. c
14. d
15. a
16. d
17. d

Chapter 20
1. c
2. d
3. b
4. a
5. b
6. d
7. a
8. (1) b, (2) b, (3) a, (4) b, (5) c
9. (1) c, (2) d, (3) e, (4) a, (5) b
10. b
11. a
12. b
13. c
14. c
15. d

Chapter 21
1. b
2. c
3. c
4. a
5. (1) d; (2) c; (3) f; (4) a; (5) b; (6) e
6. (1) d; (2) d; (3) d; (4) c; (5) c; (6) c
7. a
8. a
9. b
10. b
11. c
12. d
13. a and d
14. d
15. c
16. c
17. b
18. a
19. f

Chapter 22
1. d
2. b

3. c
4. c
5. a
6. d
7. b
8. b
9. b
10. (a) 4; (b) 1; (c) 5; (d) 3; (e) 2
11. e
12. e
13. a
14. c
15. d
16. b
17. c

Chapter 23
1. d
2. a
3. d
4. (1) d; (2) c; (3) d; (4) c; (5) e
5. d
6. a
7. d
8. e
9. b
10. a
11. a
12. a
13. a
14. c
15. b
16. c
17. d

Chapter 24
1. b
2. a
3. (1) c; (2) b; (3) g; (4) h; (5) d; (6) e; (7) f; (8) i; (9) a; (10) j; (11) b, h; (12) i
4. a
5. b
6. b
7. d
8. (1) a; (2) b; (3) a; (4) b; (5) b; (6) a
9. (1) e; (2) a; (3) a; (4) d; (5) b
10. d
11. a
12. c
13. b
14. d
15. d
16. c
17. d

Chapter 25
1. c
2. a
3. b
4. b
5. d
6. c
7. b
8. c
9. b
10. d
11. e
12. a
13. a
14. c
15. c
16. a

Word Roots, Prefixes, Suffixes, and Combining Forms

PREFIXES AND COMBINING FORMS

a-, an- *absence or lack* acardia, lack of a heart; anaerobic, in the absence of oxygen

ab- *departing from; away from* abnormal, departing from normal

acou- *hearing* acoustics, the science of sound

acr-, acro- *extreme or extremity; peak* acrodermatitis, inflammation of the skin of the extremities

ad- *to or toward* adorbital, toward the orbit

aden-, adeno- *gland* adeniform, resembling a gland in shape

amphi- *on both sides; of both kinds* amphibian, an organism capable of living in water and on land

angi- *vessel* angiitis, inflammation of a lymph vessel or blood vessel

ant-, anti- *opposed to; preventing or inhibiting* anticoagulant, a substance that prevents blood coagulation

ante- *preceding; before* antecubital, in front of the elbow

arthr-, arthro- *joint* arthropathy, any joint disease

aut-, auto- *self* autogenous, self-generated

bi- *two* bicuspid, having two cusps

bio- *life* biology, the study of life and living organisms

blast- *bud or germ* blastocyte, undifferentiated embryonic cell

broncho- *bronchus* bronchospasm, spasmodic contraction of bronchial muscle

bucco- *cheek* buccolabial, pertaining to the cheek and lip

caput- *head* decapitate, remove the head

carcin- *cancer* carcinogen, a cancer-causing agent

cardi, cardio- *heart* cardiotoxic, harmful to the heart

cephal- *head* cephalometer, an instrument for measuring the head

cerebro- *brain, especially the cerebrum* cerebrospinal, pertaining to the brain and spinal cord

chondr- *cartilage* chondrogenic, giving rise to cartilage

circum- *around* circumnuclear, surrounding the nucleus

co-, con- *together* concentric, common center, together in the center

contra- *against* contraceptive, agent preventing conception

cost- *rib* intercostal, between the ribs

crani- *skull* craniotomy, a skull operation

crypt- *hidden* cryptomenorrhea, a condition in which menstrual symptoms are experienced but no external loss of blood occurs

cyt- *cell* cytology, the study of cells

de- *undoing, reversal, loss, removal* deactivation, becoming inactive

di- *twice, double* dimorphism, having two forms

dia- *through, between* diaphragm, the wall through or between two areas

dys- *difficult, faulty, painful* dyspepsia, disturbed digestion

ec, ex, ecto- *out, outside, away from* excrete, to remove materials from the body

en-, em- *in, inside* encysted, enclosed in a cyst or capsule

entero- *intestine* enterologist, one who specializes in the study of intestinal disorders

epi- *over, above* epidermis, outer layer of skin

eu- *well* euesthesia, a normal state of the senses

exo- *outside, outer layer* exophthalmos, an abnormal protrusion of the eye from the orbit

extra- *outside, beyond* extracellular, outside the body cells of an organism

gastr- *stomach* gastrin, a hormone that influences gastric acid secretion

glosso- *tongue* glossopathy, any disease of the tongue

hema-, hemato-, hemo- *blood* hematocyst, a cyst containing blood

hemi- *half* hemiglossal, pertaining to one half of the tongue

hepat- *liver* hepatitis, inflammation of the liver

hetero- *different or other* heterosexuality, sexual desire for a person of the opposite sex

hist- *tissue* histology, the study of tissues

hom-, homo- *same* homeoplasia, formation of tissue similar to normal tissue; homocentric, having the same center

hydr-, hydro- *water* dehydration, loss of body water

hyper- *excess* hypertension, excessive tension

hypno- *sleep* hypnosis, a sleeplike state

hypo- *below, deficient* hypodermic, beneath the skin; hypokalemia, deficiency of potassium

hyster-, hystero- *uterus or womb* hysterectomy, removal of the uterus; hysterodynia, pain in the womb

im- *not* impermeable, not permitting passage, not permeable

inter- *between* intercellular, between the cells

intra- *within, inside* intracellular, inside the cell

iso- *equal, same* isothermal, equal, or same, temperature

leuko- *white* leukocyte, white blood cell

lip-, lipo- *fat, lipid* lipophage, a cell that has taken up fat in its cytoplasm

macro- *large* macromolecule, large molecule

mal- *bad, abnormal* malfunction, abnormal functioning of an organ

mamm- *breast* mammary gland, breast

mast- *breast* mastectomy, removal of a mammary gland

meningo- *membrane* meningitis, inflammation of the membranes of the brain

meso- *middle* mesoderm, middle germ layer

meta- *beyond, between, transition* metatarsus, the part of the foot between the tarsus and the phalanges

metro- *uterus* metroscope, instrument for examining the uterus

micro- *small* microscope, an instrument used to make small objects appear larger

mito- *thread, filament* mitochondria, small, filamentlike structures located in cells

mono- *single* monospasm, spasm of a single limb

morpho- *form* morphology, the study of form and structure of organisms

multi- *many* multinuclear, having several nuclei

myelo- *spinal cord, marrow* myeloblasts, cells of the bone marrow

myo- *muscle* myocardium, heart muscle

narco- *numbness* narcotic, a drug producing stupor or numbed sensations

nephro- *kidney* nephritis, inflammation of the kidney

neuro- *nerve* neurophysiology, the physiology of the nervous system

ob- *before, against* obstruction, impeding or blocking up

oculo- *eye* monocular, pertaining to one eye

odonto- *teeth* orthodontist, one who specializes in proper positioning of the teeth in relation to each other

ophthalmo- *eye* ophthalmology, the study of the eyes and related disease

ortho- *straight, direct* orthopedic, correction of deformities of the musculoskeletal system

osteo- *bone* osteodermia, bony formations in the skin

oto- *ear* otoscope, a device for examining the ear

oxy- *oxygen* oxygenation, the saturation of a substance with oxygen

pan- *all, universal* panacea, a cure-all

para- *beside, near* paraphrenitis, inflammation of tissues adjacent to the diaphragm

peri- *around* perianal, situated around the anus

phago- *eat* phagocyte, a cell that engulfs and digests particles or cells

phleb- *vein* phlebitis, inflammation of the veins

pod- *foot* podiatry, the treatment of foot disorders

poly- *multiple* polymorphism, multiple forms

post- *after, behind* posterior, places behind (a specific) part

pre-, pro- *before, ahead of* prenatal, before birth

procto- *rectum, anus* proctoscope, an instrument for examining the rectum

pseudo- *false* pseudotumor, a false tumor

psycho- *mind, psyche* psychogram, a chart of personality traits

pyo- *pus* pyocyst, a cyst that contains pus

retro- *backward, behind* retrogression, to move backward in development

sclero- *hard* sclerodermatitis, inflammatory thickening and hardening of the skin

semi- *half* semicircular, having the form of half a circle

steno- *narrow* stenocoriasis, narrowing of the pupil

sub- *beneath, under* sublingual, beneath the tongue

super- *above, upon* superior, quality or state of being above others or a part

supra- *above, upon* supracondylar, above a condyle

sym-, syn- *together, with* synapse, the region of communication between two neurons

tachy- *rapid* tachycardia, abnormally rapid heartbeat

therm- *heat* thermometer, an instrument used to measure heat

tox- *poison* antitoxic, effective against poison

trans- *across, through* transpleural, through the pleura

tri- *three* trifurcation, division into three branches

viscero- *organ, viscera* visceroinhibitory, inhibiting the movements of the viscera

SUFFIXES

-able *able to, capable of* viable, ability to live or exist

-ac *referring to* cardiac, referring to the heart

-algia *pain in a certain part* neuralgia, pain along the course of a nerve

-ary *associated with, relating to* coronary, associated with the heart

-cide *destroy or kill* germicide, an agent that kills germs

-ectomy *cutting out, surgical removal* appendectomy, cutting out of the appendix

-emia *condition of the blood* anemia, deficiency of red blood cells

-ferent *carry* efferent nerves, nerves carrying impulses away from the CNS

-fuge *driving out* vermifuge, a substance that expels worms of the intestine

-gen *an agent that initiates* pathogen, any agent that produces disease

-gram *data that are systematically recorded, a record* electrocardiogram, a recording showing action of the heart

-graph *an instrument used for recording data or writing* electrocardiograph, an instrument used to make an electrocardiogram

-ia *condition* insomnia, condition of not being able to sleep

-iatrics *medical specialty* geriatrics, the branch of medicine dealing with disease associated with old age

-itis *inflammation* gastritis, inflammation of the stomach

-logy *the study of* pathology, the study of changes in structure and function brought on by disease

-lysis *loosening or breaking down* hydrolysis, chemical decomposition of a compound into other compounds as a result of taking up water

-malacia *soft* osteomalacia, a process leading to bone softening

-mania *obsession, compulsion* erotomania, exaggeration of the sexual passions

-odyn *pain* coccygodynia, pain in the region of the coccyx

-oid *like, resembling* cuboid, shaped as a cube

-oma *tumor* lymphoma, a tumor of the lymphatic tissues

-opia *defect of the eye* myopia, nearsightedness

-ory *referring to, of* auditory, referring to hearing

-pathy *disease* osteopathy, any disease of the bone

-phobia *fear* acrophobia, fear of heights

-plasty *reconstruction of a part, plastic surgery* rhinoplasty, reconstruction of the nose through surgery

-plegia *paralysis* paraplegia, paralysis of the lower half of the body or limbs

-rrhagia *abnormal or excessive discharge* metrorrhagia, uterine hemorrhage

-rrhea *flow or discharge* diarrhea, abnormal emptying of the bowels

-scope *instrument used for examination* stethoscope, instrument used to listen to sounds of various parts of the body

-stasis *arrest, fixation* hemostasis, arrest of bleeding

-stomy *establishment of an artificial opening* enterostomy, the formation of an artificial opening into the intestine through the abdominal wall

-tomy *to cut* appendectomy, surgical removal of the appendix

-ty *conditon of, state* immunity, condition of being resistant to infection or disease

-uria *urine* polyuria, passage of an excessive amount of urine

Glossary

Pronunciation Key

The pronunciation key that is used throughout this book takes advantage of the fact that, in scientific terms, long vowels usually occur at the end of syllables, whereas short vowels usually occur at the beginning or in the middle of syllables. Therefore:

1. When vowels are *unmarked*, they have a long sound when they occur at the end of a syllable, and a short sound when they are at the beginning or in the middle of a sylla-ble. For example, in the word *kidney* (kid′ne), you auto-matically know that the middle 'i' is short and the terminal 'e' sound is long. In the word *renal* (re′nal), the 'e' is long, and the 'a' is short.

2. When a vowel sound violates the above rule, it is marked with a short or long symbol. Only those short vowels that come at the end of a syllable are marked with the short symbol, called a breve (˘); and only those long vowels that occur at the start or middle of a syllable are marked with a long symbol (¯). For example, in the word *pelvirectal* (pel″vĭ-rek′tal), all four vowels are short, but only the 'i' is marked as short, because it is the only vowel that falls at the end of a syllable. In *methane* (meth′ān), the long 'a' is marked as such because it does not fall at the end of a syllable.

3. Short 'a' is *never* marked with a breve. Instead, short 'a' at the end of a syllable is indicated by 'ah'. For example, in the word *papilla* (pah-pil′ah), each 'a' is short and pro-nounced as 'ah'.

4. In words that have more than one syllable, the syllable with the strongest accent is followed by a prime (′) mark, and syllables with the second strongest accent, where pres-ent, are indicated with a double prime (″). The unaccented syllables are followed by dashes. An example of these prin-ciples is the word *anesthetic* (an″es-thet′ik), in which thet′ is emphasized most strongly, an″ has a secondary empha-sis, and the other two syllables are not spoken with any emphasis.

Abdomen (ab-do′men)   Region of the body between the dia-phragm and the pelvis.

Abduct (ab-dukt′)   To move away from the midline of the body.

Absorption   Process by which the products of digestion pass through the lining of the alimentary canal into the blood or lymph.

Accessory digestive organs   Organs that contribute to the digestive process but are not part of the alimentary canal, including the tongue, teeth, salivary glands, pancreas, liver, and gallbladder.

Acetabulum (as″ĕ-tab′u-lum)   Cup-like cavity on lateral surface of hip bone that receives the femur.

Acetylcholine (as″ĕ-til-ko′lēn)   Chemical neurotransmitter substance released by some nerve endings.

Actin (ak′tin)   A contractile protein in cells, especially abundant in muscle cells.

Action potential   A large, transient depolarization event, including polarity reversal, that is conducted along the plasma membrane of a nerve axon or muscle cell.

Acute   Producing severe symptoms in the short term; rap-idly developing.

Adduct (ah-dukt′)   To move toward the midline of the body.

Adenohypophysis (ad″ĕ-no-hi-pof′ĭ-sis)   One of the main divisions of the pituitary gland, the other being the neu-rohypophysis; the glandular part of the pituitary.

Adenoids (ad′ĕ-noids)   The pharyngeal tonsil on the roof of the pharynx.

Adenosine triphosphate (ATP) (ah-den′o-sēn)   Molecule in cells that stores and releases chemical energy for use in body cells.

Adipose (ad′ĭ-pōs)   Fatty.

Adrenal gland (ah-dre′nal)   Hormone-secreting gland located superior to the kidney; consists of medulla and cortex areas; also called suprarenal gland.

Adrenalin (ah-dren′ah-lin)   *See* Epinephrine.

Adrenergic fibers (ad″ren-er′jik)   Nerve fibers that release norepinephrine.

Adrenocorticotropic hormone (ACTH) (ah-dre″no-kor″tĭ-ko-trop′ik)   Hormone from the anterior pituitary that influ-ences the activity of the adrenal cortex.

Adventitia (ad″ven-tish′e-ah)   Outermost layer or covering of an organ; consists of connective tissue.

Aerobic (a′er-ōb-ik)   Oxygen-requiring.

Afferent (af′er-ent)   Carrying to or toward a center; in the nervous system, *afferent* means "sensory."

Afferent neuron   Nerve cell that carries impulses toward the central nervous system; sensory neuron.

Agonist (ag′o-nist)   Muscle that bears primary responsibil-ity for causing a certain movement; also called a prime mover.

AIDS (Acquired immune deficiency syndrome)   Disease caused by the human immunodeficiency virus (HIV); symptoms include severe weight loss, swollen lymph nodes, and many opportunistic infections that lead to death.

Aldosterone (al-dos′ter-ōn)   Hormone secreted by the adre-nal cortex that stimulates reabsorption of sodium ions and water from the kidney.

Alimentary canal (al″ĭ-men′tar-e)   The digestive tube, extending from the mouth to the anus; its basic regions are the oral cavity, pharynx, esophagus, stomach, and small and large intestines.

Allantois (ah-lan'to-is)   A tubular extension of the embryonic hindgut and cloaca; becomes the urachus, a fibrous cord attached to the adult bladder.

Allergy (al'er-je)   Overzealous immune response to an otherwise harmless substance.

Alveolus (al-ve'o-lus)   (1) One of the microscopic air sacs of the lungs; (2) a spherical sac formed by the secretory cells in a gland; also called an *acinus*; (3) the socket of a tooth.

Amino acid (ah-me'no)   Organic compound containing nitrogen, carbon, hydrogen, and oxygen; building block of proteins.

Amnion (am'ne-on)   Membrane that forms a fluid-filled sac around the embryo and fetus.

Amphiarthrosis (am"fe-ar-thro'sis)   A slightly movable joint.

Anaerobic (an"a-er-o'bik)   Not requiring oxygen.

Anastomosis (ah-nas"to-mo'sis)   A union or joining of blood vessels or other tubular structures.

Androgen (an'dro-jen)   A male sex hormone, the main example of which is testosterone.

Anemia (ah-ne'me-ah)   Reduced oxygen-carrying capacity of the blood; results from too few erythrocytes or from abnormal hemoglobin.

Aneurysm (an'u-rizm)   Blood-filled dilation of a blood vessel, caused by a weakening of the vessel wall.

Angina pectoris (an-ji'nah)   Severe, suffocating chest pain caused by temporary lack of oxygen supply to heart muscle.

Antagonist (an-tag'o-nist)   Muscle that reverses, or opposes, the action of another muscle.

Anterior   The front of an organism, organ, or body part; the ventral direction.

Anterior pituitary   *See* Pars distalis.

Antibody   A protein molecule that is secreted by a plasma cell (a cell derived from an activated B lymphocyte) and that binds to an antigen in immune responses.

Antidiuretic hormone (ADH) (an"tĭ-di"u-ret'ik)   Hormone produced by the hypothalamus and released by the posterior part of the pituitary gland (pars nervosa); stimulates the kidney to reabsorb more water.

Antigen (an'tĭ-jen)   A molecule that is recognized as foreign by the immune system, activates the immune system, and reacts with immune cells or antibodies.

Anus (a'nus)   The opening at the distal end of the alimentary canal.

Aorta (a-or'tah)   Major systemic artery; arises from the left ventricle of the heart.

Apocrine gland (ap'o-krin)   A type of sweat gland in the armpit and anal-genital regions; produces a secretion containing water, salts, proteins, and lipids.

Aponeurosis (ap"o-nu-ro'sis)   Fibrous sheet connecting a muscle to the body part it moves.

Appendicular skeleton (ap"en-dik'u-lar)   Bones of the limbs and limb girdles that are attached to the axial skeleton.

Appendix (ah-pen'diks)   Worm-like tube attached to the cecum of the large intestine; technically called the *vermiform appendix*.

Aqueous humor (a'kwe-us)   Watery fluid in the anterior segment of the eye.

Arachnoid mater (ah-rak'noid ma'ter)   The web-like middle layer of the three meninges.

Areola (ah-re'o-lah)   Circular, pigmented area of skin surrounding the nipple.

Arrector pili (ah-rek'tor)   Tiny band of smooth muscle attached to each hair follicle; its contraction causes the hair to stand upright.

Arterial circle (of Willis)   A union of arteries at the base of the brain.

Arteriole (ar-te're-ōl)   A minute artery.

Arteriosclerosis (ar-te"re-o-sklĕ-ro'sis)   Hardening of the arteries: any of a number of degenerative changes in the walls of arteries leading to a decrease in their elasticity.

Artery   Vessel that carries blood away from the heart.

Arthritis (ar-thri'tis)   Inflammation of joints.

Articular capsule (ar-tik'u-lar)   The capsule of a synovial joint; consists of an outer layer of fibrous connective tissue and an inner synovial membrane.

Articulation   Joint; point where two elements of the skeleton meet.

Atherosclerosis (ath"er-o"skle-ro'sis)   Changes in the walls of large arteries involving the deposit of lipid plaques; the most common variety of arteriosclerosis.

Atlas   First cervical vertebra.

Atria (a'tre-ah)   Paired, superiorly located heart chambers that receive blood returning to the heart.

Atrioventricular bundle (AV bundle) (a"tre-o-ven-trik'u-lar)   Bundle of cardiac muscle cells that conducts impulses from the AV node to the walls of the right and left ventricles; located in the septum (wall) between the two ventricles of the heart; also called *bundle of His*.

Atrioventricular node (AV node)   Specialized mass of conducting cells located in the interatrial septum of the heart.

Atrophy (at'ro-fe)   Reduction in size or wasting away of an organ or tissue.

Auditory tube   Narrow tube of cartilage and bone that connects the superior pharynx (nasopharynx) to the middle ear.

Autonomic nervous system (aw"to-nom'ik)   General visceral motor division of the peripheral nervous system; innervates smooth and cardiac muscle, and glands.

Avascular (a-vas'ku-lar)   Having no blood supply; containing no blood vessels.

Axial skeleton   Portion of the skeleton that forms the central (longitudinal) axis of the body; includes the bones of the skull, the vertebral column, and the bony thorax.

Axilla (ak-sil'ah)   Armpit.

Axis   (1) Second cervical vertebra; (2) imaginary line about which a joint or structure rotates.

Axon   Neuron process that carries impulses away from the cell body.

B cells   Lymphocytes that oversee humoral immunity; they divide to generate plasma cells, which secrete antibodies; also called *B lymphocytes*.

Basal ganglia   *See* Basal nuclei.

Basal lamina   A thin sheet of protein that underlies an epithelium.

Basal nuclei   Areas of gray matter located deep within the white matter of the cerebral hemispheres; regulate certain aspects of movement; also called *basal ganglia*.

Basement membrane   A layer between an epithelium and the underlying connective tissue; consists of both a basal lamina and a network of reticular fibers.

Basophil (ba'so-fil)   (1) Type of white blood cell whose cytoplasmic granules stain purple with the basic dyes in blood stains; (2) a gland cell in the anterior pituitary containing cytoplasmic granules that stain with basic dyes.

Benign (be-nīn) Not malignant; not life-threatening.

Biceps (bi′seps) Two-headed, especially applied to certain muscles.

Bile Greenish fluid secreted by the liver, stored in the gallbladder, and released into the small intestine; helps start the breakdown of fats.

Bipolar neuron Neuron with just two processes, which extend from opposite sides of the cell body.

Blastocyst (blas′to-sist) Stage of early embryonic development; a hollow ball of cells; the product of cleavage.

Blood-brain barrier The feature that inhibits passage of harmful materials from the blood into brain tissues; reflects relative impermeability of brain capillaries.

Blood pressure Force exerted by blood against a unit area of the blood vessel walls; differences in blood pressure between different areas of the circulation provide the driving force for blood circulation.

Bolus (bo′lus) A rounded mass of food prepared by the mouth for swallowing.

Bone remodeling Process involving bone formation and bone destruction in response to mechanical and hormonal factors.

Bony thorax Bones that form the framework of the thorax; includes sternum, ribs, and thoracic vertebrae.

Brain stem Collectively, the midbrain, pons, and medulla of the brain.

Bronchus Air tubes of the respiratory tree between the trachea and bronchioles; bronchi enter and branch within the lungs.

Bursa A fibrous sac lined with synovial membrane and containing synovial fluid; occurs between bones and tendons (or other structures), where it acts to decrease friction during movement.

Calcaneal (Achilles) tendon Tendon that attaches the calf muscles to the heel bone.

Calcitonin (kal″sĭ-to′nin) Hormone released by the thyroid gland that promotes a decrease in calcium levels in the blood.

Calyx (ka′liks) A cup-like tributary of the pelvis of the kidney.

Cancer A malignant, invasive cellular tumor that has the capability of spreading throughout the body or body parts.

Capillary The smallest of blood vessels and the site of exchanges of molecules between the blood and the tissue fluid.

Carcinogen (kar-sin′o-jen) Cancer-causing agent.

Cardiac muscle Muscle tissue of the heart wall.

Cardiac sphincter The circular layer of smooth muscle at the junction of the esophagus and stomach; contracts to prevent reflux of stomach contents into the esophagus.

Cartilage (kar′tĭ-lij) White, semiopaque, resilient connective tissue; gristle.

Catecholamines (kat″ĕ-kol′ah-mēns″) Epinephrine, norepinephrine, and related molecules.

Caudal (kaw′dal) Literally, toward the tail; in humans, toward the inferior portion of the trunk.

Cecum (se′kum) The blind-ended pouch at the beginning of the large intestine.

Cell membrane *See* Plasma membrane.

Central nervous system (CNS) Brain and spinal cord.

Centriole (sen′tre-ōl) Barrel-shaped organelle formed of microtubules and located near the nucleus of the cell; active in cell division.

Cerebellum (ser″ĕ-bel′um) Brain region that is attached to the pons and smooths and coordinates body movements.

Cerebral aqueduct (sĕ-re′bral ak′wĕ-dukt″) The narrow cavity of the midbrain that connects the third and fourth ventricles.

Cerebral cortex The external, gray matter region of the cerebral hemispheres.

Cerebrospinal fluid (ser″ĕ-bro-spi′nal) Clear fluid that fills the cavities of the central nervous system and surrounds the CNS externally; it floats and cushions the brain and spinal cord.

Cerebrum (ser′ĕ-brum) The cerebral hemispheres (some authorities also include the diencephalon).

Cervix (ser′viks) The inferior, neck-like part of the uterus (*cervix* = neck).

Chemoreceptor (ke″mo-re-sep′tor) Receptor sensitive to chemicals in solution.

Cholesterol (ko-les′ter-ol) A steroid lipid found in animal fats as well as in the plasma membranes of cells.

Cholinergic fiber (ko″lin-er′jik) An axon whose axon terminals release the neurotransmitter acetylcholine.

Chondroblast (kon′dro-blast) Actively mitotic form of a cartilage cell.

Chondrocyte (kon′dro-sīt) Mature form of a cartilage cell.

Chorion (ko′re-on) The outermost fetal membrane; helps form the placenta; technically, it consists of the trophoblast and the extraembryonic mesoderm.

Choroid plexus (ko′roid) A capillary-rich membrane on the roof of the brain that forms the cerebrospinal fluid; technically, it consists of pia mater and ependymal cells.

Chromatin (kro′mah-tin) Strands in the cell nucleus that consist of deoxyribonucleic acid (DNA) and histone proteins.

Chromosome (kro′mo-sōm) Bar-like body of tightly coiled chromatin, visible during cell division; typical human cells have 46 chromosomes.

Chronic (kron′ik) Long-term; prolonged; not acute.

Chyme (kīm) Semifluid, creamy mass consisting of partially digested food and stomach juices.

Cilium (sil′e-um) Motile, hair-like projection from the apical surface of certain epithelial cells.

Circumduction (ser″kum-duk′shun) Movement of a body part so that it outlines a cone in space.

Cirrhosis (si-ro′sis) A chronic disease, particularly of the liver, characterized by an overgrowth of connective tissue, or fibrosis.

Cleavage An early embryonic stage consisting of rapid cell divisions without intervening growth periods; begins with a fertilized ovum and produces a blastocyst.

Cochlea (kok′le-ah) Snail-shaped chamber of the bony labyrinth in the inner ear; houses the receptor for hearing (spiral organ, or organ of Corti).

Condyle (kon′dīl) A rounded projection at the end of a bone that articulates with another bone.

Cone cell One of the two types of photoreceptor cells in the retina of the eye; provides for color vision and sharp vision.

Congenital (kon-jen′ĭ-tal) Existing at birth.

Conjunctiva (kon″junk-ti′vah) Thin, protective mucous

membrane that covers the white of the eye and the internal surface of eyelids.

Connective tissue A primary tissue; form and function vary widely, but all connective tissues contain a large amount of extracellular matrix; functions include support, holding tissue fluid, and protection from disease.

Contraction The generation of a pulling force while shortening; this ability is highly developed in muscle cells.

Contralateral (kon″trah-lat′er-al) Concerning the opposite half of the body; when nerve fibers project contralaterally, they cross over to the opposite side of the body (right to left, or vice versa).

Cornea (kor′ne-ah) Transparent anterior portion of the eyeball.

Corona radiata (kŏ-ro′nah ra-de-ah′tah) (1) Crown-like arrangement of granulosa cells around an oocyte in an ovarian follicle after the appearance of an antrum; (2) crown-like arrangement of nerve fibers radiating from the internal capsule of the brain to every part of the cerebral cortex.

Coronal plane (kŏ-ro′nal) *See* Frontal plane.

Cortex (kor′teks) Outer region of an organ.

Corticosteroids (kor″tĭ-ko-ste′roids) Steroid hormones secreted by the adrenal cortex. Examples are cortisol, aldosterone, and some sex hormones.

Cortisol (kor′tĭ-sol) A glucocorticoid hormone produced by the adrenal cortex.

Cranial nerves The 12 pairs of nerves that attach to the brain.

Cutaneous (ku-ta′ne-us) Pertaining to the skin.

Cytokinesis (si″to-ki-ne′sis) Division of the cytoplasm that occurs after the cell nucleus has divided.

Cytoplasm (si′to-plazm) The part of a cell between the plasma membrane and the nucleus; contains many organelles.

Cytotoxic T cell T lymphocyte that directly kills eukaryotic foreign cells, cancer cells, or virus-infected body cells; also called *killer T cell.*

Deep Toward the inside; inner; internal.

Defecation (def″ĕ-ka′shun) Elimination of the contents (feces) of the bowel.

Dendrite (den′drīt) Neuron process that transmits signals toward the cell body and serves as receptive region of the neuron; most dendrites branch extensively.

Deoxyribonucleic acid (DNA) (de-ok″sĭ-ri″bo-nu-kle′ik) A nucleic acid found in all living cells; carries the organism's hereditary information.

Depolarization (de-po″lar-ĭ-za″shun) Loss of a state of polarity across a cellular membrane; loss of negative charge inside the cell.

Dermis The leathery layer of skin, deep to the epidermis; composed largely of dense irregular connective tissue.

Desmosome (des′mo-sōm) A cell junction composed of two disc-shaped plaques connected across the intercellular space; the most important junction for holding epithelial cells together.

Diabetes insipidus (di″ah-be′tēz) Disease characterized by passage of a large quantity of dilute urine plus intense thirst and dehydration; caused by inadequate release of antidiuretic hormone.

Diabetes mellitus Disease caused by deficient release of, or deficient use of, insulin; characterized by an inability of the body cells to use sugars at a normal rate and by high blood sugar levels.

Diapedesis (di″ah-pĕ-de″sis) Active movement of white blood cells through the walls of capillaries and venules into the surrounding tissue.

Diaphragm (di′ah-fram) (1) Any partition or wall separating one area from another; (2) the muscular sheet that separates the thoracic cavity from the abdominopelvic cavity.

Diaphysis (di-af′ĭ-sis) Elongated shaft of a long bone.

Diarthrosis (di″ar-thro′sis) Freely movable joint; all synovial joints are diarthroses.

Diastole (di-as′to-le) Period during which the ventricles or atria of the heart relax.

Diencephalon (di″en-sef′ah-lon) The part of the forebrain between the cerebral hemispheres and the midbrain; includes the thalamus, hypothalamus, and third ventricle.

Diffusion (dĭ-fu′zhun) The spreading of particles in a gas or solution from regions of high particle concentration to regions of low concentration, with movement toward a uniform distribution of the particles.

Digestion Chemical and mechanical process of breaking down foodstuffs into molecules that can be absorbed.

Distal (dis′tal) Away from the attached end of a structure, especially a limb.

Diverticulum (di″ver-tik′u-lum) A pouch or a sac in the walls of a hollow organ or structure.

Dorsal (dor′sal) Pertaining to the back; posterior.

Duct Canal or passageway, usually tubular.

Duodenum (du″o-de′num) First part of the small intestine.

Dura mater (du′rah ma′ter) Most external and toughest of the three membranes (meninges) covering the brain and spinal cord.

Ectoderm (ek′to-derm) Embryonic germ layer that forms both the outer layer of the skin (epidermis) and nervous tissues.

Edema (ĕ-de-mah) Abnormal accumulation of tissue fluid in the loose connective tissue; causes the affected body region to swell.

Effector (ef-fek′tor) Muscle or gland capable of being activated by motor nerve endings.

Efferent (ef′er-ent) Carrying away or away from, especially a nerve fiber that carries impulses away from the central nervous system; efferent neurons are motor neurons.

Elastin (e-las′tin) Main protein in elastic fibers of connective tissues.

Embolus (embolism) (em′bo-lus) Any abnormal mass carried freely in the bloodstream; may be a blood clot, bubbles of air, mass of fat, or clumps of cells.

Embryo (em′bre-o) The developing human from week 2 through week 8 after fertilization.

Endocarditis (en″do-kar-di′tis) Inflammation of the inner lining of the heart.

Endocardium (en″do-kar′de-um) The layer that lines the inner surface of the heart wall; consists of endothelium and areolar connective tissue.

Endocrine glands (en′do-krin) Ductless glands that secrete hormones into the blood.

Endocrine system Body system consisting of glands that secrete hormones.

Endocytosis (en″do-si-to′sis) Processes by which large

molecules and particles enter cells; types are phagocytosis, pinocytosis, and receptor-mediated endocytosis.

Endoderm (en′do-derm) An embryonic germ layer that forms the lining and glands of the digestive and respiratory tubes.

Endometrium (en″do-me′tre-um) Mucous membrane lining the uterus.

Endomysium (en″do-mis′e-um) Thin connective tissue surrounding each muscle cell.

Endoplasmic reticulum (ER) (en″do-plaz′mik rĕ-tik′u-lum) A system of membranous envelopes and tubes in the cytoplasm of a cell; there are smooth and a rough varieties.

Endothelium (en″do-the′le-um) The simple squamous epithelium that lines the walls of the heart, blood vessels, and lymphatic vessels.

Enzyme (en′zīm) A protein that acts as a biological catalyst to speed up chemical reactions.

Eosinophil (e″o-sin′o-fil) Granular white blood cell whose granules readily take up a pink dye called eosin.

Epidermis (ep″ĭ-der′mis) Superficial layer of the skin, composed of a keratinized stratified squamous epithelium.

Epididymis (ep″ĭ-did′ĭ-mis) Comma-shaped structure in the scrotum adjacent to the testis; contains a duct in which the sperm mature.

Epiglottis (ep″ĭ-glot′is) A leaf-shaped piece of elastic cartilage that extends from the posterior surface of the tongue to the larynx; covers the opening of the larynx during swallowing.

Epimysium (ep″ĭ-mis′e-um) Sheath of fibrous connective tissue surrounding a muscle.

Epinephrine (ep″ĭ-nef′rin) Chief hormone produced by the adrenal medulla; also called *adrenaline*.

Epiphyseal plate (ep″ĭ-fiz′e-al) Plate of hyaline cartilage at the junction of the diaphysis (shaft) and epiphysis (end) of most bones in the growing skeleton; provides growth in the length of the bone.

Epiphysis (e-pif′ĭ-sis) The end of a long bone, attached to the shaft.

Epithelium (ep″ĭ-the′le-um) A primary tissue that covers body surfaces and lines body cavities; its cells are arranged in sheets; also forms glands.

Equilibrium The sense of balance; measures both the position and the movement of the head.

Erythrocyte (ĕ-rith′ro-sīt) Red blood cell; when mature, an erythrocyte is literally a sac of hemoglobin (oxygen-carrying protein) covered by a plasma membrane.

Estrogens (es′tro-jens) Female sex hormones.

Eustachian tube (u-sta′ke-an) *See* Auditory tube.

Exocrine glands (ek′so-krin) Glands that secrete onto body surfaces or into body cavities; except for the one-celled goblet cells, all exocrine glands have ducts.

Exocytosis (ek″so-si-to′sis) Mechanism by which substances are moved from the cell interior to the extracellular space; the main mechanism of secretion.

Expiration Act of expelling air from the lungs; exhalation.

External Outside of; superficial to.

Exteroceptor (ek″ster-o-sep′tor) Sensory end organ that responds to stimuli from the external world.

Extracellular Outside a cell.

Extracellular matrix (ma′triks) The material that lies between the cells in connective tissues; consists of fibers, ground substance, and tissue fluid.

Extrinsic (ek-strin′sik) Originating outside an organ or part.

Facet (fas′et) A smooth, nearly flat surface on a bone for articulation.

Fallopian tube (fah-lo′pe-an) *See* Uterine tube.

Fascia (fash′e-ah) Layers of fibrous connective tissue that cover and separate muscles and other structures.

Fascicle (fas′ĭ-k′l) Bundle of nerve or muscle fibers bound together by connective tissue.

Feces (fe′sēz) Material discharged from the large intestine; composed of food residue, secretions, and bacteria.

Fenestrated (fen′es-tra-ted) Pierced with one or more small openings or pores.

Fertilization Fusion of the sperm and egg nuclei.

Fetus (fe′tus) Developmental stage lasting from week 9 of development to birth.

Fibrin (fi′brin) Fibrous insoluble protein formed during blood clotting; takes the form of a fiber network.

Fibroblast (fi′bro-blast) Young, actively mitotic cell that secretes the fibers and ground substance of connective tissue proper.

Fibrocyte (fi′bro-sīt) Mature fibroblast; maintains the matrix of connective tissue proper.

Filtration Passage of a solution or suspension through a membrane or filter, with the purpose of holding back the larger particles.

Fissure (fish′er) (1) A groove or cleft; (2) the deepest depressions or inward folds on the brain.

Fixator (fik-sa′tor) Muscle that immobilizes one or more bones, allowing other muscles to act from a stable base.

Flagellum (flah-jel′um) Long, whip-like extension of the plasma membrane of some bacteria and sperm cells; propels the cell.

Follicle (fol′ĭ-k′l) (1) Spherical structure in the ovary consisting of a developing egg cell surrounded by one or more layers of follicle cells; (2) colloid-containing structure in the thyroid gland.

Follicle-stimulating hormone (FSH) Hormone secreted by the anterior pituitary that stimulates the maturation of ovarian follicles in females and the production of sperm in males.

Foramen (fo-ra′men) Hole or opening in a bone or between body cavities.

Forebrain Rostral portion of the brain, consisting of the telencephalon and diencephalon.

Formed elements Blood cells (red and white cells and platelets)

Fossa (fos′ah) A depression, often a joint surface.

Fovea (fo′ve-ah) A pit.

Frontal (coronal) plane Vertical plane that divides the body into anterior and posterior parts.

Fundus (fun′dus) The base of an organ; that part farthest from the opening of an organ.

Funiculus (fu-nik′u-lus) (1) A chord-like structure; (2) a division of the white matter in the spinal cord.

Gallbladder Sac inferior to the right lobe of the liver; it stores and concentrates bile.

Gamete (gam′ēt) Sex cell; sperm or oocyte.

Gametogenesis (gam″ĕ-to-jen′ĕ-sis) Formation of gametes.

Ganglion (gang′gle-on) Collection of neuron cell bodies outside the central nervous system.

Gap junction A passageway between two adjacent cells; formed by transmembrane proteins called connexons.

Gene One of the biological units of heredity located in chromatin; transmits hereditary information; roughly speaking, one gene codes for the manufacture of one protein.

General Pertaining to sensory inputs or motor outputs that are *widely distributed* through the body rather than localized; opposite of *special*.

Germ layers Three cellular layers (ectoderm, mesoderm, and endoderm) that represent the early specialization of cells in the embryonic body and from which all body tissues arise.

Gestation (jes-ta′shun) The period of pregnancy; averages 280 days in humans.

Gland A structure whose cells are specialized for secretion.

Glial cells (gli′al) *See* Neuroglia.

Glomerular capsule (glo-mer′u-lar) Double-walled cup forming the initial portion of the nephron in the kidney; also called *Bowman's capsule*.

Glomerulus (glo-mer′u-lus) A ball of capillaries forming part of the nephron in the kidney; forms a filtrate that will be modified into urine.

Glottis (glot′is) The opening between the two vocal cords in the larynx.

Glucagon (gloo′kah-gon) Hormone secreted by alpha cells of the pancreatic islets; raises the glucose level of blood.

Glucocorticoids (gloo″ko-kor′tĭ-koids) Hormones secreted by the cortex of the adrenal gland; they increase the concentration of glucose in the blood and aid the body in resisting long-term stress.

Glucose (gloo′kōs) The principal blood sugar; the main sugar used by cells for energy.

Glycogen (gli′ko-jen) A long chain of glucose molecules; the main form in which sugar is stored in animal cells; glycogen takes the form of dense granules in the cytoplasm.

Goblet cells Individual mucus-secreting cells of the respiratory and digestive tracts.

Gonad (go′nad) Primary reproductive organ: the testis of the male or the ovary of the female.

Gonadotropins (go-nad″o-trōp′ins) Gonad-stimulating hormones secreted by the anterior pituitary: follicle-stimulating hormone and luteinizing hormone.

Graded potential A local change in membrane potential that varies directly with the strength of the stimulus, declines with distance, and is not conducted along axons.

Gray matter Gray area of the central nervous system; contains neuron cell bodies and unmyelinated processes of neurons.

Ground substance The viscous, spongy part of the extracellular matrix of connective tissue; its large molecules attract water and hold tissue fluid.

Growth hormone Hormone that stimulates growth of the body; secreted by the anterior pituitary; also called somatotropin and somatotropic hormone (SH).

Gustation (gus-ta′shun) Taste.

Gyrus (ji′rus) A ridge on the surface of the cerebral cortex.

Hair follicle Tube-like invagination of the epidermis of the skin from which a hair grows.

Haustra (hos′trah) Pouches (sacculations) of the colon.

Heart block Impaired transmission of impulses from atria to ventricles.

Heart murmur Abnormal heart sound (usually resulting from valve problems).

Helper T cell A type of T lymphocyte that participates in the activation of other lymphocytes by secreting chemicals that stimulate newly activated lymphocytes to multiply.

Hematocrit (he-mat′o-krit) The percentage of total blood volume occupied by erythrocytes.

Hematoma (he″mah-to′mah) A mass of blood that has bled from blood vessels into the tissues.

Hematopoiesis (hem″ah-to-poi-e′sis) Blood cell formation; hemopoiesis.

Hemoglobin (he′mo-glo″bin) Oxygen-transporting protein in erythrocytes.

Hemopoiesis *See* Hematopoiesis.

Hemorrhage (hem′ŏ-rij) Bleeding.

Hepatic portal system (hĕ-pat′ik) The part of the circulation in which veins receive nutrients from capillaries in the stomach and intestines and carry these nutrients to capillaries in the liver; the liver cells then process the nutrients.

Hepatitis (hep″ah-ti′tis) Inflammation of the liver.

Hernia (her′ne-ah) Abnormal protrusion of an organ or body part through the containing wall of its cavity.

Hilus (hi′lus) A slit on the surface of an organ through which the vessels and nerves enter and leave; the spleen, lungs, kidneys, lymph nodes, and ovaries have prominent hiluses.

Histamine (his′tah-mēn) Chemical substance that increases vascular permeability in the initial stages of inflammation.

Histology (his-tol′o-je) Branch of anatomy dealing with the microscopic structure of tissues, cells, and organs.

Holocrine gland (hol′o-krin) A gland in which entire cells break up to form the secretion product; sebaceous (oil) glands of the skin are the only example.

Hormones Messenger molecules that are released by endocrine glands and travel in the blood to regulate specific body functions.

Hypertension (hi″per-ten′shun) High blood pressure.

Hypertrophy (hi-per′tro-fe) Enlargement of an organ or tissue due to an increase in the size of its cells.

Hypodermis (hi″po-der′mis) The fatty layer deep to the skin; consists of adipose and areolar connective tissue; also called the *subcutaneous layer* and *superficial fascia*.

Hypophysis (hi-pof′ĭ-sis) The pituitary gland.

Hypothalamus (hi″po-thal′ah-mus) Inferior region of the diencephalon; visceral control center of the brain.

Ileum (il′e-um) Coiled terminal part of the small intestine, located between the jejunum and the cecum of the large intestine.

Immune response Antigen-specific defenses mounted by activated lymphocytes (T and B cells).

Immunity (ĭ-mu′nĭ-te) Ability of the body to develop resistance to specific foreign agents (both living and nonliving) that can cause disease.

Immunocompetence Ability of the body's immune system to recognize specific antigens.

Infarct (in′farkt) Region of dead, deteriorating tissue resulting from a lack of blood supply.

Inferior (caudal) Below; toward the feet.

Inflammation (in″flah-ma′shun) A physiological response of the body to tissue injury; includes dilation of blood vessels and an increase in capillary permeability; indicated by redness, heat, swelling, and pain in the affected area.

Inguinal (ing′gwĭ-nal) Pertaining to the groin region.

Inner cell mass Accumulation of cells in the blastocyst from which the body of the embryo derives.

Innervation (in″er-va′shun) Supply of nerves to a body part.

Insertion Movable part or attachment of a muscle, as opposed to the muscle's origin.

Inspiration Drawing of air into the lungs; inhalation.

Insulin (in′su-lin) Hormone secreted by beta cells in pancreatic islets; it decreases blood glucose levels.

Integumentary system (in-teg″u-men′tar-e) The skin and its appendages (hairs, nails, and skin glands).

Intercalated discs (in-ter′kah-la″ted) Complex junctions that interconnect cardiac muscle cells in the wall of the heart.

Intercellular Between body cells.

Internal Deep to.

Internal capsule Band of white matter in the brain, between the basal nuclei and the thalamus.

Internal respiration Exchange of gases between blood and tissue fluid and between tissue fluid and cells.

Interneuron (in″ter-nu′ron) (1) Nerve cell that lies between a sensory neuron and a motor neuron in a reflex arc; (2) any nerve cell that is confined entirely within the central nervous system.

Interoceptor (in″ter-o-sep′tor) Nerve ending situated in a visceral organ; responds to changes and stimuli within the body's internal environment; also called *visceroceptor*.

Interstitial fluid (in″ter-stish′al) *See* Tissue fluid.

Intervertebral discs (in″ter-ver′tĕ-bral) The discs between the vertebrae of the spinal column; each consists of fibrous rings surrounding a springy core.

Intervertebral foramina (fo-ra′min-ah) Openings between the dorsal projections of adjacent vertebrae through which the spinal nerves pass.

Intracellular Within a cell.

Ion (i′on) Atom or molecule with a positive or negative electrical charge.

Ipsilateral (ip″sĭ-lat′er-al) Situated on the same side of the body; opposite of *contralateral*.

Ischemia (is-ke′me-ah) Local decrease in blood supply.

Jejunum (jĕ-joo′num) The coiled part of the small intestine that is located between the duodenum and ileum.

Joint Junction of two or more elements of the skeleton; an articulation.

Keratin (ker′ah-tin) Tension-resisting protein found in the epidermis, hair, and nails; keratin makes these structures tough and able to resist friction.

Labium (la′be-um) Lip.

Labyrinth (lab′ĭ-rinth) Bony cavities and membranes of the inner ear.

Lacrimal (lak′rĭ-mal) Pertaining to tears.

Lactation (lak-ta′shun) Production and secretion of milk.

Lacteal (lak′te-al) Lymphatic capillaries in the small intestine that take up lipids.

Lacuna (lah-ku′nah) Little depression or cavity; in bone and cartilage, each lacuna is occupied by a cell.

Lamina (lam′ĭ-nah) (1) A thin layer or flat plate; (2) the portion of a vertebra between the transverse process and the spinous process.

Larynx (lar′ingks) Cartilaginous organ located between the pharynx and trachea; contains the vocal cords; the voice box.

Lateral Away from the body midline.

Leukocyte (loo′ko-sīt) White blood cell; the five types of leukocytes are all involved in the defense against disease.

Ligament (lig′ah-ment) Band of dense regular connective tissue that connects bones.

Limbic system (lim′bik) Functional brain system involved in emotional and visceral responses; structurally, it includes a part of the cerebrum (limbic lobe) and the hypothalamus of the diencephalon.

Lumbar (lum′bar) Region of the back between the thorax and the pelvis.

Lumen (lu′men) The cavity inside a tube, blood vessel, or hollow organ.

Luteinizing hormone (LH) (loo′te-in-īz″ing) A hormone secreted by the anterior pituitary; in females, it aids maturation of follicles in the ovary and triggers ovulation; in males, it signals the interstitial cells of the testis to secrete testosterone.

Lymph (limf) The clear fluid transported by the lymphatic vessels.

Lymph node Bean-shaped lymphoid organ that filters and cleanses the lymph.

Lymphatic system (lim-fat′ik) Organ system consisting of lymphatic vessels, lymph nodes, and the lymphoid organs and tissues; drains excess tissue fluid and fights disease.

Lymphatics General term used to designate lymphatic vessels.

Lymphocyte (lim′fo-sīt) Agranular white blood cell that arises from bone marrow and becomes functionally activated in the lymphoid organs of the body; the main cell type of the immune system, each lymphocyte recognizes a specific antigen.

Lymphoid organs The organs of the lymphatic system that house lymphocytes and function in the immune response; spleen, lymph nodes, tonsils, and thymus are the main examples.

Lymphoid tissue The main tissue of the immune system; a reticular connective tissue that houses and activates many lymphocytes.

Lysosome (li′so-sōm) A membrane-bound, sac-like cytoplasmic organelle that contains a wide variety of digestive enzymes.

Macrophages (mak′ro-fāj-es) The general phagocytic cells of the body, capable of engulfing and digesting a wide variety of foreign cells, particles, and molecules; present throughout the connective tissues of the body and especially abundant in lymphoid tissues of the immune system.

Malignant (mah-lig′nant) Life-threatening; pertains to neoplasms such as cancer, that spread and lead to death.

Mammary glands (mam'ar-e) The breasts.

Mastication (mas"tĭ-ka'shun) Chewing.

Meatus (me-a'tus) A canal or opening.

Mechanoreceptor (mek"ah-no-re-sep'tor) Receptor sensitive to mechanical forces, such as touch, stretch, pressure, or vibration.

Medial Toward the midline of the body.

Median (me'de-an) In the midline of the body; midsagittal.

Mediastinum (me"de-ah-sti'num) Region of the thoracic cavity between the lungs; contains the heart, thoracic aorta, esophagus, and other structures.

Medulla (mĕ-dul'ah) Middle or internal region of certain organs.

Medulla oblongata (ob"long-gah'tah) Inferior part of the brain stem.

Meiosis (mi-o'sis) A process of nuclear division that occurs during the production of the sex cells and reduces the chromosome number by half; results in the formation of haploid (*n*) cells.

Melanin (mel'ah-nin) Dark pigment formed by cells called melanocytes; imparts color to the skin and hair.

Memory cells T and B lymphocytes that provide for immunologic memory (acquired, long-term immunity from diseases).

Meninges (mĕ-nin'jēz) Protective coverings around the brain and spinal cord; from external to internal, they are the dura mater, arachnoid mater, and pia mater.

Meningitis (men"in-ji'tis) Inflammation of the meninges.

Menstrual cycle (men'stroo-al) The changes in the female reproductive organs that occur every month (28 days, on the average).

Menstruation (men"stroo-a'shun) The periodic, cyclic discharge of blood and tissue from the lining of the female uterus in the absence of pregnancy.

Mesencephalon (mes"en-sef'ah-lon) Midbrain.

Mesenchyme (mes'eng-kīm) The type of embryonic tissue from which connective tissues and muscle tissues arise.

Mesenteries (mes'en-ter"ēz) Double-layered sheets of peritoneum that support most organs in the abdominopelvic cavity.

Mesoderm (mez'o-derm) The embryonic germ layer that gives rise to most structures in the body, including the skeleton, muscles, dermis, connective tissues, kidneys, and gonads.

Metabolic rate (mĕ-tah-bol'ik) Energy expended by the body per unit time.

Metabolism (mĕ-tab'o-lizm) Sum total of all the chemical reactions occurring in the cells of the body.

Metastasis (mĕ-tas'tah-sis) The spread of cancer from one body part or organ to another not directly connected to it.

Microvilli (mi"kro-vil-i) Immotile, cellular projections on the free surface of most epithelia; microvilli anchor sheets of mucus or increase surface area for absorption.

Micturition (mik"tu-rish'un) Urination, or voiding; emptying the bladder.

Midbrain Region of the brain stem that lies between the diencephalon and the pons.

Mineralocorticoids (min"er-al-o-kor'tĭ-koids) Steroid hormones secreted by the adrenal cortex that increase the reabsorption of sodium and water by the kidneys; the main mineralocorticoid is aldosterone.

Mitochondrion (mi"to-kon'dre-on) Cytoplasmic organelle that generates most adenosine triphosphate (ATP) for cellular activities; mitochondria are the cell's "power plants."

Mitosis (mi-to'sis) Division of the nucleus during the typical process of cell division, during which the chromosomes are distributed to the two daughter nuclei.

Mixed nerve A nerve containing fibers of both sensory and motor neurons; most nerves are mixed nerves.

Monocyte (mon'o-sīt) An agranular white blood cell, with a large nucleus that is often bent into the shape of a C; the largest of all blood cells; develops into a macrophage.

Motor neuron Nerve cell that signals muscle cells to contract or gland cells to secrete; also called *efferent neuron*.

Motor unit A motor neuron and all of the muscle cells it stimulates.

Mucosa (mu-ko'sah) *See* Mucous membrane.

Mucous membranes Moist membranes that line all tubular organs and body cavities that open to the exterior (digestive, respiratory, urinary, and reproductive tracts).

Mucus (mu'kus) A sticky, viscous fluid that covers many internal surfaces in the body; it consists of the protein mucin and a large amount of water.

Multipolar neuron (mul"tĭ-po'lar) A nerve cell that has more than two processes; most neurons are multipolar, having several dendrites and an axon.

Multipotential hematopoietic stem cell (hem"ah-to-poi-et'ik) A cell type in the bone marrow that gives rise to all formed elements of blood (blood cells).

Muscle fiber Muscle cell.

Muscle spindle Complex, spindle-shaped receptor in skeletal muscles that senses muscle stretch.

Muscle tone Continuous, low levels of contractile force produced by muscles that are not actively shortening.

Myelencephalon (mi"el-en-sef'ah-lon) Caudal part of the hindbrain; the medulla oblongata.

Myelin sheath (mi'ĕ-lin) Fatty insulating sheath that surrounds all but the thinnest nerve fibers; formed of the plasma membrane of supporting cells wrapped in concentric layers around the nerve fiber.

Myocardial infarction (mi"o-kar'de-al) Condition characterized by dead tissue areas in the myocardium of the heart; caused by interruption of blood supply to the area; also called *heart attack*.

Myocardium (mi"o-kar'de-um) Layer of the heart wall composed of cardiac muscle.

Myofibril (mi"o-fi'bril) Rod-like bundle of contractile myofilaments in the cytoplasm of a skeletal muscle cell; made of repeating segments called sarcomeres.

Myofilament (mi"o-fil'ah-ment) The contractile filaments in muscle cells; the two varieties are thick (myosin) filaments and thin (actin) filaments.

Myometrium (mi"o-me'tre-um) The thick layer of smooth muscle in the wall of the uterus.

Myosin (mi'o-sin) A contractile protein in cells, especially abundant in muscle cells.

Nares (na'rēz) The nostrils.

Necrosis (nĕ-kro'sis) Death of a cell or tissue caused by disease or injury.

Neoplasm (ne'o-plazm) *See* Tumor.

Nephron (nef'ron) A major division of the uriniferous

tubule in the kidney (the other division is the collecting tubule).

Nerve A collection of nerve fibers (long axons) in the peripheral nervous system.

Nerve fiber Any long axon of a neuron.

Neural crest Embryonic tissue derived from ectoderm that migrates widely within the embryo and gives rise to sensory neurons, all nerve ganglia, melanocytes, and other structures.

Neuroglia (nu-rog′le-ah) The supporting cells in the central nervous system; they support, protect, and insulate neurons.

Neurohypophysis (nu″ro-hi-pof′ĭ-sis) The part of the pituitary gland that derives from the brain; contains the stalk-like infundibulum and the posterior lobe of the pituitary.

Neuron (nu′ron) Cell of the nervous system specialized to generate and transmit electrical signals; a nerve cell.

Neurotransmitter (nu″ro-trans′mit-er) Chemical released by neurons that may, upon binding to receptors on neurons or effector cells, stimulate or inhibit them.

Neutrophil (nu′tro-fil) Most abundant type of white blood cell; a granulocyte specialized for destroying bacteria.

Nucleic acid (nu-kle′ik) The class of organic molecules that includes deoxyribonucleic acid (DNA) and ribonucleic acid (RNA).

Nucleolus (nu-kle′o-lus) A small, dark-staining body in the cell nucleus; represents parts of several chromosomes and manufactures the basic subunits of ribosomes.

Nucleus (nu′kle-us) (1) Control center of a cell; contains genetic material; (2) a cluster of neuron cell bodies in the brain.

Occipital (ok-sip′ĭ-tal) Pertaining to the area at the back of the head.

Occlusion (ŏ-kloo′zhun) Closure or obstruction.

Olfaction (ol-fak′shun) Smell.

Olfactory epithelium (ol-fak′to-re) A sensory receptor region in the superior lining of the nasal cavity; this epithelium contains olfactory neurons that respond to odors in the air.

Oocyte (o′o-sīt) Immature egg undergoing the process of meiosis.

Oogenesis (o″o-jen′ĕ-sis) Process of ovum (female gamete) formation.

Ophthalmic (of-thal′mik) Pertaining to the eye.

Optic (op′tik) Pertaining to the eye.

Optic chiasma (ki-az′mah) A cross-shaped structure anterior to the diencephalon of the brain, representing the point of crossover of half the axons of the optic nerve.

Organ A part of the body formed of two or more tissues and adapted to carry out a specific function; the stomach and biceps brachii muscle are examples of large organs, but many organs are smaller and simpler (sweat gland, hair follicle, muscle spindle).

Organ system A group of organs that work together to perform a vital body function; e.g., the nervous system.

Organelles (or″gah-nelz′) Small structures in the cytoplasm (ribosomes, mitochondria, and others) that perform specific functions for the cell as a whole.

Origin Attachment of a muscle that remains relatively fixed during muscular contraction.

Osmosis (oz-mo′sis) Diffusion of a solvent (water molecules) through a membrane from a dilute solution into a more concentrated one.

Ossicles (os′ĭ-klz) The three tiny bones in the middle ear: malleus (hammer), incus (anvil), and stapes (stirrup).

Ossification (os″ĭ-fĭ-ka′shun) Bone formation. *See* Osteogenesis.

Osteoblast (os′te-o-blast″) A bone-forming cell.

Osteoclast (os′te-o-klast″) Large cell that reabsorbs or breaks down the bone matrix.

Osteocyte (os′te-o-sīt″) Mature bone cell, shaped like a spider with a body and long processes, that occupies a lacuna in the bone matrix.

Osteogenesis (os″te-o-jen′ĕ-sis) Process of bone formation.

Osteon (os′te-on) Tube-shaped unit in mature, compact bone; consists of concentric layers of bone lamellae surrounding a central canal; also called *Haversian system*.

Osteoporosis (os″te-o-po-ro′sis) Age-related condition (affects many elderly women) in which bones weaken as bone reabsorption outpaces bone deposition; the weakened bones break easily.

Ovarian cycle (o-va′re-an) Monthly cycle of follicle development in the ovaries, ovulation, and formation of the corpus luteum; the menstrual cycle as it involves the ovaries.

Ovary (o′var-e) Female sex organ in which ova (eggs) are produced; the female gonad in the pelvis.

Ovulation (o″vu-la′shun) Ejection of an egg (oocyte) from the ovary.

Ovum (o′vum) (1) General meaning: the female germ cell, or egg; (2) specific meaning: female germ cell after a sperm has entered it but before the sperm nucleus and the egg nucleus have fused.

Oxytocin (ok″sĭ-to′sin) Hormone produced by the hypothalamus and released by the posterior pituitary; it stimulates contraction of the uterus during childbirth and the ejection of milk during nursing.

Palate (pal′at) Roof of the mouth.

Palpation (pal-pa′shun) Using one's fingers to feel deep organs through the skin of the body surface.

Pancreas (pan′kre-as) Tadpole-shaped gland posterior to the stomach; produces both exocrine and endocrine secretions.

Pancreatic juice Bicarbonate-rich secretion of the pancreas containing enzymes for digestion of all food categories.

Parasympathetic (par″ah-sim″pah-thet′ik) The division of the autonomic nervous system that oversees digestion, elimination, and glandular function; the resting and digesting division.

Parathyroid glands (par″ah-thi′roid) Small endocrine glands located on the posterior aspect of the thyroid gland.

Parathyroid hormone (PTH) Hormone secreted by the parathyroid glands; it increases the concentration of calcium ions in the blood.

Parenchyma (pah-reng′kĭ-mah) The functional tissue of an organ, as opposed to its supporting tissue (stroma); in a gland, the parenchyma is the secretory cells, rather than the connective-tissue capsule.

Parietal (pah-ri′ĕ-tal) Pertaining to the walls of a cavity.

Pars distalis (parz dis-tal′is) The main division of the adenohypophysis of the pituitary gland; the anterior lobe of the pituitary.

Pars nervosa (ner-vo′sah) The region of the neurohypophysis of the pituitary gland from which hormones are secreted; the posterior lobe of the pituitary.

Pectoral (pek′tor-al) Pertaining to the chest.

Pectoral girdle Bones that attach an upper limb to the axial skeleton: clavicle and scapula.

Pelvic girdle The paired hip bones that attach the lower limbs to the axial skeleton.

Pelvis (pel′vis) Inferior region of the body trunk; contains the basin-shaped, bony structure called the bony pelvis.

Pepsin (pep′sin) Protein-digesting enzyme secreted by the stomach lining.

Pericardium (per″ĭ-kar′de-um) Double-layered sac that encloses the heart and forms its superficial layer.

Perichondrium (per″ĭ-kon′dre-um) Membrane of fibrous connective tissue that covers the external surface of cartilages.

Perimysium (per″ĭ-mis′e-um) Connective tissue that surrounds and separates fascicles (bundles) of muscle fibers within a skeletal muscle.

Perineum (per″ĭ-ne′um) The region of the trunk containing the anus, vulva (females), and the posterior of the scrotum (males).

Periosteum (per″e-os′te-um) Membrane of fibrous connective tissue that covers the external surface of bones of the skeleton.

Peripheral nervous system (PNS) Portion of the nervous system consisting of nerves and ganglia that lie outside the brain and spinal cord.

Peristalsis (per″ĭ-stal′sis) Progressive, wave-like contractions that squeeze foodstuffs through the alimentary canal (or that move other substances through other body organs).

Peritoneum (per″ĭ-to-ne′um) Serous membrane that lines the interior of the abdominopelvic cavity and covers the surfaces of the organs in this cavity.

Peritonitis (per″ĭ-to-ni′tis) Infection and inflammation of the peritoneum.

Peritubular capillaries Capillaries in the kidney that surround the proximal and distal convoluted tubules; active in reabsorption.

Phagocytosis (fag″o-si-to′sis) The process by which a cell forms cytoplasmic extensions to engulf foreign particles, cells, or macromolecules and then uses lysosomes to digest these substances.

Pharynx (far′ingks) Muscular tube extending from the region posterior to the nasal cavity to the esophagus; the "throat" part of the digestive tube.

Photoreceptors (fo″to-re-sep′tors) Specialized receptor cells that respond to light energy: rod cells and cone cells.

Pia mater (pi′ah ma′ter) Most internal and most delicate of the three membranes (meninges) covering the brain and spinal cord.

Pinocytosis (pin″o-si-to′sis) The process by which cells engulf extracellular fluids.

Pituitary gland (pĭ-tu′ĭ-tār″e) Neuroendocrine gland that is situated inferior to the brain and serves a variety of functions, such as regulating the gonads, thyroid gland, adrenal cortex, lactation, and water balance; also called the hypophysis.

Placenta (plah-sen′tah) Temporary organ formed from both fetal and maternal tissues that provides nutrients and oxygen to the developing fetus, carries away fetal waste molecules, and secretes the hormones of pregnancy; shed as the afterbirth when labor is over.

Plasma (plaz′mah) The nonliving, fluid component of blood, within which the blood cells are suspended.

Plasma cell Cell formed from the division of an activated B lymphocyte; secretes antibodies.

Plasma membrane Membrane that encloses cell contents; the external, limiting membrane of the cell.

Plasmalemma (plaz″mah-lem′ah) *See* Plasma membrane.

Platelet Cell fragment found in blood; plugs small tears in blood vessels and helps initiate clotting.

Pleura (ploo′rah) Serous membrane that lines the pleural cavity in the thorax and covers the external surface of the lung.

Plexus (plek′sus) A network of converging and diverging nerves or veins.

Plica (pli′kah) A fold.

Podocytes (pod′o-sīts) Octopus-shaped epithelial cells that surround the glomerular capillaries; they help produce and maintain the basement membrane (the main filtration membrane in the kidney).

Pons The part of the brain stem between the midbrain and the medulla oblongata.

Portal system A system of vessels in which two capillary beds, rather than one, lie between the incoming artery and the outgoing vein of a body region; a portal vein lies between the two capillary beds; examples are the hepatic portal system and the hypophyseal portal system.

Posterior Toward the back; dorsal.

Posterior pituitary *See* Pars nervosa.

Postganglionic neuron (pōst″ gang-gle-on′ik) Autonomic motor neuron that has its cell body in a peripheral ganglion and projects its axon to an effector.

Preganglionic neuron (pre″ gang-gle-on′ik) Autonomic motor neuron that has its cell body in the central nervous system and projects its axon to a peripheral ganglion.

Prime mover Muscle that bears the major responsibility for a particular movement; agonist.

Process (1) Prominence or projection; (2) series of actions for a specific purpose.

Progesterone (pro-jes′tĕ-rōn) Hormone that prepares the uterus to receive the implanting embryo.

Pronation (pro-na′shun) Medial rotation of the forearm that causes the palm to face posteriorly.

Prone Refers to a body lying horizontally with the face downward.

Proprioceptor (pro″pre-o-sep′tor) Receptor that senses movement in the musculoskeletal system; more specifically, proprioceptors sense stretch in muscles, tendons, and joint capsules.

Protein (pro′tēn) A long chain of amino acids or several linked chains of amino acids; the amino acid chains have bent and folded (and often coiled) to give each protein a distinct shape.

Proximal (prok′sĭ-mal) Toward the attached end of a limb, or near the origin of a structure.

Pseudostratified (soo″do-strat′ĭ-fīd) Pertaining to an epithelium that appears to be stratified (consisting of more than one layer of cells) but is not; the cells vary in height, but all touch the base of the epithelium.

Puberty (pu′ber-te)   Period of life when reproductive maturity is reached.

Pulmonary (pul′mo-ner″e)   Pertaining to the lungs.

Pulmonary circuit   System of blood vessels that serve gas exchange in the lungs: the pulmonary arteries, capillaries, and veins.

Pulse   Rhythmic expansion and recoil of arteries resulting from the contraction of the heart; can be felt from outside the body.

Pupil   Opening in the center of the iris through which light enters the eye.

Purkinje fibers (pur-kin′je)   Modified cardiac muscle fibers of the conduction system of the heart; also called *conduction myofibers*.

Pus   Fluid product resulting from the defense response to bacterial infection; composed of dead white blood cells, bacteria, and a thin fluid.

Pyloric region (pi-lor′ik)   A funnel-shaped region of the stomach, just proximal to the pylorus.

Pylorus   The distal, ring-shaped portion of the stomach that joins the small intestine and contains the pyloric sphincter muscle.

Ramus (ra′mus)   Branch of a nerve, artery, or bone.

Raphe (ra′fe)   A seam in the midline.

Receptor   Peripheral nerve ending, or complete cell, that responds to particular types of stimulus.

Reduction   Restoring broken bone ends or dislocated bones to their original positions.

Reflex   Automatic response to a stimulus.

Relay nucleus   A nucleus in the brain whose neurons receive signals from one region of the central nervous system and relay this information to another region; within the relay nucleus, the information is organized and edited.

Renal (re′nal)   Pertaining to the kidney.

Renin (re′nin)   Hormone released by the kidneys that is involved with raising blood pressure, blood volume, and the sodium concentration in blood.

Respiratory system   Organ system that carries out gas exchange; includes the nose, pharynx, larynx, trachea, bronchi, and lungs.

Rete (re′te)   A network, often composed of nerve fibers or blood vessels.

Reticular cell (rĕ-tik′u-lar)   A fibroblast in reticular connective tissue (in bone marrow, spleen, lymph nodes, and so on).

Reticular formation   Functional system that runs through the core of the brain stem; involved in alertness, arousal, and sleep; also contains visceral centers that control heart rate, breathing rate, and vomiting; controls some body movements as well.

Reticulocyte (rĕ-tik′u-lo-sīt)   Immature or young erythrocyte.

Retina (ret′ĭ-nah)   Neural tunic of the eyeball; contains the photoreceptor cells for vision.

Rhinencephalon (ri″nen-sef′ah-lon)   The part of the cerebrum that receives and integrates olfactory (smell) impulses.

Ribonucleic acid (RNA) (ri″bo-nu-kle′ik)   Nucleic acid that contains the sugar ribose; acts in protein synthesis.

Ribosome (ri′bo-sōm)   Cytoplasmic organelle on which proteins are synthesized.

Rod cell   One of the two types of photoreceptor cells in the retina of the eye.

Rostral   Toward the superior part of the central nervous system, or toward the higher brain centers.

Rugae (roo′ge)   Elevations or ridges, as in the mucosa of the stomach.

Sacral (sa′kral)   Pertaining to the sacrum; the region in the midline of the buttocks.

Sagittal plane (saj′ĭ-tal)   A vertical plane that divides the body or a body part into right and left portions.

Sarcolemma (sar″ko-lem′ah)   The plasma membrane of a muscle cell.

Sarcomere (sar′ko-mēr)   The smallest contractile unit of skeletal and cardiac muscle; the part of a myofibril between two Z discs; contains myofilaments composed mainly of contractile proteins (actin, myosin).

Sarcoplasm (sar′ko-plazm)   The cytoplasm of a muscle cell.

Sarcoplasmic reticulum (sar′ko-plaz-mik rĕ-tik′u-lum)   Specialized smooth endoplasmic reticulum of muscle cells; stores calcium ions.

Sclera (skle′rah)   Outer fibrous tunic of the eyeball.

Scrotum (skro′tum)   The external sac that encloses the testes.

Sebaceous gland (se-ba′shus)   Gland in the skin that produces an oily secretion called sebum.

Sebum (se′bum)   The oily secretion of sebaceous glands.

Secretion (se-kre′shun)   (1) The passage of material formed by a cell to its exterior; (2) cell product that is transported to the exterior of a cell.

Section   A cut through the body (or an organ) along a particular plane; a thin slice of tissue prepared for microscopic study.

Semen (se′men)   Fluid mixture containing sperm and secretions of the male accessory reproductive glands.

Semilunar valves (sem″ĭ-lu′nar)   Valves at the base of the aorta and the pulmonary trunk that prevent blood from returning to the heart ventricles after ventricular contraction.

Seminiferous tubules (sem″ĭ-nif′er-us)   Highly convoluted tubules within the testes that form sperm.

Sensory nerve   Nerve that contains only sensory fibers.

Sensory neuron   Nerve cell that carries information received from sensory receptors; also called *afferent neuron*.

Serosa (se-ro′sah)   *See* Serous membrane.

Serous cell   Exocrine gland cell that secretes a watery product containing digestive enzymes.

Serous fluid (se′rus)   A clear, watery lubricant secreted by cells of a serous membrane.

Serous membrane   Moist, slippery membrane that lines internal body cavities (pleural, pericardial, and peritoneal cavities) and covers visceral organs within these cavities; also called *serosa*.

Serum (se′rum)   Amber-colored fluid that exudes from clotted blood plasma as the clot shrinks and then no longer contains clotting factors.

Sinoatrial (SA) node (si″no-a′tre-al)   A collection of specialized cardiac muscle cells in the superior wall of the right atrium; pacemaker of the heart.

Sinus (si′nus) (1) Mucous-membrane-lined, air-filled cavity in certain bones of the face; (2) dilated channel for the passage of blood or lymph.

Sinusoid (si′nŭ-soid) An exceptionally wide, twisted, leaky capillary; large protein molecules or whole blood cells pass easily through the walls of sinusoids.

Smooth muscle Musculature consisting of spindle-shaped, unstriped (nonstriated) muscle cells; present in the walls of most visceral organs.

Somatic (so-mat′ik) Pertaining to the region of the body that lies external to the ventral body cavity, including the skin, skeletal muscles, and the skeleton; opposite of *visceral*.

Somite (so′mīt) A mesodermal segment of the body of the embryo.

Special Pertaining to sensory inputs or motor outputs that are *localized* rather than being widespread through the body; opposite of *general*.

Special senses The senses whose receptors are confined to a small region rather than distributed widely through the body: taste, smell, vision, hearing and equilibrium.

Spermatogenesis (sper″mah-to-jen′ĕ-sis) The process by which sperm (male gametes) form in the testes; involves meiosis.

Sphincter (sfingk′ter) A muscle surrounding an opening; acts as a valve to close and open the orifice.

Spinal nerves The 31 pairs of nerves that attach to the spinal cord.

Spinal reflex A reflex mediated through the spinal cord.

Squamous (skwa′mus) Flat, plate-like; pertaining to flat epithelial cells that are wider than they are tall.

Stenosis (stĕ-no′sis) Constriction or narrowing.

Steroids (ste′roids) Group of lipid molecules containing cholesterol and some hormones.

Stimulus (stim′u-lus) An excitant or irritant; a change in the environment that evokes a response.

Striated muscle (stri′āt-ed) Muscle consisting of cross-striated (striped) muscle fibers: skeletal and cardiac muscle.

Stroke Condition in which brain tissue is deprived of its blood supply, as in blockage of a cerebral blood vessel; also called *cerebrovascular accident*.

Stroma (stro′mah) The connective tissue framework of an organ.

Subcutaneous (sub″ku-ta′ne-us) Deep to the skin.

Sulcus (sul′kus) A groove.

Superficial Located close to or on the body surface; outer; external; opposite of *deep*.

Superior Closer to the head; above.

Supination (soo″pĭ-na′shun) Lateral rotation of the forearm that causes the palm to face anteriorly.

Supine (soo′pīn) Refers to a body lying horizontally with the face upward.

Supporting cells Cells in nervous tissue that support, nourish, and insulate neurons.

Suprarenal gland (soo″prah-re′nal) The technical name for an adrenal gland.

Surfactant (ser-fak′tant) Detergent-like fluid secreted by certain cells lining the respiratory alveoli in the lungs; reduces the surface tension of water molecules, thus preventing collapse of the alveoli after each breath.

Suture (soo′cher) An immovable, fibrous joint; except at the jaw joint, all bones of the skull are united by sutures.

Sweat gland Tubular gland in the skin that secretes sweat (which cools the body).

Sympathetic division (sim″pah-thet′ik) Division of the autonomic nervous system that prepares the body to cope with danger or excitement; the fight, fright, and flight division.

Symphysis (sim′fĭ-sis) A joint in which the bones are connected by fibrocartilage.

Synapse (sin′aps) Specialized cell junction between two neurons, at which the neurons communicate.

Synaptic cleft Fluid-filled space at a synapse between neurons; also called *synaptic gap*.

Synarthrosis (sin″ar-thro′sis) Any immovable joint.

Synchondrosis (sin″kon-dro′sis) A joint in which bones are united by hyaline cartilage.

Syndesmosis (sin″des-mo′sis) A joint in which bones are united only by a ligament.

Synergist (sin′er-jist) Muscle that aids the action of a prime mover by contributing to the same movement or by stabilizing joints to prevent undesirable movements.

Synostosis (sin″os-to′sis) A completely ossified joint; a joint fused by bone.

Synovial fluid (sĭ-no′ve-al) Fluid secreted by the synovial membranes of the freely movable joints of the body; lubricates the joint surfaces and nourishes the articular cartilages.

Synovial joint Freely movable joint with a cavity and a capsule; *See* diarthrosis.

Systemic (sis-tem′ik) Pertaining to the whole body.

Systemic circuit System of blood vessels that carries oxygenated blood to the tissues throughout the body.

Systole (sis′to-le) Period during which the ventricles or the atria of the heart contract.

T cells Lymphocytes that mediate cellular immunity; include killer and helper T cells; also called *T lymphocytes*.

T tubule Extension of the muscle cell plasmalemma (sarcolemma) that protrudes deeply into the muscle cell.

Target cell A cell that is capable of responding to a hormone because it bears receptors to which the hormone can bind.

Taste buds Bulb-shaped sensory organs on and around the tongue that house the receptor cells for taste.

Telencephalon (tel″en-sef′ah-lon) Rostral division of the embryonic forebrain that develops into the cerebrum.

Tendon (ten′don) Cord of dense regular connective tissue that attaches muscle to bone.

Testis (tes′tis) Male primary sex organ that produces sperm; male gonad in the scrotum.

Testosterone (tes-tos′tĕ-rōn) *See* Androgen.

Thalamus (thal′ah-mus) An egg-shaped mass of gray matter in the diencephalon of the brain; consists of nuclei through which information is relayed to the cerebral cortex.

Thermoreceptor (ther″mo-re-sep′tor) Receptor sensitive to temperature changes.

Thoracic (tho-ras′ik) Pertaining to the chest or thorax.

Thoracic duct Large lymphatic duct that ascends anterior to the vertebral column; drains the lymph from up to three-fourths of the body (all except the body's superior right quarter).

Thorax (tho'raks) That portion of the body superior to the diaphragm and inferior to the neck.

Thymus (thi'mus) Organ of the immune system that is essential for the production of T cells (T lymphocytes); located in the anterior thorax.

Thyroid gland (thi'roid) Butterfly-shaped endocrine gland in the anterior neck; its main hormone (thyroid hormone) increases metabolic rate.

Tight junction A type of cell junction that closes off the intercellular space; also called a *zonula occludens*.

Tissue A group of similar cells (and extracellular material) that perform similar functions; primary tissues of the body are epithelial, connective, muscle, and nervous tissue.

Tissue fluid Watery fluid that, along with the molecules of ground substance, occupies the extracellular matrix of connective tissue; it is a filtrate of the blood containing all the small molecules of blood plasma; also called *interstitial fluid*.

Trabecula (trah-bek'u-lah) (1) Any one of the fibrous bands extending from the capsule to the interior of an organ; (2) a piece of the bony network in spongy bone.

Trachea (tra'ke-ah) Windpipe; the cartilage-reinforced air-tube that extends from the larynx to the bronchi.

Tract A collection of nerve fibers in the central nervous system having the same origin, destination, and function.

Transverse process One of a pair of projections that extend laterally from the neural arch of a vertebra.

Trauma (traw'mah) A wound, injury, or shock, usually caused by external forces.

Trochanter (tro-kan'ter) A large, somewhat blunt process on a bone.

Trophoblast (trof'o-blast) External layer of cells in the blastocyst (early embryo); forms the embryo's contribution to the placenta.

Tropic hormone (tro'pik) A hormone that regulates the function of another endocrine organ; tropic hormones signal endocrine glands to secrete their own hormones.

Tubercle (too'ber-k'l) A nodule or small rounded process on a bone.

Tuberosity (too"bĕ-ros'ĭ-te) A broad process on a bone, larger than a tubercle.

Tumor An abnormal growth of cells; a swelling; a neoplasm; can be cancerous.

Tunica (too'nĭ-kah) A covering or coat; a layer or membrane of tissue.

Tympanic membrane (tim-pan'ik) The eardrum located between the outer and middle ear.

Ulcer (ul'ser) Erosion of the surface of an organ or tissue, such as a peptic ulcer in the wall of the stomach or small intestine.

Umbilical cord (um-bil'ĭ-kal) A cord that attaches to the navel before birth and connects the fetus to the placenta; contains the umbilical arteries and vein.

Umbilicus (um-bil'ĭ-kus) Navel; belly button.

Unipolar neuron (u"nĭ-po'lar) A sensory neuron in which a single short process projects from the cell body but divides like a T into two long processes (central process and peripheral process).

Urea (u-re'ah) The main nitrogen-containing waste excreted in urine.

Ureter (u-re'ter) Tube that carries urine from kidney to bladder.

Urethra (u-re'thrah) Tube that carries urine from the bladder to the exterior.

Uterine tube (u'ter-in) Tube through which the ovum travels to the uterus; also called *fallopian tube* and *oviduct*.

Uterus (u'ter-us) Hollow, thick-walled pelvic organ that receives the developing embryo; site where embryo/fetus develops; the womb.

Vasa recta (vah'sah rek'tah) Capillary-like blood vessels that supply the loops of Henle and collecting tubules in the medulla of the kidney.

Vascularized (vas'ku-lar-īzd") Having a blood supply; containing blood vessels.

Vasoconstriction (vas"o-kon-strik'shun) Narrowing of blood vessels, normally through the contraction of smooth muscle cells in the vessel walls.

Vasodilation (vas"o-di-la'shun) Relaxation of smooth muscle cells in the walls of blood vessels, causing the vessels to dilate (widen).

Vasomotor fibers (vas"o-mo'tor) Sympathetic nerve fibers that regulate the contraction of smooth muscle in the walls of blood vessels, thereby regulating the diameter of the vessels.

Vein Vessel that carries blood toward the heart.

Ventilation (ven"tĭ-la'shun) Breathing; consists of inspiration and expiration.

Ventral (ven'tral) Toward the front of the body; anterior.

Ventricles (1) Paired, inferiorly located heart chambers that function as the major blood pumps; (2) fluid-filled cavities of the brain.

Venule (ven'ūl) A small vein.

Vertebral column (ver'tĕ-bral) The spine or spinal column; formed of a number of bones called vertebrae, the discs between these vertebrae, and two composite bones (sacrum and coccyx).

Vesicle (ves'ĭ-k'l) A small, liquid-filled sac; also refers to the urinary bladder.

Vesicular follicle (vĕ-sik'u-lar fol'ĭ-k'l) Mature ovarian follicle; formerly called *Graafian follicle*.

Villus (vil'us) One of many finger-like projections of the internal surface of the small intestine that tremendously increase the surface area for nutrient absorption.

Viscera (visceral organs) (vis'er-ah) The organs within the ventral body cavity, including the stomach, bladder, heart, lungs, spleen.

Visceral Pertaining to the organs and structures within the ventral body cavity and to all smooth muscle and glands throughout the body; opposite of *somatic*.

Visceral muscle Smooth muscle and cardiac muscle.

Vitamins Organic compounds required by the body in minute amounts that generally must be obtained from the diet.

Vulva (vul'vah) The external genitalia of the female.

White matter White substance of the central nervous system; contains tracts of myelinated nerve fibers.

Yolk sac Embryonic sac that stores a tiny quantity of yolk and gives rise to the lining of the digestive tube; also gives rise to the primordial germ cells and the blood cells.

Zygote (zi'gōt) Fertilized egg.

Credits

730

20.12a: © C. Overton/PhotoTake. 20.18: Kenneth Siegesmund, Ph.D, Dept. of Anatomy, Medical College of Wisconsin. "A Closer Look": Martin Rotker/ PhotoTake.

Chapter 21
21.9a: From *Tissues and Organs: A Text-Atlas of Scanning Electron Microscopy* by R. Kessel and R. Kardon. W.H. Freeman & Co. © 1979. 21.9b: From *Scanning Electron Microscopy in Biology*, by R. Kessel and C.Y. Shih, W.H. Freeman & Co. Copyright © 1974. 21.10b: © Science Photo Library / Photo Researchers. 21.3a: From *Color Atlas of Histology*, by Leslie Gartner and James Hiatt. Williams and Wilkins © 1990. 21.14: © Biophoto Associates/ Photo Researchers, Inc. 21.15d: From *Color Atlas of Histology*, by Leslie Gartner and James Hiatt. Williams and Wilkins © 1990. 21.17d: © Manfred Kage/Peter Arnold, Inc. 21.19a: © Ed Reschke/Peter Arnold, Inc. 21.19b: © Biophoto Associates/ Photo Researchers, Inc. 21.22b: © M.I. Walker/ Photo Researchers, Inc. 21.24b: Tufts Medical School, Dept.

of Anatomy. 21.25b: © Ed Reschke. "A Closer Look": Carroll H. Weiss/ © Camera M.D. Studios.

Chapter 22
22.3b: Courtesy Ruedi Thoeni, M.D., University of California, San Francisco. 22.7b: © David Phillips/ Visuals Unlimited. 22.8: © Biophoto Associates/ Photo Researchers, Inc. 22.9c: From *Tissues and Organs: A Text-Atlas of Scanning Electron Microscopy* by R. Kessel and R. Kardon. W.H. Freeman & Co. Copyright © 1979. 22.11: © Lester Bergman 1987. 22.12: © Biophoto Associates/Photo Researchers, Inc. 22.16: From *Color Atlas of Histology*, by Leslie Gartner and James Hiatt. Williams and Wilkins © 1990.

Chapter 23
23.3b: © Ed Reschke. 23.4a: From *Tissues and Organs: A Text-Atlas of Scanning Electron Microscopy* by R. Kessel and R. Kardon. W.H. Freeman & Co. Copyright © 1979. 23.5b: © Manfred Kage/Peter Arnold, Inc. 23.7a: From *Color Atlas of Histology*,

by Leslie Gartner and James Hiatt. Williams and Wilkins © 1990. 23.7b: © Biophoto Associates/ Photo Researchers, Inc. 23.13b: © Ed Reschke. 23.18a: © Dr. Michael Ross. 23.22b: © Lennart Nilsson/Boehringer Ingelheim International. 23.23b: Courtesy Dr. E.S.E. Hafez, Wayne State University, Detroit, MI. 23.23d: Photographed from the Carnegie Collection by Dr. Allan C. Enders, University of California, Davis.

Chapter 24
24.5: © Biophoto Associates/Photo Researchers, Inc. 24.8b: © Ed Reschke. 24.9b: From *Color Atlas of Histology*, by Leslie Gartner and James Hiatt. Williams and Wilkins © 1990. 24.10: © Ed Reschke. 24.11: © Don W. Fawcett/Photo Researchers, Inc. 24.12: Carolina Biological Supply Company. "A Closer Look": p. 676, Clinical Pathological Conference, *American Journal of Medicine*, 20(1956):133 with permission of Albert I. Mendelhoff, M.D.; p. 677, courtesy Dr. Charles Wilson, University of California, San Francisco.

Illustration Credits

Chapter 1
1.1, 1.2: Jeanne Koelling. 1.4, 1.5, 1.7, 1.8, 1.9a, b, 1.10: Kenneth R. Miller. 1.6a, b, c: Barbara Cousins.

Chapter 2
2.1, 2.2a, 2.5a, 2.7a, c, 2.8a, 2.12, 2.13a, 2.15b, 2.17, 2.18: Carla Simmons. 2.3, 2.4, 2.6a: Barbara Cousins. 2.9: Elizabeth Morales-Denney. 2.11b: Raychel Ciemma. 2.14, 2.16: Georg Klatt.

Chapter 3
3.1, 3.4a, b, 3.8, 3.10, 3.12: Martha Blake. 3.2, 3.3, 3.5, 3.6, 3.7, 3.11, 3.13: Barbara Cousins. 3.9: Stephanie McCann. Table 3.2: Cyndie Wooley.

Chapter 4
4.1, 4.13, 4.14: Barbara Cousins. 4.2, 4.3a, c, d, e, h, 4.4, 4.10b, 4.11, 4.12, 4.15a, b, c, e, f, g, h, i, j, k, l, 4.17a, c, 4.18, 4.19: Linda McVay. 4.3b, f: Linda McVay and Elizabeth Morales-Denney. 4.3g, 4.17b: Linda McVay and Cyndie Wooley. 4.5a, 4.15d: Carol Verbeeck; adapted from K.D. Johnson, D.L. Rayle and H.L. Wedberg, *Biology: An Introduction*, (Menlo Park, CA: Benjamin/Cummings, 1984) p. 52. © The Benjamin/Cummings Publishing Company, Inc. 4.6a, b: Georg Klatt. 4.7, 4.8, 4.9: Carla Simmons. 4.16: Cyndie Wooley.

Chapter 5
5.1, 5.2: Elizabeth Morales-Denney. 5.5a, b, c, d, 5.6, 5.8, 5.9: Linda McVay.

Chapter 6
6.1, 6.4: Laurie O'Keefe. 6.2, 6.3a, b, c, 6.6a, 6.8, 6.9, 6.11, 6.12, 6.14, Table 6.1: Barbara Cousins. 6.5a, b: Elizabeth Morales-Denney. 6.7: Georg Klatt.

Chapter 7
7.1, 7.2b, 7.4c, 7.8b, 7.14b, c, 7.16a, b, c, 7.17a, b, c, 7.18a, b, 7.20a, b, 7.21a, b, c, d, e, f, 7.22a, b, 7.23a, b, 7.26a, b, c, 7.27b, 7.28, Table 7.4: Laurie O'Keefe. 7.2a, 7.3a, b, 7.4a, b, 7.5a, 7.6a, b, 7.7a, 7.8a, c, 7.9, 7.10, 7.11, 7.12, 7.13, 7.14a, 7.15, 7.19a, b, 7.25a, 7.29a, b, c, 7.30, 7.31a, b, Table 7.3, Table 7.5: Nadine Sokol. "A Closer Look": John Karapelou. Table 7.1: Barbara Cousins.

Chapter 8
8.1, 8.2, 8.3a, c, 8.4, 8.5a, b, c, 8.5e, f, g, h, i, 8.6, 8.7, 8.8, 8.9a, b, 8.10a, c, d, 8.11a, c, d, 8.12a, b, c, d, e, 8.13, 8.14, Table 8.1: Barbara Cousins. 8.5d: Cyndie Wooley.

Chapter 9
9.1a, 9.4, 9.5b, c, d, e, 9.7, 9.8, 9.13a, b, Table 9.1, Table 9.2: Raychel Ciemma. 9.3, 9.11a, b, 9.12b, c: Barbara Cousins. 9.9, 9.14: Cyndie Wooley.

Chapter 10
10.1, 10.2, 10.4a, b, 10.5, 10.6, 10.7, 10.8a, b, d, e, 10.9a, b, 10.10, 10.11, 10.12a, b, 10.13, 10.14a, b, c, 10.15, 10.16, 10.18, 10.19, 10.20, 10.21, 10.22, 10.23: Raychel Ciemma. 10.3a, b: Barbara Cousins. 10.17a, b, c, d, 10.24a, b, c, d, e: Laurie O'Keefe.

Chapter 11
11.2, 11.5, 11.12a, b, 11.17, 11.18a, b, c: Raychel Ciemma. 11.6, 11.8, 11.9, 11.10, 11.11, 11.13a, b, c, d, e, f, 11.14b, 11.15, 11.16: Charles W. Hoffman. 11.7: John and Judy Waller.

Chapter 12
12.2, 12.3, 12.4, 12.6, 12.7a, c, 12.8, 12.10, 12.11a, b, 12.12a, b, 12.13a, 12.16, 12.17, 12.18a, b, 12.19b, 12.20a, 12.21, 12.22a, 12.24, 12.25a, 12.26a, b, 12.28, 12.29. 12.30: Stephanie McCann. 12.5, 12.27: Raychel Ciemma. 12.7b, 12.9, 12.14, 12.15, 12.18c, 12.23: Cyndie Wooley.

Chapter 13
13.1a, b, 13.3, 13.4a, : Raychel Ciemma. 13.2a, b: Charles W. Hoffman. 13.5a, b, 13.6, 13.7a, b, 13.8, 13.9a, c, d, 13.10a, b, 13.11a, b, 13.12, Table 13.2I, II, III, IV, V, VI, VIII, IX, X, XI, XII: Stephanie McCann. Table 13.1: Georg Klatt. Table 13.2VII: Cyndie Wooley.

Chapter 14
14.1, 14.2, 14.4, 14.8, 14.10: Raychel Ciemma. 14.3, 14.7, 14.9: Stephanie McCann. 14.5, 14.6a: Cyndie Wooley.

Chapter 15
15.1, 15.2, 15.5a, b, 15.6, 15.9, 15.11, 15.12, 15.16, 15.17, 15.21, 15.22a, b, c, d, 15.23, 15.24, 15.25, 15.26: Charles W. Hoffman. 15.3a, b: Elizabeth Morales-Denney. 15.7a, 15.10a, 15.18, 15.19, 15.20: Cyndie Wooley. 15.8a: Barbara Cousins. 15.14, 15.15a, "A Closer Look": Kenneth R. Miller.

Chapter 16
16.1, 16.3a, 16.5, 16.9, Table 16.1: Nadine Sokol. 16.6a, b, 16.8a, b: Barbara Cousins.

Chapter 17
17.1, 17.2a, b, c, 17.3, 17.4, 17.5a, b, c, d, 17.7, 17.8a, 17.9, 17.10, 17.11, 17.12, 17.13, 17.15, 17.16, 17.17, "A Closer Look": Barbara Cousins. 17.14a, b: Cyndie Wooley.

Chapter 18
18.1a, 18.4a, b, 18.6, 18.7, 18.8, 18.9a, 18,9c, 18.10, 18.11, 18.12, 18.13, 18.14, 18.15, 18.16, 18.18a, b, 18.19a, 18.20, 18.21, 18.22, 18.23, 18.24: Barbara Cousins. 18.3a, b: Elizabeth Morales-Denney. 18.5: Fran Milner.

Chapter 19
19.1a, b,: Barbara Cousins. 19.2a, b, 19.5, 19.8b, 19.8c: Nadine Sokol. 19.3, 19.4a, 19.10a, b: Cyndie Wooley. 19.7: Carla Simmons.

Chapter 20
20.1, 20.2, 20.3a, 20.5a, b, 20.7a, 20.8a, 20.9a, 20.10b, c, 20.11a, b, 20.14, 20.15, 20.16, 20.17, 20.19: Raychel Ciemma.

Chapter 21
21.1, 21.7b, 21.9, 21.10a: Cyndie Wooley. 21.2, 21.3, 21.4a, b, c, 21.6, 21.8, 21.11, 21.12, 21.17a, b, c, 21.22a, c, d, 21.24a: Kenneth R. Miller. 21.5a, b: Elizabeth Morales-Denney. 21.7a, 21.15a, b, c, 21.18a, c, 21.26b, c, d, 21.27: Kenneth R. Miller and Nadine Sokol. 21.14, 21.16, 21.20c, 21.21: Raychel Ciemma. 21.25a: Charles W. Hoffman.

Chapter 22
22.2, 22.3a, 22.4b, 22.6, 22.9, 22.13, 22.15a, b, 22.17, 22.18a, b, c, d: Linda McVay. 22.5a, b, 22.9b: Elizabeth Morales-Denney. 22.7a, c: Linda McVay and Elizabeth Morales-Denney. 22.10, 22.14: Raychel Ciemma.

Chapter 23
23.1, 23.2, 23.3a, 23.4b, c, 23.5, 23.6, 23.9, 23.11a, 23.11b, 23.13a, 23.14, 23.18b, 23.20, 23.21, 23.22a, 23.23a, c, 23.24a, b, c, d, e, f, 23.26, 23.27, 23.28, 23.30: Martha Blake. 23.8, 23.10, 23.12: Raychel Ciemma. 23.15: Kenneth R. Miller. 23.18, 23.19, 23.29: Darwen and Vally Hennings; from N.A. Campbell, *Biology*, (Menlo Park, CA: Benjamin/Cummings, 1987), pp. 911, 913. © The Benjamin/Cummings Publishing Company, Inc.

Chapter 24
24.1, 24.8a, 24.9a, 24.10a: Charles W. Hoffman. 24.2a, 24.2c, 24.12b: Elizabeth Morales-Denney. 24.2b: Charles W. Hoffman and Elizabeth Morales-Denney. 24.3: Fran Milner. 24.8c, 24.9c, 24.13a: Raychel Ciemma. Table 24.2: Charles W. Hoffman and Linda Harris.

Chapter 25
25.1a, b: Laurie O'Keefe. 25.2, 25.3, 25.7b, 25.9a, 25.9b, 25.12a, 25.12b, 25.13, 25.16a, 25.16b, 25.16c, 25.18b: Raychel Ciemma. 25.14: Fran Milner.

Index

Note: **Boldface** page numbers indicate definitions or major discussions; *italic* page numbers refer to tables or figures separated from accompanying text.

A bands, *222*, **223**, 224, 228, *229*
ABCD rule, 113
Abdomen, *14*, **15**, **563–564**, *565*
  arteries of, **496–498**
  auscultation of, 696
  CT scan of, *18*
  innervation of, *395*, *402*, *403*
  muscles of, *244*, *258–259*, *686*, **695–696**
  surface anatomy of, **694–696**
  veins of, *506*, *507*, **508**
Abdominal aorta, *490*, *492*, *496*, *497*
  branches of, *500*
Abdominal cavity, *12*, **13**
Abdominal hernia, **292**
Abdominopelvic cavity, *12*, *13*, **563–564**, *565*, *566*
  mesenteries in, *593*
  serous membranes in, *13*, **14**. *See also* Peritoneum
Abducens nerve (cranial nerve VI), *335*, *336*, *370*, *374*, 504
Abduction, **197–198**, 224
Abductor digiti minimi muscle
  of foot, *290*, *291*
  of hand, *274*, *275*
Abductor hallucis muscle, *290*, *291*
Abductor pollicis brevis muscle, *274*, *275*
Abductor pollicis longus muscle, *270*, *271*, *272*, *273*
Abortion, **62**, 644
Absorption, **563**
  as epithelia function, 66
  microvilli and, 73
Absorptive cells
  in large intestine, *585*, 596
  in small intestine, *579*, **580**, **581**
Accessory digestive organs, *562*, **563**
Accessory glands, male, **631–632**
Accessory hemiazygos vein, **505–506**
Accessory nerve (cranial nerve XI), *335*, *370*, *379*, **691**
Accessory pancreatic duct, *578*, **590**
Accessory sex organs, **623**
Accommodation, **425**, *426*
A cells (alpha cells), **674**, *675*
Acetabular labrum, **204**
Acetabulum, *173*, *176*, *178*
Acetylcholine (ACh), **396**, *397*
  Alzheimer's disease and, 355
  at neuromuscular junction, **367**, *368*
Acetylcholinesterase, **367–368**
Achalasia of the cardia, **407**
Achilles tendon. *See* Calcaneal tendon
Achondroplasia, **135**
Acidic stains, **15**
Acidophils, in pituitary gland, **666**
Acid phosphatase, 632
Acinar cells, **590–591**, *674*, *675*
Acinar glands, 77, *590*
Acoustic (auditory) meatus
  external, *144*, **146**, *147*, *154*
  internal, *144*, *145*, **148**, *154*, *432*

Acquired immune deficiency syndrome (AIDS), **522–523**
Acromegaly, **676**
Acromial end of clavicle, *167*, *168*
Acromioclavicular joint, *192*, **698**, 699
Acromion, **168**, *692*, *698*
Acromion process, *202*
Acrosomal reaction, **646–647**
Acrosome, **627**
ACTH, **665–666**, *668*, *678*
Actin
  in microfilaments, 33, *34*
  in muscle tissues, 91, 217
    cardiac, *229*
    skeletal, *222*, **223**, *224*, *227*
    smooth, *218*, 230
Action potentials. *See* Nerve impulses
Adam's apple, **538**, *688*, *690*
Addison's disease, **677**
Adduction, *197*, **198**, 224
Adductor brevis muscle, *277*, *278*, *288*
Adductor hallucis muscle, *290*, *291*
Adductor hiatus, *499*
Adductor longus muscle, *244*, *277*, *278*, *288*, *686*, *699*, *706*
Adductor magnus muscle, *245*, *277*, *278*, *288*, *687*
Adductor pollicis muscle, *274*, *275*
Adenine (A), in DNA, 35, *36*
Adenocarcinoma, **95**, *555*
Adenohypophysis, **664**, **665–667**, *668–669*, *676*, *678*
Adenoidectomy, **558**
Adenoids. *See* Pharyngeal tonsil
Adenoma, **95**
Adenosine triphosphate (ATP)
  mitochondria and, 28
  in thick (myosin) filaments, 223
Adhesions, **94**
ADH. *See* Antidiuretic hormone
Adipose capsule of kidney, *601*, *603*
Adipose cells. *See* Fat cells
Adipose tissue (fat tissue), **82**
  as connective tissue proper, 78, *79*
  in hypodermis, 82, *100*, 104
  sympathetic innervation and, *403*
Adluminal compartment, *627*, *629*
Adrenal cortex, **672–674**, *677*, *678–679*
Adrenal glands, *663*, **672–674**, *678*
  innervation of, **403–404**
  tumors of, hirsutism and, *107*
Adrenaline (epinephrine), **403–404**
Adrenal medulla, **403–404**, *672*, *679*
Adrenergic vs. cholinergic neurons, **396**
Adrenocorticotropic hormone (ACTH), **665–666**, *668*, *678*
Adventitia of esophagus, *567*, *574*, *575*
Adventitia of stomach, *577*
Adventitia of trachea, *541*, *542*
Adventitia of ureter, *613*
Adventitia of urinary bladder, 614

Adventitia of vagina, 642, 644
Afferent arterioles, to glomerulus, *610*, **611**
Afferent division. *See* Sensory division of PNS
Afferent lymphatic vessels, *524*, *525*
Afferent neurons. *See* Sensory neurons
Afferent transmission, *303*
Afterbirth, **652**, *653*
Aggregated lymphoid follicles. *See* Peyer's patches
Aggregate field view, 323
Aging, study of (gerontology), **45**
Agonist (prime mover), **240**
Agranulocytes, **449**, **450–451**, *452*
AIDS, **522–523**
Air-blood barrier, **543**
Alae of ilium, *173*
Alae of nose, *688*
Alae of sacrum, *162*
Alar cartilages, *535*
Alar plates, *312*, *313*
Albinism, **113**
Albumin, in plasma, 446
Aldosterone, **673**, *675*
Alimentary canal, **562–563**
  See also Digestive tube
  development of, **594–595**
  enteroendocrine cells in, 578, **582**, *675*
  histology of wall, **566–568**
  nerve plexuses in, *567*, *568*
Allantois, *594*, *614*, *617*, *618*
Alpha cells (A cells), *674*, *675*
  efferent neurons, **365**
ALS (amyotrophic lateral sclerosis), **357**
Alveolar ducts, *543*, *544*
Alveolar exocrine glands, *77*, *78*
Alveolar macrophages, **544–545**
Alveolar margins, *144*
  of mandible, **150**, *151*
  of maxillae, *150*, *151*, *153*
Alveolar nerves, *373*
Alveolar pores, **544**, *545*
Alveolar sacs, *543*, *544*
Alveoli (lung), **543–545**, *549*
  development of, 556
  in emphysema, 555, *556*
  in pulmonary circulation, *489*
Alveoli (tooth sockets), 150, *155*
Alzheimer's disease, **355**
Amacrine cell, *422*
Amino-acid-based hormones, **662**
Amino acids, ribosomes and, 29
Amnion, *52*, *53*, *56*, *650*, *651*
Amniotic cavity, *52*, *53*, *56*, *650*, *651*
Amniotic sac and cavity, *52*, *53*
Amoeboid motion, 33
Amphiarthroses, **189**, 190, 191, *192*, *193*
Ampulla
  of semicircular canal, **435**, *436*
  of ductus deferens, *623*, **629**
  of uterine tube, *634*, *638*

Amygdala/Amygdaloid nucleus, *329*, **330**, *335*, *339*, *340*
Amylin, 674
Amyloid plaques, 355
Amyotrophic lateral sclerosis, **357**
Anabolic steroids, 232, **233**
Anal columns, *583*, **584**
Anal sinuses, *583*, **584**
Anal sphincters, *260*, *261*, *583*, **584**, **585**
Anal triangle, *697*
Anal valves, *583*, **584**
Anaphase, 39, *41*
Anaplasia, **45**
Anastomoses
  portal-systemic, **510**
  vascular, **488**
Anatomical neck of humerus, **169**
Anatomical position, **7**
Anatomical snuff box, **701**
Anatomical variability, **6**
Anconeus muscle, *267*, *271*, *272*
Androgen-binding protein, **628**
Androgens, **628**
  adrenal cortex and, 673
  growth spurt and, 130
  sebum secrtion and, 108
  terminal hair growth and, 107
  testes and, **674**
Anemia, **447–448**
  aging and, 456
  sickle-cell, **456–457**
  thalassemia, **457**
Anencephaly, **356**
Anesthesia
  lumbar puncture, **345**
  pudendal nerve block, **386**
Aneurysm, 483, 485, **512**
Angina pectoris, **473**, 474
Angiogram, **513**
Angiography, digital subtraction (DSA), **18**
Angiosarcoma, **513**
Angle of rib, *165*, *166*
Angular movements, *196*, **197–198**
ANH (atrial natriuretic hormone), **675**
Ankle, muscles of, *282–289*
Ankyloglossia, **570**
Ankylosing spondylitis, **212**
Ankylosis, **211**
Annular ligament of radius, **203**
Annular ring, **417**, *418*
Annulospiral sensory endings, **364**, *365*
Annulus fibrosis, *157*
Anopsias, *372*
Anosmia, **371**, *415*
Ansa cervicalis, *379*, *380*
Antagonists, **240**
Antebrachial vein, median, *505*
Antecubital fossa. *See* Cubital fossa
Anterior, defined, *7*, *8*
Anterior arch of atlas, *159*, *160*
Anterior cardiac veins, *464*, *473*
Anterior cerebral artery, **492**, *493*